Lecture Notes in Mathe...

Editors-in-Chief:
J.-M. Morel, Cachan
B. Teissier, Paris

Advisory Board:
Camillo De Lellis (Zurich)
Mario Di Bernardo (Bristol)
Alessio Figalli (Pisa/Austin)
Davar Khoshnevisan (Salt Lake City)
Ioannis Kontoyiannis (Athens)
Gabor Lugosi (Barcelona)
Mark Podolskij (Heidelberg)
Sylvia Serfaty (Paris and NY)
Catharina Stroppel (Bonn)
Anna Wienhard (Heidelberg)

For further volumes:
http://www.springer.com/series/304

Donald A. Dawson • Andreas Greven

Spatial Fleming-Viot Models with Selection and Mutation

Donald A. Dawson
Carleton University
School of Mathematics and Statistics
Ottawa, Ontario, Canada

Andreas Greven
Department Mathematik
Universität Erlangen-Nürnberg
Mathematisches Institut
Erlangen, Germany

ISBN 978-3-319-02152-2   ISBN 978-3-319-02153-9 (eBook)
DOI 10.1007/978-3-319-02153-9
Springer Cham Heidelberg New York Dordrecht London

Lecture Notes in Mathematics ISSN print edition: 0075-8434
                              ISSN electronic edition: 1617-9692

Library of Congress Control Number: 2013954545

Mathematics Subject Classification (2010): 60J70, 92D15, 92D25, 60J80, 60J85

© Springer International Publishing Switzerland 2014
This work is subject to copyright. All rights are reserved by the Publisher, whether the whole or part of the material is concerned, specifically the rights of translation, reprinting, reuse of illustrations, recitation, broadcasting, reproduction on microfilms or in any other physical way, and transmission or information storage and retrieval, electronic adaptation, computer software, or by similar or dissimilar methodology now known or hereafter developed. Exempted from this legal reservation are brief excerpts in connection with reviews or scholarly analysis or material supplied specifically for the purpose of being entered and executed on a computer system, for exclusive use by the purchaser of the work. Duplication of this publication or parts thereof is permitted only under the provisions of the Copyright Law of the Publisher's location, in its current version, and permission for use must always be obtained from Springer. Permissions for use may be obtained through RightsLink at the Copyright Clearance Center. Violations are liable to prosecution under the respective Copyright Law.
The use of general descriptive names, registered names, trademarks, service marks, etc. in this publication does not imply, even in the absence of a specific statement, that such names are exempt from the relevant protective laws and regulations and therefore free for general use.
While the advice and information in this book are believed to be true and accurate at the date of publication, neither the authors nor the editors nor the publisher can accept any legal responsibility for any errors or omissions that may be made. The publisher makes no warranty, express or implied, with respect to the material contained herein.

Printed on acid-free paper

Springer is part of Springer Science+Business Media (www.springer.com)

Abstract

The monograph constructs a rigorous framework to analyse some phenomena in evolutionary theory of populations arising due to the combined effects of migration, selection and mutation in a spatial stochastic population model, namely the evolution towards fitter and fitter types through punctuated equilibria. The discussion is based on some new methods, in particular multiple scale analysis, nonlinear Markov processes and their entrance laws, atomic measure-valued evolutions and new forms of duality (for state-dependent mutation and multitype selection) which are used to prove ergodic theorems in this setting and are applicable for many other questions and renormalization analysis to analyse some phenomena (stasis, punctuated equilibrium, failure of naive branching approximations, biodiversity) which occur due to the combination of rare mutation, mutation, resampling, migration and selection and require mathematically to bridge between (in the limit) separating time and space scales.

Consider the following spatial multitype population model. The state of a single colony is described by a probability measure on some countable type space, the geographic space is modelled by a set of colonies and the colonies (or demes) are labelled with the countable hierarchical group $\Omega_N = Z^{\otimes \mathbb{N}}$, with Z the cyclic group of order N. This set mimics $\mathbb{Z}^2$ as $N \to \infty$. The stochastic dynamics is given by a system of interacting measure-valued diffusions and the driving mechanisms include resampling (pure genetic drift) as diffusion term, migration, selection and mutation as drift terms. Resampling is modelled in each colony by the usual Fleming–Viot diffusion. The haploid selection is based on a fitness function on the space of types, and the mutation is type-dependent and has the feature that fitter types are more stable against deleterious mutation and fitter types appear through rare mutation. The evolution starts from one basic type. The model belongs to a class of processes which have been constructed via a well-posed martingale problem by Dawson and Greven (Electron. J. Probab. **4**, paper no. 4, 181, 1999).

The goal of the monograph is to determine the long-time behaviour of the process in terms of a sequence of quasi-equilibria developing in longer and longer time scales with increasing fitness levels. This is in contrast with the more familiar single equilibrium that develops as time tends to infinity. In particular we show that the combination of mutation, selection and migration leads to the following new effects.

We show that the evolution to fitter types is governed by the immigration of rare mutants from distant locations and subsequent conquest of the colonies, which is possible due to the greater stability of the fitter mutants against further deleterious mutation which gives them a quasi-selective advantage. We describe in detail how the transition from the quasi-equilibrium on one level of fitness to the next occurs.

Since we have a spatial system this transition is taking place in different time scales. We identify five phases (corresponding to specific time-space scales) of the transitions from one quasi-equilibrium to the next, each of which we explicitly describe asymptotically.

Starting in a quasi-equilibrium of a certain level, we get some rare mutations somewhere in space creating a droplet of colonies of fitter types of exponentially growing size and total mass which later dominate everywhere in space. We describe the droplet by means of a random atomic measure and identify the limiting dynamic in terms of processes driven by Poisson random measures. This dynamic allows us to identify the random growth factor (reflecting the randomness of the rare events) and the deterministic exponential growth rate. We next determine the time scale at which the global emergence and fixation by rare mutants occurs on the level of components and blockaverages and describe the limiting dynamics on the various levels. The emergence is of a stochastic nature which makes the subsequent phase of fixation on the next higher level random. A key object for fixation is the random entrance law of a nonlinear evolution equation (McKean–Vlasov equation).

After fixation we see a neutral evolution on the higher level before, after a very long time, finally the new quasi-equilibrium on the higher level is reached.

In addition to the process describing the current population, we construct the historical process giving the information on the complete family structure of the population alive at time t. The genealogical structure of the population arising in our model depends on the migration mechanism, which determines whether we have local biodiversity or a local monoancestor situation, in other words whether the local population asymptotically includes one or countably many high level mutant families.

The method of analysis is a hierarchical renormalization scheme based on a sequence of spatial scales combined with different types of sequences of time scales, together with a limit $N \to \infty$ (in Ω_N) which corresponds to a separation of several natural time scales of the model. These separating multiscales are connected by working with entrance laws.

Due to mutation the structure of the limiting dynamics is quite different from models previously studied with renormalization techniques, and new methods are developed in this monograph to handle the present situation. We introduce and use for this purpose a new class of dual processes to derive expressions for the moment measures which allow us to treat selection and type-dependent mutation and also allows us to prove stronger ergodic theorems than previously known for diffusion models involving migration, selection, mutation and resampling.

Keywords. Atomic measure-valued processes, Biodiversity, Droplet formation, Duality, Interacting Fleming–Viot processes, Mutation, Punctuated equilibrium, Random entrance law, Random McKean–Vlasov equation, Rare mutant, Renormalization, Selection

Acknowledgements We would like to acknowledge the support of an NSERC Discovery Grant, DFG-Schwerpunkt 1033 and later DFG-NOW Forschergruppe 498, DFG-Grant Gr-876/15.1, DFG-Grant Gr-876/16.1 and D. Dawson's Max Planck Award for International Cooperation. We would also like to thank the referees for their helpful comments which helped to improve the exposition.

Contents

1	**Introduction**	1
	1.1 Ideas and Objectives	1
	1.2 Concepts and Methods of Mathematical Analysis	5
	1.3 Future Applications of Methods and Open Problems	7
	1.4 Methodologies Developed	8
	1.5 Outline of the Monograph and Different Ways to Read It	9
2	**Mean-Field Emergence and Fixation of Rare Mutants in the Fisher–Wright Model with Two Types**...........................	11
	2.1 Prelude: The Three Layers of Our Approach	11
	2.2 Formulation of the Problem and Description of Two-Type Model	13
	2.2.1 The Problem...	14
	2.2.2 The Fisher–Wright Dynamics with Two Types and Basic Objects.................................	14
	2.3 Statement of Results	18
	2.3.1 The Limiting Random McKean–Vlasov Dynamics and Entrance Laws	18
	2.3.2 The Limiting Dynamics of the Sparse Sites with Substantial Type-2 Mass	23
	2.3.3 Emergence Time, Droplet Formation, Fixation Dynamic for N-Interacting Sites: Statement of Results	27
	2.3.4 The Branching Approximation in the Fast Mixing Limit	36
	2.3.5 Some Topological Facts Concerning Atomic Measures	37
3	**Formulation of the Multitype and Multiscale Model**	39
	3.1 Main Ingredients	40
	3.2 Characterization of the Process by a Martingale Problem	42

	3.3	Detailed Specification of Selection and Mutation Mechanisms	45
	3.4	Interpretation of the Model from the Point of View of Evolutionary Theory and Connections to Related Work	50
	3.5	Extensions	53
4	**Formulation of the Main Results in the General Case**	55
	4.1	Renormalization: A Sequence of Space and Time Scales	58
	4.2	Basic Scenario and the Hierarchical Mean-Field Limit	63
		4.2.1 The Basic Scenario: Five Phases of Transition	63
		4.2.2 The Basic Limit Dynamics	67
	4.3	Ergodic Theorem for the Nonlinear McKean–Vlasov and Spatial Fleming–Viot Process	71
	4.4	Elements of Asymptotic Description: Two Interaction Chains	73
	4.5	Limit Theorems I: Quasi-equilibria and Neutral Phase	76
	4.6	Limit Theorems II: Emergence, Fixation, Droplet Formation of Rare Mutants	79
		4.6.1 Questions Concerning the Emergence and Fixation by Rare Mutants	80
		4.6.2 Characteristic Functionals for Emergence and Fixation: Limit Dynamics	85
		4.6.3 Results on Emergence, Invasion and Fixation by Rare Mutants	91
	4.7	Limit Theorems III: Local Biodiversity Versus Monotype, the Genealogy of the Mutant Population	95
		4.7.1 History of the E_{j+1}-Type Population: Concepts and Analytic Description	96
		4.7.2 Results: The Role of the Migration Parameters for the Local Bio-Diversity	99
		4.7.3 The Origin of Successful Populations	101
		4.7.4 Outlook on the Full Historical Process	102
	4.8	The Structure of the Proofs	103
5	**A Basic Tool: Dual Representations**	105
	5.1	The Dual Process and Duality Function for the Interacting System	107
	5.2	Duality Relation for Interacting Systems	116
	5.3	Two Alternative Duals for the Mutation Component	123
		5.3.1 Modified Dual for State-Independent Mutation Component	123
		5.3.2 The Dual for Mutation: Alternative Representation in the State-Dependent Case	125
		5.3.3 Pure Jump Process Dual	126
	5.4	Historical Interpretation of the Dual Process	126

5.5	Refinements of the Dual for Finite Type Space: $(\eta_t, \mathcal{F}_t^{++})_{t\geq 0}, (\eta_t, \mathcal{G}_t^{++})_{t\geq 0}$		128
	5.5.1	Ingredients	129
	5.5.2	The Refined Dual $(\eta_t, \mathcal{F}_t^{++})_{t\geq 0}$ or $(\eta_t, \mathcal{G}_t^{++})_{t\geq 0}$	135
	5.5.3	The Refined Duality Relation	135
	5.5.4	Historical Interpretation Revisited and Outlook on a Modified Dual	138
	5.5.5	Calculations with the Refined Dual: An Additional Trick	141
5.6	The Dual Representation for the McKean–Vlasov Process		142
	5.6.1	The McKean–Vlasov Dual Process	142
	5.6.2	Duality Relation for McKean–Vlasov Process	144

6 Long-Time Behaviour: Ergodicity and Non-ergodicity 147
    6.1   Non-spatial Case 148
    6.2   Spatial Case 156

7 Mean-Field Emergence and Fixation of Rare Mutants: Concepts, Strategy and a Caricature Model 161
    7.1   The Mean-Field Model and Hierarchical Mean-Field Models ... 161
        7.1.1   The Basic Scenario for the Mean-Field Model 161
    7.2   Outline of the Strategy for Sections 8–13 164

8 Methods and Proofs for the Fisher–Wright Model with Two Types 167
    8.1   Outline of the Subsections on Proofs of Results Formulated in Subsection 2.3 167
    8.2   A Warm-Up: The Case of a Single Site Model and Other Simplifications 168
        8.2.1   Emergence of Rare Mutants in the Single Site Deterministic Population Model 168
        8.2.2   Emergence of Rare Mutants in the Single Site Stochastic Population Model 169
        8.2.3   Proofs Using the Dual 169
    8.3   Proofs of Propositions 2.3, 2.6, and 2.11 173
        8.3.1   Outline of the Strategy of Proofs for the Asymptotic Analysis 173
        8.3.2   Proof of Proposition 2.3 (Properties McKean–Vlasov dynamic) 175
        8.3.3   The Structure of the Dual Process and a Crump–Mode–Jagers Process 188
        8.3.4   The Dual in the Collision-Free Regime: The Exponential Growth Rate 192
        8.3.5   The Dual in the Collision-Free Regime: Further Properties 198

	8.3.6	Dual Process in the Collision Regime: Macroscopic Emergence Proposition 2.6(a)	202
	8.3.7	Dual Process in the Collision Regime: Nonlinear Dynamics	216
	8.3.8	The Dual Process in the Collision Regime: Convergence Results	234
	8.3.9	Dual Population in the Transition Regime: Asymptotic Expansion	250
	8.3.10	Weighted Occupation Time for the Dual Process	292
	8.3.11	Proof of Proposition 2.11, Part 1: Convergence to Limiting Dynamics	295
	8.3.12	Proof of Proposition 2.11, Part 2: The Random Initial Growth Constant	297
	8.3.13	Completion of the Proof of Proposition 2.11	301
8.4	Droplet Formation: Proofs of Proposition 2.8–2.10	302
	8.4.1	Mutant Droplet Formation at Finite Time Horizon	303
	8.4.2	The Long-Term Behaviour of Limiting Droplet Dynamics (Proof of Proposition 2.10)	314
	8.4.3	Proof of Proposition 2.8	317
	8.4.4	Some Explicit Calculations	318
	8.4.5	First and Second Moments of the Droplet Growth Constant $\mathcal{W}^*$	320
	8.4.6	Asymptotically Deterministic Droplet Growth	347
8.5	Relation Between $^*\mathcal{W}$ and $\mathcal{W}^*$	358
8.6	Third Moments	366
8.7	Propagation of Chaos: Proof of Proposition 2.13	371
8.8	Extensions: Non-critical Migration, Selection and Mutation Rates	373

9 Emergence with $M \geq 2$ Lower Order Types (Phases 0,1,2) 377

9.1	Introduction		378
9.2	The $(M, 1)$ Model: Basic Objects and Notation		380
	9.2.1	Simplification for Analysis of Phases 0,1,2	380
	9.2.2	The Basic Objects	381
	9.2.3	The Intuitive Picture	382
	9.2.4	Different Mutation-Selection-Migration Parameter Regimes	383
9.3	Statement of Results: Emergence, Fixation and Droplet Formation		384
	9.3.1	Time Scale of Macroscopic Emergence	384
	9.3.2	The McKean–Vlasov Dynamics for Emergence-Fixation Process	385
	9.3.3	Droplet Formation in Early Times $t = O(1)$ and $t \to \infty$	387
	9.3.4	Exit=Entrance	390

9.4		Representations: Ordered Dual Tableau-Valued Dual	391
	9.4.1	Introduction	392
	9.4.2	Summary: Table of Dual Objects	394
	9.4.3	Warm-Up: Emergence for the Deterministic Case ($d = 0$)..	394
	9.4.4	Warm-Up: Dual Calculations with $M = 2$ for Emergence...	397
	9.4.5	The Strategy for Getting an Enriched Dual............	404
	9.4.6	Modification of Dual 1: $\tilde{\mathcal{F}}_t^{++,<}$ with Ordered Factors	405
	9.4.7	Modification of the Dual 2: $\mathcal{G}^{++,<}$ with Ordered Coupled Summands	420
	9.4.8	Extension to the McKean–Vlasov Process	425
9.5	The Marked Set-Valued Dual Process $\tilde{\mathcal{G}}^{++}$		426
	9.5.1	Dual $\tilde{\mathcal{G}}^{++}$: Examples for the Special Cases $M = 2, 3$...	427
	9.5.2	The Marked Set-Valued Duality: $\tilde{\mathcal{G}}^{++}$	434
	9.5.3	Properties 1: Set-Valued Markovian Dynamics Without Migration	438
	9.5.4	Properties 2: Set-Valued Dual with Migration..........	444
	9.5.5	Calculating with $\tilde{\mathcal{G}}^{++}$	445
9.6	The Growth of Occupied Sites in the Collision-Free Case $\tilde{\mathcal{G}}^{++}$: Basic Construction of Enriched Dynamics $\mathfrak{s}, \mathfrak{s}^*, \mathfrak{s}^{**}$		448
	9.6.1	Preparation 1: Enriched Marked Set-Valued Dynamics with Emigration	449
	9.6.2	Preparations 2: The Enriched Marked Set-Valued Markovian Dynamics $\tilde{\mathcal{G}}^{++,\ell}$	453
	9.6.3	Enriched Marked Set-Valued Dynamics: $\mathfrak{s}$	458
	9.6.4	Further Auxiliary Processes to analyse $\tilde{\mathcal{G}}^{++} : \mathfrak{s}^*, \mathfrak{s}^{**}$ and $\mathfrak{s}^{**}_{\text{enr}}$	478
9.7	Collision-Free Regime: Malthusian Parameter of $\tilde{\mathcal{G}}^{++}$		483
	9.7.1	Introduction	483
	9.7.2	Preparation: Multitype CMJ-Processes	484
	9.7.3	The CMJ Process Related to the Dual	486
	9.7.4	The Malthusian Parameter for $\vert\mathcal{G}_t^{++,\mathit{lll}}\vert$	489
	9.7.5	Formulation of the Main Results on the Marked Set-Valued Dual $\tilde{\mathcal{G}}^{++}$	494
	9.7.6	Proof of Propositions 9.40 and 9.41	496
	9.7.7	The $N \to \infty$ Limit	498
	9.7.8	Factorization Dynamics of the Set $\tilde{\mathcal{G}}^{++}$: Giant Versus Small Metafactors........................	499

9.8		The Strategy for the Collision Regime	511
	9.8.1	The Problem of Collisions	511
	9.8.2	Strategy for Subsections 9.9–9.13	512
9.9		Dual in the Collision Regime: $\tilde{\mathcal{G}}^{++,N}$ and $\mathfrak{s}^{****,N}$	514
	9.9.1	The Process $\mathfrak{s}^{****,N}$: Motivation and Basic Structure	515
	9.9.2	Dual State Description of $\mathfrak{s}^{****,N}$ in the Collision Regime	519
	9.9.3	Dynamics of $\mathfrak{s}^{****,N}$	528
	9.9.4	Clans: Their Number and Sizes, Asymptotics as $N \to \infty$	533
9.10		Empirical Process of Microstates, Clans for $\mathfrak{s}^{****,N}$: Mutation-Dominant Regime	541
	9.10.1	Empirical Process of Microstates in the Mutation-Dominance Regime	542
	9.10.2	Limiting Nonlinear Dynamics for Microstate Frequencies Under Mutation Dominance	550
	9.10.3	Convergence of the Dynamics of $(\hat{\Psi}^N(\beta^{-1}\log N + t))_{t \geq t_0}$	555
	9.10.4	Convergence of $(\hat{\Psi}^N(\beta^{-1}\log N + t)^+)_{t \in (-\infty, \infty)}$ to Entrance Law	565
9.11		The Transition Regime and Emergence: Mutation-Dominant Case	567
	9.11.1	Expansion of Dual Process as $t \to -\infty$	568
	9.11.2	Multicolour Systems	569
	9.11.3	Deterministic Regime of Droplet Growth with $M \geq 2$	573
	9.11.4	Emergence and Rare Mutation Events: Hazard Rate and Post Event Calculation	583
9.12		Emergence and Fixation: The Selection-Dominant Case	587
	9.12.1	Mesostate Branching Dynamics in the Collision-Free Case: Overview	589
	9.12.2	Mesostate Population Dynamics in the Case $M = 2, d = 0$	592
	9.12.3	Long-Time Behaviour of the MBDC-Process: $M = 2, d = 0$	605
	9.12.4	The Case $M = 2$ and $d > 0$	613
	9.12.5	Higher Moments in the Case $M = 2, d = 0$	615
	9.12.6	Extension to the Case $M > 2$: Conclusion of the Proof of Proposition 9.64	626
	9.12.7	Exponential Growth and Type Distribution of the CMJ-Mesostate Process	630

	9.12.8	Growth of $|\tilde{\mathcal{G}}_t^{++}|$: The Two Level Branching Structure..	651
	9.12.9	Empirical Process of Mesostates: Collision Case	656
	9.12.10	The Marked (Coloured) Mesostate Process	662
	9.12.11	Rare Mutations and Emergence Time: Lower Bound...	681
	9.12.12	Second Moment Calculations and L^2 Convergence of the Droplet Process	686
	9.12.13	The Random Entrance Law	687
	9.12.14	Further Perspectives: Discussion of the Non-ergodic Mutation Case	692
	9.12.15	Dependence of the Emergence Time on the Initial Measure if $\beta > \beta_{P_e}$	697
9.13	Proof of Propositions 9.1, 9.2, 9.4		698
	9.13.1	Proof of Proposition 9.1, Part 1: Emergence Time	698
	9.13.2	Proof of Proposition 9.1, Part 2: Fixation...............	702
	9.13.3	Proof of Proposition 9.2: Uniqueness for McKean–Vlasov Entrance Law	704
9.14	Proof of Propositions 9.3 and 9.5		709
	9.14.1	Proof of Proposition 9.3: Convergence to Limiting Nonlinear Dynamics........................	709
	9.14.2	Proof of Proposition 9.5	712
9.15	Asymptotic Elimination of Downward Mutations		712

10 The General (M, M)-Type Mean-Field Model: Emergence, Fixation and Droplets.. 715

10.1	The Interacting System on $\{1, \ldots, N\}$ with (M, M)-Types		716
10.2	Overview of Phases 0–3 for the (M, M) System on $\{1, \ldots, N\}$		720
10.3	Ergodic Theorem for the (M, M) Spatial and Mean-Field Systems		721
10.4	Results 1: Macroscopic Emergence, Fixation and Random Entrance Law of Higher Level Multiple Types		724
	10.4.1	The Multitype McKean–Vlasov Equation and Its Entrance Laws	725
	10.4.2	Emergence: The Multitype Case	726
	10.4.3	Fixation..	729
10.5	Results 2: Mutant Droplet Formation in (M, M)-Case		729
	10.5.1	Formation of Droplets and Their Longtime Behaviour	730
	10.5.2	The Equality of $\mathcal{W}^*$ and $^*\mathcal{W}$ (Linkage Phase 0 and Phase 1)...................................	739
10.6	Dual Calculations: Proof of Theorem 14		739

	10.7	Dual Calculations with M Higher Level Types	744
		10.7.1 Dual Calculations in the Case $(1, M)$: First Moment	747
		10.7.2 Dual Calculation for the $(1, M)$ System: Higher Moments	751
		10.7.3 The Dual Calculation in the Case of (M, M)-Types ...	760
	10.8	Proof of Proposition 10.2	764
	10.9	Proof of of Proposition 10.3	770
	10.10	Proof of Proposition 10.11	771
	10.11	Proof of Proposition 10.4	771
	10.12	Removing the Simplifying Assumption	772
	10.13	Droplets: Proofs of Propositions 10.5, 10.6, 10.7, 10.8, 10.9, 10.10	772
11	**Neutral Evolution on E_1 After Fixation (Phase 3)**		**781**
	11.1	Evolution in Neutral Equilibrium: Results	781
	11.2	Proof of Proposition 11.1	783
12	**Re-equilibration on Higher Level E_1 (Phase 4)**		**787**
	12.1	Statement of Results: From Neutral Equilibrium to Mutation-Selection on E_1	787
		12.1.1 The Finite Two-Level System: Construction	788
		12.1.2 Limit System: Construction	790
		12.1.3 Topologies on Path Space	792
		12.1.4 Statement of Three Results on Scaling Limit	794
	12.2	Proof of Proposition 12.2	796
	12.3	Convergence in the Uniform Topology	804
		12.3.1 Microscopic Viewpoint: Uniform Convergence of Blockaverages	805
		12.3.2 The Microscopic Viewpoint: Tagged Sites	808
	12.4	Proofs of Propositions 12.3 and 12.4	809
13	**Iteration of the Cycle I: Emergence and Fixation on E_2**		**811**
	13.1	How to Proceed to Higher Levels	811
	13.2	The Case of Two Levels of Spatial Organisation and Three Fitness Levels	813
		13.2.1 The (3,2)-Model and Outline	813
		13.2.2 A Two-Level System with Three Fitness Levels: Time Scale $O(N)$	815
		13.2.3 A Two-Level System with Three Fitness Levels in Time Scale $o(N \log N)$ (Phase 0)	816
		13.2.4 Droplet Formation (Transition Phase)	817
		13.2.5 A Two-Level System with Three Fitness Levels in Time Scale $N \log N$ (Phase 1)	818
		13.2.6 A Two-Level System with Three Fitness Levels in Time Scale $N \log N + t$ (Phase 2)	820

		13.2.7	A Two-Level System with Three Fitness Levels After Fixation (Phase 3)	823

          13.2.7 A Two-Level System with Three Fitness Levels After Fixation (Phase 3) 823

          13.2.8 A Three-Level System with Three Fitness Levels in Time Scale $N^2 t$ (Phase 4) 824

14 Iteration of the Cycle II: Extension to the General Multilevel Hierarchy 829

    14.1 Combination of Three Time Scales for $(j+1, j)$-Level Systems 829

    14.2 The Infinite Hierarchy in Multiple Time-Space Scales 833

    14.3 Local Genealogy and Biodiversity During Emergence and Fixation 835

          14.3.1 Proof of Theorem 11 835

          14.3.2 The Geographic Source of E_j-Valued Mutant Families: Proof of Proposition 4.14 836

15 Winding-Up: Proofs of the Theorems 3–11 839

A Tightness 841

    A.1 The Joffe–Métivier criteria for tightness of D-semimartingales 841

    A.2 Tightness criteria for continuous processes 843

B Nonlinear Semigroup Perturbations 845

References 847

Index of Notation and Tables of Basic Objects 853

Index 855

Chapter 1
Introduction

Consider a population consisting of individuals of different types and distributed in colonies located in geographic space and assume the population is locally finite. If we consider the relative frequencies of types in the different colonies, then the population in a single colony is described by a probability measure on the type space. Hence at a fixed time the essential features of the population are described by a collection of probability measures on the type space indexed by the sites of the geographic space. The time evolution of the population involves finite population resampling (often called pure genetic drift by biologists), mutation and selection in each colony and in addition migration of individuals between colonies. We shall use the large population limit leading to diffusion models, socalled interacting Fleming–Viot processes with selection and mutation, reviewed later on.

We develop here the analysis of these models under three perspectives of biological respectively methodological flavour:

- emergence from *punctuated equilibrium via rare mutation*,
- method of *hierarchical mean-field limit and renormalization*,
- *duality theory* for the genealogy of samples from the population,
- technique of *random entrance laws* for nonlinear evolutions.

We try to present the material in such a way that it can be read from all those perspectives, we comment on this as we go along.

1.1 Ideas and Objectives

Mutation and selection both play essential roles in the evolution of the population. Mutation increases genetic diversity, while selection reduces this diversity pushing it towards states concentrated on the fittest types. The balance between these two forces is influenced heavily by the geographical structure of the population and the effect of migration. Randomness enters through resampling (pure genetic drift).

The first part of the picture we focus on is a scenario (*punctuated equilibrium*) which has received considerable attention in the biological literature (see for example (Eldredge and Gould [EG1], [EG2]) and the references below). In this scenario a population remains for long periods of time in a nearly stationary state (*quasi-equilibrium or stasis*) until the emergence of one or several rare mutants of higher fitness that then take over the population relatively quickly. Possible explanations of this phenomenon based on different mechanisms include the introduction of new selectively advantageous types by rare mutation (see for example [ECL] and the trait substitution sequence of adaptive dynamics (see Champagnat, Ferrière and Méléard [CFM])), or tunnelling between local fitness maxima (see for example Newman, Cohen and Kipnis [NCK]). The explanation of the tunnelling in [NCK] was presented in the context of Wright's theory of evolution in an adaptive landscape (see [Wr1], [Wr2], [W]). In Wright's theory the long times between transitions from one quasi-equilibrium to another corresponds to "tunnelling" between one adaptive peak and another in a situation in which local stabilizing selection maintains a population near a local peak. In this setting, the Ventsel–Freidlin theory of small random perturbations of dynamical systems asserts that the actual transition, when it occurs, takes place rapidly. Thus this scenario would be consistent with the observations of paleontologists of periods of rapid evolution separated by very long periods of "stasis". This is a stochastic effect not present in the deterministic version, that is, the infinite population limit, but must be formulated in terms of a stochastic diffusion limit, which we consider here.

A second part of the picture is the role of *geographic space* and its important effects on the qualitative behaviour which continue to draw interest from mathematicians (see for example (Champagnat and Méléard [CM]), (Barton, Etheridge and Veber [BEV])). Namely, typically a large population is subdivided into small subpopulations (colonies) occupying different geographic regions. Then tunnelling between adaptive peaks or the introduction of a new more fit mutant type can occur in these subpopulations and then by spatial migration the new type can spread (for example the spread of a mutant virus) and eventually result in the take-over of the entire population.

Our goal is to introduce a class of spatial models exhibiting the above features of punctuated equilibrium starting from rare mutants in spatially structured subpopulations undergoing finite population sampling and to develop a rigorous framework for the analysis of such systems. More generally our goal is to develop tools and a framework to explore models which provide insight into the respective roles of mutation, selection and spatial migration in the emergence and spatial spread of rare mutants corresponding to a higher adaptive peak (quasi-equilibrium). The mechanism producing the rare mutants within a local colony could be as suggested by the adaptive landscape picture as above (see also Example 2 in Section 3.4). Another key feature in our formulation is the role of the fitness advantage of types based on *the relative stability against deleterious mutation*—the evolutionary importance of deleterious mutation is underlined in the literature on Muller's ratchet (see, for example [CC], [EPW]). We establish that the combined effect of *mutation, selection and migration* can result in a transition from one quasi-equilibrium to the

1.1 Ideas and Objectives

next occurring in phases (droplet formation, emergence, fixation) which take place in different time scales.

In this context it turns out that the migration together with the local stability properties, decisively enhances the speed of "progress", that is, the transition to fitter types throughout the population. In particular, we obtain rather sharp asymptotic results on the emergence of rare mutants and how they spread throughout the whole population by migration and selection. This also allows us to give a precise account on the shortcomings of using *branching approximations* in *spatial* models to study the early phase of emergence of the rare mutants.

To give a precise setting for this study we model the population dynamics according to a stochastic process arising as the diffusion limit of many individuals incorporating the basic mechanisms for a spatial population traditionally employed in genetics and evolutionary biology. In this model *random fluctuations* (see for example Ewens [Ew]) are included in contrast to other infinite population limits which are ruled by deterministic differential equations (for example, see [Bu]). More precisely the model we use arises from the particle model driven by migration of particles between colonies, and in each site by resampling of types, mutation and selection. Increase the number of particles as ε^{-1} and give them mass ε. As $\varepsilon \to 0$, a diffusion limit of interacting multitype diffusions results if the resampling rate is proportional to the number of pairs and there is weak selection, i.e. selection occurs at a rate decreasing with the inverse of the number of particles per site. Otherwise with strong selection, i.e. selection at a fixed rate, we get the deterministic limit, often referred to as infinite population limit.

These stochastic models which describe populations of possibly infinitely many types are called *interacting Fleming–Viot processes* and have been intensively studied in recent years (for example, see [EG], [D], [EK1, EK2, EK3], [Eth00], [DGV], [DG99], [DG12]). The basic mechanisms are *migration* between colonies in the diffusion limit which takes the form of deterministic mass flow between sites and then within each colony *resampling (i.e. pure genetic drift)* which gives the diffusion term, *selection* of haploid type based on a fitness function on the space of types and *mutation* between types occurring at rates given by a transition kernel on the space of types, the latter two result in a deterministic massflow between types. Some additional remarks on the relation between the model studied and evolutionary theory is given in Section 3.4.

The idea of modelling the spread of a rare advantageous mutant in a spatial population goes back to the seminal paper of Sawyer [Saw]. We will investigate this approach in our context and see that it provides some insights into the nature of the initial stage of the transition to the higher type. As a model for space we use the *hierarchical group* Ω_N (the direct sum of a countable number of copies of the cyclic group of order N, i.e. N being the alphabet length) which allows us to model approximately (as $N \to \infty$) populations in two-dimensional space and where rare mutations occur at rates of the order N^{-1}. The use of the hierarchical group goes back to the thirties of the last century and has led to mathematically important advances. We discuss this and the possibility to approximate $\mathbb{Z}^2$ in this way in detail in Section 3.

The main result of this monograph is an asymptotic (as $N \to \infty$) description of the different phases (induced by the spatial structure and migration) in which rare mutations drive the system from one *quasi-equilibrium* concentrated on types of a particular value of fitness to the next higher one. Here the system has the feature that for a long time the system remains in a quasi-equilibrium on types of a certain fitness before rapidly moving to a higher fitness value.

The transition to a new quasi-equilibrium on a higher level of fitness proceeds by first the occurrence of rare mutants of fitter types in a few colonies at sparsely located spatial sites which are rapidly taken over and form a growing droplet of fitter colonies (*droplet formation*), but a droplet which has for long time negligible intensity. From these colonized sites the new superior types invade the remaining large collection of colonies building up a positive spatial intensity (*emergence* phase), a process taking a long time but then once a "positive" intensity in space is reached very rapidly all sites are dominated by these new types due to "selection" (*fixation* phase). This process is random in two ways, there is a random time shift due to randomness in the very early stage of the first occurrence of rare mutants in the beginning of the droplet formation and there is a random frequency distribution among the different rare mutants still present at the time of fixation.

The "selection" in this picture is based on the advantage of the mutants, which at the beginning of this process is due to their *greater stability against deleterious mutation* combined with higher values of the fitness function, the latter acts only later on in much longer time scales. Then in the next phase a neutral evolution between the new (fitter) types exists which in the next phase is followed by the formation of the new quasi-equilibrium on the higher fitness level.

As a result of this effect, the evolution towards "higher level" populations proceeds very slowly for large time intervals to then tunnel suddenly to the fitter type. This tunneling occurs much faster in a system of colonies connected by migration than in systems of isolated components, where mutation to higher fitness levels must occur independently in the different components before the whole space consists of components concentrated on the higher level type. Note that this spatial spread of selectively advantageous mutants which is of a stochastic nature can be viewed as a more subtle stochastic analogue of the wave of advance described by the celebrated Fisher–Kolmogorov–Petrovsky–Piscounov equation (cf. [DMS]).

Finally we address the question of how finer properties of the migration mechanism influence this effect. For this purpose we consider an enriched version of the model, namely, the *historical process* which records the family relations between individuals and the path of descent through space. In particular we determine the number of rare mutants that are involved in producing the population with higher fitness level in a region in geographic space over which migration is effective at that time scale. We will show that this depends very crucially upon the specifics of the migration mechanism and leads to a dichotomy of *local biodiversity* versus *local monoancestor configuration* corresponding to transience and recurrence of the underlying symmetrized motion.

1.2 Concepts and Methods of Mathematical Analysis

The main mathematical tools we use and develop for the systems of interacting (spatially structured) Fleming–Viot diffusions are:

- new forms of *duality* (*set*-valued, *function*-valued)
- *multiscale* time space scalse analysis—*renormalization*
- *historical* processes

which we discuss now in more detail.

In order to give a mathematical formulation of the phenomenon introduced above we use a *hierarchically structured* spatial population model that is a special case of a general class of models which was introduced by Dawson and Greven [DG99]. In particular we impose a hierarchical structure both on the type space and on the geographic space which we will explain is quite natural. The analysis is carried out in the framework of a new type of *multiple space-time scale analysis* of probability measure-valued diffusions. The time evolution of this model is studied from a macroscopic point of view in a whole collection of space-time scales which leads to a whole hierarchy of limit dynamics.

The restriction of our method is that we assume that the different time and space scales *separate* in the limit of large times and large space scales which is of course only an approximation but allows to make precise the notion of a quasi-equilibrium as equilibrium of a limiting non-linear McKean–Vlasov dynamic. The separation of the time scales is achieved by using the socalled *hierarchical mean-field limit* that was introduced in [DG99] and amounts to letting $N \to \infty$, where N is a parameter regulating the various rates and determining the geographic space here denoted by Ω_N. However the techniques allow us to define the concept of a *quasi-equilibrium* in a mathematically rigorous fashion as equilibria of certain limiting dynamics in the various time scales.

Indeed using this technique of taking $N \to \infty$ we can produce limiting objects, which have a simple and fairly explicit description. However due to the interaction between the migration, selection, mutation and resampling mechanisms, we have to develop some new ideas to adapt the method of multi-scale analysis to resolve some delicate conceptional and mathematical problems that arise in this case. (In previous work [DG99] the qualitative analysis of the longtime behaviour did not include mutation.) For that purpose we need *new forms of duality* (suitable to treat *multitype* selection and *state-dependent* mutation). The duality we develop allows to calculate the probability that a finite sample of the population has specific genealogical distances and types. The analysis of the dual process is based on the fact that it has the structure of a two-level spatial branching process which allows us to bring the theory of generalized branching processes into play, namely, Crump–Mode–Jagers processes and processes with catastrophes. This we expand on in Sections 5, 8 and 9.

A further point is that we give a rigorously treated example of two-scale phenomena and develop as mathematical tool the notion of random nonlinear evolutions, more specifically solutions to *random McKean–Vlasov entrance laws* from time $-\infty$. The technique of random entrance laws allows to connect the *separating time scales*.

In order to describe the early occurrence of rare mutants we work with a description of the sparse set of colonies conquered by the rare mutants, which we call droplets, via *atomic measure-valued processes* which in the limit $N \to \infty$ can be described by stochastic processes driven by (inhomogeneous) Poisson random measures and in order to calculate the intensity measure we are using excursion theory.

In particular, in order to understand the transition between the limiting objects that arise in the multi-scale analysis and describe the resulting mutation-selection quasi-equilibria, we need to introduce different classes of time scales and the corresponding natural space scales. These are introduced to capture respectively *five phases* of the transition of types to the next higher level of fitness: the process in quasi-equilibrium on a certain level of fitness, the emergence of rare (fitter) mutants at sparse sites (droplet formation), the subsequent invasion of other colonies and the build-up of a positive intensity of fitter types in space and fixation on these new types, a subsequent very long neutral evolution between the new types and finally the establishment of a mutation-selection quasi-equilibrium on the higher level.

The difficulty is to identify the intermediate time scales of droplet formation, emergence and fixation to handle the five phases. However the combination of these four classes of time scales allows us to describe the combined effect of mutations, migration and selection in the hierarchical mean-field limit in a rigorous and quite transparent mathematical fashion.

The key tool to carry out the multi-scale analysis are various specifically tuned *dualities* in the spirit of [KN], but incorporating multitype selection and state-dependent mutation. For this purpose we develop a *new type* of dual process, namely a *function-valued* dual process which can be turned into a *set-valued* dual process, if we have finitely many possible types. Occasionally we also use a Feynman–Kac weight to determine the distributions of the interacting diffusions and some related nonlinear Markov processes arising as hierarchical mean-field limits.

Finally, in order to consider questions involving the genealogy, for each individual alive at time t we record not only its type and location but also its path of descent. This means that we consider the so-called *"historical process"* associated with an enrichment of our model which provides the statistics of ancestral paths of the individuals alive at time t. Questions concerning the genealogy of mutant populations, such as how many mutants gave rise to the current population, can then be answered by considering simple functionals of this process, namely the so-called *reduced historical process* which still allows to decompose the population in different families. The historical process also helps to interpret the duality as the construction of the history of a tagged finite subpopulation. However we avoid here the full development of the historical process which requires quite some technical effort. (See P. Seidel [Seidel]).

1.3 Future Applications of Methods and Open Problems

The treatment of the combined effects of selection, mutation and resampling raises many other challenging mathematical problems which we cannot develop in this monograph but which have to be taken up in the future. However the methods which are developed here can be applied to a number of these problems. We briefly mention a few possible directions of research and some potential applications.

There are two main directions for further research, on the one hand to focus on the *measure-valued process* and spatially structured populations and on the other hand on the *genealogical process* which describes the genealogical relations in the population in terms of the genealogical tree.

For example, in the range of the measure-valued process a whole new realm of questions arises if we consider mutation mechanism on the various fitness levels which are not ergodic and which lead to strong dependencies over large stretches of time introducing randomness into even the large time-space scale behaviour.

Some open problems on the measure-valued process

To illustrate we mention a few open problems.

- It is an open problem to explore the emergence and fixation in more complex spatial geometries including *euclidean lattices*, in particular $\mathbb{Z}^2$ and geographic spaces which carry *random environments* for the evolution such that fitness and mutation rates can vary with the geographic location.
- It is an open problem to explore the coevolution of interacting quasi-species and the emergence and co-evolution of new multispecies cooperative networks.

Some perspectives to study genealogical processes

One can model the genealogical process as a marked ultrametric measure space-valued process (see [GPWmetric09], [GPWmp12], [DGP11]). Then it is possible to ask detailed questions about the genealogical distances in the rare mutant families as they emerge, fixate, move neutrally or move into the next level fitness selection and mutation quasi-equilibrium. This allows to obtain insight into the finite structure of the family and subfamily structure of the current population and the connection of these structures with geographic space which can be explored in the constructed framework.

The methodology developed here carries over to the genealogical process as is shown for nonspatial models in [DGP]. The task for the genealogies is then to characterize the scaling limits of the genealogical process with martingale problems and to identify the key qualitative properties of these objects. In particular we can then get into focus the interplay between genealogical and geographical distance, the geographical structure of subfamilies defined via the genealogical distance in the different regimes, emergence, fixation, neutral stage and reequilibration on the higher level and the formation of clusters of closely related families which were sofar been studied on the level of type configurations (see [DGV], [DG96], [GHKK], [LS]).

Some open problems on the parameter range

Some open problems remain in certain parameter ranges for the relative intensity of mutation versus selection events for the models we study here. We formulate conjectures and potential concrete approaches in the proof section of this monograph.

The formulas we derive for emergence and fixation raises the question how the time to emergence depends qualitatively on the relative strength of the parameters describing migration, resampling versus mutation and selection. Due to the nonlinearity due to selection this is a complicated matter but the provided representations of the quantities by Crump–Mode–Jagers processes provide a starting point for such studies.

1.4 Methodologies Developed

The following new mathematical tools are developed in the following parts of the monograph.

1. *Duality* has been a major tool to study the dynamics of stochastically evolving populations, we review this at the beginning of Section 5. We introduce here some new dualities:

   - a positive function-valued dual replacing a signed-Feynman–Kac dual appearing earlier in the literature [D], and in a spatial context [DG99] which works for spatial models with selection and mutation. (Subsections 5.1–5.4 and 5.6),
   - a dual taking values in sums of products of indicator functions in type space which is particularly powerful for finitely many types and in deriving ergodic theorems on the longtime behaviour (Subsection 5.5),
   - a dual which is marked tableau-valued with indicator functions as entries and locations as marks, which is also very suitable studying convergence of rare mutants (Subsection 9.4),
   - a set-valued dual, again useful in studying longtime behaviour for multitype systems (Section 9.5),
   - enriched set-valued duals for carrying out our program (Section 9.6).

2. Based on the duality we develop the method to study the evolving population by means of a two-level dynamic for the dual process:

   - multitype (finite and countable) *Crump–Mode–Jagers* generalized branching process,
   - the internal dynamics of the individuals types follow a *birth–death-catastrophe dynamic*.

   This makes it possible to make use of a lot of techniques developed for branching processes, here for Fleming–Viot processes with migration-mutation-selection (subsections 9.7–9.11).

3. We develop using excursion theory and Poisson point process-driven stochastic equations as a tool to describe and analyse droplets of rare mutants (single and multitype) in forms of *atomic measure-valued processes* (Sections 8, 9, 10)
4. We develop the technique of random entrance laws for *nonlinear Markov processes* to give asymptotic descriptions of multiple scale effects (Section 8).

1.5 Outline of the Monograph and Different Ways to Read It

This monograph develops a program to study features of evolution under selection and mutation and spatial migration which we elaborate in Section 4. To carry out this program we need to develop some methods which are of broader interest which we list below together with a description of different ways yo read this monograph. To get things started, in Section 2 we give the reader an appetizing part of the theory in a simple case.

In Section 3 we shall describe the model and recall the basic uniqueness and existence results. In particular we explain the hierarchical model of mutation and selection which we use and then we explain the biological interpretation. In Section 4 we give the main results, namely in 4.1–4.5 we introduce the multiple-space time scale analysis and state the three basic rescaling theorems. We use the results to address the central questions in Subsection 4.6, to specify the questions about the combined effect of migration, mutation and selection (stasis, punctuated equilibrium, precise asymptotic description of the five phases, etc.) and in 4.7 we address the genealogical picture and the dichotomy of biodiversity versus local monotype states.

In Subsection 4.8, after the reader has gained some familiarity with the main claims, we provide some more information on the remaining sections and guidance to the reader on what can be read independently.

In Sections 5 and 6 we prepare some mathematical methods needed for the analysis. In particular Section 5 and then Subsections 9.4 and 9.5 describe the new form of the duality theory for the processes involved which is of independent interest and has other potential applications while subsection 9.6 develops a duality theory mainly relevant for studying the invasion of rare mutants. In Section 6 and Subsection 10.3 we derive using the duality tools ergodic theorems which were previously unknown for some basic spatial and non-spatial models which arise in evolutionary theory.

The Sections 7–15 contain tools and the proofs. In particular the crucial Sections 7–12 are devoted to the analysis of a N-site mean-field model which is then used in Section 13 to study finite hierarchical N^k-site models for $k = 2, 3, \cdots$ and finally $k = \infty$ by finite approximation. The Section 15 then puts everything together for the final proof. The proofs are structured in such a manner that certain restricted cases are studied first (for example 2-type case, M-lower fitness types, one higher fitness case, general (M, M)-type case) which are self-contained (Sections 8, 9, 10).

This monograph could be used in various ways (besides a top to bottom reading)

- A reader interested in the feature of successive invasions of rare mutants with selection and migration, but not in deriving proofs can read Sections 2, 3 and 4.
- A reader interested in the new duality methods in order to apply them to other applications may just read Sections 2, 3, 5 and Subsections 9.4 and 9.5, possibly 9.6.
- A reader wanting to see the key features and some key ideas of proofs may read Sections 2, 3, 4 and 8 which are self-contained.
- A reader interested in studying the longtime behaviour of stochastic populations models may focus on Sections 3, 5, 6 and part of Subsections 10.3 and 10.6.

Chapter 2
Mean-Field Emergence and Fixation of Rare Mutants in the Fisher–Wright Model with Two Types

The systematic analysis of the model is rather complex and we therefore build up the theory in levels, in order to make the main theme more visible. The present section is the overture of the whole symphony.

2.1 Prelude: The Three Layers of Our Approach

Before beginning the detailed formulation of a first simple model, we first rephrase the main themes and relate it to the structure of the monograph in order to unravel the steps we take to reduce the complexity.

The main theme is to explain the role of geographic space and migration in the evolution of a population through a chain of levels with increasing fitness through long quiet periods and short time intervals of takeover by fitter types. Furthermore we give a complete description of the five phases in which the system evolves from the emergence of new fitter types to the next quasi-equilibrium.

The basic mathematical object in modelling this is a multitype population distributed among finitely or countably many geographic sites. At each site the population undergoes mutation, selection and resampling given by Fisher–Wright dynamics. In addition there is migration between sites. We now consider for the selection a grouping of the types into levels of fitness and with rare mutation only to reach a higher level of fitness with low rates of deleterious mutations. Initially we are in a state where the population has low fitness. The idea is now that starting from a rare mutation in space a small droplet of fitter types develops, which spreads by migration and its greater fitness to eventually conquer a substantial area and the population fixates in this higher level types till even rarer mutations with slight fitness advantage play the same game on a larger time and space scale.

The objective of the monograph is to develop mathematical tools which can be used to describe the behaviour of such population systems in a whole hierarchy of space and time scales. In each space and time scale the system can be described by a quasi-equilibrium, a concept we make precise in this monograph and which corresponds to a mutation-selection equilibrium on a fixed set of types. The transition from one quasi-equilibrium to the next is a result of the introduction and emergence of advantageous new types initiated by rare mutations. We will describe in detail below these transitions in a hierarchy of space and time scales at *three levels* of description in order of increasing complexity of the methods involved.

The tools and ideas we develop or extend for the analysis include the

- introduction of a new class of *dual processes*,
- *Crump–Mode–Jagers branching* processes,
- the McKean–Vlasov limit with *random entrance laws* and
- a class of *stochastic equations* which describe the growth of rare mutant subpopulations ("droplets") as measure-valued processes.

The monograph will start with a simple scenario and level of description, in particular a more special model, to illustrate the idea on a mathematical level (Subsection 2.1) to then continue with the description of the general model and the statement of the results in their full complexity (Section 3). The proofs and development of the methods will follow the logic of treating subsequently more complex scenarios, which we explain now in more detail.

(i) In the first level of description (Subsections 2.1 and 2.2 and Section 8) we introduce the key tools in a simple setting. We begin with a population of individuals of two types $\{1, 2\}$ distributed among N demes (islands, sites) and with exchangeable diffusions with mutation and selection. We begin the population with only type 1 individuals present but allow rare mutation to an advantageous mutant type 2. We then introduce the tools listed above in this simple setting and use them to establish the emergence of type 2 in times of order $O(\log N)$ as $N \to \infty$. More precisely, we introduce the notions of microscopic emergence (small isolated subpopulations) and macroscopic emergence (presence of the mutant with positive density) and the relation between them. This analysis demonstrates the essential role of migration in the emergence since without migration the emergence occurs only in times of order $O(N)$.

From this situation we get the ideas how to formulate the general model and the picture on the hierarchical in Sections 3 and 4.

(ii) In the second level of description (Sections 9 and 10) we develop tools needed to extend these ideas to more complex populations involving many types on the lower and subsequently also upper level of fitness and in which deleterious mutations can occur leading to mutation-selection equilibria but still restricted to one spatial scale. The methodology used here involves a more complex dual process, namely a set-valued dual which allows us to compute all joint moments for multitype mutation-selection systems and to prove the ergodic theorem. In identifying the emergence rate and structure of the mutant population we now must deal with a more complex class of supercritical CMJ branching processes with countably many types and the formation mutant droplets of different types. We carry this out in two steps, first with multiple types at the lower level, Section 9, and a single advantageous type and then in Section 10 with a finite number of advantageous types.

(iii) The final level of description (Sections 11–14 is now to bring in the full impact of the combined effects of *mutation, selection, resampling* (genetic drift) and *migration* in a hierarchy of scales. The transition from one scale to the next is described in a series of *five phases* (described in detail in Subsubsection 4.2.1). This involves the hierarchical mean-field limit developed in earlier papers and the description of biodiversity and genealogy of the mutant population using the historical process. In particular we will establish that in long space and time scale the mutant population which appears locally is with high probability comprised of the descendants of successful mutant droplets which arose at some distant point in space and which spread throughout space by migration.

An important tool used throughout the monograph is the duality relation. This is first formulated in a general setting in Section 5 and then dual calculations are systematically developed at the first level of description in Subsection 9.4. The dual calculations at the second level of description are developed in two steps, first in Section 9 for the case of one advantageous type and then in Section 10 for the case of a finite number of advantageous types. Similarly the type of Crump–Mode–Jagers processes or generalizations of it are developed in these three stages. The needed McKean–Vlasov evolutions and random entrance laws are also exhibiting these three stages of complexity, from a $[0, 1]$-valued object to a multi-dimensional one passing from one to many types on the higher level (Section 10) and similarly the droplet changing from a measure to a tuple of measures (Section 10).

Developing the material in these three stages also plays a useful role in the developing the proofs of the results. For example, some of the proofs in the case $M \geq 2$ follow the same basic strategy as in the case $M = 1$ given in Section 8. In these cases this allows us to give more concise proofs in Section 9 by focussing on what is new in the more complex case. Breaking down the proofs into these stages makes them more accessible.

2.2 Formulation of the Problem and Description of Two-Type Model

We begin by considering a population of two types $\{1, 2\}$ distributed among N exchangeable colonies (also called *demes*). In Subsubsection 2.2.1 we formulate the problem of the emergence of a rare mutant in this simplified case and outline the rest of Section 2 which contains the statements of the main results in this case. Then Section 8 develops the basic tools for systems with two types and contains the proofs of the results stated in this section. Section 8 is essentially self-contained referring only to Section 5 on duality.

2.2.1 The Problem

We consider the system of N exchangeable colonies interacting by migration according to the uniform distribution (as in Wright's island model). We assume that the two types $\{1, 2\}$ satisfy:

- type 1 has fitness 0, type 2 has fitness 1 and the selection rate satisfies $s > 0$,
- the rate of mutation of type 1 to type 2 is mN^{-1}, no mutation from type 2 to 1,
- the migration rate is $c > 0$.

Let $x_2^N(i, t)$ denote the proportion of type 2 at site i, at time t and let $x_2(i, 0) = 0$ for all i. We use the notation

$$X^N(t) = \{(x_1^N(i,t), x_2^N(i,t)), i = 1, \cdots, N\}, \tag{2.1}$$

for the collection of masses of the two types at the various sites. (Note that $x_1^N(i,t) + x_2^N(i,t) = 1$.) The question is now how the type 2 *emerges* due to rare mutation and then *spreads* through space due to selection and migration.

Outline of Sections 2 and 8. We continue in the next subsubsection with writing down the simplified models precisely in terms of a system of stochastic differential equations and then the functionals used to characterize the behaviour of that process. Subsection 2.3 states the results for the two-type spatial model which are proved in Subsections 8.3 up to 8.7. In Subsection 8.8 we discuss possible extensions of the analysis to different parameter regimes. In Section 8.2 we consider the model without migration and a certain caricature to introduce some of the principal methods for the proofs of the results and also in order to develop some intuition. This gives a flavour of the techniques whose more elaborate versions involving the spatial structure are used later on in this section.

2.2.2 The Fisher–Wright Dynamics with Two Types and Basic Objects

We proceed in two steps, first giving the simplified equations and second specifying the functionals used to study the process.

Step 1 *(The two-type process)*

We next continue the analysis giving the formal set-up of a system of N-exchangeably interacting finite populations ($d > 0$) with two types $\{1, 2\}$ with fitness 0 and 1 respectively, selection rate $s > 0$ and migration rate $c > 0$.

2.2 Formulation of the Problem and Description of Two-Type Model

This means that we study the stochastic process $(X^N(t))_{t \geq 0}$, which is uniquely defined as follows:

$$X^N(t) = ((x_1^N(i,t), x_2^N(i,t)); i = 1, \ldots, N), \tag{2.2}$$

$$x_1^N(i,0) = 1, \quad i = 1, \cdots, N, \tag{2.3}$$

satisfying the wellknown SSDE:

$$dx_1^N(i,t) = c(\bar{x}_1^N(t) - x_1^N(i,t))dt - s\, x_1^N(i,t)x_2^N(i,t)dt - \frac{m}{N}x_1^N(i,t)dt$$
$$+ \sqrt{d \cdot x_1^N(i,t)x_2^N(i,t)}\, dw_1(i,t), \tag{2.4}$$

$$dx_2^N(i,t) = c(\bar{x}_2^N(t) - x_2^N(i,t))dt + s\, x_1^N(i,t)x_2^N(i,t)dt + \frac{m}{N}x_1^N(i,t)dt$$
$$+ \sqrt{d \cdot x_1^N(i,t)x_2^N(i,t)}\, dw_2(i,t), \tag{2.5}$$

where $\{(w_1(i,t))_{t \geq 0}, i = 1, \ldots, N\}$ are i.i.d. Brownian motions and w_2, w_1 are coupled via $w_2(i,t) = -w_1(i,t)$

$$\hat{x}_\ell(t) = \sum_{i=1}^{N} x_\ell^N(i,t) \text{ with } \ell = 1, 2 \tag{2.6}$$

and

$$\bar{x}_\ell(t) = \frac{1}{N}\hat{x}_\ell(t) \text{ with } \ell = 1, 2. \tag{2.7}$$

Without loss of generality we can assume for the resampling rate d that $d = 1$ and will do this except where we wish to indicate the roles of the parameters c, s, d, m. The simple collection of N one-dimensional stochastic differential equations in (2.2)–(2.5) allows us to focus first on the key features of emergence, which are already complicated enough as we shall see.

Step 2 *(Key functionals and main objectives)*

The description of the system proceeds via four objects, namely either locally by a *sample of tagged sites* or globally by the *empirical measure* over the whole spatial collection and furthermore we use as well an *atomic measure* to globally describe sparse (in space) spots of type two mass respectively the *Palm measure* to describe this locally by formalizing the concept of a typical type-2 site. The asymptotic analysis of these four objects leads to propagation of chaos results, respectively, nonlinear McKean–Vlasov limiting dynamics and atomic measure-valued processes driven by Poisson processes and formulas for Palm measures based on this.

Since it often suffices to consider the tagged site 1, i.e. $x_\ell^N(1,t)$, in the sequel to designate a tagged site we set

$$x_\ell^N(t) = x_\ell^N(1,t) \quad ; \ell = 1,2. \tag{2.8}$$

If we are interested in the *local* picture we consider for some $L \in \mathbb{N}$ the tagged sites

$$(x_\ell^N(1,t), \cdots, x_\ell^N(L,t)), \quad \ell = 1,2. \tag{2.9}$$

We abbreviate the marginal law of the type-2 mass as

$$\mu_t^N := \mathcal{L}[x_2^N(1,t)] \in \mathcal{P}([0,1]). \tag{2.10}$$

The empirical measure process gives a *global* picture and is defined by:

$$\Xi_N(t) := \frac{1}{N} \sum_{i=1}^N \delta_{(x_1^N(i,t), x_2^N(i,t))} \in \mathcal{P}(\mathcal{P}(\{1,2\})) \tag{2.11}$$

and the empirical measure process of either type is defined by

$$\Xi_N(t,\ell) := \frac{1}{N} \sum_{i=1}^N \delta_{x_\ell^N(i,t)} \in \mathcal{P}([0,1]), \, \ell = 1,2. \tag{2.12}$$

Note that since $x_1^N(i,t) + x_2^N(i,t) = 1$ it suffices in the case of two types to know one component of the pair in (2.12), the other is then determined by this condition. Then we can for two types effectively replace $\mathcal{P}(\mathcal{P}(\{1,2\}))$ by $\mathcal{P}([0,1])$ as states of the empirical measure.

Next we have to describe the sparse sites where type 2 appears. First consider N sites starting only with type 1 mass at time 0 and evolving in the time interval $[0,T_0]$. As a result of the rare mutation from type 1 to type 2 at rate $\frac{m}{N}$ mutant mass will appear and then form non-negligible colonies at a sparse set of the total of N sites, but the typical site will have mass of order $o(1)$ as $N \to \infty$. We can now again take a *global* perspective, where we describe the sparse set of colonised (by type 2) sites or a *local* perspective, where we consider a typical colonised site via the concept of the Palm measure. We introduce both viewpoints successively.

Turn to the global perspective. In order to keep track of the sparse set of sites at which nontrivial mass appears we will give a random label to each site and define the following *atomic-measure-valued process*.

We assign independent of the process a point $a(j)$ randomly in $[0,1]$ to each site $j \in \{1, \ldots, N\}$, that is, we define the collection

$$\{a(j), \quad j = 1, \cdots, N\} \text{ are i.i.d. uniform on } [0,1]. \tag{2.13}$$

2.2 Formulation of the Problem and Description of Two-Type Model

We then associate with our process and a realization of the random labels a measure-valued process on $\mathcal{P}([0, 1])$, which we denote by

$$\mathbf{J}_t^{N,m} = \sum_{j=1}^{N} x_2^N(j,t)\delta_{a(j)}. \tag{2.14}$$

This description by $\mathbf{J}_t^{N,m}$ is complemented by a local perspective, arising by zooming in on the random location where we actually find mass by studying a *typical* type-2 dominated site. In order to make precise the notion of a site as seen by a typical type-2 mass, i.e. a site seen from a randomly (among the total population in all N sites) chosen individual of type 2, we use the concept of the *Palm-distribution*, which we indicate by a hat. Define

$$\hat{\mu}_t^N = \hat{\mathcal{L}}[x_2^N(1,t)], \text{ by } \hat{\mu}_t^N(A) = \left(\int_A x\mu_t^N(dx)\right)/E[x_2^N(1,t)], \tag{2.15}$$

where $\mu_t^N = \mathcal{L}[x_2^N(1,t)]$. The law $\hat{\mu}_t^N$ describes now for an i.i.d. initial state the law of a *site, typical* for the type-2 population at time t in the limit $N \to \infty$. Namely we can think of this as picking a type-2 individual at random and then determine the site where it sits. Note that

$$\int x\hat{\mu}_t^N(dx) = \frac{E[(x_2^N(1,t))^2]}{E[x_2^N(1,t)]}. \tag{2.16}$$

The basic objective of this section is to describe the emergence of the more fit type, i.e. type 2, using the above functionals of the process. This emergence occurs in *three regimes*, two are of short duration separated by a third of long duration and arise as follows in an asymptotic description as $N \to \infty$.

Three time regimes in the emergence

1. In the first regime during times of order $O(1)$ rare mutants arise at a small set of sites (droplet) where they locally make up substantial part of the population and then develop mutant "droplets" by migration. We will see later in Subsection 8.4 in all detail that the mutant population during this stage develops as a type of measure-valued branching process. We refer to this stage as *droplet formation*.
2. The next stage, namely, the stage at which type 2 has $O(1)$ frequency at a typical (i.e. randomly chosen site) site, or equivalently at the macroscopic level has $O(1)$ density, occurs at time $O(\log N)$ and is our main interest here in this subsection and we refer to this as *emergence*.
3. The last regime at which the macroscopic density of type 2 reaches a frequency arbitrarily close to 1 is taking a further piece of time but is again only of order $O(1)$ and is referred to as *fixation*.

In the sequel we will analyse these three regimes by studying the behaviour as $N \to \infty$ of

$$\Xi_N(t), \quad \{x_\ell^N(i,t); \ell = 1, 2; i = 1, \cdots, L\} \text{ and } \mathfrak{z}_t^{N,m}, \quad \hat{\mu}_t^N, \qquad (2.17)$$

in finite times as well as in larger N-dependent time scales thus capturing each of the three regimes described above.

2.3 Statement of Results

We have three tasks, first to determine the *macroscopic emergence time scale*, that is the time scale at which the level two type appears at a typical site or equivalently has positive intensity, second, to describe the pre-emergence picture, in particular the *droplet formation* respectively the *limiting droplet dynamics* and thirdly to describe the *limiting dynamic of fixation* after the emergence with which the take-over by type 2 then occurs.

We begin the analysis with introducing the two different limit dynamics in the subsequent two subsubsections and then subsequently the emergence, fixation and convergence results are stated in a further subsubsection. We close in a fourth subsubsection with a discussion of the branching approximation for limiting values of the parameters.

2.3.1 The Limiting Random McKean–Vlasov Dynamics and Entrance Laws

We start with the initial condition $\Xi_N(0, \{1\}) = \delta_1$ (recall (2.12)), that is only type 1 appears. The objective is to establish the emergence and fixation of type 2 in times of the form $t_N = C \cdot \log N + t$, to identify the constant C, and to identify the limit of $\Xi_N(t_N, \cdot)$ as a function of t in the process of emergence and determine the dynamics in t of the fixation (takeover) which leads to the concentration of the complete population in type-2 as $N \to \infty$. In the next subsubsection we shall then discuss the behaviour at early times, i.e. times t_N with $\limsup((\log t_N)/\log N) < C$, in the stage of droplet formation.

The key ingredients in the limiting processes in times t_N for the local and global description are:

- the limiting *(nonlinear) McKean–Vlasov dynamics*,
- the *entrance law* from $-\infty$ of the McKean–Vlasov dynamic and
- *random* solutions to the McKean–Vlasov equation.

In order to define these ingredients we proceed in three steps, we recall first in Step 1 the "classical" McKean–Vlasov limit (associated with a nonlinear Markov process) before in Step 2 and Step 3 we introduce the two new objects.

2.3 Statement of Results

Step 1 Consider the above system (2.2)–(2.5) of N interacting sites with type space $\mathbb{K}$ starting at time $t = 0$ from a product measure (that is, i.i.d. initial values at the N sites). The basic McKean–Vlasov limit (cf. [DG99], Theorem 9) says that if we start initially in an i.i.d. distribution, then

$$\{\Xi_N(t)\}_{0 \leq t \leq T} \underset{N \to \infty}{\Longrightarrow} \{\mathcal{L}_t\}_{0 \leq t \leq T}, \qquad (2.18)$$

where the $\mathcal{P}(\mathcal{P}(\mathbb{K}))$-valued path $\{\mathcal{L}_t\}_{0 \leq t \leq T}$ is the law of a *nonlinear* Markov process, namely the unique weak solution of the *McKean–Vlasov equation*:

$$\frac{d\mathcal{L}_t}{dt} = (L_t^{\mathcal{L}_t})^* \mathcal{L}_t, \qquad (2.19)$$

where for $\pi \in \mathcal{P}(\mathcal{P}(\mathbb{K}))$, L^π is given by (4.28) and the $*$ indicates the adjoint of an operator mapping from a dense subspace of $C_b(E, \mathbb{R})$ into $C_b(E, \mathbb{R})$ w.r.t. the pairing of $\mathcal{P}(E)$ and $C_b(E, \mathbb{R})$ given by the integral of the function with respect to the measure. Similar equations have been studied extensively in the literature (e.g. [Gar]). The process $(\mathcal{L}_t)_{t \geq 0}$ corresponds to a nonlinear Markov process since $\mathcal{L}_t$ appears also in the expression for the generator L.

As pointed out above in (2.12), in the special case $\mathbb{K} = \{1, 2\}$, we can simplify by considering the frequency of type 2 only and by reformulating (2.19) living on $\mathcal{P}(\mathcal{P}(\{1, 2, \}))$ in terms of $\mathcal{L}_t(2) \in \mathcal{P}[0, 1]$. This we carry out now.

Namely we note that given the mean-curve

$$m(t) = \int_{[0,1]} y \, \mathcal{L}_t(2)(dy), \qquad (2.20)$$

the process $(\mathcal{L}_t(2))_{t \geq 0}$ is the *law* of the strong solution of (i.e. the unique weak solution) the SDE:

$$dy(t) = c(m(t) - y(t))dt + sy(t)(1 - y(t))dt + \sqrt{dy(t)(1 - y(t))} dw(t). \qquad (2.21)$$

Then informally $(\mathcal{L}_t)_{t \geq 0}$ corresponds to the solution of the nonlinear diffusion equation. Namely for $t > 0$, $\mathcal{L}_t(2)(\cdot)$ is absolutely continuous and for

$$\mathcal{L}_t(2)(dx) = u(t, x)dx \in \mathcal{P}([0, 1]) \qquad (2.22)$$

the evolution equation of the density $u(t, \cdot)$ is given by:

$$\frac{\partial}{\partial t} u(t, x) = -c \frac{\partial}{\partial x} \{[\int_{[0,1]} yu(t, y)dy - x]u(t, x)\}$$

$$- s \frac{\partial}{\partial x}(x(1-x)u(t, x)) + \frac{d}{2} \frac{\partial^2}{\partial x^2}(x(1-x)u(t, x)). \qquad (2.23)$$

Step 2 In order to describe the emergence and invasion process via Ξ_N, we introduce in this step and the next two extensions of the nonlinear McKean–Vlasov dynamics which describes only the limiting evolution over finite time stretches, $t \in [t_0, t_0 + T]$ given an initial condition $\mathcal{L}_{t_0}$ which we then follow as $t_0 \to -\infty$. Hence we have to define the dynamics for $t \in (-\infty, \infty)$ in terms of an *entrance law* at $-\infty$.

Since we consider the limits of systems observed in the interval $const \cdot \log N + [-\frac{T}{2}, \frac{T}{2}]$ with T any positive number, that is setting $t_0(N) = const \cdot \log N - \frac{T}{2}$, we need to identify entrance laws for the process from $-\infty$ (by considering $T \to \infty$) out of the state concentrated on type 1 with certain properties.

Definition 2.1 (Entrance law from $t = -\infty$).
We say in the two-type case that a probability measure-valued function $\mathcal{L} : \mathbb{R} \to \mathcal{P}([0, 1])$, is an entrance law at $-\infty$ starting from type 1 if $(\mathcal{L}_t)_{t \in (-\infty, \infty)}$ is such that $\mathcal{L}_t$ solves the McKean–Vlasov equation (2.19) and $\mathcal{L}_t \to \delta_1$ as $t \to -\infty$.

In the case of more than two types we work with maps $\mathcal{L} : \mathbb{R} \to \mathcal{P}(\mathcal{P}(\mathbb{K}))$ where we require as $t \to -\infty$ that $\mathcal{L}_t$ converges to measures δ_μ with $\mu \in \mathcal{P}(\mathbb{K})$ such that μ is a measure concentrated on the lower level types. □

The existence of such an object is obtained in part (c) of the proposition below.

Step 3 The usual formulation of the McKean–Vlasov limit requires that we start in an i.i.d. initial configuration. This is not sufficient for us since even though we consider systems observed in finite time stretches we do so only *after large times*, namely, after the time $const \cdot \log N$ with $N \to \infty$. Hence in our context the McKean–Vlasov limit is valid in a fixed time scale but if the system is viewed in time scales that depend on N this can (and will) break down since we only know that the initial state is exchangeable and this then leads to a random solution.

We will indeed establish that the emergence of rare mutants gives rise to "random" solutions of the McKean–Vlasov dynamics. In particular we will show that the limiting empirical measures at times of the form $C \log N + t$ are *random probability measures* on $[0, 1]$ and therefore given by exchangeable sequences of $[0, 1]$- valued truly exchangeable random variables which are *not* i.i.d., that is, the exchangeable σ-algebra is not trivial. This means that the empirical mean turns out to be a *random variable* and since this is the term driving via the migration the local evolution of a site in the McKean–Vlasov limit, the non-linearity of the evolution equation comes seriously into play. However once we condition on the exchangeable σ-algebra, we then get for the further evolution again a deterministic limiting equation for the empirical measures, namely the McKean–Vlasov equation. The reason for this is the fact that conditioned on the exchangeable σ-algebra we obtain on asymptotically (as $N \to \infty$) i.i.d. configuration to which the classical convergence theorem applies. Using the Feller property of the system, which is a direct consequence of the duality, we get our claim.

This leads to the task of identifying an entrance law in terms of a *random initial condition at time* $-\infty$. A consequence of this scenario is that when we use the duality from time T_N to $T_N + t$, we apply it to a random initial state in the limit and

2.3 Statement of Results

we therefore have to use because of the non-linearity of the evolution the appropriate formulas (recall Subsection 5.6).

The above discussion shows that we need to introduce the notion of a truly random McKean–Vlasov entrance law from $-\infty$.

Definition 2.2 (Random solution of McKean–Vlasov).

We say that the probability measure-valued process $\{\mathcal{L}(t)\}_{t \in \mathbb{R}}$ is a *random solution of the McKean–Vlasov equation* (2.19) if

- $\{\mathcal{L}_t : t \in \mathbb{R}\}$ is a.s. a solution to (2.19), that is, for every t_0 the distribution of $\{\mathcal{L}_t : t \geq t_0\}$ conditioned on $\mathcal{F}_{t_0} = \sigma\{\mathcal{L}_s : s \leq t_0\}$ is given by $\delta_{\{\mu_t\}_{t \geq t_0}}$ where μ_t is a solution of the McKean–Vlasov equation with $\mu_{t_0} = \mathcal{L}_{t_0}$,
- the time t marginal distributions of $\{\mathcal{L}_t : t \in \mathbb{R}\}$ are truly random. □

The key result of this subsubsection on the objects introduced in the previous three steps is now the following existence and uniqueness results on the solution of the McKean–Vlasov equation (2.19) and of its (random) entrance laws from $t = -\infty$:

Proposition 2.3 (McKean–Vlasov entrance law from $-\infty$).

(a) Given the initial state $\mu_0 \in \mathcal{P}([0, 1])$ there exists a unique solution

$$\mathcal{L}_t(2)(dx) = \mu_t(dx), \quad t \geq t_0 \tag{2.24}$$

to (2.19) with initial condition $\mathcal{L}_{t_0}(2) = \mu_0$.

(b) If $s > 0$ and $\int_{[0,1]} x \mu_{t_0}(dx) > 0$, then this solution satisfies:

$$\lim_{t \to \infty} \mathcal{L}_t(2)(dx) = \delta_1(dx). \tag{2.25}$$

(c) There exists a solution $(\mathcal{L}_t^{**}(2))_{t \in \mathbb{R}}$ to equation (2.19) satisfying the conditions:

$$\lim_{t \to -\infty} \mathcal{L}_t^{**}(2) = \delta_0, \tag{2.26}$$

$$\lim_{t \to \infty} \mathcal{L}_t^{**}(2) = \delta_1$$

$$\int_{[0,1]} x \mathcal{L}_0^{**}(2, dx) = \frac{1}{2}.$$

This solution is called an entrance law from $-\infty$ with mean $\frac{1}{2}$ at $t = 0$.

(d) We can obtain a solution in (c) such that:

$$\exists\, \alpha \in (0, s) \text{ and } A_0 \in (0, \infty) \text{ such that } \lim_{t \to -\infty} e^{\alpha |t|} \int_{[0,1]} x \mathcal{L}_t^{**}(2, dx) = A_0. \tag{2.27}$$

(e) The solution of (2.19) also satisfying (2.27) for prescribed A_0 is unique and if $A_0 \in (0, \infty)$ then α is necessarily uniquely determined.

For any deterministic solution

$$\mathcal{L}_t, \ t \in \mathbb{R} \tag{2.28}$$

to (2.19) with

$$0 \leq \limsup_{t \to -\infty} e^{\alpha|t|} \int_{[0,1]} x \mathcal{L}_t(2, dx) < \infty, \tag{2.29}$$

the limit $A = \lim_{t \to -\infty} e^{\alpha|t|} \int_{[0,1]} x \mathcal{L}_t(2, dx)$ exists.

If $A > 0$, then $\{\mathcal{L}_t, t \in \mathbb{R}\}$ is given by a time shift of the then unique $\{\mathcal{L}_t^{**}, t \in \mathbb{R}\}$ singled out in (2.27), i.e.

$$\mathcal{L}_t = \mathcal{L}_{t+\tau}^{**}, \quad \tau = \alpha^{-1} \log \frac{A}{A_0}. \tag{2.30}$$

For future reference we define $(\mathcal{L}_t^*)_{t \in \mathbb{R}}$ to be the unique solution satisfying

$$\lim_{t \to -\infty} e^{\alpha|t|} \int_{[0,1]} x \mathcal{L}_t^*(2, dx) = 1 \text{ for some } \alpha \in (0, s). \tag{2.31}$$

(f) Any random solution $(\tilde{\mathcal{L}}_t)_{t \in \mathbb{R}}$ to (2.19) such that

$$\limsup_{t \to -\infty} e^{\alpha|t|} E[\int_{[0,1]} x \mathcal{L}_t(2, dx)] < \infty, \quad \liminf_{t \to -\infty} e^{\alpha|t|} [\int_{[0,1]} x \mathcal{L}_t(2, dx)] > 0 \text{ a.s.}, \tag{2.32}$$

is a random time shift of $(\mathcal{L}_t^{**})_{t \in \mathbb{R}}$ (and of $(\mathcal{L}_t^*)_{t \in \mathbb{R}}$). □

Example 1. Let $\mathcal{L}_t^*$ be a solution satisfying (2.23), (2.27) and for a given value of A let τ be a true real-valued random variable. Then $\{\mathcal{L}_{t-\tau}^*\}_{t \in \mathbb{R}}$ is a truly random solution. This can also be viewed as saying that we have a solution with an exponential growth factor A which is truly random.

Remark 1. We shall derive later on a bound from above on the cardinality of our dual process which will imply that the first relation in (2.32) must be satisfied for a limiting dynamic (as $N \to \infty$) of our empirical measures Ξ_N and furthermore we get a lower bound implying that A is a.s. positive for this limiting dynamic, (see (8.147) and Subsubsection 8.3.6). Then we are able to use the identification of the solution given in (2.30) or of the one satisfying (2.31) to identify the limiting dynamic of the process of empirical measures.

2.3.2 The Limiting Dynamics of the Sparse Sites with Substantial Type-2 Mass

We next describe the limiting dynamic of the sparse set of sites which have been colonized by type 2 prior to the onset of emergence. There are two time regimes, first an initial finite time horizon of order $O(1)$, then large times $t_N \to \infty$ as $N \to \infty$ but which remain $\ll \alpha^{-1} \log N$ so that we are in the preemergence regime with still a global density of type 2 which is asymptotically zero.

It is very unlikely that a randomly chosen site has at time $O(1)$ mass of type 2 at least $\varepsilon > 0$ as $N \to \infty$. In fact an explicit calculation (see (8.703)) shows that the transition density decays like $const N^{-1}$. On the other hand the number of sites increases with N so that here we have a compensation provided we do not look at only one single tagged site but the complete population at all sites and the result is that considering all sites there is a *finite random number* of sites with substantial type 2 mass. Newly added sites in this set of sites arise from a process entering from the state 0. In fact we obtain here a Poisson distribution in the limit. This whole scenario will be made precise using entrance laws from state 0 which we derive using diffusion theory and excursion law theory.

Therefore before formulating the main results of this section we first consider entrance laws of a single site with no mutation from type 1 to type 2 which provides us with a key ingredient, the *excursion measure* $\mathbb{Q}$. We state the wellknown fact:

Lemma 2.4 (Single site: entrance and excursion laws).

(a) *Let $c > 0, d > 0, s > 0$. Then 0 is an exit boundary for the the Fisher–Wright diffusion*

$$dx(t) = -cx(t)dt + sx(t)(1-x(t))dt + \sqrt{d \cdot x(t)(1-x(t))}dw(t), \quad (2.33)$$

which then has a σ-finite entrance law from state 0 at time 0, the σ-finite excursion law

$$\mathbb{Q} = \mathbb{Q}^{c,d,s} \quad (2.34)$$

on

$$W_0 := \{w \in C([0,\infty), \mathbb{R}^+), \ w(0) = 0,$$
$$w(t) > 0 \text{ for } 0 < t < \zeta \text{ for some } \zeta \in (0,\infty)\}. \quad (2.35)$$

(b) *Moreover, denoting by P^ε the law of the process started with $w(0) = \varepsilon$ and $\varepsilon > 0$, $\mathbb{Q}$ is given by:*

$$\mathbb{Q}(\cdot) = \lim_{\varepsilon \to 0} \frac{P^\varepsilon(\cdot)}{S(\varepsilon)}, \quad (2.36)$$

where $S(\cdot)$ is the scale function of the diffusion (2.33), defined by the relation,

$$P_\varepsilon(T_\eta < \infty) = \frac{S(\varepsilon)}{S(\eta)}, \qquad 0 < \varepsilon < \eta < \infty, \tag{2.37}$$

where T_η is the first hitting time of η.

For the Fisher–Wright diffusion S is given by (cf. [RW], V28) the initial value problem:

$$S(0) = 0, \qquad \frac{dS}{dx} = \frac{e^{-2sx}}{(1-x)^{2c}}, \tag{2.38}$$

so that

$$\lim_{\varepsilon \to 0} \frac{S(\varepsilon)}{\varepsilon} = 1. \tag{2.39}$$

(c) The measure $\mathbb{Q}$ is σ-finite, namely for any $\eta > 0$, ζ as in (2.35),

$$\mathbb{Q}(\{w : \zeta(w) > \eta\}) < \infty, \tag{2.40}$$

$$\mathbb{Q}(\sup_t(w(t)) > \eta) = \mathbb{Q}(T_\eta < \infty) = \frac{1}{S(\eta)} \longrightarrow \infty \text{ as } \eta \to 0, \tag{2.41}$$

and

$$\int_0^1 x \mathbb{Q}(\sup_t(w(t)) \in dx) = \infty. \qquad \square \tag{2.42}$$

Proof. It is wellknown that 0 is an exit boundary for (2.33). In this case (a)–(c) then follows immediately from the results of Pitman and Yor [PY], Section 3. q.e.d.

We next identify the limit in distribution of $\mathbf{J}_t^{N,m}$ (defined in (2.14)) as $N \to \infty$. Let

$$\mathcal{M}_a([0,1]) = \text{ the set of finite atomic measures on } [0,1], \tag{2.43}$$

equipped with the weak atomic topology (see [EK4] and in this present work in (2.101)–(2.107) below for details on this topology) which forms a Polish space.

We will need Poisson random measures in four variables, the actual time called $s \in [0,t]$, the location parameter $a \in [0,1]$, the mutation and immigration potential called $u \in [0,\infty)$ and the path of an excursion called $w \in W_0$. This allows us to obtain the droplet dynamic $(\mathbf{J}_t^m)_{t \in [0,\infty)}$ and its properties:

Proposition 2.5 (A continuous atomic-measure-valued Markov process).
Let $\mathbf{J}_0^*(t) = \sum_i y_i(t)\delta_{a_i} \in \mathcal{M}_a([0,1])$ where the $y_i(t)$ are independent solutions of the SDE (2.33) (describing the initial droplet and its evolution) and let $N(ds, da, du, dw)$ be a Poisson random measure on (recall (2.35) for W_0)

2.3 Statement of Results

$$[0, \infty) \times [0, 1] \times [0, \infty) \times W_0, \tag{2.44}$$

with intensity measure

$$ds\, da\, du\, \mathbb{Q}(dw), \tag{2.45}$$

where $\mathbb{Q}$ is the single site excursion law defined in (2.36) in Lemma 2.4.
Then the following three properties hold.

(a) *The stochastic integral equation for $(\mathfrak{Z}_t^m)_{t \geq 0}$ a process with values in $\mathcal{M}_a([0,1])$ is given as*

$$\mathfrak{Z}_t^m = \mathfrak{Z}_0^*(t) + \int_0^t \int_{[0,1]} \int_0^{q(s,a)} \int_{W_0} w(t-s) \delta_a N(ds, da, du, dw), \quad t \geq 0, \tag{2.46}$$

where $\delta_a(\cdot) \in \mathcal{M}_1([0,1])$ and where $q(s,a)$ denotes the non-negative predictable function

$$q(s, a) := (m + c \mathfrak{Z}_{s-}^m([0,1])), \tag{2.47}$$

has a unique continuous $\mathcal{M}_a([0,1])$-valued solution, which we call

$$(\mathfrak{Z}_t^m)_{t \geq 0}. \tag{2.48}$$

(b) $(\mathfrak{Z}_t^m)_{t \geq 0}$ *is a $\mathcal{M}_a([0,1])$-valued strong Markov process.*

(c) *The process $(\mathfrak{Z}_t^m)_{t \geq 0}$ has the following properties:*

- *the mass of each atom observed from the time of its creation follows an excursion from zero generated from the excursion law $\mathbb{Q}$ (see (2.36)),*
- *new excursions are produced at time t at rate*

$$m + c \mathfrak{Z}_t^m([0,1]), \tag{2.49}$$

- *each new excursion produces an atom located at a point $a \in [0,1]$ chosen according to the uniform distribution on $[0,1]$,*
- *at each t for $\varepsilon > 0$ there are at most finitely many atoms of size $\geq \varepsilon$,*
- $t \to \mathfrak{Z}_t^m([0,1])$ *is a.s. continuous.* □

Remark 2. The process $(\mathfrak{Z}_t^m)_{t \geq 0}$ can be viewed as a continuous state analogue of the Crump–Mode–Jagers branching process with immigration (see Subsubsection 8.3.4, Step 2 for a review and more information on this type of processes). We shall see in (2.71) that the total mass grows exponentially as in a supercritical branching process. □

Remark 3. A necessary and sufficient condition for extinction of the analogue of $\hat{\mathbf{J}}_t^*$ for a general class of one-dimensional diffusions was obtained by M. Hutzenthaler ([Hu], [Hu2]). In the Fisher–Wright case with $c > 0$, $s > 0$ the fact that the probability of non-extinction is non-zero follows from the Proposition 2.10. □

Proof. (a) and (b) The existence and uniqueness of the solution to (10.88) and the strong Markov property follows as in [D-Li] and [FL].

(c) follows directly from the construction via the Poisson measure N given in (2.46).   q.e.d.

Remark 4. We can enrich the process $(\hat{\mathbf{J}}_t^m)_{t \geq 0}$ to include the genealogical information, namely which mass results from which mutation. For that purpose we have in addition to the location record the birth times of atoms due to migration (successful colonization). This means we have to split the rate at which excursions are created into mutation at the site and immigration from other sites. At time 0 we start with

$$\hat{\mathbf{J}}_0^*(t) \tag{2.50}$$

describing the initial atoms we consider to be present at time 0.

The genealogical enrichment denoted

$$(\hat{\mathbf{\mathfrak{J}}}_t^m)_{t \geq 0} \tag{2.51}$$

is a measure-valued process on a richer set E and is obtained as follows.

Let

$$E = ((\mathbb{R}_+ \times [0, 1]) \cup \emptyset)^\mathbb{N} \tag{2.52}$$

and for $y \in E$ define $\tau(y) = min\{n : y(n) = \emptyset\} - 1$.

Let $N(ds, da, du, dw)$ be a Poisson random measure on $[0, \infty) \times [0, 1] \times [0, \infty) \times W_0$ with intensity measure $ds\, da\, du\, \mathbb{Q}(dw)$ where $\mathbb{Q}$ is the single site excursion law. Then the following stochastic integral equation has a unique continuous solution, $\hat{\mathbf{\mathfrak{J}}}_t^m$:

$$\hat{\mathbf{\mathfrak{J}}}_t^m = \hat{\mathbf{\mathfrak{J}}}_0^*(t) + \int_0^t \int_E \int_{[0,1]} \int_0^{q(s,y,a)} \int_{W_0} w(t-s) \delta_{y \diamond (s,a)} N(ds, da, du, dw), \tag{2.53}$$

where (s, a) is short for $((s, a), \emptyset, \emptyset, \cdots)$,

$$q(s, y, a) = m \text{ if } \tau(y) = 0, \quad q(s, y, a) := c\hat{\mathbf{\mathfrak{J}}}_{s-}^m(y), \ \tau(y) \neq 0, \tag{2.54}$$

and

$$\diamond : E \times E \to E, \ y_1 \diamond y_2 = (y_1', y_2', \underline{\emptyset}) \quad \text{(concatenation) with },$$
$$y' = (y'(1), \cdots, y'(\tau(y'))) \quad , \underline{\emptyset} = (\emptyset, \emptyset, \cdots). \tag{2.55}$$

2.3 Statement of Results

Here $\hat{\mathfrak{Z}}_t^m$ is a measure-valued process on E and $y \diamond (s,a)$ denotes the offspring of y with birth time s and location $a \in [0,1]$. Moreover, $\hat{\mathfrak{Z}}_t^m(\{y : \tau(y) = \infty\}) = 0$ a.s.

2.3.3 Emergence Time, Droplet Formation, Fixation Dynamic for N-Interacting Sites: Statement of Results

We now have the ingredients and the background to continue the analysis and state all the results of the finite population model with two types $\mathbb{K} = \{1,2\}$ with fitness 0 and 1, respectively, $s > 0$ and $d > 0$ and with N exchangeable sites and with $c > 0$. We start with only type 1 present initially. The exposition has four parts, the *emergence*, the *preemergence* (droplet formation), the *fixation* and finally the relation between droplet formation and emergence.

Part 1: Emergence

The goal is to describe in mathematical precise form the initial formation of germs for the expression of the fitter type 2 by mutation which subsequently expand which then leads finally to the global emergence of the fitter type, and over the period of order $O(1)$ the increase of the mass of this type continues until it takes over almost the entire population (fixation). We state the results on this scenario in four main propositions.

First on emergence are Proposition 2.6 which shows that (global) emergence occurs at times of order $const \cdot \log N$ and which identifies the constant and then Propositions 2.11 and 2.13 which identify the limiting dynamics in t after global emergence of type 2, meaning we study the system observed in times $const \cdot \log N + t$ with $t \in \mathbb{R}$. The Proposition 2.8 explains the emergence behaviour by describing the very early formation of droplets of "type-2 colonies".

In addition to the four main statements mentioned above the exact properties in the early stage of droplet formation are given in Propositions 2.9 and 2.10. Namely droplet formation occurs in a random manner in the very beginning followed by a deterministic expansion of the droplet size leading to emergence on a global level.

A remarkable fact we state after the part on fixation in Proposition 2.14, which is establishing a close connection between exit behaviour in short scale and entrance behaviour in the same scale but placed much later such that both time intervals are separated by a long time stretch.

We begin with the emergence times.

Proposition 2.6 (Macroscopic emergence and fixation times).

(a) (Emergence time)
There exists a constant α with:

$$0 < \alpha < s, \tag{2.56}$$

such that if $T_N = \frac{1}{\alpha} \log N$, then for $t \in \mathbb{R}$ and asymptotically as $N \to \infty$ type 2 is present at times $T_N + t$, i.e. there exists an $\varepsilon > 0$ such that

$$\liminf_{N \to \infty} P[x_2^N(T_N + t) > \varepsilon] > 0, \tag{2.57}$$

and type 2 is not present earlier, namely for $1 > \varepsilon > 0$:

$$\lim_{t \to -\infty} \limsup_{N \to \infty} [P(x_1^N(T_N + t) < 1 - \varepsilon] = 0. \tag{2.58}$$

(b) *(Fixation time)*
After emergence the fixation occurs in times $O(1)$ as $N \to \infty$, i.e. for any $\varepsilon > 0$

$$\lim_{t \to \infty} \limsup_{N \to \infty} P[x_1^N(T_N + t) > \varepsilon] = 0. \tag{2.59}$$

(c) The constant α can be characterized as the Malthusian parameter for a Crump–Mode–Jagers branching process denoted $(\tilde{K}_t)_{t \geq 0}$ which is explicitly defined below in (8.140). The constant α can alternatively be introduced as exponential growth rate for the limiting droplet growth dynamic $\mathfrak{J}_t^m$, see (2.72) or in terms of the excursion measure of a diffusion as specified below in (8.684). □

Corollary 2.7 (Emergence and fixation times of spatial density).
The relations (2.57), (2.58) and (2.59) hold for $\bar{x}_2^N$ respectively $\bar{x}_1^N$ as well. □

Part 2: Droplet formation

This raises the question how the global emergence of type 2 actually came about and what is the role of α in the forward dynamics (instead of the view back from emergence) and why do we have a random element in the emergence. We will now demonstrate that rare mutation and subsequent selection with the help of migration produce in times $O(1)$ a cloud of sites where the type-2 is already manifest with substantial mass at a given time and this cloud then starts growing as time increases. This growth of total type-2 mass is exponential with a random factor arising as usual in the very beginning (i.e. at times growing arbitrarily slow with N) and hence in particular the growth process is given by a randomly shifted exponential. We call the growing cloud of type-2 sites a droplet (which in fact in the euclidian geographic space is literally accurate). The droplet will be described using the atomic measure $\mathfrak{J}_t^{N,m}$ introduced in (2.14) and by the Palm measure explained in (2.15).

For the purpose of making our three regime scenario precise we first need to investigate how for a finite time horizon the site of a typical type-2 mass looks like in the limit $N \to \infty$ and if we can show that such a site exhibits nontrivial type-2 mass. Then we have to see how fast the number of such sites grows and reaches size εN at time $\alpha^{-1} \log N + t(\varepsilon)$.

2.3 Statement of Results

One might expect that a growth of the droplet at exponential rate α would be the appropriate scenario. This scenario will also show that some randomness created initially, i.e. times $O(1)$ as $N \to \infty$, remains in the system up to fixation of type 2. This randomness arises since up to some finite time T there will be among the N-sites as $N \to \infty$, $O(1)$ such sites where the mass of type 2 exceeds some level $\delta > 0$. In the limit this will result in a compound Poisson number of germs for expansion. Only descendants of this early mass will make up at much later times the bulk of the type-2 mass, since we have exponential growth.

Therefore we study

- the configuration of type-2 mass on the sparse set of sites and its limiting law as $N \to \infty$ for a finite time horizon, i.e. $\mathbf{J}_t^{N,m}$ and $\hat{\mu}_t^N$ for $t \in [0, T]$,
- the configuration in a typical type-2 site as time and N tends slowly to infinity, i.e. $\hat{\mu}_{t_N}^N$, with $t_N \uparrow \infty$ but $t_N = o(\log N)$,
- the growth of the droplet as time goes to infinity but slower than the time needed for global emergence, i.e. $\mathbf{J}_{t_N}^{N,m}$ and $\mathbf{J}_{t_N}^{N,m}([0, 1])$, with t_N as above.

To verify our scenario we want to show that $\hat{\mu}_{t_N}^N$ converges for $t_N \uparrow \infty$ to a limit, say $\hat{\mu}_\infty^\infty$ which is nontrivial, i.e. $\hat{\mu}_\infty^\infty((0, 1)) = 1$. Then we want to prove that the number of sites looking like $\hat{\mu}_\infty^\infty$-realisations grows like $\mathcal{W}^* e^{\alpha^* t_N}$ for $N \to \infty$, with $t_N \to \infty, t_N = o(\log N)$ for some non-degenerate random variable $\mathcal{W}^*$ and a number $\alpha^* \in (0, \infty)$. And finally we have to identify α^* as α.

To verify that, we have to establish either that for some suitable finite and not identically zero random variable $\mathcal{W}^*$ and number α^* we have

$$\hat{x}_2^N(t_N) := \sum_{i=1}^N x_2^N(i, t_N) \sim \mathcal{W}^* e^{\alpha^* t_N} \tag{2.60}$$

as $N \to \infty$, or alternatively that we need a spatial window of size $Ne^{-\alpha^* t_N}$ to find such a type-2 colonised site with nontrivial probability in that window.

In order to explain the origin of the randomness in $\mathcal{W}^*$ we need the behaviour of the law of the random variable

$$\sum_{i=1}^N x_2^N(i, t) = \mathbf{J}_t^{N,m}([0, 1]), \quad t \in [0, T] \tag{2.61}$$

and the localization of this total mass on different sites for $N \to \infty$ in a suitable description, which is given by the full atomic measure $(\mathbf{J}_t^{N,m})_{t \geq 0}$.

In particular if we have established the above scenario, we can conclude that it requires time $(\alpha^*)^{-1} \log N + t$ for some appropriate $t \in \mathbb{R}$ to reach intensity ε of type 2 in the whole collection of sites and then we can conclude that $\alpha^* = \alpha$.

We now want to show three things to describe the droplet growth: (1) There is a limiting law for the size of the type-2 population in a typical colonized site. (2) Identify the exponential growth rate of the number of colonized type-2 sites. (3) Identify the limiting dynamic as $N \to \infty$ of $(\mathfrak{J}_t^{N,m})_{t \geq 0}$ exactly as the dynamic $(\mathfrak{J}_t^m)_{t \geq 0}$.

Addressing first point (1), (2) (for (3) see below), we shall prove the following on droplet formation. We use here the notational convention

$$a_N = \hat{o}(bf(N)) \text{ if } \limsup_{N \to \infty}(a_N/f(N)) < b. \tag{2.62}$$

Proposition 2.8 (Microscopic emergence and evolution: droplet formation).

a) *The Palm distribution stabilizes, i.e.*

$$\hat{\mu}_{t_N}^N \underset{N \to \infty}{\Longrightarrow} \hat{\mu}_\infty^\infty \text{ for } t_N \uparrow \infty \text{ with } t_N = \hat{o}(\alpha^{-1} \log N) \tag{2.63}$$

and the limit has the property

$$\hat{\mu}_\infty^\infty((0, 1)) = 1. \tag{2.64}$$

The law $\hat{\mu}_\infty^\infty$ will be identified in terms of the excursion measure in (8.635).

b) *The total type-2 mass $\mathfrak{J}_{t_N}^{N,m}([0,1])$ grows at exponential rate α^*, i.e. (2.60) holds and*

$$\alpha^* = \alpha, \text{ with } \alpha \text{ from Proposition 2.6.} \tag{2.65}$$

Furthermore we have a random growth factor:

$$\mathcal{L}\left[\hat{x}_2^N(t_N)e^{-\alpha t_N}\right] \underset{N \to \infty}{\Longrightarrow} \mathcal{L}[\mathcal{W}^*], \quad \text{for } t_N \uparrow \infty \text{ with } t_N = \hat{o}(\alpha^{-1} \log N). \tag{2.66}$$

The random variable $\mathcal{W}^*$ is non-degenerate. □

Remark 5. This result is obtained by showing that for sufficiently large t_1 and t_2:

$$\lim_{N \to \infty} P\left(|e^{-\alpha t_1}\hat{x}_2^N(t_1) - e^{\alpha t_2}\hat{x}_2^N(t_2)| > \varepsilon\right) < \varepsilon. \tag{2.67}$$

Remark 6. The mutant population making up the total population at time $\alpha^{-1}\log N + t$ of emergence dates back to a rare mutant ancestor which appeared at a time in $[0, T(\varepsilon)]$ with probability at least $1 - \varepsilon$ and $T(\varepsilon) < \infty$ for every $\varepsilon > 0$.

Remark 7. Growth in the spatial Fisher–Wright and the spatial branching model are different in nature. In the branching model we have growth of the total type-2 mass at exponential rate s, which is due to essentially a random number of families growing at that rate and hence with few sites with a large population. In the Fisher–Wright case we get an exponentially (rate $\alpha < s$) growing number of sites with a

2.3 Statement of Results

type-2 population of at least ε. Macroscopic emergence occurs once this growing droplet has volume $const \cdot N$.

In order to understand the structure of the quantity $\mathcal{W}^*$ arising from (2.66), we have to investigate the behaviour also in the earlier time horizon $t \in [0, T_0]$ of $\mathbf{J}_t^{N,m}$ as $N \to \infty$. And addressing our point (3) from above (2.62) we obtain the following convergence result.

Proposition 2.9 (Limiting droplet dynamic).
As $N \to \infty$

$$\mathcal{L}[(\mathbf{J}_t^{N,m})_{t \geq 0}] \underset{N \to \infty}{\Longrightarrow} \mathcal{L}[(\mathbf{J}_t^m)_{t \geq 0}] \tag{2.68}$$

in the sense of convergence of continuous $\mathcal{M}_a([0, 1])$-valued processes where $\mathcal{M}_a([0, 1])$ is equipped with the weak atomic topology. □

In the growth behaviour of the limit dynamics as $N \to \infty$ we obtained above, we recover the quantity α^* and $\mathcal{W}^*$ in the longtime behaviour of the limit dynamic $(\mathbf{J}_t^m)_{t \geq 0}$ in the following proposition.

Proposition 2.10 (Long-time growth behaviour of $\mathbf{J}_t^m$).
Assume that either $m > 0$ or $\mathbf{J}_0^m([0, 1]) > 0$. Then the following growth behaviour of $\mathbf{J}^m$ holds.

(a) There exists α^* such that the following limit exists

$$\lim_{t \to \infty} e^{-\alpha^* t} E[\mathbf{J}_t^m([0, 1])] \in (0, \infty). \tag{2.69}$$

We have

$$\alpha^* = \alpha, \tag{2.70}$$

where α is defined as in Proposition 2.6 (see also 8.140).

(b) Recall (2.66) for the law of $\mathcal{W}^*$. Then:

$$e^{-\alpha t} \mathbf{J}_t^m([0, 1]) \underset{t \to \infty}{\Longrightarrow} \mathcal{W}^* \text{ in probability}. \tag{2.71}$$

(c) The growth factor in the exponential is truly random:

$$0 < Var[\lim_{t \to \infty} e^{-\alpha t} \mathbf{J}_t^m([0, 1])] < \infty. \quad \square \tag{2.72}$$

Remark 8. We shall see in the proof that the random variable $\mathcal{W}^*$ reflects the growth of $\mathbf{J}_t^m([0, 1])$ in the beginning, as is the case in a supercritical branching process and hence $\mathcal{E}^* = \alpha^{-1} \log \mathcal{W}^*$ can be viewed as the random time shift of that exponential $e^{\alpha t}$ which matches the total mass of $\mathbf{J}_t^m$ for large t.

Given $\mathfrak{J}_t^m$, let $\nu_t(a,b]$, $0 < a < b \leq 1$ denote the number of atoms in the interval $(a,b]$. Then the analogue of the size distribution for the discrete CMJ process is given by

$$\frac{\nu_t(\cdot)}{\mathfrak{J}_t^m([0,1])} \in \mathcal{P}((0,1])). \tag{2.73}$$

It is reasonable to expect that there is a limiting stable size distribution in the $t \to \infty$ limit similar to that in the discrete case but we do not follow-up on this here.

Part 3: Fixation

We now understand the preemergence situation and the time of emergence. In order to continue the program to describe the dynamics of macroscopic fixation, we consider the limiting distributions of the empirical measure-valued processes in a second time scale $\alpha^{-1} \log N + t$, $t \in \mathbb{R}$. Define

$$\Xi_N^{\log,\alpha}(t) := \frac{1}{N} \sum_{i=1}^N \delta_{(x_1^N(i, \frac{\log N}{\alpha} + t), x_2^N(i, \frac{\log N}{\alpha} + t))}, \quad t \in \mathbb{R}, \quad (\Xi_N^{\log,\alpha}(t) \in \mathcal{P}(\mathcal{P}(\mathbb{K}))) \tag{2.74}$$

and then the two empirical marginals are given as

$$\Xi_N^{\log,\alpha}(t,\ell) := \frac{1}{N} \sum_{i=1}^N \delta_{x_\ell^N(i, \frac{\log N}{\alpha} + t)}, \quad \ell = 1, 2 \text{ and } t \in \mathbb{R}, \quad (\Xi_N^{\log,\alpha}(t,\ell) \in \mathcal{P}([0,1])). \tag{2.75}$$

Note that for each t and given ℓ the latter is a random measure on $[0, 1]$. Furthermore we have the representation of the empirical mean of type 2 as follows:

$$\bar{x}_2^N\left(\frac{\log N}{\alpha} + t\right) = \int_{[0,1]} x \, \Xi_N^{\log,\alpha}(t,2)(dx). \tag{2.76}$$

The third main result of this section is on the fixation process in type 2, saying that $\Xi_N^{\log,\alpha}$ converges as $N \to \infty$ and the limit can be explicitly identified as a random McKean–Vlasov entrance law starting from time $-\infty$ in type 1.

Proposition 2.11 (Asymptotic macroscopic fixation process).

(a) For each $-\infty < t < \infty$ the empirical measures converge weakly to a random measure:

$$\mathcal{L}[\{\Xi_N^{\log,\alpha}(t,\ell)\}] \underset{N\to\infty}{\Longrightarrow} \mathcal{L}[\{\mathcal{L}_t(\ell)\}] = P_t^\ell \in \mathcal{P}(\mathcal{P}([0,1])), \text{ for } \ell = 1, 2. \tag{2.77}$$

2.3 Statement of Results

In addition we have path convergence:

$$w - \lim_{N \to \infty} \mathcal{L}[(\Xi_N^{\log,\alpha}(t))_{t \in \mathbb{R}}] = P \in \mathcal{P}[C((-\infty, \infty), \mathcal{P}(\mathcal{P}(\mathbb{K})))]. \quad (2.78)$$

A realization of P is denoted $(\mathcal{L}_t)_{t \in \mathbb{R}}$ respectively its marginal processes $(\mathcal{L}_t(1))_{t \in \mathbb{R}}, (\mathcal{L}_t(2))_{t \in \mathbb{R}}$.
(b) *The laws $P_t^\ell, \ell = 1, 2$ are non-degenerate probability measures on the space $\mathcal{P}([0, 1])$. In particular*

$$E\left[\left(\int_{[0,1]} x\mathcal{L}_t(2)(dx)\right)^2\right] > \left(E[\int_{[0,1]} x\mathcal{L}_t(2)(dx)]\right)^2. \quad (2.79)$$

(c) *The process $(\mathcal{L}_t)_{t \in \mathbb{R}}$ describes the emergence and fixation dynamics, that is, for $t \in \mathbb{R}$, and $\varepsilon > 0$,*

$$\lim_{t \to -\infty} Prob[\mathcal{L}_t(2)((\varepsilon, 1]) > \varepsilon] = 0, \quad (2.80)$$

$$\lim_{t \to \infty} Prob[\mathcal{L}_t(2)([1 - \varepsilon, 1]) < 1 - \varepsilon] = 0, \quad (2.81)$$

with

$$\mathcal{L}_t(2)((0, 1)) > 0 \quad, \quad \forall t \in \mathbb{R}, \text{ a.s.}. \quad (2.82)$$

(d) *For every $t \in \mathbb{R}$, always both type 1 and type 2 are present:*

$$Prob[\mathcal{L}_t(2)(\{0\}) = 0)] = 1. \quad (2.83)$$

(e) *The limiting dynamic in (2.78) is identified as follows:*
The probability measure P in (2.78) is such that the canonical process is a random solution (recall Definition (2.2)) and entrance law from time $-\infty$ to the McKean–Vlasov equation (2.19).
(f) *The limiting dynamic in (2.78) satisfies with α as in (2.56):*

$$\mathcal{L}[e^{\alpha|t|} \int_{[0,1]} x\mathcal{L}_t(2)(dx)] \Rightarrow \mathcal{L}[^*\mathcal{W}] \text{ as } t \to -\infty, \quad (2.84)$$

and we explicitly identify the random element generating P in (2.78), namely P arises from random shift of a deterministic path:

$$P = \mathcal{L}[\tau_{*\mathcal{E}}\mathcal{L}^*] \quad, \quad {}^*\mathcal{E} = (\log^* \mathcal{W})/\alpha, \quad \tau_r \text{ is the time-shift of path by } r,$$
$$(2.85)$$

where $\mathcal{L}^*$ is the unique and deterministic entrance law of the McKean–Vlasov equation (2.19) satisfying (2.29) and with projection $\tilde{\mathcal{L}}_t(2)$ on the type 2 coordinate satisfying:

$$e^{\alpha|t|} \int_{[0,1]} x\tilde{\mathcal{L}}_t(2)(dx) \longrightarrow 1, \text{ as } t \to -\infty. \tag{2.86}$$

The random variable $^*\mathcal{W}$ satisfies

$$0 < {^*\mathcal{W}} < \infty \text{ a.s.}, \quad E[{^*\mathcal{W}}] < \infty, \quad 0 < \text{Var}(^*\mathcal{W}) < \infty. \tag{2.87}$$

(g) We have for $s_N \to \infty$ with $s_N = o(\log N)$ the approximation property for the growth behaviour of the limit dynamic by the finite N model, namely:

$$\mathcal{L}[e^{\alpha s_N} \bar{x}_2^N (\frac{\log N}{\alpha} - s_N)] \underset{N \to \infty}{\Longrightarrow} \mathcal{L}[^*\mathcal{W}]. \quad \square \tag{2.88}$$

Corollary 2.12 (Scaled total mass process convergence).
The above implies for the total mass process that:

$$\mathcal{L}[\{\bar{x}_2^N (\frac{\log N}{\alpha} + t)\}_{t \in \mathbb{R}}] \underset{N \to \infty}{\Longrightarrow} \mathcal{L}[\{\int_{[0,1]} x\mathcal{L}_t(2)(dx)\}_{t \in \mathbb{R}}], \tag{2.89}$$

and we have convergence in distribution on path space. $\quad \square$

Remark 9. The random variable $\mathcal{W}^*$ of (2.66) can be viewed as the *final value* of the microscopic development of the mutant droplets and $^*\mathcal{W}$ of (2.84) can be viewed as an *entrance value* of the macroscopic emergence of the mutant type.

Corresponding to the limit dynamics of Proposition 2.11 we also have a *stochastic propagation of chaos result*, which characterizes the asymptotic behaviour of a tagged sample of sites as follows:

Proposition 2.13 (Fixation dynamic for tagged sites).
There is a unique limiting entrance law for the process of tagged components: for fixed $L \in \mathbb{N}$, the sequence in N of laws

$$\mathcal{L}[(\{x_2^N(i, \frac{\log N}{\alpha} + t))\}_{i=1,\dots,L})_{t \in \mathbb{R}}] \tag{2.90}$$

converges as $N \to \infty$ weakly to a law

$$\mathcal{L}[(X_2(t))_{t \geq 0}] = \mathcal{L}[((x_2(1,t), \cdots, x_2(L,t))_{t \in \mathbb{R}}], \text{ with } X_2(t) \longrightarrow \underline{0} \text{ as } t \to -\infty, \tag{2.91}$$

2.3 Statement of Results

which is if restricted to $t \geq t_0$ for every $t_0 \in \mathbb{R}$ the weak solution of the SSDE:

$$dx_2(i,t) = c\left(\int_{[0,1]} x \, \mathcal{L}_t(2)(dx) - x_2(i,t)\right) dt + sx_2(i,t)(1 - x_2(i,t))dt$$
$$+ \sqrt{d \cdot x_2(i,t)(1 - x_2(i,t))} \, dw_i(t), \quad i = 1,\ldots,L, \qquad (2.92)$$

where $w_1,\ldots,w_L$ are independent standard Brownian motions which are also independent of the process $\{\mathcal{L}_t\}_{t \in \mathbb{R}}$ and the latter is given by (2.85). The behaviour at $-\infty$ in (2.91) and the equation (2.92) uniquely determine the law for a given mean curve. □

Remark 10. Note that $x_2(i,t)$ will hit 0 at random times until $\bar{x}_2(t)$ exceeds a certain level depending on the other parameters in particular c/d, only after that time where $\bar{x}_2(t)$ remains large enough the component remains a.s. occupied with positive type-2 mass for all future time. Before that time the path will have zeros at an uncountable set of time points with Lebesgue measure zero, the set of these zeros will be stochastically bounded from above. This is a consequence of standard diffusion theory (see [RW]).

Part 4: Exit and entrance, from droplet to emergence

The results above of course immediately raises the question how the macroscopic *entrance* from 0 is related to the *final* value in the smaller original time scale. We have already stated that the two exponential growth rates from these two directions are the same. But more is true also the two growth constants have the same law:

Proposition 2.14 (Relation between microscopic and macroscopic regimes).
The entrance value and the final value are equal in law:

$$\mathcal{L}[\mathcal{W}^*] = \mathcal{L}[^*\mathcal{W}]. \qquad □ \qquad (2.93)$$

Remark 11. We note that this result cannot be derived from some soft argument since we relate two time scales t and $\alpha^{-1} \log N + t$ which separate as $N \to \infty$ and indeed the limits $N \to \infty$ and $t \to \infty$ or $t \to -\infty$ cannot be interchanged in general.

Remark 12. One could go further studying the relation between final and entrance configuration and ask how the limiting Palm distribution $\hat{\mu}_\infty^\infty$, which describes the typical sparse type-2 sites is related to the limit $\mathcal{L}_{-\infty}$ as $t \to -\infty$ of the Palm distribution $\hat{\mathcal{L}}_t$ of $\mathcal{L}_t$, i.e.

$$\hat{\mathcal{L}}_t(2)(dx) = \frac{x \mathcal{L}_t(2)(dx)}{\int x \mathcal{L}_t(2)(dx)}, \qquad (2.94)$$

which describes the typical site "immediately" after emergence.

Note that this new transformed measure is now deterministic in the $t \to -\infty$ limit, since in the ratio the same random time shift is used in both numerator and denominator.

Here we have to observe that due to the exponential growth of type-2 sites the state depends on when they were colonised and that time point is typically a *finite* random time back. Define for the process given in (2.33)

$$\mu_s^*(dx) = \text{probability that an excursion of age } s \text{ has size in } dx. \tag{2.95}$$

Note that for any time the total mass $\mathbf{J}_t([0, 1])$ is finite but there are countably many non-zero atoms in $\mathbf{J}_t$. At time t let

$$\Gamma_t([a, b)) = \text{number of atoms of age in } [a, b), \quad 0 < a < b \leq 1 \tag{2.96}$$

and then define the corresponding size-biased distribution on the product space $\mathbb{R}^+ \times [0, 1]$ of age and size given by

$$\hat{\Gamma}_t(ds, dx) = \frac{\Gamma_t(ds) x \mu_s^*(dx)}{\int_0^t \Gamma_t(ds) \int_0^1 x \mu_s^*(dx) ds}. \tag{2.97}$$

By analogy with the Crump–Mode–Jagers theory, we would expect that this age and size distribution $\hat{\Gamma}_t(ds, dx)$ stabilizes as $t \to -\infty$ with stable age and size distribution $\hat{\Gamma}(ds, dx)$.

Then we would expect to have the relation:

$$\hat{\mathcal{L}}_{-\infty}(2)(dx) = \hat{\Gamma}((0, \infty), dx). \tag{2.98}$$

2.3.4 The Branching Approximation in the Fast Mixing Limit

In this subsubsection we shall note that approximation by supercritical branching occurs as $c \to \infty$. In particular, we have that as c becomes large, the growth coefficient $\alpha = \alpha(c)$ satisfies (recall $\alpha(c) < s$) that

$$\alpha(c) \longrightarrow s \text{ as } c \to \infty. \tag{2.99}$$

This means that for rapid migration the total type 2 mass growth is very close to a branching process with Malthusian parameter s and immigration rate m.

To verify this, note that in the limit $c \to \infty$, $N \to \infty$, the limiting dual process can be identified as a pure branching process in space in which each occupied site has only one particle and each particle moves to a new unoccupied site at its time of birth, respectively the initial particles do that at time 0 instantaneously. This means that in particular there is practically no coalescence taking place. In the original

2.3 Statement of Results

system this means that the rare mutant population is distributed uniformly in space and grows at the maximal speed at many sites.

2.3.5 Some Topological Facts Concerning Atomic Measures

We finally recall the definition and some facts on the *weak atomic topology with metric* ρ_a on the space of finite atomic measures $\mathcal{M}_a([0, 1])$ due to Ethier and Kurtz. Recall that we have the topology on $\mathcal{M}([0, 1])$ induced by the weak topology which is induced by the Prohorov metric ρ. Next choose a function Ψ, where $\Psi : [0, \infty) \to [0, 1]$ is continuous, nonincreasing and $\Psi(0) = 1$, $\Psi(1) = 0$. Then for $\nu, \mu \in \mathcal{M}_a([0, 1])$ one defines

$$\rho_a(\nu, \mu) := \rho(\nu, \mu) + \sup_{0 < \varepsilon \leq 1} \left| \int_{[0,1]} \int_{[0,1]} \Psi\left(\frac{|x-y|}{\varepsilon} \wedge 1\right) \mu(dx) \mu(dy) \right.$$
$$\left. - \int_{[0,1]} \int_{[0,1]} \Psi\left(\frac{|x-y|}{\varepsilon} \wedge 1\right) \nu(dx) \nu(dy) \right|. \quad (2.100)$$

We refer to ρ_a as the *Ethier–Kurtz metric*.

The space $(\mathcal{M}_a([0, 1]), \rho_a)$ is a Polish space and the topology, in other words convergence, does not depend on the choice of Ψ (the geometry of the space of course does).

The following lemma collects what we need on the relation between the weak and the weak atomic topologies, which are different.

Lemma 2.15 (Weak atomic topology and weak topology).

(a) *A sequence of random finite atomic measures on* $[0, 1]$, $(\mu_n)_{n \in \mathbb{N}}$, *converges to* μ *in the weak atomic topology if and only if*

$$\mu_n \text{ converges weakly to } \mu \quad (2.101)$$

and

$$\mu_n^*([0, 1]) \xrightarrow[n \to \infty]{} \mu^*([0, 1]), \text{ where } \mu^*([0, 1]) = \sum_{x \in [0,1]} \mu(\{x\})^2 \delta_x. \quad (2.102)$$

(b) *If the following three properties hold for* $\mu_n, \mu \in \mathcal{M}_a([0, 1])$ *(here* $\Longrightarrow$ *denotes weak convergence as* $n \to \infty$*):*

$$\mu_n \Rightarrow \mu, \text{ the ordered atom sizes converge and the set of atom locations converges,} \quad (2.103)$$

then:

$$\rho_a(\mu_n, \mu) \to 0. \quad (2.104)$$

(c) A continuous $\mathcal{M}_f$-valued process with a.s. continuous (in the weak topology) sample paths of the form $\sum_i a_i(t)\delta_{x_i(t)}$ such that $\sum a_i^2(t)$ is a.s. continuous, has also sample paths a.s. in $C([0,\infty),(\mathcal{M}_f([0,1],\rho_a))$, where ρ_a is the Ethier–Kurtz metric.

(d) Consider a sequence $\{Z_N, N \in \mathbb{N}\}$ of atomic measure-valued processes with cadlag paths in the weak atomic topology, so that Z_N has the form

$$Z_N(t) = \sum_i a_{N,i}(t)\delta_{x_{N,i}(t)} \tag{2.105}$$

for suitable functions (of t) $a_{N,i}, x_{N,i}$. Assume furthermore that

$$\{\mathcal{L}(Z_N), N \in \mathbb{N}\} in \mathcal{P}(D([0,\infty),(\mathcal{M}_f([0,1],\rho))) \text{ is relatively compact.} \tag{2.106}$$

Then the compact containment condition will hold also in $(\mathcal{M}_a([0,1]),\rho_a)$, if and only if for each $T > 0$ and $\delta > 0$ there exists $\varepsilon > 0$ such that

$$\inf_N P\left[\sup_{t\le T}\left(\int_{[0,1]}\int_{[0,1]}\Psi(\frac{|x-y|}{\varepsilon}\wedge 1)Z_N(t,dx)Z_N(t,dy) - \sum_i a_{N,i}^2(t)\right) \le \delta\right]$$
$$\ge 1 - \delta. \tag{2.107}$$

Here Ψ is as in (2.100). □

Proof. This follows from work by Ethier and Kurtz [EK4], namely (a)—Lemma 2.2, (b)—Lemma 2.5, (c)—Lemma 2.11, (d)—Remark 2.13. q.e.d.

Chapter 3
Formulation of the Multitype and Multiscale Model

We now introduce the model used in this monograph in general form. Consider a population consisting of multitype individuals divided into *colonies* (demes) that are located at sites labelled by a countable group Ω_N (modelling geographic space) and whose *types* (genotypes) belong to a countable set

$$\text{the type space} \quad \mathbb{I}. \tag{3.1}$$

The choice of the geographic space Ω_N defined below is a good caricature (for large N) of a geography around two dimensions as we shall explain later on. Formally, in this section we introduce a system with countably many $\mathcal{P}(\mathbb{I})$-valued components, each describing the *frequencies of types* within a certain colony. Here the set $\mathbb{I}$ is a hierarchically structured countable set of types describing the possible types which can successively arise via mutation starting from one basic type.

The interaction between the components comes from the *migration* inducing a drift term for each component towards the weighted average of neighboring colonies. The evolution of a single component is a diffusion with a fluctuation term representing *resampling* and a drift term representing *selection* and *mutation* (all three corresponding to the change of generation). In order to be able to use in a transparent way earlier work and important techniques from the literature we will use the framework of measure-valued diffusions and not ℓ^1-valued diffusions.

We begin by introducing in 3.1 the basic ingredients of a general class of models and then in 3.2 we characterize the processes in this class by means of a *well-posed martingale problem*. Then we get much more specific and explain in 3.3 the very specific form of the selection and mutation mechanism which we need in order to treat the questions discussed above. We finally discuss the biological interpretation of these phenomena in 3.4.

3.1 Main Ingredients

We first introduce a representation for the *hierarchically organised countable set of types* in a form adapted to the introduction of the fitness function below, namely we consider a sequence of levels of fitness, each level splitting into M-types. This leads to space of types

$$\mathbb{I} = (\mathbb{N}_0 \times \{1, 2, \ldots, M\}) \cup (0, 0) \cup (\infty, 1), \tag{3.2}$$

where $\mathbb{N}_0 = \{0\} \cup \mathbb{N}$ and M is an integer satisfying $1 < M < N$.

Remark 13. Adding the maximal point $(\infty, 1)$ allows us to work with a *compact* set of types and as a consequence a *compact Polish state space* if we embed the type set in the interval $[0, 1]$ appropriately using the relative topology (see the point (δ) below).

Then we can write:

$$\mathbb{I} = \bigcup_{j=0}^{\infty} E_j, \tag{3.3}$$

where

$$E_j = \{(j, \ell); \ell = 1, \cdots, M\}, \ j \in \mathbb{N}, \quad E_0 = \{(0, \ell), \quad \ell = 0, 1, \cdots, M\},$$
$$E_\infty = \{(\infty, 1)\}. \tag{3.4}$$

Remark 14. In the discussion below we will fix M and let $N \to \infty$. In the discussion on biodiversity in Subsection 4.7 we mention an alternative more complex type space but for the purposes of stating the main results of this monograph we use the above definition.

Next we introduce the state space of the process and the basic parameters of the stochastic evolution.

(α) The state space of a single component (describing frequencies of types) will be

$$\mathcal{P}(\mathbb{I}) = \text{set of probability measures on } \mathbb{I}. \tag{3.5}$$

The set of colonies (sites or components) will be indexed by a set Ω_N, which is countable and specified in (β) below. The *state space* $\mathcal{X}$ of the system is therefore

$$\mathcal{X} = (\mathcal{P}(\mathbb{I}))^{\Omega_N}, \tag{3.6}$$

with the product topology of the weak topology of probability measures on the compact discrete set $\mathbb{I}$. A typical element is written

$$X = (x_\xi)_{\xi \in \Omega_N} \text{ with } x_\xi \in \mathcal{P}(\mathbb{I}). \tag{3.7}$$

3.1 Main Ingredients

(β) The hierarchical group Ω_N indexing the *colonies* of the geographic space is defined by:

$$\Omega_N = \{\xi = (\xi^i)_{i \in \mathbb{N}_0} \mid \xi^i \in \mathbb{Z}, \, 0 \leq \xi^i \leq N - 1, \, \exists k_0 : \quad \xi^j = 0 \quad \forall j \geq k_0\}, \tag{3.8}$$

with group operation defined as component-wise addition modulo N. A typical element of Ω_N is denoted by $\xi = (\xi^0, \xi^1, \dots)$. Here N is a parameter with values in $\{2, 3, 4, \dots\}$.

Note that:

$$\Omega_N = \bigoplus_{i=0}^{\infty} Z_N, \quad Z_N = \{0, \cdots, N-1\} \text{ with addition mod } (N). \tag{3.9}$$

We also introduce a metric (actually an ultrametric) on Ω_N denoted by $d(\cdot, \cdot)$ and defined as

$$d(\xi, \xi') = \inf\{k \mid \xi^j = (\xi')^j \quad \forall j \geq k\}. \tag{3.10}$$

The use of Ω_N in population genetics in the mathematical literature goes back to Sawyer and Felsenstein [SF] and appeared first in the thirties to model geographically isolated subpopulations. An element $(\xi^0, \xi^1, \xi^2, \dots)$ of Ω_N can be thought of as the ξ^0-th village in the ξ^1-th county in the ξ^2-th state .... In other words we think of colonies as grouped and classified according to "extent of neighborhood inclusion". For $N \to \infty$, Ω_N is a good approximation of $\mathbb{Z}^2$ as is shown in Dawson, Gorostiza, Wakolbinger [DGW01].

(γ) The transition kernel $a(\cdot, \cdot)$ on $\Omega_N \times \Omega_N$, modelling *migration rates* has some specific properties. We shall only consider *homogeneous* transition kernels which have in addition the *exchangeability property* between colonies of equal distance. This means that $a_N(\cdot, \cdot)$ has the form:

$$a_N(\xi, \xi') = a_N(0, \xi' - \xi) \quad \forall \xi, \xi' \in \Omega_N, \tag{3.11}$$

$$a_N(0, \xi) = \sum_{k \geq j} \left(\frac{c_{k-1}}{N^{(k-1)}}\right) \frac{1}{N^k} \quad \text{if } d(0, \xi) = j \geq 1,$$

where

$$c_k > 0 \quad \forall k \in \mathbb{N}, \quad \sum_{k=0}^{\infty} \frac{c_k}{N^k} < \infty \quad \text{for all } N \geq 2. \tag{3.12}$$

The kernel $a_N(\cdot, \cdot)$ should be thought of as follows. With rate $c_{k-1}/N^{(k-1)}$ we choose a hierarchical distance k, and then each point within distance at most k

is picked with equal probability as the new location. Then (3.12) requires a finite jump rate from a given point.

Note that the scaling factor N^k in the rate $c_k N^{-k}$ represents one over the volume of the k-ball in the hierarchical metric which is the natural parameterization of the rates from the viewpoint of the potential theory of the walk. This turns out to allow us to characterize the qualitative properties of the random walk by $(c_k)_{k \in \mathbb{N}}$ independent of the parameter N and facilitates taking the limit $N \to \infty$. Different choices for the c_k correspond, from the potential theoretic point of view, to a geography analogous to 2, 2−, respectively 2+ dimensions. (See [DGW01] for more on this topic.)

It is often convenient to write the transition rates as

$$ca_N(\cdot, \cdot) \quad , \quad c \in \mathbb{R}^+ \text{ and } a_N \text{ probability transition kernel on } \mathbb{I} \times \mathbb{I}. \quad (3.13)$$

Then we can say that c is the migration rate and $a_N(\xi, \xi')$ the probability that a jump from ξ to ξ' occurs.

(*δ*) In addition to describe *mutation* and *selection* we need two further objects. Let

$$M(\cdot, \cdot) \text{ be a probability transition kernel on } \mathbb{I} \times \mathbb{I}, \quad (3.14)$$

modelling mutation probabilities from one type to another. Furthermore let

$$\chi(\cdot) \text{ be a bounded function on } \mathbb{I}, 0 \leq \chi(\cdot) \leq 1, \quad 0 = \min \chi, 1 = \sup \chi, \quad (3.15)$$

modelling relative fitness of the different types. We have set here the arbitrary minimum value of χ equal to zero and then in order to uniquely specify the parameter s (representing selective intensity) we have set the maximal value of χ equal to one. Using χ we can embed $\mathbb{I}$ into $[0, 1]$ rather naturally and use on it the relative topology induced by the euclidian topology on $[0, 1]$.

On this level of generality we can specify our process uniquely. Further explanation on selection and mutation and in particular more detailed specifications needed for our purposes here in this work will be given below in Subsection 3.3.

3.2 Characterization of the Process by a Martingale Problem

We now proceed to rigorously specify the model by formulating the appropriate martingale problem. We shall first define a general class of models and then we will specify the mechanisms we use here in detail in the next Subsection 3.3. Readers interested in the individual-based model resulting in the diffusion approximation we use are referred to [EK2] or [DGP].

The key ingredients for the martingale problem with state space E (Polish space) are a measure-determining sub-algebra $\mathcal{A}$ of $C_b(E, \mathbb{R})$, the socalled test functions,

3.2 Characterization of the Process by a Martingale Problem

typically called F here, and a linear operator on $C_b(E, \mathbb{R})$, typically called L, with domain $\mathcal{A}$ resulting in values in $C_b(E, \mathbb{R})$.

Definition 3.1 (Martingale problem).

(a) The law P on a space of E-valued path for a Polish space E, either $D([0, \infty), E)$ or $C([0, \infty), E)$, is a solution to the martingale problem for (L, ν) w.r.t. $\mathcal{A}$ if and only if

$$\left(F(X(t)) - \int_0^t (LF)(X(s)) ds \right)_{t \geq 0} \text{ is a martingale under } P \text{ for all } F \in \mathcal{A}$$

(3.16)

and

$$\mathcal{L}(X(0)) = \nu. \tag{3.17}$$

The martingale problem is called wellposed, if the finite dimensional distributions of P are uniquely determined by the property (3.16) and (3.17).

(b) In our context $E = (\mathcal{P}(\mathbb{I}))^{\Omega_N}$. □

The algebra of test functions we use here is denoted by

$$\mathcal{A} \subseteq C_b((\mathcal{P}(\mathbb{I}))^{\Omega_N}, \mathbb{R}) \tag{3.18}$$

and is defined as follows. Given a nonnegative bounded function f on $(\mathbb{I})^k$ and $\xi_1, \ldots, \xi_k \in \Omega_N$ consider the function on $(\mathcal{M}(\mathbb{I}))^{\Omega_N}$ (here $\mathcal{M}$ denotes finite measures) defined by:

$$F(x) = \int_{\mathbb{I}} \cdots \int_{\mathbb{I}} f(u_1, \ldots, u_k) x_{\xi_1}(du_1) \ldots x_{\xi_k}(du_k), \tag{3.19}$$

$$\xi_i \in \Omega_N, \quad x_\xi \in \mathcal{M}(\mathbb{I}), \quad x \in (\mathcal{M}(\mathbb{I}))^{\Omega_N}.$$

Let $\mathcal{A}$ be the algebra of functions generated by functions F of type (3.19) restricted to $(\mathcal{P}(\mathbb{I}))^{\Omega_N}$.

Our operators L will be differential operators. We define differentiation on the big space $(\mathcal{M}(\mathbb{I}))^{\Omega_N}$, even though we deal only with restrictions to $(\mathcal{P}(\mathbb{I}))^{\Omega_N}$ later on. This approach facilitates comparison arguments. Hence we differentiate the function F as follows:

$$\frac{\partial F}{\partial x_\xi}(x)[u] = \lim_{\varepsilon \to 0} \frac{F(x^{\varepsilon, \xi, u}) - F(x)}{\varepsilon}, \tag{3.20}$$

with

$$x^{\varepsilon, \xi, u} = (x'_\eta)_{\eta \in \Omega_N}, \quad x'_\eta = \begin{cases} x_\eta & \eta \neq \xi \\ x_\xi + \varepsilon \delta_u & \eta = \xi. \end{cases} \tag{3.21}$$

Correspondingly $\frac{\partial^2 F}{\partial x_\xi \partial x_\xi}(x)[u,v]$ is defined as $\frac{\partial}{\partial x_\xi}\left(\frac{\partial F}{\partial x_\xi}(x)[u]\right)(x)[v]$.
Let

$$c, s, m, d > 0 \tag{3.22}$$

be parameters that represent the relative rates of the migration, selection and mutation, respectively the resampling rate which in biological language represents the inverse effective population size, finally let $a_N(\cdot,\cdot)$ be as in (3.11). Furthermore χ and M are as in (3.14) and (3.15).

We now define a linear operator on $C_b((\mathcal{P}(\mathbb{I}))^{\Omega_N})$ with domain $\mathcal{A}$, more precisely

$$L: \mathcal{A} \longrightarrow C_b((\mathcal{P}(\mathbb{I}))^{\Omega_N}, \mathbb{R}), \tag{3.23}$$

which is the generator L of the martingale problem and is modeling migration, selection, mutation via drift terms and resampling as the stochastic term (in this order), by (here $d > 0$ and $c, s, m \geq 0$):

$$(LF)(x) = \sum_{\xi \in \Omega_N} \left[c \sum_{\xi' \in \Omega_N} \bar{a}_N(\xi, \xi') \int_{\mathbb{I}} \frac{\partial F(x)}{\partial x_\xi}(u)(x_{\xi'} - x_\xi)(du) \right. \tag{3.24}$$

$$+ s \int_{\mathbb{I}} \left\{ \frac{\partial F(x)}{\partial x_\xi}(u)\left(\chi(u) - \int_{\mathbb{I}} \chi(w) x_\xi(dw)\right)\right\} x_\xi(du)$$

$$+ m \int_{\mathbb{I}} \left\{ \int_{\mathbb{I}} \frac{\partial F(x)}{\partial x_\xi}(v) M(u, dv) - \frac{\partial F(x)}{\partial x_\xi}(u) \right\} x_\xi(du)$$

$$\left. + d \int_{\mathbb{I}} \int_{\mathbb{I}} \frac{\partial^2 F(x)}{\partial x_\xi \partial x_\xi}(u, v) Q_{x_\xi}(du, dv) \right], \quad x \in (\mathcal{P}(\mathbb{I}))^{\Omega_N},$$

where

$$Q_x(du, dv) = x(du)\delta_u(dv) - x(du)x(dv). \tag{3.25}$$

It has been proved in [DG99, 99] that a model of the type as above is well defined:

Theorem 1 (Existence and Uniqueness). *Let ν be a probability measure on $(\mathcal{P}(\mathbb{I}))^{\Omega_N}$ specifying the initial state which is independent of the evolution.*

(a) *Then the $(L; \nu)$-martingale problem w.r.t. $\mathcal{A}$, on the space $C([0, \infty)$ $(\mathcal{P}(\mathbb{I}))^{\Omega_N})$ (cf. Definition 3.1), is well-posed. For fixed value of the parameter N the resulting canonical stochastic process is denoted*

$$(X_t^N)_{t \geq 0}. \tag{3.26}$$

(b) The solution defines a strong Markov process with the Feller property. □

Remark 15. In Subsection 4.3 we shall give (Theorem 2) a general statement about equilibrium states and convergence to equilibrium.

In the sequel we shall only consider initial states which are either deterministic or random with law ν in $\mathcal{P}((\mathcal{P}(\mathbb{I}))^{\Omega_N})$ which satisfies the following two properties:

$$\nu \text{ is invariant w.r.t. spatial shift on } \Omega_N, \tag{3.27}$$

$$\nu \text{ is shift ergodic.} \tag{3.28}$$

In the discussion below the special case of monotype initial configurations is of particular importance.

3.3 Detailed Specification of Selection and Mutation Mechanisms

The formulation of specific questions about the evolutionary process in the class of models specified in Subsection 3.2, requires a specific set-up for the key ingredients, in particular selection and mutations, more precisely of the function χ and the matrix M. This we provide in this subsection and the biological interpretation is provided in the next subsection.

An important point here is the inclusion of a scaling parameter N. The scalings are chosen such that two properties hold, they are as $N \to \infty$ effectively similar to the migration mechanism which would describe a Euclidean situation around two dimensions and such that migration, mutation, selection and resampling remain in the limit $N \to \infty$ *all* of comparable strength. Hence we are in a *critical regime*. Other regimes can be tackled with the methods we develop here, but we do not follow up on this extension in the present work.

(i) *The hierarchical fitness function χ.*

Haploid *selection* drives the population towards domination by the "superior types", that is, types u at which the fitness function $\chi(\cdot)$ is a maximum. Here we will be interested in fitness functions having a special structure which we describe below.

Remark 16. The haploid fitness function χ is equivalent to a diploid fitness function $V(u, v)$ with *additive selection*, that is, $V(u, v) = \frac{1}{2}(\chi(u) + \chi(v))$. For a discussion the Fleming–Viot process with mutation, (diploid) selection and recombination see, for example, [D][1993] Chapter 10.2 and [EK3]. The general case of diploid selection as well as recombination are more complicated and are not considered in this monograph, even though many of our methods can be extended. However haploid selection is a basic mechanism in both evolutionary biology and evolutionary methods of optimization (see [H]).

46 3 Formulation of the Multitype and Multiscale Model

In order to model the adaptive landscapes as mentioned in the introduction, we will consider here a fitness function $\chi : \mathbb{I} \to [0, 1]$ corresponding to a sequence of adaptive peaks of increasing fitness and arranged in a hierarchical structure based on their fitness and stability.

Definition 3.2 (Fitness function, level-j fitness function).

(a) The function χ^N is defined for $(k, \ell) \in \mathbb{I}$ by

$$\chi^N((k, \ell)) = \left(\frac{N-1}{N}\right)\left(1 + \frac{1}{N} + \cdots + \frac{1}{N^{k-1}} + \frac{\ell}{MN^k}\right), \quad k \geq 1,$$
$$\ell = 1, \cdots, M. \tag{3.29}$$

$$\chi^N((0, \ell)) = \frac{N-1}{N}\left(\frac{\ell}{M}\right), k = 0, \quad \ell = 1, \cdots, M \tag{3.30}$$

and

$$\chi^N((0, 0)) = 0, \quad \chi^N((\infty, 1)) = 1. \tag{3.31}$$

We usually suppress N in this notation and write χ for χ^N.

(b) Define for types (j, ℓ) the level-j fitness function

$$\chi_j(j, \ell) = N^j (\chi(j, \ell) - \max_{k<j,\ell'} \chi((k, \ell'))). \quad \square \tag{3.32}$$

This parameterization of the types and fitness function induces a *hierarchical structure* on the set of types with the following properties:

(α) It can be ordered according to a sequence of fitness levels indexed by $k \in \mathbb{N}$ as well as an ordering of types at the same level by *finer distinction within levels*,

(β) The countable set of types can be ordered by fitness with the minimal type $(0, 0)$,

(γ) the fitness difference

$$\text{between type } (k-1, M) \text{ and } (k, 1) \text{ is } \frac{1}{MN^k}, \tag{3.33}$$

whereas the fitness difference

$$\text{between types } (k-1, i) \text{ and } (k-1, j) \text{ with } i \neq j \text{ is } \frac{|i-j|}{MN^{(k-1)}}, \tag{3.34}$$

$$\text{between types } (k, \ell) \text{ and } (k-1, i), \quad i \neq M \text{ is } \geq \frac{1}{MN^{(k-1)}}. \tag{3.35}$$

In this context it is useful to refer to the

$$\text{type } (k, \ell) \text{ for } \ell = 1, \cdots, M-1 \text{ and } (k-1, M) \tag{3.36}$$

3.3 Detailed Specification of Selection and Mutation Mechanisms

as a *selective level-k-type*, which is a property independent of $N > M$. Note that the fitness of a level k type is not only (strictly) greater than the fitness of any lower level type, the difference is at least $1/N^{(k-1)}$ which explains the special role of $(k-1, M)$. Furthermore a level-$(k+1)$ type will only have significantly gained against all level-k types in times of order N^k.

Remark 17. These notions are useful because types of selective level k differ in fitness by a term $\frac{const}{N^k}$ but differ from all lower level types by at least $\frac{1}{N^{(k-1)}}$ and hence for large N the difference in fitness within a selection level set is small compared to the advantage over types of lower level. This means that as the population evolves the competition between types in a level-k set can only be observed at times of order N^k at which lower selective level types have essentially disappeared and higher level ones are not yet distinguished.

Remark 18. We will later in the consideration of the role of deleterious mutation formulate a notion of "level" which removes the special role of the types (k, ℓ) with $\ell = M$ in the above set-up.

(ii) *Intra- and inter-level mutation rates*

A *mutation* occurs when an offspring has a type different from that of the parents. Let m denote the total mutation rate and let $M(u, dv)$ denote the probability distribution of the type, which results if a type u undergoes mutation.

As explained above in our model the set of types are hierarchically structured and the mutation mechanism we shall consider here will explicitly depend on this structure which provides a sequence of adaptive peaks of increasing fitness. An interpretation related to the idea of *Wright's adaptive landscape* is made explicit in Example 2 in Subsection 3.4.

The basic idea of the model is that mutation in the space of types proceeds as follows:

during the evolutionary process new types are created out of the existing ones in a random manner and the new types have three properties:

(α) they represent *small* perturbations in fitness of the types present in the system at that time
(β) the higher level types are more *stable* against (deleterious) mutation,
(γ) very *rarely* superior types arise by mutation.

We make this precise below in (3.37) by a formal definition. Since higher level types have more complex organization that results in higher stability and fitness, this means that the transitions to successive fitness levels are achieved in successively longer time scales.

Guided by the above principles we introduce the mutation rates $m(u, dv) = mM(u, dv)$ in the form of the *mutation transition matrix* $m(u, v)$ from type u to v using the hierarchical structure of our type space. Namely to define the transition rates of types in E_k among each other we use $m_{k,k}(\cdot, \cdot)$ and rates $m_{k,\ell}(\cdot, \cdot)$ between types in E_k and E_ℓ, which are independent of N in addition to a scaling factor depending on N. The latter allows us in particular to distinguish mutations and *rare mutations*. This works as follows.

Definition 3.3 (Mutation rates). We fix some positive finite $K \in (0, \infty)$. Then we set

$$
\begin{aligned}
m((k_1, j_1), (k_2, j_2)) &= \tfrac{m_{k,k}(j_1, j_2)}{N^k} && \text{if } k_1 = k_2 = k \text{ and } j_1 \in E_k, j_2 \in E_k \\
&= \tfrac{m_{k,k+1}(j_1, j_2)}{N^{k+1}} && \text{if } k_1 = k, \ k_2 = k+1 \text{ and } j_1 \in E_k, j_2 \in E_{k+1} \\
&= \tfrac{m_{k,\ell}(j_1, j_2)}{N^k} && \text{if } k_1 = k, \ k_2 = \ell < k \text{ and } j_1 \in E_k, j_2 \in E_\ell \\
&= 0 && \text{otherwise,}
\end{aligned}
\tag{3.37}
$$

where

$$ 0 \leq m_{k,\ell}(j_1, j_2) \leq K, \tag{3.38} $$

$$ m_{k,k}(j_1, j_2) > 0, \ \forall \ j_1, j_2, \quad m_{k,\ell}(j_1, j_2) > 0 \text{ for at least one } l < k \tag{3.39} $$

and

$$ m_{k,k+1}(j_1, j_2) > 0 \text{ for at least one pair } (j_1, j_2). \qquad \square \tag{3.40} $$

Remark 19. The constant K plays a similar role as the constant M we had in the fitness function, it is in particular independent of N.

We consider the case in which at each level the mutation process has only one ergodic class. Although not essential to our methods we often assume that the mutation rates between the finite set of types at each level are strictly positive and therefore we can assume that $M(u, dv)$ dominates a multiple of a probability vector independent of u.

Definition 3.4 (Decomposition with state independent mutation component). If there exists $\varrho \in \mathcal{P}(\mathbb{I})$, $\bar{m} \in (0, 1]$ with $M(u, dv) \geq \bar{m}\varrho(dv)$, then we define the probability transition kernel $\bar{M}(u, dv)$ by using the maximal $\bar{m}$ and setting

$$ M(u, dv) = \bar{m}\varrho(dv) + (1 - \bar{m})\bar{M}(u, dv). \tag{3.41} $$

Without loss of generality we can pick the uniform distribution for ϱ. $\qquad \square$

As a further simplification to keep things more transparent, we shall assume in the rest of the monograph the following *quasi-state independent* form for the mutation rates of (3.37) in the sense that only the level of the mutating type plays a role in the rate, where we only consider deleterious mutations one level down.

Definition 3.5 (Mutation rates, simplified form). Let ϱ_k be a positive distribution on E_k and let m_k^+, m_k^-, m_k be strictly positive numbers satisfying for some finite positive K:

$$ m_k^+, m_k^- \leq K \text{ for all } k \in \mathbb{N}. \tag{3.42} $$

3.3 Detailed Specification of Selection and Mutation Mechanisms

Define:

$$\begin{aligned}
m_{k,k+1}(j_1, j_2) &= m_k^+ \varrho_{k+1}(j_2), & j_1 &\in E_{k-1}, j_2 \in E_{k+1} \\
m_{k,k}(j_1, j_2) &= m_k \varrho_k(j_2), & j_1, j_2 &\in E_k \\
m_{k,k-1}(j_1, j_2) &= m_k^- \varrho_{k-1}(j_2), & j_1 &\in E_k, j_2 \in E_{k-1} \\
m_{k,\ell}(j_1, j_2) &\equiv 0 \quad \text{if } \ell \neq k-1, k, k+1, & j &\in E_k, j_2 \in E_\ell.
\end{aligned} \tag{3.43}$$

In fact we will often use for ϱ_k the uniform distribution on E_k. □

Remark 20. With these definitions above we have completely specified the model to be used in the sequel in Theorems 3–11. In Theorem 1 we stated that the corresponding martingale problem is well-posed and defines for each $N \in \mathbb{N}$ a Markov diffusion process with state space $(\mathcal{P}(\mathbb{I}))^{\Omega_N}$.

Remark 21. The parameterization of the model in terms of N has been set up so that the hierarchical mean field limit $N \to \infty$ is meaningful and leads to the separation of time scales. This means that we can effectively focus on one level of fitness if we allow for a certain time of the system to evolve and the transition from one level of fitness to the next proceeds step by step. In addition by the choice of the migration rates it is adapted as $N \to \infty$ asymptotically to a two-dimensional geographic structure.

(iii) *The concept of selection-mutation level*

A key point for the longtime behaviour is that *both* fitness levels and the mutation rates determine the longterm success of different types. More precisely, in order for a type of given fitness to survive over long time scales, it is essential that its mutation rate to types of lower fitness be small relative to the difference in fitness. In fact in the model we consider there is a hierarchy of fitness levels that could be viewed as levels of organization that arise over *successively longer time periods* and which are in addition *progressively more stable* in the sense that the rate at which they degenerate to lower order types is smaller.

We now introduce a notion of the level of a type which takes into account both selection and mutation and leads to the decomposition in (3.3) (recall (3.33) and note that $(k, 1)$ is now more stable against deleterious mutations than $(k - 1, M)$, namely by an order $N^{-(k-1)}$!):

Definition 3.6 (Selection-mutation level).
We call for $\ell = 1, \cdots, M$ (and $\ell = 0, \cdots, M$ for $k = 0$) the

$$\text{type } (k, \ell) \text{ a level-}k \text{ type.} \qquad \square \tag{3.44}$$

The rationale for this definition is that the form of $m(\cdot, \cdot)$ implies that level-k types are more stable than lower level types w.r.t. to changes due to mutation and furthermore mutation to higher level types asymptotically always involves exactly a one step increase, as N becomes large. Together with the selective level of a type we then obtain a set of types, the k-level types which are of comparable (as $N \to \infty$)

"fitness" to remain in the population under the selection-mutation dynamic and allow for a quasi-equilibrium on each level for the corresponding time scale.

Remark 22. We wish to consider the dynamics given by this system of mutation rates as N becomes large. At the fitness level denoted by k the natural time scale for selection is of the order of N^k, and consequently in order for the mutation rate to be comparable to the selection rate on that level, this rate should be of order $1/N^k$. In particular, this means that if a mutant of lower fitness is produced and the fitness difference is $1/N^{k-1}$ or more, then in time scale N^k the difference creates a drift term $O(N)$ and it immediately disappears. On the other hand if the difference is $1/N^k$ it persists in time scale N^k. Therefore mutations between the level k types change their relative frequencies and in the natural time scale for level k we will be able to prove that a mutation-selection quasi-equilibrium is reached.

3.4 Interpretation of the Model from the Point of View of Evolutionary Theory and Connections to Related Work

We now review the nature of the model we have introduced from the viewpoint of evolutionary theory. The mutation mechanism described above satisfies three qualitative requirements: (a) a type at a certain level of complexity can only mutate to types of comparable or lower levels of complexity at the natural time scale of its level, (b) mutations to lower level types occur at a rate comparable to the natural time scale of the original level so that by the selective disadvantage of this mutant it cannot be persistent in the time scale in which it arises, (c) a type can mutate to the next higher level of complexity but this occurs with a lower order rate (rare mutation).

This structure of the mutation means that the more complex a type is the more stable it is against random perturbations, more precisely, deleterious mutations. Note that in the level j-time scale lower order mutants (of level less than j), even though they occur, will be wiped out in shorter time scales. Altogether we note that the more complex types evolve at a slower rate than the simpler ones, both as far as selection and mutation goes.

One point we need to discuss a bit more is our choice of scale for mutation, selection, migration and resampling. Our choice is motivated by four properties

- corresponds for geographic space asymptotically ($N \to \infty$) to a two-dimensional situation,
- allows mutation-selection balance on each level,
- migration effects taking place on that very same scale as above and
- the evolution towards fitter types is initiated by the rare mutation.

This specifies a *critical* regime in which as $N \to \infty$ *all* components of our generator play a significant role. Other regimes would lead effectively to a deletion of one of the mechanisms in the longtime limit (compare Subsection 3.5).

To see how our "real life" migration fits into this, we recall that in [DGW01] random walks on the hierarchical group were analysed and it turns out that as $N \to \infty$ one obtains a behaviour as on $\mathbb{Z}^d$ with $d = 2, 2-0$ or $2+0$ depending on the $(c_k)_{k \in \mathbb{N}}$, that is, corresponding to the critically recurrent, recurrent and transient regimes. This scaling of migration is therefore appropriate. Then if we are looking for a regime where also all effects, namely migration, selection and mutation happen at a comparable strength and none becomes degenerate as $N \to \infty$, this then leads to our scaling. In other words we study indeed the critical regime in a natural geographic space. We shortly discuss other regimes in Subsection 8.8 once we have presented some of the methods such that the claims made are more comprehensible.

How does that fit in the context of evolutionary theory?

First, let us briefly review *Wright's theory of evolution in an adaptive landscape* (compare [W] for a modern introduction). In this model the population can be viewed as a "cloud" of individuals undergoing reproduction with mutation and selection acting and living on an "adaptive landscape" in type space in which the altitude represents the fitness level. For simplicity we can assume that initially the population consists of individuals of a simple type denoted by $(0, 0)$ and subsequently tends towards a peak in the adaptive landscape by hill-climbing. Near the adaptive peaks stabilizing selection tends to keep the population clustered near the peak. As a result many small mutations are suppressed but after a long period a rare sequence of small mutations can result in a transition to a higher level type. As a result the population is dominated by successively more complex types having a fitness advantage and greater stability, for example due to local stabilizing selection. The latter is analogous to transitions between potential wells due to large deviations that have been introduced in statistical physics and is described in Example 2 below. Recently an interesting alternative model of evolution in a very high dimensional landscape has been proposed by Gavrilets [Gav] but we do not consider this here.

Let us look closer at a more formal level, the connections of our set-up with Wright's model. Indeed the mutation matrix $m((k_1, j_1), (k_2, j_2))$ relates to the adaptive landscapes as follows. Consider a sequence of adaptive peaks, labelled by $k \in \mathbb{N}$, in the adaptive landscape in which to get to the $(k+1)$st level one must climb through the previous k levels. This resembles the situation of a *random walk in a linear random potential*. One can pass to the next highest level only by going through a valley in the landscape of the random potential of height u_k with $e^{-u_k} = \frac{m_{k,k+1}}{N^{k+1}}$, that is, $u_k = (k+1) \log N - \log m_{k,k+1} \approx (k+1) \log N$ for large N (recall $\log m_{k,k+1} \in (0, \log M)$). Now how comes selection into play? In order to dominate in the time scale in which the $k+1$-adaptive peak is reached, a fitness difference of only $\frac{1}{N^k}$ is needed. Instead of the stabilizing selection at a peak we have here in our model the stability properties against deleterious mutations which provides a stabilizing effect on fitter types and lets the process keep climbing.

Remark 23. Here we have imposed the hierarchical structure through the fitness function and have not modelled a random fitness landscape which leads asymptotically to such a linear potential. Deriving such a picture would be another very challenging task.

We discuss now an example with features of this above described scenario which have been discussed in the literature.

Example 2 (Multi-well potential on $\mathbb{R}^1$). The setting of [NCK],[KN] can lead to the mutation transition kernel that is used in our model by assuming an appropriate multiwell potential on $\mathbb{R}$ and where local stabilizing selection keeps the population concentrated near bottom of the valleys. The key ingredients are the successive mountainpass heights. If these are given by $((k+1)\log N)$ between levels k and $(k+1)$ and $(O(k \log N))$ within the k-th level, then the expected times for tunnelling (transition) to next level valleys are proportional to the exponentials of the heights of the corresponding mountain passes, that is, N^{k+1} and these times are exponentially distributed. □

Another feature of our model is the geographic structure and its approximation by the hierarchical group in the limit $N \to \infty$. This leads to:

Example 3 (Mean-field island model). In the case of a single level this *mean-field limit* as the number of colonies (demes) goes to infinity, is frequently used in statistical physics and was introduced into evolutionary theory by Kimura [Kim2] in his discussion of the development of altruistic traits. The extension of the mean-field limit to the hierarchical case was introduced by Dawson and Greven in [DG99].
 □

Example 4 (Spatially subdivided population and role of migration). The fact that in addition to mutation and selection we have migration and stochastic fluctuations due to finite population resampling, allows us to discuss the influence of communication versus isolation of subpopulations defined via the geographic space. In fact we will see later that our model allows us to discuss this problem mathematically and shows the essential role played by the interaction between colonies resulting from migration in the spread of rare mutants throughout the population. □

Example 5 (Biodiversity, bottlenecks). After leaving a punctuated equilibrium for the next one asks the question once the new state is reached how many ancestors are responsible for it. Do we have a bottleneck where only one ancestral path moves through or do we have a variety of ancestral lines persisting? This question can readily be analysed in our framework. and we will find that from a local perspective such bottlenecks can arise by migration which is not sufficiently far-reaching.
 □

3.5 Extensions

The model we have presented looks rather specific, in particular we have focused on the critical regime where migration, mutation and selection all produce effects of comparable size since this relates to the phenomenon of punctuated equilibria and stasis. However we would point out that the methods, in particular the set-valued dual developed in Sections 5 and 9, can be extended to cover more complex phenomena of interest in the theory of evolution. For example, this includes the evolution of quasi-species (a set of closely related types compared to the full spectrum of types and corresponds to the various types $1, \cdots, M$ on the different levels in our present model), mutualistic multispecies interactions and multilevel selection (e.g. [D2013]).

It should also be emphasized that the duality method which we shall develop in Section 5 extends to the case of migration on a countable graph such as $\mathbb{Z}^d$ but the asymptotic analysis in the latter case leads to interesting open questions and requires a serious strengthening of the methods. But the goal would be to exhibit the hierarchical structure in the migration on large scales in the $\mathbb{Z}^d$ model itself.

Chapter 4
Formulation of the Main Results in the General Case

Our goal is to identify the combined effect of mutation, selection and migration on the evolution of the system introduced in Section 3 as time proceeds. We do this from the point of view explained in the introduction and Subsection 3.4, i.e. *punctuated equilibrium, stasis* (quasi-equilibria) and *isolated subpopulations*. This means we remain long times in states concentrated on a certain level of fitness to then comparably quickly move to the next higher level induced by rare mutation.

The essential point which we work out in this section is that we can capture the behaviour of the system by *renormalization* using multiple space and time scales and the *hierarchical mean-field limit*, the latter means observing the spatial population in multiple space-time scales in the limit as $N \to \infty$. The necessity for multiple scales is due to the fact that the operations of taking the limits $t \to \infty$ and $N \to \infty$ do *not commute* and the transition between different types of fitness dominating occurs therefore in different phases with distinct features.

The main results we describe in this section provide a description of the evolution of the distribution of types in the population from the basic type to higher and higher levels of fitness due to rare mutations taking place in geographic areas of a corresponding size. We introduce therefore a sequence of space and time scales indexed by $j \in \mathbb{N}$ in which the population will be in a corresponding sequence of *quasi-equilibrium states* whose projection on the set of types is concentrated on E_j for $j = 0, 1, 2, \cdots$, i.e. on types of successively higher levels of fitness.

In order to describe the transition from one level E_j to the next higher one E_{j+1}, we refine this sequence of time scales by breaking down the transition from E_j-populations to E_{j+1}-populations into *five phases* involving *three different main stages* thus producing a sequence of triples of time scales. In stage 1 we see quasi-equilibria on an E_j population in stage 3 on an E_{j+1}-population and in the very delicate stage 2, where the *transition* between the two quasi-equilibria occurs, we distinguish three phases bridging from the punctuation of the evolving selection-mutation quasi-equilibrium on E_j-types by emergence of E_{j+1}-types in the emergence phase to reaching fixation on the E_{j+1}-types in the phase of fixation where the system remains for a long time in a neutral quasi-equilibrium before it reaches in the next phase the level-($j+1$) mutation-selection evolution concentrated on E_{j+1}-types.

The results will be presented in a series of theorems that describe the five different phases for every level of fitness. These stages and phases appear for *every* j at the transition of the system concentrated on types from E_j to types from E_{j+1}.

The scenario of successive invasions. The scenario of the transitions of populations concentrated on E_j, $j \in \{0,1,\cdots\}$ is occurring in steps labelled with j as follows. For each j there is a time scale such that

$$E_0 \to E_1 \to E_2 \cdots E_j \to E_{j+1} \to \cdots \qquad (4.1)$$

due to rare mutations occurring somewhere within balls with center 0

$$B_1(0), B_2(0), \cdots \text{ of size } 1, 2, \cdots, \qquad (4.2)$$

which can then expand via selection and migration in this respective ball to reach fixation. We call the transition $E_j \to E_{j+1}$, the *j-th step*.

In detail this transition of the j-th step works as follows:

- The first stage consists of one phase called *phase 0* and involves in the j-th step two features.

    1. First the j- scaling limit of the process $(X^N(t))_{t\geq 0}$ as $N \to \infty$ when the population is initially concentrated on individuals with type in E_j in the time scale of the first of the three stages. This is the time scale in which selection and mutation approach a *quasi-equilibrium* on types in E_j and in which types from E_{j+1}, i.e. of higher fitness, have not yet appeared with positive probability at a tagged site or in $B_j(0)$ respectively at a global positive density in $B_{j+1}(0)$.
    2. The second feature in phase zero is that at a sparse subset of the space rare mutation events do occur with positive probability in $B_{j+1}(0)$ in the j-time scale and we already see mutants with types from E_{j+1} dominating locally, the socalled *droplet formation* which induces heavily randomness which is becoming apparent in the far future.

- The second stage of transition consists of three phases.

    – In *phase 1* which occurs in a slightly larger time scale where we observe the j-th step emergence of a subpopulation of types in E_{j+1} that arises through (rare) mutation at rare sites within a ball of size $j+1$ forming a *growing droplet* and which subsequently spreads throughout the population by migration and selection based on fitness and stability against deleterious mutations to reach finally a size of the droplet with positive spatial intensity throughout space. This we call *emergence* in the j-th step in the sequence (4.1) of transition.

- The next *phase 2* in the j-th step which occurs after we had an emergence in the time scale we have in phase 1 and subsequently in the smaller time scale we had in phase 0, the individuals of type $j+1$ take over all but a negligible part of the population which we call *fixation* in the j-th step.

  In other words the fixation process occurs in a comparably short time after a long period needed for emergence.
- Then in the j-th step continues in a subsequent *phase 3* where the population evolves with types in E_{j+1} for a very long time on types from E_{j+1} as in a *neutral* population dynamic due to the comparably small fitness differences on the higher level.

- In the final *third stage* of the j-th step we see its *phase 4*, where the higher level types "equilibrate" in a much larger time scale into a mutation-selection *quasi-equilibrium* on the E_{j+1}-types.

Then the process starts over again to move in the $(j+1)$-step from level $j+1$ to level $j+2$, etc.

Mathematical tools. In order to exhibit this scenario we have to zoom in on the population for each $j \in \mathbb{N}_0$ in the appropriate

- time scale,
- space scale (defined in terms of the hierarchical distance).

Appropriately here means that the time and space scales have the right proportion to each other as they expand with $N \to \infty$ so that emergence and fixation can occur driven by rare mutation and subsequent action of selection and migration.

With each of the phases described above there are associated natural time scales, which we have to find. Furthermore connected with the different time scales are spatial regions of a specific range where the interaction through migration builds up a certain dependence structure over this specific range. Using the hierarchical mean-field limit $N \to \infty$ we get that all the scales *separate* and the scenario formulated above can be stated in the form of *rigorous convergence statements*.

What mathematical concepts and methods are available to verify such a scenario? Certain techniques of analysis have been worked out to treat neutral models or models with selection but without mutation. However in contrast to those cases previously analyzed in [DGV] (neutral case) and [DG99] (including selection), carrying this out for the system involving the combined action of mutation, selection and migration requires both new concepts and new tools, in particular the introduction of *three different types of scales* and new forms of limiting dynamics, certain *random nonlinear dynamics* and their entrance laws from time $-\infty$ describing global emergence of rare mutants and *atomic measure-valued processes* to describe the first appearance of rare mutants somewhere in space, the droplet formulation. This is needed in order to capture the qualitative changes now present in the system and in particular to exhibit the phenomenon of *emergence of and invasion by rare but quasi-stable mutants* which require to bridge between different time and space scales which *separate* in the limit we take here.

Outline of Section 4 The Section 4 is split in seven Subsections which group in preparatory Subsections 4.1–4.4 and the statement of the main results in Subsections 4.5–4.7.

Subsections 4.1–4.4 prepare the tools and concepts of the hierarchical mean-field limit both the "classical" objects and the new ones needed because of rare mutations. After introducing the basic ingredients of the multiple time scale analysis in Subsection 4.1 we give in Subsection 4.2 a view of the different phases in which the fitter types occur and succeed. In Subsection 4.3 we formulate the required new ergodic theorems which are of interest in their own right. The Subsection 4.4 introduces the tools to describe the spatial structure of the populations.

The Subsection 4.5 gives in three theorems basic facts on multiple space-time scale analysis and in Subsection 4.6 we apply this method to our specific questions on the *emergence* and *fixation* of rare mutants based on the results obtained in Subsection 4.5 and this is condensed in further theorems, Theorem 6–9 which is the key result of this monograph.

In order to clarify the question of founding ancestors of the population created by rare mutants we switch to a *historical* point of view in Subsection 4.7 building up to Theorem 11 on the dichotomy local *biodiversity* versus large spatial *monotype* clusters after stating in Theorem 10 existence and uniqueness of the (reduced) historical process.

4.1 Renormalization: A Sequence of Space and Time Scales

In this Subsection 4.1 we introduce the three relevant collections of time and space scales and associated renormalized systems which allows us in Subsections 4.5 and 4.6 to state the basic limit results corresponding to each of these sets of scales. In Subsubsection 4.2.1 we formulate the scenario containing our results in a nutshell but already specifying scales and all the effects quantitatively. In Subsubsection 4.2.2 we give the main ingredients for the limit dynamics. One task of course will be to describe all the arising limiting processes rigorously. This will be the content of Theorems 3–8 later on in this section.

For the moment let us fix N and consider the composition of the population at a tagged site in Ω_N as t varies from 0 to ∞. The question we wish to address is how the system approaches its final equilibrium and to understand the process in which the site is successively taken over by populations of higher and higher fitness levels. Our analysis is based on determining what happens in different times scales in the limit as $N \to \infty$. The key idea is that these time scales can be chosen to depend on N in such a way that we can observe the progression to higher and higher levels in a sequence of separating (as $N \to \infty$) scales. Furthermore the separating time scales allow us to rigorously define the notion of *quasi-equilibria* if we introduce suitable space scales, so that we can view quasi-equilibria as states in which the limiting system equilibrates if we constrain ourselves to times of observation and

4.1 Renormalization: A Sequence of Space and Time Scales

blockaverages in space of a particular order in the parameter N. The transition between the quasi-equilibria is described using the droplet of colonies occupied by rare mutants in their rescaled atomic measure description which we shall develop below (in Step 4).

In fact, as mentioned before, we need three sequences of time scales corresponding to the three main stages (and two intermediate ones for stage 2 which splits into three phases). The scales which are of a different nature in order to describe (in the presence of resampling) the combined effect of mutation, selection and migration. In addition to the time rescaling we need a rescaling of space which is of simpler nature and based on the hierarchical distance and block-averaging and finally we have to combine these operations in a meaningful way to a *time-space rescaling*. This we explain in Steps 1–3 below. In order to get a grasp on the transition between the different hierarchical levels we need a way to zoom in on the rare mutants which is done in Step 4 and the final step is to extract the different phases properly in Step 5.

Step 1 *(Time scales)*

We are interested in exhibiting the sequences of quasi-equilibria through which the evolution proceeds. This means we shall need sequences of *time scales* (the scaling parameter is always N) and the running index characterizing the different scales which will be $j \in \mathbb{N}_0$ corresponding to the levels of fitness reached. For each $j = 0, 1, 2, \ldots$ we identify the time frame during which the transition from a quasi-equilibrium on E_j to a quasi-equilibrium on E_{j+1} occurs.

Define three types of *time scales* for the j-th step in the transition, i.e. for j-equilibration, $(j+1)$-fixation and $(j+1)$-equilibration by considering for each j:

$$T(N)N^j : \quad T(N) \uparrow \infty, \text{ with } \frac{T(N)}{\log N} \to 0, \text{ as } N \to \infty, \tag{4.3}$$

$$S(N)N^j : \quad S(N) \uparrow \infty, \ S(N)N^{-1} \to 0 \text{ as } N \to \infty \text{ , with}$$
$$\liminf_{N \to \infty} \frac{S(N)}{\log N} \text{ sufficiently large,} \tag{4.4}$$

$$sN^{j+1} \text{ with } s \in (0, \infty). \tag{4.5}$$

Furthermore we will identify later on two intermediate time scales between the first two time scales which give a more detailed picture how the emergence of new types and the final fixation on these types occurs. This happens later on in Subsection 4.2 and in full detail in Subsection 4.6.

Step 2 *(Space scales and corresponding characteristic functionals)*

In addition to looking at the system in different times scales, we must specify the spatial window of observation. To do this we introduce a sequence indexed by $k \in \mathbb{N}_0$, of *spatial rescalings* (with scaling parameter N) of the spatial system as follows.

Define now the block averages over k-balls of the process $X(t) = (x_\xi(t))_{\xi \in \Omega_N}$, as

$$x_{\xi,k}(t) = N^{-k} \sum_{d(\xi',\xi) \leq k} x_{\xi'}(t), \quad \xi \in \Omega_N. \tag{4.6}$$

Then for each $k \in \mathbb{N}$, $\{x_{\xi,k}\}_{\xi \in \Omega_N}$ ia a random field which is constant on the k-balls $B_k(\xi)$. We would like to extract now a new spatially renormalised system which is of the same form as the original one, i.e. again indexed with a copy of Ω_N. We define

$$\pi_k : \Omega_N \to \Omega_N^{(k)}, \quad \pi_k((\xi^0, \xi^1, \xi^2, \cdots)) = (\xi^k, \xi^{k+1}, \xi^{k+2}, \cdots),$$

$$\xi = (\xi^0, \xi^1, \xi^2, \cdots) \in \Omega_N, \tag{4.7}$$

that is, this maps $k - balls$ in Ω_N into points in $\Omega_N^{(k)}$. We denote the hierarchical distance in $\Omega_N^{(k)}$

$$d_k(\xi, \xi') = \inf\{\ell \in \mathbb{N}_0 | \xi^m = (\xi')^m, \ \forall \ m \geq \ell\} \quad \xi, \xi' \in \Omega_N^{(k)}. \tag{4.8}$$

Then $\Omega_N^{(k)}$ is a hierarchical group isomorphic to Ω_N which we can embed in Ω_N using the mapping i_k:

$$i_k : \Omega_N^{(k)} \to \Omega_N, \tag{4.9}$$

$i_k((\xi^0, \xi^1, \xi^2, \cdots)) = (0, \cdots, 0, \xi^0, \xi^1, \xi^2, \cdots)$ with ξ^0 in the k-th position.

We then define the k-level spatially renormalized random field as follows in terms of the averages from (4.6)

$$(x_{\xi,k}(t))_{\xi \in \Omega_N^{(k)}} := (x_{i_k(\xi),k}(t))_{\xi \in \Omega_N^{(k)}}. \tag{4.10}$$

This object is again a random field indexed by a hierarchical group $\Omega_N^{(k)}$ but at the same time we can view it as the averages indexed by Ω_N observed at the thinner set $i_k(\Omega_N^{(k)})$ only.

Furthermore we define the level-k empirical measure for the level-$(k-1)$ averages:

$$\Xi_{\xi,k}^N(t) = \frac{1}{N} \sum_{\xi' \in \Omega_N^{(k-1)}, d_{k-1}(\xi,\xi') \leq 1} \delta_{x_{\xi',k-1}(t)}, \quad \xi \in \Omega_N^{(k-1)}. \tag{4.11}$$

The spatial rescalings in the form of random fields provide a sequence of spatially renormalized systems and the empirical measures for the fields allow us to abstract from the individual components the corresponding statistical properties of the collection of components.

4.1 Renormalization: A Sequence of Space and Time Scales

Step 3 *(Renormalization via multiple time-space scales)*

Next we combine the time and space scales. This provides the *multiple* space-time rescaling (indexed by $(j,k) \in \mathbb{N}_0^2$) that is used to obtain in a $\mathbb{N}_0^2$-indexed collection of *renormalised* systems for which we can define the quasi-equilibria through which the evolution process passes in the different time scales given above.

More precisely, we observe for each j the process after the time *(age)* given by one of the three time scales above in (4.3)–(4.5) for *subsequent* time spans of order tN^k, in a block of size k. Here the choices for k of interest are $k = 0, \cdots, j-1$, i.e. blocks which are small compared to the reference time and space scale N^j. The key objects of this monograph we extract from the spatial stochastic system are therefore:

Definition 4.1 (Renormalized system of index (j,k) and scaling parameter N). Fix $N \geq 2$. Then we look at the following $\mathbb{N}_0^2$-indexed collection of *space-time rescaled systems* (we display now the dependence of $X(t)$ on N by writing $X^N(t) = ((x_\xi^N(t))_{\xi \in \Omega_N})$):

$$\left\{\left\{\left(x_{\xi,k}^N(U(N)N^j + tN^k)\right)_{t \geq 0}, \; \xi \in \Omega_N\right\}, \; j,k \in \mathbb{N}\right\}, \tag{4.12}$$

with U being either S, T or sN from (4.3)–(4.5).

This collection induces for every (j,k) and U as above a random field (since the field is constant in k-balls for the original index set Ω_N)

$$\{(x_{\xi,k}^N(U(N)N^j + tN^k))_{t \geq 0}, \; \xi \in \Omega_N^{(k)}\}. \tag{4.13}$$

This recipe defines in each of the *three sets of time scales* for every index $(j,k) \in \mathbb{N}_0^2$ and every value of the scaling parameter N a new *renormalized system* again indexed by locations ξ but now with $\xi \in \Omega_N^{(k)}$ isomorphic to Ω_N and again with $\mathcal{P}(\mathbb{I})$-valued components, which is obtained by rescaling both space and time in the original system. □

Step 4 *(Droplet description of fitter types)*

Next we have in order to understand the transitions between the different regimes to provide for the j-th step a precise description of the (very small) mass of the types in E_{j+1} in times of order N^j and on the corresponding spatial scale which are j-blockaverages. This is referred to as droplet formation. For that purpose we consider the total mass in a 1-block on level-j rescaling, i.e.

$$\hat{x}_{1,j}(t) = \sum_{d_j(\xi,0) \leq 1} x_{\xi,j}(t), \quad \xi \in \Omega_N^{(j)}, \; t \in [0, Const \cdot N^j \log N]. \tag{4.14}$$

The size of this quantity *evaluated on* E_{j+1} reflects the volume of sites where the E_{j+1}-type already prevails.

A more detailed information is contained in an atomic measure-valued process given as follows. Assign to every j-block within the $(j+1)$-block which we consider and which is characterized by $\ell \in \{0,1,2,\cdots,N\}$ the j-th digit in $\xi \in \Omega_N$ a label $a_j(\ell)$, so that for each j we get a collection

$$a_j(\ell) \quad , \quad \ell \in \mathbb{N}, \tag{4.15}$$

which are i.i.d. uniformly $[0,1]$-distributed. Here we have coupled the labels on every level j for all N using the embedding of Ω_N in the set Ω_∞. Alternatively we could consider $\xi \in \Omega_N^{(j)}$ and running in the 1-ball w.r.t. d_j.

Then consider for the j-th transition step the *atomic measure-valued process* defined as

$$\mathbf{J}_{j,t}^N = \sum_\ell \sum_{i \in E_{j+1}} x_{\xi(\ell),j}^N (tN^j)(\{i\})[\delta_{a_j(\ell)} \otimes \delta_{\{i\}}] \in \mathcal{P}([0,1] \times E_{j+1}), \tag{4.16}$$

where $\xi(\ell) = (0,\cdots,\ell,0,\cdots)$, $\ell \in \{0,1,2,3,\cdots\}$ with ℓ on the j-th position. This atomic measure describes the droplet of colonies occupied by rare fitter mutants in the j-th transition step. In fact varying t, we can view this as a process with state space $\mathcal{M}_a([0,1] \times E_{j+1})$ the atomic measures on $[0,1] \times E_{j+1}$. This can be turned into a proper state space, i.e. a Polish space using the weak atomic topology of Ethier and Kurtz. (This was reviewed in detail in Section 2).

Then we obtain

Definition 4.2 (Renormalized droplets of index j). The collection of renormalized droplets at time-space scale (j,j) is given by the processes

$$\{(\mathbf{J}_{j,t}^N)_{t \geq 0}, \quad j \in \mathbb{N}\}. \quad \square \tag{4.17}$$

These are the processes describing the evolution of the droplet in the $E_j \to E_{j+1}$ transition and we can study the limit $N \to \infty$ of the droplet evolution.

Step 5 *(Quasi-equilibria)*

The final step is to make precise the idea of a quasi-equilibrium and for that purpose we work with large parameter N in order to obtain the separation of all the time-space scales introduced above. In the *hierarchical mean-field limit* of $N \to \infty$, the limiting object (certain nonlinear Markov processes) will allow us to describe asymptotically the various states (quasi-equilibria) through which the system passes on its way to the final equilibrium as time tends to infinity. (This final equilibrium would be a state where we add a more refined set E_∞ of M types of fitness 1 to the type space (instead of just type $(\infty, 1)$), so that we can define a selection-mutation equilibrium which is approached).

The *quasi-equilibrium* is defined as the equilibrium of a limiting dynamics (which we define in Subsubsection 4.2.2) and hence becomes a precise mathematical meaning. For the purpose of defining the quasi-equilibria and to obtain the transition from one quasi-equilibrium to the next we will state an ergodic theorem for such non-linear Markov processes in Subsection 4.3.

The limit $N \to \infty$ which we have to take here to get a rigorous concept of quasi-equilibrium has consequences for the geography of space, but this is such that it indeed provides some insights into the behaviour of these systems in geographical spaces around *two* dimensions as we shall explain later on precisely.

At each stage j we see three types of processes depending on whether U is S, T or sN. In addition we will need a more detailed description of the process of transition between these regimes. For that reason we will need some time to introduce these objects rigorously in the sequel.

4.2 Basic Scenario and the Hierarchical Mean-Field Limit

We continue now by first describing the basic phenomena we want to capture with our asymptotic analysis in more detail using our space-time scales we have now introduced and secondly we give the definitions needed to specify the dynamics needed for the asymptotic description in the limit $N \to \infty$.

4.2.1 The Basic Scenario: Five Phases of Transition

Having introduced the multiple time-space scales and functionals of the renormalised system we return to our scenario of the five phases. Namely we describe here the five different *phases* in which the system proceeds in the j-th step (here $j \in \mathbb{N}_0$) first to a (quasi-)stable situation on the E_j-types and then passes after longtime driven by the emergence of E_{j+1}-types mass by rare mutation quickly through three intermediate stages to a quasi-stable situation on the next level of E_{j+1}-types. The limiting dynamics occurring heuristically in the description below are defined precisely in Subsubsection 4.2.2.

The phases occurring in the j-th step of transition (as introduced in (4.1)) are as follows).

Phase 0: *Approach to quasi-equilibrium of E_j-population; germs for E_{j+1}-droplets.*

This phase exhibits two features:

(i) At the time $T(N)N^j$ a *level-j quasi-equilibrium* concentrated on E_j-types arises in the limit $N \to \infty$ as equilibrium of a limiting dynamic derived from the selection-mutation-migration mechanism for the averages in k-blocks within a tagged j-block for $k \leq j$. The quasi-equilibrium of this tagged j-block is described by a *nonlinear McKean–Vlasov equation* with selection, mutation for the limiting empirical measure $\Xi_{\xi,j}^N$ (which determines also

j-blockaverages) and certain McKean–Vlasov neutral evolutions for limiting smaller ($k < j$) k-blockaverages.

(ii) However on a sparse set in space, a subset of the tagged ($j + 1$)-block and consisting of j-blocks but whose number is $o(N)$ (as $N \to \infty$), these blocks are already dominated by fitter E_{j+1}-types which appeared through rare mutation and the subsequent selection-migration dynamic. These sparse subsets are called droplets which we described by $(\mathbf{J}^N_{j,t})_{t \geq 0}$. They will be the *germs* for the global manifestations of fitter E_{j+1}-types in the next phase. The evolution of these germs is in the limit $N \to \infty$ given by a process constructed via a *Poisson random measure* producing the excursions of the mass of fitter types somewhere in space.

This structure in (i) and (ii) arises because asymptotically no mutations to level-($j + 1$) types will occur at a tagged region of hierarchical radius j and since migration in this time scale from outside a region of hierarchical radius j is negligible no mutants from further away can play a role in the given tagged k-block. Furthermore the fluctuations of the j-blockaverage are of order 1 in this time scale and selection versus mutation reaches an quasi-equilibrium state, i.e. an equilibrium state of the McKean–Vlasov dynamic.

What is the dependence structure of the random field in the level-j quasi-equilibrium and what happens with smaller blockaverages or the components? In terms of the spatial structure in time scale $T(N)N^j$, the random field is asymptotically j-dependent, that is, the states at sites separated by hierarchical distance (strictly) greater than j are independent. This is due to the fact that at times $o(N^{j+1})$ asymptotically in the limit $N \to \infty$, the random walk does not reach components which are at a distance $j + 1$ or further.

Moreover the structure of the random field in a j-block can be then analysed at the smaller spatial scales associated with $k = 1, 2, \ldots, j - 1$ by looking at the (j, k)-renormalized systems at a tagged site with k running from 1 to $j - 1$. In the limit $N \to \infty$ this sequence is described by an *interaction Markov chain* to be defined below in Subsection 4.4.

Phase 1: *Emergence of E_{j+1}-types*

The *emergence* of level-(j+1) types in the tagged j-block (i.e. a fixed threshold of type E_{j+1}-mass is reached with positive probability as $N \to \infty$) occurs at times

$$T_j N^j + CN^j \log N = N^j (T_j + C \log N) \tag{4.18}$$

for some *constant* C depending on migration, selection and mutation parameters and a positive *random variable* T_j with values in $\mathbb{R}$. Both C and $\mathcal{L}[T_j]$ we determine via the dual process.

The time above has two distinct components.

1. The (large) part $CN^j \log N$ in (4.18) describes what occurs between time $T_j N^j$ and the time $CN^j \log N + T_j N^j$, the time interval at which time some fixed positive intensity ε of the new type can be established following a deterministic

4.2 Basic Scenario and the Hierarchical Mean-Field Limit

evolution starting from a large number of germs of the E_{j+1}-type which have developed by time $T(N)$.

2. The second part T_j arises if we wait in the *beginning*, i.e. times in $[0, T_j N^j]$, for a droplet of the rare mutants consisting of a finite random number of j-blocks. The T_j is *random* since originally at a few spots in the block of size j (i.e. N^j many sites) rare mutants occur and dominate the block (and the time at which this happens and the number of these events is random on scale N^j) which then form a growing droplet of E_{j+1}-type colonies whose growths becomes deterministic before a positive spatial intensity is reached (giving a $N^j \log N$ component of time to emergence). This initial randomness remains visible in the system after the much larger time $C \log N$ has passed.

Precisely due to rare mutations of E_{j+1} types are created *somewhere* in the ball of radius $j + 1$ around 0 before times of order N^{j+1} and they develop nontrivial intensity in some (j)-sub-blocks of the $(j+1)$-block in the limit $N \to \infty$. However by times of order $o(N^j \log N)$ they will not develop a noticeable *intensity* in the whole $(j + 1)$-block.

This stage of the development is of particular interest because it also introduces a form of *randomness* in the relative frequencies of the new fitter types in the transition from E_j to E_{j+1} populations, whose traces also remain visible much later, namely in the next phase 2.

Phase 2:   *Fixation of E_{j+1}-types.*

Suppose the E_{j+1} types have reached a positive intensity in the $(j + 1)$-block. Then after that time observe during some time in scale tN^j in which as $t \to \infty$ the fixation on the new level of fitness is reached. The *fixation* (or *take-over*) of the tagged j-block or the $(j + 1)$-blockaverage by E_{j+1} types occurs rather fast compared to phase 1 and is captured considering the time scale

$$(T_j N^j + CN^j \log N) + tN^j \quad \text{for varying } t \in \mathbb{R}, (= N^j(T_j + C \log N + t)). \tag{4.19}$$

In that time scale we can ask for two things, the dynamic of the configuration a $(j + 1)$-block or the $(j + 1)$-blockaverage or of a tagged component, i.e. a j-blockaverage and as a third item we can ask how many rare mutants are the ancestor of the population, when it finally fixates. The items in detail are:

1. The limiting dynamic of the $(j + 1)$-block empirical measure $\Xi_{0,j}^N$ is a nonlinear McKean–Vlasov process leading from a state-concentrated almost completely on E_j as $t \to \infty$ to fixation on E_{j+1}. This in principle deterministic evolution contains a random element namely it arises via a nonlinear McKean–Vlasov entrance law, arising at time $-\infty$ from mass "zero", but shifted by a *random* time in $\mathbb{R}$, namely T_j. This is due to the randomness in the occurrence of the first j-blocks colonized by rare mutants.
2. We find for a tagged j-blockaverage in the limit $N \to \infty$ an evolution, which is described by an entrance law of a Markov process in a random medium, the

latter describing the independent feeding in of the rare mutants from somewhere in the $(j+1)$-block.
3. In order to track the genealogy and spatial spread of the mutant types in that phase we will use the *historical process* and corresponding to it as a dual object a *spatial coalescent* in Section 4.7. This allows to say that the rare mutant population at fixation descends from a random number of rare mutants occurring in the j-block.

Phase 3: *Neutral evolution of E_{j+1}-types.*

After fixation a friendly *(neutral) coexistence* of all the E_{j+1}-types develops. Namely at the time scale $S(N)N^j$ the invasion and complete fixation of a region of radius $j+1$ on mutants of E_{j+1}-type has occurred.

As time varies from $S(N)N^j$ to $S(N)N^j + t_N N^j$ with $t_N \uparrow +\infty$ and $0 \le t_N = o(N)$ an *intermediate "neutral" quasi-equilibrium with immigration-emigration* on E_{j+1}-types develops in the tagged j-block. The population remains a very long time in this phase and in the limit $N \to \infty$ we obtain a stationary limiting process in s if we consider $t_N = t'_N + s$ and vary s in $\mathbb{R}$.

This is because in times of order $o(N^{j+1})$, the dynamics does not yet discriminate between types $(j+1, 1)$ and (j', ℓ) with $j' \ge j+1$, $\ell \ge 1$. Here the initial randomness from the emergence phase is still visible in the random proportion of the E_{j+1}-types.

Phase 4: *Approach to quasi-equilibrium of E_{j+1} population.*

Finally in times of order N^{j+1}, i.e. much later as we evolved in phase 3, the selection among the types of level $j+1$ starts to act and migration connects the N different j-blocks within distance $j+1$ and one observes in time scale tN^{j+1} a limiting dynamic for the $(j+1)$-blockaverages.

Later by time $T(N)N^{j+1}$ convergence to a mutation-selection *quasi-equilibrium concentrated now on level $(j+1)$-types* in the average over the $(j+1)$-block occurs. The quasi-equilibrium is in the limit $N \to \infty$, as on the previous level, obtained as the equilibrium of a nonlinear McKean–Vlasov equation but now on the type space E_{j+1} and with the level-$(j+1)$ parameters for selection, mutation parameters and distance $j+1$ migration parameters. At this stage the randomness from the transition process has been completely washed out.

Succession of levels. Once the phases 0–4 in the j-th step of transition from level-j to level-$(j+1)$ types are completed, the game starts afresh to now carry out the $(j+1)$ step of transition to the next higher level, i.e. the transition E_{j+1} to E_{j+2}, and running again through phase 0-4 and so on.

Summarized in the j-th step of transition in the time scale $T(N)N^j$ with $T(N) = o(\log N)$ we get in the limit $N \to \infty$ that the level-j block intensities of the system will approach a unique *quasi-equilibrium state which is concentrated on E_j-types* and will then be driven by rare mutation to exhibit the emergence of E_{j+1} types in time $T_j N^j + S(N)N^j$ and will then evolve towards *states concentrated on E_{j+1}-types* at times $(S(N)N^j + T_j N^j) + tN^j$ as $t \to \infty$, which evolve neutrally and in a quasi-stationary way in a j-block for times parametrised by t and given by observing at time $(S(N)N^j + T_j N^j) + (t_N + t)N^j$ till finally in times tN^{j+1} the intensities in a $(j+1)$-block will approach a *selection-mutation quasi-equilibrium* on E_{j+1}-types as $t \to \infty$.

The scenario described above will be formulated as precise convergence results in the sequel, in particular phases 0, 3, 4 are subject of Subsection 4.5, while Phase 1, 2 are the topic of Subsection 4.6. (The proofs are in Sections 8–13.)

4.2.2 The Basic Limit Dynamics

In order to precisely formulate the above objects and dynamics arising as $N \to \infty$, we introduce the following ingredients to describe the three different types of limiting dynamics of the multiple space-time scale analysis and the subsequent *hierarchical mean-field limit* $N \to \infty$. These are nonlinear Markov processes. We need first two ingredients.

We want to couple the systems for different N in a natural way by embedding them all in one bigger set. Define for that purpose

$$\Omega_\infty = \{(\xi^i)_{i \in \mathbb{N}} \mid \xi^i \in \mathbb{N}_0, \exists k \text{ with } \xi^i = 0 \quad \forall i > k\} \tag{4.20}$$

and note that (as sets):

$$\Omega_N \subseteq \Omega_{N'} \subseteq \Omega_\infty \quad \forall N \text{ and } N' \text{ with : } N, N' \in \mathbb{N}, N \leq N'. \tag{4.21}$$

Fix a set of types (here we will later use mainly $\mathbb{K} = E_j \cup E_{j+1}$ or $\mathbb{K} = E_j$)

$$\mathbb{K} = \{1, \ldots, K\} \tag{4.22}$$

and positive constants c and d, nonnegative constants m, s fitness function χ and mutation matrix M and recall $Q_z(du, dv) = z(du)\delta_u(dv) - z(du)z(dv)$ for $z \in \mathcal{P}(\mathbb{K})$.

The basic ingredients for the description of the limiting objects arising by applying the multiple space-time scale analysis and then letting $N \to \infty$ are three dynamics (see (α)-(γ) below). Namely two types of limiting dynamics and their equilibria of two (non-spatial) dynamics with state space $\mathcal{P}(\mathbb{K})$, a neutral one and one with selection-mutation, together with a third one, a random entrance law from mass zero at $t = -\infty$.

All these processes are nonlinear Markov processes, nonlinear because of the mean-field migration. (Recall a nonlinear Markov process is a process which arises typically by considering the generator of a Markov process and replacing a parameter (coefficient) by a function of the current *law* of the process). They are defined in points (α)-(γ) as follows.

(α) The *equilibrium* $\Lambda_\theta^{c,d}$ and the *equilibrium neutral Fleming–Viot process* $(\tilde{Z}_t^{\theta,c,d})_{t\geq 0}$ with immigration, emigration (with immigration source θ but without selection and mutation):

Fix $\theta \in \mathcal{P}(\mathbb{K})$ and then define the $\mathcal{P}(\mathbb{K})$-valued processes

$$(Z_t^{\theta,c,d})_{t\geq 0}, \tag{4.23}$$

as realization of the unique solution of the $(L_\theta^{c,d}, \delta_\theta)$-martingale problem on $C([0,\infty), \mathcal{P}(\mathbb{K}))$.

We define a martingale problem generator $L_\theta^{c,d}$ as a linear operator on $C_b(\mathcal{P}(\mathbb{K}), \mathbb{R})$ acting on the domain $\mathbb{F} \subseteq C_b(\mathcal{P}(\mathbb{K}), \mathbb{R})$. Here $\mathbb{F}$ is the space of functions F on $\mathcal{P}(\mathbb{K})$, which are twice continuously differentiable with bounded derivatives (recall (3.20) and (3.21) for the definition of the differential operators). The operator maps $\mathbb{F}$ into the space of bounded continuous functions on $\mathcal{P}(\mathbb{K})$ as follows:

$$\left(L_\theta^{c,d} F\right)(z) = c \int_\mathbb{K} \frac{\partial F}{\partial x}(z)[u](\theta(du) - x(du)) \tag{4.24}$$

$$+ d \int_\mathbb{K} \frac{\partial^2 F}{\partial z \partial z}(z)[u,v] Q_z(du, dv), \quad z \in \mathcal{P}(\mathbb{K}).$$

The process $Z^{\theta,c,d}$ has a unique equilibrium state and equilibrium process (compare [DGV]) denoted:

$$\Lambda_\theta^{c,d}, \ (\tilde{Z}_t^{\theta,c,d})_{t\in\mathbb{R}}. \tag{4.25}$$

(β) The *equilibrium* $\Gamma_{\rho,\chi}^{m,s,c,d}$ and the *equilibrium process* $(\tilde{Z}(t))_{t\geq 0}$ of the nonlinear Fleming–Viot process with immigration-emigration, selection and mutation.

Fix a mutation matrix M with mutation rate m and fitness function χ with selection rate s, migration rate c and with a resampling rate d.

Consider the realization of the unique solution (here we suppress all the parameters in the notation)

$$(Z(t))_{t\geq 0} \tag{4.26}$$

of the *nonlinear $(L_{M,\chi}^{\pi_t}, \delta_z)$-martingale problem* on $C([0,\infty), \mathcal{P}(\mathbb{K}))$ generated by the linear operator $L_{M,\chi}^{\pi_t}$ on $C_b(\mathcal{P}(\mathbb{K}), \mathbb{R})$ mapping from its domain $\mathbb{F}$ into $C_b(\mathcal{P}(\mathbb{K}), \mathbb{R})$ as defined below, where either $\pi_t = \mathcal{L}(Z(t))$ or we have a *random-evolution* driven by $(\pi_t)_{t\geq 0}$, in this case π_t is a random process independent of the evolution.

4.2 Basic Scenario and the Hierarchical Mean-Field Limit

Alternatively if necessary in the context, for example if we want to vary parameters, we write for Z instead of (4.26):

$$Z_{M,\chi}^{\pi_t,c,d}(t) \text{ or } Z_{M,\chi}^{\pi_t}. \tag{4.27}$$

To define $L_{M,\chi}^{\pi_t}$ let $\pi \in \mathcal{P}(\mathcal{P}(\mathbb{K}))$, and let the domain $\mathbb{F} \subseteq C_b(\mathcal{P}(\mathbb{K}), \mathbb{R})$ consist of functions $F \in C_b^2(\mathcal{P}(\mathbb{K}), \mathbb{R})$ with bounded and continuous derivatives of first and second order. Then $L_{M,\chi}^{\pi_t} = L_{M,\chi}^{\pi_t,c,d}$ maps $\mathbb{F}$ into $C_b(\mathcal{P}(\mathbb{K}), \mathbb{R})$ as follows (here we suppress the neutral parameter c and d in the notation):

$$L_{M,\chi}^{\pi}(F)(z) = c \int_{\mathbb{K}} \frac{\partial F(z)}{\partial z}(u) \left(\left[\int_{\mathcal{P}(\mathbb{K})} z' \pi(dz') \right] (du) - z(du) \right) \tag{4.28}$$

$$+ s \int_{\mathbb{K}} \left\{ \frac{\partial F(z)}{\partial z}(u) \left(\chi(u) - \int_{\mathbb{K}} \chi(w) z(dw) \right) \right\} z(du)$$

$$+ m \int_{\mathbb{K}} \left[\int_{\mathbb{K}} \frac{\partial F(z)}{\partial z}(v) M(u, dv) - \frac{\partial F(z)}{\partial z}(u) \right] z(du)$$

$$+ d \int_{\mathbb{K}} \int_{\mathbb{K}} \frac{\partial^2 F(z)}{\partial z \partial z}(u, v) Q_z(du, dv).$$

It has been established in [DG99, 99] that the $(L_{M,\chi}^{\pi_t}, \delta_z)$-martingale problem w.r.t. $C_b^2(\mathcal{P}(\mathbb{K}), \mathbb{R})$ is well-posed for the case

$$\pi_t = \mathcal{L}[Z(t)]. \tag{4.29}$$

The proof of existence and uniqueness given there carries over to the case of independent random π_t.

The marginal law of the process (an element of $\mathcal{P}(\mathcal{P}(\mathbb{K}))$) has for $t > 0$ and appropriate initial state a density with respect to the Lebesgue measure on $\mathbb{R}^{|\mathbb{K}|}$ restricted to the $(|\mathbb{K}|-1)$-dimensional simplex. For appropriate initial state for $t \geq 0$ we denote the density by:

$$u(t, x) \quad , \quad t \in [0, \infty), x \in \mathcal{P}(\mathbb{K}). \tag{4.30}$$

It satisfies the adjoint equation to (4.28), the socalled *McKean–Vlasov equation*:

$$\frac{\partial u}{\partial t} = (L_{M,\chi}^{\pi})^*(u). \tag{4.31}$$

Here for a linear operator G, G^* is the linear operator acting on the domain $D^{G^*} \subseteq L_+^1(\mathcal{P}(\mathbb{K}), \lambda^{|\mathbb{K}|})$ mapping into $L_+^1(\mathcal{P}(\mathbb{K}), \lambda^{|\mathbb{K}|})$, with λ^d the d-dimensional

Lebesgue measure restricted to Δ_{d-1} the $(d-1)$-dimensional simplex. The image G^*f, is uniquely (as L^1-element) determined by

$$< G^*f, \varphi > = < f, G\varphi > \text{ for all } \varphi \in C_b^2(\mathcal{P}(\mathbb{K}), \mathbb{R}) \qquad (4.32)$$

and D^{G^*} is the subset of $L_+^1(\mathcal{P}(\mathbb{K}), \lambda^{|\mathbb{K}|})$ where this problem has a solution.

We can formulate the equation (4.31) of course also as an equation on the level of measures, i.e. for $u(t, \cdot) \in \mathcal{P}(\mathcal{P}(\mathbb{K}))$ instead of densities. Namely we use the *mild form* of the equation, where the measure is evaluated on test functions on which L from the martingale problem acts in the obvious way.

We will show below in Subsection 4.3 under which conditions we have a unique equilibrium for this process, which we denote

$$\Gamma_{M,\chi}^{m,s,c,d} \quad \text{resp. } \Gamma_{\rho,\chi}^{m,s,c,d} \text{ if } M = 1 \otimes \rho, \quad \rho \in \mathcal{P}(\mathbb{K}). \qquad (4.33)$$

(γ) The *random nonlinear McKean–Vlasov entrance law* from $-\infty$, describing the transition from $\mathbb{K}_1$ to $\mathbb{K}_2$.

We consider here the case where we choose in (4.28) $\pi_t = \mathcal{L}[Z(t)]$ and where we are looking for a solution of (4.31) which is defined for $t \in \mathbb{R}$ with an *random* initial condition. This arises from the transition between separating time scales. We define random entrance laws now as follows.

Definition 4.3 (Random entrance law).

(a) A random entrance law P for the dynamic specified by (4.28) is the *law* of the process

$$(Z(t))_{t \in \mathbb{R}} \text{ with values in } \mathcal{P}(\mathbb{K}), \qquad (4.34)$$

which satisfies the following:
For every $t_0 \in \mathbb{R}$ and with $\mathcal{F}_t$ the σ-algebra generated by $\{(Z(s)), s \leq t\}$, the solution $\mathcal{L}_z[(Z(t))_{t \geq 0}]$ of the $(L_{M,\chi}^{\pi_t}, \delta_z)$-martingale problem defines a version of $P[(Z(t))_{t \geq 0} \in \cdot | \mathcal{F}_{t_0}, Z(t_0) = z]$.

(b) We say that the entrance law is *truly random* if the marginal laws of Z are themselves random. □

Example 6. Consider an entrance law from time $-\infty$. We see from the form of the equations (4.29) and (4.31) that a time shift will produce another entrance law from $-\infty$. Therefore if we take an entrance law from $-\infty$ and take a random variable, say $\mathcal{E}$, which is independent of the process $(Z(t))_{t \in \mathbb{R}}$ realizing the entrance law then $(Z(t + \mathcal{E}))_{t \in \mathbb{R}}$ is a truly random entrance law, if $Var(\mathcal{E}) > 0$.

We are later on particularly interested in the case where

$$\mathbb{K} = \mathbb{K}_1 \cup \mathbb{K}_2, \qquad (4.35)$$

with the following properties of the dynamic:

1. no mutation between $\mathbb{K}_1$ and $\mathbb{K}_2$, positive mutation rates on $\mathbb{K}_1$ and on $\mathbb{K}_2$ and
2. strictly higher fitness on $\mathbb{K}_2$ than on $\mathbb{K}_1$ (more precisely, fitness on $\mathbb{K}_2$ is no less than on $\mathbb{K}_1$ and there is a pair $k_2 \in \mathbb{K}_2$, $k_1 \in \mathbb{K}_1$ and k_2 has strictly higher fitness than k_1).

Typically we shall choose $\mathbb{K}_1 = E_j, \mathbb{K}_2 = E_{j+1}$ if we study the j-th step of transition from level-j to level-$(j+1)$ of fitness, i.e. types in E_j to types in E_{j+1}.
Then we consider entrance laws with the property that in probability

$$Z(t, \mathbb{K}_1) \longrightarrow 0 \text{ as } t \to \infty \text{ and } Z(t, \mathbb{K}_2) \to 0 \text{ as } t \to -\infty. \quad (4.36)$$

We will in particular show that there are solutions which satisfy, that in distribution:

$$e^{\alpha t} Z(t, \mathbb{K}_2) \xrightarrow[t \to -\infty]{} {}^*\mathcal{W}, \quad (4.37)$$

with $\alpha \in (0, \infty)$ and ${}^*\mathcal{W}$ a random variable independent of the evolution. This solution arises from one where the r.h.s. of (4.37) is 1 by a random time shift. Connecting with our examples above we would have the random time shift $\mathcal{E} = \alpha^{-1} \log^* \mathcal{W}$. These solutions are unique in the class of solutions where the l.h.s. of (4.37) is tight as we shall prove in Sections 8–10. The tightness assumption is as we shall see satisfied for the limit dynamics we are after.

4.3 Ergodic Theorem for the Nonlinear McKean–Vlasov and Spatial Fleming–Viot Process

In the formulation of our results an important role is played by the nonlinear Markov process $Z(t)$ (recall (4.28) for the definition) and its long-time behaviour. This object appears in the limit $N \to \infty$ of our renormalized processes and its equilibrium states are exactly the *quasi-equilibra* of the dynamics X^N as $N \to \infty$. Hence ergodic theorems for the limit dynamic Z are the key to having a rigorous way to establish the existence of quasi-equilibria for X^N and to obtain convergence to these states. But of course the longtime behaviour of $(X^N(t))_{t \geq 0}$ for fixed N and $t \to \infty$ is also of interest and can be studied with similar methods.

The qualitative behaviour of the socalled infinite population case for Z (deterministic system), that is, $d = 0$ is developed in [Bu]. In contrast to the deterministic case in the stochastic case $d > 0$ we must deal with the evolution of the probability law of the process $Z(t)$ and therefore we are dealing with infinite dimensional nonlinear dynamics. Here in the nonspatial case results were obtained by Ethier and Kurtz in [EK5]. In the spatial case there are results in Shiga-Uchiyama [SU] and Athreya-Swart [AS].

We turn now to the spatial and general type space and the McKean–Vlasov case. The purpose of this Section is to establish the *ergodicity* of the process $(Z(t))_{t\geq 0}$, that is, the uniqueness of the equilibrium state and the convergence as $t \to \infty$ of the probability law of $Z(t)$ to the unique equilibrium.

In the previous paper [DG99] we established uniqueness of the invariant measure for the McKean–Vlasov limiting dynamics denoted by $(Z(t))_{t\geq 0}$ under the assumption that the state-independent mutation strength $\bar{m}$ is sufficiently large compared to the selection coefficient s. However we need this result here for arbitrary positive $\bar{m}$ in order to deal with many level hierarchical models.

The intuitive idea is that for $\bar{m} > 0$, the historical path of every individual present at a late time includes at least one mutation to a new type sampled with ρ and is therefore independent of the initial state of the system. This suggests that the law of the system at late times is independent of the initial state, which excludes two invariant measures. However since the evolution of a single ancestral path depends in a nonlinear way on the state of the whole population, this conclusion is not immediate. Moreover even if the independence from the initial state is established it is still necessary to show that this also leads to convergence as $t \to \infty$ of $\mathcal{L}[Z(t)]$ towards the unique equilibrium. This "historical" argument seems difficult to carry out directly, so we will follow another route based on the duality (which on the other hand does have a historical interpretation).

Theorem 2 (Ergodicity of non-linear process and spatial process).

(a) *Consider the nonlinear mean-field process $(Z(t))_{t\geq 0}$ defined in (4.28). Here m is the mutation rate, M the mutation kernel, s the selection rate, and χ the fitness function which is assumed to be bounded.*

We also assume that the mutation kernel M is such that

$$mM \geq \bar{m}(1 \otimes \rho), \tag{4.38}$$

where ρ is a measure with support $\mathbb{K}$ and $\bar{m} > 0$. Finally assume that we have a resampling rate $d > 0$.

Then for the process $(Z(t))_{t\geq 0}$ on $\mathcal{P}(\mathbb{K})$ exists a unique equilibrium state which we denote by

$$\Gamma_{M,\chi}^{m,s,c,d}. \tag{4.39}$$

(b) *With Z as in (a) for every initial state $Z(0) \in \mathcal{P}(\mathbb{K})$:*

$$\mathcal{L}[Z(t)] \underset{t\to\infty}{\Longrightarrow} \Gamma_{M,\chi}^{m,s,c,d}. \tag{4.40}$$

(c) *Consider the spatially structured process $(X(t))_{t\geq 0}$ as defined by (3.24). Assume that*

- $x_\xi(0) \in \mathbb{I}_K$ for all $\xi \in \Omega_N$ where (recall (3.4))

$$\mathbb{I}_K = \bigcup_{j=0}^{K} E_j, \tag{4.41}$$

- $M(u, (\mathbb{I}_K)^c) = 0$ if $u \in \mathbb{I}_K$, that is mutations from $\mathbb{I}_K$ to $(\mathbb{I}_K)^c$ don't occur and
- that the mutation kernel $M(.,.)$ restricted to $\mathbb{I}_K$ has a decomposition (3.41) with $\bar{m} > 0$.

Then the process $(X(t))_{t \geq 0}$ has a unique invariant measure and is ergodic, i.e. the analogue of (4.40) holds. □

These ingredients introduced in Subsection 4.2 and the invariant measures above allow us in the next Subsection to introduce two important objects for the description of the large time-space scale behaviour.

4.4 Elements of Asymptotic Description: Two Interaction Chains

A key problem is to describe the *spatial* structure of the quasi-equilibria on the types $(E_j)_{j \in \mathbb{N}_0}$ during the transition in the steps $j \in \mathbb{N}_0$. This is accomplished by considering the space on various smaller time and space scales and leads to an object we call *interaction chain*. See [DG1, DG2, DG3], [DGV] and [DG96] for the basics on this object for neutral models.

In order to describe in our (non-neutral) situation the complete joint distribution of the sequence of blockaverages, corresponding to (4.12), in the limiting dynamics below we introduce in the presence of *rare* mutation *two different* interaction chains to cope with the new effects and to allow to describe both the preemergence and the fixation phase respectively:

- The first corresponds to the distribution at spatial scales $k = 1, \ldots, j+1$ when the quasi-equilibrium concentrated on E_j-types has been reached and E_{j+1} types have not yet appeared globally.
- The second describes the distribution at spatial scales $k = 1, \ldots, j+1$ in the (longer) time scale in which fixation of E_{j+1}-types has occurred but level $j+1$ selection is still negligible.

In other words the first object describes the situation *before emergence* of rare mutants giving fitter types, the second the situation immediately *after fixation* on the new fitter types.

The key idea is that in the natural time scale of the k sub-blocks, i.e. times $N^j + tN^k$ with t the time index, for $k \leq j - 1$ selection and mutation for level j, respectively level $j+1$, types is negligible and therefore given the distribution of the types on the spatial level j, respectively spatial level $j+1$, the system at smaller spatial scales evolves *only as neutral dynamics with migration*.

The interaction chains on level j are time-inhomogeneous Markov chains denoted

$$(M_k^j)_{k=-(j+1),\ldots,0}, \quad \text{respectively,} \ (M_k^{*,j})_{k=-(j+1),\ldots,0}, \tag{4.42}$$

with state spaces

$$\mathcal{P}(E_j), \text{ respectively } \mathcal{P}(E_{j+1}). \tag{4.43}$$

We now need to define the (i) initial laws of these Markov chains and (ii) we have to specify their transition kernels.

(i) *The initial laws of the two chains.*

For $j \geq 1$ the state M_{-j}^j denotes the equilibrium state of a McKean–Vlasov equation describing the limiting evolution of N interacting systems each which describes the empirical measure over a block of N^j sites and concentrated on E_j-types. Therefore recalling (4.33), (4.25) together with (3.43) we obtain as state the equilibrium on E_j, i.e. (cf. [DG99], Thm. 2 and (1.41)):

$$\mathcal{L}[M_{-j}^j(\cdot)] = \Gamma_{\varrho_j, \chi_j}^{m_j, s_j, c_j, d_j}, \text{ with } s_j = s\frac{d_j}{d_0}, d_j = d_0(1 + d_0 \sum_{k=0}^{j-1} c_k^{-1})^{-1}. \tag{4.44}$$

Then $M_{-(j+1)}^j$ is the limiting empirical average of the N blocks of size N^j and therefore

$$M_{-(j+1)}^j = \int_{\mathcal{P}(E_j)} \mu \, \mathcal{L}[M_{-j}^j](d\mu) = E[M_{-j}^j]. \tag{4.45}$$

This state is amended to the sequence as leftmost point.

In contrast to $M_{-(j+1)}^j$, which is deterministic, the initial state

$$M_{-(j+1)}^{*,j} \text{ is a } \textit{random} \text{ probability measure on } E_{j+1}, \tag{4.46}$$

whose law we can express in terms of its moments via a dual process or excursion measures in Section 10 and which satisfies

$$E[M_{-(j+1)}^{*,j}] = \rho_{j+1}. \tag{4.47}$$

The law of this state arises from the proportions at which the fitter types fixate after emergence and which are random.

(ii) *The transition kernels.*

The transition kernels of both the interaction chains in (4.42) will be determined at indices $-(j+1), -(j), -(j-1), \ldots, 0$ and as parameters they *only* involve

4.4 Elements of Asymptotic Description: Two Interaction Chains

resampling and migration since, recall (4.12), in time scales of order N^k with $k < j$, there is no mutation and selection taking place anymore if the system is of age $T(N)N^j$ or $S(N)N^j$ (in the limit $N \to \infty$ of course), where $S(N) \uparrow \infty$, $T(N) \uparrow \infty$ with $T(N) = o(\log N)$ and $S(N) = o(N)$ but $S(N) \log N \to \infty$. Recall the neutral nonlinear Markov process from (4.11) and its equilibrium (4.25).

Hence for the transition from time $-(j+1)$ to the state at time $-j$ it turns out we have (assuming that the mutation matrix $\{m_{j,j+1}(\cdot,\cdot)\}$ depends on ρ_{j+1} and m_j^+ as given by (3.43)):

$$\mathcal{L}[M_{-j}^j(\cdot) | M_{-(j+1)}^j] = \Lambda_{M_{-(j+1)}^j}^{c_j,d_j,(j)} \in \mathcal{P}(E_j), \tag{4.48}$$

$$\mathcal{L}[M_{-j}^{*,j}(\cdot) | M_{-(j+1)}^{*,j}] = \Lambda_{M_{-(j+1)}^{*,j}}^{c_j,d_j,(j+1)} \in \mathcal{P}(E_{j+1}). \tag{4.49}$$

We add in the notation of the equilibria here the indices $j, j+1$ to stress on which type space we are, i.e. instead of $\Lambda_\theta^{c,d}$ write $\Lambda^{c,d,(j)}$ if we are on the type space E_j and if this addendum is necessary to avoid confusion. For the transition from time $-(k+1)$ to time $-k$ (here $k < j$), i.e. on lower spatial levels $k < j$ we have a similar formula, but c_j, d_j is replaced by c_k, d_k, where

$$d_{k+1} = \frac{c_k d_k}{c_k + d_k}, \quad d_0 = d \quad \text{(cf. [DG99], Thm. 2 and (1.36))}. \tag{4.50}$$

Define therefore the kernel K_{-k}^j respectively $K_{-k}^{*,j}$ for the transition from time $-(k+1)$ to time $-k$ by the following transition kernels:

$$K_{-k}^j(\theta, d\theta') = \Lambda_\theta^{c_k,d_k,(j)}(d\theta') \text{ on } \mathcal{P}(E_j) \times \mathcal{P}(E_j) \text{ resp.} \tag{4.51}$$

$$K_{-k}^{*,j}(\theta, d\theta') = \Lambda_\theta^{c_k,d_k,(j+1)}(d\theta') \text{ on } \mathcal{P}(E_{j+1}) \times \mathcal{P}(E_{j+1}). \tag{4.52}$$

Remark 24. Note the fact that these objects are now in both cases defined based on dynamical parameters c, d etc. on spatial level j rather than on spatial level $(j+1)$ in the case of E_{j+1}-types as in [DG99]. This is due to the fact that we have now included rare mutation leading to mutation-selection induced equilibria which are migration-resampling equilibria but the input from the mean-field term occurring previously in neutral models was just the next order block average where now the relevant input is coming from mutation-selection instead and the E_{j+1}-types fixate before the migration over distance $j+1$ becomes relevant.

(iii) *Definition of interaction chains*

The ingredients from (i) and (ii) above allow us now to complete the definition of a key component of the analysis via renormalization and hierarchical mean-field limit:

Definition 4.4 (Interaction chains).

(a) The *interaction chain on level j* on $\mathcal{P}(E_j)$, denoted $(M_k^j)_{k=-(j+1),\ldots,0}$ is the time inhomogeneous Markov chain with (recall (4.44)):

transition kernel K_{-k}^j at time $-(k+1)$, $k < j$, and random initial state M_{-j}^j,

given in (4.44). (4.53)

We set

$$M_{-(j+1)}^j = E[M_{-j}^j]. \tag{4.54}$$

(b) The *interaction chain on level j after transition to E_{j+1}-types*, denoted $(M_k^{*,j})_{k=-(j+1),\ldots,0}$ is the time inhomogeneous Markov chain with (recall (4.49)):

transition kernel $K_{-k}^{*,j}$ at time $-(k+1)$, $k < j-1$,

and random initial state $M_{-(j+1)}^{*,j}$, (4.55)

where the latter arises as the random proportions of fixation and will be discussed in (4.94)–(4.97). □

This completes the list of concepts needed to formulate the basic limit results.

4.5 Limit Theorems I: Quasi-equilibria and Neutral Phase

With the concepts and the notation just introduced we can formulate our results which rigorously verify the basic scenario for the j-th step transition from types E_j to E_{j+1} described in Subsection 4.2. These identify five different phases of the evolution at each step of transition (indexed by $j \in \mathbb{N}_0$) from a level-j world of fitness to a level-$(j+1)$ world of fitness. We split the results on that scenario in two parts, Subsection 4.5 and Subsection 4.6.

Here in 4.5 we formulate the three basic limit results describing the system which has evolved for times of orders $T(N)N^j$, $S(N)N^j$ and tN^{j+1} successively, i.e. phases 0, 3 and 4 in the terminology of Subsection 4.2, while phases 1 and 2 on emergence and fixation are treated in Subsection 4.6. In each of these phases we describe the behaviour of the system in the spatial scales indexed by $k = 0, 1, \cdots, j+1$ and the corresponding natural time scales.

Remark 25 (Downward mutation and path-topology). In order to claim in the presence of downward mutations from E_{j+1} to E_j or even E_k with $k < j$ that a dynamic converges to one with values on measures on E_{j+1} in path space, we have to discuss the problem of the suitable topology. This is the Meyer–Zheng topology

4.5 Limit Theorems I: Quasi-equilibria and Neutral Phase

or the uniform topology depending on whether our time scale is less than tN^{j+1} or of this order. We discuss this in detail in Section 12, namely in Subsubsection 12.2 and 12.3.

We use the following convention: If $Z_t^{\theta,c,d}$ from (4.23) is defined for $\mathbb{K} = E_j$ we write

$$Z_t^{\theta,c,d,(j)}. \tag{4.56}$$

Then we can state a result describing the type and geographical structure of the population which has "equilibrated" on the types in E_j (recall (4.3)–(4.10), (4.53) and (4.26) for notation):

Theorem 3 (Phase 0: description of system in type E_j-quasi-equilibrium). *Assume that the system starts in the state completely occupied by the basic type. Then for every $j \in \mathbb{N}_0$:*

$$\mathcal{L}\left[\left(x_{\xi,|k|}^N(T(N)N^j)\right)_{k=-j-1,\ldots,0}\right] \underset{N\to\infty}{\Longrightarrow} \mathcal{L}\left[(M_k^j)_{k=-j-1,\ldots,0}\right]. \tag{4.57}$$

Furthermore with (recall (4.11)) $k = 0, -1, \cdots, -(j+1)$:

$$\mathcal{L}\left[\left(x_{\xi,|k|}^N(T(N)N^j + tN^{|k|})^+\right)_{t\in\mathbb{R}}\right] \underset{N\to\infty}{\Longrightarrow} \mathcal{L}\left[(\tilde{Z}_t^{M_{k-1}^j,c_{|k|},d_{|k|},(j)})_{t\in\mathbb{R}}\right]. \quad \square \tag{4.58}$$

Once the fixation on the new types from E_{j+1} has been completed (i.e. after phases 1 and 2), we have the following situation (recall also (4.55)):

Theorem 4 (Phase 3: intermediate neutral evolution after fixation on E_{j+1}). *Assume that the system starts in the state completely occupied by the basic type. Then for every time scale j:*

$$\mathcal{L}\left[\left(x_{\xi,|k|}^N(S(N)N^j)\right)_{k=-j-1,\ldots,0}\right] \underset{N\to\infty}{\Longrightarrow} \mathcal{L}\left[(M_k^{*,j})_{k=-j-1,\ldots,0}\right]. \tag{4.59}$$

Furthermore with $k = 0, -1, \cdots, -(j+1)$:

$$\mathcal{L}\left[\left(x_{\xi,|k|}^N(S(N)N^j + tN^{|k|})^+\right)_{t\in\mathbb{R}}\right] \underset{N\to\infty}{\Longrightarrow} \mathcal{L}\left[(\tilde{Z}_t^{M_{k-1}^{*,j},c_{|k|},d_{|k|},(j+1)})_{t\in\mathbb{R}}\right]. \quad \square \tag{4.60}$$

Remark 26. Note that M_k^j is concentrated on level j-types, while $M_k^{*,j}$ is concentrated on level $(j+1)$-types but note that in both cases we refer to a spatial region of size j and its subblocks of size k. This means in particular that if we observe the transitions in the various steps indexed by j, then in each such step the system has long periods of normal evolution, where a selection-mutation equilibrium builds up on the j-th level of fitness and shorter periods, where rare mutations occur which rapidly spread and take over the system leading to an $(j+1)$-th level equilibrium determined by migration and resampling and based on the relative frequency of the various mutations.

Remark 27. The dependence structure between large blockaverages at different sites arises from migration. Therefore if $t_N = o(N^{j+1})$ we have that the empirical distributions of j-blocks of balls in distance $j+1$ are independent as $N \to \infty$, i.e. for $d(\xi, \xi') \geq j+1$ (recall (9.342)):

$$[\mathcal{L}[(\Xi^N_{\xi,j}(t_N), \Xi^N_{\xi',j}(t_N)))] - (\mathcal{L}[(\Xi^N_{\xi,j}(t_N))])^{\otimes 2}] \underset{N\to\infty}{\Longrightarrow} 0. \quad (4.61)$$

Next we can ask the question how the system relaxes into the new mutation-selection quasi-equilibrium once the invasion by the rare mutants has taken place. This relaxation will take place in times of order N^{j+1} and now we observe everything in a spatial region of size $j+1$. This means we want to look at:

$$\mathcal{L}[(\Xi^N_{\xi,j}(tN^{j+1}))_{t\geq 0})], \text{ respectively } \mathcal{L}\left[\left(x^N_{\xi,j+1}(tN^{j+1})\right)_{t\geq 0}\right] \quad (4.62)$$

in the limit $N \to \infty$. This limit process is given by a dynamics which lives on the set of E_{j+1}-types. Recall Theorem 3, which describes then the behaviour and the structure of the population after the time considered here in (4.62) and in part (b) of the theorem below we connect phase 4 and phase 0 of the next transition. Recall that the intensity measure of the random measure $M^{*,j}_{-(j+1)}$ is ϱ_{j+1}.

Theorem 5 (Phase 4: approach of new type-E_{j+1} quasi-equilibrium). *Assume that the system starts in the state completely occupied by the basic type.*

(a) *Let P, with marginals denoted $(P(t), t \geq 0)$ be the solution to the $L^{\pi_l}_{M,\chi}$-martingale problem of (4.28) defined for $\mathbb{K}$ equal to E_{j+1}, with the following specification of the parameters (recall (4.50), (4.46), (3.43), (4.44), (3.32)):*

$$P(0) = \delta_{M^{*,j}_{-(j+1)}}, \quad (4.63)$$

$$c = c_{j+1}, \ d = d_{j+1}, \ M(\cdot, \cdot) = m_{j+1,j+1}, m = m_{j+1}, \ s = s_{j+1}, \ \chi = \chi_{j+1}. \quad (4.64)$$

Then for every $\ell \in \mathbb{N}_0$, with proj_ℓ denoting the mapping on measures on type space induced by the projection of the type space on E_ℓ, we have:

$$\mathcal{L}[(\text{proj}_{(j+1)} \circ \Xi^N_{\xi,j+1}(tN^{j+1}))_{t\geq 0}] \underset{N\to\infty}{\Longrightarrow} \mathcal{L}[(P(t))_{t\geq 0}], \quad (4.65)$$

$$\mathcal{L}\left[\left(\text{proj}_{(j+1)} \circ x^N_{\xi,j+1}(tN^{j+1})\right)_{t\geq 0}\right] \underset{N\to\infty}{\Longrightarrow} \mathcal{L}\left[\left(\int_{\mathcal{P}(E_{j+1})} x(P(t))(dx)\right)_{t\geq 0}\right]. \quad (4.66)$$

(b) *Furthermore, connecting with the result in Theorem 1 on the next level of fitness consider now* $t_N \to \infty$, *but* $t_N = o(\log N)$. *Then*

$$\mathcal{L}[\mathrm{proj}_{(j+1)} \circ \Xi^N_{\xi,j+1}(t_N N^{j+1})] \underset{N \to \infty}{\Longrightarrow} \delta_{\Gamma^{m,s,c,d}_{\rho,\chi}}, \qquad (4.67)$$

with the parameters as specified in (4.64). □

Remark 28. Recall from Subsection 4.3 that $P(t)$ converges from $t \to \infty$ to the equilibrium $\Gamma^{m,s,c,d}_{\rho,\chi}$.

With these results in Theorems 3–5 we have a complete set of limiting dynamics, which allows us to investigate more specific questions in the qualitative behaviour in the next step. Namely a key point is to clarify how the transition (phase 1 and 2 in the terminology of Subsection 4.2) between the regimes with states concentrated on types in E_j versus types in E_{j+1} and described in the first two theorems above actually looks like.

4.6 Limit Theorems II: Emergence, Fixation, Droplet Formation of Rare Mutants

In this section we analyze some specific features of our model motivated by questions from evolutionary biology which were outlined in the introduction. These questions involve more detailed descriptions of what happens in phase 1 and phase 2 something we did not cover in the previous theorems. In particular we obtain a detailed description of the emergence of rare mutants of types in E_{j+1} in a world of E_j-types and the process of subsequent invasion and fixation of other colonies by E_{j+1}-types.

In order to make precise what we mean by *emergence* we introduce the following

Definition 4.5 (Emergence time). Consider for a fixed geographic site ξ and small $\varepsilon > 0$ the random times $\bar{T}^{\varepsilon,j}_N$ and $T^{\varepsilon,j}_N$ defined by (recall E_j is the set of the level j-types):

$$\bar{T}^{\varepsilon,j}_N = \inf(t \,|\, x^N_{\xi,j+1}(t)(E_{j+1}) \geq \varepsilon), \qquad T^{\varepsilon,j}_N = \inf(t \,|\, x^N_{\xi,j}(t)(E_{j+1}) \geq \varepsilon).$$
(4.68)

Define the last microscopic time in a tagged j-block when the $(j+1)$-types vanish:

$$\hat{T}^{\varepsilon,j}_N = \sup(t \,|\, x^N_{\xi,j}(\bar{T}^{\varepsilon,j}_N + tN^j)(E_{j+1}) = 0). \qquad \square \qquad (4.69)$$

The times $\bar{T}^{\varepsilon,j}_N, T^{\varepsilon,j}_N$ describe the point of time at which the mutants of type in $E_{(j+1)}$ have achieved a positive spatial density of at least ε in the $(j+1)$-block around 0, respectively the time at which this happened in a single tagged

j-blockaverage in the $(j + 1)$-block. From these times on we see both in block of size $(j + 1)$ or smaller an evolution involving both E_j and $E_{(j+1)}$ types and leading eventually to fixation on the latter. The random times $\bar{T}_N^{\varepsilon,j}$, $T_N^{\varepsilon,j}$ are studied in the limit $N \to \infty$ for fixed ε and for $\varepsilon \downarrow 0$ by observing the intensity of the types in j-blocks and then in smaller tagged subblocks.

Next turn to the *fixation dynamics*. In order to formulate the questions on the qualitative behaviour, observe: Once the population of E_{j+1}-types has reached a positive intensity in the $(j + 1)$-block, it is natural to look at the spatial structure of the process in successively smaller subblocks observed over times corresponding to their natural time scale as we introduced in Subsection 4.1 but now measured *after* the emergence of the rare mutant. In our formalism this involves the evolution of the population in tagged subblocks of size $k = j, j - 1, \cdots, 1, 0$ in their respective natural time scales tN^k.

Definition 4.6 (Fixation dynamic: renormalized system in shifted time horizon). Consider the (j, k)-renormalized system in the shifted time horizon:

$$\mathcal{L}[(x_{\xi,k}^N((\bar{T}_N^{\varepsilon,j} + tN^k)^+))_{t \in \mathbb{R}}], \tag{4.70}$$

together with the sequence of laws of the empirical measures

$$\{\mathcal{L}[(\Xi_{\xi,j}^N(\bar{T}_N^{\varepsilon,j} + tN^j)^+)_{t \in \mathbb{R}}, j \in \mathbb{N}\}, \tag{4.71}$$

which captures with varying $j \in \mathbb{N}_0$ the respective fixation dynamic (in the time variable t) on next higher types and with varying k the spatial structure of this process. □

This object allows us to study the process of *fixation* in the time variable t, that is the process leading to a concentration of the probability to see types in E_{j+1} near the value 1 as $t \to \infty$. In the limit $N \to \infty$ the cases $k = j$ and $k < j$ are very different. For $k = j - 1, \cdots, 0$ we expect neutral dynamics in the natural time scales with intensities of types dictated by the $k = j$ dynamics whereas the latter involves selection leading to the fixation.

Outline Subsection 4.6 In Subsubsection 4.6.1 we formulate the *questions*, in Subsubsection 4.6.2 we introduce the required *concepts* and *limit processes* to answer the questions, and in Subsubsection 4.6.3 we state the *results* on how the emergence of and invasion by rare mutants develops up to the final take-over (fixation) in the hierarchical mean-field limit $N \to \infty$.

4.6.1 Questions Concerning the Emergence and Fixation by Rare Mutants

The last three theorems in Subsection 4.5 for the transition from E_j to E_{j+1} raise the question what can be said about the intermediate phases between phase 0 where the population is asymptotically as $N \to \infty$ composed of types in E_j and phase 3 or 4 where the population is composed of types in E_{j+1}, all described in the previous theorems. How do we move from the emergence of E_{j+1}-types to fixation on these types? This is related to another question, namely, what are the distributions of the times needed for the transition from fixation on E_j-types to fixation on $E_{(j+1)}$-types and how is the transition achieved and what are its qualitative properties.

We focus first in (α) on the points concerning the phase of preemergence (droplet formation) and emergence (macroscopic positive $E_{(j+1)}$-type density) and then in (β) on the process of fixation.

(α) *Droplet growth and various random times concerning emergence: four key questions.*

Before we formulate the questions let us reflect which qualitative features of the model are behind the phenomena of emergence and see which factors determine size and distribution of $\bar{T}_N^{\varepsilon,j}$, $T_N^{\varepsilon,j}$ as $N \to \infty$?

Some features of the emergence dynamic

The first point is that in the $(j+1)$-block in *some* j-blocks mutations to level E_{j+1} do push the state to averages above ε, even if this is not true for a *tagged* j-block, that is a typical j-block. This allows the rare mutants to occur somewhere in space that is in some random j-block with substantial intensity forming a germ which might expand. Rare mutation outside the $(j+1)$-block are irrelevant since they cannot migrate fast enough into our $(j+1)$-block to influence the emergence.

Focus on these j-blocks where rare mutants have occurred significantly. Here the next important point is the loss of mass of the E_j types due to deleterious mutation to types in E_{j-1} or from the maximal E_j-type to less fitter ones which is of order N^{-j} but in contrast the loss of E_{j+1} types to E_j or E_{j-1} types is much smaller, namely of order $N^{-(j+1)}$. The lower order mutants are then absorbed by both the E_{j+1} types *and* by the E_j-types due to selection at rates of the same order. This means that the E_{j+1} types as opposed to the E_j types do not lose mass due to deleterious mutation in the time scale $C \cdot (\log N) N^j$ (time scale of emergence) and therefore they increase their relative frequency steadily at the cost of the E_j-types in a couple of j-blocks.

Then the next step is that the number of these j-blocks with substantial E_{j+1} mass is increasing as time proceeds through migration. This is still a process involving a lot of randomness due to the resampling taking place in those blocks causing small mutant populations to collapse with a certain probability. But after a while we don't yet have positive spatial density but already a large number of E_{j+1}-blocks (blocks of order j). Hence after a while this droplet grows *deterministically* and emergence on a global scale occurs if order N such blocks of size N^j exist.

The whole process of emergence is therefore of a stochastic nature since it involves first of all some rare events of mutation and then survival or the mutants in the stochastic resampling process. Only later if we have many such colonies of higher type for the expansion process a law of large number effect can set in. Therefore we see the structure of a random entrance law of a deterministic evolution equation. Furthermore a key role is played by migration since the process involves rare mutation at distant sites and subsequent invasion. This scenario is of course what determines the structure of $\bar{T}_N^{\varepsilon,j}, T_N^{\varepsilon,j}$.

Features of emergence time.

The key observation concerning the random variables $\bar{T}_N^{\varepsilon,j}$ is that it consists of parts of a quite different nature. Namely the random time $\bar{T}_N^{\varepsilon,j}$ involves the following three components which have *different orders of magnitude* (all statements asymptotically as $N \to \infty$).

(1) During a first *short random* part of the transition "some" mass on types in E_{j+1} builds up *randomly* somewhere in a sparse subset of the interacting set of N of the j-blocks forming the $(j + 1)$-block, more precisely a small collection (that is, negligible proportion of the N subblocks) having substantial intensity on E_{j+1}. These blocks represent the *droplet* of the E_{j+1} type.

(2) Later on the size of this collection of colonies with E_{j+1}-types, then grows during a much larger time interval *deterministically* till the j-blocks colonised by types from E_{j+1} have positive intensity, i.e. make up a positive fraction of the population of the $(j + 1)$-block.

(3) Finally the macroscopic positive intensity of E_{j+1}-types has to increase to the prescribed value ε and this should also due to the law of large numbers and happens in a *deterministic* way but much faster than the previous period.

How to turn these qualitative ideas into quantitative statements?

The key question related to (1) is whether we have that recall (4.16):

$$(\mathbf{J}_{j,t}^N)_{t \geq 0} \Longrightarrow (\mathbf{J}_{j,t})_{t \geq 0}, \text{ as } N \to \infty, \tag{4.72}$$

for some atomic measure-valued process $\mathbf{J}_j$ on $[0, 1] \times E_{j+1}$ and how fast the total mass of this process grows with macroscopic time t.

Concerning (2) we note that after a *short random time* we have a fairly large droplet and then we need a *large deterministic time* of the form $CN^j \log N$ (order $CN^j \log N$, is short compared to N^{j+1} but large compared to N^j) to increase the droplet to size of order N^{j+1}, i.e. *all the way up to a "positive" intensity in the whole $(j + 1)$-block*.

Finally concerning (3) observe that some further *deterministic short time* is needed, time (order N^j) to increase this mass by migration and selective advantage in times of order N^j deterministically to an intensity ε or even an intensity close to 1. This short time is of the same order as the random time needed to get the emergence of the first j-blocks with substantial density of types in E_{j+1} going.

Hence combining the three points we expect that $\bar{T}_N^{\varepsilon,j}$ has the form:

$$CN^j \log N + (\bar{T}_\infty^\varepsilon + \bar{t}_\infty^{\varepsilon,j})N^j, \tag{4.73}$$

for a specific constant C and a non-degenerate *random* variable $\bar{T}_\infty^j$ and a *deterministic* (finite) time $\bar{t}_\infty^{\varepsilon,j}$, everything asymptotically in the limit $N \to \infty$.

4.6 Limit Theorems II: Emergence, Fixation, Droplet Formation of Rare Mutants

A similar decomposition holds for $T_N^{\varepsilon,j}$ but then instead of a deterministic $\bar{t}_\infty^{\varepsilon,j}$ we have a *random* time $t_\infty^{\varepsilon,j}$, since the j-blocks in this time scale still exhibit fluctuations.

This means that in order to understand all three parts of the emergence time we have to observe the system at the critical time scale $CN^j \log N$ and consider time parameterizations of the form

$$[CN^j \log N + tN^j] \text{ with } t \in (-\infty, \infty) \text{ as } N \to \infty. \tag{4.74}$$

Four questions.

Hence we see from the above discussion that we have to answer the following four collections of questions about $\bar{T}_N^{\varepsilon,j}, T_N^{\varepsilon,j}, \hat{T}_N^{\varepsilon,j}$ and their structure:

(o) How do we get from a random number of sites colonized by the rare mutant in times of order 1 to a growing number of such colonies? Is the growth of that droplet then deterministic later on? How fast is that growth? How long for example does it take to get $(N^j)^\delta$ with $\delta \in (0,1)$ sites where the fitter type rules? Note that such a droplet is not visible for an observer looking at a sample of tagged sites.

(i) Can we indeed split $\bar{T}_N^{\varepsilon,j}$ asymptotically in three components, that is, asymptotically decompose this time as $[(C \log N) + (\bar{T}_\infty^{\varepsilon,j} + \bar{t}_\infty^{\varepsilon,j})]N^j$ such that the elements are describing the first occurrence of E_{j+1}-types somewhere in space (corresponding to time $\bar{T}_\infty^{\varepsilon,j} N^j$) and subsequent built-up of rare mutant types E_{j+1} in time $CN^j \log N$ and then climbing up to ε in time $\bar{t}_\infty^{\varepsilon,j} N^j$? For the decomposition of $\bar{T}_\infty^{\varepsilon,j} + \bar{t}_\infty^{\varepsilon,j}$ we have to fix some very tiny reference $\tilde{\varepsilon}$ where we put $\bar{t}_\infty^{\tilde{\varepsilon},j} = 0$.

(ii) What can be said about the constant C and the law of $\bar{T}_\infty^{\varepsilon,j}$? What is the relation between $\bar{T}_N^{\varepsilon,j}$ and $T_N^{\varepsilon,j}$? Does $\bar{T}_N^{\varepsilon,j}$ properly rescaled (subtracting $CN^j \log N$ and then dividing by N^j) converge in law as $N \to \infty$ to a random limiting variable? How can we characterize the limiting laws?

(iii) What does the population look like in time $[0, \bar{T}_\infty^{\varepsilon,j} N^j]$ where the randomness sits. How can the time when the droplet $(\mathfrak{I}_{j,t}^N)_{t \geq 0}$ reaches a certain size be described by a limiting atomic measure-valued process? Then consider times $[\bar{T}_\infty^{\varepsilon,j} N^j, CN^j \log N]$. Does the latter exhibit a deterministically growing droplet of type E_{j+1} spatial j-blocks or more precisely do we get here a *Malthusian law*, with a droplet of total type-E_{j+1} mass $\mathcal{W}_j^* \exp(\beta_j t N^j)$ with random $\mathcal{W}_j^*$ and deterministic β_j? We will devote Theorem 8 to these particular questions.

The questions (o)–(ii) will be answered in Theorems 6, 7 and the questions in (iii) in Theorem 8, all in Subsubsection 4.6.3.

(β) *Emergence and fixation: three central questions on limiting dynamics*

Return to the object defined in (4.10), (4.11), (4.16), (4.70) and (4.71) which describe the type and spatial structure in the system of space-time multiscales.

We consider the limiting (as $N \to \infty$) processes of five basic statistics of our system which depend on the index indices $j = j + 1$ for the transition we consider and $k \leq j + 1$ to capture the spatial structure:

1. the global *blockaverage* (i.e. *empirical density*) in the $(j+1)$-block containing N different size-j subblocks or the *empirical measure* of the size-j blockaverages in the $(j + 1)$-block,
2. the local *blockaverage in a tagged j-block* for given empirical density in the $(j + 1)$-block,
3. the local *blockaverage in k-blocks* for $k = (0, 1, 2, \cdots, j - 1)$ for given higher level averages,
4. the global *total E_{j+1} mass* in a $(j + 1)$-block and
5. the *atomic measure-valued process* describing the E_{j+1}-droplet of j-blockaverages in the $(j + 1)$-ball in the preemergence regime.

These five objects we denoted

$$x_{\xi,j}^N, \quad \Xi_{\xi,j}^N, \quad \{x_{\xi,k}^N(\bar{T}_N^{\varepsilon,j} + tN^k), k \leq j - 1\}, \quad x_{\xi,j}^N(E_{j+1}), \quad \mathbf{J}_{j,t}^N. \quad (4.75)$$

We end up with the following three groups of questions concerning their behaviour as $N \to \infty$:

(a) Is it possible to renormalize the system by forming the level-$(j + 1)$, i.e. global intensity or more informative the empirical distribution of j-blockaverages, level-j block-averages, i.e. tagged block intensity, and rescale time measured on from $\bar{T}_N^{\varepsilon,j}$ viewed as origin in scale N^j, respectively form level-k blockaverages in this natural time scale N^k for $k \leq j - 1$ such that from a *global* perspective we get the following convergence relations:

$$\mathcal{L}\left[\left(\Xi_{\xi,j}^N(\bar{T}_N^{\varepsilon,j} + N^j t)^+\right)_{t \in \mathbb{R}}\right] \underset{N \to \infty}{\Longrightarrow} \mathcal{L}\left[(\mathcal{L}[\mathcal{Z}_\varepsilon^{*,j}(t)])_{t \in \mathbb{R}}\right], \quad (4.76)$$

$$\mathcal{L}[(x_{\xi,j+1}^N(\bar{T}_N^{\varepsilon,j} + N^j t)^+)_{t \in \mathbb{R}}] \underset{N \to \infty}{\Longrightarrow} \mathcal{L}[(\bar{\mathcal{Z}}_\varepsilon^{*,j}(t))_{t \in \mathbb{R}}], \quad (4.77)$$

and from a *local* perspective

$$\mathcal{L}\left[\left(x_{\xi,j}^N(\bar{T}_N^{\varepsilon,j} + N^j t)^+\right)_{t \in \mathbb{R}}\right] \underset{N \to \infty}{\Longrightarrow} \mathcal{L}\left[(Z_\varepsilon^{*,(j,j)}(t))_{t \in \mathbb{R}}\right], \quad (4.78)$$

$$\mathcal{L}\left[\left(x_{\xi,k}^N(\bar{T}_N^{\varepsilon,j} + N^k t)^+\right)_{t \in \mathbb{R}}\right] \underset{N \to \infty}{\Longrightarrow} \mathcal{L}\left[(Z_\varepsilon^{*,(j,k)}(t))_{t \in \mathbb{R}}\right], \quad k = j - 1, \cdots, 0, \quad (4.79)$$

with the three stochastic limiting dynamics $(\bar{\mathcal{Z}}_\varepsilon^{*,j}(t))_{t \in \mathbb{R}}$, $(Z_\varepsilon^{*,(j,j)}(t))_{t \in \mathbb{R}}$ resp. $(Z_\varepsilon^{*,(j,k)}(t))_{t \in \mathbb{R}}$ which describe the fixation of all the colonies in the block of size N^{j+1}, N^j, respectively N^k around 0 by the E_{j+1}-level types after the emergence of the rare mutant?

We write $\mathcal{Z}$ for quantities being derived from elements in $\mathcal{P}(\mathcal{P}(\mathbb{K}))$ and Z for elements in $\mathcal{P}(\mathbb{K})$ and in particular $\mathcal{Z}$ refers to a global perspective and Z to a local perspective.

(b) Can we explicitly determine the dynamic $(\mathcal{Z}_\varepsilon^{*,j}(t))_{t \in \mathbb{R}}$, $(Z_\varepsilon^{*,(j,j)}(t))_{t \in \mathbb{R}}$, respectively the somewhat different $(Z_\varepsilon^{*(j,k)}(t))_{t \geq 0}$ (here $k = j-1, \cdots, 0$)?

More specifically can we identify the three types of dynamics as follows:

– Can we write $\mathcal{Z}_\varepsilon^{*,j}(t)$ in the form

$$\mathcal{Z}_\varepsilon^{*,j}(t) = \mathcal{Z}^{*,j}(\bar{T}_\infty^{\varepsilon,j} + t), \text{ with } \bar{T}_\infty^{\varepsilon,j} = \bar{T}_\infty^j + \bar{t}_\infty^{\varepsilon,j}, \quad \bar{t}_\infty^{\varepsilon,j} \text{ deterministic} \tag{4.80}$$

and can we rewrite (4.76) in terms of an entrance law from a state with mass 0 on E_{j+1} at time $-\infty$ as follows:

$$\mathcal{L}[(\Xi_{\xi,j}^N(CN^j \log N + tN^j)^+)_{t \in \mathbb{R}}] \underset{N \to \infty}{\Longrightarrow} \mathcal{L}[(\mathcal{L}[\mathcal{Z}^{*,j}(t)])_{t \in \mathbb{R}}] \quad ? \tag{4.81}$$

– Can we describe the object $(\mathcal{Z}_\varepsilon^{*,(j,k)}(t))_{t \in \mathbb{R}}$ for $k = j$ independent of the parameter ε in terms of some $\bar{T}_\infty^{\varepsilon,j}$ as above and in terms of a random evolution for a suitably chosen Fleming–Viot process with mutation, selection and immigration-emigration where the immigration source provides the random input from the $(j+1)$-block? What is the nature of the immigration source?

– Do we obtain for $k \leq j-1$ simply *neutral* Fleming–Viot processes with immigration-emigration and immigration source constant in time?

(c) At the time of fixation of E_{j+1}-types but before mutation-selection of level $(j+1)$–types occurs what is the *relative proportion* of the different types in E_{j+1}? Is it deterministic or random? In the latter case what is its law?

In order to address all these (altogether seven circles of questions collected from (α) and (β)) we next introduce the required concepts and candidates for the limiting dynamics.

4.6.2 Characteristic Functionals for Emergence and Fixation: Limit Dynamics

We now describe, depending on the spatial scales, five classes of limiting evolutions involved in the process of emergence and fixation of types of successively higher levels. We consider the transition from E_j to E_{j+1} and let

$$\hat{E}_j := E_j \cup E_{j+1}. \tag{4.82}$$

We consider now the limiting processes for the five statistics we discussed above (4.75). All arising limit processes are of different nature. In the first and the fourth case the time scale is small compared to the one in which fluctuations of $(j+1)$-blocks are observed and we obtain random solutions to a deterministic nonlinear equation. For the cases three and four $k=j$ we have such fluctuations plus selection and mutation effects, for $k<j$ we then find neutral evolutions only. However in both cases we get stochastic evolutions.

We describe these processes in detail below in two steps, first the empirical measure of j-blockaverages in the $(j+1)$-block (statistics 2) and as a functional the global $((j+1)$-block) average (statistics 1) and then in Step 2 the two kinds of smaller blockaverages together (statistics 3). We build here on the work in Subsubsection 4.2.2.

The the fourth and fifth statistics is of a different nature. Namely we get as limit process for statistic 5 for $N \to \infty$ processes driven by Poisson random measures and involving *excursions* of certain Fleming–Viot diffusions and statistics 4 results is a functional of this process. It is possible to describe these processes explicitly but since this involves quite a bit of diffusion theory and therefore we give the lengthy details only in Section 10 and focus here on the main idea.

Step 1 *(Global functionals.)*

In this step we try to identify $\mathcal{Z}^{*,j}(t)$ from (4.80) and (4.81) and to relate this to $\mathcal{Z}_\varepsilon^{*,j}(t)$ from (4.76). First we have to identify the random measure on E_{j+1} which results from the rare mutations which occurred in the early growth of the population of level-$(j+1)$ types at a few spots in the block of size N^{j+1}. Second we have to describe the subsequent invasion and fixation process by type $(j+1)$-level types within the $(j+1)$-block, i.e. a process we get from $(\mathcal{Z}^{*,j}(t))_{t \in \mathbb{R}}$.

We begin with some observations. We first have to verify that in these time scales we obtain indeed in the limit $N \to \infty$ dynamics with state space $\hat{E}_j = E_j \cup E_{j+1}$. Note next that the fluctuations of the empirical distribution restricted to $\hat{E}_j$ in the $(j+1)$-block are asymptotically negligible in the time scales $o(N^{j+1})$ as $N \to \infty$. Therefore the evolution of $\mathcal{Z}^{*,j}(t, E_{j+1}))$ from time $CN^j \log N - t_0 N^j$ to $CN^j \log N + tN^j$, $t > -t_0$ with $t_0 < 0$ and $|t_0|$ large, will be deterministic given the state at the earlier time. This is described by a deterministic evolution, the *nonlinear McKean–Vlasov entrance law* from time $-\infty$ at zero E_j-mass denoted here:

$$(f^j(t))_{t \in \mathbb{R}}, \text{ on } \mathcal{P}(\hat{E}_j), \tag{4.83}$$

which we shall specify below. The process $(\mathcal{Z}^{*,j}(t))_{t \in \mathbb{R}}$ however carries traces of random events occurring early during the evolution and which are then preserved up to emergence and fixation. Therefore a complete description requires to combine these two effects which results in a *random entrance law* for the nonlinear McKean–Vlasov process.

4.6 Limit Theorems II: Emergence, Fixation, Droplet Formation of Rare Mutants

We need to specify next the dynamic of f^j to completely describe the situation. Begin with the ingredients. (Recall (3.32) for χ_j).

Definition 4.7 (Fixation dynamic: parameters). Consider the set

$$\mathbb{K} = \hat{E}_j = E_j \cup E_{j+1}, \tag{4.84}$$

as the basic type space. Define the fitness function $\hat{\chi}_j$ on $\hat{E}_j$ (recall (3.32)):

$$\hat{\chi}_j(v) = \max_{u \in E_j} \chi_j(u), \quad v \in E_{j+1}, \tag{4.85}$$

$$\hat{\chi}_j(v) = \chi_j(v), \quad v \in E_j.$$

Extend the mutation matrix $m_{j,j}(\cdot, \cdot)$ on E_j to a mutation matrix $\hat{m}_{j,j}(u, \cdot)$ on $\hat{E}_j$ by putting it equal to zero if either entry is in E_{j+1}. The resulting matrix we call

$$\hat{m}_{j,j}(= m_{j,j}(k, \ell) \quad ; \quad k, \ell \in \hat{E}_j). \qquad \square \tag{4.86}$$

The key process ruling the emergence of level-$(j+1)$ types can be defined as follows. Compare Proposition 4.9 below for existence, uniqueness questions.

Definition 4.8 (Fixation dynamic of E_{j+1}-types in $(j+1)$-blocks). Define the random McKean–Vlasov dynamics $(f^j(t))_{t \in \mathbb{R}}$ with values in $\mathcal{P}(\hat{E}_j)$ via the generator $L_{M,\chi}^{\pi_t}$ as in (4.28) with the parameters

$$c = c_j, d = d_j \text{ and } mM = m_j \cdot \hat{m}_{j,j}, \chi = \hat{\chi}_j, \tag{4.87}$$

as prescribed in Definition 4.7 above and for a given element $A \in \mathcal{M}(E_{j+1})$, as the unique (mild) solution of the McKean–Vlasov equation

$$\frac{\partial}{\partial t} f^j(t) = \left((L_{M,\chi}^{\pi_t})^* f^j\right)(t), \quad \pi_t = \int_{\mathcal{P}(\hat{E}_j)} \mu \, f^j(t, d\mu), \tag{4.88}$$

with the additional properties:

$$\left(f^j(t)(E_{j+1})\right) \longrightarrow 0 \text{ a.s. as } t \to -\infty, \quad (f^j(t)(E_{j+1})) \xrightarrow[t \to \infty]{} 1 \text{ as } t \to \infty, \tag{4.89}$$

$$e^{\alpha_j |t|} \left(f^j(t)(E_{j+1})\right) \longrightarrow 1 \text{ a.s. as } t \to -\infty, \text{ for some } \alpha_j \in (0, s_j), \tag{4.90}$$

$$e^{\alpha_j |t|} f^j(t)(\cdot \cap E_{j+1}) \xrightarrow[t \to -\infty]{} A \in \mathcal{M}(E_{j+1}). \qquad \square \tag{4.91}$$

Remark 29. We have the property that

$$1 > f^j(t)(E_j) > 0 \text{ for } t \in \mathbb{R} \text{ and } 1 > f^j(t)(E_{j+1}) > 0 \text{ for } t \in \mathbb{R}. \tag{4.92}$$

To describe and define the *random* entrance law $\mathcal{Z}^{*,j}$ describing the limit dynamics of the empirical measures, we use besides f^j two further basic objects, a random variable

$$^*\mathcal{E}_j \text{ with values in } (0,\infty) \tag{4.93}$$

and a random probability measure

$$\hat{\mathcal{W}}_j^* \text{ on } E_{j+1}. \tag{4.94}$$

These objects describe the *random time shift* and the *random emergence proportions*, which arise in the early stage of emergence and represent the randomness in the appearance and success of rare mutants. These quantities are obtained from a finite random measure introduced next.

In Section 8, Section 9 and Section 10 we shall show that:

Proposition 4.9 (Random data for entrance law). *If there exists a limiting dynamics as given in (4.77) then there exists a number $\beta_j \in (0,\infty)$ such that*

$$e^{\beta_j |t|} \bar{\mathcal{Z}}^{*,j}(t, \cdot \cap E_{j+1}) \Longrightarrow {}^* \mathcal{W}_j(\cdot) \text{ as } t \to -\infty, \tag{4.95}$$

for some (unnormalized) random measure $\mathcal{W}_j^*$ on E_{j+1} with

$$^*\mathcal{W}_j(A_1), {}^*\mathcal{W}_j(A_2) \text{ are independent if } A_1, A_2 \subseteq E_{j+1} \text{ with } A_1 \cap A_2 = \emptyset. \tag{4.96}$$

Here

$$^*\mathcal{W}_j \tag{4.97}$$

is a non-degenerate random variable and with law depending on the parameters c_j, d_j, s_j and m_j, ρ_j. This random variable is determined in terms of an atomic measure-valued stochastic dynamic $\mathfrak{z}_j$. See Section 9 and Section 10 for formulas.

□

Therefore we can define the (candidate for) limiting dynamics of fixation and its characteristic features as follows.

Definition 4.10 (Emergence characteristic and fixation dynamic for empirical measure). We are given the random measure $^*\mathcal{W}_j$ on E_{j+1} from (4.95).

(a) The pair

$$\left(^*\mathcal{E}_j, \frac{^*\mathcal{W}_j(\cdot)}{^*\mathcal{W}_j(E_{j+1})}\right) = (^*\mathcal{E}_j, {}^*\hat{\mathcal{W}}_j) \in \mathbb{R} \times \mathcal{P}(E_{j+1}), \tag{4.98}$$

4.6 Limit Theorems II: Emergence, Fixation, Droplet Formation of Rare Mutants

is called the emergence characteristic with the $\mathbb{R}$-valued random variable

$$^*\mathcal{E}_j = \log(^*\mathcal{W}_j(E_{j+1}))/\beta_j. \tag{4.99}$$

(b) The evolution $(\bar{Z}^{*,j}(t))_{t\in\mathbb{R}}$ is defined for a given realization of the random measure $^*\mathcal{W}_j$, in terms of the path $(f^j(t))_{t\in\mathbb{R}}$ with values in $\mathcal{P}(\hat{E}_j)$ (recall (4.91) as

$$\bar{Z}^{*,j}(t) = f^j(t+^*\mathcal{E}_j), \quad \lim_{t\to-\infty}\frac{f^j(t,\cdot\cap E_{j+1})}{f^j(t,E_{j+1})} = \frac{^*\mathcal{W}_j(\cdot)}{^*\mathcal{W}_j(E_{j+1})}. \tag{4.100}$$

(c) Moreover we set

$$\bar{t}_\infty^{\varepsilon,j} := \inf(s\,|\,f^j(s)(E_{j+1}) = \varepsilon), \tag{4.101}$$

and for $t \in \mathbb{R}$:

$$\bar{Z}_\varepsilon^{*,j}(t) := \bar{Z}^{*,j}(\bar{t}_\infty^{\varepsilon,j} -^*\mathcal{E}_j + t) \text{ so that } \bar{Z}_\varepsilon^{*,j}(0) = \varepsilon. \quad\square \tag{4.102}$$

Step 2 *(local functionals).*

In Step 1 we have described the limiting dynamics of the empirical measure over the $(j+1)$-block. Next we must determine what happens in tagged j-blocks and then what happens in tagged smaller k-blocks all the way down to single components. Hence the dynamic we want to define next in order to describe the limit of (4.78), is a *stochastic* dynamic which lives both on E_j and E_{j+1} types.

The initial state $\mathcal{L}[Z_\varepsilon^{*,(j,k)}(0)]$ is concentrated on configurations putting mass ε on E_{j+1} and this law converges as $\varepsilon \to 0$ to a law concentrated on E_j only namely the level j-equilibrium as it appears in (4.33). On the other hand its state at times tending to infinity will be in distribution equal to the law of $M_{-(j+1)}^j$, the equilibrium state of a Fleming–Viot process with immigration and emigration (recall (4.25)). What phenomena play a role here?

The first phenomenon we have to handle is that an influx of $(j+1)$-types is provided at rate c_j due to the communication with the N other j-blocks, this leads to a mean-field term.

The second phenomenon we have to handle is that level-j types (but not the level-$(j+1)$ types) have deleterious mutations at rate N^{-j} which are recovered rapidly, i.e. in time scale N^{j-1} because of the selective advantage of both level j and level $(j+1)$ over level $(j-1)$. Since level $(j+1)$ gains the ones lost by level-j at a rate proportional to its own mass we get an effective selection of level $(j+1)$ over level j. In addition the level-$(j+1)$ types grow in proportion because the maximal level-j type loses mass by mutation to level-j types not maximal which are absorbed both by level-$(j+1)$ types and the other level-j types which are superior.

Finally we have the case where $k \leq j - 1$ in its natural time scale (tN^k), in which case we simply have a neutral evolution on $E_j \cup E_{j+1}$ with immigration and emigration where immigration is from a constant source.

We need some more ingredients to handle the situation. We define (we shall prove that the limit exists) with $|_A$ denoting restriction to A:

$$\hat{\lambda}_j^\varepsilon = w - \lim_{N \to \infty} \mathcal{L}\left(x_{\xi,j}^N (\bar{T}_N^{\varepsilon,j})|_{E_j \cup E_{j+1}}\right). \tag{4.103}$$

Now we are ready to give precise formulations:

Definition 4.11 (Fixation processes for tagged j-blocks and k-blocks). Suppose we are given the process $(\mathcal{Z}_\varepsilon^{*,j}(t))_{t \in \mathbb{R}}$. Define now for $k \leq j$ the evolution of $(\mathcal{Z}_\varepsilon^{*,(j,k)}(t))_{t \in \mathbb{R}}$ conditioned on a realization of $(\pi_t)_{t \in \mathbb{R}}$ as the solution of the $L_{M,\chi}^{\pi_t}$-martingale problem defined in (4.28) with parameters

$$\pi_t = \bar{\mathcal{Z}}_\varepsilon^{*,j}(t) \text{ for } k = j, \text{ with } \bar{\mathcal{Z}}_\varepsilon^{*,j}(t) = \int_{\mathcal{P}(\hat{E}_j)} x \, [\mathcal{Z}_\varepsilon^{*,j}(t)](dx) \tag{4.104}$$

$$\pi_t \equiv Z_\varepsilon^{*,(j,k+1)}(0), \quad \forall t \in \mathbb{R}; \quad \text{for } k \leq j-1, \tag{4.105}$$

$$c = c_k, \ d = d_k, \text{ for } k \leq j, \tag{4.106}$$

$m = m_j$, mutation matrix $\hat{m}_{j,j}(\cdot,\cdot)$, $s = s_j$, fitness function $\hat{\chi}_j$, all for $k = j$

$$\tag{4.107}$$

recall Definition 4.7 for $\wedge$-quantities) and

$$m = 0, \quad \chi \equiv 0 \text{ for } k < j. \tag{4.108}$$

Here the initial state is for $k = j$ the random state $Z_\varepsilon^{*,(j,k)}(0)$ concentrated on $\hat{E}_j$ which has law $\hat{\lambda}_j^\varepsilon$ and for $k \leq j-1$ it is the respective equilibrium state. In both cases these random states are given independently of the evolution. $\square$

Remark 30. The dynamic $Z_\varepsilon^{*,(j,j)}$ is a non-equilibrium process while the $Z_\varepsilon^{*,(j,k)}$ for $k \leq j-1$ are equilibrium processes.

Remark 31. Note here for $k \leq j$ we consider types of level j and $j+1$ but everything is taking place in a spatial region of size k and hence migration and resampling parameters are c_k and d_k.

Remark 32. Since the resampling coefficients d_j depends on the $(c_k)_{k=0,\ldots,j}$ the sequence $(c_k)_{k \in \mathbb{N}}$ plays an important role. Recall that $d_{k+1} = c_k d_k/(c_k + d_k)$, which has the explicit solution $d_{k+1} = d_0/(1 + d_0 \sum_{\ell=0}^{k} c_\ell^{-1})$, (see [DGV]). This means that the role of the stochastic part depends for large j strongly on the asymptotic behaviour of the $(c_k)_{k \in \mathbb{N}}$ as $k \to \infty$. This will become very important discussing the question of biodiversity in Subsection 4.7.

Step 3 *Droplet dynamics*

The evaluation of $\mathbf{J}_t^{N,j}$ as $N \to \infty$ can be defined in terms of a stochastic equation. This stochastic equation is driven by Poisson point process and the intensity measure involves excursion laws of $\mathbb{K}$-dimensional diffusion processes. The explicit construction will be given in Section 10.5.

4.6.3 Results on Emergence, Invasion and Fixation by Rare Mutants

The limiting processes and the concepts developed in Subsubsection 4.6.2 allow us now to formulate our result on the emergence and the take over by rare mutants in two theorems, namely Theorem 6 on emergence and fixation times and Theorem 7 on the fixation dynamics. The preemergence behaviour, i.e. the droplet formation leading to the emergence is treated in Theorem 8. In Theorem 9 we finally relate the droplet growth with the emergence.

Theorem 6 (Phase 1: emergence time, emergence frequencies for E_{j+1}-types).
Assume that the system starts in the state completely occupied by the basic type $(0,0)$.

(a) *Then for $j \in \mathbb{N}_0$ the transition from E_j-types to E_{j+1}-types within the $(j+1)$, respectively j-ball, occurs at the following random times.*

With the definition 4.5(4.68) for every $\varepsilon \in (0,1)$ there exist $\mathbb{R}$-valued random variables $\bar{T}_\infty^{\varepsilon,j}$, $T_\infty^{\varepsilon,j}$ and a number β_j such that:

$$\mathcal{L}\left[\bar{T}_N^{\varepsilon,j} - \left(\frac{1}{\beta_j}(N^j \log(N)) + N^j \bar{T}_\infty^{\varepsilon,j}\right)\right] \underset{N\to\infty}{\Longrightarrow} \delta_0, \quad (4.109)$$

$$\mathcal{L}[T_N^{\varepsilon,j} - (\frac{1}{\beta_j} N^j \log N + N^j T_\infty^{\varepsilon,j})] \underset{N\to\infty}{\Longrightarrow} \delta_0. \quad (4.110)$$

The parameter β_j is explicitly determined as the Malthusian parameter of the dual process in (8.136), (9.359) or alternatively via excursion theory (10.93) and satisfies:

$$0 < \beta_j < s_j. \quad (4.111)$$

The random time $\bar{T}_\infty^{\varepsilon,j}$ is given by

$$\bar{T}_\infty^{\varepsilon,j} = -{}^*\mathcal{E}_j + \bar{t}_\infty^{\varepsilon,j} \quad (recall\ (4.100)\ and\ (4.101)). \quad (4.112)$$

The random time $T_\infty^{\varepsilon,j}$ arises as follows:

$$T_\infty^{\varepsilon,j} = \inf(t \mid \mathcal{Z}_\varepsilon^{*,(j,j)}(t, E_{j+1}) = \varepsilon) \tag{4.113}$$

and has the property

$$\mathcal{L}[T_\infty^{\varepsilon,j} \mid {}^*\mathcal{E}_j] \text{ is truly random.} \tag{4.114}$$

(b) *The law of the emergence characteristics of (4.98), (4.99) is determined by the law of $^*\mathcal{W}$. This in turn is determined as follows:*

$${}^*\mathcal{W}_j(E_{j+1}) \text{ is given in (4.132), 4.133) via an atomic measure-valued diffusion,} \tag{4.115}$$

$$\{{}^*\mathcal{W}_j(k), \quad k \in E_{j+1}\} \text{ are independent with law given} \tag{4.116}$$

via excursion theory (see Section 10.5)

and then $\hat{\mathcal{W}}_j$ in the definition of $\mathcal{Z}^{*,j}$, see (4.97), is given by

$$\hat{\mathcal{W}}_j(\cdot) = \frac{{}^*\mathcal{W}_j(\cdot)}{{}^*\mathcal{W}_j(E_{j+1})}. \qquad \square \tag{4.117}$$

Next we identify the limiting dynamics in which the fixation process is realized in which the total intensity of E_{j+1}-types increases from some small $\varepsilon > 0$ to 1. Recall Definitions 4.7–4.10 from Subsection 4.6.2.

Theorem 7 (Phase 2: the dynamics of fixation on higher level E_{j+1}).

(a) *We have for the empirical measure (of j-blocks in the $(j + 1)$-block) and empirical mean process convergence to the random McKean–Vlasov entrance law:*

$$\mathcal{L}\left[(\Xi_{\xi,j}^N(\frac{1}{\beta_j}N^j \log N + tN^j)^+)_{t \in \mathbb{R}}\right] \underset{N \to \infty}{\Longrightarrow} \mathcal{L}[(\mathcal{Z}^{*,j}(t))_{t \in \mathbb{R}}], \tag{4.118}$$

$$\mathcal{L}\left[(x_{\xi,j+1}^N(\frac{1}{\beta_j}N^j \log N + tN^j)^+)_{t \in \mathbb{R}}\right] \underset{N \to \infty}{\Longrightarrow} \mathcal{L}[(\bar{\mathcal{Z}}^{*,j}(t))_{t \in \mathbb{R}}], \tag{4.119}$$

where

$$\bar{\mathcal{Z}}^{*,j}(t) = \int_{\mathcal{P}(\hat{E}_j)} x\, [\mathcal{Z}^{*,j}(t)](dx). \tag{4.120}$$

(b) *We have convergence for a tagged j-block to a selection-mutation process in random medium:*

$$\mathcal{L}\left[(x_{\xi,j}^N(\bar{T}_N^{\varepsilon,j} + tN^j)^+)_{t \in \mathbb{R}}\right] \underset{N \to \infty}{\Longrightarrow} \mathcal{L}\left[(Z_\varepsilon^{*,(j,j)}(t))_{t \in \mathbb{R}}\right]. \tag{4.121}$$

4.6 Limit Theorems II: Emergence, Fixation, Droplet Formation of Rare Mutants

(c) After E_{j+1}-emergence the process of averages in a k-block in their natural time scale follows for $k \leq j-1$ a limiting neutral evolution on level-j and level-$(j+1)$ types:

$$\mathcal{L}\left[\left(x_{\xi,k}^N(\bar{T}_N^{\varepsilon,j}+tN^k)^+\right)_{t\in\mathbb{R}}\right] \underset{N\to\infty}{\Longrightarrow} \mathcal{L}\left[((Z_\varepsilon^{*,(j,k)}(t))_{t\in\mathbb{R}}\right], \quad k=0,1,\cdots,j-1.$$
(4.122)

If we consider in (4.122) times $\bar{T}_N^{\varepsilon,j} + sN^j + tN^k$ we obtain at the r.h.s. the same process where we replace in (4.105) $Z^{*,(j,j)}(0)$ by $Z^{*,(j,j)}(s)$. □

The final step for the $E_j \to E_{j+1}$ transition is to study the early state of the droplets of tagged j-blocks occupied by the E_{j+1}-types and to then use this to relate the two phases 0 and 1. In a first step we have to define what we mean by a typical E_{j+1}-colonised j-block. Imagine we would have a particle model with individuals. The idea is to pick a E_{j+1}-individual at random from the population of the $(j+1)$-block and then take the configuration at this j-block where the picked particle sits, which we would call then a typical E_{j+1}-populated j-block. As $N \to \infty$ this yields the *Palm distribution* w.r.t. the E_{j+1}-mass. This idea works in the continuum mass case as well. Namely define the law $\hat{\mu}_t^N$ by

$$\hat{\mu}_t^N(A) = \int_A x(E_{j+1})\mu_t(dx) \Big/ \int_{\mathcal{P}(\hat{E}_j)} x(E_{j+1})\mu_t^N(dx), \quad \mu_t^N = \mathcal{L}[x_{0,j}^N(t)].$$
(4.123)

With $\hat{\mu}_t^N$ we have the law of the typical E_{j+1}-"site", where "site" here means a j-block if we are looking out for the transition from E_j to E_{j+1}.

We want to next relate the growth rate β_j and the growth constant $^*W_j(E_{j+1})$ of the entrance law to the growth parameters in the droplet formation, i.e. we want to show that β_j actually describes the exponential growth rate of the number of j-blocks colonized at a given time by the E_{j+1}-types, measured in the time scale tN^j.

Furthermore we shall see that this droplet growth has some initial randomness arising in the very beginning (times $O(N^j)$) and is then proceeding from thereon exponentially in a completely deterministic way (in the limit $N \to \infty$), in other words in the form of a randomly shifted exponential $\exp(\alpha(t,T))$ with $T \in \mathbb{R}$ random. Emergence occurs when the droplet gains positive intensity of j-blocks in space globally and fixation if this intensity is 1. The key point below is that the droplet growth has a limiting dynamic and we can recover β_j as the growth rate of the mass in the limit droplet dynamic.

The droplet formation is now described as follows. Define (recall (4.8) for $d_j, \Omega_N^{(j)}$) with ξ' running in $\Omega_N^{(j)}$:

$$\hat{x}_{0,j+1}^N = N^{j+1} x_{0,j+1}^N = \sum_{\xi':d_j(0,\xi') \leq 1} x_{\xi',j}. \qquad (4.124)$$

Recall furthermore the atomic measure-valued process $(\mathbf{J}_{j,t}^N)_{t \geq 0}$ from (4.16). Consider t_N satisfying (pre-emergence regime)

$$\frac{t_N}{N^j} \uparrow \infty \text{ with } t_N = o(\frac{1}{\beta_j}(\log N) N^j) \text{ as } N \to \infty. \qquad (4.125)$$

Then droplets form as follows.

Theorem 8 (Between phase 0 and 1: Droplet formation with Malthusian law).

(a) We have a limiting Palm measure for a typical E_{j+1}-site:

$$\hat{\mu}_{t_N}^N \underset{N \to \infty}{\Longrightarrow} \hat{\mu}_\infty^\infty, \qquad (4.126)$$

which satisfies

$$\hat{\mu}_\infty^\infty(\{x(E_{j+1}) > 0\}) = 1, \quad \hat{\mu}_\infty^\infty(x(\hat{E}_j) = 1) = 1. \qquad (4.127)$$

(b) Furthermore the type E_{j+1}-mass grows exponentially

$$\mathcal{L}[e^{-(\beta_j)t_N}(\hat{x}_{0,j+1}^N(t_N)(E_{j+1} \cap \cdot))] \underset{N \to \infty}{\Longrightarrow} \mathcal{L}[\mathcal{W}_j^*], \qquad (4.128)$$

where $\mathcal{W}_j^*$ is a random measure on E_{j+1} with ($A \subseteq E_{j+1}, A \neq \emptyset$):

$$0 < \mathcal{W}_j^*(A) < \infty \quad , \quad Var(\mathcal{W}_j^*(A)) > 0. \qquad (4.129)$$

In particular we identify the time in which a polynomially sized E_{j+1}-droplet occurs. Namely for $\delta \in (0, 1)$:

$$\mathcal{L}[N^{-\delta j}(\hat{x}_{0,j+1}^N(\delta \frac{1}{\beta_j}(\log N) N^j)(E_{j+1})]) \underset{N \to \infty}{\Longrightarrow} \mathcal{L}[\mathcal{W}_j^*(E_{j+1})]. \qquad (4.130)$$

(c) We have furthermore a limiting dynamic for the droplet growth (see Remark 34):

$$\mathcal{L}[(\mathbf{J}_{j,t}^N)_{t \geq 0}] \underset{N \to \infty}{\Longrightarrow} \mathcal{L}[(\mathbf{J}_{j,t})_{t \geq 0}], \qquad (4.131)$$

with the growth behaviour of the limit given by

$$\mathcal{L}[e^{-\beta_j t} \mathbf{J}_{j,t}([0, 1] \times E_{j+1})] \underset{t \to \infty}{\Longrightarrow} \mathcal{L}[\mathcal{W}^*(E_{j+1})]. \quad \square \qquad (4.132)$$

4.7 Limit Theorems III: Local Biodiversity Versus Monotype, the Genealogy...

Remark 33. We conclude from this theorem that at time $\delta \left(\alpha_j N^j \log N \right)$ a "droplet" of sites with volume $N^{j+\delta}$, more precisely N^δ many j-blocks, has formed where the E_{j+1}-types have already emerged and for $\delta = 1$ we then obtain emergence in the sense of a positive intensity of E_{j+1}-types in the whole $(j+1)$-block and then also in a tagged j-block for example the one around 0.

Remark 34. We will identify in Section 10 the limiting dynamic appearing in part (c) of the Proposition in terms of a stochastic process driven by Poisson random point processes where the intensity measure involves the excursion laws of a certain multi-dimensional diffusion.

We finally relate the droplet growth and the emergence regime. We already saw that the respective exponential rates α^* and α agree. The previous two theorems raise therefore the question, whether the growth constant of the entrance law denoted $^*\mathcal{W}$ is in law equal to the growth constant of the droplet denoted $\mathcal{W}^*$. If the randomness in the emergence is indeed due to the early events when first rare mutants appear one would expect this. Note however that we are connecting here the limiting behaviour to the right and to the left in two regimes associated with time scales, which *separate* in the limit $N \to \infty$ and therefore such a conclusion is not at all immediate.

More precisely such an identity would mean that we have some randomness in the early droplet formation, i.e. at times of order $O(N^j)$, which then manifests itself again in the same way at a much later time $\alpha_j^{-1}(\log N)N^j$ when the process of emergence and fixation on a global level sets in. If we look at this by considering droplet sizes $(N^j)^\delta$ with $\delta \in [0, 1]$ then note that for $\delta < 1$ and $\delta = 1$ we are on time scales of different order of magnitude and no soft continuity arguments apply here. (In fact $\delta = 1$ is a discontinuity point since fixation occurs in a smaller time scale which shrinks in the scaling to a point!). Nevertheless we can prove for parameters s sufficiently small, respectively sufficiently large, with a gap left open at the moment (compare equations (9.401), (9.402)):

Theorem 9 ($^*\mathcal{W}$ **and** $\mathcal{W}^*$ **are equal**). *Assume that s satisfies (9.401) or (9.402). Then we have for every $j \in \mathbb{N}$ that:*

$$\mathcal{L}[\mathcal{W}_j^*] = \mathcal{L}[^*\mathcal{W}_j]. \qquad \Box \qquad (4.133)$$

4.7 Limit Theorems III: Local Biodiversity Versus Monotype, the Genealogy of the Mutant Population

In this section we enrich the perspective and consider questions related to the genealogies of the individuals in the population of very high fitness. Namely we analyse the implications of our results for the *genealogical structure* of the population in a tagged colony or a tagged block after the invasion by the rare mutants of very high level of fitness, that is, in a high level time scale. In particular we are interested in the number of founding fathers of the population observed upon fixation in a new high fitness level.

Note that the current population at that time entirely goes back to the rare mutants at the various transitions to higher fitness levels. We shall see here that *biodiversity*, meaning several different rare mutants appear as ancestors of the local current population, relies crucially on far-reaching migration while "normal" migration leads to local monotype situations given by one rare mutant populating later on every finite block.

This means our results raise a number of questions about the *genealogy* of the population described by the system $(X^N(t))_{t \geq 0}$. In particular, consider the system at times when the types in E_j dominate the system, which have earlier arisen through rare mutation, what can be said about the number of rare mutants whose descendants dominate the population later on in specific spatial areas. Do we have many or only one such rare mutant locally and how does this depend on the migration mechanism?

The first step is to see whether we can define such genealogies properly in a diffusion model and then to address the question whether in the limiting processes $(Z_\varepsilon^{*,(j,k)}(t))_{t \in \mathbb{R}}$ governing the emergence (recall Definition 4.11, assertions (4.121) and (4.122)) we can define a family decomposition of the mass into pieces of mass corresponding to a subpopulation descending from the same ancestor at time $t = -\infty$. Then in a second step we can ask whether we can determine in the process $(X^N(t))_{t \geq 0}$ asymptotically as $N \to \infty$ the *limiting number of rare mutants* which eventually produced the type-E_{j+1} population in a tagged k-block through mutation and subsequent evolution by migration, selection and resampling and whether the corresponding subpopulations resulting from these rare mutants correspond in the limit to the family decomposition of the processes $Z_\varepsilon^{*,(j,k)}$.

Outline of Subsection 4.7 In order to make the above reasoning precise the concept of the *historical process* enters which is developed in Subsection 4.7.1 including Theorem 10 on existence and uniqueness of this process. Then the historical versions of our scaling statements and the result on the *dichotomy*, monotype versus biodiversity, is stated in Subsection 4.7.2 and the locations of the ancestors of the population at fixation is studied in Subsubsection 4.7.3. Finally in Subsection 4.7.4 we briefly discuss extensions to the full historical process.

4.7.1 History of the E_{j+1}-Type Population: Concepts and Analytic Description

In order to study the genealogy of the population one can use the machinery of tree-valued processes (see [GPWmp12]) or [DGP]. We are here less ambitious and focus on the prescribed question only which allows a simpler approach where we focus on family sizes with families defined by the rare mutants.

4.7 Limit Theorems III: Local Biodiversity Versus Monotype, the Genealogy...

In order to formulate this question in a suitable framework we introduce the *historical processes* of both the interacting system $(X^N(t))_{t \geq 0}$ and of the limiting object arising in the hierarchical mean-field limit $(Z_\varepsilon^{*,(j,k)}(t))_{t \in \mathbb{R}}$. These historical processes describe the mass associated with different lines of descent. The construction of such a process for an individual based model is no problem and can be given explicitly.

The historical process in the diffusion limit is defined via a martingale problem. We specify first the state space and the generator for this martingale problem. The point is to use here the special structure of our model and the questions we want to ask, which requires only knowledge of a functional of the full historical process and the evolution of this functional is substantially simpler to describe than the full historical process. This *reduced historical process* does not provide information on the geographic position of the lines of descent in the past but only on the decomposition of the current population in *distinct families*.

The starting point for the construction of the reduced historical process is the fact that the system $(X^N(t))_{t \geq 0}$ can be embedded in a richer structure. Recall that the diffusion process arises as the small mass-many particle limit of a particle model. On the level of particle models one would simply assign *every* initial particle a different marker in addition to its type. The marks could be taken from the interval $[0, 1]$ for example. The new type space of the reduced historical process would then be

$$[0, 1] \times \mathbb{I}, \tag{4.134}$$

where only the second component influences the fitness and changes due to mutation while the first component keeps coding the founding father. The latter observation will be important for the application of duality techniques as we shall see later on.

Following this idea in the diffusion limit we define a system of interacting Fleming–Viot processes with selection and mutation with type space $[0, 1] \times \mathbb{I}$. Here the fitness function and mutation rates depend only on the second component of the type and are obtained by the natural extensions of $m(u, v)$ and $\chi(u)$ to $([0, 1] \times \mathbb{I})^2$ and $[0, 1] \times \mathbb{I}$ such that the first component experiences no mutation and selection and the weight associated with the first component changes only due to migration and resampling. This specifies the evolution mechanism of the reduced historical process as a collection of interacting Fleming–Viot processes with selection and mutation on the type space $[0, 1] \times \mathbb{I}$.

Next we have to specify the initial state. In a particle model one would simply assign to every particle initially its own type. For example by drawing independently a type from $[0, 1]$ for each individual from a uniform distribution. In the diffusion limit this would result in an initial state

$$\tau \bigotimes \delta_{(0,0)}, \tag{4.135}$$

with τ the uniform distribution on $[0, 1]$ and for $\delta_{(0,0)}$ (we recall that $(0, 0)$ is the type of lowest fitness with which everything starts).

Hence we choose in our martingale problem the initial state such that for every given value of the second component the *uniform distribution* in the first component.

A similar procedure is applied to the nonlinear Markov process $(Z(t))_{t \geq 0}$ or $(Z_\varepsilon^{*,(j,k)})_{t \geq 0}$. We omit the obvious details of this transfer.

Definition 4.12. We denote the *reduced historical process* of $(X^N(t))_{t \geq 0}$, $(Z(t))_{t \geq 0}$ and $(Z_\varepsilon^{*,(j,k)}(t))_{t \geq 0}$ by

$$(\hat{X}^N(t))_{t \geq 0}, \; (\hat{Z}(t))_{t \geq 0}, \; (\hat{Z}_\varepsilon^{*,(j,k)}(t))_{t \geq 0}. \qquad \square \qquad (4.136)$$

As a corollary of the result in [DG99], Theorem 0 therein (for part (a)) and the previous results of this monograph (for part (b)) we obtain by applying the results to the processes with the new type space that (note that we have replaced one type by uncountably many types, but these are all neutral among each other):

Theorem 10 (Historical martingale problem, hierarchical mean-field limit).

(a) The reduced historical martingale problem is well-posed.
(b) The assertions of Theorems 3–5 hold if we replace X^N by $\hat{X}^N$ and the limits by their reduced historical version and Theorem 6–7 holds if we replace $Z_\varepsilon^{*,(j,k)}$ by $\hat{Z}_\varepsilon^{*,(j,k)}$ and E_{j+1} by $[0,1] \times E_{j+1}$. $\square$

On the basis of the historical martingale problem, we can define a *family decomposition* as follows. We simply consider the masses on types agreeing in their first type-coordinate. Recall the fact ([DGV]) that for $t > 0$ the states of countably many interacting Fleming–Viot diffusions on the type space which is a continuum are in each colony *atomic* so that the family decomposition is well-defined (recall here in the first type-component the model has no selection or mutation). Recall also that the number of types with positive weight is countably infinite if the space is (countably) infinite.

Definition 4.13 (Family decomposition).

(a) We define on the type space $[0,1] \times \mathbb{I}$ the equivalence relation:

$$(u,i) \sim (v,k) \iff u = v. \qquad (4.137)$$

(b) A family cluster in a population at time t is characterized by a $\sim$ equivalence class $\vec{i}$ of $[0,1] \times \mathbb{I}$. We denote the processes of its spatial distribution of mass and the distribution of types given by restricting in each component $\hat{X}_\xi^N(t), \hat{Z}_\varepsilon^{*,j}(t), \hat{Z}_\varepsilon^{*,(j,k)}(t)$ to the equivalence class: $\vec{i}$.

$$\left(\{ \hat{X}_\xi^{N,\vec{i}}(t), \; \xi \in \Omega_N \} \right)_{t \geq 0}, \; \left(\hat{Z}^{\vec{i}}(t) \right)_{t \in \mathbb{R}}, \; \left(\hat{Z}_\varepsilon^{\vec{i},*,(j,k)}(t) \right)_{t \in \mathbb{R}}. \qquad (4.138)$$

(c) The countable collection of the processes in (4.138) indexed by $\vec{i}$ gives the family decomposition.

4.7 Limit Theorems III: Local Biodiversity Versus Monotype, the Genealogy...

(d) If we are interested in local biodiversity at a particular time t, we define the size-ordered vector of masses (notation (i) as usual for the entries of the size-ordered sequence $\vec{i}$)

$$\left(\hat{X}_\xi^{N,(i)}(t)\right)_{i\in\{1,2,\cdots\}}, \quad \left(\hat{Z}^{(i)}(t)\right)_{i\in\{1,2,\cdots\}}, \quad \left(\hat{Z}_\varepsilon^{(i),*,(j,k)}(t)\right)_{i\in\{1,2,\cdots\}}. \quad \square \tag{4.139}$$

Remark 35. An interesting observation is that we can at a later time t add a second set of colours or remove those colours from time 0 and replace them with new colours assigned to time t and then follow the genealogical relationship counted from that time on. Again we obtain a corresponding family decomposition as above.

Remark 36. The construction above will allow us to use spatial coalescent processes with birth to describe the statistics of the family clusters via the duality theory which we develop in Section 5.

4.7.2 Results: The Role of the Migration Parameters for the Local Bio-Diversity

Finally we investigate the role of the migration parameter in the scenario, which was described so far, in particular the question how many rare mutants are responsible for the current population. More precisely consider the local population in some finite set of components when the level-$(j + 1)$ types have emerged but not yet the level-$(j + 2)$ types or alternatively when the level-$j + 1$-types have fixated but the level $(j + 2)$-types have not yet appeared. The question is whether for very large j we have in the limit the following *dichotomy*:

- locally a monoancestor population,
- biodiversity (local coexistence of many rare mutant families up to the present time).

In the latter case we might expect in the limit even countably many ancestors. In formulas this dichotomy reads (convergence meant in law):

$$\left(\hat{X}_\xi^{N,(i)}(\bar{T}_N^{\varepsilon,j} + t_N N^j)\right)_{i\in\mathbb{N}} \underset{j\to\infty}{\Longrightarrow} \begin{cases} (1,0,0,\cdots) \\ \neq (1,0,0,\cdots) \end{cases}, \tag{4.140}$$

where $t_N \to \infty$, but $t_N = o(N)$. In fact we will see that the case $\neq (1,0,0,\cdots)$ results in a vector

$$(r_i)_{i\in\mathbb{N}} \text{ with } r_i > 0 \quad \forall\, i \in \mathbb{N}. \tag{4.141}$$

We approximate the situation here by first taking the limit $N \to \infty$ and only later the limit $j \to \infty$. That is, we formulate the dichotomy in terms of the limiting $\hat{Z}$ process as $N \to \infty$ that for every fixed k:

$$\left(\hat{Z}_\varepsilon^{(i),*,(j,k)}(\infty)\right)_{i \in \mathbb{N}} \underset{j \to \infty}{\Longrightarrow} \begin{cases} (1, 0, \cdots) \\ \neq (1, 0, \cdots). \end{cases} \quad (4.142)$$

The first key point is that very far reaching migration enhances the spread of the higher level type, while migration of a very recurrent nature keeps this process slow. Secondly the other important parameter is the mutation rate to lower level types since it determines the advantage of the rare mutant. Given a *fixed system of mutation rates* the goal is to determine conditions on the migration rates, $(c_\ell)_{\ell \in \mathbb{N}}$, that lead in a block of diameter k to either a single E_{j+1}-mutant type or a random countable set of E_{j+1}-mutants in the limit $j \to \infty$.

In order to carry out the analysis we have to observe for that purpose that if a fitter mutant occurs in a given family only its descendants will make up the family at much later times since the subpopulation of all the other members of the family will eventually mutate to lower order types and then disappear by selection. This mechanism was verified in Theorem 6. Therefore in order to count the number of rare mutant founders at the time that the E_{j+1} types reach mutation-selection equilibrium (and before the emergence of E_{j+2}-types) it suffices to count the number of different families in the current population after the complete fixation on E_{j+1}-types (asymptotically as $N \to \infty$ of course).

Theorem 11 (Number of rare mutant founding fathers: the dichotomy biodiversity-monoancestor). *Consider now the process* $Z^{*,(j,k)}$ *on level k in equilibrium* $(t \to \infty)$ *for varying j and in particular for $j \to \infty$. For fixed $t \in \mathbb{R}$ and k, the ℓ^1-vector of size-ordered masses of equivalence classes charged by the collection*

$$(\hat{Z}_\varepsilon^{(i),*,(j,k)}(\infty))_{i \in \mathbb{N}} \quad (4.143)$$

converges in law as $j \to \infty$ to the ℓ^1-vector:

$$\begin{array}{lll} (1, 0, 0, \cdots) & , \text{if } \sum c_\ell^{-1} = +\infty & \textit{local fixation, i.e. locally monotype} \\ (r_i)_{i \in \mathbb{N}}, r_i > 0 \;\; \forall i \, , \text{if } \sum c_\ell^{-1} < \infty & \textit{local coexistence, i.e. biodiversity.} \quad \square \end{array}$$
$$(4.144)$$

In terms of the migration process this dichotomy $\sum c_k^{-1} = \infty, < \infty$ corresponds to the usual *recurrence—transience dichotomy* for the difference walk for two independent walks corresponding each to the underlying migration mechanism. In particular this implies that for the hierarchical group with migration rates corresponding to the two-dimensional euclidean nearest neighbour random walk (which is $c_\ell \equiv c \quad \forall \ell \in \mathbb{N}$) the number of rare mutants which push the system to the next higher level is *locally due to one rare mutant* as ancestor, which means in particular that stochastic effects remain here important in selecting such an ancestor. In the hierarchical analog to the $2 + 0$ dimensional euclidean case, which would for example arise in $\mathbb{Z}^2$ by taking a symmetric random walk with infinite variance which is not in the nonnormal domain of attraction of Brownian motion, *infinitely many rare mutant families* contribute locally.

4.7.3 The Origin of Successful Populations

Consider a tagged site. The previous results imply that type E_1 types emerge and fixate at this site at times of order $O(\log N)$, types E_2 emerge and fixate at this site at times of order $O(N \log N)$, and types E_j at time $O(N^{j-1} \log N)$. A natural question which arises from Theorem 11 is on the geographical structure, namely, where does the successful population of type E_{j+1} fixating at the given site come from geographically. The idea is that the successful population at a tagged site consists of the descendant population of a successful mutation and the question is at what geographic location did this line of descendance start at time 0 and then where did these mutations take place.

For this picture it is important to recall that the emergence arises from rare mutants succeeding at a very early stage establishing large enough colonies on a small subset of the N different j-blocks in the $(j+1)$-block and then essentially grow *independent* of each other. Then we want to show that these original rare mutations come from *within a distance, at most $j+1$*. The above result now means that in the case of a recurrent migration only *one* of those rare mutants in the large block actually hits a small k-block, i.e. $k \ll j$, while many do so in the transient case.

To give a precise formulation we begin by considering the geographic locations of the founding fathers in the $(j+1)$-block leading to the emergence of E_j types in a 1-block. Given $k \in \mathbb{N}$, we decompose the $(j+1)$ block into k pieces $\{A_{N,j}(i), i = 1, \ldots, k\}$ where $A_{N,j}(i)$ contains $|A_{N,j}(i)|$ j-blocks and such that

$$A_{N,j}(1) \cup \cdots \cup A_{N,j}(k) = \{1, \ldots, N\}^{j+1}, \quad A_{N,j}(i) \cap A_{N,j}(\ell) = \emptyset, \text{ if } i \neq \ell \tag{4.145}$$

and

$$\frac{|A_{N,j}(i)|}{N} \to \frac{1}{k}, \quad i = 1, \ldots, k. \tag{4.146}$$

Recall (4.95) and let

$$(x^N_{\xi, j+1}(\frac{1}{\beta_j} N^j \log N + tN^j)^+)^i, \tag{4.147}$$

denote the mass of the E_{j+1}-type family with founding ancestors in the j-subblock $A_{N,j}(i)$ for $i = 1, \ldots, k$, at time 0. Consider the empirical distribution based on $((x^N_{\xi, j+1}(\frac{1}{\beta_j} N^j \log N + tN^j)^+)^i)_{i=1,\ldots,k}$. Then we obtain

Proposition 4.14 (Spatial origin of successful mutants of type in E_{j+1}).

(a) In the limit as $N \to \infty$ the families observed at times of order $O(N^j \log N)$ in $B_{j+1}(0)$ which have positive weight originate in $B_{j+1}(0)$.

(b) For $i = 1, \ldots, k$ we have

$$\mathcal{L}\left[\left((x^N_{\xi,j+1}(\frac{1}{\beta_j} N^j \log N + tN^j)^+)^i_{t \in \mathbb{R}}\right)_{i=1,\cdots,k}\right] \underset{N\to\infty}{\Longrightarrow} \mathcal{L}[((\tilde{Z}^{*,j}(t))^i_{t \in \mathbb{R}})_{i=1,\cdots,k}]. \tag{4.148}$$

Then

$$\mathcal{L}\left[\left(e^{\beta_j |t|}(Z^{*,j}(t, \cdot \cap E_{j+1}))^i\right)_{i=1,\cdots,k}\right] \Longrightarrow \mathcal{L}[(^*W^i_j(\cdot))_{i=1,\cdots,k}] \text{ as } t \to -\infty, \tag{4.149}$$

and $W^1_j, \ldots, W^k_j$ are i.i.d.

(c) Embed $B_{j+1}(0)$ in $[0, 1]^{j+1}$ say, by scaling $(\{0, 1, \cdots, N-1\})^{j+1}$ by $N^{-(j+1)}$. Then in the limit as $N \to \infty$ the distribution of expected mass of E_{j+1}-types based on the location of the founding father is uniform on $[0, 1]^{j+1}$.

(d) In the limit $N \to \infty$, the spatial locations of the founding fathers from which E_{j+1}-type population at a tagged site has descended are at hierarchical distance $j + 1$ from the tagged site. □

Remark 37. Letting $k \to \infty$ in Proposition 4.14, we can obtain a limiting infinitely divisible random field but we do not consider this here.

Consider now the emergence of types E_1. The identification of individual successful rare mutations as well as their locations and the resulting descendant subpopulations could be obtained by introducing the historical version of the process $\mathfrak{Y}^m_t$. This could be used to extend the analysis verifying that $^*\mathcal{W} = \mathcal{W}^*$ to prove the same relation for the different subpopulations coming from different successful rare mutations where the latter correspond to excursions of the Wright–Fisher diffusion producing the mutant droplet.

The picture for the successive emergence of types $E_1, E_2, E_3, \ldots$ could then be obtained by a multiscale analysis applied to the full historical process, which would however require the rigorous introduction of this object, for which we do not have the space here. See [Seidel08] and the brief discussion below.

4.7.4 Outlook on the Full Historical Process

In this subsubsection we sketch the possibilities of our analysis of the *full historical process* which is asymptotically described by the historical process of the interaction chains, which is based on the knowledge about the $(\mathfrak{Y}^{m,j}_t)_{t \geq 0}$ on the various levels j.

This requires some effort but would allow us to make sense of the scenario that a typical ancestral line viewed on a multiple time scale exhibits the passage through rare mutations $E_0 \to E_1 \to E_2 \to \cdots$, with the ancestral positions at these times being geographically "uniformly distributed" on the j ball if we consider the mutation from E_{j-1} to E_j. The time points on the ancestral lines are chosen as times of the order of fixation on the levels $E_j, j = 0, 1, \cdots$. Then we would want to derive a scaling theorem for these paths. However such an analysis will have to overcome some challenging mathematical problems and will be carried out in [Seidel].

This scenario becomes even more interesting if we consider more refined type spaces as follows.

Remark 38. The results above have been formulated for the simplified type space $\mathbb{I}$ and mutation kernel $M(.,.)$. However the emergence and biodiversity results can be extended to a larger class in which at each level there is more than one ergodic class of the mutation process on that level. This results in a much richer genealogy and biodiversity. For example we can look at a richer hierarchy of types by

- each level j type can give rise to distinct E_{j+1}-types - then we get a tree structure of types in which a type is associated to a sequence of mutations at successive levels; in this case biodiversity implies the presence of many distinct types and rather than just distinct families as in the present model,
- some potential mutations can be prohibited leading to multiple ergodic classes— e.g. mutation between E_{j+1}-types coming from different E_j-types; in this case the randomness inherent in the emergence stage persists in the future and the long time composition of the population is not deterministic but stochastic and very much depends on history,
- under resampling some of the resulting types disappear but this can differ from one $(j+1)$ level colony to another,

All these effects lead to a rich structure of types and in the transient case we get local coexistence of types. This would be important if as an additional mechanism recombination of rare mutants were considered.

4.8 The Structure of the Proofs

The proof of the program and the theorems stated in Section 4 begins with the development of some new techniques of duality in Section 5 and some illustrations of their use in Section 6. We then give in Section 7 (in particular 7.2, in fact it would be helpful to look at this before going into Section 4.3 in detail) a description of the key steps needed for the proofs.

The key elements needed to carry out the analysis and proofs in Sections 13–15 going through the various phases on the transition to fitter types are based on the analysis of a mean-field model with M types of lower and M types of the maximal fitness. This analysis is carried out first in Section 8 for the simpler case of $M = 1$ in a self-contained manner to then being extended to M types on the lower level in Section 9 and finally also M types at the upper level in Section 10.

Every section will state the main results in the beginning and an outline of the strategy followed in the beginning and an outline of the strategy followed by the actual execution of the proofs.

Chapter 5
A Basic Tool: Dual Representations

An important technical tool in our analysis are various new representations of the marginal distributions of the basic processes in terms of expectations under the appropriate *function-valued* respectively *set-valued* dual processes. These constructions are collected in this section and in Subsections 9.4, 9.5 and 9.6. The possible applications of these techniques go way beyond those we have in this monograph.

In general it has been realized for a long time that duals play an important role in the analysis of interacting particle systems [Lig85] and for measure-valued diffusions [D], [Eth00] and many results are obtained via this technique but for population genetics models this is of particular importance. See also [EK2] for a very general approach.

Recall that two Markov processes $(Z(t))_{t \geq 0}, (Z'(t))_{t \geq 0}$ on Polish spaces E, respectively E' are called *dual* w.r.t. to the duality function $H : E \times E' \to \mathbb{R}$, if for (deterministic) initial states Z_0 respectively Z'_0 for Z, respectively Z', if

$$E[H(Z(t), Z'_0)] = E[H(Z_0, Z'(t))] \quad , \quad \forall \ Z_0 \in E, Z'_0 \in E' \text{ and } t \geq 0. \quad (5.1)$$

This often allows us to draw conclusions on the process Z from a (simpler) process Z' on E'. Some (but not all) of these duality relations can in the case of population models be interpreted in terms of the genealogical tree of a tagged sample and this will be the case here. Depending on the application the use of *different* duals can sometimes be appropriate.

Dual process representations for Wright–Fisher processes with selection were first introduced by Shiga in [S2], [S1] and Shiga and Uchiyama [SU]. These dual representations were extended in Dawson and Greven [DG99] to spatial models with arbitrary type space, namely interacting Fleming–Viot diffusion with selection and mutation and lead to a Feynman–Kac duality. The latter can be used to establish that the martingale problem we formulated in (3.24) is indeed well-posed and to derive results on the *longtime behaviour* for sufficiently large state-independent mutation. Duals for particle models of populations with selection have been introduced first by Krone and Neuhauser in [KN97] and further duals were developed for example by Athreya and Swart [AS], Etheridge and Griffiths [EG09] and Fernhead [F]. In case of two-type and state-dependent mutation our dual is a relative of the ancestral selection graph of Krone and Neuhauser [KN97], but note that the latter does *not* have the form of a duality with another backward Markov process as in (5.1).

However in order to study the long-time behaviour in general and to study emergence we have to develop here finer dual representations both for the hierarchically interacting system on Ω_N (or any other geographical space) and for the nonlinear Markov process (McKean–Vlasov process) arising as the mean-field limit $N \to \infty$ of an N site exchangeable model. The duality we shall introduce works for

- *multitype selection*,
- *state-dependent mutation*,
- has a *historical* and *genealogical* interpretation and in fact extension to tree-valued processes (see [GPWmp12] and [DGP]) and
- *multilevel selection models* (see [D2013]).

The key point for the analysis is that we have a *function-valued* dual process, that is, in addition to a dual particle process we have a process which is function-valued but driven by a particle system.

The goal of the dual construction is to construct for a sample $\mathcal{I}$ of n-individuals from the population at a given time t the probability law of the collection

$$\{(\text{type of individual } i, \text{ genealogical distance of pairs of individuals } (k, \ell)),$$
$$\forall\ i, k, \ell \in \mathcal{I}\}. \tag{5.2}$$

See Subsubsection 5.5.4 for more on this.

In this section we first recall the Feynman–Kac duality from [DG99], next we develop the *new* dual representations for the interacting system and then at the end of the section also for the McKean–Vlasov process. This allows us in particular to establish the ergodic theorem for the mean-field limit with state-*independent* mutation component.

In particular we develop for finitely many types a further modified dual that covers state-*dependent* mutation in order to prove the results describing the transition from one quasi-equilibrium to the next. This latter dual representation is very flexible and allows for various refinements useful for specific purposes. In particular the multi-scale analysis requires some new arguments due to the interplay between selection and mutation. In the case of a finite type space we derive a refined dual which takes values in *sums of products of indicators* respectively a *set-valued dual*. Both these duals are very powerful analysing the longtime behaviour. Different arguments in our proofs require some specific features of the dual representation which requires some slight modifications of the dual mechanism that are developed here in this section systematically. These duals we use have been developed over the last decade and suit our purposes and can be generalized to tree-valued processes.

Outline of Section 5 We begin in Subsection 5.1 to introduce the duality functions and the dual process and give in Subsection 5.2 the basic duality relations

5.1 The Dual Process and Duality Function for the Interacting System

on which one can base all the modifications and refinements we need later on. In Subsection 5.3 we give special versions of the dual representation of the mutation useful for the different possible applications of the dual we have a need for. In Subsection 5.4 we discuss the historical meaning of the dual representation. In Subsection 5.5 we derive a very important refinement of the duality relation for models with finitely many types which will be the key to our analysis of the transition from level E_j to E_{j+1} later on in the proofs of our theorems. This allows also to obtain a historical interpretation of the dual which we discuss in Subsubsection 5.5.4.

In Subsection 5.6 we develop the duality theory for the McKean–Vlasov limit along the lines of Subsections 5.1–5.5.

Further dual constructions are given in Subsection 9.4 leading to a set-valued dual, respectively tableau-valued dual for the finite type model in Subsections 9.5 and 9.9. Additional calculations using these constructions are found in Subsections 10.6 and 10.7.

5.1 The Dual Process and Duality Function for the Interacting System

In this section we introduce the basic ingredients of the duality theory. In order to obtain a dual representation of a $(\mathcal{P}(\mathbb{I}))^{\Omega_N}$-valued process $(X_t)_{t \geq 0}$ we need two ingredients, (1) a *family of functions* on the state space that is measure-determining from which we construct the duality function $H(\cdot, \cdot)$ and (2) the stochastic dynamics of a *dual process* whose state space is the set which labels the family in the previous point.

(1) Duality functions

We begin with the first point. Recall that in order to determine the marginal distribution at time t of the law P_{X_0} of the system $X(t) = (x_\xi(t))_{\xi \in \Omega_N}$, with initial point $X(0) = X_0 \in (\mathcal{P}(\mathbb{I}))^{\Omega_N}$, it suffices to calculate for every fixed time t a suitable form of *mixed moments*, i.e. we have to determine the following quantities parametrized by (ξ_0, f) (the initial state of the intended dual dynamics) with ξ_0 a vector of geographic positions and a basic function f on finite products of the type space $\mathbb{I}$:

$$F_t(X_0, (\xi_0, f)) := E_{X_0}\left[\int_\mathbb{I} \cdots \int_\mathbb{I} f(u_1, \ldots, u_n) x_{\xi_0(1)}(t)(du_1) \ldots x_{\xi_0(n)}(t)(du_n)\right], \tag{5.3}$$

for all $n \in \mathbb{N}$, $\xi_0 = (\xi_0(1), \ldots, \xi_0(n)) \in (\Omega_N)^n$ and $f \in L_\infty((\mathbb{I})^n, \mathbb{R})$. This means we want to consider the bivariate function

$$H(X,(\xi,f)) = \int_{\mathbb{I}^n} f(u_1,\cdots,u_n) x_{\xi(1)}(du_1)\cdots x_{\xi(n)}(du_n) \qquad (5.4)$$

to obtain a duality. This would however only allow for a dual for the migration part. Namely introducing the dual process we shall see that we have to enrich the second argument further, to be able to obtain a Markov process as a dual, namely resampling and selection require to introduce partitions of the variables.

The class given above contains the functions which we will use for our dual representations. In fact it suffices to take smaller, more convenient sets of test functions f:

$$f(u_1,\ldots,u_n) = \prod_{i=1}^{n} f_i(u_i), \quad f_i \in L_\infty(\mathbb{I}). \qquad (5.5)$$

Since in our case $\mathbb{I}$ is countable, it would even suffice to take functions

$$f_j(u) = 1_j(u) \text{ or } f_j(u) = 1_{A_j}(u), \qquad (5.6)$$

where 1_j is the indicator function of $j \in \mathbb{I}$ and in case of finite type sets, 1_A with A some sets of types for example like $\{u|\chi(u) \geq e_k\}$ if $0 = e_0 < \cdots < e_\ell = 1$ are the fitness values, which will be a very useful choice later on.

Remark 39 *(Notational convention).* In this section we write

$$X_t \text{ instead of } X(t) \qquad (5.7)$$

on the level of the collection (but we still write $x_\xi(t)$ for components).

(2) Dual process: Spatial coalescent with births driving a function-valued process

Now we come to the second point, the dual process on a suitable state space. We observe here that due to the interaction of selection and mutation the state space of the dual process is more complicated than for most particle systems or interacting diffusions. We proceed in four steps.

Step 1 *(The state of the dual process)*

We begin this point by reviewing the dual process used in [DG99] which involves a *function-valued process* $(\mathcal{F}_t)_{t \geq 0}$ driven by a *finite ordered particle system* $(\eta_t)_{t \geq 0}$. Indeed in [DG99] the quantities on the r.h.s. of (5.3) were expressed in terms of the evolution semigroup corresponding to a

$$\text{bivariate Markovian stochastic process } (\eta_t, \mathcal{F}_t)_{t \geq 0} \qquad (5.8)$$

consisting of

5.1 The Dual Process and Duality Function for the Interacting System

- a process η involving a random set of finitely many ordered individuals,
- a function-valued process $\mathcal{F}$.

This is the dual process to be discussed in the sequel where η_t takes values in the set of *marked partitions* of a *random set of individuals* and $\mathcal{F}_t$ is a *function* defined on *products of the type set* $\mathbb{I}$. The joint evolution of these is such that the dynamic of $\mathcal{F}_t$ is driven by the process η_t which evolves autonomously.

We will use two main types of such processes which are denoted by

$$(\eta_t, \mathcal{F}_t), \text{ respectively } (\eta_t, \mathcal{F}_t^+) \text{ or its variant } (\eta_t, \mathcal{G}_t^+)_{t \geq 0}, \qquad (5.9)$$

where the first component, η_t is a particle system described formally below and the second component $\mathcal{F}_t, \mathcal{F}_t^+$, takes values in the space of functions on $\mathbb{I}^\mathbb{N}$ which are measureable, bounded and depend only on finitely many components which we can identify with

$$\bigcup_{m=1}^{\infty} L_\infty((\mathbb{I})^m), \text{ resp. } \bigcup_{m=1}^{\infty} L_\infty^+((\mathbb{I})^m). \qquad (5.10)$$

This process, called the *function-valued dual process*, will allow us in Subsections 5.2 and 5.3 to represent for every given X_0 the function F_t from (5.3) as the expectation of an appropriate functional over the evolution of $(\eta_t, \mathcal{F}_t)$ or $(\eta_t, \mathcal{F}_t^+)$ respectively $(\eta_t, \mathcal{G}_t^+)$. We construct it here in such a way (for example its order structure) that it can be refined later better for some special applications (see Sections 9 and 10).

The process $(\eta_t, \mathcal{F}_t)_{t \geq 0}$ is constructed from the following four ingredients:

- N_t is a non-decreasing $\mathbb{N}$-valued process. N_t is the number of *individuals* present in the dual process with $N_0 = n$, the number of initially tagged individuals, and with $N_t - N_0 \geq 0$ given by the number of individuals born during the interval $(0, t]$.
- $\zeta_t = \{1, \cdots, N_t\}$ is an *ordered system of individuals* where the individuals are given an assigned order and the remaining individuals are ordered by time of birth.
- $\eta_t = (\zeta_t, \pi_t, \xi_t)$ is a trivariate process consisting of the above ζ and
  - π_t: partition $(\pi_t^1, \cdots, \pi_t^{|\pi_t|})$ of ζ_t, i.e. an ordered family of subsets, where the *index* of a partition element is the smallest element (in the ordering of individuals) of the partition element.
  - $\xi_t : \pi_t \longrightarrow \Omega_N^{|\pi_t|}$, giving *locations* of the partition elements. Here ξ_0 is the vector of the prescribed space points where the initially tagged particles sit.
- $\mathcal{F}_t$ is for given $\eta_t = (\zeta_t, \pi_t, \xi_t)$ a function in $L_\infty(\mathbb{I}^{|\pi_t|})$.

Therefore the state of η is an element in the set

$$\mathcal{S} = \bigcup_{m=1}^{\infty} \{\{1, 2, \cdots, m\} \times \text{Part}^<(\{1, \cdots, m\}) \times \{\xi : \text{Part}^<(\{1, \cdots, m\}) \longrightarrow \Omega_N\}\}, \tag{5.11}$$

where $\text{Part}^<(A)$ denotes the set of ordered partitions of a set $A \subseteq \mathbb{N}$ and the set $\{1, \cdots, m\}$ is equipped with the natural order.

It is often convenient to order the partition elements according to their indices

$$\text{index}\,(\pi_t^i) = \min(k \in \mathbb{N} : k \in \pi_t^i) \tag{5.12}$$

and then assign them ordered

$$\text{labels } 1, 2, 3, \cdots, |\pi_t|. \tag{5.13}$$

Remark 40. Note that

$$N_t = |\pi_t^1| + \cdots + |\pi_t^{|\pi_t|}|. \tag{5.14}$$

Remark 41. The process η can actually be defined starting with countably many individuals located in Ω_N. This is due to the quadratic death rate mechanism at each site which implies that the number of individuals at any site will have jumped down into $\mathbb{N}$ by time t for any $t > 0$.

Remark 42. We view the configuration of $(\eta_s)_{s \leq t}$ as the description of the genealogy of the sample drawn at time t observed back at time $t - s$. We interpret $N_0 = n$ as the sample size at time t and N_s as the population at time $t - s$ which can potentially influence the type of the individuals in the sample at time t. Then $(\xi_s)_{s \leq t}$ models in law the ancestral path of the individuals which are potentially relevant for the n-sample at time t. The partition π_s describes the groups of individuals at time t which have a common ancestor at time $t - s$. See Subsection 5.4 and Subsubsection 5.5.4 for more information.

Remark 43. We can now formulate our dual process into the classical scheme of duality theory (see (5.1)). We write X' for the pair $(\eta, \mathcal{F})$,

$$E = (\mathcal{P}(\mathbb{I}))^{\Omega_N} \quad , \quad E' = \{X' \in \mathcal{S} \times (\bigcup_{m=1}^{\infty} L_\infty(\mathbb{I}^m)) \,|m = |\pi|\}. \tag{5.15}$$

If we use the standard topology on the set E, it is a Polish space.

E' can be embedded in a complete metric space which is a subset of $\tilde{\mathcal{S}}$ based on ordered partitions of $\mathbb{N}$ (rather than finite sets) and the set of functions in $L_\infty(\mathbb{I}^\mathbb{N})$ which can be approximated by functions of finitely many variables. If we want to achieve separability of E' one needs to assume more on χ and H to be able to

5.1 The Dual Process and Duality Function for the Interacting System

restrict the function space to continuous functions on $\mathbb{I}$. However one should note that the measures appearing as states of X are always at most countably supported for positive time.

Then define a map $H : E \times E' \to \mathbb{R}$ by

$$H(X, (\eta, \mathcal{F})) = \int_{\mathbb{I}} \cdots \int_{\mathbb{I}} \mathcal{F}(u_1, \cdots, u_{|\pi|}) x_{\xi(1)}(du_1) \cdots x_{\xi(|\pi_t|)}(du_m). \qquad (5.16)$$

Then the collections of bounded measureable functions

$$\{H(\cdot, (\eta, \mathcal{F})), (\eta, \mathcal{F}) \in E'\}, \quad (\{H(X, \cdot), \quad X \in (\mathcal{P}(\mathbb{I}))^{\Omega_N}\}) \qquad (5.17)$$

are measure-determining on $(E, \mathcal{B}(E))$ respectively $(E', \mathcal{B}(E'))$. The second property is not always satisfied in the duality theory and typically not needed in applications. Then $H(\cdot, \cdot)$ is called a duality function for the pair (E, E') and E and E' are legitimate state spaces for Markov processes, recall Remark 41. (See [Lig85], [D], [Eth00] and [EK2] for detailed expositions of duality theory).

We describe the dynamics of the process in two steps, first we give the autonomous dynamics of $(\eta_t)_{t \geq 0}$ and then construct $(\mathcal{F}_t)_{t \geq 0}$ respectively $(\mathcal{F}_t^+)_{t \geq 0}$ or $(\mathcal{G}_t^+)_{t \geq 0}$ given a realization of the process η. In subsequent sections we introduce various modifications of this tailored for specific purposes.

Step 2 *(Dynamics of dual particle system)*

The dynamics of $\{\eta_t\}$ is that of a pure Markov jump process with the following transition mechanisms which correspond to resampling, migration and selection in the original model (in this order):

- *Coalescence* of partition elements: any pair of partition elements which are at the same site coalesce at a constant rate; the resulting enlarged partition element is given the smaller of the indices of the coalescing elements as index and the remaining elements are reordered to preserve their original order.
- *Migration* of partition elements: for $j \in \{1, 2, \ldots, |\pi_t|\}$, the partition element j can migrate after an exponential waiting time to another site in Ω_N which means we have a jump of $\xi_t(j)$.
- *Birth* of new individuals by each partition element after an exponential waiting time: if there are currently N_t individuals then a newly born individual is given index $N_{t-} + 1$ and it forms a *partition element consisting of one individual* and this partition element is given a label $|\pi_{t-}| + 1$. Its location mark is the same as that of its parent individual at the time of birth,
- *Independence*: all the transitions and waiting times described in the previous points occur independently of each other.

Thus the process $(\eta_t)_{t \geq 0}$ has the form

$$\eta_t = (\zeta_t, \pi_t, (\xi_t(1), \xi_t(2), \cdots, \xi_t(|\pi_t|))), \qquad (5.18)$$

where $\xi_t(j) \in \Omega_N$, $j = 1, \ldots, |\pi_t|$, and π_t is an ordered partition (ordered tuple of subsets) of the current basic set of the form $\{1, 2, \ldots, N_t\}$ where N_t also grows as a random process, more precisely as a pure birth process.

We order partition elements by their smallest elements and then *label* them by $1, 2, 3, \ldots$ in increasing order. Every partition element (i.e. the one with label i) has at every time a location in Ω_N, namely the i-th partition element has location $\xi_t(i)$. (The interpretation is that we have N_t individuals which are grouped in $|\pi_t|$-subsets and each of these subsets has a position in Ω_N).

Furthermore denote by

$$\pi_t(1), \cdots, \pi_t(|\pi_t|) \tag{5.19}$$

the index of the first, second etc. partition element. In other words the map gives the index of a partition element (the smallest individual number it contains) of a label which specifies its current rank in the order.

For our concrete situation we need to specify in addition the parameters appearing in the above description, this means that we define the dual as follows.

Definition 5.1 (First component of the dual process: η_t).

(a) The initial state η_0 is of the form (ζ_0, π_0, ξ_0) with:

$$\zeta_0 = \{1, 2, \cdots, n\}, \quad \pi_0 = \{\{1\}, \cdots, \{n\}\} \tag{5.20}$$

$$\xi_0 \in (\Omega_N)^n. \tag{5.21}$$

(b) The evolution of (η_t) is defined as follows (recall (3.24) and (3.5)–(3.37) for the meaning of the parameters):

  (i) each pair of partition elements which occupy the same site in Ω_N *coalesces* during the joint occupancy of a location into one partition element after a *rate d* exponential waiting time,

  (ii) every partition element performs, independent of the other partition elements, a *continuous time random walk* on Ω_N with transition *rates* $c\bar{a}(\cdot, \cdot)$ (see (3.11)), with $\bar{a}(\xi, \xi') = a(\xi', \xi)$,

  (iii) after a rate s exponential waiting time each partition element gives *birth* to a new particle which forms a new (single particle) partition element at its location, and this new partition element is given as label $|\pi_{t-}| + 1$ and the new particle the index $N_{t-} + 1$.

All the above exponential waiting times are *independent* of each other. $\square$

Note that this process is well-defined since the total number of partition elements is stochastically dominated by an ordinary linear birth process which is well-defined for all times.

5.1 The Dual Process and Duality Function for the Interacting System

Step 3 *(Dynamics of function-valued part of dual process: $\mathcal{F}_t, \mathcal{F}_t^+, \mathcal{G}_t^+$)*

To complete the specification of the dual dynamics we want to define a bivariate process $(\eta_t, \mathcal{F}_t)$ resp. $(\eta_t, \mathcal{F}_t^+)$ associated with (3.24). Hence we now describe the evolution of the process $(\mathcal{F}_t)_{t \geq 0}$ conditioned on a realization of the process $(\eta_t)_{t \geq 0}$. For a given path of the first component $(\eta_t)_{t \geq 0}$ the second component is a *function-valued process* $(\mathcal{F}_t)_{t \geq 0}$, respectively $(\mathcal{F}_t^+)_{t \geq 0}, (\mathcal{G}_t^+)_{t \geq 0}$ with values in $\bigcup_1^\infty L_\infty(\mathbb{I}^m)$, respectively $\bigcup_1^\infty L_\infty^+(\mathbb{I}^m)$.

The evolution of $\mathcal{F}_t$, respectively $\mathcal{F}_t^+, \mathcal{G}_t^+$ starts in the state

$$\mathcal{F}_0 = \mathcal{F}_0^+ = \mathcal{G}_0^+ = f, \qquad (5.22)$$

with a bounded measureable function f of $|\pi_0|$-variables, a parameter we denoted by n, the variables running in $\mathbb{I}$.

The evolution of $(\mathcal{F}_t)_{t \geq 0}$ involves three mechanisms, corresponding to the *resampling*, the *selection* mechanism and the *mutation* mechanism. We now describe separately these mechanisms.

(i) Resampling

This transition is classical.

Definition 5.2 (Conditioned evolution of $\mathcal{F}, \mathcal{F}^+$: the coalescence mechanism).
If a coalescence of two partition elements occurs, then the corresponding variables of $\mathcal{F}_t$ are set equal to the variable indexing the partition element, (here $\hat{u}_j$ denotes an omitted variable), i.e. for $\mathcal{F}_t = g$ we have the transition

$$g(u_1, \cdots, u_i, \cdots, u_j, \cdots, u_m) \longrightarrow \hat{g}(u_1, \cdots, \hat{u}_j, \cdots, u_m) \qquad (5.23)$$
$$= g(u_1, \cdots, u_i, \cdots, u_{j-1}, u_i, u_{j+1}, \cdots, u_m),$$

so that the function changes from an element of $L_\infty(\mathbb{I}^m)$ to one of $L_\infty(\mathbb{I}^{m-1})$. □

Remark 44. We could also work with (here $\tilde{\pi}(i)$ is the label of the partition element of i)

$$\tilde{\mathcal{F}}_t(u_{\tilde{\pi}_t(1)}, \cdots, u_{\tilde{\pi}_t(N_t)}) \text{ instead of } \mathcal{F}_t(u_1, \cdots, u_{(|\pi_t|)}). \qquad (5.24)$$

This means in (5.23) we view the r.h.s. as element of $L_\infty(\mathbb{I}^m)$. Then the state space $\tilde{E}'$ is obtained by replacing in (5.15) the restriction by $m = |\zeta|$ and using the duality function $\tilde{H}(\cdot, \cdot)$ given by

$$\tilde{H}(X, (\eta, \tilde{\mathcal{F}})) = \int_{I^{|\pi|}} \tilde{\mathcal{F}}(u_{\tilde{\pi}(1)}, \cdots, u_{\tilde{\pi}(N_t)}) x_{\xi(1)}(du_1) \cdots x_{\xi(|\pi|)}(du_{|\pi|}). \qquad (5.25)$$

The form in (5.24) codes some historical information which is lost in the form in (5.16). On the other hand expressions in the duality relation often become simpler in the reduced description.

(ii) Selection dual

There are several alternate ways to incorporate the effects of selection into the dual. In particular some versions involve signed function-valued processes and others use only non-negative-valued functions. Some versions involve a Feynman–Kac factor in the representation and others do not. We now describe those versions of the dual process that will be elaborated on as required in later sections.

Definition 5.3 (Conditioned evolution of $\mathcal{F}$ and $\mathcal{F}^+, \mathcal{G}^+$: the selection mechanisms).

(i) If a birth occurs in the process (η_t) due to the partition element π_i, $i \in \{1, \ldots, |\pi_t|\}$, then for $\mathcal{F}_{t-} = g$ the following transition occurs from an element in $L_\infty((\mathbb{I})^m)$ to elements in $L_\infty((\mathbb{I})^{m+1})$:

$$g(u_1, \ldots, u_m) \longrightarrow \chi(u_i)g(u_1, \ldots, u_m) - \chi(u_{m+1})g(u_1, \ldots, u_m), \quad (5.26)$$

where the new variable is associated with the partition element of the newly born individual.

(i') For $\mathcal{F}_t^+$ the transition (5.26) is replaced by (provided that χ satisfies $0 \leq \chi \leq 1$):

$$g(u_1, \cdots, u_m) \longrightarrow \hat{g}(u_1, \ldots, u_{m+1}) = (\chi(u_i) + (1 - \chi(u_{m+1})))g(u_1, \cdots, u_m), \quad (5.27)$$

where the new variable is associated with the newly born individual. In particular

$$\|\hat{g}\|_\infty \leq 2\|g\|_\infty, \text{ and } \hat{g} \geq 0, \text{ if } g \geq 0. \quad (5.28)$$

(i'') For $(\mathcal{G}_t^+)_{t \geq 0}$ we use the following transition:

$$\begin{aligned} g(u_1, \cdots, u_m) &\longrightarrow (\chi(u_i)1(u_{m+1})g(u_1, \cdots, u_m) \\ &+ (1 - \chi(u_i))g(u_1, \cdots, u_{i-1}, u_{m+1}, u_{i+1}, \cdots, u_m), \end{aligned} \quad (5.29)$$

in which case

$$\|\hat{g}\|_\infty \leq \|g\|_\infty. \quad \square \quad (5.30)$$

Remark 45. The mechanism (i) will lead to a signed-function-valued process and requires a Feynman–Kac factor, while the mechanism (i') leads to a non-negative function-valued process and in this case NO Feynman–Kac factor is needed.

5.1 The Dual Process and Duality Function for the Interacting System

The mechanism (i'') induces the same change in the duality function H as (i') and therefore leads also to a duality function.

Remark 46. We note that the dynamic (i'') disconnects the strict parallel between the variables in the function and the individuals in the process η which corresponds to the fact that an individual in the sample may be replaced by one from outside the sample or vice versa.

(iii) Mutation dual

We now introduce some basic objects used to incorporate the mutation mechanism into the dual. Let $\mathcal{G}_t^{\mathrm{mut}}$ denote the semigroup on $\bigoplus_{n \in \mathbb{N}} L_\infty(\mathbb{I}^n)$ induced by independent copies of the mutation process whose semigroup is acting on $L_\infty(\mathbb{I})$, i.e. the process we obtain from (3.24) by letting the geographic space (i.e. the index set for the components consist of one point and putting all other coefficients d, s, r, c in (3.24) equal to 0).

We can represent $\mathcal{G}_t^{\mathrm{mut}}$ in terms of the process $\{M_t^*\}_{t \geq 0}$ which is a Markov pure jump process with jumps $L_\infty(\mathbb{I}^n) \to L_\infty(\mathbb{I}^n)$ as follows. Define M_j acting on $L_\infty(\mathbb{I}^n)$ for $j \in \{1, \cdots, n\}$ by

$$M_j g(u_1, \ldots, u_j, \ldots, u_n) = \int_\mathbb{I} g(u_1, \ldots, u_{j-1}, v, u_{j+1}, \ldots, u_n) M(u_j, dv). \tag{5.31}$$

Then consider the transitions:

$$f \to M_j f \quad \text{at rate } m \text{ for each } j = 1, \ldots, n. \tag{5.32}$$

Then

$$\mathcal{G}_t^{\mathrm{mut}}(g) = E[M_t^*(g)], \tag{5.33}$$

is a (deterministic) linear semigroup on $\bigoplus_{n \in \mathbb{N}} L_\infty(\mathbb{I}^n)$ (with exactly that domain) with generator

$$\mathcal{M}^* = m \sum_j (M_j - I). \tag{5.34}$$

This allows to define:

Definition 5.4 (Conditioned evolution of $\mathcal{F}, \mathcal{F}^+, \mathcal{G}_t^+$: the mutation mechanism). If no transition occurs in $(\eta_t)_{t \geq 0}$ then $\mathcal{F}_t$, respectively $\mathcal{F}_t^+, \mathcal{G}_t^+$ follows the deterministic evolution given by the semigroup $\mathcal{G}_t^{\mathrm{mut}}$ acting on $L_\infty(\mathbb{I}^m)$ provided that currently $\mathcal{F}_t$, respectively ($\mathcal{F}_t^+$ or $\mathcal{G}_t^+$) is a function of m variables, meaning at the last transition in (η_s) up to time t became an element of $L_\infty((\mathbb{I})^m)$. □

This completes the evolution mechanism driving the process $(\mathcal{F}_t)_{t\geq 0}$ respectively $(\mathcal{F}_t^+)_{t\geq 0}$ or $(\mathcal{G}_t^+)_{t\geq 0}$ for given η. In this fashion, the function-valued processes starting with an element $f \in L_\infty(\mathbb{I}^n)$ for some n are uniquely defined for every given path for the process η.

Step 4 *(Definition bivariate dual process)*

Combining Step 1 and Step 2 we have defined uniquely for every (ordered) set of n individuals with n locations in Ω_N and a function f_0 in $L^\infty(\mathbb{I}^n)$ a bivariate Markov process $(\eta_t, \mathcal{F}_t)_{t\geq 0}$, respectively $(\eta_t, \mathcal{F}_t^+)_{t\geq 0}$ or $(\eta_t, \mathcal{G}_t^+)_{t\geq 0}$ in which the jumps occur only at the times of jumps in the pure Markov jump process $(\eta_t)_{t\geq 0}$ and further changes in the function-valued part occur deterministically in a continuous way (action of mutation semigroup).

Remark 47. One advantage of working with the process $(\eta_t, \mathcal{F}_t)$ (or $(\eta_t, \mathcal{F}_t^+)$) over working with the original process is, that it can be explicitly constructed and only a finite number of random transitions occur in a finite time interval whereas changes occur instantaneously in countably many components of the process X in any finite time interval and are furthermore diffusive changes.

5.2 Duality Relation for Interacting Systems

Duality relations are frequently used in the theory of interacting spatial systems. The following is the basic *Feynman–Kac duality* relation that was established in [DG99] for the general class of interacting Fleming–Viot processes with selection and mutation from which we derive the *new duality relation* in Proposition 5.6 below without Feynman–Kac term which we use in this monograph.

Proposition 5.5 (Duality relation—signed with Feynman–Kac dual). *Let $(X_t)_{t\geq 0}$ be a solution of the (L, X_0)-martingale problem with L as in (3.24) and the ingredients of the form described in Section 3.1, part (3) and (4) and with $X_0 = (x_\xi)_{\xi \in \Omega_N}$. Choose $\xi_0 \in (\Omega_N)^n$ and $f \in L_\infty((\mathbb{I})^n)$ for some $n \in \mathbb{N}$. (Recall (5.10), (5.18) for notation).*

Assume that t_0 is such that:

$$E\left(\exp\left(s \int_0^{t_0} |\pi_r| dr\right)\right) < \infty. \tag{5.35}$$

Then for $0 \leq t \leq t_0$, $(\eta_t, \mathcal{F}_t)$ is the Feynman–Kac dual of (X_t), that is:

$$F_t(X_0, (\eta_0, f)) = E_{(\eta_0, \mathcal{F}_0)}\left\{\left[\exp\left(s \int_0^t |\pi_r| dr\right)\right] \cdot \right. \tag{5.36}$$

5.2 Duality Relation for Interacting Systems

$$\left[\int_{\mathbb{I}} \cdots \int_{\mathbb{I}} \mathcal{F}_t(u_1, \ldots, u_{|\pi_t|}) x_{\xi_t(1)}(du_1) \cdots x_{\xi_t(|\pi_t|)}(du_{|\pi_t|})\right]\right\},$$

where the initial state $(\eta_0, \mathcal{F}_0)$ is given by

$$\eta_0 = [\{1, \ldots, n\}, (\xi(1), \cdots, \xi(n)), (\{1\}, \{2\}, \ldots, \{n\})], \quad (5.37)$$
$$\xi(i) \in \Omega_N \text{ for } i = 1, \cdots, n; \quad n \in \mathbb{N},$$
$$\mathcal{F}_0 = f. \quad \square$$

Remark 48. In [DG99] it was shown that there exists $t_0 > 0$ for which (5.35) is satisfied.

The unpleasant features of the above dual are the exponential term and the signed function, which make it difficult to analyse for $t \to \infty$. However if we have a fitness function χ satisfying $0 \leq \chi \leq 1$ (which because of scaling properties of the dynamic really only means that we do not allow *unbounded* fitness but cover all other cases) then we can obtain the following duality relation that does *not* involve a Feynman–Kac factor and preserves the *positivity* of functions:

Proposition 5.6 (Duality relation—non-negative). *With the notation and assumptions as in Proposition 5.5 (except 5.35) we get for χ with*

$$0 \leq \chi \leq 1 \quad (5.38)$$

that for all $t \in [0, \infty)$ we have:

$$F_t(X_0, (\eta_0, f)) = E_{(\eta_0, \mathcal{F}_0^+)}\left[\int_{\mathbb{I}} \cdots \int_{\mathbb{I}} \mathcal{F}_t^+(u_1, \cdots, u_{|\pi_t|}) x_{\xi_t(1)}(du_1)\right.$$

$$\left.\cdots x_{\xi_t(|\pi_t|)}(du_{|\pi_t|})\right] \quad (5.39)$$

$$= E_{(\eta_0, \mathcal{G}_0^+)}\left[\int_{\mathbb{I}} \cdots \int_{\mathbb{I}} \mathcal{G}_t^+(u_1, \cdots, u_{|\pi_t|}) x_{\xi_t(1)}(du_1)\right.$$

$$\left.\cdots x_{\xi_t(|\pi_t|)}(du_{|\pi_t|})\right].$$

Moreover, $\mathcal{F}_t^+$ is always non-negative if $\mathcal{F}_0^+ = f \geq 0$, similarly for $\mathcal{G}_t^+$ if $\mathcal{G}_0^+ = f \geq 0$. $\square$

Remark 49. The duality relations we develop here for Ω_N work in fact for every geographic space which forms a countable group and for every migration mechanism induced by a random walk on that group. With some care we can even pass to migration given by a general Markov chain.

Remark 50. The duality relations work for general type space for example a continuum $[0, 1]$ as well, the special structure of $\mathbb{I}$ has not been used.

We note that the duality relation given above is of the general form considered in duality theory, as demonstrated by the following remark but only the state space is more complex than in the classical cases for Fisher–Wright diffusions or measure-valued processes.

Remark 51. In the language of classical duality theory as described in Remark 43 we have abbreviating $X'_t = (\eta_t, \mathcal{F}_t)$, the duality relation

$$E_{X_0}[H(X_t, X'_0)] = E_{X'_0}[H(X_0, X'_t)], \quad \forall\, t \geq 0 \tag{5.40}$$

whenever $(X_0, X'_0) \in E \times E'$. Therefore the l.h.s. which is the object of interest can be calculated in terms of the r.h.s. involving a process of a simpler nature.

For Feller processes on compact state spaces with generators G_X respectively $G_{X'}$, respectively, then the duality relation follows from the generator relation

$$(G_X H(\cdot, X'_0))(X_0) = (G_Y H(X_0, \cdot))(X'_0) \text{ for } (X_0, X'_0) \text{ in } E \times E' \tag{5.41}$$

(see e.g. [Lig85], Chap. 2). On non-compact spaces some integrability properties have to be verified in addition (see [EK2], Chap. 4).

The duality relation above can be extended as follows.

Remark 52. It is often necessary to consider moments of the time-space process when the process is observed say at two times t and $t + s$ with $s > 0$. In this case one considers the dual particle system where the dual particles corresponding to the moment at the earlier time t are activated in the dual process only at time s and the resulting dual particle system is then evaluated after evolving up to time $t + s$. This observation has first been used in the voter model and is quite useful for us later on in some calculations.

The duality relation in Proposition 5.6 can be related to the previous FK-duality in [DG99] quoted in Proposition 5.5 if we observe that the action of the selection part of the generator of X can be written in two ways, namely as (assuming that χ is bounded by 1 and in the expression χg below taking χ to be a function of *one* of the variables of g):

$$\{(\chi g - \chi \otimes g) - g\} + g = (\chi g + (1 - \chi) \otimes g) - g. \tag{5.42}$$

The first expression, that is, $\{(\chi g - \chi \otimes g) - g\} + g$ corresponds to a jump defining the transition of $\mathcal{F}_t$ (the term in $\{\}$) plus a Feynman–Kac term g as we have in the duality

5.2 Duality Relation for Interacting Systems

in (5.36). The second expression, that is, $(\chi g + (1-\chi) \otimes g) - g$ corresponds to a jump describing $\mathcal{F}_t^+$ which now yields as new function-valued state or function consisting of two summands but does not require a Feynman–Kac factor. Since integrability issues are involved we now give a detailed proof below.

Proof of Proposition 5.6. We have to do two things, (1) verify the generator relation and (2) an integrability condition to guarantee that the r.h.s. in (5.39) is always finite.

(1) *Generator relation.* Denote the generator of the process by L and of the dual process $(\eta, \mathcal{F}_t^+)_{t \geq 0}$ by L'. Then we have to verify for the duality function H that

$$(LH(\cdot, (f, \eta)))(X) = (L'H(X, \cdot))((\eta, f)). \tag{5.43}$$

In order to verify the generator relation (5.41), we have to calculate LF, recall (3.24) and for that we evaluate first the first and second order differential operators acting on F.

Fix $n \in \mathbb{N}$ and a map $\xi : \{1, \cdots, , n\} \longrightarrow \Omega_N$. We need to calculate for functions G with

$$F(X) = \int_{\mathbb{I}^n} f(u_1, \cdots, u_n) x_{\xi(1)}(du_1) \cdots, x_{\xi(n)}(du_n), \quad X \in (\mathcal{P}(\mathbb{I}))^{\Omega_N},$$

$$X = \otimes_{i \in \Omega_N} x_i \tag{5.44}$$

the first and second order derivatives (recall (3.20)–(3.21)). Note that the $\{\xi(j) : j = 1, \ldots, n\}$ are not necessarily distinct and that repetitions can occur.

For $\xi \in \Omega_N$ (cf. (3.20))

$$\frac{\partial F(X)}{\partial \mu_\xi}[v] = \sum_{\substack{\ell \in \{1, \ldots, n\} \\ \xi(\ell) = \xi}} \left(\int_{\mathbb{I}^{n-1}} f(u_1, \cdots, u_{\ell-1}, v, u_{\ell+1}, \cdots, u_n) \bigotimes_{\substack{j=1 \\ j \neq \ell}}^n x_{\xi(j)} \right), \tag{5.45}$$

$$\frac{\partial^2 F(X)}{\partial \mu_\xi \partial \mu_\xi}[v, v'] = \sum_{\substack{\ell, \ell' \in \{1, \ldots, n\}, \ell \neq \ell' \\ \xi(\ell) = \xi(\ell') = \xi}} \left(\int_{\mathbb{I}^{n-2}} f(u_1, \cdots, u_{\ell-1}, v, u_{\ell+1}, \right.$$

$$\left. \cdots, u_{\ell'-1}, v', u_{\ell'+1}, \cdots, u_n) \bigotimes_{\substack{\ell=1 \\ \ell \neq i, j}}^n x_{\xi(\ell)}(du_\ell) \right). \tag{5.46}$$

Note that if $|\{\ell : \xi(\ell) = \xi\}| = k$, then there are $\binom{k}{2}$ summands.

We can now apply this formula to our mixed moment F, recall (5.3). Recall (3.24)

$$(LF)(X) = (L^{\text{mig}} + L^{\text{sel}} + L^{\text{mut}} + L^{\text{sam}})F(X) \qquad (5.47)$$

$$= \sum_{\xi \in \Omega_N} \left[c \sum_{\xi' \in \Omega_N} a_N(\xi, \xi') \int_{\mathbb{I}} \frac{\partial F(x)}{\partial x_\xi}(u)(x_{\xi'} - x_\xi)(du) \right.$$

$$+ s \int_{\mathbb{I}} \left\{ \frac{\partial F(x)}{\partial x_\xi}(u) \left(\chi(u) - \int_{\mathbb{I}} \chi(w) x_\xi(dw) \right) \right\} x_\xi(du)$$

$$+ m \int_{\mathbb{I}} \left\{ \int_{\mathbb{I}} \frac{\partial F(x)}{\partial x_\xi}(v) M(u, dv) - \frac{\partial F(x)}{\partial x_\xi}(u) \right\} x_\xi(du)$$

$$+ d \int_{\mathbb{I}} \int_{\mathbb{I}} \frac{\partial^2 F(x)}{\partial x_\xi \partial x_\xi}(u, v) Q_{x_\xi}(du, dv) \right], \qquad x \in (\mathcal{P}(\mathbb{I}))^{\Omega_N},$$

where

$$Q_x(du, dv) = x(du)\delta_u(dv) - x(du)x(dv). \qquad (5.48)$$

We now apply this to the function $F(\cdot) = H(\cdot, (\eta, \mathcal{F}))$ (cf. (5.16)).
We first consider the action of L^{mut} using (5.44),

$$(L^{\text{mut}} H(\cdot, (\eta, \mathcal{F})))(X) \qquad (5.49)$$

$$= m \int_{\mathbb{I}^n} \left(\sum_{\ell \in \{1,\ldots,n\}} \left(\left[\int_{\mathbb{I}} f(u_1, \cdots, u_{\ell-1}, v, u_{\ell+1}, \cdots, u_n) M(u, dv) \right. \right. \right.$$

$$- f(u_1, \cdots, u_{\ell-1}, u, u_{\ell+1}, \cdots, u_n) \Big] \Big)$$

$$\bigotimes_{j=1}^{n} x_{\xi(j)}(du_j) \Big)$$

$$= H(X, (\eta, \mathcal{M}^* f)).$$

We next consider the action of L^{mig} using (5.44),

$$(L^{\text{mig}} H(\cdot, (\eta, \mathcal{F})))(X) \qquad (5.50)$$

$$= c \int_{\mathbb{I}^n} \Big(\sum_{\ell \in \{1,\ldots,n\}} \Big(\int_{\mathbb{I}} f(u_1, \cdots, u_n)$$

5.2 Duality Relation for Interacting Systems

$$\left[\sum_{\ell=1}^{n}\sum_{\xi'}\bar{a}_N(\xi(\ell),\xi')\bigotimes_{\substack{j=1\\j\neq\ell}}^{n} x_{\xi(j)}(du_j) \otimes x_{\xi'}(du_\ell) - \bigotimes_{j=1}^{n} x_{\xi(j)}(du_j)\right]\right)$$

$$= H(X, \sum_{\ell=1}^{n}\sum_{\eta'}\hat{a}_\ell(\eta,\eta')(\eta', f)) = L'^{,\mathrm{mig}} H(X,(\eta,\mathcal{F}))$$

where $\eta'_\ell = (\zeta, \pi, \xi'_\ell)$, $\xi'_\ell(j) = \xi(j)$, $j \neq \ell$, $\xi'_\ell(\ell) = \xi'$, $\hat{a}_\ell(\eta, \eta'_\ell) = a_N(\xi(\ell), \xi')$.

Consider the following function on the state space of the dual process. Fix an element $X \in (\mathcal{P}(\mathbb{I}))^{\Omega_N}$ and define for $n \in \mathbb{N}$, partition π of $\{1, \cdots, n\}$, map $\xi : \{1, \cdots, |\pi|\} \to \Omega_N$ and $f : \mathbb{I}^{|\pi|} \to \mathbb{R}^+$, define

$$G((\eta, f)) = \int_{\mathbb{I}^{|\pi|}} f(u_1, \cdots, u_{|\pi|}) x_{\xi(1)}(du_1) \cdots x_{\xi(|\pi|)}(du_{|\pi|}) \quad (5.51)$$

$$= \int_{\mathbb{I}^{|\pi|}} f(u_{\xi(1)}, \cdots, u_{\xi(|\pi|)}) \bigotimes x_\xi(du_\xi) = (H(X, \cdot))((\eta, f)).$$

We have to calculate the action of the generator L', i.e. we have to determine $(L'G)((\eta, f))$, where L' is the generator of the pure Markov jump process with a piecewise deterministic part (for the mutation) which we defined as dual process. We obtain by explicit calculation

$$(L'^{\mathrm{mig}} H(X, \cdot))((\eta, f)) = (L'G)((f, \eta)) \quad (5.52)$$

$$= \int_{\mathbb{I}^{|\pi|}} \Big[\sum_{\ell=1}^{n}\sum_{\xi'} a_N(\xi(\ell), \xi')$$

$$(f(u_{\xi(1)}, \cdots, u_{\xi(\ell-1)}, u_{\xi'}, u_{\xi(\ell+1)}, \cdots u_{\xi(|\pi|)})$$

$$- f(u_{\xi(1)}, \cdots, u_{\xi(|\pi|)}))\Big] \bigotimes x_\xi(du_\xi)$$

$$= (L^{\mathrm{mig}} H(\cdot, (\eta, f)))(X).$$

Combining both (5.50) and (5.52) we obtain the claim (5.43). The generator calculations corresponding to selection and resampling (coalescence) follow in the same way (compare [DG99]) and this completes the required generator calculation.

(2) Finite expectation

Here we have to show that

$$E_{(\eta_0, \mathcal{F}_0)}[H(X, (\eta_t, \mathcal{F}_t))] < \infty \text{ for all } t \geq 0, \quad (5.53)$$

in order to use Theorem 4.11 in [EK2] to conclude the duality relation from the generator relation we established in the previous point. We begin by verifying this for some $t_0 > 0$ which is sufficiently small.

Namely we realize first that all transitions occurring in the dual preserve the $\|\cdot\|_\infty$-norm of $\mathcal{F}_t$ except the selection transition. Here we have $g \to \chi g + (1-\chi) \otimes g$ which satisfies

$$\|\chi g + (1-\chi) \otimes g\|_\infty \leq 2\|g\|_\infty. \tag{5.54}$$

Since the number of selection events is given by a pure birth process with birth rate s we have

$$\|\mathcal{F}_t\|_\infty \leq 2^{N_t} \cdot \|\mathcal{F}_0\|_\infty. \tag{5.55}$$

Hence by explicit calculations of the Laplace-transform of N_t we have:

$$E[\|\mathcal{F}_t\|_\infty] \leq (1-p)\left(\frac{2}{1-2p}\right)^{N_0} \|\mathcal{F}_0\| < \infty, \text{ if } p = (1-e^{-st}) < \frac{1}{2}. \tag{5.56}$$

Therefore

$$E[\|\mathcal{F}_t\|_\infty] < \infty \quad , \quad \forall\, t < \frac{\log 2}{s} \text{ for all } N_0 \in \mathbb{N}. \tag{5.57}$$

We now have to extend (5.40) to all $t \geq 0$. We argue as follows.

First using Theorem 4.11 in [EK2] with $\Gamma_T = 2^{N_T}$ and $T < \frac{\log 2}{s}$, so that $E[2^{N_T}] < \infty$, we get

$$E_{X_0}[H(X_t, (\eta_0, \mathcal{F}_0))] = E_{\eta_0, \mathcal{F}_0}[H(X_0, (\eta_t, \mathcal{F}_t))], \text{ if } t \leq T, \tag{5.58}$$

and therefore the martingale problem for X is wellposed and has as solution a Markov process (in fact a Feller process). Furthermore we know that we can write (this is implied by the form of the selection transition which at each birth creates as new state a sum of two functions derived from the old function),

$$\mathcal{F}_t = F(t, (\eta_0, \mathcal{F}_0)) = \sum_{i=1}^{N_t} \mathcal{F}_{t,i}. \tag{5.59}$$

Then we know from the Markov property of the dual process and its form that

$$\mathcal{F}_{2t} = F(t, (\eta_t, \mathcal{F}_t)) = \sum_{i=1}^{N_t} F(t, (\eta_t, \mathcal{F}_{t,i})). \tag{5.60}$$

Then observe that the Markov property of $(X_t)_{t \geq 0}$ allows to calculate as follows. Using the duality for time t to time $2t$ (here $\tilde{E}$ denotes expectation over the dual dynamic)

$$\begin{aligned} E[H(X_{2t}, (\eta_0, \mathcal{F}_0))|X_0] &= E[E[H(X_{2t}, (\eta_0, \mathcal{F}_0))|X_t]|X_0] \\ &= E[\tilde{E}[H(X_t, (\eta_t, \mathcal{F}_t))|(\eta_0, \mathcal{F}_0)]X_0] \\ &= E[\tilde{E}[H(X_t, (\eta_t, \sum_{i=1}^{N_t} \mathcal{F}_{t,i}))|(\eta_0, \mathcal{F}_0)]|X_0]. \end{aligned} \quad (5.61)$$

Now we calculate the r.h.s. of the above equation as:

$$\begin{aligned} &\tilde{E}[E[H(X_t, |(\eta_t, \sum_{i=1}^{N_t} \mathcal{F}_{t,i}))]|X_0](\eta_0, \mathcal{F}_0)) \\ &= \tilde{E}\left[\sum_{i=1}^{N_t}[H(X_0, (\eta_{2t}, F(t, \mathcal{F}_{t,i})))|(\eta_0, \mathcal{F}_0)]\right] \\ &= [\tilde{E}\left[H(X_0, (\eta_{2t}, \sum_{i=1}^{N_t} F(t, (\eta_t, \mathcal{F}_{t,i})))|(\eta_0, \mathcal{F}_0)\right] \\ &= \tilde{E}[\tilde{E}[H(X_0, (\eta_{2t}, \mathcal{F}_{2t}))|(\eta_t, \mathcal{F}_t)]|(\eta_0, \mathcal{F}_0)] \\ &= \tilde{E}[H(X_0, (\eta_{2t}, \mathcal{F}_{2t}))|(\eta_0, \mathcal{F}_0)], \end{aligned}$$

where we used duality between times t and $2t$, the construction of the dual and its Markov property. Hence we get the duality relation up to time $2T$. Iteration gives the claim for all positive times. q.e.d.

5.3 Two Alternative Duals for the Mutation Component

In order to handle certain applications where mutation is a key it is sometimes useful to modify the dual representation. Here we give two modifications of the mutation induced part of the dual as well as their combined version which will be crucial later on also for the refined version of the dual which is the main tool for the renormalization analysis.

5.3.1 Modified Dual for State-Independent Mutation Component

In the case in which there is a non-zero state independent component of the mutation mechanism, that is, $\bar{m} > 0$, (recall here Definition 3.4 for the relevant notation) we can obtain a modified dual which is particularly useful when $M - \bar{m}(1 \otimes \rho) = 0$.

Definition 5.7 (Modified mutation dual). We define the *modified dual process* $(\hat{\eta}_t, \mathcal{F}_t)_{t \geq 0}$ resp. $(\hat{\eta}_t, \mathcal{F}_t^+)_{t \geq 0} (\hat{\eta}_t, \mathcal{G}_t^+)_{t \geq 0}$ for a mutation matrix M satisfying $mM \geq \bar{m} 1 \otimes \rho$ as follows.

First, we enlarge the space Ω_N to

$$\Omega_N \cup \{*\}. \tag{5.62}$$

In all the transition rates of the dual particle process η are 0. In addition the function-valued part does not change for variables associated with partition elements located on $\{*\}$.

The dynamics of the process $(\hat{\eta}_t, \mathcal{F}_t)$ resp. $(\hat{\eta}_t, \mathcal{F}_t^+)_{t \geq 0}, (\hat{\eta}_t, \mathcal{G}_t^+)_{t \geq 0}$ is now obtained by adding to the mechanism of η for partition elements still located on Ω_N (rather than $*$):

jumps of the partition elements to $\{*\}$ after exponential waiting times at rate $\bar{m}$
$$\tag{5.63}$$

independently of each other (and independent of those of all other transitions) and by changing the dynamics of $\mathcal{F}_t, \mathcal{F}_t^+, \mathcal{G}_t^+$ by

replacing mM_t^* generated by the mutation kernel M by the semigroup (5.64)
M_t^* corresponding to the transition rates $(mM - \bar{m} 1 \otimes \rho)$. □

Remark 53. Note that $\hat{\eta}_t = (\hat{\zeta}_t, \hat{\pi}_t)$ and both components have a different law than the process (ζ_t, π_t), since in particular on $*$ there is no coalescence taking place. This observation has important consequences. In the case $M - \bar{m}(1 \otimes \rho) = 0$ once a partition element reaches $\{*\}$ this element does not undergo any further change and the dual process can be easily analysed since it is eventually trapped with all locations on $\{*\}$.

With this modification the same duality relation between $(X_t)_{t \geq 0}$ and $(\hat{\eta}_t, \mathcal{F}_t)_{t \geq 0}$ and its variants holds if we enrich by an additional component associated with the site $*$, in exactly the form given in Proposition 5.5 respectively Proposition 5.6. Precisely:

Proposition 5.8 (Modified Duality for state-independent mutation jumps). Let $(X_t)_{t \geq}$ be as in Proposition 5.5 satisfying $\bar{m} > 0$ and extend it to a process on $\Omega_N \cup \{*\}$. Here the state of the original process $X(t)$ is defined in the additional state $*$ for all times as the probability measure ρ on the type space which is the measure giving the state-independent part of the mutation rates.

Let now $(\hat{\eta}_t, \mathcal{F}_t)$, respectively $(\hat{\eta}_t, \mathcal{F}_t^+), (\hat{\eta}_t, \mathcal{G}_t^+)_{t \geq 0}$ denote the modified dual processes defined in (5.62)–(5.64). Then the analogues of (5.36),(5.37) and (5.39) hold. □

5.3 Two Alternative Duals for the Mutation Component

Proof of Proposition 5.8. First we note that we can decompose the mutation semigroup in the independent superposition of the state-independent and the state-dependent part.

Next note that the expectation of the newly introduced jump occurring in our test function is exactly given by the action of the state-independent part of the mutation semigroup. In particular we have not changed the expected value of the test function switching to the new dynamic. Since the duality relation involves on both sides the expectations we get the claim. Alternatively apply (5.41) and note that the state-independent part of the generator and the part for the jump to $\{*\}$ in the dual satisfy this relation.   q.e.d.

5.3.2 The Dual for Mutation: Alternative Representation in the State-Dependent Case

The modification we now describe is useful in dealing with mutation which is not state-independent, for example in our context when we have also rare mutations from one level to the next. In that case it is useful to change the dynamics of $\mathcal{F}_t; \mathcal{F}_t^+, \mathcal{G}_t^+$ by replacing the *deterministic* function-valued evolution driven by the mutation semigroup as specified in Definition 5.4 or in Definition 5.7 by a *random* and *function-valued* jump process. The process η remains untouched.

Namely remove the deterministic evolution from (5.33) and replace it by the following Markov pure jump process in function space. Introduce the following jumps in function space for $g \in L_\infty(\mathbb{I}^k)$:

$$g(u_1, u_2, \cdots, u_k) \longrightarrow (M_i g)(u_1, u_2, \cdots, u_k), \quad k \in \mathbb{N}, \tag{5.65}$$

where M_i denotes the application of the operator M to the i-th variable of the function. These jumps occur for a function $f \in L_\infty(\mathbb{I}^k)$ independently at rate m for each $i = 1, 2, \cdots, k$. This defines (uniquely) a Markov pure jump process with values in $(\bigcup_n L_\infty(\mathbb{I}^n))$ starting from any initial state.

Consider now the situation that $\mathcal{F}_t, \mathcal{F}_t^+, \mathcal{G}_t^+$ depends on k variables. For each $i \in \{1, \cdots, k\}$ the jumps in (5.65) occur at rate m and this jump time is independent of everything else. Therefore as long as the state of $\mathcal{F}_t$, respectively $\mathcal{F}_t^+, \mathcal{G}_t^+$ is a function of k variables these random transition from (5.65) occur after exponential waiting times. We denote by

$$\hat{\mathcal{F}}_t, \hat{\mathcal{F}}_t^+, \hat{\mathcal{G}}_t^+ \tag{5.66}$$

the resulting modified function-valued processes. Note that in particular we can assign then the mutation jump occurring always with a particular partition element, which will be of some relevance for the historical interpretation.

Since the expectation of the jumps occurring in our test function is by construction given by the action of the mutation semigroup and since the duality relation only claims the identity of two expectations, we have not changed the r.h.s. of the duality relation and conclude the following.

Proposition 5.9 (Modified mutation transition). *All the previous duality relations remain valid if we replace $(\mathcal{F}_t)_{t \geq 0}$ by $(\hat{\mathcal{F}}_t)_{t \geq 0}$ or $(\mathcal{F}_t^+)_{t \geq 0}$ by $(\hat{\mathcal{F}}_t^+)_{t \geq 0}$, similarly $(\mathcal{G}_t^+)_{t \geq 0}$ by $(\hat{\mathcal{G}}_t^+)_{t \geq 0}$ and leave η untouched. The same holds for the process $(\hat{\eta}_t, \mathcal{F}_t)_{t \geq 0}, (\hat{\eta}_t, \mathcal{F}_t^+), (\hat{\eta}_t, \mathcal{G}_t^+)_{t \geq 0}$ from Definition 5.7.* □

Remark 54. Note how here the interplay between selection and mutation is reflected in this form of the duality. For example assume that we have rare mutation events, i.e. with rates $<< s$, the selection rate. Then: the rate of mutation events in the dual process is proportional to the number of partitions elements we have in the dual process and therefore rare mutations become visible in the dual process as soon as many births have occurred in the dual process due to a much higher selection rate, which then compensates via the large number of individuals a small mutation rate.

5.3.3 Pure Jump Process Dual

We can use both constructions presented in the previous two subsubsections at once:

Definition 5.10 (Pure jump process dual). In particular in combining the new representations of the state-dependent part $mM - \bar{m}1 \otimes \rho$ and the state-independent part in $\bar{m}1 \otimes \rho$ we get a pure Markov jump process

$$(\hat{\eta}_t, \hat{\mathcal{F}}_t)_{t \geq 0}, \quad (\hat{\eta}_t, \hat{\mathcal{F}}_t^+)_{t \geq 0}, (\hat{\eta}_t, \hat{\mathcal{G}}_t^+)_{t \geq 0}. \quad \square \qquad (5.67)$$

It is the dual process on which we base our refined dual in Subsection 5.5 and it is also the version best suited for the historical interpretation since it generates a marked (locations, mutation events) random graph in which we can find the marked ancestral tree of a tagged subpopulation. (Compare Subsubsection 5.5.4).

5.4 Historical Interpretation of the Dual Process

Does this analytical construction above have a heuristic meaning or is it simply an analytical trick revealing an explicitly solvable model? Both these alternative situations occur in the theory of stochastic processes. In our model, however, the duality can be interpreted in a nice way if one considers what is sometimes called the historical process. In fact this duality can be extended to reach processes including more information about the individuals past (see [DGP]).

5.4 Historical Interpretation of the Dual Process

In order to understand this, one has to remember that the diffusions we work with arise as small mass—many individual limits of particle models. In such a particle model it is possible to follow through time, the fate of individuals and their descendants and define for each individual currently alive its ancestral path leading back to its father then its grandfather and so on where this ancestral path gives the location and the type of the ancestor at every time. In fact we can from a time horizon t backward generate the collections of ancestral paths and their genealogical relation which define a tree or rather a forest of trees corresponding to different founding fathers. This way we obtain a random marked (with types and locations) forest. A more modest question would be to determine the types and the time back to the first common ancestor of the sample taken from time t (this means we only get the types at time t and not at all earlier times along the ancestral path.)

How can we study this complicated object? For each given time horizon t we can zoom in on a *finite subpopulation of tagged individuals*, tagged in the time t population and trace their history *backwards in time*. Then we can ask whether this object can be generated by a suitable stochastic process, the *backward process*. Can we hope for this process to again be Markov and time-homogeneous?

To understand this better, take the non-spatial case first. Ask the question: What are the types of a k-sample of individuals from the time t population. To answer this question generate the law of the historical evolution of this subpopulation by running a stochastic process backward. Similarly in a spatial model one can take samples at different locations. A nice case arises if this backward dynamic of the sample alone is already a Markov process, which is time-homogeneous. It is in this situation that one traditionally speaks of the existence of a dual process.

Resort first to a simpler, the *neutral* case. Due to exchangeability of individuals for the neutral case (i.e. no selection and mutation) such a Markovian backward process can be given based on a coalescent generating the family structure of the sample. A key tool to establish this in our context is the representation in terms of the lookdown process of Donnelly and Kurtz adapted to this spatial situation, (compare [GLW]). This then allows us to rigorously establish and identify the dynamics of the backward process in terms of a spatial coalescent ([GLW]). This can be extended and one can show that the genealogical trees of the neutral model evolve in such a way that their state at a fixed time t is given by the genealogical tree associated with the coalescent, see [GPWmetric09] and [GPWmp12].

In the case in which selection and mutation both depend on the type of the individual exchangeability is no longer preserved and a complicated interaction between the tagged subpopulation in the sample and the remaining population arises, which is now not equivalent in law to an autonomous evolution of the sample. Hence in order to generate the genealogy by a backward process we have to add a richer device which consists of a reservoir of possible histories in order to still obtain a Markov process driving the backward picture.

In the literature this problem has been treated first in models where the process and its dual can be specified via a random graph (graphical representation). In the case of the existence of graphical representations, such as the voter model or stochastic Lotka–Volterra models which are of that type and which appear in the models and work of Krone and Neuhauser ([KN97]) there are typically arrows between points in the random graph which might or might not be used which describe the possible action of selective forces.

In our context selection requires, as explained above, adding new individuals to our dual (sampled) population as we move backward from the time horizon which represent the *potential* insertion of a fitter type in the tagged population. Whether such a potential insertion actually takes place depends on the fitness of the "victim" in the tagged population and of the fitness of the potential intruder. This means that in order to decide the types of the k-sample of the population we have to consider for selection a growing dual population representing typical individuals drawn from the population at a certain time s back. Then we have to assign weights to these possibilities representing the probabilities with which they are realised. Therefore the ancestral lines of our sample form a subtree of the graph generated by the ordered particle process η. Among the possible choices for subtrees in this graph the various possibilities have probabilities which we can read off from the function-valued part.

In addition to the complication arising from selection, along the ancestral path mutation events have to be taken into account. For the state-independent part such an event decouples the final type from everything happening earlier, but if the mutation is state-dependent, we have again to consider the complete system of backward path but since the tagged population is small compared to the basic population a law of large number effect occurs and this can be represented through a functional dual reflecting the fact that mutations occur based on the current type and do not depend on the overall populations. However with the mechanism for the dual as described in Subsubsection 5.3.2 we can even associate with every individunal a chain of mutation events. But we do not yet have a rich enough system to be associated with the ancestral path of the individual, a path in type space. We will see in Subsection 5.5 how this possibility arises.

In the spatial context we have to sample k_1 particles from a site $\xi(1)$, k_2 from site $\xi(2), \cdots, k_m$ from site $\xi(m)$ and the ancestral paths migrate in space. Altogether we therefore get a spatial coalescent with birth and with mutation operations associated with each ancestral path which represents the historical evolution of a randomly drawn k-tuple from the time t-population of the original process. Then we can use the dual to calculate the probabilities that k-sampled individuals have specific types, a specific genealogy and paths in geographic space. We shall discuss more of this as we go along, see also the explanation of the refined dunal from a historical process perspective in Subsubsection 5.5.4.

5.5 Refinements of the Dual for Finite Type Space: $(\eta_t, \mathcal{F}_t^{++})_{t \geq 0}, (\eta_t, \mathcal{G}_t^{++})_{t \geq 0}$

In this section we focus on the case of a *finite* type space and we consider refinements of the dual $(\eta_t, \mathcal{F}_t^+)$, (resp. $(\tilde{\eta}_t, \mathcal{F}_t^+)$ or $(\eta_t, \hat{\mathcal{F}}_t^+)$) and similarly with the versions using $\mathcal{G}^+, \hat{\mathcal{G}}^+$, which we denote $(\eta_t, \mathcal{G}_t^{++})_{t \geq 0}$ which mainly require

5.5 Refinements of the Dual for Finite Type Space: $(\eta_t, \mathcal{F}_t^{++})_{t \geq 0}$, $(\eta_t, \mathcal{G}_t^{++})_{t \geq 0}$

an enrichment of the mechanism corresponding to selection and mutation in the function-valued part of the dual and which works (only) for a smaller set of functions in which the function-valued part can start, namely the function f in (5.3) has to be a *sum of products of indicators*. This set of functions however still generates the full set of functions we consider and is therefore distribution-determining for finite type space and hence suffices for a duality theory. However to guarantee that this subset of functions is *preserved* under the part of the dual dynamics corresponding to selection and mutation dynamics, we have to change the dynamics of $(\eta, \mathcal{F}^+)$ or $(\eta, \mathcal{G}^+)$.

The refined duality leads to useful relations both for type-independent and for type-dependent mutation which will be crucial in the proofs of the results stated in Section 4 and Section 6.

In particular this duality allows for a nice historical interpretation, since it generates a marked graph (marked with types and locations) of which the ancestral marked tree of a finite sample is a subtree (see Subsubsection 5.5.4).

The constructions we describe in this subsection are written for the case of a *finite* type space

$$\mathbb{I} = \{1, \ldots, K\}. \tag{5.68}$$

Remark 55. An extension to countable type space is possible, but not needed in the sequel. In particular if we assume more about the fitness functions, for example that fitness values have 1 as the only accumulation point, then we can handle also immediately the case of countably many types.

5.5.1 Ingredients

For this refined duality we use a suitably chosen smaller set of functions as state space for the function-valued process. However the key point is to also use a modified dynamic of the function-valued process $\mathcal{F}_t^+, \mathcal{G}_t^+$ (later on called $\mathcal{F}_t^{++}, \mathcal{G}_t^{++}$) which is generated by a *refinement* of the *birth process* in η_t, the particle process driving the function-valued process together with a *refinement* of the *mutation part* of the function-valued part of the dual process. The point of these modifications (leading in a way to a more complicated dynamic), will be (1) the smaller subset of test functions in preserved and (2) later on in subsections 9.4 and 9.5 we can consider the time-space process of this dual and use the additional information we build into the dynamic to define a process on a richer state space (space of tableaus representing decompositions of the set $\mathbb{I}^\mathbb{N}$) which can then be analysed better if we are concerned with the longtime behaviour.

In points (0)–(iv) we now explain step by step the needed changes in the state space and the four mechanisms in the dual to obtain the refined version.

(0) State space of function-valued part

The state space for the dual process better its function-valued part is the subset of functions on $\mathbb{I}^\mathbb{N}$ arising as follows. For $k \in \mathbb{N}$, define first the space of certain finite sums of products of at most k-indicator functions of a variable in $\mathbb{I}$ each:

$$\mathbb{F}_k := \left\{ f = \sum_{i=1}^{n} \prod_{j=1}^{k_i} 1_{B_{i,j}}(u_j), B_{i,j} \subseteq \{1, 2, \cdots, K\}, \quad n \in \mathbb{N}, k = \max_{i=1,\dots,n} k_i \right\}. \tag{5.69}$$

Here f above is viewed as a function of $\mathbb{I}^k$.

Remark 56. Note that the states in this set of functions need not satisfy $\int f d\mu^{\otimes k} \leq 1$, as would be the case if f defines a decomposition of $(\mathbb{I})^k$. Therefore we have two cases, the version of the dual based on $\mathcal{F}^+$ where this is not the case and the dual based on $\mathcal{G}^+$, where in the definition above we can impose the condition

$$\int_{\mathbb{I}^k} f d\mu^{\otimes k} \leq 1 \tag{5.70}$$

and still obtain a set of states preserved under the dynamics.

The *state space* of the function-valued component of the refined dual process is given by:

$$\mathbb{F} := \bigcup_{k=1}^{\infty} \mathbb{F}_k. \tag{5.71}$$

This function space is associated with the situation in which we have k particles in the dual particle process η. For consistency the product running from k_i to k could be filled up with $k - k_i$ indicators of the whole type space $\mathbb{I}$. The parameter n allows us to consider the evolution of functions in which there is a mechanism that replaces a product of indicators by a sum of products of indicator functions.

With each variable we associate a position in space. However this is not changed under selection, mutation or resampling. Hence in order to explain the dual mechanism for each of those we can ignore the spatial aspect for the moment.

We now explain the dual mechanisms corresponding to selection, mutation, resampling and migration in the original process step by step.

(i) Selection.

In order to describe the change of the transition in η at a birth event and the new transition in the function-valued component of the dual upon a birth event we use the following structure. Assume that there are ℓ fitness levels reaching from 0 to 1 and denote them by

$$0 = e_1 < e_2 < \cdots < e_\ell = 1. \tag{5.72}$$

5.5 Refinements of the Dual for Finite Type Space: $(\eta_t, \mathcal{F}_t^{++})_{t \geq 0}, (\eta_t, \mathcal{G}_t^{++})_{t \geq 0}$

(Note without loss of generality to assume for bounded fitness functions in assuming that $e_1 = 0$, $e_\ell = 1$.)

First change the dynamic of the first component, i.e. η, so that the birth process is replaced by a *multitype* birth process, more precisely a $(\ell - 1)$-type birth process. Births occur at total rate s but now when a birth occurs the *type of birth* ($\in \{2, \ldots, \ell\}$) is chosen i.i.d. for this event with probabilities

$$(e_i - e_{i-1})_{i=2,\cdots,\ell}. \tag{5.73}$$

We shall now introduce the jumps in the function-valued part occurring upon a birth in η of a specific type. Define therefore for each level of fitness i, with $i = 2, \cdots, \ell$ the set

$$A_i := \{j \in \mathbb{I} : \chi(j) \geq e_i\}. \tag{5.74}$$

We note that $A_{i+1} \subset A_i$.

We define for a subset $C \subseteq \{1, 2, \cdots, K\}$ of the type space the operator ψ_C on $\mathbb{F}$ as follows. Let f be a function of k-variables and define

$$\Psi_C f := \sum_{m=1}^{k} [1_C(u_m) f(u_1, \ldots, u_k) + f(u_1, \ldots, u_k)(1 - 1_C(u_{k+1})]. \tag{5.75}$$

Observe that $(\psi_C - Id)$ is the generator of a rate 1 jump process on $\mathbb{F}$. Namely denote $1_C^\ell = 1_C(u_\ell)$ and introduce for $f \in \mathbb{F}_k$ for every variable u_j, $j = 1, \cdots, k$ at rate 1 transitions:

$$f \longrightarrow 1_C^j f + f \otimes (1 - 1_C^j), \qquad j = 1, \cdots, k. \tag{5.76}$$

Then each jump from some f in $\mathbb{F}_k$ ends in $\mathbb{F}_{k+1}$.

We apply this to our context with C replaced by the sets defined in (5.74), we use the notation

$$\chi_i^j = 1_{A_i}(u_j) \tag{5.77}$$

and set for $f \in \mathbb{F}_k$,

$$(\psi_i^{j,k} f)(u_1, \cdots, u_{k+1}) = ((\chi_i^j f) \otimes 1_{\mathbb{I}}^{k+1}) + (1 - \chi_i^j) \otimes f))(u_1, \cdots, u_{k+1}). \tag{5.78}$$

Again we can define an alternative transition as follows

$$(\hat{\psi}_i^{j,k} f)(u_1, \cdots, u_{k+1}) = \chi_i^j(u_i) f(u_1, \cdots, u_n) 1(u_{k+1} \\ + (1 - \chi_i^j(u_i)) f(u_1, \cdots, u_{i-1}, u_{k+i}, u_{i+1}, \cdots, u_k).$$

Remark 57. It is easy to verify that if $0 \leq \int f d\mu^{\otimes k} \leq 1$, then $0 \leq \int \psi_i^{j,k} f d\mu^{\otimes (k+1)} \leq 2\|f\|_\infty$ for any probability measure μ, respectively $0 \leq \int \hat{\psi}_i^{j,k} f d\mu^{\otimes k+1} \leq \|f\|_\infty$.

Remark 58. Note that this definition means that if f is a sum of products that the jump occurs simultaneously in the corresponding factor in all summands.

Remark 59. A different ordering of the factors and variables in different summands and a modified dynamics will be used in Section 9 to couple the dynamics of different summands arising with every birth event.

Definition 5.11 (Selection jump). Introduce transitions (jumps in $\mathbb{F}$) for the function-valued part of the form:

$$f \longrightarrow \psi_i^{j,k} f \text{ from } \mathbb{F}_k \to \mathbb{F}_{k+1}, \qquad (5.79)$$

whenever a *birth of type i* due to partition element j in η occurred. (Recall here i is chosen with probability $(e_i - e_{i-1})$). This will replace the transition we had in $\mathcal{F}_t^+$ before. For $\mathcal{G}_t^{++}$ we use $\hat{\psi}_i^{j,k}$. □

Now the r.h.s. of (5.27) is interpreted as rate s births being of type i with probability $(e_i - e_{i-1})$ and leading to a transition given by (5.79).

The type of birth does not influence the further evolution of η, it only will change the function-valued part occurring at this transition of η_t; we therefore do not enlarge the state space of the process η to store the type assigned to the birth event.

(ii) Mutation.

We have to specify the action of mutations on functions $f \in \mathbb{F}_k$, i.e. functions of k-variables in the special case in which they are certain sums of products of indicators. We use here a refinement of the random representation of the mutation semigroup as used in Subsection 5.3.2 since the latter does not necessarily preserve indicator functions. We now construct a random *indicator-function-valued jump process* that represents the mutation semigroup $(M_t^*)_{t \geq 0}$.

Remark 60 (Set-valued dual for Markov chains). The construction we give below produces in fact for every Markov jump process on a finite state space a set-valued dual process. Let E be its state space, then we can calculate $P[Z_t = i | Z_0 = j] = P[j \in \mathcal{A}_t | \mathcal{A}_0 = \{i\}]$ for a set-valued process, i.e. values in 2^E called $(\mathcal{A}_t)_{t \geq 0}$ with jumps and rates in Definition 5.12. In particular we can calculate also its equilibrium distribution.

This process acts independently on each variable of the function $\mathcal{F}_t^{++}, \mathcal{G}_t^{++}$ (corresponding to partition elements of the dual process). That is, at random times the function $\mathcal{F}_t^{++}$ or $\hat{\mathcal{F}}_t^{++}$ for our new function-valued process in the set of sums of products of indicators is changed by a jump from a product of indicators to a new product of indicators, where

5.5 Refinements of the Dual for Finite Type Space: $(\eta_t, \mathcal{F}_t^{++})_{t \geq 0}, (\eta_t, \mathcal{G}_t^{++})_{t \geq 0}$

all factors for a *given variable* change in *every* summand at once. (5.80)

We next describe the action of this modified mutation dual acting on one variable in the argument of f (to keep the notation simple we think of f as a function of one variable with the others fixed).

The mutation semigroup driving the function-valued part of the dual process when acting on indicator functions can also be represented by an indicator-function-valued dual (random) process whose jumps are specified next. Later we extend this to the sums of indicators.

We specify for each pair (i, j) of types the jumps $f(\cdot) \to \tilde{f}(\cdot)$ corresponding to the mutation transition $i \to j$, by the following prescription. Recall $M = (m_{i,j})_{i,j,\cdots,K}$.

Definition 5.12 (Set-valued mutation jumps). This transition from type i to type j occurs at rate

$$m \cdot m_{i,j} \quad ; \quad i, j \in \{1, 2, \cdots, K\} \qquad (5.81)$$

and results in a jump (depending on whether $\ell \in \{i, j\}$ or not):

$$1_{\{j\}} \longrightarrow 1_{\{i\} \cup \{j\}} \qquad (5.82)$$
$$1_{\{i\}} \longrightarrow 0$$
$$1_{\{\ell\}} \longrightarrow 1_{\{\ell\}} \quad \ell \notin \{i, j\}$$

for $i, j \in \{1, 2, \cdots, K\}$ and $j \neq i$. □

Next extend this to $\mathbb{F}$. For this purpose let $B \subseteq \{1, \cdots, K\}$ and consider the indicator $1_B(u)$. Since this indicator can be written as

$$1_B(u) = \sum_{k \in B} 1_{\{k\}}(u), \qquad (5.83)$$

we continue the transition in (5.82) as a linear map acting on the indicators $f = 1_{\{\ell\}}(\cdot)$, with $\ell \in \{1, \ldots, K\}$.

More generally proceed as follows. The transition $f \to \tilde{f}$ associated with the parameter (i, j) is obtained by applying to f the matrix $\overline{M}$ (in (k, ℓ)) which we define as

$$\overline{M}(i, j)[k, \ell] = \begin{cases} 1 & (k, \ell) = (i, j), \quad k = i \neq \ell \\ 0 & \text{otherwise.} \end{cases} \qquad (5.84)$$

This specifies the transition occurring in one of the several variables of f.

Definition 5.13 (Refined mutation jump). We now obtain a function-valued dual process $\mathcal{F}_t^{mut}$ for the mutation semigroup acting on indicators or sums of indicators if we introduce the following collection of jumps for $f \in \mathbb{F}_k$ for each $\ell \in \{1, 2, \cdots, k\}$:

$$f(u_1, \cdots, u_\ell, \ldots, u_k) \longrightarrow \sum_v \overline{M}_\ell(i, j)[u_\ell, v] f(u_1, \cdots, v, \cdots, u_k);$$

$$i, j \in \{1, \cdots, K\}, \tag{5.85}$$

at rate

$$m\, m_{i,j}, \tag{5.86}$$

where $\overline{M}_\ell$ indicates that $\overline{M}$ is applied to the ℓth variable. □

For every variable ℓ these jumps preserve the sets $\mathbb{F}_k$ for every $k \in \mathbb{N}$ since the number of variables remains fixed. Hence the set $\mathbb{F}$ is preserved under a dynamic consisting of jumps like in (5.85). Therefore if $\mathcal{F}_0^{mut} = f \in \mathbb{F}_k$ for some k, then $\mathcal{F}_t^{mut}$ is a $\mathbb{F}$-valued process.

(iii) Resampling-Coalescence.

Here as before we identify two variables located at the same site in the corresponding factor of all summands of the element in $\mathbb{F}_k$ we are dealing with. When coalescence occurs the resulting (coalesced) element is given the lower of the indices of the coalescing elements and the remaining elements are reordered to eliminate gaps but preserve the original order. Note that this operation turns two factors $1_A(u_i), 1_B(u_j)$, $i < j$ each related to one of the coalescing variables into the indicator of the intersection of the two sets as a function of the "new" merged variable. That is upon coalescence:

$$1_A(u_i) 1_B(u_j) \longrightarrow 1_{A \cap B}(u_i), \quad 1_C(u_\ell) \longrightarrow 1_C(u_\ell), \ell \notin \{i, j\}. \tag{5.87}$$

Therefore again $\mathbb{F}$ is preserved since this operation sends

$$f \in \mathbb{F}_k \longrightarrow \tilde{f} \in \mathbb{F}_{k-1}. \tag{5.88}$$

(iv) Migration

The action of the migration is as before and has only to do with the process η_t assigning locations to the variables but not with the jumps in the function-valued part.

Remark 61. We can now proceed as in the definition of $(\hat{\eta}_t, \hat{\mathcal{F}}_t^+)$ and add a state $\{*\}$ to the geographic space where all rates are set equal to zero. Note that under this dynamic $\mathbb{F}$ is preserved.

5.5.2 The Refined Dual $(\eta_t, \mathcal{F}_t^{++})_{t \geq 0}$ or $(\eta_t, \mathcal{G}_t^{++})_{t \geq 0}$

We construct the full refined dual dynamics of the process which is denoted $(\eta_t, \mathcal{F}_t^{++})$ respectively $(\eta_t, \mathcal{G}_t^{++})_t)$ by setting:

Definition 5.14 (Refined dual $(\eta_t, \mathcal{F}_t^{++}), (\eta_t, \mathcal{G}_t^{++})$).

(a) The process $(\eta_t, \mathcal{F}_t^{++})_{t \geq 0}$ is defined as $(\eta_t, \mathcal{F}_t^{+})_{t \geq 0}$ except that births now have a random type as specified in (5.89) which changes the transition of $\mathcal{F}_t^{++}$ at a birth event as specified below and the mutation transition is replaced by a jump process for the indicators in the representation of $\mathcal{F}_t^{++}$. Similarly we proceed for $(\eta_t, \mathcal{F}_t^{++})_{t \geq 0}$. Precisely the two changes are:

   (i) The transition occurring in the second component at a birth event is modified as follows:

   if $\mathcal{F}_t^{++} \in \mathbb{F}_k$ and a birth of a type i occurs at time t from the partition element with index j we have the transition for the factor f of the i-th variable (recall (5.78)):

   $$f \to \psi_i^{j,k} f = \chi_i^j f \otimes 1 + f \otimes (1 - \chi_i^j),$$

   $$(\text{resp. for } \mathcal{G}^{++}: f \longrightarrow \chi_i^j f \otimes 1 + (1 - \chi_i^j) \otimes f), \qquad (5.89)$$

   where the factor $(1 - \chi_i^j)$ and its (new) variable corresponds to the newborn individual.

   (ii) The mutation transition in the function-valued part is replaced by jumps according to (5.85)–(5.86).

(b) Similarly we define $(\hat{\eta}_t, \hat{\mathcal{F}}_t^{++})_{t \geq 0}$ or $(\hat{\eta}_t, \hat{\mathcal{G}}_t^{++})_{t \geq 0}$ if we have state-independent mutation at rate $\bar{m}$ as the corresponding modification of $(\hat{\eta}_t, \hat{\mathcal{F}}_t^{+})_{t \geq 0}$ respectively $(\hat{\eta}_t, \hat{\mathcal{G}}_t^{+})_{t \geq 0}$. □

The key feature of the refined dual $(\eta_t, \mathcal{G}^{++})$ is the fact that the terms 1_A and $(1 - 1_A)$ which are created by births, evolve by further selection and mutation to terms being identically 0 or 1, a state which we call *resolution*. At resolution time one of the two summands generated at the birth event disappears.

5.5.3 The Refined Duality Relation

Now we can state and prove a duality relation between the process $(\eta_t, \mathcal{F}_t^{++})_{t \geq 0}$ and the interacting Fleming–Viot diffusion with mutation-selection.

Proposition 5.15 (Refined duality). *Under the assumption of finite type space we have the following duality relation between the process $(X_t)_{t \geq 0}$ and the two refined dual processes: if $|\xi| = k$ and $f \in \mathbb{F}_k$, then*

$$F_t(X_0,(\xi,f))$$
$$= E_{(\xi,f)}\left[\int_{\mathbb{I}}\cdots\int_{\mathbb{I}}\mathcal{F}_t^{++}(u_1,\cdots,u_{|\pi_t|})x_{\xi_t(1)}(du_1)\cdots x_{\xi_t(|\pi_t|)}(du_{|\pi_t|})\right] \quad (5.90)$$
$$= E_{(\xi,f)}\left[\int_{\mathbb{I}}\cdots\int_{\mathbb{I}}\mathcal{G}_t^{++}(u_1,\cdots,u_{|\pi_t|})x_{\xi_t(1)}(du_1)\cdots x_{\xi_t(|\pi_t|)}(du_{|\pi_t|})\right]$$

and analogously for state-independent mutation and $(\hat{\eta}_t,\hat{\mathcal{F}}_t^{++})_{t\geq 0}$, $(\hat{\eta},\hat{\mathcal{G}}_t^{++})_{t\geq 0}$. □

Proof of Proposition 5.15. This follows from the previous duality relations by observing that we changed two things (I) we use the dual we had previously now on a *restricted* class of test functions f, namely those in $\mathbb{F}$ which is a set of functions preserved under the dynamics. (II) we have reinterpreted this dynamic on $\mathbb{F}$ in an autonomous way, see (5.93) and (5.99), such that the *expected* jump induced in the duality function remains the same, which we now explain in detail for the selection and mutation transition where the changes in the dynamic occurred since the duality claims that certain expectations are equal it is preserved under the change.

(1) First consider the *selection* transition. We can rewrite the fitness function χ as follows. Namely

$$\chi = \sum_{i=2}^{\ell}(e_i - e_{i-1})1_{A_i}. \quad (5.91)$$

At the same time (since $e_1 = 0$, $e_\ell = 1$):

$$1 - \chi = \sum_{i=2}^{\ell}(e_i - e_{i-1})(1 - 1_{A_i}). \quad (5.92)$$

Now observe that the new state of $\mathcal{F}_t^+$ after a birth event increasing the number of basic particles from k to $k+1$ can be rewritten as follows:

$$(\chi f \otimes 1 + (1-\chi) \otimes f) = \sum_{i=2}^{\ell}[(e_i - e_{i-1})\sum_{j=1}^{k}\psi_i^{j,k} f] \quad (5.93)$$

and

$$(e_i - e_{i-1})_{i=2,\cdots,\ell} \in \mathcal{P}(\{1,\cdots,\ell\}). \quad (5.94)$$

We therefore can interpret the transition $f \longrightarrow (\chi f \otimes 1 + (1-\chi) \otimes f)$ in the process $\mathcal{F}_t^+$ now differently as superposition independent transitions defined by $\psi^{j,k}$ which gives then the transition in the refined version denoted $\mathcal{F}_t^{++}$.

5.5 Refinements of the Dual for Finite Type Space: $(\eta_t, \mathcal{F}_t^{++})_{t \geq 0}, (\eta_t, \mathcal{G}_t^{++})_{t \geq 0}$

(2) To prove this alternative representation of the *mutation* part of the dual process first note that we can write the generator of the mutation process, i.e. the jump process on type space for a fixed individual in the form of the independent superposition of mutation processes where each summand corresponds to the transition from type i to type j. This reads in formulas:

$$\mathcal{M} f(i) = \sum_{j=1}^{K} M(i, j)(f(j) - f(i)) = \sum_{k,\ell=1}^{K} \sum_{j=1}^{K} \tilde{M}(k, \ell)[i, j] f(j), \quad (5.95)$$

where

$$\{\tilde{M}(i, j), i = 1, \cdots, K; \quad j = 1, \cdots, K; i \neq j\} \quad (5.96)$$

is a collection of kernels $\tilde{M}(i, j)[\cdot, \cdot]$ given by

$$\begin{aligned}
\tilde{M}(i, j)[i, j] &= m_{i,j} \\
\tilde{M}(i, j)[i, i] &= -m_{i,j} \\
\tilde{M}(i, j)[k, \ell] &= 0, \quad \text{for } (k, \ell) \text{ different from } (i, i) \text{ or } (i, j).
\end{aligned} \quad (5.97)$$

Note that each $\tilde{M}(i, j)$ is the generator of a mutation semigroup with only mutations from type i to type j at rate $m_{i,j}$. Hence the collection above represents the collection of all possible mutation jumps at their rates.

Note that:

$$\tilde{M}(i, j)[\cdot, \cdot] = m_{i,j}(\overline{M}(i, j)[\cdot, \cdot] - \delta_{[i,i]}), \quad (5.98)$$

which means that this jump in (5.82) represents the generator $\tilde{M}(i, j)[\cdot, \cdot]$. This completes the proof of Proposition 5.15.

Remark 62. Suppose only the mutation transitions occur. One calculates that the generator of this process is $\sum_\ell \mathcal{M}_\ell$ where $\mathcal{M}_\ell$ is $\mathcal{M}$ acting on the ℓ-th component. Hence the duality representation of the mutation semigroup M_t^* acting on indicator functions is given in terms of the indicator-function-valued dual as follows:

$$M_t^* \left(\prod_{\ell=1}^{L} 1_{B_\ell}(x_\ell) \right) = E[\mathcal{F}_t^{mut}(x_1, \ldots, x_L) | \mathcal{F}_0^{mut} = \prod_{\ell=1}^{L} 1_{B_\ell}(x_\ell)]. \quad (5.99)$$

This will serve as the mutation component in the construction of an indicator-function-valued dual for the mutation-selection-migration system. As before, in the case of a sum of products of indicator functions the jumps are made simultaneously in the corresponding factors in all summands.

5.5.4 Historical Interpretation Revisited and Outlook on a Modified Dual

We can use the representation of $\mathcal{F}_t^{++}$ given in the previous points to clarify the relation between the refined duality and the selection graph of Krone and Neuhauser [KN97] and concepts in population genetics referring to the calculations of probabilities for certain genealogical relationships. Of course there are differences and specifics of our model, (1) we have multiple occupation of sites, (2) resampling only in one colony and in addition (3) we have taken a diffusion limit of many small mass particles. All this generates different features.

However if we consider an n-sample and their ancestry we are back at a discrete model and instead of arrows attached to sites specifying potential insertion of fitter types in the graphical representation we generate potential insertions attached to individuals in the sample. However our backward process is a Markov process despite the presence of selection or mutation. This we explain now in detail.

Given sets $C_1, \ldots, C_m$ the dual allows us to determine the probability that m individuals chosen randomly from the population have types in these sets and also the probability that they have a given genealogy. Already in the neutral model with mutation we have to use a function-valued dual (or for finite type space a set-valued dual) which allows to determine the probability that a given genealogy of the sample generated by the coalescent can result in a certain marking with types at time t.

Selection complicates the picture even further. The calculations involving the new factors created by selection provide tests comparing individuals with the type of other randomly chosen individuals from the population. The corresponding probabilities are obtained by summing over potential histories. To be more precise consider the following.

We can now associate with our dual particle a marked graph as follows. Starting with the tagged sample we draw from every individual an edge, till the first selection or coalescence event occurs. If a selection event occurs we place a vertex from which two new edges start, one associated with χ_i^j, the other with $(1 - \chi_i^j)$ if the birth occurs with the i-th particle and is of type j. If a coalescence occurs we join the two edges and continue with one edge. On each edge we record the current location of the corresponding individual. Therefore edges are now marked with functions χ^j or $1 - \chi^j$ and with locations.

At each edge we place at rate $m \cdot m_{i,j}$ mutation markers indicating a mutation event on the variable of the form $i \to j$.

With this construction we have generated a *random marked graph* in which we consider all paths leading from the roots (the tagged individuals) to the leaves at time t. Every such path generates a product of indicators by multiplying the function at the root and the χ-functions attached along the way. Then act with the mutation markers on the factors attached with an edge. Finally sum over all these products corresponding to path connecting the leaves to a root. This is the state of $\mathcal{F}_t^{++}$. Note that if we put $f_0 \equiv 1$, then we see that the expectation of the sum of factors arising from the birth events and undergo mutation have expectation 1.

5.5 Refinements of the Dual for Finite Type Space: $(\eta_t, \mathcal{F}_t^{++})_{t \geq 0}, (\eta_t, \mathcal{G}_t^{++})_{t \geq 0}$

Can we associate with these objects, i.e. the marked random graph the marked ancestral tree for the tagged sample meaning we know the joint law of the genealogical distances between the tagged sample together with the type and geographical location at time t? This would require that based on the random graph explained above we can decide at each branching point which way to go. If we use $\mathcal{F}^{++}$ the obstacle is that we have $\chi(u_i)$ and $1 - \chi(u_{m+1})$ which since they belong to different variables do not define a decomposition in the potential set of marked ancestral path into disjoint sets.

For the purpose of achieving this we use $\mathcal{G}^{++}$ instead. Suppose we have a birth due to particle number i of type j. Then we replace this transition by

$$f(u_i) \longrightarrow \chi^j(u_i)f(u_i)1(u_{m+1}) + (1 - \chi^j)(u_i)f(u_{m+1}). \tag{5.100}$$

The key feature is that selection introduces a *decision tree* with corresponding probabilities for the different genealogies which are possible after the interaction of the sample with the rest of the population. The selection transition $f \longrightarrow \chi f + (1 - \chi) \otimes f$ we defined for $\mathcal{F}^{++}$ (and for $\mathcal{F}^+$ of course as well) is changed in the new picture corresponding to $\mathcal{G}^{++}$ into (suppose here $f = 1_B$)

$$\begin{aligned} f &\longrightarrow 1_{A_j}(u_i)f(u_i)1(u_{m+1}) + 1_{A_j}(u_i)f(u_{m+1}) \\ &= 1_{A_j \cap B}(u_i)1(u_{m+1}) + 1_{A_j}(u_i)1_B(u_{m+1}), \end{aligned} \tag{5.101}$$

where now the two summands are alternative possibilities the ancestral path can take and their respective probabilities associated to the two new particles in each summand where one represents the preservation of the sample in the other the insertion of a superior individual in the sample.

Recall now the coupling rule of the summands (the factors follow the transitions of the individuals they are associated with). We now see that the marked ancestral graph associated with this model has now the property there is at most one path from the initial particle "root" to the "leaf" at the other end, since now $\chi^j \cdot (1 - \chi^j) = 0$ and $\chi^j + (1 - \chi^j)f \leq 1$. The $\leq$ sign appears since it might happen that not all n-path end at the leaves, which means that the tagged sample with the given type configuration and the generated genealogy is not consistent and has probability zero.

In any case we can read each summand as a possible line of ancestry and the factors generating the probabilities. In Section 9 we shall carry out and apply this construction at great length and introduce *sums of ordered factors* to parallel the genealogical relations.

Here we give an explanation how the genealogy-type structure is generated. This means that we draw n individuals from the population and record their respective type and location and the time we have to go back for every pair to find the most recent common ancestor. The joint law of this statistics we want to derive from the dual process. We start with the model (1) only resampling, (2) adding types and mutation, (3) adding selection.

(1) The genealogy of a sample of n-individuals is obtained by considering the time when pairs of two individuals of the sample first land in the same partition element. This $\binom{n}{2}$ different random times are in distribution equal to the entries in the matrix of genealogical distances with the sample.

(2) Now the individuals carry a type and for a given genealogical tree we have to calculate the probabilities that we find specific types on the individual of the sample. For a realisation of the coalescent we can, with the function-valued dual, determine the probabilities of a specific type configuration at time t given the one at time 0. Namely with every partition element at every time $s \leq t$, there is a connected factor a function on type space which changes according to the mutation dynamic. Testing with the initial state the factor turn 0 if the type is not possible and 1 if it is possible. Note that if the set-valued process has reached the full or the empty set, the process contains no further information about the time further back in the original model.

(3) Finally selection enters into the picture. The selection transition in the dual accounts for the possibility of an interaction of the sample with the rest of the population. In the neutral case this can be suppressed without changing the law since all individuals are exchangeable if they are at the same site at the moment resampling occurs. This is not true of course if we have type-based selection. Each action of the selection operation introduces two *alternative* forms of the genealogical tree of the sample by the interaction with a new randomly chosen individual from the population at the time the selection operator acts. Each realisation of all mutation transitions, coalescence events for an N-coalescent (in which the sample is embedded) k-selection transition generate 2^k-realizations of a marked random tree which are the potential genealogy-type configurations for the sample at time t. For each of those we can calculate the probability from a decision tree (whose leaves are possible genealogical trees) and where the edges carry certain factors which change according to mutation and selection.

A key point is now that all the possibilities are alternative, i.e. only one such tree is the actually realized one for the sample. Furthermore the structure is such that a finite random time back it is resolved which of the different possibilities actually occurs.

The following diagram illustrates the decision tree represented in the dual. Here the subscripts refer to the order of the factors and the selection operator has acted twice on the first factor and then the first and second factors have coalesced. We also used a coupling in which the operations are simultaneously applied to the same factor (according to the order) in each summand. The result of this history has two summands $(1 - \chi) \otimes f$ and $\chi f \otimes 1$. The indices 1,2,3 refer to the dual particle involved. This object can be viewed as a decision tree to decide which ancestral paths are possible for the tagged sample represented by f and every leaf representing a possibility.

The following picture indicates this but note that it is the *decision* tree that is depicted and *not* the genealogical tree. In that decision tree we have first two selection events (the second applies to newborn particle) and then coalescence of particle 1 and 2:

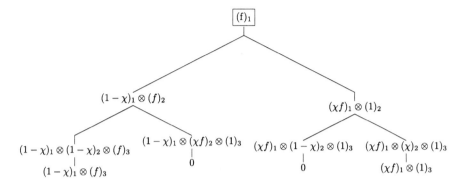

We shall return to this picture in Sections 9 and 10 where this idea is exploited.

5.5.5 *Calculations with the Refined Dual: An Additional Trick*

In this subsubsection we present an additional trick which is often helpful carrying out calculations of moments, for example to prove ergodic theorems.

In order to carry out the required calculations with the dual, i.e. calculating the r.h.s. of the equation (5.90), we now exploit more fully the additional properties of the selection and mutation mechanisms if we let them act only on indicator-functions. In some key calculations this reduces to study the expectation of a *pure product* of indicators instead of a *sum* of such products. This means we have to code the sums as additional expectations. We will have to pay for this simplification by the appearance of a Feymann–Kac term. The key fact leading to such an alternative representation is the following observation concerning the evolution of the function-valued dual with particular initial states.

To see how to proceed return to the transition in (5.76) and note that *provided that*

$$1_C^j f \equiv f \tag{5.102}$$

we can replace this transition by the new transition at rate s

$$f \longrightarrow f \otimes (1 - 1_C^j) \tag{5.103}$$

and then add a *Feynman–Kac term* corresponding to the potential:

$$s1. \tag{5.104}$$

This means that in the expression for the expectation in (5.39) on the r.h.s. the term $e^{s(t-t')}$ appears as a factor as long as the state satisfies (5.102), here t' is the time of the occurrence of the transition we have replaced.

Since we have the restriction that $1_C f = f$ in order for this relation to hold we cannot view this as defining a new dual process since we can make this replacement only for a suitable current state of the process $\mathcal{F}_t$. However it is useful to calculate the dual expression, i.e. the expectation w.r.t. $(\eta_t, \mathcal{F}_t^{++})$ in certain cases by changing the function-valued part and the duality relation as indicated. We indicate this in calculations evaluating dual expressions by writing

$$^{FK}\mathcal{F}_t^{++} \tag{5.105}$$

in this dual expression. This form of dual expression will be used in Section 6.

5.6 The Dual Representation for the McKean–Vlasov Process

The next objective is to obtain a duality relation for the *non-linear Markov process*, the McKean–Vlasov process defined in (4.28). Of course we have the duality relations for a system with N exchangeable colonies which is a special case of the duality given above in Subsections 5.1–5.3 and 5.5. This suggests the form of the dual for the McKean–Vlasov limit which is then easily verified by a moment calculation (see [DG99]), however due to the non-linearity we will have to distinguish deterministic and random initial states of the process.

5.6.1 The McKean–Vlasov Dual Process

The only change that will occur compared to the dual of the interacting process is in the evolution mechanism of the particle system η resp. $\hat{\eta}$ we used earlier, which we will therefore now replace by an evolution

$$(\tilde{\eta}_t)_{t \geq 0}. \tag{5.106}$$

The dual particle process $\tilde{\eta}$ has as site (geographic) space

$$\{0\} \cup \{1, 2, 3 \cdots\} \cup \{*\}, \tag{5.107}$$

5.6 The Dual Representation for the McKean–Vlasov Process

where site $\{0\}$ corresponds to the "tagged" site, the sites $\{1, 2, 3 \cdots\}$ are used to incorporate migration to *independent copies* of the tagged site (corresponding to the propagation of chaos property of the mean-field limit). As in Subsection 5.3 above, the "cemetery state" of $Z(t)$ in $\{*\}$ arises from the state-independent part of the mutation.

Given the initial state Z_0 of our nonlinear martingale problem, we define the "initial state" used in the duality relations as follows. The state at the site $\{*\}$ is set equal to ρ for all times $t \geq 0$ where ρ is the state-independent mutation probability as before. To the sites $0, 1, 2, \cdots$ we assign *independent copies* of our *initial state* Z_0.

Now we prescribe the dynamics of the mean-field dual.

Definition 5.16 (McKean–Vlasov dual process). We introduce the bivariate process consisting of components with values in marked partitions and a function-valued component:

$$(\tilde{\eta}_t, \mathcal{F}_t)_{t \geq 0} \text{ respectively } (\tilde{\eta}_t, \mathcal{F}_t^+)_{t \geq 0}, (\tilde{\eta}_t, \mathcal{F}_t^{++})_{t \geq 0} \text{ or } (\tilde{\eta}_t, \mathcal{G}_t^+)_{t \geq 0}, (\tilde{\eta}_t, \mathcal{G}_t^{++})_{t \geq 0}. \tag{5.108}$$

(a) The evolution of $(\tilde{\eta}_t)_{t \geq 0}$ on $\{0\} \cup \{1, 2, 3, \cdots\}$, (i.e. excluding $*$), has four different mechanisms which all act independently of each other meaning all exponential waiting times given below are independent. The mechanisms are for partition elements located on $\{0, 1, 2, \cdots\}$:

coalescence of each pair of partition element at the same site after an exponential rate d waiting time,

(5.109)

births of a new individual forming its own partition element after an exponential rate s waiting time for each existing partition element (5.110) located on, $0, 1, 2, \cdots$

jumps of the partition elements to $\{*\}$ after an exponential rate $\bar{m}$ waiting time

(5.111)

and finally a modified *migration mechanism for partition elements* due to the mean-field term, namely,

migration of each partition element located outside $\{*\}$ occurs after an exponential rate c waiting time to the unoccupied site in $\{1, 2, \cdots\}$ with the currently lowest label.

(5.112)

On the site $*$ the partition elements of the process η have no transitions at all.

(b) The dynamics of $\mathcal{F}_t$ resp. $\mathcal{F}_t^+, \hat{\mathcal{F}}_t^+, \mathcal{F}_t^{++}$ is as before for a given realization of the particle system $\tilde{\eta}$ and similarly for $\mathcal{G}_t^+, \hat{\mathcal{G}}_t^+$ and $\mathcal{G}_t^{++}$.

(c) Assume we are given an initial state $Z \in \mathcal{P}(\mathbb{K})$ and the initial state $(\tilde{\eta}_0, f_0)$ of the dual process, namely a function $f_0 \in L_+^\infty((\mathbb{K})^n)$ and $\tilde{\eta}_0$ of the form (ζ_0, π_0, ξ_0), where

$$\zeta_0 = \{1, \cdots, n\}, \pi_0 = \{\{1\}\}, \cdots, \{n\}\}, \xi_0 = (\xi_0(1), \cdots, \xi_0(n)) = (0, \cdots, 0). \tag{5.113}$$

Then we obtain the unique bivariate process $(\tilde{\eta}, \mathcal{F}_t)_{t \geq 0}$ respectively $(\tilde{\eta}, \mathcal{F}_t^+)_{t \geq 0}$.
□

5.6.2 Duality Relation for McKean–Vlasov Process

With the new rules for the dual dynamics the following duality relation holds:

Proposition 5.17 (McKean–Vlasov duality: deterministic initial state).

(a) The dual representation is given by the following equation where on the left side E denotes the expectation with respect to Z_t and on the right side E denotes expectation with respect to the process $(\tilde{\eta}_t, \mathcal{F}_t)_{t \geq 0}$, respectively $(\tilde{\eta}_t, \mathcal{F}_t^+)$:

$$\begin{aligned} &E\left[\int f(u_1, u_2, \ldots, u_n) Z_t(du_1) \cdots Z_t(du_n) | Z_0 = Z\right] \\ &= E\left[\exp(s \int_0^t \|\tilde{\pi}_r\| dr) < \mathcal{F}_t(u_1, \cdots, u_{|\tilde{\pi}_t|}), \bigotimes_{i=1}^{|\tilde{\pi}_t|} z_{\tilde{\xi}_t(i)}(du_i) > |\mathcal{F}_0 = f\right], \\ &= E\left[< \mathcal{F}_t^+(u_{\tilde{\pi}_t(1)}, \cdots, u_{\tilde{\pi}_t(|\tilde{\pi}_t|)}), \bigotimes_{i=1}^{|\tilde{\pi}_t|} z_{\tilde{\xi}_t(i)}(du_i) > |\mathcal{F}_0^+ = f\right], \end{aligned} \tag{5.114}$$

where $\|\tilde{\pi}_t\|$ is the total number of partitions excluding those in $\{*\}$ while $|\tilde{\pi}_t|$ is the usual total number of partition elements and

$$z_\eta = \begin{cases} Z & \text{if } \eta \in \{0, 1, \cdots\} \\ \rho & \eta = *. \end{cases} \tag{5.115}$$

For the first identity we have to assume $t \leq t_0$ (recall (5.35)) again.

(b) As before we can replace $\mathcal{F}_t$ by $\hat{\mathcal{F}}_t$ and $\mathcal{F}_t^+$ by $\hat{\mathcal{F}}_t^+$ defined below (5.65) or the refined process $\mathcal{F}_t^{++}$ or we can use $\mathcal{G}_t^+, \hat{\mathcal{G}}_t^+$ or $\mathcal{G}_t^{++}$ at all these instances.
□

This gives us a dual expression for *deterministic* initial states. However for a *nonlinear* Markov process Z_t the expression given a random initial state *cannot* be obtained as an integral over the expressions for fixed initial state as is the case for Markov processes, that is, the law of Z_t, conditioned on Z_0, depends both on

5.6 The Dual Representation for the McKean–Vlasov Process

$Z(0)$ and the law of $Z(0)$. Some additional considerations are therefore needed for random initial states of the process Z which we present next.

Assume we are given an initial state $Z \in \mathcal{P}(\mathbb{K})$ and the initial state of the dual process, namely a function $f \in L^\infty_+((\mathbb{K})^n)$ and $\tilde{\eta}$ of the form $\zeta_0 = \{1, \cdots, n\}$, $\pi_0 = \{\{1\}, \cdots, \{n\}\}$, $\xi_0 = \{\xi_0(1), \cdots, \xi_0(n)\} = \{0, \cdots, 0\}$.

We now consider the McKean–Vlasov system with random initial state. To determine the law of Z_t which is a random probability measure we must consider all moment measures of Z_t where the random initial state has moment measures

$$\tilde{E}\left[\int f(u_1, \cdots, u_n) Z_0(du_1) \cdots Z_0(du_n)\right]; \qquad n = 1, 2, \cdots \qquad (5.116)$$

where $\tilde{E}$ is the expectation with respect to the law of Z_0. (Recall that the moment measures $\tilde{E}[\otimes_{k=1}^n Z_0(du_k)]$ of a random probability measure are not product measures.)

We obtain in that situation:

Corollary 5.18 (McKean–Vlasov duality: random initial state). *Assume that Z_0 is random with law P_0.*

(a) *Then we obtain the duality relation for $\ell = 1$ (in 5.116):*

$$E_{P_0}\left[\int f(u_1, \cdots, u_n) Z_t(du_1) \cdots Z_t(du_n)\right] \qquad (5.117)$$

$$= \tilde{E}\left[E[<\mathcal{F}_t^+(u_1, \cdots, u_{|\tilde{\pi}_t|}), \bigotimes_{i=1}^{|\tilde{\pi}_t|} z_{\tilde{\xi}_t(i)}(du_i) > |\mathcal{F}_0^+ = f]\right],$$

where $\{z_\eta, \eta \in \{0, 1, 2, \ldots\}\}$ are as in (5.115) and $\tilde{E}$ is expectation w.r.t. P_0.

(b) *Furthermore to determine moments of higher order to handle the randomness in the initial state we consider for $\ell \in \mathbb{N}$ a dual starting with $n_1, \cdots, n_\ell$ particles at sites $1, 2, \cdots, \ell$. Since the ℓ-groups of individuals in the dual interact only by the exclusion rule in the migration rate and we have exchangeable initial distribution, we write the state of the particle system as $(\pi_t^1, \xi_t^1), \cdots, (\pi_t^\ell, \xi_t^\ell)$, giving the partitions and locations of partition elements for the individuals originally in the sites $1, 2, \cdots, \ell$ respectively.*

Then (and similarly for $\mathcal{G}^+$ and all other variants)

$$E_{P_0}\left[\prod_{k=1}^\ell \int f_k(u_1^k, \cdots, u_{n_k}^k) Z_t(du_1^k) \cdots Z_t(du_{n_k}^k)\right] \qquad (5.118)$$

$$= \tilde{E}\left[E[\prod_{k=1}^\ell <\mathcal{F}_t^+(u_1^k, \cdots, u_{|\tilde{\pi}_t^k|}^k), \bigotimes_{i=1}^{|\tilde{\pi}_t^k|} z_{\tilde{\xi}_t^k(i)}(du_i) > |\mathcal{F}_0^+ = \prod_{k=1}^\ell f_k]\right]. \qquad \square$$

Chapter 6
Long-Time Behaviour: Ergodicity and Non-ergodicity

In this section we want to illustrate the application of the duality described in Section 5 in some simpler situations. Here we use the refined dual representation to identify some effects concerning the long-time behaviour of X^N or the McKean–Vlasov process Z in a number of special cases for the parameters of the model. This illustrates some types of possible behaviours and their reflection in the structure of the dual process. The heavier arguments, in particular, the proof of the results on the long-time behaviour in the general case is postponed to Section 10.

These arguments prove in particular Theorem 2 in the case $K = 2$ but also provide some additional information which is the purpose of this section. The proof of the ergodic theorem in the general case requires additional modifications of the dual process and is deferred to Section 10.

In order to illustrate some of the ideas in a simple setting we include in Subsection 6.1 the special case of non-interacting sites, i.e. the case where there is no spatial structure and also we reprove the results for the deterministic case $d = 0$ to illustrate the application of the dual and to exhibit the changes in the behaviour of the stochastic versus the deterministic model. In Subsection 6.2 we then treat the case of interest for Theorem 2 extending the techniques utilized in Subsection 6.1.

We consider only the case of finitely many types, $\mathbb{I}$, with

$$|\mathbb{I}| = K, \text{ with } K \in \mathbb{N}, \quad K \geq 2 \text{ (often even } K = 2). \tag{6.1}$$

We note that for $|\mathbb{I}|$ finite

$$M \geq \bar{m}\rho; \; \bar{m}, \rho \text{ strictly positive} \iff M(u,v) > 0, \quad \forall (u,v) \in \mathbb{I}^2. \tag{6.2}$$

This means in particular that the model we consider can be applied to study two successive levels E_j, E_{j+1}, and the dynamics restricted to the union of these type sets.

6.1 Non-spatial Case

In this subsection we consider the non-spatial situation. We begin by considering the deterministic case ($d = 0$), often called the *infinite population model* with mutation and additive selection. In this case we note that it actually suffices to consider the non-spatial case.

We assume that type i has fitness e_i, $e_0 = 0, e_K = 1$ and selection occurs at rate s. In this case the interacting system is given by the system of ordinary differential equations

$$\frac{dx_\xi(t,i)}{dt} = sx_\xi(t,i)(e_i - \sum_{j\in\mathbb{I}} e_j x_\xi(t,j)) + \sum_{j\in\mathbb{I}} M(j,i)x_\xi(t,j)$$

$$- \sum_{j\in\mathbb{I}} M(i,j)x_\xi(t,i) \qquad (6.3)$$

$$+c \sum_{\xi'\in\Omega_N} a(\xi,\xi') \left(x_{\xi'}(t,i) - x_\xi(t,i)\right), \qquad i \in \mathbb{I}, \xi \in \Omega_N.$$

If we have a uniform initial condition

$$x_\xi(0,i) = x_{\xi'}(0,i) \geq 0, \quad \sum_{i\in\mathbb{I}} x_\xi(0,i) = 1 \qquad (6.4)$$

for all $\xi, \xi' \in \Omega_N$, $i \in \mathbb{I}$, then this property remains persists for all $t \geq 0$ and it suffices to consider a single site, that is the solution of

$$\frac{dx(t,i)}{dt} = sx(t,i)(e_i - \sum_{j\in\mathbb{I}} e_j x(t,j)) + \sum_{j\in\mathbb{I}} M(j,i)x(t,j) \qquad (6.5)$$

$$- \sum_{j\in\mathbb{I}} M(i,j)x(t,i), \quad i \in \mathbb{I}. \qquad (6.6)$$

The limiting behaviour as $t \to \infty$ of this system of ODE can be studied by standard methods of differential equations (see for example [HS], Part 6). However to illustrate the application of the dual process we give an alternative analysis using the refined dual of Subsection 5.5.

Here we first focus on the argument in case of state-independent mutation where actually $M = m(1 \otimes \rho)$, which simplifies proofs so that the main ideas concerning the effects of selection are clearer. See Subsubsection 5.5.1 for notation used here and recall in particular that A_ℓ is the set of types with maximal fitness.

Theorem 12 (Ergodic theorem: non-interacting sites). *Consider the spatial system with $c = 0$. Then the sites in X^N are independent so that it suffices to consider one site and this model we denote here by $(Y_t)_{t\geq 0}$.*

6.1 Non-spatial Case

(a) *Deterministic case, no mutation. Assume that* $d = 0$, $\bar{m} = 0$, $s > 0$, $K \geq 2$.
   If $Y_0(A_\ell) > 0$, *then*

$$Y_t(A_\ell) \to 1 \text{ as } t \to \infty. \tag{6.7}$$

(b) *Deterministic case with mutation with* $K = 2$. *Assume that* $d = 0$, $\bar{m} > 0$, $s > 0$ *and* type independent mutation *with mutation matrix*

$$M = 1 \otimes \rho. \tag{6.8}$$

Case 1: $\rho(i) > 0$, $i = 1, 2$.
Then

$$Y_t \text{ converges as } t \to \infty \text{ to a unique equilibrium point.} \tag{6.9}$$

Case 2:   No mutation to types in A_2. Assume that $\rho(1) > 0$, $\rho(2) = 0$.
If $s/m > 1$, then

$$\begin{gathered} Y_t \text{ converges as } t \to \infty \text{ to one of the different equilibrium points,} \\ \text{one corresponding to } Y_0(A_2) = 0 \text{ and others to } Y_0(A_2) > 0. \end{gathered} \tag{6.10}$$

(c) (i) *Stochastic case without mutation* ($d > 0$, $m = 0$, $s \geq 0$).
   In this case we have fixation:

$$\mathcal{L}[Y_t] \to \sum_{i=1}^{K} a(i) \delta_{\{i\}} \tag{6.11}$$

and the fixation probabilities $a(\cdot)$ depend on the initial measure.
(ii) *Stochastic case with mutation* ($d > 0$, $m > 0$, $s > 0$).
   Assume supp(ρ) = $\mathbb{I}$. Then

$$\mathcal{L}[Y_t] \text{ converges to a unique equilibrium.} \tag{6.12}$$

This is the standard mutation-selection equilibrium which can be explicitly determined as Gibbs measure (see [DG99] Theorem 10 in Section 2).
□

Remark 63. In the case of a general mutation matrix the classification of stationary measures will depend on the decomposition into ergodic classes for the mutation mechanism.

Remark 64. The proof of (b) in the case $K > 2$ is given in Theorem 14 in Section 9.

Proof of Theorem 12. Throughout this section we use the dual representation from Proposition 5.17 in Subsection 5.6 in the refined version of Subsection 5.5 in

Proposition 5.15, where we choose $\tilde{\eta}_0$ consisting of one (or n) particle at the tagged site 0 and $\mathcal{F}_0^{++} = f$, where f is a indicator function of one (or n) variables. Since $c = 0$ all individuals remain at the tagged site.

(a) Deterministic case without mutation

Here we split the argument into two different cases, treated separately. Namely we first consider two types and then later argue recursively for K types.

Special Case $\mathbb{I} = \{1, 2\}$.

Consider the dual process of the McKean–Vlasov process recalled above. In this case type 1 has fitness 0 and type 2 has fitness 1. In the infinite population case there is no coalescence of dual partition elements and consequently the factors coming from different initial elements develop completely independently. As a result it follows that

$$E[(Y_t(i))^k] = (E[Y_t(i)])^k. \tag{6.13}$$

Hence it suffices to take as initial function f of the dual a function with one variable on $\{1, 2\}$, that is start the dual with a single individual. In the case of two types it also suffices to take

$$\mathcal{F}_0^{++} = f = 1_{A_2}. \tag{6.14}$$

Recall that the dual $\mathcal{F}_t^{++}$ is given by the function-valued jump process

$$f \to f \cdot 1_{A_2} + f \otimes (1 - 1_{A_2}) \tag{6.15}$$

at rate s where 1_{A_2} is the indicator function of type 2. q.e.d.

We now note that we can calculate what the action of the selection operator is in that case. Namely

$$1_{A_2} \text{ acting on } 1_{A_2} = 1_{A_2} + 1_{A_2} \otimes (1 - 1_{A_2}), \tag{6.16}$$
$$1_{A_2} \text{ acting on } (1 - 1_{A_2}) = (1 - 1_{A_2}) \otimes (1 - 1_{A_2}).$$

Recall now the recipe from Subsection 5.5.5 how to use a Feynman–Kac term. Calculating the dual expression we work with jumps

$$1_{A_2} \otimes (1 - 1_{A_2})^{\otimes k} \to 1_{A_2} \otimes (1 - 1_{A_2})^{\otimes (k+1)} \tag{6.17}$$

with Feynman–Kac term e^{st}.

This leads to the expression

$$x(t) = E[\langle Y_t^\otimes, f \rangle] = E[\langle Y_0^{\otimes, FK}, \mathcal{F}_t^{++} \rangle] = e^{st} \sum_{n=0}^{\infty} x(1-x)^n p_n(t), \tag{6.18}$$

6.1 Non-spatial Case

where $x = Y_0(2)$, and $p_n(t)$ is the probability of n births (i.e. factors of $(1-x)$) by time t. Since this is a pure birth process with birth rate s starting with one individual, $p_n(t)$ is given by

$$p_n(t) = e^{-st}(1 - e^{-st})^n. \tag{6.19}$$

Therefore we obtain altogether the explicit expression for $(x(t))_{t\geq 0}$:

$$x(t) = E[\langle Y_t, \mathcal{F}_0^{++}\rangle] = e^{st} \sum_{n=0}^{\infty} x(1-x)^n e^{-st}(1-e^{-st})^n \tag{6.20}$$

$$= \frac{x}{1 - (1-e^{-st})(1-x)} = \frac{xe^{st}}{1 + x(e^{st} - 1)} \to 1 \text{ as } t \to \infty.$$

This is of course is a well known solution to the *logistic differential equation*.

General multitype case.

Assume that $Y_0(A_\ell) > 0$ and let $2 \leq j \leq \ell$. The using the dual representation we have

$$E[Y_t(A_j^c)] = E[\int \mathcal{F}_t^{++} dY_0^\otimes] \tag{6.21}$$

where $\mathcal{F}_t^{++}$ is the dual with $\mathcal{F}_0^{++} = f = (1 - 1_{A_j})$. Recall that the selection action on f at rate $s > 0$ and given a selection operation, with probability $(e_k - e_{k-1})$ it acts as follows:

$$f \to 1_{A_k} \cdot f + (1 - 1_{A_k}) \otimes f. \tag{6.22}$$

We note the following relations holding if we set $f = 1 - 1_{A_j}$:

$$1_{A_k} \cdot (1 - A_j) + (1 - 1_{A_k}) \otimes (1 - A_j) \tag{6.23}$$
$$= (1 - A_j) \otimes (1 - A_j), \quad \text{if } k = j$$
$$(1 - A_k) \otimes (1 - A_j) \leq 1 \otimes (1 - A_j), \quad \text{if } k > j$$
$$1_{k,\dots,j-1} + 1_{1,\dots,k-1} \otimes (1 - A_j) \leq (1 - 1_{A_j}), \quad \text{if } k < j.$$

Since the operation (6.22) is order preserving we can then iterate this operation. We then obtain

$$\mathcal{F}_t^{++} \leq (1 - A_j)^{\otimes n(t)} \tag{6.24}$$

where $\{n(t) : t \geq 0\}$ is a Poisson process with parameter $(e_j - e_{j-1})t$. Therefore substituting into (6.21) we obtain

$$E[Y_t(A_j^c)] \leq e^{-(1-Y_0(A_j^c))st(e_j-e_{j-1})} \to 0 \text{ as } t \to \infty. \tag{6.25}$$

(b) Deterministic case with mutation.

Case 1: $\rho(A_j) > 0$, for all $j = 1, \ldots, \ell$.

Here we work with $(\tilde{\eta}_t, \mathcal{F}_t^{++})_{t \geq 0}$ resp. $(\tilde{\eta}_t, {}^{FK}\mathcal{F}_t^{++})_{t \geq 0}$. Observe first that due to mutation the dual now has two sites where the second site is $\{*\}$ as introduced above. Recall that in this section we prove only the case $K = 2$.

Since the solution is deterministic it suffices to find the first moment of the mass of one of the types. Let $x = Y_0(A_2^c)$. Then it suffices to prove that $\lim_{t \to \infty} E_{Y_0}(Y_t(A_2))$ is independent of $x \in [0, 1]$. To calculate this mass we use again the dual as above using the trick of 5.5.5 (Feynman–Kac term) to calculate the dual expectation (recall (6.13)). That is, it suffices to consider:

$${}^{FK}\mathcal{F}_0^{++} = 1_{A_2}. \tag{6.26}$$

Let N_t^a denote the total number of offspring individuals created by a birth event up to time t, that have not yet jumped to $*$ at time t and K_t denotes the number of those individuals that have jumped to $\{*\}$ by time t. (Note that $N_t^a + K_t + 1 = N_t$ equals one plus the total number of births that have occurred up to time t.) Then at time t we have an expression of the form

$${}^{FK}\mathcal{F}_t^{++} = e^{s(t \wedge \tau^*)} 1_{A_2} \otimes (1 - 1_{A_2})^{\otimes N_t^a} \otimes (1 - 1_{A_2})^{\otimes K_t}, \tag{6.27}$$

where

$$\tau_* = \inf\{t : \text{initial particle is at } * \text{ at time } t\}. \tag{6.28}$$

It now suffices to prove that

$$\lim_{t \to \infty} E[\int {}^{FK}\mathcal{F}_t^{++} dY_0^{\otimes}] = \lim_{t \to \infty} E[e^{s(t \wedge \tau^*)}(1 - x) \otimes x^{\otimes N_t^a} \otimes x^{\otimes K_t}] \tag{6.29}$$

is independent of x.

Note that the Feynman–Kac term operates only until the initial 1_{A_2} jumps to $*$ (recall (6.16)) so that we have a Feynman–Kac term $e^{s(t \wedge \tau^*)}$. Also define

$$\tau^* \leq \tau^{**} := \inf\{t : N_t^a = 0, t \geq \tau^*\}. \tag{6.30}$$

Observe $N_0 = 1$, $K_0 = 0$, the birth rate is s and the jump rate to $*$ is m. The last factor in (6.27) is associated with the state $\{*\}$ the first two with the normal site, that is Y_0.

Next note that the case $m > s$ is easy because extinction occurs a.s., that is, $\tau^{**} < \infty$ and $K_{\tau^{**}} < \infty$ a.s. and hence

$${}^{FK}\mathcal{F}_t^{++} \to e^{s\tau^*} 1_{A_2} \otimes (1 - 1_{A_2})^{\otimes K_{\tau^{**}}}, \tag{6.31}$$

6.1 Non-spatial Case

where both factors are located at $\{*\}$. This random variable is for $m > s$ integrable and the result of taking the integral is

$$\rho(A_2)E[e^{s\tau^*}(1-\rho(A_2))^{K_{\tau^{**}}}] \qquad (6.32)$$

which is independent of the initial state Y_0 hence the claim follows.

We next consider the case $s > m > 0$ so that $\tau^{**} = \infty$ with positive probability and we cannot work with (6.31). To complete the proof of ergodicity it suffices to prove that the contribution of the integrated dual over the set $\{\tau^{**} > t\}$ goes to 0 as $t \to \infty$. Since $\tau^* < \infty$ a.s., the contribution of the interval over the event $\tau^* > t$ tends to 0, therefore it suffices to show that the contribution over $\{\tau^* \leq t\} \cap \{\tau^{**} > t\}$ goes to 0 as $t \to \infty$. The calculation is based on:

- the initial factor 1_{A_2} jumps to $*$ after an exponential time with parameter $m > 0$,
- during its sojourn at the initial site it serves as an immigration source of factors $(1 - 1_{A_2})$ at rate $s > 0$,
- the 1_{A_2} process at the initial site is a birth and death process N_t^a with birth rate s and death rate m,
- each death at the initial site produces a factor $(1 - 1_{A_2})$ at $*$, the number of these factors at $*$ is given by K_t.

Recall that

$$^{FK}\mathcal{F}_t^{++} = e^{s(t \wedge \tau^*)} 1_{A_2} \otimes (1 - 1_{A_2})^{\otimes N_t^a} \otimes (1 - 1_{A_2})^{\otimes K_t}. \qquad (6.33)$$

We then decompose the interval $[0, t]$ into $[0, \tau^* \wedge t]$ and $[\tau^* \wedge t, t]$. We first consider the birth and death process with immigration, $N_t^{a,IMM}$ in the first interval. By [Bai], (8.73), the probability generating function is given by

$$P(x,t) = E[x^{N_t^{a,IMM}}] = \frac{s-m}{s(1-x)e^{(s-m)t} + sx - m}. \qquad (6.34)$$

Next consider the birth process $(N_t^a)_{t \geq 0}$ of the birth due to $(1 - 1_{A_2})$-factors starting after the time s^*. The generating function of such a birth and death process is

$$P(z,t) = E[z^{N_t^a}] = \frac{z(1-\beta(t))}{1-z\beta(t)}, \qquad (6.35)$$

where $\beta(t)$ is given by (6.37) below. Namely define $\tilde{p}_k(t)$ as the probability that $N_t = k$ conditioned on non-extinction. By a standard result on birth and death processes (e.g. Bailey, Section 8.7)

$$\tilde{p}_k(t) = (1 - \beta(t))\beta(t)^{k-1}, \qquad (6.36)$$

where

$$\beta(t) = \frac{s(e^{(s-m)t} - 1)}{se^{(s-m)t} - m} \qquad (6.37)$$

154   6 Long-Time Behaviour: Ergodicity and Non-ergodicity

Recall that τ^* is an exponential random variable with parameter m that and the Feynmann–Kac factor is

$$e^{s(\tau^* \wedge t)}. \tag{6.38}$$

Then we decompose the expectation over the three events $\tau^* = \infty$, $\tau^* < \infty$, $\tau^{**} < \infty$ and $\tau^* < \infty$, $\tau^{**} = \infty$

$$\lim_{t \to \infty} E[e^{s(t \wedge \tau^*)}(1-x) x^{N_{\tau^* \wedge t}^{a,IMM}} x^{N_{t-\tau^*}^{a}} \rho(A_c^2)^{K_t}] \tag{6.39}$$

$$= \lim_{t \to \infty} e^{(s-m)t}(1-x) E[x^{N_t^{a,IMM}} (\rho(A_2^c))^{K_t}]$$

$$+ E[1(\tau^* < \infty) e^{s\tau^*} \rho(A_2) E[\rho(A_c^2)^{K_{\tau^{**}}} | N_{\tau^*-}^{a,IMM}, K_{\tau^*-}]]$$

$$+ \lim_{t \to \infty} E[1(\tau^* < t) e^{s\tau^*} \rho(A_2)$$

$$E[1(\tau^{**} = \infty) x^{N_t^{a}} \rho(A_c^2)^{K_t} | N_{\tau^*-}^{a,IMM}, K_{\tau^*-}]]$$

Using (6.34), it follows that the first summand is 0 and the second summand does not depend on x. Using (6.34) it follows that the third term is

$$\leq \int_0^t e^{(s-m)u} \cdot \frac{s-m}{s(1-x)e^{(s-m)u} + sx - m} E[1(\tau^{**}=\infty) x^{N_t^{a}} \rho(A_c^2)^{K_t} | N_u^{a,IMM}, K_u]] \tag{6.40}$$

If $x < 1$, then using (6.36) and (6.37) it follows that the limit is 0. Finally if $x = 1$ $\rho(A_2^c) < 1$, we must work with the factor $\rho(A_2)^{K_t}$. Now consider $N_{t/2}^a$. Independently each of these particles dies and produces a factor $\rho(A_2^c)$ before producing an offspring with probability $\frac{m}{m+s}$ and this happens before time t with probability $(1 - e^{-(m+s)t/2})$. Therefore the term $(\rho(A_2^c))^{K_t}$ produces an additional factor of the form const $\cdot e^{-(s-m)t/2}$. This implies that the third summand given by the contribution to the integral over $\{\tau^* \leq t\} \cap \{\tau^{**} > t\}$ goes to 0 as $t \to \infty$.

Therefore the limiting expectation $\lim_{t \to \infty} E[Y_t(A_2)]$ depends only on ρ and not Y_0. That implies that the process is ergodic.

The proof in the case $m = s$ again follows along the same lines as in the case $s > m$ but we do not give the details.

Remark 65. The main point is that when $\rho(A_2) > 0$ even when the birth and death process is supercritical, the limiting law is given by an expectation over the event that extinction occurs.

Case 2: $\rho(A_2) = 0$.

In this case we modify the argument as follows. We do not introduce ρ but note that the only changes that occur by mutation with positive rate are $1_{A_2} \to 0$ and $(1 - 1_{A_2}) \to 1$ and both with rate $m_{2,1}$. We treat the transition $(1 - 1_{A_2}) \to 1$ as a

6.1 Non-spatial Case

death of a $(1 - 1_{A_2})$ factor so again the factors $(1 - 1_{A_2})$ again form a birth and death process with immigration with birth rate s and death rate $m = m_{2,1}$ and immigration rate s.

If $x = Y_0(A_2) > 0$ and $s > m$, then the expectation at time $t > 0$ is over the event (since otherwise we get a zero factor). Then event $\{\tau_* > t\}$ has probability $e^{-m_{2,1}t}$ and the Feynman–Kac term contributes the factor

$$e^{st}. \tag{6.41}$$

We then get for the integral of (6.27) over $\{\tau_* > t\}$ the expression:

$$x \cdot e^{(s-m)t} \sum_{k=1}^{\infty} (1-x)^k \tilde{p}_k(t) \quad \text{with } x = Y_0(A_2). \tag{6.42}$$

We again use formula (6.36) to evaluate $\tilde{p}_k(t)$ for $t \to \infty$. If $s > m$ then in the case where $x = Y_0(A_2) > 0$ we obtain using (6.34)

$$\lim_{t \to \infty} E[Y_t(A_2)] = x \lim_{t \to \infty} \left(\frac{s-m}{se^{(s-m)t} - m} \right) \left(1 - \frac{s(e^{(s-m)t} - 1)(1-x)}{se^{(s-m)t} - m} \right)^{-1} = \frac{s-m}{s}. \tag{6.43}$$

On the other hand we see that if $m_{1,2} = 0$ and $Y_0(A_2) = 0$, the $Y_t(A_2) = 0$ for all $t \geq 0$. Therefore in this case we have non-ergodicity since the limit depends on whether or not $Y_0(A_2) > 0$.

By contrast if $s < m$ then the expectation $E[Y_t(A_2)]$ is bounded by $E^{(s-m)t} \to 0$ as $t \to \infty$ and ergodicity follows.

(c) Stochastic case without mutation

Proof. (i) We now observe that in the stochastic case, due to *coalescence* the number of individuals in the dual process at any time is stochastically bounded and returns to one a.s. In case of no mutation there are no jumps to $*$. Therefore in the case $K = 2$ equation (6.27) is replaced for $\mathcal{F}_0^{++} = (1 - 1_{A_2})$ by the formula:

$$\mathcal{F}_t^{++} = \otimes (1 - 1_{A_2})^{\otimes N_t}, \tag{6.44}$$

and if $\mathcal{F}_0^{++} = (1 - 1_{A_2}) \otimes 1_{A_2}$, then

$$^{FK}\mathcal{F}_t^{++} = e^{st} 1_{A_2} \otimes (1 - 1_{A_2})^{\otimes N_t} 1_{\tau_d > t}, \tag{6.45}$$

where τ_d denotes the first coalescence time. Without loss of generality we can assume that $x := Y_0(1) > 0$. But then noting that up to coalescence $N(t)$ is a pure birth process with birth rate s (recall (6.19)) and

$$\sum_n e^{st}(1-x)^n e^{-st}(1 - e^{-st})^n \leq \sum_n (1-x)^n = \frac{1}{x} \tag{6.46}$$

we have uniform integrability. Since coalescence occurs for some $t > 0$, that is $\tau_d < \infty$ with probability one, it follows that

$$\lim_{t \to \infty} E[Y_t(1) Y_t(2)] = 0. \tag{6.47}$$

This implies that the limiting law is a mixture of δ-measures on the types 1 and 2. To calculate the limiting probabilities of fixation in types 1, 2, respectively. Consider

$$E[Y_t(1)] = E[(Y_0(1))^{N_t}], \tag{6.48}$$

where now N_t is a birth and death process with birth rate sk and death rate $\frac{d}{2}k(k-1)$ when $N(t) = k$. Therefore

$$\lim_{t \to \infty} E[Y_t(1)] = E[\sum_k (Y_0(1))^k p_k], \tag{6.49}$$

where $\{p_k\}$ are the stationary probabilities of the birth and death process. This completes the proof.   q.e.d.

(ii) Stochastic case with mutation.

In the stochastic case $(d > 0)$ we must consider higher moments in order to establish the convergence in distribution. To do this let $\tilde{\eta}_0$ consist of n individuals choose $\mathcal{F}_0^{++} = f$ with functions of the form

$$f_0(x_1, \ldots, x_n) = \bigotimes_{i=1}^{n} 1_{A_{j_i}}(x_i), \quad j_i \in \{1, \ldots, \ell\}. \tag{6.50}$$

Then the distribution at time t is determined by the values of

$$E\left[\int \cdots \int f_0(x_1, \ldots, x_n) \prod_{i=1}^{n} Y_t(dx_i)\right]. \tag{6.51}$$

In the stochastic case with mutation we use again that the law at time t is determined by (6.51) and that the number of elements in the dual process is stochastically bounded and recurrent due to coalescence. Therefore with probability one it is eventually absorbed at $*$. Therefore the limit as $t \to \infty$ of $E[\prod_{j=1}^{n} Y_t(A_{i_j})]$ exists and is independent of Y_0 and the ergodicity follows.

6.2 Spatial Case

The ideas explained above allow us to incorporate the spatial structure in the case of finite type space. We can both do this for the McKean–Vlasov process Z arising as mean-field limit as well as for the spatial process X^N. We get for this interacting

6.2 Spatial Case

case the following, which essentially already proves Theorem 2 in the case $K = 2$ but contains also statements for the cases $d = 0$ or $m = 0$ which are not directly necessary for our results given in Section 4.

Recall the definition of A_k as the set of types having at least the fitness e_k if $0 = e_0 < e_1 < \cdots < e_\ell = 1$.

Theorem 13 (Ergodic theorem spatial case with finitely many types). *We assume that $|\mathbb{I}| < \infty$. Consider the McKean–Vlasov dynamic $(Z(t))_{t \geq 0}$ defined in (4.28). Assume that $c > 0, d > 0$ and $s > 0$.*

(a) *No mutation, that is, assume $m = 0$.*
   If $Z_0(A_\ell) > 0$, then $Z_t(A_\ell) \to 1$.
(b) *Assume that $K = 2$ and consider the case with mutation given by $M(u, v) = m\rho(v)$ for $u, v = 1, 2$ with $m > 0$.*

   Case 1: $\rho(A_i) > 0$, $i = 1, 2$. *Then $\mathcal{L}[Z(t)]$ converges to a unique equilibrium and the process is ergodic.*
   Case 2: $\rho(A_2) = 0$, $\rho(A_1) > 0$. *Then if s and c are sufficiently large to guarantee non-extinction of the dual process $(\eta_t)_{t \geq 0}$ outside the site $\{*\}$ with positive probability, then $\mathcal{L}[Z_t]$ converges to one of the distinct equilibria, one corresponding to $Z_0(A_2) = 0$ and the others corresponding in the case where $Z_0(A_2) > 0$ to the subsets of A_2 where Z_0 is positive. (Shiga-Uchiyama [SU])*

(c) *Interacting sites. Consider the X^N on Ω_N with $d > 0, s > 0, m > 0$ and $c > 0$. Assume we have a spatially homogeneous and shift ergodic initial condition. In this case the analogue to (b) holds.* □

Remark 66. General state-dependent mutation mechanisms and $K > 2$.

The proof of the analogue of these result for general mutation mechanisms ($M(u, v)$ arbitrary) and $K > 2$ requires the dual representations developed in Section 8. The statements and proofs are given in Section 9, Theorem 14.

Proof of Theorem 13. Here build in as a new element the spatial migration, respectively the nonlinear mean-field term in the McKean–Vlasov process in the state-independent case. The general case of state-dependent mutation $M - \bar{m}1 \otimes \rho \neq 0$ is deferred to Section 9, Theorem 14.

We evaluate the expectation in question now with $(\tilde{\eta}_t, \hat{\mathcal{F}}_t^{++})$ introduced in Subsection 5.5 in its refined form based on the dual in Subsection 5.6 and we choose $\tilde{\eta}_0$ consisting of n individuals all placed at the site $\{0\}$ and we set $\hat{\mathcal{F}}_0^{++} = f$ as above. We discuss this now in the different cases specified in the statement of the theorem.

(a) We consider here the case $\ell = 2$. Let $\mathcal{F}_0^{++} = (1 - 1_{A_2})$. At each site there are, due to the coalescence, locally a stochastically bounded number of $(1 - 1_{A_2})$ factors. Then the term $\mathcal{F}_t^{++}$ of sums of products of indicators such that every summand contains a number of factors $(1-1_{A_2})$ increasing uniformly to infinity due to migration since we assumed $c > 0$, and therefore

$$E[Z_t(A_2^c)] = E[\int \mathcal{F}_0^{++} dZ_t^\otimes] = E[\int \mathcal{F}_t^{++} dZ_0^\otimes] \to 0 \text{ as } t \to \infty. \tag{6.52}$$

This is proved for the general case $\ell \geq 2$ in ([DG99]), Section 4.

(b) *Case 1.* Assume that $\rho(A_2) > 0$.

The distribution of Z_t is determined by the values of the expectation

$$E\left[\int f(x_1,\ldots,x_n) \prod_{i=1}^n Z_t(dx_i)\right], f(x_1,\ldots,x_n) = \bigotimes_{i=1}^n 1_{A_{j_i}}(x_i),$$

$$j_i \in \{1,\ldots,\ell\}. \qquad \text{q.e.d.} \tag{6.53}$$

Consider the stopping time τ_* at which all the initial factors have reached $*$. Observe that all the remaining factors outside $*$ are of the form $(1-1_{A_2})$, i.e. do not depend on $\mathcal{F}_0^{++}$ and have the property that they all vanish for arguments in A_2 and therefore integrated w.r.t. ρ give always a contribution at most $(1-\rho(A_2)) < 1$.

On the event where *every* factor jumps to $*$ we get a value as $t \to \infty$ which is independent of the initial state since only ρ enters as integrating measure. This case occurs with probability 1 if $m \geq s$. The proof that the expectation then converges (uniform integrability) is the same as in the sequel of (6.26). However in the case $s > m$ the situation is different, here the question is what about the event of nonextinction?

Note that eventually each site is "absorbed at $*$". However if c, s are sufficiently large then the expected number of new sites produced during the lifetime of such a site is larger than one and the process is supercritical (see Section 8 for a discussion of the corresponding Crump–Mode–Jagers branching process and Nerman ([N]) for the asymptotic behaviour). In the supercritical case non-extinction (where extinction means that all terms are at $*$) can be shown to occur using the same argument as above.

Now consider the terms outside $\{*\}$ conditioned on non-extinction. We have to show that after τ_* the number of $(1-1_{A_2})$ factors increases to infinity uniformly in all summands. Using this we then have to show that

$$E[1_{\text{non-extinction}} \int \mathcal{F}_t^{++} dZ_0^\otimes] \to 0 \text{ as } t \to \infty. \tag{6.54}$$

This then completes the proof or ergodicity.

In order to show the above relation we use the same strategy as below (6.26), but the additional element now is the coalescence. The idea is that as a result of coalescence the number of dual particles is stochastically bounded at each site but no coalescence occurs between sites. Since we have a growing number of sites

6.2 Spatial Case

this means that the total number of $(1 - 1_{A_2})$ factors increases to infinity in the supercritical case. Then the contribution of summands with more that k factors decreases to 0 as $k \to \infty$.

Case 2 Assume $\rho(A_2) = 0$, $Z_0(A_2) > 0$ and both s, c sufficiently large.

Since in this case $(1 - \rho(A_2))^k = 1$, arguments similar to that in Subsection 6.1 are applicable. In particular, if s, c are sufficiently large, the probability of non-extinction is positive and $\lim_{t \to \infty} P(Z_t(A_2) > 0) > 0$ thus yielding non-ergodicity.

(c) *Case 1.* The above dual process argument used in the McKean–Vlasov process extends to the infinite site case with migration kernel. The only essential new point is to show that for sufficiently large c and s the branching coalescing random walk is supercritical, that is, spreads out and the total number of particles tends to infinity as $t \to \infty$. In the supercritical case one again shows that the contribution on the analogue of $\tau^{**} = \infty$ is zero if $\rho(A_2) > 0$ giving ergodicity and that the limit depends on $Y_0(A_2)$ if $\rho(A_2) = 0$. In the subcritical case ergodicity is immediate as above. (See the proof of Theorem 14 for details.)

Case 2. The proof of ergodicity or non-ergodicity using a different dual process has been carried out in for the case of two types in Shiga and Uchiyama [SU] in $\mathbb{Z}^d$ and by Athreya and Swart [AS] in a general countable graph.

Chapter 7
Mean-Field Emergence and Fixation of Rare Mutants: Concepts, Strategy and a Caricature Model

This Section 7 sets the stage for the proofs of our main theorems formulated in Subsection 4.5 and Subsection 4.6. In the Sections 8, 9, 12 and 13 we shall prove the basic facts stated in the framework of Section 7 needed to derive Theorem 6 to Theorem 9. This argument splits in various preparatory parts, namely the analysis of emergence and fixation (phase 1 and 3) in a mean-field model, then extension to the mean-field Fisher-Wright system first for emergence and fixation in Section 8 for two types, in Section 9 for M types on the lower level, one on the upper level, Section 10 for M types on each level, and finally the behaviour after fixation (phase 3 and 4) of the mean-field model in Sections 11 and 12. In Section 13 we treat the hierarchical finite and then infinite systems and then the combination of those results to the final argument is in Section 15. We outline the program of the Sections 7–12 on the mean-field and finite hierarchical models below in Subsection 7.2 in precise mathematical form.

7.1 The Mean-Field Model and Hierarchical Mean-Field Models

A very important building block in the understanding of the infinite hierarchical model are sequences of *finite* systems (finite in the number of geographic components and finite in the number of levels of fitness). These are so-called *mean-field* models and then building on them *finite j-level hierarchical mean-field* models.

7.1.1 The Basic Scenario for the Mean-Field Model

We start with the key concept of our analysis namely the *mean-field systems*. These are systems:

$$X^N(t) = (x^N(i,t), \quad i = 1, \cdots, N) \quad ; \quad x_i^N(t) \in \mathcal{P}(\mathbb{K}) \text{ for } i \in \{1, \cdots, N\} \quad (7.1)$$

D.A. Dawson and A. Greven, *Spatial Fleming-Viot Models with Selection and Mutation*, Lecture Notes in Mathematics 2092, DOI 10.1007/978-3-319-02153-9_7,
© Springer International Publishing Switzerland 2014

for some finite set $\mathbb{K}$ of types. This system arises from our process X in Subsection 3.2 if we replace

- the site space, i.e. the hierarchical group Ω_N, by the set $\{1, \cdots, N\}$ and
- the type space $\mathbb{I}$ by some finite set $\mathbb{K}$.

Typically we shall consider $\mathbb{K} = E_0 \cup E_1$, (or $E_1 \cup E_2$, etc.). Precisely consider the system where we consider only $B_1(0)$ in Ω_N and put $c_0 = c, c_k = 0$ for $k \geq 1$ and consider only types in $E_0 \cup E_1$. If we restrict the system now to levels of fitness at most two, we still get $2M + 1$ types, the basic one plus M on each of the next two levels.

We give now a short description how the basic scenario of the transition to the fitter types reads in the framework of the mean-field analysis, that is we consider systems for N exchangeable sites and two levels of fitness E_0 and E_1. Formally this model is obtained from our model by replacing Ω_N by $\{1, \cdots, N\}$ and letting all rates of mutation to level 3 or higher be zero so that we only see types in E_0 and in E_1 referred to as level-one and level-two types of fitness. Then the transition from a population concentrated on E_0 to one concentrated on E_1 proceeds in the following five phases.

Phase 0: (Evolution on E_0 into quasi-equilibrium, droplets)
In the time scale t, the system approaches in the limit $N \to \infty$ the McKean–Vlasov limit dynamics on E_0 and for $t = t(N) \uparrow \infty$ with $t(N) \leq C \, log \, N$ for $C \in [0, \infty)$ depending on the parameters, the law of the system approaches an equilibrium state of the McKean–Vlasov limit dynamic concentrated on individuals of types in E_0.

However on a sparse set of sites the fitter types have already completely colonized the sites and this *droplet* of size $o(N)$ will provide the germs for the occurrence of the fitter E_1-types in the next phase.

Phase 1: (Emergence of E_1-individuals)
In time scale $C \log N + t$, for suitably chosen $C \in (0, \infty)$ the E_1-individuals *emerge* and their total mass per site reaches given threshold values $\varepsilon > 0$, this arises by an exponentially growing number of sites in the droplet at which E_1-individuals have succeeded in dominating the population. At earlier times the latter sites formed a sparse set (of size $o(N)$) only.

Phase 2: (Fixation on E_1)
The population observed at times $C \log N + t$ approaches a nonlinear limit dynamic in $t \in \mathbb{R}$ as $N \to \infty$ in the form of a random entrance law from $-\infty$ and becomes concentrated on E_1 individuals *(fixation)* as both $N \to \infty$ and $t \to \infty$.

Phase 3: (Neutral evolution on E_1)
For times of order $\gg C \log N$, but $o(N)$ the population is in a *neutral* (only immigration-emigration and resampling) equilibrium on E_1-individuals as $N \to \infty$.

7.1 The Mean-Field Model and Hierarchical Mean-Field Models

Phase 4: (Mutation-selection evolution on E_1-individuals, convergence in quasi-equilibria)

Here we consider the dynamic in time scales tN which requires to consider N coupled mean-field systems since the migration steps to distance 2 now have to be considered in order to still approximate the real system in that time scale. Then we get for the collection of N blockaverages a Markovian limiting dynamic in the macroscopic time variable t and a new mutation-selection equilibrium is attained but now on E_1 individuals as $N \to \infty$ and $t \to \infty$.

In the last phase 4 we now have reached a situation which is as before only now we have a population of type in E_1 rather than in E_0 and now in larger time scales the whole cycle starts over leading to the transition from E_1 to E_2 in our spatial *two*-level model. This scheme continues this way.

Next we turn to *finite hierarchical* mean-field models of level $j \in \mathbb{N}$. These processes are the building blocks for systems with N^j components and j-level of spatial organisation with $(j+1)$-levels of fitness. These are models where space consists of a finite number, namely of N^j sites (the j-ball around 0 in Ω_N) and with types in the finite set given by the union of $E_0, E_1, \cdots, E_j$. The processes arise by connecting N hierarchical mean-field systems of spatial level $(j-1)$ and the same type space.

In other words we can find mean-field and hierarchical mean-field systems embedded in a system indexed by the hierarchical group. We restrict the spatial system to the system indexed by $B_k(0)$ and the state in each of the components to $\bigcup_{\ell=0}^{k} E_\ell$ by suppressing jumps leading out of this state space. Again this results in a *finite* system but one which is more structured than the plain mean-field system.

More formally we get in (3.11) the jump rates $c_1, c_2, \cdots = 0$ for the mean-field system, respectively $c_j, c_{j+1}, \cdots = 0$ for the j-level system and furthermore setting mutation rates from and to $\bigcup_{k=j+1}^{\infty} E_k$ equal to zero and then observe the system in $B_1(0)$ respectively $B_j(0)$ with components restricted to types in $\bigcup_{k=0}^{j} E_j$.

The key point is that above systems are *spatially finite* systems with a *finite type space*. This allows us to analyse them more easily with our dual process since we can work with the refined dual process $(\eta, \mathcal{F}^{++})$, respectively $(\eta, \tilde{\mathcal{G}}^{++})$ from Subsection 5.5 instead of the process $(\eta, \mathcal{F}^+)$. This finite hierarchical system of level j has to be analysed in the original time scale (McKean–Vlasov) and in larger time scales, more precisely in scales:

$$t, \ C \log N + t, \ tN^j, \cdots, CN^{j-1} \log N + tN, \ tN^{j+1}. \tag{7.2}$$

The mean-field model and the hierarchical j-level model mentioned above was previously constructed in [DG99]. One of the main results of that paper was to identify in the special case of *no* mutation the limiting dynamics for the model in both the original and the large time scale $tN, \cdots, tN^j$ etc. In addition, the McKean–Vlasov limit of the mean-field model was already treated for the model with mutation, however large time scales were not treated since with mutation completely new phenomena arise and new techniques are needed which are developed here.

7.2 Outline of the Strategy for Sections 8–13

In the situation *without mutation* investigated in [DG99], the proof of the theorems concerning the hierarchical mean-field limit involving the various time and space scales involved four steps:

- consideration of a size N mean-field model and verification of the McKean–Vlasov limit ($N \to \infty$) in time scale t, with a limiting dynamics which can be identified as a nonlinear Markov process,
- analysis of a size N mean-field model and proof that in the time scale tN resp. $sN + t$ there is for $N \to \infty$ a two-scale limiting dynamics which we identified explicitly,
- analysis for $N \to \infty$ of a two-level and more general j-level hierarchically structured finite systems with N^j sites and types in $E_0, E_1, \cdots, E_j$ considered from time N^j on in the time scales $tN^k, k = 0, 1, 2, \ldots, j - 1$ after forming blockaverages over the components belonging to the balls of size k,
- approximation of the infinite hierarchical system observed in time scale tN^j by the $(j + 1)$-level systems and verification of the theorems based on the $(j + 1)$-level results.

As a consequence we will also have here a four-step program, investigation of mean-field models first in slow and then in fast time scales, finite hierarchical and infinite hierarchical models, which can be naturally grouped in two parts, the mean-field analysis in Sections 7–12 and the analysis of the multilevel hierarchical structure for j and countably many levels in Section 13.

Why is that not just a modification of the earlier analysis of models with selection, migration and resampling? The combined effect of the mutation, selection and migration mechanisms, namely the emergence and take-over by rare mutants, described earlier in Section 4.6 leads already for mean-field models for $N \to \infty$ to a *new type of mean-field phenomenon* and as a result we must insert four additional steps in this Section 7–12 which are the following:

- analysis of the "random times" at which rare mutants of higher fitness arise *(emergence)*,
- the time scale in which the higher level mutants spread throughout the system *(fixation)*,
- the time scale at which the rare mutants after the fixation follows a *neutral evolution*,
- the time scale in which the population reaches a new *mutation-selection equilibrium* at the *higher level*.

7.2 Outline of the Strategy for Sections 8–13

We will do this extension of the analysis to include rare mutation in this Section 7, the subsequent Sections 8 to 12, first for phase 0, 1, 2 for

- one-level size N mean-field Fisher–Wright models with two fitness levels and one type on each level in Section 8,
- in Section 9 with M types on the lower level and one on the upper level,
- in Section 10 with M types on both levels.

Then we consider phases 3 and 4 in such models in Section 11 and 12 respectively. Finally we extend the results in Section 13 to

- two-level and more generally j-level (spatially) finite hierarchically structured systems with a finite hierarchy of $(j+1)$ fitness levels,
- infinite hierarchical system, both for space and fitness observed in order j space-time scales.

This structure already shows that the main work is in understanding the problem for two fitness levels, the extension to more levels is less tough. One crucial observation in carrying out this analysis is that the mutation part of the generator is linear in the configuration, which means that it commutes with the operation of building the blockaverages.

Chapter 8
Methods and Proofs for the Fisher–Wright Model with Two Types

The proofs of the results formulated in Section 2 require precise information on the laws of a tagged site (colony) and the total mass. These laws are determined by the collection of joint moments which we can calculate using the dual process $(\eta_t, \mathcal{F}_t^+)_{t \geq 0}$. To carry out the analysis of the limiting behaviour as $N \to \infty$ we must develop the asymptotic analysis of the dual process and this is the main objective of this section. Section 8 therefore introduces the principal tools, namely the analysis of the dual particle system in the so-called pre-collision and post-collision regimes (referring to asymptotically self-avoiding or non-self-avoiding migration of the complete dual population) using ideas from the theory of Crump–Mode–Jagers branching processes and the dynamical law of large numbers, that form together the basis for the proofs of the main results.

8.1 Outline of the Subsections on Proofs of Results Formulated in Subsection 2.3

We begin in Subsection 8.2 with a warm-up exhibiting some of the methods in simpler cases. (This is not needed for the rigorous argument). Then we continue with the formal proofs, namely proving

- in Subsection 8.3 assertions related to the large time scale, in particular Propositions 2.3, 2.6 and 2.11,
- in Subsection 8.4 assertions concerning the small time scale and droplet formations, in particular Propositions 2.5, 2.8 and 2.9,
- in Subsections 8.5 and 8.6 assertions concerning the connection between the two time scales, in particular Proposition 2.14,
- in Subsections 8.7–8.8 additional assertions where different methods are used, in particular Proposition 2.13 in Subsection 8.7 and extensions in Subsection 8.8.

8.2 A Warm-Up: The Case of a Single Site Model and Other Simplifications

Before we start proving all the statements made in Subsection 2.3 we look at some simpler cases and already give in the arguments a flavour of the methods used later on. This subsection may be skipped from a purely logical point of view.

In the *absence of migration*, $c = 0$, the individual sites evolve independently. In this case it suffices to study a system

$$X(t) = (x_1(t), x_2(t)) \tag{8.1}$$

at a single isolated site, for example use $x_1(t) = x_1(1, t), x_2(t) = x_2(1, t)$. Recall (2.2)–(2.7).

We begin by showing in this case that emergence in $\log N$ time scale does occur in the infinite population, that is, deterministic case, i.e. $d = 0$ but *does not* occur in the finite population case, that is, random case, where $d > 0$. This demonstrates the effect of the bounded component (mass restriction) and the important role played by space and migration.

The proof of this result gives us the opportunity to introduce some basic ideas and methods in a simple setting. We consider here successively the deterministic and stochastic case and collect the proofs with the dual calculations in the third Subsubsection 8.2.3.

8.2.1 Emergence of Rare Mutants in the Single Site Deterministic Population Model

We first consider the deterministic, i.e. $d = 0$ (often called infinite population) model without migration, i.e. $c = 0$. In the case of small mutation rate, $\frac{m}{N}$, we will show with two different methods that asymptotically as $N \to \infty$ emergence occurs at times $\frac{\log N}{s} + t, t \in \mathbb{R}$ with fixation as $t \to \infty$:

Lemma 8.1. *Assume that the mutation rate is $\frac{m}{N}$, and $d = 0$, $s > 0$, $c = 0$. Then we have:*

$$\lim_{N \to \infty} x_2\left(\frac{\log N}{s} + t\right) = 1 - \frac{s}{s + me^{st}} \begin{cases} \to 1, \text{ as } t \to \infty \\ \to 0, \text{ as } t \to -\infty. \end{cases} \qquad \square \tag{8.2}$$

Remark 67. This means that in the deterministic case we have emergence and fixation in the "critical" time scale $(s^{-1} \log N) + t$.

8.2.2 Emergence of Rare Mutants in the Single Site Stochastic Population Model

The above analysis can easily be extended to the stochastic case (often called finite population), that is, $d > 0$, but now we get a qualitatively different conclusion. We again take mutation rate $\frac{m}{N}$. Namely we observe that

Lemma 8.2. *Let $d > 0$ and $c = 0$, then*

$$\mathcal{L}[x_2^N(a_N)] \underset{N \to \infty}{\Longrightarrow} \delta_0 \text{ whenever } a_N/N \to 0 \text{ as } N \to \infty. \tag{8.3}$$

Therefore emergence and takeover does not occur in time scale $\log N$ if there is no migration, that is, $c = 0$. □

Remark 68. This means that the "finite population sampling", i.e. $d > 0$, seriously slows down the emergence of rare advantageous mutants. This effect is removed as soon as we have migration, which is therefore the key mechanism for the emergence if $d > 0$.

8.2.3 Proofs Using the Dual

Proof of Lemma 8.1. The Lemma 8.1 is now proved in two ways with and without duality. We consider first an auxiliary object $x_2(t)$ for which if we consider the case where the parameter m is replaced by $\frac{m}{N}$ we get our x_2^N.

(i) Analytic approach The quantity $x_2(t)$ satisfies the ODE

$$\frac{dx_2}{dt} = m(1 - x_2) + sx_2(1 - x_2); \quad x_2(0) = 0 \tag{8.4}$$

with solution

$$x_2(t) = \frac{m(e^{(m+s)t} - 1)}{s + me^{(m+s)t}}. \tag{8.5}$$

We can now study this explicit solution to the ODE for our choices of the parameters and the appropriate time scales where t is replaced by $\frac{1}{s} \log N + t$. (See point (iii) below).

(ii) Duality approach. As a prelude to the analysis of the stochastic case we show how the formula (8.5) can be obtained using the *dual representation* with the modified dual process and with the Feynman–Kac representation explained in the sequel of (5.103).

We assume that $Y_0 = \delta_1$, $s > 0$ and we choose $f_0 = 1_{\{1\}}$. The mutation part of the dual consists of the simple transition corresponding to a mutation from type 1 to type 2. This means (recall (5.82)) for the dual that we have jumps $1_{\{2\}} \to 1$, $1_{\{1\}} \to 0$ both at rate m.

Let $\{p_k(t)\}_{k \in \mathbb{Z}_+}$ be the distribution of a pure birth process N_t starting with one individual (and each individual independently gives birth at rate s at time t) and let

$$\tau = \text{time of first mutation jump}. \tag{8.6}$$

Let $\tilde{p}_k(t)$ denote the probability that there were $(k-1)$ births in $[0,t]$ and that no mutation jump has occurred.

Then the random state of $^{FK}\mathcal{F}_t^+$ at time t is given by:

$$^{FK}\mathcal{F}_t^+ = (1_{\{1\}})^{\otimes k} 1_{\{\tau > t\}}, \tag{8.7}$$

with probability $\tilde{p}_k(t)$ and 0 with probability $1 - \sum_{k=1}^\infty \tilde{p}_k(t)$. Then

$$x_1(t) = E[< X(t), 1_{\{1\}} >] = E[\int {}^{FK}\mathcal{F}_0^+ dX_t] = E[\int {}^{FK}\mathcal{F}_t^+ dX_0^{\otimes |\pi_t|}] = P[\tau > t]. \tag{8.8}$$

We have:

$$P(\tau > t) = \sum_{k=1}^\infty \tilde{p}_k(t). \tag{8.9}$$

Noting that the rate of a jump to zero is mn when there are n factors it follows that the collection $\{\tilde{p}_n(t), n \in \mathbb{N}\}$ satisfies the system of ODE's:

$$\frac{d\tilde{p}_n(t)}{dt} = (n-1)s\,\tilde{p}_{n-1}(t) - n(s+m)\tilde{p}_n(t) \tag{8.10}$$

$$\tilde{p}_1(0) = 1.$$

To determine the $\{\tilde{p}_k(t), \ k \in \mathbb{N}\}$ we introduce the Laplace transform:

$$R(\theta, t) = \sum_{n=1}^\infty e^{n\theta} \tilde{p}_n(t). \tag{8.11}$$

Then using (8.10) it can be verified that $R(\cdot, \cdot)$ satisfies

$$R(\theta, 0) = e^\theta \tag{8.12}$$

$$\frac{dR(\theta, t)}{dt} = \sum_{n=1}^\infty [se^{n\theta}(n-1)\tilde{p}_{n-1}(t) - (s+m)\sum ne^{n\theta}\tilde{p}_n(t)]$$

$$= se^\theta \frac{\partial}{\partial \theta} R(\theta, t) - (s+m) \frac{\partial}{\partial \theta} R(\theta, t)$$

$$= (se^\theta - (s+m)) \frac{\partial}{\partial \theta} R(\theta, t).$$

8.2 A Warm-Up: The Case of a Single Site Model and Other Simplifications

The solution to this PDE is obtained by the method of characteristics as follows. In symbolic notation:

$$\frac{dt}{1} = \frac{d\theta}{-(se^\theta - (s+m))} = \frac{dR}{0}, \tag{8.13}$$

so that we obtain the characteristic curve

$$e^{(s+m)t}(s - (s+m)e^{-\theta}) = \text{constant}. \tag{8.14}$$

Hence the general solution is

$$R(\theta, t) = \Psi(e^{(s+m)t}(s - (s+m)e^{-\theta})), \tag{8.15}$$

where Ψ is an arbitrary function. Using the initial condition

$$e^\theta = \Psi(s - (s+m)e^{-\theta}) \tag{8.16}$$

we get that

$$\Psi(u) = \frac{s+m}{s-u}. \tag{8.17}$$

It follows that

$$R(\theta, t) = \frac{s+m}{s - e^{(s+m)t}(s - (s+m)e^{-\theta})}. \tag{8.18}$$

Therefore

$$1 - x_2(t) = \sum_{n=1}^{\infty} \tilde{p}_n(t) = R(0, t) = \frac{s+m}{s + me^{(s+m)t}}. \tag{8.19}$$

(iii) Large N asymptotic. We now consider the emergence of a rare mutant with m replaced by $\frac{m}{N}$ and consider the time scale $(C \log N) + t$, $t \in \mathbb{R}$. We get

$$1 - x_2(C \log N + t) = \frac{s + \frac{m}{N}}{s + \frac{m}{N} e^{(s+\frac{m}{N})(C \log N + t)}} = \frac{s + \frac{m}{N}}{s + \frac{m}{N} e^{(s+\frac{m}{N})t} N^{C(s+\frac{m}{N})}}, \tag{8.20}$$

so that $1 - x_2(C \log N + t) \to 1$ or 0 for all $t \in \mathbb{R}$ depending on $C < \frac{1}{s}$ or $C > \frac{1}{s}$. In other words the rare mutant emerges and takes over in time scale $s^{-1} \log N$ as $N \to \infty$. We shall derive more precise results later in Subsection 8.4 for example in (8.647). Finally, for $C = \frac{1}{s}$,

$$1-x_2(C\log N+t) = \frac{s+\frac{m}{N}}{s+me^{(s+\frac{m}{N})t}N^{\frac{m}{sN}}} \to \frac{s}{s+me^{st}}, \quad \text{as } N \to \infty. \quad \text{q.e.d.} \tag{8.21}$$

Proof of Lemma 8.2. We consider the dual process $(\eta_t^N, {}^{FK}\mathcal{F}_t^+)_{t \geq 0}$ and define

$$\tau_N = \inf\{t : {}^{FK}\mathcal{F}_t^+ \equiv 0\}. \tag{8.22}$$

For the deterministic case, $d = 0$, the random time τ_N was of order $s^{-1} \log N$, but this now changes to a much larger order of magnitude. This is made precise in the following Lemma that together with the dual identity completes the proof of Lemma 8.2.   q.e.d.

Lemma 8.3. *Let $d > 0$ and $c = 0$.*
If $\lim \frac{a_N}{N} \to 0$, then

$$P(\tau_N < a_N) \to 0 \text{ as } N \to \infty \text{ for all } t > 0. \qquad \square \tag{8.23}$$

Proof of Lemma 8.3. Start the dual $\mathcal{F}^+$ in $1_{\{1\}}$. In this case up to the time of the first mutation the number of factors $1_{\{1\}}$ in (8.7) is equal to or greater than 1. Therefore for each N, $P(\tau_N < \infty) = 1$.

This implies that for every *fixed* N

$$x_2^N(t) \to 1, \text{ a.s. as } t \to \infty. \tag{8.24}$$

On the other hand the number of factors is stochastically bounded due to the coalescence (which occurs at quadratic rate as soon as $d > 0$). To verify this we can calculate the expected number of factors and show that it remains bounded by obtaining an upper bound given by the solution of an ODE as follows.

Consider the birth and death process $\Pi_t^{(n)} = \Pi_t^{N,(n,1)}$ representing the number of factors of which $\mathcal{F}_t^+$ consists of at time t if we are starting at $\Pi_0^{(n)} = n$ at one site. The dynamics of this object follows immediately from the evolution of $(\mathcal{F}_t^+)_{t \geq 0}$. The forward Kolmogorov equations of the process $\Pi_t^{(n)}$ are therefore given by

$$p'_{ij}(t) = s(j-1)p_{i,j-1}(t) + \frac{d}{2}(j+1)jp_{i,j+1}(t) - (sj + \frac{d}{2}j(j-1))p_{ij}(t), \tag{8.25}$$

$$p_{n,n}(0) = 1. \tag{8.26}$$

Letting $E[\Pi_t^{(n)}]^k = m_k(t) = \sum_{j=0}^{\infty} j^k p_{n,j}(t)$, we obtain with $\hat{s} = s + \frac{d}{2}$:

$$\frac{dm_1(t)}{dt} = \hat{s}m_1(t) - \frac{d}{2}m_2(t) \leq \hat{s}m_1(t) - \frac{d}{2}m_1^2(t), \quad m_1(0) = n. \tag{8.27}$$

8.3 Proofs of Propositions 2.3, 2.6, and 2.11   173

Therefore

$$E[\Pi_t^{(n)}] \leq \frac{2\hat{s}ne^{\hat{s}t}}{nd(e^{\hat{s}t}-1)+2\hat{s}}, \quad (8.28)$$

so that

$$E[\int_0^t \Pi_u^{(n)} du] \leq \frac{2}{d}\log(\frac{2\hat{s}}{d}+n(e^{\hat{s}t}-1)). \quad (8.29)$$

This implies that if we let $1 > \varepsilon > 0$ and choose α so that $1 - e^{-\alpha} = \varepsilon$, then using (8.29) we get:

$$P[\tau_N < a_N] \leq \varepsilon + P[\frac{m}{N}\int_0^{a_N}\Pi_s^{(n)}ds > \alpha] \leq \varepsilon + \frac{2m}{d\cdot\alpha}\frac{\text{const}+[\log n + \hat{s}a_N]}{N} \quad (8.30)$$

Hence $\lim_{N\to\infty} P(\tau_N < a_N) \leq \varepsilon$ for arbitrary $\varepsilon > 0$ and the proof is complete. q.e.d.

8.3 Proofs of Propositions 2.3, 2.6, and 2.11

In this section we consider the large time scale $\alpha^{-1}\log N + t$, $t \in \mathbb{R}$ and prove Proposition 2.3 on properties of the limit dynamics in this *emergence and fixation time scale* and the convergence results stated in Subsubsection 2.3.3, in particular, on the time needed for emergence and the form of the process of fixation, i.e. Propositions 2.6, 2.11. (Recall the remaining assertions requiring analysis in different time scales, in particular Propositions 2.8 and 2.10 on droplet formation are proved later in Subsection 8.4 and Proposition 2.13 on tagged sample convergence which uses different methods will be proved in Subsection 8.7.)

8.3.1 Outline of the Strategy of Proofs for the Asymptotic Analysis

The proofs for the large time scale proceeds in altogether thirteen subsubsections. Therefore we describe here in 8.3.1 first the general strategy and list at the end what happens in the various subsubsections.

The proofs of the results on the $N \to \infty$ asymptotics for the process in large time scales involve calculating the mean of $x_1^N(t)$ or $x_2^N(t)$, respectively the higher moments, and hence depend in an essential way on an asymptotic analysis of the dual process $(\eta_t, \mathcal{F}_t^+)$ which we introduced in Section 5. Recall the formula

(5.39) giving moments in terms of the dual and recall Subsections 5.1–5.3 for the ingredients of the dual.

In the case of two types the required dual calculation involves a system of particles (corresponding to factors) located at sites in $\{1, \ldots, N\}$ with linear birth rate and quadratic death rate at each site and migration between sites. Assuming that the initial population consists of only one individual and the initial function f is the indicator of type 1 at site 1, then we get a growing string of factors of such indicators, until the rare mutation operator (at rate $\frac{m}{N}$) acts on it at which time it and therefore the product becomes zero. Hence the main point is to prove that a mutation event somewhere in the string will occur in the $\log N$ time scale. For this to be the case we have to prove that the total number of factors reaches $O(N)$ in that time scale.

Consider now the dual particle system η starting with only finitely many particles. As long as the number of occupied sites remains $o(N)$ the migration of individuals leads in the limit $N \to \infty$ always to an unoccupied site. More precisely, in this regime the proportion of sites at which a collision occurs, that is a migration to a previously occupied site, is negligible. For that reason the collection of *occupied sites* can be viewed (asymptotically as $N \to \infty$) as a *Crump–Mode–Jagers branching process* (a notion we shall recall later on). This process grows exponentially and the identification of the emergence time corresponds to the Malthusian parameter α of this CMJ-branching process.

In order to investigate the dynamics in the next phase between emergence and fixation, i.e. in times $C \log N + t$ with $t \in \mathbb{R}$ the new time parameter, we must on the side of the dual particle system now consider the role of *collisions* of dual particles once the number of occupied sites is of order $O(N)$ and the proportion of occupied sites at which collisions occur is asymptotically non-zero. This can be handled by identifying a dynamical law of large numbers described by a *nonlinear evolution equation*. Nonlinear due to the immigration term at a site, resulting from immigration from other sites and which in the limit depends on their law. On the side of the McKean–Vlasov limit, this makes the evolution of the frequency of type 2 at single site depend on a global signal, namely the current spatial intensity of type 2.

The delicate issue of the *transition* between these two regimes with and without collision must be addressed to fully describe the emergence and to show that this leads to the McKean–Vlasov dynamics with random initial condition. For that purpose the system is considered in the time scale $\alpha^{-1} \log N + t$, with $t \in \mathbb{R}$ the macroscopic time parameter. The key element in the analysis is the determination of the leading term in an asymptotic expansion of the type-2 mass in time scale $\alpha^{-1} \log N + t$ with t varying, once we have taken $N \to \infty$, as $t \to -\infty$. On the side of the dual process this will correspond to an expansion of the normalised number of occupied sites as $N \to \infty$ and then $t \to \infty$ in the above time scale. However in the discussion on droplet formation later on in subsection 8.4 it is also necessary to keep track of the collisions since we need higher-order terms in the expansion of the total type-2 mass and these terms involve collisions of dual particles. This will be dealt with in Subsubsection 8.4.5 and the sequel.

8.3 Proofs of Propositions 2.3, 2.6, and 2.11

To carry out this program we must study the behaviour of the dual particle system in two time regimes, namely,

$$s(N) + t \text{ where } s(N) \to \infty \text{ but } s(N) = o(\log N) \tag{8.31}$$

and for suitably chosen α,

$$\frac{\log N}{\alpha} + t, \quad -\infty < t < \infty, \tag{8.32}$$

and their asymptotics as $N \to \infty$. The first we call the *collision-free regime* or *pre-emergence regime*, the second the *collision regime*.

A key step in the dual process analysis is a coupling of the pre-emergence and collision regimes in order to understand the *transition* between the two regimes and to be able to consider the time scale

$$\frac{1}{\alpha} \log N + t_N \text{ with } t_N \to -\infty, \quad t_N = o(\log N). \tag{8.33}$$

Outline of Subsection 8.3.

This subsection focusses on the proofs of Proposition 2.3 (Subsubsection 8.3.1) on the limiting McKean–Vlasov dynamics, Proposition 2.6 (Subsubsections 8.3.1–8.3.6) and Proposition 2.11 on emergence respectively convergence to the limiting dynamics of fixation (Subsubsections 8.3.7–8.3.13).

The Subsubsection 8.3.3 is devoted to the proof of Proposition 2.3, which specifies the limiting laws in short and large time scales.

In Subsubsections 8.3.4–8.3.6 we begin the proof of Proposition 2.6 on the order of magnitude of the emergence time. In a first step we exhibit the branching structure for the number of sites in 8.3.3 and then analyse the collision free, respectively the collision regime, 8.3.4 to 8.3.6.

The proof of Proposition 2.11 on the limit dynamics is more involved and is broken in six preparatory steps which are entirely concerned with the dual process at times $\frac{1}{\alpha} \log N + t$ and a final argument returning to our process, each of these is in one of the separate Subsubsections 8.3.7–8.3.13. Namely, we study in Subsubsections 8.3.7, 8.3.8, 8.3.9 and 8.3.10 the structure of the dual process more carefully, show in 8.3.11 convergence of the process to the limiting dynamics and examine in Subsubsection 8.3.12 the random growth constant in the emergence. Then we are able in Subsubsection 8.3.13 to complete the proof of Proposition 2.11 on the limiting dynamics of fixation.

8.3.2 Proof of Proposition 2.3 (Properties McKean–Vlasov dynamic)

In order to prove the results on the McKean–Vlasov dynamic recall the duality in Subsubsections 5.2–5.6. We now go through the different parts of the proposition to prove, but here the hard part is assertions d and e.

(a) This is the wellknown McKean–Vlasov limit for the interacting Fleming–Viot model with migration and selection (compare, for example, [DG99], Theorem 9).

(b) We want to show that $E[x_1(t)] \to 0$ as $t \to \infty$ if $E[x_1(0)] < 1$. This follows by a simple application of the mean-field dual $(\tilde{\eta}_t, \mathcal{F}_t^+)_{t \geq 0}$ started with $\mathcal{F}_0^+ = 1_{\{1\}}$ and η_0 having one particle located at site 0. As time proceeds in the dual process the selection operator keeps being applied. But the ℓ-fold application of the selection operator, which now in the two-type case is simply obtained as $f \to f \otimes 1_{\{1\}}$, produces as surviving terms for $\mathcal{F}_t^+$ a string of the form (note coalescence only changes here the number of factors belonging to distinct variables)

$$\Pi_{i=1}^{\ell(t)} 1_{\{1\}}(u_{\xi_i(t)}) \tag{8.34}$$

and since the total number of factors diverges as soon as the migration rate c is positive, i.e. $\ell(t) \to \infty$, the result follows from the assumption $E[x_1(0)] < 1$.

(c) For existence it suffices to obtain a particular solution $(\mathcal{L}_t^*(2))_{t \in \mathbb{R}}$ where $\mathcal{L}_0^*(2) \in \mathcal{P}([0,1])$. Take the solution $(\tilde{\mathcal{L}}_t^n(2))_{t \geq 0}$ with $\tilde{\mathcal{L}}_0^n(2) = \mu_n$ with

$$a_n := \int_0^1 x \mu_n(dx), a_n \to 0 \text{ as } n \to \infty. \tag{8.35}$$

In particular we can take here $\mu_n = \delta_{a_n}$. Choose now $(a_n)_{n \in \mathbb{N}}$ with $0 < a_n < \frac{1}{2}$ and $a_n \to 0$ as $n \to \infty$. Then first of all for every fixed n

$$\lim_{t \to \infty} \tilde{\mathcal{L}}_t^n(2) = \delta_1 \tag{8.36}$$

and there exists $r(a_n)$ such that $\int_0^1 x \tilde{\mathcal{L}}_{r(a_n)}^n(2) dx = \frac{1}{2}$. In fact by duality we see that $r(a)$ is strictly increasing and continuous in a.

Then define

$$(\mathcal{L}_t^n(2))_{t \geq -r(a_n)} = (\tilde{\mathcal{L}}_{t+r(a_n)}^n(2))_{t \geq -r(a_n)}. \tag{8.37}$$

Since for any $t > 0$,

$$\lim_{n \to \infty} \tilde{\mathcal{L}}_t^n(2) = \delta_0, \tag{8.38}$$

it follows again by duality that $\lim_{n \to \infty} r(a_n) = \infty$. The value at $t = -r(a_n)$ tends to δ_0 as $n \to \infty$.

We note that if for a subsequence $(n_k)_{k \in \mathbb{N}}$ the sequence $(\mathcal{L}_s^{n_k})_{k \in \mathbb{N}}$ converges as $k \to \infty$ for some $s \in \mathbb{R}$, then it converges for all times $t \geq s$, as we immediately read off from the duality which gives the Feller property (continuity in the initial state) of the McKean–Vlasov process. Therefore by considering for example $s = -k, k \in \mathbb{N}$ we can obtain a further subsequence converging for all times $t \in \mathbb{R}$.

8.3 Proofs of Propositions 2.3, 2.6, and 2.11

Now we take such a convergent subsequence of $(\mathcal{L}^n(2))_{n \in \mathbb{N}}$ and define $\mathcal{L}^*(2)$ as the $(n \to \infty)$ limit. By construction $\int x \tilde{\mathcal{L}}_0^n(2)(dx) = \frac{1}{2}$ and therefore $\mathcal{L}_t^*(2) \to \delta_1$ as $t \to \infty$. It remains therefore to argue that $\mathcal{L}^*(2)$ is a solution and is converging to δ_0 as $t \to -\infty$.

The fact that we obtain a solution of the equation is a consequence of the construction together with the continuity in the initial state which is immediate using the duality. Namely for every n we have a solution for $t \geq -\tau(a_n)$. Then using duality we see that the limit is a solution as well, since it satisfies the duality relation between times $s < t; s, t \in \mathbb{R}$.

The convergence to δ_0 as $t \to -\infty$ follows by contradiction as follows. Assume we converge to a strictly positive mean. Then applying the duality to calculate the mean at time 0 starting at times $t \downarrow -\infty$ we see, that this mean converges to one and therefore we obtain convergence to δ_1 (by (8.36)) at time 0. This contradicts the assumption that the mean is $\frac{1}{2}$ at time 0. This completes the proof of existence of a solution on $(-\infty, \infty)$ satisfying (2.26).

(d, e) Here we proceed in three steps, first we establish existence of solutions whose mean curve has the claimed growth behaviour at $-\infty$ and then we show uniqueness of the mean curve with this growth behaviour and subsequently the uniqueness of the solution for given mean curve.

Step 1 *Growth of mean-curves*

The first step is to consider the growth of solutions of the McKean–Vlasov equation, $\mathcal{L}_t(2)$, at $-\infty$. We will show two things,

- first there exists a solution with mean curve $m(\cdot)$ which behaves like $m(t) \sim Ae^{\alpha t}$ for $t \to -\infty$ with $A \in (0, \infty)$,
- secondly that all solutions satisfying the growth condition posed in (d) have mean curves satisfying $m(t) \sim Ae^{\alpha t}$ as $t \to -\infty$ for some $A \in [0, \infty)$.

In fact we shall combine these two points in our proof.

First assume that $(\mathcal{L}_t(2))_{t \in \mathbb{R}}$ is a solution (and entrance law) with mean-curve $(m(t))_{t \in \mathbb{R}}$ which satisfies:

$$m(t) > 0, \qquad 0 < \liminf_{t \to -\infty} e^{-\alpha t} m(t) < \infty. \tag{8.39}$$

We want to show that $e^{-\alpha t} m(t)$ converges to some $A \in \mathbb{R}^+$.

We can chose a sequence t_n with $t_n \to -\infty$ so that for some $A \in (0, \infty)$ we have

$$e^{\alpha|t_n|} m(t_n) = e^{\alpha|t_n|} \int_0^1 x \mathcal{L}_{t_n}(2)(dx) \to A \text{ as } n \to \infty. \tag{8.40}$$

We now abbreviate

$$\mu_{t_n} := \mathcal{L}_{t_n}(2). \tag{8.41}$$

What we know about μ_{t_n} is its mean and from this we have to draw all conclusions on the convergence, namely that the limit A does not depend on the choice of $(t_n)_{n\in\mathbb{N}}$.

On the other hand we can start with such a measure μ_{t_n} as in (8.41) and try to construct a solution with a mean-curve $(m(t))_{t\in\mathbb{R}}$, $m(t) = (A + o(1))e^{\alpha t}$ as $t \to -\infty$ as the limit point of a sequence of solutions starting at time t_n in μ_{t_n}. A solution starting in μ_{t_n} at time t_n is denoted:

$$(\tilde{\mathcal{L}}_t^n)_{t\geq t_n}. \tag{8.42}$$

Then we know, that if $(\mathcal{L}_t(2))_{t\in\mathbb{R}}$ is an entrance law with $\mathcal{L}_{t_n} = \mu_{t_n}$, then by uniqueness of the solution starting at t_n we must have

$$\mathcal{L}_t(2) = \tilde{\mathcal{L}}_t^n(2), \text{ for } t \geq t_n. \tag{8.43}$$

Therefore our strategy now is twofold. First we verify that in fact the mean curve $(m(t))_{t\in\mathbb{R}}$ of $(\mathcal{L}_t(2))_{t\in\mathbb{R}}$ has the form:

$$m(t) = \exp(\alpha(t + \frac{1}{\alpha}\log A)) + o(e^{\alpha t}). \tag{8.44}$$

Then we argue that in fact the $(\tilde{\mathcal{L}}_t^n)_{t\geq t_n}$ with mean curves satisfying (8.40) converge to an entrance law with mean curve satisfying (8.44). We begin with the first point and come to the second one (which is then simpler by what we have prepared for the first point) after (8.89).

In order to carry this program out we must consider the different possibilities we have for $\mu_{t_n} \in \mathcal{P}([0,1])$ with mean curve $m(t_n) \sim \frac{A}{e(n)}$ where we use the abbreviation

$$e(n) = e^{\alpha|t_n|}. \tag{8.45}$$

This is somewhat complicated by the possible interplay of probabilities and sizes in producing a given mean.

We begin by considering two special cases that illustrate the main idea. Let

$$\mu_{t_n} = (1 - \frac{p_1}{e(n)})\delta_{\frac{a_1}{e(n)}} + \frac{p_1}{e(n)}\delta_{a_2}, \tag{8.46}$$

so that $A = a_1 + p_1 a_2$. We can distinguish now different cases. The first case arises in our example if (small probability for an order one value a_2)

$$p_1 = 0, a_1 > 0, \tag{8.47}$$

in which case the mean curve satisfies

$$e(n) \int x \mu_{t_n} dx \longrightarrow a_1, \tag{8.48}$$

8.3 Proofs of Propositions 2.3, 2.6, and 2.11

the second case arises if (large probability for very small values)

$$p_1 = 1, \ a_1 = 0, a_2 > 0, \tag{8.49}$$

in which case the mean-curve satisfies

$$e(n) \int x \mu_{t_n} dx \longrightarrow \text{const} \tag{8.50}$$

given in (8.62).

We shall now consider these two cases each with a distinct flavour which then allows us, since they catch the key features, to see how we can treat the *general* case, which we call case 3 and for which we need to use an additional trick.

A key tool in all cases is the duality relation from Subsection 5.6 and the fact that the dual process consists of a collection of birth and death processes acting independently so that with generating functions we can analyse the dual expectation. The number of such birth and death processes we have to consider grows with time due to the emigration of dual particles from sites occupied with more than one particle. We study this behaviour and the composition of the collection in Subsubsections 8.3.3 and 8.3.4 in detail.

Case 1

Here we first treat the deterministic case $\mu_{t_n} = \delta_{\frac{a_1}{e(n)}}$. We will use the dual representation for the McKean–Vlasov limit (see Subsection 5.6) together with (8.144) (giving below $W_n \Rightarrow W$) to compute the expected value of x_2 mass at time t, i.e. the mean of $\mathcal{L}_t(2)$. Note that then we consider the time stretch $|t_n| + t$ from the time point t_n where we have information about the state.

Observe that we know that the dual process at time T is a collection of K_T many independent birth and death processes $(\zeta(t))_{t \geq 0}$, which have evolved for a time T. We shall show in Subsections 8.3.3 and 8.3.4 that in the large time limit $T \to \infty$ they satisfy

$$K_T \sim W e^{\alpha T} \text{ a.s. as } T \to \infty \tag{8.51}$$

and the *stable age distribution*, i.e. the empirical distribution of the states of the collection enriched by the age (time back till the site became colonised) converges as $T \to \infty$ to $\mathcal{U}(\infty, \cdot)$. Therefore the state of the empirical measure of the birth and death processes at times $T + t$ is given by the stable size distribution, which arises as

$$\int \mathcal{U}(\infty, ds) v_s, \tag{8.52}$$

where v_s is the law of the state of a single birth and death process at time s starting with one particle at an occupied site.

We now fix some time $t > t_n$ and suppress the t-dependence in the notation. We need the generating function

$$G_n(z) = E[z^{\zeta(t-t_n,i)}] \tag{8.53}$$

of the state of the birth and death process at a randomly chosen site i. To calculate we introduce the generating function

$$G(z) = E[z^\zeta], \quad 0 < z \leq 1, \tag{8.54}$$

where ζ represents the number of dual factors at a dual site under the stable size distribution and $(\zeta(i), i \in \mathbb{N})$ is an i.i.d. collection of such variables. We know that

$$G_n \longrightarrow G \quad \text{as } n \to \infty, \tag{8.55}$$

$$G'_n \longrightarrow G' \quad \text{on the unit circle as } n \to \infty. \tag{8.56}$$

We observe first that we can calculate $\int_0^1 (1-x)\tilde{\mathcal{L}}_t^n(dx)$ using the duality for the McKean–Vlasov process from Subsection 5.6. Then noting that $G'(1) = B = \frac{\alpha+\gamma}{c}$ and using (8.51) (see also (8.166)):

$$\int_0^1 (1-x)\tilde{\mathcal{L}}_t^n(dx) = E\left[\prod_{i=1}^{K_{t-t_n}} (1-\frac{a_1}{e(n)})^{\zeta(t-t_n,i)}\right] = E\left[\prod_{i=1}^{K_{t-t_n}} G_n(1-\frac{a_1}{e(n)})\right] \tag{8.57}$$

$$= E\left[\left(G(1-\frac{a_1}{e(n)})\right)^{W_n e(n) e^{\alpha t}(1+o(1))}\right], \text{ as } n \to \infty$$

$$\xrightarrow[n\to\infty]{} E[e^{-Wa_1 B e^{\alpha t}}]$$

$$\sim 1 - E[W]a_1 B e^{\alpha t} + O(e^{2\alpha t}), \text{ as } t \to -\infty.$$

From the first to the second line we have approximated G_n by G and used at the same time the convergence to the stable age distribution. To go from the second to the third line we use that $W_n \longrightarrow W$ as $n \to \infty$.

Therefore as $t \to -\infty$:

$$m(t) \sim [E[W]a_1 B]e^{\alpha t} + O(e^{2\alpha t}). \tag{8.58}$$

Case 2

In case 2 the number of dual sites occupied in the dual process which have the value a_2 at time t_n, is denoted $Z_n(t, a_2)$, and is $\text{Bin}(W_n e(n) e^{\alpha t}, \frac{1}{e(n)})$ distributed. Hence we get

8.3 Proofs of Propositions 2.3, 2.6, and 2.11

$$\int_0^1 (1-x)\tilde{\mathcal{L}}_t^n(dx) = E[\prod_{i=1,\ldots,K_{t-t_n}} (1-a_2)^{\zeta(t-t_n,i)}]. \tag{8.59}$$

The r.h.s. of (8.59) is approximated as above as in case 1 (see (8.57)) for $n \to \infty$ by

$$E\left[(G(1-a_2))^{Z_n(t,a_2)}\right]. \tag{8.60}$$

Then $Z_n(t, a_2)$ converges as $n \to \infty$ in distribution to a Poisson distribution with parameter $We^{\alpha t}$.

For the calculation it is convenient to *condition on* W and read expectations as expectations for given W and write $\tilde{E}$ for this conditional expectation suppressing the condition in the notation. We conclude that as $n \to \infty$ and then $t \to -\infty$:

$$\tilde{E}[\int_0^1 (1-x)\tilde{\mathcal{L}}_t^n(dx)] \longrightarrow e^{[G(1-a_2)-1]We^{\alpha t}}, \quad \text{as } n \to \infty \tag{8.61}$$

$$\sim 1 - [(1-G(1-a_2))]We^{\alpha t} + O(e^{2\alpha t}), \quad \text{as } t \to -\infty.$$

Hence as $n \to \infty$ and then for $t \to -\infty$ we get:

$$E[\int_0^1 x\tilde{\mathcal{L}}_t^n(dx)] \sim [(1-G(1-a_2))]E(W)]e^{\alpha t} + O(e^{2\alpha t}). \tag{8.62}$$

In other words also in this case $m(t) = Const \cdot e^{\alpha t} + O(e^{2\alpha t})$.

Case 3

We now turn to the general case. The key idea is to focus on the components where we do have type-2 mass even as $t_n \to -\infty$, which is achieved using the Palm measure. Consider the family $\{\hat{\mu}_{t_n}\}_{n \in \mathbb{N}}$ of Palm measures on $[0, 1]$:

$$\hat{\mu}_{t_n} := \frac{x\mu_{t_n}(dx)}{\int_0^1 x\mu_{t_n}(dx)} \tag{8.63}$$

and denote the normalizing constant again as in cases 1, 2 by

$$e(n) = (\int_0^1 x\mu_{t_n}(dx))^{-1}. \tag{8.64}$$

Since the family $\{\hat{\mu}_{t_n}\}_{n \in \mathbb{N}}$ is tight (by construction the measures are supported on $[0, 1]$) we can find a subsequence and this sequence we denote again by $(t_n)_{n \in \mathbb{N}}$ such that $t_n \to -\infty$ and $\hat{\mu}_{t_n}$ converges. This means we can find a measure $\hat{\mu}$ on $[0, 1]$ (with $\hat{\mu}(\{0\}) = 0$) such that

$$\frac{x\mu_{t_n}(dx)}{\int_0^1 x\mu_{t_n}(dx)} \Longrightarrow a\delta_0 + \hat{\mu}, \quad \text{as } n \to \infty. \tag{8.65}$$

Moreover there exists a scale s_n, with $s_n > 0$ and $s_n \to 0$ such that

$$\frac{\int_0^{s_n} x\mu_{t_n}(dx)}{\int_0^1 x\mu_{t_n}(dx)} \to a. \tag{8.66}$$

We shall study the contributions due to the part $a\delta_0$ and due to $\hat{\mu}$ first separately in (i) and (ii) below and then join things to get the total effect in (iii).

(i) We begin with the effect from $a\delta_0$. For this purpose consider the measures $(\nu_n)_{n \in \mathbb{N}}$ on $[0, 1]$ defined by setting for every interval (c, d), with $0 \leq c < d \leq 1$:

$$\nu_n((c, d)) = \frac{\int_{ds_n}^{cs_n} x\mu_{t_n}(dx)}{\int_0^{s_n} x\mu_{t_n}(dx)}. \tag{8.67}$$

This allows us to consider the contribution due to the terms leading to $a\delta_0$ in the expectation in the duality relation. Define

$$(\tilde{\nu}_n)_{n \in \mathbb{N}} \tag{8.68}$$

as the empirical measure from a sample of size $e^{W_n \alpha(t-t_n)}$ sampled from ν_n. Now we consider the state of the dual process at time $t - t_n$ and acting on the state at time t_n but considering only the contributions from sites with values in the interval s_n, which means that we can represent the state at time by an empirical measure as follows. As $n \to \infty$ we have:

$$\int_0^1 (1-x) \tilde{\mathcal{L}}_t^n(dx) \sim E\left[\exp\left(W_n e^{\alpha t} \int_0^1 \log G(1 - ys_n) \tilde{\nu}_n(dy)\right)\right]. \tag{8.69}$$

By (8.127) $G''(1) < \infty$ and therefore the error term in (8.69) is $O(s_n)$. Then taking the limit as $n \to \infty$ we get using relation (8.66) that:

$$\begin{aligned} E\left[\exp\left(W_n e^{\alpha t} \int_0^1 \log G(1 - ys_n) \tilde{\nu}_n(dy)\right)\right] & \\ \to E\left[\exp\left(-G'(1) W e^{\alpha t} a\right)\right] & \quad \text{as } n \to \infty \\ \sim 1 - G'(1) E[W] e^{\alpha t} a & \quad \text{as } t \to -\infty. \end{aligned} \tag{8.70}$$

Hence in this case we get if there are only contributions from $a\delta_0$, i.e. $\hat{\mu} \equiv 0$, that:

$$m(t) = (G'(1) E[W] a) e^{\alpha t} + O(e^{2\alpha t}), \tag{8.71}$$

i.e. $A = G'(1) E[W] a$.

8.3 Proofs of Propositions 2.3, 2.6, and 2.11

(ii) Next we consider the contribution if they are only coming from $\hat{\mu}$. We have to "return" here from the Palm $\hat{\mu}$ to a measure μ which need not be a probability measure since x^{-1} diverges at $x = 0$. Note that for any $\varepsilon > 0$ the set of finite measures given by

$$\{e(n)(1_{x>\varepsilon})\mu_{t_n}(dx)\}_{n \in \mathbb{N}}, \tag{8.72}$$

is tight in the space of finite measures on $\mathcal{M}([0, 1])$ and we can choose a convergent subsequence. We can then obtain a convergent sequence, we call again $(t_n)_{n \in \mathbb{N}}$ such that:

$$e(n)(1_{x>\varepsilon})\mu_{t_n}(dx) \Rightarrow \mu, \text{ as } n \to \infty \tag{8.73}$$

where μ is a measure on $(0, 1]$ with

$$\int_0^1 x\mu(dx) < \infty, \text{ i.e. in particular } \mu([\varepsilon, 1]) < \infty \text{ for all } \varepsilon > 0. \tag{8.74}$$

We denote

$$\mu_\varepsilon = \text{ restriction of } \mu \text{ to } [\varepsilon, 1]. \tag{8.75}$$

By the dual representation we then have under the assumption

$$\mu_\varepsilon \text{ is discrete with atoms at } x_1, x_2, \cdots, x_M, \tag{8.76}$$

that with $x_{g_n(i)}$ being the value at site i at time t_n the representation:

$$\int_0^1 (1-x)\tilde{\mathcal{L}}^n(dx) = E[\prod_{i=1,\ldots,K_{t-t_n}} (1 - x_{g_n(i)})^{\zeta(t-t_n,i)}]. \tag{8.77}$$

Denote by $Z_n(t, x_j)$ the number of dual sites occupied in the dual process which have at time t_n the value x_j. The r.h.s. of (8.77) we approximate again for $n \to \infty$ by

$$E\left[\prod_j G(1-x_j)^{Z_n(t,x_j)}\right]. \tag{8.78}$$

We know

$$\mathcal{L}[(Z_n(t, x_j))_{j=1,\cdots,M}] = \text{Mult}\left(W_n e(n)e^{\alpha t}, \left(\frac{\mu(x_j)}{e(n)}\right)_{j=1,\cdots,M}\right). \tag{8.79}$$

The assumption (8.76) is removed as follows. For a general μ_ε, we consider the random measure

$$Z_n(t,dx) \quad \text{(the number of dual sites at which } x_2 \in dx) \tag{8.80}$$

which converges as $n \to \infty$ in distribution to an *inhomogeneous Poisson random measure* with the intensity measure

$$\mu(dx)We^{\alpha t}. \tag{8.81}$$

For the calculation it is convenient to condition on W and read expectations as expectations for given W and write $\tilde{E}$ for this object. We conclude that as $n \to \infty$ and then $t \to -\infty$ with the same reasoning as in the other cases:

$$\tilde{E}[\int_0^1 (1-x)\tilde{\mathcal{L}}_t^n(dx)] \to e^{\int_0^1 (G(1-x)-1)\mu(dx)We^{\alpha t}}, \quad \text{as } n \to \infty \tag{8.82}$$

$$\sim 1 - [\int (1-G(1-x))\mu(dx)]We^{\alpha t} + O(e^{2\alpha t}),$$

as $t \to -\infty$.

Hence taking first $n \to \infty$ we then get for $t \to -\infty$ the expansion:

$$E[\int_0^1 x\tilde{\mathcal{L}}_t^n(dx)] \sim [\int_0^1 (1-G(1-x))\mu(dx)]E[W]e^{\alpha t} + O(e^{2\alpha t}), \tag{8.83}$$

which then implies that the mean curve satisfies if $a = 0$, that

$$m(t) = Ae^{\alpha t} + O(e^{2\alpha t}), \tag{8.84}$$

with

$$A = E[W]\int_0^1 (1-G(1-x))\mu(dx). \tag{8.85}$$

(iii) Finally we have to use the results in the two different cases where $\hat{\mu} \equiv 0$ respectively $a = 0$, to get the answer in full generality. Combining the expressions for the two contributions we obtain as $n \to \infty$ and then considering the expansion as $t \to -\infty$ that

$$\int_0^1 (1-x)\tilde{\mathcal{L}}_t^n(dx) \xrightarrow[n \to \infty]{} E\left[e^{\int(G(1-x)-1)\mu(dx)We^{\alpha t}} e^{-WG'(1)e^{\alpha t}a}\right] + O(e^{2\alpha t}), \tag{8.86}$$

and as a consequence for $t \to -\infty$ the r.h.s. is asymptotically given by

$$\sim \left([\int_0^1 [G(1-x)-1]\mu(dx) + aG'(1)]E[W]\right)e^{\alpha t} + O(e^{2\alpha t}) = A \cdot e^{\alpha t} + O(e^{2\alpha t}) \tag{8.87}$$

8.3 Proofs of Propositions 2.3, 2.6, and 2.11

with

$$A = \left(\int_0^1 [G(1-x) - 1]\mu(dx) + aG'(1) \right) E[W]. \tag{8.88}$$

Finally we have to exclude the possibility that the choice of different subsequences in (8.40), (8.65), (8.73) (and hence of different values for A) would give a different mean curve. However this would lead to a contradiction since this would give two values to the same quantities. Therefore the resulting mean curve is independent of the choice of subsequence $(t_n)_{n \in \mathbb{N}}$ of starting times.

This means that we have proved that an entrance law with at most exponentially growing mean has the property that the mean curve has asymptotically as $t \to -\infty$ the form

$$m(t) = A \cdot e^{\alpha t} + O(e^{2\alpha t}), \tag{8.89}$$

with A given by the formula (8.88) which completes the first of the two points we specified around (8.44).

We next need to argue that in fact such an entrance law exists by showing that actually $\tilde{\mathcal{L}}^n$ (from (8.42)) converges along a subsequence. Due to the continuity of $\mathcal{L}_t$ in the initial state at time t_0, the tightness is straightforward. We have to show that first of all the limit is a solution and secondly the mean curve has the desired asymptotics. To show that we obtain in the limit a solution we argue as in the proof of part (c) of the proposition. It remains to verify the asymptotics of the mean curve.

For this purpose we use the same line of argument as above for the first moment of $x_2(t)$ to show that also all higher moments of $x_2(t)$ converge as $n \to \infty$ for every fixed t, so that we have convergence of the processes using the Feller property. For that we observe that we then simply have to start the dual process with k-initial particles and then we carry out the same calculations, since the CMJ-theory works also for the CMJ-process started with a different internal state of the starting site. We omit the straightforward modification.

Step 2 *Uniqueness of solution with the $t \to -\infty$ asymptotics*

We next show that if the mean curves of two entrance laws differ at most by $o(e^{\alpha t})$ as $t \to -\infty$, then they are identical. This then allows to conclude that if we prescribe the value A then there is exactly one entrance law with this asymptotics as $t \to -\infty$.

For that purpose we consider two solutions $(\mathcal{L}_t^1)_{t \in \mathbb{R}}, (\mathcal{L}_t^2)_{t \in \mathbb{R}}$ such that the mean curves $(m^\ell(t))_{t \in \mathbb{R}}$ satisfy for $\ell = 1, 2$ that $m^\ell(t) \sim A \exp(\alpha t)$ as $t \to -\infty$. We consider a sequence $(t_n)_{m \in \mathbb{N}}$ with $t_n \to -\infty$ as $n \to \infty$ and we fix a time point $t \in \mathbb{R}$. Then we define the two processes we apply the duality relation at time t

$$(\tilde{\mathcal{L}}_s^{n,1})_{s \geq t_n} \quad , \quad (\tilde{\mathcal{L}}_s^{n,2})_{s \geq t_n}, \tag{8.90}$$

which have at time t_n the initial distributions given by

$$\mathcal{L}^1_{t_n} \text{ resp. } \mathcal{L}^2_{t_n}. \tag{8.91}$$

We apply the duality relation at time t. Our goal is to show that the mean curve satisfies

$$m^1(t) = m^2(t), \quad \forall\, t \in \mathbb{R}, \tag{8.92}$$

by proving that for every $t \in \mathbb{R}$

$$\tilde{m}^{n,1}(t) = \tilde{m}^{n,2}(t) + o(1) \text{ as } n \to \infty. \tag{8.93}$$

We first verify this in Case 1. Let $\tilde{\mathcal{L}}^{n,1}$, $\tilde{\mathcal{L}}^{n,2}$ be two solutions as in case 1 but such that $m^1(t_n) - m^2(t_n) = o(\frac{1}{e(n)})$. Then by (8.57),

$$\left| \int_0^1 (1-x) [\tilde{\mathcal{L}}^{n,1}_t(dx) - \tilde{\mathcal{L}}^{n,2}_t(dx)] \right| \tag{8.94}$$

$$\leq E\left[\left| \prod_{i=1}^{K_{t-t_n}} (1 - \frac{a_1}{e(n)} + o(\frac{1}{e(n)}))^{\zeta(t-t_n,i)} - \prod_{i=1}^{K_{t-t_n}} (1 - \frac{a_1}{e(n)})^{\zeta(t-t_n,i)} \right| \right]$$

$$= E\left[\left| \prod_{i=1}^{K_{t-t_n}} G_n(1 - \frac{a_1}{e(n)} + o(\frac{1}{e(n)})) - \prod_{i=1}^{K_{t-t_n}} G_n(1 - \frac{a_1}{e(n)}) \right| \right]$$

$$\leq E[G'(1) W_n e(n) e^{\alpha t} o(\frac{1}{e(n)})] \to 0 \text{ as } n \to \infty,$$

where we have first used that we can split the expectation in the duality relation into one over the initial state and then over the dual dynamic, furthermore for $z \in [0,1]$, $G'_n(z) \leq G'_n(1) \leq G'(1)$ (the latter by coupling). Hence we have $m^1 = m^2$ in this case.

In case 2 there is $o\left(\frac{1}{e(n)}\right)$ change in the mean curve resulting from the change of a_2 to $a_2 + o\left(\frac{1}{e(n)}\right)$. Moreover given a change of the frequency $\frac{p_1}{e(n)}$ to $\frac{p_1}{e(n)} + o\left(\frac{1}{e(n)}\right)$ we have $Z_n(t, a_2)$ is $\text{Bin}\left(W_n e(n) e^{\alpha t}, \frac{1}{e(n)} + o\left(\frac{1}{e(n)}\right)\right)$ and the limiting Poisson distribution is not changed by the $o\left(\frac{1}{e(n)}\right)$ perturbation.

Again, the general case follows by a combination of the arguments of cases 1 and 2 exactly as in case 3 in Step 1.

Step 3 *Uniqueness given the mean curve*

To complete the proof for uniqueness of the entrance law with specified behaviour as $t \to -\infty$, it remains to show that for a given mean curve $(m(t))_{t \in \mathbb{R}}$

8.3 Proofs of Propositions 2.3, 2.6, and 2.11

there is a unique solution $(\mathcal{L}_t^*)_{t\in\mathbb{R}}$ to the McKean–Vlasov equation having this mean curve. Since the McKean–Vlasov dynamics is unique for $t \geq t_0$ it suffices to prove uniqueness of $\mathcal{L}_{t_0}^*$ for arbitrary t_0.

Now consider the tagged component in the McKean–Vlasov system $(x_2^\infty(1,t))_{t\geq t_0}$ assuming that the mean curve $(m(t))_{t\in\mathbb{R}}$ satisfies

$$\limsup_{t\to-\infty} m(t) = 0. \tag{8.95}$$

Then $\mathcal{L}_t^*$ is the law of a tagged component conditioned on the realization of the mean curve and this conditioned law is a weak solution of the wellknown SDE

$$dx_2^\infty(t) = c(m(t)-x_2^\infty(t))dt + sx_2^\infty(t)(1-x_2^\infty(t))dt + \sqrt{x_2^\infty(t)(1-x_2^\infty(t))}dw(t), \tag{8.96}$$

(cf. (2.21)) which has a unique weak solution given an *initial* value at time t_0. However we must prove the existence and uniqueness of a solution with time running in $\mathbb{R}$ and satisfying

$$\lim_{t\to-\infty} x_2^\infty(1,t) = 0. \tag{8.97}$$

We construct a "minimal solution" $\tilde{x}_2^\infty(1,t)$ for this problem by starting with a sequence of solutions $x_{(n)}(t)$ of (8.96) (driven by independent Brownian motions $w_n(t)$) starting with value $x^n(t_n) = 0$ at times $(t_n)_{n\in\mathbb{N}}$ with $t_n \to -\infty$. We then construct a coupled version $\{\hat{x}^m : m \geq m_0\}$ by coupling paths x^m and x^n, $n \leq m$ at a time where they collide. It can be verified that the coupled system $\{(\hat{x}^n(t))_{t\in\mathbb{R},n\in\mathbb{N}}\}$ is a stochastically monotone increasing sequence as $n \to \infty$ with limit $\hat{x}_2^\infty(1,t)$. If we assume *existence* of a solution of (8.96) with time index $\mathbb{R}$ and mean curve $m(t)$ we then know it will be stochastically bounded below by $\hat{x}_2^\infty$.

Next we need uniqueness in law of the solution and we shall show that it must agree with $\hat{x}_2^\infty$, which then automatically must also be a solution. We observe that any solution to (2.92) must have zeros at $-|t|$ for arbitrarily large $|t|$ (cf. classical property of Wright–Fisher diffusions [S1]). If it has a zero at $t^* > t_n$, then we can couple it with $\hat{x}_n$ and therefore we can couple the diffusion with the minimal one therefore the considered diffusion must agree with a version of $\hat{x}_2^\infty$ after the coupling time. Combined with the uniqueness property of the McKean–Vlasov equation this also proves uniqueness of $\mathcal{L}_t^*$, $t \geq t_0$. Since t_0 is arbitrary, we obtain the *uniqueness* for all $t \in \mathbb{R}$. This completes the proof of (e) of the Proposition 2.3.

(f) Finally, we consider a *random* solution $\{\mathcal{L}_t : t \in \mathbb{R}\}$ to the McKean–Vlasov equation. Since $\mathcal{L}_t$ is a.s. a solution and since by assumption it satisfies a.s. the growth condition with $A \in (0,\infty)$, it follows that it is given a.s. by a (random) time shift of the standard solution.

8.3.3 The Structure of the Dual Process and a Crump–Mode–Jagers Process

The key tool for the proof of the emergence results is the dual representation for the system of N exchangeably interacting sites, namely, the dual process $(\eta_t, \mathcal{F}_t^+)$ (note the dependence on N is suppressed in the notation) in the form it takes if we have two types only (recall Section 5).

Since in the case of two types, $\mathbb{K} = \{1, 2\}$ the two frequencies of type one and two add up to 1, it suffices to determine the law of $x_1(\frac{\log N}{\alpha} + t)$ and $\Xi_N(\frac{\log N}{\alpha}) + t, \{1\})$ in the limit $N \to \infty$. This will, as we shall see in Subsubsection 8.3.11, involve computing (ℓ, k)-moments, i.e. the product of k-th moments at ℓ-different sites for the N-site system and for the random McKean–Vlasov limit. The calculation is carried out using the dual process started with k particles at each of ℓ different sites. We denote the number of particles in the dual process at time t with such an initial condition by

$$\Pi_t^{N,k,\ell}, \text{ we often write } \Pi_t^N \text{ if } (k, \ell) \text{ are fixed.} \tag{8.98}$$

First, to prove that the mass of the inferior type at a tagged site goes to zero in time scale $\alpha^{-1} \log N + t$ as $N \to \infty$ and then $t \to \infty$, it suffices to work with $E[x_1(\cdot)]$. The calculation of this expectation involves only one initial particle for the dual process. To carry out the calculation we assume that $x_1(0, i) = 1$ for $i = 1, \cdots, N$ and take the initial element for the dual process

$$\mathcal{F}_0^+ = 1_{\{1\}}(u_{i_0}), \quad \eta_0 = \delta_{i_0}, \tag{8.99}$$

i.e. one factor at a tagged site in $i_0 \in \{1, \ldots, N\}$. Note that because of migration the law of this process depends indeed heavily on N if t is large enough.

First, *on the event* that *no* rare mutation occurs up to time t, the process $\mathcal{F}_t^+$ has the form

$$\mathcal{F}_t^+ = \prod_{k=1}^{\Pi_t^N} 1_{\{1\}}(u_{i_k}), \text{ where } i_k = i_k(t) \text{ is the location at time } t \text{ of the } k\text{th factor.} \tag{8.100}$$

If we start with k factors at each of ℓ sites the structure is the same.

The number of factors at each site form a system of dependent birth and death processes with immigration, i.e. a process with state space $(\mathbb{N}_0)^N$, denoted

$$(\zeta^N(t))_{t \geq 0} \text{ with } \zeta^N(t) = \{\zeta^N(t, i), i = 1, \cdots, N\}. \tag{8.101}$$

In other words

$$\Pi_t^{N,k,\ell} = \sum_{i=1}^{N} \zeta^N(t, i), \tag{8.102}$$

and $\zeta^N(0)$ has exactly ℓ-non zero components each containing exactly k factors.

8.3 Proofs of Propositions 2.3, 2.6, and 2.11

Next we describe the dynamic of ζ^N. Consider first a single component i. If the current state of ζ_t^N at a site i is $x \in \mathbb{N}_0$, then

the immigration rate of one new individual to this site is $c(N^{-1}(\Pi_t^{N,k,\ell} - x))$, where every individual at the other sites has the same chance to be selected for migration to i, (8.103)

$$\text{the emigration rate for one individual at this site is } c\frac{N-1}{N}x, \quad (8.104)$$

the individual moves to a randomly chosen different site,

the death rate, i.e. rate for the death of one individual at this site is $(d/2)x(x-1)$, and (8.105)

$$\text{the birth rate for one new individual is } \quad sx. \quad (8.106)$$

The dependence of the birth and death processes $(\zeta(t,i))_{t \geq 0}$, $i = 1, \cdots, N$ at different sites arises from the migration transition in (8.103), namely the immigration is coupled with the emigration at another site.

Note that because of the exchangeability of X^N and the initial state $X^N(0)$ it suffices for the dual ζ^N to keep track of

- the number of occupied sites,
- the number of individuals, i.e. factors, at each occupied site,

in order to specify the state of $\mathcal{F}_t^+$ to the extent needed to calculate the integral in the duality relation.

We now introduce the notation needed to describe the dual population. First the process

$$\{K_t^N\}_{t \geq 0} \quad (8.107)$$

that counts the number of *occupied sites* at time t. Secondly the process recording the *age and size* of each site (age is counted from the time of "birth", that is, the time when the site first became occupied the last time). Note that each site is uniquely identified by its birth time. Equivalently, due to exchangeability it suffices to keep track of the number of occupied sites K_t^N and the empirical age and size distribution at time t, a measure on $\mathbb{R}^+ \times \mathbb{N}$ denoted

$$\Psi^N(t, ds, dy), \quad (8.108)$$

where $\Psi^N(t, (a,b], y)$ denotes the number of sites in which the birth time, that is, the time at which the initial immigrant arrives, lies in the time interval $(a,b]$ and the current size is $y \in \mathbb{N}$.

We will use the abbreviation

$$\Psi^N(t,ds) = \Psi^N(t,ds,\mathbb{N}). \tag{8.109}$$

Consider next for a fixed $t \in \mathbb{R}^+$ the collection of processes

$$\{(\zeta_s^N(u))_{u \geq s}, \ s \in \text{supp } \Psi^N(t,\cdot,\mathbb{N})\}, \tag{8.110}$$

which denotes the size of the population at a site having birthtime s, that is, one particle arrives at time s. In the absence of collisions these processes for different birth times are independent birth and death processes and their distributions depend only on their current age. However in the presence of *collisions* these are *coupled* (by migration) and no longer independent.

The number of occupied sites fluctuates due to migration to *new* sites respectively jumps from sites with one particle to other occupied sites. Note that the probability of a jump of the latter type is non-negligible only if the number of occupied sites is comparable to the total number of sites. Hence up to the time that the number of occupied sites reaches $O(N)$ the number of occupied sites is (asymptotically) non-decreasing and the birth and death processes at different sites are asymptotically independent.

We obtain an upper bound for the growth of $(K_t^N)_{t \geq 0}$ if we suppress all collisions and assume that always a new site is occupied. In order to construct this process, we enlarge the geographic space from $\{1,\cdots,N\}$ to $\mathbb{N}$ and drop in (8.103) all jumps to occupied sites replacing them by jumps to a new site, namely the free one with the smallest label. We refer to this process as the *McKean–Vlasov dual* since it arises by taking in our model of N exchangeable sites the limit as $N \to \infty$ of the dual process. This process is denoted (if we want to stress how it arises we add the superscript ∞):

$$(\zeta_t)_{t \geq 0} = (\zeta_t(1), \zeta_t(2), \cdots)_{t \geq 0} \tag{8.111}$$

and the process of the total number of individual by

$$(\Pi_t^{(k,\ell)})_{t \geq 0}, \tag{8.112}$$

and ζ has the following markovian dynamic.

If we have at a site $i \in \mathbb{N}$ occupied with $x \in \mathbb{N}$ individuals then we have

$$\text{one death at rate } \frac{d}{2}x(x-1), \tag{8.113}$$

$$\text{one birth at rate } sx \tag{8.114}$$

8.3 Proofs of Propositions 2.3, 2.6, and 2.11

and if j is the smallest index with $\zeta_t(j) = 0$ then

$$\text{at rate } c\zeta(i) \text{ an individual moves from } i \text{ to } j. \tag{8.115}$$

Since in the duality relation we only need the occupation numbers of occupied sites, we can *ignore migration steps of single factors* to another unoccupied site. Therefore in the migration rate in (8.115) we take

$$c\zeta_t(i)1_{(x \geq 2)} \text{ instead of } c\zeta_t(i)1_{(x \geq 1)}. \tag{8.116}$$

This way we obtain, completely analogous to (8.107)–(8.110) new processes which we denote

$$(K_t)_{t \geq 0}, (\Psi(t, ds, dj))_{t \geq 0}, \{(\zeta_s(u))_{u \geq s}, s \in \mathbb{R}^+\}. \tag{8.117}$$

Then we know that for every $t \in \mathbb{R}^+$ the collection

$$\{(\zeta_s(u))_{u \geq s}, s \in \text{supp } \Psi(t, \cdot, \mathbb{N})\} \tag{8.118}$$

consists of *independent* birth and death processes.

We can show that

$$K_t^N \leq K_t, \quad \text{stochastically for all } t \geq 0 \text{ and } N \in \mathbb{N}. \tag{8.119}$$

Namely on the r.h.s. we suppress collisions and at each jump occupies a *new* site. Since after a collision we can have coalescence with one of the particles already at that site, a coupling argument shows that we have fewer particles as in the model with collisions, in fact strictly fewer with positive probability.

Then the dual $(\eta_t, \mathcal{F}_t^+)$ on the N-site model evolves exhibiting the following features. The dual population consists of an increasing number, Π_t^N, of $(1_{\{1\}})$ factors (due to birth events related to selection). Each factor $(1_{\{1\}})$ (and therefore the product) can jump to 0 by rare mutation (recall (5.85)) at rate $\frac{m}{N}$ (because the factor jumps to $(1_{\{2\}})$, which becomes 0 in the dual expression for $E[x_1^N(t)]$ by (8.100) since $x_2^N(0) = 0$). Therefore asymptotically as $N \to \infty$ to have non-zero probability of such a jump to zero to occur, we need $\Pi_t^N \sim O(N)$. This means that asymptotically the probability that a mutation event occurs becomes positive only for sufficiently large times such that the number of $(1_{\{1\}})$ factors in $\mathcal{F}_t^+$ reaches $O(N)$.

In order to see the meaning of this behaviour for the original process recall the *first moment duality relation* (assuming $x_1^N(0) = 1$) rewritten in terms of the quantities introduced (8.107)–(8.110):

$$E[x_1^N(t)] = E[\exp(-\frac{m}{N} \int_0^t \Pi_u^{N,1,1} du)]$$

$$= E\left[\exp\left(-\frac{m}{N} \int_0^t \left(\int_0^u \zeta_v^N(u) \Psi^N(u, dv)\right) du\right)\right]. \tag{8.120}$$

In order to prove Proposition 2.6, (2.58), we therefore need to show that asymptotically the probability that a mutation event occurs by time $T_N + t$ goes to 0 as $N \to \infty$ and $t \to -\infty$.

A complication in the analysis of the dual process arises in that if a migrant lands at a previously occupied site, that is, a *collision occurs*, then the different birth and death processes are not independent due to coalescence with its quadratic rate and we must take this into account when K_t^N reaches $O(N)$ since there is then a non-negligible probability of a collision.

We will therefore study in Subsubsections 8.3.4 and 8.3.5 first the collision-free regime of the dual dynamic and in Subsubsection 8.3.2 its consequences for the process and then in Subsubsections 8.3.6–8.3.10 the regime with collisions.

8.3.4 The Dual in the Collision-Free Regime: The Exponential Growth Rate

Due to coalescence (if $d > 0$) the total number of $(1_{\{1\}})$ factors is locally stochastically bounded but due to migration the number of occupied sites can grow and with it the total number of factors. We need to find α such that the number of factors is at time $\alpha^{-1} \log N + t$ of order N for $t \geq t_0$ and N large and the integrated lifetime of all factors up to that time is $O(N)$ as $N \to \infty$, with smaller and smaller number of factors as $t \to -\infty$ and a diverging number as $t \to \infty$. We therefore will fix a candidate α and then show lower and upper bound on the number of factors. We will introduce this α as the exponential growth rates of the number factors as $N \to \infty$ as follows.

If $c > 0$ and $N = \infty$ (i.e. "mean-field migration"), then the number of factors (i.e. Π_t^∞) would grow exponentially fast as we prove below. We will now first work with this scenario of mean-field migration in Step 1 providing a candidate for α and an upper bound. In the next subsubsection we return to the effect of collisions, which can arise if $N < \infty$ and we have migration to a previously occupied site to get a lower bound.

In order to verify (2.58), first note that by Jensen and (8.120)

$$E[x_1^N(t)] \geq \exp\left(-\frac{m}{N} E\left[\int_0^t [\int_0^u \zeta_v^N(u)\Psi^N(u,dv)]du\right]\right). \tag{8.121}$$

In the next three steps we analyse first $\zeta_v^N(u)$ given K_τ^N for τ from 0 up to u to get an estimate for the growth rate of the integral in the r.h.s. in terms of the process K^N of (8.121) and then in the second step we focus on the growth of K_u^N in u and in third step will then be simple, we combine the results to get that the α we have chosen is an upper bound for the exponential growth rate (in t) of the double-integral in the r.h.s. of (8.121).

8.3 Proofs of Propositions 2.3, 2.6, and 2.11

Step 1 *Sufficient condition for (2.58)*

We next show in order to control the r.h.s. of (8.121) that

$$\limsup_{N \to \infty} E[\zeta_v^N(u) | \{K_s^N \leq N : s \leq u\}] \leq L \qquad (8.122)$$

and L is non-random, i.e. a constant. Given α and recalling (8.121) in order to establish the upper bound on the emergence time, it suffices once we have (8.122) to show that:

$$\lim_{t \to \infty} \lim_{N \to \infty} \frac{1}{N} \int_0^{\frac{\log N}{\alpha} - t} E[K_s^N] du = 0, \qquad (8.123)$$

which we will do in the Step 2.

To prove (8.122) consider the collection of auxiliary processes (here $\iota \in \mathbb{R}^+$)

$$(\zeta_{0,t}(\iota))_{t \geq 0}, \qquad (8.124)$$

which as it turns out bounds our process from above in a site colonized at time 0 at time t for proper choice of the parameter ι. It has the dynamic of a birth and death process with *linear birth rate sk, quadratic death rate $\frac{d}{2}k(k-1)$, emigration* at rate $ck1_{(k \geq 2)}$ and *immigration* of a particle at rate ι. The parameter ι we choose later. The case $\iota = 0$ corresponds to the McKlean–Vlasov dual, i.e. the dual in the collision-free regime. The process has a unique equilibrium state

$$\zeta_{0,\infty}(\iota), \qquad (8.125)$$

and is ergodic (with exponential convergence to equilibrium).

A standard coupling argument can then be used to show that for all $t \geq 0$ and every s the random variable (recall we start in one particle)

$$\zeta_{s,t}(\iota) \text{ is stochastically dominated by } \zeta_{0,\infty}(\iota). \qquad (8.126)$$

Moreover writing down the differential equation for the second moment of $\zeta_{0,t}$ it is easy to verify that for every $\iota \geq 0$:

$$E[(\zeta_{0,\infty}(\iota))^2] < \infty. \qquad (8.127)$$

We have (using a standard coupling argument) in stochastic order:

$$\zeta_{0,\infty}(0) \leq \zeta_{0,\infty}(\iota). \qquad (8.128)$$

We next consider the birth and death process in which the parameter ι is time-dependent denoted by ι_t. The associated mean-field birth and death process is then obtained by setting ι_t to be given by the *current mean* of the process at time t. (Note that the mean-field birth and death process is of McKean–Vlasov type but should not be confused with the McKean–Vlasov dual corresponding to the collision-free regime). This process $\zeta_{0,t}(\{m_u\}_{u \leq t})$ arises as the limit $N \to \infty$ of the process with migration between the N sites according to the uniform distribution and starting one particle at each site. The mean-field birth and death process with migration rate c then has a unique equilibrium and it can be verified that the fixed point equation

$$\iota_* = E[\zeta_{0,\infty}(c\iota_*)] \tag{8.129}$$

has a finite solution which gives the expectation for the equilibrium of this mean-field process and that in fact determines the equilibrium.

To show this we consider the nonlinear Kolmogorov equations for the process $(\zeta_{0,t}(c\iota))_{t \geq 0}$ which are spelled out in all detail later on in (8.267) where we set in (8.267) $\alpha(t)u(t) \equiv c\iota$. More precisely let $p_{k,\ell}(t) = p_{k,\ell}(c\iota, t)$ denote the transition probabilities and set

$$m_t(\iota) = \sum_{j=2}^{\infty} j p_{1,j}(c\iota, t), \tag{8.130}$$

$$m(\iota) = \lim_{t \to \infty} \sum_{j=2}^{\infty} j p_{1,j}(c\iota, t). \tag{8.131}$$

To look for equilibria we look for fixed points $m(\iota) = \iota$. Using coupling we can show that $m(\iota)$ is monotone increasing, continuous, ultimately sublinear in ι and $\lim_{\iota \to 0} m(\iota) > 0$ since the smallest state is 1 if we start in 1. Therefore there exists a unique

$$\text{largest solution } \iota^* \text{ of the equation } m(\iota) = \iota \text{ with } m_t \leq m(\iota^*). \tag{8.132}$$

If we insert this ι^* as parameter the mean remains constant and the equilibrium of the mean-field birth and death process, is the unique equilibrium of the Markov process $(\zeta_{0,t}(\iota^*))_{t \geq 0}$.

We see again by coupling that for $\iota \leq \iota^*$:

$$\zeta_{0,\infty}(0) \leq \zeta_{0,\infty}(\iota) \leq \zeta_{0,\infty}(\iota^*). \tag{8.133}$$

We now show that this provides the bound required in (8.122). Recall that our dual is obtained by starting one particle at one randomly chosen site in $\{1, \ldots, N\}$ with the other sites empty. Then the key point is that by coupling it follows that $\zeta_0^N(t)$ is dominated in the stochastic order by the system denoted by $\tilde{\zeta}_{0,t}^N$ started with one particle at each site at time 0. Let $\tilde{\zeta}_{0,}^N(j)$ denote the number of particles at site $j \in \{1, \ldots, N\}$ in this exchangeable system.

8.3 Proofs of Propositions 2.3, 2.6, and 2.11

Note that at a tagged site 1 in the exchangeable system of N sites the number of particles is given by

$$\tilde{\zeta}_{0,t}^N(1) = \zeta_{0,t}((c\iota_u^N)_{u \leq t}) \text{ with } \iota_u^N = \frac{1}{N} \sum_{j=1}^N \tilde{\zeta}_{0,u}^N(j), \qquad (8.134)$$

A standard mean-field limit argument then shows that $(\tilde{\zeta}_{0,t}^N(1), \iota_t^N)$ converges in law as $N \to \infty$ to $(\zeta_{0,t}(\{cm_u\}_{u \leq t})(1), m(t))$, where $m(t) = E(\zeta_{0,t}(\{cm_u\}_{u \leq t})(1))$. Therefore using (8.133) and the construction of ι^* we see that:

$$\lim_{N \to \infty} E[\zeta_0^N(t)] = E[\zeta_{0,t}(\{cm_u\}_{u \leq t})] \leq E[\zeta_{0,\infty}(\iota_*)] = \iota^* \qquad (8.135)$$

and this proves (8.122).

Step 2 *A Crump–Mode–Jagers process gives a Malthusian parameter α.*

To show (8.123) and to obtain a candidate for α we can use the collision-free mean-field dual Π_t in particular its ingredients K_t and ζ_t.

The process K in (8.117) has some important structural properties. Recall the process counts the number of occupied sites, each of which has an internal state, which consists of the number of particles at these sites. Each of the particles exceeding the first particle can migrate and produce a new occupied site. Whenever a new site is colonized then an independent copy of the basic one site process starts and evolves independently of the rest.

To analyse this object we recall the concept and some basic results on *supercritical Crump–Mode–Jagers branching processes* and in particular its *Malthusian parameter α*. Such a process is defined by the following properties.

The process counts the number of individuals in a branching population whose dynamics is as follows. Individuals can die or give birth to new individuals based on the following ingredients:

- individuals have a lifetime (possibly infinite),
- for each individual an independent realization of a point process $\xi(t)$ starting at the birth time specifying the times at which the individual gives birth to new individuals,
- different individuals act independently,
- the process of birth times is not concentrated on a lattice.

This is exactly the structure of $(K_t)_{t \geq 0}$ we described in Subsubsection 8.3.3.

Let $(K_t)_{t \geq 0}$ be a process with the described structure. The corresponding *Malthusian parameter*, $\alpha > 0$ is obtained as the unique solution of

$$\int_0^\infty e^{-\alpha t} \mu(dt) = 1 \text{ where } \mu([0, t]) = E[\xi([0, t])], \qquad (8.136)$$

with

$$\xi(t) \text{ counting the number of births of a single individual up to time } t, \quad (8.137)$$

(see for example [J92, N] Equation (1.4)).

In our case we have

$$\mu([0,t]) = c \int_0^t E[\zeta_0(s) 1_{(\zeta_0(s) \geq 2)}] ds. \quad (8.138)$$

Remark 69. Note that in our model particles at singly occupied sites do not migrate. We can suppress the emigration step if only one particle is left, since in that case everything would simply start afresh at some other site. We have done this so that in the pre-collision regime the number of occupied sites equals the number of migrating particles. In this case the rate of creation of new sites due to migrations from a given site is given by $\zeta_0(s) 1_{\zeta_0(s) \geq 2}$ and the death rate when $\zeta_0(s) = 1$ is 0.

However we can also consider the original model in which particles at singly occupied sites do migrate. Then sites have *finite lifetimes* and the rate of production of new sites by a fixed site at time s is given by $\tilde{\zeta}_0(s)$ where now $\tilde{\zeta}_0(s)$ has death rate c when $\tilde{\zeta}_0(s) = 1$. Then we have instead of (8.138) the equation

$$\mu([0,t]) = c \int_0^t E[\tilde{\zeta}_0(s)] ds. \quad (8.139)$$

However the Malthusian parameter describing the exponential growth of the number of sites occupied by at least one particle is the same in both cases as is shown by explicit calculation.

We therefore define α by (recall for ζ_0 (8.117) and above):

$$c \int_0^\infty e^{-\alpha s} E[\zeta_0(s) 1_{(\zeta_0(s) \geq 2)}] ds = 1. \quad (8.140)$$

The first thing we want to know is that in our case

$$\alpha \in (0, s) \quad \text{if} \quad d > 0 \text{ and } \alpha \in (0, s] \text{ if } d \geq 0. \quad (8.141)$$

In the case of our dual process $K_t \leq N_t$ where N_t is a rate s pure birth process and therefore $\alpha \leq s$.

To show that the Malthusian parameter satisfies:

$$\alpha > 0 \quad (8.142)$$

note first that when $d > 0$ still no deaths (in the number of occupied sites) occur and that a lower bound to the growth process is obtained by considering only those

8.3 Proofs of Propositions 2.3, 2.6, and 2.11

migrants which come from the birth of a particle at sites containing sofar only one particle and which then migrates before coalescence or the next birth occurs. This process of the number of such special sites is a classical birth and death process with in state k birth rate $s(\frac{c}{c+s+d})k$ and death rate zero which clearly has a positive Malthusian parameter (if $c > 0$) and hence (8.142) holds.

Next to see that if $d > 0$ then $\alpha < s$, note that a non-zero proportion of the new particles generated by a pure birth process at rate s dies due to coalescence and therefore $\alpha < s$ in this case.

If we know the Malthusian parameter, we need to know that it is actually equal to the almost sure growth rate of the population. It is known for a CMJ-process $(K_t)_{t \geq 0}$ that (Proposition 1.1 and Theorem 5.4 in [N]) the following basic growth theorem holds. If

$$E[X \log(X \vee 1)] < \infty, \tag{8.143}$$

where here $X = \int_0^\infty e^{-\alpha t} d\xi(t)$, then

$$\lim_{t \to \infty} \frac{K_t}{e^{\alpha t}} = W, \text{ a.s. and in } L_1, \tag{8.144}$$

where W is a random variable which has two important properties, namely

$$W > 0, \ a.s. \text{ and } E[W] < \infty. \tag{8.145}$$

We now have to check that (8.143) holds in our case.

We know using (8.140) that $\int_0^\infty u e^{-\alpha u} \mu(du) < \infty$, since we can bound the birth rate by the one in which the number of particles at each site is given by the equilibrium state $\mathcal{L}(\zeta_\infty)$, which has a finite mean. This also implies that assumption (8.143) holds. Namely, in equilibrium we can estimate

$$E[X^2] \leq \int_0^\infty \int_0^\infty e^{-\alpha(s+t)} E[d\xi(s) d\xi(t)] < \infty, \tag{8.146}$$

by using Cauchy–Schwartz and using (8.127).

Remark 70. In the above we considered the dual $\mathcal{F}^+$ starting with one factor (particle) at one initial site. We can also consider the case with k initial particles at one site. In this case the same exponential growth occurs since (8.138) still holds but the corresponding random growth factor $W^{(k)}$ in (8.144) now depends on k. Similarly we can start with k particles at each of ℓ different sites and again get exponential growth with the same Malthusian parameter α but now with random variable $W^{(k,\ell)}$. This will be frequently used in the sequel of this section.

Step 3 *Completion of argument*

To complete the argument we note that (8.123) follows from (8.144) (which implies that $K_t = e^{-\alpha t} W_t$ with $W_t \to W$) and (8.119) since

$$E(\int_0^{\alpha^{-1} \log N - t} K_s^N \, ds) \leq \text{Const} \cdot \frac{1}{\alpha} E(W) e^{\log N - \alpha t}. \tag{8.147}$$

We saw earlier that (8.123) is sufficient for (2.58) and the proof is complete that emergence does not happen before time $\alpha^{-1} \log N$ as $N \to \infty$.

8.3.5 The Dual in the Collision-Free Regime: Further Properties

We shall need for the analysis of the regime with collisions some further properties concerning the dual in the collision-free regime. First, we obtain further properties of the growth constant W, more precisely its moments, and then the stable age distribution which among other things allows to give an explicit expression for the growth rate α. The random variable W is defined in (8.144) in terms of the process $(K_t)_{t \geq 0}$ introduced in (8.117), that is, we consider the dual process without collision.

Step 1 *Higher moments of W.*

In the sequel we will perform moment calculations which use:

Lemma 8.4 (Higher Moments of W). *For all n,*

$$E[W^n] < \infty. \qquad \square \tag{8.148}$$

Proof of Lemma 8.4. By [BD], Theorem 1, W has a finite nth moment if and only if

$$E\left[\left(\int_0^\infty e^{-\alpha s} d\xi(s)\right)^n\right] < \infty, \tag{8.149}$$

where $\xi(t)$ denotes the number of particles that have emigrated by time t from a fixed site starting with 1 particle. We will verify this condition for the second moment.

$$E\left[\int_0^\infty e^{-\alpha s} d\xi(s)\right]^2 = 2E\left[\int_0^\infty \int_s^\infty e^{-\alpha s} e^{-\alpha(s+s')} d\xi(s') d\xi(s)\right] \tag{8.150}$$

$$= c^2 E\left[\int_0^\infty \int_s^\infty e^{-\alpha(s+s')} \zeta_0(s) E[\zeta_0(s')|\zeta_0(s)] ds' ds\right]$$

$$= c^2 E\left[\int_0^\infty \int_s^\infty e^{-\alpha s} \zeta_0(s) \zeta_0(s') ds' ds\right].$$

8.3 Proofs of Propositions 2.3, 2.6, and 2.11

Recall that here ζ_0 is a birth and death process with birth rate sk and death rate $ck1_{k\geq 2} + \frac{1}{2}dk(k-1)$. But then by explicit calculation $\sup_s E(\zeta_0(s)^2) < \infty$, and therefore

$$2E\left[\int_0^\infty \int_s^\infty e^{-\alpha s}\zeta_0(s)\zeta_0(s')ds'ds\right] < \infty. \qquad (8.151)$$

This completes the proof that $E(W^2) < \infty$.

Recalling that $\zeta_0(t)$ is for all t stochastically dominated by the equilibrium distribution $(p_k)_{k\in\mathbb{N}}$ of this birth and death process and that the latter satisfies for some $C < \infty$,

$$p_k \leq \frac{C^k}{k!}, \qquad \text{for all } k = 1, 2, \cdots \qquad (8.152)$$

it follows that for every $n \in \mathbb{N}$,

$$\sup_s E[(\zeta_0(s))^n] < \infty. \qquad (8.153)$$

Then a similar argument to the second moment case argument given above verifies that nth moments of W are finite.   q.e.d.

Remark 71. Alternatively to the above we can argue as follows. In the pre-collision regime we can also represent the particle process as a branching Markov chain, more specifically a *branching birth and death process* (linear birth rate and quadratic death rate), that is a counting measure-valued process, on $\mathbb{N}$,

$$\Lambda_t \in \mathcal{M}(\mathbb{N}), \qquad (8.154)$$

where $\Lambda_t(k)$ counts the number of sites occupied by exactly k partition elements. In other words the individuals of this new process Λ correspond to the occupied sites of the dual particle system and the location of individuals is now the size of the population at the sites of the dual particle system.

The dynamics of $(\Lambda_t)_{t\geq 0}$ are therefore given by

- The branching. Individuals branch independently with a location-dependent branching rate, namely, an individual at k dies at rate ck and produces two new individuals with new locations, namely one individual at $k-1$ and a second one at 1. Since sites of size 0 that would be produced if a site of size 1 dies would not contribute to the production of new sites, they can be ignored. For this reason we suppress these deaths and set the death rate at k to be $ck1_{k\neq 1}$.
- The motion: the motion between locations is given by a Markov process $(\tilde{\zeta}(t))_{t\geq 0}$ on $\mathbb{N} \cup \{\infty\}$, namely, a birth and death process $(\zeta(t))_{t\geq 0}$, where the birth rate is sk and the death rate $dk(k-1)/2$.

In other words the CMJ process can be embedded in the branching birth and death process and the number of individuals in that process is the same as the number of occupied sites in our original CMJ-process.

We work with the probability generating function, M_t, of branching birth and death process Λ_t and use this to compute the distribution of the number of occupied sites as follows.

Define the generating function:

$$M_t(k,z) = E_{\delta_k}[z^{\Lambda_t(1_{\{\mathbb{N}\}})}] = \sum_{j=1}^{\infty} p(t,k,j)z^j, \quad 0 < z \le 1, \tag{8.155}$$

where we write $\Lambda_t(f) = \sum_{i=1}^{\infty} f(i)\Lambda_t(k)$.

Then M_t satisfies the standard functional equation

$$M_t(k,z) = E_k[e^{-c\int_0^t \tilde{\zeta}(s))ds}]z$$
$$+ E_k\left[\int_0^t e^{-c\int_0^s \tilde{\zeta}(u)du} \cdot c\tilde{\zeta}(s) \cdot M_{t-s}(\tilde{\zeta}(s)-1,z)M_{t-s}(1,z)ds\right], \tag{8.156}$$

with $M_t(0,z) \equiv 1$. Here $\tilde{\zeta}$ is a birth and death process with birth rate sk and death rate $\frac{1}{2}dk(k-1)$. The above equation can be solved recursively for the coefficients of z^j.

This relation can also be used to obtain moment formulas by evaluating the appropriate derivatives at $z = 1$.

Consider the semigroup $(T_t)_{t \ge 0}$ of the migration defined by

$$T_t(f)(k) = E_k[e^{-c\int_0^t \tilde{\zeta}(s)ds} f(\tilde{\zeta}(t))], \quad k \in \mathbb{N} \tag{8.157}$$

and let

$$f_\ell(j) := j^\ell, \quad \ell = 1,2,3,\ldots. \tag{8.158}$$

Let for $\ell = 1, 2, 3, \cdots$ and $k \in \mathbb{N}$:

$$m_\ell(t,k) = E_k((\Lambda_t(1_{\{\mathbb{N}\}}))^\ell). \tag{8.159}$$

Then the following moment relations hold:

$$m_1(t,k) = T_t(f_0)(k) + c\int_0^t T_s(f_1(\cdot)[m_1(t-s,1) + m_1(t-s,\cdot-1)])(k)ds. \tag{8.160}$$

8.3 Proofs of Propositions 2.3, 2.6, and 2.11

$$m_2(t,k) = T_t(f_0)(k) + c \int_0^t (T_s(f_1(\cdot) \cdot [(m_2(t-s,\cdot-1) + m_2(t-s,1)) \quad (8.161)$$

$$+ (m_1(t-s,1) \cdot m_1(t-s,\cdot-1))])(k) ds.$$

Step 2 *Stable age and size distribution*

Return to the McKean–Vlasov version of the dual process (which is the dual in the collision-free regime in the limit $N \to \infty$). If we consider for every time t for each individual (i.e. occupied site in our case) currently alive its age and size, then we can introduce a random probability measure, the normalized empirical age and size distribution of the current population, which we denote by

$$\mathcal{U}(t, du, j) = \frac{\Psi(t, du, j)}{K_t}. \quad (8.162)$$

The marginal random measure $\mathcal{U}(t, du, \mathbb{N})$ converges in law (we use the weak topology on measures) as $t \to \infty$ to a *stable age distribution*

$$\mathcal{U}(\infty, du, \mathbb{N}) \text{ on } [0, \infty), \quad (8.163)$$

according to Corollary 6.4 in [N], if condition 6.1 therein holds. The condition 6.1 in [N] or (3.1) in [JN] requires that (with μ as in (8.136)):

$$\int_0^\infty e^{-\beta t} \mu(dt) < \infty \quad \text{for some } \beta \geq 0. \quad (8.164)$$

The condition 2.1 in [N] holds for any $\beta > 0$ since the local (at one site) population of the dual McKean–Vlasov particle system $(\tilde{\eta}_t)_{t \geq 0}$ given by $(\zeta_{0,t})_{t \geq 0}$ goes to a finite mean equilibrium.

Since the distribution of size at a site depends only on the age of the site, it follows that $\mathcal{U}(t, du, \cdot)$ converges to a *stable age and size* distribution, i.e.

$$\mathcal{U}(t, \cdot, \cdot) \Longrightarrow \mathcal{U}(\infty, \cdot, \cdot), \text{ as } t \to \infty \text{ in law.} \quad (8.165)$$

We can now also calculate α and the asymptotic density of the total number of individuals B as follows using a law of large number effect for the collection of independent birth and death processes at the occupied sites. Namely the frequencies of specific internal states stabilize in the stable size distribution, while the actual numbers diverge with the order of K_{t_N} as $N \to \infty$. Given site i let $\tau_i \geq 0$ denote the time at which a migrant (or initial particle) first occupies it. Noting that we can verify Condition 5.1 in [N] we have that

$$\lim_{t\to\infty} \frac{1}{K_t} \sum_{i=1}^{K_t} \zeta_{\tau_i}(t-\tau_i) = \int_0^\infty E[\zeta_0(u)]\mathcal{U}(\infty, du) = B \text{ (a constant)}, a.s.,$$
(8.166)

by Corollary 5.5 of [N]. The constant B in (8.166) is in our case given by the average number of particles per occupied site and the growth rate α arises from this quantity neglecting single occupation. Namely define

$$\alpha = c \sum_{j=2}^\infty j\mathcal{U}(\infty, [0,\infty), j) < \infty, \qquad \gamma = c\,\mathcal{U}(\infty, [0,\infty), 1). \quad (8.167)$$

Then

$$B = \frac{\alpha + \gamma}{c}. \quad (8.168)$$

Furthermore the average birth rate of new sites (by arrival of a migrant at an unoccupied site) at time t (in the process in the McKean–Vlasov dual) is equal to

$$\alpha = c\,B - \gamma. \quad (8.169)$$

8.3.6 Dual Process in the Collision Regime: Macroscopic Emergence Proposition 2.6(a)

We have obtained in Subsubsection 8.3.4 a lower bound on the emergence time by an upper bound on the number of dual particles. Now we need an upper bound on the emergence time via a lower bound on the number of dual particles.

In order to prove (2.59) return to the dual particle system specified below (8.98). Then after the lower bound for the emergence time in Subsubsection 8.3.4, where we ignored collisions in the dual particle system, we derive here a lower bound on the growth of the number of occupied sites in this system starting with one occupied site incorporating the effect of *collisions*. We proceed in five steps, first we consider lower bounds on the number of particles in the dual process, then refine this in a second step by constructing a multicolour particle system which is an enrichment of the dual particle system, then in a third step we prepare the estimation of the difference between collision-free and dual system, and in the fourth step we turn this into an upper bound for the emergence time. In Step 5 we then show the convergence of the hitting times for reaching $\lfloor \varepsilon N \rfloor$ dual particles.

Step 1: Number of sites occupied by the dual process: preparation

Here we must take into account the effect of collisions and show that (8.140) indeed gives the correct α which describes the growth of the number of factors in the dual process.

If we have collisions, that is, migration to occupied sites, we have two effects. (1) We have *interaction* between sites. (2) If a particle migrates from a site containing only one particle to an occupied site, this results in a *decrease* by one in the number of occupied sites. For these two reasons the process counting the number of occupied sites, denoted by K_t^N, is no longer approximated by a nondecreasing Crump–Mode–Jagers process (which is based on independence of internal states) when it reaches size $O(N)$ and at this moment it is also no longer non-decreasing since now singly occupied sites can disappear by a jump to an occupied site.

To handle these two problems the key idea is to carry out the analysis of the dynamics separately in three time intervals defined as follows. Set first

$$\tau_{\log N} = \inf\{t : K_t = \lfloor \log N \rfloor\}, \quad \tau^N(\varepsilon) = \inf\{t : K_t^N = \lfloor \varepsilon N \rfloor\}. \quad (8.170)$$

Then define the three time intervals as:

$$[0, \tau_{\log N}),$$

$$[\tau_{\log N}, \tau_{\log N} + \frac{1}{\alpha}(\log N - \log \log N)), \quad (8.171)$$

$$[\tau_{\log N} + \frac{1}{\alpha}(\log N - \log \log N), \tau^N(\varepsilon)].$$

What happens in the *first interval*? Note (since $\tau_{\log N} \ll \tau^N(\varepsilon)$ for $N \to \infty$) that asymptotically as $N \to \infty$, in a single migration step the probability of a collision is $O(N^{-1} \log N)$. In particular over a time horizon of length $o(N/\log N)$ the probability to observe a collision ever goes to zero as $N \to \infty$. Hence looking at the formulas we obtained for α and with the properties of exponentially growing populations (occupation measure in current size) we can replace our dual with the mean-field dual without collisions so that with (8.144) we can conclude as $N \to \infty$:

$$\tau_{\log N} = \frac{\log \log N}{\alpha} - \frac{\log W}{\alpha} + o(1) \quad (8.172)$$

and we know furthermore that this process has (asymptotically) reached the stable age distribution at time $\tau_{\log N}$ according to the analysis of Subsubsection 8.3.5, Step 2 therein.

Now consider the *second interval*. From the previous upper bound result still no collisions occur comparable to the dual population in this interval (the collision probability is now at most $O((\log N)^{-1})$ and hence at most only finite many collisions may occur over the whole time interval, essentially at the end and we can assume that asymptotically as $N \to \infty$ we can replace our dual process again by the collision free mean-field dual at the left end point but now in the *whole time span* the stable age distribution is in effect. Hence we can in the second interval continue working with the mean-field dual without collisions and we can even assume the stable age distribution in effect.

Returning to our dual process we conclude (recall (8.171)) that at the end of the second interval we have as $N \to \infty$:

$$K^N_{\tau_{\log N} + \frac{1}{\alpha}(\log N - \log \log N)} \sim W \frac{N}{\log N}. \tag{8.173}$$

Remark 72. Note that this means that $\tau^N_{s_N} - \tau^N_{\log N}$, with $s_N \gg \log \log N$ but $s_N \ll \log N$, is asymptotically as $N \to \infty$ deterministic.

We now consider the *third interval*, i.e. the regime in which by (8.173) above

$$K^N_t \geq W \frac{N}{\log N}. \tag{8.174}$$

In this interval two new effects must be considered, namely,

(A) a decrease in the number of occupied sites when a lone particle at an occupied site jumps to another currently occupied site
(B) a particle moves from a site occupied by more than one particle and jumps to another occupied site.

We note that effect (A) also increases the age distribution of occupied sites since in a site the process is stochastically increasing and the young sites can disappear therefore easier (and hence the age of the total population becomes stochastically larger). Effect (B) tends to increase the size of occupied sites of a given age compared to the case without collision. Therefore the result of (A) and (B) is to tend to increase the average size of occupied sites compared to a system without collision (recall the older sites are stochastically larger), but decreases the number of occupied sites.

We note that larger sites increase the number of migration steps. Altogether this means that we expect that the rate of growth of K^N_t denoted $\beta_N(t)$ (i.e. $\beta_N(t) = (K^N_{t+\Delta t} - K^N_t)/K^N_t$) satisfies in the third time interval (for $N \geq \log \log N$) if the present state of K^N_t is k:

$$\beta_N(\cdot) \geq \alpha(1 - \frac{k}{N}). \tag{8.175}$$

In order to control these effects precisely we use the technique of coupling constructed from an enriched, i.e. multicolour particle system, which we introduce next.

8.3 Proofs of Propositions 2.3, 2.6, and 2.11

Step 2: A multicolour particle system

To examine rigorously the growth in the time span, in which the number of occupied sites reaches $O(N)$ we construct a coupled system of *multicoloured particles* such that we can compare the new system with collisions effectively with the simpler one without collisions, and are able to carry out estimates on the difference. This way we can obtain the lower bound on the number of occupied sites and the total number of particles.

First an informal description. We shall have *white, black* and *red* particles. We start with white particles only. In intervals 1 and 2 the process of the white particles grows as before, i.e. without collisions. However as soon as collisions occur we consider a *modified system*. The colouring of the system allows us to keep track of events like (A) and (B) given above. For example in the case of (A) we will mark the lost site by placing there a black particle and the particle that jumped and collided with other particles on the new location is marked by giving it a red colour. We handle the second effect (B) by using bounds from below on the rate of founding of new occupied sites.

Formally proceed as follows. The multicolour comparison system has black, white and red particles. It has white and red particles located at the sites $\{1, 2, \cdots, N\}$ and black particles located at a site in $\mathbb{N}$ where $\{1, 2, \cdots, N\}$ and $\mathbb{N}$ are disjoint finite and countable sets respectively. In other words the geographic space of this new system is

$$\{1, 2, \cdots, N\} + (\mathbb{N}), \tag{8.176}$$

and the state space is

$$(\{1, 2, \cdots, N\} + (\mathbb{N}))^{\mathbb{N}_0^3}. \tag{8.177}$$

The initial state is given by having only white particles, which are located at sites in $\{1, 2, \cdots, N\}$ such that occupation numbers are exchangeable on this part of the geographic space.

The dynamics of the new system is markovian more precisely it is a pure Markov jump process and we specify the transitions and their rates. Instead of writing the generator we describe this more intuitively in words. This runs as follows:

- *white* particles at a site follow the same local dynamics as the dual particle system η as far as birth (of white particles) and coalescence (of white particles) goes, changes occur for migration.

  Let k denote the number of sites having currently at least one white particle. Each migrating white particle moves with probability $1 - \frac{k}{N}$ to a new site in $\{1, 2, \cdots, N\}$ which prior to this event did not contain any white particles and with probability $\frac{k}{N}$ changes to a black particle now located at a new unoccupied site in ℕ and at the same time also a red particle is produced at an occupied site in $\{1, 2, \cdots, N\}$ chosen at random among the k occupied sites.

Hence a migrating white particle produces:

$$\text{a new site occupied with a white particle with probability } (1 - \frac{k}{N}), \quad (8.178)$$

and two particles,

$$\text{one black and one red, with probability } \frac{k}{N}, \quad (8.179)$$

the red particle is placed at a randomly chosen occupied site in $\{1, 2, \cdots, N\}$, the black one at the smallest free site in ℕ.

$$(8.180)$$

- *Red* particles have the same dynamics as the dual particle system η on $\{1, 2, \cdots, N\}$ (newborn particles are also red) and in addition when a red and white at the same site coalesce the outcome is always white.
- *Black* particles follow the same dynamics as the white except that migrating black particles move on ℕ and always go to a new, so far unoccupied site in ℕ.

This means that no collisions (by migration!) occur among white particles and furthermore by the collision convention the white particles are not influenced by the presence of red particles, nor are they influenced by the black particles. Hence the key observation about the new system is that:

- The number of occupied sites in the *union* of the *black and white* particles follows the dynamics of the process without collisions, i.e. the number of sites they occupy is a version of $(K_t)_{t \geq 0}$,
- the number of occupied sites in the *union* of the *white and red* particles follows the exact dual dynamics, i.e. the number of sites they occupy is producing a version of $(K_t^N)_{t \geq 0}$.
- The process of white and black particles per site follow the dynamics of ζ and the one of white and red particles per site that of ζ^N.
- We have a coupling of ζ and ζ^N given by the embedding in the multicolour system and the difference process $\zeta_t - \zeta_t^N$ can be represented as the number of black minus the number of red particles at the various sites. In particular also K, K^N and $K - K^N$ can be represented in terms of the multicolour system.

This construction allows to give upper and lower bounds on the time to reach with the dual process $\lfloor \varepsilon N \rfloor$ occupied sites, which then in turn gives bounds on the number of factors in the dual process.

8.3 Proofs of Propositions 2.3, 2.6, and 2.11

Since we know from the last subsubsection how the growth of the population of both black and white particles works, we use this as an upper bound on the number of white particles (conditioned on W).

We denote the *number of sites occupied only by black particles* (recall "only" is by the construction not a constraint), respectively the number of sites occupied by either *white or black particles* at time t by

$$\bar{Z}_t^N, \text{ respectively } \hat{K}_t. \tag{8.181}$$

Note that $\hat{K}$ is a version of K.

We shall get below an upper bound on the number of sites occupied by black particles and then we can use this upper bound to get a *lower bound* on the number of sites occupied by white particles and therefore a lower bound for the number of factors of the dual (red plus white particles).

Let

$$\tilde{\tau}^N(\varepsilon) = \inf\{t : \hat{K}_t \geq \varepsilon N\}. \tag{8.182}$$

This means in particular that

$$\tilde{\tau}^N(\cdot) \text{ is nondecreasing.} \tag{8.183}$$

(Note the difference between $\tilde{\tau}^N(\varepsilon)$ and $\tau^N(\varepsilon)$ is that we use in the first case white and *black* particles and in the second case white and *red* particles. In addition $\hat{K}$ is a version of K.)

We just saw above that we have to estimate the number of black particles to compare the dual and the dual of the McKean–Vlasov limit which differ by the difference of black and red particles. The latter is estimated above stochastically by the number of black particles. This holds since if we associate the black and red upon their creation, we see that the red ones have an extra chance to disappear once they coalesce with a white particles or collide with red particles from another black–red creation event.

Step 3: Estimating the number of black particles.

An upper bound on the number of black sites (recall black particles sit on $\mathbb{N}$, while white and red sit on $\{1, \cdots, N\}$) at time $\tilde{\tau}^N(\varepsilon)$, is obtained as follows. Bound the number of migration steps in the model by a Crump–Mode–Jaegers process which is in distribution given by the process of black and white particles. This means we construct up to terms of order $o(N)$ a stochastic upper bound for the number of black particles by realizing *independently* a Poisson stream of potential collision events according to the law of a driving CMJ-process generating potential new sites and with a collision probability given by the current number of white sites divided by N.

Therefore an upper bound on the production rates of new black populations is obtained by integrating the production rate of new sites generated by the process of white and black particles at a given time (which is for large times converging to α by (8.137)). Furthermore note that in state k (meaning the number of sites with at least one white particle) with a migration step of a white particle, the probability of a collision and hence the probability to become black is k/N. This gives at time s with the CMJ-theory a production rate of collisions which is stochastically bounded by

$$\alpha \frac{W_s e^{\alpha s}}{N} W_s e^{\alpha s}. \tag{8.184}$$

Hence we get a stochastic bound on the intensity of migration steps resulting in collisions of the form $W_s^2 \alpha (e^{\alpha s}/N) e^{\alpha s}$. This expression therefore bounds the rate of creation of *founders* of new black families of particles.

We can replace for $s \geq \tau_{\log N}$, the quantity W_s^2 in the limit $N \to \infty$ by W^2, due to the convergence theorem for CMJ-processes. Therefore we can estimate the rate at which new black families are created asymptotically as $N \to \infty$ and work with the expression

$$W^2 \alpha e^{2\alpha s}/N \text{ for } s \geq \tau_{\log N}. \tag{8.185}$$

The contribution to the stream of production of collisions arising before time $\tau_{\log N}$ goes to zero in probability as $N \to \infty$ and therefore the probability that black families are founded before time $\tau_{\log N}$ tends to zero as $N \to \infty$, more precisely at rate $(\log N)/N$.

In principle the time we consider, i.e. from $\tau_{\log N}$ to $\tilde{\tau}_N(\varepsilon)$ seems random on first sight, however it is asymptotically deterministic. Namely $\tilde{\tau}_N(\varepsilon) - \tau_{\log N}$ becomes deterministic in the sense that

$$(\tilde{\tau}_N(\varepsilon) - \tau_{\log N}) - (\frac{1}{\alpha} \log(\frac{\varepsilon N}{W}) - \frac{1}{\alpha} \log(\frac{\log N}{W})) \xrightarrow[N \to \infty]{} 0, \text{ a.s.} \tag{8.186}$$

so that conditioned on W as $N \to \infty$, $\tilde{\tau}_N(\varepsilon) - \tau_{\log N}$ can be replaced by a deterministic quantity. This allows us to generate as upper bound on the number of black particles the following process.

1. Realise a Poisson point process with intensity measure

$$(W^2 \alpha e^{2\alpha s}/N) ds. \tag{8.187}$$

2. Then let from each point of this Poisson point process evolve independent families of black particles with the usual dynamics but independent of the Poisson point process.

Next observe that a new black founding particle generated by the Poisson point process has a *descendant black population* which forms by construction a Crump–Mode–Jagers process with growth rate α, i.e. it grows like $\tilde{W}_{(\tilde{\tau}^N(\varepsilon)-s)} e^{\alpha(\tilde{\tau}^N(\varepsilon)-s)}$, if s

8.3 Proofs of Propositions 2.3, 2.6, and 2.11

is the birth time of the black particle and provided we observe up to the final time $\tilde{\tau}^N(\varepsilon)$. Here $\{\tilde{W}_s\}_{s \in \mathbb{R}}$ are independent copies of W.

The newly founded black populations created at the Poisson point process jump times observed at time $\tilde{\tau}_N(\varepsilon)$ have sizes

$$\{\tilde{W}_{(\tilde{\tau}_N(\varepsilon)-s), N} \exp(\alpha(\tilde{\tau}_N(\varepsilon) - s)), \quad s \in \{s_1^N, s_2^N, \ldots, s_{M(N)}^N\}\}, \tag{8.188}$$

(we will suppress the subscript N for s_i^N in formulas below) and are independently distributed conditioned on the process of birth-times and $\tilde{\tau}_N(\varepsilon)$. For these copies we have a law of large numbers acting. The claim is more precisely that if we take the expectation $\tilde{E}$ over $\{\tilde{W}_{\tilde{\tau}_N(\varepsilon)-s_i, N}, \quad i \in \mathbb{N}\}$ then as $N \to \infty$ we have the following law of large numbers effect:

$$\sum_{i=1}^{M(N)} \tilde{W}_{\tilde{\tau}_N(\varepsilon)-s_i, N} \exp(\alpha(\tilde{\tau}_N(\varepsilon) - s_{i,N}))$$

$$\sim \int_0^{\tilde{\tau}_N(\varepsilon)} \alpha W^2 (e^{2\alpha s}/N) \tilde{E}[\tilde{W}_{(\tilde{\tau}_N(\varepsilon)-s), N}] e^{\alpha(\tilde{\tau}_N(\varepsilon)-s)} ds \tag{8.189}$$

and the r.h.s. is asymptotically as $N \to \infty$ equal to

$$E[W] \int_0^{\tilde{\tau}_N(\varepsilon)} \alpha W^2 (e^{2\alpha s}/N) e^{\alpha(\tilde{\tau}_N(\varepsilon)-s)} ds, \tag{8.190}$$

using the convergence theorem for the Crump–Mode–Jagers process.

Then combining (8.188) and (8.189) together with (8.190) we obtain a *mean conditioned on W* of the l.h.s. of (8.189) which is asymptotically as $N \to \infty$ equal to:

$$\left[\int_{\tau_{\log N}}^{\tilde{\tau}^N(\varepsilon)} E[W] \alpha W^2 e^{\alpha \tilde{\tau}_N(\varepsilon)} \cdot e^{\alpha s} ds \right] \sim E[W] W^2 \frac{1}{N} \left(\frac{1}{W^2} \varepsilon^2 N^2 \right) = E[W] \varepsilon^2 N. \tag{8.191}$$

To justify the law of large numbers given in (8.189), we observe that the number of birth events of black populations goes to zero for times t_N with

$$t_N - \frac{\log N}{2\alpha} \underset{N \to \infty}{\Longrightarrow} -\infty \tag{8.192}$$

and for these times the order of magnitude of a descending population is at most of order $\sqrt{N} = o(N)$. Therefore we get a diverging number of contributions which are all $o(N)$.

More precisely, the law of large numbers is verified by showing that

$$\tilde{V}ar\left[\sum \tilde{W}_{(\tilde{\tau}_N(\varepsilon)-s_i),N} e^{\alpha(\tilde{\tau}_N(\varepsilon)-s_i)}\right] = O(\frac{N^2 \log N}{N^3}) = O(\frac{\log N}{N}). \quad (8.193)$$

Here we use the explicit representation and the fact that $Var(W_s) < \infty$ and $Var(W_s) \to Var(W)$ as $s \to \infty$ together with the asymptotics of $\tilde{\tau}_N(\varepsilon)$. This concludes the proof for the law of large numbers.

Hence we can bound the number of black particles by the asymptotic growth in N of its expectation over the randomness in the process of growth, i.e. the $\tilde{W}_{s_i^N}$ in the black population.

This reasoning gives the following asymptotic stochastic upper bound on the number of black sites, namely the expression:

$$W^2 \int_{\tau_{\log N}}^{\tilde{\tau}^N(\varepsilon)} \alpha \frac{e^{\alpha s}}{N} e^{\alpha s} E[\tilde{W}_s] e^{\alpha(\tilde{\tau}^N(\varepsilon)-s)} ds \lesssim E[W] N \varepsilon^2, \text{ as } N \to \infty, \quad (8.194)$$

using the upper bound on $\tilde{\tau}^N(\varepsilon)$ implied by (8.144).

Therefore we conclude that the *proportion of black sites* among all sites in the black and white system at time $\tilde{\tau}^N(\varepsilon)$ is at most (in the limit $N \to \infty$) equal to

$$\frac{E[W]}{W}\varepsilon. \quad (8.195)$$

In particular by letting $\varepsilon \to 0$ this relative frequency goes to 0.

In particular this would imply that the calculation below of $\tau^N(\varepsilon)$ (the time for the number of white plus red sites to reach ε) will give a stochastic lower bound for

$$\tau^N(\varepsilon - const \cdot \varepsilon^2), \quad (8.196)$$

if we condition on W.

Step 4: Upper bound on emergence time

Recalling (8.175) we see that given the current state k of K_t^N, the time of the next birth in K_t^N is a positive random variable denoted τ_k for which we are waiting for the next migration step of a white or red particle to an unoccupied site. This waiting time is *bounded above* by the waiting time of a white particle migration step to an unoccupied site.

Observe that due to the uniform migration distribution, the property to jump to an unoccupied site is an experiment independent of everything else and the jump is successful with probability $1 - k/N$ if k is the current number of occupied sites. Therefore we can view the formation of new occupied colonies as a *pruning* of the point process of white particle migration steps.

8.3 Proofs of Propositions 2.3, 2.6, and 2.11

What can we say about the point process of the attempted migration steps by white particles? We note first that this migration step is *given all current internal states* the infimum of k, independent exponential waiting times, each of which is the minimum of exponential waiting times for the internal transitions like birth, death and emigration. Once an internal transition other than migration occurs, the experiment starts over with this new internal states. The rates of all this exponential waiting times depend on the internal state in the corresponding site. It is best to view this as trials consisting of migration steps arising from k waiting times built from exponentials according to the internal states, waiting for one which leads to a new colony.

We know from the CMJ-theory that the number of migration steps in the white plus black system, with k occupied at time t, in the time interval $[t, t + \Delta t]$ behaves asymptotically as $\alpha k \Delta t$ as $k \to \infty$ a.s. and in L_1 as long as k has not reached too large values. We now use the observation that the relative frequency of black colonies is asymptotically as $N \to \infty$ negligible so that asymptotically the white migration steps are as frequent as the ones of the white and black system as long as k is large but k/N remains small.

When the current number of occupied sites is k the waiting time for the next colonization has mean *bounded above* by that of the waiting time for the birth of white site. To obtain information on this random time we note that the next migration of a particle is given by an exponential random clock which occurs with a hazard function given by the rate $c \sum_{i=1}^{k} \zeta_i(\cdot)$ and the probability that a migration produces a new white site is $1 - \frac{k}{N}$ (compare (8.175)). To get an upperbound on the time $\tau_{\log N}$ we note that $\zeta_i(\cdot) \geq 1$ and note that the number of migration events needed to obtain a new white site is geometric with mean $(1 - k/N)^{-1}$.

Recall that $\tau_{\log N}$ is given by (8.172). Then recalling the convergence to the stable age and size distribution for $t \to \infty$ for $k \to \infty$ and therefore $\frac{c}{k} \sum_{i=1}^{k} \zeta_i(\cdot) \sim \alpha$, we argue below in detail that the expected time, given the internal states of the occupied sites to obtain the next white site for $\log N \leq k \leq \varepsilon' N$, behaves for $N \to \infty$ as follows:

$$\frac{1}{k(\frac{1}{k}\sum_{i=1}^{k} 1_{\zeta_i > 1}\zeta_i(\cdot))(1 - \frac{k}{N})} \sim \frac{1}{k\alpha(1 - \frac{k}{N})} = \frac{1}{\alpha k} + \frac{1}{\alpha N}\frac{1}{1 - \frac{k}{N}}$$

$$\leq \frac{1}{\alpha k} + \frac{1}{\alpha N(1 - \varepsilon')}. \quad (8.197)$$

To make more precise the first step, note first that

$$\lim_{N \to \infty} \left(\frac{1}{K^N_{\tau_{\log N}+j}} \sum_{i=1}^{K^N_{\tau_{\log N}+j}} 1_{\zeta_i > 1}\zeta_i(\cdot) \right) = \lim_{N \to \infty} \left(\sum_{k=2}^{\infty} k U^N(\tau_{\log N+j}, k) \right). \quad (8.198)$$

We know from the CMJ-theory that:

$$\left| \lim_{N \to \infty} \sum_{k=2}^{\infty} k U^N(\tau_{\log N+j}, k) - \sum_{k=2}^{\infty} k U(\infty, k) \right| = 0. \tag{8.199}$$

Therefore

$$\left| \lim_{N \to \infty} \sum_{k=2}^{\infty} k U^N(\tau_{\log N+j}, k) - \alpha \right| = 0 \quad \text{uniformly in } j. \tag{8.200}$$

Secondly we now have to deal with the fact that the quantity we handled above appears in the denominator and is random, so that taking *expectation* needs some care. Note that $\sum_{k=2}^{\infty} k U^N(\tau_{\log N} + j), k) \geq 1 - U^N(\tau_{\log N+j}, 1)$ and

$$\lim_{N \to \infty} (1 - U^N(\tau_{\log N+j}, 1)) > \frac{1}{2}(1 - U(\infty, 1)) > 0. \tag{8.201}$$

We treat below the event $(1 - U^N(\tau_{\log N+j}, 1)) \leq a$ for small a as a rare event and obtain a large deviation bound for it allowing us to proceed for $N \to \infty$ with assuming we are in the complement.

We break the details of the argument down in two parts, namely, the probability that more than $\frac{\log N}{2}$ of the individuals have age less than or equal to B, that is, $\frac{1}{2} \log N$ particles produce $\frac{1}{2} \log N$ new particles in time $\leq B$. We choose B so that the expected number of particles produced by one site in time B is $\leq \frac{1}{3}$. The second part is to calculate that more than a fraction $(1 - a)$ of the sites of age $> B$ have only a single particle where $a > 0$ is chosen so that if a site is of age $\geq B$ then $P(\zeta \geq 2) > 2a$. Since the events that $\frac{\log N}{2}$ particles produce $\frac{\log N}{2}$ particles in time B and the event that more than a fraction $(1 - a)$ sites of age B have $\zeta = 1$ are both "rare" events, we can verify the exponential bound:

$$P[(1 - U^N(\tau_{\log N+j}, 1)) \leq a] \leq e^{-C(\log N+j)} \text{ for large } N, \text{ for some } C > 0. \tag{8.202}$$

We now take expectations *conditioned* to be on the good event,

$$E_*[\cdot] = E[\cdot | (1 - U^N(\tau_{\log N+j}, 1)) > a \text{ for } j = 1, \ldots, \varepsilon N - \log N]. \tag{8.203}$$

We can then conclude (by bounded convergence) that:

$$\lim_{N \to \infty} E_*\left[\frac{1}{\sum_{k=2}^{\infty} k U^N(\tau_{\log N+j}, k)}\right] \leq \frac{1}{\alpha}, \quad \text{uniformly in } j. \tag{8.204}$$

8.3 Proofs of Propositions 2.3, 2.6, and 2.11

This means that we now can conclude with T_k denoting the waiting time for the next migration step if we have k occupied sites satisfies:

$$E_*[T_k] \leq \frac{1}{\alpha k} + \frac{1}{\alpha N(1-\varepsilon')}, \quad k \in [\log N, \varepsilon' N]. \tag{8.205}$$

Remark 73. For the case considered in Section 8.3.4 it is convenient to note the following simple generalization. Consider a system of k independent exponential clocks $C_1, \ldots, C_k$ with exponential waiting times with parameters $c_1, \ldots, c_k$. Then

$$P(\min_i C_i \leq \frac{x}{k}) = (1 - \prod_i P(C_i > \frac{x}{k})) \tag{8.206}$$

$$= (1 - \prod_i e^{-c_i x/k}) = 1 - e^{-(\frac{1}{k}\sum_{i=1}^k c_i)x} \quad \text{as } k \to \infty.$$

In our case $c_i = \zeta_i \in \mathbb{N}$ and this leads to $1 - e^{-(\sum_k k \cdot U^N(\tau_{\log N+j},k))}$, and therefore mean $\frac{1}{k \cdot \sum_i i \cdot U^N(\tau_{\log N+j},i)}$ and second moment $2\left(\frac{1}{k \cdot \sum_i i \cdot U^N(\tau_{\log N+j},i)}\right)^2$ and so the variance is $\left(\frac{1}{k \cdot \sum_i i \cdot U^N(\tau_{\log N+j},i)}\right)^2$.

In order to conclude the argument we next choose $\varepsilon > 0$. Then for sufficiently large N, the mean time, $E_*[\tau^N(\varepsilon)]$, for K_t^N required to reach εN (cf. (8.205)) satisfies

$$E_*[\tau^N(\varepsilon)] \leq E[\tau_{\log N}] + \frac{1}{\alpha}[\frac{1}{\log N} + \cdots + \frac{\log N}{N}] \tag{8.207}$$

$$+ \frac{1}{\alpha}[\frac{\log N}{N} + \cdots + \frac{1}{\varepsilon N}] + \frac{\varepsilon}{\alpha(1-\varepsilon)}.$$

Hence

$$E_*[\tau^N(\varepsilon)|W] \leq \frac{\log N}{\alpha} + c(\varepsilon), \tag{8.208}$$

where $c(\varepsilon)$ is a constant depending on ε and W.

We also have from the CMJ-theory the lower bound

$$E[\tau^N(\varepsilon)|W] \geq E[\tilde{\tau}^N(\varepsilon)|W] = \alpha^{-1}\log N - \left(\frac{|\log \varepsilon'| + \log W}{\alpha}\right). \tag{8.209}$$

Therefore it follows that $\tau^N(\varepsilon)$, conditioned on W, is of the form $\alpha^{-1} \log N + O(1)$, as claimed. We also see that the difference between $\tau^N(\varepsilon)$ and $\tilde{\tau}^N(\varepsilon)$ is a random variable satisfying

$$E_*[|\tau^N(\varepsilon) - \tilde{\tau}^N(\varepsilon)|] < \infty. \tag{8.210}$$

We can also estimate now the time till fixation as follows. The coupling construction now implies that if we choose ε small enough, then at time $\alpha^{-1} \log N$ we will reach εN dual particles which will then in finite random time produce a mutation jump so that we get indeed (2.59) and thus completing the argument. Namely since

$$\Pi_t^N \geq \hat{K}_t - \bar{Z}_t^N, \quad \bar{Z}_t^N \leq \text{r.h.s. (8.194)}, \tag{8.211}$$

after time $\tilde{\tau}^N(\varepsilon)$ there is a mutation rate at least

$$m(\varepsilon - E[W]\varepsilon^2)^+ \tag{8.212}$$

uniformly in large N and therefore the probability (recall $\tau^N(\varepsilon) - \tilde{\tau}^N(\varepsilon) = O(1)$ as $N \to \infty$) that a mutation does not occur before $\tau^N(\varepsilon) + t$ goes to zero as $t \to \infty$ uniformly in N if we choose ε small enough. This completes the proof of (2.59).

Step 5: Extension: convergence of hitting times

However we can get even more mileage from the above argument to make it clear that the randomness in the evolution of spatially macroscopic variables is created only in the very early stages of the growth of the dual particle system.

We next estimate the variance of $\tau^N(\varepsilon) - \tau^N_{\log N}$, a random variable of which we know from the analysis of Step 4 above that it is asymptotically close to $\tilde{\tau}^N(\varepsilon) - \tilde{\tau}^N_{\log N}$ as $N \to \infty$ and $\varepsilon \to 0$. If we now condition on all the internal states of all occupied sites we can calculate the variance of the waiting time T_k for the next migration step explicitly since we know it is exponential. Hence we can, denoting by $\tilde{V}ar(\cdot)$ this conditional variance, estimate $\tilde{V}ar(\tau^N(\varepsilon) - \tau^N_{\log N})$. Conditioned on W and restricted to the complement of the rare event introduced in Step 4 (which is asymptotically negligible (8.202)) we can verify by explicit calculation (following the same argument as for the means in (8.207) now for the variances, in this case we get the sum of squares of the terms we had in the previous mean calculation) that with $\tilde{V}ar_*$ denoting this conditional variance *restricted to the complement of the rare event*, we get the bound

$$\limsup_{N \to \infty} \tilde{V}ar_*[(\tau^N(\varepsilon) - \tau_{\log N})] \leq \limsup_{N \to \infty} \left(\text{const} \cdot \left(\frac{1}{\log N} - \frac{1}{\varepsilon N} \right) \right) = 0. \tag{8.213}$$

Now we note that the term in the limsup on the r.h.s. is independent of the internal states and we conclude that conditional on W the (restricted) variance of the difference between $\tau^N(\varepsilon)$ and $\tau_{\log N}$ goes to zero and hence that is, the variance of the sum conditioned on W of the waiting times (with means given in (8.197)) in the second and third time intervals given in (8.171) goes to 0 as $N \to \infty$ since the rare event (and hence the restriction) become negligible as $N \to \infty$. The analogous result also holds for $\tilde{\tau}^N(\varepsilon) - \tilde{\tau}_{\log N}$.

8.3 Proofs of Propositions 2.3, 2.6, and 2.11

This means that the growth in the second and third time interval we have specified in (8.171) is *deterministic* and the randomness occurred earlier in the first time interval.

Note that during the first time interval as $N \to \infty$ the expected number of collisions is $O(\frac{(\log N)^2}{N})$ so that no collisions occur with probability tending to 1. Therefore as $N \to \infty$,

$$\tilde{\tau}^N_{\log N} - \tau^N_{\log N} \xrightarrow[N \to \infty]{} 0 \quad \text{in probability.} \tag{8.214}$$

Together with the fact that as $N \to \infty$ the limiting probability of the rare event is 0 it follows that there exists a deterministic sequences, namely

$$(\tilde{\delta}^\varepsilon_N = E[\tilde{\tau}^N_{\varepsilon N} - \tilde{\tau}^N_{\log N}])_{N \in \mathbb{N}}, (\delta^\varepsilon_N = E_*[\tau^N_{\varepsilon N} - \tau^N_{\log N}])_{N \in \mathbb{N}} \tag{8.215}$$

such that for $\eta > 0$,

$$P[|\tilde{\tau}^N(\varepsilon) - \tilde{\tau}^N_{\log N} - \tilde{\delta}^\varepsilon_N| > \eta] \to 0, \tag{8.216}$$

and

$$P[|\tau^N(\varepsilon) - \tau^N_{\log N} - \delta^\varepsilon_N| > \eta] \to 0. \tag{8.217}$$

We now collect the above estimates to prove the following.

Proposition 8.5 (Convergence of hitting times of level ε). *There exists a non-degenerate $\mathbb{R}$-valued random variable $\tilde{\tau}(\varepsilon)$ such that*

$$\tilde{\tau}^N(\varepsilon) - \frac{\log N}{\alpha} \Rightarrow \tilde{\tau}(\varepsilon), \text{ as } N \to \infty, \tag{8.218}$$

$$\tau^N(\varepsilon) - E_*[\tau^N(\varepsilon)] \text{ converges in distribution} \tag{8.219}$$

and there exists a constants $\{C(\varepsilon)\}_{\varepsilon > 0}$ such that $\lim_{\varepsilon \to 0} C(\varepsilon) = 0$ and

$$\lim_{N \to \infty} P\left(\frac{\log N + \log \varepsilon - \log W}{\alpha} - C(\varepsilon) < \tau^N(\varepsilon) \right. \tag{8.220}$$
$$\left. < \frac{\log N + \log \varepsilon - \log W}{\alpha} + 2C(\varepsilon) \Big| W\right) = 1. \quad \square$$

Proof. We begin with some preparatory points. Note that asymptotically as $N \to \infty$ by (8.144),

$$\tilde{\tau}^N_{\log N} - \frac{\log \log N - \log W_N}{\alpha} = o(1) \tag{8.221}$$

where $W_N \to W$ a.s., and by (8.214) it follows that

$$\tau^N_{\log N} - \frac{\log \log N - \log W_N}{\alpha} = o(1). \tag{8.222}$$

But by (8.216), (8.217),

$$P[|(\tilde{\tau}^N(\varepsilon) - \tilde{\tau}_{\log N} - \tilde{\delta}^\varepsilon_N) - (\tau^N(\varepsilon) - \tau^N_{\log N} - \delta^\varepsilon_N)| > 2\eta] \to 0. \tag{8.223}$$

But then (8.210) implies that

$$\sup_N |\tilde{\delta}^\varepsilon_N - \delta^\varepsilon_N| < \infty. \tag{8.224}$$

We also have that $\tilde{\delta}^\varepsilon_N - (\frac{\log(N\varepsilon) - \log\log N}{\alpha}) \to 0$. Therefore as $N \to \infty$:

$$\tilde{\tau}^N(\varepsilon) - \frac{\log N}{\alpha} \to \frac{\log \varepsilon - \log W}{\alpha}. \tag{8.225}$$

Having finished the preparation we already see that (8.225) implies the weak convergence claimed in (8.218). Now we come to the proof of (8.219) and (8.220).

The relation (8.220) then follows from using (8.233) and inserting (8.222), (8.221) together with (8.224) and the fact that $\sup_N |\tilde{\delta}^\varepsilon_N - \delta^\varepsilon_N|$ tends to zero as $\varepsilon \to 0$. In order to see now (8.219) we have to use that given W the $\tau^N(\varepsilon) - \tau^N_{\log N}$ is as $N \to \infty$ asymptotically deterministic as we saw above. Then the claim follows with (8.222). q.e.d.

8.3.7 Dual Process in the Collision Regime: Nonlinear Dynamics

In order to obtain more detailed results on the emergence regime and on the process of fixation via the dual process we will need more precise information about the growth of the number of sites occupied by the dual process, respectively the states at these sites, once we reach the regime where *collisions* play a role.

The behaviour in this regime is qualitatively very different from the collision-free situation and is the result of three properties of the evolution mechanism:

1. Randomness enters in early stages of the growth of the dual process, i.e. times in $[0, O(1)]$, here the dual behaves as a CMJ branching process described by *linear* (i.e. Markov) but *random* dynamics,

8.3 Proofs of Propositions 2.3, 2.6, and 2.11

2. in the time regime

$$[s(N), \frac{\log N}{s(N)}), \text{ with } s(N) \to \infty, \ s(N) = o(\log N) \qquad (8.226)$$

the law of large numbers is in effect leading to *deterministic* but still *linear* dynamics

3. in the regime ($s(N)$ as above)

$$[\frac{\log N}{s(N)}, \frac{\log N}{\alpha} + t) \qquad (8.227)$$

the dynamics is *deterministic* but the evolution equation becomes *nonlinear* due to collisions.

The problem is now that these regimes separate as $N \to \infty$ and we have to connect them through a careful analysis. To integrate all three phases into a *deterministic nonlinear* dynamic with *random* initial condition, the full dual process for the finite system with N exchangeable sites is analysed.

Indeed in this and the next subsubsections we show that in the limit $N \to \infty$ the initial asymptotically collision-free evolution stage in (1) and (2) produces *a random initial condition* for the nonlinear dynamics arising in (3) where the randomness arises from stage (1). A key step in linking the random and nonlinear aspects is the analysis of the evolution in the time interval in (8.227) in the limit $N \to \infty$, $t \to -\infty$.

In this subsection we consider the time interval

$$\frac{1}{\alpha}[\log N - \log \log N, \log N + T) \qquad (8.228)$$

in which the dynamics are deterministic. Recall that

$$u^N(\frac{1}{\alpha}(\log N - \log \log N)) = W_N \frac{N}{\log N} = W \frac{N}{\log N} + o(1) \text{ as } N \to \infty, \qquad (8.229)$$

since $W_N \Rightarrow W$ by the equation (8.144).

We have to describe the dual process in the above time interval in the limit $N \to \infty$ and to obtain a limit dynamic which allows to draw the needed conclusions for the original process. We proceed in three steps. It turns out that the main properties of the dual process needed for that purpose can be captured in a triple of functionals of this process which we introduce in Step 1. In Step 2 we formulate the corresponding limiting (for $N \to \infty$) objects which allows us to finally state and prove in Subsubsection 8.3.8 the key convergence relation for the dual process. Step 3 proves that the defined evolutions have the desired properties.

Step 1 *Some functionals of the dual in the time span up to fixation*

First, we observe that to determine the quantity of interest for emergence namely

$$\frac{\Pi^N_{T_N+t}}{N}, \text{ for } T_N = \frac{\log N}{\alpha} \tag{8.230}$$

as a function of t we need to know the number of sites occupied at time $T_N + t$ and relative frequencies of the occupation numbers of these sites. Since the occupation number of a site depends on the age of the site, we will keep track of the number of occupied sites and the relative frequencies of both age and occupation number for these sites. As $N \to \infty$ a law of large number argument is then used to obtain the desired information.

We consider the number of sites occupied at time t denoted K^N_t and the corresponding measure-valued process on $[0, \infty) \times \mathbb{N}_0$ giving the unnormalized number of sites of a certain age u in $[a, b)$ and occupation size j:

$$\Psi^N(t, [a,b), j)) = \int_{(t-b)}^{(t-a)} 1_{(K^N_u > K^N_{u-})} 1_{(\zeta^N_u(t)=j)} dK^N_u, \tag{8.231}$$

where $\zeta^N_u(t)$ denotes the occupation number at time t of the site born at time u, that is, a site first occupied the last time at time u, which is therefore at time t exactly of age $t - u$.

The *normalized empirical age and size distribution* among the occupied sites is defined as:

$$U^N(t, [a,b), j) = \frac{1}{K^N_t} \Psi^N(t, [a,b), j), \quad t \geq 0, \ j \in \{1, 2, 3, \dots\}. \tag{8.232}$$

Thus for $(\zeta^N_t)_{t \geq 0}$ we have a growing number K^N_t of different interacting birth and death processes which each have in state k linear birth rate sk, linear death rate $c(1 - \frac{1}{N})k$, quadratic death rate $dk(k-1)$. In addition immigration of one additional element into a site of size k occurs with rate

$$c\left(\frac{K^N_t c^{-1}(\alpha_N(t) + \gamma_N(t)) - k}{N}\right), \tag{8.233}$$

where $\alpha_N(t)$ is c times the mean size of sites with more than one particle, γ_N is c times the frequency of singletons; these will be defined precisely below in (8.238). Note that $U^N(t, dx, j)$ is a pure atomic measure which evolves with jumps of size $\frac{1}{K^N_t}$ one for each migration event to a yet unoccupied site, or $-\frac{1}{K^N_t}$ when a singly occupied site migrates and a collision occurs.

Remark 74. Note that the $\zeta^N_u(\cdot)$ interact if we are in a regime where collisions occur, that is at times after emergence has occurred which happens at times of order $\frac{\log N}{\alpha}$. The marginal distribution of age is the analogue of the empirical age

8.3 Proofs of Propositions 2.3, 2.6, and 2.11

distribution in (8.162). However now we take into account possible collisions, that is, a migration to a previously occupied site when we consider the system of N sites with symmetric migration among these sites and in this case the distribution of $\zeta_u^N(t)$ may depend not only on the age $t - u$ but also on t due to changes in the collision rates.

Set for convenience to describe the pair given by the number of sites and size-age distribution:

$$u^N(t) := K_t^N. \tag{8.234}$$

We have obtained with (8.232), (8.234) now a pair which is $\mathbb{N} \times \mathcal{P}([0, \infty) \times \mathbb{N})$-valued, denoted

$$(u^N(t), U^N(t, \cdot, \cdot))_{t \geq 0} \tag{8.235}$$

and which completely describes our dual particle system up to permutations of sites. We can recover the functional Π_t^N as

$$\Pi_t^N = u^N(t) \sum_{j=1}^{\infty} j U^N(t, [0, t], j). \tag{8.236}$$

However since we work with exchangeable initial states for our original population process, the pair (u^N, U^N) provides a sufficiently complete description of the dual process to allow us to calculate the laws of our original processes.

In the interval $\alpha^{-1}[\log N - \log \log N, \log N + T]$ the process $(u^N(t))_{t \geq 0}$ increases by one, respectively decreases by one, at rates

$$\alpha_N(t)(1 - \frac{u^N(t)}{N})u^N(t), \text{ respectively } \gamma_N(t)\frac{(u^N(t))^2}{N}, \tag{8.237}$$

where $\alpha_N(t), \gamma_N(t)$ are defined:

$$\alpha_N(t) = c \int_0^t \sum_{j=2}^{\infty} j U^N(t, ds, j) ds, \quad \gamma_N(t) = c \int_0^t U^N(t, ds, 1). \tag{8.238}$$

These rates of change of $U^N(t, \cdot, \cdot)$ follow directly from the dynamics of the dual particle system η.

We note that if we start with k particles at each of ℓ different sites we obtain similar quantities, which we denote by

$$(u_t^{N,k,\ell}, U_t^{N,k,\ell})_{t \geq 0}. \tag{8.239}$$

The dynamics of the system

$$(u^N(t), U^N(t, ds, dx)) \tag{8.240}$$

evolves in two main regimes corresponding to times $o(\log N)$ and those of the form $\alpha^{-1} \log N + t$, with $t = O(1)$. They are asymptotically as $N \to \infty$ *linear* and *random* for times of order $o(\log N)$ as we have demonstrated in the previous subsubsection with the randomness captured by the random variable W given in (8.144) respectively (8.229) above. On the other hand the dynamics become *deterministic* and *nonlinear* in the post-emergence regime, that is, for times of the form $(T_N + t)_{t \in (-\infty, \infty)}$ where $T_N = \frac{\log N}{\alpha}$. This latter case involves the law of large numbers in the limit $N \to \infty$ and collisions play a decisive role. We will introduce the limiting dynamics in Step 2. We will later in Subsubsection 8.3.9 show how the two regimes are linked and that the first produces a random initial condition for the second, the nonlinear regime.

Step 2 *The limiting dynamics for dual in collision regime*

The second step is to define for $k, \ell \in \mathbb{N}$ candidates

$$(\Pi_t^{k,\ell}, u^{k,\ell}(t), U^{k,\ell}(t, ds, dx))_{t \in \mathbb{R}} \text{ and } (\zeta(t))_{t \in \mathbb{R}} \tag{8.241}$$

for the limiting objects as $N \to \infty$ related to (here $T_N = \alpha^{-1} \log N$)

$$\left(\frac{\Pi_{T_N+t}^{N,k,\ell}}{N}, \frac{u_{T_N+t}^{N,k,\ell}}{N}, U^{N,k,\ell}(T_N + t, ds, dx) \right), \quad -\infty < t < \infty,$$
$$\{\zeta^{N,(k,\ell)}(T_N + t, i), \quad i = 1, \cdots, N\}, \quad -\infty < t < \infty, \tag{8.242}$$

when the dual process initially has k *particles* at each of ℓ *distinct sites* at time $t = 0$.

We now turn to the limiting objects as $N \to \infty$ which we denote

$$(\Pi_t^{k,\ell}, u(t), U(t))_{t \in \mathbb{R}}, \quad (\zeta^{k,\ell}(t))_{t \in \mathbb{R}}. \tag{8.243}$$

We remark that it will turn out that conditioned on $(\Pi_{t_0}^{k,\ell}, u^{k,\ell}(t_0), U^{k,\ell}(t_0, ds, dx))$, the dynamics of $(\Pi_t^{k,\ell}, u^{k,\ell}(t), U^{k,\ell}(t, ds, dx))_{t \geq t_0}$ is independent of k, ℓ. For this reason we often suppress the superscripts and later indicate how the initial conditions for $k, \ell > 1$ are determined.

Since

$$\Pi_t^{k,\ell} = u^{k,\ell}(t) \cdot \sum_{j=1}^{\infty} j \cdot U^{k,\ell}(t, \mathbb{R}^+, j), \tag{8.244}$$

and since the process ζ will depend on this pair from the triple in (8.241) it suffices to specify the pair $(u(t), U(t))_{t \in \mathbb{R}}$. This is done by first establishing that they satisfy a pair of deterministic equations and then determine the entrance law at $-\infty$ as follows.

8.3 Proofs of Propositions 2.3, 2.6, and 2.11

We first define two ingredients (parallel to (8.238)):

$$\alpha(t) = \alpha(U(t)) := c \int_0^\infty \sum_{j=2}^\infty j U(t, ds, j), \quad \gamma(t) = \gamma(U(t)) := c \int_0^\infty U(t, ds, 1). \tag{8.245}$$

As before we first consider the evolution equation for times $t \geq t_0$ and then we try to characterize an entrance law by considering $t_0 \to -\infty$. For the moment, for notational convenience let $t_0 = 0$.

Then to specify

$$(u, U) = (u(t), U(t))_{t \geq 0}, \text{ with } (u(t), U(t)) \in \mathbb{R}_+ \times \mathcal{M}_1(\mathbb{R}_+ \times \mathbb{N}), \tag{8.246}$$

we introduce the (coupled) system of nonlinear evolution equations:

$$\frac{du(t)}{dt} = \alpha(t)(1 - u(t))u(t) - \gamma(t)u^2(t), \tag{8.247}$$

$$\frac{\partial U(t, dv, j)}{\partial t} \tag{8.248}$$

$$= -\frac{\partial U(t, dv, j)}{\partial v}$$

$$+ s(j-1) 1_{j \neq 1} U(t, dv, j-1) - sj U(t, dv, j)$$

$$+ \frac{d}{2}(j+1) j U(t, dv, j+1) - \frac{d}{2} j(j-1) 1_{j \neq 1} U(t, dv, j)$$

$$+ c(j+1) U(t, dv, j+1) - cj U(t, dv, j) 1_{j \neq 1}$$

$$- c u(t) U(t, dv, 1) 1_{j=1}$$

$$+ u(t)(\alpha(t) + \gamma(t))[1_{j \neq 1} U(t, dv, j-1) - U(t, dv, j)]$$

$$+ ((1 - u(t))\alpha(t)) 1_{j=1} \cdot \delta_0(dv)$$

$$- \left(\alpha(t)(1 - u(t)) - \gamma(t) u(t)\right) \cdot U(t, dv, j).$$

Note that if $U(0, \cdot, \cdot)$ is a probability measure on $\mathbb{R}_+ \times \mathbb{N}$, then this property is preserved under the evolution. (Recall here that α and γ contain the constant c of migration.)

We note that the right side equation (8.248) is only defined in terms of $\mathcal{M}_{\text{fin}}(\mathbb{R}^+ \times \mathbb{N})$ elements, if we restrict to measures for which $\int_0^\infty \sum_j j^2 \cdot U(dv, j) < \infty$, so we have to restrict to a subset of $\mathcal{M}_1(\mathbb{R}^+ \times \mathbb{N})$. Define ν as the measure on $\mathbb{N}$ given by

$$\nu(j) = 1 + j^2. \tag{8.249}$$

and let $L^1(\nu, \mathbb{N})$ denote the corresponding space of integrable functions.

We can write these equations in compact form as

$$\frac{d}{dt}(u(t), U(t, \cdot, \cdot)) = \vec{G}^*(u(t), U(t, \cdot, \cdot)), \qquad (8.250)$$

with the *nonlinear* operator

$$\vec{G}^* \text{ mapping } \mathbb{R}_+ \otimes L^1_+(\mathbb{R}_+ \times \mathbb{N}, \nu) \to \mathbb{R}_+ \otimes \mathbb{F}, \qquad (8.251)$$

with $\mathbb{F}$ denoting the class of finite measurable functions f on $\mathbb{R}_+ \times \mathbb{N}$.

If $u(0) = u_0 > 0$, we also work with the equivalent $\mathcal{M}_{\text{fin}}(\mathbb{R}_+ \times \mathbb{N})$-valued process

$$\hat{\Psi}_t := u(t) \cdot U(t) \qquad (8.252)$$

which solves the nonlinear systems

$$\frac{d\hat{\Psi}_t}{dt} = G^*(\hat{\Psi}_t) \qquad (8.253)$$

equivalent to (8.250) (note that it is easy to verify that if $u(t_0) = \hat{\Psi}_{t_0}(\mathbb{R}_+ \times \mathbb{N}) > 0$, then $u(t) > 0$ for all $t \geq t_0$).

We think of this as a forward equation for a deterministic measure-valued process. In order to formulate the corresponding Markov semigroup generator we consider functions $F_{g,f}$ on $(\mathbb{R}_+ \otimes L^1_+(\mathbb{R}_+ \times \mathbb{N}, \nu))$ of the form

$$F_{g,f}(\hat{\Psi}_0) = g\left(\sum_{j=1}^\infty \int_0^\infty f(r, j) \hat{\Psi}_0(dr, j)\right) \qquad (8.254)$$

$g : \mathbb{R} \to \mathbb{R}$, $f : [0, \infty) \times \mathbb{N} \to \mathbb{R}$, g has bounded first and second derivatives,

f bounded and has bounded first derivatives in the first variable.

The algebra of functions, which contains all functions of the form (8.254) is denoted

$$D_0(\mathbf{G}) \qquad (8.255)$$

and is a separating set in $C_b(\mathcal{M}_{\text{fin}}([0, \infty) \times \mathbb{N}_0))$ and is therefore *distribution-determining for laws* on $\mathcal{M}_{\text{fin}}([0, \infty) \times \mathbb{N}_0)$.

Remark 75. In fact since we are concerned with $\hat{\Psi}$ which are sub-probability measures we note that the integrals are bounded, we can include unbounded functions g defined on $[0, 1]$ and in $C^2([0, 1], \mathbb{R})$. In particular, we can consider $g(x) = x$ and $g(x) = x^2$.

8.3 Proofs of Propositions 2.3, 2.6, and 2.11

Then we introduce the semigroup

$$V_t F_{g,f}(\hat{\Psi}_0) := F_{g,f}(\hat{\Psi}(t)), \tag{8.256}$$

where $\hat{\Psi}(t) = u(t)U(t)$ and $(u(t), U(t))$ is the solution of the pair (8.247), (8.248) with initial condition $\hat{\Psi}_0 = u_0 \cdot U_0$. Let

$$\hat{\Psi}(f) := \left(\sum_{j=1}^{\infty} \int_0^{\infty} f(r,j) \hat{\Psi}_0(dr,j) \right). \tag{8.257}$$

Then with $F_{g,f}$ as in (8.254) we set

$$\begin{aligned}
\mathbf{G}F_{g,f}(\hat{\Psi}) &= g'\left(\hat{\Psi}(f)\right) \left(\hat{\Psi}(G^{\hat{\Psi}} f(r,j))\right) \\
&= g'\left(\hat{\Psi}(f)\right) \int_0^{\infty}\int_1^{\infty} \tfrac{\partial f(r,x)}{\partial r} \hat{\Psi}(dr,dx) \\
&\quad + g'(\hat{\Psi}(f)) \int_0^{\infty}\int_1^{\infty} sx \left[(f(r,x+1) - f(r,x))\right] \hat{\Psi}(dr,dx) \\
&\quad + g'(\hat{\Psi}(f)) \int_0^{\infty}\int_1^{\infty} \tfrac{d}{2} x(x-1) \left[f(r,x-1) - f(r,x)\right] \hat{\Psi}(dr,dx) \\
&\quad + g'(\hat{\Psi}(f)) \cdot c(1 - \hat{\Psi}(1)) \Big(\int_0^{\infty}\int_1^{\infty} [f(0,1) + f(r,x-1) - f(r,x)] \\
&\qquad x 1_{x \neq 1} \hat{\Psi}(dr,dx) \Big) \\
&\quad + g'(\hat{\Psi}(f)) \hat{\Psi}(1) c \int_0^{\infty}\int_1^{\infty} [f(r,x+1) - f(r,x))] \, x \hat{\Psi}(dr,dx) \\
&\quad + g'(\hat{\Psi}(f)) c (\int_0^{\infty}\int_1^{\infty} x \hat{\Psi}(dr,dx)) \Big(\int_0^{\infty}\int_1^{\infty} [1_{\tilde{x} \neq 1} f(\tilde{r}, \tilde{x} - 1) \\
&\qquad - f(\tilde{r}, \tilde{x}))] \hat{\Psi}(d\tilde{r}, d\tilde{x}) \Big) \\
&= g'(\hat{\Psi}(f)) \cdot \int_0^{\infty}\int_1^{\infty} G^{\hat{\Psi}} f(r,x) \hat{\Psi}(dr,dx),
\end{aligned} \tag{8.258}$$

with the obvious extension to $D_0(\mathbf{G})$.

We can then consider the martingale problem associated with $(\mathbf{G}, D_0(\mathbf{G}))$, that is, to determine probability measures, P, on $C([0,\infty), \mathcal{M}_{\text{fin}}(\mathbb{R}_+ \times \mathbb{N}) \cap \mathbb{B})$ (where $\mathbb{B}$ is defined below in (8.261)) such that

$$F_{g,f}(\hat{\Psi}(t)) - \int_0^t (\mathbf{G} F_{g,f})(\hat{\Psi}(s)) ds \tag{8.259}$$

is a P-martingale for all $F \in D_0(\mathbf{G})$.

Since $\mathbf{G}$ is a *first order* operator any solution to this martingale problem is deterministic. To verify this consider $g(x) = x^2$ and $g(x) = x$ and show that $\mathbf{G}(F_{x^2}) - 2 F_x \mathbf{G} F_x = 0$. This holds by inspection of (8.258). This implies that the variance of the marginal distribution $\hat{\Psi}_t(f)$ for a solution to the martingale problem is 0.

Then taking next $g(x) = x$ we have that any solution satisfies

$$\hat{\Psi}_t(f) - \int_0^t \hat{\Psi}_s(G^{\hat{\Psi}_s} f) ds = \hat{\Psi}_t(f) - \int_0^t G^* \hat{\Psi}_s(f) ds. \tag{8.260}$$

Taking $f \equiv 1$ and then indicator functions $f = 1_j$ we obtain our nonlinear system. In other words any solution to the martingale problem is a deterministic trajectory satisfying the nonlinear dynamics (8.247), (8.248). This form will play a role later on when we consider the convergence.

In order to discuss the existence and uniqueness of solutions to these equations we introduce a norm and reformulate the equations as a nonlinear evolution equation in a Banach space, namely,

$$\hat{\mathbb{B}} := \mathbb{R} \otimes L^1(\mathbb{R}_+ \times \mathbb{N}, ds \times \nu) \tag{8.261}$$

which we furnish with the norm

$$\|\hat{\Psi}\| = \|(u, (a_i(\cdot))_{i \in \mathbb{N}})\| = |u| + \|(a_i(\cdot))_{i \in \mathbb{N}}\|_1, \tag{8.262}$$

where $u = \hat{\Psi}(\mathbb{R} \times \mathbb{N})$, $a_i(\cdot) = \hat{\Psi}(\cdot \times \{i\})$ and

$$\|(a_i(\cdot))_{i \in \mathbb{N}}\|_1 := \sum_{j=1}^{\infty} (1 + j^2) |a_j(\cdot)|_{\text{var}}, \tag{8.263}$$

where var denotes the total variation norm for measures on $\mathbb{R}_+$. On this space the r.h.s. of the equation (8.248) is well-defined.

We note that the Markov property is satisfied by the equivalent systems $(u(t), U(t, \mathbb{R}, \cdot))$ and $\hat{\Psi}(t) = u(t) \cdot U(t, \mathbb{R}, \cdot)$. Since this all that is needed for the dual representation we often suppress the ages. In this case we use

$$\mathbb{B} := \mathbb{R} \otimes L^1(\mathbb{N}, \nu) \tag{8.264}$$

and the simpler norm

$$\|(u, (a_i)_{i \in \mathbb{N}})\| = |u| + \|(a_i)_{i \in \mathbb{N}}\|_1, \quad \text{where} \quad \|(a_i)_{i \in \mathbb{N}}\|_1 := \sum_{j=1}^{\infty} (1 + j^2) |a_j|. \tag{8.265}$$

Existence of a solution to this system will follow from the convergence result below in Subsubsection 8.3.8 (Proposition 8.8). The question of uniqueness given the initial point $(u(t_0), U(t_0))$ is considered in Step 3 below.

Return now to the finite N-system and imagine the collection of occupied sites in our dual particle process at a specific time t represented in the form $\zeta^N(t, i), i = 1, \ldots, N_t$. To complete the description we must specify what happens at a single tagged site for times $r \geq t$, i.e. to $\zeta^N(r, i), i \in \mathbb{N}$ in the limit $N \to \infty$.

8.3 Proofs of Propositions 2.3, 2.6, and 2.11

Given the function $(u(t), U(t))$ this will be given by the time-inhomogeneous birth, death and immigration process

$$\{\zeta(t,i), t \geq 0\}, \text{ with transition probabilities } \{p_{k,j}(t), (k,j) \in \mathbb{N}^2, \quad t \geq 0\}, \tag{8.266}$$

given by the *forward* nonlinear Kolmogorov equation starting with one particle at site i at time 0, i.e., $p_{1,1}(0) = 1$ and which reads as follows:

$$p'_{1,j}(t) = (s(j-1) + \alpha(t)u(t))p_{1,j-1}(t) + \frac{d}{2}(j+1)jp_{1,j+1}(t) - ((s+c)j$$

$$+ \frac{d}{2}j(j-1))p_{1,j}(t) \tag{8.267}$$

for $j \neq 0, 1$,

$$p'_{1,0}(t) = cu(t)p_{1,1}(t) - \alpha(t)u(t)p_{1,0}(t), \tag{8.268}$$

$$p'_{1,1}(t) = d \cdot p_{1,2}(t) - ((s + \alpha(t)u(t)))p_{1,1}(t), \tag{8.269}$$

where

$$u(t) = 1 - p_{1,0}(t), \quad \alpha(t)u(t) = c\sum_{j=2}^{\infty} jp_{1,j}(t). \tag{8.270}$$

This process $\zeta(t,i)$ will describe the evolution of a tagged site i in the limiting system in the collision regime.

Next, in order to prepare for the study of the nonlinear system (8.247),(8.248) we consider a related set of equations for the non-collision regime obtained by setting $u \equiv 0$, namely,

$$\tag{8.271}$$

$$\frac{\partial \tilde{U}(t, dv, j)}{\partial t} = -\frac{\partial \tilde{U}(t, dv, j)}{\partial v}$$

$$+ s(j-1)1_{j \neq 1}\tilde{U}(t, dv, j-1) - sj\tilde{U}(t, j)$$

$$+ \frac{d}{2}(j+1)j\tilde{U}(t, dv, j+1) - \frac{d}{2}j(j-1)1_{j \neq 1}\tilde{U}(t, dv, j)$$

$$+ c(j+1)\tilde{U}(t, dv, j+1) - cj\tilde{U}(t, dv, j)1_{j \neq 1}$$

$$+ \tilde{\alpha}(t)[\delta_0 \times 1_{j=1} - \tilde{U}(t, dv, j)],$$

where $\tilde{\alpha}(t) := c\sum_{\ell=2}^{\infty} \int_0^{\infty} \ell \cdot \tilde{U}(t, dv, \ell)$.

Since the dual representation expressions only involve the marginals $U(t, \mathbb{R}_+, j)$ and the jump rates depend on j but not the age of an occupied site it suffices to work with these marginal distributions we denote by $U(t, j)$. We then work with the corresponding versions of (8.247),(8.248) and (8.271) and use the norm (8.265).

We observe next that the solution to the nonlinear system (8.271) can be obtained as

$$\tilde{U}(t, dv, j) = \frac{V(t, dv, j)}{\sum_j \int_0^\infty V(t, dv, j)} \qquad (8.272)$$

where $\{V(t, \cdot, \cdot)\}$ solve the *linear* system is obtained by deleting the last term in (8.288). The resulting linear equations coincide with the mean equations for the CMJ process and hence the convergence of $U(t, \cdot, \cdot)$ as $t \to \infty$ to the stable size and age distribution follows from the CMJ theory. In addition we have that $\alpha(t) \to \alpha$ as $t \to \infty$, the Malthusian parameter of the CMJ process.

We define a linear operator given by the infinite matrix

$$Q^{a,*} \text{ on } L_+^1(\mathbb{N}, \nu) \subseteq \mathcal{M}_{\text{fin}}(\mathbb{N}), \qquad (8.273)$$

obtained by first integrating over ages setting $\alpha(t) = a$ on the right side of the (8.271). Note that the equation resulting this way from (8.271) can be viewed as the forward equations for a Markov chain Z on $\mathbb{N}$ with transition matrix $Q^{a,*}$. Let

$$(S^a)_{t \geq 0} = \text{ the dual semigroup on } \mathcal{M}_1(\mathbb{N}) \text{ of this time-homogeneous Markov chain on } \mathbb{N}. \qquad (8.274)$$

To verify that the semigroup S_t^α corresponding to $Q^{\alpha,*}$ is strongly continuous on $(\mathbb{B}, \|\cdot\|)$, note that for $\mu \in \mathcal{M}_1(\mathbb{N})$,

$$\lim_{t \to 0} \|S_t \mu - \mu\| = \lim_{t \to 0} [E[Z^2(t)] - E[Z^2(0)]] = 0, \qquad (8.275)$$

follows from elementary properties of the Markov chain $(Z(t))_{t \geq 0}$.

Note that the generator of S^α is an unbounded operator but has domain containing

$$D_0 := \{(a_i)_{i \in \mathbb{N}} \in \mathbb{B} : \sum_{j=1}^\infty (1 + j^4)|a_j| < \infty\} \qquad (8.276)$$

and in fact this is a core for the semigroup ([EK2], Ch. 1, Prop. 3.3), since the birth and death process with linear birth and quadratic death rates has moments of all orders at positive times. (Note that the corresponding Markov process on $\mathbb{N}$ has an entrance law starting at $+\infty$. This property is wellknown for the Kingman coalescent but extends to our birth and death process provided $d > 0$ and is easily extended to include an additional immigration term.)

8.3 Proofs of Propositions 2.3, 2.6, and 2.11

The Markov chain associated with $Q^{a,*}$ has a unique equilibrium state $U^a(\infty)$. Let $m(a)$ denote the mean equilibrium value

$$m(a) = \sum_{j=2}^{\infty} j \cdot U^a(\infty, j). \tag{8.277}$$

The equation $m(a) = a$ has a unique positive solution and it is equal to α. Moreover S_t^α satisfies

$$\lim_{t\to\infty} (S_t^\alpha U(0)) = \mathcal{U}(\infty), \quad \limsup_{t\to\infty} |\alpha(t) - \alpha| = 0, \tag{8.278}$$

where $\mathcal{U}(\infty) = U^\alpha(\infty)$ is the unique equilibrium with mean α and is the stable size distribution for the CMJ process.

Step 3 *Uniqueness and existence results*

We next show that the equations we gave in Step 2 uniquely determine the objects needed for our convergence results.

Lemma 8.6 (Uniqueness of the pair (u, U) and of ζ).

(a) *The pair of equations (8.247)–(8.248), for $(u(t), U(t, \mathbb{R}_+, \cdot))$ given an initial state from $\mathbb{R}^+ \times \mathcal{M}_1(\mathbb{N})$ satisfying*

$$(u(t_0), U(t_0, \cdot)) \in [0, 1] \otimes \mathcal{M}_1(\mathbb{N}) \cap L^1(\mathbb{N}, \nu) \tag{8.279}$$

at time t_0, has a unique solution $(u(t), U(t))_{t \geq t_0}$ with values in $[0, 1] \otimes (\mathcal{M}_1(\mathbb{N}) \cap L_+^1(\mathbb{N}, \nu))$.

(b) *Given (u, U) the process $(\zeta(t))_{t \geq t_0}$ is uniquely determined by (8.267).*

(c) *There exists a solution (u, U) with time parameter $t \in \mathbb{R}$ for every $A \in (0, \infty)$ with values in $\mathbb{B} = \mathbb{R} \otimes L^1(\mathbb{N}, \nu)$, such that*

$$u(t)e^{-\alpha t} \to A \text{ as } t \to -\infty, \tag{8.280}$$

$$U(t) \xrightarrow[t \to -\infty]{} \mathcal{U}(\infty), \quad \text{and } \limsup_{t \to -\infty} e^{-\alpha t}|\alpha(t) - \alpha| = B < \infty. \tag{8.281}$$

Here $\mathcal{U}(\infty)$ is the stable age and size distribution of the CMJ-process corresponding to the particle process $(K_t, \zeta_t)_{t \geq 0}$ given by the McKean–Vlasov dual process η, which was defined in (8.165).

(d) *Given any solution (u, U) of equations (8.247)–(8.248) integrating out the age (given in (8.286)–(8.287)) for $t \in \mathbb{R}$ with values in the space $\mathbb{B} := \mathbb{R} \otimes L^1(\mathbb{N}, \nu)$ satisfying*

$$u(t) \geq 0, \quad \limsup_{t \to -\infty} e^{-\alpha t} u(t) < \infty, \quad t \to \|U(t)\| \text{ is bounded}, \tag{8.282}$$

has the following property of u

$$A = \lim_{t \to -\infty} e^{-\alpha t} u(t) \qquad (8.283)$$

exists and the solution of (8.247)–(8.248) satisfying (8.283) for given A is unique. Furthermore U(t) and α(t) satisfy

$$U(t) \Longrightarrow \mathcal{U}(\infty) \text{ as } t \to -\infty, \quad \text{and } \limsup_{t \to -\infty} e^{-\alpha t} |\alpha(t) - \alpha| < \infty. \qquad (8.284)$$

(e) *The solution of the system including the age distribution, that is, (8.247) and (8.248) has a unique solution.* □

The following is immediate from (8.279) and the discussion in (8.271)–(8.260).

Corollary 8.7. *The $(G, D_0(G))$ martingale problem is well-posed.*

Remark 76. Note that looking at the form of the equation we see that a solution indexed by $\mathbb{R}$ remains a solution if we make a time shift. This corresponds to the different possible values for the growth constant A in (8.280). In particular the entrance law from 0 at time $-\infty$ is unique up to the time shift.

Proof of Lemma 8.6. (a) First define

$$b(t) = 1 + \frac{\gamma(t)}{\alpha(t)} \qquad (8.285)$$

and we express $\alpha(t)$ and $\gamma(t)$ in terms of $U(t)$ and and consider $L^1(\nu, \mathbb{N})$ as a basic space for the analysis of the component $U(t, \cdot)$. Then we obtain a system of coupled differential equations for the pair (u, U):

$$\frac{du(t)}{dt} = \alpha(t)(1 - b(t)u(t))u(t), \qquad (8.286)$$

$$\frac{\partial U(t, j)}{\partial t} = +s(j-1)1_{j \neq 1} U(t, j-1) - sj U(t, j)$$

$$+ \frac{d}{2}(j+1) j U(t, j+1) - \frac{d}{2} j(j-1))1_{j \neq 1} U(t, j)$$

$$+ c(j+1) U(t, j+1) - cj U(t, j) 1_{j \neq 1}$$

$$+ [\alpha(t) - \alpha(t) u(t) - \gamma(t) u(t))] 1_{j=1} \qquad (8.287)$$

$$+ u(t)(\alpha(t) + \gamma(t))[U(t, j-1) 1_{j \neq 1} - U(t, j)]$$

$$- \Big(\alpha(t)(1 - u(t)) - \gamma(t) u(t)\Big) \cdot U(t, j).$$

We now consider the related nonlinear system obtained by setting $u \equiv 0$ in (8.287), namely,

8.3 Proofs of Propositions 2.3, 2.6, and 2.11

$$\frac{\partial \tilde{U}(t,j)}{\partial t} = +s(j-1)1_{j \neq 1}\tilde{U}(t, j-1) - sj U(t, j)$$

$$+ \frac{d}{2}(j+1)j\tilde{U}(t, j+1) - \frac{d}{2}j(j-1))1_{j \neq 1}\tilde{U}(t,j)$$

$$+ c(j+1)\tilde{U}(t, j+1) - cj\tilde{U}(t,j)1_{j \neq 1}$$

$$+ \alpha(t)1_{j=1} - \alpha(t) \cdot \tilde{U}(t, j), \tag{8.288}$$

where $\alpha(t) = c \sum_{\ell=2}^{\infty} \ell \cdot \tilde{U}(t, \ell)$.   q.e.d.

We will consider the system (8.287) as a nonlinear perturbation of the linear strongly continuous semigroup S^α on the Banach space $\mathbb{B} = L^1(\nu, \mathbb{N})$ given in (8.274).

Then we can rewrite (8.286) and (8.287) in the form (G^* stands for adjoint of G)

$$\frac{d}{dt}(u(t), U(t, \cdot)) = \vec{G}^*(u(t), U(t, \cdot)) = (0, U(t, \cdot)Q^{\alpha,*}) + F(u(t), U(t, \cdot)), \tag{8.289}$$

where the infinite matrix $Q^{\alpha,*}$ is the infinitesimal generator (Q-matrix) of the Markov semigroup of S^α and denote by $U^{\alpha,*}(t)$ the state if S^α acts on measures and $F : \mathbb{B} \longrightarrow \mathbb{B}$ is a (locally) Lipschitz continuous function on $\mathbb{B}$, namely,

$$F(u, U) = \big(\alpha(U)u - (\gamma(U) + \alpha(U))u^2, ((\alpha - \alpha(U))$$
$$+ u(\alpha(U) + \gamma(U)))Z(U)\big), \tag{8.290}$$

where $Z(U) = (Z_j)_{j \in \mathbb{N}}$, with $Z_j = U(j-1), j \neq 1, Z_1 = -1$.

Note that F involves a third degree polynomial in u and $\alpha(t)$. Moreover $U \to \alpha(U)$ is a linear function and differentiable on $\mathbb{B}$. Therefore we can conclude that F is C^2 with first and second derivatives uniformly bounded on bounded subsets of $\mathbb{B}$. Therefore we can use the result of Marsden [MA], Theorem 4.17 (see Appendix B) to conclude that there is a unique flow $(u(t), U(t))$ satisfying the equations and in addition the mapping $t \to (u(t), U(t))$ and the mapping $(u_{t_0}, U_{t_0}) \to (u(t), U(t))$ are Lipschitz in $\mathbb{B}$.

Remark 77 (Alternate proof of (a)). We write the equation (8.243) as an a system of ODE in the form

$$B'(t) = \vec{G}^*(B(t)) = L(B(t)) + N(B(t)), \tag{8.291}$$

where L is a linear operator (generator of a Markov process) and N is a nonlinear operator on the Banach space $\mathbb{B}$. The latter is locally Lipschitz but not globally Lipschitz. Furthermore the operator L maps $\mathbb{B}$ into finite well-defined functions on $\mathbb{N}$, but *not* into $\mathbb{B}$. Only if we restrict to the smaller sets of configurations $\tilde{\mathbb{B}}$, i.e. configurations b satisfying $b \in L^1(\mathbb{N}, \nu)$ and $\sum j^4 |(b(j)| < \infty$ we obtain

$Lb \in \mathbb{B}$. This subset of configurations forms a dense subset in $L^1(\mathbb{N}, \nu)$. This means we cannot view this as a standard evolution equation in the Banach space $\mathbb{B}$ and have to bring in some extra information to make the usual arguments work. Namely we have to show that if we start in the set $\tilde{\mathbb{B}}$ we remain there; we get an evolution in the Banach space $\mathbb{B}$. The extra information needed comes from the probabilistic interpretation.

Indeed the evolution equation

$$\tilde{B}'(t) = Q^*(\tilde{B}(t)) \tag{8.292}$$

has a nice dual Markov semigroup (backward equation) with which we can work. Namely we know that the corresponding Markov process on $\mathbb{N}$ has an entrance law starting at $+\infty$. This property is wellknown for the Kingman coalescent but extends to our birth and death process provided $d > 0$. This is easily extended furthermore to an additional immigration term.

Hence the evolution (forward equation) can be started with "$\tilde{B}(0) = \delta_\infty$" and for all solution we obtain a smaller path in stochastic order, in fact the entrance law from infinity has the property that in particular $\sum_j \tilde{B}(t, j)(1 + j^4) < \infty$, for $t > 0$.

Therefore we can solve the evolution equation (8.292) in $\mathbb{B}$ uniquely. This has to be extended to the nonlinear equations where again starting somewhere in $\mathbb{B}$ after any positive time we reach the stronger summability condition.

However if we truncate L to L^k where $L^k x = 1_{\{j \leq k\}} Lx$, we can view (8.243) as a nonlinear evolution equation in a subset of the Banach space $\mathbb{R} \otimes L^1(\mathbb{N}, \nu)$ (recall (8.262)). This means now we are in the set-up of *ODE's in Banach spaces* and can use the classical theory of existence and uniqueness for such equations to obtain unique solutions B^k.

Then we return to the original equation observing that for the solution B^k of the truncated problem we have for initial states with $\sum_i B(0, i)(i)^4 < \infty$ as $k \to \infty$

$$(B^k(t))' \longrightarrow B'(t), \quad B^k(t) \longrightarrow B(t), \quad t \geq 0, \tag{8.293}$$

using the bound in stochastic order from the entrance law as pointed out above.

If we have existence of the solution in $R^+ \otimes L^1_+(\mathbb{N}, \nu)$ for all t (where $L^1_+(\mathbb{N}, \nu)$ denotes the set of sequences in $L^1(\mathbb{N}, \nu)$ having non-negative components) it suffices to show local uniqueness.

Turn first to the existence problem. We shall obtain in Proposition 8.8 that we have convergence of the finite N systems to a solution of the equations and we need for the existence to know that it is in our Banach space. It can easily be verified that the solution obtained in the limit must lie in $\mathbb{R}^+ \times L^1_+(\mathbb{N}, \nu)$ and has no explosions using a priori estimates. (Recall that for $\zeta^N(t, i)$ the second moments (in fact *all* moments) are finite over finite time intervals since this is true for the process without collision which provides upper bounds on u^N, U^N independently of N.) Using the fact that second moments remain bounded in bounded intervals the proof is standard.

8.3 Proofs of Propositions 2.3, 2.6, and 2.11

Therefore a solution arising as a limit for $N \to \infty$ lies in the subspace of $\mathbb{R}^+ \times \mathcal{M}_1(\mathbb{N})$ where $\|\cdot\|$ is finite. This gives existence of a solution with values in the specified Banach space.

We turn to the uniqueness. The form of the equation now immediately gives the local Lipschitz property of F. Therefore the solution is locally unique by the uniqueness result for nonlinear differential equations in Banach spaces (e.g. [Paz], Chapt. 6, Theorem 1.2) applied to the truncated problem and together with the approximation property we get uniqueness of our equation.

(b) It is standard to check that ζ is for given (u, U) and for (8.267)–(8.269) a nonexplosive time-inhomogeneous Markov process which is uniquely determined by these equations. We have to show that the self-consistency relation (8.270) can be satisfied. This will follow from our convergence result for (u^N, U^N) below in Proposition 8.8, Part (e).

(c) To prove existence of an entrance law with the prescribed properties, we construct a solution as a limit of the sequence $(u_n(t), U_n(t))_{t \in \mathbb{R}}, n \in \mathbb{N}$ defined as follows:

$$u_n(t) = Ae^{\alpha t_n} \text{ for } t \leq t_n \tag{8.294}$$

$$U_n(t) = \mathcal{U}(\infty) \text{ for } t \leq t_n$$

(u_n, U_n) satisfy (8.286) and (8.287) for $t \geq t_n$.

Standard arguments imply the tightness of the sequence $\{(u_n(t), U_n(t))_{t \in \mathbb{R}}, n \in \mathbb{N}\}$ and that any limit point (u, U) satisfies (8.286) and (8.287). We now have to show that the limit points satisfy the conditions on the behaviour as $t \to -\infty$.

We observe that the solution to the equation for $u \equiv 0, U \equiv \mathcal{U}(\infty)$ gives a stationary solution of the coupled pairs of equations. We know that the growth by $Ae^{\alpha t}, t \in \mathbb{R}$ is an upper bound for the u-part of the limit, since this is the collision-free solution of the equation and therefore $\limsup_{t \to \infty}(e^{-\beta t} u(t)) \leq A < \infty$. We therefore must verify that the limit point matches exactly that growth and starts from $\mathcal{U}(\infty)$ at time $t = -\infty$. We note that (u, U) must then agree with a stationary solution for $u \equiv 0$, which is unique by standard Markov chain theory and hence the point at $t = -\infty$ must be $\mathcal{U}(\infty)$. We next argue that a limit point (u, U) actually satisfies $e^{-\beta t} u(t)$ converges to A as $t \to -\infty$. We shall see below that this would actually follow if $u(t)e^{-\beta t}$ has for $t \to -\infty$ A as limes superior. If we know that the solution with this property is unique, then our limit point in the construction must then have the desired properties and in particular we must actually have convergence of (u_n, U_n) as $u \to \infty$. This however we shall show below in (d).

(d) Let (u, U) be a solution satisfying (8.282). Then there exists a sequence $t_n \to -\infty$ such that as $n \to \infty$ the limit of $(u(t_n), U(t_n))$ exists.

Now consider the nonlinear system $(u(t), U(t))$. We must verify that as $t \to -\infty$ we actually converge to the case where $u \equiv 0$ and $\alpha(t) \equiv \alpha$. First note that by (8.132) we can assume that $\alpha(t)$ is bounded for all t. By tightness we can find a sequence $(t_n)_{n \in \mathbb{N}}$, $t_n \to -\infty$ as $n \to \infty$ and $A \in (0, \infty)$ such that as $n \to \infty$, $u(t_n)e^{-\alpha t_n} \to A$. Since by assumption $\limsup_{t \to -\infty} e^{-\alpha t}u(t) < \infty$ it follows from the equations that $|\alpha(t) - \alpha| \leq \text{const} \cdot e^{\alpha t}$. Then any such solution must satisfy

$$u(t) - Ae^{\alpha t} = o(e^{\alpha t}). \tag{8.295}$$

We next note that given $u(t), \alpha(t), \gamma(t)$, the equation for $U(t)$ is linear and has a unique solution by the standard Markov chain theory. Therefore it suffices to prove that $u(t), \alpha(t), \gamma(t)$ as functions on $\mathbb{R}$ are unique given the asymptotics behaviour of u as $t \to -\infty$. Remember that $\alpha(t)$ and $y(t)$ are functions $U(t)$.

We next consider the leading asymptotic term for U. Let $(S^\alpha(t))_{t \geq 0}$ the semigroup defined in (8.274) and $S^{\alpha,*}$ is action on measure, $U^{\alpha,*}$ its path, $U^{\alpha,*}(\infty)$ its equilibrium measure. Furthermore let $L(\cdot) = \alpha(\cdot) + \gamma(\cdot)$. Let $t_n, t_m \to -\infty$ so that $t_n - t_m \to \infty$. Then by (8.243)

$$U(t_n) = \lim_{t_m \to -\infty} \left[S^{\alpha,*}(t_n - t_m)U(t_m) + \int_{t_m}^{t_n} S^{\alpha,*}(t_n - s)(\alpha - \alpha(s))Z(U(s))ds \right.$$

$$\left. + \int_{t_m}^{t_n} u(s) S^{\alpha,*}(t_n - s) L(s) ds \right] \tag{8.296}$$

$$= \lim_{t_m \to -\infty} \left[S^{\alpha,*}(t_n - t_m)U(t_m) \right] + \int_{-\infty}^{t_n} S^{\alpha,*}(t_n - s)(\alpha - \alpha(s))Z(\mathcal{U}(\infty))ds$$

$$+ \int_{-\infty}^{t_n} Ae^{\alpha s} S^\alpha(t_n - s)(\alpha + \gamma)Z(U^{\alpha,*})ds + o(e^{\alpha t_n})$$

$$= \mathcal{U}(\infty) + \int_{-\infty}^{t_n} S^{\alpha,*}(t_n - s)(\alpha - \alpha(s))Z(\mathcal{U}(\infty))ds$$

$$+ \int_{-\infty}^{t_n} Ae^{\alpha s} S^{\alpha,*}(t_n - s)(\alpha + \gamma)Z(U^{\alpha,*}(\infty))ds + o(e^{\alpha t_n}).$$

Therefore we have

$$\lim_{t_n \to -\infty} U(t_n) = \mathcal{U}(\infty). \tag{8.297}$$

Note that then also $\alpha(t_n) \to \alpha$ and $\gamma(t_n) \to \gamma$ as $n \to \infty$, since these are continuum functionals of $U(t_n)$ and we have the integrability properties.

Finally, assume that there are two solutions. We shall show below that if two such solutions (u_1, U_1) and (u_2, U_2) were distinct a contradiction would result. Hence in particular in (c) we must have convergence as claimed.

Let

$$(v(t), V(t)) := (u_1(t), U_1(t)) - (u_2(t), U_2(t)),$$
$$\hat{\alpha}(t) = \alpha_1(t) - \alpha_2(t), \quad \hat{\gamma}(t) = \gamma_1(t) - \gamma_2(t). \tag{8.298}$$

8.3 Proofs of Propositions 2.3, 2.6, and 2.11

From the above (i.e. (8.295) and using $(\alpha - \alpha(s)) = o(e^{\alpha s})$ in (8.296) and estimating therein the third term by explicit bound) we know that

$$|v(t_n)| = o(e^{\alpha t_n}) \quad \text{and} \quad \|V(t_n)\| = o(e^{\alpha t_n}). \tag{8.299}$$

Then

$$\frac{dv(t)}{dt} = \alpha v(t) + e^{\alpha t}\hat{\alpha}(t) - (\alpha + \gamma)e^{\alpha t}v(t) - e^{2\alpha t}(\hat{\alpha} + \hat{\gamma}), \tag{8.300}$$

$$\frac{dV(t)}{dt} = V(t)Q^{\alpha,*} - \alpha V(t) + A(\alpha + \gamma)e^{\alpha t}Z((V(t))) + (\alpha + \gamma) \cdot v(t)Z(U^{\alpha,*})$$
$$+ e^{\alpha t}((\hat{\alpha} + \hat{\gamma} + Z(V(t))),$$

$$\frac{d\hat{\alpha}(t)}{dt} = -\alpha\hat{\alpha}(t) + A(\alpha + \gamma)e^{\alpha t}\alpha(Z(V(t))) + (\alpha + \gamma) \cdot v(t)\alpha(Z(U^{\alpha,*}))$$
$$+ o(e^{\alpha t}),$$

where $\hat{\alpha}(t) = \alpha(V(t))$.

We then obtain by inspection from (8.300) that:

$$\frac{d(|v(t)| + \|V(t)\|)}{dt} \leq \alpha \cdot (1 + \text{const} \cdot e^{\alpha t})(|v(t)| + \|V(t)\|). \tag{8.301}$$

Then use the v equation and (8.299), more precisely we use the V equation to get an expression for $\hat{\alpha}(t)$ and $\hat{\gamma}(t)$ as a linear function of v. to conclude $v(t) \equiv 0$ as follows. Namely we conclude using Gronwall's inequality,

$$|v(t_n)| + \|V(t_n)\| \leq o(e^{\alpha t_m})e^{\alpha(t_n - t_m)}. \tag{8.302}$$

Letting $t_m \to -\infty$ we obtain

$$|v(t_n)| + \|V(t_n)\| = 0. \tag{8.303}$$

This completes the proof of uniqueness of $(u(t), U(t))$.

(e) It remains to show that not only $U(t, \mathbb{R}^+, \cdot)$ is unique but that this holds also for $U(t, \cdot, \cdot)$. Note that our result already implies that $u(\cdot)$ is uniquely determined. Note also $\alpha(t)$ and $\gamma(t)$ depend only on U integrated over the age and are continuous functions of t and are therefore also uniquely determined. Therefore $u(\cdot), \alpha(\cdot), \gamma(\cdot)$ are uniquely determined. Hence we can insert the unique objects as *external* (time-inhomogeneous) input into the equation. Then the equation (8.248) for given u, α, γ is the forward equation of a time inhomogeneous Markov process with state space $M_1([0, \infty) \times \mathbb{N})$ restricted to elements in $\mathbb{R} \otimes L^1(\mathbb{N}, \nu)$, which has a unique solution by standard arguments.

This completes the proof of Lemma 8.6.

8.3.8 The Dual Process in the Collision Regime: Convergence Results

Now we are ready to state the limit theorem for the growth dynamics of the dual population in the critical time scale where collisions are essential, with the key result for the application to the original process given in e) of Proposition 8.8 below, see in particular (8.320). As usual we denote by $\Longrightarrow$ convergence in law.

Consider the dual process which starts with k particles at each of ℓ distinct sites and denote the corresponding functionals of the dual process by $(\Pi_s^{N,k,\ell}, u_s^{N,k,\ell}, U_s^{N,k,\ell})_{s \geq 0}$. In order to consider them in the time scale in which fixation occurs, we introduce:

$$\tilde{u}^{N,k,\ell}(t) = u^{N,k,\ell}((\frac{\log N}{\alpha} + t) \vee 0), \ t \in (-\infty, T], \quad \tilde{u}^{N,k,\ell}(-\frac{\log N}{\alpha}) = \ell \tag{8.304}$$

$$\tilde{U}^{N,k,\ell}(t) = U^{N,k,\ell}((\frac{\log N}{\alpha} + t) \vee 0), \ t \in (-\infty, T], \quad \tilde{U}^{N,k,\ell}(-\frac{\log N}{\alpha}) = \delta_{(k,0)}. \tag{8.305}$$

Furthermore we need the standard solution of the nonlinear system (8.247) and (8.248), (see Lemma 8.6 part (c)) denoted

$$(u^*(t), U^*(t, \cdot, \cdot)))_{t \geq t_0}, \tag{8.306}$$

which is the solution satisfying (recall $\mathcal{U}(\infty)$ was the stable age and size distribution)

$$\lim_{t \to -\infty} e^{-\alpha t} u^*(t) = 1, \quad \lim_{t \to -\infty} U^*(t) = \mathcal{U}(\infty). \tag{8.307}$$

Finally we have to specify three time scales. Let t_0 stand for some element of $\mathbb{R}$ and let $t_0(N)$ and $s(N)$ be as follows:

$$t_0(N) = s(N) - \frac{\log N}{\alpha}, \tag{8.308}$$

with

$$s(N) \to \infty, s(N) = o(\log N). \tag{8.309}$$

Recall the time shift in (8.304) and (8.305).

The behaviour of the dual particle system is asymptotically as $N \to \infty$ as follows.

8.3 Proofs of Propositions 2.3, 2.6, and 2.11

Proposition 8.8 (Growth of dual population in the critical time scale in the $N \to \infty$ limit).

(a) *Assume that for some $t_0 \in \mathbb{R}$ as $N \to \infty$, $(\frac{1}{N}\tilde{u}^{N,k,\ell}(t_0), \tilde{U}^{N,k,\ell}(t_0))$ converges in law to the pair $(u(t_0), U(t_0))$ automatically contained in $[0, \infty) \times L_1(\mathbb{N}, \nu)$. Then as $N \to \infty$*

$$\mathcal{L}\left[(\frac{1}{N}\tilde{u}^{N,k,\ell}(t), \tilde{U}^{N,k,\ell}(t, \cdot, \cdot))\right)_{t \geq t_0}\right] \Longrightarrow \mathcal{L}\left[(u(t), U(t, \cdot, \cdot)))_{t \geq t_0}\right], \quad (8.310)$$

in law on pathspace, where the r.h.s. is supported on the solution of the nonlinear system (8.247) and (8.248) corresponding to the initial state $(u(t_0), U(t_0))$. (Note that the mechanism of the limit dynamics does not depend on k or ℓ, but the state at t_0 will. That is why we suppress k, ℓ on the r.h.s.)

(b) *The collection $\{(\tilde{u}^{N,k,\ell}, \tilde{U}^{N,k,\ell}), N \in \mathbb{N}\}$ can be constructed on a common probability space such that for $t_0(N)$ as in (8.308) we have:*

$$(e^{-\alpha s(N)}\tilde{u}^{N,k,\ell}(t_0(N)), \tilde{U}^{N,k,\ell}(t_0(N))) \to (W^{k,\ell}, \mathcal{U}(\infty)), \text{ a.s., as } N \to \infty. \quad (8.311)$$

(c) *The scaled occupation density converges (recall (8.304) and (8.305) and (8.307)):*

$$\frac{\tilde{u}^{N,k,\ell}}{N}(t) \underset{N \to \infty}{\Longrightarrow} u^{k,\ell}(t), \; t \in (-\infty, T], \quad (8.312)$$

in the sense that for $\varepsilon > 0$ (with $t_0(N)$ as in (8.308)),

$$P\left(\sup_{t_0(N) \leq t \leq T} e^{-\alpha t}|\frac{\tilde{u}^{N,k,\ell}}{N}(t) - u^{k,\ell}(t)| > \varepsilon\right) \to 0 \text{ as } N \to \infty. \quad (8.313)$$

Also the age-size distribution of the dual converges:

$$\tilde{U}^{N,k,\ell}(t, \cdot, \cdot) \underset{N \to \infty}{\Longrightarrow} U^{k,\ell}(t, \cdot, \cdot), \; t \in (-\infty, T], \quad (8.314)$$

in the sense that for $\eta > 0$, with $\|\cdot\|$ denoting the variational norm,

$$\lim_{N \to \infty} P\left(\sup_{t_0(N) \leq t \leq T} \|\tilde{U}^{N,k,\ell}(t, \cdot, \cdot) - U^{k,\ell}(t, \cdot, \cdot)\| > \eta\right) = 0. \quad (8.315)$$

(d) *The limits $(u^{k,\ell}(t), U^{k,\ell}(t, \cdot, \cdot)))_{t \geq t_0}$, can be represented as the unique solution of the nonlinear system (8.247) and (8.248) satisfying*

$$\lim_{t \to -\infty} e^{-\alpha t} u^{k,\ell}(t) = W^{k,\ell}, \quad \lim_{t \to -\infty} U^{k,\ell}(t) = \mathcal{U}(\infty), \quad (8.316)$$

with $W^{k,\ell}$ having the law of the random variable appearing as the scaling (by $e^{-\alpha t}$) limit for $t \to \infty$ of the CMJ-process $K_t^{k,\ell}$ started with k particles at each of ℓ sites.

These solutions are random time shifts of the standard solution (8.306), (8.307), namely:

$$u^{k,\ell}(t) = u^*(t + \frac{\log W^{k,\ell}}{\alpha}), \quad U^{k,\ell}(t) = U^*(t + \frac{\log W^{k,\ell}}{\alpha}). \qquad (8.317)$$

(e) Let $(\Pi_u^{N,k,\ell})_{u \geq 0}$ denote the number of dual particles and $T_N = \alpha^{-1} \log N$ and $t_0(N), s(N)$ as in (8.308). Then for each $k \in \mathbb{N}, \ell \in \mathbb{N}$ we have:

$$\mathcal{L}[\{N^{-1} \Pi_{T_N+t}^{N,k,\ell}, \, t_0(N) < t < \infty\}] \underset{N \to \infty}{\Longrightarrow} \mathcal{L}[\{\frac{u^{k,\ell}(t)(\alpha(t) + \gamma(t))}{c}\}_{t \in \mathbb{R}}], \qquad (8.318)$$

$$\mathcal{L}[(N^{-1} \Pi_u^{N,k,\ell}, \, 0 \leq u \leq s(N))] \underset{N \to \infty}{\Longrightarrow} \delta_{\underline{0}}, \qquad (8.319)$$

with $\underline{0}$ the function on $\mathbb{R}^+$, which is identically zero.

Furthermore for every $t \in \mathbb{R}$, there exists (a deterministic) $v_{k,\ell}(t) \in \mathcal{P}([0, \infty))$ such that

$$\mathcal{L}[\frac{1}{N} \int_0^{T_N+t} \Pi_u^{N,k,\ell} du] \underset{N \to \infty}{\Longrightarrow} v_{k,\ell}(t). \qquad \square \qquad (8.320)$$

Remark 78. Note that the dual at time T describes the genealogy of a typical sample from the original population drawn at the time T. Here time runs backwards, in particular the evolution of the dual with total time horizon $T = \alpha^{-1} \log N + t$ in the time described by varying t corresponds in the genealogy of the tagged sample to the early moments, while the emergence phenomena of the dual population (arising collisions) reflect in the original population later times, where law of large number effects rule the further expansion of the fitter type on macroscopic scale (i.e. observing the complete space).

In the case in which the rare mutant has succeeded our result tells us that we have a situation where the mutations which occur very early in the typical ancestral path have generated $O(N)$ possible choices for a fitter type to occur and to then prevail at the observation time at a fixed observation site. The randomness enters since these very early mutations during the first moments of the evolution have in the limit $N \to \infty$ a Poisson structure. This produces a random time shift of emergence and takeover which then follows a deterministic track. This is the global picture, if we look at it locally we find that also a collection of tagged sites follows a random evolution in small time scale.

8.3 Proofs of Propositions 2.3, 2.6, and 2.11

Proof of Proposition 8.8. We begin with some preparations needed for all of the proofs of the different claims of the proposition, this is part 0, followed by three further parts explained at the end of part 0.

Part 0 (Preparation)

For the proof we represent the process differently, namely we rewrite the problem in measure-valued form. We will need the following functional of the process Π_t^N, denoted $\hat{\Pi}_t^N$ respectively $\hat{\Pi}_t^{N,k}$, $\hat{\Pi}_t^{N,(k,\ell)}$ if we indicate the initial state in the notation (here $s_i = s_i^N$):

$$\hat{\Pi}_t^N = \sum_{i=1}^{K_t^N} \zeta_{s_i}^N(t) 1(\zeta_{s_i}^N(t) > 1), \tag{8.321}$$

which counts at time t only those dual particles which are not the only ones at their site, i.e. they are the ones which can generate new occupied sites. Here $\zeta_s^N(t)$ is the single site birth and death process starting at time s, the time where the site was first occupied, with mean-field immigration at time t given by $c\frac{\Pi_t^{N,1}}{N} - o(1)$ as $N \to \infty$ from the other sites.

Note that we have now made the N-dependence explicit. Also note that by definition of U^N, u^N and $\hat{\Pi}^N$ we have

$$\hat{\Pi}_t^{N,1} = u^N(t) \int_0^\infty \sum_{j=2}^\infty j U^N(t, ds, j). \tag{8.322}$$

To continue we introduce a new random object from which we can read off (u^N, U^N) by taking suitable functions. Namely we consider $\Psi^N(t, ds, dx)$ the $\mathcal{M}_{\text{fin}}([0, \infty) \times \mathbb{N}_0)$- valued random variable (finite measures on $[0, \infty) \times \mathbb{N}_0$) defined by

$$\Psi^N(t, ds, dx) = \sum_{s_i \leq t} \delta_{(t-s_i, \zeta_{s_i}^N(t))}, \tag{8.323}$$

where $\{s_i\}$ (recall $s_i = s_i^N$) denote the times of birth of new sites and i runs from 1 to K_t^N (this set of times depends on N). Note that here *empty sites are not counted*. We know from the CMJ-theory that $\Psi^N(t, \cdot, \cdot)$ is bounded by $W^* e^{\alpha t}$. Since we are interested in times $\alpha^{-1} \log N + r$ we introduce the scaled object:

$$\hat{\Psi}_t^N = \frac{\Psi_t^N(\cdot, \cdot)}{N}. \tag{8.324}$$

Note that for fixed N this is an element of $\mathcal{M}_{\text{fin}}([0, \infty) \times \mathbb{N}_0)$.

Next note that indeed we obtain our pair (u^N, U^N) as a functional of Ψ^N. Observe:

$$u^N(t) = \Psi_t^N([0, \infty) \times \mathbb{N}), \quad U^N(t, \cdot, \cdot) = \frac{\Psi_t^N(\cdot, \cdot)}{u^N(t)} = \frac{\hat{\Psi}_t^N(\cdot, \cdot)}{u^N(t)/N} \tag{8.325}$$

and the stochastic integral below gives:

$$\Pi_t^N = \int_0^t \zeta_u^N(t) dK_u^N = \int_0^t \int_0^\infty x \Psi_t^N(ds, dx). \tag{8.326}$$

Therefore from (8.325) we obtain the convergence of $(N^{-1}u^N, U^N)$ in the time scale $\alpha^{-1} \log N + t$ with $t \in \mathbb{R}$ if we show the convergence of $\hat{\Psi}^N$. This will also allow us later on to argue that the mass and time scaled process

$$\bar{\Pi}^N(t) = N^{-1} \Pi^N_{\frac{\log N}{\alpha} + t} \tag{8.327}$$

converges as $N \to \infty$ to a limit.

The proof is broken into three main parts. The first two parts concern the pair (u, U) the third one the quantity $N^{-1}\Pi^{N,k,\ell}_{T_N + t}$. The parts can be outlined as follows (where $(a), \cdots$ refers to the parts of the proposition):

1. Assuming the marginals of $(N^{-1}\tilde{u}^N, \tilde{U}^N)$ converge in distribution at $t = t_0$, prove (a) that the limiting evolution during $t \geq t_0$ follows the deterministic nonlinear dynamics specified in the system of coupled equations (8.247), (8.248) and that pathwise convergence holds.
2. Then prove (b) that the marginal distributions converge for $t_0 = t_0(N) = \frac{1}{\alpha} \log N + s, s \in \mathbb{R}$ and show (c), (d) on the corresponding entrance behaviour.
3. Show the convergence of the process $\{N^{-1}\Pi^{N,k,\ell}_{T_N + t}, t \in \mathbb{R}\}$ and of its time integrals to complete the proof of (e).

Part 1 (Convergence of dynamics of $\hat{\Psi}^N$)

Let $P^N \in \mathcal{P}(D([t_0, \infty), \mathcal{M}_{\text{fin}}(\mathbb{R}_+ \times \mathbb{N})))$ denote the law of the Markov process

$$\{\hat{\Psi}^N_{T_N + t}\}_{t \geq t_0} \text{ (recall (8.324)), with } T_N = \frac{\log N}{\alpha}. \tag{8.328}$$

In this part we assume that the marginal distributions $P^N_{t_0}$ converge as $N \to \infty$. Our next goal is then to show that the processes converge in law on path space, that is, $P^N \Rightarrow P$ as $N \to \infty$ where P is the law of a deterministic dynamics.

Recall that we have defined the candidate for the limiting object by a nonlinear equation, which we can view as forward equation for a nonlinear Markov process on $\mathbb{R}_+ \times \mathbb{N}$. This means we want to show that if we consider the marginal laws, that is, elements P^N_t of $\mathcal{P}(\mathcal{M}_{\text{fin}}(\mathbb{R}_+ \times \mathbb{N}))$ they converge to an element $\delta_{\hat{\Psi}_t}$ with $\hat{\Psi}_t \in \mathcal{M}_{\text{fin}}(\mathbb{R}_+ \times \mathbb{N})$. We will now take the viewpoint of backward equations and their generators, that is equations for expectations of measures rather than actual weights.

In order to do so in this present Part 1, we proceed as follows.

- we obtain moment bounds on $\hat{\Psi}^N_{T_N + t}$ uniform for $N \in \mathbb{N}$ and $t_0 \leq t \leq t_0 + T$,
- we formulate the martingale problem for $\hat{\Psi}^N$ with generator $\mathbf{G}_N$ given by a linear operator on the space $C_b(\mathcal{M}_{\text{fin}}(\mathbb{R}_+ \times \mathbb{N}_0)))$ with common domain $\mathcal{D}$ given by a certain dense subset of $C_b(\mathcal{M}_{\text{fin}}(\mathbb{R}_+ \times \mathbb{N}_0))$,

8.3 Proofs of Propositions 2.3, 2.6, and 2.11

- we verify the tightness of the laws P^N,
- we prove the weak convergence of P^N to P by showing that $\mathbf{G}_N$ converges pointwise to $\mathbf{G}$ on these functions in the domain $\mathcal{D}$ where $\mathbf{G}$ is the operator defined in (8.258),
- verify that limit points of $(P^N)_{N \in \mathbb{N}}$ are characterized as solutions to the $\mathbf{G}$ martingale problem,
- conclude the convergence since we have proved that the $\mathbf{G}$ martingale problem has a unique solution given by the nonlinear system (8.247), (8.248),
- the convergence for fixed t_0 is obtained later in Part 2.

Abbreviate for bounded measureable test functions f on $[0, \infty) \times \mathbb{N}_0$

$$\hat{\Psi}(f) = \int_0^\infty \int_\infty^\infty f(r, x) \hat{\Psi}(dr, dx) \tag{8.329}$$

and in particular $\hat{\Psi}(1) = \hat{\Psi}([0, \infty) \times (\mathbb{N}_0))$.

The process $\hat{\Psi}^N$ is a Markov process with state space $\mathcal{M}_{\text{fin}}([0, \infty) \times \mathbb{N})$. We calculate the action of the generator $\mathbf{G}_N$ acting on the domain $\mathcal{D}$ of functions $F_{g,f}$ of the form (8.254) as follows. Let $\hat{\Psi} \in \mathcal{M}_{\text{fin}}([0, \infty) \times \mathbb{N})$. Then:

$$(\mathbf{G}_N F_{g,f})(\hat{\Psi})$$
$$= c(1 - \hat{\Psi}(1)) \left(\int_0^\infty \int_0^\infty N \left[g(\hat{\Psi}(f) + \tfrac{f(0,1) + f(r,x-1) - f(r,x)}{N}) - g(\hat{\Psi}(f)) \right] \right.$$
$$\left. \times 1_{x \neq 1} \hat{\Psi}(dr, dx) \right)$$
$$+ \int_0^\infty \int_1^\infty g'\left(\hat{\Psi}(f)\right) \tfrac{\partial f(r,x)}{\partial r} \hat{\Psi}(dr, dx)$$
$$+ \int_0^\infty \int_1^\infty sxN \left[g(\hat{\Psi}(f) + \tfrac{f(r,x+1) - f(r,x)}{N}) - g(\hat{\Psi}(f)) \right] \hat{\Psi}(dr, dx)$$
$$+ \int_0^\infty \int_1^\infty \tfrac{d}{2} x(x-1) N \left[g(\hat{\Psi}(f) + \tfrac{f(r,x-1) - f(r,x)}{N}) - g(\hat{\Psi}(f)) \right] \hat{\Psi}(dr, dx)$$
$$+ c\hat{\Psi}(1) \int_0^\infty \int_1^\infty \int_0^\infty \int_1^\infty N \left[g(\hat{\Psi}(f) + \tfrac{f(\tilde{r}, \tilde{x}+1) - f(\tilde{r}, \tilde{x})}{N} + \tfrac{1_{x \neq 1} f(r, x-1) - f(r,x)}{N}) \right.$$
$$\left. - g(\hat{\Psi}(f)) \right] x \hat{\Psi}(dr, dx) \tfrac{\hat{\Psi}(d\tilde{r}, d\tilde{x})}{\hat{\Psi}(1)}.$$

(8.330)

Note that in order for the right side to be well-defined we require $\int x^2 \hat{\Psi}(\mathbb{R}, dx) < \infty$. This condition is automatically satisfied if $\hat{\Psi} \in \mathcal{M}_{\text{fin}}([0, \infty) \times \mathbb{N}_0) \cap \mathbb{B}$. This means that $\mathbf{G}_N F_{g,f}$ is a welldefined function on $\mathbb{B}$, where it is even continuous. Note however $C_b(\mathbb{B}, \mathbb{R})$ is *not* mapped in $C_b(\mathbb{B}, \mathbb{R})$. This means we now have to investigate next that under our evolution we stay with $\hat{\Psi}$ being $\mathbb{B}$. Therefore we turn next to bounds on the empirical moments of the number of particles at the occupied sites.

We now establish some a priori bounds which will guarantee that the empirical moments remain stochastically bounded over the considered time interval. By definition we have

$$\int_0^\infty x^m \hat{\Psi}^N_{T_N+t}(\mathbb{R}, dx) = \frac{1}{N} \sum_{i=1}^{u^N(T_N+t)} \sum_{k=1}^\infty k^m 1_k(\zeta^N_{s_i}(T_N+t)). \tag{8.331}$$

Then we have with s_i denoting the time of birth of the i-th colonized site (recall $s_i = s_i^N$) and letting S be the birth time of a randomly picked site:

$$E\left[\int_0^\infty x^m \hat{\Psi}^N_{T_N+t}(\mathbb{R}, dx)\right] = \frac{1}{N} E\left[\sum_{i=1}^{u^N(T_N+t)} \sum_{k=1}^\infty k^m 1[(\zeta^N_{s_i}(T_N+t) = k)]\right] \tag{8.332}$$

$$= E\left[\frac{1}{N} \sum_{i=1}^{u^N(T_N+t)} (\zeta^N_{s_i})^m (T_N+t)\right]$$

$$\leq E\left[(\zeta^N_S(T_N+t))^m\right]$$

and by Cauchy–Schwarz applied to the m-th process of $\zeta^N_{s_i}$ we get

$$E\left[\left(\int_0^\infty x^m \hat{\Psi}^N_{T_N+t}(\mathbb{R}, dx)\right)^2\right] \tag{8.333}$$

$$\leq \left(E[\frac{1}{N} \sum_{i=1}^{u^N(T_N+t)} (\zeta^N_{s_i}(T_N+t))^m]\right)^2 + E[\frac{1}{N} \sum_{i=1}^{u^N(T_N+t)} (\zeta^N_{s_i}(T_N+t))^{2m}].$$

We can obtain an upper bound for $\int_0^\infty x^m \hat{\Psi}^N_{T_N+t}(\mathbb{R}, dx)$ by comparing with the collision free case (recall construction in Step 2 of Subsubsection 8.3.6 that is, the total number of particles is bounded by the collision free case) to obtain

$$\lim_{L \to \infty} P\left[\sup_{t_0 \leq t \leq t_0+T} \int_0^\infty x \hat{\Psi}^N_{T_N+t}(\mathbb{R}, dx) > L\right] \tag{8.334}$$

$$\leq \lim_{L \to \infty} P\left[\frac{1}{N} \sum_{i=1}^{u_N(T_N+t)} \sup_{t_0 \leq t \leq t_0+T} \zeta^N_{s_i}(T_N+t) > L\right]$$

$$\leq \lim_{L \to \infty} \frac{E\left[\sup_{t_0 \leq t \leq t_0+T}(\zeta_s(t))^2\right]}{L^2} = 0,$$

using the independence of the $\{\zeta_s\}$ in the collision-free case and the martingale problem together with a standard martingale inequality for the birth and death process $\zeta_s(\cdot)$.

8.3 Proofs of Propositions 2.3, 2.6, and 2.11

We have seen that the first empirical moment bound (8.334) holds by comparison with the collision-free regime. We now have to establish analogous results for higher empirical moments and incorporating collisions which involves additional immigration from the other colonies at each occupied site. Let $K > 0$. Let

$$\zeta_{s_i}^{N,K}(t) \tag{8.335}$$

denote the birth and death process with immigration at rate K which starts at time s_i. This process is ergodic and has a unique equilibrium. Let $m_n(K)$ denotes the corresponding n-th equilibrium moment.

Remark 79. We used above the fact that the birth and death process with linear birth and quadratic death rates has moments of all orders at positive times. This can be verified by comparing with a subcritical branching process with immigration. Moreover the corresponding Markov process on $\mathbb{N}$ has an entrance law starting at $+\infty$. This property is wellknown for the Kingman coalescent but extends to our birth and death process provided $d > 0$ and is easily extended to include an additional immigration term.

Then up to the time

$$\tau_K := \inf\{t : \int_0^\infty x\hat{\Psi}_{T_N+t}(\mathbb{R}, dx) \geq K\} \tag{8.336}$$

we can verify with a stochastic monotonicity argument that we get a bound uniformly in N:

$$E[(\zeta_{s_i}^{N,K}(\cdot))^n] \leq m_n(K) < \infty, \quad \text{for } n, N \in \mathbb{N}. \tag{8.337}$$

Now recall that (writing $F_{x,f}$ if $g(x) = x$)

$$M_t^N := \hat{\Psi}_{T_N+t}^N(f) - \hat{\Psi}_{T_N+t_0}^N(f) - \int_{T_N+t_0}^{T_N+t} G_N F_{x,f}(\hat{\Psi}_{T_N+s}^N(\mathbb{R}, dx)) ds \tag{8.338}$$

is a martingale. Using the above moment inequalities, (see (8.333)), we can show that this is an L^2 martingale and we can then use a martingale inequality to show that for f with finite support

$$\lim_{L \to \infty} \sup_N P(\sup_{t_0 \leq t \leq (t_0+T) \wedge \tau_K} \hat{\Psi}_{T_N+t}^N(f) > L) \tag{8.339}$$

$$\leq \lim_{L \to \infty} \sup_N \frac{4}{L^2} \left\{ E\left[\left(\hat{\Psi}_{T_N+t_0}^N(|f|) + \int_{T_N+t_0}^{T_N+T} |G_N F_{x,f}(\hat{\Psi}_{T_N+s}^N(\mathbb{R}, dx))| ds \right)^2 \right] \right\}$$

$$+ \lim_{L \to \infty} \sup_N \frac{4}{L^2} E[(M_{(t_0+T) \wedge \tau_K}^N)^2]$$

$$= 0.$$

For every $n \geq 2$ we can now choose a sequence $(f_m)_{m \in \mathbb{N}}$ with finite support in $C(\mathbb{R}, \mathbb{R})$ and let $f_m(x) \uparrow x^n$ verify that for given K

$$\limsup_{L \to \infty} \sup_N P\left[\sup_{t_0 \leq t \leq (t_0+T) \wedge \tau_K} \int_0^\infty x^n \hat{\Psi}^N_{T_N+t}(\mathbb{R}, dx) > L\right] = 0. \tag{8.340}$$

We next note that equation (8.334) implies that

$$\tau_K^N \uparrow \infty \text{ as } K \to \infty \tag{8.341}$$

uniformly in N. This together with (8.340) implies that the evolutions $\{\hat{\Psi}^N_t, \ t \in [T_N + t_0, T_N + t_0 + T]\}$, $N \in \mathbb{N}$ and their weak limit points are concentrated on *path with values in* $\mathbb{B}$.

We can conclude now with noting that $\mathbf{G}_N(F_{g,f})$ is a finite and even continuous function on $\mathbb{B}$, even though $\mathbf{G}_N$ does not map functions in its domain $\subseteq C_b(\mathbb{B}, \mathbb{R})$ into $C_b(\mathbb{B}, \mathbb{R})$ but only into the set $C(\mathbb{B}, \mathbb{R})$ of *continuous functions* on $\mathbb{B}$. Therefore we will have to show first that for the starting points of the evolution we work with, namely the processes $\{\hat{\Psi}^N(t), t \in [t_0, T]\}$ and the limit points of their laws are concentrated on states satisfying this extra condition to be in $\mathbb{B}$. This is now immediate from the a priori bound (8.340).

We now verify first tightness in path space and then in a further step convergence of the distribution.

Tightness. We verify the *tightness* in path space that is the tightness of the laws

$$\{P^N = \mathcal{L}[(\hat{\Psi}^N_t)_{t \geq t_0}]; N \in \mathbb{N}\} \text{ as } N \to \infty, \tag{8.342}$$

assuming convergence of the marginals at time t_0. Here we deal with a sequence of probability measure-valued (on a Polish space) jump processes. With the help of Jakubowski's criterion, (see Theorem 3.6.4 in [D]) we obtain that it suffices to consider the tightness of laws of real valued semimartingales, namely of $\{\mathcal{L}[(F_{g,f}(\hat{\Psi}^N_t))_{t \geq 0}]$, admissible f and $g\}$. In fact $g(x) = x$ and $g(x) = x^2$ (recall Remark 75) alone suffice according to that theorem and for these two functions g and all f with $0 \leq f \leq 1$ we shall now verify the criterion.

Recall first the a priori bound (8.340). The tightness of the laws of $F_{g,f} \circ \hat{\Psi}^N$ is obtained using the Joffe–Métivier criterion (see Appendix A.1) in the form of Corollary 3.6.7. in [D], applied to $F_{g,f}$ for $g(x) = x$ and $g(x) = x^2$. The bound (8.340) verifies the conditions therein. This means that now $\{\mathcal{L}[(\Psi^N_{T_N+t})_{t \geq t_0}], N \in \mathbb{N}\}$ has limit points in the set of laws on Skorohod space.

Let $P^N_{\hat{\Psi}^N_0}$ denote the law of $\{\hat{\Psi}^N_0\}_{t \geq t_0}$ on the Skorohod space $D([0, \infty), \mathcal{M}_{\text{fin}}(\mathbb{R}_+ \times \mathbb{N}) \cap \mathbb{B})$ with initial values $\hat{\Psi}^N_{t_0} \Rightarrow \hat{\Psi}_{t_0} \in \mathcal{M}_{\text{fin}}(\mathbb{R}_+ \times \mathbb{N}) \cap \mathbb{B}$. By tightness we can choose a convergent subsequence. Since the largest jumps decrease to 0 the limit measure is automatically concentrated on $C([0, \infty), \mathcal{M}_{\text{fin}}(\mathbb{R}_+ \times \mathbb{N}) \cap \mathbb{B})$ (see [EK2],

8.3 Proofs of Propositions 2.3, 2.6, and 2.11

Chap. 3, Theorem 10.2). Convergence in law on the Skorohod space follows if any limit point satisfies a martingale problem with a unique solution.

Convergence. Next we turn to the convergence in f.d.d. by establishing *generator convergence*. For this purpose we consider $\mathcal{M}_{fin}(\mathbb{R}_+ \times \mathbb{N}) \cap \mathbb{B}$ as the state space for the processes $\hat{\Psi}_N(\cdot)$ and the *function space* $C_b(\mathcal{M}_{fin}(\mathbb{R}_+ \times \mathbb{N}) \cap \mathbb{B}, \mathbb{R})$ as the basic space for our semigroup action. Recall the algebra of functions $D_0(\mathbf{G})$ given in (8.258). The operator $\mathbf{G}_N$ is a *linear map* from $D_0(\mathbf{G})$ into $C(\mathcal{M}_{fin}(\mathbb{R}_+ \times \mathbb{N}) \cap \mathbb{B}, \mathbb{R})$.

We consider the convergence for two choices for the "initial state" at some time $t_0 = t_0(N)$, namely we assume that $\hat{\Psi}_{t_0}^N$ converges as $N \to \infty$, but either

$$\hat{\Psi}_{t_0}^N(1) \xrightarrow[N \to \infty]{} 0 \quad \text{or} \quad \hat{\Psi}_{t_0}^N(1) \xrightarrow[N \to \infty]{} x > 0. \tag{8.343}$$

In the first case we would be back in the linear regime (no collisions) we discussed in Subsubsection 8.3.7 and which we have identified in terms of the CMJ process. Next we consider the limit corresponding to initial points at some time t_0 such that $\hat{\Psi}^N(t_0)$ does not converge to the zero measure.

Let $f \geq 0$, $\|f\|_\infty \leq 1$, $\hat{\Psi}(1) \leq 1$, g with $g|_{([0,1],\mathbb{R})} \in C_b^2([0,1], \mathbb{R})$ and $F_{g,f}$ as in (8.254). Recall the definition of the operator $\mathbf{G}$ on $D_0(\mathbf{G})$ given in (8.258). Then there exists a constant (const) depending only on c, d, s such that:

$$|\mathbf{G}_N(F_{g,f})(\hat{\Psi}) - \mathbf{G}(F_{g,f})(\hat{\Psi})| \leq \frac{\text{const}}{N} \|g'' 1_{[0,1]}\|_\infty \int_0^\infty \int_0^\infty x^2 \hat{\Psi}(dr, dx)$$

$$\leq \frac{\text{const}}{N} \|\hat{\Psi}\|_\mathbb{B}. \tag{8.344}$$

We get that for all $F_{g,f}$ of the form in (8.254) with the above extra restrictions that

$$\mathbf{G}_N(F_{g,f})(\hat{\Psi}) \xrightarrow[N \to \infty]{} \mathbf{G}(F_{g,f})(\hat{\Psi}), \tag{8.345}$$

provided that the argument $\hat{\Psi}$ satisfies

$$\int_0^\infty \int_0^\infty x^2 \hat{\Psi}(dr, dx) < \infty, \tag{8.346}$$

which is the case for $\hat{\Psi} \in \mathbb{B}$. More precisely, considering the functions $\mathbf{G}_N(F_g), \mathbf{G}(F_g)$ on any finite ball in $\mathbb{B}$, we have

$$\mathbf{G}_N(F_{g,f}) \longrightarrow \mathbf{G}(F_{g,f}), \quad N \to \infty \tag{8.347}$$

pointwise uniformly on such balls.

From (8.347) and the convergence of the initial laws we want to conclude that weak limit points of $\mathcal{L}[(\hat{\Psi}_t^N)_{t \geq t_0}]$ conditioned on a particular initial state at time t_0

are δ-measures on the space of $\mathcal{M}_{\text{fin}}(\mathbb{R}^+ \times \mathbb{N})$-valued path which must satisfy the evolution equation

$$\frac{d}{dt}(\hat{\Psi}_t) = G^*\hat{\Psi}_t \quad , \quad t \geq t_0. \tag{8.348}$$

For this we have to show

1. the variance process of $<\hat{\Psi}^N, f>$ converges to zero, and
2. $\hat{\Psi}^N$ remains in large enough balls with high probability such that (8.347) can be applied,
3. making the choice $g(x) = x$ in (8.254), to verify that the path solves equation (8.348),
4. to prove convergence it then suffices to show that the **G**-martingale problem is wellposed.

For the point (1) consider $g(x) = x^2$ and $g(x) = x$ and show that

$$\lim_{N \to \infty} [\mathbf{G}_N(F_{x^2,f}) - 2F_{x,f}\mathbf{G}_N F_{x,f}] = 0. \tag{8.349}$$

Note that replacing G_N by G the relation says that G is a *first order* operator and this is read off from the form of G directly. This can be verified also directly using (8.258) and (8.344). This implies that the variance of a solution to the limiting martingale problem is 0. Since the higher moments of $\langle \hat{\Psi}^N, f \rangle$ are bounded this implies (1). Point (2) is given by the a priori estimate (8.340).

For point (3) we take $g(x) = x$ and conclude that the limit point is given by the unique trajectory in $\mathbb{B}$ which satisfies

$$\int_0^\infty \int_0^\infty f(r,x)\hat{\Psi}_t(dr,dx) - \int_0^t \int_0^\infty \int_0^\infty G^{\hat{\Psi}_s} f(r,x)\hat{\Psi}_s(dr,dx)ds = 0, \tag{8.350}$$

that is, the nonlinear system (8.253) (recall Lemma 8.6). In other words, any limit point is a solution to the **G** martingale problem. For point (4), we have established the uniqueness and properties of the solution to the **G** martingale problem in Lemma 8.6 and Corollary 8.7.

To complete the proof of convergence, given $\hat{\Psi}_{t_0}$ and T, find $K_0 > 0$ such that the solution of the nonlinear system stays in the interior of the ball of radius K_0 during the interval $[t_0, t_0 + T]$ and let $K \geq 2K_0$, say. Then it follows from the above that we have weak convergence up to time $(t_0 + T) \wedge \tau_K$ for every $K > 0$ and therefore for $\varepsilon > 0$

$$\lim_{N \to \infty} P[\sup_{t_0 \leq t \leq t_0+T} \|\hat{\Psi}_t^N - \hat{\Psi}_t\| > \varepsilon] = 0. \tag{8.351}$$

This completes the proof of part (a) of the proposition.   q.e.d.

8.3 Proofs of Propositions 2.3, 2.6, and 2.11

Part 2 Convergence of t_0-marginals and properties of (u^N, U^N)

Our goal is to prove part (b) and (c) of the proposition under the assumptions of convergence at a fixed time $\alpha^{-1} \log N + t_0$.

Proof of (1). For that purpose we prove first the convergence of the marginals at times $\alpha^{-1} \log N + t_0$, starting with "$t_0 = -\infty$", (8.311), and then later consider $t_0 \in \mathbb{R}$ to finally come to the convergence of the path.

Proof of (8.311). Here we have a time scale where as $N \to \infty$ the collisions become negligible since the observation time diverges sublogarithmically as function of N. Using the multicolour particle system (see Subsubsection 8.3.6) we can build all variables on one probability space. From this construction it follows that collisions become negligible and we have the claimed convergence from the one for the collision-free system proved in Subsubsection 8.3.4, see (8.144), and Subsubsection 8.3.5 see (8.165) via the CMJ-theory. This proves part (b) of the proposition. q.e.d.

Proof of (c). We now have to prove the convergence result for the marginal of $(\hat{\Psi}^N_t)_{t \in \mathbb{R}}$ at times $t = \alpha^{-1} \log N + t_0$, where we can choose $t_0 \in \mathbb{R}$ as small as we want since we know that the dynamic converges to a continuous limit dynamic. This means we have to show (8.312), (8.313) and (8.314). Finally we have to relate the limit of $(N^{-1}\tilde{u}^N, \tilde{U}^N)$ as $N \to \infty$ to the system of nonlinear equations to then prove (8.316) and (8.317).

Proof of (8.312) and (8.313). We recall that we proved in the previous part the convergence of $N^{-1}\tilde{u}^N$ pathwise if we consider path for $t \geq t_0$ and if we have convergence in law at $t = t_0$. Therefore it suffices now to show that $N^{-1}\tilde{u}^N(t_0)$ converges (in law) as $N \to \infty$.

To do this we first note that the time-scaled and normalized (i.e. scaled by N^{-1}) number of sites occupied at time $\alpha^{-1} \log N + t_0$, denoted $N^{-1}\tilde{u}^N$, does not go to zero as $N \to \infty$ nor does it become unbounded. This was proved in Proposition 2.6. This proves tightness of the one-dimensional marginal distributions at times $\alpha^{-1} \log N + t$ and that limit points are non-degenerate.

The tightness gives us the existence of limiting laws for $(N^{-1}\tilde{u}^N, \tilde{U}^N)$ along subsequences for time $\alpha^{-1} \log N + s$ for fixed s. Since $N^{-1}\tilde{u}^N \leq We^{\alpha t}$ any limit point must satisfy

$$\limsup_{t \to -\infty} e^{-\alpha t} u(t) < \infty \text{ and } \|U(t)\| \text{ bounded.} \tag{8.352}$$

Therefore by Lemma 8.6(d) any limit point corresponds to a solution to the nonlinear system given by (8.280) and (8.281) and any two such solutions are (random) *time shifts* of each other. Hence we have to show that each pair of limit points belongs to the same time shift from the standard solution of the system with growth constant 1.

Recall the definitions of $\tau^N(\varepsilon)$ (which is the first time where $u_s^{N,k,\ell}$ reaches $\lfloor \varepsilon N \rfloor$ for all $\varepsilon \in (0,1)$), $\tilde{\tau}^N(\varepsilon)$ from (8.182), (8.183). Now assume that there are two limit points and let the corresponding limiting passage times be denoted $\tau^1(\varepsilon)$ and $\tau^2(\varepsilon)$ such that $\tau^{N_{n_i},i}(\varepsilon) \to \tau^i(\varepsilon)$, $i = 1,2$ in law. But by (8.220), $P[|\tau^{N_{n_2},2}(\varepsilon) - \tau^{N_{n_1},1}(\varepsilon)| > 2C(\varepsilon)] \to 0$. But since $C(\varepsilon) \to 0$ as $\varepsilon \to 0$, the time shift must be arbitrarily small, which means equal to zero and therefore the two limit points coincide. This immediately proves that $\tau^N(\varepsilon)$ converges for any $\varepsilon > 0$.

We now elaborate on this argument. To show convergence of the marginal distributions it suffices to prove the convergence in law of the *normalized* first passage times $\tau^N_{\text{norm}}(\varepsilon) := \tau^N(\varepsilon) - (\alpha^{-1} \log N)$. We give two arguments which verify that convergence of the $\tau^N_{\text{norm}}(\varepsilon)$ is sufficient.

(i) Given the value $u(s)$ we know exactly the first passage time for all $\varepsilon > u(s)$ (due to the *unique* and also *deterministic* evolution from any initial point). Indeed using the fact that the path converges (for convergent sequences at a fixed time), the collection is monotone non-decreasing and cadlag, we know that the whole collection (in ε) of first passage time converges in path space. The uniqueness of the limit points follows since two different limiting distributions at times $\alpha^{-1} \log N + s$ for fixed s would not be compatible with the uniqueness in distribution of the limiting first passage times, since the limiting dynamics $(u(t))_{t \in \mathbb{R}}$ is monotone for sufficiently small (i.e. negative t) so that it is uniquely determined by the first passage times in that range and hence everywhere.

(ii) Alternatively we can use the fact that there is according to Lemma 8.6 up to a time shift a unique entrance law from 0 at $t \to -\infty$ which is compatible with the growth conditions of a limit of that quantity as $N \to \infty$. Therefore by one first passage time this random time shift is determined and hence it suffices to prove convergence in law of one first passage time concluding the second argument.

Having proved the reduction of the convergence statement to one on τ^N_{norm}, we have to show next that for one $\varepsilon \in (0,1)$ and therefore by the above for every such ε:

$$\mathcal{L}[\tau^N(\varepsilon) - \frac{\log N}{\alpha}] \underset{N \to \infty}{\Longrightarrow} \mathcal{L}[\tau(\varepsilon)], \tag{8.353}$$

with

$$\tau(\varepsilon) = \inf(t \in \mathbb{R} | u(t) = \varepsilon), \tag{8.354}$$

where u is the (unique) entrance law with $e^{\alpha|t|} u(t) \underset{t \to -\infty}{\longrightarrow} W$ (recall Lemma 8.6).

We have two tasks, to identify the limit on the r.h.s. of (8.353) as the quantity defined in (8.354) this however we showed in part 1 where we established in particular the convergence of the dynamic for the frequency of occupied sites, and secondly we have to show the convergence in (8.353). So only the latter remains.

8.3 Proofs of Propositions 2.3, 2.6, and 2.11

We use our results on the corresponding times $\tilde{\tau}^N(\varepsilon)$, assuming no collisions occur, namely we see that it follows from Proposition 8.5 that $\tilde{\tau}^N(\varepsilon)$ converges to a limit $\tilde{\tau}(\varepsilon)$ defined in terms of the *collision-free* process. Hence it remains to bridge from $\tau^N(\varepsilon)$ to $\tilde{\tau}^N(\varepsilon)$.

To relate $\tilde{\tau}^N(\varepsilon)$ and $\tau^N(\varepsilon)$ we will use that according to the above arguments arbitrarily small $\varepsilon > 0$ and we can work with the multicolour particle system which we introduced in the proof of Proposition 2.6 below (8.175).

Consider $\varepsilon \downarrow 0$. We have the fact from the multicolour particles system (recall (8.194) and (8.196)) that for sufficiently small ε asymptotically as $N \to \infty$ in probability

$$\tilde{\tau}(\varepsilon) \leq \tau^N(\varepsilon) \leq \tilde{\tau}(\tilde{\varepsilon}), \tag{8.355}$$

where $\tilde{\varepsilon}$ solves

$$\tilde{\varepsilon} - \text{Const}\, \tilde{\varepsilon}^2 = \varepsilon, \quad \tilde{\varepsilon} > 0. \tag{8.356}$$

Note that the Const is less than 1, so that there is a unique $\tilde{\varepsilon} > \varepsilon$ such that

$$\tilde{\varepsilon} - \varepsilon = O(\tilde{\varepsilon}^2). \tag{8.357}$$

In (8.355) the right inequality holds for ε small enough and N *sufficiently large*, while the l.h.s. holds even for every N.

Denote by $\tau_{\text{norm}}^N(\varepsilon)$ the normalised $\tau^N(\varepsilon)$, i.e. $\tau^N(\varepsilon) - \alpha^{-1} \log N$. We know that $\tilde{\tau}^N(\varepsilon) \longrightarrow \tilde{\tau}(\varepsilon)$ as $N \to \infty$ if we use the coupled collection of processes we constructed in the proof of equation (8.312). Hence for ε_0 sufficiently small realizations of two different limit points $\tau_{\text{norm}}^{\infty,1}(\varepsilon_0), \tau_{\text{norm}}^{\infty,2}(\varepsilon_0)$ satisfy on a suitable probability space:

$$\tilde{\tau}(\varepsilon) \leq \tau_{\text{norm}}^{\infty,1}(\varepsilon), \tau_{\text{norm}}^{\infty,2}(\varepsilon) \leq \tilde{\tau}(\tilde{\varepsilon}). \tag{8.358}$$

Therefore $|\tau_{\text{norm}}^{\infty,1}(\varepsilon) - \tau_{\text{norm}}^{\infty,2}(\varepsilon)| \leq |\tilde{\tau}(\varepsilon) - \tilde{\tau}(\tilde{\varepsilon})|$. We furthermore know that $\tilde{\tau}(\varepsilon)$ is given by $(\log \varepsilon - \log W)/\alpha$ so that the l.h.s. and the r.h.s. of (8.358) differ by a *deterministic* quantity, which equals $const \cdot \tilde{\varepsilon}_0$ as $\tilde{\varepsilon}_0 \downarrow 0$. As $N \to \infty$ we have a limiting evolution between $\tau^N(\tilde{\varepsilon}_0)$ and $\tau^N(\tilde{\varepsilon})$ which is deterministic. This evolution is a *random shift* of some deterministic curve. Then it is not possible to have two limit points for the law $\tau^N(\tilde{\varepsilon})$, since their difference are shifted versions of a given curve would translate into the same difference but now bounded above by arbitrarily small $\tilde{\varepsilon}_0$ and therefore equal to 0. Hence all limit points are equal and $\tau_{\text{norm}}^N(\varepsilon) = (\tau^N(\varepsilon) - \alpha^{-1} \log N)$ must converge as $N \to \infty$ as claimed and we are done. This concludes the argument for the convergence of the marginals at time $\alpha^{-1} \log N + t_0$.

Proof of (8.314). We first need some preparation, involving the collision-free regime. Turn to the dual particle process and recall that at time $\alpha^{-1} \log \log N$ asymptotically as $N \to \infty$ there are $W \log N$ occupied sites and that at time

$\alpha^{-1} \log \log N$ the empirical age and size distribution is given by the CMJ stable age and size distribution $\mathcal{U}(\infty, \cdot, \cdot)$, since we are in the non-collision regime due to $\log \log N \ll \alpha^{-1} \log N$. Therefore at time $\tau_{\log N}$ the age distribution is (asymptotically as $N \to \infty$) given by the stable age distribution. Moreover for each realization of W there is a time shift of the limiting dynamics which is a function of W. Since the evolution from this point $\tau_{\log N}$ on is asymptotically deterministic we conclude that the age and size distribution U^N at time $\alpha^{-1} \log N - t_N + t$, with $t_N \uparrow \infty$ but $t_N \ll \alpha^{-1} \log N$, converges to a time shift of the one determined by equation (8.247) and (8.248) with initial condition given by $u = 0$ and U given by the stable age distribution. We now have to bridge from times $\alpha^{-1} \log N - t_N + t$ to $\alpha^{-1} \log N + t$, where collisions play a role. This runs as follows.

For a given approximation parameter $\varepsilon > 0$ we choose a time horizon $\alpha^{-1} \log N + T(\varepsilon)$ such that $(K_t^N, \zeta_t^N)_{t \geq 0}$ is, as $N \to \infty$, ε-approximated by the collision-free process $(K_t, \zeta_t)_{t \geq 0}$ up to time $\alpha^{-1} \log N + T(\varepsilon)$ if we choose ε sufficiently small and this implies the approximation for the functionals U^N by $\mathcal{U}$ as $N \to \infty$ for ε sufficiently small. More precisely we require that we can find a coupling between U^N and $\mathcal{U}$ such that ($\|\cdot\|$ denotes variational distance)

$$\limsup_{N \to \infty} P \left(\sup_{t \leq \frac{\log N}{\alpha} + T(\varepsilon)} \|U^N(t, \cdot, \cdot) - \mathcal{U}(t, \cdot, \cdot)\| > \varepsilon \right) \to 0 \text{ as } \varepsilon \to 0. \quad (8.359)$$

In other words we have to show that by the collisions up to time $\alpha^{-1} \log N + T(\varepsilon)$ at most difference ε develops between U^N and $\mathcal{U}$ for very large N.

To show this note that we can bound the total effect of collisions occurring up to time $\alpha^{-1} \log N + T(\varepsilon)$. Namely we return to the multicolour particle system defined below (8.175), which shows that we can relate the difference of the system without and with collision through the black and red particles. In order to show that $\|U^N(t, \cdot, \cdot) - \mathcal{U}(t, \cdot, \cdot)\| \leq \varepsilon$ it suffices to show that the proportion or red and black particles is $o(\varepsilon)$ in probability as $N \to \infty$. But this was verified for the number of black sites with $T(\varepsilon) = \tilde{\tau}(\varepsilon)$ in (8.194), which immediately by using that ζ is stochastically bounded gives the claim for the number of particles. Since the number of descendants of red particles grows at a slower rate as the black and new red founding particles occur at the same rate as black ones then the claim follows. This completes the proof of Part (c).

Proof of (d): (8.316) and (8.317). Again we first turn to the collision-free process. We approximate (recall $\mathcal{U}$ refers to the collision-free CMJ-process) $\mathcal{U}(t, \cdot, \cdot)$ by $\mathcal{U}(\infty, \cdot, \cdot)$. By [N] (Theorem 6.3) we know that $\mathcal{U}(r, \cdot, \cdot)$ converges a.s. as $r \to +\infty$ to the stable age and size distribution $\mathcal{U}(\infty, \cdot, \cdot)$ of the Crump–Mode–Jagers process (see (8.163)) and hence

$$\mathcal{U}(\alpha^{-1} \log N + t, \cdot, \cdot) \xrightarrow[N \to \infty]{} \mathcal{U}(\infty, \cdot, \cdot), \text{ a.s..} \quad (8.360)$$

8.3 Proofs of Propositions 2.3, 2.6, and 2.11

Now take collisions into account and consider the limit of $U^N(\frac{\log N}{\alpha} + t, \cdot, \cdot)$ as $N \to \infty$, $t \to -\infty$. Combining (8.359) and (8.360) it follows that for $\eta > 0$ there exists $\tau(\eta)$ such that

$$\limsup_{N \to \infty} P\left(\sup_{t \leq \tau(\eta)} \|U^N(\frac{\log N}{\alpha} + t), \cdot, \cdot) - \mathcal{U}(\infty, \cdot, \cdot)\| > 2\eta\right) < \eta. \qquad (8.361)$$

Therefore

$$\lim_{t \to -\infty} U(t, \cdot, \cdot) = \mathcal{U}(\infty, \cdot, \cdot). \qquad (8.362)$$

Finally we have to verify that indeed $(e^{-\alpha t} \frac{\tilde{u}^N(t)}{N})_{t \in \mathbb{R}}$ converges to a limit $(e^{-\alpha t} u(t))_{t \in \mathbb{R}}$ such that $e^{-\alpha t} u(t) \to W^{k,\ell}$ as $t \to -\infty$, where $W^{k,\ell}$ arises as the growth constant of the collision-free process. This was already essentially done in the proof of (8.311), if we observe that the difference between the collision-free process and $\tilde{u}^N(t)$ is bounded by the number of black sites, whose intensity vanishes as $t \to -\infty$, since

$$\limsup_{N \to \infty} P[\tau^N(\varepsilon) \leq \alpha^{-1} \log N + t] \to 0 \text{ as } t \to -\infty \qquad (8.363)$$

and hence the intensity of black sites is $O(\varepsilon^2)$ at time $\tau^N(\varepsilon)$ for N large.

This completes the proof of part (d). q.e.d.

Part 3: Occupation density of dual particle system

Proof of (e)

Note first that (8.319) follows from what we proved for the collision-free regime, in particular the fact that $\sup(e^{-\alpha t} K_t | t > 0) < \infty$.

(8.320) Next note that it suffices to prove the convergence in (8.318) to obtain (8.320) using (8.319). Namely we then have pointwise convergence on a suitable probability space and an upper bound is given by the collision-free process with single sites in equilibrium, which provides an integrable integrand and is independent of N and indeed using Proposition 8.8, part (b) we see that (8.320) holds.

(8.318) Using Proposition 8.8 part (c) we see that for (8.318) we have to reduce the problem concerning $u^N U^N$ and uU to what we showed there for u^N and u respectively U^N and U. Namely each of them converges weakly. However since

$$(u(t), U(t, \cdot)) \longrightarrow u(t)U(t, \cdot) \qquad (8.364)$$

is a continuous mapping this follows from the convergence properties we proved for $(N^{-1}\tilde{u}^N, \tilde{U}^N)$ as $N \to \infty$.

This completes the proof of Proposition 8.8.

Remark 80. Consider the normalized first passage times $\bar{\tau}^N(\varepsilon)$, where $\bar{\tau}^N(\varepsilon)$ is the first time where $\Pi_s^{N,k,\ell}$ (rather than $K_s^{N,k,\ell}$) reaches $\lfloor \varepsilon N \rfloor$.
Then one can show that for every $\varepsilon \in (0, 1)$:

$$\mathcal{L}[\bar{\tau}^N(\varepsilon) - \frac{\log N}{\alpha}] \underset{N \to \infty}{\Longrightarrow} \mathcal{L}[\bar{\tau}(\varepsilon)], \tag{8.365}$$

with

$$\bar{\tau}(\varepsilon) = \inf(t \in \mathbb{R} | u(t) U(t, \mathbb{R}, \mathbb{N}) = \varepsilon), \tag{8.366}$$

and u is the entrance law with $\lim_{t \to \infty} e^{\alpha |t|} u(t) = W$.

8.3.9 Dual Population in the Transition Regime: Asymptotic Expansion

The purpose of this subsubsection is to use and extend the techniques we developed above to later be able to connect the early times (studied later in Subsection 8.4) and the late times studied in Subsection 8.3.8 above.

Return first to our original population model. On the side of the original process the limit in the first time scale, i.e. of times $O(1)$ as $N \to \infty$, is described by the branching process $(\mathbf{J}_t^m)_{t \geq 0}$ and in the second time range, i.e. after times $\alpha^{-1} \log N$ by the deterministic McKean–Vlasov equation but with random initial condition. For both these limits we have derived (respectively for $\mathbf{J}_t^m$ we will in Subsection 8.4) exponential growth at the same rate α and for the first we had obtained the growth constant $\mathcal{W}^*$ and for the latter the growth constant $^*\mathcal{W}$. The objective below is to show that we can consider the limit simultaneously in both time scales in such a way that the final random value arising from the branching process describing the number of type-2 sites at small times provides the initial value of the McKean–Vlasov equation describing the evolution from emergence on till fixation meaning that in fact $^*\mathcal{W}$ and $\mathcal{W}^*$ have the same distribution.

This section provides the analysis of the dual needed for this purpose. The completion of the argument is obtained in the following subsubsection which is devoted to the limiting branching process and the transfer back of properties of the dual to the original process.

Recall the definition of u^N, U^N in (8.234)–(8.238). In this section we deal with the relation of the two basic limits as $N \to \infty$ of the functional of the dual population given by $(u^N(t), U^N(t))$, namely, the limit in the two different time scales

$$t_N = o(T_N) \text{ and } t_N = T_N + t, \text{ where } T_N := \frac{\log N}{\alpha} \text{ and } t \in \mathbb{R}, \tag{8.367}$$

as well as the *transition* between these two time scales which *separate* in the limit $N \to \infty$.

8.3 Proofs of Propositions 2.3, 2.6, and 2.11

The main tool developed in this subsection is the analysis of the dual population in this two-time scale context. For that purpose we make use of two techniques: (1) *Nonlinear evolution equations* for the dual particle systems and their enrichments, (2) *couplings* of particle systems to estimate and control the effects of collisions in the dual process by comparing it with collision-free systems, i.e. the dual of the McKean–Vlasov process.

Heuristically we can approximate u^N/N by $\tilde{u}^N$ which are the solutions of the ODE

$$\frac{d\tilde{u}^N(t)}{dt} = \alpha_N(t)\tilde{u}^N(t)(1 - \frac{b_N(t)}{N}\tilde{u}^N(t)), \quad t \geq t_0(N), \tag{8.368}$$

$$b_N(t) = 1 + \frac{\gamma_N(t)}{\alpha_N(t)}, \quad \tilde{u}^N(t_0(N)) = u^N(t_0(N)) = W_N e^{\alpha t_0(N)}, \tag{8.369}$$

which allows to get some feeling for the behaviour based on explicit calculation for the case where α_N and γ_N are constant and explicit formulas for the solution of the ODE (8.368) and (8.369) can be given.

There are two regimes depending on which of the time scales in (8.367) is used: the first (linear) regime where $N^{-1}u^N(t_N) \to 0$ as $N \to \infty$ together with $\alpha_N(t_N) \to \alpha$ and $\gamma_N(t_N) \to \gamma$ and the second (nonlinear) regime where $u^N(t_N) = O(N)$ and where $\alpha_N(t_N), \gamma_N(t_N)$ differ from α, γ.

We have proved in Subsubsection 8.3.3 that for times $t_N \to -\infty$ the number of sites occupied by the dual process at times $\frac{\log N}{\alpha} + t_N$ and multiplied by N^{-1} converges to zero in probability. On the other hand we saw in Subsubsection 8.3.7 that assuming that in law

$$\frac{1}{N}u^N(\frac{\log N}{\alpha} + t_0) \underset{N \to \infty}{\Longrightarrow} u(t_0), \tag{8.370}$$

then in law on path space

$$\{\frac{1}{N}u^N(\frac{\log N}{\alpha} + t)\}_{t \geq t_0} \underset{N \to \infty}{\Longrightarrow} \{u(t)\}_{t \geq t_0} \tag{8.371}$$

and given the state at time t the limiting dual dynamics $u(t)$ is deterministic and nonlinear, namely, $(u(t), U(t))_{t \geq t_0}$ is for all $t_0 \in \mathbb{R}$ the solution to the system (8.247), (8.248) where in the latter quantity we integrate out the age. Moreover we proved that this system has a unique solution satisfying

$$\lim_{t \to -\infty} e^{\alpha|t|}u(t) = W, \quad \lim_{t \to -\infty} U(t) = \mathcal{U}(\infty), \tag{8.372}$$

where $\mathcal{U}(\infty)$ is the stable size distribution of the CMJ process induced by the collection of the occupied sites of the McKean–Vlasov dual process.

In order to later on relate $^*W$ and W^* we next focus on $^*W$ and try to get more information on its law. This requires, as we shall see later, when we return to the original process from the dual to obtain in (8.372) the higher order terms as $t \to -\infty$ for the limiting equation (as $N \to \infty$) as well as in the approximation as $N \to \infty$ of this behaviour. This means we want to write for $t \to -\infty$ the limiting (as $N \to \infty$) intensity of $N^{-1}\tilde{u}^N$, denoted u:

$$u(t) = We^{\alpha t} - \text{Const} \cdot W^2 e^{2\alpha t} + o(e^{2\alpha t}) \qquad (8.373)$$

and determine the constant, but moreover we want to consider for $N \to \infty, t \to -\infty$ the quantity

$$u^N\left(\frac{\log N}{\alpha} + t\right) = W_N(t)e^{\alpha t} \cdot N + C(t, N) \qquad (8.374)$$

and to estimate the order in both N and t of the correction term $C(t, N)$. The latter is equivalent to determining up to which order the expansion in (8.373) is approximated by the finite N-system. First we turn to the question in (8.374) and then to (8.373) for each point of view formulating a separate proposition.

We therefore refine (8.372) by providing speed of convergence, in t and uniformity, in N, results.

Proposition 8.9 (Order of approximation of entrance law in N and t).

(a) Let $T_N = \frac{\log N}{\alpha}$ and $0 < \delta < 1$. Then $(u^N(t), U^N(t))$ and $(u(t), U(t))$ have the following three properties:
We can couple the $(u^N(t), U^N(t))_{t \geq 0}$ with a fixed CMJ process $(K_t, \mathcal{U}(t))_{t \geq 0}$ such that

$$e^{-\alpha t} K_t \longrightarrow W \text{ a.s. as } t \to \infty, \qquad (8.375)$$

$$\lim_{t \to -\infty} \limsup_{N \to \infty} P\left(|e^{-\alpha t}\frac{1}{N}u^N(T_N + t) - W| > e^{\delta \alpha t}\right) = 0 \qquad (8.376)$$

and

$$\lim_{t \to -\infty} \lim_{N \to \infty} \left(\frac{e^{-\alpha t}}{N}u^N(T_N + t), U^N(T_N + t)\right) = (W, \mathcal{U}(\infty)). \qquad (8.377)$$

(b) Given W the system (8.247), (8.248) has unique solution $(u(t), U(t))$ with $e^{-\alpha t}u(t) \to W$ as $t \to -\infty$ which can be realized together with $\{(u^N, U^N), N \in \mathbb{N}\}$ on one probability space such that the process

$$(e^{-\alpha t} N^{-1} u^N(T_N + t))_{t \in \mathbb{R}} \qquad (8.378)$$

can for every $\delta \in (0, 1)$ be approximated by $(e^{-\alpha t}u(t))_{t \in \mathbb{R}}$ with "initial condition" W, up to order $e^{\alpha \delta t}$ for $t \in \mathbb{R}$ uniformly in N. □

8.3 Proofs of Propositions 2.3, 2.6, and 2.11

We now have established a precise relation between (u^N, U^N) and the limiting entrance law (u, U). We now return to the solution (u, U) to (8.247), (8.248) and $\alpha(t)$, $\gamma(t)$ as defined in equation (8.245). The purpose of the rest of this section is to identify the asymptotic behaviour of this nonlinear system as $t \to -\infty$ up to higher order terms, i.e. we want to identify in particular the correction of order $e^{2\alpha t}$ as $t \to -\infty$ which is due to the occurrence of collisions by migration of the dual individuals which by subsequent coalescence of the collided individuals can change the behaviour of the limiting system.

Proposition 8.10 (Transition between linear and nonlinear regime limit dynamics (u, U)).

(a) The pair (u, U) and the functionals α and γ arising as the limit in (8.312)–(8.314) via (8.245) satisfy (here we suppress the k, ℓ):

$$(\alpha(t) + \gamma(t))u(t) \leq (\alpha + \gamma)We^{\alpha t}, \tag{8.379}$$

$$\lim_{t \to -\infty} \alpha(t) = \alpha, \quad \lim_{t \to -\infty} U(t) = \mathcal{U}(\infty), \quad \lim_{t \to -\infty} \gamma(t) = \gamma, \tag{8.380}$$

$$u(t) = We^{\alpha t} - \kappa W^2 e^{2\alpha t} + O(e^{3\alpha t}) \quad \text{as } t \to -\infty, \text{ for some constant } \kappa > 0 \tag{8.381}$$

and furthermore as $t \to -\infty$ the functions $\alpha(\cdot)$ and $U(\cdot)$ satisfy that

$$\alpha(t) = \alpha + \tilde{\alpha}_0 e^{\alpha t} + O(e^{2\alpha t}), \quad U(t) = \mathcal{U}(\infty) + \tilde{U}_0 e^{\alpha t} + O(e^{2\alpha t}), \tag{8.382}$$

$$\gamma(t) = \gamma + \tilde{\gamma}_0 e^{\alpha t} + O(e^{2\alpha t}), \tag{8.383}$$

where $\tilde{\alpha}_0$ is a positive number $\tilde{U}_0$ is an $\mathbb{N}$-vector and $\tilde{\gamma}_0 = \tilde{U}_0(1)$.

Using a multicolor system construction, we will see below that all the randomness is given by the (non-degenerate positive) random variable W, namely

$$\tilde{\alpha}_0 \text{ is explicitly given by (8.569), as } Const_1 \cdot W, \tag{8.384}$$

$\tilde{U}_0$ is explicitly given by (8.575) and has the form $W \cdot \vec{const}$, $\tilde{\gamma}_0 = Const_2 \cdot W$.
$\tag{8.385}$

(b) The total number of particles satisfies

$$(\alpha(t) + \gamma(t))u(t) = (\alpha + \gamma)We^{\alpha t} - \kappa^* W^2 e^{2\alpha t} + O(e^{3\alpha t}) \quad \text{as } t \to -\infty, \tag{8.386}$$

for some constant $\kappa^* > 0$.

More precisely with

$$\kappa^* W^2 = (\tilde{\alpha}_0 + \tilde{\gamma}_0)W + \kappa W^2 = W^2(Const_1 + Const_2 + \kappa), \quad (8.387)$$

we have that

$$\limsup_{t\to-\infty} |(\alpha(t)+\gamma(t))u(t)-(\alpha+\gamma)We^{\alpha t} - \kappa^* W^2 e^{2\alpha t}|\cdot e^{-3\alpha t} < \infty. \quad (8.388)$$

(c) Furthermore $U(t,\cdot,\cdot)$ is uniformly continuous at $t = -\infty$ and as $t \to -\infty$

$$\|U(t) - \mathcal{U}(\infty) - \tilde{U}_0 e^{\alpha t}\|_{\mathrm{var}} = O(e^{2\alpha t}). \square \quad (8.389)$$

Remark 81. Note that the results above show that all the relevant randomness in the dual process sits indeed in the random variable W.

The proof proceeds in four steps. The proof of Propositions 8.9 and 8.10 are based on an enriched version of the dual particle system η and a reformulation of the nonlinear system (8.247) and (8.248) which are given in Steps 1 and 2 below which are then followed by the Steps 3 and 4 giving the proof of the two propositions using these tools. In Step 1 we focus on the finite N system in a time regime where collisions due to migration become essential, while the Step 2 develops the analysis of the limiting system as $N \to \infty$ in this time regime. The main focus here is on the behaviour as the effect of collisions becomes small and to investigate its precise asymptotics in this regime.

Step 1: Modified and enriched coloured particle system

Our goal is to realize on one probability space the CMJ-process and the dual particle system in such a way that we can identify in a simple way the difference between the two dynamics with a higher degree of accuracy than before by using three colours white, black and red. In particular we can identify the part of the population involved in one or in more than one collision (more than two etc.). This joint probability space can be explicitly constructed easily, based on collections of Poisson processes. Since no measure theoretic subtleties occur, we do not write out this construction in all its lengthy detail and just spell out the evolution rules.

In order to identify the contributions at time $\alpha^{-1}\log N + t$ of order $e^{\alpha t}, e^{2\alpha t}, e^{3\alpha t}$, we want to expand the number of occupied sites at time $\alpha^{-1}\log N + t$, using quantities $W_N(t)$ defined as $e^{-\alpha t} K_t^N$ and then analog higher-order objects, in the form:

$$u^N(T_N + t) \sim W_N(t)e^{\alpha t} - W_N'(t)e^{2\alpha t} + W_N''(t)e^{3\alpha t}, \quad (8.390)$$

with $T_N = \alpha^{-1}\log N$ and $N^{-1}W_N(t), N^{-2}W_N'(t), N^{-3}W_N''(t)$ converging as $N \to \infty, t \to -\infty$, more precisely,

8.3 Proofs of Propositions 2.3, 2.6, and 2.11

$$\lim_{t \to -\infty} \lim_{N \to \infty} [e^{-\alpha t} N^{-1} u_N(T_N + t)] = W$$
$$\lim_{t \to -\infty} \lim_{N \to \infty} [e^{-2\alpha t} N^{-2} [(WNe^{\alpha t} - u^N(T_N + t))]i] = \kappa W^2 \quad ,$$
$$\lim_{t \to -\infty} \lim_{N \to \infty} [e^{3\alpha t} N^{-3} [u_N(T_N + t)) - WNe^{\alpha t} - \kappa N^2 W^2 e^{2\alpha t}]] < \infty, \tag{8.391}$$

where the limits are needed in probability and in L^1. The latter convergence follows from the convergence in probability and the finiteness of the moments of W (see Lemma 8.4).

We use the dual particle system to obtain information on the asymptotics as $t \to -\infty$ of the pair (u, U) and also to identify their random initial condition. However to carry this out we must enrich the *dual particle system* to a *multicolour* particle system on an *enriched geographic space*.

Since we want to read off more properties than in the emergence argument based on the mean where we had white, black and red particles we shall need now some more colours which allow us to record particles which have been involved in one, two, etc. collisions and with associated colours their counterpart in a collision-free system.

Furthermore in order to realize the CMJ and the dual on one probability space we have to couple the evolution of certain colours where one belongs to the CMJ-part and the other to the dual particle system part. This will also require us to refine the geographic space $\{1, \cdots, N\} + \mathbb{N}$ (where the black particles were collision-free and moving on $\mathbb{N}$) we had before for the white, red, black particles system by adding a further copy $\mathbb{N}$ (or more if we need more accuracy).

More precisely we now introduce a *multicolour particle system* that is obtained by modifying and enriching the coloured particle system defined by (8.176)–(8.181) by introducing (1) new colours and (2) new type of sites. The number of new colours and sites necessary depends on how precisely we want to control the behaviour as $t \to -\infty$ meaning up to which higher order terms as $t \to -\infty$ we want to go.

We first need a new colour (green) which allows us to control exactly (i.e. not just estimating from above) the difference between the dual particle system η and our old white–black collision-free particle system. Recall that in the old system we created a red–black pair upon a collision such that the further evolution of black remains collision-free and the red follows the true mechanism of the dual particle system. A real difference in the total number of particles between dual and CMJ will however occur only if the red particle coalesces with a white particle but no such loss of an individual occurs in the CMJ-process which now will have one particle more due to this transition. To mark this we will use the colour *green* to mark the disappearing red particle in the dual.

To allow a fine comparison between the white–red and the white–black system we will use coupling techniques which allow us to estimate not only the effect of *collisions* (and subsequent coalescents) but also *recollisions*, the latter will generate parallel to the red–black construction two new colours *purple* (in the dual system) and *blue* (in the collision-free comparison system). For this purpose we have to modify the evolution rules when red particles collide or red particles collide with white particles. For the bookkeeping of the effects from these events we use the new further colours.

The white particles in the old process also have the same dynamics as the white particles in our new formulation but in addition to white, red, and black particles the *green, purple* and *blue* particles can have now locations in

$$\{1, 2, \cdots, N\} + \mathbb{N} + \mathbb{N}, \tag{8.392}$$

which we will explain as we describe the evolution rules.

In order to achieve all these goals of a higher order expansion the modified dynamics should satisfy the following requirements:

- for each N the union of black plus white particles is equivalent to the particle system without collisions and can be defined on a common probability space by a single CMJ process,
- the white plus red plus purple particles give a version of $(u^N(t), U^N(t, \cdot, \cdot))$,
- the red plus purple particles give a refinement of the red particle system in the white–black–red process used earlier, i.e. *both together* have the same dynamics as the set of red particles in (8.3.6),
- the green particles describe the loss due to coalescence after collision, i.e. after a particle (white, red or purple) migrated to an occupied site on $\{1, \cdots, N\}$ (by a white, red or purple particles) and then coalesced with such particle at this site,
- the black particles are placed and evolve on the second copy of $\mathbb{N}$ and blue particles are placed and evolve on the first copy of $\mathbb{N}$.
- We can match certain pairs of red and black respectively purple and blue particles when they are created to better compare the CMJ-part and the one with the dual dynamics.
- The number of the sites occupied by red, green and blue particles has the same law as the number of sites occupied by black particles.

For each N the evolution rules below define a *Markov pure jump process* which describes a growing population of individuals carrying a *location* and a *colour* and in addition the information which pairs of individuals are *coupled*.

Let I_n be a copy of $\{1, \cdots, n\}$. Denote by

$$C = \{\text{ white, black, red, purple, blue, green}\}, \tag{8.393}$$

Then the state space is given by the union over n of the set of maps

$$I_n \longrightarrow C \times (\{1, \cdots, N\} + \mathbb{N} + \mathbb{N}) \times (I_n^2). \tag{8.394}$$

If we just count the number particles with a certain colour and location we get again a Markov pure jump process, specifying a multitype particle system on $\{1, 2, \ldots, N\} + \mathbb{N} + \mathbb{N}$. The state space of this system is given by:

$$(\mathbb{N}_0)^{(\{\{1,2,\ldots,N\}+\mathbb{N}+\mathbb{N}\}\times C)}. \tag{8.395}$$

We shall work with both processes.

8.3 Proofs of Propositions 2.3, 2.6, and 2.11

In order to satisfy the properties specified above, the dynamics of the enriched coloured particle systems has the following dynamics:

- Given that k is the number of sites in $\{1, 2, \ldots, N\}$ which are occupied, and a white particle migrates to another site in $\{1, \cdots, N\}$ the outcome is given as in (8.178)–(8.180) (a *red–black pair* is created) but we record that these particles form a red–black pair. White particles give birth to white particles and two white particles coalesce to produce one white particle.
- Black particles evolve as before but live on the second copy of $\mathbb{N}$.
- Red particles give birth to red particles, two red particles at the same site coalesce to one *red particle*.
- When a red particle coalesces with a white particle (at the same site) the outcome is a white particle and a *green* particle at this site. Moreover we then couple the green particle with the black particle that had been associated with the involved red particle.
- Given that the current number of sites occupied by white, red, purple or green particles on $\{1, \cdots, N\}$ is ℓ, when a red particle migrates it moves to an empty site in $\{1, \ldots, N\}$ with probability $1 - \frac{\ell}{N}$ and with probability $\frac{\ell}{N}$ it dies and produces a *blue–purple pair* where the blue particle is placed at the first empty site in the first copy of $\mathbb{N}$ and the purple particle is placed at a randomly chosen occupied site on $\{1, 2, \cdots, N\}$. We also associate the blue particle with the black particle that had been associated with the involved red particle.
- Blue particles have the same dynamics as black particles.
- Purple particles have the same dynamics as the dual particle systems η on $\{1, \ldots, N\}$, in particular they migrate to a randomly chosen site in $\{1, \ldots, N\}$.
- Green particles give birth to green particles and two green particles at one site coalesce giving one green particle. Green particles on $\{1, 2, \cdots, N\}$ migrate to empty sites on $\{1, \cdots, N\}$ with probability $1 - \frac{\ell}{N}$ and with probability $\frac{\ell}{N}$ to the first empty site in the first copy of $\mathbb{N}$ where ℓ is the number of occupied sites on $\{1, \cdots, N\}$. Green particles on the first copy of $\mathbb{N}$ migrate to the first empty site.
- When a green and red particle coalesce they produce a just red particle.

Remark 82. The reason for this rule is as follows. The red and green pair correspond to two red particles one of which has coalesced with a white particle. The green particle corresponds to a particle loss due to collision and coalescence. First note that the green particle which coalesces with a red one must have been created at that site. Hence both red and green must descend from the same collision of a particle arriving at the site and hence the coupled black particles share the same site. The red and green particle are coupled with a black or blue particle each. The two black particles can coalesce only if they descend from the same black particle after arriving at the current site. In this case the number of red and green, respectively black and blue is reduced both by one and the original loss of black–white versus white–red is cancelled. When the green and red coalesce this corresponds to the coalescence of two red particles that would have been at a site having no white particles. Then the coalescence reduces the number of red particles to one. This means that the potential loss is canceled out and in both cases we end up with a single red particle.

Green and white particles do not coalesce.
- We couple for a newly created red–black pair the birth, coalescence with offspring up to the time of the first collision of a red particle with a site occupied by red or white particles. We then continue by coupling for all future times the associated black particle with the blue particle (from the blue–purple pair) created at the collision time.
- When red and purple particles coalesce they produce a red particle.
- When a purple particle coalesces with a white particle the outcome is a white and a green particle at the site.

Note that these rules lead to a system with the six desired properties we had listed below (8.392).

The state space of this system is given by:

$$(\{\{1, 2, \ldots, N\} + \mathbb{N} + \mathbb{N}\} \times C)^{\mathbb{N}_0}. \tag{8.396}$$

We shall single out specific subsystems comprised of particles of certain colours, for example, the *WRP-system* (white, red, purple) or the *WRGB-system* (white, red, green, blue) which have state-spaces:

$$\mathbb{N}_0^{(\{\{1,2,\ldots,N\}+\mathbb{N}+\mathbb{N}\}\times\{W,R,P,\})}, \quad \mathbb{N}_0^{(\{\{1,2,\ldots,N\}+\mathbb{N}+\mathbb{N}\}\times\{W,R,G,B\})}, \tag{8.397}$$

respectively. If we need more information, then we have the analogue of (8.394).

Observe that white, red, and purple particles are located only in $\{1, \ldots, N\}$, blue particles are located only in the first copy of $\mathbb{N}$ and green particles can be in either $\{1, \ldots, N\}$ or first copy of $\mathbb{N}$ and all black particles live in the second copy of $\mathbb{N}$. It is important to note that the transitions of green particles do not depend on whether they are located in $\{1, \cdots, N\}$ or in the first copy of $\mathbb{N}$.

We note that the difference between the number of red plus purple particles and the number of black particles is due to the coalescence of red or purple particles with white particles. Each such loss is compensated by the creation of a green particle and we note that green particles never migrate to an occupied site.

By the construction above we define on a common probability space five *coupled* particle systems,

- a CMJ-process which is the union of black and white particles which give a version of $(K_t, U(t))_{t \geq 0}$,
- the white, red and purple particles generate a version of $(u^N(t), U^N(t))$,
- a CMJ process which is the union of white, red, green and blue particles and which generates a version of the pair $(K_t, U(t))_{t \geq 0}$,
- a subset of this above system, the white system generating the pair $(K_{(W),t}^N, U_{(W)}^N(t, \cdot))$, a CMJ-process which is the union of black and white particles which give a version of $(K_t, U(t))_{t \geq 0}$,

8.3 Proofs of Propositions 2.3, 2.6, and 2.11

- a *coupled* subsystem, on the one hand the system of red and purple and on the other hand a subset of the black particles. This coupling is induced by the convention how purple–blue and red–black pairs evolve.

As before in the analysis of (u, U) it suffices for our purposes to study a functional of the multicolour particle system by observing at sites only how many particles of the various colours occur and counting the number of sites of a specific colour configuration. Our functional of a state is given by the counting process

$$\Psi_t^N(i_W, i_R, i_P, i_G, i_B, i_{BL}), \tag{8.398}$$

denoting the number of sites containing i_W white, i_R red, i_P purple, i_G green, i_B blue and i_{BL} black particles at time t.

The process $(\Psi_t^N)_{t \geq 0}$ evolves itself as a *pure jump-strong Markov process* with state space the counting measures on $C^{\mathbb{N}}$, the set of colour configurations, due to the fact that the dynamic only depends on the vector of the numbers of particles of the various colours at a site. Similarly we define pure Markov jump processes

$$\Psi_{W,t}^N(i_W), \quad \Psi_{WRP,t}^N(i_W, i_R, i_P), \quad \Psi_{WRGB,t}^N(i_W, i_R, i_G, i_B). \tag{8.399}$$

Next we want to define the coloured versions of the (u, U) system and of certain subsystems. We therefore define the number of sites exhibiting certain colours:

$$K_t = \sum_{i_i, \dots, i_{BL}} 1_{i_W + i_{BL} \geq 1} \Psi_t^N(i_W, i_R, i_P, i_G, i_B, i_{BL}), \tag{8.400}$$

$$K_{WRP,t}^N = \sum_{i_W, i_R, i_P} 1_{i_W + i_R + i_P \geq 1} \Psi_{WRP,t}^N(i_W, i_R, i_P), \tag{8.401}$$

$$K_{WRGB,t}^N = \sum_{i_W, i_R, i_G, i_B} 1_{i_W + i_R + i_G + i_B \geq 1} \Psi_{WRGB,t}^N(i_W, i_R, i_G, i_B) \tag{8.402}$$

and occasionally we make use of

$$K_{C,t}^N = \sum_{i_L : L \in C} 1_{\sum_{L \in C} i_L \geq 1} \Psi_{C,t}^N(i_C), \quad C \subseteq \{W, R, P, G, B, BL\}, \quad i_C = (i_L)_{L \in C}. \tag{8.403}$$

Now we can define the coloured relatives of (u^N, U^N) due to our construction all on one probability space, setting

$$(u(t), u_W^N(t), u_{WRP}^N(t), u_{WRGB}^N(t)) = (K_t, K_{W,t}^N, K_{WRP,t}^N, K_{WRGB,t}^N), \tag{8.404}$$

$$U_{W,t}^N(i_W) = \frac{\Psi_{W,t}^N(i_W)}{K_{W,t}^N},$$

$$U^N_{WRP,t}(i_W, i_R, i_P) = \frac{\Psi^N_{WRP,t}(i_W, i_R, i_P)}{K^N_{WRP,t}}, \quad (8.405)$$

$$U^N_{WRGB,t}(i_W, i_R, i_G, i_B) = \frac{\Psi^N_{WRGB,t}(i_W, i_R, i_G, i_B)}{K^N_{WRGB,t}}.$$

Lemma 8.11 (Properties of enriched coloured particle system).

(a) *Ignoring colour the process $(u^N_{WRP,t}, U^N_{WRP}(t,\cdot))_{t\geq 0}$ has the same law as $((u^N_t, U^N_t))_{t\geq 0}$ given by (8.234)–(8.238).*

(b) *Ignoring colour, the total number of particles or occupied sites in the WRGB-system has the same distribution as the number of particles respectively occupied sites in W-BL-system and therefore we can identify the total number of occupied sites with the total number of individuals of a CMJ process which we denote by $(K_t)_{t\geq 0}$.*

(c) *The total number of particles (or sites) in the WRP-system is less than that in the WRB-system and larger that in the WR-system a.s.*

(d) *Consider the coloured particle system at times*

$$T_N + t, \quad T_N = \alpha^{-1} \log N. \quad (8.406)$$

The total number of purple or blue particles is $O(e^{3\alpha t})$, more precisely, we have uniformly in N,

$$\sum_{i_W, i_R, i_P} i_P \cdot \Psi^N_{WRP, T_N+t}(i_W, i_R, i_P) \quad (8.407)$$

$$\leq \sum_{i_W, i_R, i_G, i_B} i_B \cdot \Psi^N_{WRGB, T_N+t}(i_W, i_R, i_G, i_B) \leq const \cdot NW_* e^{3\alpha t},$$

where

$$W_* = \sup_t (e^{-\alpha t} K_t) < \infty, \text{ a.s. .} \quad (8.408)$$

(e) *We consider again a time horizon as in (d). Define $W_N(t) = N^{-1} e^{-\alpha t} u^N(T_N + t)$. The process counting the number of red particles produced by collisions of white particles is bounded above by an inhomogeneous Poisson process with rate*

$$N e^{2\alpha t} (W^2_N(t))^2 \quad (8.409)$$

where

$$\lim_{N\to\infty} W_N(t) = W \text{ for each } t \in \mathbb{R}. \quad \square \quad (8.410)$$

8.3 Proofs of Propositions 2.3, 2.6, and 2.11

Corollary 8.12. *The normalized (by N^{-1}) number of red respectively purple particles at time $T_N + t$ conditioned on (W, W_*) is uniformly in N*

$$O(e^{2\alpha t}), \text{ respectively } O(e^{3\alpha t}), \text{ as } t \to -\infty. \qquad \Box \qquad (8.411)$$

Remark 83. By introducing further colours, for example, following the first collision of a purple particle with a site occupied by some other colour introducing a new pair of colours analogous to purple and blue, etc., we could obtain upper and lower bounds with errors of order $O(e^{k\alpha t})$ for any $k \in \mathbb{N}$.

Proof of Lemma 8.11. (a) follows by observing that combined number of white, red and purple particles at a site behaves exactly like the dual particle process of typical site and when they migrate they can have collisions according to the same rules as the dual particle system. This follows since when a white particle migrates it collides with the correct probability and then becomes red and when a red particle migrates it can collide and become purple and when a purple particle migrates it can collide again following the same rule (and remains purple).

(b) Since the green particles compensate for the loss of red particles due to coalescence after collision of a white with another white particle and the blue particles arise upon collision of red with red or white and they evolve as the black particles, the total number of particles is as in the W-BL system.

(c) follows since (i) the blue particles are produced in one to one correspondence with the purple particles and (ii) the purple particles can suffer loss due to collision and coalescence but this does not occur with the blue particles.

(d), (e) We have proved earlier in this section using the CMJ-theory that the normalized rate for the number of white particles

$$\frac{1}{N} u_W^N (T_N + t) = O(e^{\alpha t}), \qquad (8.412)$$

if we condition on $\{W, W_*\}$ since for all N this number is smaller than the number of the white and black particles together, which is then asymptotically $W_N(t) N e^{\alpha t}$, with $W_N(t) \to W$ as $N \to \infty$ by the CMJ-theory. When white particles migrate they collide with another white particle with probability $u_W^N(T_N + t)/N$. Moreover white migrants are at times before $T_N + t$ produced at rate at most $W_N(t) N \alpha e^{\alpha t}$. Therefore the normalized rate for the number of red particles produced conditioned on $\{W, W_*\}$ at time $T_N + t$ is $O(Ne^{2\alpha t})$.

Finally a purple or blue particle occurs when a red particle collides with a white or red and therefore the rate of this event is of order

$$O(Ne^{2\alpha t}) \times O(e^{\alpha t}) = O(Ne^{3\alpha t}), \qquad (8.413)$$

if we condition again on $\{W, W_*\}$.

This completes the proof of Lemma 8.11 which will be a key tool in the further analysis of the dual process.   q.e.d.

Remark 84. Since the WRGB system can be identified with the W-BL(by assigning the union of the red and green particles at a white site to an empty site in the first copy of $\mathbb{N}$) a system we will primarily work with the former and ignore the BL system. We then have both upper and lower bounds for the WRP system given by the WRB, WR systems, respectively and this will be our primary object for the analysis of the $t \to -\infty$ asymptotics. This will provide upper and lower bounds with error of $O(e^{3\alpha t})$ and therefore determine the second order asymptotics as $t \to -\infty$.

Step 2: Reformulation of the nonlinear (u, U), $(u_{(C)}, U_{(C)})$ equations

This step has two parts, first rewriting the (u, U) equation and then a second part where we do this for the multicolour version of this equation.

Part 1   We will start by bringing the equations of the limit dynamic (u, U) in a form suitable for the purpose of the proof of Propositions 8.9 and 8.10. Recall that (u, U) solves the following system of differential equations in the Banach space $(\mathbb{R} \times L_1(\mathbb{N}, \nu), \|\cdot\|)$ (remember for the first line that U is a normalized quantity):

$$\frac{du(t)}{dt} = \alpha(t)(1 - u(t))u(t) - \gamma(t)u^2(t), \quad t \geq t_0, \tag{8.414}$$

$$\begin{aligned}\frac{\partial U(t, j)}{\partial t} =\ & +s(j-1)1_{j \neq 1} U(t, j-1) - sjU(t, j) \\ & + \frac{d}{2}(j+1)jU(t, j+1) - \frac{d}{2}j(j-1))1_{j \neq 1}U(t, j) \\ & + c(j+1)U(t, j+1) - cjU(t, j)1_{j \neq 1} \\ & + \alpha(t)1_{j=1} \\ & + u(t)(\alpha(t) + \gamma(t))[U(t, j-1)1_{j \neq 1} - U(t, j)] \\ & - u(t)(\alpha(t) + \gamma(t))1_{j=1} \\ & - \Big(\alpha(t)(1 - u(t)) - \gamma(t)u(t)\Big) \cdot U(t, j).\end{aligned} \tag{8.415}$$

Remark 85. We want to interpret these evolution equations (8.414), (8.415) by a particle system of the type of the mean-field dual but now making more explicit the role of collision and in particular the first collisions of particles. This will allow us to analyse the behaviour as $t \to -\infty$ of the nonlinear evolution equation above. The analysis will be based on the fact that $u(t)$ arises as the limit of $N^{-1}u^N(T_N + t)$ and similarly $U(t)$ as the limit of $U^N(T_N + t)$ where $T_N = \alpha^{-1}\log N$. This will allow us in the second part of this Step 2 to introduce enrichments of the solution of the nonlinear equations through coloured particle systems. Then by taking the

8.3 Proofs of Propositions 2.3, 2.6, and 2.11

$N \to \infty$ limit of the normalized coloured particle systems we obtain a coloured limiting evolution and by that information about the nonlinear original system.

We next rewrite the equation (8.415) in a form suitable for the analysis of the solution viewed as a *perturbation* of the linear (collision-free and hence $u \equiv 0$, $\alpha(t) \equiv \alpha, \gamma(t) = \gamma$) equation in the limit as $t \to -\infty$. We shall see below in Lemma 8.14, that with $u(t) \equiv 0$, the system (8.415) has the stable age distribution $\mathcal{U}(\infty, \mathbb{R}, \cdot)$ of the CMJ as equilibrium and $\alpha = c \sum_{k=2}^{\infty} k \mathcal{U}(\infty, \mathbb{R}, k)$. Hence we should organize the r.h.s. of (8.415) in such a way that we isolate the linear part on the one hand and the nonlinear perturbations of this linear part on the other hand.

We define for every parameter $a \in (0, \infty)$ (for which we shall later choose the value α) the triple of $\mathbb{N} \times \mathbb{N}$-matrices

$$(Q_0^a, Q_1, L), \tag{8.416}$$

by the equations:

$$Q_0^a = Q_0^0 - aI + a1_{j=1} = (q_{j,k})_{j,k \in \mathbb{N}} \left(= (q_{j,k}(a))_{j,k \in \mathbb{N}}\right), \tag{8.417}$$

where

$$q_{12} = s, \quad q_{1,1} = -s$$
$$q_{2,3} = 2s, \quad q_{2,1} = d + 2c + a,$$
$$q_{j,j+1} = sj, \quad q_{j,j-1} = \frac{d}{2} j(j-1) + cj, \quad q_{j1} = a, \quad j \neq 1, 2$$
$$q_{jj} = -sj - cj - \frac{d}{2} j(j-1) - a, \quad j \neq 1,$$
$$Q_1 = (\tilde{q}_{jk})_{j,k \in \mathbb{N}}, \text{ where}$$
$$\tilde{q}_{jj} = -1, \tag{8.418}$$
$$\tilde{q}_{jk} = 0 \quad j \neq k,$$

and finally

$$L = (\ell_{jk}), \quad \ell_{jj} = 0, \text{ and for } j \neq 1, \ \ell_{j-1,j} = 1. \tag{8.419}$$

Note that the matrix Q_0^a is for the forward equation but we consider $\bar{Q}_0^a$ for the backward equation. Then for each $a > 0$:

$$\bar{Q}_0^a \text{ generates a semigroup } (S^a(t))_{t \geq 0} \text{ on } L_\infty(\mathbb{N}) \tag{8.420}$$

corresponding to a unique (pure jump) Markov process on $\mathbb{N}$. This process is as follows. The matrix Q_0^a defines a Markov process on $\mathbb{N}$. For $a = 0$ we obtain the birth and death process which corresponds to a colony where we have birth rate s for each particle, coalescence of two particles at rate d and emigration of a particle at rate c, except when there is only one particle. For positive a the process is put at rate a in the state with only one particle.

Furthermore we abbreviate

$$\tilde{\alpha}(t) = c \sum_{k=2}^{\infty} k U(t, k) - \alpha. \tag{8.421}$$

With these four ingredients equation (8.415) finally reads:

Lemma 8.13 (Rewritten U-equation).

$$\frac{\partial U(t)}{\partial t} = U(t) Q_0^\alpha \tag{8.422}$$
$$+ \tilde{\alpha}(t)[U(t) Q_1 + 1_{j=1}] + u(t)(\alpha(t) + \gamma(t))[U(t) L - 1_{j=1}]. \qquad \square$$

Remark 86. Note that our equations for (u, U) are forward equations and hence all the operators Q_0^a, Q_1, L act "from the right" on $U(t)$.

Remark 87. Note that we can consider the equation (8.422) also for arbitrary values $a \in (0, \infty)$. Denote the solutions as

$$(U^a(t))_{t \in \mathbb{R}}. \tag{8.423}$$

We shall see below how we can characterize U among those by a self-consistency property.

We can later use the following information about the semigroup S^a to characterize the growth rate α of $(u(t))_t \in \mathbb{R}$ in the entrance law (u, U) as $t \to -\infty$.

Lemma 8.14 (Representation of α). *Consider the evolution equation for $U^a(t)$ in the regime in which $(u(t))_{t \in \mathbb{R}}$ is identically zero and $\alpha(t) \equiv a$. Then for given $a > 0$ there is a unique equilibrium state (positive eigenvector)*

$$\{q_j^*(a)\}_{j \in \mathbb{N}} \text{ for the operator } Q_0^a (\text{ equalling } Q_0^0 - aI + a1_{j=1}). \tag{8.424}$$

Then α is uniquely determined as the fixed point defined by the self-consistency equation

$$\alpha = c \sum_{j=2}^{\infty} j q_j^*(\alpha). \tag{8.425}$$

Also,

$$\{q_j^*(\alpha), j = 1, 2, \cdots\} = \{\mathcal{U}(\infty, \mathbb{R}, j), j = 1, 2, \cdots\} = (p_1, p_2, \ldots), \tag{8.426}$$

the r.h.s. being the stable size distribution of the CMJ process. $\qquad \square$

8.3 Proofs of Propositions 2.3, 2.6, and 2.11

Proof. The existence of the unique equilibrium $q_j^*(a)$ follows from standard Markov chain theory, the question is whether the self-consistency equation (8.425) has a solution. Note for this purpose first that $q_j^*(0)$ is the equilibrium for the single site birth and death process appearing in the McKean–Vlasov dual process for one component and observe that $\sum_{j=2}^{\infty} j q_j^*(0) > 0$. We define:

$$F : a \longrightarrow c \sum_{j=2}^{\infty} j q_j^*(a). \tag{8.427}$$

Then we are left to show that the fixed point equation $\alpha = F(\alpha)$ has a solution. We saw above $F(0) > 0$. We claim next that the function is monotone decreasing in a, converging to 0 as $a \to \infty$ and continuous.

To verify the monotonicity, let $a_2 > a_1 > 0$. Recall that the chain for $a \equiv 0$ starting from 1 is stochastically increasing to its equilibrium. Now consider the two Markov chains for a_1, a_2 as a sequence of independent excursions away from 1 which end when a jump to 1 at rate a occurs but otherwise follow the Q_0^0 dynamic. To compare the average height over the excursions, we consider two such excursion lengths given by coupled exponentials $(a_2)^{-1}\mathcal{E}$, $(a_1)^{-1}\mathcal{E}$, respectively. Then we observe by a simple coupling argument that the heights of the a_1-excursion at times $(a_2)^{-1}\mathcal{E} < t \leq (a_1)^{-1}\mathcal{E}$ stochastically dominate the height at time $(a_2)^{-1}\mathcal{E}$. Therefore the average height over an excursion for the a_1-chain is stochastically greater than or equal to that for the a_2-chain.

The continuity follows by noting that the a_2-excursions from zero converge to the a_1 excursion from 0 if $a_2 \downarrow a_1$.

Noting finally that $\sum_{j=2}^{\infty} j q_j^*(a)$ converges to 0 when $a \to \infty$, we obtain the existence and uniqueness of a solution of (8.425) which we read as a fixed point of the equation $\alpha = F(\alpha)$. q.e.d.

Part 2. We now turn to the second part of Step 2, where we give the limiting dynamics of the multicolour enrichment as represented by $(u_{(WRPGB)}^N, U_{(WRPGB)}^N)$. We also need systems of other subsets of colours in our arguments and therefore we denote the limiting objects analog to the noncoloured system by

$$(u_{(C)}, U_{(C)}), \tag{8.428}$$

with

$$C = \{W, R, P, G, B\} \text{ or some other colour subset of } \{W, R, P, G, BL, B\}, \tag{8.429}$$

where u_C is a positive real number and $U_{(C)}$ is a measure on

$$(\mathbb{N}_0)^C. \tag{8.430}$$

This set-up means that we consider the processes of *occupied sites*, occupied with the various colour combinations. The dynamics, in the $N \to \infty$ limit, of the WRP-system (or the WRGB-system) at time $T_N + t$ is given by an enrichment $(u_{(WRP)}(t), U_{(WRP)}(t))$ of the nonlinear system (8.414), (8.415). Namely the latter is recovered as follows:

$$u(t) = u_{(WRP)}(t), \quad U(t,k) = \sum_{\{(i_W, i_R, i_P): i_W + i_R + i_P = k\}} U_{(WRP)}(t; i_W, i_R, i_P). \tag{8.431}$$

In order to write down the evolution equation for the quantities from (8.427), we need the changes occurring in the underlying multicolour system and its various subsystems which then induce the changes in the measure on these configurations. For this purpose we need operators associated with the various possible transitions, their rate and their form which are associated with a particular current state. This state is given by a tuple of the form $(i_W, i_R, i_P, i_B, i_G, i_{BL})$ or a tuple for a smaller set of colours which gives the number of particles of the various colours at a site. Therefore transitions and transition rates are specified by matrices of the form

$$Q_{\underline{i},\underline{j}} \quad , \quad \text{with } \underline{i}, \underline{j} \in (\mathbb{N}_0)^C \quad , \quad C \subseteq \{W, R, P, B, G, BL\}. \tag{8.432}$$

We can now distinguish two different groups of transitions those which concern only particles of *one colour* and then there are transitions where particles of different *colours interact*. We specify these transitions and the corresponding operators in (8.433) and then in (8.435), (8.437).

For the first type the one colour operators let

$$Q^{0,i} = \text{operator } Q_0^0 \text{ applied to the ith colour occupation numbers - see (8.417)}$$
$$\tag{8.433}$$

corresponding to birth, coalescence and emigration

(as long as there are more than one particle) of the ith colour individuals.

Next we introduce the appropriate *inter-type coalescence* operators (for versions with formulas for the r.h.s., see (8.438) and sequel below) which describe the changes in the limiting frequency measure U based on changes in the occupation numbers which correspond to actions of the coloured particles. There are essentially two types of intertype-transitions, (1) changes which occur at a site with rates depending on the state at this site and (2) the creation of a new site at rates depending on the state at the founder site leading to the migration operators.

The two groups of matrices corresponding to the coalescence, respectively migration operator, are the following matrices

$$Q^A = (Q_{\underline{i},\underline{j}}^A), \quad \underline{i}, \underline{j} \in \mathbb{N}_0^C \text{ with } A \subseteq C \times C \text{ or } A \subseteq \{a \to b | a, b \in C\} \tag{8.434}$$

8.3 Proofs of Propositions 2.3, 2.6, and 2.11

running through the list of names of the various transitions are, beginning with the *coalescence* operators:

$$Q^{WR} = \text{coalescence of white and red at same site} \quad (8.435)$$
$$\text{yielding a white and green pair at this site,}$$

$$Q^{WP} = \text{coalescence of white and purple at same site}$$
$$\text{yielding a white and blue pair at this site,}$$

$$Q^{PR} = \text{coalescence of red and purple at same site}$$
$$\text{yielding a red at this site,}$$

$$Q^{RG} = \text{coalescence of green and red at same site}$$
$$\text{yielding a red at this site} \quad (8.436)$$
$$= \text{coalescence of purple and blue at same site}$$
$$\text{yielding a purple at this site,}$$

$$Q^{WG} = Q^{WB} = 0.$$

and the *migration* operators, which are effectively creation operators for new occupied sites, which in particular are then sites occupied by initially only one particle. For sites occupied by *only* one colour we can talk of green (G), blue (B), purple (P), red (R), white (W) sites without ambiguity. However for multicolours, a WR site denotes a site having at least one white or one red particle, etc.

$$Q^{W \to W} = \text{creation of white site at an empty site in } \{1, \ldots, N\}, \quad (8.437)$$
$$Q^{W \to R} = \text{creation of red site at randomly chosen occupied site in } \{1, \ldots, N\},$$
$$Q^{R \to R} = \text{creation of red at empty site in } \{1, \ldots, N\},$$
$$Q^{R \to P.B} = \text{creation of purple–blue pair}$$
$$\text{with purple at randomly chosen occupied site in } \{1, \ldots, N\}$$
$$\text{and blue at the first empty site in the first copy of } \mathbb{N},$$
$$Q^{P \to P} = \text{creation of purple site at randomly chosen site in } \{1, \ldots, N\},$$
$$Q^{G \to G} = \text{creation of green site at first empty site in the first copy of } \mathbb{N},$$
$$Q^{B \to B} = \text{creation of blue site at first empty site in the first copy of } \mathbb{N}.$$

The verbal description on the r.h.s above corresponds to the following expressions which we spell out in two examples:

$$Q^{W \to W}_{(i_W, i_R, i_G, i_B), (1,0,0,0)} = c \cdot 1_{i_W + i_R \neq 1} \, i_W, \quad (8.438)$$

$$Q_{(i_W, i_R, i_G, i_B),(i'_W, i'_R+1, i'_G, i'_B)}^{W \to R} = c \cdot 1_{i_W + i_R \neq 1} i_W U_{WRGB}(t, i'_W, i'_R, i'_G, i'_B), \quad \text{etc..}$$
(8.439)

We now develop in the detail the system of equations for $U_{WRGB}(t)$ since the former provides the necessary upper and lower bounds.

The pair (u, U) arises as the limit as $N \to \infty$ of the particle system (u^N, U^N) in times

$$T_N + t \text{ where } T_N := \frac{\log N}{\alpha}.$$
(8.440)

The proofs are based on first taking the limit as $N \to \infty$ of the $(u^N_{(WRGB)}, U^N_{(WRGB)})$ system and then identifying the order of the terms corresponding to the different colours in the limit as $t \to -\infty$ based on the structure of the particle system.

Now we set for the rescaled processes (misusing earlier-notation)

$$\tilde{u}^N_{WRGB}(t) = N^{-1} u^N_{WRGB}(T_N + t),$$

$$\tilde{U}^N_{WRGB}(t, i_W, i_R, i_G, i_B) = U^N_{WRGB}(T_N + t),$$
(8.441)

$$\tilde{u}^N_{WRP}(t) = N^{-1} u^N_{WRP}(T_N + t),$$

$$\tilde{U}^N_{WRP}(t, i_W, i_R, i_P) = U^N_{WRP}(T_N + t, i_W, i_R, i_G, i_B).$$
(8.442)

Then let $N \to \infty$ and we again obtain, the existence of the limit of the rescaled system (u^N, U^N), the existence of the coloured versions. The limiting objects are denoted:

$$(u_{WRGB}(t), U_{WRGB}(t, \cdot, \cdot, \cdot)), \quad (u_{WRP}(t), U_{WRP}(t, \cdot, \cdot, \cdot)),$$
(8.443)

which will satisfy a system of equations which we give below in (8.454), and (8.455) respectively in (8.460). We omit the details of the convergence proof here, which are straightforward modifications of the argument given in the proof of Proposition 8.8.

Remark 88. In order to give an intuitive picture, even after taking the limit $N \to \infty$, we shall still refer to this nonlinear system in terms of coloured particles, but of course the quantities in question are now continuous quantities that arise in the limit as $N \to \infty$ of the normalized quantities corresponding the well defined particle system given by the above pure jump process.

Note first the following facts about the limiting evolution which are important for our purposes and which follow from the corresponding finite-N properties:

- since the green particles compensate for the loss of any red particles due to coalescence, the number of red plus green particles at a given site has an evolution which is independent of the white particles (but not the distribution of the relative proportions of red and green),

8.3 Proofs of Propositions 2.3, 2.6, and 2.11

- the red plus green plus blue particle system has the same dynamics as the black particle system and since red and black founding particles are created at the same time both systems have the same distribution,
- the particle system comprised of white, red and purple particles has the same distribution as the dual particle system.

Recalling that the WBL-system is a CMJ-system and the RGB process can be identified with the set of black particles, it follows that $(K_{(W)}^N(t) + K_{(RGB)}^N(t))_{t \geq 0}$ is less than a CMJ process with Malthusian parameter α but the difference is of order $o(e^{\alpha t})$ and we have for $t \in \mathbb{R}$, that the intensity of the occupied sites in the various coloured systems satisfies:

$$u_{WR}(t) \leq u_{WRP}(t) \leq u_{WRB}(t) \leq u_{WRGB} \tag{8.444}$$

$$\leq u_W(t) + u_{RGB}(t) = \lim_{N \to \infty} \left[\frac{1}{N}(u_W^N(T_N + t) + u_{RGB}^N(T_N + t)) \right] \leq We^{\alpha t}.$$

The intensity of individuals in the various coloured systems satisfy that

$$\sum_{i_W, i_R, i_G, i_B} (i_W + i_R) u_{WRGB}(t) U_{WRGB}(t, i_W, i_R, i_G, i_B) \tag{8.445}$$

$$\leq \sum_{i_W, i_R, i_P} (i_W + i_R + i_P) u_{WRP}(t) U_{WRP}(t, i_W, i_R, i_P)$$

$$\leq \sum_{i_W, i_R, i_G, i_B} (i_W + i_R + i_B) u_{WRGB}(t) U_{WRBG}(t, i_W, i_R, i_G, i_B).$$

As a result of these inequalities we note that the $WRGB$ system provides *lower* and *upper* bounds for the "number" of occupied sites and "total number of particles" for the WRP system which corresponds to the dual particle system of interest. We note that the difference between the upper and lower bounds for the total number of particles (which corresponds to the number of blue particles) is of order $O(e^{3\alpha t})$. Since each occupied site must contain at least one particle this provides bounds for the number of occupied sites with the same order of accuracy. For this reason we focus on the WRGB system and then read off the required estimates for the WRP system up to this order of a accuracy.

We now write down the equations for the WRGB-system, i.e. for the corresponding pair (u, U). We need the following abbreviations where all sums are over i_W, i_R, i_G and i_B:

$$u_{(WR)}(t) = u_{(WRGB)}(t) \cdot \left(\sum 1_{i_W + i_R \geq 1} U(t; i_W, i_R, i_G, i_B) \right), \tag{8.446}$$

$$u_{(WRB)}(t) = u_{(WRGB)}(t) \cdot \left(\sum 1_{i_W + i_R + i_B \geq 1} U(t; i_W, i_R, i_G, i_B) \right), \tag{8.447}$$

$$\alpha_W(t) = c \sum i_W \cdot 1_{i_W + i_R + i_G \geq 2} U_{(WRGB)}(t; i_W, i_R, i_G, i_B), \tag{8.448}$$

$$\alpha_R(t) = c \sum i_R \cdot 1_{i_R+i_W+i_G \geq 2} U_{(WRGB)}(t; i_W, i_R, i_G, i_B), \tag{8.449}$$

$$\alpha_G(t) = c \sum i_G \cdot 1_{i_G \geq 2} U_{(WRGB)}(t; i_W, i_R, i_G, i_B), \tag{8.450}$$

$$\alpha_B(t) = c \sum i_B \cdot 1_{i_B \geq 2} U_{(WRGB)}(t; i_W, i_R, i_G, i_B), \tag{8.451}$$

$$\gamma_W(t) = c 1_{i_W+i_R+i_G=1} 1_{i_W=1} U_{(WRGB)}(t; 1, 0, 0, 0),$$

$$\text{similarly for } \gamma_R(t), \gamma_G(t), \gamma_B(t). \tag{8.452}$$

Then we can write down the equation for $(u_{(C)}, U_{(C)})$ as coloured version of the (u, U)-equation. Again as in the latter case we work in the very same Banach space $(\mathbb{R} \times L_1(\mathbb{N}, \nu), \|\cdot\|)$ without mentioning this explicitly below.

Consider the WRGB-system described by the

$$(u_{(WRGB)}(t), U_{(WRGB)}(t, \cdot)), t \in \mathbb{R}) \tag{8.453}$$

satisfying the following nonlinear equation:

$$\frac{\partial u_{(WRGB)}(t)}{\partial t} = (\alpha_W(t) + \alpha_R \alpha_R(t)) u_{(WRGB)}(t)(1 - u_{(WR)}(t)) \tag{8.454}$$

$$+ \alpha_G(t) u_{(WRGB)}(t) + \alpha_B(t) u_{(WRGB)}(t)$$

$$- (\gamma_W(t) + \gamma_R(t)) u_{(WR)}(t) u_{(WRGB)}(t).$$

$$\frac{\partial U_{(WRGB)}(t)}{\partial t} = U_{(WRGB)}(t) Q^{(WRGB)}(t), \tag{8.455}$$

where

$$Q^{(WRGB)}(t) = \sum_{i=W,R,G,B} Q^{0,i} + Q^{(WR)} + Q^{RG} \tag{8.456}$$

$$+ (1 - u_{(WR)}(t))[Q^{W \to W} + Q^{R \to R}] + u_{(WR)}(t)[Q^{W \to R} + Q^{R \to B}],$$

$$- \alpha(t) I,$$

with initial condition at time $t = -\infty$ given by

$$\lim_{t \to -\infty} u_{(WRGB)}(t) = 0, \tag{8.457}$$

$$\lim_{t \to -\infty} U_{WRGB}(t, i_W, i_R, i_G, i_B) = \mathcal{U}_\infty(i_W) \quad \text{if } i_R = i_G = i_B = 0 \tag{8.458}$$

$$= 0, \quad \text{otherwise,}$$

where $\mathcal{U}_\infty$ is the stable size distribution of the CMJ process.

If we write for a better understanding of (8.455) this equation out pointwise, we obtain (we suppress on the r.h.s. the subscript WRGB in U):

8.3 Proofs of Propositions 2.3, 2.6, and 2.11

$$\frac{\partial U_{(WRGB)}}{\partial t}(t; i_W, i_R, i_G, i_B) \tag{8.459}$$

$$= +\frac{d}{2}i_W(i_R + 1)U(t; i_W, i_R + 1, i_G - 1, i_B) - \frac{d}{2}i_W i_R U(t; i_W, i_R, i_G - 1, i_B)$$

$$+\frac{d}{2}(i_W + i_R)(i_G + 1)U(t; i_W, i_R, i_G + 1, i_B)$$

$$-\frac{d}{2}(i_W + i_R)i_G 1_{i_R \ne 1} U(t; i_W, i_R + 1, i_G - 1, i_B)$$

$$+\alpha_W(t)(1 - u_{(WR)}(t))1_{(i_W=1, i_R=0, i_G=0, i_B=0)}$$

$$+\alpha_R(t)(1 - u_{(WR)}(t))1_{(i_W=0, i_R=1, i_G=0, i_B=0)}$$

$$+(\alpha_W(t) + \gamma_W(t))u_{(WR)}(t)(U(t; i_W, i_R + 1, i_G - 1, i_B)$$

$$-U(t; _W, i_R + 1, i_G - 1, i_B)$$

$$+(\alpha_G(t) + \sum_{i+j \ge 1} U(t; i_W, i_R + 1, i_G - 1, i_B))1_{(0,0,1,0)}$$

$$+(\alpha_B(t) + \sum_j j U(t; i_W, i_R + 1, i_G - 1, i_B)u_{(WR)}(t))1_{(0,0,0,1)}$$

$$-(\alpha(t)(1 - u_{WR}(t)) - \gamma(t)u_{WR}(t)) \cdot U_{(WRGB)}(t; i_W, i_R + 1, i_G - 1, i_B).$$

The process $(u_{(WRP)}, U_{(WRP)})$ satisfies for $u_{(WRP)} = u$ our original equation, and the quantity $U_{(WRP)}(t)$ satisfies a similar set of equations as above with B replaced by P but in this case we must add the terms corresponding to the migration of a purple particle to a site occupied by white, red or purple. Also when a purple coalesces with red at the same site (which occurs with rate d) it produces a red–green pair and when it coalesces with a white at the same site it produces a white–green pair. The equation for $U_{(WRP)}$ has therefore the form

$$\frac{\partial U_{(WRP)}(t)}{\partial t} = U_{(WRP)}(t) Q^{WRP}(t), \tag{8.460}$$

where

$$Q^{WRP}(t) = \sum_{i=W,R,P} Q^{0,i} + Q^{WR} + Q^{WP} + Q^{PR} \tag{8.461}$$

$$+c(1 - u(t))[Q^{W \to W} + W^{R \to R}] + cu(t)[Q^{W \to R} + Q^{R \to P}]$$

$$+cQ^{P \to P} - \alpha(t)I.$$

This now completes the set of limiting ($\mathbb{N} \to \infty$) equations for the coloured particle system at time $T_N + t$.

For the further analysis it is crucial to observe the following two facts:

- for calculating the asymptotics of *first* moments of the McKean–Vlasov system it suffices to isolate dual process effects which in the limit $N \to \infty$ are of the order $e^{\alpha t}$ as $t \to -\infty$ and ignore error terms of order $e^{2\alpha t}$ and higher,
- for calculating the asymptotics of *second* moments of the McKean–Vlasov system it suffices to isolate effects which are in the limit $N \to \infty$ of the order $e^{2\alpha t}$ as $t \to -\infty$ and ignore error terms of order $e^{3\alpha t}$ and higher.

It is easily verified that if we *condition* on the growth constant W and on W_*, we can verify that as $t \to -\infty$, we have the following estimates on the "number of sites" occupied by certain colour combinations:

$$\sum_{i_W, i_R, i_G, i_B} 1_{i_R + i_G + i_B \geq 1} U_{(WRGB)}(t; i_W, i_R, i_G, i_B) = O(e^{\alpha t}), \tag{8.462}$$

$$\sum_{i_W, i_R, i_G, i_B} 1_{i_B \geq 1} U_{(WRGB)}(t; i_W, i_R, i_G, i_B) = O(e^{2\alpha t}) \tag{8.463}$$

and therefore

$$u_{RGB}(t) = O(e^{2\alpha t}) \text{ and } u_B(t) = O(e^{3\alpha t}). \tag{8.464}$$

We conclude showing that as $t \to -\infty$ this difference between the red plus white particle system (with WRGB dynamics) and the true dual (WRP) is of order $o(e^{2\alpha t})$ once we have taken the limit $N \to \infty$. This difference is the difference between the system of *purple* particles and the system of *blue* particles.

Again we condition on W and W_*. We first note that the intensity (after taking $N \to \infty$) of red particles is $O(e^{2\alpha t})$ and therefore the rate of production for collisions of red particles with other occupied sites by white, red or purple is $O(e^{3\alpha t})$. Hence we have as $t \to -\infty$

$$u_P(t) = O(e^{3\alpha t}), \quad u(t) - u_{(WRGB)}(t) = O(e^{3\alpha t}). \tag{8.465}$$

The sites occupied by red, green or blue particles correspond to black sites and therefore bound the number of occupied sites eventually lost (compared to CMJ) due to collisions. What is the meaning of the three components? The population of green particles represent asymptotically as $t \to -\infty$ the total number of particles lost (in the dual particle system compared with the collision-free CMJ-system) due to collisions up to order $o(e^{2\alpha t})$.

Remark 89. It is important to remember that we have seen above in (8.444), (8.445) that we can obtain upper and lower bounds for the exact dual using *white plus red* for the lower bound and *white plus red plus blue* for the upper bound and that by (8.465) we know that the difference of lower and upper bound is $O(e^{3\alpha t}) = o(e^{2\alpha t})$. Hence suffices for the purpose of first and second moment calculations for the random McKean–Vlasov entrance law to work with the WRGB-system on the dual side.

Step 3:

Proof of Proposition 8.9.

(a) We define the coupling using the multicolour system where white and black gives the CMJ-process and WRP the dual particle system. The evolution rules define the processes in a standard way for all N on one probability space using Poisson stream for all potential sites, colours and transitions.

Recall that the total number of sites occupied by black plus white particles can be identified with a CMJ process K_t and our processes for each N can be coupled based on a single realization of this CMJ process using the multicolour system. Let on this common probability space

$$W := \lim_{t \to \infty} e^{-\alpha t} K_t. \tag{8.466}$$

Then the existence of the limit in (8.466) and therefore assertion (8.375) follows immediately from the Crump–Mode–Jagers theory.

In order to obtain the assertion (8.376) we shall estimate below the *mean* and *variance* of

$$e^{-\alpha t} N^{-1} u^N (T_N + t) \qquad \text{(recall } T_N = \alpha^{-1} \log N\text{)} \tag{8.467}$$

by bounding the mean and variance of the normalized number of black particles (equivalently, red plus green plus blue particles) and finally put this together to get the claim of the proposition.

Part 1: Bounds on the mean.

Define here

$$W_t = e^{-\alpha t} K_t, \quad W_{N,t} = W_{T_N + t}, \quad W = W_\infty. \tag{8.468}$$

Recall that by CMJ theory ([N]), and finiteness of $E[W^2]$,

$$\bar{w} = \sup_t E[W_t] < \infty, \quad \overline{w^2} = \sup_t E[W_t^2] < \infty. \tag{8.469}$$

We condition now on the path $(K_t)_{t \geq 0}$ (and then in particular W is given) and give a conditional upper bound for the number of black particles produced by time $T_N + t$. This bound is given by considering the following upper bound for the rate of collisions of white particles at time s which is

$$(\alpha_N(s) + \gamma_N(s)) K_s^N \frac{K_s^N}{N}, \tag{8.470}$$

where $\alpha_N(\cdot)$ and $\gamma_N(\cdot)$ are defined as in (8.238). Recall that $\alpha_N(s) \to \alpha(s)$, $\gamma_N(s) \to \gamma(s)$ and $(\alpha(s) + \gamma(s))K_s \leq (\alpha + \gamma)W_s e^{\alpha s}$ as $N \to \infty$. We get a stochastic upper bound of the quantity in (8.470) by taking the random process

$$(\alpha + \gamma) W_s e^{\alpha s} \frac{K_s^N}{N}. \tag{8.471}$$

Now assume that we realize a Poisson point process on $[0, \infty)$ with intensity measure given by the density given in (8.470). Now we use this Poisson point process to generate the founders of the black families. Then we let a black cloud grow descending from one black ancestor born at time s up to time $(T_N + t - s)$ and this evolution is independent of the Poisson point process. This black cloud born at time s grows till time $T_N + t$ as

$$\tilde{W}_{N,s}(t) e^{\alpha(t-s)} \tag{8.472}$$

and is given by an independent copy of the CMJ process starting with one particle at time s. With this object we obtain a stochastic upper bound on the number of black particles in the original dual process.

Since at every time s of birth of a black particle we get a cloud independent of all other clouds and also independent of $(K_s)_{s \geq 0}$, we get an upper bound for the expected number of black particles *given a realization of* $(K_s)_{s \geq 0}$:

$$\int_0^{T_N+t} E[\tilde{W}_{N,t-s}] e^{\alpha(T_N+t-s)} (\alpha_N(s) + \gamma_N(s)) K_s^N \frac{K_s^N}{N} ds. \tag{8.473}$$

Furthermore with the Crump–Mode–Jagers theory applied to $(K_s, \mathcal{U}(s))$ we get setting $D = E[\tilde{W}] \frac{\alpha+\gamma}{\alpha} \in (0, \infty)$ that the quantity in (8.473) can be bounded above asymptotically by (recall (8.468) for $W_{N,s}$):

$$\frac{1}{N} \int_0^{T_N+t} e^{\alpha(T_N+t-s)} \alpha D W_s^2 e^{2\alpha s} ds \quad \text{as } N \to \infty. \tag{8.474}$$

The *mean* of the expression in (8.474) is bounded by

$$\overline{w^2} \alpha D \frac{1}{N} \int_0^{T_N+t} e^{\alpha(T_N+t-s)} e^{2\alpha s} ds. \tag{8.475}$$

This quantity in turn is equal to

$$N \overline{w^2} e^{2\alpha t} (1 - e^{-\alpha(T_N+t)}) = N D \overline{w^2} e^{2\alpha t} (1 - e^{-\alpha t} \frac{1}{N}). \tag{8.476}$$

8.3 Proofs of Propositions 2.3, 2.6, and 2.11

Therefore we obtain the upper and lower bound:

$$\bar{w}e^{\alpha t} \geq E[\frac{1}{N}u^N(T_N + t)] \geq E[W_{N,t}]e^{\alpha t} - \overline{w^2}e^{2\alpha t}(1 - O(\frac{1}{N})), \quad (8.477)$$

where the last expression is an upper bound on the expected number of black particles.

Therefore we get the final bound for the mean of the normalized number of sites:

$$0 \geq E[e^{-\alpha t}N^{-1}u^N(T_N + t) - W_{N,t}] \geq c_N \cdot e^{\alpha t}(1 - O(\frac{1}{N})) \quad (8.478)$$

where $\sup_N(|c_N|) < \infty$.

Part 2: Analysis of the variance

To complete the result we now have to estimate

$$Var[e^{-\alpha t}N^{-1}u^N(T_N + t) - W_{N,t}]. \quad (8.479)$$

We will verify as a first step that the variance of the *normalized* number of black particles converges to 0 as $N \to \infty$ and then we will return to the dual particle system. Recall that in our calculations we condition on $(K_t)_{t \geq 0}$.

The first step is now to condition again on a realisation of a process which is a stochastic upper bound on the number of white particles (compare part 1). This bound is given by a CMJ-process. Therefore we observe first that given this number the evolution of the black clouds once they are founded are all independent. Therefore we obtain an *upper bound* on the normalized variance of the black particles if we use a path of the CMJ-process, which dominates the white population.

Note that therefore the *birth of new black clouds* of particles as $N \to \infty$ (which arises upon collision) can by (8.474) be bounded by a time inhomogeneous Poisson process with intensity

$$\left(\alpha \frac{W_{N,t}^2 e^{2\alpha t}}{N}\right) ds, \quad (8.480)$$

where we condition on a realisation of $W(t)$, $W_{N,t} = W(T_N + t)$, or alternatively on the path $(K_s)_{s \geq 0}$.

Define L_s^r as (for given r in the variable s) the Laplace transform of $e^{-\alpha(r-s)}K_{r-s}$ (starting with one particle).

The Laplace transform L of the total number of black particles at time r then is given by

$$L^r(\lambda) = \exp\left(-\int_0^r D\frac{\alpha(K_s^N)^2}{N}(1 - L_s^r(\lambda e^{\alpha(r-s)}))ds\right). \quad (8.481)$$

Note that by construction $L_s^r = L_{r-s}$. Furthermore we have that:

$$\sup_{s \geq 0} Var[e^{-\alpha s} K_s] = \sup_{s \geq 0} L_s''(0) < \infty. \tag{8.482}$$

Now we apply this to $r = T_N + t$ and conclude that conditioned on W (this is indicated by $\sim$ on E, Var, etc.):

$$\tilde{V}ar[e^{-\alpha t} N^{-1} \cdot (\text{\# of black particles at time } T_N + t)] \tag{8.483}$$

is bounded by (Const means here a function of W only)

$$\frac{e^{-2\alpha t}}{N^2} \int_0^{T_N+t} DW^2 \alpha e^{2\alpha(T_N+t-s)} \frac{e^{2\alpha s}}{N} L_s''(0) ds \leq \text{Const} \frac{N^2(\log N)}{N^3}. \tag{8.484}$$

The r.h.s. is independent of t. Hence *uniformly* in $t \in \mathbb{R}$, as $N \to \infty$:

$$\tilde{V}ar[e^{-\alpha t} \frac{1}{N}(\text{\# black particles at time } T_N + t)] = o(1). \tag{8.485}$$

Hence *conditioned on* W the number of black and white particles normalized by $Ne^{\alpha t}$ is *deterministic* in the limit $N \to \infty$. Hence by (8.485) the density of white particles alone and the normalized difference between the number of black and white particles is deterministic in the $N \to \infty$ limit.

Part 3 Conclusion of argument

Since the dual lies between the set of CMJ particles and the set of white particles, conditioned on the variable W, in the limit $N^{-1} e^{-\alpha t} u^N(T_N + t)$ lies between *two deterministic curves*, both converging to the same constant as $t \to -\infty$, namely W. Using (8.477) and (8.478) we conclude inequality (8.376).

The assertion (8.377) is proved as follows. Since the first component was just treated, we focus on the second, i.e. U^N. Note that the total variation distance between $U^N(T_N + t)$ and $\mathcal{U}(T_N + t)$ is bounded by the normalized number of black particles. Then combining (8.473) with (8.476) we have

$$\lim_{t \to -\infty} \|U^N(T_N + t) - \mathcal{U}(T_N + t)\| = 0. \tag{8.486}$$

(b) is an immediate consequence of the analysis above.　　q.e.d.

Step 4:

Proof of Proposition 8.10. We proceed in six parts. Four parts prove the various claims and two parts are needed to prove some key lemmata at the end of the argument.

8.3 Proofs of Propositions 2.3, 2.6, and 2.11

Part 1: Proof of (8.379)

To verify (8.379) we use that the $u(t), \alpha(t), \gamma(t)$ arise as the limit $N \to \infty$ of $\frac{u^N(t)}{N}, \alpha^N(t), \gamma^N(t)$ for which we can use the representation by the multicolour particle system. For each N the inequality

$$(\alpha_N(t) + \gamma_N(t))u^N(t) \leq (\gamma + \alpha)N W_{N,t} e^{\alpha t} \tag{8.487}$$

follows since the left side counts the number of particles in a subset of the set of particles on the right side (the difference is induced by the green particles in the multicolour construction). Furthermore we know that $W_{N,t} \to W$, a.s.. Hence the inequalities in (8.487) are therefore inherited in the limit as $N \to \infty$ and give

$$(\alpha(t) + \gamma(t))u(t) \leq (\alpha + \gamma)W e^{\alpha t}. \tag{8.488}$$

Part 2: Proof of (8.381), (8.386)

To obtain (8.381), (8.386) we use the coloured particle system introduced in Step 1 of this subsection. We show first that κ and κ^* are strictly positive and then later on we identify these numbers by a more detailed analysis.

Recall that the total number of white, red, green and blue particles at time $T_N + t$ equals the total number, $W_{N,t} N e^{\alpha t}$ in the CMJ process and the collection of white, red and purple particles is a version of the actual dual particle system η^N. We have shown that if we consider these particle systems at time $T_N + t$ and let $N \to \infty$ we obtain a limit dynamic. The analogous statement holds for the limiting dynamic as $N \to \infty$, now with t as the time variable for the multicolour system. The techniques are the same as used in Subsubsection 8.3.8 and we do not write out here the details again. We are now interested in the expansion as $t \to -\infty$ in this multicolour limit dynamics.

In contrast to the proof of Proposition 8.9 the argument here now involves the green and blue particles.

The purpose of the green particles is to keep track of the particles *lost due to collisions* of white with red and purple particles (lost by coalescence). The purpose of the blue particles is to obtain with the WRB-system an upper bound for the total number of particles in the WRP-system by suppressing the loss of particles that could occur if a red or purple particle migrates to an occupied site. We observe that therefore that since the number of blue particles is $O(e^{3\alpha t})$ (recall (8.464)), then in order to identify the $e^{2\alpha t}$ term in the total number of dual particles as $t \to -\infty$, it is sufficient to control the asymptotics as $t \to -\infty$ of the green particles. Therefore we now have to study the green population, for which we first need more information about the red particles.

The rate at which white particles migrate and collide with occupied sites thus creating a red particle is given and estimated as follows (recall (8.488)):

$$(\alpha_W(t) + \gamma_W(t))u_W^2(t) \leq (\alpha_W(t) + \gamma_W(t))u^2(t) \leq W^2(\alpha + \gamma)e^{2\alpha t}. \tag{8.489}$$

At the particle level, once a red particle is created it begins to develop a growing cloud of red sites which grows with exponential rate $\leq \alpha$. Using the latter and (8.489) we obtain

$$\sum_{i_W, i_R, i_P} 1_{i_R \geq 1} U_{WRP}(t, i_W, i_R, i_P) = O(e^{2\alpha t}). \tag{8.490}$$

Return now to the production of the green particles. We will show first that the number of green particles and sites with green particles is of order $e^{2\alpha t}$ as $t \to -\infty$.

We can assume that when a red particle is created on an occupied site the number of white particles is given by the stable age distribution by (8.458). (We will verify below in (8.559) that in fact the error in (8.458) is $O(e^{\alpha t})$.) We now want to study the production of sites with green particles. We introduce the concept of special sites for this purpose.

We briefly return to the finite N-system. We call occupied sites (by a white particle) in the coloured particle system

$$\text{"special sites" at time } s \tag{8.491}$$

if s lies between the (random) time when a first red particle arrives at this site until it contains no red, purple or green particles. This random time is a.s. finite.

We transfer this concept to the $N \to \infty$ limit at time $T_N + t$. This means that we incorporate the additional mark in the measure-valued description.

During the lifetime of a special site a special site can produce red, purple and green migrants. Special sites are created at rate (recall for the first equality sign that α, γ are the limits of $\alpha(t), \gamma(t)$ as $t \to -\infty$, i.e. the ones characterizing the collision-free regime and then use (8.464) to get $o(1)O(e^{2\alpha t})$ as bound for the difference of both sides):

$$(\alpha_W(t) + \gamma_W(t))u_W^2(t) = (\alpha + \gamma)u^2(t) + o(e^{2\alpha t}) = W^2(\alpha + \gamma)e^{2\alpha t} + o(e^{2\alpha t}) \tag{8.492}$$

and therefore by integration from $-\infty$ to t the number of special sites created up to time t is bounded below by

$$W^2 \frac{\alpha + \gamma}{2\alpha} e^{2\alpha t} + o(e^{2\alpha t}). \tag{8.493}$$

Then green particles at a special site are produced at rate

$$d \sum_{i_W, i_R, i_B} i_W(i_R + i_P) U_{WRP}(t; i_W, i_R, i_P) \geq d \sum_{i_W, i_R, i_P} 1_{i_W(i_R + i_P) \geq 1} U_{WRP}(t; i_W, i_R, i_P). \tag{8.494}$$

8.3 Proofs of Propositions 2.3, 2.6, and 2.11

Since $i_W(i_R + i_P) \geq 1$ automatically at a special site we use (8.492) and (8.493), to get that the expected number of green particles at time t is bounded below by

$$d \cdot W^2 \frac{\alpha + \gamma}{2\alpha} e^{2\alpha t} + o(e^{2\alpha t}). \tag{8.495}$$

Provided the limit exists (see below), this implies that (recall the green particles describe the loss of the WRP-system compared to the W-BL-system)

$$\lim_{t \to -\infty} e^{-2\alpha t}[(\alpha + \gamma)We^{\alpha t} - (\alpha(t) + \gamma(t))u(t)] = \kappa^* > 0. \tag{8.496}$$

In order to now get also information on κ, we need information on sites rather than on particle numbers. In order to get that the number of sites with only green particles is of order $e^{2\alpha t}$ as $t \to -\infty$, we argue as follows. Since there is a positive probability that a green particles migrates before coalescing with a red or white, this implies (if the limit exists, see below) that

$$\lim_{t \to -\infty} e^{-2\alpha t}[(We^{\alpha t} - u(t)] = \kappa > 0. \tag{8.497}$$

Hence we now know (provided that the limits taken exist) that:

$$\kappa, \kappa^* > 0. \tag{8.498}$$

Next, in order to identify the constant κ, κ^* we need to obtain the 2nd order asymptotics. It remains therefore to determine the actual value of κ, κ^* using this information.

We first consider the production of blue particles. Since there are $O(e^{2\alpha t})$ special sites and a blue particle is created only when a migrant comes from a special site and hits the set of size $u(t)$ of occupied sites, it follows that the number of blue particles produced is of order $O(e^{3\alpha t})$ and hence indeed the *blue* particles and therefore also the *purple* particles play *no role* determining the $e^{2\alpha t}$-term. In particular the production of green particles by the loss of purple particles is of order $O(e^{3\alpha t})$ and can be omitted.

We first obtain an expression for κ^*. Since the green particles represent the particles lost in the WRP-system due to coalescence of red or purple with white particles, this is obtained by considering the growth of the green particles in more detail. We note that the loss of purple particles is of smaller order than the loss of the red particles as mentioned above. As a result, in order to identify the constant κ or κ^* in the expressions for $u(t)$ (number of sites) or $(\alpha(t) + \gamma(t))u(t)$ (number of particles) we can work with the WRGB-system instead of the WRP-system. Hence in carrying out the analysis using the WRGB-system we obtain a lower bound for the number of particles lost but as shown above the resulting error is $O(e^{3\alpha t})$.

Once a new green particle is produced by a white-red coalescence at a special site we are interested in the number of green particles that migrate before the end of the life time of the special site. As noted above we can assume that when the red arrives the number of white particles is given by the stable size distribution. The number of green particles at the special site and the process of producing green migrants can then be obtained by the analysis of a *modified birth and death* process where we now have two types, namely, red and green with birth and death rules inherited from the WRGB dynamics. In terms of the limit process this involves the forward Kolmogorov equations for this modified birth and death process which serves as a source of migrants for the green particle system which then evolves by the CMJ dynamics.

To summarize, in order to obtain the distribution of green mass up to an error term of order $O(e^{3\alpha t})$ we consider

- the WRGB-system instead of the WRP-system, ignore blue particles
- the production of special sites by red–white collision,
- the production of green particles and green migrants at a special site,
- a CMJ process with immigration, with immigration source given by the green migrants from special sites.

Recall from the derivation of a lower bound on κ above that in the *WRGB*-system red always migrates to a new (unoccupied) site or otherwise a purple–blue pair is created so that the number of *white plus red and blue* particles gives an *upper bound* to the original interacting dual particle system in the limit $N \to \infty$. The number of the *green particles* produced only by white–red coalescence (omitting purple–white coalescence) gives up to an error of order $O(e^{3\alpha t})$ a *lower bound* to $W(\alpha + \gamma)e^{\alpha t} - (\alpha(t) + \gamma(t))u(t)$.

We make the following definitions in order to turn bounds into precise asymptotics. We say below "expected", to distinguish from the usual expected value, when we calculate quantities of the form $\sum i_G U(t, i_W, i_R, i_G, i_B, i_P)$, where the sum is overall $i_A, A \in C$. Let

$$g_1(t, s) \tag{8.499}$$

be the "expected" number of green particles at time t at a special site created at time s. Note that a special site has a finite lifetime (with finite expected value) since due to coalescence it will revert to a single white particle at some finite random time after its creation. Therefore the function $g_1(t, s) = g_1(t - s)$ is bounded.

Let

$$g_2(s, r)dr \tag{8.500}$$

be the rate of production of green migrants at time r at a special site created at time, i.e. the c times the "expected" number of green particles at special sites. This function is also bounded as above.

Let

$$g_3(r, t) \leq \text{const} \cdot e^{\alpha(t-r)} \tag{8.501}$$

8.3 Proofs of Propositions 2.3, 2.6, and 2.11

be the "expected" number of green *sites* produced at time t from a founder at time r and finally

$$g_3^*(r,t) \leq \text{const} \cdot e^{\alpha(t-r)} \tag{8.502}$$

is the "expected" number of green *particles* produced at time t from a founder at time r. These four functions determine the numbers κ and κ^* uniquely as we shall see next.

Now the creation of the first green particles at a site and then subsequently green sites occurs from a red–white coalescence at a site. At such an event a green particle arises and if there is no green particle yet a new green site is created at this moment. From these founders now a cloud of green particles and sites develops. The evolution of these clouds is independent of the further development of the number of white particles and white sites. Therefore conditioned on W we get that the *"expected" number of green particles at time t*, denoted κ_t^* is asymptotically as $t \to -\infty$

$$W^2 \kappa^* e^{2\alpha t} + O(e^{3\alpha t}), \tag{8.503}$$

where $\kappa^* = \lim_{t \to -\infty} \kappa_t^*$ and

$$\kappa_t^* = e^{-2\alpha t} \int_{-\infty}^{t} (\alpha + \gamma) e^{2\alpha s} \left[g_1(t,s) + \left(\int_s^t g_2(s,r) g_3^*(r,t) dr \right) \right] ds + O(e^{\alpha t}). \tag{8.504}$$

Since g_1 and g_2 are bounded, the integral is finite. Observe that $g_1(t,s) = \tilde{g}_1(t-s)$ and $g_2(s,r) = \tilde{g}_2(r-s)$, $g_3^*(r,t) = \tilde{g}_3^*(t-r)$ by construction of the dynamic of the green particles, which do not coalesce with white or red particles. Therefore the first term in (8.504) is independent of t. The second part is given by an integral depending again only on $t - s$ and hence the complete term again does not depend on t. This implies the convergence of κ_t^* as $t \to -\infty$.

We now turn to the identification of κ that is we turn from particle numbers to number of sites. As in the identification of κ^* we can show that there exists κ such that

$$u_{WR}(t) = W e^{\alpha t} - \kappa W^2 e^{2\alpha t} + O(e^{3\alpha t}). \tag{8.505}$$

We obtain the constant κ by counting *green sites* (i.e. sites which are not also occupied by only red or white) instead of green particles. Conditioning on W we get the *"expected" number of green sites*, at time t is asymptotically as $t \to -\infty$:

$$W^2 \kappa e^{2\alpha t} + O(e^{3\alpha t}), \tag{8.506}$$

where $\kappa = \lim_{t \to -\infty} \kappa_t$ with

$$\kappa_t = e^{-2\alpha t} \int_{-\infty}^{t} (\alpha + \gamma) e^{2\alpha s} \left[\left(\int_s^t g_2(s,r) g_3(r,t) dr \right) \right] ds + O(e^{3\alpha t}). \tag{8.507}$$

The existence of the limit follows as above.

Part 3: Proof of (8.380), (8.382), (8.389)

Next we return to our original nonlinear equation, which we now relate with the quantities of our multicolour particle system. We will need to collect in (8.509)–(8.514) some facts on this system used in the proof.

We first note the relation between the original dual and the WRP system and the possibility to replace them with the WRGB-system for the asymptotic as $t \to -\infty$, namely as $t \to -\infty$:

$$u(t) = u_{\text{WRP}}(t) = u_{\text{WR}}(t) + O(e^{3\alpha t}), \tag{8.508}$$

$$U(t,k) = \sum_{i_W, i_R, i_P} 1_{i_W + i_R + i_P = k} U_{\text{WRP}}(t, i_W, i_R, i_P) \tag{8.509}$$

$$= \frac{u_{\text{WRGB}}(t)}{u(t)} \sum_{i_W, i_R, i_G, i_B} 1_{i_W + i_R = k} U_{\text{WRGB}}(t; i_W, i_R, i_G, i_B) + o(1).$$

Here the first factor arises since the normalization (by the number of sites) for the WRP system is $u(t)$ and it is $u_{\text{WRGB}}(t)$ for the WRGB system. We also recall that by construction

$$u_{\text{WR}} \leq u = u_{\text{WRP}} \leq u_{\text{WRB}}. \tag{8.510}$$

Recall furthermore that the sites occupied by black particles are in one-to-one correspondence with the sites occupied by the red, green and blue particles and the number of white plus black sites is $W_t e^{\alpha t}$ with $W_t \to W$ as $t \to \infty$. Therefore the difference in the number of occupied sites in the (W-BL)-system and the WRGB-system arises from white particles sharing a site with the coloured particles, but those sites are represented as two sites in the (W-BL)-system. Hence

$$\frac{u_{\text{WRGB}}(t)}{W e^{\alpha t}} = \frac{\sum_{i_W, i_R, i_G, i_B} [1_{i_W + i_R + i_G + i_B \geq 1}] U_{\text{WRGB}}(t, i_W, i_R, i_G, i_B)}{\sum_{i_W, i_R, i_G, i_B} [1_{i_W \geq 1} + 1_{i_R + i_G + i_B \geq 1}] U_{\text{WRGB}}(t, i_W, i_R, i_G, i_B)}. \tag{8.511}$$

Furthermore the estimates on the number of green, red and purple particles in the part 2 of our argument for Proposition 8.10 imply as well

$$\frac{u_{\text{WRGB}}(t)}{u_{\text{WRP}}(t)} = 1 + \kappa W e^{\alpha t} + O(e^{2\alpha t}) \tag{8.512}$$

and

$$\alpha(t) = \frac{u_{\text{WRGB}}(t)}{u_{\text{WRP}}(t)} (\alpha_W(t) + \alpha_R(t)) + O(e^{2\alpha t}), \tag{8.513}$$

8.3 Proofs of Propositions 2.3, 2.6, and 2.11

$$\gamma(t) = \frac{u_{\text{WRGB}}(t)}{u_{\text{WRP}}(t)} \sum_{i_W, i_R, i_G, i_B} (1_{i_R+i_W=1} U_{\text{WRGB}}(t, i_W, i_R, i_G, i_B)) + O(e^{2\alpha t}). \tag{8.514}$$

Using the equations (8.508)–(8.514) we shall show below that (conditioned on W) we have the following approximation relations of $t \to -\infty$ for u, α, γ and U, which finish the proof of (8.380), (8.382) and (8.386):

$$|u(t) - W e^{\alpha t}| = O(e^{2\alpha t}), \tag{8.515}$$

$$|\alpha(t) - \alpha| = O(e^{\alpha t}), \tag{8.516}$$

$$|\gamma(t) - \gamma| = O(e^{\alpha t}), \tag{8.517}$$

$$\sum_{k=1}^{\infty} |\sum_{i_W, i_R, i_G, i_B} 1_{i_R+i_W=k} U_{\text{WRGB}}(t, i_W, i_R, i_G, i_B) - \mathcal{U}(\infty, k))| = O(e^{\alpha t}). \tag{8.518}$$

It therefore remains now to verify (8.515)–(8.518) in the sequel.

The bound (8.515) follows from the fact that the difference between the W-BL-system giving $W e^{\alpha t}$ and the WRP-system giving $u(t)$ is bounded by the RGB-system which satisfies $u_{RGB}(t) = O(e^{2\alpha t})$.

Turn next to the proof of (8.516). First note that the bound of the WRP-system (for which $\alpha(t)$ is the rate of colonization of new sites by multiple occupied sites) from below and above by the WR-system respectively the WRB-particles from the WRGB-system yields by dividing the inequality by $u_{WRP}(= u)$:

$$\frac{u_{\text{WRGB}}(t)}{u_{\text{WRP}}(t)} (\alpha_W(t) + \alpha_R(t)) \leq \alpha(t) \tag{8.519}$$

$$\leq \frac{u_{\text{WRGB}}(t)}{u_{\text{WRP}}(t)} (\alpha_W(t) + \alpha_R(t) + \alpha_B(t)) \tag{8.520}$$

The difference between the first and third expressions in (8.519) is $O(e^{2\alpha t})$ (bounding $\alpha_B(t)$ by $O(e^{2\alpha t})$ and using (8.512)). Therefore we get using the representation of $\alpha(t)$ by the WRP-system (compare (8.513)):

$$\alpha(t) = \sum_{i_W, i_R, i_G, i_B} [1_{i_W>1}(i_W) + 1_{i_W=1, i_R>0}(i_W + i_R) + 1_{i_W=0, i_R>1}(i_R)]$$

$$\cdot U_{\text{WRGB}}(t, i_W, i_R, i_G, i_B) \cdot \frac{u_{\text{WRGB}}(t)}{u_{\text{WRP}}(t)} + O(e^{2\alpha t}).$$

We can obtain the expression for α, which is suitable for the comparison with $\alpha(t)$, as follows. This constant α arises from the (W-BL)-system. Furthermore the RGB-particles are in correspondence with the black particles, only sit always on the copy of $\mathbb{N}$ instead of $\mathbb{N}$ or $\{1, \cdots, N\}$. (Note that we are interested here on particle numbers!) Therefore since $W - BL$ is a CMJ-system in the stable age-type distribution

$$\begin{aligned}
\alpha &= u_{WRGB}(t)(We^{\alpha t})^{-1} \\
&\quad (\sum_{i_W, i_R, i_G, i_B} [1_{i_W>1} i_W + 1_{i_W=0} 1_{i_R+i_G+i_B>1}(i_R + i_G + i_B) \\
&\qquad + 1_{i_W=1, i_R+i_G+i_B>0}(i_W + i_R + i_G + i_B) \\
&\qquad + 1_{i_W>1}(i_R + i_G + i_B)] U_{WRGB}(t, i_W, i_R, i_G, i_B)).
\end{aligned} \tag{8.521}$$

Therefore we can represent $\alpha(t) = \alpha + \tilde{\alpha}(t)$ and by combining (8.521) and (8.521) together with the fact that blue particles are $O(e^{3\alpha t})$, we get an expression for $\tilde{\alpha}(t)$ which together with (8.512) results as $t \to -\infty$ in

$$\alpha(t) = \alpha + O(e^{\alpha t}), \tag{8.522}$$

which proves (8.516). (See Lemma 8.15 below for the characterization of $\tilde{\alpha}(t)$.)

Similarly we can proceed with the remaining claims (8.517) and (8.518). This completes the proof of (8.380), (8.382) and (8.389) as pointed out below (8.514).

Part 4: Proof of (8.384) and (8.385)

For this purpose we must investigate the difference between $\alpha(t)$, $U(t)$, $\gamma(t)$ and α, $\mathcal{U}(\infty)$, γ, respectively more accurately than in the bounds above. Namely we need an expression with error terms of order $O(e^{3\alpha t})$ resp. $O(e^{2\alpha t})$ in (8.515) resp. (8.516)–(8.518). For that purpose we use again the multicolour representation. We begin by establishing the order of these differences rather than only upper bounds. This is based on the coloured particle system. From the above discussion we expect that $\tilde{\alpha}(t) = \alpha(t) - \alpha \sim \tilde{\alpha}_0 e^{\alpha t}$ for some $\tilde{\alpha}_0 > 0$ which we have to identify.

Recall that the intensity of green particles at time t corresponds to

$$(\alpha + \gamma)e^{\alpha t} - (\alpha(t) + \gamma(t))u(t) \tag{8.523}$$

and the number of green particles is given W of order $O(e^{2\alpha t})$ so that conditioned on W we have

$$(\alpha + \gamma)We^{\alpha t} - (\alpha(t) + \gamma(t))u(t) \geq \delta e^{2\alpha t}, \quad -\infty < t \leq t_0, \tag{8.524}$$

for some positive constant δ.

We want to sharpen this inequality above to a precise second order expansion. To do this and to thereby identifying $\tilde{\alpha}_0$, $\tilde{\gamma}_0$ we now return to the analytical study of the nonlinear system (8.414), (8.415) and prove the following.

Lemma 8.15 (Expansion of $\alpha(t)$ and $\gamma(t)$).

(a) Let

$$\tilde{\alpha}(t) := \alpha(t) - \alpha, \quad \tilde{\gamma}(t) = \gamma(t) - \gamma. \tag{8.525}$$

8.3 Proofs of Propositions 2.3, 2.6, and 2.11

Then as $t \to -\infty$ we have

$$\tilde{\alpha}(t) = \tilde{\alpha}_0 e^{\alpha t} + O(e^{2\alpha t}), \tag{8.526}$$

$\tilde{\alpha}_0 > 0$ is given explicitly by (8.569) and is of the form $\vec{Const} \cdot W$. (8.527)

Moreover,

$$\tilde{U}(t) = U(t) - \mathcal{U}(\infty) = \tilde{U}_0 e^{\alpha t} + O(e^{2\alpha t}), \tag{8.528}$$

where $\tilde{U}_0$ is given explicitly in (8.575) and is of the form $Const \cdot W$. Then $\tilde{\gamma}_0 = \tilde{U}_0(1)$.

(b) Define u^* as the solution to the nonlinear equation (8.414) with initial condition at $-\infty$

$$\lim_{t \to -\infty} e^{-\alpha t} u^*(t) = 1. \tag{8.529}$$

Then as $t \to -\infty$ we have the second order expansion:

$$u^*(t) = e^{\alpha t} - (b - \frac{\tilde{\alpha}_0^*}{\alpha})e^{2\alpha t} + O(e^{3\alpha t}), \tag{8.530}$$

where $\tilde{\alpha}_0^*$ is obtained from $\tilde{\alpha}_0$ by setting $W = 1$ in the formula (8.569). □

With these results we obtain immediately (8.386). This would complete the Proof of Proposition 8.10.   q.e.d.

It remains to prove Lemma 8.15. To prove this we need some preparation we do in the next part.

Part 5: Statement of proof of Lemma 8.16

We and formulate a statement that gives an explicit representation of u^* (and hence of the limit of u^N suitably shifted in terms of $\alpha(\cdot)$ which is exact up to the third order error terms:

Lemma 8.16 (Identification of u in terms of $\hat{u}$). *Consider the solution to the ODE*

$$\frac{du^*(t)}{dt} = \alpha(t)[u^*(t) - b(t)(u^*(t))^2], \quad -\infty < t < \infty, \tag{8.531}$$

with boundary condition at $-\infty$ given by $\lim_{t \to -\infty} e^{-\alpha t} u^*(t) = 1$ *and with the abbreviations*

$$b(t) = 1 + \frac{\gamma(t)}{\alpha(t)}. \tag{8.532}$$

(Recall that the general case is obtained by a time shift by $\frac{\log W}{\alpha}$ if 1 is replaced by W and $\alpha(\cdot), b(\cdot)$ are shifted accordingly.)

Define with $b = 1 + (\gamma/\alpha)$ the function $\hat{u}$ on $\mathbb{R}$:

$$\hat{u}(t) = \frac{e^{\alpha t} e^{\int_{-\infty}^{t}(\alpha(s)-\alpha)ds}}{1 + be^{\alpha t} e^{\int_{-\infty}^{t}(\alpha(s)-\alpha)ds}}, \quad -\infty < t < \infty. \tag{8.533}$$

(a) Then

$$|u^*(t) - \hat{u}(t)| = O(e^{3\alpha t}). \tag{8.534}$$

The function $\hat{u}$ satisfies (here again $b = 1 + \frac{\gamma}{\alpha}$)

$$\hat{u}(t) \sim \frac{e^{\alpha t} e^{\int_{-\infty}^{t}(\alpha(s)-\alpha)ds}}{1 + be^{\alpha t}} \quad \text{as } t \to -\infty. \tag{8.535}$$

(b) We have for $T_N = \alpha^{-1} \log N$ that in distribution:

$$\frac{1}{N} u^N(T_N + t) \underset{N \to \infty}{\Longrightarrow} u^*(t + \frac{\log W}{\alpha}) = \hat{u}(t + \frac{\log W}{\alpha}) + O(e^{3\alpha t}). \quad \square \tag{8.536}$$

Part 6:

Proof of Lemma 8.16.

(a) We start with the following observation. Since by the coloured particle calculation we obtained earlier in this proof, see (8.516), (8.517),

$$|\gamma(t) - \gamma| + |\alpha(t) - \alpha| \leq \text{const} \cdot e^{\alpha t} \tag{8.537}$$

and $\alpha > 0$, then

$$|b(t) - b| = |b(t) - (1 + \frac{\gamma}{\alpha})| \leq \text{const} \cdot e^{\alpha t}. \tag{8.538}$$

When $b(\cdot)$ is not constant we cannot obtain a closed form solution of (8.531). However we can obtain an approximation that describes the asymptotics as $t \to -\infty$ accurate up to terms of order $O(e^{2\alpha t})$ as follows.

Let

$$(\hat{u}_{t_0}(t))_{t \in \mathbb{R}} \tag{8.539}$$

denote the solution of the modified equation (8.531) where we replace our function $b(t)$ by the constant b and put the initial condition u_{t_0} at time t_0.

8.3 Proofs of Propositions 2.3, 2.6, and 2.11

The solution of the modified equation is given by the formula

$$\hat{u}_{t_0}(t) = \frac{\hat{u}_{t_0}(t_0)e^{\int_{t_0}^t \alpha(s)ds}}{1 + b\hat{u}_{t_0}(t_0)(e^{\int_{t_0}^t \alpha(s)ds} - 1)}, \quad t \geq t_0. \tag{8.540}$$

Now let

$$u_{t_0} = e^{\alpha t_0}, \tag{8.541}$$

so that

$$\hat{u}_{t_0}(t) = \frac{e^{\alpha t} e^{\int_{t_0}^t (\alpha(s) - \alpha)ds}}{1 + be^{\alpha t} e^{\int_{t_0}^t (\alpha(s) - \alpha)ds} - be^{\alpha t_0}}, \quad t \geq t_0. \tag{8.542}$$

We then get with (8.535) that

$$\hat{u}(t) = \lim_{t_0 \to -\infty} \hat{u}_{t_0}(t) = \frac{e^{\alpha t} e^{\int_{-\infty}^t (\alpha(s) - \alpha)ds}}{1 + be^{\alpha t} e^{\int_{-\infty}^t (\alpha(s) - \alpha)ds}}, \quad -\infty < t < \infty, \tag{8.543}$$

noting that the integrals are well defined by (8.537). Hence $\hat{u}$ satisfies a differential equation ((8.531) with $b(t) \equiv b$) which we now use to estimate $\hat{u} - u^*$.

Let $v(t) := (\hat{u}(t) - u^*(t))$. Then $v(t)$ satisfies

$$\frac{dv(t)}{dt} = \tilde{\alpha}(t)v(t) - b(\hat{u}(t) + u^*(t))v(t) + (b(t) - b)(u^*(t))^2. \tag{8.544}$$

Using (8.537), (8.538) and the fact that $u^*(t) \leq \text{const} \cdot e^{\alpha t}$ we then obtain that

$$\frac{dv(t)}{dt} \leq \text{const} \cdot e^{\alpha t} v(t) + O(e^{3\alpha t}). \tag{8.545}$$

Therefore $|v(t)| \leq \text{const} \cdot e^{3\alpha t}$ and hence

$$|u^*(t) - \hat{u}(t)| \leq \text{const} \cdot e^{3\alpha t} \text{ as } t \to -\infty. \tag{8.546}$$

We conclude with an estimate for the integral term in $\hat{u}$

$$(\alpha(t) + \gamma(t))\hat{u}(t) \tag{8.547}$$

$$= \frac{(\alpha(t) + \gamma(t))e^{\alpha t} e^{\int_{-\infty}^t (\alpha(s) - \alpha)ds}}{1 + b(e^{\alpha t} e^{\int_{-\infty}^t (\alpha(s) - \alpha)ds})}.$$

Furthermore by (8.524),

$$(\alpha(t) + \gamma(t))u^*(t) \leq (\alpha + \gamma)e^{\alpha t}. \tag{8.548}$$

Then solving in (8.547) for the $e^{\int}$-term we obtain with (8.534) for $t \to -\infty$ the relation:

$$I(t) = e^{\int_{-\infty}^{t}(\alpha(s)-\alpha)ds} \leq \frac{1}{1-be^{\alpha t}} = 1 + be^{\alpha t} + o(e^{\alpha t}). \tag{8.549}$$

(b) The second assertion follows from part (a). The first assertion follows from (8.312).

This completes the proof of Lemma 8.16. q.e.d.

Part 6: Proof of Lemma 8.15

We separately show the different claimed relations in points (1)–(3) for $\tilde{\alpha}, \tilde{\gamma}$ and then u^*.

(1) Relation (8.526) and (8.527)

Here our strategy is to express $\tilde{\alpha}(\cdot)$ in terms of the rewritten (u, U) using the Markov process generated by Q_0^α and the operators Q_1 and L. Using (8.422), and (8.572) we can rewrite the nonlinear system defining (u, U) in the following form. Let $U(t), Q_0^a, Q_1, L$ be defined as in (8.415), (8.417), (8.418) and define ν by

$$\nu(j) = j 1_{j>1}. \tag{8.550}$$

Furthermore recall that by setting $a = \alpha$ we get the following:

$$(S_t^\alpha)_{t \geq 0} = \text{semigroup with generator } Q^\alpha = Q_0^\alpha + \alpha 1_{j=1}. \tag{8.551}$$

This semigroup has a unique entrance law from $t \to -\infty$, since it is standard to verify that S^α defines an ergodic Markov process.

Next let $(U^\alpha(t))_{t \in \mathbb{R}}$ solve the (forward) equation for generator Q^α, i.e.

$$\frac{\partial}{\partial t}U^\alpha(t) = U^\alpha(t)Q^\alpha, \quad t \in \mathbb{R}, \tag{8.552}$$

and define the difference process $(\tilde{U}(t))_{t \in \mathbb{R}}$ (we now suppress the superscript α in $\tilde{U}$):

$$\tilde{U}(t) = U(t) - U^\alpha(t). \tag{8.553}$$

Then

$$\frac{\partial \tilde{U}(t)}{\partial t} = \tilde{U}(t)Q_0^\alpha + \tilde{\alpha}(t)[U(t)Q_1 + 1_{j=1}] + u(t)(\alpha(t) + \gamma(t))[U(t)L - 1_{j=1}]. \tag{8.554}$$

8.3 Proofs of Propositions 2.3, 2.6, and 2.11

We can now represent due to the definition of $\alpha(\cdot), U, U^\alpha, \tilde{\alpha}(\cdot)$ the function $\alpha(t)$ as

$$\alpha(t) - \alpha = <c\tilde{U}(t), \nu>. \tag{8.555}$$

Therefore (recall (8.550)) by the formula of partial integration for semigroups (for ν see (8.550) with the semigroup S^α of U^α as the reference semigroup and then U as the wanted object:

$$\tilde{\alpha}(t) = \alpha(t) - \alpha = \langle c\tilde{U}(t), \nu \rangle \tag{8.556}$$

$$= \int_{-\infty}^{t} c[\tilde{\alpha}(s)\langle (U(s)Q_1 + 1_{j=1})S^\alpha_{t-s}, \nu \rangle$$

$$+u(s)(\alpha + \gamma)\langle (U(s)L - 1_{j=1})S^\alpha_{t-s}, \nu \rangle]ds.$$

Next set

$$\mathcal{U}(\infty) = (p_1, p_2, \ldots) \tag{8.557}$$

and note that by explicit calculation (see (8.418), (8.419)):

$$\mathcal{U}(\infty)Q_1 + 1_{j=1} = (1-p_1, -p_2, -p_3, \ldots), \quad \mathcal{U}(\infty)L - 1_{j=1} = (-1, p_1, p_2, \ldots). \tag{8.558}$$

Note that for $t \to -\infty$:

$$\lim_{t \to -\infty} U(t) = \mathcal{U}(\infty), \quad \|\mathcal{U}(\infty) - U(t)\|_1 = O(e^{\alpha t}), \tag{8.559}$$

where $\mathcal{U}(\infty)$ is the stable size distribution of the McKean–Vlasow dual process where the norm $\|\cdot\|_1$ is as in (8.262). Hence we get from (8.556) that

$$\tilde{\alpha}(t) = \int_{-\infty}^{t} c[\tilde{\alpha}(s)\langle (\mathcal{U}(\infty)Q_1 + 1_{j=1})S^\alpha_{t-s}, \nu \rangle$$

$$+u(s)(\alpha + \gamma)\langle (\mathcal{U}(\infty)L - 1_{j=1})S^\alpha_{t-s}, \nu \rangle]ds$$

$$+O(e^{2\alpha t}).$$

Next note that (recall (8.550) and (8.558), (8.557) and the fact that the Markov process for S^α increases from the initial value 1 stochastically to its equilibrium) we have the relations:

$$-\alpha < \langle (\mathcal{U}(\infty)Q_1+1_{j=1})S^\alpha_t, \nu \rangle < 0 \text{ and } \langle (\mathcal{U}(\infty)L-1_{j=1})S^\alpha_t, \nu \rangle > 0. \tag{8.560}$$

Recall furthermore that

$$\langle \mathcal{U}(\infty)S^\alpha_t, \nu \rangle = \frac{\alpha}{c} \text{ for all } t \geq 0. \tag{8.561}$$

Let now (use (8.560) for positivity):

$$A_1 = c \int_0^\infty \langle (\mathcal{U}(\infty)L - 1_{j=1})e^{-\alpha s} S_s^\alpha, v \rangle ds > 0, \tag{8.562}$$

$$A_2 = -c \int_0^\infty e^{-\alpha s} \langle (\mathcal{U}(\infty)Q_1 + 1_{j=1}) S_s^\alpha, v \rangle ds > 0. \tag{8.563}$$

Now multiply through both sides of (8.556) by $\alpha e^{-\alpha t}$ and set

$$\hat{\alpha}(t) = e^{-\alpha t} \tilde{\alpha}(t), \quad \hat{S}_t^\alpha = e^{-\alpha t} S_t^\alpha. \tag{8.564}$$

We obtain the equation:

$$\hat{\alpha}(t) = c \int_{-\infty}^t [\hat{\alpha}(s) \langle (\mathcal{U}(\infty)Q_1 + 1_{j=1}) \hat{S}_{t-s}^\alpha, v \rangle \tag{8.565}$$
$$+ \hat{u}(s)(\alpha + \gamma) \langle (\mathcal{U}(\infty)L - 1_{j=1}) \hat{S}_{t-s}^\alpha, v \rangle] ds.$$

This is a *renewal-type equation* of the form (choosing suitable positive functions f, g, recall (8.560)):

$$\hat{\alpha} = \hat{\alpha} * (-f) + \hat{u} * g, \tag{8.566}$$

where the integral over f over $\mathbb{R}$ is equal to A_2, which is less than 1 due to relation (8.561) and (8.564), furthermore $\hat{u}(t) \to W$ as $t \to -\infty$ and g is an integrable positive function with integral A_1.

We now claim that as $t \to -\infty$

$$\hat{\alpha}(t) \longrightarrow C \text{ and } |\hat{\alpha}(t) - C| = O(e^{\alpha t}), \tag{8.567}$$

where C is calculated as usual in renewal theory.

Then the solution (uniqueness is verified below) to (8.566) is given by

$$\tilde{\alpha}(t) = \tilde{\alpha}_0 e^{\alpha t} + O(e^{2\alpha t}), \tag{8.568}$$

where $\tilde{\alpha}_0 > 0$ is given by

$$\tilde{\alpha}_0 = \frac{A_1}{1 + A_2} W. \tag{8.569}$$

The fact that the errorterm in (8.568) is of the form $O(e^{2\alpha t})$ follows from (8.560). Moreover (8.568) is the unique solution of order $O(e^{\alpha t})$ to (8.556). To verify the uniqueness we argue as follows.

8.3 Proofs of Propositions 2.3, 2.6, and 2.11

The difference of two solutions h_1, h_2 must solve according to (8.566) that

$$(h_1 - h_2) = (h_1 - h_2) * (-f). \tag{8.570}$$

We can then verify that either $h_1 \equiv h_2$ or

$$|(h_1 - h_2)(t)| < \sup_{(-\infty,0)} |h_1(t) - h_2(t)| \tag{8.571}$$

and therefore indeed $h_1 - h_2 \equiv 0$.

This completes the proof of (8.526) and (8.527).

Remark 90. We have with (8.245) and (8.287) the following ODE for $\alpha(t)$ as function of (u, U):

$$\frac{\partial \alpha(t)}{\partial t} = \frac{\partial}{\partial t}[c \sum_{j=2}^{\infty} jU(t,j)] = c \sum_{j=2}^{\infty} j(\frac{\partial}{\partial t}U(t,j)) \tag{8.572}$$

$$= c \sum_{j=2}^{\infty} j \Big\{ [\alpha(t)(1-u(t)) - \gamma(t)u(t)] 1_{j=1}$$

$$+ u(t)(\alpha(t) + \gamma(t))[1_{j \neq 1} U(t, j-1) - U(t, j)]$$

$$+ c(j+1)U(t, j+1) - cjU(t, j)1_{j \neq 1}$$

$$+ s(j-1)1_{j \neq 1} U(t, j-1) - sjU(t, j)$$

$$+ \frac{d}{2}(j+1)jU(t, j+1) - \frac{d}{2}j(j-1))1_{j \neq 1} U(t, j)$$

$$- \Big(\alpha(t)(1-b(t)u(t))\Big) \cdot U(t, j)\Big\}.$$

Collecting terms we obtain:

$$\frac{\partial \alpha(t)}{\partial t} = (s - \alpha(t) + \frac{d}{2})(\alpha(t) + \gamma(t)) - \frac{cd}{2} \sum_{j=1}^{\infty} j^2 U(t, j) \tag{8.573}$$

$$+ u(t)(\alpha(t) + \gamma(t))^2 + cu(t)(\alpha(t) + \gamma(t)).$$

Hence we can determine α as solution of an ODE once we are given (u, U).

(2) Relations (8.528) and (8.385).

Recall (8.554) which yields

$$\frac{\partial \tilde{U}(t)}{\partial t} = \tilde{U}(t)Q_0^\alpha + \tilde{\alpha}(t)[\mathcal{U}(\infty)Q_1 + 1_{j=1}] + u(t)(\alpha+\gamma)[\mathcal{U}(\infty)L - 1_{j=1}] + O(e^{2\alpha t}). \tag{8.574}$$

Then using the expression (8.568) for $\tilde{\alpha}(t)$ and $u(t) = We^{\alpha t} + O(e^{2\alpha t})$ we obtain

$$\tilde{U}_0 = \int_{-\infty}^0 e^{\alpha s} S_{-s}^\alpha V ds \tag{8.575}$$

where $V = W(\tilde{\alpha}_0^*[\mathcal{U}(\infty)Q_1 + 1_{j=1}] + (\alpha + \gamma)[\mathcal{U}(\infty)L - 1_{j=1}])$. Recalling that S_t^α is a semigroup on $\mathcal{P}(\mathbb{N})$, it follows that the integral is well-defined.

(3) Relation (8.530).

With the knowledge we have now the expression (8.530) then follows by (8.534) by substitution with (8.568) in (8.535).

8.3.10 Weighted Occupation Time for the Dual Process

In this subsubsection we focus on the behaviour of the quantity $\Xi_N(\alpha^{-1}\log N + t, 2)$ in the limit $N \to \infty$ as a function of t. We know that this limit is $\mathcal{L}_t(2)$ which is a random probability measure on $[0, 1]$, in order to identify its distribution it suffices to compute the moments via duality and (8.320). Recall also that the weighted occupation time of the dual determines the probability that no mutation occurred changing the value at the tagged site from type 1 to 2. Why, what do we have to prove about the dual process? This we now explain first.

Consider $T_N = \alpha^{-1}\log N$ and $t_0(N) \uparrow \infty$, $t_0(N) = o(\log N)$. Recall that

$$\Pi_s^{N,k,l} = \sum_{i_W, i_R, i_P} (i_W + i_R + i_P)\Psi_s^N(i_W, i_R, i_P), \tag{8.576}$$

where $\Pi_s^{N,k,l}$ is given by the dual particle system started with k particles at each of ℓ sites.

Then we know that we only have to replace W by $W^{(k,\ell)}$ and otherwise we get the same equation as we had for $k = \ell = 1$. Hence we have

$$\int_{t_0(N)}^{(t+T_N)} [N^{-1}\Pi_s^{N,k,\ell}]ds \sim \frac{1}{c}\int_{-\infty}^t (\alpha^{(k,\ell)}(s) + \gamma^{(k,\ell)}(s))u^{(k,\ell)}(s)ds, \text{ as } N \to \infty \tag{8.577}$$

and we need the r.h.s. in the limit as $t \to -\infty$. Integrating the expression given by (8.386) we get the following expansion:

$$\int_{-\infty}^t (\alpha(k,\ell)(s) + \gamma(k,\ell)(s))u(k,\ell)(s)ds$$
$$= W^{(k,\ell)}\frac{(\alpha + \gamma)}{\alpha}e^{\alpha t} - \frac{\kappa^*}{2\alpha}(W^{(k,\ell)})^2 e^{2\alpha t} + O(e^{3\alpha t}). \tag{8.578}$$

8.3 Proofs of Propositions 2.3, 2.6, and 2.11

Recall that in terms of the *normalized* solution of the equation for the dual (u^*, U^*), (i.e. $e^{\alpha t} u^*(t) \to 1$ as $t \to -\infty$) we have by the above on the one hand

$$\mathcal{L}\left\{[\tfrac{1}{N} \int_{t_0(N)}^{T_N+t} \Pi_u^{N,k,\ell} du]\right\} \underset{N\to\infty}{\Longrightarrow} \mathcal{L}\left\{\int_{-\infty}^{t+\frac{\log W^{k,\ell}}{\alpha}} \tfrac{1}{c}(\alpha(s)+\gamma(s))u^*(s)ds\right\} \quad (8.579)$$
$$=: \nu_{k,\ell}(t) \in \mathcal{P}([0,\infty))$$

and on the other hand in terms of the original process we have (cf. (8.120), (8.597)),

$$E\left(\left[\int_0^1 x^k \mathcal{L}_t(2, dx)\right]^\ell\right) = \int_0^\infty e^{-my} \nu_{k,\ell}(t, dy) \text{ for all } k, \ell \in \mathbb{N}. \quad (8.580)$$

Note that the random time shift by $\frac{\log W^{k,\ell}}{\alpha}$ plays an essential role. The interplay between this shift and the nonlinearity of $u(t)$ determines the distribution of $\mathcal{L}_t(2)$. To determine this effect we study the properties of the r.h.s. in the two equations above. In particular we want to show that $\mathcal{L}_t(2)$ is neither δ_0 nor δ_1 and we want to show in fact that it is truly random and has its mass on $(0, 1)$. This can be translated into properties of the r.h.s. of (8.580) which we will study using the dual.

We next collect some key facts needed to calculate the probability of mutation jumps as $N \to \infty$ which appear on the r.h.s. of (8.580).

Lemma 8.17 (Properties limiting dual occupation density).

(a) *If $c > 0$ the limit object $\nu_{k,\ell}$ satisfies:*

$$\nu_{k,\ell}(t, (0, \infty)) = 1, \quad (8.581)$$

and

$$\lim_{n\to\infty} \nu_{n,1}(t, [K, \infty)) = 1, \text{ for all } K. \quad (8.582)$$

(b) *Denote by $<$ strict stochastic order of probability measures. Then for $c > 0, d > 0$:*

$$\nu_{k+1,1} > \nu_{k,1} \quad (8.583)$$
$$\nu_{1,2}(t) < \nu_{1,1}(t) \star \nu_{1,1}(t), \quad (8.584)$$

where $\star$ denotes convolution.

(c) *Let $n_0 \in \mathbb{N}$ and set*

$$t_{n_0}(\varepsilon, N) := \inf\{t : \Pi_t^{N,n_0,1} = \lfloor \varepsilon N \rfloor\} - \frac{1}{\alpha} \log N. \quad (8.585)$$

Then every weak limit point (in law) as $N \to \infty$, denoted $t_{n_0}(\varepsilon)$ satisfies:

$$t_{n_0}(\varepsilon) \to -\infty \text{ in probability as } n_0 \to \infty. \tag{8.586}$$

Furthermore

$$t_{n_0}(\varepsilon) \longrightarrow -\infty \text{ in probability as } \varepsilon \to 0. \quad \square \tag{8.587}$$

Proof of Lemma 8.17.

Proof of (a). The claim $v_{k,\ell}(t, \{0\}) = 0$ follows from (8.381) since $W^{k,\ell} > 0$ a.s. and $v_{k,\ell}(t, (\{\infty\})) = 0$ since for $T_N = \alpha^{-1} \log N$ and every $t \in \mathbb{R}$ we have:

$$\limsup_{N \to \infty} E\left[\frac{1}{N} \int_0^{T_N + t} \Pi_u^{N,k,\ell} du\right] < \infty. \tag{8.588}$$

Therefore $v_{k,\ell}$ is a probability measure concentrated on $(0, \infty)$.

The relation (8.582) will follow from the argument given for the proof of (8.586) in (c).

Proof of (b). The first part of (b) is immediate by a coupling argument realising the $(k+1)$ and k particle system on one probability space. The second part follows from the fact two typical individuals picked among the descendants of the two populations interact through coalescence with positive probability before they jump since once they both occupy $O(N)$ sites, so that we have a fraction of both populations overlapping. This proves (b).

Proof of (c). The main idea in the proof of (c) is that, asymptotically as $n_0 \to \infty$, starting n_0 particles at a tagged site the number of particles at this site decreases to $O(1)$ in time 1 but during this time still produces $O(\log n_0)$ migrants as we shall see below. Therefore if the number of initial particles, n_0 increases to infinity, then an increasing number of populations start growing like $W_i e^{\alpha t}$ and they do so independently. Hence writing the total population as

$$\sum_{i=1}^{m(n_0)} e^{\alpha(t + \alpha^{-1} \log W_i)}, \tag{8.589}$$

we see that if $m(n_0)$ diverges as $n_0 \to \infty$, then the total number of particles reaches $O(N)$ at a time $(\frac{1}{\alpha} \log N + t_{n_0})$ where t_{n_0} decreases as n_0 increases to $+\infty$. Hence it remains to give the formal argument for the fact that $m(n_0)$ is of order $\log n_0$ as $n_0 \to \infty$, which runs as follows.

Consider Kingman's coalescent $(C_t)_{t \geq 0}$ starting with countably many particles. Consider the number of particles in a spatial Kingman coalescent which jump before coalescing at the starting site. This is the given via the rate of divergence of the entrance law of Kingman's coalescence from 0. Hence we need the rate of divergence

$$\int_{\delta}^{1} |C_t| dt \sim |\log \delta| \text{ as } \delta \to 0, \tag{8.590}$$

which follows since the rate of divergence is wellknown to be δ^{-1}. We connect this to an initial state with n_0 particles which is obtained by restricting the entrance law of Kingman's coalescent. We then see that the number of migration jumps diverges as n_0 tends to infinity.

To get (8.587) we use that $t_{n_0}(\varepsilon, N)$ can be bounded from above by the time where $\Pi^{\infty, n_0, 1}$ reached first $\lfloor \varepsilon N \rfloor$. By the analysis of (K_t, ζ_t) we know that for this it suffices to show that $\tilde{\tau}^{N, n_0}(\varepsilon) \longrightarrow -\infty$ as $\varepsilon \to 0$, which follows from the exponential growth of K_t immediately.

This completes the proof of the lemma.   q.e.d.

8.3.11 Proof of Proposition 2.11, Part 1: Convergence to Limiting Dynamics

In the proof of Proposition 2.6 it sufficed to work with the expected value $E[x_1^N(\frac{1}{\alpha} \log N + t)]$ (since $0 \leq x_1 \leq 1$). However in order to prove Proposition 2.11 it is necessary to work as well with higher moments in order to identify not only the time of emergence but to determine the dynamic of fixation. The strategy is to carry out the following two steps next:

1. show that weak convergence of $(\Xi_N^{\log, \alpha}(t))_{t \geq t_0}$ to the limiting deterministic McKean–Vlasov dynamics for $t \geq t_0$ follows from the weak convergence of the marginal distributions at t_0,
2. prove that the one dimensional marginals $(\Xi_N^{\log, \alpha}(t_0))$ converge weakly to a random (not deterministic) probability measure on $[0, 1]$.

Step 1   *Reduction to convergence of one-dimensional marginals.*

In order to prove that $(\Xi_N^{\log, \alpha}(t))_{t \in \mathbb{R}}$ converges weakly to a random solution of the McKean–Vlasov equation we argue first that it suffices to prove that the one-dimensional marginals converge. Namely if one has that, then one can use the Skorohod representation to get a.s. convergence on some joint probability space. Then however we can use that (using duality to obtain a Feller property)

$$\Xi_N^{\log, \alpha}(t_0) \to \mu_0, \tag{8.591}$$

implies $\{\Xi_N^{\log, \alpha}(t)\}_{t_0 \leq t \leq t_0 + T}$ converges to a solution of the McKean Vlasov equation (2.19) with initial value μ_0.

This follows by noting that we claim here a standard McKean–Vlasov limit of an exchangeable system satisfying the martingale problem (2.4), (2.5) and with

initial empirical measures converging. The proof is a modification of the standard proof of the McKean–Vlasov limit (for details see proof of Theorem 9, [DG99]).
Step 2   Convergence of marginals: reformulation in terms of the dual.
We now prove the required convergence of the one-dimensional marginal distributions. To do this we recall first that moments determine probability measures on $[0, 1]$ and that therefore the collection of "moments of moments" of a random measure denoted X on $[0, 1]$ given by

$$E[\prod_{i=1}^{m} \int x^{k_i} X(dx)], \quad k_i, m \in \mathbb{N}, \qquad (8.592)$$

determine the law. Compactness is automatic so therefore it suffices to verify that for all $k, \ell \in \mathbb{N}$ the (k, ℓ) moments of the mass of type 1

$$m_{k,\ell}^N(t) = E([\int_{[0,1]} x^k \, \Xi_N^{\log,\alpha}(t, 1)(dx)]^\ell), \quad \forall k, \ell \in \mathbb{N}, \qquad (8.593)$$

converge as $N \to \infty$ to conclude weak convergence of the one-dimensional marginal distributions.

In order to analyse the empirical measure and its functionals we return to the original system on the site space $\{1, 2, \cdots, N\}$ and express the quantity through moments of observables of the system. These moments are obtained in terms of the dual process by considering the dual with initial function generated by taking the k-product given by $1_{\{1\}} \otimes \cdots \otimes 1_{\{1\}}$ at ℓ not necessarily distinct sites.

This representation can be simplified a bit. Namely note that for bounded exchangeable random variables $(X_1, \cdots, X_N)$ one has:

$$\lim_{N \to \infty} E[(\frac{1}{N} \sum_{i=1}^{N} X_i^k)^\ell] = E[\prod_{i_1 \neq i_2 \neq \cdots \neq i_\ell} X_{i_j}^k]. \qquad (8.594)$$

Therefore to compute $\lim_{N \to \infty} E[\int_{[0,1]} x^k \, \Xi_N^{\log,\alpha}(t, 1)(dx)]^\ell$ we use the dual process $(\eta_t, \mathcal{F}_t^+)_{t \geq 0}$ where the particle system η starts in configuration η_0 with k-particles at *each of ℓ distinct sites* $i = 1, \cdots, \ell$ and the function-valued part starts with

$$\mathcal{F}_0^+ = \bigotimes_{j=1}^{\ell} \bigotimes_{i=1}^{k} (1_{\{1\}}). \qquad (8.595)$$

Let

$$\Pi_u^{N,k,\ell} \qquad (8.596)$$

denote the resulting number of dual particles at time u.

8.3 Proofs of Propositions 2.3, 2.6, and 2.11

Recall that the mutation jump to type 2 (from type 1) is given by $1_{\{1\}} \to 0$. Recall furthermore that the selection operator $1_{A_2} = 1_{\{2\}}$ preserves the product form (note for every state f before the first mutation $1 \to 2$, $f1_{(2)} \equiv 0$) and just generates another factor $1_{\{1\}}$ with a new variable. Therefore the dual $\mathcal{F}_t^+$ (before the first rare mutation $1 \to 2$ occurs) is a product of factors $1_{\{1\}}$ and this product integrated with respect to $\mu_0^\otimes$ with $\mu_0(\{1\}) = 1$ is 0 or 1 depending on whether or not a mutation jump has occurred. Therefore

$$m_{k,\ell}^N(t) := P[\text{no mutation jump } 1 \to 2 \text{ occurred by time } T_N + t] \quad (8.597)$$

$$= E[\exp(-\frac{m}{N} \int_0^{T_N+t} \Pi_u^{N,k,\ell} du)].$$

Hence in order to prove that the following limit exists:

$$m_{k,\ell}(t) = \lim_{N \to \infty} m_{k,\ell}^N(t), \quad (8.598)$$

it suffices to prove that the following holds:

$$\mathcal{L}\left[\frac{1}{N} \int_0^{T_N+t} \Pi_u^{N,k,\ell} du\right] \underset{N \to \infty}{\Longrightarrow} \nu_{k,\ell}(t) \in \mathcal{P}([0, \infty]),$$

$$\nu_{k,\ell}(t)((0, \infty)) = 1, \text{ for every } k, \ell \in \mathbb{N} \quad (8.599)$$

and then we automatically have as well the formula

$$m_{k,\ell}(t) = \int_0^\infty e^{-my} \nu_{k,\ell}(t)(dy). \quad (8.600)$$

The existence of the limit (8.599) is a result on the dual process which we proved in Proposition 8.8, part (8.320). This will be used in the next section to prove that we have convergence to a random McKean–Vlasov limiting dynamic which is truly random by showing that the limiting variance of the empirical mean is strictly positive.

8.3.12 Proof of Proposition 2.11, Part 2: The Random Initial Growth Constant

In the previous sections we have studied the asymptotics of the dual process in terms of the Crump–Mode–Jagers branching process and the pair (u, U) following a certain nonlinear equation. In this section we use these results to establish that the limiting empirical distribution of types is given by the McKean–Vlasov dynamics with random initial condition at time $-\infty$. Recall that in order to describe emergence we assume that initially only type 1 is present and we wish to determine the distribution of $\bar{x}_2(t) := \lim_{N\to\infty} \bar{x}_2^N(T_N + t)$ and its behaviour as $t \to -\infty$ subsequently.

We have so far established that $\Xi_N^{\log,\alpha}(t,2)$ converges to $\mathcal{L}_t(2)$ which is a solution to the McKean–Vlasov dynamics and that

$$\int_0^1 x\mathcal{L}_t(2)(dx) \to 0 \text{ as } t \to -\infty. \tag{8.601}$$

Therefore by Proposition 2.3 (see proof in Section 8.3.2), Proposition 2.6 and (2.77) the following limits exist in distribution and satisfy

$$\lim_{t\to-\infty} e^{\alpha|t|} \int_0^1 x\mathcal{L}_t(dx) = \lim_{t\to-\infty} \lim_{N\to\infty} e^{\alpha|t|}\bar{x}_2^N(\frac{\log N}{\alpha} + t) =^* \mathcal{W}, \tag{8.602}$$

so that $\mathcal{L}_t$ is a random shift of $\mathcal{L}^*$ (cf. (2.31)), namely, $\mathcal{L}^*_{t+\frac{\log *\mathcal{W}}{\alpha}}$ and

$$\mathcal{L}[^*\mathcal{W}] = \lim_{t\to-\infty} \mathcal{L}[e^{\alpha|t|}\int_0^1 x\mathcal{L}_t(dx)]. \tag{8.603}$$

We next consider the first and second moments of $^*\mathcal{W}$.

Proposition 8.18 (Moments of $^*\mathcal{W}$). *The first and second moments of $^*\mathcal{W}$ satisfy:*

$$E[^*\mathcal{W}] = m^* E[W], \text{ with } m^* = \frac{mb}{c}, \quad b = (1+\frac{\gamma}{\alpha}), \tag{8.604}$$

$$Var\left[\int_0^1 x\mathcal{L}_t(2)(dx)\right] = O(e^{2\alpha t}), \tag{8.605}$$

and

$$Var[^*\mathcal{W}] = m^*\frac{\kappa^*}{2\alpha}(E[W])^2. \quad \square \tag{8.606}$$

Proof of Proposition 8.18. Recall the equation (8.578) from Subsubsection 8.3.10 and note that this approximates $(u(t))_{t\in\mathbb{R}}$ for $t \to -\infty$ up to terms $O(e^{3\alpha t})$:

$$\int_{-\infty}^t \frac{1}{c}(\alpha(s) + \gamma(s))u(s)ds = \frac{1}{c}\left(\frac{(\alpha+\gamma)}{\alpha}We^{\alpha t} - \frac{\kappa^*}{2\alpha}W^2e^{2\alpha t}\right)$$
$$+ O(e^{3\alpha t}), \quad \kappa^* > 0. \tag{8.607}$$

This follows from Proposition 8.10—see Lemma 8.15 for the proof that $\kappa^* > 0$.

8.3 Proofs of Propositions 2.3, 2.6, and 2.11

Therefore

$$\exp\left(-\frac{m}{c}\int_{-\infty}^{t}(\alpha(s)+\gamma(s))u(s)ds\right) \quad (8.608)$$

$$= \exp\left(-m^*\left(We^{\alpha t}-\frac{\kappa^*}{2\alpha}W^2e^{2\alpha t}\right)+O(e^{3\alpha t})\right),$$

where $m^* = \frac{mb}{c}$.

An immediate consequence of this relation is that (recall (8.598) for a definition):

$$m_{1,1}(t) = 1 - m^*E[W]e^{\alpha t} + O(e^{2\alpha t}), \text{ for } t \to -\infty \quad (8.609)$$

and therefore

$$\lim_{t\to-\infty} e^{\alpha|t|}E[\int_0^1 x\mathcal{L}_t(2)(dx)] = m^*E[W]. \quad (8.610)$$

Now consider the second moment of the type two mass at time t which we denote $m^{(2)}(t)$. Here we consider the dual process with one particle at each of two distinct sites. The corresponding growing clouds have independent random growth constants W_1 and W_2. We use the formula

$$m^{(2)}(t) := E[\int x\mathcal{L}_t(2)(dx)]^2 = E[\int x\mathcal{L}_t(1)(dx)]^2 \quad (8.611)$$

$$-2E[\int x\mathcal{L}_t(1)(dx)] + 1 = m_{1,2}(t) - 2m_{1,1}(t) + 1.$$

In order to calculate $m_{1,2}(t), m_{1,1}(t)$ we expand (8.608) to terms of order $o(e^{2\alpha t})$ for $\tilde{W} = W_1 + W_2$ respectively W_1 where W_1, W_2 are independent copies of W, to obtain

$$\exp\left(-m^*\left(b\tilde{W}e^{\alpha t}-\frac{\kappa^*}{2\alpha}\tilde{W}^2e^{2\alpha t}\right)+O(e^{3\alpha t})\right) \quad (8.612)$$

$$= 1 - m^*\left(b\tilde{W}e^{\alpha t}-\frac{\kappa^*}{2\alpha}\tilde{W}^2e^{2\alpha t}\right) + \frac{1}{2}\left(m^*b\tilde{W}e^{\alpha t}\right)^2 + O(e^{3\alpha t}).$$

Therefore

$$m_{1,2}(t) = 1 - m^*bE[W_1+W_2]e^{\alpha t} + m^*\frac{\kappa^*}{2\alpha}E[W_1+W_2]^2e^{2\alpha t} \quad (8.613)$$

$$+ \frac{1}{2}(m^*b)^2 E[W_1+W_2]^2 e^{2\alpha t} + O(e^{3\alpha t}).$$

Combining this with (8.609) and (8.611), we see that:

$$m^{(2)}(t) = m^* \frac{\kappa^*}{2\alpha} E[W_1]E[W_2]e^{2\alpha t} \qquad (8.614)$$
$$+ \frac{1}{2}(m^*b)^2 E[W_1]E[W_2]e^{2\alpha t} + O(e^{3\alpha t}).$$

Note that the coefficients of the higher order terms involve higher moments of W and cancel provided that the latter are finite. Hence to justify the cancellation it is necessary to verify that the higher moments of W are finite. However this was established in Lemma 8.4 in Subsubsection 8.3.5.

Inserting (8.609) and (8.614) in $Var[\bar{x}_2(t)] = m^2(t) - (E[\bar{x}_2(t)])^2$ we get:

$$Var[\bar{x}_2(t)] = m^* \frac{\kappa^*}{2\alpha} b(E[W])^2 e^{2\alpha t} + O(e^{3\alpha t}). \qquad (8.615)$$

If we can show that

$$E[(\bar{x}_2(t))^3] = O(e^{3\alpha t}), \qquad (8.616)$$

we get via normalizing by $e^{\alpha t}$ and letting $t \to -\infty$ indeed:

$$Var[^*\mathcal{W}] = m^* \frac{\kappa^*}{2\alpha} b(E[W])^2. \qquad (8.617)$$

Therefore we get (8.606) once we have the bound on the supremum over time of the third moment of $\bar{x}_2(t)$, which follows from Lemma 8.32 in Subsection 8.6.

Here we can use the fact that the collision-free regime gives a stochastic upper bound of the finite N system and then that in fact according to (8.617) we can bound the third moment of the scaled variable in t. q.e.d.

Remark 91. The randomness expressed in $^*\mathcal{W}$ arises as a result of the fact that the mutation jump occurs in this time scale at an exponential waiting time (in t) together with the nonlinearity in (8.247), (8.248). The random variable W governing the initial growth of the dual population (which describes the limiting behaviour of the CMJ process) arise from the initial birth events in the dual process and has its main influence due to the nonlinearity of the evolution in t. However W will enter only through its mean in the law of $^*\mathcal{W}$ due to a law of large number effects.

The moments of $^*\mathcal{W}$ are determined in terms of the random variable W arising from the growth of K_t in the dual process. However as we have seen in the second moment calculation above the determination of the coefficients of $e^{k\alpha t}$ depend on an analysis of an asymptotic expansion of the nonlinear system and in particular $u(t)$ in orders up to k. Although we do not attempt to carry this out here, we show that with respect to Laplace transform order the law of $^*\mathcal{W}$ lies between the case in which the collisions are suppressed (deterministic case) and a modified system in which the correction involving $\int (\alpha(s) - \alpha)ds$ is suppressed, that is, replacing

$\alpha(s)$ by α. This illustrates the role of the collisions (nonlinearity) in producing the randomness in $^*W$ and allows us to obtain bounds for the expected time to reach a small level ε (cf. [L1]). This we pursue further in Subsection 8.5.

8.3.13 Completion of the Proof of Proposition 2.11

We now collect all the pieces needed to prove all the assertions of the proposition, we proceed stepwise.

Step 1   Completion of the Proof of Proposition 2.11(a)
This was proved in Subsubsection 8.3.11 which heavily used 8.3.10.

Step 2   Completion of the proof of the Proposition 2.11(b)
In order to verify that the limiting marginal random probability measure $\mathcal{L}_t(j)$ has for $j = 1, 2$ really a nontrivial distribution and is not just deterministic, it suffices to show that

$$E([\int x\mathcal{L}_t(1)(dx)]^2) > (E([\int x\mathcal{L}_t(1)(dx)]))^2. \tag{8.618}$$

The quantity $\int x\mathcal{L}_t(1)(dx)$ arises as the limit in distribution of $\bar{x}_1^N(t)$, the empirical mean mass of type 1. Note that the empirical mean mass process $(\bar{x}_\ell^N(t); \ell = 1, 2)_{t \geq 0}$ is a random process which is not Markov! To see this look at the second moment of mean mass, it involves the covariance between two sites! More generally the dual representation of the kth moment of the empirical mean is given by starting k particles at k distinct sites.

Note that no coalescence occurs until time $O(\log N)$ and then the two clouds descending from the two different initial dual particles *interact* nontrivially. Then it is easy to conclude that the limiting variance of $\bar{x}_2^N(T_N + t)$ as $N \to \infty$ is not zero using (8.584).

Step 3   Completion of the proof of Proposition 2.11 (c).
The relations (2.80), (2.81) follow from the dual representation (8.580) of $\int_0^1 x^k \mathcal{L}_t(2) dx$ from the fact that $v_{k,\ell}(t, \cdot) \Longrightarrow \delta_0$ as $t \to -\infty$ and $v_{k,\ell}(t, \cdot) \Longrightarrow \delta_\infty$ as $t \to \infty$. This follows from combining (8.318) and (8.320) and then using (8.316) to get the first claim and using $u^{k,\ell}(t) \to \infty$ as $t \to \infty$ to get the second claim. For the proof of (2.83) we will use the general fact that

$$P(\mathcal{L}_t(1)(\{1\}) = 1) = \lim_{k \to \infty} m_{k,1}(t). \tag{8.619}$$

The result then follows from Lemma 8.17 equation (8.582).

Step 4   Completion of the proof of Proposition 2.11 (e)
The convergence of the empirical measure processes to a random solution of the McKean–Vlasov equation follows from Proposition 8.8 together with Proposition 2.11 (b).

Step 5   *Proof of Proposition 2.11(f)*
The claims (2.84)–(2.86) was proved in Subsubsection 8.3.12 in (8.602) and (8.603). The assertion (2.87) follows from Proposition 8.18.
Step 6   *Proof of Proposition 2.11 (g)*
For the assertion (2.88) we refer to a calculation we do later, see (8.895).

All these steps complete the proof of Proposition 2.11.

Remark 92. The same approach can be carried out on Ω_L for fixed L or $\mathbb{Z}^d$. Namely the dual can be viewed as follows. Start with one particle, then the particle system (factors) can be viewed as a stochastic Fisher-KPP equation (at least until the time of the first mutation). In particular in $\mathbb{Z}^1$ we expect linear growth according to a travelling wave solution (see [CD]) and not exponential growth. In this case emergence occurs in time $O(\sqrt{N})$.

8.4 Droplet Formation: Proofs of Proposition 2.8–2.10

In this section we assume that $c > 0$, $m > 0$, $s > 0$, and examine the process in which rare mutants first appear in a finite time horizon in the population at the microscopic level, that is only at some rare sites they appear at a substantial level, then later on after large times these rare sites develop into growing *mutant droplets*, that is, a growing collection of sites occupied by the mutant type but still being of a total size $o(N)$. Then finally in even much larger times this leads to emergence at the macroscopic level once we have $O(N)$ sites in the droplet.

We show that in the microscopic growth regime of the droplet the type-2 mass is described by a population growth process with Malthusian parameter α and random factor $\mathcal{W}^*$. In the next Section 8.5 we use this structure to investigate the relation between $^*\mathcal{W}$ and $\mathcal{W}^*$. Recall that $^*\mathcal{W}$ arose in the context of emergence and fixation by considering time $\alpha^{-1} \log N + t$ and letting first $N \to \infty$ and then $t \to -\infty$ whereas $\mathcal{W}^*$ arises from the droplet growth at times t and by letting first $N \to \infty$ and then $t \to \infty$.

To carry out the analysis for the droplet growth we consider the following four time regimes for our population model:

$$[0, T_0), \text{ where } 0 < T_0 < \infty, \tag{8.620}$$

$$[T_0, T_N), \text{ where } T_N \to \infty, \text{ as } N \to \infty \text{ and } T_N = o(\log N), \tag{8.621}$$

$$[T_N, \frac{\delta \log N}{\alpha} + t], \text{ where } 0 < \delta < 1, \ t \in \mathbb{R}, \tag{8.622}$$

$$[\frac{\delta \log N}{\alpha}, \frac{\log N}{\alpha} + t], \text{ where } 0 < \delta < 1, \ t \in \mathbb{R}. \tag{8.623}$$

8.4 Droplet Formation: Proofs of Proposition 2.8–2.10

The first two time regimes are needed in proving the Propositions 2.8–2.10, the two remaining ones are necessary to prepare the stage for Subsection 8.5 where we shall relate $^*W$ and W^*.

Outline of Subsection 8.4

We now give a description of the evolution through the stages corresponding to the time intervals given in (8.620)–(8.623).

The first step is to examine the sparse set of sites at which the mutant population is of order $O(1)$ in a fixed finite time interval and then to describe this in terms of a Poisson approximation in the limit $N \to \infty$. This happens in Subsubsection 8.4.1 and proves Proposition 2.9.

In Subsubsection 8.4.2 we analyse the consequences for the longtime properties and prove Proposition 2.10 and in Subsubsection 8.4.3 the Proposition 2.8.

Subsequently in 8.4.4 we recall some related explicit calculations and in 8.4.5 and 8.4.6 we continue with the time intervals (8.621)–(8.623) to exhibit the law and properties of W^*.

8.4.1 Mutant Droplet Formation at Finite Time Horizon

Recall the definition of $(\mathfrak{J}_t^{N,m})_{t\geq 0}$ in (2.14). Furthermore we recall the abbreviation:

$$\hat{x}_2^N(t) = \sum_{j=1}^N x_2^N(j,t) \tag{8.624}$$

for the total mass of type 2 in the whole population of N sites. We have to prove the convergence of $(\mathfrak{J}_t^{N,m})_{t\geq 0}$ as $N \to \infty$ to $(\mathfrak{J}_t^m)_{t\geq 0}$ and then we have to derive the properties of this limit as $t \to \infty$, which we had stated as three propositions which we now prove successively, but not in order, we close with Proposition 2.8.

Proof of Proposition 2.9. The strategy to prove the convergence in distribution of $(\mathfrak{J}_t^{N,m})_{t\geq 0}$ as $N \to \infty$ is to proceed in steps as follows. We first consider only the contributions of mutation at a typical site where we have in reality two type of rare events, (1) the *immigration* of mass of type two from other sites and (2) the building up of this mass via *rare mutation* at this site. Indeed if we isolate the two effects and first suppress the immigration of type-2 mass, this simplification leads to N-independent sites. We therefore first look at one site, then at an independent collection of N sites and then finally build in the effect of immigration (destroying the independence) from other sites. In other words we work with the simplification in Step 1 and Step 2 and return in Step 3 to the original model. In Step 1 and Step 2 each the main point is condensed in a Lemma.

Step 1 *(Palm distribution of a single site dynamic)*

Recall Lemma 2.4 and the definition (2.36) of the single site excursion measure $\mathbb{Q}$ for the process without mutation. We must now consider the analogue for the process with mutation. Consider the *single site* dynamics (with only emigration but no immigration), which is given (misusing notation):

$$dx_2^N(1,t) = -cx_2^N(1,t)dt + s\, x_1^N(1,t)x_2^N(1,t)dt + \frac{m}{N}(1-x_2^N(1,t))dt \quad (8.625)$$

$$+ \sqrt{d\cdot(1-x_2^N(1,t))x_2^N(1,t)}\, dw_2(i,t),$$

$$x_2^N(1,0) = \frac{a}{N}.$$

We use size-biasing to focus on the set of sites at which mutant mass appears in a finite time interval and use the excursion law of the process without mutation (with law P) to express the excursions of the process with mutation (with law $\tilde{P}$). We prove first as key tool the following. Let

$$Q \in \mathcal{M}(C(\mathbb{R}^+,[0,1])) \quad (8.626)$$

be defined as the excursion law of the process in (2.33) (which is (8.625) with $m=0$ which then does not depend on N), compare also the more general case in (9.34) in Section 9. Furthermore let

$$\tilde{P}^\varepsilon = \mathcal{L}[(x_2^\varepsilon(t))_{t\geq 0}] = \mathcal{L}[(x_2^\varepsilon(1,t))_{t\geq 0}] \text{ and } x_2^\varepsilon \text{ solves (8.625) with } x_2^\varepsilon(0)$$

$$= \frac{a}{N} = 0, \frac{m}{N} = \varepsilon. \quad (8.627)$$

Furthermore let

$$W_{\text{ex}} := \{w \in C([0,\infty),\mathbb{R}^+),\ w(t)=0 \text{ for } t<\zeta_b \text{ and } t>\zeta_d,\ w(t)>0 \quad (8.628)$$

on the interval $\zeta_b < t < \zeta_d$ for some $\zeta_b < \zeta_d \in (0,\infty)\}$.

Note that then ζ_b and ζ_d are unique functions of the element $w \in W_{\text{ex}}$. We shall use these functionals

$$\zeta_b, \zeta_d : W_{\text{ex}} \longrightarrow (0,\infty) \quad (8.629)$$

below.

Next observe that every continuous path starting at 0 and having an unbounded set of zeros we can write uniquely as a sum of elements of W_{ex} with disjoint intervals of positivity and with every point x not a zero of the path we can associate a unique excursion between

$$\zeta_b^x \text{ and } \zeta_d^x. \quad (8.630)$$

8.4 Droplet Formation: Proofs of Proposition 2.8–2.10

Lemma 8.19 (Single site excursion law and Palm distribution). *Consider the evolution of (8.625) and let $a = 0$, $d > 0$, $c \geq 0$, $t_0 > 0$. Then the following properties hold:*

(a) *Fix $t_0 > 0$ and $K \in \mathbb{N}$. For $k = 1, \cdots, K$ let $x_{2,k}^\varepsilon$ denote the solution of (8.625) but with the mutation term replaced by the time dependent mutation $\varepsilon 1_{(\frac{k-1}{K}t_0, \frac{k}{K}t_0]}$. Set $I_k := x_{2,k}^\varepsilon(\frac{k}{K}t_0)$ and $A_k(a_3)$ denote the event that the excursion trajectory satisfies (recall (8.628)) $\zeta_b^x \in (\frac{k-1}{K}t_0, \frac{k}{K}t_0]$ and $\zeta_d^x - \frac{k}{K} > a_3$ with $x = \frac{k}{K}t_0$.*

Then the I_k are identically distributed random variables with means $\frac{\varepsilon}{K}t_0$ and variance asymptotically of the form $\frac{const\cdot\varepsilon}{K}$ as $K \to \infty$.

Then the probability of more than one such excursion is

$$\lim_{\varepsilon \to 0} \tilde{P}^\varepsilon(\sum_{k=1}^K 1(A_k) > 1 | \sum_{k=1}^K 1(A_k) \geq 1) = 0. \tag{8.631}$$

This remains true for A_k^* where the latter are defined as A_k but for the solution x_2^ε of (8.625) with $\frac{m}{N} \equiv \varepsilon$.

(b) *The following family of measures $\tilde{\mathbb{Q}}((a_1, a_2), a_3, \cdot)$ and a measure $\mathbb{Q}$ on W_{ex} exists. Choose $a_1, a_2 \in [0, \infty), a_1 < a_2, a_3 > 0$ and $A \in \mathcal{B}(C([0, \infty), \mathbb{R}^+))$. Then*

$$\tilde{\mathbb{Q}}((a_1,a_2),a_3,A)) =: \tilde{\mathbb{Q}}(\zeta_b(w) \in (a_1,a_2), \zeta_d - \zeta_b > a_3, w \in A) \tag{8.632}$$

$$= \lim_{\varepsilon \to 0} \frac{\tilde{P}^\varepsilon(\exists x \in (a_1,b_1)\zeta_b^x(w) \in (a_1,a_2), \zeta_d^x - \zeta_b > a_3, w \in A)}{\varepsilon}.$$

The measures $\tilde{\mathbb{Q}}$ on W_{ex} can be represented as ($\mathbb{Q}$ as in (2.36)),

$$\tilde{\mathbb{Q}}((a_1,a_2), a_3, A) = \int_{a_1}^{a_2} \mathbb{Q}(\zeta > a_3, w(\cdot - s)) \in A)ds. \tag{8.633}$$

(c) *Now consider (8.625) which has mutation rate $\varepsilon = \frac{m}{N}$ and put $a = 0$. Then for $t_0 > 0$*

$$\lim_{N \to \infty} N \cdot E[x_2^N(1,t_0)] > 0. \tag{8.634}$$

(d) *Let $\hat{\mu}_t^N$ be as defined in (2.15), that is, the Palm distribution of $x_2^N(1,t)$ in (8.625) with $a = 0$. The following limit exists:*

$$\hat{\mu}_t^\infty(dx) = \lim_{N \to \infty} \hat{\mu}_t^N(dx) = \frac{x \int_0^t \tilde{\mathbb{Q}}(\zeta_b \in ds, \zeta_d > t, w(t) \in dx)ds}{\int_0^\infty \int_0^t x\tilde{\mathbb{Q}}(\zeta_b \in ds, \zeta_d > t, w(t) \in dx)ds}.$$

$$\tag{8.635}$$

Hence the Palm distribution $\hat{\mu}_t^N$ is the law of a random variable of order $O(1)$ which is also asymptotically non-degenerate as $N \to \infty$.

Furthermore as a consequence of (8.634) and (8.635) the first moment of the Palm distribution of $x_2^N(1, t_0)$ has mean satisfying:

$$\lim_{N \to \infty} \hat{E}[x_2^N(1, t_0)] = \lim_{N \to \infty} \frac{E[(x_2^N(1, t_0))^2]}{E[(x_2^N(1, t_0))]} \in (0, \infty). \qquad \square \qquad (8.636)$$

Proof of Lemma 8.19. We prove separately the parts (a), (b) and then (c), (d) of the Lemma.

(a) and (b).

To prove (a) and (b) we begin by approximating the single site process $\{x_2^\varepsilon(t) : 0 \leq t \leq t_0\}$ with law $\tilde{P}^\varepsilon$ (which satisfies (8.625) with $\frac{m}{N} = \varepsilon$) by a sequence of processes, where we replace the mutation term (induced by the rare mutation) in different time intervals by mutation terms at a grid of discrete time points getting finer and finer. In order to keep track of the mutations at different times we split the type 1 into different types. Namely let $K \in \mathbb{N}$ and consider a $K + 1$ type Wright–Fisher diffusion with law

$$\tilde{P}^{\varepsilon, K}, \qquad (8.637)$$

which is starting with type 0 having fitness 0 and the other types having fitness $1 > 0$ (selection at rate s) and with mutation from type 0 to type k at rate ε only during the interval $(\frac{(k-1)t_0}{K}, \frac{k t_0}{K}]$.

We first consider a simpler process, which is built up from K-independent processes. Namely consider independent processes $x_{2,k}^\varepsilon$ representing the masses of types $k = 1, \ldots, K$ in which in each time interval we suppress mutation of all types except some k. Then $x_{2,k}^\varepsilon(\cdot)$ is a solution of (8.625) with mutation term $\varepsilon 1_{(\frac{k-1}{K}t_0, \frac{k}{K}t_0]}(\cdot)$. Let $I_k(\varepsilon) := x_{2,k}^\varepsilon(1, \frac{k}{K}t_0)$. Let $A_k(a_3)$ denote the event that the continuing trajectory satisfies $\zeta_d - \frac{k}{K} > a_3$. Note that the $I_k(\varepsilon)$ are identically distributed random variables with means $\frac{\varepsilon}{K}t_0$ and variance asymptotically of the form $\frac{\varepsilon t_0^2}{2K^2} + o(\frac{\varepsilon}{K^2})$.

Note that as $\varepsilon \to 0$.

$$\frac{1}{\varepsilon} P_{I_k(\varepsilon)}(A_k(a_3)) = \frac{I_k(\varepsilon)}{\varepsilon} \mathbb{Q}(\zeta > a_3) + o(I_k(\varepsilon)). \qquad (8.638)$$

Therefore

$$\sum_{k=1}^{K} 1(\frac{k t_0}{K} \in (a_1, a_2)) \frac{1}{\varepsilon} P_{I_k(\varepsilon)}(A_k(a_3)) \qquad (8.639)$$

$$= \frac{1}{K} \sum_{k=1}^{K} 1(\frac{k t_0}{K} \in (a_1, a_2)) \frac{K I_k(\varepsilon)}{\varepsilon} \mathbb{Q}(\zeta > a_3) + o(\varepsilon) \text{ as } \varepsilon \to 0.$$

8.4 Droplet Formation: Proofs of Proposition 2.8–2.10

Moreover by independence, $Prob(A_k \cap A_{k'}) = P_{I_k(\varepsilon)}(A_k) P_{I_{k'}(e)}(A_{k'}) = O(\varepsilon^2)$, $k \neq k'$, and therefore the probability of more than one such excursion in the limit $\varepsilon \to 0$ is equal to zero.

$$\lim_{\varepsilon \to 0} \frac{1}{K^2} \sum_{k,k'} \frac{1}{\varepsilon} P_{I_k(\varepsilon)}(A_k) \cdot P_{I_{k'}(\varepsilon)}(A_{k'}) = 0. \tag{8.640}$$

Then letting $K \to \infty$ we get using the continuity of $t \to \mathbb{Q}(w(t) \in \cdot)$ that the following limit exists

$$\tilde{\mathbb{Q}}(\zeta_b \in (a_1, a_2), \zeta_d - \zeta_b > a_3, w(t_0) > x) \tag{8.641}$$

$$= \lim_{\varepsilon \to 0} \lim_{K \to \infty} \left[\frac{1}{K} \sum_k 1\left(\frac{kt_0}{K} \in (a_1, a_2)\right) \frac{KI_k(\varepsilon)}{\varepsilon} \frac{1}{I_k(\varepsilon)} P_{I_k(\varepsilon)}(A_k \cap \{x(t_0) > x\}) \right]$$

$$= \lim_{\varepsilon \to 0} \left[\lim_{K \to \infty} \frac{1}{K} \sum_k 1\left(\frac{kt_0}{K} \in (a_1, a_2)\right) \frac{KI_k(\varepsilon)}{\varepsilon} \mathbb{Q}(\zeta > a_3, w(t - \frac{kt_0}{K}) > x) + o(\varepsilon) \right]$$

$$= \int_{a_1}^{a_2} \mathbb{Q}(\zeta > a_3, w(t_0 - s) > x) ds.$$

We claim that the analogous result is valid for the multitype model, that is, letting A_k^* denote the presence of an excursion of type k of length greater that a_3 and having at time t_0 mass bigger than x, starting in (a_1, a_2)

$$\lim_{\varepsilon \to 0} \tilde{P}^{\varepsilon,K}\left(\bigcup_{k=1}^K A_k^*\right) = \int_{a_1}^{a_2} \mathbb{Q}(\zeta > a_3, w(t_0 - s) > x) ds. \tag{8.642}$$

To verify that this remains true with dependence between the types as they arise in a multitype Fisher–Wright diffusion, we return to the law $\tilde{P}^{\varepsilon,K}$. The result then follows from (8.641) by noting that

$$\tilde{P}^{\varepsilon,K}(A_k^* \cap A_{k'}^*) \leq P_{I_k(\varepsilon)}(A_k) \cdot P_{I_{k'}(\varepsilon)}(A_{k'}). \tag{8.643}$$

The latter holds due to the competition between excursions from different subintervals.

(c) and (d)

The proof of the convergence of $\hat{\mu}_t^N$ as $N \to \infty$ proceeds by showing that for every $k \in \mathbb{N}$ we have:

$$\lim_{N \to \infty} N \cdot E([x_2^N(i,t)]^k) = m \lim_{\delta \downarrow 0} \int_0^1 x^k \int_0^t \tilde{\mathbb{Q}}(\zeta_b \in ds, \zeta_d - \zeta_b > \delta, w(\cdot - s) \in dx) \tag{8.644}$$

and that the limiting variance is positive.

We first use the dual process to the process defined in equation (8.625) to compute the first and second moments. Since we discuss the limit $N \to \infty$ over a finite time horizon with finitely many initial particles, we get in the limit no immigration term, but only emigration at the migration rate.

We warn the reader at this point that we use the same notation for the dual process of this single site diffusion as for the one of our interacting system. Let $\Pi_t^{(j)}$ denote the number of factors in (8.7) at time t starting with j factors at time 0 where $\Pi_t^{(j)}$ is the birth and death process with birth and death rates

$$sk \text{ and } ck + \frac{d}{2}k(k-1) \text{ respectively.} \tag{8.645}$$

If we consider finite N we have additional immigration at rate $N^{-1}(\Pi_t^{(1),N} - k)^+$ if at the site considered we have k particles. This means that as $N \to \infty$ and over a finite time horizon with probability tending to 1 at rate $O(N^{-1})$ no immigration jumps occur.

Then for fixed t we have that:

$$E[x_2^N(i,t)] = 1 - E[x_1^N(i,t)], \tag{8.646}$$

$$E[x_1^N(i,t)] = E[\exp(-\frac{m}{N} \int_0^t \Pi_u^{(1),N} du)]$$

$$\sim E[\exp(-\frac{m}{N} \int_0^t \Pi_u^{(1)} du)], \quad \text{as } N \to \infty \tag{8.647}$$

$$\sim 1 - \frac{m}{N} E[\int_0^t \Pi_u^{(1)} du], \quad \text{as } N \to \infty$$

$$\sim 1 - \frac{mc_1 t}{N}, \quad \text{as } N \to \infty,$$

where $c_1 = c_1(t) \in (0, \infty)$ is constant in N. Hence

$$\lim_{N \to \infty} N \cdot E[x_2^N(i,t)] = mc_1 = m \int_0^t E[\Pi_u^{(1)}] ds. \tag{8.648}$$

For fixed t, as $N \to \infty$ we calculate the second moment:

$$E[(x_1^N(i,t))^2] = E[\exp(-\frac{m}{N} \int_0^t \Pi_u^{(2),N} du)]$$
$$\sim E[\exp(-\frac{m}{N} \int_0^t \Pi_u^{(2)} du)] \tag{8.649}$$
$$\sim 1 - \frac{mc_2(t)}{N},$$

8.4 Droplet Formation: Proofs of Proposition 2.8–2.10

for some $c_2 = c_2(t) \in (0, \infty)$. Hence

$$E[x_2^N(i,t)^2] = 1 - 2E[x_1^N(i,t)] + E[x_1(i,t)^2]$$
$$\sim 1 - 2E[\exp(-\tfrac{m}{N}\int_0^t \Pi_u^{(1)} du)] + E[\exp(-\tfrac{m}{N}\int_0^t \Pi_u^{(2)} du)] \quad (8.650)$$
$$\sim 2\tfrac{mc_1 t}{N} - \tfrac{mc_2 t}{N}, \quad \text{as } N \to \infty.$$

Therefore

$$\lim_{N \to \infty} N \cdot E[x_2^N(i,t)^2] = 2mc_1(t) - mc_2(t) = m[2\int_0^t E[\Pi_u^{(1)}] du - \int_0^t E[\Pi_u^{(2)}] du]. \quad (8.651)$$

Therefore the l.h.s. of (8.636) equals

$$\frac{(2c_1(t) - c_2(t))}{c_1(t)}. \quad (8.652)$$

This limit is larger than 0 since (for every t)

$$c_2(t) < 2c_1(t) \quad (8.653)$$

holds because of the positive (uniformly in N) probability of coalescence between the two initial particles of the two populations before a migration step of either one. This implies that every limit point of the sequence of laws of $x_2^N(t)$ (solving (8.625)) has positive variance.

Similarly we introduce

$$c_k(t) = \int_0^t \Pi_u^{(k)} du \quad (8.654)$$

and we can establish writing

$$E[(x_2^N(i,t))^k] = \sum_{j=0}^k \binom{k}{j} (-1)^j E[(x_1^N(i,t))^j] \quad (8.655)$$

and then using

$$E[(x_1^N(i,t))^j] = E[\exp(-\tfrac{m}{N}\int_0^t \Pi_u^{(j)} du] \quad (8.656)$$

the convergence of all k-th moments as for $k = 1, 2$. This then gives the weak convergence of the Palm distribution $\hat{\mu}_t^N$ as $N \to \infty$.

Finally we have to verify the relation (8.635), that is, the limiting Palm distribution is represented in terms of the excursion measure.

Note that for every $k \in \mathbb{N}$ we know that:

$$\int_0^1 x^k \tilde{\mathbb{Q}}_t(dx) \tag{8.657}$$

$$= \int_0^t \int_0^1 x^k \mathbb{Q}(x_2(t-s) \in dx) ds = \int_0^t \int_0^1 x^k (\lim_{\varepsilon \to 0} \frac{1}{\varepsilon} \tilde{P}^\varepsilon[x_2(t-s) \in dx) ds$$

$$= \int_0^t \lim_{\varepsilon \to 0} \frac{1}{\varepsilon} E_\varepsilon[(x_2(t-s))^k] ds.$$

Using the duality we calculate then (with $\tilde{E}_\varepsilon$ denoting the expectation with respect to $\tilde{P}_\varepsilon$)

$$\lim_{\varepsilon \to 0} \frac{1}{\varepsilon} \tilde{E}_\varepsilon[x_2(u)] = \lim_{\varepsilon \to 0} \frac{1}{\varepsilon} E[1 - (1-\varepsilon)^{\Pi_u^{(1)}}] = E[\Pi_u^{(1)})]. \tag{8.658}$$

$$\lim_{\varepsilon \to 0} \frac{1}{\varepsilon} \tilde{E}_\varepsilon[x_2(u)^2] = \lim_{\varepsilon \to 0} \frac{1}{\varepsilon} \left(1 - 2\tilde{E}_\varepsilon[x_1(u)] + \tilde{E}_\varepsilon[x_1(u)^2]\right) \tag{8.659}$$

$$= \lim_{\varepsilon \to 0} \frac{1}{\varepsilon} \left(1 - 2E[1 - (1-\varepsilon)^{\Pi_u^{(1)}}] + E[1 - (1-\varepsilon)^{\Pi_u^{(2)}}]\right)$$

$$= 2E[\Pi_u^{(1)}] - E[\Pi_u^{(2)}].$$

Substituting this (with $k = 2$) in (8.657) and comparing with (8.648) and (8.649) we verify (8.644) for $k = 1, 2$ and (8.636). Similar calculations can be used to verify the claim for all $k \in \mathbb{N}$. q.e.d.

Step 2 *(Compound Poisson limit)*

Consider a collection of processes where each component of the system is still as in (8.625) in which we ignore the effects of immigration into sites and only *emigration* is still accounted for, namely,

$$dx_2^N(i,t) = -cx_2^N(i,t)dt + s x_1^N(i,t)x_2^N(i,t)dt + \frac{m}{N}(1 - x_2^N(i,t))dt \tag{8.660}$$

$$+ \sqrt{d \cdot (1 - x_2^N(i,t))x_2^N(i,t)}\, dw_2(i,t),$$

$$x_2^N(i,0) = \frac{a}{N}, \quad i = 1, 2, \cdots, N.$$

Note that this is a system of N independent diffusion processes. Furthermore we calculate for the total mass process

8.4 Droplet Formation: Proofs of Proposition 2.8–2.10

$$d\hat{x}_2^N(t) \leq ((-c+s)\hat{x}_2^N(t) + \frac{m}{N})dt + dM_t, \qquad (8.661)$$

with a martingale $(M_t)_{t \geq 0}$, where

$$\langle M \rangle_t \leq d \cdot \hat{x}_2^N(t). \qquad (8.662)$$

Therefore we can bound by a submartingale and if we start with only type 1, i.e. *if $a = 0$*, then by submartingale inequalities we get:

$$P[\sup_{t \leq T} |x_2^N(i,t)| > \varepsilon] \leq \frac{\text{constant}}{N}. \qquad (8.663)$$

We return now to the study of the independent collection of diffusions we introduced in the beginning of this step. Consider the atomic measure-valued process

$$(\tilde{\mathfrak{z}}_t^{N,m})_{t \geq 0}, \qquad (8.664)$$

which is defined as the analogue of $(\mathfrak{z}_t^{N,m})_{t \geq 0}$ but with only excursions arising from mutation, that is ignoring immigration at each site (emigration is still accounted for) which was defined in (8.654). Similarly let

$$\{\tilde{x}_2^N(i,t), i \in \mathbb{N}\} \qquad (8.665)$$

be the corresponding collection of independent diffusions. Then we carry out the limit $N \to \infty$.

Lemma 8.20 (Droplet growth in absence of immigration). *At time t_0, asymptotically as $N \to \infty$ we have the following three properties for the dynamics introduced above in (8.664), (8.665).*

(a) *There is a Poisson number of sites j at which $\tilde{x}_2^N(j, t_0) > \varepsilon$ where the parameter is*

$$m \int_0^{t_0} \mathbb{Q}(w(t_0 - s) > \varepsilon) ds. \qquad (8.666)$$

(b) *At time t_0, in the limit $N \to \infty$, there are a countable number of sites contributing to a total mutant mass of order $O(1)$.*
(c) *For each $t > 0$ the states of $\tilde{\mathfrak{z}}_t^{N,m}$ converge in the sense of the weak atomic topology to*

$$\tilde{\mathfrak{z}}_t^m \text{ on } [0,1], \qquad (8.667)$$

with atomic random measure specified by the two requirements

$$\text{atom locations i.i.d. uniform on } [0, 1] \tag{8.668}$$

and atom masses given by a Poisson random measure on $[0, 1]$ with intensity measure

$$m\nu_t(dx), \text{ where } \nu_t(dx) = \int_0^t \mathbb{Q}(w : w(t-s) \in dx) ds. \tag{8.669}$$

The Palm measure $\hat{\nu}_t$ of ν_t satisfies (recall $\hat{\mu}_t^\infty$ from Lemma 8.19 (d)):

$$\hat{\nu}_t = \hat{\mu}_t^\infty. \qquad \square \tag{8.670}$$

Proof of Lemma 8.20. (a) It follows from Lemma 8.19 (a) that

$$\lim_{N \to \infty} N \cdot P[x_2^N(i,t) > \varepsilon] = m\tilde{\mathbb{Q}}_t((\varepsilon, 1]). \tag{8.671}$$

Therefore in the collection of N sites it follows that asymptotically as $N \to \infty$ the number of sites j with $\tilde{x}_2^N(j, t_0) > \varepsilon$ (recall that in this step we suppress immigration and we are considering only mass originating at this site) is Poisson with parameter $m\tilde{\mathbb{Q}}_t((\varepsilon, 1])$.

(b) Since $\nu_t((0, \infty)) = \infty$ and $\nu_t((\delta, \infty)) < \infty$ for $\delta > 0$, there are countably many atoms in the limit as $N \to \infty$.

(c) Consider the sequence obtained by size-ordering the atom sizes in $\tilde{\mathfrak{z}}_t^{N,m}$. Note that the limiting point process on $[\varepsilon, 1]$ is given by a Poisson number with i.i.d. sizes and the distribution of the atoms is given by $\frac{\tilde{\mathbb{Q}}_t(dx)}{\tilde{\mathbb{Q}}_t([\varepsilon, 1])}$. It follows that the order statistics also converge and therefore the joint distribution of the largest k atoms converge as $N \to \infty$.

To verify that the limit is pure atomic we note that the expected mass of the union of sites of size smaller than ε converges to 0 as $\varepsilon \to 0$ uniformly in N by the explicit calculations in the next section (cf. (8.712)). The convergence of $\tilde{\mathfrak{z}}_t^{N,m}$ in the weak atomic topology then follows by Lemma 2.15. q.e.d.

Step 3 *(Completion of proof of convergence)*

Here we have to incorporate the migration of mass between sites in particular the *immigration* to a site from all the other sites. Note that the immigration rate at fixed site is as the rare mutation of order N^{-1} as long as the total mass is $O(1)$ as $N \to \infty$ and therefore we get here indeed an effect comparable to the one we found in Step 1 and 2. We can show directly that the first two moments of the total mass $\hat{x}_2^N(t) = \mathfrak{z}_t^{N,m}([0, 1])$ converge to the first two moments of $\hat{x}_2(t) = \mathfrak{z}^m(t)([0, 1])$. These moment calculations are carried out later on in Subsubsection 8.687, in particular the first moment result is given by (8.867) and the second moment is given by (9.600). This gives the tightness of the marginal distributions.

8.4 Droplet Formation: Proofs of Proposition 2.8–2.10

Consider the sequence of time grids $\{\frac{\ell}{2^k}, \ell = 0, 1, 2, \ldots\}$ with index k and with width 2^{-k}. Then the migration of mass between time $\frac{\ell}{2^k}$ and $\frac{\ell+1}{2^k}$ and the mutation in this interval will be replaced by immigration from other sites and by rare mutation at the *end* of the interval. We then define using this idea an approximate evolution in discrete time by an approximate recursion scheme of atomic measures on $[0, 1]$.

Let $T_t(x)$ denote the evolution of atomic measures, where each atom follows the single site Fisher–Wright diffusion with emigration and the initial atomic measure is x. Furthermore let $Y_\ell^{N,k}$ be an atomic measure with a countable set of new atoms such that the atoms in $Y_\ell^{N,k}$ are produced as above in Step 1 if we observe the process at time 2^{-k} but now the intensity of production of a new atom is instead of just m in the previous steps now given by:

$$m + c\hat{x}_2^N(\frac{\ell}{2^k}), \tag{8.672}$$

where $\hat{x}_2^N$ is given in (8.624). Then define:

$$\tilde{X}_2^N(\frac{\ell+1}{2^k}) = T_{\frac{1}{2^k}} \tilde{X}_2^N(\frac{\ell}{2^k}) + Y_\ell^{N,k}, \quad k \in \mathbb{N}, \tag{8.673}$$

which defines, for each N and a fixed parameter $k \in \mathbb{N}$, a piecewise constant atomic measure-valued process which we denote by

$$(\mathbf{J}_t^{N,k,m})_{t \geq 0}. \tag{8.674}$$

We will now first investigate the weak convergence of this process, in the parameters k, N tending to infinity. First we focus on the weak convergence w.r.t. the topology of the weak convergence of finite measures on the state space and then we pass to the (stronger) one w.r.t. the weak atomic topology on the state space.

Consider first $N \to \infty$. For given k and t, as $N \to \infty$, the random variable $\mathbf{J}_t^{N,k,m}$, converges weakly even in $\mathcal{M}_a([0, 1])$ to $\mathbf{J}_t^{\infty,k,m}$ as proven in Lemma 8.20 (c). We can then obtain the convergence of the finite dimensional distributions of $(\mathbf{J}_t^{N,k,m})_{t \geq 0}$ to those of $(\mathbf{J}_t^{\infty,k,m})_{t \geq 0}$. Tightness and convergence is obtained as we shall argue at the end of this proof so that we have as $N \to \infty$ weak convergence of processes to

$$(\mathbf{J}_t^{\infty,k,m})_{t \geq 0}. \tag{8.675}$$

Next we let $k \to \infty$ and consider the weak topology on the state space. Note first that the random variable in (8.675) is by construction *stochastically increasing in k*, since passing from k to $k+1$ we pick up additional contributions from newly formed atoms. We can then show that as $k \to \infty$ the $\mathbf{J}_t^{\infty,k,m}$ converge in distribution to $\mathbf{J}_t^m$. Similarly we can take the limit $k \to \infty$ in $\mathbf{J}_t^{N,k,m}$ to get $\mathbf{J}_t^{N,m}$.

We can then use the triangle inequality for the Prohorov metric $\rho(\mathbf{J}_t^{N,m}, \mathbf{J}_t^m)$ to get our result. Namely from the weak convergence of $\mathbf{J}_t^{N,k,m}$ to $\mathbf{J}_t^{\infty,k,m}$ as $N \to \infty$, the convergence as $k \to \infty$ of $\mathbf{J}_t^{N,k,m}$ to $\mathbf{J}_t^{N,m}$ as well as the convergence as $k \to \infty$ of $\mathbf{J}_t^{\infty,k,m}$ to $\mathbf{J}_t^m$. Using then the Markov property and a standard argument we can verify the convergence of the finite dimensional distributions of $\mathbf{J}_t^{N,m}$ to the finite dimensional distributions of $\mathbf{J}_t^m$.

It now remains to check for $(\mathbf{J}^{N,m})_{t\geq 0}$ as $N \to \infty$ the tightness condition of the laws and then convergence actually holds also in $C([0,\infty), (\mathcal{M}_a([0,1]), \rho_a))$. We first prove tightness with respect to the *weak topology* on the state space, i.e. we first prove tightness in the path space $C([0,\infty), \mathcal{M}_f([0,1]))$.

We write the state of the process as

$$\mathbf{J}_t^{N,m} = \sum_{i=1}^N a_i^N(t) \delta_{x_i^N(t)}. \tag{8.676}$$

Then it suffices to show that the laws of

$$\{\sum_{i=1}^N f(a_i^N(t), x_i^N(t)), \quad N \in \mathbb{N}\} \tag{8.677}$$

are tight for a bounded continuous function f on $[0,1] \times [0,1]$. But the $x_i^N(t)$ do not change in time and the $a_i^N(t)$ are semimartingales with bounded characteristics and are therefore tight by the Joffe–Métivier criterion.

To get to the convergence in path space based on the *weak atomic topology* on the state space of the process, we next note that since the joint distributions of the ordered atom sizes at fixed times converge, the locations are constant in time (as long as they are charged), we have convergence of the finite dimensional distributions in the weak atomic topology. The verification of the condition for tightness in $C([0,\infty), \mathcal{M}_a([0,1]))$ (2.107) follows as in the proof of Theorem 3.2 in [EK4].   q.e.d.

8.4.2 The Long-Term Behaviour of Limiting Droplet Dynamics (Proof of Proposition 2.10)

We now investigate the behaviour of the limiting ($N \to \infty$) droplet dynamic, where again the dual process is the key tool, now the dual process of McKean–Vlasov dynamics and certain subcritical Fisher–Wright diffusions.

Proof of Proposition 2.10. We prove separately the three parts of the proposition.

(a) Let $(\mathbf{J}_t^m)_{t\geq 0}$ be defined as in Propositions 2.5. We begin by deriving a formula for the first moment

8.4 Droplet Formation: Proofs of Proposition 2.8–2.10

$$m(t) = E[\mathfrak{Z}_t^m([0,1])], \tag{8.678}$$

given $\mathfrak{Z}_0^m(\cdot)$ with $\mathfrak{Z}_0^m([0,1]) < \infty$, which is accessible to an asymptotic analysis.

First we introduce a key ingredient in the formula for $m(\cdot)$ and we obtain its relation to the exponential growth rates α^* respectively α.

Let

$$f(t) = \int_{W_0} w(t) \mathbb{Q}(dw) = \lim_{\varepsilon \to 0} \frac{1}{\varepsilon} E_\varepsilon[\hat{x}_2(t)], \tag{8.679}$$

where $\hat{x}_2(\cdot)$ denote the solution of (2.33) and E_ε refer to the initial state ε. We now obtain an expression for (8.679).

To compute the first moment of $\hat{x}_2(t)$ we use the dual representation for the moments of the SDE (2.33). The dual process to be used then is $(\eta_t, \mathcal{F}_t^+)_{t \geq 0}$ with one initial factor $1_1(\cdot)$. The dual particle process $(\eta_t)_{t \geq 0}$ then is effectively a birth and death process denoted $(\tilde{D}_0(t))_{t \geq 0}$ on $\mathbb{N}_0$, with a dynamic given by the evolution rule that the process can jump up or down by 1 (birth or death) and the rates in state k are given by:

$$\text{birth rate } sk, \text{ death rate } ck + \frac{d}{2}k(k-1) \tag{8.680}$$

and

$$\text{initial state } \tilde{D}_0 = 1. \tag{8.681}$$

Then the dual expression is given by

$$E_\varepsilon[\hat{x}_2(t)] = 1 - E[(1-\varepsilon)^{\tilde{D}_0(t)}]. \tag{8.682}$$

Therefore

$$f(t) = \int_{W_0} w(t) \mathbb{Q}(dw) = \lim_{\varepsilon \to 0} \frac{1}{\varepsilon} E_\varepsilon[\hat{x}_2(t)] = \lim_{\varepsilon \to 0} \frac{E[1-(1-\varepsilon)^{\tilde{D}_0(t)}]}{\varepsilon} = E[\tilde{D}_0(t)]. \tag{8.683}$$

Now let α^* be chosen so that

$$c \int_0^\infty e^{-\alpha^* r} f(r) dr = c \int_0^\infty e^{-\alpha^* r} E[\tilde{D}_0(r)] dr = 1. \tag{8.684}$$

Recalling that by (8.139), (recall the death rate of $D_0(t)$ is zero if $k-1$)

$$\int_0^\infty e^{-\alpha^* r} E[\tilde{D}_0(r)] dr = \int_0^\infty e^{-\alpha^* r} E[D_0(r) 1_{D_0(r) \geq 2}] dr, \tag{8.685}$$

it follows that $\alpha^* = \alpha$ with α as defined in (8.136), (8.138).

Now we are ready to write down the final equation for $m(\cdot)$. Consider

$$\mathfrak{Z}_0^*(t) = (\sum y_i(t)\delta_{a_i})_{t\geq 0}, \tag{8.686}$$

where $y_i(t)$, $\sum y_i(0) < \infty$, are those realizations of the (independent) Fisher–Wright diffusions (2.33) which represent the atoms that were present at time 0 without immigration. Furthermore assume that $\sum_i y_i(0) < \infty$. Then we consider the process $(\mathfrak{Z}_t^m)_{t\geq 0}$ defined in Proposition 2.5 with $\mathfrak{Z}^*$ therein as given in (8.686).

By taking expectations in the stochastic equation defining $\mathfrak{Z}^m$, namely (2.46), we get if we define

$$g(t) = E[\sum_i y_i(t)], \tag{8.687}$$

the renewal equation

$$m(t) = E[\mathfrak{Z}_t^m([0,1])] = E[\sum_i y_i(t)]$$

$$+ \int_0^t E[(m + c\mathfrak{Z}_r^m([0,1]))] \int_{W_0} w(t-r)\mathbb{Q}(dw)]dr \tag{8.688}$$

$$= g(t) + m\int_0^t f(r)dr + \int_0^t cm(r)f(t-r)dr.$$

The last step is now to analyse the growth behaviour of $(m(t))_{t\geq 0}$ using renewal theory. Multiplying through equation (8.688) by $e^{-\alpha^* t}$ we get an equation for $(e^{-\alpha^* t}m(t))_{t\geq 0}$ in terms of $(e^{-\alpha^* t}f(t))$ and $(e^{-\alpha^* t}g(t))$ as follows:

$$e^{-\alpha^* t}m(t) = e^{-\alpha^* t}[(g(t)+m\int_0^t f(r)dr)] + c\int_0^t (e^{-\alpha^* r}m(r))(e^{-\alpha^*(t-r)}f(t-r))dr. \tag{8.689}$$

This equation in $e^{-\alpha^* t}f(t)$ and $e^{-\alpha^* t}g(t)$ has the form of a renewal equation.

We want to apply now a *renewal theorem* to this equation. Define R by

$$R = c\int_0^\infty re^{-\alpha^* r}f(r)dr. \tag{8.690}$$

Since $f(t)$ and $g(t)$ are continuous and converge to 0 exponentially fast, it can be verified that $a(u) = e^{-\alpha^* u}\left[g(u) + m\int_0^u f(r)dr\right]$ is directly Riemann integrable. Therefore by the renewal theorem ([KT], Theorem 5.1) we obtain from (8.689) that:

$$\lim_{t\to\infty} e^{-\alpha^* t}m(t) = \frac{1}{R}\int_0^\infty e^{-\alpha^* u}\left[g(u) + m\int_0^u f(r)dr\right]du \in (0,\infty). \tag{8.691}$$

8.4 Droplet Formation: Proofs of Proposition 2.8–2.10   317

Hence $e^{-\alpha^* t} E[\mathfrak{Z}^m(t)]$ converges as $t \to \infty$ to the r.h.s. of (8.691) which concludes the proof of part (a) of the proposition.

(b) Let

$$\hat{x}_2(t) = \mathfrak{Z}_t^m([0, 1]). \tag{8.692}$$

The convergence in distribution of $e^{-\alpha t} \hat{x}_2(t)$ as $t \to \infty$ follows since by (a) the laws of $\{e^{-\alpha t} \hat{x}_2(t), \ t \geq 0\}$ form a tight family and we have to exclude only that several limit points exist. We know that for any $\tau > 0$

$$\lim_{t \to \infty} E([e^{-\alpha(t+\tau)} \hat{x}_2(t+\tau) - e^{-\alpha t} \hat{x}_2(t)]^2) = 0, \tag{8.693}$$

which is proved in Proposition 8.27. (Recall here that over a finite time horizon $[0, \tau]$ the mutation can be ignored and hence the proposition applies). We have to strengthen this statement (8.693) to

$$\lim_{t \to \infty} (\sup_{\tau > 0} E[e^{-\alpha(t+\tau)} \hat{x}_2(t+\tau) - e^{-\alpha t} \hat{x}_2(t))^2]) = 0. \tag{8.694}$$

This extension is provided in Corollary 8.28 of Subsubsection 8.4.6 where it is proved using moment calculations. This implies that there exists $\mathcal{W}^*$ such that

$$\hat{x}_2(t) \longrightarrow \mathcal{W}^* \text{ in } L^2. \tag{8.695}$$

The non-degeneracy of the limit $\mathcal{W}^*$ follows from (c).

(c) This claim on the variance of $\mathcal{W}^*$ follows from Proposition 8.21 in Subsubsection 8.4.5.   q.e.d.

Remark 93. The proof of (c) demonstrates that the randomness arises from the early rare mutation events.

8.4.3 Proof of Proposition 2.8

This proof builds mainly on the previous subsubsection and on calculations we shall carry out in Subsection 8.5.

(a) This follows from Proposition 2.9
(b) Recall that we know already that $e^{-\alpha t} \hat{x}_2^N(t)$ converges in law as $N \to \infty$ to a limit $e^{-\alpha t} \hat{x}_2(t)$ and hence by Skorohod embedding and our moment bounds we have convergence on some L^2 space, furthermore $e^{-\alpha t} \hat{x}_2(t)$, converges in L^2 as $t \to \infty$ to a limit $\mathcal{W}^*$ as proved in (8.695) above. Here we shall show that

$$\lim_{t \to \infty} \lim_{N \to \infty} E[(e^{-\alpha t} \hat{x}_2^N(t) - e^{-\alpha t_N} \hat{x}_2^N(t_N))^2] = 0 \tag{8.696}$$

and then we use (2.71) to get the claim. The above relation follows via Corollary 8.31 which is proven by moment calculation which requires subtle coupling arguments which we shall develop in Subsubsections 8.4.5 and 8.4.6.

8.4.4 Some Explicit Calculations

The entrance law from 0 for the type-2 mass can also be derived using explicit formulas for densities of the involved diffusions both for Fisher–Wright and branching models. We collect this in the following two remarks.

Remark 94. Consider the Fisher–Wright given by the SDE

$$dx(t) = \frac{m}{N}[1 - x(t)] + c[\frac{y(t)}{N} - x(t)]dt$$
$$+ sx(t)(1 - x(t))dt + \sqrt{d \cdot x(t)(1 - x(t))}dw(t), \quad (8.697)$$
$$x(0) = p,$$

where $y(t) = \hat{x}_2^N(t)$ is inserted as an external signal. By Kimura [Kim1] if $m = c = s = 0$, then there is no N-dependence and the density of the solution at time t is

$$f(p, x; t)$$
$$= \sum_{i=1}^{\infty} p(1-p)i(i+1)(2i+1)F(1-i, i+2, 2, p) \quad (8.698)$$
$$\cdot F(1-i, i+2, 2, x)e^{i(i+1)dt/2}$$
$$= 6p(1-p)e^{-dt} + 30p(1-p)(1-2p)(1-2x)e^{-3dt} + \dots,$$

where $F(\alpha, \beta, \gamma; x)$ is the hypergeometric function.

Consider now the neutral Fisher–Wright diffusion, $s = 0$ and constant $y(t) = m$, $c > 0$, $d = 1$. Crow–Kimura (see [GRD], Table 3.4, or Kimura [Kim1], (6.2)) obtained the density at time t

$$\tilde{f}^N(y, x, t) = x^{\mu-1}(1-x)^{\nu-\mu-1}$$
$$\times \sum_{i=0}^{\infty} \frac{(\nu+2i-1)\Gamma(\nu+i-1)\Gamma(\nu-\mu+i)}{i!\Gamma^2(\nu-\mu)\Gamma(\mu+i)} \times \tilde{F}_i(1-x)\tilde{F}_i(1-y)\exp(-\lambda_i t),$$
$$(8.699)$$

where the initial value of the process is y and

$$\mu = \frac{2m}{N}, \nu = 2(\frac{m}{N} + c), \lambda_i = i[2(\frac{m}{N} + c) + (i-1)]/2, \quad (8.700)$$

$$\tilde{F}_i(x) = F(\nu + i - 1, -i; \nu - \mu; x), \quad (8.701)$$

8.4 Droplet Formation: Proofs of Proposition 2.8–2.10

where F is the socalled hyper-geometric function that is that solution of the equation

$$x(1-x)F'' + [\gamma - (\alpha + \beta + 1)x]F'' - \alpha\beta F = 0, \qquad (8.702)$$

which is finite at $x = 0$.

Setting $y = 0$ and taking $N \to \infty$, then for every $t > 0$ and $x \in [0, 1]$, taking $\tilde{f}^N(x, t) = \tilde{f}^N(0, x, t)$, we get that there exists a function $\tilde{f}^\infty(x, t)$ which is for every t the density of a σ-finite measure such that:

$$N\tilde{f}^N(x,t) \longrightarrow \tilde{f}^\infty(x,t). \qquad (8.703)$$

The limit defines an entrance type law. Note that $\tilde{f}^N$ has the same behaviour near zero as that of the branching density below in (8.711).

Consider now the case $s > 0$. Then selection term then leads to a new law which is absolutely continuous w.r.t. the neutral one and the density is given by the Girsanov factor

$$L_t = \exp\left[\frac{s}{d}\int_0^t \sqrt{d \cdot x(s)(1-x(s))}dw(t) - \frac{s^2}{2d^2}\int_0^t d \cdot x(s)(1-x(s))ds\right]. \qquad (8.704)$$

Therefore the selection term does not change the asymptotics since it involves a bounded Girsanov density (uniform in N).

Alternatively to the above argument we can take the Kimura solution [Kim1](4.10) of the forward equation and take the limit $\frac{\phi(p,x;t)}{p}$ as $p \to 0$. □

Remark 95. For comparison, consider the branching SDE

$$d\tilde{x}^N(t) = \frac{m}{N} + \tilde{s}\tilde{x}^N(t)dt + \sqrt{d\tilde{x}^N(t)}dw(t). \qquad (8.705)$$

For $m = 0$, a simple calculation using Itô's lemma shows that

$$\exp\{-\tilde{x}^N(t-s) \cdot v_N(s)\}, \quad 0 \le s \le t, \qquad (8.706)$$

is a martingale provided that $v_N(\cdot)$ satisfies the equation

$$\frac{dv(t)}{dt} = \tilde{s}v_N(t) - \frac{d}{2}v_N^2(t), \quad v_N(0) = \lambda > 0. \qquad (8.707)$$

The solution of (8.707) is

$$v_N(t) = \frac{2\lambda\tilde{s}e^{\tilde{s}t}}{d\lambda(e^{\tilde{s}t} - 1) + 2\tilde{s}}. \qquad (8.708)$$

Then if $\tilde{x}^N(0) = 0$ and the immigration is given by $\frac{2m}{Nd}$, and $d = 2$ we have

$$E[e^{-\lambda \tilde{x}^N(t)}] = \exp\left(\frac{m}{N}\int_0^t v_N(s)ds\right) \qquad (8.709)$$

$$= \left(\frac{\tilde{s}}{\lambda(e^{\tilde{s}t}-1)+\tilde{s}}\right)^{\frac{m}{N}}. \qquad (8.710)$$

Then we obtain the Gamma density

$$f^N(x,t) = \frac{const}{N} x^{\frac{m}{N}-1} e^{-x/e^{\tilde{s}t}}, \quad x \geq 0, \qquad (8.711)$$

and Nf^N converges to the density $x^{-1}e^{-x/e^{\tilde{s}t}}$ on $\mathbb{R}^+$ of a σ-finite measure. It follows that

$$N \cdot \int_0^\varepsilon x f^N(x,t)dx \to 0 \text{ as } \varepsilon \to 0 \text{ uniformly in } N. \qquad \square \qquad (8.712)$$

8.4.5 First and Second Moments of the Droplet Growth Constant $\mathcal{W}^*$

Continuing the study of the droplet growth we now turn to times in $[T_0, T_N]$ as given in (8.621) and (8.622). In the sequel we compute the limits as $N \to \infty$ of the first two moments of the total type-2 mass in the population denoted $\hat{x}_2^N(t_N)$, for $t_N \to \infty$, $t_N = o(\log N)$. The present calculation allows in particular to show the existence of $\mathcal{W}^*$ and to determine its moments.

As before we use the dual representation to compute moments. For example, recall that

$$E[x_1^N(i,t_N)] = E[\exp(-\frac{m}{N}\int_0^{t_N} \Pi_s^{N,1} ds)], \qquad (8.713)$$

where $\Pi_t^{N,1}$ is of the total number of particles in the dual cloud in $\{1, \ldots, N\}$ starting with one factor $1_{\{1\}}$ at site i at time $t = 0$. As developed earlier this cloud is described by a CMJ process and has the form

$$\Pi_t^{N,1} = \sum_{j=1}^{W_N(t)e^{\alpha t}} \zeta_j^N(t), \qquad (8.714)$$

8.4 Droplet Formation: Proofs of Proposition 2.8–2.10

where $W_N(t) \to W(t)$ as $N \to \infty$ and $W(t) \to W$ a.s. as $t \to \infty$ and the processes at the different sites satisfy $\mathcal{L}[(\zeta_j^N(t))_{t \geq 0}] \Longrightarrow \mathcal{L}[(\zeta_j(t))_{t \geq 0}]$ as $N \to \infty$, where the $(\zeta_j(t))_{t \geq 0}$ for $j \in \mathbb{N}$ are all evolving independently and are given by the birth and quadratic death process at an occupied site with the same dynamic but starting from one particle at the random time, when the site is first occupied. The dependence between the different occupied sites arises through these random times even in the $N \to \infty$ limits.

If we consider second moments we have to start with two particles and then get the analog of (8.713) but in the representation of the r.h.s. of this formula given in (8.714), $W_N(t)$ and ζ^N have to be replaced accordingly and in the limit $N \to \infty$ the $W = W^{1,1}$ is replaced by $W^{1,2}$.

We have noted above that if $t_N = o(\log N)$ then the probability that a collision between dual particles due to a migration step occurs up to time t_N goes to 0 as $N \to \infty$. The new element that arises in this subsubsection is the fact that in the computation of the higher moments of $\hat{x}_2^N(t_N)$, these collision events must be taken into account. The reason for this is that we will compute moments of the mutant mass $\hat{x}_2^N(t)$ in terms of the moments of $\hat{x}_1^N(t)$. We see below from the formula (8.722) that in the expansion of k-th moments of the total mass powers of N appear, which forces us to consider in the expansions of these moments as well as higher order terms in N^{-1}. Hence we have to analyse more carefully the overlaps in the dual clouds corresponding to different ancestors which have probabilities which go to zero but not fast enough to be discarded.

The key tool we work with is the coloured particle systems (WRGB) introduced in Subsubsection 8.3.9.

The main result of whose proof this subsubsection is devoted to is:

Proposition 8.21 (Limiting growth constant of droplet). Let $t_N \to \infty$ and

$$\limsup_{N \to \infty} \frac{t_N}{\log N} < \frac{1}{2\alpha}. \qquad (8.715)$$

(a) The family

$$\{e^{-\alpha t_N} \hat{x}_2^N(t_N) : N \in \mathbb{N}\} \text{ is tight}, \qquad (8.716)$$

with non-degenerate limit points $\mathcal{W}^*$ with

$$0 < E[\mathcal{W}^*] < \infty, \quad 0 < \mathrm{Var}[\mathcal{W}^*] < \infty. \qquad (8.717)$$

(b) Furthermore:

$$\lim_{N \to \infty} e^{-\alpha t_N} E[\hat{x}_2^N(t_N)] = m^* E[W_1], \qquad (8.718)$$

$$\lim_{N \to \infty} e^{-2\alpha t_N} E[(\hat{x}_2^N(t_N))^2] = (m^*)^2 (EW_1)^2 + 2m^*(EW_1)^2 \kappa_2, \qquad (8.719)$$

where α is defined as in (8.140), $\kappa_2 > 0$ (which is given precisely in (8.751)) and

$$m^* = \frac{m(\alpha + \gamma)}{c}, \text{ with } b = 1 + \frac{\gamma}{\alpha}. \tag{8.720}$$

(c) The third moment satisfies

$$\limsup_{N \to \infty} E[e^{-3\alpha t_N}(\hat{x}_2^N(t_N))^3] < \infty, \tag{8.721}$$

so that the first and second moments of W^* are given by the formulas on the r.h.s. of (8.718) and (8.719). □

The proof of the main result is given in the rest of this subsubsection which has five parts, namely we require four tools to finally carry out the proof.

1. First we give a preparatory lemma 8.22 that gives asymptotic expressions for the moments of $\hat{x}_2^N(t_N)$ in terms of the moments of $\hat{x}_1^N(t_N)$.
2. In order to calculate the moments of $\hat{x}_1^N(t_N)$ we must determine the growth of dual clouds starting with one or two initial factors. This is stated in Proposition 8.24 and is then proved.
3. Here we introduce a multicolour system to analyse the two dual clouds and their interaction starting both from one particle (at different sites).
4. Then we present in Lemma 8.26 the tools to obtain the asymptotics of the expectations of the exponentials involving the occupation times of the dual clouds.
5. Finally we complete the proof of the above main result, the Proposition 8.21, using the dual expressions at the end of the section and the results from the previous Parts 1–4.

Part 1 (Moment formulas)

Lemma 8.22 (k-th moment formulas).

(a) We have the k-th moment formula for the total type-2 mass:

$$\bar{m}_k^{N,2}(t_N) := E[(\hat{x}_2^N(t_N))^k] = E\left[\left(\sum_{i=1}^N x_2^N(i, t_N)\right)^k\right]$$

$$= \sum_{\ell=1}^k \binom{k}{\ell}(-1)^\ell N^{k-\ell}\left(\bar{m}_\ell^{N,1}(t_N)\right), \quad k = 1, 2, \cdots \tag{8.722}$$

where $\bar{m}_\ell^{N,1}$ is the ℓ-th moment for the total type-1 mass.

(b) The term on the r.h.s. of (8.722) is calculated as follows. Let $q = (q_1, \cdots, q_j)$ and let $\Pi^{N,q}$ be the dual starting with q_i-particles at site $i = 1, \cdots, j$. Define

8.4 Droplet Formation: Proofs of Proposition 2.8–2.10

$$\mu_{q_1,\ldots,q_j;q}(t_N) = E\left[\exp\left(-\frac{m}{N}\int_0^{t_N} \Pi_s^{N,(q_1,\cdots,q_j)} ds\right)\right]. \quad (8.723)$$

Then

$$\bar{m}_q^{N,1}(t_N) = \sum_{j=1}^q \binom{N}{j} \sum_{q_1+\ldots+q_j=q} \frac{q!}{q_1!\ldots q_j!} \mu_{q_1,\ldots,q_j;q}(t_N). \quad \square \quad (8.724)$$

Proof of Lemma 8.22.

(a) First recall that

$$\hat{x}_1^N(t) = \sum_{i=1}^N x_1^N(i,t), \quad \hat{x}_2^N(t) = \sum_{i=1}^N x_2^N(i,t) = N - \hat{x}_1^N(t) \quad (8.725)$$

and therefore

$$\bar{m}_k^{N,2}(t_N) := E[(\hat{x}_2^N(t_N))^k] = E\left[\left(\sum_{i=1}^N x_2^N(i,t_N)\right)^k\right]$$

$$= \sum_{\ell=1}^k \binom{k}{\ell}(-1)^\ell N^{k-\ell} E\left[\left(\sum_{i=1}^N x_1^N(i,t_N)\right)^\ell\right], \quad k=1,2,\cdots \quad (8.726)$$

Recalling the definition of $\bar{m}_\ell^{N,1}(t)$ the claim follows.

(b) Moreover we can express the moments of the mass of type 1 appearing on the r.h.s. of (8.726) by expanding according to the *multiplicity* of the occupation of sites in the mixed moment expression, namely:

$$\bar{m}_k^{N,1}(t_N) = E\left[\left(\sum_{i=1}^N x_1^N(i,t)\right)^k\right] \quad (8.727)$$

$$= \sum_{i_1 \neq i_1 \neq \cdots \neq i_k} E\left[x_1^N(i_1,t)\ldots x_1^N(i_k,t)\right] + \cdots + \sum_{i=1}^N E[(x_1^N(i,t))^k].$$

The moments on the r.h.s. above can be computed as follows. Define

$$\mu_{q_1,\ldots,q_j;q}(t_N) = E\left[\prod_{\ell=1}^j (x_1^N(i_\ell, t_N))^{q_\ell}\right], \quad i_\ell \text{ distinct}, \quad \sum_{\ell=1}^j q_\ell = q. \quad (8.728)$$

Then

$$\bar{m}_q^{N,1}(t_N) = \sum_{j=1}^q \binom{N}{j} \sum_{q_1+\ldots+q_j=q} \frac{q!}{q_1!\ldots q_j!} \mu_{q_1,\ldots,q_j;q}(t_N). \quad (8.729)$$

Next use the duality relation to express $\mu_{q_1,\cdots,q_j;q}$. Define

$$\Pi_{t_N}^{(q_1,\cdots,q_j),N}, \tag{8.730}$$

as the total number of dual particles at time t_N starting the particle component η_t of the dual process $(\eta_t, \mathcal{F}_t^{++})$ with

$$q_i \text{ particles at distinct sites } i = 1,\cdots,j. \tag{8.731}$$

Then write

$$\mu_{q_1,\ldots,q_j;q}(t_N) = E\left[\exp\left(-\frac{m}{N}\int_0^{t_N} \Pi_s^{N,(q_1,\cdots,q_j)} ds\right)\right]. \quad \text{q.e.d.} \tag{8.732}$$

Part 2 *(Growth of dual clouds)*

Looking at the r.h.s. of formula (8.722)–(8.724) we see that for calculating higher moments we have to study the asymptotic (as $N \to \infty$) behaviour of the dual system starting with *more* than one particle. Of particular importance is to estimate the *overlap* in the dual particle population of different families due to different ancestors at the starting time of the evolution since this will allow to control the terms of order N^{-k} which are needed to compensate for the $\binom{N}{k}$ term on the r.h.s. of (8.722). This is addressed in the form needed for the calculation of first and second moments in the following definition and proposition.

Definition 8.23 (Dual clouds for moment calculations).

(a) To calculate first and second moments at one time we calculate the dual particle number asymptotically as follows if we start with one or two particles. We choose W_a as:
- $W_a = W_1$ where $W_1 = W^{1,1} = W$ defined by (8.144) starting with one $1_{\{1\}}$ factor at time 0
- $W_a = W^{1,2} = W_1 + W_2$ where W_1, W_2 are independent copies of the random variable W arising from one factor at each of two different sites,
- $W_a = W^{2,1}$ which is the corresponding random variable when the CMJ process is started with the two factors $1_{\{1\}} \otimes 1_{\{1\}}$ at time 0 at the same site.
- Similarly we write $W_a(\cdot)$ for the function $(W_a(t))_{t\geq 0}$ arising from $e^{-\alpha t} K_t^a$.

Let ξ be a time-inhomogeneous Poisson process on $[0,\infty)$ with intensity measure given by

$$\frac{(\alpha+\gamma)}{c}\frac{(W_a(s))^2 e^{2\alpha s}}{N} ds. \tag{8.733}$$

Note ξ depends on the chosen a but only through W_a.

8.4 Droplet Formation: Proofs of Proposition 2.8–2.10

Let
$$\tilde{g}(u,s) \tag{8.734}$$

(which evolves independently of the parameter N) be the size of that green cloud of sites of age u produced by one collision (creating a red particle) at time s. Then $\tilde{g}(t,s)$ has a law depending only on $t-s$, namely

$$\tilde{g}(t-s) = e^{\alpha(t-s)} W^g(t-s), \tag{8.735}$$

where $W^g(t-s) \to W^g$ as $t-s \to \infty$ and W^g is defined by this relation.

Furthermore introduce for a realization of ξ a process $\mathfrak{g}$ depending on the parameters N and whose law depends on the choice of a but only through $W_a(\cdot)$ and which describes the size of the complete green population at time t:

$$\mathfrak{g}(W_a(\cdot), N, t) = \int_0^t \tilde{g}(t,s)\xi(ds). \tag{8.736}$$

Analogously to $\tilde{g}, \mathfrak{g}$ we define the pair

$$\tilde{g}_{\text{tot}}, \mathfrak{g}_{\text{tot}} \tag{8.737}$$

as the number of particles in the green cloud, instead of the number of sites.

(b) To compute the joint second moment at two different times we consider

$$W_a = W_{j_1} + W_{j_2,s}, \tag{8.738}$$

which is the corresponding random variable when the CMJ process is started with the factor $1_{\{1\}}$ located at j_1 at time 0 and an additional factor $1_{\{1\}}$ is added at location j_2 at time $s > 0$.

Let $\xi^{(1,2)}$ be an inhomogeneous Poisson process on $[0,\infty)$ with intensity measure

$$\left[(\alpha + \gamma) \frac{W_1(u) W_2(u) e^{2\alpha u}}{N} \right] du. \tag{8.739}$$

Then with

$$t = t_N, \tag{8.740}$$

we consider time $t + s$. Furthermore set (recall (8.740))

$$\mathfrak{g}_{(1,2)}(W_1(\cdot), W_2(\cdot), N, t, t+s) = \int_s^t \tilde{g}(t_N, u) \xi^{(1,2)}(du). \tag{8.741}$$

We denote by $K_t^{N,(j_1,0),(j_2,s)}$ the number of occupied sites in the dual particle system starting with a particle at j_1 at time 0 and one at time j_2 at time s. □

Now we can state the main result on the growth of the dual clouds.

Proposition 8.24 (Dual clouds). *Let*

$$t = t_N, \text{ where } t_N \to \infty, \quad t_N = o\left(\frac{\log N}{2\alpha}\right), \tag{8.742}$$

more precisely,

$$2\alpha t_N - \log N \to -\infty \text{ as } N \to \infty. \tag{8.743}$$

Then consider the growth of the dual cloud in various initial states indicated by the subscript a.

(a) We have

$$K_t^{N,(a)} = W_a(t)e^{\alpha t} - \mathfrak{g}(W_a(\cdot), N, t) + \mathcal{E}_a(N, t), \tag{8.744}$$

$$\Pi_t^{N,(a)} = \frac{1}{c}(W_a)(t))(\alpha + \gamma)e^{\alpha t} - \mathfrak{g}_{\text{tot}}(W_a(\cdot), N, t) + \mathcal{E}_{a,\text{tot}}(N, t), \tag{8.745}$$

where we have for some $\kappa_2 \in (0, \infty)$

$$\lim_{N \to \infty} Ne^{-2\alpha t_N} E[\mathfrak{g}_{\text{tot}}(W_a(\cdot), N, t_N)] = \kappa_2 E[(W_a)^2] \tag{8.746}$$

and the error term $\mathcal{E}_a$ or $\mathcal{E}_{a,\text{tot}}$ satisfies that

$$\lim_{N \to \infty} E[Ne^{-2\alpha t_N}|\mathcal{E}_a(N, t_N)|] = 0. \tag{8.747}$$

(b) We have (as $N \to \infty$, recall (8.740)):

$$K_{t+s}^{N,(j_1,0),(j_2,s)} \sim (W_1 e^{\alpha(t+s)} + W_2 e^{\alpha t}) - \mathfrak{g}(W_1(\cdot), W_2(\cdot), N, t, t+s) + O\left(\frac{e^{3\alpha t}}{N^2}\right), \tag{8.748}$$

and

$$\Pi_{t+s}^{N,((j_1,0),(j_2,s))} = \frac{\alpha + \gamma}{c}(W_1 e^{\alpha(t+s)} + W_2 e^{\alpha t})$$

$$- \mathfrak{g}_{\text{tot}}(W_1(\cdot), W_2(\cdot), N, t, t+s) + O\left(\frac{e^{3\alpha t}}{N^2}\right), \tag{8.749}$$

8.4 Droplet Formation: Proofs of Proposition 2.8–2.10

where $\mathfrak{g}_{tot}(W_1(\cdot), W_2(\cdot), N, t, t+s)$ satisfies the two conditions

$$\mathfrak{g}_{tot}(W^{1,2}(\cdot), N, t, t+s) = \mathfrak{g}_{tot}(W_1(\cdot), N, t+s) + \mathfrak{g}_{tot}(W_2(\cdot), N, t) \\ + \mathfrak{g}_{(1,2),tot}(W_1(\cdot), W_2(\cdot), N, s, t), \quad (8.750)$$

and

$$\lim_{N\to\infty} N \cdot e^{-\alpha(2t_N+s)} E[\mathfrak{g}_{(1,2),tot}(W_1(\cdot), W_2(\cdot), N, t_N, t_N+s)] = 2\kappa_2 E[W_1]E[W_2]. \quad (8.751)$$

Similarly such relations hold for $\mathfrak{g}$ instead of $\mathfrak{g}_{tot}$ with κ_2 replaced by the appropriate different positive constant.

(c) *Now consider the case in which $N \to \infty$ but t is independent of N (that is, we no longer take $t_N \to \infty$). Then we do not replace $W_i(\cdot)$ by W_i and $\alpha(t), \gamma(t)$ by α, γ. Then in (8.748), (8.749) $W_1 \cdot (\alpha + \gamma)$ is replaced by $W_1(t+s)(\alpha(t+s) + \gamma(t+s))$ and $W_2 \cdot (\alpha + \gamma)$ is replaced by $W_2(t)(\alpha(t) + \gamma(t))$.* □

Remark 96. Note that in the expansion $(W_1 + W_2)^2 = W_1^2 + 2W_1W_2 + W_2^2$ we can associate the term $2W_1W_2$ with collisions of a particle from one cloud with sites occupied by the other cloud. In order to asymptotically evaluate these collision terms we could keep track of the white, red, green, blue particles from the two sites by labelling them, for example, by

$$(W_-, R_-, G_-, B_-), (W_+, R_+, G_+, B_+), \quad (8.752)$$

respectively, and then considering the collisions of W_+ with W_-, etc.

Remark 97. Note that the existence of κ_2 follows exactly in the same way as the argument leading up to (8.507) (proof of existence of κ) in Subsubsection 8.3.9. The only difference here is that it involves the growth of the number of green sites arising from either self collisions or between collisions of two different clouds.

Proof of Proposition 8.24. (a) In order to justify (8.745) we now reconsider the development of the dual particle system on $\{1, \ldots, N\}$ starting from one factor $1_{\{1\}}$ at location 1. We have used above the fact that for the range $0 \le t \le t_N$ where $\lim_{N\to\infty} \alpha \frac{t_N}{\log N} < 1$, the particle system is described asymptotically as $N \to \infty$ in first order by a CMJ process with Malthusian parameter α. Correction terms are analysed using the multicolour particles system.

Moreover if we follow two families starting with one factor $1_{\{1\}}$ at locations 1, 2 respectively, then the probability that they collide, that is occupy the same site in $\{1, \ldots, N\}$, in this time regime goes to 0 as $N \to \infty$. This is essentially equivalent to the statement that the probability that a green particle is produced in the WRGB system in this time regime goes to zero as $N \to \infty$. However for finite N these probabilities are not zero and we now determine them by expansion in terms of powers of N^{-k}.

Recall that the system of white and black particles is a CMJ process as described above in which the number of occupied sites has the form

$$W(t)e^{\alpha t} \tag{8.753}$$

where $\lim_{t\to\infty} W(t) = W$ and this black and white system serves as an *upper bound* for the system of occupied sites in the dual process. Hence if at time t a migrant is produced by this collection following the dual dynamics, it moves to a randomly chosen point in $\{1,\ldots,N\}$. Then the probability that it hits an occupied site is therefore at most:

$$\frac{W(t)e^{\alpha t}}{N}. \tag{8.754}$$

Therefore a upper bound for the process of collisions is given by an inhomogeneous Poisson process $\xi(s)$ with rate

$$\frac{W^2(t)e^{2\alpha t}}{N} \tag{8.755}$$

and number of sites that have been hit this way in the time interval $[0,t]$ is at most

$$\int_0^t \frac{W^2(s)e^{2\alpha s}}{N}ds \leq \frac{1}{N}(\sup_t W(t))^2 \cdot \frac{1}{2\alpha}e^{2\alpha t} = \frac{1}{N}(W^*)^2\frac{1}{2\alpha}e^{2\alpha t}. \tag{8.756}$$

Recall that $W^* = \sup_{t\geq 0} W(t) < \infty$ a.s. For bounding second moments we can also show that $\sup_{t\geq 0} E[(W(t))^2] < \infty$. In particular the number of sites at which collisions take place by time t_N is $o(N)$, i.e. has spatial intensity zero as $N \to \infty$.

Moreover the expected number of sites to be hit more than once in the time interval $[0,t]$ goes to 0 as $N \to \infty$ at the order N^{-2}. This order holds also for red–red collision by migration in the multicolour particle representation.

Next we recall that when a white particle hits an occupied site it produces a red particle at this occupied site. Subsequently if the red particle is removed by coalescence with a white particle, then a green particle is produced. Due to the coalescence process eventually no red particles will remain at this site. Also due to migration eventually this site will revert to a pure white site. During the "lifetime" of such a multicolour site there is a positive probability that one or more green particles will migrate and then produce a growing cloud. (Recall that the lifetime has finite exponential moment.)

Once such a new site containing white and red particles is formed, a green particle is created by coalescence unless either the white or the red disappears before coalescence. The number of green particles to migrate from such a special site is random with probability law that depends on the number $\zeta(s)$

8.4 Droplet Formation: Proofs of Proposition 2.8–2.10

of the occupying white particles at the time s of arrival of the red particle (asymptotically as $N \to \infty$ no further red particle will arrive at this site before the coalescence or migration step of the red particle or disappearance of all whites at the site).

In other words there is a positive probability that the green particle will produce green migrants before it coalesces with a red particle (such that only one red particle remains and the green particle disappears (recall Remark 82)). We denote the probability $p^N(s)$ that at least one green migrant is produced during the lifetime of a special site due to a collision at time s satisfies

$$p^N(s) \longrightarrow p(s) \text{ as } N \to \infty, \quad p(s) \longrightarrow p > 0 \text{ as } s \to \infty. \tag{8.757}$$

After the migration step the resulting green family, if non-empty, grows according to the CMJ process. Combining the two facts we see that the number of green sites this family occupies at time t, denoted $\tilde{g}(t, s)$, satisfies:

$$\tilde{g}(t, s) = W^g(t, s) e^{\alpha(t-s)}, \tag{8.758}$$

and $\lim_{t \to \infty} W^g(t, s) = W^g_s$ exists and converges to W^g as $s \to \infty$. Note that the event $\{W^g = 0\}$ occurs with the probability that a collision does not create a green migrant (see Remark 98 below) and otherwise

$$W^g(t, s) := \text{ is the random growth constant for all the sites occupied by} \tag{8.759}$$

green migrants from a collision at a site at time s in the

interval $[s, t]$.

Also $W^g(t, r)$ is independent of $W(\cdot)$ and $W^g(t, r), W^g(t, s)$ are independent for $r \neq s$. The analogous quantity $W^g_{\text{tot}}(t, r)$ is defined to be the number of green particles produced in $[r, t]$ by a collision at time r.

Define the analog of $W^g(t, s)$ in the case where the white sites are in the stable age distribution as

$$W^{g,*}(t, s). \tag{8.760}$$

If we assume that the white population has reached the stable size distribution (for the number of white particles at an occupied site) at time s, then the probability law of the corresponding process $W^{g,*}(t, s)$ depends only on $t - s$.

Remark 98. We note that at times $t \geq s_N$ with $s_N \to \infty$ the distribution of ζ in the CMJ process population approaches the stable age distribution as $N \to \infty$ uniformly in $t \geq s_N$.

Then combining (8.756) and (8.758) we get that

$$E[\mathfrak{g}(W(\cdot), N, t)] \leq \int_0^t E[(W(s))^2] \frac{e^{2\alpha s}}{N} E[W^g(t,s)] e^{\alpha(t-s)} ds \leq \text{const} \cdot \frac{e^{2\alpha t} - e^{\alpha t}}{N}. \tag{8.761}$$

Provided the first inequality could be *modified* such that it becomes up to terms of order $o(N^{-1})$ an *equality*, the following limit exists

$$\lim_{N \to \infty} N e^{-2\alpha t_N} E[\mathfrak{g}(W(\cdot), N, t_N)] = \kappa_2 E[W^2], \quad \kappa_2 = \int_0^\infty e^{-\alpha u} E[W^{g,*}(u)] du. \tag{8.762}$$

We note that the probability of collision in $[0, t_N]$ tends to 0 as $N \to \infty$ therefore we determine the probability of such a collision, the distribution of the collision time given that a collision occurs and the resulting number of green sites produced by this one collision. Given that a collision occurs the site at which it occurs is obtained by choosing a randomly occupied site.

In order to verify that such a modification exists and that the assumption on the error in the first inequality in (8.761) we have to (1) estimate how much we overestimate the collision rate by using the CMJ process (of black and white particles) instead of the real dual (i.e. (W, R, P) particles) and (2) in the application of W^g we need to control the error introduced by assuming that collision events involve sites with the stable age distribution for the white particles and replacing $\int_0^{t_N} e^{-\alpha(t_N - s)} E[W^g((t_N, s)] ds$ as $N \to \infty$ by the second part of the equation (8.762).

1. We again note that the over estimate of the number of green particles corresponds to events having two or more collisions and this occurs with rate $\frac{e^{3\alpha t}}{N^2}$ and this correction as well as the exclusion of the purple particles produces a higher order error term.
2. Consider the early collisions before reaching the stable age distribution. Noting that

$$\int_0^{s_N} e^{2\alpha s} e^{\alpha(t_N - s)} ds = \text{Const} \cdot e^{\alpha(t_N + s_N)} \tag{8.763}$$

therefore these contribute only a term increasing as $e^{\alpha t_N + s_N}$ whereas the latter ones grow at a rate $e^{2\alpha t_N}$. If $s_N \to \infty$, $s_N = o(t_N)$, then the early collisions form a negligible contribution. But then as $N \to \infty$, the white population approaches the stable age distribution by time s_N.

Note that for the late ones we cannot assume that $W^g(t,s)$ takes on its limiting value and for this reason we obtain the integral in the definition of κ_2. This proves Part (a) of Proposition 8.24.

(b) We now consider (8.749) when we begin with one $1_{\{1\}}$ factor at site j_1 at time 0 and one factor $1_{\{1\}}$ at $j_2 \neq j_1$ at time $s > 0$ and the total number of occupied sites at time t is what we called $\Pi_t^{N,(1,2)}$. This can be handled exactly as in (a).

8.4 Droplet Formation: Proofs of Proposition 2.8–2.10

However we want to keep track of the subset of sites occupied by migrants coming from collisions between the two different families. For this reason we constructed the multicolour system.

Finally we have to show that the bound (8.794) gives actually asymptotically the correct answer. Following the argument in part (a) we note that the upper bound on the $G_{1,2}$ particles gives us a lower bound on the white particles and this in turn gives us by (8.792) with a bootstrapping a lower bound on the number of $G_{1,2}$ particles. It then follows that the overestimate of the number of $G_{1,2}$ particles is no larger than const $\cdot \frac{e^{3\alpha t_N}}{N^2}$.

The analogous arguments also yields the claim (8.751) for $\mathfrak{g}_{(1,2),\text{tot}}$.

(c) Other than including the time dependence for $W_i(\cdot)$ the proof follows the same lines as above. In particular the first two terms in (8.748) now correspond to the CMJ process obtained when $N = \infty$ and the limiting expected value of the correction term now has the form

$$\lim_{N\to\infty} N \cdot e^{-\alpha(2t+s)} E[\mathfrak{g}_{(1,2),\text{tot}}(W_1(\cdot), W_2(\cdot), N, t, t+s)] = 2\kappa_2(t) E[W_1(t)] E[W_2(t)]. \tag{8.764}$$

This completes the proof of Proposition 8.24.   q.e.d.

Part 3   *(Two dual clouds interacting)*

We want to compare the dual populations (note the different brackets and the $\sim$ indicating the populations instead of just the population sizes)

$$\tilde{\Pi}^{N,\{j_1\}}, \tilde{\Pi}^{N,\{j_2\}}, \tilde{\Pi}^{N,\{j_1,j_2\}} \tag{8.765}$$

corresponding to the descendants of the particles at j_1, j_2, respectively evolving independently and the resulting population starting with both initial particles and evolving jointly. In order to carry out this comparison we construct a *coupling*, i.e. construct all three populations on one probability space in such a way that the respective laws are preserved. At the same time we want to be able to compare these populations with the respective collision-free versions evolving as CMJ-processes.

Construction of enriched multicolour system.

To construct this coupling we use again the WRGB-system and we construct the particle system starting with 1 particle at both sites j_1 and j_2, recall here the constructions from subsubsection 8.3.9. We now modify this coloured WRGB-particle system by marking the offspring of the initial particles at j_1, j_2 by using an enriched colour system, i.e. using the colours

$$\{W_1, R_1, G_1, B_1, P_1, W_2, R_2, G_2, B_2, P_2\}. \tag{8.766}$$

together with

$$\{W_1^*, W_2^*, R_1^*, R_2^*, G_1^*, G_2^*, B_1^*, B_2^*, P_1^*, P_2^*, R_{1,2}, G_{1,2}, B_{1,2}, P_{1,2}\}. \tag{8.767}$$

The second set of colours is reserved for particles involved in a collision between the two different families or their descendants.

The new features are

- When a W_1 particle hits a W_2 particle the result is a W_1^* and $R_{1,2}$ pair as well as the original W_2 particle at this site. A W_1^* particle does not coalesce with the W_2 particle.
- If the $R_{1,2}$ particle coalesces with the W_2 particle, the result is the pair consisting of a W_2 particle and a $G_{1,2}$ particle.
- The above holds with 1 and 2 interchanged.
- If W_1^* hits a W_1 particle it changes to a R_1^* particle.

With this expanded colour set, we now have to define a suitable multicolour dynamic. We define this system more precisely after outlining the goals of this construction. This is done so that the (W_i, W_i^*, R_i, G_i), $i = 1, 2$ describe the families in the absence of interactions between them. The particles $(R_{1,2}, G_{1,2})$ correspond to the particle families resulting from the interaction between the two families and $(W_1, W_2, R_1, R_2, G_1, G_2, R_{1,2}, G_{1,2})$ describe the interacting system. As before we want to couple the interacting system with a CMJ process.

Recall that the covariance of the type-1 mass in two locations if we write it in the form $E[XY] - E[X]E[Y]$ consists in the dual representation of a cloud of dual particles evolving jointly from two initial particles at different locations and two independently evolving clouds starting from one particle each. Hence the idea is that the covariance of $x_1^N(i, t+s)$ and $x_1^N(i, t)$ is determined by the particles in

$$\{\tilde{\Pi}^{N,\{1\}} \cup \tilde{\Pi}^{N,\{2\}}\} \setminus \tilde{\Pi}^{N,\{1,2\}}, \tag{8.768}$$

where $\tilde{\Pi}^{N,\{1\}}, \tilde{\Pi}^{N,\{2\}}$ denotes the two independently evolving dual particle systems and $\tilde{\Pi}^{\{1,2\}}$ the jointly evolving ones. The set $\tilde{\Pi}^{N,\{1\}} \cup \tilde{\Pi}^{N,\{2\}} \setminus \tilde{\Pi}^{N,\{1,2\}}$ corresponds to the $G_{1,2}$ green particles.

We list now the precise evolution rules of the new coloured particle system. In the evolution of the populations arising from the two initial particles as before particles *migrate, coalesce* and *give birth* to new particles at the rates as in the dual. One modification is required, namely, in this case we allow singletons of one family to migrate to sites occupied by the other family and also singletons of one family in the presence of one or more particles of the other type can migrate. Here are the precise rules.

(0) The rules for the particles and colours appearing in (8.766) are exactly as before as long as the two populations (i.e. particles with a colour with one index either 1 or 2) do not interact or occupy the same site. If particles of the two families occupy the same site singletons of one family can now migrate.

Furthermore we have the four additional rules concerning the colours newly introduced in (8.767):

(1) When a white particle from family 1 and a white particle from family 2 coalesce the result is one W_1^*, one W_2^* particle and a $R_{1,2}$ particle. The W_ℓ^* particles

8.4 Droplet Formation: Proofs of Proposition 2.8–2.10

evolve according to the rules of the dual particle system, but ignoring further collisions with the respective other population and upon collision with the own population a R_ℓ^*-particle is created replacing the W_ℓ^*-particle. Similarly proceed with the other $*$-colours.

(2) $R_{1,2}$ particles that coalesce with W_1 (respectively W_2) produces a W_1 (respectively W_2) and $G_{1,2}$ pair.

Two $R_{1,2}$ particles that hit an occupied site produce blue–purple pairs $(B_{1,2}, P_{1,2})$ where as before the $B_{1,2}$ particle is placed on the first empty site in the first copy of $\mathbb{N}$ continuing according to the collision-free dynamic and the $P_{1,2}$ particle behaves as a typical dual particle for $\tilde{\Pi}^{N,\{1,2\}}$.

(3) Coalescence of $G_{1,2}$ and $R_{1,2}$ particles produce $R_{1,2}$ particles (cf. Remark 82).

(4) If a $R_{1,2}$ or $G_{1,2}$ particle more generally every colour in (8.767) gives birth the new particle carries the same type.

These set of rules uniquely define a pure jump Markov process on the state space describing the individuals with their colour and location and with the coupling information (recall (8.394) for how to formalize this).

The key properties of the new coloured dynamics are the following. On the same probability space we have the various systems we consider. Namely

$$\tilde{\Pi}^{N,\{i\}} = \text{union of the } W_i, W_i^*, R_i, P_i, R_i^*, P_i^*, \text{ for } i = 1, 2 \qquad (8.769)$$

The new system also includes our dual starting with two particles, namely, (by construction)

$$\tilde{\Pi}^{N,\{1,2\}} = \text{the union of the } W_1, W_2, R_1, R_2, P_1, P_2, R_{1,2}, G_{1,2} \text{ particles.} \qquad (8.770)$$

Furthermore we can again identify a CMJ-process,

$$\text{union of } (W_1, W_2, R_1, R_2, G_1, G_2, G_{1,2}, B_1, B_2, R_{1,2}, B_{1,2})\text{-particles} \hat{=} \text{CMJ process.} \qquad (8.771)$$

Remark 99. To compute higher moments a *complete* bookkeeping of the difference would require more and more different colours whenever such a collision with subsequent coalescence occurs. However although these collisions do occur, we note that they do not contribute to the calculation up to terms of order N^{-2}. Therefore we again ignore these higher order terms instead of introducing further colours.

In the sequel we shall retain the W_1^* and W_2^* particles but not the $R_\ell^*, B_\ell^*, P_\ell^*$ particles since the order in N of these sets of particles are of too small to play a role for our purposes.

Lemma 8.25 (Negligible colours). *Let $\mathcal{N}_t^\ell$ denote the number of R_ℓ^*, B_ℓ^* or P_ℓ^* particles in the system ℓ. Then*

$$E[\mathcal{N}_t^\ell] = O(\frac{e^{3\alpha t}}{N^2}). \qquad \square \qquad (8.772)$$

Proof of Lemma 8.25. This follows since these correspond to families resulting from two or more collisions and these have order $O(\frac{e^{3\alpha t}}{N^2})$. q.e.d.

If we count only the occupation numbers by the various colours the state space is given by

$$\mathcal{M}_c[(\{W_1, W_1^*, R_1, G_1, B_1\} \cup$$

$$\{W_2, W_2^*, R_2, G_2, B_2\} \cup \{R_{1,2}, G_{1,2}, B_{1,2}, P_{1,2}\}) \times (\{\{1, 2, \ldots, N\} + \mathbb{N} + \mathbb{N}\})] \tag{8.773}$$

(where $\mathcal{M}_c$ denotes the set of counting measures).

This concludes the construction of a coupling via a new enriched enriched multicolour particle system.

Consequences of the coupling

Using the arguments of part (2) above we can conclude that:

$$\Pi_t^{N,\{1\}} + \Pi_t^{N,\{2\}} \sim \frac{1}{c}((W' + W''))(\alpha + \gamma)e^{\alpha t} - [\mathfrak{g}_{\text{tot}}((W'(\cdot)), N, t)$$
$$+ \mathfrak{g}_{\text{tot}}((W''(\cdot)), N, t)] + \mathcal{E}_{1,\text{tot}}(N, t) + \mathcal{E}_{2,\text{tot}}(N, t), \tag{8.774}$$

$$\Pi_t^{N,\{1,2\}} \sim \frac{1}{c}((W' + W''))(\alpha + \gamma)e^{\alpha t} - \mathfrak{g}_{\text{tot}}((W'(\cdot) + W''(\cdot)), N, t) + \mathcal{E}_{1,2;\text{tot}}(N, t), \tag{8.775}$$

where

$$\mathcal{E}_{\ell,\text{tot}}(N, t) = o(N^{-1}e^{2\alpha t_N}) \text{ as } N \to \infty, \quad \ell = 1, 2 \text{ or } (1, 2), \tag{8.776}$$

$$\mathfrak{g}_{\text{tot}}((W'(\cdot) + W''(\cdot)), N, t) = \int_0^{t_N} \tilde{g}_{\text{tot}}(t_N, s) \xi^{(1,2)}(ds), \tag{8.777}$$

with $\xi^{(1,2)}$ a time inhomogeneous Poisson process on $[0, \infty)$ with intensity measure

$$\frac{((W'(\cdot)(s) + W''(\cdot)(s)))^2 e^{2\alpha s}}{N} ds \tag{8.778}$$

and $\tilde{g}_{\text{tot}}(t, s)$ is the size of the green cloud at time t produced at time s by a *newly* created red particle at a white site.

We know furthermore that

$$E[\mathfrak{g}_{\text{tot}}((W' + W'')(\cdot), N, t)] = \kappa_2 E[(W' + W'')^2] \frac{(e^{2\alpha t} - e^{\alpha t})}{N}. \tag{8.779}$$

8.4 Droplet Formation: Proofs of Proposition 2.8–2.10

Then combining all these properties (8.741) follows by comparing the expression for $\mathfrak{g}_{\text{tot}}((W'+W'')(\cdot), N, t)$ with $\mathfrak{g}_{\text{tot}}(W'(\cdot), N, t) + \mathfrak{g}_{\text{tot}}(W''(\cdot), N, t)$.

Therefore the process corresponding to the quantity

$$\Pi_t^{N,1} + \Pi_t^{N,2} - \Pi_t^{N,\{1,2\}}, \quad t \geq 0 \tag{8.780}$$

has an upper bound which is generated by the Poisson process on $[0, \infty)$ with intensity measure

$$\left(\frac{(2W'W'')e^{2\alpha s}}{N} + O(e^{3\alpha s} N^{-2}) \right) ds. \tag{8.781}$$

This corresponds to the difference term (i.e. lost particles due to collision and coalescence between the two families) given by the $G_{1,2}$-particles. It is bounded above by allowing the rate of production of collisions leading to the production of $G_{1,2}$ particles by the CMJ-process which excludes collisions. We can then obtain a lower bound by reducing the white families by the respective upper bounds for the sets of green particles as well as the blue particles. This replaces the white family $W(t)e^{\alpha t}$ by $W(t)e^{\alpha t} - O(\frac{e^{2\alpha t}}{N})$. Using this we can show that the result is an error term which is of order $O(\frac{e^{3\alpha t}}{N^2})$.

We now want a more detailed description in which we keep track explicitly of the collisions of the two families and the resulting $G_{1,2}$ families. We therefore consider

$$K_t^{N,\{1,2\}}, \Psi_t^N(i_{W_1}, i_{W_2}, i_{R_1}, i_{R_2}, i_{R_{1,2}}, i_{G_1}, i_{G_2}, i_{G_{1,2}}, i_{B_1}, i_{B_2}, i_{B_{1,2}}, i_{B_{1,2}}), \tag{8.782}$$

which denotes the number of occupied sites at time t, respectively the number of sites having i_{W_1} white particles from the first population, i_{W_2} white particles from the second population, etc. Similarly we can define systems for subsets of colours as we did before. We also consider the pair (u^N, U^N) describing the new coloured particle system and a new corresponding limiting system (u, U) for $N \to \infty$. For the $(W_1, W_2, R_1, R_2, G_1, G_2, R_{1,2}, G_{1,2}, B_1, B_2)$-system we modify the equation for U_{WRGB}, namely, (8.459) to include the contribution to the dynamics of collisions between W_1 and W_2 particles. Similar we can proceed with other colour combinations.

Then we obtain the following representation formula. Here we use the convention to write

$$U_t^N(i_{W_j}, i_{W_j^*}, i_{R_j}, i_{P_j}), U_t^N(i_{W_1}, i_{W_2}, i_{R_{1,2}}), \cdots \tag{8.783}$$

for $U_t^N(\cdot)$ all other colours not appearing in the argument summed out. Then:

$$\Pi_t^{N,\{j\}} = u^N(t)$$

$$\cdot \sum_{i_{W_j}, i_{W_j^*}, i_{R_j}, \cdots, i_{P_j^*}} (i_{W_j} + i_{W_j^*} + i_{R_j} + i_{P_j} + i_{R_j^*} + i_{P_j^*})$$

$$U_t^N(i_{W_j}, i_{W_j^*}, i_{R_j}, i_{P_j}, i_{R_j^*}, i_{P_j^*}),$$

$$j = 1, 2, \qquad (8.784)$$

$$\Pi_t^{N,\{1,2\}} = u^N(t) \sum_{i_{W_1},\cdots,i_{P_{1,2}}} (i_{W_1} + i_{W_2} + i_{R_1} + i_{R_2} + i_{R_{1,2}} + i_{P_1} + i_{P_2} + i_{P_{1,2}})$$

$$U_t^N(i_{W_1}, i_{W_2}, i_{R_1}, i_{R_2}, i_{R_{1,2}}, i_{P_1}, i_{P_2}, i_{P_{1,2}}), \qquad (8.785)$$

where the sum is over $i_{W_1}, i_{W_2}, i_{R_1}, i_{R_2}, i_{R_{1,2}}, i_{P_1}, i_{P_2}, i_{P_{1,2}}$.

We now focus on the are sites at which a collision occurs between a W_1 particle and a W_2 site (or vice versa) which can then produce migrating $R_{1,2}$ or $G_{1,2}$-particles. Therefore we define as *"special"* sites at time t, those sites at which

one or more $R_{1,2}$ particles are present at time t and *also* $\qquad$ (8.786)

a W_1 or W_2-particle is present at time t.

Note that special sites have a finite lifetime and can produce migrating $R_{1,2}, G_{1,2}$ particles during their lifetime. Denote the number of special sites by:

$$K_t^{N,*} = \left(\sum_{i_{W_1},i_{W_2},i_{R_{1,2}}} (1_{i_{R_{1,2}} \geq 1} \cdot (1_{i_{W_1} \neq 0} \vee 1_{i_{W_2} \neq 0}) U_t^N(i_{W_1}, i_{W_2}, i_{R_{1,2}}) \right) \cdot u^N(t). \qquad (8.787)$$

In the calculation of second moments we can ignore errors of size $o(N^{-2})$, we can work with the system of $W_1, W_2, R_1, R_2, R_{1,2}$ and ignore the purple particles i.e. the colours $P_1, P_2, P_{1,2}$. Therefore it suffices to determine the number of sites occupied by $W_1, W_2, R_1, R_2, R_{1,2}$ particles which estimates $K_{t+s}^{N,(j_1,0),(j_2,s)}$ up to errors of order $O(\frac{e^{3\alpha t}}{N^2})$. This results in the expansion given in (8.748) where in (8.750) the three $\mathfrak{g}_{\text{tot}}$ terms correspond to the sites occupied by only G_1, G_2 and $G_{1,2}$ particles. It remain to estimate these terms. The G_1, G_2 terms are similar to those considered earlier so we focus on the $G_{1,2}$ sites and particles. We will see below that key to the determining the asymptotics of the covariance at two sites as $N \to \infty$ is to analyse the number of $G_{1,2}$-particles asymptotically as $N \to \infty$. We have now all the tools to start the estimation.

Estimates on the number of $G_{1,2}$-particles

In order to study the asymptotics of the *number of $G_{1,2}$ particles* at time t_N, in sublogarithmic time scales which we denote this number by:

$$\Pi_{t_N}^{N,G_{1,2}} \text{ with } t_N \text{ as in (8.742).} \qquad (8.788)$$

We next recall that *independent* systems of dual particle systems with starting particles in site j_1 at time 0 and site j_2, at time $s > 0$ with $j_1 \neq j_2$, produce

8.4 Droplet Formation: Proofs of Proposition 2.8–2.10

clouds as $N \to \infty$ with independent stable size distributions and the number of sites of the two dual populations is denoted by

$$K_t^{N,\{1\}}, K_t^{N,\{2\}}. \tag{8.789}$$

We then have by the CMJ-theory and the K_t for K_t^N approximation $N \to \infty$,

$$K_{t_N}^{N,\{1\}} \sim W' e^{\alpha t_N}, \quad K_{t_N}^{N,\{2\}} \sim W'' e^{\alpha t_N}. \tag{8.790}$$

To identify the correction terms needed for second moment calculations we need to consider the dual particle system starting with two particles one at site 1, one at site 2 and estimate the effects of the *collisions between the two clouds*. Using our embedding in the coloured particle system this can involve a collision in which a W_1 particle can migrate to a site already occupied by a W_2 particle (or vice versa). Once a collision occurs, this produces a W_1^*, $R_{1,2}$ pair. The $R_{1,2}$ particle can move to another site before it coalesces with one of the W_2 particles of the other colour. If it coalesces the W_2 particle remains an a new $G_{1,2}$-particle is added at this site. The $G_{1,2}$ particles can then reproduce and coalesce with other particles. This way they can produce more $G_{1,2}$ particles or migrate to produce a new $G_{1,2}$ site. The system of $G_{1,2}$ particles after the creation of a $G_{1,2}$-particle has the property that after foundation a $G_{1,2}$-family does not anymore depend on the other colours, therefore the $G_{1,2}$-system evolves as a copy of the basic one type *CMJ particle system* with *immigration* given by a randomly fluctuating source which is independent of the CMJ-process. The creation of new $R_{1,2}$ particles at time s is determined by the number of (W_1, W_2)-pairs at time s.

Suppose we are given the evolution of the W_1, W_2-particles. Then the events of collisions between W_1 and W_2 particles is given by a Poisson process and when a collision occurs at a site there is a positive probability of a coalescence and therefore the production of a green particle. This then produces a growing population of $G_{1,2}$ descendants according to the CMJ-theory leading to the analogue expression to (8.761) which we had for W_1, W_2-particles.

If we replace the system of white particles of the two types by the *independent* site-1 and site-2 system, we obtain an *upper* bound for the mean number of $G_{1,2}$ particles given the growth constants W_1 and W_2 of the two clouds, so that at time $t = t_N$ provided that

$$t_N = o(\frac{\log N}{\alpha}), \tag{8.791}$$

by just calculating the mean.

This results in the following asymptotic population size for the $G_{1,2}$ particles which is then again given by the behaviour of the conditional mean (given the W_1, W_2 populations):

$$\frac{c}{N} \int_0^{t_N} E[W] e^{\alpha(t_N - u)} (\Pi_u^{N,\{1\}} K_u^{N,\{2\}} + \Pi_u^{N,\{2\}} K_u^{N,\{1\}}) p^N(u) du + o(N^{-1}). \tag{8.792}$$

The real system requires a correction term due to the interaction of the two clouds leading to a reduction of the two independent white systems by the $G_{1,2}$ but which is of lower order in N than the W_1, W_2-particle numbers and their effect, i.e. $o(N)$. Hence from (8.792) we get the asymptotic upper bound for the mean number of $G_{1,2}$ particles as $N \to \infty$ if we condition on the growth constants W' and W'' of the two clouds. As bound we get an expression which is asymptotically ($N \to \infty$) equal to:

$$\sim \frac{\kappa_2^N}{N} W' W'' e^{2\alpha t_N}, \tag{8.793}$$

and which differs from the real system by an error term of order $o(\frac{e^{2\alpha t_N}}{N})$.

In fact we can obtain

$$\lim_{N \to \infty} N e^{-2\alpha t_N} E[\mathfrak{g}_{(1,2)}(W(\cdot), N, t_N)] = \kappa_2 E[W'] \cdot E[W''],$$

$$\kappa_2 = \int_0^\infty e^{-\alpha u} E[W^{g,*}(u)] du. \tag{8.794}$$

exactly the same way as for (8.762).

Remark 100. In the longer time regime $\frac{\log N}{\alpha} + t$ the growing cloud of green particles $G_{1,2}$ satisfies a law of large number effect similar as in (8.485), that is, conditioned on W_1, W_2, the variance of the normalized number of $G_{1,2}$-particles goes to zero as $N \to \infty$.

Part 4 *(Random exponentials)*

We next present the facts needed about the expansion of random exponentials in the following lemma.

Lemma 8.26 (Asymptotics of random exponentials).

(a) Let X be a nonnegative random variable satisfying $E[X^k] < \infty$ for all $k \in \mathbb{N}$. Let $f(\varepsilon, X)$ be such that

$$f(\varepsilon, X) \geq 0, \quad f(\varepsilon, X) \sim \varepsilon[X - \varepsilon \kappa X^2 + O(\varepsilon^2) X^3], \tag{8.795}$$

that is,

$$\sup_{0 < \varepsilon \leq 1} \frac{|f(\varepsilon, x) - \varepsilon X + \varepsilon^2 X^2|}{\varepsilon^3} < \infty. \tag{8.796}$$

Consider

$$E[1 - e^{-f(\varepsilon, X)}]. \tag{8.797}$$

8.4 Droplet Formation: Proofs of Proposition 2.8–2.10

Then the following two relations hold:

$$\lim_{\varepsilon \to 0} \frac{E[1 - e^{-f(\varepsilon, X)}]}{\varepsilon} = E[X], \tag{8.798}$$

$$\lim_{\varepsilon \to 0} \frac{E[1 - e^{-f(\varepsilon, X)} - \varepsilon X]}{\varepsilon^2} = (1 + 2\kappa) E[X^2]. \tag{8.799}$$

(b) *Consider now random variables X, Y with $E[Y^k + X^k] < \infty$ for all $k \in \mathbb{N}$. Let Y_ε be a superposition of a Poisson number of independent copies of a random variable Y with intensity εX^2. Assume now that $f(\varepsilon, X, Y) \geq 0$ and that we have:*

$$f(\varepsilon, X, Y) \sim \varepsilon[X - Y_\varepsilon]. \tag{8.800}$$

Then (8.798) and (8.799) are satisfied with

$f(\varepsilon, X)$ *replaced by* $f(\varepsilon, X, Y)$ *and* $1 + 2\kappa$ *by* $1 + EY$. □ (8.801)

Proof. (a) Since the limits (8.798) and (8.799) are not affected by the $O(\varepsilon^2)$ term in (8.795) we can assume that it is a constant without loss of generality. Then by the mean value theorem,

$$\frac{1 - e^{-\varepsilon X + \varepsilon^2 X^2 - \varepsilon^3 X^3}}{\varepsilon} = D(X, \varepsilon^*) \text{ where } D(X, \varepsilon) = \frac{d}{d\varepsilon}[1 - e^{-\varepsilon X + \varepsilon^2 \kappa X^2 - \varepsilon^3 X^3}] \tag{8.802}$$

and where $0 < \varepsilon^* \leq \varepsilon$ and $\lim_{\varepsilon \to 0} D(X, \varepsilon) = X$. Therefore we have for $c > 0$ that

$$E\left[\frac{1 - e^{-\varepsilon X + \varepsilon^2 \kappa X^2 - \varepsilon^3 X^3}}{\varepsilon}\right] = E[D(X, \varepsilon^*) 1_{(X \leq \frac{c}{\varepsilon})}] + \frac{\text{const}}{\varepsilon} \cdot P[X > \frac{c}{\varepsilon}]. \tag{8.803}$$

Then

$$D(X, \varepsilon) = \frac{d}{d\varepsilon}[1 - e^{-\varepsilon X + \varepsilon^2 \kappa X^2 - \varepsilon^3 X^3}] = [(X - 2\varepsilon^2 \kappa X^2 + 3\varepsilon^2 X^3) e^{-\varepsilon X + \varepsilon^2 \kappa X^2 - \varepsilon^3 X^3}]. \tag{8.804}$$

Note that

$$E[\sup_{0 < \varepsilon \leq 1, 0 \leq X \leq \frac{c}{\varepsilon}} |D(X, \varepsilon)|] \leq \text{const} E[X + X^2 + X^3] < \infty. \tag{8.805}$$

Therefore by dominated convergence for every $c > 0$:

$$\lim_{\varepsilon \to 0} E[1_{(X \leq \frac{c}{\varepsilon})} \cdot D(X, \varepsilon)] = E[X]. \tag{8.806}$$

Since $E[X^k] < \infty$ for $k \in \mathbb{N}$ we have

$$P[X > \frac{c}{\varepsilon}] \leq E[X^k] \cdot (\frac{\varepsilon}{c})^k. \tag{8.807}$$

Combining (8.806), (8.807) we conclude from (8.803) the claim (8.798).
Similarly for the second order expansion write,

$$\frac{1}{\varepsilon}[\frac{1 - e^{-\varepsilon X + \varepsilon^2 \kappa X^2 - \varepsilon^3 X^3}}{\varepsilon} - X] = \frac{1}{\varepsilon}[D(X, \varepsilon^*) - D(X, 0)] = D^2(X, \varepsilon^{**}), \tag{8.808}$$

where $D^2(X, \varepsilon) = \frac{d^2}{d\varepsilon^2}[1 - e^{-\varepsilon X + \varepsilon^2 \kappa X^2 - \varepsilon^3 X^3}]$ and

$$\lim_{\varepsilon \to 0}[D^2(X, \varepsilon)] = X^2 + 2\kappa X^2 = (1 + 2\kappa)X^2. \tag{8.809}$$

As above we then conclude the claim (8.799).
(b) We have

$$1 - e^{-f(\varepsilon, X, Y)} = 1 - e^{-\varepsilon X + \varepsilon Y} 1_{Y \neq 0} - e^{-\varepsilon X} 1_{Y = 0} \tag{8.810}$$

and

$$\sup_{0 < \varepsilon \leq 1, Y < \frac{1}{\varepsilon}} |\frac{d}{d\varepsilon}(1 - e^{-f(\varepsilon, X, Y)})| \leq \text{const} \cdot (X + Y). \tag{8.811}$$

Therefore the relation is now obtained as follows:

$$\lim_{\varepsilon \to 0} \frac{E[1 - e^{-f(\varepsilon, X, Y)}] - \varepsilon E[X]}{\varepsilon} \tag{8.812}$$

$$= \lim_{\varepsilon \to 0} \left[\frac{Er[((1 - e^{-\varepsilon X}]) + \varepsilon E[(1 - e^{-\varepsilon X^2})]E[Y 1_{0 < Y \leq \frac{1}{\varepsilon}}] - \varepsilon E[X]}{\varepsilon^2} \right]$$

$$= \lim_{\varepsilon \to 0} \left[\frac{\varepsilon^2 E[X^2] + \varepsilon^2 E[X^2]E[Y] + P[Y > \frac{1}{\varepsilon}]}{\varepsilon^2} \right] = E[X^2](1 + E[Y]). \quad \text{q.e.d.}$$

Part 5: *(Proof of Proposition 8.21)*

To prove (a) (8.716) and (8.717) it suffices to show that

$$0 < \lim_{N \to \infty} E[e^{-\alpha t_N} \hat{x}_2^N(t_N)] < \infty, \quad 0 < \lim_{N \to \infty} \text{Var}[e^{-\alpha t_N} \hat{x}_2^N(t_N)] < \infty, \tag{8.813}$$

$$\limsup_{N \to \infty} E[(e^{-\alpha t_N} \hat{x}_2^N(t_N))^3] < \infty, \tag{8.814}$$

8.4 Droplet Formation: Proofs of Proposition 2.8–2.10

so that the family $\{\mathcal{L}(\hat{x}_2^N(t_N)) : N \in \mathbb{N}\}$ is tight and their limit points are non-degenerate and random since they have finite non-zero mean and variance. If we can identify the limits in (8.813), then we can obtain (8.718) and (8.719) via (8.814).

The proof proceeds in steps, using duality representation of general moments in order to obtain first, second and then third moments of $\hat{x}_2^N(t_N)$. It follows from Lemma 8.22 that we must consider *correction terms* of order $\frac{1}{N^k}$ in the calculation of kth moments of $\hat{x}_1^N(t)$ in order to later derive the asymptotics of the moments of $\hat{x}_2^N$ using the representation of formula (8.722).

Recall that the basic dual relation

$$E[(x_2^N(i,t))^a] = E\left[1 - \exp\left(-\frac{m}{N}\int_0^{t_N} \Pi_s^{N,a} ds\right)\right] \qquad (8.815)$$

where $\Pi_s^{N,a}$ denotes the total number of dual particles if we start with $a \in \mathbb{N}$ particles at site $i \in \{1,\ldots,N\}$. We have obtained a representation of this set of particles in terms of a decomposition into white, red and purple particles embedded in a multi-colour particle system of white, red, purple, blue and green particles (WRGPB-system). We have also coupled this system to a related CMJ process which in particular provides an upper bound, which was the WRGB-system. We now look at this system in more detail in the time interval $[0, t_N]$ where t_N satisfies the hypothesis (8.715). We first recall that the number of occupied sites in the CMJ has the form

$$\Pi_s^{CMJ,a} = W_a(s)e^{\alpha s}, \quad \lim_{s\to\infty} W_a^s = W_a, \text{ a.s.} \qquad (8.816)$$

Moreover the set of white and red particles in given by removing a set of green and blue particles from the CMJ process and we can bound the number of purple particles by the set of blue particles. Therefore we can represent the dual population by removing the green and blue particles with an error given by the difference between the blue and purple particles. Note that the blue and purple particles result from the event of *2 or more collisions* and therefore the error term can be obtained by determining this probability. Moreover we have seen that the collision process is given by a Poisson process with intensity bounded by $\alpha(s)W_a(s)e^{\alpha s}\frac{W_a(s)e^{\alpha s}}{N}$. Therefore the probability of a collision in $[0, t_N]$ is $O(\frac{e^{2\alpha t_N}}{N})$ and this goes to 0 as $N \to \infty$ as $N \to \infty$ under hypothesis (8.715). Let # denote the number of collisions in $[0, t_N]$. Moreover the probability of the event two or more collisions is

$$P(\# \geq 2) \leq Z_1\left(\frac{e^{4\alpha t_N}}{N^2}\right), \qquad (8.817)$$

where Z_1 is a random variable having finite moments. Conditioned on the presence of a collision at time $s \in [0, t_N]$ the resulting set of green particles is given by $W_{\text{tot}}^g(t_N, s)e^{\alpha(t_N-s)}$ where $W_{\text{tot}}^g(t_N, s)$ is a nonnegative random variable with all moments and bounded by an independent copy of the CMJ process.

Then using Taylor's formula with remainder we obtain

$$E\left[1 - \exp\left(-\frac{m}{N}\int_0^{t_N} \Pi^{N,a}(s)ds\right) - \left(\frac{m}{N}\left[\int_0^{t_N} \Pi^{N,a}(s)ds\right]\right)\right] \quad (8.818)$$

$$= -\frac{m^2}{2N^2}E\left[\left(\int_0^{t_N} \Pi^{N,a}(s)ds\right)^2\right] + E[\mathcal{E}_2(N)]$$

$$\mathcal{E}_2(N) \leq \frac{1}{6}\left(-\frac{m}{N}\left[\int_0^{t_N} \Pi^{CMJ,a}(s)ds\right]\right)^3 \leq \frac{m^3}{6N^3}Z_2 e^{3\alpha t_N}, \quad (8.819)$$

where Z_2 has finite moments. Calculate first the expectation of the occupation integral.

$$E\left[\frac{m}{N}\left[\int_0^{t_N} \Pi^{N,a}(s)ds\right]\right] \quad (8.820)$$

$$= \frac{m}{N}E\left[\int_0^{t_N} \frac{(\alpha(s) + \gamma(s))}{c} W_a(s)e^{\alpha s} ds\right.$$

$$\left. -\frac{1}{N}\int_0^{t_N} \frac{(\alpha(s) + \gamma(s))}{c}(W_a(s))^2 W_{g,\text{tot}}(t_N, s)ds\right]$$

$$+ E[\mathcal{E}_1(N)]$$

$$\mathcal{E}_1(N) \leq Z_1 \frac{e^{4\alpha t_N}}{N^3}.$$

Here the second term comes from integrating over the event $\{\# = 1\}$ and corresponds to a green family produced by a single collision. To estimate the first term on the r.h.s. of (8.818) we calculate:

$$E\left[\frac{m}{N}\left(\int_0^{t_N} \Pi^{N,a}(s)ds\right)^2\right] \quad (8.821)$$

$$= \frac{m^2}{N^2}E\left[\int_0^{t_N} \frac{(\alpha(s) + \gamma(s))}{c} W_a(s)e^{\alpha s} ds\right]^2 + E(\mathcal{E}_3(N))$$

$$\mathcal{E}_3(N) \leq Z_3 \frac{e^{3\alpha t_N}}{N^3}.$$

8.4 Droplet Formation: Proofs of Proposition 2.8–2.10

We are now ready to obtain study the first and second moments of $\hat{x}_2(t_N)$. Consider a sequence $s_N \to \infty$ $0 \le t_N - s_N \to \infty$. Then using the basic properties of the CMJ process (8.144) and (8.165) we have $W_a(s) \to W_a, \alpha(s) \to \alpha$ and $\gamma(s) \to \gamma$ (a.s. and in L^1) as $s \to \infty$. Using this we obtain

$$\lim_{N \to \infty} N e^{-\alpha t_N} E\left[\int_0^{t_N} \Pi_s^{N,a} ds\right] \tag{8.822}$$

$$= \lim_{N \to \infty} N e^{-\alpha t_N} E\left[\int_0^{t_N} \frac{\alpha(s) + \gamma(s)}{c} W_a(s) e^{\alpha s} ds\right]$$

$$= \lim_{N \to \infty} N e^{-\alpha t_N} \left(E\left[\int_0^{s_N} \frac{\alpha(s) + \gamma(s)}{c} W_a(s) e^{\alpha s} ds\right] \right.$$

$$\left. + E\left[\int_{s_N}^{t_N} \frac{\alpha(s) + \gamma(s)}{c} W_a(s) e^{\alpha s} ds\right]\right)$$

$$= \frac{(\alpha + \gamma)}{c} W_a.$$

Therefore

$$\lim_{N \to \infty} N e^{-\alpha t_N}\left(1 - E[\exp(-\frac{m}{N}\int_0^{t_N} \Pi_s^{N,a} ds)]\right) \tag{8.823}$$

$$= \lim_{N \to \infty} N e^{-\alpha t_N} E\left[m \int_0^{t_N} \Pi_s^{N,a} ds\right]$$

$$= m^* E[W_a] \tag{8.824}$$

where $m^* = \frac{\alpha + \gamma}{\alpha c}$.

In the same way we obtain

$$\lim_{N \to \infty} N^2 e^{-2\alpha t_N}\left(1 - E\left[\exp\left(-\frac{m}{N}\int_0^{t_N} \Pi_s^{N,a} ds\right) - \frac{m}{N}\int_0^{t_N} \Pi_s^{N,a} ds\right]\right) \tag{8.825}$$

$$= \kappa_2 E[(W_a)^2] + \frac{(m^*)^2}{2}(E[W_a])^2.$$

Step 1   First moment of $\hat{x}_2(t_N)$

Note that by exchangeability

$$E[\hat{x}_1^N(t)] = NE[x_1^N(t)]. \tag{8.826}$$

Then

$$E[\hat{x}_2^N(t)] = N - NE[x_1^N(t)]. \tag{8.827}$$

Using the above calculations we have

$$E[\hat{x}_2^N(t_N)] = NE[x_2^N(t_N)] = NE\left[1 - \exp\left(-\frac{m}{N}\int_0^{t_N} \Pi_s^{N,a}\,ds\right)\right]. \tag{8.828}$$

Therefore using (8.823), we obtain

$$\lim_{N\to\infty} E[e^{-\alpha t_N} x_2^N(t_N)] = m^* E[W_a]. \tag{8.829}$$

Step 2   Second moment of $\hat{x}_2(t_N)$

We will focus next on the calculation of the second moment. We follow the method used above but leave out some details.

To compute the second moment of $\hat{x}_2^N(t)$ note that

$$\begin{aligned}E[(\hat{x}_2^N(t))^2] &= N^2 - 2NE[\hat{x}_1^N(t)] + E[(\hat{x}_1^N(t))^2] \\ &= N^2 - 2N^2 E[x_1^N(i,t)] + E[(\hat{x}_1^N(t))^2]\end{aligned} \tag{8.830}$$

and with $i \neq j$

$$E[(\hat{x}_1^N(t))^2] = NE[(x_1^N(i,t))^2] + N(N-1)E[x_1^N(i,t)x_1^N(j,t)]. \tag{8.831}$$

Since the formula contains $O(N^2)$ factors, in order to compute the r.h.s. up to order 1, as $N \to \infty$, it is necessary to include corrections of order $\frac{1}{N^2}$ in both $E[x_1^N(i,t)]$ and $E[x_1^N(i,t)x_1^N(j,t)]$, etc. The first moment expansion we have done in Step 1. This means that we have to consider corrections due to possible *intersections* of different clouds (clouds meaning the descendants of one of the initial individuals.)

Consider two clouds starting with one factor at different sites in the dual population which occupy the following number of sites respectively

$$W_{1,N} e^{\alpha t_N} \text{ and } W_{2,N} e^{\alpha t_N}, \text{ as } N \to \infty, \tag{8.832}$$

8.4 Droplet Formation: Proofs of Proposition 2.8–2.10

with t_N as in (8.715) and where $W_{i,N} = W_i(t_N) \to W_i$ as $N \to \infty$ for $i = 1, 2$. Then the rate at which the first cloud produces particles that migrate onto one of the sites occupied by the second cloud and symmetrically is as $N \to \infty$:

$$\frac{2(\alpha+\gamma)^2}{c} \frac{W_{1,N} W_{2,N} e^{2\alpha t_N}}{N} + o\left(\frac{e^{2\alpha t_N}}{N}\right) = \frac{2(\alpha+\gamma)^2}{c} \frac{W_1 W_2 e^{2\alpha t_N}}{N} + o\left(\frac{e^{2\alpha t_N}}{N}\right). \tag{8.833}$$

If the two clouds have an intersection, coalescence can reduce the number of particles compared to $(W_1 + W_2)(\alpha + \gamma)e^{\alpha t_N}$ and this produces a *correction term* corresponding to the $G_{1,2}$ particles. We have analysed this correction term in all detail (in the proof of Proposition 8.24) using a coloured particle system but here using that construction and its consequences, recall (8.745)–(8.747), we get a result on the behaviour of the number of dual particles and can perform moment calculations. We use these results to calculate covariances for $i \neq j$, namely, we use the dual and expand the time integral using (8.749) with $s = 0$ to represent $\Pi_u^{N,(1,2)}$ for $u \in [0, t_N]$. We estimate $E[\int_0^{t_N} \mathfrak{g}_{\text{tot}}(W^{1,2}(\cdot), N, u) du]$ by $\frac{\kappa_2}{2c\alpha} E[(W^{1,2})^2] \left(\frac{e^{2\alpha t_N} - 2e^{\alpha t_N}}{N^2}\right)$ and (8.747) to bound the error term.

If t_N satisfies the above growth assumption then as $N \to \infty$ we can expand the exponential and again verify that we can take expectation in this expansion, as in (8.829). (Recall Lemma 8.26 for the justification of these expansions.) Then using again (8.815) we calculate as follows.

$$E[x_1^N(i, t_N) x_1^N(j, t_N)]$$
$$= E\left[e^{-\frac{m^*}{N}\left[(W_1 + W_2)(\alpha+\gamma)(e^{\alpha t_N} - 1) - \int_0^{t_N} \mathfrak{g}_{\text{tot}}((W_1+W_2)(\cdot), N, u) - e^{\alpha t_N} du + o(N^{-1})\right]}\right]$$
$$= 1 - \frac{m^*}{N} E[(W_1 + W_2)](\alpha + \gamma)(e^{\alpha t_N} - 1) + \frac{m\kappa_2}{2c\alpha} E[(W^{1,2})^2] \frac{e^{2\alpha t_N}}{N^2}$$
$$+ \frac{(m^*)^2}{2N^2} (E[W_1 + W_2])^2 e^{2\alpha t_N} + o\left(\frac{e^{2\alpha t_N}}{N^2}\right). \tag{8.834}$$

We have as $N \to \infty$ for $i \neq j$:

$$E[x_1^N(i, t_N) x_1^N(j, t_N)] \tag{8.835}$$
$$\sim 1 - \frac{m^*}{N} E[W_1 + W_2] e^{\alpha t_N}$$
$$+ \frac{m\kappa_2}{2c\alpha} E[(W^{1,2})^2] \frac{e^{2\alpha t_N}}{N^2} + \frac{(m^*)^2}{2N^2} (E[W_1 + W_2])^2 e^{2\alpha t_N} + o\left(\frac{e^{2\alpha t_N}}{N^2}\right),$$

where $W^{2,1}$ is the growth constant if we start with *two* particles at *one* site. Similarly we get

$$E[(x_1^N(i,t_N))^2] \tag{8.836}$$

$$\sim 1 - \frac{m^*}{N} E[W^{2,1}] e^{\alpha t_N}$$

$$+ \frac{m\kappa_2}{2c\alpha} E[(W^{2,1})^2] \frac{e^{2\alpha t_N}}{N^2} + \frac{(m^*)^2}{2N^2} (E[W^{2,1}])^2 e^{2\alpha t_N} + o\left(\frac{e^{2\alpha t_N}}{N^2}\right).$$

Furthermore note that

$$\text{Var}\left(\hat{x}_2^N(i,t_N)\right) = \sum_{i=1}^N \text{Var}(x_1^N(i,t_N)) + \sum_{i \neq j}^N \text{Cov}(x_1^N(i,t_N), x_1^N(j,t_N)). \tag{8.837}$$

Then by (8.831) combined with (8.815) and (8.834) the total mass of type 2 satisfies for $N \to \infty$ by expanding the exponential (below $i \neq j$):

$$\begin{aligned}
E[(\hat{x}_2^N(t_N))^2] &= N^2 - 2NE[\hat{x}_1^N(t_N)] + E[(\hat{x}_1^N(t_N))^2] \\
&= N^2 - 2N^2 E[x_1^N(i,t)] + NE[(x_1^N(i,t_N))^2] + N(N-1) \\
&\quad E[x_1^N(i,t_N)x_1^N(j,t_N)] \\
&= -N^2 + 2Nm^* E[W_1] e^{\alpha t_N} + N(1 - \frac{m^*}{N} E[W^{(2,1)}] e^{\alpha t_N}) \\
&\quad - m^2 E[(W_1)^2] e^{2\alpha t_N} \\
&\quad + N(N-1)(1 - \frac{m^*}{N} 2E[W_1] e^{\alpha t_N} + \frac{(E[(W_1+W_2)^2]\kappa_2 m^* e^{2\alpha t_N})}{N^2}) \\
&\quad + \frac{1}{2} N(N-1) \frac{m^{*2}}{N^2} E[(W_1+W_2)^2] e^{2\alpha t_N}.
\end{aligned} \tag{8.838}$$

Therefore we get

$$(\text{r.h.s. (8.838)}) \tag{8.839}$$
$$= 2E[W_1 W_2] \kappa_2 m^* e^{2\alpha t_N} - m^* E[W^{(2,1)}] e^{\alpha t_N}$$
$$+ m^{*2} (E[W_1])^2 e^{2\alpha t_N} + o(1), \text{ as } N \to \infty.$$

This implies

$$\lim_{N \to \infty} (e^{-2\alpha t_N} E[(\hat{x}_2^N(t_N))^2]) = (2\kappa_2 m^* + m^{*2})(E[W_1])^2, \tag{8.840}$$

$$\lim_{N \to \infty} (e^{-2\alpha t_N} Var[(\hat{x}_2^N(t_N))]) = 2\kappa_2 m^* (E[W_1])^2. \tag{8.841}$$

We now consider the case with two times t_N and $t_N + t$. Following the same steps as above we can show that

$$\lim_{N \to \infty} (e^{-2\alpha t_N} E[(\hat{x}_2^N(t_N))\hat{x}_2^N(t_N+t)]) = (2\kappa_2 m^* + m^{*2})(E[W_1])^2 e^{\alpha t}. \tag{8.842}$$

Step 3 (Third moments)

We can proceed again as before, expand the $x_2^N(t)$ in terms of $x_1^N(t)$ and apply the dual representation again. Here however we only need an upper bound and therefore we can bound the total number of dual particles from above by the ones of the collision-free process. Then the claim follows again easily by using (8.148).

This completes the proof of Proposition 8.21.

8.4.6 Asymptotically Deterministic Droplet Growth

We now return to the analysis of the beginning of the evolution, where the droplet growth starts. Over time periods which are large but remain much smaller than $\frac{\log N}{\alpha}$ as $N \to \infty$ the total mass of type 2 is expected to grow deterministically as an exponential with rate α and constant (random) factor. The random factor developed over a short time period at the very beginning of the evolution. This leads to the idea to focus here in a first step on the case where after some time t_0 *we turn off the mutation*. The point is that the mutation which enters later than some large time t_0 is not relevant for the growth of the droplet anymore if we choose t_0 sufficiently large which we show in the second step.

In this section we therefore compute the moments of $\hat{x}_2^N(t)$ conditioned on the configuration at a fixed time t_0, that is, given a specific realization of the random configuration $x_2^N(i, t_0)$, $i = 1, \ldots, N$, then identify the asymptotics as first $N \to \infty$ and then $t \to \infty$. In the Corollary 8.28 and Corollary 8.30 and then later also in the next subsubsection in Proposition 8.29 we will then consider the joint limit of $N \to \infty, t \to \infty$, for the quantity $e^{-\alpha t} \hat{x}_2^N(t)$, more precisely $e^{-\alpha t_N} \hat{x}_2^N(t_N)$ for $N \to \infty$ with $t_N \uparrow \infty$ and $\limsup_{N \to \infty}(t_N / \log N) < \alpha$. Those quantities converge in probability to a random limit.

In order to mimic the situation, we consider a process, where the rare mutation is turned of after time t_0 for times $t \geq 0$, we consider a process starting in finite mass and $m = 0$ for all times.

Proposition 8.27 (Deterministic regime of droplet growth). *Assume that at some time $t_0 \geq 0$ the distributions of $\{\tilde{x}_2^N(t_0, i) : i = 1, \ldots, N\}$ for the system (2.4), (2.5) are converging to the one of $\{\tilde{x}_2(t_0, i), i \in \mathbb{N}\}$ with $\sum_i \tilde{x}_2(t_0, i) < \infty$ a.s.*

Moreover we assume that for every N and for the limit configuration we have that $\forall \varepsilon > 0$ there is a finite random set (we suppress the dependence on N in the notation):

$$\mathcal{I}(\varepsilon) = \mathcal{I}(\varepsilon, N) \text{ of } k = k(\varepsilon, N) \text{ sites}, \tag{8.843}$$

such that for every $N \geq 2$:

$$\sum_{i \in (\mathcal{I}(\varepsilon))^c} \tilde{x}_2^N(t_0, i) < \varepsilon. \tag{8.844}$$

Assume

$$m = 0 \quad \text{for } t \geq t_0. \tag{8.845}$$

We set (recall (8.287) for $\mathcal{U}(t,\cdot)$ and (8.284), (8.163) for $\mathcal{U}(\infty,\cdot)$):

$$g(t,x) = (1 - \sum_t^\infty (1-x)^\ell \mathcal{U}(t,\ell)), \quad g(\infty,x) = (1 - \sum_{\ell=1}^\infty (1-x)^\ell \mathcal{U}(\infty,\ell)). \tag{8.846}$$

Then the following four convergence properties hold:

(a)

$$\lim_{t \to \infty} e^{-\alpha t} E[\mathbf{J}_t^{m,t_0}([0,1])] = \lim_{t \to \infty} e^{-\alpha t} \lim_{N \to \infty} E[\hat{x}_2^N(t)] \tag{8.847}$$

$$= E[W] \sum_{i=1}^\infty g(\infty, \tilde{x}_2(t_0, i)),$$

(b)

$$\lim_{t \to \infty} \lim_{N \to \infty} (E[e^{-\alpha(t+s)} \hat{x}_2^N(t+s)] - E[e^{-\alpha t} \hat{x}_2^N(t)]) = 0, \quad \forall s \in \mathbb{R}, \tag{8.848}$$

(c)

$$\lim_{t \to \infty} \lim_{N \to \infty} e^{-\alpha(s+2t)} \mathrm{Cov}(\hat{x}_2^N(t+s), \hat{x}_2^N(t))$$

$$= (E[W])^2 \left[2\kappa_2 \sum_{i=1}^k g(\infty, \tilde{x}(t_0, i)) \right], \quad \forall s \in \mathbb{R}, \tag{8.849}$$

and

(d)

$$\lim_{t \to \infty} \lim_{N \to \infty} E\left[([e^{-\alpha(t+s)} \hat{x}_2^N(t+s)] - e^{-\alpha t} E[\hat{x}_2^N(t)])^2 \right] = 0, \quad \forall s \geq 0. \tag{8.850}$$

(e) *Now assume that $m > 0$ for all $t \geq t_0$ (recall (8.845)). Then the conclusions of (a)–(d) remain valid.* □

We next verify that we can compute the simultaneous limits $N \to \infty, t \to \infty$ for times $t_N(\beta, t) = (\beta/\alpha) \log N + t, 0 < \beta < 1$.

Corollary 8.28 (Droplets of size N^β grow deterministically in β). *Consider again the case where $m = 0$ and the initial state is as in Proposition 8.27. Choose β with $0 < \beta < 1$ and let*

8.4 Droplet Formation: Proofs of Proposition 2.8–2.10

$$t_N(\beta, t) = \frac{\beta(\log N)}{\alpha} + t. \tag{8.851}$$

We have

$$\lim_{t \to \infty} \lim_{N \to \infty} E\left[\left(e^{-\alpha t} \hat{x}_2^N(t) - e^{-\alpha t_N(\beta,0)} \hat{x}_2^N(t_N(\beta,0))\right)^2\right] = 0. \quad \square \tag{8.852}$$

This result shows that the randomness in the droplet growth is generated in the very beginning at times of order $O(1)$ as $N \to \infty$. In particular we can then show that, conditioned on the total mass of type 2 at time $t_N(\beta_1, 0)$, the total mass at a later time $t_N(\beta_2, 0)$, $1 > \beta_2 > \beta_1 \geq 0$ is *deterministic* in the limit $N \to \infty$.

Remark 101. The limiting total droplet mass dynamic as $N \to \infty$ becomes deterministic in the following sense. Assume that $\{x_2^N(i, t_0), i = 1, \ldots, N\}$ is a measure with total mass N^β and supported on a set of size aN^β for some $a \in (0, \infty)$ and $\beta \in (0, 1)$. Then we can show that for $t > t_0$ we have:

$$\frac{Var[\hat{x}_2^N(t)]}{(E[\hat{x}_2^N(t)])^2} \to 0 \tag{8.853}$$

as $N \to \infty$.

Remark 102. Given a fixed initial measure $\{\tilde{x}(t_0, i), i \in \mathbb{N}\}$ as in the statement of Proposition 8.27 we can also prove that as $t \to \infty$

$$\lim_{N \to \infty} E[\sum_i (x_2^N(i,t))^2] = O(\lim_{N \to \infty} E[\sum_i (x_2^N(i,t))]). \tag{8.854}$$

This is analogous to the Palm picture (8.636) and implies that the mass of type 2 is clumped on a set of sites on which the mass is almost one.

Proof of Proposition 8.27. The proof proceeds by using the dual representation of moments and the multicolour particle system introduced in Subsubsection 8.4.5 to analyse the expressions.

(a) We determine the contribution to the first moment coming only from the initial mass in $\mathcal{I}(\varepsilon)$. The result then follows by letting $\varepsilon \to 0$.

Without loss of generality we use (recall $k = k(\varepsilon, N), \mathcal{I}(\varepsilon) = \mathcal{I}(\varepsilon, N)$)

$$\mathcal{I}(\varepsilon) = \{1, \cdots, k\}. \tag{8.855}$$

Note that since we are now not working with a spatially homogeneous initial measure, we must modify the dual. In particular we eliminate the rule that single particles at a site do not move. In the calculation of $E[\hat{x}_2^N(t)]$ we compute

$$E[\hat{x}_2^N(t)] = \sum_{j=1}^{N} E[x_2^N(j,t)]. \tag{8.856}$$

To compute $E[x_1^N(j,t)]$ we start the dual with one particle at site $j \in \{1,\ldots,N\}$. However the number of particles at this site will return to 1 after at most a finite random time and therefore after some finite random time the single particle will jump to a randomly chosen site and site j will become empty. Therefore for large times t the value of $E[x_1^N(j,t)]$ is independent of j (even if it is in $\mathcal{I}(\varepsilon) = \{1,\ldots,k\}$).

We calculate the expectation of type 1 in a site j using the dual starting with one particle at site j.

$$E[x_1^N(t,j)] = E\left[\prod_{i=1}^{K_t^N} (1 - \tilde{x}_2^N(t_0, i))^{\zeta^N(t,i)}\right]. \tag{8.857}$$

We first consider the initial configuration where we have

$$\tilde{x}_2^N(t_0, i) = 0 \text{ for every } i \notin \mathcal{I}(\varepsilon). \tag{8.858}$$

Recall that here $U^N(t, \ell)$ is the frequency of occupied sites that have ℓ particles.

Hence noting that there will be a $Bin(K_t^N, \frac{k}{N})$ number of hits of the k-set $\mathcal{I}(\varepsilon)$ by the dual cloud and the frequency of sites in the cloud with ℓ-sites is given by $U^N(t, \ell)$ and the contribution of such a site by $(1 - \tilde{x}_2^N(t_0, i))^\ell$, we get for sites $j \notin \{1, \cdots, k\}$ that:

$$E[x_1^N(t,j)] = E\Bigg[(1 - \frac{k}{N})^{K_t^N}$$

$$+ (1 - \frac{k}{N})^{K_t^N - 1} \frac{K_t^N}{N} \sum_{i=1}^{k} \sum_{\ell=1}^{\infty} U^N(t, \ell)(1 - \tilde{x}_2^N(t_0, i))^\ell$$

$$+ (1 - \frac{k}{N})^{K_t^N - 2} \frac{K_t^N(K_t^N - 1)}{2N^2}$$

$$\times \sum_{i_1, i_2 = 1}^{k} \sum_{\ell_1, \ell_2 = 1}^{\infty} U^N(t, \ell_1) U^N(t, \ell_2)(1 - \tilde{x}_2^N(t_0, i_1))^{\ell_1}$$

$$(1 - \tilde{x}_2^N(t_0, i_2))^{\ell_2}\Bigg]$$

$$+ o(N^{-2}). \tag{8.859}$$

8.4 Droplet Formation: Proofs of Proposition 2.8–2.10

Note that as $N \to \infty$, $U^N(t, \ell) \to U(t, \ell)$, with $U(t, \cdot)$ the size distribution of the CMJ process at time t.

We rewrite the formula (8.859) as follows. Let

$$g^N(t,x) = 1 - \sum_{\ell=1}^{\infty} (1-x)^\ell U^N(t, \ell) \tag{8.860}$$

and

$$g^{N,2}(t,x) = 1 - \sum_{\ell=1}^{\infty} (1-x)^\ell (U^N(t)*U^N(t))(\ell). \tag{8.861}$$

Note that $g^N, g^{N,2}$ do depend as well on the initial state of the dual process via $U^N = U^{N,(1,1)}$, this dependence we suppress here in the notation.

Then:

$$\tag{8.862}$$

$$E[x_1^N(t, j)]$$

$$= E\Bigg[1 - \frac{K_t^N}{N} \sum_{i=1}^{k} g^N(t, \tilde{x}_2^N(t_0, i))$$

$$+ \frac{K_t^N(K_t^N - 1)}{2N^2} \Bigg\{\sum\sum_{i_1 \neq i_2}^{k} g^N(t, \tilde{x}_2^N(t_0, i_1))g^N(t, \tilde{x}_2^N(t_0, i_2))$$

$$+ \sum_{i=1}^{k} g^{N,2}(t, \tilde{x}_2^N(t_0, i))\Bigg\}\Bigg] + o(N^{-2}).$$

Note that because of the convergence of the dual particle system to the collision-free McKean–Vlasov dual we have:

$$g^N(t,x) \to 1 - \sum_{\ell=1}^{\infty} (1-x)^\ell U(t, \ell) \text{ as } N \to \infty, \tag{8.863}$$

$$g^{N,2}(t,x) \longrightarrow 1 - \sum_{\ell=1}^{\infty} (1-x)^\ell U^{*2}(t, \ell) \text{ as } N \to \infty. \tag{8.864}$$

Therefore we can conclude from equation (8.862) that for $i \in \{1, 2, \cdots, k\}$:

$$E[x_2^N(t, i)] = E\Bigg[\frac{K_t^N}{N} \sum_{i=1}^{k} g^N(t, \tilde{x}_2^N(t_0, i))\Bigg] + o(N^{-1}). \tag{8.865}$$

Consequently we calculate:

$$E[\hat{x}_2^N(t)] = E\left[K_t^N \sum_{i=1}^{k} g^N(t, \tilde{x}_2^N(t_0, i))\right] + o(1). \qquad (8.866)$$

Then by $K_t^N \to K_t$ as $N \to \infty$, (8.863) and dominated convergence we get:

$$\lim_{N \to \infty} E[\hat{x}_2^N(t)] = E\left[K_t \sum_{i=1}^{k} g(t, \tilde{x}_2(t_0, i))\right] \qquad (8.867)$$

where we define $g(t, x)$ using (8.863) by

$$g(t, x) = 1 - \sum_{\ell=1}^{\infty}(1 - x)^{\ell} \mathcal{U}(t, \ell), \qquad (8.868)$$

together with the fact that $K_t, \mathcal{U}(t, \cdot)$ can be identified as the number of occupied sites and the size distribution for the McKean–Vlasov limit dual process obtained above in Subsubsection 8.3.4 (Step 2).

Finally putting everything together we have:

$$\lim_{t \to \infty} (e^{-\alpha t} \lim_{N \to \infty} E[\hat{x}_2^N(t)]) = E[W] \sum_{i=1}^{k} g(\infty, \tilde{x}(t_0, i)). \qquad (8.869)$$

Finally we have to remove the restriction on the initial state being 0 outside $\mathcal{I}(\varepsilon, N)$. We note that the system where $\tilde{x}^N(0)$ is different from zero only at $\mathcal{I}(\varepsilon)$ and the true system differ at time t at most by $Const \cdot \varepsilon e^{\alpha t}$ and hence we get letting $\varepsilon \to 0$ (in $\mathcal{I}(\varepsilon)$) finally the claim of part (a).

(b) The claim on the first moment is immediate from (a).
(c) We now consider the second moment calculation again using the dual particle system. As before in the proof of part (a) we consider first the system arising by setting the type 2 mass equal to zero outside the set $\mathcal{I}(\varepsilon)$ and then later we shall let $\varepsilon \to 0$ to obtain our claim for the general case with the very same argument as in part (a).

First note that

$$E[\hat{x}_2^N(t+s)\hat{x}_2^N(t)] = E[\hat{x}_2^N(t+s)]E[\hat{x}_2^N(t)] + \text{cov}(\hat{x}_2^N(t+s), \hat{x}_2^N(t)) \qquad (8.870)$$

$$= E[\hat{x}_2^N(t+s)]E[\hat{x}_2^N(t)] + \sum_{i=1}^{N} \text{Cov}(x_1^N(t+s, i), (x_1^N(t, i)))$$

$$+ \sum_{j \neq \ell}^{N} \text{Cov}(x_1^N(t+s, j)), x_1^N(t, \ell)). \qquad \text{q.e.d.}$$

8.4 Droplet Formation: Proofs of Proposition 2.8–2.10

Since the last term of (8.870) has $N(N-1)$ summands it is necessary to consider terms of order $O(\frac{1}{N^2})$ in the computation of $\text{Cov}(x_1^N(t+s, j)), x_1^N(t, \ell))$. Therefore the calculation is organized to identify the terms up to those of order $o(N^2)$ as $N \to \infty$, and then to calculate the contributing terms of $O(N^{-2})$.

To handle the calculation of $\text{Cov}(x_1^N(t+s, j), x_1^N(t, \ell))$ both for $j \neq \ell$ and for $j = \ell = i$ we use the dual process. We introduce as "initial condition" for the dual particle system:

one initial dual particle at j_1 at time 0 and a second dual particle at j_2 at time s
(8.871)

and then consider the dual cloud resulting from these at time $t + s$.

Since we shall fix times 0 and s where the j_1, respectively j_2, cloud start evolving we abbreviate the number of sites occupied by the cloud at time u and similarly the process of the frequency distribution of sizes of sites by

$$K_u^{N, j_1, j_2} = K_u^{N, (j_1, 0), (j_2, s)} \text{ and } U^{N, j_1, j_2} = U^{N, (j_1, 0), (j_2, s)}. \tag{8.872}$$

Precisely this runs as follows. There will be a $Bin(K_t^{N, j_1, j_2}, \frac{k}{N})$ number of hits of the k-set $\mathcal{I}(\varepsilon)$, hence as $N \to \infty$ we obtain:

(8.873)

$$E[x_1^N(t+s, j_1) \cdot x_1^N(t, j_2)]$$

$$= E[\prod_{\ell=1}^{K_{t+s}^{N, j_1, j_2}} (1 - \tilde{x}_2^N(t_0, \ell))^{\zeta_\ell^{(t+s)}}]$$

$$= E\Bigg[(1 - \frac{k}{N})^{K_{t+s}^{N, j_1, j_2}} + \frac{K_{t+s}^{N, j_1, j_2}}{N}(1 - \frac{k}{N})^{K_{t+s}^{N, j_1, j_2} - 1}$$

$$\sum_{i=1}^{k} \sum_{\ell=1}^{\infty} U_{t+s}^{N, j_1, j_2}(\ell)(1 - \tilde{x}_2^N(t_0, i))^\ell$$

$$+ (1 - \frac{k}{N})^{K_{t+s}^{N, j_1, j_2} - 2} \frac{K_{t+s}^{N, j_1, j_2}(K_t^{N, j_1, j_2} - 1)}{2N^2}$$

$$\sum_{i_1, i_2 = 1}^{k} \sum_{\ell_1, \ell_2 = 1}^{\infty} U^{N, j_1, j_2}(t+s, \ell_1) U^{N, j_1, j_2}(t+s, \ell_2)$$

$$(1 - \tilde{x}_2^N(t_0, i_1))^{\ell_1}(1 - \tilde{x}_2^N(t_0, i_2))^{\ell_2}\Bigg] + o(N^{-2}).$$

Therefore if we now use the function $g^N(t,x)$ defined in (8.860) but now for the initial state specified by the space-time points $(j_1,0), (j_2,s)$ (note $g^N = g^{N,(j_1,0)}$, respectively $= g^{N,(j_2,0)}$), we can rewrite the above expression. However there is a small problem with the time index, since our two clouds now have a time delay s therefore depending on which of the two clouds has the hit of $\mathcal{I}(\varepsilon)$ we have to use t or $t+s$ as the argument in our function $g^N(\cdot, x)$. Here we use therefore the *notation* $t(\cdot)$ which indicates that actually the time parameter here is either t or $t+s$ and depends on which of the two clouds is involved. However this abuse of notation is harmless since in the arguments below both the times t and $t+s$ will go to ∞.

We get, with the above conventions, from the formula (8.873) that:

$$E[x_1^N(t+s, j_1) x_1^N(t, j_2)] \tag{8.874}$$

$$= 1 - E\left[\frac{K_{t+s}^{N,j_1,j_2}}{N} \sum_{i=1}^{k} g^N(t(\cdot), \tilde{x}_2^N(t_0, i))\right]$$

$$+ E\left[\frac{K_{t+s}^{N,j_1,j_2}(K_{t+s}^{N,j_1,j_2} - 1)}{2N^2} \sum_{i_1 \neq i_2=1}^{k} g^N(t(\cdot), \tilde{x}_2^N(t_0, i_1)) g^N(t(\cdot), \tilde{x}_2^N(t_0, i_2))\right]$$

$$+ o(N^{-2}).$$

Now we continue and calculate the covariances of $x_1^N(t+s, j_1)$ and $x^N(t, j_2)$ based on (8.874) and the first moment calculations. For this purpose we can consider the dual systems starting in $(j_1, 0)$ respectively (j_2, s) and evolving *independently* and the one starting with a factor at each time-space point $(j_1, 0)$ and (j_2, s) but evolving together and with *interaction*. As we saw we can couple all these evolutions via a multicolour particles dynamic. This now allows to calculate with the first moment formulas obtained in (8.857)–(8.868) and using the notation

$$s_m = 0 \text{ if } m = 1 \text{ and } s_m = s \text{ if } m = 2, \tag{8.875}$$

and we obtain the covariance formula

$$\mathrm{Cov}(x_1^N(t+s, j_1) x_1^N(t, j_2)) \tag{8.876}$$

$$= 1 - E\left[\frac{K_{t+s}^{N,j_1,j_2}}{N} \sum_{i=1}^{k} g^N(t(\cdot), \tilde{x}_2^N(0, i))\right]$$

$$+ E\left[\frac{K_{t+s}^{N,j_1,j_2}(K_{t+s}^{N,j_1,j_2} - 1)}{2N^2} \sum_{i_1 \neq i_2=1}^{k} g^N(t(\cdot), \tilde{x}_2^N(t_0, i_1)) g^N(t(\cdot), \tilde{x}_2^N(0, i_2))\right]$$

$$- \prod_{m=1}^{2} E\left[1 - \frac{K^{N,j_m}(t+s_m)}{N} \sum_{i_m=1}^{k} g^N(t+s_m, \tilde{x}_2^N(t_0, i_m))\right]$$

8.4 Droplet Formation: Proofs of Proposition 2.8–2.10

$$+ \frac{K_{t+s_m}^{N,j_m}(K_{t+s_m}^{N,j_m}-1)}{2N^2} \sum_{i_{m,1}=1}^{k} \sum_{i_{m,2}=1}^{k} g^N(t+s_m, x_{i_{m,1}}) g(t+s_m, x_{i_{m,2}}) \Bigg]$$

$$+ o(N^{-2}).$$

We now have to distinguish two cases, namely $j_1 \neq j_2$ and $j_1 = j_2$ and we argue separately in the two cases.

From (8.876), using (8.865) and Proposition 8.24, we get that if $j_1 \neq j_2$, then

$$\mathrm{Cov}(x_2^N(t+s, j_1), x_2^N(t, j_2)) \tag{8.877}$$

$$= \frac{1}{N^2} \left\{ E[N(K_{t+s}^{N,j_1} + K_t^{N,j_2} - K_{t+s,t}^{N,j_1,j_2}) \sum_{i=1}^{\cdot k} g^N(t(\cdot), \tilde{x}_2^N(t_0, i))] \right\} + o(N^{-2}).$$

The expression for $j_1 = j_2$ is a slight modification of this, since either we have as first step a coalescence and are left with one particle or we get first a migration event and get a $j_1 \neq j_2$ case, or finally we get a birth event first. Since we have the time delay s of course coalescence can act only with this delay.

Therefore combining both cases up to terms of order $o(\frac{1}{N^2})$ as $N \to \infty$,

$$\mathrm{Cov}(\hat{x}_2^N(t+s), \hat{x}_2^N(t)) \tag{8.878}$$

$$= \left\{ E[N(K_{t+s}^{N,j_1} + K_t^{N,j_2} - K_{t+s,t}^{N,j_1,j_2}) \sum_{i=1}^{k} g^N(t, \tilde{x}_2^N(t_0, i))] \right\} + o(1).$$

Recall that $g^N(t, \tilde{x}_2^N(t_0, i_1)) \to g(t, \tilde{x}_2(t_0, i_1))$ as $N \to \infty$. Therefore the main point is to identify the behaviour of

$$N(K_{t+s}^{N,j_1} + K_{t+s}^{N,j_2} - K_{t+s}^{N,j_1,j_2}), \text{ as } N \to \infty, \tag{8.879}$$

which is N times the difference between two independent dual populations each starting in j_1 and j_2 and one combined interacting system. In order to evaluate this we must determine the effect of the collisions of the dual populations.

In order to analyse the dual representation of the r.h.s. of equation (8.870) rewritten in the form (8.878) we will again work with the enriched coloured particle system which allows us to identify the contribution, on the one hand of the two evolving independent clouds, and on the other hand the effects of the interaction between the two clouds by collision and subsequent coalescence.

The construction of the coloured particle system (observing $W_1, W_2, R_1, R_2, R_{1,2}$, $G_1, G_2, G_{1,2}$ particles which are relevant for the accuracy needed here) was given above in the sequel of (8.765). Recall that the $G_{1,2}$ particles are produced in the coloured particle system from collision of W_1 and W_2 particles, that is, W_1 and W_2 particles at the same site which then suffer coalescence. Such an event reduces the number of occupied sites in the interacting clouds by 1 and then creates a $G_{1,2}$

family. Then the number of occupied sites in the W_1, W_2 interacting system K_t^{N,j_1,j_2} is given by (8.748).

First we observe that if we evaluate the duality relation (recall the form of the initial state we consider here) we see that the contributing terms of order $O(\frac{1}{N^2})$ arise from either double or single hits of the dual particle cloud of a point in $\mathcal{I}(\varepsilon)$.

We note that in general there are two cases which must be handled differently, namely, the case in which the difference in (8.879) is based on a Poisson collision event with mean of order $O(\frac{1}{N})$ as in Lemma 8.26 (b) and the other in which there is a law of large numbers effect as in(8.485) and Lemma 8.26(a). However in both cases the expected value used in the calculations below have the same form. In the present case for fixed t we are in the Poisson case with vanishing rate so that at most one Poisson event contributes to the limit.

Using Proposition 8.24 (8.748) we se that the key quantity in (8.879) satisfies as $N \to \infty$ that:

$$E[N(K_{t+s}^{N,j_1} + K_t^{N,j_2} - K_{t+s,t}^{N,j_1,j_2})] \sim N \cdot E[\mathfrak{g}_{(1,2)}(W'(\cdot), W''(\cdot), N, s, t)] \quad (8.880)$$

where $\mathfrak{g}_{(1,2)}(W'(\cdot), W''(\cdot), N, s, t)$ is given by (8.741) and has mean given in (8.751). Hence we conclude with (8.751) that the limit of the r.h.s. of (8.880) is given by

$$2\kappa_2(E[W])^2 e^{\alpha(2t+s)}. \quad (8.881)$$

Remark 103. Here we have noted that using the coupling introduced above the number of *single hits* of the k sites in $\mathcal{I}(\varepsilon)$ coming from either one of the two interacting clouds, can be obtained by deducting from the single hits of the "non-interacting W_1 and W_2 particles" the single hits by $G_{1,2}$ particles in the interacting system. Moreover the event of a *double hit*, that two occupied sites in the dual cloud coincide with one of the $k = k(\varepsilon)$ sites, can occur from either two W_1 sites, two W_2 sites or one W_1 and one W_2 site. (Since the number of $G_{1,2}$ sites or including sites at which both W_1 and W_2 particles coexist are of order $O(\frac{1}{N})$, the contribution of double hits involving $G_{1,2}$ sites is of higher order and can be ignored.)

Taking the limit as $N \to \infty$ in (8.878), using (8.881), that leads to

$$\mathrm{Cov}(\hat{x}_2^N(t+s), \hat{x}_2^N(t)) \xrightarrow[N \to \infty]{}$$
$$2\kappa_2 e^{\alpha(2t+s)} E[(W'(t+s)W''(t) \sum_{i=1}^{k} g(t, \tilde{x}_2(t_0, i)]. \quad (8.882)$$

Recalling (8.868), (8.763) and the fact that $\mathcal{U}(t, k) \to \mathcal{U}(\infty, k)$ as $t \to \infty$, we get next as $t \to \infty$:

$$\text{r.h.s. of } (8.882) \sim 2e^{\alpha(2t+s)} E\left[\kappa_2 W'W'' \sum_{i=1}^{k} g(t, \tilde{x}_2(t_0, i))\right] \quad (8.883)$$

8.4 Droplet Formation: Proofs of Proposition 2.8–2.10

and hence

$$\left| e^{-(2t+s)}\mathrm{Cov}(\hat{x}_2^N(t+s), \hat{x}_2^N(t)) - 2(E[W])^2 \left[\kappa_2 \sum_{i=1}^{k} g(\infty, \tilde{x}_2(t_0, i)) \right] \right| \to 0, \tag{8.884}$$

as $t \to \infty$ uniformly in s. To verify the uniformity we work with the expression for the expected number of $G_{1,2}$ sites (in the $N \to \infty$ limit dynamics), recall here (8.759), to get:

$$E\left[\int_s^{t+s} W'(u) W''(u-s) e^{\alpha(u-s)} e^{\alpha u} W^g(t+s, u) e^{\alpha(t+s-u)} du \right]. \tag{8.885}$$

Then using that for the $N \to \infty$ limiting dynamics always $W(u) \to W$ as $u \to \infty$ and showing that the contribution from the early collisions (in $[0, s+s_0(t)]$, $s_0(t) = o(t)$) form a negligible contribution as in the argument following (8.763) we obtain

$$e^{-\alpha(2t+s)} E\left[\int_s^{t+s} W_1(u) W_2(u-s) e^{\alpha(u-s)} e^{\alpha u} W^g(t+s, u) e^{\alpha(t+s-u)} du \right] \tag{8.886}$$

$$-(E[W])^2 \left[\kappa_2 \sum_{i=1}^{k} g(\infty, \tilde{x}_2(t_0, i)) \right] \le \mathrm{const} \cdot e^{-\alpha t} \text{ as } t \to \infty.$$

This proves (8.849) and completes the proof of (c).

(d) follows immediately from (c) and (a).

(e) To verify this note let $\hat{x}_2^{N,m,[t_0,\infty)}(t)$ denote the contribution to the population resulting from rare mutations that occur after t_0. Then

$$E[e^{-\alpha t} \lim_{N \to \infty} \hat{x}_2^{N,m,[t_0,\infty)}(t)] \le \frac{m}{\alpha} e^{-\alpha t_0}. \tag{8.887}$$

Since the above results are valid for any t_0, we can let $t_0 \to \infty$, to conclude that they remain true if we do not turn off the rare mutation at time t_0.

Remark 104. In the above calculations we have included double hits—that is two clouds including the same site and that conditioned on a hit the number of particles at that site is given by the size distribution and therefore converges to the stable size distribution as $t \to \infty$.

Proof of Corollary 8.28. Again we use the dual representation of the moments and analyse the dual based on the enriched coloured particle system. We consider separately two cases $0 < \beta < 1/2$ and $1/2 \le \beta < 1$, which correspond to the distinction whether two clouds of dual particles descending from the two initial particles do not meet as $N \to \infty$ in the $(\beta/\alpha) \log N$ time scale with a nontrivial probability or whether they do next.

The proof in the case $0 < \beta < 1/2$ (where the clouds do not meet asymptotically) follows the same lines as for Proposition 8.27 only that now in (8.835) the limits $N \to \infty$ and $s \to \infty$ are carried out at once. Hence the additional step is to verify that the error terms remain negligible when s is replaced by s_N with

$$s_N = t_N(\beta) - t. \tag{8.888}$$

Now we have to analyse the first and second moments in this time scale.

To prove the claims we first note that for $\beta < 1$ the probability that the particle system hits a given site still goes to 0 as $N \to \infty$. Therefore with (8.866) we still can apply the CMJ-theorem on exponential growth which takes care of the *first moment* term asymptotic.

Moreover, the remainder term in the calculation of the *second order term s*, i.e. of

$$E[e^{-\alpha t} x_1^N(t, j_1) e^{-\alpha t_N(\beta)} x_1^N(t_N(\beta), j_2)] \tag{8.889}$$

has the order

$$O\left(\frac{e^{\alpha t_N(\beta)}}{N^3}\right) = o(N^{-2}). \tag{8.890}$$

This gives the claim.

The additional consideration in case $\frac{1}{2} < \beta < 1$ is that there is a positive probability that two clusters will collide in this time scale. However the question is whether the collisions with their effects can change the occupation integral appearing in the dual representation in the exponential. We see that since the number of those sites at which collisions occur among the total of the N sites is $o(N)$ we do not need to consider higher level occupation. q.e.d.

8.5 Relation Between *$\mathcal{W}$ and $\mathcal{W}^*$

Let us briefly review the two limiting regimes we have considered. In the first regime we have considered the limiting droplet process $\mathfrak{Z}_t^m$ which is a $\mathcal{M}_a([0, 1])$-valued Markov process. We have also proved that there exists $\alpha^* > 0$ such that

$$\{e^{-\alpha^* t} \mathfrak{Z}_t^m([0, 1])\}_{t \geq 0} \text{ is tight} \tag{8.891}$$

and determined the limiting first moment and variance of $e^{-\alpha^* t} \mathfrak{Z}_t^m([0, 1])$ as $t \to \infty$. This has been obtained by considering the process $\mathfrak{Z}_t^{N,m}$ for times in $[0, t_N]$ with $t_N \to \infty$, $t_N - \frac{\log N}{\alpha^*} \to -\infty$ and letting $N \to \infty$. Moreover by (8.850) it follows that $\mathfrak{Z}_t^m([0, 1])$ is Cauchy as $t \to \infty$ and therefore has a limit $\mathcal{W}^*$ in L^2 as $t \to \infty$.

8.5 Relation Between *W and W*

On the other hand there exists $\alpha > 0$ such that the empirical process $\int_0^1 x \Xi_N^{\log,\alpha}(t,dx)$ in time window $\{\frac{\log N}{\alpha}+t\}_{t\in\mathbb{R}}$ converges to $\mathcal{L}_t$, and the limit $*W = \lim_{t\to-\infty} e^{-\alpha t} \int_0^1 x\mathcal{L}_t(dx)$ exists and is random.

Next recall that we have established that the two exponential growth rates α and α^* in the two time regimes $t_N \uparrow \infty$, $t_N = o(\log N)$ as $N \to \infty$ respectively $t_N = \frac{1}{\alpha}\log N + t$ as $N \to \infty$, $t \to -\infty$ are the same, so that in that respect limits interchange.

It is then natural to conjecture that also the distributions of $*W$ and W^* are the same even though there is the following obstacle. Define

$$t_N(\beta,t) = \frac{\beta}{\alpha}\log N + t, \text{ with } 0 < \beta < 1. \tag{8.892}$$

Consider the law $\mathcal{L}[\bar{x}_2^N(t_N(\beta,t))]$ and note that for fixed t there is a discontinuity at $\beta = 1$ since then collisions of the dual occur at order $O(1)$ and we know that as function of β we jump from type-2 mass zero for $\beta < 1$ to a positive value at $\beta = 1$. Hence the matter to relate $*W$ and W^* is very subtle.

In order to investigate the relation between the laws of $*W$ and W^* we next consider the relations between the dual process at times corresponding to the droplet formation and the dual in the time scale corresponding to the macroscopic emergence. In this direction we verify first that the first two moments of W^* and $*W$ agree and then use the techniques of proof to conclude with L_2-arguments that the two random variables have asymptotically as $N \to \infty$, $t \to +\infty$ respectively $N \to \infty$, $t \to -\infty$ L_2-distance zero.

Recall the notation $t_N(\beta,t) = \beta\alpha^{-1}\log N + t$, $t_N = \alpha^{-1}\log N$, which we shall use in the sequel.

Proposition 8.29 (Relation between *W and W*).

(a) For $0 < \beta < 1$,

$$\lim_{t\to-\infty}\lim_{N\to\infty} E\left[\left(e^{-\alpha t_N(\beta,0)}\hat{x}_2^N(t_N(\beta,0)) - e^{-\alpha(\frac{\log N}{\alpha}+t)}\hat{x}_2^N(\frac{\log N}{\alpha}+t))\right)^2\right] = 0. \tag{8.893}$$

(b) We have the relation:

$$\mathcal{L}[*W] = \mathcal{L}[W^*]. \quad \square \tag{8.894}$$

Remark 105. We can summarize the results in (8.893) as follows. Consider for $0 < \beta < 1$, with $\bar{x}_2^N(t) := \frac{\hat{x}_2^N(t)}{N}$, the quantities:

$$\{\hat{x}_2^N(t): 0 \le t \le \frac{\beta\log N}{\alpha}\}, \quad \{\bar{x}_2^N(\frac{\log N}{\alpha}+t): \frac{(\beta-1)\log N}{\alpha} \le t \le 0\}. \tag{8.895}$$

Then we have shown so far that these two processes converge in law to

$$\{\hat{x}_2(t) : 0 \leq t < \infty\}, \quad \{\bar{x}_2(t) : -\infty < t \leq 0\}, \text{ respectively.} \tag{8.896}$$

We can, by the classical embedding theorem of weakly convergent distributions on a Polish space, construct both sequences and their limits on one probability space $(\Omega, \mathcal{A}, P)$ preserving the law of each random variable in question and such that we have a.s. convergence in the supnorm on compact time intervals. On this probability space we consider $L^2(\Omega, \mathcal{A}, P)$.

Then we have:

$$\mathcal{W}^* = L^2 - \lim_{t \to \infty} e^{-\alpha t} \hat{x}_2(t) = L^2 - \lim_{t \to -\infty} e^{-\alpha t} \bar{x}_2(t) = {}^* \mathcal{W}. \tag{8.897}$$

Corollary 8.30 (Equality of moments for growth constant).

$$E[({}^*\mathcal{W})^k] = E[(\mathcal{W}^*)^k], \quad k = 1, 2, \ldots. \quad \square \tag{8.898}$$

Proof of Corollary 8.30. This follows from Proposition 8.29. q.e.d.

Remark 106. We can calculate the moments of $^*\mathcal{W}$ and $\mathcal{W}^*$ by explicit calculation. However since we have not proved that all moments are finite we cannot use this to prove the equality in law (8.894)).

Proof of Proposition 8.29. We start with the rough idea. We have calculated $Var[\mathcal{W}^*]$ in (8.841) which involved the time scales smaller than $\alpha^{-1} \log N$. An expression for $Var[^*\mathcal{W}]$ was derived in (8.617). We see that $Var[\mathcal{W}^*] = Var[^*\mathcal{W}]$. We shall now prove (8.893) which will give (8.894). The point is that the moment calculations now have to include time scales up to $\alpha^{-1} \log N$ and must therefore different from before treat collisions of the dual cloud for a given point occurring with positive probability.

We proceed in five steps. We again start by considering the case without mutation but start instead with positive but finite initial type-2 mass. We calculate in Step 1 first moments, in Step 2 second moments at the given time, in Step 3 we calculate the mixed moment between the two times specified in the assertion under our assumption. In Step 4 of our arguments we include mutation. In Step 5 we conclude the proof.

Step 1 We begin with the *first moment* calculation. Consider the calculation of the expected value using the dual particle system starting with one particle at j and recall $\mathcal{I}(\varepsilon)$ from (8.855). We will assume that at time 0 we have an initial distribution as in Proposition 8.27 and that furthermore $m = 0$. As above there will then be a $Bin(K^N_{t_N(1,t)}, \frac{k}{N})$ distributed number of hits of the k-set $\mathcal{I}(\varepsilon)$ by the cloud of dual particles. To apply the Poisson approximation recall that by (8.744)–(8.747) we have:

$$K^{N,(a)}_t = W_a(t) e^{\alpha t} - \mathfrak{g}(W_a(\cdot), N, t) + \mathcal{E}_a(N, t), \tag{8.899}$$

8.5 Relation Between $*\mathcal{W}$ and $\mathcal{W}*$

where the error term satisfies

$$E[\mathcal{E}_a(N,t)] \leq \text{const} \frac{e^{2\alpha t}}{N} \cdot \sup_u E[W_g(u)]. \tag{8.900}$$

Therefore the number of hits of the k-set converges as $N \to \infty$ weakly to a *Poisson distribution with parameter* $We^{\alpha t} - \kappa_2 W^2 e^{2\alpha t} + O(e^{3\alpha t})$. Therefore from (8.873) we get looking at 0,1 or 2 hits of $\mathcal{I}(\varepsilon)$ and denoting by

$$\{\xi_\ell, \quad \ell = 1, \cdots, K^N_{t_N(1,t)}\} \tag{8.901}$$

the states of the sites occupied by the dual at time $t_N(1,t)$ and by K^N_t, U^N_t the number of occupied sites and the size distribution for the finite N dual (nonlinear) system studied above:

$$E[x_1^N(t_N(1,t), j)] = E\left[\prod_{\ell=1}^{K^N_{t_N(1,t)}} (1 - x_2^N(t_0, \xi_\ell))^{\zeta^N(t_N(1,t),\xi_\ell)}\right] \tag{8.902}$$

$$= E\left[(1 - \frac{k}{N})^{K^N_{t_N(1,t)}}\right.$$

$$+ (1 - \frac{k}{N})^{K^N_{t_N(1,t)}-1} \frac{K^N_{t_N(1,t)}}{N} \sum_{i=1}^{k} \sum_{\ell=1}^{\infty}$$

$$U^N(t_N(1,t), \ell)(1 - \tilde{x}_2^N(t_0, i))^\ell$$

$$+ (1 - \frac{k}{N})^{K^N_{t_N(1,t)}-2} \frac{K^N_{t_N(1,t)}(K^N_{t_N(1,t)} - 1)}{2N^2}$$

$$\sum_{i_1, i_2=1}^{k} \sum_{\ell_1, \ell_2=1}^{\infty} U^N(t_N(1,t), \ell_1) U^N(t_N(1,t), \ell_2)$$

$$\left. (1 - \tilde{x}_2^N(t_0, i_1))^{\ell_1}(1 - \tilde{x}_2^N(t_0, i_2))^{\ell_2}\right]$$

$$+ o(N^{-2}).$$

We rewrite this in the form

$$E[x_1^N(t_N(1,t), j)] \tag{8.903}$$

$$= E\left[1 - (W_N(t_N)e^{\alpha t} + O(W_N^2(t_N)e^{2\alpha t})) \sum_{i=1}^{k} g^N(t_N(1,t), \tilde{x}_2^N(t_0, i))\right.$$

$$+ \frac{W_N(t_N)e^{2\alpha t}}{2} \sum_{i_1=1}^{k} \sum_{i_2=1}^{k} g^N(t_N(1,t), x_{i_1}) g^N(t_N(1,t), x_{i_2}) \Bigg] + o(N^{-2}),$$

with

$$g^N(t_N(1,t), x) = 1 - \sum_{\ell=1}^{\infty} (1-x)^\ell U^N(t_N(1,t), \ell). \tag{8.904}$$

Therefore as $N \to \infty$, we have (recall that $W_N(t_N) \to W$ a.s. as $N \to \infty$ and $W_N(t_N)$ are uniformly square integrable):

$$E[x_2^N(t_N(1,t), j)] = E\Bigg[(We^{\alpha t} + O(W^2 e^{2\alpha t})) \sum_{i=1}^{k} g^N(t_N(1,t), \tilde{x}_2^N(t_0, i))\Bigg]$$
$$+ o(N^{-2}). \tag{8.905}$$

Next we need to know more about $g^N(t_N(1,t), \tilde{x}_2^N(t_0, i))$ as $N \to \infty$ and then $t \to -\infty$. Recall that $g^N(t_N(1,t), x) = 1 - \sum_{\ell=1}^{\infty}(1-x)^\ell U^N(t, \ell)$ (see 8.860), $\lim_{N \to \infty} U^N(t_N(1,t), \cdot) = U(t, \cdot)$ (see 8.52) and $\lim_{t \to -\infty} U(t, \cdot, \cdot) = \mathcal{U}(\infty, \cdot, \cdot)$ (see 8.362). Therefore

$$\lim_{t \to -\infty} \lim_{N \to \infty} g^N(t_N(1,t), \tilde{x}_2^N(t_0, i)) = g(\infty, \tilde{x}_2(t_0, i)) = 1 - \sum_{\ell=1}^{\infty}(1 - \tilde{x}_2(t_0, i))^\ell \mathcal{U}(\infty, \ell). \tag{8.906}$$

This allows to calculate next as $N \to \infty, t \to -\infty$:

$$E[\hat{x}_2^N(t_N(1,t))] = E\Bigg[WNe^{\alpha t} \sum_{i=1}^{k} g^N(t_N(1,t), \tilde{x}_2^N(t_0, i))\Bigg] + o(e^{\alpha t} N^{-1}). \tag{8.907}$$

Then

$$\lim_{t \to -\infty} \lim_{N \to \infty} E[e^{-\alpha t_N(1,t)} \hat{x}_2^N(t_N(1,t))] = E\Bigg[W \sum_{i=1}^{k} g(\infty, \tilde{x}_2(t_0, i))\Bigg]. \tag{8.908}$$

Finally putting everything together results in

$$\lim_{t \to -\infty} \lim_{N \to \infty} E[e^{-\alpha t_N(1,t)}(\hat{x}_2^N(t_N(1,t)))] = E[W] \sum_{i=1}^{k} g(\infty, \tilde{x}(t_0, i)). \tag{8.909}$$

8.5 Relation Between *W and W*

Step 2 We now consider the *second moments*. Consider the time points $t_N(\beta_1, 0)$ and $t_N(\beta_2, t)$, $\beta_1 < \beta_2$. The case $\beta < 2$ follows as in Proposition 8.27. The case $\beta = 1$ requires more careful analysis, we carry out now. Let $0 < \beta \leq 1$, then

$$E[x_1^N(t_N(\beta, 0), j_1) \cdot x_1^N(t_N(1, t), j_2)] \tag{8.910}$$

$$= E\left[\prod_{j=1}^{K_{t_N(1,t)}^{N,j_1,j_2} \zeta(t_N(1,t))} \prod_{\ell=1}^{\infty} (1 - \tilde{x}_2^N(t_0, j))^\ell\right]$$

$$= E\left[(1 - \frac{k}{N})^{K_{t_N(1,t)}^{N,j_1,j_2}} + \frac{K_{t_N(1,t)}^{N,j_1,j_2}}{N}(1 - \frac{k}{N})^{K_{t_N(1,t)}^{N,j_1,j_2}-1}\right.$$

$$\sum_{i=1}^{k} \sum_{\ell=1}^{\infty} U_t^N(\ell)(1 - \tilde{x}_2^N(t_0, i))^\ell$$

$$+ (1 - \frac{k}{N})^{K_{t_N(1,t)}^{N,j_1,j_2}-2} \frac{K_{t_N(1,t)}^{N,j_1,j_2}(K_{t_N(1,t)}^{N,j_1,j_2} - 1)}{2N^2}$$

$$\left.\sum_{i_1,i_2=1}^{k} \sum_{\ell_1,\ell_2=1}^{\infty} U^N(t, \ell_1)U^N(t, \ell_2)(1 - \tilde{x}_2^N(t_0, i_1))^{\ell_1}(1 - \tilde{x}_2^N(t_0, i_2))^{\ell_2}\right]$$

$$+ o(N^2).$$

Proceeding as in the proof of Proposition 8.27 in the case where $j_1 \neq j_2$ we get:

$$\text{Cov}(x_2^N(t_N(\beta, 0), j_1), x_2^N(t_N(1, t), j_2))$$

$$= \frac{1}{N^2}\left\{E\left[N\left(K_{t_N(1,t)}^{N,j_1} + K_{t_N(\beta,0)}^{N,j_2} - K_{t_N(1,t),t_N(\beta,0)}^{N,j_1,j_2}\right)\sum_{i=1}^{k} g^N(t_N(\cdot), \tilde{x}_2^N(t_0, i))\right]\right\}$$

$$+ o(N^{-2}). \tag{8.911}$$

Step 3 The next point is to prove the following claim on second moments for $N \to \infty$ and then as $t \to -\infty$. For $0 < \beta \leq \beta_2 \leq 1$,

$$\lim_{t \to -\infty} \lim_{N \to \infty} E\left[\left(e^{-\alpha t_N(\beta,0)} \hat{x}_2^N(t_N(\beta, 0)) \cdot e^{-\alpha t_N(\beta_2,t)} \hat{x}_2^N(t_N(\beta_2, t))\right)\right] = \text{const.}, \tag{8.912}$$

with a constant independent of β, β_2.

In order to verify (8.912) we must now include new considerations that arise since $e^{\alpha t_N(1,t)} = Ne^{\alpha t}$ leading to a positive particle density and hence collisions. Recall our representation in terms of the *coloured particle system* and a careful analysis of the number of W_1, W_2 and finally $G_{1,2}$ particles. We begin with the W_1 and W_2 particles.

The key point is that in the analysis of the κ_2 term one needs to include slower growth of the W_1 and W_2 particles due to the nonlinear term. In particular the populations of white particles are asymptotically as $N \to \infty$ of the form $NF(t)$ and the function $F(t)$ can be expanded as $t \to -\infty$ in terms of k-th powers ($k = 1, 2, \cdots$) of $e^{\alpha t}$. Hence the number of particles in the two white clouds, the one starting in j_1 and the one starting in j_2 are each asymptotically as $N \to \infty$ behaving as

$$We^{\alpha t_N(\gamma,t)} - \frac{1}{N}\kappa_2 W^2 e^{2\alpha t_N(\gamma,t)} \sim WNe^{\alpha t} - \kappa_2 W^2 Ne^{2\alpha t} = Ne^{\alpha t}W(1 - \kappa_2 We^{\alpha t}), \tag{8.913}$$

where we take two independent realisations of W for each cloud and with $\gamma = 1-\beta$. The system with start of a particle at j_1 at time 0 and j_2 at time $t_N(\beta, 0)$ evolves similarly but now an additional correction is needed because of the collisions between the two clouds. This effect is represented by the $G_{1,2}$ particles.

Hence the important term for the covariance calculation is the one involving the $G_{1,2}$ particles. The number of $G_{1,2}$ particles is taking again $N \to \infty$ and then afterwards $t \to -\infty$ of order

$$O(W'W''\kappa_2 e^{2\alpha t}). \tag{8.914}$$

The self-intersections of the W_1 and the $G_{1,2}$ particles produce only a higher order correction to the number and size distribution of the $G_{1,2}$ particles. As a result in the limit as $t \to -\infty$ the correction terms are all of higher order.

From together with (8.911, 8.913) and (8.914) we can verify that the covariance term is asymptotically the same as (8.849), that is,

$$\lim_{t \to -\infty} \lim_{N \to \infty} \mathrm{cov}(e^{-\alpha(t_N(1,t))}\hat{x}_2^N(t_N(1,t)), e^{-\alpha t_N(\beta,0)}\hat{x}_2^N(t_N(\beta,0))) \tag{8.915}$$

$$= 2(E[W])^2 \left\{ \kappa_2 \sum_{i=1}^{k} g(\infty, \tilde{x}_2(t_0, i)) \right\}.$$

and we get the claim in (8.912).

Step 4 We are now ready to conclude the argument for the two assertions of the proposition. First we have to show the initial masses we used in the calculation above can in fact arise as configuration at time t_0. This follows from (2.68), where we show convergence of the $\mathbf{J}^{m,N}$ to $\mathbf{J}^{m}$ (recall (2.103)). Second we have to prove that the case we treated can be replaced by the masses which arise from mutation. For any fixed t_0 these masses satisfy (8.843), (8.844) and the above argument gives therefore the result for mutation turned of after time t_0. We now have to argue that this is not relevant for large t_0.

8.5 Relation Between *W and W*

The idea is that the contribution to W^* due to mutation after t_0 is negligible for large t_0, more precisely, the expected contribution is $O(e^{-\alpha t_0}) \to 0$ as $t_0 \to \infty$. The result then follows by letting $t_0 \to \infty$ in Corollary 8.31 below.

Step 5 Next we have to go through the list of claims.

(a) This follows from (8.915) above, (8.852) together with (8.849) and (8.912) for $\beta = \beta_2 = 1$ since we see that both the variances of the two random variables and their covariance all converge to the same value.

(b) Follows from (a) by using the construction of Remark 105 ($W^*,^* W$) both on one probability space and noting that with $t_N(0, t) = t$

$$e^{-\alpha t_N(1,t)} \hat{x}_2^N(t_N(1,t)) = e^{-t} \bar{x}_2^N(t_N(1,t)), \tag{8.916}$$

$$W^* = \lim_{t \to \infty} \lim_{N \to \infty} e^{-\alpha t_N(0,t)} \hat{x}_2^N(t_N(0,t)) \tag{8.917}$$

and

$$^*W = \lim_{t \to -\infty} \lim_{N \to \infty} e^{-t} \bar{x}_2^N(t_N(1,t)). \tag{8.918}$$

This completes the proof of Proposition 8.29 if we could show that in (8.918) we can replace $t_N(0, t)$ and $t \to \infty$ by t_N with $t_N \to \infty$ but $t_N = o(\log N)$. Hence the proof of Proposition 8.29 follows from the next corollary. q.e.d.

Corollary 8.31. *Consider* $t_N \uparrow \infty$ *with* $t_N = o(\log N)$. *Then:*

$$\lim_{t \to \infty} \lim_{N \to \infty} E[(e^{-\alpha t} \hat{x}_2^N(t) - e^{-\alpha t_N} \hat{x}_2^N(t_N))^2] = 0 \quad \square \tag{8.919}$$

Proof of Corollary 8.31. If we consider the case in which we exclude mutation but insert initially some finite mass (even for $N \to \infty$) then the result follows due to (8.852). What we want is however the statement where initially we have no type-2 mass but where we have mutation from type 1 to type 2 at rate mN^{-1}. The idea now is to use the fact that only early mutation counts, that is, mutation after some late time has a negligible effect. We can use this idea together with coupling to prove the result with mutation.

We consider solutions of our basic equation for mutation rate m and mutation rate 0. We denote the solutions by

$$(X_t^{N,m})_{t \geq 0} \text{ and } (X_t^{N,0})_{t \geq 0}. \tag{8.920}$$

We can define these two processes on one probability space so that the following relation holds:

$$\hat{x}_2^{N,0}(t) \leq \hat{x}_2^{N,m}(t), \tag{8.921}$$

with

$$E[\hat{x}_2^{N,m}(t) - \hat{x}_2^{N,0}(t)] \leq Const \cdot m \int_0^t e^{\alpha s} ds. \qquad (8.922)$$

We now apply this to compare the process with mutation turned off after time t_0 with the original process with mutation running on $[0, \infty)$. We know $\hat{x}_2^{N,0}(t_0) = \hat{x}_2^{N,m}(t_0) = O(e^{\alpha t_0})$) and with (8.921) and (8.922) we have

$$0 \leq E[e^{-\alpha(t_0+t)}\hat{x}_2^{N,m}(t_0+t) - e^{-\alpha(t+t_0)}\hat{x}_2^{N,0}(t_0+t)] \leq Const \cdot m e^{-\alpha t_0}. \qquad (8.923)$$

Letting $t_0 \to \infty$ we get that we can approximate the systems with mutations by ones where mutation is turned off after t_0 uniformly in N. Then the claim follows by using the result for $O(1)$ initial mass of type 2 and no mutation. q.e.d.

8.6 Third Moments

We expect that the random variables $\mathcal{W}^*(^*\mathcal{W})$ have moments of all orders which determine the distribution. We do not verify this here but in this section we verify that the third moment of the rare mutant type is finite given if we start with total mass of type 1 only using the same method as that used for the first and second moments which for convenience we review here.

Lemma 8.32 (First, second and third moments of $\mathcal{W}^*$). *Consider times $t_N \to \infty$ with $t_N = o(\log N)$. The first, second and third moment of the rescaled total type two mass satisfies:*

$$\lim_{N \to \infty} e^{-\alpha t_N} E[(\hat{x}_2^N(t_N))] = m^* E[W], \qquad (8.924)$$

$$\lim_{N \to \infty} e^{-2\alpha t_N} E[(\hat{x}_2^N(t_N))^2] = (m^*)^2 (E[W])^2 + 2m^* \kappa_2 (E[W])^2 \qquad (8.925)$$

and for some $0 < \kappa_3 < \infty$, $m^* = \frac{1}{c}(1 + \frac{\gamma}{\alpha})m$ (as in (8.604)), κ_2 given by (8.794)

$$\lim_{N \to \infty} e^{-3\alpha t_N} E[(\hat{x}_2^N(t_N))^3] = (m^*)^3 (E[W])^3 + 6(m^*)^2 \kappa_2 (E[W])^3$$
$$+ 6m^* \kappa_3 (E[W])^3. \quad \square \qquad (8.926)$$

Proof of Lemma 8.32. The proof will be based on duality and as a prerequisite we need (in order to determine the moment asymptotics) to evaluate the expression which represents the moments in terms of the dual particle system. We therefore carry out the proof in *two steps*, first an expansion of dual expression and then the moment calculation.

8.6 Third Moments

Step 1 *(Expansion of $\Pi^N(t)$ up to terms of order $O(\frac{1}{N^3})$)*

In this subsection we extend the coloured particle system in order to derive and expansion of $\Pi_t^{N,a}$ including terms of order $\frac{1}{N}$ and $\frac{1}{N^2}$ which are needed in the calculation of the second and third moments of $\hat{x}_1^N(t)$.

Lemma 8.33 (Expansion of $\Pi_{t_N}^{N,a}$). *Consider the dual particle system starting with one, two or three particles. To obtain a unified expression we state the result using W_a.*

The corresponding CMJ random growth constants (recall (8.144) $W_a = W_1$ with one initial particle, $W_a = W_1 + W_2$ if two particles start at different sites, $W_a = W^{(2)}$ if two particles start at the same site, $W_a = W_1 + W_2 + W_3$ is three particles start at different sites and $W_a = W_1^{(2)} + W_2$ if two particles start at one site and one particle at a different site.

Then there exists $\kappa_3 > 0$ such that

$$\Pi_{t_N}^{N,a} \sim \frac{1}{c}(\alpha + \gamma)\left[(W_a)e^{\alpha t_N} - \frac{2\kappa_2}{N}(W_a^2)\frac{e^{2\alpha t_N}}{N}\right. \tag{8.927}$$
$$\left. - \frac{3\kappa_3}{N^2}(W_a^3)e^{3\alpha t_N} + O(\frac{e^{4\alpha t_N}}{N^3})\right].$$

Then we get for the time-integral and the exponential of it:

$$H_a^N(t) := \frac{m}{N}\int_0^{t_N}\Pi_s^N ds \sim \frac{m^*}{N}(W_a)e^{\alpha t_N} - \frac{m^*\kappa_2}{N^2}(W_a^2)e^{2\alpha t_N} \tag{8.928}$$
$$- \frac{m^*\kappa_3}{N^3}W_a^3 e^{3\alpha t_N} + O(\frac{e^{4\alpha t_N}}{N^4}),$$

$$e^{-H_a^N(t)} = 1 - H_a^N(t) + \frac{1}{2}(H_a^N(t))^2 - \frac{1}{6}(H_a^N(t))^3 + O((H_a^N(t))^4). \qquad \square$$
$$\tag{8.929}$$

Proof of Lemma 8.33. We use an extension of the multicolour system defined so far in Subsubsection 8.3.9 (Step 1) therein and in the above Subsection 8.5 in order to identify $\Pi_{t_N}^{N,a}$ up to terms of order $e^{3\alpha t_N}$ with an error term of order $O(\frac{e^{4\alpha t_N}}{N^3})$. To obtain this we must introduce new colours.

Recall in the expansion to order $e^{2\alpha t}$ we worked with the system of white, black, red, green, purple and blue coloured particle system and that the dual system is given by the white, red and purple particles. The green particles are produced by coalescence of a red and white particle and a green site is produced by the migration of a green particle. These represent particles which are lost due to one collision (followed by coalescence). The white particles grow as $W_a\frac{\alpha+\gamma}{c}e^{\alpha t}$ and the green particles are produced at rate $W_a^2 e^{2\alpha s}$ at time s. Since the green families grow with exponential rate α the total number of green particles grows like $W_a^2 \kappa_2 e^{2\alpha t}$. The number of purple–blue particles is of order $O(e^{3\alpha t})$ and the lower bound is given by the white and red particles obtained by subtracting the number of green particles. The upper bound is obtained by adding the blue particles.

In order to identify the *next order term* we must include the purple particles (and exclude the blue particles) and follow what happens when the purple particles coalesce or migrate. This requires to introduce two further colours, namely *pink* and *yellow*. When a purple particle migrates and hits a white, red or purple site we now produce a pink–yellow pair. There are $O(e^{4\alpha t})$ pink (yellow) particles where now the pink particles provide the dual and the yellow particles now behave as the blue particles did before in obtaining the error term.

We note that the purple particles can coalesce with a white (or red) particle and when this happens a white (or red) and green (to distinguish these we label them G_2) pair of particles is produced at the site where coalescence occurs. Green (G_2) particles are produced at rate const $\cdot W_a^3 e^{3\alpha t}$.

The two particle systems we have to compare, the dual one and the collision-free one, are now given as follows.

- The dual is given by the *white, red, purple* and *pink particles* and
- the collision free by *white, red, green* ($G \cup G_2$), *purple* and *yellow*.

The modified rules are now:

- If a purple migrates and hits a site occupied by white or red it dies and creates a pink–yellow pair which have subsequently coupled times of birth, migration,
- the yellow particle is placed in the first unoccupied site in the first copy of $\mathbb{N}$,
- if a purple particles coalesces with a white or red particle it produces a green G_2 particle
- the pink particles behaves like a standard dual particle.

This will allow us to approximate the dual with an error of order as: $N \to \infty$:

$$O(N^{-3} e^{4\alpha t_N}). \tag{8.930}$$

Now we have to represent the quantity $\Pi_t^{N,a}$ in terms of the numbers of particles of the various colours. Here we follow the procedure explained in Subsubsection 8.4.5, Part 2 therein. Then we obtain the claimed expansion of (8.927). Recall that $\kappa_2 W^2 e^{2\alpha t}$ denotes the term corresponding to the growth of the green G particles in subsubsection 8.4.5. In the same way the $\kappa_3 W^3 e^{3\alpha t}$ in (8.927) term corresponds to the set of G_2 particles, that is, the particles lost due to particles involved in exactly two collisions.   q.e.d.

The assertions (8.928) and (8.929) follow from (8.927) by explicitly calculating the integral and then inserting the expansion in the Taylor expansion of the exponential around 0. q.e.d.

8.6 Third Moments

Step 2 *(Moment calculation)*

First recall the dual identity

$$E[(x_1^N(i,t))^k] = E\left[e^{-H_k^N(t)}\right], \qquad (8.931)$$

where H_k^N is given by (8.928) and where W_a is the CMJ random growth constant (recall (8.144)) obtained by starting the dual with k particles at site i. We obtain a similar expression for $E[x_1^N(i,t)x_1^N(j,t)]$, where now we start one particle at i and one particle at j, etc.

We now compute the first three moments following the same method. Note that

$$E[(\hat{x}_2^N(t))] = E[N - \hat{x}_1(t)] = N - NE[x_1(t)] \qquad (8.932)$$

$$= N - NE[(1 - m^* \frac{W_1}{N} e^{\alpha t} + O(\frac{1}{N^2}))].$$

Then replacing t by t_N we get

$$\lim_{N\to\infty} e^{-\alpha t_N} E[(\hat{x}_2^N(t_N))] = \lim_{N\to\infty} e^{-\alpha t_N} N[\frac{m^* E[W]}{N} e^{\alpha t_N} + O(\frac{1}{N^2})] \qquad (8.933)$$

and (8.924) follows provided that $t_N = o(\log N)$.

We now consider the second and third moments:

$$E[(\hat{x}_1^N(t))^2] \qquad (8.934)$$
$$= NE[(x_1^N(i,t))^2] + N(N-1)E[(x_1^N(i,t))x_1^N(j,t)], \quad i \neq j$$

and for $i, j, k \in \{1, \cdots, N\}$ distinct

$$E[(\hat{x}_1^N(t))^3] = NE[(x_1^N(i,t))^3] + 3N(N-1)E[(x_1^N(i,t))^2 x_1^N(j,t)]$$
$$+ N(N-1)(N-2)E[x_1^N(i,t)x_1^N(j,t)x_1^N(k,t)]. \qquad (8.935)$$

We first illustrate the method of calculation by calculating in detail the second moment. We substitute (8.928 and 8.929) in (8.940) keeping track of only those terms that do not go to 0 as $N \to \infty$. We get:

$$E[(\hat{x}_2^N(t))^2] = E[N - \hat{x}_1^N(t)]^2$$
$$= N^2 - 2NE[\hat{x}_1^N(t)] + E[\hat{x}_1^N(t))^2]$$
$$= N^2 - 2N^2 E[x_1^N(t)] + NE[(x_1^N(t))^2] + N(N-1)E[x_1^N(i,t)x_1^N(j,t)]. \qquad (8.936)$$

We now insert for the appearing moments of x_1^N the formula (8.931) and then use the expansion of the exponential from (8.929). We first compute the terms coming from the $1 - H^N(t)$ in the expansion of the exponential. This gives

$$N^2 - 2N^2 E\left[1 - \frac{m^*}{N} W e^{\alpha t} + \frac{m^* \kappa_2}{N^2} W^2 e^{2\alpha t} + O\left(\frac{e^{3\alpha t}}{N^3}\right)\right] \tag{8.937}$$

$$+ NE\left[1 - \frac{m^*}{N} W^{(2)} e^{\alpha t} + \frac{m\kappa_2}{N^2} W^2 e^{2\alpha t} + O\left(\frac{e^{3\alpha t}}{N^3}\right)\right]$$

$$+ N(N-1) E\left[1 - \frac{m^*}{N}(W_1 + W_2) e^{\alpha t} + \frac{m^* \kappa_2}{N^2}(W_1 + W_2)^2 e^{2\alpha t}\right]$$

$$= N^2(1 - 2 + 1)$$

$$+ NE\left[(2m^* W e^{\alpha t} + 1 - 1 - m^*(W_1 + W_2) e^{\alpha t}) + O\left(\frac{e^{3\alpha t}}{N^3}\right)\right]$$

$$+ E\left[-2m^* \kappa_2 W^2 e^{2\alpha t} - m^* W^{(2)} e^{\alpha t} + m^* \kappa_2 (W_1 + W_2)^2 e^{2\alpha t} + O\left(\frac{e^{3\alpha t}}{N^3}\right)\right]$$

$$= 2m^* \kappa_2 (E[W])^2 e^{2\alpha t} + m^* E[W_1 + W_2] e^{\alpha t} - m^* E[W^{(2)}] e^{\alpha t} + O\left(\frac{e^{3\alpha t}}{N}\right).$$

We next note that the $\frac{1}{2}(H^N(t))^2$ gives

$$(m^*)^2 (E[W])^2 e^{2\alpha t}. \tag{8.938}$$

Here W_1, W_2 are independent random variable coming from the CMJ limit starting with particles at two disjoint sites and $W^{(2)}$ is the random variable coming from the CMJ limit starting with two particles at the same site.

Combining all these terms we obtain

$$E[(\hat{x}_2^N(t))^2] \tag{8.939}$$

$$= 2(E[W])^2 m\kappa_2 e^{2\alpha t} + 2m^* E[W] e^{\alpha t} - m^* E[W^{(2)}] e^{\alpha t} + (m^*)^2 (E[W])^2 e^{2\alpha t}$$

$$+ O\left(\frac{e^{3\alpha t}}{N}\right).$$

Letting $t = t_N$, multiplying by both sides $e^{-2\alpha t_N}$ and taking the limit $N \to \infty$, we obtain (8.925) provided that $t_N = o(\log N)$.

We now follow the same method for the third moment but omit writing out the numerous intermediate terms which cancel and get:

$$E[(\hat{x}_2^N(t))^3] \tag{8.940}$$

$$= N^3 - 3N^2 E[\hat{x}_1^N(t)] + 3NE[(\hat{x}_1^N(t))^2] - E[(\hat{x}_1^N(t))^3]$$

$$= N^3 - 3N^3 E[(x_1^N(t))] + 3N^2 E[(x_1^N(t))^2] + 3N^2(N-1)E[x_1^N(i,t)x_1^N(j,t)]$$
$$- NE[(x_1^N)^3] - 3N(N-1)E[(x_1^N(i))^2 x_1(j)]$$
$$- (N^3 - 3N^2 + 2N)E[x_1^N(i)x_1^N(j)x_1^N(k)].$$

The proof proceeds by substituting (8.928) and (8.929) via (8.931) in (8.940). One can verify by direct algebraic calculations that the resulting coefficients of N^3, N^2, N^1 are zero. The $O(1)$ terms involves powers of $e^{\alpha t}, e^{2\alpha t}$ and $e^{3\alpha t}$. To prove the Lemma it suffices to identify the coefficient of $e^{3\alpha t}$. The coefficient involving $(m^*)^3$ comes from the term $\frac{N^3}{6} \cdot (H_N(t))^3$ in the expansion of the exponential. The κ_2 term comes from the term $\frac{N^3}{2} \cdot (H_N(t))^2$ and the κ_3 term comes from the term $N^3 H_N(t)$ in the expansion of the exponential.

To illustrate this structure we consider the contribution coming from the term

$$\frac{N^3}{2}(H_a^N)^2 = -(m^*)^2 \kappa_2 e^{3\alpha t} (W_a)^3. \tag{8.941}$$

We must apply this to each term in the following expression

$$N^3[-3E(x_1^N(t)) + 3E[x_1^N(i,t)x_1^N(i,t)] - E[x_1^N(i,t)x_1^N(j,t)x_1^N(k,t)] \tag{8.942}$$

The corresponding terms with a $(W_a)^3$ result in

$$-3E[W_1^3]3E[(W_1 + W_2)^3] - E[(W_1 + W_2 + W_3)^3] = -6E[W_1]E[W_2]E[W_3]. \tag{8.943}$$

Note that the coefficients of $e^{\alpha t}, e^{2\alpha t}$ depend on higher moments of W. This leads to

$$E[(\hat{x}_2^N(t))^3] = m^* \kappa_3 (E[W])^3 e^{3\alpha t} + 6(m^*)^2 \kappa_2 (E[W])^3 e^{3\alpha t} + 6(m^*)^3 (E[W])^3 e^{3\alpha t}$$
$$+ \text{const} \cdot e^{2\alpha t} + \text{const} \cdot e^{\alpha t} + O\left(\frac{e^{\alpha 4 t}}{N}\right)$$

Letting $t = t_N$, multiplying by $e^{-3\alpha t_N}$ and taking the limit $N \to \infty$ we obtain (8.926) provided that $t_N = o(\log N)$.   q.e.d.

8.7 Propagation of Chaos: Proof of Proposition 2.13

Here we prove Proposition 2.13. We proceed in three steps.

Step 1 Fix $t_0 > -\infty$ and tagged sites $1, \ldots, L$. We can again prove for $T_N = \frac{1}{\alpha} \log N$ that $\{x_2^N(1, T_N + t_0), \ldots, x_2^N(L, T_N + t_0)\}$ converges in distribution as

$N \to \infty$ using the dual representation to show that all joint moments converge. Here of course it is equivalent to do the same for $x_1(i, T_N + t_0)$ which is more convenient for calculation. In particular we argue as below (8.595) using the dual process, except that now we start k_i-particles at site $i, i = 1, \ldots, L$ to calculate the mixed higher moments of $x_1^N(1, T_N + t_0), \cdots, x_1^N(L, T_N + t_0)$, i.e.

$$E[\prod_{j=1}^{L}(x_1^N(j, T_N + t_0))^{k_j}], \quad k_1, k_2, \cdots, k_L \in \mathbb{N}, \tag{8.944}$$

in terms of

$$E[\exp(-\frac{m}{N} \int_0^{T_N+t} \Pi_s^{N,(k_1,\cdots,k_L)} ds)] \tag{8.945}$$

using formula (8.722) and (8.723) and denoting with the superscript $(k_1, \cdots, k_L)$ the initial position.

Then we obtain convergence as $N \to \infty$ from Proposition 8.8 part (e). (Note the fact that the k_i here are possibly different does not change the argument). Therefore we have proved the convergence of the marginal distribution at time t_0 of the tagged size-L sample.

Step 2 Next we verify that the limiting dynamics for $t \geq t_0$ is as specified in (2.92). To do this we first review some results proved earlier. Using the results from Step 1 and Skorohod representation of weakly converging laws on a Polish space we can assume that $\Xi_N^{\log,\alpha}(t_0, 2) \xrightarrow[N \to \infty]{} \mathcal{L}_{t_0}(2)$, $a.s.$. Note that conditioned on $\mathcal{L}_{t_0}(2)$, the path $\{\mathcal{L}_t(2)\}_{t \geq t_0}$ is deterministic.

Consider for $i \in \{1, \ldots, L\}$ the system:

$$dx_2^N(i,t) = c(\bar{x}_2^N(t) - x_2^N(i,t))dt - s\, x_2^N(i,t)(1 - x_2^N(i,t))dt \tag{8.946}$$

$$- \frac{m}{N} x_2^N(i,t)dt$$

$$+ \sqrt{d \cdot x_2^N(i,t)(1 - x_2^N(i,t))} dw_2(i,t).$$

Conditioned on $\mathcal{L}_{t_0}(2)$, we know already for $N \to \infty$ (on our Skorohod probability space):

$$\bar{x}_2^N(t) \to \int x \mathcal{L}_t(2, dx), \quad a.s., \quad \text{for } t \geq t_0. \tag{8.947}$$

Since we know from Step 1 that

$$\mathcal{L}[\{x_2^N(i, t_0), \ i = 1, 2, \cdots, L\}] \underset{N \to \infty}{\Longrightarrow} \mathcal{L}[\{x_2^\infty(i, t_0), \ i = 1, \ldots, L\}], \tag{8.948}$$

then a standard coupling argument yields

$$\mathcal{L}[\{x_2^N(i,t)\}_{i=1,...,L;\,t\geq t_0}] \underset{N\to\infty}{\Longrightarrow} \mathcal{L}[\{x_2^\infty(i,t)_{i=1,...,L;\,t\geq t_0}\}], \qquad (8.949)$$

where $x_2^\infty(\cdot,\cdot)$ satisfies for $i=1,\ldots L$, $t \geq t_0$.

$$dx_2^\infty(i,t) = c(\int x\mathcal{L}_t(t,dx) - x_2^\infty(i,t))dt - s\,x_2^\infty(i,t)(1-x_2^\infty(i,t))dt$$
$$\qquad\qquad (8.950)$$
$$+\sqrt{d\cdot x_2^\infty(i,t)(1-x_2^\infty(i,t))}\,dw_2(i,t).$$

Step 3   Now we need to show that we have for the equation (8.950) a unique solution for $t \in \mathbb{R}$.

We can prove that, given $(\int x\mathcal{L}_t(2,dx))_{t\in\mathbb{R}}$, which converges to 0 as $t \to -\infty$, (8.950) has a unique solution on $(-\infty, t_0]$ as follows.

We can construct a minimal solution by considering a sequence (in the parameter n) of solutions corresponding to starting the process at time t_n. Then consider the tagged site SDE driven by this mean curve starting at 0 at time $t_n \to -\infty$. The sequence of solutions forms a stochastically monotone increasing sequence as $n \to \infty$ (use coupling). Then observe that any solution $x(t)$ that satisfies $x(t) \to 0$ as $t \to -\infty$ must have zeros (cf. classical result on Wright–Fisher diffusions). Then by coupling at a zero we get that it must agree with the minimal solution. This gives uniqueness for a given mean path $(\mathcal{L}_t(2))_{t\in\mathbb{R}}$. Note that this also implies the uniqueness of $(\mathcal{L}_t)_{t\in\mathbb{R}}$ since we have already established the uniqueness of the limiting mean curve.

This completes the proof.

8.8 Extensions: Non-critical Migration, Selection and Mutation Rates

The hierarchical mean field analysis allows us to investigate the emergence times and behaviour for a wide range of scenarios involving different parameter ranges for mutation rates, migration rates and relative fitness of the different levels. Our main focus in this work is the *critical case* in which mutation, migration and fitness all play a comparable role and for this reason we have chosen the parameterization from Section 3 (cf. $\beta_1 = 0$, $\beta_2 = 0$, $\beta_3 = 1$ in the listing below). However the basic tools and analysis we develop here can be adapted to other scenarios. Although we will not carry out the analysis in detail we now briefly indicate the main features of the different cases.

For example, consider the following "general" parametrization of the migration, selection and mutation rates:

$$c_{N,k} = \frac{c}{N^{(k-1)(1-\beta_1)}}, \quad 0 \leq \beta_1 < 1, \tag{8.951}$$

$$s_N = \frac{s}{N^{\beta_2}}, \quad \beta_2 \geq 0, \tag{8.952}$$

$$m_N = \frac{m}{N^{\beta_3}}, \quad \beta_3 \geq 0. \tag{8.953}$$

The different ranges of the migration parameters have the following interpretation. The value $\beta_1 = 0$ corresponds to euclidian space with dimension 2. The values $0 < \beta_1 < 1$ correspond to dimensions $d > 2$. We do not consider this case in this monograph but for a detailed discussion of the relation of random walks on the hierarchical group to random walks on $\mathbb{Z}^d$ and for further references see [DGW01].

We have discussed the case $\beta_1 = \beta_2 = 0$ and $\beta_3 = 1$ in Section 8 and we will now briefly discuss the case (1) where $\beta_1 = 0$ but $\beta_2 > 0$ and (2) the cases $\beta_1 = \beta_2 = 0$ but either $\beta_3 < 1$ or $\beta_3 > 1$, which are the other most interesting cases.

(1) In order to understand the behaviour $\beta_1 = 0$, $\beta_2 > 0$ consider the behaviour of the dual process. Then the dual process has the property that most occupied sites have only one factor (until the number of factors is $O(N)$) and the number of factors grows like $e^{\frac{st}{N^{\beta_2}}}$. In speeded-up time scale $N^{\beta_2} t$ the factors move quickly with possibility of coalescence each time a pair collides—see discussion in Step 1 of the proof of Proposition 12.2. If $\beta_3 \leq 1$, then heuristically, emergence will occur when

$$e^{\frac{st}{N^{\beta_2}}} = O(N^{\beta_3}) \tag{8.954}$$

that is,

$$t = O(N^{\beta_2} \frac{\beta_3}{s} \log N). \tag{8.955}$$

This means that emergence occurs much later than in the critical regime considered in our work.

(2) We now restrict our attention to the case $\beta_1 = \beta_2 = 0$ and discuss the effect of different mutation rates. The case of *faster mutation*, $0 < \beta_3 < 1$, has the effect that emergence can occur before collisions become non-negligible. In particular, if the mutation rate is $\frac{m}{N^{\beta_3}}$ with $\beta_3 < 1$ we get emergence at time $\beta_3 \alpha \log N$ and the *emergence is deterministic*, i.e. $\mathrm{Var}(\bar{x}_2^N(\alpha \beta_3 \log N)) \to 0$. In this case if we look at the dual process there are $O(N^{1-\beta_3})$ occupied sites at time $O(1)$ and therefore we have this number of droplets of size N^{β_3} at time $\alpha \beta_3 \log N$. Therefore we have a law of large numbers and deterministic emergence dynamics.

8.8 Extensions: Non-critical Migration, Selection and Mutation Rates

On the other hand, *slower mutation*, $\beta_3 > 1$, means that the "wave of advance" of the dual system must move to distance 2 or higher depending on the value of $\beta_3 > 1$ (cf. see discussion in Section 12) in order to produce the number of dual particles needed to guarantee emergence.

Chapter 9
Emergence with $M \geq 2$ Lower Order Types (Phases 0,1,2)

We now have a good understanding how emergence occurs in the Fleming–Viot model with *two* fitness levels with one type at each level and fitness difference 1 between the two types (Section 8). It remains to determine in this Section 9 and the next Section 10 respectively what is the effect of having

- $M \geq 2$ types on the lower level,
- $M \geq 2$ types on the upper level.

Both of these cases are going to exhibit some new phenomenon and challenges not present in the two-type model. What are the effects of having more than two types precisely?

First turn to the higher level. It turns out that having more types at the higher level has no effect on the emergence time but the random state after emergence is different, namely we obtain again as for two types an entrance law which is the random shift of a deterministic one, but for more than one type on the upper level in addition the *relative proportions* of the different types of rare mutants are now also *random* after their emergence which is a new effect and we need a technique to determine the law of the random frequencies. The case with M higher level types, will be discussed in Section 10.

It is however the first case, $M \geq 2$ types at the lower level which gives rise to the even more serious mathematical challenges. The analysis with more than one type at the lower level is substantially more complex and results indeed in a *slower emergence*, i.e. the constant in front of $\log N$ is larger, compared to one type on the lower level. This is reflected in a more complicated structure of the dual process, where dual particles now will have an internal structure changing due to selection, mutation and coalescence leading to interactions between geographic sites due to the internal states which forms potentially expanding spatial clusters.

In the case $M \geq 2$ lower level types and one higher level type there are $M^2 + M$ mutation parameters and M selection parameters. It is natural to expect that the emergence-fixation behavior can depend on these parameters. In particular we identify different parameter regimes and develop tools applicable to these regimes.

The first regime is called the mutation-dominant regime in which assuming sufficiently rapid mutation the program parallel to the case of $M = 1$ can again be carried out. The remaining cases, the selection-dominant regime, require a more complex approach with a number of additional complications.

The whole analysis in the $M \geq 2$ case involves the introduction and study of three new mathematical objects. (1) A *set-valued dual* which is used not only for the purpose of this section but also is applicable to more general spatial models with mutation and selection. (2) The application of this set-valued dual to the *island model* as the number of islands $N \to \infty$ in the selection dominant regime which is a CMJ branching system in which the individuals now have *multitype* internal dynamics, a new object called a *mesostate* with dynamics given by a multilevel birth, death, catastrophe process. We obtain a number of properties of this *branching mesostate process* but in addition pose some *open problems and conjectures* concerning their behaviour. (3) A branching population of mesostates with collisions which is again described in the $N \to \infty$ limit by a *nonlinear evolution equation for the distribution on the space of mesostates*.

9.1 Introduction

The slower emergence taking place with many types at the lower level is due to a complex scenario of competition between the fitter E_1 types and the E_0 types where both mutation and selection play an important role and which is reflected in a different structure of the dual process. We will see that the form of the results for the case $M \geq 2$ on the lower level is the same as before but what changes is that the dual representation of the emergence constant in front of $\log N$ now has a very different form. The reason being that, we now have in the dual representation *multitype* CMJ-processes. The different types correspond to the internal states of the dual particles which relates to the fact that the interplay of mutation and selection on the level E_0 which now plays a role in the E_1 versus E_0 competition. Correspondingly (1) the growth rate of the process is a much more subtle matter and (2) we get more complex nonlinear equations in the collision regime for the dual.

One effect of this more complex structure is that we have two regimes of behaviour, to which we have to adapt the analysis. This is the case of *mutation dominance* versus *selection dominance*, which requires to develop different tools for the proof in Subsections 9.10, 9.11 respectively 9.12 for the two regimes. We explain more about this in Subsubsection 9.7.8.

This complex situation will be addressed in this section via a new tool, namely a duality where the *form* of the function-valued part becomes part of the state information, namely the factors and summands will form an array producing a *tableau-valued process* and will induce a *set-valued dual*, based on the dual from Section 5, namely the Markovian dynamic $(\eta, \mathcal{G}^{++})$. For this purpose, as an intermediate step we introduce an *ordered* and *enriched* (incorporating historical information) version of $(\eta, \mathcal{F}^{++})$ respectively $(\eta, \mathcal{G}^{++})$ called $(\eta', \mathcal{F}^{++,<})$,

9.1 Introduction

respectively $(\eta', \mathcal{G}^{++,<})$ which will also later on play a role in the required enrichments of the set-valued dual.

Once we have the new ordered dual we shall see that this process will drive a *set-valued dual process* $\tilde{\mathcal{G}}^{++}$, which arises as a functional of the former. However we are still able to give a *Markovian* dynamic for this process. This latter dual object is a very flexible tool and is closely related to the genealogical process of the population. This allows in particular to prove ergodic theorems for such processes.

A bit more subtle is to study the emergence after rare mutation and here we need to find an underlying branching structure or in other words the underlying CMJ-processes. We construct therefore processes $\mathfrak{s}, \mathfrak{s}^*, \mathfrak{s}^{**}$ related to $\tilde{\mathcal{G}}^{++}$ (in fact enrichments respectively enriched modifications based on the process $(\eta', \mathcal{G}^{++,<})$) which unravel the various CMJ-structures related to the underlying set-valued dual and which allow us to carry out the analysis of the growth of the dual analog to Section 8 but in the more complex case of M-types ($M \geq 2$) on the lower level of fitness.

Due to the complex notation necessary to deal with the dual for more than two types on the lower level we first consider shortly below in the beginning of Subsection 9.4 the problem of the slowdown of emergence in the dual picture in two examples for $M = 2, M = 3$ lower level types and then subsequently develop systematically the full picture in the general case $M \geq 2$.

Outline of Section 9. In Subsubsection 9.2 we introduce the model and notation for M types on the lower and one on the upper level and formulate the key results in Subsection 9.3, which we then prove in Subsections 9.4–9.14 without deleterious mutation from the higher level and in Subsection 9.15, with deleterious mutation.

In Subsection 9.2 we give the $(M + 1)$-type model and a simplified version with no deleterious mutation for the fittest type. Next we give in Subsection 9.3 the results. Namely in Subsubsection 9.3.1 we state the results on emergence, in Subsubsection 9.3.2 on the limiting fixation dynamic and in Subsubsection 9.3.3 on the early droplet formation preceding emergence while Subsubsection 9.3.4 relates droplets and emergence, which are related to two-time scales of different order.

Then we develop the *new* duality methods in Subsections 9.4 and 9.5 and in Subsection 9.6 we develop some aspects further which are tuned to apply it to the emergence and fixation analysis. We begin in Subsubsection 9.4.1 with an introduction in the problem and give in Subsubsection 9.4.2 a table of where to find the different objects and construction and carry out as a warm-up the duality proof for the emergence result in the deterministic case. In Subsubsections 9.4.4–9.4.8 we develop the dual process with values in *marked tableaus of factors*, organized according to variables and summands. In Subsection 9.5 we develop from this a new Markovian *set-valued* process and a duality relation for our process and show how this process is driven by a Markov process $\mathfrak{s}$ which is suitable to analyse the growth behaviour. In Subsection 9.6, we then systematically expand and refine the duality theory further to be able to handle emergence and the limiting nonlinear evolution equations efficiently.

In Subsections 9.7–9.12 we apply the duality techniques and analyse the $N \to \infty$ limit of the dual process in the different time regimes with Subsection 9.7 devoted to the collision-free case (i.e. space is $\mathbb{N}$). In Subsection 9.8 then we describe the problems and strategy for Subsections 9.8–9.11. In Subsection 9.9 the case of the finite space $\{1, \cdots, N\}$ is treated. Subsections 9.10 and 9.11, respectively Subsection 9.12, analyse the objects as $N \to \infty$ in the two different regimes (mutation versus selection dominance). In Subsection 9.13 and Subsection 9.14 we then prove the results stated in Subsection 9.3. In Subsection 9.15 we finally remove the assumption of no deleterious mutation for the advantageous type.

9.2 The $(M, 1)$ Model: Basic Objects and Notation

We consider the case with types $\{1, \ldots, M\}$ at the lower level and type $\{M + 1\}$ at the higher level. As before we consider the system

$$X^N(t) = ((x_1^N(i,t), \ldots, x_M^N(i,t), x_{M+1}^N(i,t)); i = 1, \ldots, N) \in (\Delta_M)^N. \quad (9.1)$$

The mutation rates are given by

$$m_{i,j} : i \to j, \quad i, j = 1, \ldots, M \quad (9.2)$$

$$\frac{m_{M+1,\text{up}}}{N} : i \to M+1 \quad i = 1, \ldots, M$$

$$\frac{m_{M+1,\text{down}}}{N} : M+1 \to i, \quad i = 1, \ldots, M$$

and the fitness of types $(1, \ldots, M)$ and, $M + 1$ are

$$(0, e_2, \ldots, e_M) \text{ and } 1, \quad \text{respectively}, \quad (9.3)$$

where $e_1 = 0 < e_2 < e_3 < \cdots < e_M = \frac{N-1}{N}$ and the selection rate is s.

9.2.1 Simplification for Analysis of Phases 0,1,2

If we are only interested in phases 0,1,2, i.e. up to a fixation on the higher level, one can simplify the model in two further ways (for selection and for mutation) without changing any of the results. We next explain this.

To see how this simplification works with selection observe that (9.3) means that in the level zero time scale (i.e. $N^0 t$), there is a fitness difference of size $e_{i+1} - e_i$ between individuals of types $i, i + 1$ with $i = 1, \cdots, M - 1$ and of size $\frac{1}{N}$ between type M and $M + 1$. However the fact that types M and $M + 1$ have a fitness difference of order $O(\frac{1}{N})$ plays no role in the emergence which occurs in times of order $o(N)$, while higher order selection effects occur at times $O(N)$ only.

9.2 The (M, 1) Model: Basic Objects and Notation

Therefore to simplify the essential argument in this subsection we assume that both type M and type $M + 1$ have fitness 1.

Next to the simplification of the mutation part. Note here that also the slow mutation of type $M + 1$ to types $1, \ldots, M$ plays no role in the emergence since the latter already occurs in times $o(N)$ and for simplicity of exposition in the statements of our results we assume that the deleterious mutation from the upper type is zero.

We summarize this simplification, useful for phases 0,1,2 only, in

$$m_{M+1,\text{down}} = 0,$$
$$e_M = e_{M+1} = 1. \tag{9.4}$$

In Subsection 9.15 we explain in detail how to remove these assumptions.

9.2.2 The Basic Objects

Next we introduce again the key objects and notation with which we describe the system.

If not otherwise stated, throughout this section we assume that the initial configuration is given by

$$x_1^N(i,0) = 1, \quad \forall\, i \in \{1, 2, \cdots, N\}, \tag{9.5}$$

i.e. all the mass is on the *lowest type*. We will explicitly indicate in separate remarks where we need to take different initial conditions and formulate the appropriate results. Of particular interest is to work with the low level equilibrium.

For each type j define the empirical measure as well as the total empirical measure for all types:

$$\Xi_N(t,j) := \frac{1}{N} \sum_{i=1}^{N} \delta_{x_j^N(i,t)}, \quad j = 1, \ldots, M+1;$$

$$\Xi_N(t) := \frac{1}{N} \sum_{i=1}^{N} \delta_{(x_1^N(i,t),\ldots,x_M^N(i,t),x_{M+1}^N(i,t))}. \tag{9.6}$$

Given $\beta > 0$ we then define the time-shifted empirical measure

$$\Xi_N^{\log,\beta}(t) = \frac{1}{N} \left(\sum_{i=1}^{N} \delta_{(x_j^N(i,\frac{\log N}{\beta}+t); j=1,\ldots,M+1)} \right) \in \mathcal{P}(\mathcal{P}(\{1, \ldots, M+1\})). \tag{9.7}$$

Again we denote the empirical mean vector of the $M+1$ types in the new time window by

$$\bar{x}_j^N(\frac{\log N}{\beta}+t) = \int_{[0,1]} x_j \left(\Xi_N^{\log,\beta}(t)\right)(dx), \quad j=1,\ldots,M+1. \qquad (9.8)$$

In the early stage we consider the sparse subset of the N sites at which type $M+1$ appears. Even though we assumed that initially there is no type $M+1$ mass at any site, as a result of rare mutations from types $\{1,\ldots,M\}$ to $M+1$ at rates $\frac{m_{M+1,\mathrm{up}}}{N}$, mutant mass of type $M+1$ will appear. As in the previous section, in order to keep track of the sparse set of sites at which substantial mass appears, we will give a random label to each site and define an atomic measure-valued process as follows.

We assign to each site $j \in \{1,\ldots,N\}$ independently of the process a point $a(j)$ randomly in $[0,1]$ for *all* time, i.e.

$$\{a(j), j \in \mathbb{N}\} \text{ are i.i.d. with the uniform distribution on } [0,1]. \qquad (9.9)$$

We then associate with our process and a realization of the random labels a measure-valued process on $[0,1]$, which we denote by (the superscript 1 indicating one type on the upper level)

$$\mathfrak{z}_t^{N,m,1} = \sum_{j=1}^N \delta_{a(j)} x_{M+1}^N(j,t). \qquad (9.10)$$

This description is complemented by studying the type $M+1$ mass at a typical site or the type $M+1$ mass at a site at which a randomly chosen type $M+1$ individual lives, that is in our context, the Palm measure of a single site.

9.2.3 The Intuitive Picture

As in the two-type model we have two main tasks: (1) Show that emergence of type $M+1$ occurs in times $\frac{1}{\beta}\log N + t$ as $N \to \infty$ for some $\beta \in (0,s)$ (with in general $\beta < \alpha$) which we have to identify, (2) obtain a nice limiting dynamic describing the fixation of type $M+1$ in the macroscopic time variable t and (3) to show a propagation of chaos result for the dynamic in t. The method to handle point (3) propagation of chaos for tagged sites remains unchanged from that used for two types, while (1) and (2) need some serious modification which we explain next.

Intuitively a key feature is that on short time scales a mutation-selection equilibrium among types 1 up to M develops before type $M+1$ comes into play through rare mutation. This means that when type $M+1$ spreads to new sites it faces not just the lowest level type with fitness 0 (as in the two-type model), but low level types $\{1, \ldots, M-1\}$ with fitness *between* 0 and 1. In addition we have the type M which also has fitness 1. Both these facts *reduce* the speed of spread of the high level type $M + 1$. Therefore we can think of the populations with types $\{1, \ldots M\}$ as having for $N \to \infty$ an

$$\text{effective fitness } \tilde{e}, \tag{9.11}$$

where $\tilde{e}$ is the expected fitness of of the types $\{1, \ldots, M\}$ in the selection-mutation equilibrium on the lower-level types $\{1, 2, \cdots, M\}$ arising by restricting the dynamics to this set and removing all upward mutation.

Then heuristically we can think of this as a system of two types with the higher level having an effective fitness advantage of only $1 - \tilde{e}$ (instead of 1 in Section 8) and the analysis of the previous section would predict emergence of the higher level type in $\log N$ time scale, more precisely, in times $\frac{1}{\beta} \log N + t$ with $\beta \in (0, s \cdot \tilde{e})$ and in particular since $\tilde{e} < 1$ we expect the β to be strictly smaller than the α of the two-type case.

This scenario is reflected in the dual process by the replacement of dual individuals by individuals with an *internal structure* which is driven by the *mutation and selection dynamic on E_0* and produces new effects in the growth behaviour of the dual population.

The objective of this section is to confirm this and to explain how this arises from a *refinement* of the dual process which leads to a replacement of the *CMJ-process* given by the McKean–Vlasov dual we used for two types by a *multitype* CMJ-process which then gives the strictly positive exponential growth rate $\beta < \alpha$. We will also see that a CMJ-process with *countably* many types comes into play. We shall explain in detail the more complex structure of the dual in the multitype case in Subsubsection 9.4.1.

9.2.4 Different Mutation-Selection-Migration Parameter Regimes

A natural question is to what extent the longtime behaviour of the system depends on the relative sizes of the selection, mutation and migration rates. One important dichotomy is based on whether or not the mutation matrix on E_0 is irreducible. Our main results on emergence are formulated under the assumption that the mutation matrix on E_0 is irreducible (ergodic). In the case of nonergodic mutation we will exhibit in Subsubsections 9.12.14–9.12.15 some completely new qualitative features of the long time behaviour of the Fleming–Viot process itself but will not explore the extension to this case in depth.

Another distinction of regimes arises from the relative strength of mutation versus selection. This arises as follows from the properties of the dual process.

The growth of the particles in the dual process is driven by the selection rate s which determines the number of birth events. We already saw that state-independent mutation can reduce this growth. We will see here in general that the complexity of the dual can be reduced by sufficiently high mutation rate between the types from E_0 if the mutation on E_0 is ergodic (irreducible). The effect of the high mutation rate (compared to the selection rate) is that we are not faced with large random networks of sites in which the evolution of the internal states of the dual by mutation-selection-coalescence is coupled. This means that for the analysis of our dual process we have to deal with two regimes which require different approaches when it comes to the analysis of emergence.

We will therefore introduce in Subsection 9.7.8 the distinction of *selection* and *mutation-dominant* parameter ranges. In the mutation-dominant case we will prove all our claims in all detail, in the selection-dominant case we will do so only for $M = 2$, the case $M > 2$ is presented with less detail. The discussion of the parameter ranges belonging to neither one is kept short, since no replacement of the L^2-methods is available at this time. Since this would be needed to prove that $^*\mathcal{W} = \mathcal{W}^*$ (see Proposition 9.5), this identity in the intermediate range remains an open problem.

9.3 Statement of Results: Emergence, Fixation and Droplet Formation

In this section we formulate the results on *emergence, fixation* and *droplet formation* in three respective subsubsections for the case of $(M, 1)$-model with M types at the lower level and one type at the higher level which illustrates one of the two principal new effects which occur with multiple types.

9.3.1 Time Scale of Macroscopic Emergence

We now begin by determining the time of emergence. Define the hitting time of intensity ε of the type $M + 1$,

$$\bar{T}_\varepsilon^N = \inf_{t \geq 0}(\bar{x}_{M+1}^N(t) = \varepsilon). \tag{9.12}$$

Then the $N \to \infty$ asymptotics of $\bar{T}_\varepsilon^N$ is as follows.

Proposition 9.1 (Emergence time with M lower level types). *Assume that parameters now satisfy: $d > 0$, $s > 0$, $c > 0$ and*

$$m_{i,j} > 0, \ i, j \in \{1, \ldots, M\}, \ m_{M+1,\mathrm{up}} > 0 \ but \ m_{M+1,\mathrm{down}} = 0. \tag{9.13}$$

9.3 Statement of Results: Emergence, Fixation and Droplet Formation

(a) *There exists a constant $\beta \in (0, \alpha)$ (with α as in Proposition 2.6) such that if $t_N = \frac{1}{\beta} \log N$, $t \in \mathbb{R}$ and $\varepsilon \in (0, 1)$ then at a tagged site:*

$$\liminf_{N \to \infty} P[x_{M+1}^N(t_N + t) > \varepsilon] > 0, \quad (9.14)$$

and for any $\varepsilon \in (0, 1)$:

$$\lim_{t \to -\infty} \limsup_{N \to \infty} P[\sum_{i=1}^{M} x_i^N(t_N + t) < 1 - \varepsilon] = 0, \quad (9.15)$$

$$\lim_{t \to \infty} \limsup_{N \to \infty} P[\sum_{i=1}^{M} x_i^N(t_N + t) > \varepsilon] = 0. \quad (9.16)$$

(b) *The constant β is given in terms of the dual process of the McKean–Vlasov process. More precisely, by the Malthusian parameter (9.359) of the multitype CMJ-process defined in Proposition 9.13.*

(c) *We have for $\varepsilon \in (0, 1)$:*

$$\mathcal{L}[\bar{T}_\varepsilon^N - (\frac{1}{\beta} \log N)] \underset{N \to \infty}{\Longrightarrow} \nu_\varepsilon, \quad \nu_\varepsilon \in \mathcal{P}((-\infty, \infty)), \quad (9.17)$$

where ν_ε is nontrivial (not a point mass). □

9.3.2 The McKean–Vlasov Dynamics for Emergence-Fixation Process

Again we obtain for Ξ^N observed at time $\beta^{-1} \log N + t$, with $t \in \mathbb{R}$ a limiting evolution as $N \to \infty$ characterized as a random entrance law of a *differential equation in a Banach space*, the so-called *McKean–Vlasov equation* (recall (4.28)) which for the present $(M, 1)$ model is given by:

$$\frac{d}{dt}(\mathcal{L}_t) = (L^{\mathcal{L}_t})^*(\mathcal{L}_t), \quad t \in \mathbb{R}. \quad (9.18)$$

Here $\mathcal{L}_t \in \mathcal{P}(\Delta_M)$ a.s., with Δ_M the set of probability measures on $\{1, \cdots, M+1\}$. Recall the notion the entrance law from $-\infty$ which we introduced in Subsection 2.3.1. Then we have

Proposition 9.2 (McKean–Vlasov equation, random entrance law).

(a) *(Existence-Uniqueness) Given $t_0 \in \mathbb{R}$ and $\mathcal{L}_{t_0} \in \mathcal{P}(\Delta_M)$, there exists a unique weak solution to the equation (9.18) for $t \geq t_0$.*

(b) (Mutation-Selection Equilibrium) If $\int_{\Delta_M} \mu(M+1)\mathcal{L}_{t_0}(d\mu) = 0$, then as $t \to \infty$,

$$\mathcal{L}_t \Rightarrow \mathcal{L}_{equil}, \text{ with } \mathcal{L}_{equil}|_{\{1,\ldots,M\}} = P_{equil} \in \mathcal{P}(\mathcal{P}(\Delta_{M-1})) \qquad (9.19)$$

and the latter is the mutation-selection equilibrium on $\{1, \cdots, M\}$ arising for the process defined in (9.1) if we set $m_{M+1,up} = 0$.

(c) (Fixation) If $\int_{\Delta_M} \mu(M+1) d\mathcal{L}_{t_0}(d\mu) > 0$, then

$$\int_{\Delta_M} \mu(\{M+1\})\mathcal{L}_t(d\mu) \to 1 \text{ as } t \to \infty. \qquad (9.20)$$

(d) (Entrance law from $-\infty$) The assertions of Proposition 2.3 (c-f) hold if we replace x_2 by x_{M+1} and α by β from (9.17) (see also Proposition 9.13(e)). In addition $(\mathcal{L}_t)_{t \in \mathbb{R}}$ (not only $\mathcal{L}_t(M+1)_{t \in \mathbb{R}}$) is unique. □

Remark 107. This means that by (d) above again we have a unique entrance law. from $-\infty$ with decay of the mass of type $(M+1)$ given by $e^{\beta t}$ as $t \to -\infty$.

We next state the result on the fixation occurring after the emergence.

Proposition 9.3 (Convergence and limiting fixation dynamics). *Assume that the interacting systems has the initial condition (9.5). Then the following properties hold.*

(a) *For each $-\infty < t < +\infty$ the empirical measures converge weakly to a non-degenerate random measure:*

$$\mathcal{L}[\Xi_N^{\log,\beta}(t)] \underset{N \to \infty}{\Longrightarrow} \mathcal{L}[\mathcal{L}_t] \in \mathcal{P}(\mathcal{P}(\Delta_M)). \qquad (9.21)$$

(b) *In addition we have convergence in path space and identify the limit as follows:*

$$\mathcal{L}[(\Xi_N^{\log,\beta}(t))_{t \in \mathbb{R}}] \underset{N \to \infty}{\Longrightarrow} P \text{ on } C((-\infty,\infty), \mathcal{P}(\Delta_M)), \qquad (9.22)$$

where the probability measure P is such that the canonical process, which we denote by $(\mathcal{L}_t)_{t \in \mathbb{R}}$, is a random entrance law to the McKean–Vlasov equation (9.18).

(c) *For $t \in \mathbb{R}$, denote the "marginals" of $\mathcal{L}_t$ as $(\mathcal{L}_t(1), \ldots, \mathcal{L}_t(M), \mathcal{L}_t(M+1))$, with $\mathcal{L}_t(j)$ being the probability measure on $[0,1]$ corresponding to type j obtained by projecting the M-simplex on its j-th coordinate.*
Then $(\mathcal{L}_t)_{t \in \mathbb{R}}$ is an emergence dynamic, that is:

$$\lim_{t \to -\infty} P[\mathcal{L}_t(\{\mu : \mu(M+1) \in (\varepsilon, 1]\}) > \varepsilon] = 0, \qquad \forall \varepsilon \in (0,1), \quad (9.23)$$

$$\lim_{t \to \infty} P[\mathcal{L}_t(\{\mu : \mu(M+1) \in [1-\varepsilon, 1]\}) < 1-\varepsilon] = 0, \qquad \forall \varepsilon \in (0,1).$$
$$\qquad (9.24)$$

9.3 Statement of Results: Emergence, Fixation and Droplet Formation

(d) Moreover

$$P[\mathcal{L}_t(\{\mu : \mu(M+1) = 0\})] = 0 \text{ for all } t \in \mathbb{R}. \tag{9.25}$$

(e) There exists a $\mathbb{R}_+$-valued random variable $^*W$ such that in distribution:

$$\lim_{t \to -\infty} e^{\beta|t|} \int_{\Delta_M} \mu(M+1)\mathcal{L}_t(d\mu) = {}^*W \text{ and } 0 < {}^*W < \infty \text{ a.s.}. \tag{9.26}$$

(f) The limit dynamics $(\mathcal{L}_t)_{t \in \mathbb{R}}$ is obtained as random shift of the standard solution, namely:

$$(\mathcal{L}_t)_{t \in \mathbb{R}} \stackrel{(d)}{=} (\tilde{\mathcal{L}}_{t+{}^*\mathcal{E}})_{t \in \mathbb{R}}, \quad {}^*\mathcal{E} = \beta^{-1} \log {}^*W, \tag{9.27}$$

where $\stackrel{(d)}{=}$ denotes equality in law and $(\tilde{\mathcal{L}}_t)_{t \in \mathbb{R}}$ is the unique solution of (9.18) with

$$e^{\beta|t|} \int_{\Delta_M} \mu(M+1)\tilde{\mathcal{L}}_t(d\mu) \longrightarrow 1 \text{ as } t \to -\infty. \qquad \square \tag{9.28}$$

Remark 108. If we start with a mass distribution $(x_1, \ldots, x_M, z)$ with $z = 0$ and $\sum_{i=1}^M x_i = 1$ instead of $(1, \ldots, 0, 0)$, then the law of $^*W$ will depend on $(x_1, \ldots, x_M)$ decisively, since the randomness in the evolution comes from the very beginning of the evolution of our system, more precisely the droplet formation. The latter will be different for different initial type distributions in the set $\{1, \cdots, M\}$. However the emergence scale $\beta^{-1} \log N$ will remain.

Remark 109. As before we can also study tagged sites and we obtain the analog of Proposition 2.13 now for the case $(M, 1)$, i.e. for given mean curve the Fisher-Wright diffusion with immigration-emigration is replaced by the corresponding multitype Fisher-Wright diffusion. We don't state this here explicitly since both the statements and the proof works exactly as before. $\square$

9.3.3 Droplet Formation in Early Times $t = O(1)$ and $t \to \infty$

The next objective is to clarify the origin of the emergence of a positive density of type-$(M+1)$ mass. This requires the description of the growth of the small subpopulations of type $M+1$ individuals we call *droplets* which live on a small subset of the geographic space ($o(N)$-sites for $N \to \infty$). For this purpose we begin by introducing the limiting dynamics of the droplet $(\mathfrak{J}_t^{N,m,1})_{t \geq 0}$. The key ingredients are again *excursions*, this time the excursions of the diffusion process in Δ_M which describes the single site dynamic without immigration of the $(M+1)$-type population. Here excursions still only refer to one type, namely $M + 1$, but now starting from the *different* possible configurations of the M other masses.

Example 7. In the case $M = 2$ the basic diffusion is given by the (weak) solution to the system of SDE's

$$dx_3(t) = -cx_3(t)dt + sx_3(t)x_1(t)dt \qquad (9.29)$$
$$+ \sqrt{x_3(t)x_1(t)}dw_3(t) + \sqrt{x_3(t)x_2(t)}dw_2 + \frac{m_{3,up}}{N}(x_1(t) + x_2(t))dt$$

$$dx_1(t) = -cx_1(t)dt + (m_{2,1}x_2(t) - m_{1,2}x_1(t))dt - sx_1(t)(x_2(t) + x_3(t))dt$$
$$+ \sqrt{x_1(t)x_2(t)}dw_1(t) - \sqrt{x_1(t)x_3(t)}dw_3(t) - \frac{m_{3,up}}{N}x_1(t)dt$$

$$dx_2(t) = -cx_2(t)dt + (m_{1,2}x_1(t) - m_{2,1}x_2(t))dt + sx_2(t)x_1(t)dt$$
$$- \sqrt{x_1(t)x_2(t)}dw_1(t) - \sqrt{x_2(t)x(t)_3}dw_2(t) - \frac{m_{3,up}}{N}x_2(t)dt, \qquad (9.30)$$

where w_1, w_2, w_3 are independent standard Brownian motions. Note that if $x_3(0) = 0$, then $x_3(t) = 0$ for all $t \geq 0$, if we let $m_{3,up} = 0$.

We now consider excursions for type $M + 1$ for the $(M + 1)$-type version of the diffusion (9.29) that now plays the same role as (2.33) does in the two type case. An important difference is that we must not only specify the birth time of an excursion but also the proportions $x_1, \ldots, x_M$ of types $1, \ldots, M$ at the birth time of the excursion.

An excursion is therefore given by a continuous path in the simplex Δ_M (here Δ_M is the simplex for $M + 1$ types) beginning from the point $(x_1, \ldots, x_M, 0)$. Let therefore

$$\Delta_{M,0} = \{\mathbf{x} \in \Delta_M, x_{M+1} = 0\}. \qquad (9.31)$$

The set of paths we need is denoted

$$W_0 = \{w \in C([0, \infty), \Delta_M) | w(0) \in \Delta_{M,0},$$
$$\exists \zeta \in (0, \infty) \text{ with } w(r) \in \Delta_{M,0} \text{ for } r \geq \zeta < \infty,$$
$$w(r) \notin \Delta_{M,0} \text{ for } r \in (0, \zeta)\}. \qquad (9.32)$$

Denote the law of the process started with $w(0) = ((1 - \varepsilon)(x_1, \ldots, x_M), \varepsilon)$ with $(x_1, \cdots, x_M) \in \Delta_{M-1}$ and $1 > \varepsilon > 0$ by

$$P^\varepsilon_{x_1,\ldots,x_M} \text{ for } (x_1, \ldots, x_M, 0) \in \Delta_{M,0}. \qquad (9.33)$$

9.3 Statement of Results: Emergence, Fixation and Droplet Formation

We then introduce the corresponding excursion law $\mathbb{Q}_{x_1,\ldots,x_M}$. The idea is to again obtain this as follows:

$$\mathbb{Q}_{x_1,\ldots,x_M}(\cdot) = \lim_{\varepsilon \to 0} \frac{P^\varepsilon_{x_1,\ldots,x_M}(\cdot)}{\varepsilon}. \tag{9.34}$$

The construction is carried out by reducing this to the two type case using a martingale change of measure - see Proposition 10.5 for details.

Let

$$P_{\text{equil}} \tag{9.35}$$

be the mutation-selection equilibrium for the McKean–Vlasov M-type system with types $\{1, \cdots, M\}$ which have fitness $e_1, \cdots, e_M$ and mutation rates $m_{i,j}$.

Then we characterize the limiting measure-valued process of $\mathbf{J}^{N,m,1}$ (where $m = m_{M+1,\text{up}}$) as follows:

Let $N(ds, dx_1, \ldots dx_M, da, du, dw)$ be a Poisson random measure on (recall (9.32) for W_0)

$$[0, \infty) \times \Delta_{M,0} \times [0, 1] \times [0, \infty) \times W_0, \tag{9.36}$$

with intensity measure

$$ds \, P_{\text{equil}}(dx_1, \ldots, dx_M) \, da \, du \, \mathbb{Q}_{x_1,\ldots,x_M}(dw). \tag{9.37}$$

Furthermore let $(\mathbf{J}^0_*(u))_{u \geq 0}$ be given by

$$\sum_i y(i, u) \delta_{a_i}, \tag{9.38}$$

where $\{y(i, \cdot) \in C([0, \infty), \Delta_M), \; i = 1, 2, \ldots\}$ are independent solutions of the M-type version of the system (9.29), (9.30) with the initial conditions

$$y(i, 0) = (x_1(i, 0), \ldots, x_M(i, 0), x_{M+1}(i, 0)) \text{ and } x_{M+1}(i) > 0, \sum_i x_{M+1}(i, 0) < \infty. \tag{9.39}$$

Proposition 9.4 (Convergence of $\mathbf{J}^{N,m,1}$, characterization properties of limiting measure-valued droplet process).

(a) Consider with the ingredients (9.36)-(9.39) the stochastic integral equation for $(\mathbf{J}^{m,1}_t)_{t \geq 0}$ is given, for $t \geq 0$, as

$$\mathbf{J}_t^{m,1} = \mathbf{J}_*^0(t) +$$

$$\int_0^t \int_{\Delta_{M,0}} \int_{[0,1]} \int_0^{q(s,a)} \int_{W_0} w_{x_1,\ldots,x_M}(t-s) \quad (9.40)$$

$$\times \delta_a N(ds, dx_1, \ldots, dx_M, da, du, dw),$$

where $q(s,a)$ denotes the non-negative predictable function

$$q(s,a) := (m + c \mathbf{J}_{s-}^{m,1}([0,1])), \quad (9.41)$$

with

$$m = m_{M+1,up}. \quad (9.42)$$

The equation (9.40) has a unique continuous solution, which we call

$$(\mathbf{J}_t^{m,1})_{t \geq 0}. \quad (9.43)$$

(b) The solution $(\mathbf{J}_t^{m,1})_{t \geq 0}$ is a $\mathcal{M}_a([0,1])$-valued strong Markov process, where $\mathcal{M}_a$ denotes the set of atomic measures and is topologized with the weak atomic topology (for the latter see Section 8, (2.100)).

(c) Given $\mathbf{J}_*^{0,N}(0) \Longrightarrow \mathbf{J}_*^0(0)$ in law as $N \to \infty$, the process $\mathbf{J}^{m,1}$ is given by the limit in law:

$$\mathcal{L}[(\mathbf{J}_t^{N,m,1})_{t \geq 0}] \Longrightarrow \mathcal{L}[(\mathbf{J}_t^{m,1})_{t \geq 0}] \quad \text{as } N \to \infty. \quad (9.44)$$

(d) We have for $\mathbf{J}^{m,1}$ the following longtime behaviour:

$$\mathcal{L}[e^{-\beta t} \mathbf{J}_t^{m,1}([0,1])] \underset{t \to \infty}{\Longrightarrow} \mathcal{L}[\mathcal{W}^*], \quad 0 < \mathcal{W}^* < \infty \text{ a.s.} \quad \square \quad (9.45)$$

Since we can immediately obtain the generalization to the (M, M) case Proposition 10.8, the proof of this result is postponed to Section 10.

Remark 110. In (9.37) we could equally work with another distribution on the lower types than P_{equil}, but then everything is time inhomogeneous, which essentially involves only a notational change. The equilibrium measure as initial law will also be needed later in Chapters 10–13.

9.3.4 Exit=Entrance

Now again the question is whether the entrance and the exit behaviour agree. We already know we have the same exponential growth rate β for both. It remains to show that $^*\mathcal{W}$ and $\mathcal{W}^*$ have the same law.

The statement below is proved in detail here only for the case of mutation dominance and selection dominance, see (9.388) where $\hat{\lambda}$ is introduced as the exponential tail parameter of the so-called decoupling time,

$$2s < \hat{\lambda}, \text{ with } \hat{\lambda} = \hat{\lambda}(c,d,s,m) > 0, \text{ respectively } s > \varsigma\hat{\lambda}. \tag{9.46}$$

The selection-dominant case is more complex and we will give a detailed outline but not include complete details in Subsection 9.12. This identity for the intermediate parameter range remains open but we conjecture that this result are true in all ergodic cases.

Proposition 9.5 (Exit equals entrance). *Assume that we are in the mutation-dominant or selection-dominant case. If we define $\mathcal{W}^*$ by (2.66) but replacing x_2 by x_{M+1} and with underlying dynamics $(\mathbf{1}_t^{N,m,1})_{t \geq 0}$ defined by (9.43), then*

$$\mathcal{L}[^*\mathcal{W}] = \mathcal{L}[\mathcal{W}^*]. \quad \square \tag{9.47}$$

9.4 Dual Representations: Ordered Dual and Tableau-Valued Dual

The Sections 9.4–9.6 will be devoted to a new duality theory for processes with selection and mutation and we briefly outline the purpose and form of this theory.

The principal tool in the study of the $(M,1)$ system is the dual process $(\eta_t, \mathcal{F}_t^{++})_{t \geq 0}$ or its variants $(\eta_t, \mathcal{G}_t^{++})$ and the dual representation (5.90) in Proposition 5.15. Recall that the function-valued part $\mathcal{F}_t^{++}$ lives in the state space $\mathbb{F}$, that is, finite sums of products of indicator functions. We now develop the analysis of this dual by enriching it to an ordered dual $(\eta', \mathcal{F}^{++,<})$ in Subsection 9.4.6 and then we obtain again parallel to the change from $\mathcal{F}_t^{++}$ to $\mathcal{G}_t^{++}$ in Section 5 an *ordered* coupled dual $(\eta', \mathcal{G}^{++,<})$ in Subsection 9.4.7. The enrichment builds some historical information into the state, in particular the order in which birth occurred and from which rank they occurred.

This new ordered dual process drives a functional with Markovian evolution our key dual process, namely a *set-valued process* $(\tilde{\mathcal{G}}_t^{++})_{t \geq 0}$, which allows us to analyse the longtime behaviour and which we introduce and study in Subsection 9.5. A reader only interested in the duality theory can move now directly to Subsection 9.4.6, 9.4.7 and to Subsection 9.5.

The constructions with the dual which we carry out in this Subsection 9.4 apply for general finite type models as well as the set-valued dual we construct in Subsection 9.5. In Subsection 9.6 we use the more specific features of the model to construct *enrichments* and *refinements* of the set-valued dual process which lead to Markov processes $\mathfrak{s}, \mathfrak{s}^*, \mathfrak{s}^{**}, \mathfrak{s}^{****,N}$. The latter are tailored for our present purposes, namely to study emergence and fixation since they are related to CMJ-branching processes, but do not contain some of the information contained in the ordered dual $(\eta', \mathcal{G}^{++,<})$ which is not needed.

9.4.1 Introduction

In order to first give the big picture of the role of the dual we recall the main steps in the proof with $M = 1$. Although this dual is given by a sum of products of factors, in the case $M = 1$, choosing the appropriate initial dual factor we were able to reduce the calculation to a single summand being a *product of indicator functions*, which would then turn to zero as soon as rare mutation acts on one of the factors. More precisely, the resulting single summand involved a product of factors located at a growing number of sites which would then turn zero as soon as rare mutation acts on one of the factors. Therefore we studied effectively a population of factors with locations in space each of which can jump to zero by rare mutation and otherwise is driven by migration of factors, birth of new factors and by death of factors by coalescence. *Therefore for $M = 1$ the dual can be identified with a spatial particle model*, being a collection (indexed by geographic space) of birth and death processes coupled by migration between sites.

The key point used to establish emergence was that before a positive particle intensity is reached the growing number of occupied sites could be identified with a *CMJ branching process* and the critical time scale was determined by the Malthusian parameter associated to this CMJ process which allowed us to determine at what time we would have $O(N)$ factors and hence $O(1)$ probability to become zero by rare mutation. More precisely, the proof of emergence is based on the idea that, up to the first rare mutation, $\mathcal{F}_t^{++}$ is in some sense bounded above by a product of a random number of factors which are indicators on the lower type and this number grows at some exponential rate.

In the case $M > 1$ it is not possible to work with a single summand process of one product of factors, as we shall demonstrate in Subsubsection 9.4.4 below. This causes three problems.

(1) It is therefore not possible to stick with a particle model, but the particles have to be given an *internal structure* changing by mutation-selection on E_0, whose own evolution we have to take into account and
(2) furthermore the evolution of different particles internal states are *coupled* since the evolution of internal states of a particle will depend on the complete configuration at the site and in addition to the parent sites and their parents, etc. Tis leads in addition to migration, to a coupling of the states at *different sites*.
(3) A further complication is that now in order to become zero the factor has to pass after a rare mutation through a whole *chain* of changes of internal states first, instead of just jumping to the zero state.

Therefore we shall have to work with a *multitype version* of the particle model from Section 8 where however the evolution mechanism depended on the total configuration at a site and therefore by migration the internal states of individuals at different sites become dependent. The key point allowing us to proceed with the multitype version is that we have for finitely many particles at a site only finitely

9.4 Dual Representations: Ordered Dual and Tableau-Valued Dual

many types and the total number of particles is stochastically dominated by the old particle system.

The objective of this subsection is therefore to develop a further refinement of the dual and methods to work with it. The key idea is to build in *historical information* and to keep track of the factors and summands as they appear successively by the selection operator. Note that it was already implicit in the description of $(\eta_t, \mathcal{F}_t^{++})$ that when a particle gives birth or two particles coalesce, then this occurs simultaneously in all summands since the function-valued part was driven by the transition in η. We also coupled in Section 5 the transitions of the factors differently in the various summands and by modifying the transition upon birth and selection passing to $(\eta_t, \mathcal{G}_t^{++})$.

The key idea in this Subsection 9.4 is that since the dual expectation involves only the expectation of the sum of the summands only the marginal distribution of the summands is involved. This means that we can introduce an enrichment of the dynamics of the summands and products in the $(\eta_t, \mathcal{F}_t^{++})_{t \geq 0}$ process which will lead to a new dual process $(\eta'_t, \mathcal{F}_t^{++,<})$ respectively $(\eta'_t, \mathcal{G}_t^{++,<})_{t \geq 0}$ in which the *form* of the function-valued part as an *ordered array of summands of products of factors* and this description is part of the state of the process. These arrays of factors in summands of products we call *tableaus*. This form of the dual is of general applicability and allows conveniently for moment calculations.

The tableau-valued process also allows to produce in Subsection 9.5 a new (marked) *set-valued dual process* $\{\tilde{\mathcal{G}}_t^{++}\}_{t \geq 0}$ where the state is a subset of $\prod_{j \in S} \mathbb{I}^{\mathbb{N}}$, where S is the geographic space. This form of the dual process is very convenient to prove *ergodic theorems* and is of general applicability for a broad range of questions.

If we study the emergence of rare mutants, another observation becomes important. The evolution of $\tilde{\mathcal{G}}^{++}$ can also be thought of as driven by a richer Markov process $(\mathfrak{s}_t)_{t \geq 0}$ based on some of the information contained in the time-space process of $(\eta', \mathcal{G}^{++,<})$ and which is the main objective of the further analysis in Subsections 9.7–9.12, since it allows to analyse the emergence and fixation behaviour.

In the case of our mean-field migration process this will allow us to again determine the long-term behaviour via a CMJ branching process but now with *multitype* branching (Subsection 9.7). As before the analysis of the emergence time requires an estimate of the speed at which the dual particle system grows but now combined with an analysis of the composition of $\tilde{\mathcal{G}}^{++}$.

In order to study the fixation regime we have to analyse the nonlinear evolution equations for the full function-valued dual process which does not reduce, as in Section 8 to a dual particle system and hence these equations are now more complex.

However the point is that we can view all the new objects we obtain in the analysis as a "splitting" of a particle in different internal states, so that the descriptions by number of occupied sites and empirical measures of particle numbers as internal states of sites from Section 8 turn into occupied sites with internal states given by occupation vectors of multitype particles. Hence measure-valued processes arising from empirical measures over the population result in measures on a set which is again countable, so that many arguments do not change in essence but only in notational complexity since the set on which the measures live are of a more complex structure. However depending on parameters a problem arises from the fact that the internal states of "particles" at different sites may change simultaneously, as we shall see in Section 9.6. Therefore the size of the *random network* of sites at which such simultaneous changes occur, becomes an important parameter and gives rise to a distinction of mutation versus selection-dominant regimes leading to small respectively fairly large such random networks.

Because of the structural similarities on the one hand and some significant changes just pointed out we proceed here by focussing precisely on those new effects which are different in the $M > 1$ case compared to the $M = 1$ case.

Remark 111. The dual representation that will be developed below is of independent interest in the study of mutation-selection systems with multiple types. It could also be adapted to time dependent and/or random selection.

Outline of Subsection 9.4. We shall give in Subsubsection 9.4.2 a table of orientation to the objects we shall construct in Section 9, Subsubsections 9.4.3 and 9.4.4 serve as warm-up and treat the deterministic case respectively exemplify the problems with some concrete calculations. In Subsubsection 9.4.5 we summarize the conclusions from the examples and give the strategy. In Subsubsection 9.4.6 we construct $(\eta', \mathcal{F}^{++,<})$, then in 9.4.7 $(\eta', \mathcal{G}^{++,<})$ and in 9.4.8 we extend this to the dual of the McKean–Vlasov process.

9.4.2 Summary: Table of Dual Objects

In order to better find the various dual constructions developed and used during the Section 9 and 10 we add here a table of everything to come and everything used from earlier sections to which the reader can return to Table 9.1. The first part of the table comprises objects which can be constructed for general spatial processes on countable abelian groups, the second part uses the special choice of the space we consider later on for our mean-field analysis. Our dual is of the form $(\eta_t, \mathcal{H}_t)_{t \geq 0}$ with always the same underlying particle system η (or some historical enrichment η'_t) but with various possibilities for the process $\mathcal{H}$ which we list now in Table 9.1. The associated state spaces are listed in Table 9.2.

9.4.3 Warm-Up: Emergence for the Deterministic Case $(d = 0)$

As in the two type case, before considering the stochastic case it is instructive to first consider the deterministic case $d = 0$ where there already appears a slowdown of the emergence compared with the case of one type on each level ($M = 1$). For simplicity we consider in this subsubsection only the case $M = 2$.

9.4 Dual Representations: Ordered Dual and Tableau-Valued Dual 395

Table 9.1 Dual objects

Name	S	State space	Dynamics	Appl.	Section
$\mathcal{F}$	Countable set	Function-valued	Signed, FK term	Wellposedness	Subsections 5.1–5.3
$\mathcal{F}^+$	Countable set	Function-valued	Nonnegative, no FK term	Wellposedness	Subsections 5.1–5.3
$\mathcal{F}^{++}$	Countable set	Sum of indicators	Free summands	Wellposedness	Subsection 5.5
$\mathcal{G}^{++}$	Countable set	Sum of indicators	Newly coupled summands	Wellposedness	Subsection 5.5
$\mathcal{F}^{++,\check{}}$	Countable set	Sum of indicators, order	Ordered factors/summands	Longtime behaviour	Subsection 9.4
$\mathcal{G}^{++,\check{}}$	Countable set	Sum of indicators, order	Ordered factors/summands	Longtime behaviour	Subsection 9.4
$\tilde{\mathcal{G}}^{++}$	Countable set	I-valued, $I \subset (\mathcal{I}^{\mathbb{N}})^S$	Marked set-valued	Longtime behaviour	Subsection 9.5
$\tilde{\mathcal{G}}^{++,\ell}$	$\mathbb{N}$	Enriched I	Options I-III ($\ell = I - III$)	dual if $X_0 = \mu^\otimes$, $\ell = I, II$	Subsection 9.6
$\mathfrak{s}$	$\mathbb{N}$	$\mathfrak{S}$	McKean–Vlasov	Dual if $X_0 = \mu^\otimes$	Subsection 9.6
$\mathfrak{s}^*$	$\mathbb{N}$	$\mathfrak{S}^*$	Non-Markov	Auxiliary	Subsection 9.6
$\mathfrak{s}^{**}$	$\mathbb{N}$	$\mathfrak{S}^{**}$	h-transformed	Malthusian par.	Subsection 9.6
$\mathfrak{s}^{****,N}$	$\{1,\ldots,N\}$	$\mathfrak{S}_N^{****}$	Ghost clan dynamics	Evolution eq.	Subsection 9.9

Table 9.2 State spaces

Notation	Defn.	Section
$\mathbb{I} = \mathbb{I}_M$	$\{1,\ldots,M,M+1\}$	Section 9
$\mathcal{I}$	Algebra of subsets of $\mathbb{I}$	Section 9
$\mathcal{T}$	(9.176)	Section 9
$\mathcal{T}^*$	(9.226)	Section 9
$\mathcal{T}^{**}$	(9.229), (9.231)	Section 9
Υ	(9.237)	Section 9
$\mathfrak{T}$	(9.242)	Section 9
$\mathfrak{T}^{**}$	(9.242)	Section 9
$\mathfrak{T}^{****}$	(9.423)	Section 9
$\mathfrak{S}_N^{****}$	(9.437), (13.35)	Section 9

Again we will see that emergence for $d = 0$ arises at time $\frac{1}{\tilde{s}} \log N + t$ and with

$$\tilde{s} = ps \tag{9.48}$$

the effective selective advantage being $(1 - p)s$, where p is the equilibrium proportion of type 1 in the type 1,2 system, which means that below in (9.51) we put $m_{3,\text{up}} = 0 = m_{3,\text{down}}$ and start with $(x, y, 0)$ and look for the constant solution $(\bar{x}, \bar{y}, 0)$ getting $p = \bar{x}$. The value $\tilde{s}$ is bigger than the value in the stochastic case once again as was the case for the two-type case. We see however the slowdown from s to ps due to more than one type on the lower level.

In the deterministic case migration is irrelevant for the evolution. To see this simply pass to the dual and note that without coalescence the location of particles does not influence the transitions of the dual process. The objective is to prove the analogue of Prop. 6.1 for the case of 3 types (two types at level 1 and one type at level 2). The proof again uses the dual representation which in this case has a more complex structure than in the two type case.

Let μ be the initial state of the process which is a collection of N measures on the type space $\{1, 2, 3\}$ of which we can form the product measure denoted $\mu^\otimes$. The corresponding object at time t is $\mu_t^\otimes$. We use the notation 100, 010, 001 for the indicators of the three types and for the set of the two types 1 and 2 the symbol 110 etc.

Then set (note that both the various functions y_i and $\mathcal{F}_t^{++}$ depend on N without exhibiting this in the notation)

$$y_{100}(t) := E_{(100)}[\int \mathcal{F}_t^{++} d\mu^\otimes]. \tag{9.49}$$

Since $\mu(\{3\}) = 0$, we have $y_{001}(0) = 0$. Also note that $y_{110} = y_{100} + y_{010}$.

Start the dual $\mathcal{F}_t^{++}$ at $\mathcal{F}_0^{++} = (100)$ (we sometimes also use the short notation $1_{\{1\}}$ for the indicator of type 1) or with two particles $(100) \otimes (100)$, etc. Then by independence (because there is no coalescence) we have

9.4 Dual Representations: Ordered Dual and Tableau-Valued Dual

$$E_{(100)\otimes(100)}[\int \mathcal{F}_t^{++} d\mu^\otimes] = (E_{(100)}[\int \mathcal{F}_t^{++} d\mu^\otimes])^2, \text{ etc.} \quad (9.50)$$

Using this fact, the moment equations become closed and we can write down the backward equations for the first moments. In order to simplify, recall that we take type 2 also of fitness 1. Then with this assumption we obtain the system of ODE for the proportions of type 1, 2, 3, respectively:

$$\frac{dy_{100}}{dt} = -m_{12} y_{100} + m_{21} y_{010} + s(y_{100}^2 - y_{100}) - \frac{m_{3,\text{up}}}{N} y_{100} + \frac{m_{3,\text{down}}}{N} (y_{101} - y_{100})$$

$$\frac{dy_{010}}{dt} = m_{12} y_{100} - m_{21} y_{010} + s y_{010} y_{100} - \frac{m_{3,\text{up}}}{N} y_{010} + \frac{m_{3,\text{down}}}{N} (y_{011} - y_{010})$$

$$\frac{dy_{001}}{dt} = \frac{m_{3,\text{up}}}{N} y_{110} - \frac{m_{3,\text{down}}}{N} y_{001} + s(y_{001} y_{100}). \quad (9.51)$$

Then it can be verified from these equations that if $m_{3,\text{up}} > 0$, $m_{3,\text{down}} = 0$ then $y_{001}(\cdot)$ reaches $O(1)$ in time $\frac{\log N}{\tilde{s}}$ and

$$\lim_{t \to \infty} \lim_{N \to \infty} y_{001}(\frac{\log N}{\tilde{s}} + t) = 1, \quad \tilde{s} = \frac{m_{12} s}{m_{12} + m_{21}}. \quad (9.52)$$

Moreover, even if $m_{3,\text{down}} > 0$, then the equilibrium value of y_{001} still goes to 1 as $N \to \infty$ in that scale.

9.4.4 Warm-Up: Dual Calculations with $M = 2$ for Emergence

In order to introduce some of the new issues involved in the proofs and some of the tools required in the multitype case we begin by considering as example the special case $M = 2$ and present some preliminary calculations with the dual $(\eta_t, \mathcal{F}_t^{++})_{t \geq 0}$ which will then explain why and how we have to develop a modification of the dual.

We assume that the initial state μ of the process satisfies:

$$\mu(i, \{3\}) = 0, \; i = 1, \ldots, N, \quad \mathcal{L}[\Xi_N(0)] = \delta_\mu$$

where $\mu = (b, 1 - b, 0)$ for some $b \in [0, 1]$. \quad (9.53)

The first question is to determine the rate of decay of the total mass of the types 1 and 2 in the original process. We therefore study the behaviour of the dual process with the appropriate initial state.

Before we come to the details of the proof let us explain what is new in the behaviour of the dual process compared to the $M = 1$ case. Before in the dual process we had in $\mathcal{F}_t^{++}, \mathcal{G}_t^{++}$ factors $1_{\{1\}}$ and $1_{\{2\}}$ only, but now we have $1_{\{1\}} + 1_{\{2\}}$

and $1_{\{3\}}$ as the analogous quantities which reflects that in the forward evolution the *dynamics between type 1 and 2* enters the picture in a decisive way. In the backward evolution this requires us to work with a dual process that is given by *sums of products of indicators* instead of just a string of factors, each an indicator.

We begin in this subsection in three steps. In Step 1 we describe the structure of the dual in the $M = 2$ case and develop comparison systems. We use these comparisons to obtain upper and lower bounds on the constant β, i.e. $0 < \beta < \alpha$, for the emergence time $\frac{1}{\beta} \log N + t$. We first get a bound from above in Step 2 and then a bound from below in Step 3. The calculations used to obtain these bounds are very useful in the sequel (in particular the ideas in Step 3 will reappear in Section 10) but cannot be readily sharpened to give the precise time of emergence as we shall see. This happens only in Subsection 9.7 where we refine the analysis to get the exact time scale, i.e. we identify β as a growth constant in a new dual process, but this requires quite some preparation in this and the next subsection.

Step 1   *The evolution of the dual: some explicit calculations*

We begin by doing some moment calculations using the dual representation based on the refined dual $(\eta_t, \mathcal{F}_t^{++})_{t \geq 0}$ of Subsection 5.5. In the case $M = 1$ this object could be simplified with $\mathcal{F}_0^{++} = 1_{\{1\}}$, since then $\mathcal{F}^{++}$ is a product of indicators. The new point for $M = 2$ is that in this case we must work with a dual process that involves not only products of indicator functions but also sums of products.

Notational Conventions

We can view the system $X^N(t)$ as a collection $\{x_i^N(\ell, t); i = 1, 2, 3\}_{\ell=1,\ldots,N}$ and for each t view $x^N(\ell, t)$ as a *probability measure* on the types $\{1, 2, 3\}$. We use the following notation:

$$(x^N(t))^\otimes = x^N(1, t) \otimes \cdots \otimes x^N(N, t). \tag{9.54}$$

If we consider a tagged site we drop the index ℓ of the location since we can w.l.o.g. consider $\ell = 1$. Then denoting by $\int$ the integral over the measure in (9.54) we get using duality:

$$E[\int \mathcal{F}_0^{++} d(x^N(t))^\otimes] = E[\int \mathcal{F}_t^{++} d(x^N(0))^\otimes] \tag{9.55}$$

$$= \int E[\mathcal{F}_t^{++}] d(\mu^\otimes).$$

In order to compute the first moments of $x_1^N(t) + x_2^N(t)$ or of $x_3^N(t)$ at a tagged site we take the initial conditions

$$\mathcal{F}_0^{++} = 1_{\{1\}} + 1_{\{2\}}, \text{ placed at site 1 or } \mathcal{F}_0^{++} = 1_{\{3\}} \text{ placed at site 1.} \tag{9.56}$$

9.4 Dual Representations: Ordered Dual and Tableau-Valued Dual

It is useful to encode the indicator functions 1_A of a subset A of $\{1,2,3\}$ in the form

$$(000), (100), \cdots, (111) \tag{9.57}$$

writing a one or zero if the corresponding type belongs to A or not. We also use the identifications $(011) = (010) + (001)$, $(110) = (100) + (010)$. When needed, we denote the location of factors by writing

$$(110)_\ell \quad , \quad \ell \in \{1, \cdots, N\}. \tag{9.58}$$

Often we will only write $(110)_1, (110)_2$, etc. to indicate that we have different locations since, due to exchangeability of the law at fixed times, the actual site is not relevant for us. Whenever the location plays no role we simply suppress the index in the notation. □

We now describe the dynamics of the dual $(\mathcal{F}_t^{++})_{t \geq 0}$ in this notation. We have to consider (1) births in η associated with the action of the selection operator on $\mathcal{F}_t^{++}$, (2) mutation, (3) coalescence acting on $\mathcal{F}_t^{++}$ and (4) changing η by migration. We now go through this list, starting with the most complicated transition.

(1) The possible transitions of the function-valued part $\mathcal{F}_t^{++}$ in the dual process $(\eta_t, \mathcal{F}_t^{++})$ due to a *birth event in η combined with the action of the selection operator on $\mathcal{F}_t^{++}$* are as follows:

$$\begin{aligned} (001) &\to (001) + (001) \otimes (100) \quad \text{rate } s \\ (110) &\to (010) + (110) \otimes (100) \quad \text{rate } s \\ (100) &\to (100) \otimes (100) \quad \text{rate } s \\ (010) &\to (010) + (010) \otimes (100) \quad \text{rate } s. \end{aligned} \tag{9.59}$$

Remark 112. Note that as indicated in Section 9.2 assumption (9.4) we consider only $1_{A_2} = 1_{(011)}$ (instead of 1_{A_2} and $1_{A_3} = 1_{(001)}$). Here the selection operator 1_{A_3} which would occur with rate $O(N^{-1})$ if we would aply (9.3) can be neglected since it does not play any role in the emergence. The point is that it introduces up to the time we have N dual particles only $O(1)$ factors which could change under rare mutation and hence that even with N dual particles it could only affect $O(1)$ factors under upward mutation.

(2) The *mutation* jumps result in the following transitions of $\mathcal{F}_t^{++}$:

$$\begin{aligned} (100) &\to (000) \quad \text{rate } m_{12} \\ (100) &\to (110) \quad \text{rate } m_{21} \\ (010) &\to (110) \quad \text{rate } m_{12} \\ (010) &\to (000) \quad \text{rate } m_{21} \end{aligned} \tag{9.60}$$

$$(100) \text{ or } (010) \to (000) \quad \text{rate } \frac{m_{3,\text{up}}}{N}$$
$$(001) \to (011) \text{ or } (101) \quad \text{rate } \frac{m_{3,\text{up}}}{N}$$
$$(010) \to (011) \quad \text{rate } \frac{m_{3,\text{down}}}{N}$$
$$(100) \to (101) \quad \text{rate } \frac{m_{3,\text{down}}}{N}.$$

Note that for the mutations $1 \to 2$ and $2 \to 1$ we have:

$$(110) \longrightarrow (110). \tag{9.61}$$

(3) The *coalescence* of factors of indicators corresponding to two variables results in a single factor corresponding to the intersection of the sets, such as

$$(110) \otimes (100) \to (100) \text{ etc. }, \tag{9.62}$$

but again

$$(110) \otimes (110) \longrightarrow (110) \tag{9.63}$$

and only the number of such factors is affected by coalescence.

(4) In addition to selection and mutation the process $\mathcal{F}_t^{++}$ can change the location of factors since individuals in the process η are *migrating* between sites in $\{1, \ldots, N\}$. Hence we have

$$(***)_k \longrightarrow (***)_\ell, \ k \neq \ell, \text{ at rate } \frac{c}{N-1}. \tag{9.64}$$

Remark 113. The process of factors exhibits traps and quasi-traps. Note that for the evolution of the dual process

$$(111) \text{ and } (000) \text{ are traps} \tag{9.65}$$

and

$$(110) \text{ and } (001) \text{ are "quasi-traps" under mutation and coalescence,} \tag{9.66}$$

where quasi-traps are factors that survive (i.e. do not change to (000)) up to an exponentially distributed random time with mean $O(N)$. We will see later that even under selection (110) factors proceed to traps or return to this state after finite time. Therefore it makes sense to call (110) a quasi-trap.

Step 2 *Lower bound for emergence time*

Recall that we are working in the case that the parameters satisfy: $d > 0$, $s > 0$, $c > 0$, $m_1 \neq 0, m_2 \neq 0, m_{3,\text{up}} > 0, m_{3,\text{down}} = 0$ and we assume that

9.4 Dual Representations: Ordered Dual and Tableau-Valued Dual

$$\mathcal{F}_0^{++} = (110). \tag{9.67}$$

One can show that eventually one reaches (111) with probability 1. To verify that this happens in time $const \cdot \log N$ one needs exponential growth in the number of factors.

We look for the first upward mutation to type 3 from type 1 or 2. We can bound the dual expression for the expectation of the total mass of type 1 and 2, i.e. the dual with the initial state (110) from below to obtain a *lower bound on the emergence time*.

Lemma 9.6 (Lower bound system for emergence time). *Under selection, more precisely selection acting based on 1_{A_2} (denote the action by Ψ) we can estimate the dual starting in $\mathcal{F}_0^{++} = (110)$ from below (and hence the one starting in (001) from above) using for each action of the selection operator the bound:*

$$\Psi(110) \geq (110) \otimes (110), \tag{9.68}$$

and emergence of type 3 does not occur at times $\gamma \log N$ with $\gamma < \frac{1}{\alpha}$. □

Proof of Lemma 9.6. We calculate:

$$\Psi(110) = (010) + (110) \otimes (100) = (101) \otimes (111) + (110) \otimes (100)$$
$$= (010) \otimes (111) + (110) \otimes (110) - (110) \otimes (010)$$
$$= (110) \otimes (110) + (010) \otimes (001)$$
$$\geq (110) \otimes (110). \tag{9.69}$$

Note that in (9.69) the r.h.s. is exactly what we would get from the selection operator if both types 1 and 2 had fitness level 0 and type 3 fitness level 1 which would then be exactly the same as the previous analysis with $K = 2$ if we consider the dual up to the first appearance of an upward mutation and start it in the indicator (110). Therefore we get a lower bound for the expected mass of type 1 and 2 and hence an *upper bound on the expected mass of type* 3 and hence by comparison with the $M = 1$ case we know that emergence does not occur in time $\gamma \log N$ with $\gamma < \frac{1}{\alpha}$. Hence we can conclude that if emergence occurs by time $\beta^{-1} \log N$, then

$$\beta \leq \alpha. \quad \text{q.e.d.} \tag{9.70}$$

Step 3 *Upper bound for emergence time*

We next find a lower bound for $E[x_3^N(t)]$ in order to establish emergence for some $\log N$ time scale. Recall that to make things a little simpler here we had assumed that both types 2 and 3 have fitness 1 (instead of $1 - N^{-1}$ and 1) but this provides a lower bound since type 3 looses a bit of its advantage.

We now derive a further comparison that will yield a lower bound on the growth of the rare mutant population and hence an *upper bound on the emergence time*. To do this consider the dual with a different initial state as above, namely

$$\mathcal{F}_0^{++} = (001). \tag{9.71}$$

Now the dual expression gives zero until we have the first mutation jump to type 3 on the initial factor. However the development of the dual prior to this plays an important role since the evaluation of the actual value of the integral in (9.55) will be heavily dependent on this. That is, although we must wait quite long until such a jump occurs, we must study how the dual expression has grown by the selection mechanism due to the order 1 advantage of type 2 and 3 over type 1. Note that (110) changes to (000) (through upward mutation) only at rate N^{-1}. However before the rare mutation jump occurs, the growth of the number of summands of such factors can compensate for small probabilities for the mutation jump in a certain time window.

Recall that the selection operator acting with 1_{A_2} on (001), migration and mutation (from type 2 to type 1) yields jumps of the following form with positive rate $\bar{s} \in (0, s)$ (recall the first jump below has rate s):

$$(001)_1 \to (001)_1 + (001)_1 \otimes (100)_1$$
$$\to (001)_1 + (001)_1 \otimes (100)_2 \to (001)_1 \otimes (1_1 + (110)_2), \tag{9.72}$$

where the subscripts $1, 2 \in \{1, \ldots, N\}$ indicate two different sites and we include only the production of terms where the factor (100) moves to a new site before any other transition occurs at this site and the last change $(100) \to (110)$ due to mutation from type 2 to type 1 occurs after a finite time after arriving at the new site. (In fact in general the mutation mechanism will kill off some factors by mutation from 1 to 2 and change others to (110) by mutation from 2 to 1). Nevertheless since no negative terms appear in (9.83) and hence in the r.h.s. of (9.55), this yields the following lower bound for the expected mass of type 3.

Lemma 9.7 (Upper bound system for emergence time).

(a) *Consider* $\mathcal{F}_0^{++} = (001)$. *Then (ignoring the migration steps leading to "collisions") there exists* $\bar{s} \in (0, s)$ *with:*

$$E[\mathcal{F}_t^{++}] \geq E[(001)_1 \otimes (1 + (110)_2) \cdots \otimes (1 + (110)_{k(t)})], \tag{9.73}$$

where $k(t)$ is a Poisson process with rate $\bar{s}$.

(b) *Assume that $x_3^N(0) = 0$, a.s. Then*

$$\lim_{N \to \infty} E\left[x_3^N\left(\frac{\log N}{\bar{s}} + t\right)\right] > 0. \qquad \square \tag{9.74}$$

9.4 Dual Representations: Ordered Dual and Tableau-Valued Dual

Proof of Lemma 9.7. (a) Note that the first transition in (9.72) occurs at times given by a Poisson process with rate s. Then the second and third transitions in (9.72) occur in a time interval of length 1 with positive probability. Therefore a lower bound on the production of the factors $(1+(110))$ at distinct sites is determined by a Poisson process with some rate in $(0, s)$.

(b) We next claim that even including collisions, the number of factors $(1+(110))$ on the right hand side of (9.73) at time $\frac{\log N}{\bar{s}}+t$ is $O(N)$ and therefore we get a factor of the form $\geq C2^N$ for some $C > 0$ for all large N. The precise proof of this is as follows.

First we note that a rare mutation from either type 1 or type 2 to type 3 will have two effects - first and most important it changes the first factor on the r.h.s. of (9.73) namely (001) to (011) or (101) thus making the integral of $\mathcal{F}_t^{++}$ with the initial state (recall (9.53)) non zero.

In addition, it can kill off some terms in the other factors by changing, for example, (100) to (000). However this is negligible until the number of these factors is of order N. In fact this process continues as the number of factors grows and eventually there is an equilibrium with $O(N)$ factors but we do not use this fact at this time.

Each migration step in the sequence of events in (9.72) without the last step produces a factor $(1+(100))$ which gives after taking the expectation over the initial state the factor $(1+a)$ where a is the expected mass of type 1 initially. The whole sequence produces the factor 2 for initial states concentrated on type 1 or 2. The whole sequence of steps as given in (9.72) produces therefore the factor $(1+(110))$ and after the transition 1 or 2 to 3 this contributes a factor 2 to the expectation. Let $n(t)$ denote the number of such factors $(1+(110))$ produced.

Note that as long as we can ignore rare mutations of the factors $(1+(110))$ (which results in $1+(110) \to 1$) the production of these factors is bounded below by a Poisson process at some rate $\bar{s}$ as in (9.73). However we must take into account the fact that these factors $(1+(110))$ are only quasi-traps, i.e. they can change in times of $O(N)$ back to 1. This is due to the fact that the factors (110) produce a growing number of additional (110) factors by selection and then the whole product will have a positive death rate as soon as the number of such factors reaches N. Namely death occurs if a mutation to type 3 happens with subsequent mutations of type 1 or 2 (recall fourth line of (9.68)).

Therefore $(n(t))_{t\geq 0}$ can be viewed as an immigration-death process with constant immigration rate $\bar{s}$ (recall (9.72)) and *age dependent death rate*. Then the mean at time t is as $N \to \infty$ given as:

$$E[n(t)] \sim \bar{s}t, \qquad (9.75)$$

provided that $t = o(N)$, since then two conditions are satisfied, namely new sites are available and hence no collisions occur and in addition no rare mutation, i.e. death, takes place.

Since $a = \mu(110) = 1$, we have effectively at time $t_N = \tau + \frac{\log N}{\bar{s}}$, with $\tau \in \mathbb{R}$:

$$E[\int \mathcal{F}_{t_N}^{++} d\mu^{\otimes}] \geq E[(1+1)^{n(t_N)} 1_{A(t)}] = \frac{1}{2} E[2^{n(t_N)} 1_{A(t)}], \qquad (9.76)$$

where $A(t)$ is the event that a rare mutation has changed the initial factor $(001) \to (101)$ or (011) (each with probability $\frac{1}{2}$) and this event has rate $\frac{m_{3,\mathrm{up}}}{N}$ *independent* of what happens to all the other factors.

On the other hand we know that

$$E[2^{n(t_N)}] = \sum_{k=0}^{\infty} \frac{2^k (\bar{s} t_N)^k}{k!} e^{-\bar{s} t_N} = e^{(\bar{s} t_N)} = O(N). \qquad (9.77)$$

Therefore since (9.75) is valid for times $t = \frac{\log N}{\bar{s}} + \tau$ with τ such that $\tau \ll 0$, we obtain the lower bound establishing that $\beta > 0$ by having proved now:

$$E[x_3^N (\frac{\log N}{\bar{s}} + \tau)] = O(1) \text{ as } N \to \infty. \qquad \text{q.e.d.} \qquad (9.78)$$

Remark 114. It turns out that it is easier to complete this analysis and to identify the *exact* logarithmic time scale using the dual with initial value (110) and this will be carried out in the next subsubsection.

9.4.5 *The Strategy for Getting an Enriched Dual*

In the next two subsubsections we shall define in two steps a new enriched dual process which even though more complex can be better controlled in many situations and allows us furthermore to also construct auxiliary processes we need for our analysis of emergence and fixation.

This dual provides the first step in the analysis necessary for the proof of emergence of type $M + 1$ in time scale of order $\log N$, and to identify the correct constant. This requires that we refine and generalize the above analysis of the dual process $\mathcal{F}_t^{++}$ to compute the moments of $(x_1^N(t) + \cdots + x_M^N(t))$ to show that the first moment for $t_N = \beta^{-1} \log N + \tau$ is close to 0 for τ large and close to 1 for τ small, (i.e. $\ll 0$).

For this purpose we work with

$$\mathcal{F}_0^{++} = (1, \ldots, 1, 0). \qquad (9.79)$$

The goal is now to bring the dual into a form such that we can read off two facts:

- There exists $\beta \in (0, s)$ such that $\mathcal{F}_t^{++}$ is equivalent (w.r.t. the value of the expectation in the duality relation) to an expression which has $O(N)$ factors at time $\beta^{-1} \log N + T$, with T a $\mathbb{R}$-valued random variable.
- Since we assume (9.53), the integral of $\mathcal{F}_t^{++}$ with $\mu^\otimes$ is 1 before the first rare upward mutation event and is strictly less than 1 with positive probability afterward.

Together these points imply emergence at times $\beta^{-1} \log N + T$ with $T \in \mathbb{R}$.

As explained in Subsubsection 9.4.4 in Step 1, in the $M = 2$ case, the dual process is given by a *sum* of products of indicator functions. The number of factors follows a birth and death process given by $|\eta_t|$, and the number of summands increases by one whenever we have a birth event in η and hence it appears that we get a very complex expression. To simplify this expression we took advantage of the fact that factors can become trivial, i.e. 0 or 1, but more importantly of *cancelation effects* and arrived at $(\eta, \mathcal{G}^{++})$ which is also crucial here. We shall introduce a simpler structure for this object in order to keep track of its evolution.

To achieve this simplification we shall in the next subsubsection introduce an enriched version of $\mathcal{F}_t^{++}$ denoted by $\mathcal{F}_t^{++,<}$ which contains some *historical* information (in particular on the order in appearance of the action of the selection operator up to time t) and introduces a *new order*. This will allow in the subsequent subsubsection to implement the new idea for the process $\mathcal{G}^{++}$, which is enriched to $\mathcal{G}^{++,<}$ in Subsubsection 9.4.7 and in the next Subsubsection 9.4.6 we focus in constructing the process $\mathcal{F}_t^{++,<}$ needed for that purpose.

9.4.6 Modification of Dual 1: $\tilde{\mathcal{F}}_t^{++,<}$ with Ordered Factors

In this subsubsection we carry out the modification of the dual $\mathcal{F}_t^{++}$ constructed in Subsection 5.5 involving an enrichment (see Remark 115). In the case of our (M, M)-model we introduce here a simplified version applying for the case of M lower level types and one higher level type $M + 1$ with only up-mutation to the higher level. (The simplified dual for the (M, M) system is developed in the Section 10.)

Remark 115. That enrichments of dual processes again give duality relations as a general principle. For this we recall the general form of a duality (see (5.1)) and note that if we have for X a dual process Y with a duality function $H(\cdot, \cdot)$ and if we can introduce a new process Y^* on a new state space $E'^{,*}$ such that there exists a map

$$\kappa : E'^{,*} \to E' \text{ such that } \mathcal{L}[\kappa(Y_t^*)] = \mathcal{L}[Y_t]; \qquad (9.80)$$

then we have a duality between X and Y^* with duality function $H^* : E \times E'^{,*} \to \mathbb{R}$ by setting

$$H^*(\cdot, \cdot) = H(\cdot, \kappa(\cdot)). \qquad (9.81)$$

Remark 116. Typically such enrichments arise from a duality relation of the new dual to an enriched original process. We no not use this fact here and hence do not prove here this new relation.

We proceed in six steps.

Step 1: *Preparation: further observations on the nature of $\mathcal{F}_t^{++}$.*

As a preliminary step to the construction of the modified dual process that we shall use in the case $M \geq 2$ we now look in more detail at the structure of the function-valued part $\mathcal{F}_t^{++}$ of the dual process that appears on the r.h.s. of (9.55) before the first rare mutation occurs. For that purpose we view $\mathcal{F}_t^{++}$ not only as a function on products of the type space but we explicitly work with its form as a *sum* of *products of indicators* where new factors or new summands occur in a certain *order* in time, in other words, both this specific form of the dual state and the historical information will become part of the state, see (9.84) below. The historical information about the process η is the information about the complete ancestral relations in the particle system η. This will be, even though looking more complicated at first sight, easier to evaluate.

Example 8. Consider the special case $M = 2$. Define the abbreviations:

$$f_1 = (110), f_2 = (100), f_3 = (010), f_4 = (000), \tag{9.82}$$
$$f_5 = (111), f_6 = (001), f_7 = (011), f_8 = (101).$$

Let $\mathbb{F} = (f_1, \ldots f_8)$ denote the set of possible factors.

Then starting with $\mathcal{F}_0^{++} = (110)$ and applying selection and the lower level mutation operations we have an object of the form:

$$\mathcal{F}_t^{++} = \sum_{i}^{\tilde{N}_t} \bigotimes_{j=1}^{N} f_1^{\otimes k_1(i,j,t)} f_2^{\otimes k_2(i,j,t)} f_3^{\otimes k_3(i,j,t)} \cdots f_8^{\otimes k_8(i,j,t)} := \sum_i^{\tilde{N}_t} \mathcal{F}_{i,t}^{++}, \tag{9.83}$$

where $\tilde{N}_t$ denotes the (random) number of summands and

$$k_1, k_2, k_3, \cdots, k_8 : \{1, \cdots, \tilde{N}_t\} \times \{1, 2, \cdots, N\} \times [0, \infty) \longrightarrow \mathbb{N}_0, \tag{9.84}$$

are appropriate (random) functions which depend on t, for example, $k_1(i, j, t)$ is the number of $f_1 = (110)$ factors in the ith summand at time t located at $j \in \{1, \ldots, N\}$. We can encode all information by coding $\mathcal{F}_t^{++}$ in the form of a different type of object (misusing notation since we refrain from using a new letter), since we go even further below introducing $\mathcal{F}^{++,<}$, where we explicitly spell out the dynamics:

$$\mathcal{F}_t^{++} = (\tilde{N}_t, \{(k_1(i, j, t), \cdots, k_8(i, j, t)); i \in \{1, \cdots, \tilde{N}_t\}, j \in \{1, \cdots, N\}\}). \tag{9.85}$$

9.4 Dual Representations: Ordered Dual and Tableau-Valued Dual

We can perform calculations by expressing the dual expectation in terms of a suitably chosen population of factors, which gives the dynamic of the $(k_1(i, j, t), \cdots, k_4(i, j, t))$. In particular we now consider the population of summands and the dynamic of the exponents in each summand. However we can give a more refined description by noting that whenever a birth event in η occurs and the selection operator acts on $\mathcal{F}_t^{++}$, we can get new summands. We can therefore record the information on the order at which variables appeared in the expression and we can also order the summands by splitting at each selection event a summand in two (ordered) summands getting a matrix with entries among $\{f_1, \cdots, f_8\}$.

After the first step starting with f_1 we get the tableau

$$f_1 \, 1_A \quad 1 \atop f_1 \quad\ 1 - 1_A, \qquad (9.86)$$

if the selection operator 1_A acts on the function f_1.

Remark 117. Note we could have if our present state f satisfies $1_A f \equiv 0$ that

$$f \longrightarrow 1_A f + (1 - 1_A) \otimes f = (1 - 1_A) \otimes f \qquad (9.87)$$

and even so a birth occurs we get no new summand in $\mathcal{F}^{++}$ nor $\mathcal{G}^{++}$.

Step 2: *(The basic idea: passing to tableaus)*

The key trick is to order factors and introduce factors 1. This is based on the following observation. We can also keep track of some historical information in order to deal with the sum on the r.h.s. of (9.83). In particular we introduce an *ordering of the factors* that allows us to *associate factors in different summands*. First note that a product of indicators can change into a *sum* of products of indicators only by the selection mechanism and all other transitions preserve the product structure of the indicators of a summand in the $\mathcal{F}_t^{++}$ respectively $\mathcal{G}_t^{++}$ dynamic.

Recall next that selection operates by $1_A \cdot f + (1 - 1_A) \otimes f$ with $1 - 1_A$ corresponding to a new variable. Observe first that we can write without changing integrals but having the same number of variables for both summands as

$$f \to 1_A \cdot f \otimes 1 + f \otimes (1 - 1_A), \qquad (9.88)$$

where with $M = 2$ types we get $1_A = (011)$. Then it is useful to place the new factors 1 and $(1 - 1_A)$ not at the end of the product but next to the factor giving birth.

Remark 118. Recall that in the $\mathcal{G}^{++}$ we have *reordered* the factors so that in all summands $1_A \cdot f$ and $(1 - 1_A)$ always are associated with the *same* variable, we call this having the same *rank*. Then we can couple the operations on different summands in a different way so that both are operated on simultaneously by the mutation, selection and migration operators acting at a given rank (and not particle). This will be exploited again in the next subsubsection.

From (9.88) we see that we get a sum of two terms now involving a new variable corresponding to the particle added by the birth event in the dual particle system. As time goes on this produces a *binary tree* and each path in this tree starting from the root and ending in a leaf corresponds uniquely to a summand which consists of an ordered product of factors.

In general and more precisely we make the resulting two summands comparable in the sense that they have the same number of variables in both summands in (9.148), but without changing the value. To do this we write this action of the selection operator as follows. Take a function $f = f^3 \otimes f^1 \otimes f^2$, where f^1 is a function of *one* variable, the one on which selection acts and f^2, f^3 of the remaining variables ordered in the order of the corresponding partition elements in the ordered particle system. Write:

$$f \to f^3 \otimes (1_A \cdot f^1) \otimes 1 \otimes f^2 + f^3 \otimes f^1 \otimes (1 - 1_A) \otimes f^2, \qquad (9.89)$$

where 1 stands for (111), in general $(11 \cdots 1)$ with M ones, which integrates to 1 for any probability measure. This means that we have inserted a new variable in *both* summands. This allows us then to associate factors in the two summands after the selection jump one to one starting from the initial factor and ending with the last born variable.

To formalize the above ideas we need two changes.

(1) We want to view the function $\mathcal{F}_t^{++}$ not just as a function but to add as part of its description some additional information on its *form*. Namely, we want to consider a sum of *products of indicators* where each factor is associated with a particular particle in η and the particles are *ordered* in a certain way related to their appearance as selection acts. Note that this involves historical information on the evolution of the dual $(\eta, \mathcal{F}^{++})$.
(2) This means we can enrich the $\mathcal{F}_t^{++}$ to a marked *tableau* whose rows correspond to summands and columns to factors (which are indicators) each of which is a function of one variable.

For this purpose we need to *order the factors* and to introduce *new factors of 1* where necessary in order that all the summands in the expression for $\mathcal{F}_t^{++}$ consist of an equal number of factors. This leads to a collection of factors which are ordered, carry a location and are organized in summands. This object we will call $\mathcal{F}_t^{++,<}$. In the next two steps we formalize this procedure and we introduce this new process rigorously.

Step 3: *State space of the ordered dual $\mathcal{F}_t^{++,<}$*

Next we introduce the ingredients necessary to construct the *modified dual*, $\mathcal{F}_t^{++,<}$ formally. This means that we add to the particle system a further function assigning each particle a rank, an element in $\mathbb{N}$ which defines automatically an additional order relation. (Note that we use here a *different* order than in Subsections 5.1–5.3.)

9.4 Dual Representations: Ordered Dual and Tableau-Valued Dual

Start by observing how selection acts in order to see how the order must be set up. An initial set of factors $f^1 \otimes f^2 \cdots \otimes f^n$ each corresponding to *one* single variable is given a linear order (from left to right). The new particles appear by birth naturally ordered in time. However we now associate with a factor a *rank* which introduces then a *new order* by rank. The rank is defined dynamically as follows.

Definition 9.8 (Order relation among factors, ranks, transition under selection).

(a) We start with giving rank $1, \cdots, N_0$ to the N_0 initial particles and the associated factors. New factors are created by the selection operator as follows.

Suppose there is a birth by selection operating on the j-th particle in η and the corresponding factor is in the ℓ-th rank and m denotes the current number of partition elements. That is, selection hits the factor with rank ℓ and produces the transition of $f^1 \otimes \cdots \otimes f^m$ to:

$$f^1 \otimes \cdots f^{\ell-1} \otimes (f^\ell 1_A) \otimes 1 \otimes f^{\ell+1} \otimes \cdots \otimes f^m$$
$$+ f^1 \otimes \cdots f^{\ell-1} \otimes f^\ell \otimes (1 - 1_A) \otimes f^{\ell+1} \otimes \cdots \otimes f^m. \quad (9.90)$$

In order to work with an ordered object we use the rule that

- the additional factor 1 is placed at the $(\ell + 1)$-th position directly to the right of the factor on which the operator is acting producing $f^\ell 1_A$ on the ℓ-th position,
- the remaining factors are shifted one unit to the right,
- the factor 1 and $1 - 1_A$ is put at the $(\ell + 1)$-th position.

(b) We denote this *order* of factors of one variable by

$$<, \text{ that is, } f^1 < f^2 \quad (9.91)$$

means that the factor f^1 lies to the left of the factor f^2 in this order.

(c) We shall relabel all factors counting left to right and we assign the factors the

$$\text{ranks } 1, 2, 3, \cdots \quad (9.92)$$

with the natural order on $\mathbb{N}$. □

Note that in this labelling the *order in rank* is inherited. Therefore in our example in (9.83) each of the $\sum_{\ell=1}^{8} k_\ell(i, j, t)$ many factors has a rank in the linear order. This means that for each summand the factors are assigned ordered by their ranks. This leads then to the following *formal* description.

Definition 9.9 (State description of the enriched ordered dual $(\eta', \mathcal{F}_t^{++,<})$).

(a) The state of the ordered dual has the form of an enriched ordered particle system together with a collection of maps defined on the particle system, where the collection of maps is given by

$$\{\varphi_i(k)(t); \quad i = 1, \cdots, \tilde{N}_t; \quad k = 1, \cdots, N_t\}, \tag{9.93}$$

$$\varphi_i : \quad \{1, \cdots, N_t\} \longrightarrow \mathbb{F} \times \{1, \cdots, N\} \times \mathbb{N}, \tag{9.94}$$

where rows corresponding to the index i represent summands, and columns corresponding to the index k represent the factors of a given rank in different summands. Furthermore by the φ_i every column is assigned the **type of factor** in position (i, k) in the array as well as a **location** in $\{1, \cdots, N\}$ and assigning the **rank** to dual particles corresponding to columns.

A row of the tableau can be split into a product of a set of sub-products of factors, one sub-product for each occupied location.

We denote (a factor is a function $\mathbb{I} \to \mathbb{R}^+$)

$$\mathbb{F} = \text{the set of possible factors.} \tag{9.95}$$

(b) Then the state at time t can be uniquely described by a collection of

$$\tilde{N}_t \text{ summands (rows) of } N_t \text{ factors of one variable (columns)} \tag{9.96}$$

and the state is denoted

$$\{\varphi_i, \quad i \in \{1, \cdots, \tilde{N}_t\}\}, \tag{9.97}$$

where each summand is associated to a map $\varphi_i (= \varphi_i(t))$, specifying separately for each summand for each factor the *type* of the factor, the *location* in geographic space and finally *rank* in the order called $<$.

We write

$$\varphi_i(k)(t) = (f_k^i(t), j_k(t), \ell_k(t)) \quad , \quad k = 1, \cdots, N_t, \tag{9.98}$$

with the constraints that

$$f_k^i = f_{k'}^i \text{ and } j_k = j_{k'} \text{ if } \ell_k = \ell_{k'}. \tag{9.99}$$

We set

$$\mathcal{F}_t^{++,<} = \{f_k^i(t) : k = 1, \cdots, N_t, \ i = 1, \ldots, \tilde{N}_t\}. \tag{9.100}$$

9.4 Dual Representations: Ordered Dual and Tableau-Valued Dual

(c) At time 0 we set $\tilde{N}_0 = 1$ and the individuals $\{1, \ldots, N_0\}$ are assigned an initial index and the particles $\{1, \ldots, N_t - N_0\}$ are indexed in order of their birth times, the first birth assigned index $N_0 + 1$, etc. In contrast to the index the rank assigned to a particle changes dynamically due to coalescence or other births (recall also (9.90) to be described by the dynamics below after (9.107)).

(d) The set of particles assigned the same rank corresponds to the partition elements defined in Subsection 5.1. As in Subsection 5.1 the set of partition elements at time t is denoted

$$\pi_t = (\pi_t(1), \cdots, \pi_t(|\pi_t|)) \qquad (9.101)$$

and the partition elements have the form

$$\pi_t(\ell) = \{k : j_k = j, \ell_k = \ell\}, \quad \ell = 1, \ldots, |\pi_t|. \qquad (9.102)$$

This defines a mapping

$$\xi_t : \{1, \ldots, |\pi_t|\} \to \{1, \ldots, N\}, \qquad (9.103)$$

such that the partition element corresponding to rank ℓ is located at site

$$\xi_t(\ell) = j. \qquad (9.104)$$

Then π_t defines a mapping

$$\pi_t : \{1, \ldots, N_t\} \to \{1, \ldots, |\pi_t|\}, \qquad (9.105)$$

where the partition elements are now *indexed* by the smallest index of the particles they contain and ξ_t defines a mapping $(\pi_t(1), \cdots, \pi_t(|\pi|)) \to \Omega$ (giving the locations of partition elements).

We define the enriched (from η) particle system

$$\eta'_t = (N_t, \pi_t, \xi_t, \Re_t), \qquad (9.106)$$

that is, η'_t is given by the set of particles $\{1, \cdots, N_t\}$, their partition structure π_t, locations ξ_t together with the list of current ranks $\Re_t$ of partition elements. □

Step 4: *Dynamics of ordered dual process* $(\eta', \mathcal{F}_t^{++,<})$.

In order to clarify the ideas we start in **(i)** with an informal description and then we formalize this in **(ii)** writing down a Markov jump process by specifying the state, transitions and their rates.

(i) The point will be to place the factors in the summands of $\mathcal{F}_t^{++,<}$ in a particular order useful for analysing the resulting expressions. We describe now informally the transitions of the dual ordered in this way.

The dual evolution can be described as follows. First the particle system η. Starting with N_0 initial dual particles each located at one of the sites in $\{1, \ldots, N\}$ at time 0, new particles are created by a pure birth process with birth rate sk when the current number of dual particles is k. More precisely, independently each particle gives birth at rate s. Each particle is located at one of the points in $\{1, \ldots, N\}$ and is assigned in addition a *rank*. Then the enriched evolution with the transitions by (1) coalescence, (2) mutation and (3) birth (selection) is as follows.

(1) *Coalescence* (two particles at the same location) *effectively* decreases the number of factors even though here we keep formally the number of factors and introduce an additional factor 1. We make the convention that upon coalescence of two factors we replace the one with lower rank by the product of the two factors and the one with upper rank becomes (111) (this is the analogue of the wellknown "look-down" process). In formulas, at the coalescence of individuals i' and k' in the particle system η which corresponds in a particular summand to factors with ranks i and k (in the labelling (9.92)) we get for the function-valued part the transition given by

$$f^1 \otimes \cdots \otimes f^i \otimes \cdots \otimes f^k \otimes \cdots \longrightarrow f^1 \otimes \cdots \otimes (f^i f^k) \otimes \cdots \otimes 1 \otimes \cdots , \tag{9.107}$$

with each factor we associate a rank and the new factor 1 gets the same rank which introduces then a new order by rank. The rank is defined as in a selection operation the newly created factor.

(2) *Mutations* act as before on factors corresponding to each variable independently and we note that in particular acting on (111) or (000) mutation has no effect.

(3) *Selection* creates a new particle in the dual particle system η and at the same time the transition given in (9.88) occurs. *Ranks* change at the same time as follows. The offspring of a particle of rank k is assigned the rank $k+1$ and the ranks of all particles with ranks $\ell \geq k+1$ are reassigned rank $\ell + 1$. We denote the resulting number of particles in the dual system η_t at time t by N_t where $N_t - N_0$ denotes the number of births. Moreover at a birth time the offspring of a particle is located at the same site as the parent.

The rank allows us to consider for a selection transition the new factors now next to each other, i.e. we have if the factor f gives birth now after the transition the sum of the two factors:

$$1_A \cdot f \otimes 1, \quad f \otimes (1 - 1_A). \tag{9.108}$$

(Recall 1_A and $1 - 1_A$ are connected with different variables).

To complete the description of the dual we must define the corresponding function $\mathcal{F}_t^{++,<}$ which is a function of N_t variables but has the form of a sum of *ordered* products of factors. The transitions of $\mathcal{F}_t^{++,<}$ are as listed above.

We can now represent the dual function $\mathcal{F}_t^{++}$ via our new and *richer* state. By construction we have:

9.4 Dual Representations: Ordered Dual and Tableau-Valued Dual

Lemma 9.10 (Representation via enriched process).
Consider the dual process $(\eta'_t, \mathcal{F}_t^{++,<})_{t\geq 0}$. Let pr denote the map $(N_t, \pi_t, \zeta_t, \mathfrak{R}_t)$ $\to (N_t, \pi_t, \xi_t)$ associating with η' the triple η arising by ignoring the rank. Given the collection of maps associated with $(\eta', \mathcal{F}^{++}, <)$ as $f_k^i,: k \leq N$, $i = 1, \ldots, \tilde{N}\}$, we define for this state the pair:

$$(\eta, \mathcal{F}^{++}) = \left(pr(\eta'), \sum_{i=1}^{\tilde{N}} \prod_{j\in S} \prod_{\ell=1}^{|\pi|} (1_{\xi(\ell)=j} \left(f_\ell^i(u_\ell)\right))\right). \tag{9.109}$$

Then we obtain a version of $(\eta_t, \mathcal{F}_t^{++})_{t\geq 0}$ on the probability space of $(\eta', \mathcal{F}^{++,<})$. □

Remark 119. Note that we can also construct a version of $(\eta'_t, \mathcal{F}_t^{++,<})$ from the path $(\eta_s, \mathcal{F}_s^{++})_{s\leq t}$.

It is often convenient to order the summands in the tableau in a specific way, i.e. instead of a set as in (9.97) we want to use a tuple. Recall that the action of selection (corresponding to a birth event for η_t) on a function f at a given location and rank is defined by (9.108). This replaces a summand by two summands (one of which can be 0) in $\mathcal{F}^{++,<}$ and we must keep track of the summands.

In the examples we have arranged our $\mathcal{F}_t^{++}$ in form of tableaus of indicator functions meaning that we have also given the rows a certain order. The *tableau* provides a nice way to to represent the state of $\tilde{\mathcal{F}}_t^{++,<}$ which is an array of strings of factors called *rows*. The *columns* in this array have a nice structure and correspond to factors with the same rank. For that purpose we can write at each selection event the new rows directly under the parent row to get a nice to work with tableau. (See (9.155) for an example). Via the particle system we can identify the location of the columns.

We formalize this procedure now.

Definition 9.11. (Ordering the rows)

(a) In order to keep track of the summands and produce a convenient *visualization as a tableau* we adopt the following convention for ordering the summands. Starting with a single factor f operated on by selection as in (9.108) we produce the ordered rows

$$\begin{array}{cc} \chi_A \cdot f & 1 \\ f & (1 - \chi_A). \end{array} \tag{9.110}$$

In general

- The operation of selection at a rank is applied successively to each row starting at the top and then moving down to each original row. When selection acts on the rank at a row it produces an additional row immediately

below the row on which it acts. If selection produces a row with a zero factor this row is removed.
- Also mutation and coalescence can produce a row with a zero factor which is then removed but otherwise they do not change the order of the rows.

(b) This means that we want to think of the

$$\text{state of } (\eta', \mathcal{F}^{++,<}) \text{ as a } tableau \qquad (9.111)$$

where *rows* are ordered dynamically according to (a) and correspond to summands and *columns* to variables and *marks* on the entries indicate the location of the factor.

(c) For convenience we denote the tableau of ordered rows and columns of indicator functions (of one variable in each column) associated to the particle system η'_t and to the object $\mathcal{F}_t^{++,<}$ as

$$\mathfrak{t}((\eta'_t, \mathcal{F}_t^{++,<})), \qquad (9.112)$$

where a *column* corresponds to a *variable* which are ordered according to the rank in the tableau from left to right and a *row* to a particular *summand* which we order according to the convention in (9.110). □

The ordered summands (rows) allow to put the entries 1 in convenient positions for calculation and estimates, as we shall see later on working with this object.

Definition 9.12 (Tableau-valued process).
The evolution of the Markov jump process $(\eta'_t, \mathcal{F}_t^{++,<})_{t \geq 0}$ induces a *marked tableau-valued* pure jump process driven by an autonomous particle process η'_t:

$$(\mathfrak{t}((\eta'_t, \mathcal{F}_t^{++,<})))_{t \geq 0}. \qquad \square \qquad (9.113)$$

The description of the dynamic we give below will indeed show that the marked tableau-valued process from (9.113) is *Markov*. The information which we lose in the tableau description is which particles which were born up to the present time are in one partition element. This information is not needed to evaluate the expectation in the duality relation.

Remark 120. A natural way to keep track of the summands is as the set of leaves in a tree. When selection event occurs at a vertex two additional edges and vertices are produced in the tree. The dual representation will then involve an expectation over the summands indexed by the set of leaves in the tree. This means that only the marginal distributions of the summands are involved, not the joint distribution of the summands. One possible choice for the dynamic is that different summands evolve independently. But then we must have a separate birth process (associated to the selection events) for each summand. Instead in the construction below we have a single birth process but this process is applied simultaneously to all summands following the dynamics described below.

9.4 Dual Representations: Ordered Dual and Tableau-Valued Dual

(ii) Based on the ideas introduced above we now formally construct the dual process $(\eta', \mathcal{F}_t^{++,<}))$ in the case of M types at the lower level and one type at the higher level by specifying transitions and transition rates.

We next define the *dynamic of the ordered dual* first in η' and then in $\mathcal{F}^{++,<}$, however this is only the explicit version of what we described in point (i).

We specify the transitions that result in an increase or decrease of the number of factors, namely, selection and coalescence at one site. This will produce an enriched version of the birth and death process used in the case $M = 1$, $\widetilde{N}_t$. We now describe the transitions in the dynamic step by step, first for η' in (1) and (2) and then for $\mathcal{F}^{++,<}$ in (3).

(1) The first transition of the quadruple $(N_t, \pi_t, \xi_t, \mathfrak{R}_t)$ of processes is due to *birth* and is defined as follows. N_t takes values in $\mathbb{N}$ and is a pure birth process with birth rate at time t

$$s \cdot |\pi_t| \tag{9.114}$$

and π_t is taking values in the partitions of the ordered set $\{1, 2, \cdots, N_t\}$. $|\pi_t|$ is the number of partition elements. If the number of elements increases by one the new element N_t forms its own partition element. Each partition element with smallest element i has a rank $\mathfrak{R}_t(i)$. Each new individual produced when there are n partition elements is the offspring of a randomly chosen rank ℓ which is chosen with the uniform measure on $1, \ldots, n$. At the moment a new individual is born to the rank ℓ the individual (and its partition element) gets rank $\ell + 1$ and all larger ranks are increased by one.

(2) The second transition of $(N_t, \pi_t, \xi_t, \mathfrak{R}_t)$ is *death*. Deaths can occur at each site due to coalescence and migration and occur at rate

$$d \binom{m}{2} 1_{m \geq 2} + cm 1_{m \geq 1}, \quad (1_{m>1} \text{ for } N = \infty \text{ in the latter}). \tag{9.115}$$

When a death occurs, that is a rank is removed from a site, it is a migration with probability

$$\frac{c 1_{m \geq 1}}{cm 1_{m \geq 1} + d \binom{m}{2}}, \quad (1_{m>1} \text{ for } N = \infty \text{ with the above convention}). \tag{9.116}$$

The rank of a migrant is chosen with uniform measure on $1, \ldots, |\pi_t|$. Then the map ξ changes and assigns the partition element with the chosen rank instead of the current site a new one chosen uniformly among $\{1, \cdots, N\}$, respectively for $N = \infty$ the smallest site not yet occupied.

If a death is not a migration, it is a coalescence and partitions and ranks change. On coalescence of ranks ℓ and $\ell' > \ell$ which are located at the same site the rank ℓ'

is removed, more precisely the partition element at rank ℓ now includes all particles that had been in the two partition elements with ranks ℓ and ℓ' and the ranks larger than ℓ' are reduced by 1.

Remark 121. (a) The ranks (ℓ, ℓ'), $\ell' > \ell$ involved in a *coalescence* are chosen with uniform measure on the set of $m(m-1)/2$ pairs of particles. Note that when $m = 1$ deaths due to coalescence do not occur.

(b) Also we can adopt the convention that deaths at a site where $m = 1$ due to migration do not occur in the case of infinitely many sites and initial measures of the form $\mu^{\otimes \mathbb{N}}$, that is, case we adopt the convention that ranks at singly occupied sites do not migrate. Note that we cannot use this convention in the collision regime.

(3) Turn now to the transitions of $\mathcal{F}^{++,<}$.

When a birth occurs in η', then a selection transition occurs for the selection operator 1_A with $A = \{M + 1\}$. We first give the description of the transition that results from selection operator associated to 1_A acting on a rank ℓ at a site j. We first consider the action on just the i-th summand, namely that this summand is modified and a new summand is produced and we describe below how to label the summands and the newly created ones.

The original summand is modified as follows:

$$\{\varphi_i(k) = (f_k^i, j_k, \ell_k)\}_{k=1,...,N_*}, \text{ as original state.}$$
number of particles changes:
$$N_* \to N_* + 1,$$
function change to: \hfill (9.117)
$$\varphi_i(k) = (f_k^i, j_k, \ell_k), \quad \text{if } \ell_k < \ell,$$
$$\varphi_i(N_* + 1) = (1, j_k, \ell + 1),$$
$$\varphi_i(k) = (f_k^i, j_k, \ell_k + 1), \quad \text{if } \ell_k > \ell,$$
$$\varphi_i(k) = (1_A f_k^i, j_k, \ell_k), \quad \text{if } \ell_k = \ell.$$

Moreover for every summand a new offspring summand $\varphi_{i'}$ (if we use the special order of summands from Definition 9.11, $i' = i + 1$ and the number of all other higher summands is shifted by one) is produced which is defined as follows:

$$\{\varphi_{i'}(k) = (f_k^{i'}, j_k', \ell_k')\}_{k=1,...,N_*+1}$$
is defined by
$$\varphi_{i'}(k) = (f_k^i, j_k, \ell_k), \quad \text{if } \ell_k < \ell,$$
$$\varphi_{i'}(k) = (f_k^i, j_k, \ell_k + 1), \quad \text{if } \ell_k > \ell, \hfill (9.118)$$
$$\varphi_{i'}(k) = (f_k^i, j_k, \ell_k), \text{ if } \ell_k = \ell$$
$$\varphi_{i'}(N_* + 1) = (1 - 1_A, j_k, \ell + 1), \text{ if } \ell_k = \ell.$$

The transition due to coalescence of ranks ℓ and $\ell' > \ell$ at a site j is given as follows:

9.4 Dual Representations: Ordered Dual and Tableau-Valued Dual

$\{\varphi_i(k) = (f_k^i, j_k, \ell_k)\}_{k=1,\ldots,N_*}$
changes to
$$\begin{aligned}
\varphi_i(k) &= (f_k^i, j_k, \ell_k), & \text{if } \ell_k < \ell, \\
\varphi_i(k) &= (f_\ell^i \cdot f_{\ell'}, j_k, \ell_k), & \text{if } \ell_k = \ell, \\
\varphi_i(k) &= (f_k^i, j_k, \ell_k), & \text{if } \ell < \ell_k < \ell', \\
\varphi_i(k) &= (f_k^i, j_k, \ell_k - 1), & \text{if } \ell_k > \ell', \\
\varphi_i(k) &= (f_\ell^i \cdot f_{\ell'}, j_k, \ell), & \text{if } \ell_k = \ell'.
\end{aligned} \quad (9.119)$$

Next we consider the transition that occurs if rank ℓ migrates to a new site say j':

$$\begin{aligned}
&\{\varphi_i(k) = (f_k^i, j_k, \ell_k)\}_{k=1,\ldots,N_*} \\
&\text{changes to} \\
&\varphi_i(k) = (f_k^i, j_k, \ell_k), \quad \text{if } \ell_k \neq \ell, \\
&\varphi_i(k) = (f_k^i, j', \ell), \quad \text{if } \ell_k = \ell.
\end{aligned} \quad (9.120)$$

Finally the mutation acts again on every rank k independently and a mutation jump

$$(f_k^i)_{i=1,\cdots,\tilde{N}} \longrightarrow (f_k^i)'_{i=1,\cdots,\tilde{N}} \quad (9.121)$$

takes places and the mutation generator M_k acts on every f_k^i simultaneously.

Remark 122. The collection of summands has a natural tree structure which is described as follows. We begin with

$$\tilde{N}_0 = 1 \text{ and give this index } 1. \quad (9.122)$$

We index

the offspring of a summand of index i by $i1, i2, i3, \ldots$ $\quad (9.123)$

which then defines a natural tree structure. In this way we see that after $N_t - N_0 = n$ births, there are 2^n summands. However we note that some summands can be 0 so that this provides an upper bound on the set of non-zero summands.

Remark 123. Assume we can work with the collision-free regime ($N = \infty$). We can construct a richer labelling system that incorporates both the *birth order* and *rank* in the linear order, as follows.

It can be encoded starting with the founder at 0. The history of transitions up to time t and therefore the state at time t is determined by a sequence $\{0 \to, \ldots\}$ of transitions, where each transition is given by one of

$$k \to, k \leftarrow \ell \text{ (where } \ell > k\text{) and } k \uparrow. \quad (9.124)$$

The first corresponds to a selection event and produces an offspring at $k+1$, the second a coalescence (of ℓ with k) and the third an emigration, i.e. migration to an new unoccupied site. Note that the founder position 0 never migrates. We denote the position of the kth particle born in the linear order at time t by $v(k,t)$. We can have $k \leftarrow \ell$ only if $v(\ell) > k$, that we, have a *lookdown*. The relative order of the particles does not change but when a lookdown occurs we denote the outcome as the "compound particle" $[k, \ell]$, etc.

For example

$$0, 0 \to, 0 \to, 1 \to, 0 \to, 1 \to, 2 \leftarrow 3 \tag{9.125}$$

would result in (the labels according to birth times)

$$0, 01, 021, 0213, 04213, 042153, 04[2,3]15. \tag{9.126}$$

Note that in the special case with no coalescence this labelling above identifies the tree associated to a branching process and in the special case with no births identifies the coalescent graph (via the lookdown process). Therefore this can be viewed as a coding of the branching coalescing (or selection-mutation) graph.

The following observation for the enriched process is important. Since the birth and death process is recurrent, when the number of particles at a given location returns to one, the sequence can be deleted and relabelled 0 and the process begins again.

Step 5: *(Summarizing the construction)*

We can summarize the construction so far as follows. We have modified the $(\eta_t, \mathcal{F}_t^{++})_{t \geq 0}$ to a new process $(\eta_t', \mathcal{F}_t^{++,<})_{t \geq 0}$ where $(\eta_t')_{t \geq 0}$ is described by the enriched (by information on the *rank*) birth and death process (branching coalescing particle system) and the function-valued part $(\mathcal{F}_t^{++,<})_{t \geq 0}$ and the associated tableau of indicators is given by

$$\mathcal{F}_t^{++,<} = \sum_{i=1}^{\widetilde{N}_t} \mathcal{F}_{i,t}^{++,<} \quad , \quad \mathfrak{t}(\eta_t', \mathcal{F}_t^{++,>}), \tag{9.127}$$

where $\widetilde{N}_t$ denotes the number of summands and the (possibly also ordered, (recall (9.110)) summands are each an ordered products of factors.

We have for the modified process according to the Remark 119 the duality relation:

Proposition 9.13 (Modified duality).

We have $x^\otimes$ abbreviating the initial product measure state.

$$E[\int \mathcal{F}_t^{++} dx^\otimes] = E[\int \mathcal{F}_t^{++,<} dx^\otimes]. \quad \square \tag{9.128}$$

9.4 Dual Representations: Ordered Dual and Tableau-Valued Dual

We can therefore in the sequel study the evolution of marked tableaus of indicator functions, which allows for convenient calculations.

Step 6: *Simplifying the ordered dual to* $(\eta', \tilde{\mathcal{F}}_t^{++,<})$.

Our main application of the dual is to determine the emergence time scale. For this reason we make a few simplifications that are applicable for the calculations of the moments of the type $M + 1$ when we begin with an initial measure on $(\Delta_M)^N$ of the form

$$\mu^{\otimes N} \text{ with } \mu(M + 1) = 0. \tag{9.129}$$

Note that for the moment we have included summands that when integrated against $\mu^{\otimes N}$ yield 0, hence the dynamic can be further simplified if we are interested only in the integral on the r.h.s. of (9.55). We observe that a summand which contains a factor

$$(0, \ldots, 0, *), \text{ with } * = 0 \text{ or } 1 \tag{9.130}$$

becomes 0 when integrated w.r.t. μ since we start the original process with all the mass on types $1, \cdots, M$ so that a summand with such a factor would not contribute taking the expectation. In addition such a factor can only change and create summands without this property by upward mutation to type $M + 1$ which requires times of order N to happen at a fixed rank but presently we consider the evolution before such large times. Therefore we can and shall agree to prune the summands in the dual process and hence prune the list of rows of our tableau by setting

$$\tilde{\mathcal{F}}_{t,i}^{++,<} = 0, \quad \text{summand } i \text{ contains a factor } (0, \ldots, 0, *) \tag{9.131}$$
$$\tilde{\mathcal{F}}_{t,i}^{++,<} = \mathcal{F}_{t,i}^{++,<}, \quad \text{otherwise,}$$

respectively removing a row containing $(), \cdots, 0, *)$ from the tableau.

With this convention we have to deal only with summands which have a contribution in the integral in (9.55).

Similarly the evaluation of ranks for which $f = (1, \ldots, 1, 1)$ in all summands does not change the integral on the r.h.s. of (9.55). However for the moment we do keep factors which are effectively one, i.e. $(1, \ldots, 1, 1)$ since below we shall see that this will help doing the bookkeeping once later on we couple summands.

Furthermore note that the factor $(0, \ldots, 0, 1)$ can only gain further 1''s by a downward mutation and at the moment we consider $m_{3,\text{down}} = 0$. (Even for $m_{3,\text{down}} > 0$ this would at best become visible if the dual process reaches a state where we have $O(N)$ summands with such a factor).

Lemma 9.14 (Simplified duality with ordered factors).
For initial measures satisfying $\mu(M+1) = 0$, we have the dual representation in terms of the simplified dual:

$$E[\int \mathcal{F}_t^{++} d(\mu^\otimes)] = E[\int \tilde{\mathcal{F}}_t^{++,<} d(\mu^\otimes)]. \qquad \square \qquad (9.132)$$

The main idea for the analysis of $(\eta', \tilde{\mathcal{F}}^{++,<})$ will be to modify the specification of the dynamic in such a way that the duality remains true for the new modified function-valued dynamic as well as getting cancelations of terms by coupling the summands for different i in the representation in (9.83) by obtaining sums combining to give factors 1. What is in the way is that upon selection 1_A and $1 - 1_A$ have different ranks. We defined in Section 5 a new process denoted by $(\mathcal{G}_t^{++})_{t \geq 0}$ and we define the enriched $(\mathcal{G}_t^{++,<})_{t \geq 0}$ precisely in the next Subsubsection 9.4.7 by specifying the mutation and selection transitions such that they occur simultaneously at the same rank in each summand.

9.4.7 Modification of the Dual 2: $\mathcal{G}^{++,<}$ with Ordered Coupled Summands

In the dual $(\eta', \mathcal{F}^{++,<})$ factors and summands are ordered and part of the state and summands are coupled, since transitions in the particle system η' induce transitions in $\mathcal{F}_t^{++,<}$, which occur for certain factors in *all* the summands. The strategy of this section is to to use the *new coupling* between summands used by $\mathcal{G}^{++}$ and introduced in Section 5, that took advantage of the structure of this dual which introduces *cancellations* which will make combinations of some summands in one variable equal to 1. This involved changing the transition of the function-valued part at selection to $f \to f 1_A \otimes 1 + (1 - 1_A) \otimes f$ (instead of $f 1_A \otimes 1 + f \otimes (1 - 1_A)$). Then everything we did in Subsubsection 9.4.6 can be transferred (a bit more detail is provided in Step 1 below). This generates using the process $\mathcal{G}^{++}$ in combination with the enriched particle system η' and the order introduced for $(\eta'_t \mathcal{F}_t^{++,<})$ to arrive at

$$(\eta'_t, \mathcal{G}^{++,<}). \qquad (9.133)$$

Remark 124. Note that $\mathcal{G}_t^{++,<}$ have the same type of state as $\mathcal{F}_t^{++,<}$ but the dynamic is now changed.

Remark 125. A key point is that with this new coupling the different (i.e. not repetitions of another one) non-zero summands always correspond to *disjoint* events. Also note that some summands can correspond to the empty event which does not contribute in the duality.

9.4 Dual Representations: Ordered Dual and Tableau-Valued Dual

This construction will be the basis for the second key idea that will be taken up in the next Subsection 9.5, which is to introduce another dual object that allows us to represent the process with coupled summands as a nice *set-valued process* which allows to study the longtime behaviour. In our application we can also calculate moments and with enrichments of it we evaluate the Malthusian parameter β that determines the critical time scale.

The objective is to use the basic duality formula

$$E[\int \mathcal{F}_0^{++} d(x(t))^\otimes] = E[\int \mathcal{F}_t^{++} d(x(0))^\otimes] = E\left[\int \mathcal{F}_t^{++,<} d(\mu^\otimes)\right], \quad (9.134)$$

where $\mathcal{F}_t^{++}$ is the refined dual process of Subsection 5.5 and to rewrite the r.h.s. further, in the sense that for every initial state $x^\otimes$:

$$E\left[\int \mathcal{F}_t^{++,<} d(x^\otimes)\right] = E\left[\int \mathcal{G}_t^{++,<} d(x^\otimes)\right]. \quad (9.135)$$

Remark 126. Note that the new process $(\mathcal{G}_t^{++,<})_{t \geq 0}$ can be viewed as a process, where the birth, coalescence and migration transitions are now jumps associated with the ranks (recall (9.127)). Recall that the *coupling* of summands in (9.146) was given by the evolution rule that every transition occuring for an individual in η, say k of some particular rank say ℓ occurs simultaneously in each summand that is for every $\tilde{\mathcal{F}}_{i,t}^{++,<}$ in (9.127). This induces a coupling of the rows in the tableau. Alternatively we can use η' as the driving process by permuting for each individual in η' the position in the ranking to the factor it is associated with.

Recall that rows which contain factors $(0 \ldots 0)$ play no role in the duality we do not need them in any calculation and we can completely remove them which we did in passing from $\mathcal{F}_t^{++,<}$ to $\tilde{\mathcal{F}}_t^{++,<}$. Similarly a column which has only factors $(1 \cdots 1)$ plays no role in the duality expression. We call these ranks *inactive*.

Definition 9.15 (Non-zero summands and active ranks).

(a) If we include only summands which are not 0, we number these as

$$1, \ldots, \tilde{N}_t^*. \quad (9.136)$$

(b) Columns (ranks) in the tableau which contain only $(11 \cdots 1)$ are called inactive. In particular sites which are only colonized by inactive rows are called inactive sites. □

Definition 9.16 (The coupled dual process $(\eta', \mathcal{G}^{++,<})$ and $\mathfrak{t}(\eta', \mathcal{G}^{++,<})$).

We shall denote the process arising from $(\eta', \mathfrak{t}((\eta', \tilde{\mathcal{G}}^{++,<})))$ by introducing coupled summands via the rule (9.145) and by removing inactive ranks and zero rows by

$$(\eta'_t, \mathcal{G}_t^{++,<})_{t \in \mathbb{R}}. \quad (9.137)$$

The part $\mathcal{G}_t^{++,<}$ of the process consists of the collection of summands of marked factors

$$\{\mathcal{G}_{i,t}^{++} : i = 1, \ldots, \tilde{N}_t^*\}, \tag{9.138}$$

each consisting of N_t^* active ranks and the associated factors with their location, thus specifying again a *marked tableau* of indicator functions, which if the rows are ordered as in (9.110), is denoted:

$$\mathfrak{t}(\eta', \mathcal{G}^{++,<}). \qquad \square \tag{9.139}$$

We have the identity (which we prove in Step 1 or alternatively Step 2 below):

Proposition 9.17 (Tableau-valued duality with coupled summands).
The process $(\eta', \mathcal{G}^{++,<})$ gives the same expectation as the dual $(\eta, \mathcal{F}^{++,<})$, namely:

$$E\left[\int \left(\sum_{i=1}^{\tilde{N}_t} \mathcal{F}_{i,t}^{++,<}\right) d(\mu^\otimes)\right] = E\left[\int \sum_{i=1}^{\tilde{N}_t^*} \mathcal{G}_{i,t}^{++,<} d(\mu^\otimes)\right]. \qquad \square \tag{9.140}$$

The statement can be proven immediately viewing the process as enrichment of $\mathcal{G}^{++}$ and recalling that the latter satisfies the duality (see Step 1 below in the proof). We give an independent argument based on $\mathcal{F}^{++,<}$ in Step 2 below.

A key property of the new dual is that it allows us in a more transparent way to keep track of how far our expression deviates from a product, which we summarize in the next two remarks.

Remark 127. Recall from above that the sum can be organised with the help of a tree, a splitting occurring with each operation of the selection operator. Consider at time $t + s$ the two summands corresponding to each of the two subtrees starting at a birth time s with selection operator $(1_A f \otimes 1 f)$, respectively $f \otimes (1 - 1_A)$ (where the order of the factors is as indicated left to right). We have changed the dynamic such that we couple the summands such that effectively we work instead of this with

$$(1_A f) \otimes 1 \text{ and } (1 - 1_A) \otimes f, \tag{9.141}$$

which is then a *decomposition in complementary terms* since $1_A + (1 - 1_A) = 1$. Furthermore if mutation acts on the partition A the decompositions $(\emptyset, \mathbb{I})$ respectively $(\mathbb{I}, \emptyset)$ are traps and if the mutation rates are strictly positive these traps are actually reached. If the trap is reached (we call this resolution) then only one of the two summands in (9.141) remains.

Remark 128. We have a look at the behaviour of sites contributing in the duality.

9.4 Dual Representations: Ordered Dual and Tableau-Valued Dual

In the two-type case we saw new sites being colonized by particles was always equivalent of a contribution of this site to the dual expression. This is not the case for the case of $M > 1$ and becomes very transparent with the new dual.

Consider the following sequence of transitions

$$\begin{aligned} & f \\ & \to \\ & f \cdot 1_B \otimes 1 + \\ & 1_{B^c} \otimes f \\ & \to \\ & 1_A 1_B \cdot f \otimes 1 \otimes 1 + \\ & 1_{A^c} \otimes 1_B \cdot f \otimes 1 + \\ & 1_A 1_{B^c} \otimes f \otimes 1 + \\ & 1_{A^c} \otimes 1_{B^c} \otimes f \otimes 1. \end{aligned} \quad (9.142)$$

If next the second rank migrates and then the first rank evolves by mutation via the jump $1_A 1_B \to 1$ (hence $1_A 1_{B^c}, 1_{A^c} 1_B, 1_{A^c} \to 0$), then the newly occupied site becomes inactive ($=1$).

Hence we see that factors which have colonized a new site need not lead to a permanent colonization of that site as active site which is in contrast to the case $M = 1$.

Note that the jump to 1 of a rank to the right of a migrating rank does not have this effect. For example consider

$$\begin{aligned} & f \to f \cdot 1_B \otimes 1 + 1_{B^c} \otimes f \\ & \to \\ & 1_A 1_B \cdot f \otimes 1 \otimes 1 + \\ & 1_{A^c} \otimes 1_B \cdot f \otimes 1 + \\ & 1_A 1_{B^c} \otimes f \otimes 1 + \\ & 1_{A^c} \otimes 1_{B^c} \otimes f \end{aligned} \quad (9.143)$$

Now first let the first column migrate and then the second column resolve $1_B \to 1$. We then obtain

$$\begin{aligned} & 1_A 1_B \cdot f \otimes 1 \otimes 1 + \\ & 1_{A^c} \otimes f \otimes 1 \otimes 1 + \\ & 1_A 1_{B^c} \otimes f \otimes 1 \otimes \\ & = \\ & 1_A 1_B \cdot f \otimes 1 \otimes 1 \otimes 1 + \\ & 1_{B^c} \otimes f \otimes 1 \otimes 1 \otimes 1 \end{aligned} \quad (9.144)$$

but where now the first and second columns are located at different sites. However the first column is not removed.

Proof of Proposition 9.17.

Step 1 *(Construction of the new couplings of summands in the ordered dual)*

We obtain the *new* coupling of the summands on the level of the tableau-valued process by changing the rule of the selection transition, parallel to the change in Section 5 from $\mathcal{F}^{++}$ to $\mathcal{G}^{++}$, as follows. The action of selection operator acting on a particle with rank ℓ and associated factor f and location j produces now at j and the ranks ℓ and $\ell+1$ the factors induced by

$$f \longrightarrow 1_A f \otimes 1 + (1 - 1_A) \otimes f. \tag{9.145}$$

This means that now the factors 1_A and $(1 - 1_A)$ have the *same* variable, i.e. they are in the same column and the same rank. In particular this means that under the further action of the migration, coalescence or the selection operator we apply this simultaneously to the $(1 - 1_A)$ and 1_A in the product $(1 - 1_A) \otimes f$ and $1_A \cdot f \otimes 1$, that is, to every factor in the corresponding rank (in the linear order). This includes the transitions of these two factors in the two summands arising from coalescence, migration, selection and mutation, both normal and rare but for the moment we do not consider the rare mutation.

Note that at this moment the variables of 1_A and $(1 - 1_A)$ are associated by this modified mechanism, even though originally they were associated with different individuals in η. The point is that taking the expectation of the sum w.r.t the dual dynamics we have *not* changed the value of the expectation since the expectation depends only on the marginals corresponding to the summands and not the joint law. This means that we permute the order of the individuals in the η-process depending on the summand, in order to make the cancellation effects explicitly visible.

Step 2 *(Alternative Proof of Proposition 9.17).*

In order to prove Proposition 9.17 we recall how we rewrote the state of $\tilde{\mathcal{F}}_t^{++,<}$ (compare (9.127) and (9.148) below). Consider the finite sum decomposition from (9.127)

$$\tilde{\mathcal{F}}_t^{++,<} = \sum_{i=1}^{\tilde{N}_t} \tilde{\mathcal{F}}_{i,t}^{++,<}, \quad \mu^\otimes - a.s., \tag{9.146}$$

where each $\tilde{\mathcal{F}}_{i,t}^{++}$ is an (ordered) product of indicator functions.

Given a realization of the particle system, we will build a version of $\tilde{\mathcal{F}}_t^{++,<}$. Then in (9.134) we can write for the r.h.s.:

$$E[\int \{\tilde{\mathcal{F}}_t^{++,<}\} d(\mu^\otimes)] = \hat{E}\left[\int \{E_{\tilde{N}_t} \tilde{\mathcal{F}}_t^{++}\} d(\mu^\otimes)\right], \tag{9.147}$$

9.4 Dual Representations: Ordered Dual and Tableau-Valued Dual

where $\hat{E}$ denotes the expectation over the number of summands process $(\tilde{N}_t)_{t \geq 0}$ and $E_{N_t} = E(\cdot | N_s : 0 \leq s \leq t)$ the respective conditional expectation with respect to the mutation, coalescence and migration dynamics.

As in the case $M = 2$ we have that $\mathcal{F}_t^{++,<}$ is decomposed into a sum and then

$$E \int \left\{ \sum_{i=1}^{\tilde{N}_t} \mathcal{F}_{i,t}^{++,<} \right\} d(\mu^\otimes) = \hat{E} \left[\int \{E_{\tilde{N}_t} \left[\sum_{i=1}^{\tilde{N}_t^*} \mathcal{F}_{i,t}^{++,<} \right] \} d(\mu^\otimes) \right]. \quad (9.148)$$

We note that we have the following upper bound on the number of non-zero summands $\tilde{N}_t^* \leq \tilde{N}_t \leq 2^{N_t}$ (since selection acting on a rank can replace each summand by at most two non-zero summands). We note that the sum over $\tilde{N}_t$ or $\tilde{N}_t^*$ results in the same integrand for $\int$.

Therefore given $\{\tilde{N}_s : 0 \leq s \leq t\}$ we have to keep track of $\{\mathcal{F}_{i,s}^{++,<} : 1 \leq i \leq \tilde{N}_s^*, 0 \leq s \leq t\}$ and the way they combine to form the sum. On the r.h.s. of (9.148) we see that the conditional expectation depends only on the marginal distribution of the vector of summands and *not* on the joint distribution for the given birth process. This is where the coupling will come in.

Since in the dual representation we compute the expected value of the sum this depends only on the marginal distributions of the dynamics (conditioned on their starting points) of the different summands after time s (recall (9.148). Therefore we can *couple* the evolution of the dynamics of the two summands differently and still preserve the duality relation.

Observe here that the different transitions of the dual $(\eta'_t, \mathcal{F}_t^{++,<})_{t \in \mathbb{R}^+}$ occur independently for each individual in the dual process and the dynamic of the dual system is Markov. Therefore if our coupling of summands is constructed by coupling transitions of factors in $\{\mathcal{F}_{t,i}^{++,<}, \ i = 1, \cdots, N_t^*\}$, which correspond to different leaves i in the binary tree generated by the dual population the basic duality relation (9.140) *will be satisfied* automatically.

9.4.8 Extension to the McKean–Vlasov Process

The construction we have carried out above can be immediately extended to the McKean–Vlasov process following the device of Subsection 5.5 on the dual for the McKean–Vlasov process and obtain the objects and relations for the case $N \to \infty$. In order to distinguish the cases we denote the coupled dual for the mean-field process respectively the McKean–Vlasov process differently as:

$$\mathcal{G}^{++,N,<}, \text{ resp. } \mathcal{G}^{++,<}. \quad (9.149)$$

These two objects we have to analyse for $N \to \infty$ and $t \to \infty$ in the sequel.

9.5 The Marked Set-Valued Dual Process $\tilde{\mathcal{G}}^{++}$

The next step is to obtain a different coding of the dual process $(\eta', \mathcal{G}^{++,<})$ to be better able to study the longtime behaviour. Namely we now associate with the Markov process $(\eta', \mathcal{G}^{++,<})$, respectively the marked tableau-valued process, *marked set-valued states* which leads to a functional $\tilde{\mathcal{G}}^{++}$ which is marked set-valued. Here set-valued refers to sets which are product sets in the sense that we form a product over the locations and for each site we have a product of a (random) number of copies of the type set $\mathbb{I}$. The information we lose coding the dual state this way is the information on the *global rank* of the factors (columns) in the tableau, which is on the one hand a useful simplification of the dual, but also limits its power to deal with some questions.

The new dynamic is *not* anymore *a priori* Markov if we are in a spatial situation (for a single site process the Markov property remains automatically). We show however that the evolution of the functional is in fact still *Markov*. We obtain then a dual representation of our process with $\tilde{\mathcal{G}}^{++}$.

In fact we can construct a richer Markov process $\mathfrak{s}$ that can be related to a *Crump-Mode-Jagers* structure, a process we shall call $\mathfrak{s}^{**}$.

The Subsections 9.5–9.9 introduce (in Subsection 9.5) and then later on analyse a set-valued dual process which we derive as functional of the process $(\eta', \mathcal{G}^{++,<})$.

This set-valued dual has monotonicity properties and corresponds to the genealogical structure which allows a detailed analysis. The key task is to obtain from the dual $(\eta', \mathcal{G}^{++,<})$ both for the case N finite and $N = \infty$ a Markovian and transparently behaving *set-valued dynamics* which we call $\tilde{\mathcal{G}}^{++,N}, \tilde{\mathcal{G}}^{++}$ (i.e. the ˜ indicates this new dynamics) in the case of finite, respectively infinite N. Then we obtain here a new dual but in fact this construction works for every countable geographic space not just these special geographic spaces $\{1, \cdots, N\}$ and $\mathbb{N}$.

The reader only interested in the duality theory may just read Subsubsections 9.5.2–9.5.4, which contain the part of the construction of wider applicability which subsubsection 9.5.1 explains and motivates from the point of view of studying emergence and fixation.

This Subsection 9.5 constructs and lays the basis for the analysis of the set-valued dual needed for proving ergodic theorems on the one hand and studying emergence and fixation on the other hand in Subsections 9.6–9.12. For the latter purpose an enrichment of the process is needed. Indeed the analysis of the further longtime behaviour and the behaviour as $N \to \infty$ requires a new construction and representation of the objects in (9.149) and of $\tilde{\mathcal{G}}^{++}$.

Subsections 9.6 and 9.7 carry out the study of emergence and fixation first for the simpler case $N = \infty$. We want to recover in this case a CMJ-process as we had in Section 8, but we need to preserve the duality relation. We will see that indeed not all of this can be achieved. This will lead us to three dynamics closely related to $\tilde{\mathcal{G}}^{++}$, two of which are duals and one is of auxiliary nature but has a CMJ-structure. Once we have carried out in Subsections 9.6 and 9.7 the $N = \infty$ case, which is

9.5 The Marked Set-Valued Dual Process $\tilde{\mathcal{G}}^{++}$

somewhat simpler we come to the case of $N < \infty$ in Subsection 9.9 where we introduce the necessary modifications to handle collisions from migration which occur on the geographic space $\{1, 2, \cdots, N\}$.

9.5.1 Dual $\tilde{\mathcal{G}}^{++}$: Examples for the Special Cases $M = 2, 3$

To understand how the enriched process of Subsubsection 9.4.7 which took the form of a marked tableau of indicator functions of one variable might in a useful (for our question of emergence and fixation) may be represented as marked set-valued process we associate with the tableau a subset of $(\mathbb{I}_M^{\mathbb{N}})^S$ with S the geographic space, we consider the example of $M = 2, 3$ and calculate what form the transitions take in $\mathcal{G}_t^{++,<}$ and indicate how we can represent the resulting process which we call $\tilde{\mathcal{G}}^{++}$ again in form of a functional of a Markov process and use this process as the new dual. These examples also suggest the concept of *resolved* and *unresolved* factors, which is a key to evaluate the dual expression as time becomes large.

The Case M = 2

In the case $M = 2$ we construct a Markov process with dynamics looking as follows. There are a finite number of types of factors, namely (100), (010), (110), (111). Starting with a factor of type (110) at a tagged site, new factors and a new summand are produced once a birth in η' occurs by the action of the selection operator. Factors can move to a new site, can coalesce or change due to mutation. The mutation can result in the removal of a summand as we shall see below, a phenomenon not occurring with one type on the lower level. We shall show below that the factor-valued process associated with a given site is recurrent and returns either to the state (110) after a finite random time or reaches (000) and the whole summand is deleted. During this sojourn some factors emigrate to new sites and the founding factor at the new site now (in contrast to the case $M = 1$ in Section 3) may become inactive and thus a *new site can be lost again*. The essential question is the rate of the *successful emigration* (establishing a new permanent site) which governs the growth of the number of the total number of (active) factors.

Here are the transitions in detail. We write an $*$ if either the value 0 or 1 can be inserted.

- **Selection** We first consider selection. Recall that the basic selection operator in $\mathcal{G}^{++}$ is given by

$$f \to (1_A \cdot f) \otimes (111) + (1 - 1_A) \otimes f \qquad (9.150)$$

where $1_A = (011)$ and the terms $1_A \cdot f$ and $(1 - 1_A)$ have the same rank in the order so that all other operations are performed simultaneously on them.

Since $1_A, 1-1_A$ are of the form (011) respectively (100) they are not traps under the mutation chain (different from (110) disregarding rare mutation) and can be changed to (00∗) or (11∗), which are quasi-traps. We will therefore introduce the following concept:

$$(01*), (10*) \text{ are called } \textit{unresolved} \text{ factors.} \tag{9.151}$$

- **Mutation** We now include the action of mutation at rate $(m_{12} + m_{21})$ which can *resolve* factors, i.e. a transition of the factors (011) and (100), which are produced if selection acts, to the quasi-trap (110) or to the quasi-trap (001) occurs:

$$(100) \to (110), \ (010) \to (000), (011) \to (001) \text{ with probability } \frac{m_{21}}{m_{12}+m_{21}}, \tag{9.152}$$

or

$$(100) \to (000), \ (010) \to (110), (011) \to (111) \text{ with probability } \frac{m_{12}}{m_{12}+m_{21}}, \tag{9.153}$$

(thanks to the coupling we use). This means that one summand is zero (or (001) and contributes 0 in the dual expression if we start with zero mass on type 3) and the other has only the factor (110), which is a quasi-trap. Factors where the mutation has occurred are called "*resolved* factors". These factors are important because they don't undergo further changes by mutation (except rare mutation), where with unresolved factors in a row we do not know yet if this summand remains or will become zero due to the action of mutation.
- **Coalescence** Now consider the coalescence of ranks k_1 and k_2. If a rank k_2 looks-down to a position $k_1 < k_2$, then the column k_2 is removed and the rows in which the combination (010), (100) or (100), (010) occur are removed (remember that here we are taking the product of indicator functions).
- **Migration** Any factor can undergo migration, that is, move to a new site, but factors of the same rank do this together. For the moment we consider the case of infinitely many sites. In this case a rank always moves to an unoccupied site (but note that the rank of the particles it contains does not change). The case with finitely many sites in which a rank can move to an occupied site will be dealt with in Subsubsection 9.9.

The importance of the location is that factors at different sites do not coalesce.

We now make a number of observations being crucial in the sequel. We observe first that (100) or (101) respectively (010) or (011), respectively (110) or (111) are for the purpose of emergence equivalent since in the duality relation they give the same contribution and they can change in the last component only by rare mutation so that it suffices here to discuss either of the two.

9.5 The Marked Set-Valued Dual Process $\tilde{\mathcal{G}}^{++}$

Note that migrating factors can be one of three types (referred to as $*$-types)

$$(110), (100)^*, (010)^*, \qquad (9.154)$$

where for example $(100)^*$ denotes a factor that is currently given as factor (100) later on the resolution $(100) \to (110)$ occurs, which makes the summand with (010) equal to zero and the one with (100) (which became (110)) acts as the original factor (110). Hence the $*$ highlights that the factor has not yet resolved in (110) or (000) and we call $(100)^*, (010)^*$ then $*$-*types*.

Note that we can decide on the final type of a factor already at the time of migration with probabilities $(\frac{m_{2,1}}{m_{1,2}+m_{2,1}})$, $(\frac{m_{1,2}}{m_{1,2}+m_{2,1}})$. Note furthermore that the choice of the final type also determines the fate of the particles to the right of that rank. This means the $*$-type of a given factor is together with the two-traps now a four-state Markov chain.

Note that it is possible for a newly occupied site to be deleted if the founding rank is deleted due to resolution of an ancestral position which turns the column into (111) (recall (9.142) in Remark 128). In this case the descendant newborn factors then have no effect anymore leading to $1_A \otimes 1 + (1 - 1_A) \otimes 1 = 1 \otimes 1$ upon selection. Hence we have to keep track of the ancestral relations in order to determine which factors need to be deleted!

How do these transitions combine? Consider an example.

Example 9. Three successive applications of the selection operator $1_A = (011)$ to a factor f, i.e. successively to rank 1 can be described in the form of a tableau with 8 rows and four columns. Deleting now the zero rows arising from $1_A^1, (1 - 1_A^1)$, etc. we end up with the simplified tableau of only four rows as follows:

$$\begin{array}{llll} 1_A^1 \cdot f & \otimes 1 & \otimes 1 & \otimes 1 \\ (1 - 1_A^1) & \otimes 1_A^2 \cdot f & \otimes 1 & \otimes 1 \\ (1 - 1_A^1) & \otimes (1 - 1_A^2) & \otimes 1_A^3 \cdot f & \otimes 1 \\ (1 - 1_A^1) & \otimes (1 - 1_A^2) & \otimes (1 - 1_A^3) & \otimes f. \end{array} \qquad (9.155)$$

Here the first column corresponds to the initial position of f and the "offspring" of a selection operation is placed immediately to its right. Each column consists of indicator functions which induces a partition of the type space and the partition elements evolve via the mutation process but in such a way that the evolving partition elements at all time remain disjoint.

Applied to the case $f = (110)$, $1_A = (011)$, (9.155) yields the following *tableau* of indicators corresponding to subsets and inducing a partition of $\{1, 2, 3\}^4$:

$$\begin{array}{llll} (010) \otimes 1 & \otimes 1 & \otimes 1 \\ (100) \otimes (010) & \otimes 1 & \otimes 1 \\ (100) \otimes (100) & \otimes (010) & \otimes 1 \\ (100) \otimes (100) & \otimes (100) & \otimes (110). \end{array} \qquad (9.156)$$

Consider the case in which at the second rank in (9.156), we have (100) → (000), (010) → (110), by mutation. Then there are only two summands and the third and fourth ranks are inactive (and therefore columns are deleted). Hence we are then left by a 2 × 2-tableau instead of a 4 × 4. This deletion of formerly active ranks due to mutation is the reason that the growth is slower than in the two type case.

Associated Disjoint Decomposition of Product of Type Set

In the *tableau* of (9.156) each row corresponds to one summand in the dual $\mathcal{G}^{++}$ and each column defines a decomposition of the set of the M low level types. The collection of columns defines a decomposition of the 4-fold product of the set of low-level types. The number of factors is a pure birth process with rate s as in the two-type case.

More generally each state of the $(k' \times k)$-tableau induces a *decomposition of the k'-th product of the set of low level types*, i.e. $\{1,2\}^{k'}$ and thereby also of the set $\{1,2,3\}^{k'}$ by just putting no constraint on the element 3. The collection of *rows* of the resulting tableau corresponds for (9.156) to the decomposition of $\{1,2\}^4$ given by the elements $(x_1, x_2, x_3, x_4) \in \{1,2\}^4$:

$$\begin{aligned}&\{x_1 \in \{2\}\}\\&\cup \{x_1 \in \{1\}\} \cap \{x_2 \in \{2\}\}\\&\cup \{x_1 \in \{1\}\} \cap \{x_2 \in \{1\}\} \cap \{x_3 \in \{2\}\}\\&\cup \{x_1 \in \{1\}\} \cap \{x_2 \in \{1\}\} \cap \{x_3 \in \{1\}\} \cap \{x_4 \in \{1,2\}\}.\end{aligned} \quad (9.157)$$

The Case M = 3

We next consider the effect of having $M > 2$ lower level types using the case $M = 3$ exhibiting the key features. The main new feature is that with $M = 2$ we had only one selection operator, since we had two fitness levels. In general $M - 1$ of the M lower level types have different fitness levels from 1. Having more such levels results in more possible ways for unresolved factors to reach their final resolved state in a quasi-trap called above $*$-types. Hence the essential difference between the case $M = 2$ and $M = 3$ is the *resolution mechanism* for the dual process which we shall describe in the case $M = 3$.

Suppose the fitness of the four types is

$$0, a, 1, 1 \text{ with } a \in (0,1). \quad (9.158)$$

Recall the upper level type has downward mutation at rate N^{-1} while the three others have mutation rates $O(1)$.

9.5 The Marked Set-Valued Dual Process $\tilde{\mathcal{G}}^{++}$

We need to analyse the dual starting with one particle and with:

$$\mathcal{G}_0^{++,<} = (1110). \tag{9.159}$$

Most of the things said for $M = 2$ remain but what we have to study in addition is what happens after the creation of an unresolved factor by the action of a selection operator, i.e. a factor not of the form $(111*)$ or $(000*)$ which are the quasi-traps, i.e. traps under mutation, excluding rare mutation. On this unresolved factors various actions can take place, mutations, further selection, coalescence and migration. The new feature is that the selection can act with different selection operators at one rank and can be combined with intermediate mutation steps. *Consequently the process of resolution is no longer a - two-states plus two-traps- Markov chain* (with states $(10*)$ or $(01*)$, but requires paths of intermediate steps in further states.

We focus on this feature now and look at the transitions of resolved and unresolved factors step by step.

- **Coalescence and Migration.** This is exactly as what we described in the case $M = 2$.
- **Selection and Mutation.** We focus first on the selection operator followed by subsequent mutations. We first note that we have two selection operators corresponding to the sets of types

$$A_2 \quad (\text{fitness} \geq a) \text{ and } A_3 \quad (\text{fitness} \geq 1), \tag{9.160}$$

which leads at rate s to the multiplication by (0111) resp. (1000) and by (0011) respectively (1100). Recall the choice between A_2 and A_3 occurs with probability a respectively $(1 - a)$.

In particular we have for the selection operator with $A = A_3$:

$$(1110) \to (0010) \otimes (1111) + (1100) \otimes (1110) \quad \text{at rates } s \cdot (1 - a). \tag{9.161}$$

After a finite random time due to the action of mutation this yields either $(1110) \otimes (1110)$ (mutation from type 3 to type 1 or 2) or $(1110) \otimes (1111)$ (mutation from type 2 to type 3 and then type 1 to 3) and this selection event is then in either case *resolved* since we arrive in a quasi-trap under mutation.

If we use the selection operator $A = A_2$ we have the transition:

$$(1110) \longrightarrow (0110) \otimes (1111) + (1000) \otimes (1110) \quad \text{at rates } s \cdot a. \tag{9.162}$$

In other words, selection occurs at rate s and then the selection operator (0111) is chosen with probability a and the selection operator (0011) is chosen with probability $1 - a$.

Using the coupling between summands in $\mathcal{G}^{++}$, eventually after a finite random time and several subsequent mutation steps (for example $3 \to 1, 2 \to 1$ for the (0110) factor) the part of the sum due to the considered first action of the selection operator collapses by mutations to either (0000) or alternatively to (1110) ⊗ (1111) or (1110) ⊗ (1110), and the factor on the r.h.s. of (9.162) is resolved.

What makes this dynamic difficult to handle? The problem is to determine what happens between creation of an unresolved factor and its resolution by subsequent mutation steps. As before when a position migrates we can decide in advance, i.e. at creation, which state will occur at resolution and with which probability. But now to determine the probabilities of resolution in one of the possible states, namely, the two quasi-traps (111∗) and (000∗), we need to follow the *intermediate* states which were not present for $M = 2$.

To do this we must work with a finite state Markov chain conditioned to be absorbed by one of the two absorbing states (111∗), (000∗). Having first chosen this outcome we can then secondly replace the mutation Markov chain by its *h-transform* corresponding to the condition to reach the appropriate absorbing point. The overall effect of this does not change the law of the resolution process.

For example the r.h.s. in (9.162) can resolve to

$$(1110) \otimes (1111) + (0000) \text{ mutations } 1 \to 3 \text{ or } 1 \to 2. \tag{9.163}$$

Similarly r.h.s. of (9.162) can resolve to

$$(000*) \otimes (1111) + (1110) \otimes (1110) \text{ by mutations } 2 \to 1, 3 \to 1, \text{ etc..} \tag{9.164}$$

The length of such paths to resolution is in principle *unbounded* but the path is eventually absorbed in (000∗) or (111∗) and can pass only through a *finite* number of intermediate states.

In other words between birth or coalescence events each rank involves the dynamics of a partition of the set of types $\{1, 2, 3\}$. Moreover each partition element is associated to a subset of the set of columns of the tableau of factors in the different summands. The dynamics of the partition associated with a rank is given by a Markov chain induced by the mutation process. Note that the number of partition elements can only increase by a selection operation and can only decrease by coalescence in each case by one element. Otherwise the partition process eventually ends in a quasi-trap and stays there until the next selection event at this rank.

Note that a founding particle is not influenced by coalescence at that new rank and its behaviour is given by a Markov chain involving mutation and selection at that rank by the look-down principle in carrying out the coalescence (this uses of course the symmetry between individuals under the resampling mechanism). We can use this as in the $M = 1$ case to obtain a simpler form of the sum. Namely we can determine its resolved value in advance and then follow the h-transformed process as in the $M = 2$ case. As before we distinguish for each of the selection operators two versions associated with the two principal states for absorption but now instead of making the choice with probabilities $m_{1,2}/(m_{1,2}+m_{2,1})$ respectively $m_{2,1}/(m_{1,2}+m_{2,1})$ we now use the two absorption probabilities. The states of the

9.5 The Marked Set-Valued Dual Process $\tilde{\mathcal{G}}^{++}$

process before resolution is then described by the *h-transform* corresponding to the absorbing state chosen.

Remark 129. In the new formulation of the dual process with coupled summands namely $(\eta', \mathcal{G}^{++})$ it is important to code the *new site* according to whether we have a factor which survives or one which can be deleted after some random time. In particular in the regime where we have no collision we can do this easily since a site colonized will never be hit by another migrant ever, so that these two events do not depend on the evolution at other sites. For this purpose we shall later introduce classes of permanent, removed, and transient sites. A site is called permanent at the moment it is clear that it survives, it's called removed once it becomes deleted and transient after the time of birth, until it will eventually be deleted. Alternatively using the *h*-transform we can immediately at the time of colonization decide to assign permanent or transient to the site.

Example 10 (coalescence of selection factors). In order to get a better feeling for the process, we discuss a particular effect. Since there are more types we must also consider the result of coalescence between two different partitions. As we has seen above in the three type case it is necessary to determine the outcome of collisions of unresolved factors of different types.

Now consider the case of selection operators corresponding to $A = (0111)$, $B = (0011)$ acting on two different ranks, say $i < j$, both with factors corresponding to (1110):

$$(1110) \to_{1_A} \begin{matrix}(0110) \otimes (1111) \\ (1000) \otimes (1110),\end{matrix} \quad (1110) \to_{1_B} \begin{matrix}(0010) \otimes (1111) \\ (1100) \otimes (1110).\end{matrix} \quad (9.165)$$

Then after coalescence of the first columns and after insertion of two ranks at ranks i and $j + 1$ we get the non-zero rows at rank i

$$\begin{aligned} 1_A 1_B &= (0110)(0010) = (0010), \\ (1 - 1_A)(1 - 1_B) &= (1000)(1100) = (1000) \\ 1_A(1 - 1_B) &= (0111)(1100) = (0100). \end{aligned} \quad (9.166)$$

The following (see (9.167)) illustrates the effect of coalescing two columns followed by a mutation. Consider the tableau obtained starting with (1110) and then evolving as follows. Let the selection operator 1_B act on column 1 twice and then the selection operator 1_A acting on the third resulting column. This yields the tableau on the left in (9.167). Then the coalescence of the 3rd and 1st columns results in the tableau on the right of (9.167).

$$\begin{vmatrix} 0010 & 1111 & 1111 & 1111 \\ 1100 & 0010 & 1111 & 1111 \\ 1100 & 1100 & 0110 & 1111 \\ 1100 & 1100 & 1000 & 0110 \end{vmatrix} \mapsto \begin{vmatrix} 0010 & 1111 & 1111 & 1111 \\ 1100 & 0010 & 1111 & 1111 \\ 0100 & 1100 & 1111 & 1111 \\ 1000 & 1100 & 1111 & 0110 \end{vmatrix} \quad (9.167)$$

Then mutation in the first column from 1 to 2 removes the last row and mutation in the first row from 2 to 1 removes the third row, resulting in the total change to the tableau of active ranks given by:

$$\begin{vmatrix} 0010 & 1111 \\ 1100 & 0010 \end{vmatrix} \tag{9.168}$$

Return to tableau (9.167). After resolution of the first column this column would consist either of (1110) factors or (0000) factors. For example if the next transition is a mutation $1 \to 2$ or $2 \to 1$ in column 1, then we know either the third or fourth rows can be deleted as shown above. If we first have the mutations $1 \to 3, 2 \to 3$ rows 2–4 are deleted, etc.

We note that very long strings on the way to resolution are very unlikely since the mutation rates between the three lower types are strictly positive and all other rates bounded above by some constant, so that the quasi-traps for this mechanism are reached in finite time. We also note that at each step we have a finite number of possible outcomes. Therefore by Markov chain theory the waiting time needed to reach resolution and reach a quasi-trap has therefore *finite mean*.

Conclusion from Examples $M = 2, 3$

We can summarize the message of the two examples for the case of M-types on the lower level as follows. Again as for $M = 1$ we can simplify the dual expression but instead of analysing strings of factors we now work with *tableaus* of factors. In addition we have to distinguish *resolved* and *unresolved* factors and hence resolved and unresolved sites. In comparison with the case $M = 2$, for the general case $M \geq 3$ there are two new features in the resolution of the selection factor:
(i) resolution does not occur at the first action of mutation, and
(ii) there are a finite number of intermediate states, more precisely ($2^M - 2$) (here we have M lower level types) that can be reached by mutation before resolution.

Nevertheless the observations on the form of $\mathcal{G}^{++,<}$ in the two examples does suggest that our *Markovian dynamics of marked tableaus* introduced in Subsection 9.4, corresponds to a *decomposition of subsets of* $\{1, \cdots, M + 1\}^{\mathbb{N}}$ at each geographic location. This we will now formally introduce in Subsections 9.5.2–9.6.2 for the case of general M.

9.5.2 The Marked Set-Valued Duality: $\tilde{\mathcal{G}}^{++}$

We are now ready to set up formally the functional of $(\eta', \mathcal{G}^{++,<})$ given by a marked *set*-valued Markov process and in particular specify its state space. We make some definitions in a form, that it can be used also for the finite N process and in fact the

9.5 The Marked Set-Valued Dual Process $\tilde{\mathcal{G}}^{++}$

process on any geographic space, which is countable. All the duality constructions we are carrying out in Subsubsections 9.5.2–9.5.4 work for every finite type space.

Let

$$\mathbb{I} = \mathbb{I}_M := \{1, \ldots, M, M+1\}, \tag{9.169}$$

$$\mathcal{I} := \text{the algebra of subsets of } \mathbb{I}. \tag{9.170}$$

Let the geographic space be S with

$$S = \{1, \ldots, N\} \text{ or } S = \mathbb{N}. \tag{9.171}$$

In this Subsection 9.5 we will not include in the notation whether $S = \mathbb{N}$ or $S = \{1, \cdots, N\}$, later we write $\tilde{\mathcal{G}}^{++}$ respectively $\tilde{\mathcal{G}}^{++,N}$ for the two cases.

The states of $(\eta', \mathcal{G}^{++,<})$ can be coded as *marked tableaus* consisting of columns of factors of one variable, each marked by a location, where a row defines then the product of factors. This defines a set-valued state

$$\tilde{\mathcal{G}}_t^{++}, \tag{9.172}$$

as follows. Consider the subtableau associated with one and the same mark. The rows of the subtableau (recall (9.109)) correspond to the indicator function of subsets of $(\mathbb{I}_M)^m$ for some m and the product over the marks gives then a subset of $\prod_{j \in S} (\mathbb{I}_M)^{m_j}$ such that *distinct rows* (recall rows can be identical) correspond to *disjoint subsets*. Therefore the sum of the rows of the complete tableau is the *indicator function* of a subset of $\prod_{j \in S} (\mathbb{I}_M)^{m_j}$. It is often convenient to associate with this subset a subset of $\prod_{j \in S} (\mathbb{I}_M)^{\mathbb{N}}$ by considering

$$\prod_{j \in S} (A_j \times \mathbb{I}^{\mathbb{N}}) \quad , \quad A_j \subseteq (\mathbb{I}_M)^{m_j}. \tag{9.173}$$

This allows us to work with rows corresponding to subsets of one and the same set.

Definition 9.18 (Marked set-valued process of $\tilde{\mathcal{G}}^{++}$).

(a) The rows of the marked tableau $t(\eta', \mathcal{G}^{++,<})$ define disjoint subsets of

$$\prod_{j \in S} (\mathbb{I}_M)^{m_j} \text{ for some } \{m_j\}_{j \in S}, \text{ with all } m_j < \infty. \tag{9.174}$$

The union of these disjoint subsets is a subset of $\prod_{j \in S} (\mathbb{I}_M)^{m_j}$ whose indicator function is given by the sum of the rows (where the factors in the row are multiplied corresponding to taking the intersection of the sets associated with the different ranks) of the tableau.

(b) The set in (9.174) defines a new process (the state space will be formally introduced below) of subsets of $\prod_{j \in S} (\mathbb{I}_M)^{\mathbb{N}}$ (recall (9.173)) denoted:

$$(\tilde{\mathcal{G}}_t^{++})_{t \geq 0}. \qquad \square \qquad (9.175)$$

Remark 130. The subset of $\prod_{j \in S}(\mathbb{I}_M)^{m_j}$ does not characterize the state of $(\eta', \mathcal{G}^{++,<})$ since in general the (global) ranks cannot be reconstructed from ranked subtableaus corresponding to different sites in general (for more than one site).

Remark 131. Note that since the map giving $\tilde{\mathcal{G}}^{++}$ from the Markov process $(\eta', \mathcal{G}^{++,<})$ is not bijective (at least if space contains more than one site, see above remark), the former need not be a priori Markov in a spatial situation. However we will verify the Markov property below in Proposition 9.21.

We now introduce some needed notation:

$$\mathcal{T}_m := \{ \text{ the algebra of subsets of } (\mathbb{I}_M)^m \}, \qquad (9.176)$$

$$\mathcal{T} := \text{ subsets of } (\mathbb{I}_M)^{\mathbb{N}} \text{ of the form } A \times (\mathbb{I}_M)^{\mathbb{N}} \text{ with } A \in \mathcal{T}_m \text{ for some } m, \qquad (9.177)$$

$$\mathcal{T} \subset \hat{\mathcal{T}} := \sigma - \text{algebra of subsets of } (\mathbb{I}_M)^{\mathbb{N}}. \qquad (9.178)$$

The state space of $\tilde{\mathcal{G}}^{++}$ is contained in the σ-algebra

$$\mathsf{I}_* = \bigotimes_{j \in S} \mathcal{I}_j, \text{ where for each } j, \ \mathcal{I}_j = \hat{\mathcal{T}}. \qquad (9.179)$$

Given a non-empty set $G \in \mathsf{I}_*$ let (recall we remove inactive ranks in $\mathcal{G}^{++}$ and hence in $\tilde{\mathcal{G}}^{++}$ we get only finitely many factors not equal to $\mathbb{I}_M$)

$$|G| := \min\{ j : \exists S_j = \{s_1, \ldots, s_j\} \subset S : G = G_j \otimes ((\mathbb{I}_M)^{\mathbb{N}})^{S \setminus S_j}$$

$$\text{with } G_j \in \bigotimes_{i \in S_j} \mathcal{T} \}$$

$$:= +\infty \text{ if no such } S_j \text{ exists} . \qquad (9.180)$$

If $|G| = j < \infty$, the *support* of G, supp(G) is defined to be the set S_j that appears in (9.180).

Then we define

Definition 9.19 (State space of $\tilde{\mathcal{G}}^{++}$).
The state space of $(\tilde{\mathcal{G}}_t^{++})_{t \geq 0}$ is defined as countable algebra of sets

$$\mathsf{I} = \mathsf{I}_S := \{ G \in \mathsf{I}_* : |G| < \infty \}. \qquad \square \qquad (9.181)$$

9.5 The Marked Set-Valued Dual Process $\tilde{\mathcal{G}}^{++}$

Given the finite type spatial Fleming–Viot process with selection $X_t^S = \{x^N(t,j)\}_{j \in S} \in (\mathcal{P}(\mathbb{I}_M))^S$, we define the expression

$$\hat{X}_t^S := \prod_{j \in S} (x(t,j))^{\otimes \mathbb{N}} \in (\mathcal{P}((\mathbb{I}_M)^{\mathbb{N}}))^S,$$

with $S =$ the countable geographic space. (9.182)

The following dual representation is satisfied because of the very construction of the process $\tilde{\mathcal{G}}^{++}$ as functional of the dual $\mathcal{G}^{++,<}$:

Proposition 9.20 (Set-valued duality relation).
The process $\hat{X}$ and $\tilde{\mathcal{G}}^{++}$ are dual:

$$E_{X_0}\left[<\hat{X}_t, \tilde{\mathcal{G}}_0^{++}>\right] = E_{\tilde{\mathcal{G}}_0^{++}}\left[<\hat{X}_0, \tilde{\mathcal{G}}_t^{++}>\right], \; \forall \; \tilde{\mathcal{G}}_0^{++} \in I \text{ and } X_0 \in (\mathcal{P}(\mathbb{I}))^S. \quad \square$$
(9.183)

The point in the sequel will be to formulate a Markovian dynamics which generates the process $\tilde{\mathcal{G}}^{++}$ (thus proving the Markov property) and doing so in a transparent way. This turns the above duality relation in the usual pairing of two *Markov* processes given by a duality relation, where the r.h.s. can be conveniently calculated. The Markovian dynamics of the process $\tilde{\mathcal{G}}_t^{++}$ is constructed in Sub-subsections 9.5.3–9.5.4 first without migration, then incorporating the latter. Then indeed we can prove in Subsubsection 9.5.4:

Proposition 9.21 (Markov property of $\tilde{\mathcal{G}}^{++}$).
The process $(\tilde{\mathcal{G}}_t^{++})_{t \geq 0}$ is Markov. $\quad \square$

In order to analyse the behaviour of the r.h.s. in (9.183) for our questions on emergence and fixation we have to classify factors of the geographic sites as follows.

Definition 9.22 (Occupied, inactive and resolved sites).

(a) A factor is called resolved if its indicator has the form $(1, \cdots, 1, *)$ or $(0, \cdots, 0, *)$.
(b) A site for a state in I_S is occupied if there is at least one factor with indicator not equal to $(1, \ldots, 1, *)$ at the site. An occupied site is called *resolved* if it contains only one type of factor, namely with indicator $(1, \ldots, 1, *)$, or *inactive* if it contains only factors with indicators $(1, 1, \cdots, 1)$. $\quad \square$

Definition 9.23 (Tableau of sets and corresponding process).
Following the procedure in Subsubsection 9.4.6 we can write the state of sets in terms of a disjoint union of subsets, where in the union we order the subsets and we can define a process

$$\tilde{\mathfrak{t}}(\eta_t', \mathcal{G}_t^{++}) \tag{9.184}$$

with values in tableaus of subsets, which is Markov and is obtained by replacing indicators in $\mathfrak{t}(\eta', \mathcal{G}^{++,<})$ by the corresponding subset of $\mathbb{I}$. $\square$

We now have to prove Proposition 9.21. For this purpose we have to find a mechanism which turns $\tilde{\mathcal{G}}^{++}$ into a *Markov* process. On the way we perform some constructions and prove some results relevant for the later analysis of the longterm behaviour. As a preparation for this analysis of the set-valued dual dynamic $\tilde{\mathcal{G}}_t^{++}$ we begin by focussing on a single factor with subsets of $\mathbb{I}_M$ (corresponding in $\mathcal{G}^{++,<}$ to a single column (rank)) since in this case $\tilde{\mathcal{G}}^{++}$ is in fact automatically a Markov process. The difficulty arises due to migration.

9.5.3 Properties 1: Set-Valued Markovian Dynamics Without Migration

First we analyse the behaviour of $\tilde{\mathcal{G}}^{++}$ under the action of the mutation transition at a single factor at a single location. Later we consider also selection and coalescence, in particular the interaction of mutation with selection.

Where we had before in Section 8 a collection of birth and death processes with values in $\mathbb{N}$, we now get a collection of dynamics on another countable set, which can be viewed as subsets of $\mathbb{I}_M^{\mathbb{N}}$. Acting on each column we have a dynamic of subsets of $\mathbb{I}_M$ if j is the number of factors $\mathbb{I}_M^j$. We study here these refined dynamics and begin by looking at the dynamics at a given factor $\mathbb{I}_M$ (i.e. rank), first under the effect of mutation, then only under selection and finally under the combination of both. At last coalescence is added in a straightforward way.

At a single location we deal with subsets of $\mathbb{I}_M^{\mathbb{N}}$ as in (9.177). Our subsets typically do not have product form but are disjoint unions of subsets of products of $\mathbb{I}_M$. So the disjoint union induces a partition of $\mathbb{I}_M$ in each factor of $\mathbb{N}$. Therefore we will study the dynamics of these partitions of $\mathbb{I}_M$ as well. We introduce migration in the next subsubsection.

(1) Mutation. Consider the mutation mechanism in the dual process $(\tilde{\mathcal{G}}_t^{++})$, ignoring for the moment all other transitions. This induces transitions on indicators of subsets of $\mathbb{I}_M$ - see (9.152), (9.153), (9.163), (9.164) for examples. In general, translating the transitions of indicator functions into transitions on sets we obtain for $j \in \mathbb{I}_M$:

$$A \to A \cup \{j\} \text{ at rate } \sum_{\ell \in A} m_{j,\ell}, \quad A \to A \setminus \{j\} \text{ at rate } \sum_{\ell \in A^c} m_{j,\ell}. \qquad (9.185)$$

This defines a *Markov process* on subsets of $\mathbb{I}_M$ which we denote by (here m stands for the mutation matrix $(m_{i,j})_{i,j=1,\cdots,n}$):

$$(\zeta_t^m)_{t \geq 0}. \qquad (9.186)$$

9.5 The Marked Set-Valued Dual Process $\tilde{\mathcal{G}}^{++}$

The above process describes the evolution of a set of types which have mutated to a given type, say i or to a given set of types, say A. We want to define now a dynamic on a whole *partition* of $\mathbb{I}_M$, i.e. we vary the type i respectively the sets A. Given a *partition of* $\mathbb{I}_M$ we can let the mechanism in (9.185) act simultaneously on *all* the indicator functions corresponding to the elements of a given partition.

Note that for our model order $O(1)$ transitions do not change the $(M+1)$ component so that effectively this is described by a Markov process on the set of partitions of $\{1,\ldots,M\}$ with transition and transition rates given for every $k = 1, \cdots, n$ by:

$$(A_1,\ldots,A_n) \to (A_1,\ldots,A_k\backslash\{j\},\ldots,A_\ell \cup \{j\},\ldots,A_n) \text{ at rate } \sum_{j' \in A_\ell} m_{j,j'}. \tag{9.187}$$

We obtain a process:

$$(\tilde{\zeta}_t^m)_{t \geq 0}. \tag{9.188}$$

Then we can prove

Lemma 9.24 (The set-valued mutation dynamics and h-transform).

(a) *The Markov process $(\tilde{\zeta}^m(t))_{t \geq 0}$ is a pure Markov jump process with state space given by the set of partitions of $\{1,\ldots,M\}$. The partition*

$$(\mathbb{I}_M\backslash\{M+1\},\emptyset) \text{ is a trap} \tag{9.189}$$

Assume that the mutation matrix is strictly positive for all types $i,j \in \mathbb{I}_M\backslash\{M+1\}$. Then starting with an initial partition, exactly one initial partition element will grow to $\mathbb{I}_M\backslash\{M+1\}$ and all others will go to the empty set.

(b) *Given the initial partition $(\{1\},\ldots,\{M\}))$ of $(\mathbb{I}_M\backslash\{M+1\})$ consider the process*

$$(\tilde{\zeta}_t^m(1),\ldots,\tilde{\zeta}_t^m(M)), \tag{9.190}$$

where the $\{\tilde{\zeta}_t^m(i), i = 1, \cdots, M\}$ are possibly empty disjoint subsets with union $\mathbb{I}_M\backslash\{M+1\})$. The quantity $\tilde{\zeta}_t^m(i)$ denotes the set of initial types that have mutated to type i. This is a Markov process.

If the $m_{k,\ell} > 0$ for $k,\ell \in \{1,\ldots,M\}$, then for every $k \in \mathbb{N}, 1 \leq j \leq k$ the k-tuples

$$A_j^* := (\emptyset,\ldots,\emptyset,\mathbb{I}_M\backslash\{M+1\},\emptyset,\ldots,\emptyset), \text{ with } \mathbb{I}_M\backslash\{M+1\} \text{ at } j\text{-th position}, \tag{9.191}$$

are absorbing points for the the $\tilde{\zeta}^m_t$ process and with probability one the process reaches an absorbing point at a finite time.

(c) Given $1 \leq j \leq k$ the process conditioned to hit the absorbing point A^*_j is given by the h-transform corresponding to h_j which is the Markov process with transition rates

$$q^{h_j}_{\zeta_1,\zeta_2} = \frac{h_j(\zeta_2)}{h_j(\zeta_1)} q_{\zeta_1,\zeta_2} \quad \text{if } h_j(\zeta_1) \neq 0, \qquad (9.192)$$

where

$$h_j((B_1,\ldots,B_k)) = P_{(B_1,\ldots,B_k)}(\{\tilde{\zeta} \text{ hits } A^*_j \text{ eventually}\}). \qquad (9.193)$$

Denote the law of the h-transformed process by P^h. Then the law P of $\tilde{\zeta}^m$ is given by

$$P = \sum_j h_j P^{h_j}. \quad \square \qquad (9.194)$$

Proof of Lemma 9.24.

(a) Note first that in a mutation transition two types i, j are involved and hence it suffices to consider the effect for a set A and A^c. Consider therefore the initial condition

$$\psi(0) = (A, A^c), A \subset \mathbb{I}_M \backslash \{M+1\}. \qquad (9.195)$$

Then a mutation from type i to type j makes no change if $i, j \in A$ or $i, j \in A^c$. However if $i \in A$, $j \in A^c$, then after the transition at time τ,

$$\tilde{\zeta}_{\tau+} = (A\backslash i, A^c \cup i). \qquad (9.196)$$

Therefore after each mutation, we have a partition therefore resulting in a partition-valued Markov process.

The process continues until it reaches

$$\tilde{\zeta}^m = (\emptyset, \mathbb{I}_M \backslash \{M+1\}) \text{ or } \tilde{\zeta}^m = (\mathbb{I}_M \backslash \{M+1\}, \emptyset). \qquad (9.197)$$

Therefore given any partition exactly one partition elements grows to $\mathbb{I}_M \backslash \{M+1\}$ and all the other go to the empty set since we have assumed positive rates on $\mathbb{I}_M \backslash \{M+1\}$.

We can next consider the enriched process where we start with a complete partition of $\mathbb{I}_M$. When type i changes to type j, i is moved to the partition element containing j. This process continues until there is only one non-empty partition element (these are the traps). Note that the number of different non-empty

9.5 The Marked Set-Valued Dual Process $\tilde{\mathcal{G}}^{++}$

partition elements is non-increasing and the process eventually hits a trap with probability one.

(b) and (c) Relations (9.192) and (9.194) follow as a special case of Doob's h-transform of a Markov process ([RW], III.45). q.e.d.

This dynamic of *partitions* of the set $\mathbb{I}_M$ can be immediately generalized to a dynamics of *tuples of sets* which are pairwise either disjoint or repetition of the other subset. (Note here the union can be strictly contained in the type set). Given a k disjoint subsets with union in $\{1,\ldots,M\}$ the process from (9.185) induces a Markov jump process with values in k-tuples of disjoint possibly empty subsets with union in $\{1,\ldots,M\}$.

By collecting the types not in the union of the tuple in a $(k+1)$-element we obtain a partition and can use the previous result.

This means we can next define uniquely the *evolution of a column* of a *tableau* our object discussed earlier, but here we view it as a tableau of subsets of $\mathbb{I}_M$ and we let it evolve under mutation since the indicators of sets, say A, B, in a column satisfy either $A \cap B = \emptyset$ or $A = B$. Given a specified rank in a tableau at a site the above mechanism induces the evolution of the column at this rank where of course in the case of repetition of sets they follow all the *same* evolution.

If we now consider j-different columns, the mutation acts independently on each of these components. Therefore we have now all what we need to handle the mutation part of the evolution of the set-valued tableau. Misusing notation we use for this process $\tilde{\zeta}^m$.

(2) Selection. Consider the set-valued state. Given a factors k, the action of selection on this factor produces a new rank (birth) placed at $k+1$ and the rank $k+\ell$ is moved to rank $k+\ell+1$ for $\ell = 1,2,\ldots$.

Now consider the action of a selection operator on an indicator function at a factor at a site. Recall that selection acts on each factors with rate s and when this occurs the selection operators corresponding to A_i, $i = 2,\ldots,M$, is chosen with probabilities $(e_i - e_{i-1})$ where $\{e_i\}$ are defined in (9.3).

Then the action of the chosen selection operators 1_{A_ℓ}, $\ell \in \{2,\cdots,M\}$ act by multiplication on the active elements of the factor and intersects with A_ℓ^c on the new elements of the factor as specified by (9.117), (9.118). On the level of a set-valued dynamics, if $B = \cup_{i=1}^m \cap_{k=1}^n B_{i,k} \in \mathcal{I}^n$, then selection 1_{A_ℓ} acting on factor $j \in \{1,\ldots,n\}$ produces a transition to the disjoint union of ordered products

$$\left(\bigcup_{i=1}^m \left(\prod_{k=1}^{j-1} B_{i,k} \times (B_{i,j} \cap A_\ell) \times \mathbb{I} \times \prod_{k=j+1}^n B_{i,k} \right) \right) \qquad (9.198)$$

$$\cup \left(\bigcup_{i=1}^m \left(\prod_{k=1}^{j-1} B_{i,k} \times (A_\ell^c) \times \prod_{k=j}^n B_{i,k} \right) \right) \in \mathcal{I}^{n+1}.$$

We can represent such a subset in a more transparent way in a reduced tableau consisting of subsets of $\mathbb{I}_M$ as entries, where the columns form a tuple of subsets of $\mathbb{I}_M$ taken from a partition of $\mathbb{I}_M$ and the rows define disjoint *nonempty* subsets of $\mathbb{I}_M^{n+1}$.

Remark 132. Note that the corresponding tableau can be rewritten taking advantage of the simplifications $1_{A_i} + 1_{A_i^c} = \mathbb{I}$, $1_A 1_C = 0$ if A and C are disjoint, which may reduce the number of rows, but of course not the number of columns.

Remark 133. Consider the result of two selection operations acting on the same rank - we obtain four summands

$$(f 1_{A \cap B}) \otimes 1 \otimes 1, \quad (1-1_A) 1_B \otimes f \otimes 1, \quad (1-1_B) \otimes 1_A f \otimes 1, \quad (1-1_B) \otimes (1-1_A) \otimes f. \tag{9.199}$$

corresponding to the decomposition of $\mathbb{I}_M$ in each factor and hence of $(I_M)^3$. If $f \equiv 1_{\mathbb{I}}$ the decomposition is induced by:
$A \cap B$, $A^c \cap B$, B^c in the first factor, respectively A^c and A in second factor. This gives four elements in $\mathbb{I}_M$. The further evolution under mutation then determines which of the four decomposition elements takes over, that is, which of these will grow to $\mathbb{I}_M \backslash \{M+1\}$. For general f we get more decomposition elements. Note that if $B = A$, then two of the four terms become empty sets.

(3) Combined Effect of Selection and Mutation. The effect of selection is to produce new nontrivial factors (real subsets of $\mathbb{I}_M$). Since mutation can change factors, it can change such a factor into $\mathbb{I}_M$ or $\emptyset$ - when this happens the factor becomes either trivial or the whole state becomes $\emptyset$. In both cases the size $|\mathcal{G}|$ of the state $\mathcal{G}$ does not increase.

Example 11. We can then think of the selection process as a "proposal" process and then mutation as acceptance-rejection process as follows (in the case $M = 2$, i.e. first and selection corresponding to 1_A with $A = \{2, 3\}$ acts and next mutation acting with the two possible outcomes corresponding to $1 \to 2$ or $2 \to 1$):

$$\begin{aligned} \{1,2\} &\longrightarrow \{2\} \times \{1,2,3\} \cup \{2\} \times \{1,2,3\} \\ \{1,2\} &\longrightarrow \{1,2\} \times \{1,2,3\} \cup \{1,2\} \times \{1,2\}. \end{aligned} \tag{9.200}$$

In the first case the new rank produced by selection is removed (rejected) by mutation, in the second it is accepted.

(4) Coalescence. Sets of the tableau at rank ℓ and $\ell' > \ell$ which are at the same site *coalesce* by a look-down process: the rank ℓ' is removed and the factor ℓ is modified by taking the intersections of the corresponding partition elements - see (9.119).

As a result in the representation as irreducible tableau some rows and columns can disappear. For example, in the case $M = 2$ the row with (100) in the ℓ'th rank

9.5 The Marked Set-Valued Dual Process $\tilde{\mathcal{G}}^{++}$

and (010) in the ℓ' th rank is removed on coalescence so that one row and one column is removed.

(5) Set-Valued Markovian Dynamic. Now we have defined all three mechanisms occuring in a single site model and we can combine this now to a set-valued evolution in $\mathcal{T}$. We denote the corresponding set-valued dynamics (respectively tableau-valued) combining mutation (rate matrix m), selection (rate s), coalescence (rate d) as

$$(\zeta_t^{m,s,d})_{t \geq 0}, \quad (\tilde{\zeta}_t^{m,s,d})_{t \geq 0}. \tag{9.201}$$

The corresponding jump rates are denoted

$$\{q_{\zeta,\zeta'}^{m,s,d}\}_{\zeta,\zeta' \in \mathcal{T}}. \tag{9.202}$$

By construction the evolution is corresponding to the evolution of the set-valued process as induced by $(\eta', \mathcal{G}^{++,<})$. Starting with a single factor or finite number of factors, for example $(1, \ldots, 1, 0)^{\otimes k}$, this determines the evolution at the given site in our indicator-function-tableau-valued process $\mathcal{G}^{++,<}$. This proves in particular Proposition 9.21 for the case of a geographic space with one single site.

The set-valued process has the following properties.

Lemma 9.25 (Local set-valued dynamics with no migration).

(a) *Consider the Markov process $(\zeta_t^{m,s,d})_{t \geq 0}$ with $d > 0$, $m > 0$. This process is positive recurrent and returns to states $A \subseteq \mathbb{I}_M^\mathbb{N}$ where at most the first factor is different from $\mathbb{I}_M$.*

(b) *In particular for our concrete model of M lower and one upper type the process is positive recurrent and ergodic, namely returns to the state $(1, 1, \ldots, 1, 0)$ (corresponding to $(\mathbb{I}_M \setminus \{M+1\}) \times \mathbb{I}_M^\mathbb{N}$) after a recurrence time called the site resolution time τ which has a distribution satisfying for some $\lambda > 0$ and k denoting the initial number of variables*

$$E_k[e^{\lambda \tau}] < \infty. \quad \square \tag{9.203}$$

Proof of Lemma 9.25. We first note that the number of active ranks at a site undergoes a birth and death process with linear birth rate and quadratic death rate. Therefore the process will reach a state with only one active rank with probability one. But then there is a positive probability that the mutation process at this rank will resolve, that is, reach a trap or quasi-trap, before the next selection event.

To prove the existence of a finite exponential moment, note that due to the quadratic death term there exists k_0 such that for $k \geq k_0$ the birth and death rates are dominated by a subcritical linear birth and death process. Therefore if τ_{k_0-1} is the time to hit $k_0 - 1$, then there exists some $\lambda' > 0$ such that for $k \geq k_0$,

$$E_k[e^{\lambda' \tau_{k_0-1}}] < \infty. \tag{9.204}$$

We also note that (by recurrence) for each $k \leq k_0$, there is a positive probability of hitting 1 before $k + 1$. This means that the process returns to k at most a finite number of times before hitting 1 and this random number has a geometric distribution. Therefore starting at $k \leq k_0$ the number of jumps before hitting 1 is the sum of a finite number of geometric random variables. Since the time intervals between jumps are exponentially distributed, this implies that τ_1 has a finite exponential moment. Combining this with (9.204) we see that the hitting time of 1 has a finite exponential moment starting from any point.

Noting that when the number of active ranks is reduced to 1 this corresponds to a column containing members of a partition of $\mathbb{I}_M \setminus \{M + 1\}$ and therefore has union some $A \subset \mathbb{I}_M \setminus \{M + 1\}$. If this equals $\mathbb{I}_M \setminus \{M + 1\}$ then $\tau = \tau_1$. Otherwise there is a positive probability that this rank will reach a trap $\mathbb{I}_M \setminus \{M + 1\}$ or $\emptyset$ before the next selection event. This will then occur after a geometric number of trials. This implies the result.   q.e.d.

9.5.4 Properties 2: Set-Valued Dual with Migration

We have sofar shown that we have for the non-spatial process a set-valued dual process with a Markovian dynamic. We now turn to the spatial case, when migration can occur. A transition due to mutation, selection and coalescence only effects the site at which the transition event occurs. Moreover if $\tilde{\mathcal{G}}_t^{++}$ *factors* between sites, then this property is *preserved* under these three types of transitions. Recall that we denote by $|\tilde{\mathcal{G}}_t^{++}|$ the number of sites with a factor of $\tilde{\mathcal{G}}_t^{++}$, see (9.180) for a formal definition.

We now introduce the transition due to *migration* which is a transition changing both the originating (parent) site and the site to which the migrant individual moves. Recall that in $\mathcal{G}^{++,<}$ this amounts to a column changing its mark from say i to j corresponding to a migration step from i to j. More specifically at rate $ca(i, j)$ a migration event from a specific factor at an *occupied* site i to a site j takes place.

There are two possibilities on migration at a time t, namely in the first case (1) j is an unoccupied site thus yielding $|\tilde{\mathcal{G}}_t^{++}| = |\tilde{\mathcal{G}}_{t-}^{++}| + 1$, or (2) j is an occupied site in which case we have $|\tilde{\mathcal{G}}_t^{++}| = |\tilde{\mathcal{G}}_{t-}^{++}|$.

We note

For $S = \mathbb{N}$ only case 1 occurs, for $S = \{1, \cdots, N\}$ both cases do occur.   (9.205)

In particular for the dual of the McKean–Vlasov process the distinction of the two cases is irrelevant.

In case (1) the factor is removed at site i and all factors to the right of the migrant are shifted one to the left and at site j this migrant is placed as the first factor.

In case (2) we distinguish two subcases.

9.5 The Marked Set-Valued Dual Process $\tilde{\mathcal{G}}^{++}$

If prior to the migration the state of $\tilde{\mathcal{G}}_t^{++}$ factored at these two involved sites, then at the new site we add the individual at the first inactive rank at the new site and have the product form of the new individual and the old state at the new site.

If $\tilde{\mathcal{G}}_t^{++}$ does not factor at the site where the migrant appears, the situation is more complicated. This situation will be dealt with in Subsubsection 9.9.1–9.9.3 since here in the sequel we consider only the case $S = \mathbb{N}$ with no collision.

Altogether transitions for the set-valued process are well-defined by these finite rates and well-defined transitions.

We denote the process and the resulting transition rates in the case $S = \mathbb{N}$ as (with E denoting the state space):

$$(\tilde{\zeta}_t^{m,s,d,c})_{t \geq 0}, \quad \{q_{\zeta,\zeta'}^{m,s,d,c}\}_{\zeta,\zeta' \in E}. \tag{9.206}$$

Proof of Proposition 9.21 (Markov property).

We have seen that the transitions that occur in the function-valued dual $\mathcal{G}_t^{++}$ (restricted to indicator functions) translate into the transitions on the corresponding set. Given any set $G \in \mathsf{I}$, the set of possible transitions $G \to G'$ due to selection, mutation, coalescence and migration are well defined and the transition rate is finite. Therefore these transitions define *uniquely* a continuous time Markov chain $\tilde{\mathcal{G}}_t^{++}$ on the countable space I. Furthermore by construction of the transitions and their rate as read off from those induced by $(\eta', \mathcal{G}^{++,<})$ we conclude that the set-valued functional of $(\eta', \mathcal{G}^{++,<})$ we defined and the process $\tilde{\mathcal{G}}^{++}$ driven by the above dynamic have the same law. Hence $\mathcal{G}^{++}$ is Markov.   q.e.d.

9.5.5 Calculating with $\tilde{\mathcal{G}}^{++}$

We first note that to carry out actual calculations it is sometimes appropriate to work with sets but it is also often convenient to represent sets by their indicators. For example (9.200) then reads:

$$(110) \to (010) \otimes 1 + (010) \otimes 1 \to (110) \text{ or } (110) \otimes (110). \tag{9.207}$$

Also we can then work with irreducible tableaus of indicator factors to represent subsets, see (9.198) for example.

Remark 134. Note that the decomposition of the indicator function into the sum of products of indicator functions is preserved but the constraint on which other individuals the migrating individual can coalesce with changes.

We demonstrate the effects of migration with some examples. However one should have in mind that the objects we deal with should be viewed as subsets. The art of using the duality is to use the right representation at the right moment.

446   9 Emergence with $M \geq 2$ Lower Order Types (Phases 0,1,2)

In order to make the effect of migration and its interplay with mutation more transparent, we give some examples for the main effect that new occupied sites are created by migration, but as a result of mutation the parent site the offspring site can be deleted at a random time or the site can become permanently occupied.

Example 12 (The effects of migration for $\tilde{\mathcal{G}}^{++}$).

We consider the case where we have two types of low fitness and one of higher fitness. Let us start with (110) and assume that 4 births occur due to selection successively acting at the current ranks 1,2,3,4 and assume that we simplify the tableau by carrying out as much sums as possible (so that each time here only one row is added), so we get:

$$\begin{pmatrix} (010)_1 \\ (100)_1 \ (010)_1 \\ (100)_1 \ (100)_1 \ (010)_1 \\ (100)_1 \ (100)_1 \ (100)_1 \ (010)_1 \\ (100)_1 \ (100)_1 \ (100)_1 \ (100)_1 \ (110)_1 \end{pmatrix} \qquad (9.208)$$

Here the missing entries are (111).

If the individual at the third (local) rank below migrates to an empty site and then there are two selection operations at the new site (denoted by subscript 2).

The symbol $\odot$ indicates that sites are coupled and 2* indicates that this rank is actually a copy of the rank 2 at the second site, later in Subsection 9.6 we shall call this a "ghost rank" coupled to the new site 2.

We obtain (representing the a set by its indicator function with each row given by a product of indicator functions and the different rows being the indicator function of disjoint sets):

$$\begin{pmatrix} (010)_1 \\ (100)_1 \ (010)_1 \\ (100)_1 \ (100)_1 \ (010)_{2*} \\ (100)_1 \ (100)_1 \ (100)_{2*} \ (010)_1 \\ (100)_1 \ (100)_1 \ (100)_{2*} \ (100)_1 \ (110)_1 \end{pmatrix} \odot \begin{pmatrix} (010)_2 \\ (100)_2 \ (010)_2 \\ (100)_2 \ (100)_2 \ (010)_2 \end{pmatrix} \qquad (9.209)$$

More precisely, (9.209) corresponds to the multisite tableau:

$$(010)_1$$
$$(100)_1 \ (010)_1$$
$$(100)_1 \ (100)_1 \ (010)_2$$
$$(100)_1 \ (100)_1 \ (100)_2 \ (010)_2 \qquad (9.210)$$
$$(100)_1 \ (100)_1 \ (100)_2 \ (100)_2 \ (010)_2$$
$$(100)_1 \ (100)_1 \ (100)_2 \ (100)_2 \ (100)_2 \ (010)_1$$
$$(100)_1 \ (100)_1 \ (100)_2 \ (100)_2 \ (100)_2 \ (100)_1 \ (110)_1.$$

9.5 The Marked Set-Valued Dual Process $\tilde{\mathcal{G}}^{++}$

If a mutation $1 \to 2$ occurs in either column 1 or column 2 (turning (100) to (000) and (010) to (110)), then the new site is removed, i.e. becomes (111) and therefore inactive.

On the other hand if the mutation $2 \to 1$ occurs (turning (100) to (110) and (010) to (000)) in both column 1 and column 2, and the last two columns coalesce with one of the first two columns, then the new site is active and is decoupled from the parent site, i.e. the two sites factor.

One additional complication arises if one of the last two columns migrates before this happens. Then the future of this migrant site depends on the site 2 in the same way. This results in:

$$\begin{pmatrix} (010)_1 & & & & \\ (100)_1 & (010)_1 & & & \\ (100)_1 & (100)_1 & (010)_{2*} & & \\ (100)_1 & (100)_1 & (100)_{2*} & (010)_{3*} & \\ (100)_1 & (100)_1 & (100)_{2*} & (100)_{3*} & (110)_1 \end{pmatrix} \odot \begin{pmatrix} (010)_2 & & \\ (100)_2 & (010)_2 & \\ (100)_2 & (100)_2 & (010)_2 \end{pmatrix} \odot \begin{pmatrix} (010)_3 \\ (100)_3 \end{pmatrix}, \quad (9.211)$$

corresponding to the multisite tableau

$$\begin{aligned}
&(010)_1 \\
&(100)_1 \ (010)_1 \\
&(100)_1 \ (100)_1 \ (010)_2 \\
&(100)_1 \ (100)_1 \ (100)_2 \ (010)_2 \\
&(100)_1 \ (100)_1 \ (100)_2 \ (100)_2 \ (010)_2 \\
&(100)_1 \ (100)_1 \ (100)_2 \ (100)_2 \ (100)_2 \ (010)_3 \\
&(100)_1 \ (100)_1 \ (100)_2 \ (100)_2 \ (100)_2 \ (100)_3 \ (110)_1.
\end{aligned} \quad (9.212)$$

Now two scenarios can occur for the further evolution.

If the mutation $2 \to 1$ occurs in both column 1 and column 2, and the last two columns coalesce with one of the first two columns, then the new first site is active. Then if the mutation $1 \to 2$ occurs in the first column at the second site, then the third site is removed. On the other hand if mutations $1 \to 2$ occur at all three columns of the second site then the third site remains active and is decoupled, i.e. factors from the rest of the expression.

Outlook on Subsections 9.6–9.12 The set-valued dual process $\tilde{\mathcal{G}}_t^{++}$ is the basic tool to be used in the remainder of this Section 9. As we have seen above if this begins with $(11\cdots 10)$, i.e. the subset $\{1, 2, \cdots, M\}$ of $\mathbb{I}_M = \{1, \cdots, M+1\}$, then the support $|\tilde{\mathcal{G}}_t^{++}|$ can increase as t increases. As in the case $M = 1$ the analysis of the exponential growth of $|\tilde{\mathcal{G}}_t^{++}|$ in this case plays a key role in the proof of the results on emergence. However there are a number of complications that arise when $M > 1$ which did not appear when $M = 1$. The first is the observation above that one or more newly occupied sites can be *removed* due to a mutation at the parent site. This means that the dynamics of the *parent* and *offspring sites* are *linked* and in fact simultaneous transitions can occur at two or more sites. In particular the joint event at these sites cannot be factored into a product of events at the individual sites.

In order to manage these complexities and prove the main results on emergence using the set-valued dual a number of techniques will be developed. One strategy will enable us to identify the Malthusian parameter describing the growth of $|\tilde{\mathcal{G}}_t^{++}|$, and another will lead to the McKean–Vlasov limit of the set-valued dual and the corresponding nonlinear deterministic equations with random entrance law.

9.6 The Growth of Occupied Sites in the Collision-Free Case $\tilde{\mathcal{G}}^{++}$: Basic Construction of Enriched Dynamics $\mathfrak{s}, \mathfrak{s}^*, \mathfrak{s}^{**}$

The duality constructions we have so far are sufficient to resolve *well-posedness* of martingale problems and to prove *ergodic theorems*. However in this subsection we want to further develop the dual in order to study the *emergence and fixation* problem after a rare mutation. For this purpose we have to control the expansion of the set of active sites of the dual, that is, to determine the number of sites occupied by $\tilde{\mathcal{G}}^{++}$ with nonempty strict subsets of $(\mathbb{I}_M)^{\mathbb{N}}$. For this we need some modified and richer dynamics to unravel CMJ-structures we had in Section 7 also here in the multitype case, which then have to be analysed.

The main problem here is the *dependence* between the evolution of some factors at different occupied sites, which can result in *losses* of active occupied sites. This is *not* just the interaction by migration, but nonlocal effects of mutation and selection operations at the parent site of a migrant which can effect the migrants new site. These effects stem from the fact that the evolution of the tableau of the ranks at the parent site and at a new site do *not factor* and hence do *not* effectively evolve independently under the action of mutation or selection operators and coalescence, in contrast to the particle model where once a particle moved away from a site it was not influenced directly by the selection and mutation transitions at its parent site at the moment of the transition since at any time the state of the dual factored.

We therefore have to set up a new object which is an enriched version of $\tilde{\mathcal{G}}^{++}$ which will be called $\mathfrak{s}$ (together with two relatives $\mathfrak{s}^*, \mathfrak{s}^{**}$, where $\mathfrak{s}^{**}$ is the desired CMJ process) and which allows us to effectively *monitor these dependencies* between sites effectively using some historical information, what we shall call ghost copies of migrating ranks in their internal states at the old site. These enriched versions are obtained from modifications $\tilde{\mathcal{G}}^{++,\ell}$ for $\ell = I, II, III$ of $\tilde{\mathcal{G}}^{++}$, but where $\ell = I, II$ remain *dual* and $\ell = I, III$ are *Markov* and $\ell = III$ is a CMJ process and all do *not* require to keep the very complex information about all *global ranks* as we have to carry around with us in $(\eta', \mathcal{G}^{++,<})$, we then have only local ranks.

The link between these $\mathfrak{s}, \mathfrak{s}^*, \mathfrak{s}^{**}$ and the set-valued dual process $\tilde{\mathcal{G}}^{++}$ is given via enriched set-valued states we call $\tilde{\mathcal{G}}^{++,\ell}, \ell = I, II, III$ which are *functionals* of $\mathfrak{s}, \mathfrak{s}^*, \mathfrak{s}^{**}$ but the latter can be read off from the historical process of the $\tilde{\mathcal{G}}^{++,I}$. In fact we will explain first in 9.6.1 the enrichment of $\tilde{\mathcal{G}}^{++}$ we need and in 9.6.2 the modifications of the evolution rules needed for $\tilde{\mathcal{G}}^{++,\ell}$, i.e. we explain what transitions will occur in these objects $\mathcal{G}^{++,\ell}$, since this will allow to obtain an intuitive understanding for the heavy technical machinery needed to construct $\mathfrak{s}, \mathfrak{s}^*, \mathfrak{s}^{**}$.

The constructions are all somewhat easier in the case of the collision-free process, on which we focus first and which is crucial for emergence. For fixation we then have to cope with a general system with migration and including the possibility of collision which we do later in Subsection 9.9.

9.6.1 Preparation 1: Enriched Marked Set-Valued Dynamics with Emigration

So far we have a set-valued dual process as functional of $(\eta', \mathcal{G}^{++,<})$, namely $\tilde{\mathcal{G}}^{++}$ for which we have also specified a Markovian dynamic generating a version of this process (see (9.213)). However the growth of occupied sites is now more subtle. Both migration and dependencies between many sites complicate the picture, when we want to analyse the growth of the number of occupied sites in the corresponding process (set-valued, respectively tableau-valued, more precisely tableau of subsets of $\mathbb{I}_M$)

$$(\zeta_t^{m,s,d,c})_{t \geq 0}, \text{ resp. } (\tilde{\zeta}_t^{m,s,d,c})_{t \geq 0}, \tag{9.213}$$

with positive migration rate c which we discussed in Subsubsection 9.5.4. This occurs due to the fact that two sites may evolve with heavy dependencies. In the dual $\langle \eta', \mathcal{G}^{++,<} \rangle$ we can control this using the *global rank*, however this object is too complicated and contains more than the needed information. The point is to enrich $\tilde{\mathcal{G}}^{++}$ with just the needed information. The complications occur in the case $S = \{1, \cdots, N\}$ and to a lesser degree $S = \mathbb{N}$ (so that ranks migrate out but no immigrants arrive at an active site).

We first describe the problem more precisely and then show how to proceed with an enriched state description of $\tilde{\mathcal{G}}^{++}$ or equivalently the driving Markov processes given in (9.213) involving (1) *ghost* copies and (2) *local ranks* to solve it. This will be still considerably simpler than keeping track of global ranks in $\langle \eta', \mathcal{G}^{++,<} \rangle$ itself.

Migration involves two sites, the parent site and a new site which in the case $S = \mathbb{N}$ is the first unoccupied integer. We have noted earlier (Subsubsection 9.5.1) in the case $M = 2$ that an interaction due to a mutation and selection operation at the parent site can take place between the parent site and new site. The device introduced in the special case $M = 2$, namely deciding upon occurrence in which quasi-trap a factor resolves, does not extend to the general case $M \geq 2$.

Let us first indicate the nature of the interaction between parent and new site in $< \eta', \mathcal{G}^{++,<} >$ and its reflection in $\tilde{\mathcal{G}}^{++}$. There are two effects and one important conservation fact:

- due to a resolution transition at the parent site the state at the new site can revert to $(1, \ldots, 1, *)$ so that the site is again inactive and is removed from the list of active sites (see Example (12), (9.210)) from Subsection 9.5.5.
- A transition at the new site occupied by a migrant can remove ranks in the parent site that correspond to ranks that were produced by the action of selection acting on ranks to the right of the rank of the migrant (at the migration time). This can result in the removal of a rank at the parent site. (see again Example (12) (9.210) and consider $1 \to 2$ mutation at rank 1 at site 2 which removes ranks 3 and 4 at site 1).
- Note the conservation property: the number of active sites after a migration and resolution is at least equal to the number before the migration event.

Remark 135. A consequence of this interaction is that the dynamics depends on the ranks both at the parent and the new site. It is sometimes useful to distinguish between local and global ranks. The global ranks introduced up to now change only due to a birth (selection) and can be deleted due to coalescence. The global rank does not change due to migration.

At a given site we can give columns at that site a local rank with order consistent with the global rank. Note that the global rank of a column at a site can change due to a transition at another site but the local rank does not change in this case. It is exactly the information on the global rank which cannot be reconstructed from $\tilde{\mathcal{G}}_t^{++}$.

Therefore in order to obtain a richer Markov process which in particular allows us to represent $\tilde{\mathcal{G}}^{++}$, we need to incorporate enough information on the ranking at both the parent site and the new colonized site to be able to work without the global rank. Such a trick will be the introduction of two objects (1) a *ghost copy* coded as an element of a set Υ to be introduced below and (2) a *local* rank. The required extension of the state space will be constructed formally in all detail in Subsection 9.6.2, Step 2 and below will be specified only verbally.

We now have to *enrich the state description* to be able to introduce the *ghost copy* and a *local* rank. This construction is for the collision-free system, i.e. $S = \mathbb{N}$ only and does not work if we have returns to a site or migration to occupied sites.

We now specify the change of state when a migration event occurs at global rank ℓ at a given site. When this rank migrates, then

- a new site is chosen at random and the corresponding partition (or column) is assigned as the first rank at this new site for the *local rank* (note in $(\eta', \mathcal{G}^{++,<})$ it keeps its global rank),
- a *ghost copy* of this column (rank) is maintained at the parent site in a set Υ to be described in more detail later and with subsequent dynamics that mirrors the dynamics at the new site,
- at the resolution time of the founding rank either the new site is removed or it factors and becomes independent of the parent site and then in both cases the ghost copy is removed at the parent site,

9.6 The Growth of Occupied Sites in the Collision-Free Case... 451

- due to the mirror dynamics the ranks in Υ are subject to mutation, selection, coalescence (within Υ) and migration but *do not undergo coalescence with the other ranks at the parent site*.

We must keep track of both the local dynamics at the parent site and also the global change due to the creation of an additional active site. The latter can either be permanent or else be transient, that is, have a finite lifetime. For the local dynamics at the parent site we must include the ghost dynamics. This leads to a new enriched *"microstate space"* compared to $\tilde{\mathcal{G}}^{++}$ since we have information on local ranks and we have the ghost copies.

Focus now on the evolution of the parent site with its ghost copies. The microstate space $\mathfrak{T}$ will be defined precisely in the next subsubsection. We denote the enriched (by the ghost copies and local ranks) version of the process $(\tilde{\zeta}_t^{m,s,d,c})_{t\geq 0}$ from (9.206) for $S = \mathbb{N}$ as

$$(\hat{\zeta}_t^{m,s,d,c})_{t\geq 0}. \tag{9.214}$$

Remark 136. To summarize the above

- A migrated rank and its offspring can no longer coalesce with the ranks at the parent site.
- However ranks to the right of the migrated rank ℓ (and its offspring) at the parent site can possibly be deleted due to the mutation trajectory at these ranks.
- The migrated position ℓ can also be removed by resolution at $j < \ell$ at the parent site (see Remark 128).

Now we have a situation that the evolution at different sites is potentially connected till the time of decoupling of the parent site from the new site. This we have handled by introducing the ghost copies and we want to represent this dynamic now in a different form which is more convenient for our later analysis. For this purpose we add some information to the state of $\hat{\xi}^{m,s,d,c}$ namely on whether the migrants are removed or decouple as permanent sites in a particular state of their initial rank, so for a migrant better its ghost copy this then has the form of an element in $\{UPe, URe\} \times \mathcal{T}$. We then know that our new dynamics "represents" the old one.

We define the following time-inhomogeneous markovian mechanisms. If the first migration event occurs we decide with the appropriate probabilities to give the migrated rank the mark *UPe* or *URe* (ultimately permanent, respectively ultimately removed) and in case of *UPe* τ^{**} the current partition and the index indicating which partition element takes over when the migrant rank resolves in $(1, 1, \ldots, 1, 0)$. Then the process continues on the events *UPe*, respectively *URe* in the following way till the next migration event.

- UPe: take the h-transformed transitions, with h being the function which assigns to the present state of $\hat{\xi}^{m,s,d,c}$ at the migration step, the probability that the

migrant becomes permanent and if its first rank resolves in $(1, 1, \ldots, 1, 0)$ it is the subset τ^{**} of $\mathbb{I}_M$ that takes over.
- URe: We take the h'-transformed evolution, h' being the probability that the migrant becomes eventually removed as a function of the state at the migration step.

At the next migration event we continue in this way but now the h-transforms are taken w.r.t. the event prescribing the fate of the two migrant ranks on the then four possible events. This way we continue and obtain a dynamic

$$(\hat{\xi}_t^{*,m,s,d,c})_{t \geq 0} \tag{9.215}$$

which is Markov if we store the current restrictions posed on the migrants up to the current time.

Lemma 9.26 (Local set-valued dynamics with migration, h-transform).

(a) *Consider the Markov process $(\hat{\xi}_t^{m,s,d,c})_{t \geq 0}$ with $c, d > 0$, $m > 0$ (i.e. m has strictly positive entries) in the case $S = \mathbb{N}$. Consider the process $\hat{\xi}^{*,m,s,d,c}$ constructed above and the projection on the state space of $\hat{\xi}^{m,s,d,c}$ denotes $proj(\hat{\xi}_t^{*,m,s,d,c})$. Then*

$$\mathcal{L}[(proj(\hat{\xi}_t^{*,m,s,d,c}))_{t \geq 0}] = \mathcal{L}[((\hat{\xi}_t^{m,s,d,c}))_{t \geq 0}] \tag{9.216}$$

(b) *Given that a migration took place at rank ℓ in the parent site the dynamics at that site then follow the appropriate h-transform dynamics consistent with the selected outcome. Consider the event in which the new site becomes permanent with h-transform dynamics at the new site as in (a). Then the respective h-transform dynamics at the ghost ranks corresponding to the new site and the h-transform dynamics at ranks of the parent site to the left of the migrant rank ℓ are independent.*

(c) *The special tableau corresponding to exactly one column and partition $(\mathbb{I} \setminus \{M+1\}, \emptyset)$ is recurrent under the dynamic $(\xi_t^{m,s,d,c})_{t \geq 0}$.* □

Proof.

(a). This is done in Lemma 9.24, in fact this is a general Markov chain property.

(b) To verify the independence note that the determination of the permanence/recurrence of the new site depends only on the ranks to the left of the rank of the migrant site. Furthermore, due to the structure of the factors of the tableau a lookdown from a rank to the right of the migrant particle to the ranks to the left of the migrant particle do not affect the probability of the outcomes for the survival of the migrant rank.

For example, in Example 12 assume that column 3 migrates and then column 4 coalesces with column 1. We are left with

9.6 The Growth of Occupied Sites in the Collision-Free Case... 453

$$\begin{vmatrix} 0010 & 1111 & (1111) \\ 1100 & 0010 & (1111) \\ 1100 & 1100 & (0110). \end{vmatrix} \tag{9.217}$$

The probability that the migrant site is not removed is the probability that the mutation process leads to the takeover of 1100 at resolution which coincides with the probability before the coalescence.

Therefore transitions at the new site do not affect the probability that the new site is or is not removed which therefore depends only on the dynamics at the parent site at ranks to the left of the rank of the migrant rank. Then the required independence of the h-transformed laws follows, since the condition is that the new site is not removed.

(c) This follows from Lemma 9.25.   q.e.d.

Remark 137. (a) Note that the outcome in (a) of which tagged partition element takes over does determine whether a later migrant from the parent site from a rank to the right of ℓ is removed at the resolution time of the new site.

(b) Note that the resolution time of a new site resulting from a migration can occur before the resolution time at the parent site.

(c) Note that when a migrant site is removed at the resolution time of the parent site all of its descendants produced by migrants from the site are removed at the same time.

9.6.2 Preparations 2: The Enriched Marked Set-Valued Markovian Dynamics $\tilde{\mathcal{G}}^{++,\ell}$

We discuss now three "variants" of the enriched dual introduced above in (9.214) and the sequel and which allow to generate the latter as a functional but is better for the bookkeeping of the number of active sites. Namely in this section we focus on the global picture which includes describing the dynamics of the number of *active* sites and the resulting jump process $\tilde{\mathcal{G}}^{++}$ with state space I which we defined as functional of $(\eta', \mathcal{G}^{++,<})$ and for which we showed how to generate it by a collection of autonomous Markov processes $\hat{\zeta}^{m,s,c,d}$. The interaction between a newly colonized site and the parent site destroys the CMJ-structure for occupied sites we had in Section 8. We will now show that we can nevertheless relate the process in a manageable way to some multitype CMJ-dynamics with finitely many types which allows us to determine the growth rate in the number of active sites.

In order to keep track of the interaction between parent site and new site and to unravel this CMJ-structure we use the enriched process from above. We give below first in a formal manner the Markovian dynamic which we now call

$$\tilde{\mathcal{G}}^{++,I}, \tag{9.218}$$

to distinguish it better from two related dynamics to be introduced later on and which generates via a functional (projection on collection of local states) a version for the process of collections $\hat{\zeta}^{m,s,c,d}$ from (9.214) and $\tilde{\mathcal{G}}^{++}$. We construct two further related auxiliary dynamics needed to analyse $\tilde{\mathcal{G}}^{++}$ via multitype CMJ-processes with *countable*, respectively, *finitely* many types. Those processes will be called $\tilde{\mathcal{G}}^{++,\ell}$, $\ell = II, III$.

These new processes contain a set-valued component and the information at which site factors sit and also enough information on the evolution of ranks via the ghost copies to allow to determine what happens till the resolution of factors, after which the newly created and the parent site factors and continue independently. The key point for the construction is that in the case $S = \mathbb{N}$ we have only *emigration*, that is, immigrants never arrive at an occupied site.

To continue constructing the dynamics of $\tilde{\mathcal{G}}^{++,\ell}$, $\ell = I, II, III$ we follow the following principles.

- *Mutation, selection* and *coalescence:* The action at a single site is as described in Subsubsection 9.5.3 and is the same for all the dynamics we introduce.
- *Migration:* In order to deal with the interaction between the parent and offspring sites that occur after emigration from a (parent) site to a new site we introduce *three options* I-III for the transition at the migration step.

The constructions needed are now discussed step by step below.

In order to do actual calculations with the set-valued process in particular to obtain nice representations of these processes (as motivated in Subsubsection 9.6.1) the task will be to construct the set-valued processes corresponding to $\tilde{\mathcal{G}}^{++,\ell}$ as a *functional* of an enriched *Markov* process $\mathfrak{s}, \mathfrak{s}^*, \mathfrak{s}^{**}$ incorporating further information which we construct in Subsubsection 9.6.3 after giving below the set-valued dynamic.

Assumption 1. *Since the main application of this dual is to determine the emergence time scale we assume that the initial measure of the interacting process is*

$$X(0) = \mu^{\otimes \mathbb{N}}, \text{ where } \mu \text{ is concentrated on } \{1, \ldots, M\} \quad (9.219)$$

and the initial state of the dual is a finite product of factors $(1, \ldots, 1, 0)$, *possibly at different initial sites. The reader should note that the resulting dual expressions in the case of Option II dynamics below are valid only under this assumption.*

Remark 138. We note that the coding by subsets of $\mathbb{I}_M^\mathbb{N}$ sets need not carry the information on the locations of the factors. It suffices to know which subsets of $\mathbb{I}_M^\mathbb{N}$ belong to different sites, since when we create an emigrant, due to the assumption made in this section (9.219) that the measure against which the dual is integrated has the form $\mu^{\otimes \mathbb{N}}$ its actual position is not important.

The constructions needed are now discussed step by step below.
The three dynamics $\tilde{\mathcal{G}}^{++,\ell}$, $\ell \in I, II, III$.
We first briefly review the basic terminology.

9.6 The Growth of Occupied Sites in the Collision-Free Case...

Terminology

- An *occupied site* is one at which the state is not $\mathbb{I}_M^{\mathbb{N}}$.
- At each site an *active rank* is one at which the column is not identically $\mathbb{I}_M$, and
- an *column* at an active rank at a site has an internal state which is determined by a tuple of sets taken from a partition at that rank in the tableau.

We introduce *three* different dynamics, each having features allowing for some specific applications in our proofs.

Dynamics Option I -Instantaneous migration.
Recall that at any occupied site any rank can undergo migration. In this option I if a column at a rank undergoes a migration two things happen: (1) this column is immediately placed at a new site and is assigned the local rank 1 there. (2) A *ghost copy* of this column is kept at the parent site and local rank till its resolution and this ghost rank does not coalesce with ranks at the parent site. The dynamics of the ghost copy is identical (that is, completely coupled) to that of the corresponding migrant at the new site.

This uniquely defines (with the above) the dynamics. This way at each site we have *all* the information for the further evolution at this site which becomes autonomous this way. By construction we have the following fact.

Lemma 9.27 (Identification of $\tilde{\mathcal{G}}^{++}$).
The dynamic just defined defines a pure jump process $(\tilde{\mathcal{G}}_t^{++,I})_{t \geq 0}$ for which we can consider the states with ghost copies removed. Then $(\tilde{\mathcal{G}}_t^{++})_{t \geq 0}$ defined in (9.175) is a version of the resulting process. □

As we have explained above, after a migration step of a rank the subsequent dynamics at the parent site and the new migrant site are not independent, in particular the two sites *cannot* be factored. For example, it is possible for a newly occupied site to be deleted if the founding position is deleted due to resolution of an ancestral position.

However the key point is that after a *decoupling time* (see below (9.29)), the parent site and migrant site have *product form*, i.e. the set $(\subset (\mathbb{I}_M)^{\mathbb{N}} \times (\mathbb{I}_M)^{\mathbb{N}})$ associated with the two sites is a *product set* and the subsequent dynamics of the two factors are *independent*.

At the decoupling time it is possible that the migrant site reverts to $(1, \cdots, 1, 1)$ and the site is therefore removed from the set of occupied sites. The sites that are not removed in this way form an increasing set of permanent active sites. (These sites are permanent since in this case we apply the convention that singly occupied sites do not migrate.)

This defines a set-valued dynamics $\tilde{\mathcal{G}}_t^{++,I}$ that provides the required dual representation (and does not require Assumption 1). However the set of occupied sites is *not* a CMJ process due to the coupled dynamics of the parent site and sites occupied by offspring of the parent site.

Dynamics Option II: Delayed migration.

In this option when a column at a rank undergoes migration it does not immediately move to a new site, only coalescence with other ranks at the site stops, but the evolution of the migrant column at a potential new site is encoded by the state (at the parent site) in Υ (see below), that is, the microstate of this site includes the state at a potential new site. (Recall that the location determines which initial measure with which to integrate and which ranks can coalesce.) The creation of the new site *occurs only at a decoupling time* and the associated global state is then given by the *set-product* of the *local states at the permanent active sites*. This defines the process $\tilde{\mathcal{G}}^{++,II}$.

This dynamic produces a CMJ process in which simultaneous multiple births can occur (due to already resolved potential migrants at the new migrant site at its resolution time) and since the number of such multitype births is not bounded the *set of potential types is infinite* (the types correspond to the possible internal states of the new sites, which due to the unbounded number of possible ranks is countably infinite).

Dynamics Option III: h-transform dynamics.

The task is now to reduce the infinite number of types appearing above in dynamics Option II to a finite one. Recall the examples from Subsubsection 9.5.1, which suggests how to proceed.

The event as to whether a migrating rank is permanent (Pe) or temporary (Tr) (i.e. removed at a later random time), and if permanent, the event specifying which partition elements take over are chosen according the appropriate probabilities which are measurable functions of the current state. The subsequent time evolution at the site is given by the corresponding h-transform. On the event that a migrant site is permanent we start the assigned column with *h-transform dynamics* specifying which partition element takes over at resolution. This determines which additional migrants from the parent site (to the right) can be permanent (recall Example (12)).

There are three facts about this dynamics which are important for us. (1) The set of *permanent sites* forms an auxiliary process $\tilde{\mathcal{G}}_t^{++,III}$, which is *not* a dual process since it *excludes the temporary sites*. (2) However at any given time the combined *temporary and permanent sites* form a version of the dual. (3) The *set of permanent sites* form a *CMJ process* with *finitely* many types corresponding to the specification as to which partition element for the migration rank takes over.

In point (2) we proceed as follows. On the event that a migrant site is temporary we start a *temporary tree* following Type I dynamics. This tree is removed at its root at a random time when the parent site h-transform process hits the appropriate state (the distribution of the random time is also a measurable function of the state of the parent site).

We will show later on that the process $\tilde{\mathcal{G}}^{++,III}$ still determines the growth rate for the total number of *all* active sites. For that purpose the transient sites are incorporated as descendants of the permanent sites with finite lifetimes as described above.

Duality

The three options for migration described above determine dynamics of *enriched* (by ghost copies, respectively, by migration information) set-valued processes denoted

$$\{\mathcal{G}^{++,\ell}\}, \quad \ell = I, II, III. \tag{9.220}$$

The key result is the following representation for the integral in the duality relation (9.183) with the (marked) set-valued dual process $\tilde{\mathcal{G}}^{++}$ in terms of the enriched marked set-valued dual processes $\tilde{\mathcal{G}}^{++,\ell}$, $\ell = I, II$, which we can analyse using the dynamic III. Denote by the maps

$$s_\ell, \ell = I, II \tag{9.221}$$

the reduction of the state to a set, i.e. ignoring the ghost copy ($\ell = I$), respectively passing to the set product of the sites and potential sites ($\ell = II$). Recall the $\widehat{}$ notation of (9.182).

Proposition 9.28 (Duality modified).
  Assume that $X(0) = \mu^{\otimes \mathbb{N}}$, i.e. $\hat{X}(0) = (\mu)^{\otimes \mathbb{N}}$, $\mu \in \mathcal{M}_1(\mathbb{I}_M)$. Then $(\tilde{\mathcal{G}}_t^{++,\ell})_{t \geq 0}$, $\ell = I, II$ determine marked set-valued dual processes $s_\ell \circ \tilde{\mathcal{G}}^{++,\ell}$, $\ell = I, II$ with state space state space I and the following duality relationship is satisfied (recall (9.182):

$$E\left[\int \tilde{\mathcal{G}}_t^{++} d(\hat{X}(0))\right] = E\left[\int s_\ell \circ \tilde{\mathcal{G}}_t^{++,\ell} d(\hat{X}(0))\right]. \tag{9.222}$$

For $\ell = 1$ the above holds for general initial states. □

Proof of Proposition 9.28. For $\ell = I$ this follows combining Proposition 9.20 and Lemma 9.27.
  Now consider $\ell = II$. We observe that $\mathcal{G}_t^{++,II}$ defines a function-valued process that is also given by a sum of indicator functions of disjoint subsets. The only difference now is that we integrate the factors in Υ with respect to the measure μ at the parent site. Since we have assumed in this case that the initial measure is $\mu^{\mathbb{N}}$, this produces the same integral as we did with $\mathcal{G}_t^{++,I}$. q.e.d.

Remark 139. The above dynamics involving mutation, selection, coalescence and migration determine the enriched set-valued process $\tilde{\mathcal{G}}_t^{++,\ell}$, $\ell = I, II$. As we have indicated the set of occupied sites in $\tilde{\mathcal{G}}_t^{++,II}$ is a CMJ process but it has potentially infinitely many types. Some results on supercritical branching processes with infinitely many types are known, but we could not find in the literature the required results on the Malthusian parameter and stable age-type distribution in the needed generality of CMJ processes. For this reason, here we introduce a related process $\mathcal{G}_t^{++,\ell,III}$, which is a finitely many type CMJ process. We then need some additional considerations to relate the growth of the process $\mathcal{G}_t^{++,I}$ to that of $\mathcal{G}_t^{++,III}$. We return to the question of countable many type branching processes in subsubsection 9.12.

9.6.3 Enriched Marked Set-Valued Dynamics: $\mathfrak{s}$

The process $\tilde{\mathcal{G}}^{++,\ell}$, $\ell = I, II$ contains enough information to be Markov and to calculate the expectation we need in the duality relation (recall Propositions 9.21 and 9.28). However in order to control the behaviour of the expectation in the duality relation, we do not have sufficient information on the occupied sites. We now fill this gap.

The deficit of $\tilde{\mathcal{G}}^{++}$ respectively $\tilde{\mathcal{G}}^{++,\ell}$ is that we do not not have the information on the global ranks anymore, which was what simplified the dual but a little more information than contained in local ranks and ghost copies is still needed for some purposes such as for the analysis of the longtime behaviour and in order to carry out some concrete calculations. Therefore some additional information has to be incorporated.

This requires three processes, first a process $\mathfrak{s}$ which is Markov but does not store the full information on all the global ranks, which we had in $(\eta', \mathcal{G}^{++,<})$ but which contains more information about the migration steps than contained in $\tilde{\mathcal{G}}^{++}$ alone or in the states of $\mathcal{G}^{++,\ell}$ with $\ell = I, II$. This is done by incorporating information arising in the *time-space process* of $\mathcal{G}^{++,\ell}$ into the current state of the new process. Depending on whether we use Option II or Option I dynamics we have processes $\mathfrak{s}$ and $\mathfrak{s}_I$ here.

Furthermore as above with the dynamics I, which was complemented by dynamics II,III, we need also here some auxiliary processes, which are easier to analyse as far as the growth of active sites goes and from which we can draw conclusions for dynamics I using perturbation analysis via coupling techniques. The auxiliary processes will be called $\mathfrak{s}^*, \mathfrak{s}^{**}$ (parallel to a non-Markov version of $\mathcal{G}^{++,II}$ or in the Markov case to $\tilde{\mathcal{G}}^{++,III}$).

We deal first in Step 1 with the process $\mathfrak{s}$ based on Option II and then in Step 2 the one based on Option 1.

Step 1: The Markov Process $\mathfrak{s}(t)$: Construction.

The process $\tilde{\mathcal{G}}^{++,II}$ provides a dual to the multitype process $X(t)$ under the Assumption 1. In this step we also assume that $S = \mathbb{N}$ so we do not need to consider collisions. However it is not yet in the optimal form for the analysis of its growth behaviour in space and for this purpose we need a more effective bookkeeping for the active sites and their current subtableau by using historical

9.6 The Growth of Occupied Sites in the Collision-Free Case...

migration information. In this step we therefore construct a representation of the set-valued dual object $\tilde{\mathcal{G}}^{++}$ and $\tilde{\mathcal{G}}^{++,\ell}$ as a functional of the *enriched* process, called

$$\mathfrak{s} = (\mathfrak{s}(t))_{t \geq 0} \qquad (9.223)$$

and in Subsubsection 9.6.4 two related processes $\mathfrak{s}^*$, $\mathfrak{s}^{**}$ which have the advantage that they help to keep better track of the number of sites which are occupied by active ranks at a given time and have a countable type, respectively finite type, CMJ-structure which is crucial in the Subsubsection 9.7 to calculate the Malthusian parameter.

The process $(\mathfrak{s}(t))_{t \geq 0}$ will be based on a specific coding of the current state of the dual process $(\eta'_t, \mathcal{G}_t^{\tilde{+}+,<})_{t \geq 0}$ to obtain an *enriched* set-valued state. This is done in terms of a

collection of labels taken from a copy of $\mathbb{N}$, here denoted $\mathbb{N}_\ell$, (9.224)

which then indexes the subtableaus corresponding to sites where the tableau does not factor, as they are successively created by a migration event (note this may not be the time of the actual move depending which option of dynamics we use) up to the subsequent decoupling event. At time 0 we give the labels in the order generated by the ordered particle system on which η' is based. In other words the labels index the *irreducible factors* in the complete tableau $\mathsf{t}(\eta', \mathcal{G}^{++,<})$.

Each label is assigned a state called a

mesostate of a label (9.225)

which collects the information of *microstates* at the potential sites (subtableaus at a site) which evolve together since they do not yet factor and which are in the Option-II dynamic not yet moved to the new site. The whole collection of labels together with their mesostates then is the description of the global *(macrostate)* state of the process. The precise definition of a mesostate is given below in (9.288) and this state evolves according to a mesodynamic driven by the mechanism we have seen in the description of the process $\tilde{\mathcal{G}}^{++,\ell}$.

The *global* dynamics describes the *creation* of new labels at a decoupling time at one of the existing labels (this point addresses now the fact that in contrast to the $M = 1$ case not every migration step results in a permanently occupied new site). Otherwise the dynamics of different labels are *independent*. The site assigned to a newly created label is called the *founding site*.

Remark 140. The reader may ask himself why in addition to sites we need labels. The concept of labels allows us to follow the evolution of a dependent set of sites (including the microstates of the occupied sites) which is very crucial for Option-I dynamics.

We next proceed in this Step 1 focussing on the option II dynamic in parts (0)-(v), first in (0) introducing elements for the *description* of a state, in (i) the migration information, in (ii) the *state space*, in (iii) the *mesodynamic* at a fixed label (option II), in (iv) the *global* dynamic of all factors which then allows us to define the process. Then we represent in (v) $\tilde{\mathcal{G}}^{++,II}$ by the process $\mathfrak{s}$ and then in (vi) specify properties. (Recall only in Step 2 below we discuss option I for the dynamics).

(0) Ingredients 1: basic objects

Here we shall give indices to different copies of $\mathbb{N}$ if we want to distinguish them.

The ingredients from which we build the state of $\mathfrak{s}$ are grouped into two classes:

- The set of potential *geographic sites* $\mathbb{N}_g = \{1, 2, 3, \dots\}$. All but a finite number of sites are empty (the label is trivial) at time 0 and we keep track of sites that are currently occupied or have been visited before.
- The *internal state at a site*: The state at a fixed site plays the role of the state of the birth and death process in the case of $M = 1$. However in the case $M \geq 2$ the state space has a more complex structure describing the (1) *tableau* of subsets and (2) *migration information* associated with this label.

The internal state of a site to be introduced later on in point (i) is based on the following objects:

- an integer n denoting the *number* of "active local ranks" among the total number of ranks created,
- each *column* at a local rank in the tableau is a vector with entries given by elements of a partition of $\mathbb{I}_M$ such that entries are disjoint or repetitions of earlier entries,
- a *list of the subset of the local ranks* $\{1, \dots, n\}$ that have migrated together with the site to which migration of that local rank takes place, (note that the name of the site to which they migrated is not needed for the McKean–Vlasov dual where no collisions occur and where we consider only exchangeable states),
- the *-type* of a column at an active local rank in a tableau, where the *-type codes the entries in the column as

$$(B_1, \dots, B_\ell), \tag{9.226}$$

which is an ordered tuple of elements belonging to a partition of $\{1, \dots, M\}$ into disjoint subsets (i.e. the B_i are disjoint or repetition of another);
- we denote:

$\mathcal{T}^*$ the set of *-types = ℓ-tuple taken from disjoint subsets of $\{1, \dots, M\}$,
$$\tag{9.227}$$

the special element of $\mathcal{T}^*$ given by the 1-tuple consisting of the single set $\{1, \dots, M\}$ is denoted

$$\ddagger; \tag{9.228}$$

9.6 The Growth of Occupied Sites in the Collision-Free Case...

- we introduce $**$-*types* which will be particularly important with Option-III dynamics and which have the form

$$\tau^{**} = (\tau_1^*, \tau_2^*), \qquad (9.229)$$

where $\tau_1^* \in \mathcal{T}^*$ and τ_2^* (recall (9.226)):

$$\tau_2^* \in \{1, \ldots, \ell\}, \qquad (9.230)$$

where the latter gives the marking of *one* special partition element (which will later represent the one taking over at resolution, compare example in (9.162)–(9.164) and Lemma 9.24 and which determines whether or not ranks to the right of this factor in a tableau at the parent site can produce permanent migrants),
- we denote by

$$\mathcal{T}^{**} \text{ the set of } **\text{-types.} \qquad (9.231)$$

Remark 141. We will talk of sums of indicator functions as they appear above, disjoint unions of subsets, or of tableaus as introduced above depending on whichever is more suggestive at the moment. Not all possible tableau can appear under the dynamic. We can consider the restricted set that can be reached by a sequence of the selection, coalescence and mutation operations.

(i) Ingredients 2: migration information

We begin by specifying the second key element of the state. We have to code a tableau of factors (subsets of $\mathbb{I}_M$) marked with their *locations* in geographic space in a way suitable for the description of the dynamics and the growth behaviour in space.

For the form of the tableau of factors recall the definition of $\mathcal{T}_n$ and $\mathcal{T}$ from (9.176). Note that for $\tau \in \mathcal{T}$, we can identify a set $A \in \mathcal{T}_{n(\tau)}$ such that

$$\tau = A \otimes \mathbb{I}_M^{\mathbb{N}}, \qquad (9.232)$$

which is the set describing the factors of the active ranks in the various summands (rows of the tableau). Recall that a column consists of the factors associated with one rank in the various summands at a given rank which are given by elements from a partition of the set $\mathbb{I}_M$, i.e. an element of $\mathcal{T}^*$, and the complete tableau is then given by a decomposition of the set $(\mathbb{I}_M)^{n(\tau)}$. Let

$$n : \mathcal{T} \to \mathbb{N} \qquad (9.233)$$

$$n(\tau) := \min\{m : \tau \in \mathcal{T}_m \otimes (\mathbb{I}_M)^{\mathbb{N}}\},$$

denote the number of active ranks for each $\tau \in \mathcal{T}$ with $n(\tau) < \infty$.

We now have to describe the structure related to space and migration in the case $S = \mathbb{N}$, where migration means *emigration*. Recall that we want to encode the information needed to exhibit the dependencies between factors corresponding to different sites.

Let us now focus on one label. Imagine that we only consider the evolution of the population of factors attached to this label. To describe migration we have to account for the fact that ranks marked for migration can generate additional ranks by selection and which are then marked for migration and furthermore new emigration steps keep occuring to the already emigrated ranks. We therefore code the migration information using a *hierarchy of intervals* in the set which is indexing the local ranks, i.e. the ranks marked with the same location.

In order to record migrant ranks from a label at the founding site indexed by $\{1, 2, \cdots, n\}$ we use intervals marked with the destination (recall below $n = n(\tau)$ from (9.233)), i.e. we collect the *ranks assigned to the same destination* (recall by birth events such number can be bigger than one) and set $\mathcal{Y}_{n,0}$ the tuple of "subintervals" of rank 1 and each of its descendants that have not migrated, that is each containing one point. Next consider the subintervals of ranks after their *first* move from our tagged site and their descendants which do not move:

$$\mathcal{Y}_{n,1} = \text{the tuple of disjoint subintervals in } \{1, \ldots, n\}. \qquad (9.234)$$

On this first level a new subinterval is added at each migration event of a rank *not* appearing in a previously formed subinterval.

In the same way, at each migration event that occurs at a rank in one of the subintervals in $\mathcal{Y}_{n,1}$ a subinterval of this subinterval is formed, these are the ones making a second migration step or they are descendants of the one-step migrant making their first jump. The tuple of these tuples of subintervals is denoted $\mathcal{Y}_{n,2}$.

In the same way at each migration event that occurs in one of the subintervals in $\mathcal{Y}_{n,\ell}$ a subinterval of this subinterval is formed. The offspring of an element in $\mathcal{Y}_{n,\ell}$ is always added in the subinterval of the parent. The set of these subintervals is denoted $\mathcal{Y}_{n,\ell+1}$. That is,

$$\mathcal{Y}_{n,\ell+1} = \text{the tuple of disjoint subintervals each of which} \qquad (9.235)$$
$$\text{is contained in subinterval in } \mathcal{Y}_{n,\ell}.$$

This bookkeeping results in a *tuple of nested subintervals of* $\{1, 2, \cdots, n\}$ and we set

$$\Upsilon_n \text{ is the collection of all possible tuples}$$
$$\text{of nested subintervals of } \{1, \cdots, n\} \qquad (9.236)$$

and

$$\Upsilon = \cup_n \Upsilon_n. \qquad (9.237)$$

9.6 The Growth of Occupied Sites in the Collision-Free Case...

Elements have the form

$$v = (1, \ldots, [\ldots [\ldots]_{1,1}]_1, \ldots, [\ldots [\ldots [\ldots]_{\ell,1,1} \ldots]_{\ell,1} \ldots]_\ell, \ldots, n) \in \Upsilon_n, \tag{9.238}$$

for example (allowing for the moment collisions)

$$(1, 2, [3, [4, 5]_{1,1}, 6]_1, 7, 8, [9, 10]_2, [11, 12]_3), \tag{9.239}$$

which means ranks 1,2,7,8 are at the founding site and

$$\mathcal{Y}_{n,0} = ([1], [2], [7], [8]), \quad \mathcal{Y}_{n,1} = ([3, 6], [9, 10], [11, 12]) \tag{9.240}$$

and 3 and 6 are at the same site having moved once or descending from 3 and not moving, 4 made a second move and 5 as well or it descended from 4,7 and 8 are at the founding site, 9 and 10 are at a different site where 9 moved to and 10 either moved there as well or is child of 9 and finally 11 moved in a migration step from the founding site and gave birth to 12. We note that in the collision-free case never two ranks move to the same site so that above 6 is a descendent of 3, etc.

It is often useful to extract the following information from v. Namely (recall (9.233)) and introduce the set of migrants and their descendants at each step and consider the set of all possibilities for all steps taken.

Fix a label and a $v \in \Upsilon_{n(\tau)}$. Then denote the set of all sequences of the form

$$\mathcal{Y} := \{v(1) \supset v(2) \supset \cdots \supset v(\ell), \; v(i) \text{ entry of } \mathcal{Y}_{n(\tau),i}, \; i = 1, \ldots, \ell\}. \tag{9.241}$$

(ii) State space

We next consider the *state space for Option II dynamics*. We distinguish the microstates associated with sites, the mesostates associated with labels and the macrostate (global state). We first specify the state space of a mesostate at a founding site and then the global state containing the information on all the mesostates at the occupied sites. The information on the micro-states of a site associated with the founding site can be read of from the migration information.

The state space of the *meso-state* at label $j \in \mathbb{N}_\ell$ is

$$\mathfrak{T} := \{\mathfrak{t} = (\tau^*, \tau, v) : \tau^* \in T^*, \quad \tau \in \mathcal{T} \text{ with } n(\tau) < \infty, \quad v \in \Upsilon\}, \tag{9.242}$$

τ^* is the $*$-type of the label at the moment of its creation. The element τ (recall (9.177)) codes the tableau associated with the label j. Finally v specifies the corresponding coding for the various ranks at this label which have been designated as potential migrants of ℓth order, $\ell = 1, 2, \ldots$ having the same destinations.

If we are in need to write the state in the form of a product over *all* of $\mathbb{N}_\ell$, then the mesostate at the not yet created label j is indicated as

$$(\ddagger, (\mathbb{I}_M)^\mathbb{N}, \emptyset). \tag{9.243}$$

Let the set of possible paths in the countable set $\mathfrak{T}$, endowed with the discrete topology, be denoted by

$$\Omega_* := D([0, \infty), \mathfrak{T}). \tag{9.244}$$

The states at different sites, which do not factor and descend from one founding site then form a mesostate.

Remark 142. We note that we can write elements in this mesostate space as a marked tableau, namely, as a tableau of rows and columns, where the row given a string of factors in a summand and the columns the corresponding factors at a given rank marked by a potential location. Recall that rows are ordered according to rank and columns evolve under selection by splitting a column by first writing the 1_A and below the $(1 - 1_A)$-term.

We now define the *global state space* based on the space $\mathbb{N}_\ell$ as

$$\mathfrak{G} := \{\mathfrak{s} \in (\mathfrak{T})^{\mathbb{N}_\ell} : \mathfrak{s}_j = (\ddagger, (\mathbb{I}_M)^\mathbb{N}, \emptyset) \quad \text{a.a. } j\}, \tag{9.245}$$

i.e.

$$\mathfrak{s}_j \text{ is the projection of } \mathfrak{s} \text{ on label } j. \tag{9.246}$$

We write for the j-th factor

$$\mathfrak{s}_j = (\tau^*, \tau, v)_j = (\tau_j^*, \tau_j, v_j) \quad , \quad j \in \mathbb{N}_\ell. \tag{9.247}$$

For $\mathfrak{s} \in \mathfrak{G}$, let

$$|\mathfrak{s}| := \sum_{j \in \mathbb{N}_\ell} 1(\mathfrak{s}_j \neq (\ddagger, (\mathbb{I}_M)^\mathbb{N}, \emptyset)). \tag{9.248}$$

We have now specified the space where the states of a Markov jump process with Option II dynamics, $(\mathfrak{s}(t))_{t \geq 0}$ will obtain their value.

The label collects the n nontrivial factors associated (recall we do *not* move a rank before decoupling) to a finite number of geographic sites such that it is irreducible, i.e. cannot be written as a product of subsets of $\mathbb{I}^\mathbb{N}$ and correspond to a mesostate. This distinction allows to simplify the dual of the McKean–Vlasov process where no collisions occur and achieve a factorization in independent evolutions, each of which however involves potentially (and typically) several sites and thereby constituting a nontrivial mesostate.

9.6 The Growth of Occupied Sites in the Collision-Free Case... 465

(iii) Mesodynamics 1: transitions at single sites

Given $\mathfrak{s} \in (\mathfrak{T})^{\mathbb{N}_\ell}$ the possible transitions of $(\mathfrak{s}(t))_{t \geq 0}$ and their rates are of two different types. We have to distinguish the changes occuring for a rank only at one designated site (called local transition), which we discuss in this point and the global transitions occuring for ranks (of a given label) at several designated sites which we treat in point (iv).

The local transitions work as follows:

$$\text{transitions occur at the different labels } \textit{independently} \text{ of each other.} \qquad (9.249)$$

Now focus on the transition for a single nontrivial label $\mathfrak{s}_n$, $n \in \mathbb{N}_\ell$ which gives a process in $\mathfrak{T}$. We first adopt the rule that

$$\text{all transitions below occur only if we have an } \textit{active} \text{ rank.} \qquad (9.250)$$

Then the four local transitions below take place (ranks here are always local ranks):

- *Selection*

  - Each of the active ranks $1, \ldots, n(\tau)$ can give birth to a new rank hence creating new factors due to the immediate action of the selection operator. When a birth occurs, $n(\tau) \to n(\tau) + 1$. The births occur for each rank independently at rate s.
  - If the selection acts on a rank i, then the new rank is inserted immediately to the right of i and the previous indices $i + 1, \ldots$ are shifted one place to the right.
  - Given a birth event the selection operators $\Psi_i f = 1_{A_i} \cdot f \otimes 1 + (1 - 1_{A_i}) \otimes f$ are chosen with probabilities $p_i = e_i - e_{i-1}$, $i = 2, \ldots, M$. Recall that $A_i = \{i, \ldots, M + 1\}$.
  - All lists $\mathcal{Y}_{n,j}$ are updated after the insertion of a rank and the shift of ranks due to the selection operation (i.e. boundaries of subintervals are shifted accordingly).

- *Coalescence*

  - Coalescence takes place only among the ranks not included in $\mathcal{Y}_{n,1}$ or among ranks in an interval in $\mathcal{Y}_{n,1}$ but not in $\mathcal{Y}_{n,2}$, etc.,
  - for each pair $k_2 > k_1$ for which coalescence is allowed, rank k_2 looks-down to a rank $k_1 < k_2$ at rate d, and coalesces,
  - if at time t_0 position k_2 looks-down to a position $k_1 < k_2$ and coalesces, then the column k_2 is removed and the new factor with rank k_1 are given by the common refinement (at the coalescence time) of the partition elements in each row at ranks k_1, k_2. (Remember that upon coalescence we are taking the product of indicator functions). All ranks to the right of k_2 are also shifted down by one.

- *Mutation*
    - Between selection events at a given active site, mutation acts on the set of elements starting with the current column. The time development of the partition elements is given by the process $\tilde{\zeta}(t)$ as specified above and in the Lemma 9.24.
    - Due to mutation of a factor some rows and positions to the right of the rank where the mutation took place can be deleted (i.e. only left with factors $1_{\mathbb{I}_M}$).
- Migration (local effects for option II migration)
    - At each *label* each rank except the founder position independently of each other can be targeted for migration, that is, become *potential migrants* with rate c. Once a potential migrant its dynamics is the same as the dynamics described above. (Note here that using the look-down dynamic for coalescence the exceptional role of the founder in the migration step can be made, since it does never look down to another factor and due to the exchangeable initial state this does not influence the way the founding factor is integrated.)
    - At such a migration event as above a new subinterval is created, namely when a position, k in $\mathcal{Y}_{n,j}$ is selected for migration, a new subinterval $[k,k]$ is added to $\mathcal{Y}_{n,j+1}$

The above mechanism shall be what is needed to describe the evolution of the process $(\mathfrak{s}(t))_{t\geq 0}$ at all labels between the random time where global transitions occur, the socalled decoupling times which we discuss next.

(iv) Mesodynamics 2: global transitions at decoupling times - Option II dynamics

Before we come to the global transitions, we need a further concept. As we already pointed out in the examples discussed in Subsubsection 9.5.1 an important concept is that of the *resolution* of a rank and the consequential *decoupling* of the factor from the other ones leading to the creation of one or more new irreducible tableau, that is, new labels.

Remark 143. The decoupling time plays an essential role in the global description of the set-valued dual. Namely, at such a decoupling time at label j we have the factorization of the set $\tilde{\mathcal{G}}^{++}$ describing the new state as $\tilde{\mathcal{G}}^{++} = A \times (\mathbb{I}_M \setminus \{M+1\}) \times B$, where $A \subset (\mathbb{I}^{\mathbb{N}})^{L\setminus\{j\}}$, where $L \subseteq \mathbb{N}_\ell$ is the set of nontrivial labels. If the state just prior to the decoupling time was $A \times B'$, then B is a set corresponding to the ranks in B' which are not the founding one. (See Example 13 below for some simple cases.)

Definition 9.29 (Resolution and decoupling).

We start the dual for the $(M,1)$-model in the factor $(11\cdots 10)$ and suppress downward and upward mutation. (Now all factors have a 0 at position $M+1$ throughout).

9.6 The Growth of Occupied Sites in the Collision-Free Case...

(a) A rank (column of the tableau) is said to be *resolved* by mutation when a mutation produces a partition of $\{1,\ldots,M\}$ with only one non-empty element, that is the factor $(1,\ldots,1,*)$ results, which then remains until again selection takes place at this rank. (See Lemma 9.24.)

(b) Now consider the dynamics of label j, called $\zeta(t)$ at a founding site. Let

$$\zeta(0) := (\zeta(0,1), \zeta(0,2), \zeta(0,3)) = (\tau^*, (\mathbb{I}_M \setminus \{M+1\})^{\otimes m}, v) \in \mathfrak{T}, \quad (9.251)$$

and

T_1 is the time of the first birth (due to selection) at this label and site.

(9.252)

The associated *decoupling time* of the label at this site, denoted T_{R^*}, is defined by

$$T_{R^*} := \inf\{t > T_1 : \zeta(t,2) = (\mathbb{I}_M \setminus \{M+1\})\} \quad \text{if } m = 1 \quad (9.253)$$
$$:= \inf\{t > 0 : \zeta(t,2) = (\mathbb{I}_M \setminus \{M+1\})\} \quad \text{if } m > 1.$$

If $m = 1$, this is the time at which an excursion from $(\mathbb{I}_M \setminus \{M+1\})$ is completed where the excursion begins with the first selection event at the label and site.

The state at the given site jumps at time T_{R^*}, so that (also see (9.256)).

$$\zeta(T_{R^*}-) = (\tau^*, \tau, v), \quad \text{implies } \zeta(T_{R^*}) = (\ddagger, (\mathbb{I}_M \setminus \{M+1\}), \varnothing). \quad \square$$
(9.254)

The global transitions occur at the decoupling times of the various labels. For the *Option II dynamics* there will be a global jump at a decoupling time, namely, new labels are created and occupied by the level 1 elements of Υ and new sites are occupied. In addition any level two elements of Υ that are resolved at this time are also removed and new sites become occupied, etc.

Fix a label j and a corresponding founding site. We now consider the decoupling event at this label j occuring with founding site corresponding to j, say at time T_{R^*}. Recall that at time $T_{R^*}-$, the state ζ at label j has the form:

$$\zeta(T_{R^*}-) = (\tau^*, \tau, v). \quad (9.255)$$

Then at time T_{R^*} two things happen.

First at site j a new label is created and the old label at this site jumps to the state

$$\zeta(T_{R^*}) = (\ddagger, (\mathbb{I}_M \setminus \{M+1\}), \varnothing). \quad (9.256)$$

Secondly if in (9.255) we have $v \neq \varnothing$, then one or more new sites become occupied with irreducible tableaus at time T_{R^*}, i.e. new labels are created at these sites. In this way a *random number of offspring* can be produced at a decoupling time. This works in detail as follows.

If the entries $[\ldots]_1, \ldots, [\ldots]_k$ in $\mathcal{Y}_{n(\tau),1}$ are resolved but the entry $[\ldots]_{k+1}$ in $\mathcal{Y}_{n(\tau),1}$ is not resolved then $k+1$ new sites become occupied with irreducible tableaus as founding sites. The states at the site j_i coming from $[\ldots]_i, i = 1, \ldots, k$ have the form

$$\zeta_i(T_{R^*}) = (\ddagger, \tau_i, v_i), \tag{9.257}$$

where (τ_i, v_i) is given by the tableau in $[\ldots]_i$ and $\mathcal{Y}_{n(\tau_i),\ell}$ is given by the sequence of nested subintervals in $\mathcal{Y}_{n(\tau),\ell+1}$ contained in $[\ldots]_i$. Moreover at site j_{k+1}

$$\zeta_{k+1}(T_{R^*}) = (\tau^*_{k+1}, \tau_{k+1}, v_{k+1}), \tag{9.258}$$

with τ^*_{k+1} given by the partition in the first column of the tableau in $[\ldots]_{k+1}$, τ_{k+1} given as by the tableau in $([k+1], [k+2], \ldots)$ and v_{k+1} is given as for $i \leq k$ but with $[\ldots]_{k+2}, [\ldots]_{k+3}, \ldots$ added to $\mathcal{Y}_{n(\tau_{k+1}),1}$.

Example 13. We give a simple illustration for $m = 2$. We start with (010) at site 1. Here the subscripts 1,2,3 indicate different potential locations. The ∗ denotes the fact that this is a potential migrant to a site, not yet moved.

Then after 2 selection operations and the migration of the second column we have

$$\begin{array}{l} (010)_1 \\ (100)_1 \ (010)_{2*} \\ (100)_1 \ (100)_{2*} \ (010)_1 \end{array} \tag{9.259}$$

Then after one selection operation on the second column and then migration of the third column we have

$$\begin{array}{l} (010)_1 \\ (100)_1 \ (010)_{2*} \\ (100)_1 \ (100)_{2*} \ (010)_{3*} \\ (100)_1 \ (100)_2 \ (100)_{3*} \ (010)_1. \end{array} \tag{9.260}$$

Finally after a selection operation on the third column we have

$$\begin{array}{l} (010)_1 \\ (100)_1 \ (010)_{2*} \\ (100)_1 \ (100)_{2*} \ (010)_{3*} \\ (100)_1 \ (100)_{2*} \ (100)_{3*} \ (010)_{3*} \\ (100)_1 \ (100)_{2*} \ (100)_{3*} \ (100)_{3*} \ (010)_1. \end{array} \tag{9.261}$$

9.6 The Growth of Occupied Sites in the Collision-Free Case...

Then after 2 selection operations and the migration of the second column we have

$$\begin{array}{l}(010)_1\\(100)_1\ (010)_{2*}\\(100)_1\ (100)_{2*}\ (010)_1.\end{array} \quad (9.262)$$

Then after one selection operation on the second column and then migration of the third column we have

$$\begin{array}{l}(010)_1\\(100)_1\ (010)_{2*}\\(100)_1\ (100)_{2*}\ (010)_{3*}\\(100)_1\ (100)_2\ (100)_{3*}\ (010)_1.\end{array} \quad (9.263)$$

Finally after a selection operation on the third column we have

$$\begin{array}{l}(010)_1\\(100)_1\ (010)_{2*}\\(100)_1\ (100)_{2*}\ (010)_{3*}\\(100)_1\ (100)_{2*}\ (100)_{3*}\ (010)_{3*}\\(100)_1\ (100)_{2*}\ (100)_{3*}\ (100)_{3*}\ (010)_1.\end{array} \quad (9.264)$$

Then after 2 selection operations and the migration of the second column we have

$$\begin{array}{l}(010)_1\\(100)_1\ (010)_{2*}\\(100)_1\ (100)_{2*}\ (010)_1.\end{array} \quad (9.265)$$

Then after one selection operation on the second column and then migration of the third column we have

$$\begin{array}{l}(010)_1\\(100)_1\ (010)_{2*}\\(100)_1\ (100)_{2*}\ (010)_{3*}\\(100)_1\ (100)_2\ (100)_{3*}\ (010)_1.\end{array} \quad (9.266)$$

Finally after a selection operation on the third column we have

$$\begin{array}{l}(010)_1\\(100)_1\ (010)_{2*}\\(100)_1\ (100)_{2*}\ (010)_{3*}\\(100)_1\ (100)_{2*}\ (100)_{3*}\ (010)_{3*}\\(100)_1\ (100)_{2*}\ (100)_{3*}\ (100)_{3*}\ (010)_1.\end{array} \quad (9.267)$$

Now if we have next mutations on the second and third column, resolving these and then a mutation in the second one, we obtain at that latter moment a creation of new labels 2 and 3 at sites 2 and 3.

Then after 2 selection operations and the migration of the second column we have

$$\begin{array}{l}(010)_1 \\ (100)_1 \ (010)_{2*} \\ (100)_1 \ (100)_{2*} \ (010)_1.\end{array} \qquad (9.268)$$

Then after one selection operation on the second column and then migration of the third column we have

$$\begin{array}{l}(010)_1 \\ (100)_1 \ (010)_{2*} \\ (100)_1 \ (100)_{2*} \ (010)_{3*} \\ (100)_1 \ (100)_2 \ (100)_{3*} \ (010)_1.\end{array} \qquad (9.269)$$

Finally after a selection operation on the third column we have

$$\begin{array}{l}(010)_1 \\ (100)_1 \ (010)_{2*} \\ (100)_1 \ (100)_{2*} \ (010)_{3*} \\ (100)_1 \ (100)_{2*} \ (100)_{3*} \ (010)_{3*} \\ (100)_1 \ (100)_{2*} \ (100)_{3*} \ (100)_{3*} \ (010)_1.\end{array} \qquad (9.270)$$

In addition coalescence can occur only between columns with the same subscript. At this stage we have $\mathcal{Y}_{5,1} = [2,3,4]_1$, $\mathcal{Y}_{5,2} = [3,4]_{1,1} \subset [2,3,4]_1$.

If we have mutation $2 \to 1$ in the second column, then coalescence of the 5 and 1 columns, and the decoupling mutation $2 \to 1$ in the 1st column, then at this decoupling time we produce 2 new sites

$$(110)_2 \text{ and } \begin{array}{l}(010)_3 \\ (100)_3 \ (010)_3.\end{array} \qquad (9.271)$$

(v) The process $\mathfrak{s}$ using Option II dynamics

We observe that the decoupling transition at the involved sites can be reformulated as follows. Consider such a state at which we can reach a quasi-trap at the next local transition. Then the decoupling transition can be viewed as a jump after an exponential waiting time occuring in the special set of states just singled out.

We can therefore define $\mathfrak{s}$ formally as follows.

9.6 The Growth of Occupied Sites in the Collision-Free Case... 471

Definition 9.30 (The process $\mathfrak{s}$ - Option II dynamics).
The process
$$(\mathfrak{s}(t))_{t\geq 0}, \tag{9.272}$$
is defined to be a Markov pure jump process with state space $\mathfrak{S}$, that is in particular with trajectories in
$$\Omega_{**} = D([0,\infty), \mathfrak{S}). \tag{9.273}$$

The process dynamics is given by the transitions corresponding to (1) local transitions due to selection, mutation, coalescence as described in point (ii) above, (2) migration as specified above in point (iii) and (3) the conventions we made above in point (iv) on the global transitions resulting from decoupling events. □

From the process above we now obtain $\tilde{\mathcal{G}}^{++}$ as a functional as follows. Let

$$(\mathfrak{s}_j)_{j\in\mathbb{N}} \text{ be the sequence of the projections of } \mathfrak{s} \text{ on the } j\text{-th label}, \tag{9.274}$$

i.e. the type of the j-label, the current ranks with their columns and the migration information for these ranks. Then let

$$\tilde{F}(\mathfrak{s}_j) = \tau_j \in \mathcal{T} \quad \text{if } \mathfrak{s}_j = (\tau_j^*, \tau_j, v_j) \in \mathfrak{T}, \tag{9.275}$$

denote the subset of $(\mathbb{I}_M)^n$ defined by the ranks specified by the second component of an element $\mathfrak{s}_j \in \mathfrak{T}$.

Lemma 9.31 (Indicator-valued process $(\tilde{\mathcal{G}}_t^{++})_{t\geq 0}$ as functional of $\mathfrak{s}$).
Given $\mathfrak{s}(t)$ we now recover a version the set-valued dual process denoted $(\tilde{\mathcal{G}}_t^{++,II})_{t\geq 0}$ on the same probability space by setting:

$$\tilde{\mathcal{G}}_t^{++,II} = \bigotimes_{j=1}^{\infty} \tilde{F}(\tilde{\mathfrak{s}}_j(t)), \quad t \geq 0, \tag{9.276}$$

where

$$\tilde{\mathfrak{s}}_j = \begin{cases} \mathfrak{s}_j & \text{if } j \text{ is an active label} \\ (\ddagger, \mathbb{I}_M^{\mathbb{N}}, \emptyset), & \text{otherwise}. \end{cases} \quad \Box \tag{9.277}$$

Proof of Lemma 9.31. By inspection, namely, compare the list of transitions and rates as defined in the Subsection 9.5 for $\tilde{\mathcal{G}}$ and Subsubsection 9.6.3 for $\mathfrak{s}$ after projecting the latter. q.e.d.

Remark 144. Suppose we are in the finite N case but in the preemergence regime. When we consider the interactive case with geographic space $\{1, \ldots, N\}$

replacing $\mathbb{N}_g$, then, when a particle migrates it moves to a new site which can be assumed asymptotically to be empty (in the pre-emergence regime). In principle a migrating particle moves to each existing occupied site with probability $\frac{1}{N}$ and otherwise moves to an empty site which without loss of generality can be taken as the $\min\{j : \mathfrak{s}(j) = (\mathbb{I}_M)^\mathbb{N}\}$ if we work with exchangeable initial states.

(vi) Properties of $\mathfrak{s}$

We collect now some facts about the process $\mathfrak{s}$. Starting with

$$\mathfrak{s}(0) = (1, \ddagger, (\mathbb{I}_M \setminus \{M+1\}) \otimes (\mathbb{I}_M)^\mathbb{N}, \emptyset) \times (\ddagger, \mathbb{I}_M^\mathbb{N}, \emptyset) \cdots , \qquad (9.278)$$

let T_1 be the time of the first birth event. Let $\mathfrak{s}_1(t, 1)$ be the collection of all columns with label one which are located (recall in this dynamic we have delayed migration moves) at the site 1. Then let

$$n_0(t) = n(\mathfrak{s}_1(t, 1)) - \sum_{\xi \in \mathcal{Y}_{n(\mathfrak{s}_1(t,1)),1}} |\xi|, \qquad (9.279)$$

be the current number of ranks with label 1 which remain to be assigned to site 1 (i.e. no delayed migration step is up to now assigned to the rank) and define

$$T_R := \inf\{t > T_1 : |n_0(t)| = 1\}, \qquad (9.280)$$

the time this collection returns to the state where it consists of only one active rank. Then the *decoupling time* of resolution of this site

$$T_{R^*} = \inf\{t \geq T_1 : |n_0(t)| = 1, \ A(t) = (\mathbb{I}_M \setminus \{M+1\})\}, \qquad (9.281)$$

where with $n(t)$ the number of active ranks of the label still assigned site 1, $A(t)$ is the associated subset of $(\mathbb{I}_M)^{n(t)}$ for a collection of $n(t)$ factors.

Proposition 9.32 (Properties of $(\mathfrak{s}(t))_{t \geq 0}$).
 Consider now $\mathfrak{s}$ *(driven by the option-II dynamic) defined above.*

(a) *Starting with the factor* $(1, \ldots, 1, 0)$ *at a tagged site, this tagged site never becomes empty before the time of the first rare mutation.*
(b) *We have for T_{R^*} as defined in (9.281),*

$$E[T_{R^*}] < \infty. \qquad (9.282)$$

(c) *There exists a $\lambda > 0$ such that with T_{R^*} as in (9.281) we have:*

$$E[e^{\lambda T_{R^*}}] < \infty. \qquad \square \qquad (9.283)$$

9.6 The Growth of Occupied Sites in the Collision-Free Case... 473

Proof. (a) Note that if the type $M + 1$ component of a factor is zero, then the only transition that can change it to 1 is the rare mutation transition and every unresolved factor lies between $(1, \cdots 10)$ and $(1 \cdots 10)^{\otimes k}$ for some $k \in \mathbb{N}$.

(b) The process $n(t)$ can be coupled with the $M = 1$ birth and death process. Namely using a multi-colour system with normal particles and *pink* particles, such that both colours together give the $M = 1$ birth and death process. For this purpose create a particle of a different colour, e.g. pink, when a rank is deleted due to the mutation-selection operation and with the convention that when a pink particle coalesces with a regular particle the regular particle remains and the pink particle is removed. Since the original birth and death process is positive recurrent this means that $E[T_R] < \infty$.

We note that on each visit to $n(t) = 1$ there is a positive probability of mutation to $(1, \ldots, 1, 0)$ before the next selection event. Therefore at most a geometrically distributed number of return to $(1, \ldots, 1, 0)$ are needed for T_{R^*} to be reached before the next return and (9.282) follows.

(c) Furthermore we get (9.283) using that $dk^2 > ck$ for $k \geq k_0$ with k_0 sufficiently large so that there is a minimal positive "drift" to the left if $k \geq k_0$ (see proof of Lemma 9.25 for details).   q.e.d.

Remark 145. We see from the above that the process $\mathfrak{s}$ will consist of an increasing number of occupied sites and after the decoupling time from its parents, we can observe a unique label arising from it, the mesostate assigned to this label will approach an equilibrium distribution on $\mathfrak{T}$ exponentially fast.

Step 2. The Case of Option I Dynamics: $\tilde{\mathcal{G}}^{++,I}, \mathfrak{s}_I, \mathfrak{s}_{\text{enr},I}$

We discuss now also the Option-I dynamics since it is also a process which is in duality (as is the option-II dynamics with exchangeable initial measures) to our Fleming–Viot process and will be needed later on and is the version which can be adapted to the finite $S = \{1, \cdots, N\}$ case. We use here two versions $\mathfrak{s}_I$ and the very useful, later on slightly enriched one $\mathfrak{s}_{I,\text{enr}}$.

(1) The process $\mathfrak{s}_I$

For Option I we *immediately* move the migrants but we also keep ghost copies at the founding site replacing the migration information we stored with $\tilde{\mathcal{G}}^{++,II}$ respectively $\mathfrak{s}$. It is natural to view the process $\mathfrak{s}_I$ as a process with basic objects given by multisite irreducible tableau and we now keep track of the status of the configuration at both the founding and migrant sites and only store the ghost copy at the founding site. Nevertheless this requires to set up a new state space and corresponding notation.

Remark 146. Note that again we have the concept of microstate, mesostate and global state, but however these are now new objects. We refrain from introducing different names for these objects since they are as we shall see enrichments of what we had before.

In order to define the appropriate state spaces we denote by $\mathbb{N}^g$ a copy of $\mathbb{N}$ which enumerates the set of geographic sites. We shall now make use of the fact that every active label corresponds to a unique founding site from where migrants move to related new sites, which are not yet decoupled from the founding rank.

Definition 9.33 (Rooted tree coding migration steps).

(a) The information on the successive migration steps is encoded in a tree with vertices given by sites and edges corresponding to a migration step from one site to the next.
(b) The set of such trees will be denoted by $\mathcal{E}$.

$$\mathcal{E} := \{ \text{ finite rooted trees with vertices in } \mathbb{N}_g \}. \tag{9.284}$$

An element of $\mathcal{E}$ is given as (here $V \subseteq \mathbb{N}_g$)

$$(v^*, (V, E)) \quad , \quad v^* \in V \quad , \quad E \subseteq \{\{v_1, v_2\} | v_1, v_2 \in V\}, \tag{9.285}$$

specifying *root, vertices and edges* of the rooted tree. $\square$

Now we want to associate with every label and the migration information the tree of successive migration steps and the location of its founding site.

Definition 9.34 (The map Geo).

(a) We define a map

$$Geo : \mathbb{N}_\ell \times \Upsilon \longrightarrow (\mathbb{N}_g, \mathcal{E}), \tag{9.286}$$

which associates to each label and migration information the *founding* site (the site at which the local rank 1 sits), the *finite set* of sites that have been assigned to it forming the set of vertices V and finally a tree from the set $\mathcal{E}$ by specifying the set of edges. Namely if ($\upsilon(1) \in \mathcal{Y}_{n,1}, \ldots, \upsilon(\ell) \in \mathcal{Y}_{n,\ell}$) is a sequence of tuples of nested subintervals of subintervals in Υ_n, then Geo assigns sites $j_1, \ldots, j_\ell \in \mathbb{N}_g$ to these and an edge between j_i and j_{i+1} and this then assigns to the tree-structured subintervals in v a tree with vertices in $\mathbb{N}_g$ and with root at j. A subinterval in $\mathcal{Y}_{n,\ell}$ is connected by an edge in $\mathcal{E}$ to each of its subintervals in $\mathcal{Y}_{n,\ell+1}$.
(b) We denote the range of Geo, that is, the set of vertices as $\mathfrak{R}$, that is, given the pair $(j, v) \in \mathbb{N}_\ell \times \Upsilon$

$$\mathfrak{R}(Geo)(j, v) \subset \mathbb{N}_g = \{ \text{ set of vertices in } Geo(j, v)\}. \quad \square \tag{9.287}$$

Now we can specify the state space of the mesostate processes.

Definition 9.35 (State space for mesostates).
The state space of the *mesostate* at label $j \in \mathbb{N}_\ell$ is the countable set (recall below (9.177, 9.227, 9.233, 9.236)):

$$\hat{\mathfrak{T}} := \{\hat{\mathfrak{t}} = (\tau, v, \tilde{v}) : \tau \in \mathcal{T} \text{ with } n(\tau) < \infty, \ v \in \Upsilon, \tilde{v} = Geo(j, v)\}, \tag{9.288}$$

9.6 The Growth of Occupied Sites in the Collision-Free Case... 475

where $\hat{t}$ is a marked multisite tableau, $\tau \in \mathcal{T}$ is an irreducible tableau, marks v defined as in (9.238) and with marks $\tilde{v}$ given by the tree specified by $Geo(j, v)$ as in (9.286). Furthermore $(\tau, v) \in \mathcal{T} \times \Upsilon_{n(\tau)}$ specifies the corresponding coding for the various ranks at this label and which ranks have been designated as potential migrants of ℓth order, $\ell = 1, 2, \ldots$ together with their destinations. □

We call a label j

$$\text{inactive if } \hat{t}_j \text{ has as first three components } (j, (\mathbb{I}_M)^{\mathbb{N}}, \emptyset), \text{ active otherwise.} \tag{9.289}$$

There is a natural equivalence relation on $\hat{\mathfrak{T}}$ defined by

$$\hat{t}_1 \equiv \hat{t}_2 \tag{9.290}$$

if $\hat{t}_1$ is transformed into $\hat{t}_2$ by a permutation on $\mathbb{N}_g$.

Definition 9.36 (Micro states at a site).
Given a label j with mesostate $\hat{t}_j$ we can define the

$$\textit{microstate} \text{ at each site in } \mathfrak{R}(Geo)(j, v). \tag{9.291}$$

More precisely, consider the following tuple of sets of ranks $(\bar{v}(i))_{i=1,\cdots,\tau(j)}$, $v(i)$ entry of $\mathcal{Y}_{\tau(j),i}$, $\bar{v}(i)$ the ranks in $v(i)$, which describe all ranks associated with the sites after $1, 2, \cdots$ migration steps. Then the microstate at site k as coded in $Geo(j, v)$ is obtained by taking the columns in the tableau τ which belong to ranks associated with one site and defining a subtableau of columns restricted to the ranks $\bar{v}(i)$ in a subinterval $v(i)$ which have as destination the site k in question. □

We do not repeat each step in the construction of the dynamic which arises replacing Option-II by Option-I but otherwise is again induced by the evolution of $(\eta', \mathcal{G}^{++,\ell})$. Compare also Remark 148 below.

Definition 9.37 (The process $\mathfrak{s}_I$).
The process $(\mathfrak{s}_I(t))_{t \geq 0}$ has as state space

$$\hat{\mathfrak{S}} := (\hat{\mathfrak{T}})^{\mathbb{N}_\ell} \tag{9.292}$$

and evolves as Markov pure jump process as $\mathfrak{s}$ did but now with Option-I dynamics, instead of Option-II dynamics. □

(2) *The process* $\mathfrak{s}_{\text{enr}}$. We also consider a modification of the process $\mathfrak{s}_I$ by enriching its state by adding *flags* to each of the sites. We call the new process

$$\mathfrak{s}_{\text{enr},I}. \tag{9.293}$$

Then we are also able to use the Option-I dynamic (instead of the option-II dynamic as in Step 1) together with this richer version (containing the information on the flags at sites) of the process $\mathfrak{s}$. In fact we point out in Remark 147 below that we can construct both processes on one probability space.

We now construct the process $\mathfrak{s}_{\text{enr},I}$. In the case of Option I dynamics we must keep track of the interaction between the site occupied by a new migrant and the parent site up to the analogue of the decoupling time. In particular we focus on what happens at those transition times which involve simultaneous changes to 2 sites, namely the removal of a potential migrant from the parent site and creation of a new permanent site. If we run a dynamic using the local transitions, we can exhibit certain features of the state at a site which will then determine the creation of new labels. We will use the following concepts.

We associate *flags* to sites to characterize their *status in the process of resolution*. There will be the following flags

$$\{Pe, Re, Tr, Un\} \tag{9.294}$$

and if necessary we can use a refinement of Tr (however this requires looking into the *future*):

$$\{UPe, URe\}, \tag{9.295}$$

which we now first explain informally.

We can at a migration event at a (parent) site identify a new site and place a ghost copy of the new state at this parent site (and couple the dynamics at both sites). Such a site is assigned the flag (*Tr*) *transient*. If the interval is ultimately removed by resolution it is denoted as *URe* and it changes to *removed* (Re) at the time of resolution, which is an instantaneous state which immediately disappears (as well as all its descendants that are also automatically (*Re*)). If the site ultimately becomes permanent after resolution of all ranks to the left of its ghost copy, it is denoted as *UPe* and it changes to *permanent* (*Pe*) at the resolution time. The *Pe* sites are immortal (except due to a rare mutation). At the very moment where we change to *Pe*, the new site creates a new label. A site $j \in \mathbb{N}_g$ which is not the founding site of a label gets the flag Un. Sometimes we refer to transient sites as *Tr* but when the status at resolution is important we use *UPe* or *URe* where we have to remember that this requires information from the future.

Given a label j consider the set of sites consisting of the founding *Pe* site and the attached set of undecoupled migrant sites flagged Tr defined by the mesostate. This object is called a

direct clan founded by the *Pe* site of the label j. (9.296)

The dynamics of such direct clans will be considered in subsection 9.12. In this subsection we work with the individual sites involved.

We next introduce the global transition that occurs when a new label is created.

9.6 The Growth of Occupied Sites in the Collision-Free Case...

- A migrant becomes *permanent* when the ranks to the left of the migrant position at the parent site are resolved at time T_R (at the parent site) and the flag of the associated site is changed from Tr (or in the finer description of Tr by UPe) to Pe but it still belongs to the label associated with the parent site. Then consider the *decoupling time* for this new Pe site, namely the first time $T'_R \geq T_R$ at which the attached rank is also resolved. At the decoupling time a new label is created. Then the further evolution of the new label is independent of the parent label and is identical (in law) to the one we started at time 0.
- At the first time a migrant is permanent *and* resolved we assign a unique site to the label (associated to the permanently occupied Pe site) referred to as founding site.
- At the new site the first interval $[\ldots]_1$ of $\mathcal{Y}_{n(\tau^*),1}$ at the parent site forms the non-migrated positions at the new site and the remaining elements to the right form the subintervals of $\mathcal{Y}_{n(\tau^*),1}$ at the new site. These factors at the new site (and their descendants) are assigned the next label in $\mathbb{N}_\ell$.
- In addition the ranks in $\mathcal{Y}_{n(\tau),1}$ are removed from the parent site.
- The subsequent evolution at the parent site and newly created site are independent and they have now different labels.

Note that our construction defines a map from the set of sites

$$\mathbb{N}_g \longrightarrow \{Pe, Tr, Re, Un\}; \quad Tr = (UPe \cup URe), \tag{9.297}$$

which associates with every geographic site the current flag of the site.

Remark 147. The reason we introduce the flags Pe, Re and Tr at geographic sites is to keep track of the combined effects of mutation, coalescence and migration as follows. There is a positive probability that a particle targeted for migration will be eliminated by a resolution (due to the mutation process) of a rank to its left. Moreover ranks to the right of the founder position cannot be preassigned a $**$-type because this could lead to an inconsistency if a look-down occurs. Therefore a newly formed site is assigned the flag UPe which changes to Pe when all ranks to the left of the rank of the founder in the parent site are resolved. Note however that including the information coded by Υ at the set of Pe sites contains a complete description and can be used directly to obtain the dual (product) representation if $X_0 = \mu^\otimes$. Similarly the set of Pe sites (ignoring the information coded by Υ) together with the set of Tr sites also gives a complete representation of the dual.

The system $\mathfrak{s}_{\text{enr},I}$ just constructed and $\mathfrak{s}_I$ constructed before can be *coupled*, that is realized on one probability space as follows.

Remark 148. To obtain the system $\mathfrak{s}_{\text{enr},I}$ with Option I dynamics on the same probability space as $\mathfrak{s}(t)$ we use a specific construction. We consider a version of the system on a copy of the geographic space coupled to a given realization of the process $\mathfrak{s}(t)$. The mutation-selection- coalescence process in the new system mirrors that in $\mathfrak{s}(t)$. When a rank is designated for migration in $\mathfrak{s}(t)$ we immediately create a new site for the new system and then simply mirror the dynamics at the first interval in $\mathcal{Y}_{n,1}$ in the original process. We can assign flags as above so that the creation of the new site for the $\mathfrak{s}(t)$ process coincides with the change of a *UPe* rank to a *Pe* rank and if a rank in $\cup_i \mathcal{Y}_{n,i}$ is removed by the mutation process, then the flag is changed from *URe* to *Re*.

Remark 149. To establish emergence we start with label 1 and associate with site 1 the interval $\{1, 2, \cdots, n\} \setminus \Upsilon_{n,1}$, with $[\ldots]_1 \in \Upsilon_{n,1}$ the site 2, etc., continuing with working this way until label 1 resolves. The sites flagged *Pe* are assigned in order of the times at which they become permanent. We continue this process until we reach a label from where on the component two to four of the state is $(\ddagger, (I_M)^{\mathbb{N}_\ell}, \emptyset)$.

To obtain moments we start the system with a finite number of occupied sites. To establish emergence we start the system with one occupied site with the factor $(11 \cdots 10)$, which means with initial state given by set $(\mathbb{I}_M \setminus \{M+1\}) \otimes (\mathbb{I}_M)^{\mathbb{N}}$, that is,

$$\mathfrak{s}(0) \text{ has one active label } (\ddagger, (\mathbb{I}_M \setminus \{M+1\}) \otimes (\mathbb{I})^{\mathbb{N}}, \emptyset). \tag{9.298}$$

9.6.4 Further Auxiliary Processes to analyse $\tilde{\mathcal{G}}^{++}$: $\mathfrak{s}^*, \mathfrak{s}^{**}$ and $\mathfrak{s}^{**}_{\text{enr}}$

The major complication in the case $M \geq 2$ results from the fact that selection factors resolve only after finite random times and then reach a quasi-trap under the selection-mutation process (excluding rare mutation). It is only then do they function as the particles in the $M = 1$ case and remain in the system until rare mutation. Therefore during the time following the action of the selection operation till its resolution the expressions in the dual are not so nicely accessible to calculation and estimation as in the case $M = 1$ in Section 8, since new site and parent site do not factor and are dependent in their evolution. This results in the somewhat complex nature of the process $\mathfrak{s}(t)$ or $\mathfrak{s}_I$ that includes unresolved factors which are selected to migrate. Its advantage is that it is a *time-homogeneous Markov* process, which can be used as a *dual*.

To facilitate the analysis of the growth of occupied sites despite this phenomenon we need to relate our process to one with a CMJ-structure. It is therefore convenient for our analysis to also consider the following two modifications of the process $\mathfrak{s}(t)$ together with an enrichment of the second choice which lose some of the good points of $\mathfrak{s}$ respectively $\mathfrak{s}_I, \mathfrak{s}_{\text{enr},I}$ but gain other nice features, namely allowing to isolate finite-type CMJ-processes.

The new processes are denoted as

$$\mathfrak{s}^*(t), \quad \mathfrak{s}^{**}(t), \quad \mathfrak{s}^{**}_{\text{enr}}(t). \tag{9.299}$$

9.6 The Growth of Occupied Sites in the Collision-Free Case...

We begin with an informal description. The three processes are distinguished as to how the evolution from creation of a new rank up to its resolution is handled.

(1) The process $\mathfrak{s}^*(t)$ eliminates the "uncertainty" (of what happens at resolution) by looking into the future to see what in fact happens at the time of resolution. (2) The process $\mathfrak{s}^{**}(t)$ eliminates the "uncertainty" by making the decision at the time of birth of a factor with the appropriate probabilities and then using the appropriate h-transformed process for the further evolution up to the resolution time of the ultimately permanent sites and drops migrants which are not ultimately permanent immediately. (3) To get $\mathfrak{s}^{**}_{\text{enr}}$, we can now also (as we showed in the examples we discussed in Subsection 9.5.1, where $M = 2, 3$) define a process where a selection factor is assigned its values upon resolution immediately at creation, but we keep the information on the factors we have dropped since they become zero at resolution and store this information.

The first process is a *non*-Markov process. The process $\mathfrak{s}^{**}, \mathfrak{s}^{**}_{\text{enr}}$ is *Markov* but as it will turn out it is *time-inhomogeneous*, but gains the *finite-type CMJ-property*, giving a powerful tool. This process is not a dual since the *URe* factors are missing. The process $\mathfrak{s}^{**}$ will be extremely important once we study the growth of the number of factors and of occupied sites.

Next we give in (1)-(3) below the precise and formal description of the two new processes and then the enrichment of the latter. We work here on a probability space where we also have constructed both $\mathfrak{s}$ and $\mathfrak{s}_{\text{enr},I}$ as pointed out above.

1. The process $\mathfrak{s}^*(t)$.

Consider the càdlàg process $(\mathfrak{s}_{\text{enr},I}(\cdot))_{t \geq 0}$ constructed in Step 1 and let $(\mathcal{F}^{\mathfrak{s}}_t)_{t \geq 0}$ be the natural filtration. Consider a label $j \in \mathbb{N}_\ell$ with state

$$\mathfrak{s}_j \in \mathfrak{S} \tag{9.300}$$

Recall that *Geo* induces a 1-1 mapping between the *Tr* sites which result from migration events at label j and the nested set of subintervals $\mathcal{Y}_{\tau_j}$.

Consider the set of *Tr* sites with associated τ^* given by the partition at their position at the time they are targeted to migrate. Then the event that such a site becomes permanent (*Pe*) or removed (Re) at the decoupling time and the element τ_2^* in τ^{**} (recall (9.229), this is the $**$-type) that takes over when this rank is resolved are measurable functions with respect to an *enlarged* filtration $(\mathcal{F}^{\mathfrak{s},*}_t)_{t \geq 0}$ including the information on the outcome at resolution of the currently unresolved factors (see (9.229–9.231) for definitions). Recall that we have proved that the time at which both are determined is a.s. finite. For each *Tr* site j and $t \geq$ birth time of j, this defines a map:

$$\tilde{H}_j(t) : v_j \longrightarrow \{Pe, Re\} \times \mathcal{T}^{**}, \tag{9.301}$$

which is measurable with respect to $(\mathcal{F}^{\mathfrak{s},*}_t)_{t \geq 0}$. Moreover the *distribution* of (the finite-valued random variable) $\tilde{H}_j(t)$ is measurable with respect to $\sigma(\mathfrak{s}(t))$.

The process $\mathfrak{s}^*$ is defined by (recall the convention (9.243))

$$\mathfrak{s}^*(t) = \{(\mathfrak{s}_i(j,t), \tilde{H}_j(t)), j \in \mathbb{N}_g \; i \in \mathbb{N}_\ell\}, \quad t \geq 0. \tag{9.302}$$

Since $\mathfrak{s}^*(t)$ involves future information (namely what happens at the times of resolution), it is *not* a Markov process.

2. *The process $\mathfrak{s}^{**}(t)$ - Option III dynamics.*

We now introduce a modified version of $\mathfrak{s}(t)$ based on Option III dynamics that leads to a (finite) *multitype CMJ process* for which we then determine the Malthusian parameter in the next subsection. In this version we anticipate two properties at the moment that a rank is selected for migration at a label and creates a new label, namely, (1) whether this is an ultimately removed *URe* or ultimately permanent migrant *UPe* and (2) the state in which the founding migrant factor arrived together with the information which partition element takes over if it resolves at the new site. An *UPe* site becomes *Pe* and a *URe* site becomes *Re* at the time of the decoupling from the parent. Recall the example $M = 2, 3$ in Subsubsection 9.5.1 here for this idea.

The dynamics of $\mathfrak{s}^{**}(t)$ are hence formally defined as follows. At migration events a new label is created and the *choice of the outcomes* $O^{\text{mig}}(\subseteq \{Pe, Re\} \times \mathcal{T}^{**})$ upon resolution (recall (9.301)) are made based on the current state with the respective probabilities and the future dynamics are then the *conditioned dynamics using the h-transforms* as given by the previous Lemma 9.26 which relates it to the original dynamic. The outcomes we coded by the complement of the taboo set, is the initial taboo set H of the new label at the moment it is created at the decoupling time. As more and more migrants appear, more and more taboos are created and we describe the total taboo by the subset of points in the state space we should not visit.

To formalize this we introduce the enriched mesostate space associated with label j. It is denoted $\mathfrak{T}^{**}$ and which is defined as follows (recall (9.229), (9.232), (9.233), (9.236)):

$$\mathfrak{T}^{**} := \{(\tau^{**}, \tau, v, H(\tau))| \; (\tau^{**}, \tau, v) \in \mathfrak{T}, \quad H(\tau) \in \mathfrak{H}(n(\tau))\}, \tag{9.303}$$

where (a taboo state is the complement of the one we want to reach)

$\mathfrak{H}(n) =$ set of possible *taboo states* for the mutation-selection-coalescence process of our label $\hat{\zeta}^{m,s,c,d}$ of (9.201) with values in $\mathcal{T}_n \otimes \mathbb{I}_M^{\mathbb{N}}$.
$$\tag{9.304}$$

Here again τ^{**} is the initial $**$-type, τ is the current indicator of a subset of $(\mathbb{I}_M)^{\mathbb{N}}$, v the list of migrants with their destination and in addition we now include the element H, which is the taboo event which together with the selection-mutation-coalescence-migration dynamics determines the local dynamics for the label considered. Note that $\mathfrak{T}^{**}$ is again a *countable* state space.

9.6 The Growth of Occupied Sites in the Collision-Free Case... 481

The Markov jump process, $\mathfrak{s}^{**}(t)$, has global state space (almost all entries are "trivial")

$$\mathfrak{S}^{**} := \{\mathfrak{s}^{**} \in (\mathfrak{T}^{**})^{\mathbb{N}_\ell} : (\mathfrak{s}^{**})_j = (\ddagger, (\mathbb{I}_M)^{\mathbb{N}}, \emptyset, \diamond)\ a.a.\ j\}. \tag{9.305}$$

For $\mathfrak{s}^{**}(\cdot)$ we immediately *delete all transient sites and ranks that eventually become removed and immediately place the designated ultimately permanent migrants at new sites*. This is implemented by choosing these outcomes of the resolution with their probabilities (these probabilities are adapted to the natural filtration of the $\mathfrak{s}(t)$ process) and then modifying the dynamics with the appropriate h-transform (recall (9.192)) on the event where this is the final outcome in order to condition that the newly designated permanent sites are not killed by mutation at the parent site. This can be coded by specifying the appropriate *taboo event*, H (which forms part of the state description of $\mathfrak{s}^{**}$).

In more detail this means given $\mathfrak{s}_j^{**}(t) = (\tau^{**}, \tau, v, H)$ the dynamics at the new label j are as follows.

- the dynamics at the future new label follow the mutation selection-coalescence dynamics h-transformed as required by τ^{**} which defines the initial taboo for the future label,
- between migration events the transitions at the remaining ranks occur following the transition mechanism of $\hat{\zeta}^{m,s,d}$, but h-transformed according to the current value of H,
- if one of the required absorbing sets specified by H is reached the corresponding taboo event is removed from H,
- at a migration event at rank i in τ the value *UPe* or *URe* is selected according to the the probability that it would be ultimately *Pe* or *Re* and then follows the h-transformed dynamic w.r.t. current value of H, a new label will be created later at decoupling at the new site and in the mean time the appropriate taboo states are added via the ghost copy if *UPe* is chosen and if *URe* is chosen the rank is immediately deleted.

3. The enrichment $\mathfrak{s}_{\text{enr}}^{**}$.

Recall the process $\mathfrak{s}_{\text{enr},I}$ and its construction with flags. On this probability space we can also construct $\mathfrak{s}^{**}$ and suppose now we have done that. Then we see that in the process $\mathfrak{s}^{**}$ we consider only the *Pe* and *UPe* sites and we lose information about the excluded *URe* factors. We can remedy this and construct an enriched version of this process in which the *URe* sites are included in the state of the process and follow the corresponding h-transformed dynamics (clearly this can be constructed on the same probability space as $\mathfrak{s}^{**}(t)$). Then we add to each *UPe* or *Pe* site the *Tr* migrants (with their factors) produced at this site that have not yet been killed up to the present time. We denote this enriched process by

$$(\mathfrak{s}_{\text{enr}}^{**}(t))_{t \geq 0}. \tag{9.306}$$

Using the construction of the process $\mathfrak{s}^{**}_{\text{enr}}(\cdot)$ and $\mathfrak{s}_{\text{enr},I}(\cdot)$ on one and the same probability space, we include the flags in this enriched process with the following map

$$\mathbb{N}_g \longrightarrow Pe, UPe, URe, Re, Un, \qquad (9.307)$$

which associates with every geographic site the *current flag of the site*.

We shall show later on that the *URe* sites play in fact no role in the growth rate of the set of occupied sites and determination of the Malthusian parameter but they are necessary for the description of the state of the complete configuration needed for the duality relation at a fixed time.

Remark 150. The identification with a CMJ process is based on the following heuristics. During its lifetime a given site will produce migrants to new sites and the new sites will be of finitely many types (namely one type for each partition of $\mathbb{I}_M$) and which are permanent (*Pe*). The probability of production of such migrants at a given occupied site in a given time increment $[t, t + \Delta t)$ depends only on the type of the initial migrant and the age and type of the parent site, that is, the initial type and time passed since arrival of the initial immigrant. The development of such a site given its initial type is independent of the other occupied sites assuming that collisions do not occur and the rank is not deleted. Therefore the collection of occupied permanent (*Pe*) sites forms a CMJ process.

We again introduce here the flags we used in the construction of $\mathfrak{s}_{\text{enr},I}$. Note that in $\mathfrak{s}^{**}$ we can simplify the flagging and use the flags

$$\tilde{Pe} \text{ to denote } UPe \text{ or } Pe \text{ sites }, \tilde{Re} \text{ to denote } URe \text{ or } Re \text{ sites}, \qquad (9.308)$$

since we have decided on the final fate of removed or permanent. Then conditioned on the choices selected, the dynamics at the parent site and new site are *conditionally independent*. In this modification, at a migration time a migrating position is assigned the flag permanent or removed. To each such *Pe* migrant a new site is assigned and a copy of the migrant interval is attached. There are no transient sites. When a migrant interval is assigned the flag *Re* no new site is created and this interval is deleted after a finite lifetime on resolution and produces no descendants after the completion of their lifetime. For the process $\mathfrak{s}^{**}$ after the creation of the new label it evolves independently of the old label and their evolution.

We have now defined $\tilde{\mathcal{G}}^{++}$ as functional of the process $\mathfrak{s}$ and various forms of its driving processes $\mathfrak{s}^*, \mathfrak{s}^{**}, \mathfrak{s}^{**}_{\text{enr}}$ which we need in our further analysis of the growth of occupied sites.

Remark 151. We use the three versions of the modified dual process based on $\mathfrak{s}, \mathfrak{s}^*, \mathfrak{s}^{**}$ as appropriate. Two choices are important. (1) Using $\mathfrak{s}$ provides the dual representation for moment calculations at a fixed time. (2) In determining the long time growth in the number of occupied sites, that is, for determining the Malthusian parameter and the emergence time scale, it is more convenient to use

$\mathfrak{s}^{**}$ and $\mathfrak{s}^{**}_{\mathrm{enr}}$, since we can then focus on sites which are not removed later due to the mutation-selection-coalescence resolution process. The process $\mathfrak{s}^*$ is of auxiliary nature.

9.7 Collision-Free Regime: Malthusian Parameter of $\tilde{\mathcal{G}}^{++}$

The goal of this subsection is to verify again for our dual process $\tilde{\mathcal{G}}^{++}$ that its relative $\tilde{\mathcal{G}}^{++,III}$, respectively $\mathfrak{s}^{**}$ is a *multitype CMJ-process* in order to determine the time to emergence via the Malthusian parameter of $\tilde{\mathcal{G}}^{++}$ derived from the one of the related CMJ process $\mathfrak{s}^{**}$.

We describe the problem in Subsubsection 9.7.1, recall the multitype CMJ-theory in Subsubsection 9.7.2 and then carry out the analysis in Subsubsections 9.7.3–9.7.8.

9.7.1 Introduction

The probability of emergence is determined, as we shall show, by the probability of a rare mutation to occur which in turn depends on the *number of active ranks* in the process $\mathfrak{s}(t)$ and therefore on the number of Pe occupied sites in $\mathfrak{s}^*_{\mathrm{enr},I}$ together with the typical number of active ranks per site and their typical factors. In this subsection as in section 8 we first ignore collisions and consider the dual $\tilde{\mathcal{G}}^{++}_t$ for the case $S = \mathbb{N}$ as introduced in the previous subsections and determine the growth of the total number of active ranks at all occupied sites in the collision-free dual process.

We first show that as in the case $M = 1$ we can identify a CMJ branching process related to $\mathfrak{s}_{\mathrm{enr},I}$ but we have to allow this process to be *multitype* (with finitely many types). This will be based on $\mathfrak{s}^{**}$ which does however *not* include transient sites that are eventually removed. However we will show that adding back these transient sites does not effect the exponential growth rate of the active ranks of $\tilde{\mathcal{G}}^{++}$.

We next develop in detail the multitype CMJ process, i.e. the branching process of occupied sites and determine first the *Malthusian parameter* β in the pre-collision regime and then the corresponding stable *size-type-age distribution*. Then we have to pass to the factors. The behaviour of this object is in many respects similar to the corresponding one in the case of $M = 1$. At each site the number of active factors remains bounded due to coalescence and due to Proposition 9.32 the kind of factors associated with the site stabilize in some equilibrium distribution and hence the exponential growth rate in the total number of factors is due to the exponentially increasing number of occupied sites.

More precisely, the set of $\tilde{P}e$ sites (see (9.308)) in the the process $\mathfrak{s}^{**}$ can be represented as a CMJ branching process, which again satisfies the conditions needed to establish growth with a Malthusian parameter β which we have shown in the beginning in (9.70) must be bounded by α (and in fact is strictly smaller). Since the number of factors at a site is stochastically bounded, the growth in the number of factors is indeed determined by the growth rate of the number of occupied sites as it was before in Section 8 for $M = 1$.

What exactly is then the difference from the CMJ-process in the case $M = 1$? The point process of production of a migrant, that is, new sites coming from a given site is similar to the two type case except that now there is the *potential loss of* $(1, \ldots, 1, 0)$ *factors* at the old site (due to mutation) as described above in Subsubsection 9.5.3 which results in the removal of a site created upon a migration step of a rank and hence gives a smaller growth rate β (i.e. $\beta < \alpha$) since this can lead to the situation in which either no additional site comes about or at least the production rate from that site becomes smaller.

In other words in order to determine the growth of the number of sites we must consider individuals of *types* indexed by partitions of $\mathbb{I}_M$ together with the mark that specifies the partition element which takes over at resolution. Hence we need now for $M \geq 2$ a CMJ-processes with finitely many possible types (in fact the number is a combinatorial function of M).

9.7.2 Preparation: Multitype CMJ-Processes

In this subsubsection we review some basic concepts and results on *multitype* CMJ-processes. For more details than provided here we refer to [Do2].

A multitype CMJ process describes a *population of individuals with types* in a *finite* type set K. Starting at their time of birth an individual of type $i \in K$ can give birth to individuals of type $j \in K$ according to a point process counting the birth times of descendents of type j up to time t after the birth of the considered individual:

$N_{ij}(t)$ for $t \in (0, L_i)$ where $L_i \leq \infty$ is the random lifetime of the type-i-individual. (9.309)

The collections of the above processes

$$\{(N_{i,j}^\iota(t))_{t \in [0, L_\iota)}\}_{\iota \in \mathbb{N}} \qquad (9.310)$$

for the different individuals ι being member of the population from their birthtime on counted in order of appearance, evolve independent of each other.

Often the process N_{ij} is generated by an evolving internal state of an individual, as we already saw in Section 8 where this was the birth and death process given by the process $\eta_t(k)$ at locations k, where the locations corresponded to individuals in the CMJ. Now the internal state has become more complex.

We define the multitype CMJ-process

$$Z = (Z(t))_{t \in \mathbb{R}}, \qquad (9.311)$$

as the vector of the numbers of individuals of the different types at time t.

9.7 Collision-Free Regime: Malthusian Parameter of $\tilde{\mathcal{G}}^{++}$

In order to study the process we need some characteristic quantities. Starting with an initial population of *one individual of type i*, let

$$Z_{ij}(t) = \text{the number of individuals alive at time } t \text{ of type } j, \quad i, j \in K. \quad (9.312)$$

In order to determine the longtime behaviour we need the following parameters:

$$\mathbf{n}(s) = (n_{i,j}(s))_{i,j \in K}, \quad (9.313)$$

$$n_{ij}(r) = \int_0^\infty e^{-rt} d(E[N_{ij}([0,t])]), \quad s \in (0, \infty), \quad i, j \in K, \quad (9.314)$$

$$\mu_{ij}(r) = \int_0^\infty t e^{-rt} d(E[N_{ij}([0,t])]), \quad i, j \in K, \quad (9.315)$$

$$\|\mathbf{n}(r)\| = \sup_{i,j} n_{ij}(r). \quad (9.316)$$

Assume that

$$\text{the matrix } \mathbf{n} \text{ is } \textit{strictly} \text{ positive.} \quad (9.317)$$

Then the matrix $\mathbf{n}$ has a

maximal eigenvalue ρ, which is *positive* and *simple*. (9.318)

Note that if we consider $\mathbf{n}$ not necessarily *strictly* positive, but which is irreducible, then a power of n satisfies this condition and then one can proceed similarly.

By the Perron-Frobenius theorem for a matrix $\mathbf{n}$, which is strictly positive and has maximal eigenvalue ρ we know for the entries $n_{i,j}^n$ for the n-th power that

$$n_{ij}^n \sim \rho^n \text{ as } n \to \infty. \quad (9.319)$$

The process Z is *supercritical* if for some $r > 0$:

$$\text{either } \|\mathbf{n}(r)\| < \infty \text{ and } \rho(r) > 1 \text{ or if } \|\mathbf{n}(r)\| = \infty. \quad (9.320)$$

In order to find the generalization of α from the scalar case ($M = 1$) in the multitype case, we consider the following situation. Suppose that Z satisfies (1) it is supercritical and (2) there exists a real number β such that $\mathbf{n}(\beta)$ has finite entries and a simple, maximal eigenvalue 1. Then we conclude that indeed β is the *Malthusian parameter of Z* which as we shall see next determines the growth of the entries of the mean of $Z_{i,j}$ from (9.312).

In order for β to be also the *exponential growth* rate for the mean, we need the condition

$$\max_{i,j} \mu_{ij}(\beta) < \infty, \quad (9.321)$$

which then implies that

$$E[Z_{ij}(t)] \sim w_{ij}e^{\beta t}, \quad \text{where } w_{ij} > 0. \tag{9.322}$$

To get a.s. statements let

$$\mathbf{W}(t) := e^{-\beta t}\mathbf{Z}(t) \tag{9.323}$$

and let

$$Y_{ij} = \int_0^\infty e^{-\beta u} dN_{ij}(u). \tag{9.324}$$

If in addition:

$$\sup_{i,j} E[Y_{i,j}|\log Y_{ij}|] < \infty, \tag{9.325}$$

then

$$\mathbf{W}(t) \to \mathbf{W}, \text{ a.s. as } t \to \infty, \tag{9.326}$$

with $\mathbf{W}$ not identically zero. This means that β is indeed the growth rate of the process $\mathbf{Z}$ a.s. and in L_1.

The related convergence results of the age-multitype distribution are included in the work of Nerman [N, N2] Theorem 6.7, Doney [Do2], and Jagers and Nerman [JN2].

9.7.3 The CMJ Process Related to the Dual

We have constructed a marked set-valued dual. The task now is to determine its structure in different time scales in the same way as we did for the case $M = 1$. The main objective of Subsection 9.7 is to use CMJ theory to determine (1) the exponential growth rate β in the number of active sites (2) a random time shift of the form $\frac{\log(\overline{W})}{\beta}$ to produce $\mathbf{W}(\cdot)$ from a standard solution of the linear (collision-free) evolution equation with $|\mathbf{W}| = 1$ and (3) the empirical distribution of internal states of the occupied sites. Points (1)–(3) are used to prove Propositions 9.1, 9.2, 9.3.

In Section 8 we applied the CMJ-theory and we want to substitute this now for the $(M, 1)$-case by the theory described in Subsection 9.7.2. However since $\tilde{\mathcal{G}}^{++,I}$ is not a CMJ-process, we have to turn to the processes $\tilde{\mathcal{G}}^{++,\ell}$, $\ell = II, III$ and the processes $\mathfrak{s}, \mathfrak{s}^{**}$ which are used to define those or their enrichments. Here for $\ell = II$ we only get a CMJ-process with countably many types for the theory of the growth behaviour is involves additional difficulties that we shall address in

9.7 Collision-Free Regime: Malthusian Parameter of $\tilde{\mathcal{G}}^{++}$

subsubsection 9.12.7 and is essential for the selection-dominant case. Here, in order to work with finitely many types, we turn to $\ell = III$ which is sufficient for the analysis of the mutation-dominant case.

To apply the knowledge described in Subsection 9.7.2 for multitype CMJ-processes we use the auxiliary processes $\tilde{\mathcal{G}}^{++,III}$, $\mathfrak{s}^{**}$, which we introduced in Subsection 9.6.2 and 9.6.3. We begin here with developing the theory in the $N = \infty$ case to obtain the growth rate β (recall 9.180) the growth constant $\overline{W}$ and the limiting empirical distribution of the internal states at the different sites for $\tilde{\mathcal{G}}^{++,III}$.

Define now for $j \in \mathbb{N}_g$ (recall (9.276) and note that in both dynamics new sites are colonized only after decoupling *such that the labels correspond to sites*)

$$\tilde{\mathcal{G}}_{t,j}^{++,II} = \tilde{F}(\mathfrak{s}_j^*(t)), \quad \tilde{\mathcal{G}}_{t,j}^{++,III} = \tilde{F}(\mathfrak{s}_j^{**}(t)). \tag{9.327}$$

We also define

$$\bar{K}_t := |\mathfrak{s}^{**}(t)| \tag{9.328}$$

and

$$\bar{U}(t, \cdot) := \frac{\sum_j 1(\tilde{\mathcal{G}}_{t,j}^{++,III} = \cdot)}{\bar{K}_t}. \tag{9.329}$$

We now determine the growth of $|\tilde{\mathcal{G}}_t^{++,III}|$, that is the number of sites at which the mesostate is a strict subset of $\mathbb{I}^\mathbb{N}$.

The first step is to identify a (finite) multitype CMJ-process in this structure. For this purpose we note that a newly occupied site with type III dynamics has a type which describes the internal state, namely, τ^{**}. To recall the relevant features consider a rank marked to migrate. At the time of resolution of the rank the various factors at this rank in the different rows have the property that one of the partition elements of $\mathbb{I}_M$ associated with this rank takes over and the others become empty. Recall that in the previous subsection this information together with the tuple of partition elements was denoted

$$* *\text{-type} \in \mathcal{T}^{**}. \tag{9.330}$$

Important for the CMJ-structure is the following observation. The choice of the partition element that survives determines which ranks *to the right of the migrant rank* at the parent site, if any, are deleted when the new site resolves. As the $\mathfrak{s}^{**}$ dynamics at the new site follow the appropriate h-transform, then the exact path followed to resolution determines the time at which the ranks at the parent site are deleted. However these ranks (to be deleted at resolution) play no role in the future evolution of the other factors at the parent site.

In general we can therefore immediately delete such ranks at the parent site already at the migration time since these ranks play no role for the long-time growth rate of the number of occupied sites since they and all their descendants have finite expected lifetime.

To illustrate this phenomenon consider again the tableau (recall the subscripts denote sites and $*$ the ghost status).

$$\begin{pmatrix} (010)_1 \\ (100)_1 \ (010)_1 \\ (100)_1 \ (100)_1 \ (010)_{1*} \\ (100)_1 \ (100)_1 \ (100)_{1*} \ (010)_1 \\ (100)_1 \ (100)_1 \ (100)_{1*} \ (100)_1 \ (110)_1 \end{pmatrix} \qquad (9.331)$$

If the third column is selected to migrate and the eventual resolution of this partition is $(010)_{1*} \to (110)_{1*}$, then the fourth and fifth columns *as well as their descendants* are removed at the resolution time of $(010)_{1*}$. Since this occurs after a finite time the effect of removing these columns at the migration time of $(010)_{1*}$ has no effect on the total population after that resolution time.

Once this is done the evolution under the h-transform dynamics at the new site and the remaining factors at the parent site evolve independently by Lemma 9.26, part (c). Therefore the rate of production of migrants of each type by a single site process depends only on the current state of the process.

The key result is:

Proposition 9.38 (Collision-free dual and a multitype CMJ-process).
Consider the process $\mathfrak{s}^*$ or $\mathfrak{s}^{**}$ defined above (recall they are collision-free). Observe the set of permanent sites (Pe) in $\mathfrak{s}^*$ or equivalently the set of ultimately permanent or permanent sites in $\mathfrak{s}^{**}$ (respectively Pe or UPe in $\mathfrak{s}^{**}_{\mathrm{enr}}(t)$).

Then this collection of occupied sites marked with their $**$-types forms a supercritical multitype CMJ branching process. □

Proof of Proposition 9.38. During excursions at a given site that begin and end at the state $(1, \ldots, 1, 0)$ the site produces migrants that migrate to unoccupied sites. Here we focus on the new sites which are Pe in $\mathfrak{s}^*(t)$ or Pe, UPe in $\mathfrak{s}^{**}(t)$ (or Pe or UPe in $\mathfrak{s}^{**}_{\mathrm{enr}}(t)$). When a position is selected for migration, the corresponding column in the tableau is placed at the new site the column consists of sets selected from a partition of $\{1, \ldots, M\}$ and in particular there are only finitely many possibilities for this internal state. One of the partition elements takes over under the action of mutation and selection at the new site.

Focus on $\mathfrak{s}^{**}$. We want to show with these observations that the collection of Pe and UPe sites with their $**$-type is a multitype CMJ process, where the sites correspond to what is called *individuals* in the CMJ-theory and the *types* in the CMJ-theory correspoind to the $**$-type. If $\mathfrak{s}(0)$ is given by (9.278), then $\mathfrak{s}^{**}(t)$ determines a population of occupied sites with types indexed by partitions of $\mathbb{I}_M$. Note that Pe and UPe sites are immortal and that they produce new Pe or UPe sites

9.7 Collision-Free Regime: Malthusian Parameter of $\tilde{\mathcal{G}}^{++}$

with finite expected times between production of new sites. Moreover the production mechanisms for new *Pe* sites (for $\mathfrak{s}^{**}(t)$) at different *Pe* sites are independent. In particular, the production at a *Pe* offspring site, conditioned on its $**$-type is independent of the future evolution of the parent site and the production of further *Pe* sites by the parent site (following the appropriate h-transform) since after a migration step the further evolution depends only on the $**$-type of the migrant and not the precise trajectory.

To summarize, during their lifetimes *Pe* sites produce new offspring sites of these types and the new sites evolve in $\mathfrak{s}^{**}(t)$ independently after birth. Therefore it follows that this system has the structure of a CMJ process.

However since the *Pe* sites can be of finitely many different types, namely, the possible $**$-types, this is a multitype CMJ process. Moreover, since the lifetime of each *Pe* site itself is infinite and the creation rate is positive, the process is supercritical. This completes the proof because the *Pe* sites in $\mathfrak{s}^*$ have the same distribution as the *Pe* sites in $\mathfrak{s}^{**}$.   q.e.d.

9.7.4 The Malthusian Parameter for $|\mathcal{G}_t^{++,III}|$

The reason for the introduction of $\mathfrak{s}^{**}(t)$ and $\tilde{\mathcal{G}}_t^{++,III}$ was to exploit the theory of *multitype CMJ processes* to determine the critical time scale for the $M \geq 2$ type cases in the same way as we used CMJ processes in the two-type case, i.e. $M = 1$. We now state the main results in this direction. Of course later we have to pass from $\tilde{\mathcal{G}}^{++,III}$ to $\tilde{\mathcal{G}}^{++,II}$ which is a *dual* in order to get the information about emergence.

Connection to CMJ-process The discussion of the multitype CMJ-theory and the fact that we can exhibit a CMJ-structure in our process $\mathfrak{s}^{**}$ as we explained above in Proposition 9.38 allows us to now apply this theory. To verify that the properties needed for the asymptotic CMJ-theory, namely (9.321), (9.325), are fulfilled in our multitype context, note that the in our case the processes "$N_{ij}(t)$" correspond to the production of new occupied sites, occupied by migrants of different types from a given site. These processes have jump rates proportional to the number of active ranks at the parent site. However the number of such active ranks is dominated by a birth and death process with linear birth and quadratic death rate as in the case $M = 1$. Since starting with one particle the distribution of the number of active ranks at the site can be bounded by the equilibrium distribution of the birth and death process, it follows that the number of such migrations in a bounded interval, say of length 1, has a bounded second moment. The conditions (9.321) and (9.325) then follow (even with β replaced by any $r > 0$) by a simple calculation. We can then apply the results including (9.326) to determine the growth of the number of *Pe* sites $\bar{K}$ (replacing the number of occupied sites K of the $M = 1$ case). In addition the normalized size and age distribution $\mathcal{U}$ in the $M = 1$ case is replaced by a normalized *size-age-tyPe distribution* $\bar{\mathcal{U}}$.

We denote the quantities replacing $(K,\mathcal{U})$ from (8.117), (8.162) in the $M=1$ case now in the $M \geq 2$ case by

$$((\bar{K}_t)_{t \geq 0}, (\bar{U}^{III})(t,\cdot,\cdot,\cdot))_{t \geq 0}, \tag{9.332}$$

where $\bar{K}_t := |\mathfrak{s}^{**}(t)|$ and the normalized empirical site-age-type distribution $\bar{U}^{III}(t,\cdot)$ will be introduced below in (9.341).

For each $k \in \mathbb{N}$ the *birth time* of the k-th site is

$$\mathfrak{t}(k) = \inf(t : \mathfrak{s}^{**}(t) \text{ has } k\text{-occupied sites}) \tag{9.333}$$

and the $**$-*type* of the k-th site is (recall at time $\mathfrak{t}(k)$ we give out the new label) at the time of its creation:

$$\mathfrak{f}(k) = (\mathfrak{s}^{**}((\mathfrak{t}(k)))_k. \tag{9.334}$$

The finite *set of types* and the set of *internal states* at labels is denoted by:

$$\mathcal{T}^{**} \text{ and } \mathfrak{T}^{**}. \tag{9.335}$$

In order to match the details of the set-up of multitype CMJ-processes described in (9.309)–(9.311) to $\mathfrak{s}^{**}$, observe the following. The "individuals" in the CMJ process are now $\tilde{P}e$ sites and such a site has various possibilities for its internal state. Consider a $\tilde{P}e$ site with label i. Then since this is immortal the lifetime $L_i = \infty$. The type of this "individual" is $\tau_i^{**} \in \mathcal{T}^{**}$. The production of offspring of types τ^{**} is given by the point process which we denote by $N_{\tau_i^{**}, \tau^{**}}(t)$ which is completely determined by the Markov process which denotes the internal state at the label specified by the site i at time t, denoted

$$\zeta(t,i) \in \mathfrak{T}^{**}, \ t \geq 0, \tag{9.336}$$

where the Markov property follows since the instantaneous rate of production of particles of type τ^{**} is a function of the current state $\zeta(t,i)$. Moreover, since the lifetimes of permanent sites are infinite, the process of the population generated by the site i is supercritical.

In order to determine the Malthusian parameter we must describe the growth of the total number of $\tilde{P}e$ sites as well as the relative proportions of the types in $\mathcal{T}^{**}$ and as before the age distribution.

Consider the process $\mathfrak{s}^{**}$ and define the following characteristic quantities. Consider an occupied site k. Starting at its birth time $\mathfrak{t}(k)$ the factor first located at site k gives rise to the $\mathfrak{T}^{**}$-valued process

$$\zeta(t,k), \quad t \geq \mathfrak{t}(k). \tag{9.337}$$

9.7 Collision-Free Regime: Malthusian Parameter of $\tilde{\mathcal{G}}^{++}$

The number of $\tilde{P}e$ sites of $**$-type τ^{**} with tableau $\zeta \in \mathcal{T}$ at time t is given by:

$$\bar{\Psi}_t(\tau^{**},\zeta) = \sum_{k=1}^{\infty} 1_{\mathfrak{t}(k)\le t} 1_{\mathfrak{f}(k)=\tau^{**}} 1_{\zeta(t,k)=\zeta}, \qquad (9.338)$$

where $\mathfrak{f}(k)$ is the $**$-type of the site k. Define the number of τ^{**}-type sites and total numbers of occupied sites at time t by:

$$\bar{K}_t(\tau^{**}) = \sum_{\zeta\in\mathfrak{T}_{**}} \bar{\Psi}_t(\tau^{**},\zeta), \qquad \bar{K}_t = \sum_{\tau^{**}\in T^{**}} \bar{K}_t(\tau^{**}). \qquad (9.339)$$

Let (recall (9.335))

$$\mathcal{N} = \text{space of counting measures on } [0,\infty) \times T^{**} \times \mathfrak{T}^{**}. \qquad (9.340)$$

For $0 \le a < b$, $\tau^{**} \in T^{**}$, $\zeta \in \mathfrak{T}$ define the enrichment of $\bar{\Psi}_t$ from (9.338), $\bar{\Psi}_t \in \mathcal{N}$ by:

$$\bar{\Psi}_t([a,b),\tau^{**},\zeta) = \sum_{k=1}^{\bar{K}_t} \left(1(\mathfrak{t}(k) \in [t-b, t-a)) \cdot 1(\mathfrak{f}(k) = \tau^{**}) \cdot 1\,(\zeta_k(t) = \zeta)\right), \qquad (9.341)$$

respectively its normalized version as:

$$\bar{\mathcal{U}}^{III}(t;[a,b),\tau^{**},\zeta) = \frac{\bar{\Psi}_t([a,b),\tau^{**},\zeta)}{\bar{K}_t}. \qquad (9.342)$$

We need furthermore the following ingredient. Define (recall (9.226)–(9.231))

two $**$-types are *equivalent* if they belong to the same partition of $\{1,\cdots,M\}$. $\qquad (9.343)$

The number of such equivalence classes is finite. Now consider a matrix

$\mathbf{n}(r)$ with entries indexed by the equivalence classes. $\qquad (9.344)$

Define first the ingredients of this matrix.
Recall $\zeta(t,k)$ was the internal state at the k-th created site. Let

$$\tilde{\zeta}(t,\tau_1^{**}) = \text{the state in } \mathfrak{T}^{**} \text{ at a site of } **\text{-type } \tau_1^{**} \text{ at time } t \qquad (9.345)$$

and

$$n(\zeta) = \#\text{ of active ranks in } \zeta \in \mathfrak{T}^{**}, \qquad (9.346)$$

$\tau_2^{**}(\ell)$ denotes the ℓth component in $\mathcal{T}^{**}$ elements with $\ell = 1, 2$ (9.347)

and

$\tilde{\zeta}_j(t, \tau_1^{**})$ denotes the jth rank in ζ from (9.345). (9.348)

Now for every $r \in (0, \infty)$ define the entries of the matrix $\mathbf{n}(r)$ indexed with the equivalence classes of two $**$-types τ_1^{**}, τ_2^{**} as:

$$n_{\tau_1^{**}, \tau_2^{**}}(r) \tag{9.349}$$

$$:= c \int_0^\infty e^{-rt} E \left[\sum_{j=2}^{n(\zeta)} (1_{\tilde{\zeta}_j(t,\tau_1^{**})=\tau_2^{**}(1)} \cdot 1_j \text{ is permanent} \cdot 1_{\tau_2^{**}(2) \text{ is selected}}) \right] dt.$$

(9.350)

Then we can state the following about the longtime behaviour of of the process $\mathfrak{s}^{**}$ which then immediately allows to say something about the longtime behaviour of $\tilde{\mathcal{G}}^{++,III}$.

Proposition 9.39 (The process $\tilde{\mathcal{G}}^{**,III}$ with M lower level types before first rare mutation).

Consider the process $\mathfrak{s}^{**}$ and the induced version of $\tilde{\mathcal{G}}^{++,III}$.

(a) If $\mathfrak{s}^{**}(0)$ is given by (9.278) and $\mu(\{E_0\}) = 1$, then for all t smaller than the time of the first rare mutation jump in the $\tilde{\mathcal{G}}^{**,III}$ process we have:

$$\int \tilde{\mathcal{G}}_{td}^{++,III}(\mu^\otimes) = 1. \tag{9.351}$$

(b) The system

$$\{\bar{K}_t, \{\bar{\zeta}(t,k) : k = 1, \ldots, \bar{K}_t\}\} \text{ is a supercritical multitype CMJ process}, \tag{9.352}$$

and satisfies (9.321) and (9.325).
There exists $\beta > 0$ such that

$\mathbf{n}(\beta)$ has finite entries and a simple maximal eigenvalue 1. (9.353)

(c) There exists a non-degenerate random variable $\overline{W}$ and positive vector $\mathbf{v}$ such that

$$e^{-\beta t} \bar{\Psi}_t \to \overline{W} \mathbf{v}, \quad a.s. \text{ as } t \to \infty, \tag{9.354}$$

$$e^{-\beta t} \bar{K}_t \longrightarrow \overline{W}, \quad a.s. \text{ as } t \to \infty. \tag{9.355}$$

9.7 Collision-Free Regime: Malthusian Parameter of $\tilde{\mathcal{G}}^{++}$

(d) Then starting with one rank corresponding to the factor ($\mathbb{I}_M \setminus \{M+1\}$), the mean number of sites of type-** grows at rate β:

$$E[\bar{K}_t(\tau^{**})] \sim w_{\tau^{**}} e^{\beta t}, \qquad \text{for } \tau^{**} \in T^{**}, \tag{9.356}$$

that is, β is the Malthusian parameter for $\tilde{\mathcal{G}}^{++,III}$.

(e) The Malthusian parameter β can also be calculated using the stable age-type-internal state distribution which exists and is defined as:

$$\bar{\mathcal{U}}^{III}(\infty; ds, \tau^{**}, \zeta) = \lim_{t \to \infty} \bar{\mathcal{U}}^{III}(t; ds, \tau^{**}, \zeta). \tag{9.357}$$

Let

$r_{(\tau^{**},\zeta),\tilde{\tau}^{**}}(v) =$ the transition rate for a $\tilde{P}e$ site of type τ^{**} and internal state ζ and age v to produce a $\tilde{P}e$ migrant of type $\tilde{\tau}^{**}$.

$$\tag{9.358}$$

The Malthusian parameter β satisfies (recall (9.322)):

$$\beta = c \sum_{\tau^{**}, \tilde{\tau}^{**} \in T^{**}} \sum_{\zeta \in \mathfrak{Z}^{**}} \int_{-\infty}^{0} r_{(\tau^{**},\zeta),\tilde{\tau}^{**}}(v) \bar{\mathcal{U}}^{III}(\infty; dv, \tau^{**}, \zeta). \tag{9.359}$$

(f) The parameter β satisfies for $M \geq 2$:

$$\beta < \alpha, \tag{9.360}$$

where α is the Malthusian parameter in the case $|E_0| = |E_1| = 1$. □

Note that the numerical value of β will depend on all the parameters which determine the probabilities in which trap the resolved factors end up, which are of course different for different numbers of types on the lower level and which of course also depend on the mutation distribution and the relative fitness on the lower level. In other words β is now in a subtle way depending on the details of the dynamic on the lower level and *all* its parameters.

Proof of Proposition 9.39. (a) follows immediately since with no rare mutation the transition rates to $(*, \ldots, *, 1)$ from $(*, \ldots, *, 0)$ are zero.

(b) To identify the system $\{\bar{K}_t(\cdot), \{\zeta(t,k)\} : k = 1, \ldots, \bar{K}_t\}$ as a multitype CMJ we identify occupied sites as the individuals in the population. As defined above the type of a site is determined by the type of the founding particle. Furthermore during the lifetime of the site it produces new sites when ranks migrate from the site and thus producing a new site. Furthermore the production mechanisms at different sites are independent. Therefore this has the structure of a CMJ process with types T^{**}

and internal states $\mathfrak{T}^{**}$. Since occupied sites never become unoccupied and there is a positive rate of production of new sites, the process is supercritical.

The proof of (9.321) and (9.325) is a result of the positive recurrence of the birth and death process at a site and is essentially the same as that for the $M = 1$ case. Briefly, since the local production rate at a site converges to a finite mean stationary process (as $t \to \infty$) the condition (9.321) is satisfied. Similarly we can verify that $E[Y_{ij}^2] < \infty$ and therefore (9.325) is satisfied with the same argument as in Section 8.

Next we show (9.353). We first note that $\mathbf{n}(v)$ is a strictly positive matrix. To verify this note that starting with any partition there is a positive probability (due to mutation and selection) of producing $(1, \ldots, 1, 0)^{\otimes k}$ between any two migration times. From these ranks we can then obtain by selection and mutation the product of partitions of the form $((1, \ldots, 1, 0 \ldots, 0), (0, \ldots, 0, 1 \ldots, 1))$. Then noting that coalescence of the product of two partitions gives their mutual refinement it follows that we can obtain any partition of $\mathbb{I}_M \setminus \{M+1\}$. The existence of β then follows from the Perron-Frobenius theorem.

(c), (d) The statements (9.356) and (9.354) follow by [Do2], Proposition 3 and Theorem 2.

(e) The existence of the stable age distribution $\bar{\mathcal{U}}_\infty$ follows from the multitype version of the results of Nerman [N,JN2] Theorem 6.7. The relation (9.359) follows as in the $M = 1$ case.

(f) The fact that $\beta \leq \alpha$ follows by coupling with the case in which $e_i = e_M$, $i = 1, \ldots, M$ which is then equivalent to the case $M = 1$. The strict inequality follows if $e_1 < e_M$ since then a positive proportion of ranks created by selection are removed by mutation.   q.e.d.

9.7.5 Formulation of the Main Results on the Marked Set-Valued Dual $\tilde{\mathcal{G}}^{++}$

We have shown in Proposition 9.39 exponential growth in the number of occupied sites in $\tilde{\mathcal{G}}^{++,III}$ with Malthusian parameter $\beta \in (0, \alpha)$. However this process is not a dual process. Hence in order to prove that β is the Malthusian parameter of our original dual system we have (1) to return to the *dual* process $\tilde{\mathcal{G}}^{++,II}$ and then (2) we have to pass from numbers of occupied sites to the actual growth of the *number of active ranks*.

These issues will be addressed in two propositions which combined give the first step for the desired conclusion, namely we then can control the number of sites as $t \to \infty$ in $\tilde{\mathcal{G}}^{++,I}$ and $\tilde{\mathcal{G}}^{++,II}$ and the number of ranks in $\tilde{\mathcal{G}}^{++,III}$. What then still *remains* is to control the asymptotics of the *number of ranks* in the dual processes, not only for $\mathfrak{s}^{**}$. This problem we address in Subsubsection 9.12.11.

9.7 Collision-Free Regime: Malthusian Parameter of $\tilde{\mathcal{G}}^{++}$

The key results about these dual processes which we can prove with our techniques from this subsection are the following. (Recall that $|\tilde{\mathcal{G}}^{++,II}|$ is the number of occupied *sites* in $\tilde{\mathcal{G}}^{++,II}$.)

Proposition 9.40 (Asymptotic structure of $\tilde{\mathcal{G}}^{++}$).
Consider again the processes, $\mathfrak{s}, \mathfrak{s}^{**}$ starting with $|\mathfrak{s}(0)| = |\mathfrak{s}^{**}(0)| < \infty$.

(a) *The Malthusian parameter for the growth of both of the total number of $\tilde{Pe}$ sites (Pe and UPe) and the total number of Pe, UPe and URe sites (permanent, ultimately permanent with ultimately removed) at time t is given by β, i.e. we have for $t \to \infty$ that*

$$e^{-\beta t}|\mathfrak{s}_t^{**}| \longrightarrow \overline{W}, \quad \lim_{t\to\infty} \frac{1}{t}\log(e^{-\beta t}|\mathfrak{s}_t|) = 0. \quad (9.361)$$

(b) *The number of ranks at an occupied site in $\mathfrak{s}$ and $\mathfrak{s}^{**}$ is a stochastically bounded random variable and is stochastically bounded by the number in the $M = 1$ case.*

(c) *The total number of active ranks in the process $\mathfrak{s}^{**}(t)$ denoted R_t satisfies*

$$e^{-\beta t}R_t \underset{t\to\infty}{\longrightarrow} \overline{W}^{\text{rank}} = \text{const} \cdot \overline{W}, \quad \text{a.s. as } t\to\infty \quad (9.362)$$

with $L = |T^{**}|$ and $\overline{W}, \mathbf{v}$ are as in (9.354),

$$0 < \overline{W}^{\text{rank}} = \sum_{i=1}^{L}(\text{Expected equilibrium no. of type } i\text{-ranks per}$$

$$\text{site in } \mathfrak{s}^{**}) \cdot \overline{W}v^i < \infty, \; a.s. \qquad \square \quad (9.363)$$

Proposition 9.41 (Growth of dual occupied sites with $M \geq 2$, collision-free).
We consider now on one probability space $\mathfrak{s}, \mathfrak{s}^*, \mathfrak{s}^{**}$ and $\tilde{\mathcal{G}}^{++,\ell}, \ell = I, II, III$ with $\mathfrak{s}(0) = \mathfrak{s}^*(0) = \mathfrak{s}^{**}(0)$ with $|\mathfrak{s}(0)| < \infty$ and $S = \mathbb{N}$.

(a) *Then there exists $\beta > 0$ and random variable $W^{III} > 0$ such that*

$$\lim_{t\to\infty} e^{-\beta t}|\mathfrak{s}^{**}(t)| = W^{III}, \quad (9.364)$$

where

$$W^{III} = \overline{W} \text{ and } \overline{W} \text{ is given by (9.355)}, \overline{W} > 0 \quad a.s. \quad (9.365)$$

(b) *There exists a random variable $W^{II} > 0$ such that*

$$\lim_{t\to\infty} e^{-\beta t}|\tilde{\mathcal{G}}_t^{++,II}| = W^{II}, \quad (9.366)$$

$$W^{II} = C_{II}\overline{W}, \quad C_{II} > 0 \quad (9.367)$$

where $\overline{W}$ is the random variable given by (9.355) and C_{II} a (deterministic) constant.

(c) There exists a random variable $W^I > 0$ such that,

$$\lim_{t \to \infty} e^{-\beta t} |\tilde{\mathcal{G}}_t^{++,I}| = W^I \tag{9.368}$$

where

$$W^I = C_I \overline{W}, \quad C_I > 0 \tag{9.369}$$

with $\overline{W}$ given by (9.355) and C_I a (deterministic) constant. .

(d) The following limit exists and defines the stable age - type-internal state distribution:

$$\lim_{t \to \infty} \bar{\mathcal{U}}^{III}(t, \cdot, \cdot, \cdot) = \bar{\mathcal{U}}^{III}(\infty; \cdot, \cdot, \cdot) \in \mathcal{P}([0, \infty) \times \mathcal{T}^{**} \times \mathfrak{T}^{**}). \tag{9.370}$$

(e) Recall the definition of $\tilde{\mathcal{G}}_j^{++,II}$ in (9.366). Then the set-valued process has a product form factors indexed by sites, i.e.

$$\tilde{\mathcal{G}}_t^{++,II} = \prod_{j \in \mathbb{N}_g} \tilde{\mathcal{G}}_{t,j}^{++,II}. \qquad \square \tag{9.371}$$

Remark 152. We note that relation (9.368) allows us to conclude as in Section 8 that the number of sites at which we have a dual factor which can undergo rare mutation is again of the form $W^I e^{\beta t}$ only that now β is strictly less than α since new sites may be removed later on upon resolution of the founding rank.

9.7.6 Proof of Propositions 9.40 and 9.41

Proof of Proposition 9.40.

(a) We have shown above in Proposition 9.39 that the Malthusian parameter for the set $\tilde{P}e$ of UPe and Pe sites is β. Let β_2 the Malthusian parameter for all sites ($\tilde{P}e \cup URe$). We want to show $\beta = \beta_2$ and for that it suffices to show $\beta_2 \leq \beta$.

The proof of this inequality is by contradiction, so we assume that $\beta_2 > \beta$. By coupling we have $\beta_2 \leq \alpha$ so that it is finite. Next note that the resolution time has a finite moment generating function and exponential tail, say index λ. (Recall Proposition 9.32). To see this note that the killing of a URe particle occurs when a mutation process hits the boundary. We note that before resolution, that is, $1_A \uparrow 1_{\mathbb{I} \setminus \{M+1\}}$, the process visits each of the finitely many possible intermediate cases at most a geometrically distributed number of times and each of the times at which a jump $1_{A^*} \to 1_{A^{**}}$ occurs is exponentially distributed.

9.7 Collision-Free Regime: Malthusian Parameter of $\tilde{\mathcal{G}}^{++}$

Now consider times t and $2t$. Then the total number of particles at time $2t$ is given by

$$const \cdot e^{\beta t} e^{\beta_2 t} + const \cdot e^{\beta_2 t} e^{-\lambda t} e^{\beta_2 t} \neq const \cdot e^{2\beta_2 t}, \quad \text{with } \lambda > 0, \quad (9.372)$$

so that we have a contradiction. Namely the first term gives a bound on the total number of particles at time $2t$ that are offspring of permanent particles at time t. The second term given the offspring of particles at time $2t$ that are offspring of particles at time t that eventually turn Re - the proportion of such particles at time t that produce descendants at time $2t$ is $e^{-\lambda t}$ (recall that all descendants of a Re particle are removed when it is removed on resolution).

(b) This is immediately clear from the construction.

(c) Here we have to argue in a similar way as in the $M = 1$ case in Section 8. Note that the age-type-internal state distribution converges to the stable age-type-internal state distribution and the number of active ranks arises by integrating or summing over age, internal state and type. Then from (9.354) and (9.357) we have

$$e^{-\beta t} R_t = e^{-\beta t} \bar{K}_t \sum_{\tau^{**} \in T^{**}} \sum_{\zeta \in \mathfrak{T}^{**}} \int_0^\infty n(\zeta) \bar{\mathcal{U}}(t; ds, \tau^{**}, \zeta) \quad (9.373)$$

$$\rightarrow \sum_{\tau^{**} \in T^{**}} W_{\tau^{**}}^{\text{rank}} \sum_{\zeta \in \mathfrak{T}^{**}} \int_0^\infty n(\zeta) \bar{\mathcal{U}}^{III}(\infty; ds, \tau^{**}, \zeta), \text{ as } t \rightarrow \infty$$

where $n(\zeta)$ denotes the number of active ranks in $\zeta \in \mathfrak{T}^{**}$. q.e.d.

Remark 153. In Proposition 9.40 we have shown that the exponential growth rate of UPe, Pe and Re is not larger than β. It remains to determine if there can be a subexponential difference in the growth rate of all sites and the growth rate of the UPe, Pe sites we want. This cannot occur in the mutation dominant case discussed below in Subsubsection 9.7.8. For a discussion of the other case, the selection-dominant case, we refer to Subsubsection 9.12.7.

Proof of Proposition 9.41.

(a) follows from Proposition 9.39 together with part (a) of Proposition 9.40.
(b) Note that the sites in the support of $\tilde{\mathcal{G}}_t^{++,II}$ coincide with the Pe sites in $\tilde{\mathcal{G}}_t^{++,III}$. Then (b) follows from (a) where C_{II} is the limiting proportion of Pe sites in $\tilde{\mathcal{G}}_t^{++,III}$, given by

$$C_{II} = \lim_{T \to \infty} \lim_{t \to \infty} e^{-\beta T} \int_0^{t-T} \int_{T^{**} \times \mathfrak{T}^{**}} n_{Pe}(t, T, v, \tau^{**}, \zeta) U(t-T, dv, \tau^{**}, \zeta) \quad (9.374)$$

where $n_{Pe}(t, T, v, \tau^{**}, \zeta)$ denote the expected number of Pe sites at time t produced by a UPe or Pe site at time $t - T$ of age $v \in [0, t - T]$ and type (τ^{**}, ζ).

(c) Note that every site in $\tilde{\mathcal{G}}_t^{++,I}$ corresponds to either a *Pe*, *UPe* site or a *URe* offspring of one of the former. These are included in $\mathfrak{s}_{\text{enr}}^{**}(t)$ and can be obtained by looking at the *URe* trees originating from each of the *Pe* or *UPe* sites starting during this lifetime. These trees of finite lifetime have a law depending only on the type of the site upon birth and the present state at time t is obtained by cutting it at time $t - v$, if v is the birth time of the site. Let

$$n_{URe}(t, v, \tau^{**}, \zeta) \tag{9.375}$$

denote the expected number of *URe* sites at time t produced by a *UPe*, *Pe* site of age v and type (τ^{**}, ζ). Then we obtain

$$C_I = 1 + \lim_{t \to \infty} \int_0^t \int_{\mathcal{T}^{**} \times \mathfrak{T}^{**}} n_{URe}(t, v, \tau^{**}, \zeta) U(t, dv, \tau^{**}, \zeta).$$

(d) This follows from the results of Nerman [N2] on the limiting stable age distribution for a multitype CMJ process.

(e) This follows since at the decoupling time for option II dynamics, the tableaus at the parent and new site factor since we delayed the move to the new site up to the decoupling time. Therefore the factorization holds by construction. q.e.d.

9.7.7 The $N \to \infty$ Limit

We can show that as $N \to \infty$ the dual system $(\tilde{\mathcal{G}}_t^{++,N})_{t \geq 0}$ converges in fact to the collision-free dual $(\tilde{\mathcal{G}}_t^{++})_{t \geq 0}$, which we have defined in Subsubsections 9.5.3 and 9.5.4. The key point is the following.

Proposition 9.42 (Exponential growth of dual sites for $M \geq 2$ as $N \to \infty$).

(a) For $t \in [0, T)$ we have convergence to the collision-free ($S = \mathbb{N}$) system. Namely for $\ell = I, II, III$, we have

$$\mathcal{L}[\overline{K}_t^N] = \mathcal{L}[|\tilde{\mathcal{G}}_t^{++,\ell,N}|] \Longrightarrow \mathcal{L}[|\tilde{\mathcal{G}}_t^{++,\ell}|], \text{ as } N \to \infty, \tag{9.376}$$

and for $\ell = I$ or II,

$$\mathcal{L}[\hat{X}_0^N(\tilde{\mathcal{G}}_t^{++,\ell,N})] \underset{N \to \infty}{\Longrightarrow} \mathcal{L}[\hat{X}_0(\tilde{\mathcal{G}}_t^{++,\ell})], \text{ for } X^N = X|_{\{1,\cdots,N\}}, \quad X \in (\mathcal{P}(\mathbb{I}))^{\mathbb{N}}. \tag{9.377}$$

(b) Assume that

$$t_N \uparrow +\infty, t_N = o(\log N). \tag{9.378}$$

9.7 Collision-Free Regime: Malthusian Parameter of $\tilde{\mathcal{G}}^{++}$

Then with C^1 a constant as given in (9.368)

$$e^{-\beta t_N} \bar{K}_{t_N}^N \Longrightarrow C^1 \overline{W}, \text{ as } N \to \infty, \tag{9.379}$$

where $\overline{W}$ has the distribution given by (recall (9.355), (9.328))

$$\mathcal{L}[\overline{W}] = \mathcal{L}[\lim_{t \to \infty} e^{-\beta t} \bar{K}(t)]. \tag{9.380}$$

Furthermore let R_t^N be the number of active ranks in $\tilde{\mathcal{G}}_t^{++,N}$, then we get

$$e^{-\beta t_N} R_{t_N}^N \underset{N \to \infty}{\sim} e^{-\beta t_N} R_{t_N}^\infty. \quad \square \tag{9.381}$$

Proof of Proposition 9.42. The proof here is essentially the same as in the case $M = 1$.

(a) is a simple consequence of the fact that as $N \to \infty$ a proportion of the occupied sites involved in a collision in a fixed time interval is asymptotically negligible.
(b) follows from the basic properties of collision-free processes (Proposition 9.39) and the fact that for $t_N = o(\log N)$ the difference between the collision-free process and $\bar{K}_{t_N}^N$ is asymptotically negligible.   q.e.d.

Corollary 9.43. *The number of $Pe \cup UPe$ sites at time t grows as $e^{\beta t + \log \overline{W}}$.* $\square$

Proof of Corollary 9.43. First note that every occupied site at time $t + T$ is an offspring of a Pe site at time t or a UPe or URe site that has not yet decoupled from the parent site after time T. Therefore if we consider the set of all sites including those with flags Pe, UPe or URe as specified by $\mathfrak{s}_{enr}^{**}(t)$ for each Pe site at time t we can consider the set of UPe and URe sites which correspond to migrants from this Pe site. From the above we have that the number of Pe or UPe sites at time t grows as $\cdot e^{\beta t + \log \overline{W}}$.

9.7.8 Factorization Dynamics of the Set $\tilde{\mathcal{G}}^{++}$: Giant Versus Small Metafactors

The main difference from Section 8 in the analysis of emergence and fixation is that the dual in the case $M > 1$ does not have the form of a product *indexed by the sites* but now to get a product form the factors are *indexed by labels*. This holds for $\tilde{\mathcal{G}}^{++,I}$ but also for $\tilde{\mathcal{G}}^{++,II}$ if we consider the dynamic involving more than one site (recall from a dynamical point of view a "site" is characterized by the set of ranks which can coalesce). Hence for the study of the effect of rare mutation on emergence and fixation it is important to decompose $\tilde{\mathcal{G}}^{++}$ into *metafactors* which are comprised of irreducible multisite tableau, that is, tableau over random sets of sites over which the state does not factor further into products of subsets of $\mathbb{I}_M^{n_i}$. That is, we refer to an *irreducible tableau* as *metafactor*. We have seen in the previous subsubsections that the metafactors in the collision-free case *correspond to the labels with their mesostates*.

Then two questions arise: (1) what is the rate of growth of the *number of such metafactors* for large large times t, and (2) how many sites and associated active ranks are included in the single metafactors. These points are addressed in this Subsubsection for $\tilde{\mathcal{G}}^{++}$ and Subsubsection 9.9.4 for $\tilde{\mathcal{G}}^{++,N}$ the process with collisions.

We next note that the number of metafactors (labels) corresponds to the number of *Pe* sites. Since the form of a newly created labels depend on the mesostate of the parent label we get a CMJ-process of labels with a *countable* number of types.

Two scenarios for the size and number of metafactors in $\tilde{\mathcal{G}}^{++}$ are possible. Namely

- a regime in which the number of metafactors grows at a slower exponential rate than the total number of active ranks or at least differ by a subexponential factor so that the size of the largest metafactor has occupied sites which are not necessarily of negligible frequency among all occupied sites,
- a regime in which the number of metafactors and the total number of active ranks are of the same order of magnitude.

In the first case large networks of sites must develop, that is, sufficiently rapidly growing sets of sites occupied by one label (metafactor) develop so that the asymptotic number of ranks and of factors do not grow at the same exponential rate. However we will see in Lemma 9.45 part (a) that, provided that $m > 0$, a *positive fraction of sites* in $\tilde{\mathcal{G}}^{++}$ at times $\beta^{-1} \log N + t$ will be in labels with number of sites *bounded by some constant uniformly in N* and by choosing the constant can be made *arbitrarily close to one*. In the second case the metafactors involve a random number of sites such that even the largest one has a negligible proportion among all occupied ranks.

The distinction between the two regimes is of particular importance for the evolution towards fixation since it is crucial in determining the nonlinear evolution equations describing limiting empirical distribution of local states. In one regime more complicated expressions arise if the clans of connected sites grow since then simultaneous transitions at a large random number of sites have to be taken into account in the limiting dynamic. In the latter case we are forced to include in the *state description* information about these networks of connected sites.

The answer to the question (1) and (2) above will lead in particular to two regimes for the behaviour of these metafactors depending for fixed parameters c and d on the relative sizes of the selection parameters s and $e_2, \ldots, e_M$ and mutation parameters $m = ((m_{i,j}); i, j \in \{1, \cdots, M\})$.

We know (Proposition 9.40 (c)) that the number of active ranks in $\tilde{\mathcal{G}}^{++}$ grows like $W^{\mathrm{rank}} \exp(\beta t)$ as $t \to \infty$. We can now ask the question into how many metafactors $\tilde{\mathcal{G}}_t^{++}$ can be decomposed such that each metafactor is supported by

9.7 Collision-Free Regime: Malthusian Parameter of $\tilde{\mathcal{G}}^{++}$

a *random number of sites*, and how this quantity grows in the limit $t \to \infty$. Since the selection operator acts at rate s, while the mutation operator acts at rates $m = ((m_{i,j}); i, j = 1, \cdots, M)$ driving the new factors to resolution, the growth of the number of metafactors and the length of the latter are determined by the relation between m and s.

More precisely: For given parameters c and $d > 0$ there is a number of ranks per occupied site which is stochastically bounded and is monotone increasing in s. If the minimal mutation rate m increases, these ranks resolve at an increasing minimal rate. Therefore the tail of the number of unresolved sites decreases more rapidly with increasing minimal rate of mutation and becomes flatter with increasing s.

It is of interest to consider the time scale $\beta^{-1} \log N + t$, even though working only with the collision-free process $\tilde{\mathcal{G}}^{++}$. We shall see as a consequence that for s sufficiently small compared to the mutation rates $\{m_{i,j}; i, j = 1, \cdots, M\}$ the length of metafactors is a finite random variable with second-moments and at least the proportion of metafactors of size N^a with $a > 0$ remains negligible as $N \to \infty$. On the other hand for sufficiently large s it can occur at least for non-ergodic mutation on E_0 that at time $\beta^{-1} \log N$ we have metafactors with a length of N^a for some $a \in (0, 1)$. We explain this in detail later on in this subsubsection.

We next give an answer to the basic question on the growth behaviour of metafactors.

Proposition 9.44 (Exponential growth of the number of metafactors in $\tilde{\mathcal{G}}^{++}$).

(a) There is an enriched version of the set-valued process $\tilde{\mathcal{G}}_t^{++}$ that can be factored into F_t many irreducible factors of subsets each supported on a random number of sites. Then there exists a rate β_{Pe}, a random variable W_{Pe} such that:

$$F_t \sim W_{Pe} e^{\beta_{Pe} t} \text{ as } t \to \infty \text{ a.s., } 0 < W_{Pe} < \infty, \qquad (9.382)$$

where

$$0 < \beta_{Pe} \leq \beta, \qquad (9.383)$$

that is, β_{Pe} is the Malthusian parameter for the set of Pe sites.
(b) The cases $\beta_{Pe} < \beta$ and $\beta_{Pe} = \beta$ are both possible depending on the parameters c, d, m, s. □

Proof of Proposition 9.44. The proof of Proposition 9.44 (a) is deferred to Subsubsection 9.12.7, Lemma 9.75.

Part (b) is verified by exhibiting examples for both scenarios. We have given an example with $\beta = \beta_{Pe}$ above and will give an example with $\beta > \beta_{Pe}$ below (Example 20 in Subsubsection 9.12.3). q.e.d.

Remark 154. The subsets of sites which are offspring of different Pe sites yield random sets that are completely decoupled and therefore are product sets. The system of permanent sites alone is a CMJ-process with a rate $\beta_{Pe} \leq \beta$. Finally since the decoupling time of any Pe set is a finite random variable it produces at most a finite random number of sites between decoupling times. (See Subsection 9.11 for details.) Below we give an example (see Example 20) with $\beta_{Pe} < \beta$.

We will now turn to the question of the size of the irreducible metafactors (that is, mesostates). In particular we first identify a parameter regime which implies $\beta_{Pe} = \beta$ and will later given an example where $\beta_{Pe} < \beta$.

Consider the set-valued dual $\tilde{\mathcal{G}}_t^{++}$. Let $\zeta(t)$ be the state of a given generic label, then let

$$\mathfrak{l}(\zeta(t)) \text{ denote the length (number of ranks) of } \zeta \text{ at time } t. \tag{9.384}$$

If $\mathfrak{l}(\zeta) > 1$, let

$$T(\zeta) := \inf\{t > 0 : \mathfrak{l}(\zeta_t) = 1\}. \tag{9.385}$$

Starting at the *birth time of an unresolved tableau*, let

$$T_{\text{dec}} = T_{R^*} \tag{9.386}$$

where T_{R^*} is the decoupling time defined in (9.281). At the time T_{dec} we set $\mathfrak{l}(\zeta) = 1$.

We now consider the question of the behaviour of the first and second moments of $\mathfrak{l}(\zeta(t \wedge T_{\text{dec}}))$ and prove at the end of this section after some preparation the following.

Consider the parameters

$$c, d, m = (m_{i,j})_{i,j=1,\cdots,k}, \ s \text{ and } e = (e_2, \ldots, e_M) \tag{9.387}$$

and define

$$\hat{\lambda} = \hat{\lambda}(c, d, m, s, e) := \sup\{\lambda : E[e^{\lambda T_{\text{dec}}}] < \infty\}. \tag{9.388}$$

Remark 155. In the case $M = 2$ we have $\hat{\lambda} = m_{12} + m_{21}$.

Lemma 9.45 (Moments of irreducible tableau length for $\tilde{\mathcal{G}}^{++}$).

(a) *Assume that $\hat{\lambda} > 0$. Then the family*

$$(\mathfrak{l}(\zeta(t \wedge T_{\text{dec}}-)))_{t \geq 0} \text{ is tight} \tag{9.389}$$

and there exists $\varepsilon > 0$ such that

$$E[(\mathfrak{l}(\zeta(T_{\text{dec}}-))^{\varepsilon}] < \infty. \tag{9.390}$$

9.7 Collision-Free Regime: Malthusian Parameter of $\tilde{\mathcal{G}}^{++}$

In particular the fraction of sites which are in labels which are at times $\beta^{-1} \log N + t$ bounded in N is arbitrarily close to one.

(b) Assume that $\hat{\lambda} > s$. Then

$$\lim_{t \to \infty} E[\mathfrak{l}(\zeta(t \wedge T_{\text{dec}}-))] < \infty. \tag{9.391}$$

If we assume that $\hat{\lambda} > 2s$, resp. $\hat{\lambda} > 3s$. Then

$$\lim_{t \to \infty} E[(\mathfrak{l}(\zeta(t \wedge T_{\text{dec}}-)))^2)] < \infty \quad , \text{resp.} \quad \lim_{t \to \infty} E[(\mathfrak{l}(\zeta(t \wedge T_{\text{dec}}-)))^3)] < \infty. \tag{9.392}$$

(c) Under the assumption $\hat{\lambda} > s$,

$$\beta_{Pe} = \beta. \tag{9.393}$$

(d) *Under the assumption $\hat{\lambda} > s$, at resolution the expected number of different (irreducible) offspring tableau is finite.* □

Remark 156. Under the assumption that $m_{ij} > 0$ for all $i, j \in \{1, \ldots, M\}$, the decoupling time T_{dec} has a finite exponential moment - recall Lemma 9.25.

Proof of Lemma 9.45. (a) Recall the analysis in the sequel of (9.411). Note that $\mathfrak{l}(\zeta(t))$ is dominated by a Yule process with birth rate s and resolution at rank 1 occurs at an exponential time with parameter m. Therefore we can for our estimate certainly work with the process $\tilde{\zeta}$ where we suppress mutation at the higher ranks. Then recall the Yule process at time t has geometric distribution with parameter $(1 - e^{-st})$.

$$P[\mathfrak{l}(\tilde{\zeta}(T_{\text{dec}}-) \geq k] = \int_0^\infty m e^{-mt} (1 - e^{-st})^{(k-1)} dt \tag{9.394}$$

$$= \frac{m}{s} \int_0^1 u^{\frac{m-s}{s}} (1-u)^{k-1} du$$

and this converges to 0 as $k \to \infty$.

Moreover

$$P[\mathfrak{l}(\tilde{\zeta}(T_{\text{dec}}-) = k] = \frac{m}{s} \int_0^1 u^{\frac{m}{s}} (1-u)^{k-1} du. \tag{9.395}$$

Then

$$\sum_k k^\varepsilon P[\mathfrak{l}(\tilde{\zeta}(T_{\text{dec}}-) = k] = \frac{m}{s} \int_0^1 u^{\frac{m}{s}} \sum_k k^\varepsilon (1-u)^{k-1} du. \tag{9.396}$$

Noting that we can bound:

$$\sum_k k^\varepsilon (1-u)^k \le Const \cdot \int_0^\infty x^\varepsilon e^{x \log(1-u)} dx = \frac{1}{(|\log(1-u)|)^{1+\varepsilon}} \Gamma(\varepsilon+1), \quad (9.397)$$

and $\int_0^1 u^{\frac{m}{s}} \frac{1}{(|\log(1-u)|)^{1+\varepsilon}} du < \infty$ if $\varepsilon < \frac{m}{s}$, (9.398)

(9.390) follows.

(b) We prove this for the case $M = 2$. The general case follows in a similar way. Comparing to the Yule process, if $m > 2s$ then the second moment at time $t \wedge T_{\text{dec}}$ satisfies:

$$\sup_t E[(\iota(\tilde{\zeta}(t \wedge T_{\text{dec}})-)^2] \le \sup_t \{\frac{m}{m-2s} e^{2st} e^{-mt}\} < \infty. \quad (9.399)$$

The first moment argument is the same.

(c) Since the tableau length has bounded first moments, the expected number of occupied sites cannot grow faster than the expected number of tableau.

(d) The expected number of offspring is bounded by const $\cdot \iota(\tilde{\zeta}(T-))$. q.e.d.

In order to consider all parameter values we introduce the following concept of mutation versus selection dominance leaving our intermediate regime where arguments become more subtle. We first define *mutation-dominance* which will be assumed in Subsections 9.9 and 9.10.

Definition 9.46 (Mutation dominance condition).

Consider $\tilde{\mathcal{G}}^{++}$ and assume that the decoupling time of a site has an exponential moment, say with parameter

$$\hat{\lambda} = \hat{\lambda}(c,d,m,s,e). \quad (9.400)$$

Then the condition

$$\hat{\lambda} > 2s, \quad (9.401)$$

guarantees the boundedness of first and second moments of the number of sites associated with a tagged metafactor. In this case we say that the pair $(\hat{\lambda}, s)$ respectively parameters (c,d,m,s,e) satisfy the *mutation-dominance* condition. □

Now consider the case $d = 0$ where s exceeds the total mutation rate m which means that even the first moment of the birth process with rate s can be infinite at the decoupling time. This leads to the more subtle notion of *selection-dominance* which will be dealt with in detail in Subsection 9.11. The main point is that in the ergodic case where all mutation rates $\{m_{ij} : i \ne j, i,j \in \{1,\ldots,M\}\}$ are positive

9.7 Collision-Free Regime: Malthusian Parameter of $\tilde{\mathcal{G}}^{++}$

the growth of the size of the mesostate is controlled by *mutation acting not only at the initial rank but on an increasing set of ranks produced by the selection process*. In the case $d > 0$, similar considerations are required but now involve the growth of the number of occupied sites in a mesostate. In this case the number of factors can grow so fast that most of them are not resolved by mutation at the founding site and hence large clans of sites with irreducible tableaus could form as time increases.

Definition 9.47 (Selection-dominance condition).

Consider the selection rate s which gives the birth rate of the particle system η'_t associated with $\tilde{\mathcal{G}}^{++}$.

If

$$\frac{s}{m} > \varsigma \text{ (if } M = 2\text{) or } \frac{s}{\lambda} > \varsigma \text{ (if } M > 2\text{),} \tag{9.402}$$

(where ς is defined in Proposition 9.65) we say the pair (s, m), respectively the parameters (c, d, m, s, e), satisfy the *selection-dominance* condition. □

Remark 157. Note that there are some remaining cases, for example $M = 2$, $\frac{s}{\varsigma} \leq m \leq 2s$ which don't satisfy either condition. It turns out that these can also be handled by modifications of the methods of Subsection 9.11 but will not be developed here.

In our case where mutation on E_0 is ergodic we conjecture that the distinction does not affect the emergence, fixation or limiting behaviour, only requires different analytical tools. However if we do not have this ergodicity, dramatic changes in the behaviour occur.

Remark 158 (Initial growth regimes).

We have looked above at the two regimes at the level of the dual process, we can demonstrate the behaviour also on the level of the original process as follows. Recall that the deterministic case corresponds to the system of ordinary differential equations:

$$\frac{dx_3}{dt} = sx_3x_1 + r(x_1 + x_2), \quad x_3(0) = 0, \ r > 0,$$

$$\frac{dx_1}{dt} = m(p_{21}x_2 - p_{12}x_1) - sx_1(x_2 + x_3) - rx_1, \quad 0 < x_1(0) \neq x_1(eq),$$

$$\frac{dx_2}{dt} = m(p_{12}x_1 - p_{21}x_2) + sx_1x_2 - rx_1,, \quad x_2(0) = b, \tag{9.403}$$

with $a + b + c = 1$, r the rare mutation rate.

Note that $x_1(t)$ approaches the equilibrium value $x_1(eq) = \frac{p_{21}}{p_{12}+p_{21}}$ as

$$|x_1(t) - x_1(eq)| \sim e^{-mt}. \tag{9.404}$$

On the other hand the growth of x_3 initially goes as

$$x_3(t) = \frac{r}{sa}(e^{ast} - 1). \tag{9.405}$$

If for $m > 0$ fixed but s is sufficiently large, then $x_3(t)$ can reach a fixed $\varepsilon > 0$ before $x_1(t)$ gets close to $x_1(eq)$.

Therefore we have two regimes

- $s \ll m$ where the system on the E_0-types approaches the lower level (on E_0) equilibrium faster than the emergent type grows.
- $s \gg m$ where emergence rate depends on the initial proportions of the E_0 types.

These regimes are analogous to the dichotomy at the dual level between bounded and unbounded mean length of the irreducible subtableau which correspond to the mutation-dominant and selection-dominant regimes. Note however if $r = \frac{1}{N}$ but $m > 0$ as in our ergodic case, as $N \to \infty$ types 1 and 2 do approach equilibrium before $x_3(t)$ reaches a positive ε so that the emergence *rate* is not affected.

Example 14. To illustrate the two regimes further, take the case $s > \tilde{m}(= m_{1,2} + m_{2,1})$, $d = 0$ with three types, i.e. $M = 2$. Consider the tableau generated by (110) until decoupling which occurs with rate m. Once a second generation individual is born it grows as $e^{\beta t}$. For example consider the array:

$$\begin{array}{l} (010) \\ (100)\ (000) \\ (100)\ (110)\ (000) \\ (100)\ (110)\ (110)\ (010) \\ (100)\ (110)\ (110)\ (100)\ (010) \\ (100)\ (110)\ (110)\ (100)\ (100)\ (110). \end{array} \tag{9.406}$$

We see that each (110) factor can produce a compound cloud of active individuals that grows at rate s if m_{12} is zero so that no truncations of the tableau occur. Therefore if $s > \tilde{m}(= m_{1,2} + m_{2,1})$ and we only allow mutation at the first rank the expected size of $\zeta((\tilde{t} \wedge T_{\text{dec}})_-)$ at time $\tilde{t} = \frac{\log N}{s} + t$ is given by

$$\int_0^{(\frac{\log N}{s}+t)} e^{(s-(m_{12}+m_{21}))u} du = N^{1-\frac{m_{12}+m_{21}}{s}} e^{(s-(m_{12}+m_{21}))t} \tag{9.407}$$

and

$$\int_0^\infty e^{(s-m)u} du = \infty. \tag{9.408}$$

9.7 Collision-Free Regime: Malthusian Parameter of $\tilde{\mathcal{G}}^{++}$

On the other hand if $m_{12} > 0$ and we allow mutation to occur at any rank, then the tableau can be truncated at any rank and the question is more subtle. □

The example suggests that the the mean length of the tableau generated from a single factor (110) before decoupling can grow with a power of $N^a, a \in (0, 1)$ for a certain parameter range in which however the mutation on E_0 is not ergodic ($m_{12} = 0$). However if $m_{12} > 0$ the situation is more subtle and will be considered in more detail in subsection 9.11. Using the case considered in the proof of Corollary 10.1 we can obtain an example in which the mean length diverges in the case $m_{1,2} = 0$, $m_{2,1} > 0$. This contrasts with the case $m_{1,2} > 0$, $m_{2,1} = 0$ - see Lemma 9.48 (b).

Remark 159. To describe the global picture we keep track of the number of irreducible tableau and the empirical distribution of the tableau (given by a probability measure on the set of tableau of finite length). We expect that as $t \to \infty$ we obtain the Malthusian parameter β_{P_e} for the number of irreducible tableau and β for the sum of the lengths of the tableau, that is, the total number of active ranks. In the mutation-dominant regime we obtain the analogue of the stable type distribution. We can then obtain the analogue of the limiting nonlinear (u, U)-equations.

In subsection 9.10 we consider the convergence of the empirical distribution of local states for the spatial case with coalescence and migration. In order to prove the existence of the limiting nonlinear equations for the empirical measures of micro-states we need the analogue of the second moments of the tableau size to be uniformly bounded. We will again encounter the two emergence regimes described above and discuss in detail the mutation-dominant case. The analysis of the selection-dominant case requires the empirical distribution of mesostates, this will be introduced in subsubsection 9.12.7.

To prepare the proofs concerning the factorization properties of $\tilde{\mathcal{G}}^{++,N}$, we consider first the case $d = 0$.

Factorization in the Case $d = 0$

In order to introduce some of the ideas which will be used to analyse the selection-dominant case in a simpler setting here we consider the deterministic case $d = 0$. Since we can embed the case with $d > 0$ into this by coupling the particle systems, in this way we get from the analysis of the $d = 0$-case analysis some tools for the $d > 0$-case.

Since for $d = 0$ coalescence plays no role the location of individuals plays no role and without loss of generality the problem can be reduced to the single site case. In the deterministic case the set-valued dual can be decomposed into a product of one or more tableaus each located in a interval in $\mathbb{N}$. An *irreducible tableau of length* $k \in \mathbb{N}$ is given by a subset of $(\mathbb{I}_M \backslash \{M+1\})^k$ which cannot be factored into the product of two tableau.

The set-valued dual dynamics can be described as follows. The action of selection on an irreducible tableau of length k produces a tableau of length $k + 1$. On the other hand the action of mutation can result in replacing an irreducible tableau by the product of two or more irreducible tableau with the property that the sum of the lengths of the resulting irreducible tableau is less that or equal the length of the original irreducible tableau. The number of irreducible tableau is nondecreasing and we are interested in the growth of the number and the lengths of the various tableau.

We will see that an important property of a tableau of length ≥ 2 is its life cycle, that is the period of time between its birth (selection acting on the *founding individual*) and the next resolution time of the founding individual at which the length of the tableau reduces to 1. During its lifetime the tableau undergoes Markovian internal dynamics. At the end of its lifetime the tableau gives rise to a random number ≥ 1 of offspring tableau. Therefore at the global level we can view this as a CMJ branching process with countably many types, namely the set of possible tableau.

In order to introduce the key ideas we consider the case

$$M = 2, \text{ with parameters } s, \tilde{m} = (m_{12} + m_{21}) \text{ and } p_{12} = \frac{m_{12}}{(m_{12} + m_{21})},$$
$$p_{21} = 1 - p_{12}. \tag{9.409}$$

Denote the tableau-valued process by ζ_t and set $\zeta_0 = (110)$. The state space for the tableau process is the set of all tableau of finite length. If selection acts (with rate s) on ζ_0 we obtain the tableau of length 2:

$$\begin{array}{ll} (010) & (111) \\ (100) & (010). \end{array} \tag{9.410}$$

The tableau can then *grow by further selection events* and also be *truncated by mutation events*. When the first column resolves with $(010) \to (110)$, then we return to the original state (110). If the first column resolves with $(010) \to (000)$, then we we can factor the tableau into (110) times a new tableau. If the second column has resolved in the same way we have two factors $(110) \otimes (110)$, etc.

Lemma 9.48 (Irreducible tableau length: deterministic case for three types).
Assume here that $d = 0$ and that we have $|E_0| = 2$ and $|E_1| = 1$.

(a) If $\tilde{m} > 0$, then the irreducible tableau process $\{\zeta_t\}_{t \geq 0}$ is a positive recurrent countable state Markov chain.
(b) If $m_{12} > 0$ and $m_{21} = 0$, then $l(\zeta_{t \wedge T_{\text{dec}}})$ is stochastically bounded by a random variable with a geometrically decreasing tail. □

Remark 160. In fact, in case (b) we can determine the probabilities inductively which is given by a geometric with parameter $\frac{m_{21}}{s+m_{21}}$.

9.7 Collision-Free Regime: Malthusian Parameter of $\tilde{\mathcal{G}}^{++}$

Proof of Lemma 9.48.

(a) The Markov property follows since each transition occurs after an exponentially distributed random time. Eventually the founding individual resolves (010) → (110) or (100) → (110) and the cycle is complete and the process begins again. In the first case we return to (110) and in the second to (110) ⊗ ζ^* where ζ^* is an offspring tableau. The resolution time is exponential with mean $\frac{1}{m}$ and the recurrence time is the sum of an exponential of mean $\frac{1}{s}$ and an exponential of mean $\frac{1}{m}$. Therefore it is positive recurrent.

(b) Assume that $m_{12} > 0$. We first consider a tableau produced by selection acting during $[0, t]$ starting with (110) at time 0:

$$\begin{array}{l}(010)\\(100)\ (010)\\(100)\ (100)\ (010)\\(100)\ (100)\ (100)\ (010)\\(100)\ (100)\ (100)\ (100)\ (010)\\(100)\ (100)\ (100)\ (100)\ (100)\ (110)\end{array} \qquad (9.411)$$

The number of columns in this tableau is a Yule process starting in 1 at splitting rate s observed at time t and hence is a random variable with moment generating function

$$G(\theta, t) = \frac{1}{1 - e^{st}(1 - e^{-\theta})}, \quad \theta < \log\left(\frac{1}{1 - e^{-st}}\right), \qquad (9.412)$$

and probability mass function (cf. [Bai] (8.13), (8.15))

$$p_n(t) = e^{-st}(1 - e^{-st})^{(n-1)}, \quad n \geq 1, \qquad (9.413)$$

that is, *geometric* with parameter $1 - e^{-st}$.

In the process $(\mathfrak{l}(\zeta(t)))_{t \geq 0}$ we have in addition transitions due to the mutation $1 \to 2$, which at the first rank induces the resolution and leads to the return of the length process to one. However this mutation may also occur at the higher ranks. The transition (010) → (110) only occurs after an exponential time after birth with parameter m_{12}. Note that if the current population size is n, then at the next birth, the parent is chosen at random so that each existing individual has probability of $\frac{1}{n}$ of being the parent. This means that the successive columns in the above tableau have random ages and the probability that they have undergone the mutation event depends on their age.

We will distinguish now two events on which we separately bound the length. The event in question is defined by having a sufficient number of ranks of at least age 1 and its complement. We therefore next consider the probability that the age is at least 1. Consider $(n_j(t))_{j \in \mathbb{N}}$ the number of ranks descending by time t from a rank labelled j, by the Yule process at rate s. To this collection we want to apply a large deviation estimate. Given that $n(t) = k$, the probability that

$$P(\frac{n(t)}{n(t+1)} \geq \varepsilon e^{-s}|n(t) = k) = P(\sum_{j=1}^{k} n_j(1) \leq \frac{ke^s - \varepsilon}{\varepsilon}) \quad (9.414)$$

$$= 1 - P(\sum_{j=1}^{k} n_j(1) \geq \frac{ke^s - \varepsilon}{\varepsilon}) \geq 1 - e^{-I(\varepsilon)k}.$$

for some $I(\varepsilon) > 0$ if $\varepsilon < 1$.

Given that the population at time t, $n(t) = k$, and the proportion of the population that has age a at least 1 is at least εe^{-s}, the probability that any individual has undergone the mutation event by time t is bounded below by $\varepsilon e^{-s}(1 - e^{-m_{12}})$. Starting from the left we look for the position of the first individual which has undergone the transition (010) to (110). From the above it follows that the distribution of this position can be dominated by a geometric distribution with parameter $\varepsilon e^{-s}(1 - e^{-m_{12}})$. But the rank at which this transition occurs truncates the remaining tableau and this determines the length, that is, $l(\zeta(t))$ is dominated by a fixed geometric distribution uniformly in t. Let $P(n, a)$ denote the equilibrium probability that the length is n and the proportion of the n ranks which have age a.

Therefore either (i) the length is k for some $k \geq k_0$ and the proportion of age ≥ 1 is $\geq \varepsilon e^{-s}$ or (ii) the length is bounded by a geometric with parameter independent of k, since the probability of (i) which is the complement of the event in (9.414) is $\leq e^{-I(\varepsilon)k}$. Therefore

$$P(n(t) = k) \quad (9.415)$$
$$= P(n(t) = k, a(t) \geq \varepsilon e^{-s}) + P(n(t) = k, a(t) \leq \varepsilon e^{-s})$$
$$\leq (1 - \varepsilon e^{-s}(1 - e - m_{12}))^k + e^{-I(\varepsilon)k}.$$

Together these imply that the tail of the distribution of the length is stochastically dominated by a geometric r.v..   q.e.d.

Remark 161. In fact, in case (b) we can determine the probabilities inductively which is given by a geometric with parameter $\frac{m_{21}}{s+m_{21}}$.

Remark 162. The proof of (b) is based on the truncation effect of a $1 \to 2$ mutation at any of the ranks. Such a truncation is called a *catastrophe* transition and will be developed in a more general setting Subsection 9.11.

Remark 163. The number of occupied sites and occupation numbers are different for the dual process or $(\eta', \mathcal{G}^{++})$ and the set-valued dual process $\tilde{\mathcal{G}}^{++}$, since in $\tilde{\mathcal{G}}^{++}$ factors can be killed by combination of selection and mutation and or coalescence, but not in the particle system η' of the dual system, where only coalescence reduces the number of individuals per site. This means that the number of occupied sites and factors per site is stochastically smaller in $\tilde{\mathcal{G}}_t^{++}$ than in the particle system η' of $(\eta', \mathcal{F}^{++})$ or $(\eta', \mathcal{G}_t^{++})$ (in fact that is the purpose in introducing $\tilde{\mathcal{G}}_t^{++}$). Therefore we consider in the multitype case only the process $\tilde{\mathcal{G}}^{++}$ and its variants.

Growth Regimes with $M \geq 2$ and $d > 0$

We now return to the general case with $d > 0$ and $M \geq 2$. When $d > 0$ the expected size of a single site tableau remains bounded due to the quadratic death rate due to coalescence. However in the collision-free regime there is no coalescence between factors originating from different sites and therefore the lifetime of a multisite tableau is determined only by the selection and mutation processes at the founding site. Therefore we are in the same situation as in the $d = 0$ case but now where occupied sites play the role of ranks and we must again identify the regime in which the mean length of the tableau remains bounded in the same way as with $d = 0$.

9.8 The Strategy for the Collision Regime

9.8.1 The Problem of Collisions

As before the main tool in establishing emergence, in particular precise results rather than just bounds, is the dual process $(\tilde{\mathcal{G}}_t^{++,N})_{t \geq 0}$. For small times $t_N \ll \beta^{-1} \log N$ we can approximate $\tilde{\mathcal{G}}^{++,N}$ by $\tilde{\mathcal{G}}^{++}$, recall Subsubsection 9.7.7 and then use the results of Section 9.7 of $\tilde{\mathcal{G}}^{++}$. We have determined the Malthusian parameter for $\tilde{\mathcal{G}}^{++}$, the dual of the McKean–Vlasov process.

However to prove emergence the dual calculations must be carried out on $\{1, \ldots, N\}$ on a time scale where *collisions* occur, that is, migration steps to an occupied site occur with positive probability even in the limit as $N \to \infty$. Of course $\tilde{\mathcal{G}}^{++,N}$ arises immediately again as the functional of $(\eta', \mathcal{G}^{++,N,<})$ so this is no problem. However we now have to adapt the definitions of $\tilde{\mathcal{G}}^{++,\ell}$, $\mathfrak{s}$, etc. to the case with collisions by migration steps to occupied sites. Here some problems arise and these problems play also a role for the analysis of the fixation dynamics where additional critical points arise. Therefore we start now with the basic construction for the analysis carried out in the subsequent Subsections 9.10–9.13.

In this Subsection 9.9 we pursue two points. We first point out the problems in the analysis of $\tilde{\mathcal{G}}^{++,N}$ in the collision regime treated subsequently in Subsections 9.9–9.13, secondly in order to prevent the problems arising from collisions we introduce instead of $\mathfrak{s}^N$ or its variants $\mathfrak{s}^{*,N}, \mathfrak{s}^{**,N}$ an approximate (by repressing *multiple collisions*) new object $\mathfrak{s}^{****,N}$ to study the mutation-dominant case, which serves our purposes and for the selection-dominant case we will introduce $\mathfrak{s}_{l,\text{enr}}^N$.

9.8.2 Strategy for Subsections 9.9–9.13

The tasks. The multitype system *with collisions* is rather complicated since not only the ranks at different sites undergo *dependent* dynamics but even different labels which we defined in the collision-free case and which then gave a factorization into irreducible factors of tableaus which evolved independently and always arose from a migration to a new site after successful decoupling do not do this anymore. Namely, they can now become dependent by collision-coalescence events. This means we cannot simply proceed as in Section 8, with the modifications we carried out so far in Subsections 9.6 and 9.7.

However we can identify the key steps needed to complete the proofs of the analogous results to the ones for the two type case in Section 8. There this complete program was carried out in great detail for the 2-type case in Subsubsections 8.3.7–8.3.12. In this section we identify the changes needed to carry out the same program in the general case of M types on the lower level but we do not duplicate all the detailed arguments.

For times $t_N = o(\log N)$ the dual on $\{1,\ldots,N\}$ started with a fixed finite number of factors can be well-approximated by the dual on $\mathbb{N}$ since we do not see collisions of ranks due to migration steps in the limit $N \to \infty$ (recall Proposition 9.42). A consequence of this is the independent evolution of different labels. When a positive proportion of the finite set of sites $\{1,\ldots,N\}$ are occupied the *independence* of labels (as would be the case with $\mathfrak{s}^{**,N}$ when $t = o(\log N)$) is no longer valid and the CMJ branching picture breaks down. The reason is that at a newly occupied site we can no longer decide first whether it becomes permanent and then take h-transforms since different migrants could *collide* and this would produce inconsistencies. Hence we now have to control the effect of these collisions in the dual $\tilde{\mathcal{G}}^{++,N}$, and hence also in all the enriched versions $\mathfrak{s}^N, \mathfrak{s}^{*,N}, \mathfrak{s}^{**,N}$, which we obtain if we allow collisions, i.e. immigration at an occupied site.

Because of this problem we shall introduce in Subsection 9.9.1 a new enriched version but excluding what we call *self-collisions* of the set-valued dual $\tilde{\mathcal{G}}^{++,N}$, namely $\mathfrak{s}^{****,N}$ which provides a good enough approximation for $N \to \infty$ for the mutation-dominant case and similarly introduce $\mathfrak{s}^N_{l,\text{enr}}$ for the selection-dominant case. The distinction i that in one case we work with *sites* and *microstates* and in the second case with *labels* and *mesostates*.

Next, in order to study the process of emergence and fixation we have to study the evolution equations in the emergence and fixation regime where now we have to control the number of those sites with factors whose evolution remains linked even asymptotically as $N \to \infty$. This we carry out in Subsection 9.10.

The main new difficulty that arises in the study of even simple collisions is that when a migrant moves from a site i to site j the dynamics at the two sites are *coupled* during a (finite) transient period so that mutation-selection-coalescence steps at one site can have impact on the other one and this influence goes both ways. In particular, then two labels can get coupled and by further collisions three

9.8 The Strategy for the Collision Regime

founding sites and their labels become coupled, ..., etc., in other words the situation becomes very complex.

In dealing with the asymptotics of simple collisions as we approach the collision free limit (cf. $t \to -\infty$ in the two type case) we can track the pairs involved. Once multiple collisions are important the interactions become difficult to track but in the limit as $N \to \infty$ we again obtain nonlinear dynamics. We first focus on the dynamics where we are considering only *simple collisions*.

In the collision regime a simpler analysis which can only be carried out in the case of mutation dominance is developed in Subsections 9.10 and 9.11. The selection-dominant case requires a more elaborate method is required since mesostates must be considered instead of microstates associated to sites, this is addressed in Section 9.12.

The strategy The following are the key steps we develop below in the two respective regimes:

In the *mutation-dominant* regime we develop methods which allow us to prove both the emergence results and the relation between W^* and $^*W$. This proceeds as follows.

- We will follow the effect of simple collisions to establish emergence at time $\beta^{-1} \log N + t$ and we show multiple collisions do not change this scale. We also establish the relation between W^* and $^*W$ in the same way.
- We can establish fixation for the McKean–Vlasov dynamics with a positive initial mass $\geq \varepsilon > 0$ of the superior type at all sites (or at a positive proportion of sites) since this involves only short time periods where multiple collisions do not play a role.
- We can identify the deterministic mean-field dual dynamics in the $N \to \infty$ limit because in this case we can eliminate the pair-dependence described above by decoupling since in this case we can assume that a migrant goes to a site chosen at random and therefore for the empirical distribution of the complete dual population of tableaus only the probabilities for the various states at a site are relevant, not the particular dynamics of a specific site.
- For fixed t we determine the limiting moments in the time scale $(\frac{\log N}{\beta} + t)$ using the mean-field dual dynamics with random initial condition obtained from the longtime limit of the multitype CMJ process.

Next the *selection-dominant regime*

- we establish emergence using the h-transform dynamics in the pre-collision regime together with an analogue of the coloured particle system.
- We then identify the limiting empirical distribution of mesostates.

The first regime is treated in this subsection and Subsections 9.9, 9.10 and 9.11, the second regime is then treated in Subsection 9.12.

The objective of this section is to justify the analogues of (8.607), Proposition 8.24 and Proposition 8.27 and Proposition 8.29 and to explain that we can follow a very similar logic except that we have to now incorporate the more complicated internal structures at the site. We will focus on indicating how these modifications work and remain rather brief where we can follow the logic developed in Section 8.

In the proof of emergence in time scale $\frac{\log N}{\beta}$ we can handle this by isolating the effects of single collisions to obtain an upper bound on the transition time.

9.9 Dual in the Collision Regime: $\tilde{\mathcal{G}}^{++,N}$ and $\mathfrak{s}^{****,N}$

The Program for the Mutation-Dominant Case. We will follow here very closely the program of Section 8 which is an advantage if a description based on sites is possible rather than mesostates. In the previous subsection we worked with the McKean–Vlasov dual process $\tilde{\mathcal{G}}^{++}$, i.e. $N = \infty$ with no migration to already occupied sites, and under this assumption we identified the CMJ process and its Malthusian parameter denoted by β. As in the case $M = 1$ this is an essential part of the development but we must now return to the case in which the geographic space is $S = \{1, \ldots, N\}$, $N < \infty$ and determine the dual dynamics in the nonlinear regime with collisions. This means that we must keep track (1) of *collisions* and (2) note that *singletons* (i.e. only one active rank at a site) can now move and collide.

As described above the basic dual process $\tilde{\mathcal{G}}^{++}$ is a set-valued process with values in $\prod_{j \in S} \mathbb{I}^{\mathbb{N}}$ and the moments of the mutation-selection-migration system are obtained by integrating with respect to $\mu^{\otimes S}$. With $M \geq 2$ we have seen that there is a more complex structure than in the case $M = 1$ and therefore the empirical distribution of the states at the different sites is no longer sufficient to characterize the dynamics of these empirical distributions, namely we have the following obstacles:

- the sets are not necessarily the product of sets over different sites,
- a transition at a site can result in changes at a number of other sites (e.g. elimination of migrants at other sites)
- the presence of collisions can result in linked sets of sites at which simultaneous changes can occur.

In the non-collision regime we have handled the simultaneous changes at several sites in Subsection 9.7 using h-transformed versions of the dynamics. However with N fixed and in the presence of collisions we can no longer do this (because of possible inconsistencies if collisions occur) and must introduce new techniques. On the other hand we will show that in the $N \to \infty$ limit, chaos re-emerges in the mutation dominant regime (recall Subsection 9.7.8) and we can again work with the limiting empirical distribution of microstates which is a more straightforward generalization of the technique of Section 8.

In the following subsections we first introduce the state space description of the finite N system with collisions (Subsections 9.9.1–9.9.4). Having the constraints from the three items above in mind we can follow the same program as for the 2-type

9.9 Dual in the Collision Regime: $\tilde{\mathcal{G}}^{++,N}$ and $\mathfrak{s}^{****,N}$

system but now involving the dual where *particles* are replaced by columns hence by *multitype objects*. Therefore it is now much more complicated than in Section 8 to control the effect of a *collision*. Therefore the essential new point is to identify the result of a *simple* collision on the state at a particular site and to obtain an upper bound on the effect of *second order collisions* to justify their exclusion.

Therefore we are facing the task as in Section 8 to *quantify the effect of the collisions*. There we used a *multicolour particle system* to couple the CMJ process and the real finite N dual, they consisted of the black and white, respectively, white and red particles of a three-colour system. With further colours green, blue, purple we tracked the difference between the two populations, more precisely between the red and black particle populations. This allowed us to give sharp estimates for the $N \to \infty$ behaviour. The same principle must be applied now to the ranks in $\tilde{\mathcal{G}}^{++}$ and $\tilde{\mathcal{G}}^{++,N}$ instead of the particles in order to obtain these sharp estimates. We then apply the constructions from the present Subsection 9.9, (1) in Subsection 9.10 to obtain the nonlinear evolution-equations as the limiting dynamics as $N \to \infty$ and then (2) in Subsection 9.11 to derive the first order asymptotics as $t \to -\infty$ to determine the law of $^*\vec{\mathcal{W}}$.

More precisely, for point (2) above, we shall carry out the analysis with multicolour systems in Subsection 9.11 to study the transition regime between the collision-free and the collision case. Indeed we recall that for most purposes, for example emergence, the calculation of first and second moment measures suffices. One key point is that again in order to carry out the second moment calculations using the dual $\mathcal{G}_t^{++,N}$ with $M \geq 2$ in the transition regime ($\frac{\log N}{\beta} + t$, $t \to -\infty$) it suffices to obtain the *first order correction* to the *CMJ process* which depends only on simple collisions (i.e. multiple collisions involving the same particle can be suppressed).

In Subsection 9.12 we shall then indicate how one can carry out this program in Subsection 9.10 and 9.11 if one deals with empirical distribution of mesostates instead of the empirical distribution of microstates.

9.9.1 The Process $\mathfrak{s}^{****,N}$: Motivation and Basic Structure

We must now identify the key points important to construct $\mathfrak{s}^{****,N}$, which replaces the finite N versions of $\mathfrak{s}, \mathfrak{s}^*$ and $\mathfrak{s}^{**}$ processes from the collision-free case which we can no longer use (we comment on this more below), but which is as $N \to \infty$ easier to handle than the analog of $\mathfrak{s}, \mathfrak{s}^N$. (Recall $\mathfrak{s}^{*,N}, \mathfrak{s}^{**,N}$ cannot be consistently defined as before because of the dependency between sites. For this reason we will also have *no* analogue of $\tilde{\mathcal{G}}^{++,\ell}$ for $\ell = II, III$ and N finite.)

The additional difficulty compared to the two type case and which we have seen in the previous section is that a combination of selection, resampling and mutation together with migration with possible collisions results in a dual system in which there can be *simultaneous jumps* at two or more sites.

We now establish an approach to the dual which allows us to generalize techniques from Section 8. We start by clarifying some terminology needed for that purpose. We enrich the set-valued state in such a way that it contains the information on the dependence between sites in the related parent site and newly colonized sites. In order to make the link to Section 8, where we had branching-coalescing individuals populating sites in space, we want to think of the corresponding population of dual objects here as

- A set of *columns* indexed by the rank but having an internal state.
- The state of a *column at a rank* corresponds to a tuple of ordered elements from a partition of $\mathbb{I}_M$ with possible repetition of elements.
- The rank is assigned a *site* in geographic space and a *local rank* at the site inducing an order.
- We refer to a rank as *migrant* if we remove a rank from a site and place it somewhere else.
- We say a rank changes its *internal state* if the sets in the associated tuple of sets change due to mutation, selection or coalescence.
- We say a rank is *born*, when another one undergoes a selection transition and a new rank (column) is created.

With this view of the dual, we will later be able to adopt the techniques of the multicolour particle systems from Section 8, to systems of individuals where the internal state is enriched by a colour (white, black, red, green, etc.). This will be essential to proving statements on emergence.

In order to elaborate on the problems arising in working with these objects, we return to $\mathcal{G}^{++}$ for a moment. What has changed from the $M = 1$ case (besides α being replaced by the smaller value β already in the collision-free regime) is that now the process

$$\{(n(\bar{\zeta}_s^N(t,k)), t \geq s), \quad k = 1, 2, \cdots, \bar{K}_t^N\}, \tag{9.416}$$

which describes the number of local ranks (representing factors) at time t at the various sites k, which was first colonized at time s does not provide the complete information since each site has an internal structure contained only in $\{\bar{\zeta}_s^N(t,k), k = 1, \cdots, \bar{K}_t^N\}$ itself and this internal structure is induced by the factors which are now not just 1_{E_0}.

The new feature compared to the process $\mathfrak{s}$ on $\mathbb{N}$ is that a potential migrant from one label can be assigned a site in $\{1, \ldots, N\}$ which is *already occupied*. We must then decide how to handle the *local rank of the incoming column* since this now generates dependencies if coalescence occurs. We must consider both cases of permanent and transient elements (depending on the future at the parent site of the migrant).

To get an idea how the transitions work, consider the following. Consider the time at which a rank is targeted for migration at a founding site i.

9.9 Dual in the Collision Regime: $\widetilde{\mathcal{G}}^{++,N}$ and $\mathfrak{s}^{****,N}$

- Then a site i' is selected at random in $\{1, \ldots, N\}$ to which it should migrate.
- Now two cases have to be distinguished:
  - If the site is *un*occupied we proceed as in the collision-free case.
  - If the site i' is *occupied*, i.e. it appears already in the list of potential migrants in some other founding site, say j, then the migrant gets a *local* rank in the label corresponding to the founding site j by putting it at that point of the list corresponding to the site i' at the rightmost position and shifting all other ranks to the right by one.

At the arrival time at i' the tableau corresponding to the founding site j and i still *factor* if it does not already contain an earlier migrant from the tableau of the founding site. That is, the product form does not hold if there already is a migrant at the new site from an unresolved excursion from the same parent site. The product form also disappears as soon as we have coalescence of the migrant with a rank from label j.

However the probability that two or more migrants move from a parent site to the same site is $o(N^{-1})$ and we will show that these do not contribute to the limiting quantities we consider when $N \to \infty$.

We note that in a collision event there can be one label involved or two different labels, we can have a self-collision or a collision. The latter ones are much more likely as $N \to \infty$, except if labels grow to size of order N.

Remark 164. Note that in contrast to the case of migration for a tableau at label ℓ to an empty site (with label ℓ') we cannot assign a $**$-type or $*$-trajectory in advance since when a coalescence involving the migrant occurs this could lead to incompatibilities with such prescriptions at other ranks at label ℓ'. The reason being that the collision at one site followed by coalescence results in a coupling between the internal dynamics at two (or more) sites during the time interval between collision and resolution of the factors present at the same site.

This means that after a collision and subsequent coalescence the evolution at former label ℓ' has an influence on (former) label ℓ until label ℓ' decouples and the dynamics at ℓ and ℓ' are not independent during the transient period. This implies that networks of ranks of a random size develop which evolve with coupled dynamics and hence whole networks of sites become coupled in their evolution which without collision would factor in different labels and the network of sites would have a tree structure.

This requires that we refine our states and state spaces. Recall that we had before for the collision-free dual dynamics on $\mathbb{N}$, microstates of the tableau at single sites, mesostates corresponding to labels which could involve several sites not yet factored (however the different labels do factor) and we have the (global) macrostate describing the state of all occupied active sites. We contrast the changes occuring from $N = \infty$ to $N < \infty$ in a table where we abbreviate $S = \{1, 2, \cdots, N\}$. Corresponding to the collision-free case (on the left) we change (on the right) to different objects which we are going to explain step by step below but which we summarize here for the reader to return to as he follows the partly very complex arguments and which are as follows:

Collision-free: $\tilde{\mathcal{G}}^{++}$	With collision: $\tilde{\mathcal{G}}^{++,N}$
Label $\longleftrightarrow$ irreducible tableau, (called mesostates).	Labels become connected and form compound labels, a single label is family component of irreducible factor.
Geo: label $\to \mathbb{N}^g \times$ Graph(2^S, arrows)	Geo: label $\to \bigcup_{n=1}^{\infty} (S)^n \times$ Graph(2^S, arrows)
(i, A) founder i and his direct clan A.	Clan: connected subset of $\{1, \cdots, N\}$
Column $\longleftrightarrow$ state at a rank.	Remains
Label associated with mesostate	Remains
Mesostate: tableau + migration info.	Mesostate: tableau + migration info + coalescence info
Single root for tree of sites	Multiple roots

We will return to discuss the mesostates in the collision regime in Subsection 9.12. In the remainder of this subsection we will concentrate on the *micro*state description related to a site.

It turns out that it is useful to represent the information contained in the system by first considering the "local state" as seen and is visible from a particular site but sufficiently detailed so that the information concerning the clan associated with a label can be reconstructed from the microstates at the sites it contains. To do this we collect at a site the information on the *tableau* on *departing* and *arriving ranks* together with *lists of coalescence*. This we call a

microstate of label k at site k^* : $(\tau_{k^*}, \Lambda_A(k, k^*), \Lambda_D(k, k^*), \mathcal{C}_A(k, k^*))$.
(9.417)

The mesostate description will then have to include in Subsection 9.12 the information on the clan structure and connections between unfactorized sites.

Expected Effects in $N \to \infty$ Limit in the Collision Regime

We begin by reviewing the previous discussions in view of the effects expected for $N \to \infty$ and which result in some simplifications for the dependence structure and simplifying in the evolution of empirical measures of microstates.

In the case of finite N but in the pre-emergence region we can replace $\tilde{\mathcal{G}}^{++,N}$ by $\tilde{\mathcal{G}}^{++}$. We described $\tilde{\mathcal{G}}^{++}$ using $\mathfrak{s}$ involving both *UPe* and *Pe* sites. The *Pe* sites turn out to be the key object but it is important to note the distinction between *Pe* sites, that is sites that are resolved and do not die in the absence of rare mutation and on the other hand the *UPe*, that is, ultimately *Pe* sites not yet decoupled from the parent site, that is, sites that are either *UPe* (or *Tr*) but become *Pe* on resolution. As pointed out above the indicator function that is 1 if a *Tr* site becomes *Pe* is not adapted to the natural filtration. For this reason in the dual representation at a fixed time we must consider both *Tr* and *Pe* sites (recall (9.276)). However in the definition of $(\mathfrak{s}^{**})_{t \geq 0}$

9.9 Dual in the Collision Regime: $\tilde{\mathcal{G}}^{++,N}$ and $\mathfrak{s}^{****,N}$

we only register *Pe* sites since *Re* particles play no role in the overall growth of occupied sites, in the calculation of the Malthusian parameter, (recall (9.338)) only the $\tilde{Pe}$ sites of $\mathfrak{s}^{**}$ drive the growth of the number of occupied sites. However the actual value of the expectation in the duality expression depends also on the *URe* sites that are eventually removed but whose removal has not yet occurred.

What are now the effects of taking $N \to \infty$ in the time regime $\beta^{-1} \log N + t$, $t \in \mathbb{R}$? The point is that asymptotically as $N \to \infty$

- we do *not* see collisions between two specified sites,
- but collisions between a tagged site and a growing finite random subset of the other sites can occur.

Hence what enters in the limit for the situation at a tagged site is the *immigration* (arrival) of a factor from another occupied site *sampled from the empirical law*. Similarly for *emigration* we only have the two cases: either we get a collision with a site sampled from the empirical measure or we arrive at a new site. Therefore in the $N \to \infty$ limit in the collision regime we have to consider in addition the effect of the *immigration from sites sampled from the empirical measure*.

We need some further observations concerning the status of sites. The following is important for the evolution of *Pe* factors. Since a factor that becomes *Re* has only a transient effect which disappears on resolution the only question is the rate of arrival of ultimately *Pe* factors (i.e. particles which are not killed on resolution at the parent site). The key point is that upon coalescence *Pe* has precedence over Re, so that the evolution of *Pe* sites are not disturbed.

In the formulation of the nonlinear dynamics we shall give in Subsection 9.11, we include the union $\tilde{Pe}$ of the *Pe* and ultimately *UPe* sites. This can be done since the probability that a migrating *Tr* site is ultimately *Pe* (or *Re*) is adapted to the natural filtration.

Remark 165. The following observations will be relevant below for taking the mean-field limit in the mutation-dominant collision regime:

- The probability of a new singly occupied site with the various **-types is based on the production of such sites from all other sites where we incorporate the probability of the site ending *Pe*, not *Re*. In other words in the mean field limit we do not need to tag the migrating position with its parent that will determine whether or not it becomes *Pe* - i.e. we decouple the pair dependence.
- The internal state space for the mean field limit is (countable) (see the definition of $\mathfrak{T}^{**}$ below).

9.9.2 Dual State Description of $\mathfrak{s}^{****,N}$ in the Collision Regime

Recall the definition of $\tilde{\mathcal{G}}^{++,N}$ remains the same, it is the functional of $\mathcal{G}^{++,N}$ taken as before but now from $(\eta', \mathcal{G}^{++,N})$ in relation to (9.175). The Markovian dynamics of $\tilde{\mathcal{G}}^{++,N}$ is then the same as before except that,

at a migration event the new location is chosen at random among $\{1,\cdots,N\}$,
(9.418)

instead of the first free site in $\mathbb{N}$. Accordingly the state space has a different index for the geographic space S which is now $\{1,\cdots,N\}$.

Now the distinction. The task we face is to define the basic dual objects for finite N needed to analyse $\tilde{\mathcal{G}}_t^{++,N}$:

$$(\mathfrak{s}^{****,N}(t))_{t\geq 0}, \quad ((\bar{K}_t^N)_{t\geq 0}, (\bar{\mathcal{U}}^N)(t,\cdot,\cdot))_{t\geq 0}) \text{ and}$$
$$((\bar{\zeta}_s^N)(t,k))_{t\geq s}, \quad k \in \{1,\cdots,N\}.$$
(9.419)

In order to analyse this *collision regime* we introduce as a first easy step an enriched description of the microstate space in terms of local ranks and ghost local ranks as follows.

We simultaneously move a column at a local rank targeted for migration to the designated site with the assignment of the *next free local rank* there and place at the original site a *ghost column* at the original local rank, which then becomes a *ghost rank*. The ghost and associated active rank are given *completely coupled dynamics*. As before we call the old site the *parent site*.
(9.420)

A ghost local rank can no longer coalesce with other local ranks at this site or be affected by mutation or selection operations associated with other individuals (or ghost individuals) at this site.
(9.421)

For each ghost rank we must keep track of the location and local rank at that location of the corresponding (linked) active rank.

Now we come to the real challenge posed by the collisions. The point is that during the time the parent site resolves the migrant local rank can be hit by the selection operator and then any of the resulting new offspring ranks can also migrate and so on. This means that there can be more than two sites that can make a common transition when the "founding" parent site resolves. That is, *such linked individuals make coupled simultaneous transitions*. This simultaneous transition process is carried out at the global level (driven by the site at which the active representative is living). Note that such a connection or coupling may be created if at a site we have ranks which do not factorize yet and one of them is assigned a migration step.

The influence of the parent site is guaranteed to have disappeared at the moment the parent site reduces to one rank and has resolved, which needs at most a finite random number of returns to one rank, which at the parent site are geometrically distributed. Hence the critical time period for a site is between a first migration event up to the next return to just one rank and subsequent resolution. Therefore it

9.9 Dual in the Collision Regime: $\tilde{\mathcal{G}}^{++,N}$ and $\mathfrak{s}^{****,N}$

will be convenient to work with these excursions from the first migrant after return to 1 rank till the next return to 1 rank. (Recall we do not move single ranks if they migrate to an empty site.)

To keep track of the simultaneous transitions we introduce a secondary object which includes data on the sites linked to the same active rank and allows to define a new process

$$\mathfrak{s}^{****,N}. \tag{9.422}$$

We begin here in this subsection by specifying the *state space*. Precisely we discuss in the present subsubsection (1) decoupling events, (2) local state space and (3) global state space. In Subsubsection 9.9.3 we then continue by giving the *dynamic* of $\mathfrak{s}^{****,N}$.

(1) Decoupling Events

We begin defining the notion of a *(de)coupling event* (recall the Subsubsection 9.7.8 on factorization) and a basic lemma on this.

Definition 9.49 (Coupling-decoupling of a parent site).
A coupling is initiated between a parent site and a site to which a rank at that site migrates at the migration time or descendants (by birth of new ranks via action of the selection operator) of him migrate to.

A decoupling event for a given parent site occurs at the next time at which the number of active ranks at that site decreases to 1 and subsequently resolves. □

We now recall from Section 8 that the immigration rate at a tagged site can be stochastically bounded by the rate of immigration arising from a site in the McKean–Vlasov equilibrium state. Namely the rate of immigration to a tagged site depends on the average of the occupation numbers of the total population. Since we start with a single dual particle, we can bound this stochastically by a random variable which converges as $N \to \infty$ to a number θ^*, see the arguments following (8.127).

This motivates to state the following:

Lemma 9.50 (Bounds on immigration rates).
Consider $\tilde{\mathcal{G}}^{++,N}$ for fixed N or the McKean–Vlasov limit under the assumption that the rate of arrival of immigrants from other sites is bounded by some constant. Then

(a) *The random times between decoupling events are finite and satisfy (9.203).*
(b) *During an excursion (the path between the coupling and decoupling events) the number of births at the parent site due to selection events is a.s. finite.*
(c) *During each excursions in (b), the number of migration events is a.s. finite.* □

Proof of Lemma 9.50.

(a) Noting that emigration adds a linear death rate to a quadratic death rate due to coalescence while immigration is like a constant birth rate. Then the result follows using from the argument in Lemma 9.25.
(b) This follows from (a) and the fact that the number of selection events is dominated by a linear birth process.
(c) This follows from (b) since the number of migrants cannot exceed the number of births.   q.e.d.

(2) State Space of Enriched Microstates

We introduce the state space of the *enriched* microstates at the sites associated with a label and hence a specific mesostate. Here a new label is always created when a migrant yo a new site decouples from the parent site. We define next the state space describing the enriched microstates of sites k^* at a label k:

$$(\mathbb{R} \cup \{*\}) \times \mathfrak{T}^{****}, \text{ with } \mathfrak{T}^{****} = \{(\tau, \Lambda_A, \Lambda_D, \mathcal{C}_A)\} \quad (9.423)$$

with ingredients of the r.h.s. explained below.

The state at a site $k \in \mathbb{N}_g$ is given by the following objects

- a pair $(a, \mathfrak{t}^{****})$
  - where $a \in [0, \infty) \cup \{*\}$ is the age of a site and takes the value $*$ if the site is unoccupied and
  - $\mathfrak{t}^{****}$ is an element of the countable set $\mathfrak{T}^{****}$.

An element $\mathfrak{t}^{****} \in \mathfrak{T}^{****}$ combines the following objects from the lists below which refer to the considered site k^*

$$\mathfrak{t}^{****} = (\tau_k, \Lambda_D(k,k), \Lambda_A(k,k^*), \mathcal{C}_A(k,k^*)): \quad (9.424)$$

- a tableau (with rows given by the indicator functions of disjoint subsets of $(\mathbb{I}_M)^\mathbb{N}$ which are of the form $A_j \times I_M^\mathbb{N}$ and $A_j \in \mathcal{T}_j$) defining a set $\tau \in \mathcal{T}$,
- two lists Λ_D (Departures) and Λ_A (Arrivals) which we explain below in points (1), (2),
- $\mathcal{C}_A$ specifying the coalescences between the ranks in Λ_A which we explain below in (3).

(1) The list

$$\Lambda_D = (\Lambda_D(k,k^*), \quad k \in \mathbb{N}_\ell, \quad k^* \in \mathbb{N}_g) \quad (9.425)$$

gives the *sites* and *ranks* at label k that are occupied by migrants that have *departed from* site k^* and are not yet decoupled from the ranks at this site.

9.9 Dual in the Collision Regime: $\tilde{\mathcal{G}}^{++,N}$ and $\mathfrak{s}^{****,N}$

The list Λ_D has the form (recall (9.234)–(9.237)):

$$\Lambda_D(k, k^*) = \{\{[\ell_i^1, \ell_i^2]\}, i \in \{1, 2, \ldots, |\mathcal{Y}_{|\tau|,1}|\}; \tilde{\xi}_{k^*}\}, \tag{9.426}$$

$$\tilde{\xi}_{k^*} : \mathcal{Y}_{|\tau|,1} \times \{1, 2, \ldots, |\mathcal{Y}_{|\tau|,1}|\} \to \{1, \ldots, N\} \times \mathbb{N},$$

more precisely $\mathcal{Y}_{|\tau|,1}$ is the tuple of subintervals describing the migrants (one-step) and their descendants not yet decoupled from the founding rank of the label $\tilde{\xi}_{k^*}$ maps the first rank of the different subintervals of the tuple (indexed by the position in the tuple) in $\mathcal{Y}_{|\tau|,1}$ to the migrant (new) sites and local ranks. Here $\mathcal{Y}_{|\tau|,1}$ (defined as in (9.234)) is a tuple of subintervals of the set of $|\tau|$ active ranks at the site k^*, one subinterval for each undecoupled migrant from this site. (The development of the corresponding interval is completely coupled to dynamics at the site to which it migrates and associated further migrations from that site - recall the hierarchy Υ from (9.234)–(9.237).

(2) The lists

$$\Lambda_A = (\Lambda_A(k, k^*), \quad k \in \mathbb{N}_\ell, \quad k^* \in \mathbb{N}_g) \tag{9.427}$$

of sites and ranks of the immigrants to site k^* listed in the label k and still not decoupled from their parent site, that is, the set of parent sites and current local ranks of unresolved immigrants:

$$\Lambda_A(k, k^*) = \{([\ell_i^1, \ell_i^2], (t_i, r_i)); \xi_{k^*}(i), \text{flag}(i), i = 1, \ldots, m_A(k^*)\}, \tag{9.428}$$

where in $\Lambda_A(k, k^*)$, $m_A(k^*)$ is the number of immigrants (arriving migrants) at site k^*, $[\ell_i^1, \ell_i^2]$ is the subinterval of local ranks occupied by the sub-tableau generated by the ith immigrant family at site k^*, (t_i, r_i) denotes the subtableau and current rank at the parent site of the individual that immigrated, ξ_{k^*} is a map from $\{1, \ldots, m_A(k^*)\}$ to $\{1, \ldots, N\} \times \mathbb{N}$ (giving the parent site and local rank of migrant at the parent site), more precisely, ξ_{k^*} maps the founding rank of the families in Λ_A to the parent sites and ranks and the flag(i) is either *Pe* or *Tr*.

Assuming that the arrivals come from different sites, at the arrival time the new migrant individual begins a tableau which factorizes, that is, up to the time of a coalescence between the arrivals already there at the arrival time the resulting tableau at the site is a product of the old tableau and the new tableau.

(3) The collection of lists of coalescences $\mathcal{C}_A$ in the labels at certain sites. Here we need elements (j, r_j) where j is from the index set of the objects in the list $\Lambda_A(k, k^*)$ for which we write shortly $j \in \Lambda_A(k, k^*)$. The needed collection consists of

$$\mathcal{C}_A = \{\mathcal{C}_A(k, k^*), \quad k \in \mathbb{N}_\ell, \quad k^* \in \mathbb{N}_g\}; \tag{9.429}$$

with

$$\mathfrak{C}_A(k,k^*) := \{((j,r_j),(j',r_{j'})); j,j' \in \Lambda_A(k,k^*), j \neq j', r_j \in I_{j,j'}, r_{j'} \in I_{j,j'}\}. \tag{9.430}$$

Here the ingredients are as follows:

$$j \text{ and } j' \in \{1,\ldots,m_A(k^*)\} \text{ denote two immigrant families at site } k^*, \tag{9.431}$$

and

$$I_{\ell,m} \text{ is the set of local ranks in the } \ell\text{-th immigrant family} \tag{9.432}$$

which have coalesced with a rank from the m-th immigrant family sofar

This list collects the set of coalescences among the individuals from label k at site k^* listed in Λ_A and which are incorporated into the resulting compound tableau as illustrated in (9.454) below. Here $(j,r_j), (j',r_{j'})$ indicates that local rank r_j in $[\ell_j^1, \ell_j^2]$ and $r_{j'}$ in $[\ell_{j'}^1, \ell_{j'}^2]$ have coalesced (and therefore have completely coupled dynamics).

(3) Global State Space

In the mutation-domination regime we will base the description of the global state on the microstates (in thee selection-dominat case this will be the messtates). Hence the global state (macrostate) includes the

$$\textit{enriched microstate } \mathfrak{t}^{****}(k^*) \text{ at all sites } k^* \in \{1,\ldots,N\}. \tag{9.433}$$

However in order to better keep track of sites at which simultaneous transitions can occur, we introduce the relevant information as a separate component of the global state, namely we also include a new object in the state description, namely, the

collection of (maximal connected) linked ghost clans (or simply *clans*). (9.434)

Such a clan is a *random graph* (undirected) with the vertices given by the sites, i.e. $V \subseteq \{1,\cdots,N\}$ and edges drawn between sites if the sites are not yet decoupled. We shall see below that the picture has to be refined and that we have to also keep track through which ranks at the sites they are linked, so that the vertices become $V \subseteq \{1,\cdots,N\} \times \mathbb{N}$.

The picture is however even a bit more complicated. A transition at a site can result in a simultaneous transition at another site in two ways:

- *(forward influence)* Consider a migration event from a given rank at a given site and the site to which it migrates. A mutation event at the parent site to the left

9.9 Dual in the Collision Regime: $\tilde{\mathcal{G}}^{++,N}$ and $\mathfrak{s}^{****,N}$

of the migrant rank at the parent site can convert its internal state to the identity factor and can therefore result in the removal of the migrant site.

- *(backward influence)* Now consider the possible transitions at the migrant site, a mutation there can remove ranks to the right of the migrant rank at the parent site, more precisely, convert them to the identity factor.

In order to keep track of all the possibilities we introduce a *directed graph* with a system of *forward arrows* and *backward arrows*. The vertices in the graph are given by the occupied sites and the forward arrows (directed edge) corresponds to the first item, i.e. corresponds to a migration step, the backward arrows to the second item, which point opposite to the migration jump. Then we get in particular two types of clans defined by the directed graph:

- a (maximal) collection of sites connected *only* by using forward arrows, called a *direct clan*,
- a (maximal) collections of sites connected by *only* using backward arrows called a *reverse clan*.

Direct clans already played a role in the non-collision case, so what is the new effect due to collisions? Transitions due to backward arrows within a direct clan can occur but what turns out to be important are backward arrows that simultaneously effect two or more clans. This does not occur in the collision-free regime but can occur when *collisions* take place. Collisions can *link* different direct clans if two active individuals from different clans meet at a site and then coalesce. In this case a transition at this coalesced individual can result in simultaneous changes at the two parent sites. These effects are described by the reverse clans.

The fact that now simultaneous changes at several sites of a clan can occur will have an impact on the asymptotic analysis of the process $\mathfrak{s}^{****,N}$. In particular we will need to bound the number of clans at which a simultaneous transition can occur in the emergence regime (see Lemma 9.52 (a)). In order to keep track of of these effects we use the directed random graph.

A clan consists of a number of sites which are not yet decoupled (recall Definition 9.49), that is, a collection of sites at which a simultaneous transition can occur is given in Γ^N. The set $\hat{\Gamma}^N$ is the set of states of clans which can occur and which we shall define in detail only further below (subsequent to (9.439)).

Hence to define the global state space introduce as further ingredients the information about the set of all clans which appear in $\tilde{\mathcal{G}}^{++,N}$ and with it the corresponding state space for the dynamics $\mathfrak{s}^{****,N}$:

$$\Gamma^N = \{\text{ collection of maximal ghost clans}\}, \qquad (9.435)$$

$\hat{\Gamma}^N$ the set of potential collections Γ^N \hfill (9.436)

$= $ set of collections of finite *directed* graphs with vertices in $\{1, \ldots, N\} \times \mathbb{N}$.

The role of directed graphs which occur here will be explained below.

In order to define states of $\mathfrak{s}^{****,N}$ in the nonlinear regime, we assign age $*$ to unoccupied sites and the age of a site is the time that has passed since the arrival of a migrant to the empty site). Then the *global state space* of $\mathfrak{s}^{****,N}$ is given by

$$\mathfrak{S}_N^{****} = (([0,\infty) \cup \{*\}) \times \mathfrak{T}^{****})^{N_g} \times \hat{\Gamma}^N. \tag{9.437}$$

The first component of the state space is a Polish spaces as product space, the second component is equipped with the discrete topology, so that our complete state space is a Polish space.

We also define the projection extracting the part of the state we need in the duality relation:

$$pr: \mathfrak{S}_N^{****} \to (\mathfrak{T}^{****})^{N_g}. \tag{9.438}$$

Note that with this map the global state uniquely defines a subset of $((\mathbb{I}_M)^\mathbb{N})^S$ with $S = \{1, \ldots, N\}$, which we can then use in the duality relation.

The directed random graphs Γ^N: We now have to explain the component $\hat{\Gamma}^N$ describing the clans more formally, in particular a

marked directed random graph (9.439)

which allows us to keep track of the sites and ranks that are currently coupled. Note that the currently, at time t say, connected ranks and sites are connected due to certain events taking place during the time in $[0, t]$. This means we must build a random graph containing this historical information and then marks to read off the current time-t relations.

The directed random graph is built as follows:

- *vertices* correspond to the pairs (site, local rank), i.e. they are given by

$$\{(k, \ell) \in \{1, \ldots, N\} \times \mathbb{N} : (\mathbf{t}^{****}(k))_1 \neq (^\ell \mathbf{t}^{****}(k))_1\}, \tag{9.440}$$

$(^\ell \mathbf{t}^{****}(k))_1$ arises by setting all rows of the ℓth column equal to $(1, \ldots, 1)$, i.e. we consider the set of active ranks at occupied sites, (here $(\mathbf{t}^{****}(k))_1$ denotes the first component of the 4-tuple, that is, $\tau \in \mathcal{T}$.)
- *(forward arrows)* there is a directed edge

$$(k_1, j_1) \to (k_2, j_2), \text{if} \exists\, i = i(j_2) \in \{1, \ldots, m_A(k_2)\} \text{and } j_2 = \ell_i^1 \text{ (recall (9.428))} \tag{9.441}$$

with (ξ is defined below (9.426)

$$\xi_{k_2}(j_2) = (k_1, j_1) \tag{9.442}$$

and where j_1 must be an initial point of a subinterval contained in $\mathcal{Y}_{n(\tau_{k_1}),1}$.

9.9 Dual in the Collision Regime: $\tilde{\mathcal{G}}^{++,N}$ and $\mathfrak{s}^{****,N}$

This indicates that a migrant originally located at k_1 (at ghost rank currently given by j_1) migrated to k_2 at the rank currently given by j_2 and belongs to the *ith* immigrant family at k_2;
- *(backward arrows)* there is a directed edge

$$(k_1, j_1) \leftarrowtail (k_2, j_2) \text{ if } \exists\, i = i(j_1) \text{ entry of } \mathcal{Y}_{|\tau|,1}(k_1),$$
$$j_1 = \ell_i^1 \text{ and } \tilde{\xi}_{k_1}(\mathcal{Y}_{|\tau|,1}, i) = (k_2, j_2), \tag{9.443}$$

this indicates that a migrant originally located at k_1 and at rank currently given by j_1 migrated to k_2 at the rank currently given by j_2 and belongs to the *ith* emigrant family at k_1;
- A *clan*

$$(\gamma^N, \varrho) \tag{9.444}$$

consists of two components, first γ^N is a *maximal connected directed subgraph* of the above directed graph and second a map (marks)

$$\varrho : \text{vertices of } \gamma^N \to \{1, \ldots, N\} \times \mathbb{N}, \tag{9.445}$$

which assigns the vertices of the graph to the appropriate sites and ranks. We can then read off which columns in the total tableau belong to an *irreducible factor* and which locations are inhabited or connected by ranks from this factor.
Although this directed graph is defined by a function on $(\mathfrak{T}^{****})^{\{1,\ldots,N\}}$, for convenience in determining transitions that occur simultaneously at more than one site we have included it as a component in the definition of the global state space.

Terminology

- An *active* vertex of a clan is a vertex which is a leaf, that is, has an ingoing edge but no outgoing edge.
- A *founder* is a vertex which is a root, that is, can have an outgoing edge but no ingoing edge, that is it is an original vertex or one that is decoupled from its parent and therefore at a *Pe* site.
- The set of descendant vertices not yet decoupled from the parent site is a *direct clan*.
- A *reverse clan* is a collection of sites connected via backward (ancestral) arrows.
- The *ancestral depth* of a vertex is the number of sites traced back to its founder.
- A *cycle* can occur when *self collision* occurs, that is either a descendant of a site migrates back to the founder site or two ranks migrate from one site and meet together at a second site.

9.9.3 Dynamics of $\mathfrak{s}^{****,N}$

We now have to specify the Markovian dynamics of $\mathfrak{s}^{****,N}$. Of course we just take the transitions in $(\eta', \mathcal{G}^{++,<})$ and translate what the changes mean to obtain the list of transitions and rate with which we define $\mathfrak{s}^{****,N}$ formally. We consider the transitions occuring in the tableaus τ at the different labels due to the selection operator, mutation and coalescence and then in the lists $\Lambda_A, \Lambda_D, \mathfrak{C}_A$ on the one hand and in the clan structure Γ^N on the other hand.

We consider separately the two different types of transitions. We first in (1)–(3) introduce the three operators corresponding to the action of selection, coalescence, and mutation acting on a rank which is *not* in the corresponding (to this label and site) $\Lambda_A \cup \Lambda_D$, then in (4) we discuss what happens to those ranks listed in $\Lambda_A \cup \Lambda_D$, subsequently in (5) we consider *migration* and finally in (6) we prescribe how *clans* are created and how the evolution of the clans works, respectively we describe the effects of the earlier defined dynamics on clans.

We now introduce in (1)–(3) transitions which occur for local ranks at a given fixed site at specified rates which are

$$\text{at the current time } t \text{ not listed in } \Lambda_A \cup \Lambda_D. \tag{9.446}$$

This means in particular the transitions we consider now act on the state at one site only. We denote for this purpose the tableau at this site by

$$\zeta \text{ and the } j\text{-th column by } \zeta(j). \tag{9.447}$$

1. Transitions due to *birth* (i.e. selection) acting on a local rank j at some site k^*:

   Due to birth events the following jump occurs,

   $$\zeta \to \Phi^{\text{sel}}_{j,A_i} \zeta, \text{ with rate } s(e_i - e_{i-1}), \tag{9.448}$$

   and the operator $\Phi^{\text{sel}}_{j,A_i}$ is defined on the tableau of factors in the different summands i.e. $\Phi^{\text{sel}}_{j,A_i} : \mathcal{T} \to \mathcal{T}$ and changes this object in the following way:

   replace column $\zeta(j)$ and add a new column at position $j+1$ so that we obtain now in place of each row the 2 new rows (indicting only the two changing entries):

   $$\begin{array}{ll} \zeta(j) \cdot 1_{A_i} & 1 \\ (1 - 1_{A_i}) & \zeta(j), \end{array} \tag{9.449}$$

   where the operation acts simultaneously on each row of $\zeta(j)$.

2. Transitions due to *coalescence* of ranks j_1 and j_2:

   Due to coalescence the following jump occurs

9.9 Dual in the Collision Regime: $\widetilde{\mathcal{G}}^{++,N}$ and $\mathfrak{s}^{****,N}$

$$\zeta \to \Phi^{coal}_{j_1,j_2}\zeta, \quad j_2 > j_1, \quad \text{at rate } 2d, \qquad (9.450)$$

where the operator $\Phi^{coal}_{j_1,j_2}$ acts on the tableau ζ by

- removing the j_2 column,
- the columns $j_2 + 1,\ldots$ are shifting down by one position and
- the column j_1 is replaced by taking the products of the indicator functions of the sets in the corresponding rows of columns j_1 and j_2 and
- removing rows containing a 0 factor.

Here the waiting times for coalescence for all pairs $\{(j_2, j_1), j_2 > j_1\}$ are independent.

3. *Transitions due to mutation*:

Mutation from type k to type ℓ at rank j induces the jump

$$\zeta \to \Phi^{mut}_{j,(k,\ell)}\zeta, \quad \text{with rate } m_{k\ell}, \qquad (9.451)$$

where $\Phi^{mut}_{j,(k,\ell)}$ changes the j-th column of the tableau ζ by changing each component by (9.185). When due to this transition a rank is resolved in $(11\cdots 1)$ it is removed from the list of active ranks.

This completes the jumps involving the tableau at one site and acting on the ranks there which satisfy that they are not listed in $\Lambda_A \cup \Lambda_D$.

4. *Transitions for listed ranks*:

We now turn to transitions involving more than one site at a single time point due to the links between sites.

The first point is to give the dynamics of the ranks listed in Λ_A or Λ_D for the specific site.

- The ranks at the migrant site then develop as above in (1)–(3) (the action of of selection, coalescence, mutation) and this dynamics is *mirrored* at the parent site in Λ_D of the parent site.
- In addition the dynamics at the parent site is mirrored in the list Λ_A.

Furthermore the following changes can occur to the list Λ_A and $\mathfrak{C}_A$ by the transition (1)–(3):

- An item is removed from the list Λ_A at the decoupling time of the parent site and leaves behind the resulting immigrant family (if it is not removed) which is now decoupled from the parent site. (Note that since there is a positive probability, uniformly bounded below by some $\varepsilon > 0$ for all possible partitions of $\mathbb{I}_M$, that any such migrant site results in the removal of the migrant, at most a geometrically distributed number can be involved and therefore the time until this decoupling event has an exponential moment.)

- Coalescence can introduce dependence between different members of Λ_A. These are tracked by the collection $\mathfrak{C}_A$. When a rank $r_j \in \Lambda_A$ coalesces with a rank $r_{j'} \in \Lambda_A$ then the pair $((j, r_j), (j', r_{j'}))$ is added to the corresponding element of the collection $\mathfrak{C}_A$.

5. *Migration*

When two ranks migrate from possibly different sites and then collide, that is, migrate to the same site, it is possible that they will coalesce at this new site. We cannot make both of them immortal, because once they coalesce before anything else happened at most one of them can be "immortal", that is, ultimately produce a permanent site, thus leading to a potential loss. In addition we then have a larger clan given by the union of the two coalescing clans. Therefore the dynamics used before has to be modified to capture these effects. In particular when this occurs the two clans merge which means that simultaneous transitions can now occur in both the original clans. The formal rules are as follows.

Transitions due to *migration* from a site i to a site k.

At a migration event taking place in a label ℓ (which we do not explicitly exhibit in the notation below) of an individual at local rank j with jump from site i to site k defines the transition:

$$\zeta \to \Phi^{\mathrm{mig}}_{(i,k),j}\zeta, \text{ with rate } c. \qquad (9.452)$$

Each active individual can migrate at rate $c\,(1 - N^{-1})$ to one of the other $(N-1)$ sites chosen at random. There are two cases we have to distinguish for the transition:

Case 1. The new site k is *un*occupied.

In this case the new site is assigned the migration time as birth time, that is

- the birth time changes: $* \to$ *time of the migration* and
- the migrant is added to the corresponding list $\Lambda_A(\ell, k)$.

Case 2. The new site is *occupied*.

Here we have to distinguish two subcases.

Subcase 2a The parent site and new site are *decoupled* at the migration time (i.e. no simultaneous transitions can occur at these two sites).

Then the immigrant combines its tableau with the existing tableau at the new site as a *product* of the two tableaus (c.f. (9.454), occupying with its ranks the first free *local ranks* at this site.

In addition

- a ghost copy (member of $\mathcal{Y}_{|\tau|,1}$) remains at the parent site (if it is not the only rank at this site)
- the migrant (together with the parent state and rank of migrant at the parent site) is added to the list in the collection Λ_A with rank given by the first unoccupied local rank and proceeds to form a migrant family.

Subcase 2b The parent site and new site are *not* decoupled at the migration time.

9.9 Dual in the Collision Regime: $\tilde{\mathcal{G}}^{++,N}$ and $\mathfrak{s}^{****,N}$

If, for example, an individual has already migrated from the parent site to the same new site the situation is more complex. We call this a *self collision*. The resulting transitions are read off as before from the evolution of the process $(\eta', \mathcal{G}^{++})$, but they are very complicated to describe and would require us to store additional information.

Fortunately in the limit as $N \to \infty$ these self collisions are *negligible* even in the case of positive density of occupied sites (since the intensity of self-collisions would still be order N^{-1}) and therefore we do not elaborate here on the detailed description of the transitions. In the analysis below these terms will be *included in an error term* that goes to 0 as $N \to \infty$.

6. *Clan dynamics and simultaneous transitions*

We now have to describe two transitions here, (1) the creation of *new clans* and (2) the evolution of the *existing clans*.

The *creation* of clans proceeds as follows:

- when an active rank migrates from a site the pair including the parent and migrant (sites, local rank) forms what we call a clan involving 2 sites,
- if a rank then migrates from the migrant site we then have a clan involving three sites,
- if before migrating the migrant produced an offspring rank then it can also migrate and then produce two active individuals (ordered by the time of migration).

The *dynamics* of the clans after creation are determined by the following:

- as a consequence of the rules given so far we have when one of the operators (suppressing the subscript)

$$\Phi^{\text{sel}}, \Phi^{\text{coal}}, \Phi^{\text{mut}} \qquad (9.453)$$

acts on a rank at an active vertex (leaf in the graph) it does so *simultaneously* at all ghost copies at their respective sites and they make the same transitions, that is, the same change in $\mathcal{T}$ occurs as if the transition due to one of the above operators occurred at the ghost rank,
- if all those local ranks in the originating site (root of the clan) have undergone resolution which are ancestors of the original migrant, then there are two possible outcomes: (1) the entire linked clan is removed or (2) a new clan is created with the "son" as the founder, depending on whether at the originating site the ranks to the left resolve (1) in $(0 \cdots 0*)$ or (2) in $(1 \cdots 1*)$,
- when a rank, which is not a ghost resolves, the appropriate action at the parent site is carried out (e.g. including the removal of appropriate descendant ranks to the right of the ghost) and also on its descendants at other sites in the clan if the clan has one root and if the clan has more than one root then the removal is carried out only on the edges involving descendants of that root,
- clans merge when active ranks of two clan meet at a site and coalesce.

As a consequence of the rules we have given above for transitions of the type 6, simultaneous transitions can occur at multiple sites as follows.

- *Direct clan*: when a transition at a *founder* results in decoupling then the descendants in the clan are either removed or the next level creates one or more new labels.
- *Reverse clan*: up to its resolution time transitions at an *active* rank can result in changes at the ancestral positions - for example removal and this can also result in removal of offspring of the active rank as it moved through other sites; this also affects other clans which have merged and formed part of this reverse clan.

The list (1)–(6) of the different types of transitions given above completes the list of transitions and their rates in the dynamics of the *Markov pure jump process* $\varsigma^{****,N}$, which is therefore welldefined for every fixed choice of N. We demonstrate the mechanism in Example 15.

Remark 166. Observe the following features of the structure of clans arising from certain transitions. Clans can *merge* by migration of the active vertex to the same site followed by a coalescence or by coalescence of two founder vertices. In other words, two linked sets can be joined by a coalescence, that is, when two *active* vertices from different clans coalesce at a site. When this occurs we take first as usual the corresponding mutual refinement and then the two active vertices are replaced by this single active vertex and forming a single connected component. In this case we can then have two incoming and two outgoing edges from the same individual.

Simultaneous transitions can occur at some or all members of a clan when a transition occurs at either a founder or an active vertex in a clan. There are potentially $O(N)$ clans, one associated to each occupied site.

In the Subsubsection 9.9.4 we shall investigate the structure and dynamics of the network of clans further.

Remark 167. The (complicated) construction above has far more applications than we exploit here. The dual obtained above is an interacting system which allows us to compute the moments of the interacting Fisher-Wright system with mutation and selection. If N is fixed, we start the dynamics on types in E_0 and we suppress the rare mutation, then in the limit $t \to \infty$ this system converges to an equilibrium which determines the moments of the equilibrium for the interacting Fisher-Wright system with types in E_0 and N sites. We do not develop this in detail here.

Example 15 (Finite N-dynamics). Here we illustrate the effect of coalescence on the product of two sub-tableaus at the same given site. We indicate with subscripts a respectively b the sub-tableaus (which are the factors of the tableau we consider) to which the columns belong.

We start with the tableau of the form:

$$\begin{pmatrix} (010)_a & & & \\ (100)_a & (010)_a & & \\ (100)_a & (100)_a & (010)_a & \\ (100)_a & (100)_a & (100)_a & (011)_a \end{pmatrix} \otimes \begin{pmatrix} (010)_b & & \\ (100)_b & (010)_b & \\ (100)_b & (100)_b & (010)_b \end{pmatrix} \qquad (9.454)$$

9.9 Dual in the Collision Regime: $\tilde{\mathcal{G}}^{++,N}$ and $\mathfrak{s}^{****,N}$

Here we fill up with the (111) terms on the upper part and consider then the 7 columns and multiply the sums out (recall rows are summands consisting of products of the factors in the row) to obtain the $(4 \times 3) \times 7$ tableau.

This corresponds to the full 12×7 tableau at this site given next:

$$\begin{pmatrix} (010)_a & (111)_a & (111)_a & (111)_a & (010)_b & (111)_b & (111)_b \\ (010)_a & (111)_a & (111)_a & (111)_a & (100)_b & (010)_b & (111)_b \\ (010)_a & (111)_a & (111)_a & (111)_a & (100)_b & (100)_b & (010)_b \\ (100)_a & (010)_a & (111)_a & (111)_a & (010)_b & (111)_b & (111)_b \\ (100)_a & (010)_a & (111)_a & (111)_a & (100)_b & (010)_b & (111)_b \\ (100)_a & (010)_a & (111)_a & (111)_a & (100)_b & (100)_b & (010)_b \\ (100)_a & (100)_a & (010)_a & (111)_a & (010)_b & (111)_b & (111)_b \\ (100)_a & (100)_a & (010)_a & (111)_a & (100)_b & (010)_b & (111)_b \\ (100)_a & (100)_a & (010)_a & (111)_a & (100)_b & (100)_b & (010)_b \\ (100)_a & (100)_a & (100)_a & (011)_a & (010)_b & (111)_b & (111)_b \\ (100)_a & (100)_a & (100)_a & (011)_a & (100)_b & (010)_b & (111)_b \\ (100)_a & (100)_a & (100)_a & (011)_a & (100)_b & (100)_b & (010)_b \end{pmatrix} \quad (9.455)$$

(since the rows correspond to disjoint events, we are free to choose the ordering so that we start with the first row of the first tableau and then produce the product of this factor with each of the rows of the second tableau, etc.)

Now assume that the 2nd rank of the first factor and 2nd rank of the second factor coalesce. Then the rows 6,8,11 contain a (000) and are deleted and the 6-th column turns into a (111) column which is inactive. We then obtain the following tableau by deleting 0-rows and inactive columns:

$$\begin{pmatrix} (010)_a & (111)_{ab} & (111)_a & (111)_a & (010)_b & (111)_b \\ (010)_a & (010)_{ab} & (111)_a & (111)_a & (100)_b & (111)_b \\ (010)_a & (100)_{ab} & (111)_a & (111)_a & (100)_b & (010)_b \\ (100)_a & (010)_{ab} & (111)_a & (111)_a & (010)_b & (111)_b \\ (100)_a & (010)_{ab} & (111)_a & (111)_a & (100)_b & (111)_b \\ (100)_a & (100)_{ab} & (010)_a & (111)_a & (010)_b & (111)_b \\ (100)_a & (100)_{ab} & (010)_a & (111)_a & (100)_b & (010)_b \\ (100)_a & (100)_{ab} & (100)_a & (011)_a & (010)_b & (111)_b \\ (100)_a & (100)_{ab} & (100)_a & (011)_a & (100)_b & (010)_b \end{pmatrix}, \quad (9.456)$$

which can no longer be factored since we have 9 summands and 4 a- or ab-columns, but only two columns in the b-group.

9.9.4 Clans: Their Number and Sizes, Asymptotics as $N \to \infty$

Our next objective is to identify the structure of the set-valued dual with $S = \{1, \ldots, N\}$ in the limit as $N \to \infty$ in the time scale $(\frac{\log N}{\beta} + t)$ for fixed t and in the limit as $t \to -\infty$. Structure here means, the properties of the number and length of (compound) mesostates corresponding to irreducible factors in Proposition 9.40. Recall that we proved for times $t = t_N$ of order $o(\log N)$ a factorization of the set-valued process into $O(e^{\beta \rho_e t})$ many irreducible metafactors each corresponding to the permanent founder sites and linked clans of sites associated to Pe sites. A key quantity is the following:

Definition 9.51 (Size of direct clan).

a) We define the *size* of a (direct) clan γ^N as

$$|\gamma^N| = |\pi_1 \circ \rho(\gamma^N)|, \tag{9.457}$$

i.e. as the number of sites which appear as first components in the vertices of the (direct) clan γ^N.

b) The frequency of γ^N is defined as

$$|\gamma^N|N^{-1}. \qquad \square \tag{9.458}$$

Before proceeding with the collision regime we now review the description in the case $S = \mathbb{N}$ developed in Subsubsection 9.7.8. In this case we have described the set-valued process in terms of a product of a growing number of subsets of $\mathbb{I}_M^\mathbb{N}$, each given by a tableau. There we identified two regimes, the *mutation-dominated* and the *selection-dominated* regimes. We verified that in the mutation-dominated regime the first and second moments of the length of the tableau of any age are uniformly bounded and we can introduce then the empirical distribution of the microstates of the different sites. The resulting system of direct clans forms the basis for the next step in our discussion. Again in the mutation-dominated regime given by (9.401) the size of the direct clans have bounded first and second moments. In this case we complete the analysis using the *empirical distribution of the microstates at individual sites* along the same lines as in the case $M = 1$. In the selection-dominant case this must be replaced by the *empirical distribution of mesostates* we treat in Section 9.12.

In the following lemma we collect some basic properties for times of observation $o(\log N)$, i.e. we are collision-free asymptotically as $N \to \infty$.

Lemma 9.52 (Migrant production and growth of direct clans in collision-free regime).

Let T_{R*} denote the decoupling time of a single site. Let $\hat{\lambda}$ denote the parameter in the dominating exponential tail of T_{R*}.

(a) Let $N(t)$ denote the number of migrants from a fixed site during a bounded interval $[0, t]$. This random variable has an exponential moment and furthermore there exists $L = L(\theta)$ such that:

$$E[e^{\theta N(t)}] \leq e^{Lt}, \quad \theta \in [0, C) \text{ for some } C > 0. \tag{9.459}$$

Moreover for $\hat{\lambda} > L$,

$$\sup_t E[e^{\theta N(t \wedge T_{R*}-)}] \leq \int_0^\infty e^{-(\hat{\lambda}-L)t} dt < \infty. \tag{9.460}$$

9.9 Dual in the Collision Regime: $\tilde{\mathcal{G}}^{++,N}$ and $\mathfrak{s}^{****,N}$

(b) The total number of migrant individuals produced during a decoupling cycle has finite first and second moments.

(c) Assume we are in the mutation dominance regime, i.e. that $\hat{\lambda}$ satisfies (9.401). Consider a direct clan $\mathcal{C}$ initiated at a fixed founding vertex.
   Then for any t the distribution of the size of the direct clan satisfies

$$E[|\mathcal{C}((t \wedge T_{R^*})-)|^2] < \infty. \tag{9.461}$$

(d) For times $t = o(\log N)$, asymptotically as $N \to \infty$:

   (1) Consider the frequencies of sites of directed clans among all active occupied sites. The maximal frequency at time t_N tends to zero in probability as $N \to \infty$. All but at most a negligible proportion of the active occupied sites belong to connected components of the directed graph that are direct clans and

   (2) two direct clans sampled without replacement from all direct clans at time t are not linked. (These correspond to Tr sites attached to Pe sites as described in subsection 9.7.)   □

Proof of Lemma 9.52.

(a) Let $|\tau(t)\backslash \mathcal{Y}|$ denotes the number of local ranks not contained in one of the subintervals in $\mathcal{Y}$. The migration process at a site has Poisson rate

$$cn(t) = c|\tau(t)\backslash\mathcal{Y}|, \tag{9.462}$$

Recall that $n(\cdot)$ is stochastically bounded by a birth and death process with linear birth rate and quadratic death rate and starting in equilibrium. The exponential moment of the number of migrants, $N(t)$ is then bounded by

$$E[e^{\theta N(t)}] \leq E\left[\exp\left((e^\theta - 1)\int_0^t cn(s)ds\right)\right] \leq e^{Lt}, \text{ for some } L = L(\theta) > 0. \tag{9.463}$$

The result then follows from [EK5], Lemma 2.1.

(b) The expected number of migrants during a decoupling interval is bounded as follows:

$$E\left[\int_0^{T_{R^*}} c \cdot n(t)dt\right] = \text{const} \cdot E[T_{R^*}] \leq \frac{\text{const}}{\hat{\lambda}}. \tag{9.464}$$

Similarly, the second moment can be obtained using the uniform boundedness of the second moments of $n(t)$.

(c) Consider the *direct clan*, that is the set of sites occupied by a migrant clan. During the lifetime the clan can add active particles at sites in the clan and add new sites by selection respectively migration. Since the migration and selection rates (c, s) are finite, this means that the direct clan is stochastically dominated by a CMJ branching process. Moreover the size of the clan can be dominated by the length of the analogous tableau in the deterministic case, i.e. where $d = 0$ (since coalescence only reduces the size) and the result follows by the result for the deterministic case in Subsubsection 9.7.8.

(d) Since $t_N = o(\log N)$ by the above CMJ-bound a clan cannot reach positive intensity. Therefore in particular also the claim follows since the probability of a collision of clans is then $o(N^{-1})$. q.e.d.

Above we were in the collision-free regime. As we enter the collision regime this picture must be modified to include the *interaction* of the *direct clans*. For times of the form $\frac{\log N}{\beta} + t$ we have a system of ranks associated with the *sites of a clan* (i.e. sites which appear as first coordinate in the vertices of a clan) such that migration steps hit other occupied sites. When there is a collision between clans followed by a coalescence of two of their collided ranks, a change at the rank of an active leaf of the direct clan creates a change at vertices at the appropriate ranks at the ancestral sites. The resulting system of connected sets of sites is described by a complex dynamic of the *clan coalescence-fragmentation process*.

For the analysis of our transition regime we focus on the early stages of coalescence involving only sets formed by a small number of clan coalescence events. In the proof of the convergence theorem below we need to determine the set of vertices and the number of sites changed simultaneously when the rank corresponding to the active vertex is subject to a transition. Since we are here concerned with transition regime it suffices to prove this when the proportion of occupied sites in less than some given $u^* > 0$.

In our assumptions in the subsequent lemma on simultaneous changes we work with the fact that the number of occupied sites in $\tilde{\mathcal{G}}^{++,N}$ can be bounded above by the one in $\tilde{\mathcal{G}}^{++}$. We call

$$\text{CMJ-majorant of a direct clan} \tag{9.465}$$

this stochastic upper bound.

The required bounds on clan sizes and sets of sites at which a simultaneous transition can occur in the collision regime is the following.

Lemma 9.53 (Collision regime: Simultaneous transition sets).

Let T_u^N denote the first time when the proportion of occupied sites in $\{1, \ldots, N\}$ is $u \in (0, 1)$. Given $u_0 \in (0, 1)$, and $t_0 > 0$ consider the time interval of times s given by

$$s \in [0, T_{u_0}^N + t_0]. \tag{9.466}$$

Given u_0, t_0 let u_* be chosen so that

$$t_0, u_0 \text{ such that } e^{\beta t_0} u_0 < u^*. \tag{9.467}$$

9.9 Dual in the Collision Regime: $\tilde{\mathcal{G}}^{++,N}$ and $\mathfrak{s}^{****,N}$

Let

$$|\tilde{\gamma}^N|_{u_0,t_0} \tag{9.468}$$

denote the size of the direct clan initiated by a migrant from a randomly chosen occupied site among those which are between successive decoupling events during the time interval specified in (9.466).

(a) Reverse clans and simultaneous transitions.

Let $\mathfrak{n}^N$ denote the random number of direct clans in $\{1,\ldots,N\}$ involved in a simultaneous transition triggered by a transition at an active site of a reverse clan in the time interval (9.466).

(1) In the general case one can choose u_0 sufficiently small so that

$$\lim_{N\to\infty} E[(\mathfrak{n}^N)^2] < \infty. \tag{9.469}$$

(2) Under condition (9.401) the total number of sites at which a simultaneous transition can occur due to a transition at an active site of a reverse clan in the time interval (9.466) has a stochastic bound with finite second moment.

(b) Connected components in the mutation-dominance regime.

Let $p_k = p_k(t_0)$ denote the probability that the CMJ-majorant of a direct clan contains k sites up to the reference time t_0 (with the clan founded at time 0). Assume that the parameters (c,d,s,m) are such that

$$\sum_k p_k e^{\lambda k} < \infty \text{ for some } \lambda > 0. \tag{9.470}$$

Then there exists $u^* > 0$ such that if we choose

$$u_0, t_0 \text{ such that } e^{\beta t_0} u_0 + t_0 < u^*, \tag{9.471}$$

under (9.470) the law of $|\tilde{\gamma}^N|_{u_0,t_0}$ has an exponential moment uniform in N, that is, there exists $\kappa > 0$ such that

$$E[e^{\kappa|\tilde{\gamma}^N|_{u_0,t_0}}] < K < \infty \quad \text{for all } N \in \mathbb{N}. \tag{9.472}$$

If the condition (9.470) is weakened to

$$\sum_k k^3 p_k < \infty, \tag{9.473}$$

(which is implied by the mutation-dominance condition (9.392)), then (9.472) is replaced by

$$\sup_N E[|\gamma^N|^2_{u_0,t_0}] < \infty. \tag{9.474}$$

(c) Clan sizes

Assume the parameters c, d, s, m are such that (9.470) holds and let

$$\Gamma_{t_0, u_0} \tag{9.475}$$

be the collection of clans we can observe in our time interval as described in (9.460) and satisfying (9.467).
Then there exists $C > 0$ such that

$$\limsup_{N \to \infty} P^N[\max\{|\gamma^N|, \gamma^N \in \Gamma^N_{t_0,u_0}\} > C \log N] = 0. \tag{9.476}$$

If we assume only (9.473), then for $\varepsilon > 0$,

$$\limsup_{N \to \infty} P^N[\max\{|\gamma^N|, \gamma^N \in \Gamma^N_{t,u}\} > CN^{\frac{1}{2}+\varepsilon}] = 0. \quad \square \tag{9.477}$$

Remark 168. The case described in (c) equation (9.476) corresponds to the basic result in the theory of Erdős-Rényi graphs in the regime before the emergence of the giant component occurs.

One can now ask whether the result of (a) is true for all $u_0 < 1$ or whether we can get a simultaneous transition at $O(N)$ sites for some u larger than u_0. We conjecture that the latter does not happen. However since for our purposes it suffices to describe the evolution up to T_u for some $u > 0$ and so we do not address this question here.

Proof of Lemma 9.53.

(a) We want to bound the number of simultaneous transitions within a linked clan. Here backward and forward links have to be considered to obtain the affected sites. We first have to control (1) the possible founder sites playing a role in our transition and (2) the affected sites in the corresponding direct clans.

(1) Consider a fixed tagged leaf in a direct clan. A reverse clan associated to this vertex is given by a backward-connected set of vertices starting with the leaf and ending in one or more founder sites. More than one founder site can be involved in the case of collisions, namely if the concerned leaf had previously been at a site which was hit by another migrant rank and coalescence between these active ranks had occurred.

Note that this exhibits a branching structure and a branch occurs if during its excursion the path followed by this rank occupies a site at which another migrant hits and then this new migrant coalesces with the tagged rank.

Since we are considering the time interval (9.466), the number of occupied sites in $\{1, \ldots, N\}$ is not larger than u^*N. In addition each occupied site i at time t independently produces migrants at rate $cn_i(t)$ given by (9.462), (where $t \leq t_0$) and recall from Section 8 that $n_i(\cdot)$ can be bounded by a birth and death process

9.9 Dual in the Collision Regime: $\tilde{\mathcal{G}}^{++,N}$ and $\mathfrak{s}^{****,N}$

with linear birth rate, quadratic death rate and immigration in equilibrium where the immigration rate is a random variable converging to a constant and all moments converge. (cf. (9.463),(9.464))

Therefore each vertex in the path in the reverse clan could have been hit with this rate and then coalescence could have produced a linkage. The reverse clan then has the structure of a binary branching process. The law of the number of branches at a vertex in a time period $[t, t + \Delta t]$ in given by a $Pois(\frac{c}{N} \sum_{i=1}^{N} \int_{t}^{t+\Delta t} u(s) n_i(s) ds)$ and therefore has first moment bounded by $b(u^*) \Delta t$ where $b(u^*) = \text{const} \cdot u^*$ by (9.464).

The branching tree in reverse time has branching rate bounded by $b(u^*)$ and terminates at the resolution time of the active individual. Therefore we can choose $u^* > 0$ sufficiently small so that the second moment of the number of branches in the reverse clan is bounded by (recall that $\hat{\lambda}$ is a finite exponential moment of the resolution time at the tagged leaf)

$$\int_{0}^{\infty} e^{2b(u^*)t} \hat{\lambda} e^{-\hat{\lambda} t} < \infty \qquad (9.478)$$

(uniformly in N). Since $\mathfrak{n}^N$ corresponds to the number of vertices in the branching tree, the result follows.

(2) The set of other vertices which are affected by a transition is comprised of the direct clan together with the set of sites which have previously sent migrants which coalesced with this rank. This implies that under condition (9.401), a change at the rank at a leaf can result in changes at a random number of sites with distribution having a second moment bounded due to (9.461).

(b) In order to consider the connected components of the directed graph we note that in addition to the direct clan additional sites can belong to a connected set if a direct migrant clan from another site collides with the direct clan from the original tagged site and subsequently the two corresponding ranks coalesce or the direct clan hits a direct clan which hits the tagged one etc. So we focus next on direct connections only. We shall first consider the direct connections, then the second order one, etc.

We begin with some preparation. We obtain a stochastic upper bound for the size of a direct clan by using the CMJ-bound on it over the considered time interval. This results in a distribution which is independent of u_0 and only depends on the considered time span.

Since we have assumed (9.473) and hence $\sum k p_k < \infty$, the probability that a randomly chosen site belongs to the tagged direct clan is $O(\frac{1}{N})$. A direct tagged ghost clan of size ℓ can be hit by a direct ghost clan of size k starting at another site with probability bounded above by $\frac{\ell k}{N}$ since each of the sites in the clan can coincide with a site in the tagged clan. If the two active ranks which are occupying the same site coalesce during their mutual sojourn at the site, then they would form a single label leading to a *compound clan*. This produces a collection of sites at which a simultaneous transition could occur. (For example, a mutation (at a rank located at a leaf) in the clan can eliminate ranks at each of the sites in the clan.)

We now consider the connections of the tagged clan via direct collisions and subsequent coalescence. Consider the formation of a new direct clans at the set of occupied sites. The rate at which direct clans of size k are produced is $uNp_k i$ if u is the current intensity. Let

$$\{Y_i, i = 1, \ldots, \lfloor uN \rfloor\} \tag{9.479}$$

denote the sizes of the direct clans attached to the $\lfloor uN \rfloor$ occupied sites when the current proportion of occupied sites is u. The probability that a collision-coalescence occurs between one of these Y_i and a direct clan of size ℓ at a tagged occupied site can be bounded above by $\frac{r\ell Y_i}{N}$ for some $r \in (0, 1)$ (since here we need a coalescence to happen). Let $\tilde{Y}_i$ denote the indicator of the the event that this occurs.

We then want to obtain an upper bound for the exponential moment of

$$\hat{Y} = \text{total number of sites in set of direct clans that merge with given direct clan,} \tag{9.480}$$

which is given by:

$$E[e^{\lambda \hat{Y}} | Y_0 = \ell] = E[e^{\lambda \sum_{i=1}^{\lfloor uN \rfloor} \tilde{Y}_i Y_i} | Y_0 = \ell]. \tag{9.481}$$

Since the events that the direct clans fuse with the tagged direct clan are independent we have

$$E[e^{\lambda \hat{Y}} | Y_0 = \ell] \tag{9.482}$$

$$\leq E\left[\prod_{i=1}^{\lfloor uN \rfloor} \left(1 + \frac{r\ell Y_i}{N}(e^{\lambda Y_i} - 1)\right)\right] = \prod_{i=1}^{\lfloor uN \rfloor} E\left[\left(1 + \frac{r\ell Y_i}{N}(e^{\lambda Y_i} - 1)\right)\right]$$

$$= \left(1 + \sum_k \frac{r\ell k}{N} p_k (e^{\lambda k} - 1)\right)^{\lfloor uN \rfloor}$$

$$\longrightarrow e^{r\ell \sum k p_k u (e^{\lambda k} - 1)}, \text{ as } N \to \infty.$$

Since we assumed that $\sum k p_k e^{\lambda k} < \infty$ for some $\lambda > 0$, it follows that the number of sites with direct clans that link to the tagged direct clan has an exponential moment and the mean number of sites in the resulting clan is $ru\ell \sum k^2 p_k$.

We have now completed the first step in obtaining the clan associated to a given site during an excursion in which the largest direct connected set linked to the tagged site involving either 0 or 1 single coalescence event. We must now consider *second order connections*, namely direct clan that do not hit the initial direct clan but hit (and coalesce) one of the other direct clans that does hit the initial direct clan. In other words we now consider the clan involving one and two coalescence events. This can be iterated for higher-order connections using the same idea. We now consider the sequence of these random variables corresponding to i coalescence events given a sequence denoted

$$(\hat{Y}_i)_{i=1,2,\dots}. \tag{9.483}$$

We obtain a branching process having the offspring distribution function given by the law of $\hat{Y}$. Since the mean offspring size is $\ell r u (\sum_k k^2 p_k)$, there then exists $u_* > 0$ such that for $u < u_*$, this branching process is *subcritical*. The the resulting connected set of sites corresponds to the total progeny $(\hat{Y}_1 + \hat{Y}_2 + \dots)$ which has an exponential bound (see e.g. [H], 13.2).

To prove (9.474) we carry out the same calculations using the Laplace transform and obtain the second moment by differentiating twice.

(c) This follows from (9.472) by the exponential Chebyshev inequality. To verify this let γ_i^N denote the largest clan which includes site i. Then by (b)

$$P^N[[\max\{|\gamma^N|, \gamma^N \in \Gamma_{t_0,u_0}^N\} > C \log N] \tag{9.484}$$
$$= P^N[\cup_{i=1}^N |\gamma_i^N| > C \log N]$$
$$\leq \sum_{i=1}^N P^N[|\gamma_i^N| > C \log N]$$
$$\leq \frac{NE[e^{\kappa|\gamma^N|}]}{e^{C\kappa \log N}} \leq NKe^{-C\kappa \log N}.$$

The result then follows by taking C sufficiently large and noting that the number of different linked clans at a fixed time that have been initiated at a tagged site also has an exponential moment.

Similarly, (9.477) follows from (9.474) and Chebyshev's inequality. q.e.d.

9.10 Empirical Process of Microstates, Clans for $\mathfrak{s}^{****,N}$: Mutation-Dominant Regime

Recall that in the case $M = 1$ we considered the empirical distribution of microstates (which were simply the particle numbers at the various sites) and identified the limiting nonlinear dynamics in time window $\{\frac{\log N}{\alpha} + t\}_{t \in \mathbb{R}}$. In the case $M > 1$, this becomes more complex since particles carry now an *internal state*.

In this section we formulate and prove the convergence results for the dual process as $N \to \infty$ in the collision regime for the range of parameters

$$c, d, s, (m_{i,j}) \text{ in the } mutation\text{-}dominance \text{ } parameter \text{ } regime \tag{9.485}$$

for times of the form $\beta^{-1} \log N + t$ with $t_0 \leq t < t_1$ and later in Subsection 9.11.1 we shall consider the behaviour as $t_0 \downarrow -\infty$. The selection-dominant case is discussed in Subsubsection 9.12.

In this subsection we will primarily work with the process $\mathfrak{s}^{****,N}$ (which of course determines the dual but contains more information) rather than $\tilde{\mathcal{G}}^{++,N}$ itself. We first introduce for $\mathfrak{s}^{****,N}$ the *empirical process of microstates* and its limit (as $N \to \infty$) process with its properties and finally prove the convergence ($N \to \infty$) in the time window $\{\beta^{-1} \log N + t\}_{t \in \mathbb{R}}$ to a random entrance law associated with an appropriate McKean–Vlasov equation. The point here will be to extract the information on the enriched microstate needed for the N evolution in the $N \to \infty$ limit. Hence our goal now is to write down the *nonlinear evolution equations* governing the dual for the emergence-fixation regime.

9.10.1 Empirical Process of Microstates in the Mutation-Dominance Regime

We introduce here the two key objects the microstates of $\mathfrak{s}^{****,N}$ and the empirical process of microstates of $\mathfrak{s}^{****,N}$. The point is now that in the limit $N \to \infty$ we need not work with all the information contained in the *enriched* microstate needed to define the evolution of $\mathfrak{s}^{****,N}$ as a Markov process.

We proceed in two steps, first we introduce the description of the microstates and the corresponding state spaces for our process, in the second step we introduce the quantities describing the evolution of these states based on the empirical measure and density of microstates. The resulting limit dynamics are described in the next subsection.

Step 1 *(State description for empirical measure of microstates)*

We recall that the state of the process $\mathfrak{s}^{****,N}$ is specified by the set of mesostates corresponding to the labels (including the lists $\Lambda_A, \Lambda_D, \mathfrak{C}_A$) and the list of sets of (ranks, sites) with their connections as specified in a random graph. This system is Markov. Recall that the state space for $\mathfrak{s}^{****,N}$ is $\mathfrak{S}_N^{****}$.

However as we have noted above, as a result of migrations and *collisions*, the state at a single label (corresponding to an irreducible tableau) of $\tilde{\mathcal{G}}^{++,N}$ does not have product form, that is cannot be written in the form $\prod_{j \in S} \zeta^N(j,t)$ with $\zeta^N(j,t)$ the internal states (i.e. columns of the attached ranks) at site j at a given time t, and in addition simultaneous changes of the internal states at many different sites can occur, namely for many sites in a label, when a rank at a leaf of the random graph resolves. In fact mesostates at different labels can now be coupled and factorize only as $N \to \infty$. In particular *irreducible factors may now consist of several labels*.

Recall that in Section 8 we formulated the limiting deterministic evolution equation for the *empirical measure* of internal states over the collection of occupied sites. We want to do something similar here. We consider a label and the sites

9.10 Empirical Process of Microstates, Clans for $\mathfrak{s}^{****,N}$: Mutation-Dominant Regime

associated with it. We now have to *describe the information on the mesostate of that label which concerns only a specific site*, the microstate so that we can describe the evolution of this microstate effectively and in such a way that in the limit $N \to \infty$ we obtain a *Markov* process for the evolution of the *empirical measure of microstates*.

In order to formulate the empirical process over the sites of this process, we must first determine a state space which is based *only* on information on the microstates associated with each of the sites and not on *joint information* involving two or more sites. For fixed N, and projection to the age and state at each site (this map denoted pr) the resulting image empirical process of the process $pr(\mathfrak{s}^{****,N})$ is no longer Markov due to the coupling of different sites of a label (or even between different labels) as long as they do not factor. This is the key problem.

However the key idea is that the *Markov property reemerges in the $N \to \infty$ limit*. In order to exhibit this effect we now isolate the information on the mesostate that is needed to define autonomously the evolution of the empirical measure of microstates for finite N and which then as $N \to \infty$ becomes a functional of the empirical distribution of microstates.

We now first make a couple of modifications that are needed to define the *empirical process of microstates*. The required space

$$\bar{\mathfrak{T}} \text{ of possible microstates} \tag{9.486}$$

arises from the complete state description by an element in $\mathfrak{S}_N^{****}$ by dropping meso- and global state information as much as possible, but retaining enough information for the evolution equation of the empirical measures. This is necessarily somewhat complex. The state space of (9.486) will be specified in (9.495) below after introduction of all the needed ingredients.

We shall need for the evolution of the empirical measure of the microstates (0) the tableau at this site and furthermore two pieces of information, (i) description of the *immigration* into the microstate itself and the *emigration* from it and (ii) information about all the parent sites of the ranks at a site. Then we can combine all this in (iii) to get the needed state space.

We must include detailed information concerning the (i) *arrival of immigrants* at a site, that is, information about the state of the parent site of the immigrants at the migration time and (ii) the *subsequent evolution of the parent site* up to the decoupling time. For the description of occupied and active sites we need the following four different lists of objects given in (9.487) to (9.492) and below. Begin with the point (i).

(i) Information on internal state at site (microstate)

We abbreviate in the sequel $\Lambda_A(k, k^*)$, etc. by $\Lambda_A(k^*)$ since the label plays no role. Begin with the immigration information. For $\Lambda_A(k^*)$ we remove the information on the label k in $\Lambda_A(k, k^*)$ concerning the location of sites different from k^*. Similarly we proceed with the collection of lists $\Lambda_D, \mathfrak{C}_A$.

First, at a site k^* consider the ranks which have immigrated indexed by $j \in \mathbb{N}$, more precisely, indexed by $1, 2, \ldots$ in the order of arrival and furthermore the ranks belonging to their families, i.e. the ranks of their descendants created upon the action of the selection operator:

$-\mathsf{t}_j \in \mathcal{T} = $ current associated *tableau* of the jth immigrant rank $\in \Lambda_A(k^*)$

(i.e., descendant family of immigrant rank)

$-\mathfrak{C}_A(k^*) = \{(j, r_j; j', r_{j'})\} = $ set of pairs of coalesced ranks

from different immigrant families j, j'

− the age of the site is defined to be the time since the arrival of the (9.487)

first immigrant family at the site.

Finally individuals are migrating out of the site k^* and this information is collected in:

$v_{k^*} \in \Upsilon$ which gives the description of individuals that have emigrated

from site k^*

as in the definition of s (see (9.237)). (9.488)

This information is contained in $\Lambda_D(k^*)$,

(ii) Information on parent state (microstate):

Next consider the *parent state* information for a site $k^*\mathrm{m}$ which in fact may include information on parent of parent sites, etc.. This is coded by the objects (here i indexes the immigrants)

$$\{\vec{\mathsf{t}}_i : i \in \Lambda_A(j^*) \text{ for the site } j^* \in \{1, \ldots, N\}\}, \quad (9.489)$$

with $\vec{\mathsf{t}}_i$ taken from a set $\tilde{\mathfrak{T}}_{\mathrm{par}}$ introduced below.

The parent information is needed to determine whether a migrant family has or has not been removed by a resolution at the parent site k_{par}. This depends on the rank at the parent site that was responsible for initiating the migrant family and on the tableau at the parent site to the left of this rank. For example in the case in which the second rank of

9.10 Empirical Process of Microstates, Clans for $\mathfrak{s}^{****,N}$: Mutation-Dominant Regime

$$\begin{pmatrix} (010) & & & & \\ (100) & (110) & & & \\ (100) & (100) & (110) & & \\ (100) & (100) & (100) & (110) & \\ (100) & (100) & (100) & (100) & (110) \end{pmatrix} \quad (9.490)$$

migrates and produces a migrant family only a mutation at rank 1 can remove the migrant family and ranks $3, 4, 5$ at the parent site are not needed. In turn whether or not the migrant family is removed in this way can depend on a migrant family at the parent site k_{par}, and so on. Thus we must search back to the ancestral site $k_{\text{anc0}} \in \bar{\tilde{\mathfrak{T}}}_{\text{anc0}}$ at which $\vec{t}_{\text{anc0}} = \emptyset$. Here

$$\bar{\tilde{\mathfrak{T}}}_{\text{anc0}} \quad (9.491)$$

denotes the set of such states associated with exactly one label, that is, the offspring of a founding migrant (having a resolved parent site) at a given site.

The complete parent state information has then the form:

$$\vec{t}_i = (\bar{\tilde{t}}, \ell, \tau^*) \quad \text{for an } \textit{un}\text{resolved family} \quad (9.492)$$

$$\vec{t}_j = \emptyset \quad \text{if the family is resolved,}$$

where the rank $\ell \in \mathbb{N}$ and $\tilde{t}$ is the substate associated to the tableau to the left of this rank in the state $\bar{t} \in \bar{\mathfrak{T}}$ at the parent site, this includes the tableau to the left of ℓ, the partition at rank ℓ and the corresponding subsets of Λ_A, Λ_D and $\mathfrak{C}_A$. and $\tau^* \in T^*$ is the partition at rank ℓ. We denote by

$$\bar{\tilde{\mathfrak{T}}}_{\text{par}} \quad (9.493)$$

the image of $\bar{\tilde{\mathfrak{T}}} \times \mathbb{N} \cup \{\emptyset\}$ defined by (9.492). Note that in general $\vec{t}_j$ can in turn refer back to its parent sites and individuals, that is, $\Lambda_A(j)$ can contain a number of migrants which remain unresolved which *again* can refer back to their parent sites. For any time t this chain of back references is finite. We have to be careful once we let t together with N tend to infinity.

Remark 169 (Self-collisions can be ignored).

In principle for finite N, the information contained in (9.487) and (9.492) would not be enough to determine the further evolution if further migrants would arrive from the same decoupling cycle at a fixed site. This is the case because since we have then no information on their connection.

However since in the limit $N \to \infty$, the set of such self-intersections occurs in at most $O(1)$ sites, these do not contribute to the limiting empirical measure and therefore we can ignore these for our limiting nonlinear equations and for emergence and include them in our finite N equations as error terms in N. More rigorously speaking we can define a Markovian dynamics which suppresses this dependence and of which we know from the argument just given that asymptotically we get the same limit dynamic.

(iii) State space

We assume here that no further interaction with the parent sites of the immigrants occurs and note that this assumption is valid in the $N \to \infty$ limit (See Remark 169). Then for the remaining sites the information we extracted in (i) and (ii) together with the development of the family initiated by the migrating rank is then sufficient to determine the future evolution at the tagged site. Similarly in the $N \to \infty$ limit we can assume that the future evolution of a migrant rank from a tagged site (given by a member $v \in \mathcal{Y}_{\vec{\imath}}$) can be viewed as exogenous to the evolution at other ranks at the tagged site.

To define *countable* state space we shall need a condition on the possible states. We require the

finite depth condition (9.494)

for each i either $\vec{t}_i$ is resolved

or it is the ℓth generation descendent of a resolved family for some $\ell \in \mathbb{N}$.

Definition 9.54 (State space of microstate process).

(a) We define the *state space* of a tagged occupied active site k^* in a label k, the space of *microstates*, as follows:

$$\bar{\mathfrak{T}} := \left\{ \bar{\mathfrak{t}} = \left(\{(\vec{t}_i; t_i) : \vec{t}_i \in \bar{\mathfrak{T}}_{\mathrm{par}}, t_i \in \mathcal{T}, i \in \Lambda_A \}, \right. \right. \tag{9.495}$$

$$\Lambda_A(k,k^*), \Lambda_D(k,k^*), \mathfrak{C}(k,k^*),$$

and the finite depth condition (9.494) is satisfied $)\}$

where $\Lambda_A(k,k^*), \Lambda_D(k,k^*)$ are defined as in (9.428), respectively, (9.426).

Now the mappings ξ, respectively $\tilde{\xi}$, as well as as well as the information on the label k to which k^* belongs **are omitted**.

The state just defined contains the information on microstates important for the study of the limiting empirical measure in the $N \to \infty$ limit but doesn't contain information on the clan structure.

(b) A site is said to be *inactive* (unoccupied) if the corresponding factor is $\mathbb{I}_M^\mathbb{N}$, we then write its state as $((\emptyset, \mathbb{I}_M^\mathbb{N}), \emptyset, \emptyset, \emptyset)$ and its age is set to be $\dagger$. When an arrival takes place at an unoccupied site its age is changed from $\dagger$ to 0 at the arrival time. □

Remark 170. Note that the self-reference involved in the definitions (9.492) and (9.495) is resolved under the finite depth hypothesis.

9.10 Empirical Process of Microstates, Clans for $\mathfrak{s}^{****,N}$: Mutation-Dominant Regime 547

Table 9.3 Basic state spaces mutation-dominant regime

State space	Process	Eqn. no.
$\mathfrak{S}_N^{****}$	$\mathfrak{s}^{****,N}$	(9.437)
$\mathbb{R}_+ \times \mathfrak{T}^{****}$	Enriched microstate process	(9.423)
$\mathbb{R}_+ \times \bar{\mathfrak{T}}$	Microstate process	(9.495)
$\mathcal{M}_{\text{fin}}([0,\infty) \times \bar{\mathfrak{T}})$	$\bar{\Psi}^N$	(9.499)

The space $\bar{\mathfrak{T}}$ together with the age of the site will serve as the state space for a microstate at a site for the purpose of defining the empirical process and is denoted

$$(\mathbb{R} \cup \{\dagger\}) \times \bar{\mathfrak{T}}. \tag{9.496}$$

The second component is equipped (as countable set) with the discrete topology and $\mathbb{R} \cup \{\dagger\}$ with the Euclidian topology and augmented with an isolated point. Hence the state space is a Polish space.

Remark 171. Here are some observations before we continue. For finite N for the process $\mathfrak{s}^{****,N}$ the dynamics of $\vec{\mathfrak{t}}_j$ mirrors the dynamics at the parent site. In the limit $N \to \infty$ the dynamics is given by an *independent* copy of the dynamics starting with the configuration at the parent site. Similarly, in the $N \to \infty$ limit the dynamics of the ranks in contained in the parent state information are given by running the basic process independently of the other ranks at the tagged site up to the decoupling time. This will be made more precise below.

We note that the structure mentioned above reveals that a microstate contains a *reduced microstate* which contains the information which is relevant in the $N \to \infty$ limit, where we can in particular drop the links to the parent site. For convenience we summarize the different spaces involved and indicate where their definitions appear in the following Table 9.3.

Step 2 *Empirical quantities*

We now introduce the empirical quantities. First, let

$$\bar{K}_t^N = \sum_{k^*=1}^N 1\left(\mathfrak{t}_j(t) \neq \mathbb{I}_M^\mathbb{N} \text{ for some } j \in \Lambda_A(k^*)\right) = \text{ the number of occupied}$$

sites at time t. \hfill (9.497)

Let $\mathcal{M}_{\text{fin}}([0,\infty) \times \bar{\mathfrak{T}})$ (recall 9.495)) denote the space of finite measures on $[0,\infty) \times \bar{\mathfrak{T}}$, that is, the ages and microstates of occupied sites. We now consider the mapping from the global state of $\mathfrak{s}^{****,N}$ to the collection of a microstates

$$pr : \mathfrak{S}_N^{****} \to ([0,\infty) \times \bar{\mathfrak{T}})^N, \tag{9.498}$$

which gives the current age and microstates at each site *but excluding information as to the* state *and locations of the linked sites.*

In order to verify that the process $pr : \mathfrak{S}_N^{****}$ can be realized on the space $\tilde{\mathfrak{T}}$ the key point is that in the backward view each backward reference requires a migration time. Since each earlier parent site has a finite decoupling time, the probability that there are ℓ backward references can be dominated by a geometric random number.

Next we introduce the empirical process $\bar{\Psi}^N(t, ds, d\bar{\mathfrak{t}})$ given by the $\mathcal{M}_{\text{fin}}([0, \infty) \times \tilde{\mathfrak{T}})$- valued random variable:

$$\bar{\Psi}^N(t, ds, d\bar{\mathfrak{t}}) = \sum_{k^*=1}^{N} 1\left(t_j(t) \neq \mathbb{I}^{\mathbb{N}} \text{ for some } j \in \Lambda_A(k^*)\right) \delta_{(prs^{****,N}(t))(j)}(ds, d\bar{\mathfrak{t}}).$$
(9.499)

We introduce the corresponding scaled object:

$$\hat{\Psi}^N(t) = \frac{\bar{\Psi}^N(t, \cdot, \cdot)}{N}.$$
(9.500)

Note that for fixed N this is indeed an element of $\mathcal{M}_{\text{fin}}([0, \infty) \times \tilde{\mathfrak{T}})$.

Next we introduce the pair $(\bar{u}^N, \bar{U}^N)$ as a functional of $\bar{\Psi}^N$:

$$\bar{u}^N(t) = \frac{\bar{K}_t^N}{N} = \bar{\Psi}^N(t, [0, \infty) \times \tilde{\mathfrak{T}}), \quad \bar{U}^N(t, \cdot, \cdot) = \frac{\bar{\Psi}^N(t, \cdot, \cdot)}{\bar{u}^N(t)}.$$
(9.501)

As in the $M = 1$ case, we often work with the marginal

$$\bar{U}^N(t, \bar{\mathfrak{t}}) = \bar{U}^N(t, \mathbb{R}_+, \bar{\mathfrak{t}}).$$
(9.502)

Recall that without the additional information on the sets of sites and the random graph which are attached to one label the process $(\bar{\Psi}^N(t))_{t \geq 0}$ is *not Markov*. However we will prove below that in the "mean-field" limit $N \to \infty$ do we recover a Markovian dynamics.

Remark 172. The key ideas in the simplification obtained in the mean-field limit below are the following.

(1) Consider a typical site. In the limit $N \to \infty$ when a new migrant site arrives, then the parent site can be viewed as a typical occupied site of a randomly chosen rank (i.e. it is chosen with weights depending on the numbers of active ranks at these sites) and the chosen site can be assumed to be decoupled from the considered site. This means that no further immigrant will arrive from the same parent site before decoupling of the first immigrant from that site. In terms of clans this means that the clans do not have *self-intersections*. The latter is a consequence of the fact that the proportion of the N sites that can appear in a

9.10 Empirical Process of Microstates, Clans for $\mathfrak{s}^{****,N}$: Mutation-Dominant Regime

clan tends to zero in the $N \to \infty$ limit. This is the *propagation of chaos* effect for this system.

(2) When a migrant leaves a tagged site at the decoupling time, then the site to which it migrates (and subsequent associated clan sites) can be viewed as randomly chosen and its future evolution is independent of the evolution at the tagged site.

(3) In taking the limit $N \to \infty$ we can assume that the evolution of the parent site states of two or more immigrant families at a tagged site are independent and therefore their lifetimes and resolution states are independent. We can do this because (as for self interactions) the probability that two (or any bounded number of) migrating sites interact, that is, receive unresolved migrants from the same interval between decoupling events at a fixed site, is $O(\frac{1}{N})$. This means that in a fixed time interval this occurs in at most $O(1)$ of the N sites and these do not contribute to the limiting empirical distribution.

(4) The state space $\bar{\bar{\mathfrak{T}}}$ allows us for the above reasons to include the necessary information to determine the future evolution of the considered site in the $N \to \infty$ limit.

As indicated above in the limit $N \to \infty$ there are two types of transitions: (1) internal transitions at a permanent site and (2) the transition that occurs when a decoupling event happens at a site. The former are the result of selection, mutation, coalescence and migration. The latter can occur due to coalescence or mutation involving the leading active rank.

We must spell out the possible immigration events that can occur at an occupied site. We note that the probability that a given migrant from an occupied site will be removed at the decoupling time and the lifetime is a measurable function of the state at the parent site at the time of migration. At a decoupling event one or more sites can become permanent. Note that a permanent site which is a concept from the *collision-free* systems is not necessarily immortal under the dynamic of the finite N-system including collisions - a permanent site can be deleted if it migrates and collides with another permanent site.

Note that the links prior to decoupling between the affected sites are contained in the *linked ghost clan* information but *not* in the empirical measure of microstates. In particular if a migrant rank is removed at the parent site this must be removed at the other occupied site. At a decoupling event simultaneous changes occur at the parent site and one or more linked sites, in particular the immigrant and its descendants at the new site are removed from the list Λ_A and the corresponding member is removed from Λ_D at the parent site.

Remark 173. An alternate approach is to work with the empirical distribution of the mesostate including information as to coupling with other microstates which however works only in the selection-dominant case (see Subsection 9.12).

9.10.2 Limiting Nonlinear Dynamics for Microstate Frequencies Under Mutation Dominance

We now want to formulate the nonlinear equations for a pair $(\bar{u}, \bar{U})$ arising in the limit $N \to \infty$ from $(\frac{\bar{u}^N}{N}, \bar{U}^N)$ (recall (9.501)) and (9.502). This again involves the law of a tagged site in the $N \to \infty$ limit and a propagation of chaos. One point here is that this limit dynamic is given by a nonlinear McKean–Vlasov evolution on $\mathcal{M}([0, \infty) \times \tilde{\mathfrak{T}})$ and the linked ghost clan structure no longer plays a role. Namely the idea here is that if the clans are of order $o(N)$ then we have a propagation of chaos in that a colliding migrant can be viewed as colliding with a site randomly chosen from the set of occupied sites and the evolution at these two sites is independent after the collision. Then in the limit dynamics we do *not* need to keep track of ghost ranks in the sense that they can be mimicked by a randomly chosen rank according to the empirical measure. Therefore in the limit $N \to \infty$ the empirical measures of microstates evolve *autonomously* without reference to richer structures and hence this limiting dynamic of a tagged site has state space $\tilde{\mathfrak{T}}$ defined by (9.495).

To prepare the formulation of the limit dynamic we rephrase some key effects of the dynamics. We focus on a *tagged site*. Then we will have arrivals of factors from a *randomly chosen site and rank*. In the limit the corresponding parent site dynamics is independent of our tagged site. By convention, arrivals are placed at the first unoccupied local rank. Recall that the arrivals are independent and therefore at the arrival time the new rank and the tableau existing at that site factor (provided they are not migrants from the same site - note that as $N \to \infty$ the probability that a migrant ever returns to the originating site is $O(\frac{1}{N})$ and therefore is negligible).

The number of occupied sites can change via the following three different transitions:

- The number of occupied sites can *increase* since a new site can be occupied by the arrival of a immigrant from an existing occupied site
- The number of occupied sites can *decrease* since a site can die if all the immigrant families at that site are removed by resolution at their parent site(s).
- The number of occupied sites can *decrease* since a site can die if a single active rank at that site migrates and collides with an existing occupied site.

On the other hand remember that our state space allows us to store the following information. For each occupied site we specify the (1) age (time since arrival of the last founding immigrant), (2) immigrant family information and (3) the collision events. Recall that a site is *Pe* if it is occupied by at least one resolved migrant. A collision event is specified by the *arrival time, parent state information* and nontrivial *partition of* $\{1, \ldots, M+1\}$. Recall that the distribution of the lifetime and probability of a survival/removal event at the lifetime is a measurable function of the state at the parent site at the time of migration. This should allow in the $N \to \infty$ limit a *Markovian evolution* of the empirical measure of microstates.

9.10 Empirical Process of Microstates, Clans for s****,^{N}: Mutation-Dominant Regime

Before writing down the equations we describe the $N \to \infty$ limit dynamics informally. Assume that we have for the limiting dynamic a current value of $u \in (0, 1)$ the density of occupied sites and age-type distribution $\mu \in \mathcal{M}_{\text{fin}}([0, \infty) \times \bar{\mathfrak{T}})$, i.e. $\bar{u}(t) = u, \bar{U}(t, \cdot) = \mu(\cdot)$. Then we have:

- arrivals: with rate cu migrants and active ranks are chosen from sites using the empirical measure but weighted with number of active ranks at each site; we then run the migrant up to its decoupling time,
- departures: rate c per active rank - then run to its decoupling time,
- mutation-selection-coalescence acting on each of the set of active ranks at a site,
- at the site decoupling time, there is a single active rank at the site with factor $(11 \cdots 10)$.

We now focus on the *production of migrants* and the ingredients to describe it. A *migrant* will have the form $\vec{t} \in \bar{\mathfrak{T}}_{\text{par}}$, as in (9.492) describing the internal state of the parent site, at the migration time.

The microstate $\bar{t}$ at a site produces migrants from that site of type $\vec{t} = (\tilde{t}, \ell, \tau^*)$ of the form (9.492) at rate

$$\beta_{\vec{t}}(\bar{t}) = c \sum_i 1(\bar{t}_i = \tilde{t}, \bar{t}_i(\ell) = \tau^*), \qquad (9.503)$$

where i runs over all ranks at the site. The total rate of migration from a site with microstate $\bar{t}$ is then given by

$$\beta(\bar{t}) = \sum_{\vec{t}} \beta_{\vec{t}}(\bar{t}). \qquad (9.504)$$

The rate of production for migrants of type $\vec{t}$ at a fixed time t in the population is:

$$\beta_{\vec{t}}(t) = \int_{\mathcal{M}(\bar{\mathfrak{T}})} \beta_{\vec{t}}(\bar{t}) \, \bar{U}(t, d\bar{t}). \qquad (9.505)$$

Here $\beta_{\vec{t}}(t)$ is the arrival rate at a tagged site of migrants $\vec{t}$. Note that at the arrival time we have *the product form* between old and new tableaus and this family is added to the list Λ_A.

Remark 174. Note that these production rates include production of temporary sites that are ultimately removed.

For the limiting equation satisfied by the pair $(\bar{u}, \bar{U})$ we need the following ingredients. Recall $\bar{\beta}(t)$ is given by (9.504) and define the aggregated (over all ages and microstates of sites) production rate of migrants in the total population at time t, $t \in \mathbb{R}$ as

$$\bar{\beta}(t) = \int_{\mathcal{P}(\bar{\mathfrak{T}})} \int_0^\infty \beta(\bar{t}) \bar{U}(t, ds, d\bar{t}), \quad t \in \mathbb{R} \qquad (9.506)$$

and introduce furthermore the rates at time t related to the removal of sites given by

$$\bar{\gamma}_2(t) = c \cdot \text{probability (a site has a single active rank)}, \quad t \in \mathbb{R} \quad (9.507)$$

$$\bar{\gamma}_1(t) = \text{rate of removal of an active site due to resolution}, \quad t \in \mathbb{R}. \quad (9.508)$$

Then the $(\bar{u}, \bar{U})$-equation reads (the terms are explained below):

$$\frac{\partial \bar{u}}{\partial t} = \bar{\beta}(t)\bar{u}(t)(1 - \bar{u}(t)) - \bar{\gamma}_1(t)\bar{u}(t) - \bar{\gamma}_2(t)\bar{u}^2(t), \quad t \in \mathbb{R} \quad (9.509)$$

$$\frac{\partial \bar{U}(t, ds, \vec{\mathfrak{t}})}{\partial t} = -\frac{\partial \bar{U}(t, ds, \vec{\mathfrak{t}})}{\partial s} \quad (9.510)$$
$$+ \sum_{\vec{\mathfrak{t}}' \in \vec{\mathfrak{X}}} \bar{U}(t, ds, \vec{\mathfrak{t}}') Q^0_{\vec{\mathfrak{t}}', \vec{\mathfrak{t}}}$$
$$- \bar{\beta}(\vec{\mathfrak{t}}) \bar{U}(t, ds, \vec{\mathfrak{t}}) + \beta_{\vec{\mathfrak{t}}}(t) 1((s, \vec{\mathfrak{t}}) = (0, \vec{\mathfrak{t}}))$$
$$+ \bar{u}(t) \left[\beta_{(\vec{\mathfrak{t}})}(t) \bar{U}(t, ds, \vec{\mathfrak{t}}') 1\left((s, \vec{\mathfrak{t}}) = (s, \vec{\mathfrak{t}}') \otimes \vec{\mathfrak{t}}\right) \right]$$
$$- [\bar{\beta}(t)(1 - \bar{u}(t)) - \bar{\gamma}_1(t) - \bar{\gamma}_2(t)\bar{u}(t)] \bar{U}(t, ds, \vec{\mathfrak{t}}), \quad t \in \mathbb{R}.$$

Explanation of terms of (9.510).

(1) The first term on the right hand side of (9.510) indicates that the age of the site (time since current founder arrived at the site) increases at the same constant rate.
(2) Here $\vec{\mathfrak{t}}, \vec{\mathfrak{t}}' \in \vec{\mathfrak{X}}$ and $Q^0_{\vec{\mathfrak{t}}', \vec{\mathfrak{t}}}$ collects the transitions due to selection, mutation and coalescence at a tagged site.

An important point is that the dynamics of the $\vec{\mathfrak{t}}$ (the limit of the parent site dynamics) is an independent copy of the one site dynamics - we can suppress the possibilities of collisions here because they do not effect the lifetime or resolution state. Note that the migration operator cuts off coalescence of the resulting migrant block (which is added to Λ_D) - which develops using the same basic mechanisms (further migrations induce a nested structure given by Υ) with the other ranks.
(3) The third line corresponds to the migration of a migrant $\vec{\mathfrak{t}}$ to an unoccupied site.
(4) The fourth line in (9.510) corresponds to the collision of a migrant with a site with state $\vec{\mathfrak{t}}'$ and we obtain the product set $\vec{\mathfrak{t}}' \otimes \vec{\mathfrak{t}}$ (we have the product of the tableaus and this migrant family is added to the list Λ_A).
(5) The fifth line comes from the changes in the normalizing term.

We must define the sense in which equations (9.509), (9.510) have solutions and if they are unique. In order to do so we proceed as in Section 8 and we introduce an appropriate *Banach space* in order to formulate this in the setting of *nonlinear semigroups*. Consider the linear space

9.10 Empirical Process of Microstates, Clans for $\mathfrak{s}^{****,N}$: Mutation-Dominant Regime

$$\mathbb{B} := \mathbb{R} \times \mathcal{M}_{f,\pm}([0,\infty) \times \tilde{\mathfrak{T}}) \text{ with the norm } \|\cdot\|_{\mathbb{B}}, \qquad (9.511)$$

$$\|(u,\mu) - (u',\mu')\|_{\mathbb{B}} = |u - u'| + \sum_{j=1}^{\infty} \left\{ \left(\sum_{\{\bar{\mathfrak{t}} \in \tilde{\mathfrak{T}} : |\bar{\mathfrak{t}}| = j\}} |\mu(.,\bar{\mathfrak{t}}) - \mu'(.,\bar{\mathfrak{t}})|_{\text{var}} \right) (1+j)^2 \right\}, \qquad (9.512)$$

where $|\cdot|_{\text{var}}$ denotes total variation of signed measures on $\mathbb{R}$ and $|\bar{\mathfrak{t}}| = n(\bar{\mathfrak{t}})$ denotes the number of active ranks in the state in $\tilde{\mathfrak{T}}$.

As a preparation for our existence and uniqueness result for the nonlinear equation we need:

Lemma 9.55. *(Related linear equation)*
The linear equation obtained from (9.510) by setting $\bar{u}(t) \equiv 0$ and removing the last (normalization) line corresponds to the mean equations for a countably-many-type CMJ branching process with type space $\tilde{\mathfrak{T}}$. Let $\bar{V}^0(t), t \in \mathbb{R}$, denote the solution of these equations. Then there exists a Malthusian parameter β and stationary type distribution $\bar{U}^0(\infty)$ such that as $t \to \infty$:

$$\frac{\bar{V}^0(t, \bar{\mathfrak{t}})}{\bar{V}^0(t, \tilde{\mathfrak{T}})} \to \bar{U}^0(\infty). \qquad \square \qquad (9.513)$$

Proof of Lemma 9.55. Note that the linear system corresponds exactly to the mean equations for the CMJ process with type space $\tilde{\mathfrak{T}}$ determined by $\tilde{\mathcal{G}}_t^{++}$. In this case individuals can be removed (by resolution at an ancestor). However individuals with states in $\tilde{\mathfrak{T}}_{\text{anc0}}$ are immortal and therefore the process is supercritical.

We claim that starting from a single rank with factor $(1, \cdots, 0)$ we converge to a limit distribution for age-type frequency, which is obtained as a fixed point equation. Then the existence of the Malthusian parameter β is given by the Perron-Frobenius Theorem for positive countable matrices (see Subsubsection 9.12.7), and $\bar{U}^0(\infty)$ is obtained as the solution of a fixed point equation as in Lemma 9.75. In Subsubsection 9.12.7 we will give an extensive and complete argument for the corresponding problem for mesostates. Here the arguments would follow the same line but the keypoint there is to check the so-called positive-1-recurrence condition for $\bar{U}^o(t)$ for some t is simpler since in the present regime we have good bounds on the microstates and their tightness. In particular we know that the the second moment of the number of sites per label is bounded, say by some $L < \infty$, in the mutation-dominant regime without collisions which allows us to use the argument of the proof of Lemma 9.75 - see Remark 197.

Remark 175. Note that $\bar{U}^0(\infty)$ is related to but different from $\bar{\mathcal{U}}^{III}(\infty)$, where $\bar{\mathcal{U}}^{III}(\infty)$ is stable age-type-internal state distribution of the multitype CMJ process arising from type *III* dynamics because the latter included only *Pe* sites, and the present object includes both *Pe* and *Tr* sites. Here we consider a CMJ-process where the individuals correspond to the occupied sites and the types are given by $\bar{\mathfrak{t}}$ which contains the tableau at the site plus the parent site information.

We now return to the nonlinear system which arises in the time window $\left(\frac{\log N}{\beta} + t\right)_{t \geq -\frac{\log N}{\beta}}$ in the limit as $N \to \infty$.

Proposition 9.56 (Wellposedness of nonlinear evolution equation).

(a) The nonlinear equations (9.509), (9.510) for the time index in $[0, T]$ have a unique solution and the state at time $t \in [0, T]$ depends continuously on the initial value at time 0.

(b) The nonlinear system (9.509) and (9.510) has for given $A > 0$ a unique solution satisfying

$$\lim_{t \to -\infty} e^{-\beta t} \bar{u}(t) = A, \quad \lim_{t \to -\infty} \bar{U}(t) = \bar{U}^0(\infty), \tag{9.514}$$

where β and $\bar{U}^0(\infty)$ are given by Lemma 9.55. □

Proof of Proposition 9.56. (a) The proof that $(\bar{u}, \bar{U})$ are uniquely determined through the system of coupled equations (9.509) and (9.510) and condition (9.514) can be established as in Section 8 in the two-type case, more precisely in Subsection 8.3.7.

To get continuity in the initial state we note that the evolution equation (9.509), (9.510) can be viewed as a nonlinear perturbation (described by a vector field) of a linear semigroup on $\mathbb{B}$. Since $\bar{\beta}(\bar{t}), \beta_{((t^*,\ell),\tau^*),\bar{t}}$ grows at most linearly with the number of ranks (ranks at the considered site in t for $\bar{t}$ contains (t_i^*, t_i), i labelling the different arriving migrants, we can easily verify that the vector field is C^2 with first and second derivatives uniformly bounded in norm on bounded sets. To verify this we again note that (9.509), (9.510) can be rewritten as

$$\frac{d}{dt}(u(t), U(t)) = (0, U(t)\tilde{Q}) + F(u, U) \tag{9.515}$$

where $\tilde{Q}$ generates a linear semigroup on $\mathbb{B}$, $F(u, U)$ is given by a third degree polynomial in u and $\beta(U)$ where $\beta(U)$ is linear and differentiable on $\mathbb{B}$. We can then apply Marsden [MA] Theorem 4.17 (see Appendix B) to verify that the solution of equations (9.509), (9.510) defines a strongly continuous nonlinear semigroup.

(b) The proof follows along the same lines as Lemma 8.6 as follows.

First note that for any solution and $t \in \mathbb{R}$ we have

$$\|(\bar{u}(t), \bar{U}(t)) - (Ae^{\beta t}, \bar{U}^{**}(t))\|_{\mathbb{B}} \leq C_1 e^{2\beta t} \text{ for some } 0 < C_1 < \infty, \tag{9.516}$$

where $\bar{U}^{**}(t)$ is the solution of the following linear equation:

$$\frac{\partial \bar{U}^{**}(t, ds, \bar{t})}{\partial t} = -\frac{\partial \bar{U}^{**}(t, ds, \bar{t})}{\partial s} \tag{9.517}$$

9.10 Empirical Process of Microstates, Clans for $\mathfrak{s}^{****,N}$: Mutation-Dominant Regime

$$+ \sum_{\vec{t}' \in \vec{\mathfrak{T}}} \bar{U}^{**}(t, ds, \vec{t}') Q^0_{\vec{t}',\vec{t}} - \bar{\beta}(\vec{t}) \bar{U}^{**}(t, ds, \vec{t}) + \beta_{\vec{t}}(t) 1((s,\vec{t}) = (0,(\vec{t}))$$

$$+ A e^{\beta t} \left[\beta_{\vec{t}}(-\infty) \bar{U}^{**}(-\infty, ds, \vec{t}') 1 \left((s,\vec{t}) = (s, \vec{t}' \otimes \vec{t}) \right) \right]$$

$$+ (\bar{\beta}(-\infty) + \bar{\gamma}_2(-\infty)) A e^{\beta t} \bar{U}^{**}(-\infty, ds, \vec{t}) - (\bar{\beta}(-\infty) + \bar{\gamma}_1(\infty)) \bar{U}^{**}(t, ds, \vec{t})$$

$$- (\bar{\beta}^*(t) + \bar{\gamma}_1^*(t)) \bar{U}^{**}(-\infty, ds, \vec{t}),$$

$$\bar{\beta}^*(\vec{t}) = \int \int \beta(\mathfrak{t})[\bar{U}^{**}(t, ds, d\vec{t}) - U^{**}(-\infty, ds, d\vec{t})], \qquad (9.518)$$

$$\bar{\gamma}_1^*(\vec{t}) = \int \int \gamma_1(\mathfrak{t})[\bar{U}^{**}(t, ds, d\vec{t}) - U^{**}(-\infty, ds, d\vec{t})],$$

with

$$\beta(\vec{\mathfrak{t}}), \gamma_1(\vec{\mathfrak{t}}) \qquad (9.519)$$

are the production rates of new sites in the CMJ-process respectively the removal rates in type $\vec{\mathfrak{t}}$.

Then (9.516) implies that for any two solutions we have

$$\|(\bar{u}_1(t_0), \bar{U}_1(t_0)) - (\bar{u}_2(t_0), \bar{U}_2(t_0))\|_{\mathbb{B}} \leq 2 C_1 e^{2\beta t_0} \qquad (9.520)$$

But then we can solve the equation uniquely for $t \geq t_0$ with the initial state at $t = t_0$ given by $(\bar{u}_\ell, t_0), \bar{U}_\ell(t_0))$, $\ell = 1, 2$ and estimate the growth of the distance by $C_2 e^{\beta(t-t_0)}$ with appropriate $C_2 < \infty$ depending on the two chosen initial states at time t_0. This implies altogether that:

$$\|(\bar{u}_1(t), \bar{U}_1(t)) - (\bar{u}_2(t), \bar{U}_2(t))\|_{\mathbb{B}} \leq C_2 e^{\beta(t-t_0)} \cdot 2C_1 e^{2\beta t_0} = 2 C_1 C_2 e^{\beta t} e^{\beta t_0},$$

$$\text{for } t \geq t_0 \in \mathbb{R}. \qquad (9.521)$$

Uniqueness follows by taking $t_0 \to -\infty$.   q.e.d.

9.10.3 Convergence of the Dynamics of $(\hat{\Psi}^N(\beta^{-1} \log N + t))_{t \geq t_0}$

As in the case $M = 1$ the time of the first rare mutation transition in the dual $\tilde{\mathcal{G}}^{++,N}$ depends on the growth of the number of active ranks. In particular we need to determine now for $M > 1$ the hazard rate for the rare mutation which requires to write down the analogue of the stochastic integral

$$\Pi_t^N = \int_0^t |\zeta^N(t)| dK_u^N, \qquad (9.522)$$

we had in the case $M = 1$. However due to the more complex structure in the case $M > 1$ we need more information which is contained in $(\bar{u}^N, \bar{U}^N)$ and the limiting equations for (u, U). We will develop this in this and the next subsubsections and return in Section 9.13 to the hazard function and its behaviour as $N \to \infty$.

Remark 176. We now give some intuitive ideas behind the proof of the mean-field limit as $N \to \infty$. The key problem are the simultaneous transitions at several sites. However also in the limit a simplification emerges in that we can consider the dynamics of the parent sites of migrants to a tagged site as independent.

To understand this consider the event that a migrant family is removed or becomes permanent at a time t due to a mutation at the parent site. This involves a simultaneous transition at these two sites and therefore the rate of such transitions depends on the state at the parent site. In order to capture this in the empirical process description we included a copy of the appropriate parent site dynamics as part of the description, that is, $\vec{t}$ which is the part of the state $\bar{t}$ containing the parent site information (recall (9.492)). In the $\mathfrak{s}^{****,N}$ description the dynamics of the tableau t at the tagged site is coupled with the dynamics at the parent site.

Focus on this joint transition at tagged site and parent site. In the empirical process description since we do not have the linkage information, we must consider the set of sites consistent with this transition of $\vec{t}$, denoted:

$$\mathcal{A}(\vec{t}). \qquad (9.523)$$

Since all sites in $\mathcal{A}(\vec{t})$ have the same transition rate, then the distribution of the actual linked parent is uniform on this set of sites. Moreover this set of sites is exchangeable.

Note that (assuming no self interaction) the dynamics of $\vec{t}$ for a particular migrating rank depends only on the subsequent selection, mutation, coalescence and migration to the left of the migrating rank and the distribution of the next transition time τ_1 and transition probabilities $\mathfrak{t} \to \mathfrak{t}'$ are measurable with respect to the σ-algebra generated by the process at the parent site. (The probability of self interaction, that is, the collision of a migrant with the parent site or second migrant from the same parent site is asymptotically negligible in the $N \to \infty$ limit.)

Then due to exchangeability and the fact

that at most $o(N)$ sites are linked to any of the $O(N)$ active sites

under consideration, $\qquad (9.524)$

it follows that as $N \to \infty$ the limiting empirical distribution of the next transition time and transition of $\vec{t}$ is given by the actual limiting law of this random objects, i.e.:

$$P(\tau_1 \in ds, \vec{t} \to \vec{t}'). \qquad (9.525)$$

9.10 Empirical Process of Microstates, Clans for $\mathfrak{s}^{****,N}$: Mutation-Dominant Regime   557

To illustrate the law of large numbers effect in this context recall that for $u < u^*$ (given in Lemma 9.52) we have small clans only and then note that for fixed immigrant rank j for all but a set of k which is of order $o(N)$, we know that:

$$E[1_{\tau_{1,j} \leq s, \vec{\mathfrak{t}}_j \to \vec{\mathfrak{t}}'_j} \cdot 1_{\tau_{1,k} \leq s, \vec{\mathfrak{t}}_k \to \vec{\mathfrak{t}}'_k}] = E[1_{\tau_{1,j} \leq s, \vec{\mathfrak{t}}_j \to \vec{\mathfrak{t}}'_j}] \cdot E[1_{\tau_{1,k} \leq s, \vec{\mathfrak{t}}_k \to \vec{\mathfrak{t}}'_k}]. \qquad (9.526)$$

This implies that we have

$$\mathrm{Var}[\frac{1}{|\mathcal{A}(\vec{\mathfrak{t}})|} \sum_{j=1}^{|\mathcal{A}(\vec{\mathfrak{t}})|} 1_{\tau_{1,j} \leq s, \vec{\mathfrak{t}}_j \to \vec{\mathfrak{t}}'_j}] \to 0 \text{ as } N \to \infty, \qquad (9.527)$$

and this implies that (9.525) is the limit of the empirical distribution of the given event.

We now prove that if we know that for some t_0 at times $\beta^{-1} \log N + t_0$ we have convergence in law of $(\bar{u}^N, \bar{U}^N)$, then the subsequent dynamic converges as $N \to \infty$ to a limiting one at least up to some time horizon $\beta^{-1} \log N + t_0 + t_1$ up to which we can guarantee small clan sizes. This will be the key to later on establish the convergence of $(\bar{u}^N, \bar{U}^N)$ to an entrance law from time $-\infty$ we show in the next subsubsection:

Remark 177. We formulate the following proposition which gives convergence of the empirical process of microstates to the the nonlinear system only in the "low density" region, that is up to the point the density of occupied sites reaches the value u^* obtained in Lemma 9.52. The existence and uniqueness of the nonlinear system has been proved without the restriction. The problem is to exclude that a linked clan appears which is of order $O(N)$ with positive probability (giant component appears in the terminology of random graphs) so that in the limit dynamic we would have to add an extra term. As indicated in Remark 168 we conjecture that we can actually take $u^* = 1$ but do not prove this in this monograph. This limited version is adequate for our purposes since we use this result only in the transition regime in the proof of the equality of $^*\mathcal{W}$ and $\mathcal{W}^*$.

Proposition 9.57 (Convergence of dynamics for intensity, empirical measure of microstates).
Assume here that we are in the mutation-dominant regime according to Definition 9.46. Consider $(\frac{\bar{u}^N(t)}{N}, \bar{U}^N(t))$ and $\hat{\Psi}^N(t) = \frac{\bar{u}^N(t)}{N} \cdot \bar{U}^N(t)$. Let $t_1 > t_0$ be such that for given $\bar{u}(t_0)$ we have

$$e^{\beta(t_1 - t_0)} \bar{u}(t_0) < u^*, \qquad (9.528)$$

where u^* is defined as in Lemma 9.52.

(a) Let $T_N = \frac{\log N}{\beta}$. Assume that we have at $T_N + t_0$ convergence, i.e. we have:

$$\mathcal{L}\left[\left(\frac{\bar{u}^N(T_N + t_0)}{N}, \bar{U}^N(T_N + t_0)\right)\right] \Longrightarrow \mathcal{L}\left[(\bar{u}(t_0), \bar{U}(t_0, \cdot, \cdot))\right],$$
in law as $N \to \infty$, (9.529)

with $\bar{u}(t_0) > 0$.
Then as $N \to \infty$:

$$\mathcal{L}\left[\left(\frac{\bar{u}^N(T_N + t)}{N}, \bar{U}^N(T_N + t)\right)_{t_0 \le t \le t_1}\right] \Longrightarrow \mathcal{L}\left[(\bar{u}(t), \bar{U}(t, \cdot, \cdot))_{t_0 \le t \le t_1}\right].$$
(9.530)

(b) *The limiting pair of a.s. all realizations in (a), denoted*

$$(\bar{u}(t), \bar{U}(t, \cdot))_{t \in [t_0, t_1]} \qquad (9.531)$$

satisfies the nonlinear system of equations given in (9.509) and (9.510). □

Proof of Proposition 9.57. We assume that the marginal distributions at t_0 converge. Since the scaled object $\hat{\Psi}^N$ of (9.500) is not necessarily Markov we must view it as a function $\ell(\cdot)$ of the Markov process $\tilde{\mathcal{G}}^{++,N}$, which is coded by the Markov process $\mathfrak{s}^{****,N}$ (recall $\tilde{\mathcal{G}}^{++,N} = \tilde{\pi}(\mathfrak{s}^{****,N})$).

Our strategy now is to show, assuming convergence at time $T_N + t_0$, that the stochastic process

$$\left(\hat{\Psi}^N_{(T_N + t)}\right)_{t \ge t_0}, \qquad (9.532)$$

has for $N \to \infty$ a limiting deterministic dynamics. In order to do so we explicitly calculate first in Step 1 the generator G_N on functionals of $\hat{\Psi}^N$ which operates on the space $\mathfrak{S}^{****}_N$ and show in Step 2 that $G_N F \to GF$ on a collection of functions Γ which is separating in $\mathcal{M}_{\text{fin}}([0, \infty) \times \vec{\mathfrak{T}})$ and in Step 3 that G is the generator of a unique nonlinear evolution on $\mathcal{M}_{\text{fin}}([0, \infty) \times \vec{\mathfrak{T}})$.

We can verify tightness in path space of $\{\hat{\Psi}^N\}_{N \in \mathbb{N}}$ in a similar way to the case $M = 1$. We will then show that any weak limit point satisfies the G-martingale problem since $G_N F \to GF$ and that the G-*martingale problem* corresponds to the pair of equations (9.509) and (9.510). The weak convergence of the empirical processes then follows from the uniqueness of solutions to these equations.

Step 1 *(Martingale problem generator G_N)*

We first determine the appropriate algebra of test functions and secondly we calculate the action of G_N on such functions.

9.10 Empirical Process of Microstates, Clans for $\mathfrak{s}^{****,N}$: Mutation-Dominant Regime

(1) Consider functions F_g of the form

$$F_g(\hat{\Psi}) = g\left(\int_0^\infty \int_{\bar{\mathfrak{T}}} f(r,\bar{\mathfrak{t}})\hat{\Psi}(dr,d\bar{\mathfrak{t}})\right), \tag{9.533}$$

where $f : [0,\infty) \times \bar{\mathfrak{T}} \to \mathbb{R}$ and $g : \mathbb{R} \to \mathbb{R}$ are measurable and bounded, g has bounded first and second derivatives and f has bounded first derivatives in the first variable. We view this function as a function on $\mathfrak{S}_N^{****}$, even though it does not depend on the clan structure, namely,

$$F_g(\hat{\Psi}^N) = \tilde{F}_g(\mathfrak{s}^{****,N}). \tag{9.534}$$

The question is as to whether this statistics becomes in the limit $N \to \infty$ sufficient to determine the further evolution so that the clan structure becomes irrelevant and whether we can even reduce $\vec{\mathfrak{t}}$ to $\bar{\mathfrak{t}}$.

Observe first, that the algebra of functions generated by F_g is dense in $C_b(\mathcal{M}_1([0,\infty) \times \bar{\mathfrak{T}}), \mathbb{R})$ and is therefore distribution-determining for laws on $\mathcal{M}_1([0,\infty) \times \bar{\mathfrak{T}})$. We now consider this smaller space and show that the limiting evolution indeed lives on that space by showing that in the limit $N \to \infty$ the generator of $\mathfrak{s}^{****,N}$ applied to functions of the form (9.533) does not depend on the clan structure.

Given a configuration in $\mathfrak{s}_N^{****} \in \mathfrak{S}_N^{****}$ consider the limit:

$$(G_N F_g(\hat{\Psi}_0^N))(\mathfrak{s}^{****,N}) = \lim_{t \downarrow 0} \frac{1}{t} E[F_g(\hat{\Psi}_t^N) - F_g(\hat{\Psi}_0^N)|\hat{\Psi}_0^N] = h(\mathfrak{s}^{****,N}). \tag{9.535}$$

The right side or (9.535) can be written in terms of the possible transitions at each site and rank and requiring that such transitions also occur simultaneously at all ghost ranks linked to the given site and rank according to the evolution rules of $\mathfrak{s}^{****,N}$.

A key point for the limiting form of $G^N F_g$ as $N \to \infty$ is, that *over bounded time intervals every set of* sites appearing in a clan *is no larger than* $o(N)$ since otherwise the clan would be appearing in the limit expression. Recall that we have verified in the mutation-dominant regime even that, the largest connected component is at most $O(N^{\frac{1}{2}+\varepsilon})$ for some $\varepsilon > 0$ (see (9.477)).

Introduce now the notation $\hat{\Psi}(f)$ as:

$$\hat{\Psi}(f) = \int_0^\infty \int_{\bar{\mathfrak{T}}} f(r,\bar{\mathfrak{t}})\hat{\Psi}(dr,d\bar{\mathfrak{t}}) \tag{9.536}$$

and in particular $\hat{\Psi}(1) = \hat{\Psi}([0,\infty) \times \bar{\mathfrak{T}})$.

(2) The next step is now to calculate the action of G_N on F_g of the form in (9.533). In order to do this we first recall the set of possible transitions of $\mathfrak{s}^{****,N}$ and the transitions they induce on the empirical measure, recall the microstate $\bar{\mathfrak{t}}$ contains $(\mathfrak{t}^*, \mathfrak{t})$ with parent site, actual site information:

[D_r] *Aging:* - the ages of all sites increase at the same constant rate. The age of a site at time t: is $r = t - s$ where s is the arrival time of the first migrant.
[$\Phi^{\text{sel}}, \Phi^{\text{mut}}_{j,j'}, \Phi^{\text{coal}}_{i,i'}$] *mutation, selection, coalescence* acting at an active site:

- these transitions do not create any new active sites,
- at the active site such a change will decrease the empirical measure of the original type and add one to the type resulting from the action of the appropriate operator,
- the changes are reflected in the set of linked sites,
- these transitions can also result in the removal, that is, change the microstate to $\mathbb{I}^N_M$ (i.e. to RE status), of one or more of sites including the linked sites by resolution due to mutation. In particular, at the decoupling time of a site linked to a tagged site, it is removed from the list $\Lambda_A, \mathcal{Y}$.

[Imm]*Migration,* immigration into a site and emigration from a site

- This can create new active sites linked to their parent sites with birth time given by this transition time. At the same time a ghost site is created at the parent site.
- This can result in a collision with an existing active site. The resulting configuration at the arrival site is given by the product of the existing configuration and the migrating one. By convention the local rank assigned is the first unoccupied local rank and the coalescence is the usual lookdown mechanism.

We must also define the action of *migration* on $v \in \Upsilon$. If j is at nested level m the action of migration on the corresponding active site and rank results in moving the ghost rank to the $(m+1)$-st level in Υ. Note that a mutation action on a site can remove a linked site thus decreasing the number of occupied sites.

We now return to the main step. Let $\hat{\Psi} \in \mathcal{M}([0, \infty) \times \bar{\mathfrak{T}})$. Recall that a microstate component $\bar{\mathfrak{t}}$ in $\bar{\mathfrak{T}}$ contains the pair $(\vec{\mathfrak{t}}_i, \mathfrak{t}_i)$ for the immigrated ranks. Then letting $\Phi^S_{j,i}$, etc. denoting the action of the jth selection operator on the ith rank $\mathfrak{t}_i$ in the microstate $\bar{\mathfrak{t}}$, let $|\cdot|$ denote the number of nontrivial ranks in $\mathfrak{t}$ the collection of $\mathfrak{t}_i$. Furthermore define

$$(\bar{\mathfrak{t}}', \bar{i}') \sim (\bar{\mathfrak{t}}, i) \text{ if and only if there is a } \textit{directed path} \text{ from } (\bar{\mathfrak{t}}, i) \text{ to } (\bar{\mathfrak{t}}', \bar{i}'). \quad (9.537)$$

Then we introduce the following abbreviations:

- let $\setminus \ell$ mean that the rank ℓ is removed from the set $\mathfrak{t}$ in $\bar{\mathfrak{t}}$ (we use $\bar{\mathfrak{t}} \setminus \ell$ or $\mathfrak{t} \setminus \ell$ whatsoever is more convenient) and changes in $\Lambda_A, \Lambda_D, \mathcal{C}_A$ are made accordingly,

9.10 Empirical Process of Microstates, Clans for $\mathfrak{s}^{****,N}$: Mutation-Dominant Regime

- $\mathfrak{t} \perp \mathfrak{t}'$ for $\mathfrak{t}, \mathfrak{t}' \in \bar{\mathfrak{T}}$ means that $\mathfrak{t}'$ does not contain an unresolved factor from the same parent site as $\mathfrak{t}$, which is also from the same excursion between decoupling times,
- $(\bar{\mathfrak{t}}', i') \sim (\bar{\mathfrak{t}}, i)$ denotes elements of $\bar{\mathfrak{T}}$ and ranks that are listed as ghosts linked to the active rank i from the same parent sites as $\bar{\mathfrak{t}}$ and $\sum_{(\bar{\mathfrak{t}}', i') \sim (\bar{\mathfrak{t}}, i)}$ is the sum over these,
- $\mathfrak{t} \setminus \mathfrak{t}_j$ means that the rank j is moved from the active ranks to the ghost ranks in $\bar{\mathfrak{t}}$ and the lists Λ_A, Λ_D are changed accordingly.

Now note that the number of sites at which a simultaneous transition occurs and this is given by the size of the connected sets,

$$\mathfrak{n}(\bar{\mathfrak{t}}, i) = |\{(\bar{\mathfrak{t}}', i') : (\bar{\mathfrak{t}}', i') \sim (\bar{\mathfrak{t}}, i)\}|. \tag{9.538}$$

Connected sets of this form are contained in a connected set of a clan.
Then we write out the action of G_N on the jump process $\hat{\psi}(\mathfrak{s}_t^{****,N})$ explicitly using the Taylor expansion of the function g with the second order correction term as:

$$G_N F_g(\hat{\Psi})(\mathfrak{s}^{****,N}) =$$

$$+ \int_0^\infty \int_{\bar{\mathfrak{T}}} g'\left(\hat{\Psi}(f)\right) D_r f(r, \bar{\mathfrak{t}}) \hat{\Psi}(dr, d\bar{\mathfrak{t}})$$

$$+ \sum_{j=2}^M \int_0^\infty \int_{\bar{\mathfrak{T}}} s(e_j - e_{j-1}) N \sum_{i=1}^{|\bar{\mathfrak{t}}|}$$

$$\left[g(\hat{\Psi}(f)) + \sum_{(\bar{\mathfrak{t}}', i') \sim (\bar{\mathfrak{t}}, i)} \frac{f(r, \Phi_{j,i}^{\text{sel}} \bar{\mathfrak{t}}') - f(r, \bar{\mathfrak{t}}')}{N} \right) - g(\hat{\Psi}(f)) \right] \hat{\Psi}(dr, d\bar{\mathfrak{t}})$$

$$+ \sum_{j \neq j' = 1}^M \int_0^\infty \int_{\bar{\mathfrak{T}}} m_{j,j'} N \sum_{i=1}^{|\bar{\mathfrak{t}}|}$$

$$\left[g(\hat{\Psi}(f)) + \sum_{(\bar{\mathfrak{t}}', i') \sim (\bar{\mathfrak{t}}, i)} \frac{f(r, \Phi_{(j,j'),i}^{\text{mut}} \bar{\mathfrak{t}}') - f(r, \bar{\mathfrak{t}}')}{N} \right) - g(\hat{\Psi}(f)) \right] \hat{\Psi}(dr, d\bar{\mathfrak{t}})$$

$$+ \int_0^\infty \int_{\bar{\mathfrak{T}}} d \cdot N \sum_{j=2}^{|\bar{\mathfrak{t}}|} \sum_{i=1}^{|\bar{\mathfrak{t}}|} \sum_{i' > i} 1(i, i' \text{ active ranks})$$

$$\left[g(\hat{\Psi}(f)) + \sum_{(\bar{\mathfrak{t}}', i') \sim (\bar{\mathfrak{t}}, i)} \frac{f(r, \Phi_{i,i'}^{\text{coal}} \bar{\mathfrak{t}}') - f(r, \bar{\mathfrak{t}}')}{N} \right) - g(\hat{\Psi}(f)) \right] \hat{\Psi}(dr, d\bar{\mathfrak{t}})$$

$$+ c(1 - \hat{\Psi}(1)) \left(\int_0^\infty \int_{\bar{\mathfrak{T}}} N \sum_{\ell, \tau^*} \right.$$

$$\left[g(\hat{\Psi}(f)) + \frac{f(0, (\vec{t}, \ell, \tau^*)) + f(r, \vec{t}\backslash \ell) - f(r, \vec{t})}{N}\right) - g(\hat{\Psi}(f))\right]$$

$$1(t(\ell) = \tau^*)1_{|t|\neq 1}\hat{\Psi}(dr, d\vec{t})\bigg)$$

$$+c\hat{\Psi}(1)\int_0^\infty \int_{\bar{\mathfrak{T}}} N\sum_{\ell,\tau^*}$$

$$\left[g(\hat{\Psi}(f)) + \frac{f(r', \vec{t}' \otimes (t, \ell, \tau^*)) - f(r', \vec{t}')}{N} + \frac{f(r, \vec{t}\backslash \ell) - f(r, \vec{t})}{N}\right) - g(\hat{\Psi}(f))\right]$$

$$1(\vec{t} \perp \vec{t}')1(t(\ell) = \tau^*)\hat{\Psi}(dr, d\vec{t})\frac{\hat{\Psi}(dr', d\vec{t}')}{\hat{\Psi}(1)}$$

$$+o(N). \tag{9.539}$$

Explanation of terms in (9.539).

The $o(N)$ terms denotes the summand corresponding to the contributions from the last term except with $\vec{t} \not\perp \vec{t}'$, that is, this is an error term corresponding to the sparse set of sites involved in self interaction.

The interior sums in the expression in the fifth row correspond to the occurrence of simultaneous jumps at multiple sites.

Step 2 *(Limiting martingale problem generator, generator convergence)*

We consider the convergence for the "initial state" at some N-dependent initial time of the form $T_N + t_0$, namely

$$\hat{\Psi}^N_{T_N+t_0}(1) \xrightarrow[N\to\infty]{} x \geq 0. \tag{9.540}$$

Note that if $\Psi^N_{T_N+t_0}(1) \xrightarrow[N\to\infty]{} 0$, then $\hat{\Psi}^N_{T_N+t_0+t}(1) \xrightarrow[N\to\infty]{} 0$ for $0 \leq t \leq T < \infty$.

We now consider the case $x > 0$ and specify a corresponding generator G for a limiting problem. With F_g as in (9.533) we let the operator G acting on F_g by only taking into account the measure on $[0, \infty) \times \bar{\mathfrak{T}}$ and ignoring the clan component and we set (here Φ^{sel}, Φ^{mut}, Φ^{coal} are the selection, mutation and coalescence operators) and the operations are projected in the obvious way from $\vec{t}$ to $\vec{t}$:

$GF_g(\hat{\Psi}) =$

$$+g'\left(\hat{\Psi}(f)\right)\int_0^\infty \int_{\bar{\mathfrak{T}}} D_r f(r, \vec{t})\hat{\Psi}(dr, d\vec{t})$$

$$+g'(\hat{\Psi}(f))\int_0^\infty \int_{\bar{\mathfrak{T}}} \left[((\Phi^{\text{sel}} + \Phi^{\text{mut}} + \Phi^{\text{coal}})f(r, \vec{t}))\right]\hat{\Psi}(dr, d\vec{t})$$

9.10 Empirical Process of Microstates, Clans for $\mathfrak{s}^{****,N}$: Mutation-Dominant Regime 563

$$+g'(\hat{\Psi}(f)) \cdot c(1 - \hat{\Psi}(1))$$

$$\left(\int_0^\infty \int_{\tilde{\mathfrak{T}}} \sum_{\ell, \tau^*} [f(0, ((\bar{\mathfrak{t}}, \ell, \tau^*)) + f(r, \vec{\mathfrak{t}}\backslash_\ell) - f(r, \vec{\mathfrak{t}})] 1(\zeta(\ell) = \tau^*)) 1_{|\bar{\mathfrak{t}}| \neq 1} \hat{\Psi}(dr, d\vec{\mathfrak{t}}) \right)$$

$$+g'(\hat{\Psi}(f))\hat{\Psi}(1)c \int_0^\infty \int_{\tilde{\mathfrak{T}}} \sum_\ell \left[f(r, \vec{\mathfrak{t}}\backslash_\ell) - f(r, \vec{\mathfrak{t}}) \right] \hat{\Psi}(dr, d\vec{\mathfrak{t}})$$

$$+g'(\hat{\Psi}(f))c(\int_0^\infty \int_{\tilde{\mathfrak{T}}} \sum_{\ell, \tau^*} 1(\zeta(\ell) = \tau^*))\hat{\Psi}(dr, d\vec{\mathfrak{t}}))$$

$$\left(\int_0^\infty \int_{\tilde{\mathfrak{T}}} \left[f(r', \vec{\mathfrak{t}}' \otimes (\bar{\mathfrak{t}}, \ell, \tau^*)) - f(r', \vec{\mathfrak{t}}) \right] \hat{\Psi}(dr', d\vec{\mathfrak{t}}') \right).$$

(9.541)

Our task is now to compare the action of G_N and G on a given function F_g in a point with the empirical distribution part of the state of $\mathfrak{s}^{****,N}$ given by $\hat{\Psi}$. We can now expand g up to second order terms to compare G_N and G. Only the error term in the equation remains.

We can control the coefficient of g'' in the expansion (9.539) in terms of the empirical second moment of the size of the connected set of clans.

Since the number of summands where simultaneous transitions occur is given by the number $\mathfrak{n}(\bar{\mathfrak{t}}, i)$ if we are in the state $\bar{\mathfrak{t}}$ and here rank i is changed, the empirical second moment of the resulting change is

$$|J_N| \leq \text{Const} \cdot \int_{\tilde{\mathfrak{T}}} \int_0^\infty \sum_{i=1}^{|\bar{\mathfrak{t}}|} (\mathfrak{n}(\bar{\mathfrak{t}}, i))^2 \hat{\Psi}(dr, d\vec{\mathfrak{t}}). \quad (9.542)$$

Therefore the coefficient g'' tends to 0 in the limit provided this empirical moment remains bounded in probability as $N \to \infty$.

More precisely, given the empirical measure $\hat{\Psi}$ based on $\mathfrak{s}^{****,N}$ we have

$$(|G_N(F_g)(\hat{\Psi}) - G(F_g)(\hat{\Psi})|) \leq \quad (9.543)$$

$$\frac{\text{const}}{N} \|g''\| \left[\int_{\tilde{\mathfrak{T}}} \int_0^\infty \sum_{i=1}^{|\bar{\mathfrak{t}}|} (\mathfrak{n}(\bar{\mathfrak{t}}, i))^2 \hat{\Psi}(dr, d\vec{\mathfrak{t}}) \right].$$

Consider $\hat{\Psi}$ satisfying that

$$\int_{\tilde{\mathfrak{T}}} (\mathfrak{n}(\bar{\mathfrak{t}},i))^2 \hat{\Psi}(dr,d\bar{\mathfrak{t}}) < \infty. \tag{9.544}$$

For any F_g of the form in (9.533), we have the following convergence restricting to states of $\mathfrak{s}^{****,N}$ such that the corresponding $\hat{\Psi}$ is satisfying (9.544) (more precisely in the sense that the time integrals from t_0 to $t_0 + t_1$ converge):

$$G_N(F_g)(\hat{\Psi}) \xrightarrow[N\to\infty]{} G(F_g)(\hat{\Psi}). \tag{9.545}$$

Therefore we can control the coefficient of g'' in the expansion (9.539) in terms of the empirical second moment of the size of the connected set of clans. Therefore the coefficient g'' tends to 0 in the limit if we can show that the empirical second moment remains stochastically bounded.

Now we restrict our attention to the mutation-dominated regime so that (9.473) is satisfied by Lemma (9.52)(c), since the second moment of the sizes of linked clusters remain $o(N)$ during $(-\infty, T]$ as $N \to \infty$. Using (9.461) in Lemma 9.52, we obtain in fact that:

$$\sup_N \int_{t_0}^{t_1} E\left[\int_0^\infty \int_{\tilde{\mathfrak{T}}} \sum_{i=1}^{|\bar{\mathfrak{t}}|} (\mathfrak{n}(\bar{\mathfrak{t}},i))^2 \hat{\Psi}^N_{\beta^{-1}\log N + s}(dr,d\bar{\mathfrak{t}})\right] ds < \infty. \tag{9.546}$$

As in Section 8 this can be used to conclude again that in the mutation-dominant case the empirical measure process (of microstates) converges weakly to a solution of the G martingale problem. (In the selection-dominant regime we must work with the empirical measure of mesostates - see Subsection 9.12.)

Step 3 *(Convergence to deterministic evolution equation)*

We now show that any solution to the G martingale problem corresponds to a solution to the nonlinear system (9.509), (9.510). We note G is a differential operator, in fact G is a *first order* differential operator and the solution $(V_t)_{t \geq t_0}$ of the evolution equation

$$\frac{d}{dt} V_t = G V_t, \tag{9.547}$$

corresponds therefore to a *deterministic* process. The forward equation that corresponds to the solution of the G martingale problem is exactly the equation for $(\bar{u}(t), \bar{U}(t,\cdot,\cdot))$. We observe that for the test function $f \equiv 1$ we get the equation for u and given this information we conclude that the normalized object satisfies the U-equation. Therefore uniqueness of the G-martingale problem follows from the uniqueness (see Proposition 9.56) for the system of coupled equations (9.510) and (9.509). The convergence result then follows from a version of the Kurtz convergence theorem (see e.g. [EK2], Chapt. 4, Section 8). See for the detailed argument we gave in Section 8, (8.343)–(8.351) which can be carried out again, we leave things here to the reader. q.e.d.

9.10.4 Convergence of $(\hat{\Psi}^N (\beta^{-1} \log N + t)^+)_{t \in (-\infty, \infty)}$ to Entrance Law

The convergence of $(\bar{u}^N, \bar{U}^N)$ in the time window $\{\frac{\log N}{\beta} + t\}_{t \in [t_0, \infty)}$ follows from Proposition 9.57 together with the convergence of the marginal distribution $\{\hat{\Psi}^N(t_0)\}$ which is included in the next proposition.

Proposition 9.58 (The convergence $N \to \infty$ for dual in the nonlinear regime with $M \geq 2$).

Assume that we are in the regime of mutation dominance.

(a) For fixed $t_0 \in (-\infty, \infty)$ we have that:

$$\frac{\bar{K}^N_{\frac{\log N}{\beta} + t_0}}{N} \to \bar{u}(t_0), \text{ in law as } N \to \infty, \tag{9.548}$$

and

$$\bar{U}^N(\frac{\log N}{\beta} + t_0) \underset{N \to \infty}{\Longrightarrow} \bar{U}(t_0), \text{ in law as } N \to \infty, \tag{9.549}$$

where $(\bar{U}(t_0, \cdot)) \in \mathcal{P}(\bar{\mathfrak{T}})$.

(b) As $N \to \infty$,

$$\mathcal{L}\left[\left(\frac{1}{N}\bar{u}^N\left(\left(\frac{\log N}{\beta} + t\right)^+\right)\right), \bar{U}^N\left(\left(\frac{\log N}{\beta} + t\right)^+\right)_{t \in \mathbb{R}}\right]$$
$$\Longrightarrow \mathcal{L}\left[(\bar{u}(t), \bar{U}(t, \cdot, \cdot))_{t \in \mathbb{R}}\right], \tag{9.550}$$

where $(\bar{u}(t), \bar{U}(t, \cdot, \cdot))_{t \in \mathbb{R}}$ is the unique solution to the equations (9.509), (9.510) which satisfies

$$\lim_{t \to -\infty} [e^{\beta |t|} \bar{u}(t)] = W^I, \quad \lim_{t \to -\infty} \bar{U}(t) = \bar{U}^0(\infty), \tag{9.551}$$

with $\bar{U}^0(\infty)$ the stable type distribution given by Lemma 9.55 and W^I given by (9.368). □

Remark 178. Applying this with the appropriate configuration of initial factors can be used to prove the convergence of the marginal distributions of the process $\Xi_N^{\log \beta}(t)$ as $N \to \infty$.

Proof of Proposition 9.58. The proof follows the same main steps as that of Proposition 8.8 in the case $M = 1$.

(a) We first prove that the marginal distributions converge for $t_0(N) = \frac{1}{\beta} \log N + t_0$. This is again done as in Section 8 by first using Proposition 9.57 and by then showing that the hitting times $\tau^N(\varepsilon)$ of the intensity of occupied sites to reach ε converge in law. But now in addition we have to show that the relative proportions of the types of factors at these times converge in law.

For this purpose we have to study the times at which new sites are occupied, once we enter the collision regime. This works exactly as in the two-type case, see Subsubsection 8.3.6, but we have to build in now in addition the multitype structure of the sites. In other words instead of a growth process we have a *multitype* growth process. The good point is that this is taking place in the deterministic regime of the evolution again.

The structure of the waiting times for the next colonization, i.e. to move from k to $k + 1$ colonized sites is for each type of site created as before, but now depends on the relative proportions of the different types of factors (resolved or unresolved with a specified result once resolved) in the state of k occupied sites, where k is large for example of order $\log N$ (i.e. after times of order $\log \log N$), so that we have the stable age-type-internal state distribution in effect. However, as soon as we enter the regime with collisions, the age-type-internal state distribution is also evolving. But this we control with the evolution equation for $\bar{U}^N(\beta^{-1} \log N + t)$ and we know from above that for that mechanism we have a limiting dynamic $\bar{U}(t)$ for $t \in \mathbb{R}$, in fact we have this for the pairs $(\bar{u}^N, \bar{U}^N), (\bar{u}, \bar{U})$. Therefore whenever we are in a point $k(k \in [\log N, \varepsilon N])$ we have waiting times of different types but the proportion of the different types is given as $N \to \infty$ by $\bar{U}(t)$ and as $\varepsilon \to 0$ by $\bar{\mathcal{U}}^0(\infty)$ (recall Lemma 9.55).

The convergence of $\tau^N(\varepsilon)$ rests now with the convergence of the time by which we reach $k = \log N$, by showing that the remaining time to reach $k = \lfloor \varepsilon N \rfloor$ is as $N \to \infty$ deterministic. But again we argue as before to get a waiting time for the next step which is $\frac{1}{\beta} \frac{1}{k}$, instead of $\frac{1}{\alpha} \frac{1}{k}$. Of course we must again justify this following the arguments in (8.197)–(8.212) but do not repeat these steps here.

(b) We must show (recall that combining part (a) with Proposition 9.57, we know that we have convergence to a random entrance law indexed by $\mathbb{R}$ as $N \to \infty$ already) that

$$(\frac{1}{N} e^{-\beta t} \bar{u}^N(\frac{\log N}{\beta} + t), \bar{U}^N(\frac{\log N}{\beta} + t))_{t \in \mathbb{R}} \Longrightarrow (W^I, \bar{U}^0(\infty)) \text{ as}$$
$$N \to \infty, \ t \to -\infty, \tag{9.552}$$

where W^I is given by (9.368). But any weak limit point of the left hand side of (9.550) must be a solution of (9.509), (9.510) and by the particle picture we shall derive in Subsection 9.11 an expansion as $t \to -\infty$ for the limit (see below Proposition 9.59) which in particular implies that this solution must satisfy

$$\lim_{t \to -\infty} e^{-\beta t} \bar{u}(t) = W^I, \qquad \lim_{t \to -\infty} \bar{U}(t) = \bar{U}^0(\infty). \tag{9.553}$$

The result then follows since by Proposition 9.56 the system has a unique solution with entrance behaviour given by (9.553).   q.e.d.

9.11 The Transition Regime and Emergence: Mutation-Dominant Case

We continue here under the **assumption** of *mutation dominance* with the analysis of $\mathfrak{s}_{I,\text{enr}}$ and $\mathfrak{s}^{****,N}$ in the time regime $\beta^{-1} \log N + t \cdot (N \to \infty, t \to -\infty)$, i.e. the transition regime from droplet growth to emergence and then the *emergence* at times $\beta^{-1} \log N + t$.

In this subsection we proceed in four subsubsections in the first three items dealing with the transition to emergence and in the last discussing the actual emergence: (1) we derive the $t \to -\infty$ asymptotics of the set-valued dual process arising with $N \to \infty$ limit, (2) introduce for that purpose the *multi-colour system*, (3) study the asymptotics of the *droplet growth* and finally (4) calculate the *hazard function* for rare mutation acting on the dual process in the $N \to \infty$ limit to be able to obtain fixation. Before we begin we describe the task and the changes compared to Section 8 on the $(M, M) = (1, 1)$-case.

We have now in the case $(M, 1)$ described the growth of the droplet process on the one hand (recall the proofs are in Section 10 since we deal directly with the general (M, M)-case and on the other we have developed the set-valued dual in the collision regime and obtained the limiting nonlinear equations for the limiting empirical measure of microstates of the dual process. In addition we have identified the entrance laws of these nonlinear equations satisfying (9.514) in Proposition 9.56. This parallels the development of Section 8 but has involved significant differences both for the droplet process and in particular the dual process. Having now at hand the basic tools the remainder of the program is to prove the main results on emergence and fixation (in the mutation dominance regime) which follows rather closely the development in the special case $(M, M) = (1, 1)$, as we explain next.

In order to proceed to the proofs of the main results it remains to consider the asymptotics in the transition regime, i.e. determining the behaviour as $t \to -\infty$ in the limit $N \to \infty$ and asymptotically as $N \to \infty$ in the finite N expression, and in particular we have to develop the tools to finally show that $\mathcal{L}[^*\mathcal{W}] = \mathcal{L}[\mathcal{W}^*]$.

This is parallel to a method developed in Subsubsections 8.3.9–8.3.13, 8.4.2-8.4.6 and Subsection 8.5. We will again use a *coloured system* to investigate the dual process asymptotics in the limit as $t \to -\infty$. Our plan is to carry out this program below following the same basic strategy as in Section 8, taking advantage of the detailed development of Section 8 and primarily focussing on the additional complexities involved in the multitype system due to the replacement of particles by ranks with their column, which we can view as *particles with internal state*.

In particular we recall that (see Proposition 8.29) the essential step in proving that $^*\mathcal{W} = \mathcal{W}^*$ is the asymptotic analysis of the second moment

$$E\left[\left(\sum_{i \in E_0} x_i(\frac{\log N}{\beta} + t)\right)\left(\sum_{i \in E_0} x_i(\frac{\log N}{\beta} + t + s)\right)\right] \qquad (9.554)$$

as $N \to \infty$ and $t \to -\infty$.

The expression above can be expressed by the dual process starting with the factors $(1, \ldots, 1, 0)$ at two sites at two different times and we see that it becomes less than 1 due to a rare mutation in one of the ranks of the cloud. In contrast to the case $M = 1$ the occurrence of a single rare mutation does not necessarily lead to a jump to zero in the dual expression (but does become less than one if we integrate with respect to an initial measure with support equal to E_0). However once a rare mutation occurs there is a positive probability that the jump to 0 will then occur due to mutations at the level E_0, however there is also a positive probability that the effect of the rare mutation is reversed by a mutation at level E_0. As a result the main contribution does come from the event that the *first* rare mutation is in fact successful. The dynamics $(\bar{u}, \bar{U})$ allows us to compute the hazard rate for the rare mutations to occur for some individual. In spite of the complexity the key point of the proof of the required second moment result is the same.

9.11.1 *Expansion of Dual Process as* $t \to -\infty$

The next step is to formulate in the next two Propositions the following analogue of Proposition 8.10 and to indicate the necessary modification to the proofs necessary in the case $M > 1$. Recall here that we have established the convergence to the limiting evolution equation only under the assumption of mutation-dominance, which we assume in Subsection 9.11. This convergence would in fact *not* hold if the clan size could reach $O(N)$ in the collision time scale and in fact if the maximal clan size would exceed $\sqrt{N}$ we run into technical problems.

Proposition 9.59 (Transition between linear and nonlinear regime limit dynamics $(\bar{u}, \bar{U})$).

Assume that we are in the mutation-dominance regime.

(a) *The pair $(\bar{u}, \bar{U})$ starting with k-factors at ℓ-different sites with factors $(11 \cdots 10)$ satisfies (here we suppress the k, ℓ specifying the initial particle configuration in the constants in the expansion) (recall (9.551)):*

9.11 The Transition Regime and Emergence: Mutation-Dominant Case

$$\bar{u}(t) = W^I e^{\beta t} - \bar{\kappa}(W^I)^2 e^{2\beta t} + O(e^{3\beta t}) \quad \text{as } t \to -\infty,$$
for some constant $\bar{\kappa} > 0$. (9.555)

(b) Furthermore $\bar{U}(t, \cdot, \cdot)$ is uniformly continuous at $t = -\infty$ and moreover as $t \to -\infty$:

$$\|\bar{U}(t) - \bar{U}^0(\infty)\|_{\text{var}} = O(e^{\beta t}). \qquad \square \qquad (9.556)$$

Corollary 9.60 (Time scale $\frac{\log N}{\beta} + t, t \to -\infty$).
Define $\tau_\varepsilon^N = \inf(t \| \bar{K}_t^N | = \varepsilon N)$. Then we can construct $\bar{u}(t)$ and $\bar{K}^N$ on one probability space such that for $N \geq N_0(\varepsilon)$

$$E[|(N\bar{u}(t) - \bar{K}_t^N)| \cdot 1_{t \leq \tau_\varepsilon^N}] \leq C\varepsilon^2 N. \qquad \square \qquad (9.557)$$

We also need the following analogue of Proposition 8.24 on the asymptotics of $t \to -\infty, N \to \infty$ simultaneously.

Proposition 9.61 (Expansion for space-time dual).
Assume that we are in the mutation dominance regime. Consider the case in which we start the process with one factor $(1 \cdots 10)$ at two distinct points j_1 and j_2 in $\{1, \ldots, N\}$ at time 0 and time $s > 0$, respectively. We consider the times:

$$t_N = (T_N + t)^+, \quad T_N = \frac{\log N}{\beta}, \quad t \in \mathbb{R}. \qquad (9.558)$$

We get the analogue of (8.748), that is as $N \to \infty$ for some $\bar{\kappa}_2 > 0$:

$$\bar{K}_{t_N+s}^{N, j_1, j_2} = (W_1^I e^{\beta(t_N+s)} + W_2^I e^{\beta t_N}) - \bar{\kappa}_2 W_1^I W_2^I \frac{e^{\beta(2t_N+s)}}{N^2} + o(e^{\beta(2t_N+s)}/N^2). \qquad \square$$
(9.559)

The relation (9.559) is used to identify the $O(\frac{1}{N^2})$ terms.

9.11.2 Multicolour Systems

The first step for the proof of the propositions above is to introduce the *multicolour version of* $\tilde{G}^{++,N}, \mathfrak{s}^{****,N}$. The set of colours

$$C := \{ \text{white, black, red, green, purple and blue} \} = \{W, B, R, G, P, BL\}, \qquad (9.560)$$

is used to mark *ranks* as we have done in Section 8 with particles. We introduce a function c which assigns a colour to each active rank in a microstate $\bar{t}$ at an occupied site $j \in \{1, \ldots, N\}$:

$$c_j : \{1,\ldots,|t|\} \to \{W, B, R, G, P, BL\}. \tag{9.561}$$

As before, the colours $\{B, R, G, P, BL\}$ arise only by *collision or coalescence* events. We can then enrich the microstate space by including the colour of each rank and introduce the analogues of the transition rules of Section 8 replacing particles by ranks. Note that for the time regime where collision do not occur, so that all ranks individuals stay white and we have the process $\mathfrak{s}$, while at later times when collisions do occur white and black have then the same law as the collision-free process. On the other hand in the fully nonlinear regime described by $\mathfrak{s}^{****,N}$, we would have to introduce a countable set of colours. The objective here is to analyse the emergence regime as in Section 8 by identifying sites that have been involved in only one or two collisions using the above colours.

The key point is that at time $T_N + t$, after $N \to \infty$ as $t \to -\infty$, the total intensity of occupied sites is of order $O(e^{\beta t})$, the number of red or green sites is of order $O(e^{2\beta t})$ and the number of blue or purple sites is of order $O(e^{3\beta t})$ and further colours would occur with masses $O(e^{4\beta t})$, which again as in Section 8 is not relevant for our purposes.

We can now determine the influence that collisions have on the system. To carry this out we introduce the modified local and global state spaces.

(i) The local and global state spaces

The *mesostate space* will be a modification of $\mathfrak{T}^{****,N}$, the finite N-version of $\mathfrak{T}^{****}$ (see equation (9.423)), denoted

$$\mathfrak{T}^{****,N,C} \tag{9.562}$$

including the maps c_j (see (9.561)), in which some ranks can be assigned a colour B, R, G, P, BL. The other ranks are assigned W (white). Here the ranks and its ghost copy carry always the same colour.

A site is assigned a red tag if it becomes occupied by a red rank or ghost rank, a purple tag if it is occupied by (at least) a purple rank and a blue tag if it is occupied by (at least) a blue rank. It is tagged green if it is occupied by a green migrant, i.e. green tags occur only at sites to which a green particle *migrates*.

The *global state space* partitions the set of sites into four subsets

$$\{1,\ldots,N\} = N_1 \cup N_2 \cup N_3 \cup N_4 \cup N_5, \tag{9.563}$$

where

- N_1 are the occupied sites that have not been involved in a collision (white sites)
- $N_2 = N_{2,A} \cup N_{2,D}$ are the occupied sites that have been involved in exactly one collision (red sites); $N_{2,A}$ are sites at which migrants have arrived from other sites and $N_{2,D}$ are sites from which migrants have left and collided,

9.11 The Transition Regime and Emergence: Mutation-Dominant Case

- N_3 = purple sites, upper bound for the occupied sites that have been involved in more than one collision,
- N_4 = green sites,
- N_5 = the unoccupied sites.

In the transition regime calculations we do not track the particular states of sites in N_3 but only bound the number of such sites.

(ii) Dynamics

The dynamics of the coloured system is analogous to that in the case $M = 1$ but we must note that in the calculations using the set-valued dual with $M > 2$ the differences are (lower order refers to rates of $O(1)$ as $N \to \infty$)

- lower order selection can produce births leading to new migrant sites as before,
- but some of these additional new sites can be removed by lower order mutation,
- the above two effects combined lead to the Malthusian parameter β (rather than α).

The rules of evolution for the multicolour system that carry over immediately are as follows:

- all ranks that have never been involved in a collision are white,
- a white ranks that migrates and hits an occupied site becomes red and a black rank on a separate copy of $\mathbb{N}$ is produced on the first free site,
- when a red rank coalesces with a white rank a red-green pair is produced,
- when a red rank migrates and hits an occupied site it produces a blue-purple pair of ranks with the blue one placed on an extra copy of $\mathbb{N}$.

The remaining rules are also as the above completely as in Section 8 and are not repeated here.

The additional rules occuring in our more complex state space are as follows:

- if a white rank migrates and hits an occupied site the resulting immigrant family is assigned the colour red and the corresponding ghost rank in Υ is also assigned the colour red,
- if a red member of an immigrant family coalesces with a white rank the result is a white rank at the former position of the white rank plus a green rank at the former position of the red rank,
- if a red individual migrates and hits an occupied site, it produces a blue-purple pair of ranks. The purple rank is placed at the new location and the blue rank is placed at the first empty site in a copy of $\mathbb{N}$, and the ghost rank is coloured purple.
- green ranks migrate to an empty site in $\{1, \ldots, N\}$ with probability $1 - \frac{\ell}{N}$ and to the first empty site in a copy of $\mathbb{N}$ with probability $\frac{\ell}{N}$ where ℓ denotes the number of occupied sites,

- the dynamics at sites in $(N_1, N_2, N_3, N_4, N_5)$ is as described above up to a migration time,
- when a migration event at rank k_1 at a site j occurs, a new site is chosen at random,
    - if $j \in N_1$ and if the new site is in N_5 we proceed as before,
    - if $j \in N_2$ and if the new site is in N_2, then both sites are moved to N_3,
    - if j_D and the new site j_A are both in N_1 then we proceed as in $\mathfrak{s}^{****.N}$ and in addition the corresponding objects are red and both sites are moved into N_2,
    - if a coalescence occurs between a red rank and white rank a white-green pair is produced at that site,
    - when a green individual migrates a new site is occupied and is assigned colour green.

The key *new feature* compared to Section 8 ($M = 1$) is that a red or green site can be *eliminated by resolution* at the parent site.

We need the following splitting of $\bar{u}$ into $\bar{u}_i, i = 1, \cdots, 5$, "the colour-induced" intensities:

$$\bar{u}_i^N(t) = \frac{\bar{K}_i^N(T_N + t)}{N}, \quad i = 1, \ldots, 5 \quad , T_N = \frac{\log N}{\beta}, \tag{9.564}$$

where $\bar{K}_i^N(t)$ denotes the number of sites in N_i.

Proof of Proposition 9.59. The proof is obtained essentially as in Section 8. The main difference is now twofold, the ranks not only have colours but also can take on a finite number of possible internal states (partitions or subpartitions) and that ranks can be removed by resolution. As we have shown this results in growth at rate β rather than α. However noting that the basic growth rate is now β, the arguments concerning the order of the total numbers of white, red, green and blue particles can be extended to the number of white, red, ... ranks by replacing $e^{k\alpha t}$-terms by $e^{k\beta t}$-terms for $k = 1, 2, 3, 4$. Recall that we have proved that $\bar{u}^N(T_N + t)$ (in fact the pair $(\bar{u}^N, \bar{U}^N)$) converges in law as $N \to \infty$ and then $t \to -\infty$. A similar argument carries over for the coloured version, so that $((\bar{u}_1^N, \cdots, \bar{u}_5^N), \bar{U}^N)$ converge in law evaluated at time $T_N + t$.

The main steps of the proof are then as follows:

-
$$\lim_{t \to -\infty} e^{-\beta t} \lim_{N \to \infty} \frac{\bar{u}_1^N(T_N + t)}{N} = \bar{W} \in (0, \infty). \tag{9.565}$$

This follows by noting that by the CMJ result (9.565) is satisfied for the set of white plus black.
- The number of individuals involved in one or more collisions is bounded above by the red and purple population and hence bounded (for one respectively two collisions) by a term of order

9.11 The Transition Regime and Emergence: Mutation-Dominant Case

$$O(Ne^{2\beta t}) \text{ as } N \to \infty \text{ and } t \to -\infty, \tag{9.566}$$

$$\lim_{t \to -\infty} e^{-3\beta t} \limsup_{N \to \infty} \frac{\bar{u}_3^N(T_N + t)}{N} < \infty. \tag{9.567}$$

This is based on the upper bound for the number of blue ranks which follows since the rate at which a red rank collides with an occupied site is bounded above by $1/N$ times the number of white and green sites and the number of red ranks is of order $\frac{1}{N}e^{2\beta(T_N+t)}$ as we saw above.

- The above argument also gives:

$$\lim_{t \to -\infty} e^{-2\beta t} \lim_{N \to \infty} \frac{\bar{u}_2^N(T_N + t)}{N} > 0. \tag{9.568}$$

-

$$\lim_{t \to -\infty} e^{-2\beta t} \lim_{N \to \infty} \frac{\bar{u}_4^N(T_N + t)}{N} = \kappa \in (0, \infty), \tag{9.569}$$

which follows as in (8.504).

Finally $\bar{U}^N(t,\cdot,\cdot)$ is uniformly continuous at $t = -\infty$ where $\bar{U}^N(t,\cdot,\cdot)$ denotes the normalized measure on $\mathfrak{T}^E$ (ignoring colour)

$$\limsup_{t \to -\infty} e^{-\beta t} \lim_{N \to \infty} \|\bar{U}^N(T_N + t) - \bar{U}^0(\infty)\|_{\text{var}} < \infty. \tag{9.570}$$

Remark 179. Note that $\bar{U}^N$ is the empirical measure of local states at all occupied sites and not just $UPe \cup Pe$ sites and therefore corresponds to U^1. Also see Remark 175.

Proof of Proposition 9.61. This follows the same basic steps as in the proof of Proposition 8.24 except that now the coloured dual particle system is started with one factor $(1,\ldots,1,0)$ at site j_1 time 0 and another factor $(1,\ldots,1,0)$ at site j_2 at time s. q.e.d.

9.11.3 Deterministic Regime of Droplet Growth with $M \geq 2$

We now return to the original system of interacting Fleming–Viot diffusions and use the expansion Subsubsection 9.11.2. We consider the analysis of the linkage between the droplet growth process and the McKean–Vlasov deterministic dynamics for the empirical measure of the dual microstates and obtain the proof of (9.47) in Proposition 9.5. This analysis follows the same lines as that in Section 8. We now outline the overall plan of the argument and the additional complication in the $M > 2$ case.

In the absence of rare mutation, the interacting Fleming–Viot diffusions converge in distribution to the mutation-selection equilibrium on the types in E_0. Even during the period before this equilibrium is reached some rare mutants can be produced.

These rare mutants initiate small droplets of E_1-types which grow as described by the process $\mathfrak{J}_t^m$ (see Proposition 9.4). As in the case of supercritical branching with immigration, the main contribution to the mass at large times comes from the early mutations. For this reason as in Section 8 we consider the growth of the mutant droplet that forms by some time t_0 (conditioned on the event that this does not suffer extinction). Recall that we have described this in terms of the excursion process and verified that the mass is concentrated on a small number of sites inhabited by larger excursions. We shall prove below that this undergoes Malthusian growth with Malthusian parameter β instead of α as in Section 8.

We then proceed as in Proposition 8.27 and we derive the behaviour of the second order moments of the type-2 $(M+1)$ mass in the complete population taking the limit $N \to \infty$, then scale by $e^{-\beta t}$ and consider $t \to +\infty$. This in turn by expressing the behaviour in terms of again the dual process requires now to obtain the asymptotic for $N, t \to \infty$ of

$$\frac{K_t^N}{N} \text{ and } \frac{K_t^N(K_t^N - 1)}{N^2}, U^N(t, \cdot) \qquad (9.571)$$

given that the dual starts with 1_{E_0} at one, two sites, respectively. In the calculations with the dual the exact distribution on the lower order type now enters our expressions. The additional complication of the $M > 2$ case is an *additional correction* (compared to the $M = 1$ case) to the terms containing now instead of $g^N(t, x)$ we had in Section 8 a more complicated argument and become $g^N(t, \mu)$, μ describing the initial type distribution, taking hereby care of the multitype structure of the state generating the droplet, which carries over to the limiting expression appearing in the next proposition below. Here we use the idea that if we turn of mutation after time t_0, then due to the convergence of $(\mathfrak{J}_t^{m,N})_{t \leq t_0}$ to $(\mathfrak{J}_t^m)_{t \leq t_0}$ the states at time t_0 do converge as $N \to \infty$ a limit in law. On the other hand turning of mutation after time t_0 does not affect the growth of the droplet if we let $t_0 \to \infty$.

Proposition 9.62 (Deterministic regime of droplet growth).

Consider the system (9.1) and assume that at some time $t_0 \geq 0$ the site type frequencies at time t_0 satisfy:

$$\mu_{t_0}^N(i) = \{\tilde{x}_\ell^N(t_0, i)\}_{\ell \in E_0 \cup E_1} \in \mathcal{P}(E_0 \cup E_1), \ i = 1, \ldots, N, \qquad (9.572)$$

converge in law for $N \to \infty$ to

$$\tilde{\mu}_{t_0}(i) = \{\tilde{x}_\ell(t_0, i)\}_{\ell \in E_0 \cup E_1}, \ i \in \mathbb{N} \text{ with } \sum_i \sum_{\ell \in E_1} \tilde{x}_\ell(t_0, i) < \infty \quad a.s. \qquad (9.573)$$

9.11 The Transition Regime and Emergence: Mutation-Dominant Case

Moreover we assume that for every N and for the limit configuration we have that $\forall \, \varepsilon > 0$ there is a finite random set (we suppress the dependence on N in the notation):

$$\mathcal{I}(\varepsilon) = \mathcal{I}(\varepsilon, N) \text{ of } k = k(\varepsilon, N) \text{ sites,} \qquad (9.574)$$

such that for every $N \geq 2$:

$$\sum_{i \in (\mathcal{I}(\varepsilon))^c} \sum_{\ell \in E_1} \tilde{x}_\ell^N(t_0, i) < \varepsilon. \qquad (9.575)$$

Assume that mutation acts only on $[0, t_0]$, i.e. the mutation rate $m(t)$ satisfies:

$$m(t) = m \text{ for } t \leq t_0, \, m(t) = 0 \quad \text{for } t > t_0. \qquad (9.576)$$

Then the following four convergence properties (a)-(d) hold:

(a) *Consider the droplet process with $\mathfrak{X}_{t_0}^0 = \sum_{i=1}^k \delta_{a_i} \tilde{\mu}_{t_0}(i)$ where $a_i = \hat{\mu}_{t_0}(i)|_{E_1}$. Then*

$$\lim_{t \to \infty} e^{-\beta t} E[\mathfrak{X}_t^0([0,1])] = \lim_{t \to \infty} e^{-\beta t} \lim_{N \to \infty} E[\sum_{\ell \in E_1} \hat{x}_\ell^N(t)] \qquad (9.577)$$

$$= E[\bar{W}_{t_0}] \sum_{i=1}^\infty g(\infty, \tilde{\mu}_{t_0}(i)).$$

(b) *Again the expectations below depend on t_0 but satisfy always:*

$$\lim_{t \to \infty} \lim_{N \to \infty} (E[e^{-\beta(t+s)} \hat{x}_{E_1}^N(t+s)] - E[e^{-\beta t} \hat{x}_{E_1}^N(t)]) = 0, \quad \forall s \in \mathbb{R}. \qquad (9.578)$$

(c) *As above for every $t_0 > 0$ we have:*

$$\lim_{t \to \infty} \lim_{N \to \infty} E[e^{-\beta(s+2t)} \hat{x}_{E_1}^N(t+s) \hat{x}_{E_1}^N(t)]$$
$$= (E[\bar{W}_{t_0}])^2 \left[2\bar{\kappa}_2 \sum_{i=1}^k g(\infty, \tilde{\mu}_{t_0}(i)) \right], \quad \forall s \in \mathbb{R}, \qquad (9.579)$$

(d) *For every $t_0 \geq 0$:*

$$\lim_{t \to \infty} \lim_{N \to \infty} E\left[([e^{-\beta(t+s)} \hat{x}_{E_1}^N(t+s)] - e^{-\beta t} E[\hat{x}_{E_1}^N(t)])^2 \right] = 0, \quad \forall \, s \geq 0. \qquad (9.580)$$

(e) *Next replace (9.576) and assume that $m > 0$ (constant) for all t and that $\hat{x}_\ell^N(0, i) = 0$ for all $i \in \{1, \cdots, N\}$ and $\ell \in E_1$. Then the conclusions* of (a)–(d) remain valid and $\tilde{\mu}_{t_0}$ is replaced by a frequency vector satisfies $x_\ell(0, i) = 0$, $\forall\, i \in \mathbb{N}$ and W_{t_0} by $\bar{W}$. □

Proof of Proposition 9.62. The proof follows the same lines as in Section 8. (a) is the analogue of (8.867). (c) is the analogue of (8.849).

We briefly indicate the modification required to the proofs in the case $M > 1$.
(a,b) We begin with the proof of the first moment calculation and consider the initial configuration where we have

$$\tilde{x}_\ell^N(t_0, i) = 0 \text{ for every } i \notin \mathcal{I}(\varepsilon) \text{ and } \ell \in E_1. \tag{9.581}$$

Recall that here $U^N(t, \bar{\mathfrak{t}})$ is the frequency that an occupied site has at time t microstate $\bar{\mathfrak{t}}$, see (9.501) for $U(t, \cdot)$ and (9.551) for $U^0(\infty, \cdot)$.

Denote with

$$\mu_{t_0}^\otimes(\bar{\mathfrak{t}}) \in \mathcal{M}_1((\mathbb{I}_M \setminus \{M+1\})^\mathbb{N}), \mu_{t_0}^\otimes(i, \bar{\mathfrak{t}}) \in \mathcal{M}_1((\mathbb{I}_M \setminus \{M+1\})^\mathbb{N}), \quad \bar{\mathfrak{t}} \in \bar{\mathfrak{T}},$$
$$i \in \{1, \cdots, N\}, \tag{9.582}$$

the former is (product) measure of the event (subset of $\mathbb{I}_M^\mathbb{N}$) associated to $\bar{\mathfrak{t}} \in \bar{\mathfrak{T}}$. For the latter recall that the configurations $\bar{\mathfrak{t}}$ are defined at a specific site but contains the parent state information and hence the resulting set of involved sites may consist of a finite random number of other sites (i.e. Λ_A, Λ_D). Similarly

$$\mu_{t_0}^\otimes(i, \bar{\mathfrak{t}}) \tag{9.583}$$

is the corresponding measure associated with site i and microstate $\bar{\mathfrak{t}}$ at the time t_0 when the mutation $E_0 \to E_1$ stops to act.

Hence noting that there will be a $Bin(K_t^N, \frac{k}{N})$ number of hits of the k-set $\mathcal{I}(\varepsilon)$ by the dual cloud, we get the expression (here $t > t_0$):

$$E[x_{E_0}^N(t, j)] = E\left[(1 - \frac{k}{N})^{\bar{K}_t^N} + (1 - \frac{k}{N})^{\bar{K}_t^N - 1} \frac{\bar{K}_t^N}{N} \sum_{i=1}^{k} \sum_{\bar{\mathfrak{t}} \in \mathfrak{T}^E} \mu_{t_0}^\otimes(i, \bar{\mathfrak{t}}) \bar{U}^N(t, \bar{\mathfrak{t}})\right.$$

$$+ (1 - \frac{k}{N})^{\bar{K}_t^N - 2} \frac{\bar{K}_t^N(\bar{K}_t^N - 1)}{2N^2}$$

$$\left.\times \sum_{i_1, i_2=1}^{k} \sum_{\bar{\mathfrak{t}}_1, \bar{\mathfrak{t}}_2} \mu_{t_0}^\otimes(i_1, \bar{\mathfrak{t}}_1) \mu_{t_0}^\otimes(i_2, \bar{\mathfrak{t}}_2) \bar{U}^N(t, \bar{\mathfrak{t}}_1) \bar{U}^N(t, \bar{\mathfrak{t}}_2)\right]$$

$$+ o(N^{-2}). \tag{9.584}$$

Note that as $N \to \infty$, $\bar{U}^N(t, \bar{\mathfrak{t}}) \to \bar{U}(t, \bar{\mathfrak{t}})$, with $\bar{U}(t, \cdot)$ the microstate distribution of the McKean–Vlasov dual process at time t.

We rewrite the formula (9.584) as follows. Define

9.11 The Transition Regime and Emergence: Mutation-Dominant Case

$$g^N(t,\mu) := 1 - \sum_{\bar{i} \in \mathfrak{T}^E} \mu^\otimes(\bar{t}) \bar{U}^N(t,\bar{t}), \tag{9.585}$$

$$g(t,\mu) := \left(1 - \sum_{\bar{i} \in \mathfrak{T}^E} \mu^\otimes(\bar{t}) \bar{U}(t,\bar{t})\right), \quad g(\infty,\mu) := \left(1 - \sum_{\bar{i} \in \mathfrak{T}^E} \mu^\otimes(\bar{t}) \bar{U}^0(\infty,\bar{t})\right) \tag{9.586}$$

and (here we do not need the $N = \infty, t = \infty$ versions):

$$g^{N,2}(t,\mu) = 1 - \sum_{\bar{t}_1,\bar{t}_2} \mu \otimes (\bar{t}_1) \mu \otimes (\bar{t}_2) U^N(t,\bar{t}_1) U^N(t,\bar{t}_2). \tag{9.587}$$

Note that g^N, $g^{N,2}$ depend on the initial state of the dual process via $U^N = U^{N,(1,1)}$.

Remark 180. The function $g(\infty,\mu)$ is obtained by integrating with respect to independent copies of the selection-mutation equilibrium on E_0 at the other sites referred to by the propagation of chaos property of the N-interacting Fleming–Viot diffusions as $N \to \infty$. Note however that this excludes the case in which ℓ is one of the sites occupied by the droplet and one or more sites in $\Lambda_A(\ell) \cup \Lambda_D(\ell)$ is also one of the $k(\varepsilon, N)$ sites occupied by the droplet. Therefore there is an additional correction. However this is a correction of order $\frac{\bar{K}_t^N}{N^2}$ and is therefore negligible with respect to the other terms (9.571) in the limit as $\bar{K}_t^N \to \infty$ as $N \to \infty$.

Then we can rewrite (9.584). Here $\mu_{t_0}(i)$ is the state (of the measure-valued process) at time t_0 at site i:

$$E[x_{E_0}^N(t,j)]$$

$$= E\left[1 - \frac{K_t^N}{N} \sum_{i=1}^k g^N(t,\mu_{t_0}(i))\right.$$

$$+ \frac{K_t^N(K_t^N - 1)}{2N^2} \left(\sum_{i_2=1}^k \sum_{i_1 \neq i_2}^k g^N(t,\mu_{t_0}(i_1)) g^N(t,\mu_{t_0}(i_2))\right)$$

$$\left. + \sum_{i=1}^k g^{N,2}(t,\mu_{t_0}(i))\right] + o(N^{-2}). \tag{9.588}$$

We then proceed as in (8.867).

(c) We now consider the second moment calculation again using the dual particle system. As before in the proof of part (a) we consider first the system arising by setting the type (M+1) mass equal to zero outside the set $\mathcal{I}(\varepsilon)$ and then later we shall let $\varepsilon \to 0$ to obtain our claim for the general case with the same argument as in part (a).

First note that

$$E[\hat{x}_{M+1}^N(t+s)\hat{x}_{M+1}^N(t)] \quad (9.589)$$
$$= E[\hat{x}_{M+1}^N(t+s)]E[\hat{x}_{M+1}^N(t)] + Cov[\hat{x}_{M+1}^N(t+s), \hat{x}_{M+1}^N(t)]$$
$$= E[\hat{x}_{M+1}^N(t+s)]E[\hat{x}_{M+1}^N(t)] + \sum_{i=1}^N Cov[x_{E_0}^N(t+s,i), x_{E_0}^N(t,i)]$$
$$+ \sum_{j \neq k}^N Cov[x_{E_0}^N(t+s,j)), x_{E_0}^N(t,k)].$$

As in the $M = 1$ case it is necessary to consider terms of order $O(\frac{1}{N^2})$ in the computation of $Cov[x_{E_0}^N(t+s,j)], x_{E_0}^N(t,k))$. Therefore the calculation is organized to identify the terms up to those of order $o(N^2)$ as $N \to \infty$, and then to calculate the contributing terms of $O(N^{-2})$.

To handle the calculation of $Cov(x_{E_0}^N(t+s,j), x_{E_0}^N(t,k))$ both for $j \neq k$ and for $j = k = i$ we use the dual process. We introduce as "initial condition" for the dual particle system:

one initial dual particle at j_1 at time 0 and a second dual particle at j_2 at time s (9.590)

and choose as associated function 1_{E_0} in both cases. Then consider the dual cloud resulting from these at time $t + s$.

Since we shall fix times 0 and s where the j_1 respectively j_2 cloud start evolving we abbreviate the number of sites occupied by the cloud at time u and similarly the process of the frequency distribution of sizes of sites by

$$\bar{K}_u^{N,j_1,j_2} = \bar{K}_u^{N,(j_1,0),(j_2,s)} \text{ and } \bar{U}^{N,j_1,j_2} = \bar{U}^{N,(j_1,0),(j_2,s)}. \quad (9.591)$$

The second moment calculation now runs as follows. There will be a $Bin(\bar{K}_t^{N,j_1,j_2}, \frac{k}{N})$ (more precisely $Bin(\bar{K}_t^{N,j_1,j_2} - 2, \frac{k}{N})$) number of hits of the k-set $\mathcal{I}(\varepsilon)$ for $j_1, j_2 \neq \{1, 2, \cdots, k\}$, hence as $N \to \infty$ we obtain:

$$E[x_{E_0}^N(t+s, j_1) \cdot x_{E_0}^N(t, j_2)]$$
$$= E\left[(1 - \frac{k}{N})^{\bar{K}_{t+s}^{N,j_1,j_2}} + \frac{K_{t+s}^{N,j_1,j_2}}{N}(1 - \frac{k}{N})^{\bar{K}_{t+s}^{N,j_1,j_2}-1} \sum_{i=1}^k \sum_{\bar{i} \in \mathcal{I}^E}^\infty \mu_{t_0}(i, \bar{t})\bar{U}_{t+s}^{N,j_1,j_2}(\bar{t})\right.$$
$$+ (1 - \frac{k}{N})^{\bar{K}_{t+s}^{N,j_1,j_2}-2} \frac{\bar{K}_{t+s}^{N,j_1,j_2}(\bar{K}_t^{N,j_1,j_2} - 1)}{2N^2}$$

9.11 The Transition Regime and Emergence: Mutation-Dominant Case

$$\sum_{i_1,i_2=1}^{k}\sum_{\bar{t}_1,\bar{t}_2\in\mathfrak{T}}\bar{U}^{N,j_1,j_2}(t+s,\bar{t}_1)\bar{U}^{N,j_1,j_2}(t,\bar{t}_2)\mu_{t_0}^{\otimes}(i_1,\bar{t}_1)\mu_{t_0}^{\otimes}(i_2,\bar{t}_2)\Bigg]$$

$$+o(N^{-2}). \tag{9.592}$$

Therefore if we now use the function $g^N(t,x)$ defined in (8.860) and modified to $g^N(t,\mu)$ as in (9.582), but now for the initial state specified by the space-time points $(j_1,0),(j_2,s)$ (note $g^N = g^{N,(j_1,0)}$, respectively $= g^{N,(j_2,0)}$), we can rewrite the above expression. However as in Section 8 there is a small problem with the time index, since our two clouds now have a time delay s therefore depending on which of the two clouds has the hit of $\mathcal{I}(\varepsilon)$ we have to use t or $t+s$ as the argument in our function $g^N(\cdot,x)$. Here we use therefore the *notation* $t(\cdot)$ which indicates that actually the time parameter here is either t or $t+s$ and depends on which of the two clouds is involved. However this abuse of notation is harmless since in the arguments below both the times t and $t+s$ will go to ∞.

We get, with the above conventions, from the formula (9.592) that:

$$E[x_{E_0}^N(t+s,j_1)x_{E_0}^N(t,j_2)] \tag{9.593}$$

$$= 1 - E\left[\frac{\bar{K}_{t+s}^{N,j_1,j_2}}{N}\sum_{i=1}^{k}g^N(t+s,\mu_{t_0}(i))\right]$$

$$+ E\left[\frac{\bar{K}_{t+s}^{N,j_1,j_2}(\bar{K}_{t+s}^{N,j_1,j_2}-1)}{2N^2}\sum_{i_1\neq i_2=1}^{k}g^N(t+s,\mu_{t_0}(i_1))g^N(t+s,\mu_{t_0}(i_2))\right]$$

$$+o(N^{-2}).$$

Now we continue and calculate the covariances of $x_{E_0}^N(t+s,j_1)$ and $x_{E_0}^N(t,j_2)$ based on (9.593) and the first moment calculations. For this purpose we can consider the dual systems starting in $(j_1,0)$ respectively (j_2,s) and evolving *independently* and the one starting with a factor at each time-space point $(j_1,0)$ and (j_2,s) but evolving together and with *interaction*. As we saw we can couple all these evolutions via a multicolour particles dynamic. Using the notation of the above first and second moment calculations we get with setting below

$$s_m = 0 \text{ if } m = 1 \text{ and } s_m = s \text{ if } m = 2, \tag{9.594}$$

the formula

$$\text{Cov}[x_1^N(t+s,j_1)x_1^N(t,j_2)] \tag{9.595}$$

$$= 1 - E\left[\frac{\bar{K}_{t+s}^{N,j_1,j_2}}{N}\sum_{i=1}^{k}g^N(t(\cdot),\mu_{t_0}(i))\right]$$

$$+E\left[\frac{\bar{K}^{N,j_1,j_2}_{t+s}(\bar{K}^{N,j_1,j_2}_{t+s}-1)}{2N^2}\sum_{i_1\neq i_2=1}^{k}g^N(t(\cdot),\mu_{t_0}(i_1))g^N(t(\cdot),\mu_{t_0}(i_2))\right]$$

$$-\prod_{m=1}^{2}E\left[1-\frac{\bar{K}^{N,j_m}(t+s_m)}{N}\sum_{i_m=1}^{k}g^N(t+s_m,\mu_{t_0}(i_m))\right.$$

$$\left.+\frac{\bar{K}^{N,j_m}_{t+s_m}(\bar{K}^{N,j_m}_{t+s_m}-1)}{2N^2}\sum_{i_{m,1}=1}^{k}\sum_{i_{m,2}=1}^{k}g^N(t+s_m,\mu_{t_0}(i_{m,1}))g^N((t+s_m,\mu_{t_0}(i_{m,2}))\right]$$

$$+o(N^{-2}).$$

We now have to distinguish two cases, namely $j_1 \neq j_2$ and $j_1 = j_2$ and we argue separately in the two cases.

Case $j_1 \neq j_2$. With (9.595) we get using (9.588) that:

$$\text{Cov}[x^N_{M+1}(t+s,j_1), x^N_{M+1}(t,j_2)] \tag{9.596}$$

$$= \frac{1}{N^2}\left\{E[N(\bar{K}^{N,j_1}_{t+s}+\bar{K}^{N,j_2}_t-\bar{K}^{N,j_1,j_2}_{t+s,t})\sum_{i=1}^{k}g^N(t(\cdot),\mu_{t_0}(i))]\right\}+o(N^{-2}).$$

Case $j_1 = j_2$ The expression here is a slight modification of the above, since either we have as first step a coalescence and are left with one particle or we get first a migration event and get a $j_1 \neq j_2$ case, or finally we get a birth event first. Since we have the time delay s of course coalescence can act only with this delay. Therefore up to terms of order $o(\frac{1}{N^2})$ as $N \to \infty$,

$$\text{Cov}[\hat{x}^N_{M+1}(t+s), \hat{x}^N_{M+1}(t)] \tag{9.597}$$

$$= \left\{E[N(\bar{K}^{N,j_1}_{t+s}+\bar{K}^{N,j_2}_t-\bar{K}^{N,j_1,j_2}_{t+s,t})\sum_{i=1}^{k}g^N(t,\mu_{t_0}(i))]\right\}+o(1).$$

Recall that $g^N(t,\mu_{t_0}(i_1)) \to g(t,\mu_{t_0}(i_1))$ as $N \to \infty$. Therefore the main point is to identify the behaviour of

$$N(\bar{K}^{N,j_1}_{t+s}+\bar{K}^{N,j_2}_{t+s}-K^{N,j_1,j_2}_{t+s}), \text{ as } N \to \infty, \tag{9.598}$$

which is N times the difference between two independent dual populations starting in j_1 and j_2 each with a factor 1_{E_0} in both cases and one combined system which is interacting through the collisions and to estimate the difference we must take into account the collisions of the dual populations.

In order to analyse the dual representation of the r.h.s. of equation (9.589) rewritten in the form (9.597) we will follow closely the arguments given in Section 8 for this type of analysis and we again work with the construction of the *enriched*

9.11 The Transition Regime and Emergence: Mutation-Dominant Case

coloured system which allows us to identify the contribution of on the one hand the two evolving independent clouds and on the other hand the effects of the interaction between the two clouds by collision and subsequent coalescence. This means every coloured rank carries a "tag" whether it is descending from the j_1 or the j_2 rank. In particular each colour is now of either of the two types and for the colour green we have to distinguish $G_1, G_2, G_{1,2}$ depending on whether two type-1, two type-2 or one type-1 and 1 type-2 rank interacted. (Compare Section 8 for the details of the construction and the rules of the dynamic.)

We now consider the system of *coloured ranks* (observing $W_1, W_2, R_1, R_2, R_{1,2}$, $G_1, G_2, G_{1,2}$ ranks which are relevant for the accuracy needed here). Recall that the $G_{1,2}$ ranks are produced in the coloured ranks system from collision of W_1 and W_2 ranks, that is W_1 and W_2 ranks at the same sites which then suffer coalescence. Such an event reduces the number of occupied sites in the interacting clouds by 1 and then creates a $G_{1,2}$ family. Then the number of occupied sites in the system of interacting populations is denoted shortly $\bar{K}_t^{N,j_1,j_2}$ analog to (8.872).

First we observe that if we evaluate the duality relation (recall the form of the initial state we consider here) we see that the contributing terms of order $O(\frac{1}{N^2})$ arise from either double or single hits of the dual cloud of ranks of a point in $\mathcal{I}(\varepsilon)$.

We note that in general there are two cases which must be handled differently, namely, the case in which the difference in (9.598) is based on a Poisson collision event with mean of order $O(\frac{1}{N})$ as studied in Section 8 in Lemma 8.26 (b) and the other in which there is a law of large numbers effect as exhibited in Section 8 in (8.485) and Lemma 8.26(a). However in both cases the expected value used in the calculations below have the same form. In the present case for fixed t we are in the Poisson case with vanishing rate so that at most one Poisson event is involved.

Using (9.559) we see that the key quantity in (9.598) satisfies as $N \to \infty$ that:

$$E[N(\bar{K}_{t+s}^{N,j_1} + K_t^{N,j_2} - \bar{K}_{t+s,t}^{N,j_1,j_2})] \sim N \cdot E[\mathfrak{g}_{(1,2)}(W'(\cdot), W''(\cdot), N, s, t)] \quad (9.599)$$

where $\mathfrak{g}_{(1,2)}(W'(\cdot), W''(\cdot), N, s, t)$ is given by the analogue of (8.741).

Remark 181. Here we have noted that using the coupling introduced above the number of *single hits* of the k sites in $\mathcal{I}(\varepsilon)$ coming from either one of the two interacting clouds, can be obtained by deducting from the single hits of the "non-interacting W_1 and W_2 ranks" the single hits by $G_{1,2}$ ranks in the interacting system. Moreover the event of a *double hit*, that two occupied sites in the dual cloud coincide with one of the $k = k(\varepsilon)$ sites, can occur from either two W_1 sites, two W_2 sites or one W_1 and one W_2 site. (Since the number of $G_{1,2}$ sites or including sites at which both W_1 and W_2 ranks coexist are of order $O(\frac{1}{N})$, the contribution of double hits involving $G_{1,2}$ sites is of higher order and can be ignored.)

Taking the limit as $N \to \infty$ leads to

$$\mathrm{Cov}[\hat{x}^N_{M+1}(t+s), \hat{x}^N_{M+1}(t)] \xrightarrow[N\to\infty]{} 2\kappa_2 e^{\beta(2t+s)} E[(W'(t+s)W''(t)\sum_{i=1}^{k} g(t, \mu_{t_0}(i))]. \tag{9.600}$$

Recalling (8.868), (8.763) and the fact that $\mathcal{U}(t,k) \to \mathcal{U}(\infty, k)$ as $t \to \infty$, we get next as $t \to \infty$:

r.h.s. of (9.600) $\sim 2e^{\beta(2t+s)} E\left[\kappa_2 W'W'' \sum_{i=1}^{k} g(t, \mu_{t_0}(i))\right]$ and

$$\left| e^{-(2t+s)} \mathrm{Cov}[\hat{x}^N_{M+1}(t+s), \hat{x}^N_{M+1}(t)] - 2(E[W])^2 \left[\kappa_2 \sum_{i=1}^{k} g(\infty, \mu_{t_0}(i))\right] \right| \to 0, \tag{9.601}$$

uniformly in s. To verify the uniformity we work with the expression for the expected number of $G_{1,2}$ sites:

$$E\left[\int_s^{t+s} W'(u)W''(u-s)e^{\beta(u-s)}e^{\beta u}W_g(t+s,u)e^{\beta(t+s-u)}du\right]. \tag{9.602}$$

Then using $W_i(u) \to W_i$ as $u \to \infty$ and showing that the contribution from the early collisions (in $[0, s + s_0(t)]$, $s_0(t) = o(t)$) form a negligible contribution as we agreed in Section 8 in (8.763) and hence we obtain

$$\left| e^{-\beta(2t+s)} E\left[\int_s^{t+s} W_i'(u)W''(u-s)e^{\beta(u-s)}e^{\beta u}W_g(t+s,u)e^{\beta(t+s-u)}du\right] \right.$$
$$\left. - (E[W])^2 \left[\kappa_2 \sum_{i=1}^{k} g(\infty, \mu_{t_0}(i))\right] \right| \leq \mathrm{const} \cdot e^{-\beta t} \text{ as } t \to \infty. \tag{9.603}$$

This proves (9.579) and completes the proof of (c).

(d) The result follows from an explicit calculation as in Section 8 based on (a)–(c).

(e) Here we have to use that the mutant mass entering after a large time t_{late} is as $N \to \infty, t \to \infty$ negligible (in the sense of convergence in L_2) compared to the one entering before. To verify this denote the subpopulation produced by mutation acting only during the time interval $[k, k+1]$ by $\sum_{\ell \in E_1} \hat{x}^{N,k}_\ell(t)$. Then we have

$$E[(W_{n+1} - W_n)^2] = \lim_{t\to\infty} \lim_{N\to\infty} E[(e^{-\beta t} \sum_{\ell \in E_1} \hat{x}^{N,k}_\ell(t))^2] \leq \mathrm{const} \cdot e^{-2\beta n}. \tag{9.604}$$

This implies that W_{t_0} is Cauchy in L^2 and there exists a limit $\bar{W}$ such that $W_{t_0} \to \bar{W}$ in L^2 as $t_0 \to \infty$. Then the argument is finished as in Section 8.

9.11 The Transition Regime and Emergence: Mutation-Dominant Case

Remark 182. We use here that the asymptotic distribution of k tagged sites are given by i.i.d. random tableau where the sets (tableau) are chosen using $U(t, \cdot)$ obtained from the nonlinear equation. This can be seen as follows. Consider the local dynamics (with emigration) at a finite set of k sites as $N \to \infty$. The probability that any two of the sites belong to same direct clan of size $O(1)$ converges to 0 and the connected sets of clans in the low density regime have the same property. The local transitions are independent but they both have immigrants from the same source given by the empirical measure. In the limit $N \to \infty$ the source is deterministic and given by the nonlinear dynamics. Also the probability that there are any migrations among the fixed k sites converges to 0 as $N \to \infty$.

9.11.4 Emergence and Rare Mutation Events: Hazard Rate and Post Event Calculation

We want to show now emergence at time $\beta^{-1} \log N + t$ and hence we want that the expectation of type $M+1$ at a tagged site will be less than 1 at this time of emergence but also that it is as $N \to \infty$ strictly positive. We first consider the dual starting from $\tilde{\mathcal{G}}^{++,N}$ with one single factor $(1, \ldots, 1, 0)$. To obtain information on emergence we must first determine the random time for the *first occurrence of a rare mutation event* for the dual process.

We note that in contrast to the case $M=1$ such a rare mutation event does not necessarily result in a zero factor and hence gives a jump to 0 of the dual expression for the first moment, which in turn makes the expectation of the dual expression smaller than 1. In the case $M \geq 2$ the first occurrence of a zero factor involves a rare mutation followed by one or more rate one mutations.

Example 16. Consider $M=2$ and the tableau

010	111	111
100	010	111
100	100	110,

which changes by rare mutation ($2 \to 3$) in the rank (column) 2 to

010	111	111
100	100	110

and then to $\boxed{000}$

by mutation $2 \to 1$ in column 1 and $1 \to 2$ in column 2.

Note however immediately after the rare mutation the integrated value is $x_2 + x_1^2 < 1$ if the initial probability is $\mu = (x_1, x_2, 0)$ and $x_1 \neq 1$, if $x_1 = 1$ the value is strictly less than one on the event that the two mutations occur as described above. We learn from this example in particular that the integrated value of the dual expression at the time of rare mutation depends on the initial state on the type distribution on lower level E_0.

In order for the first rare mutation event to occur there must be $O(N)$ (active) ranks which we have shown occurs in $\tilde{\mathcal{G}}^{++,N}$ at times $\beta^{-1} \log N + t$ with t

arbitrarily small ($<< 0$), if we ignore collisions and replace the dual $\tilde{\mathcal{G}}^{++,N}$ by the process $\tilde{\mathcal{G}}^{++}$. Next consider the system when there are $\bar{K}_t^N$ permanent sites. Then each of these indicator functions can be bounded above by $(1,\ldots,1,0)$ and therefore the rate of jumps to zero is at least proportional to the number of occupied sites, namely a rare mutation from type $i \in \{1,\ldots,M\}$ to type $M+1$ leads to $(1,\ldots,0,\ldots,1,0)$ (the 0 in the i-th position which followed by the mutations at rate $O(1)$ can lead to $(0,\ldots,0)$ after a finite random time τ after the rare mutation and then the integrated value is $\int \tilde{\mathcal{G}}_{\tau+}^{++,N} d\mu^\otimes = 0$.

For example, if the resolved factor $(1,\ldots,1,0)$, jumps by rare mutation to $(01\cdots10)$, then we can have the sequence of mutations:

$$(01,\ldots,1,0) \xrightarrow[\tilde{m}]{} (01\cdots01\cdots0) \longrightarrow \cdots \xrightarrow[\Pi_{j=2}^M m_{j1}]{} (0,\ldots,0,0), \qquad (9.605)$$

where $\tilde{m}$ is $m_{j,1}$ with $j \in \{2,\cdots,M\}$ at the j-th arrow. Note that there is a positive probability that this sequence of transitions occurs in an interval of finite length and *before* the next selection operation.

Moreover, transition to $(0,\ldots,0,0)$ can occur even if a selection operator acts. For example in the case $M = 2$, if a selection transition after the rare and before a regular mutation acting on (110) we obtain

$$(110) \xrightarrow[\text{rare}]{} (100) \xrightarrow[s]{} (100) \otimes (100) \xrightarrow[m_1]{} (000). \qquad (9.606)$$

This argument can be continued with further selection operators acting. Therefore the occurrence of the rare mutation is exactly the time needed to have emergence.

There are three key points of difference from the case $M = 1$

(i) since each active rank at a site can undergo rare mutation the time that this first occurs depends on the integral over time of a function of the current *microstate* in each occupied site; given the occurrence of a rare mutation the associated rank is randomly chosen from the set of all active ranks,
(ii) the probability that after a rare mutation further $O(1)$ mutations result in a zero factor and the distribution of the time until this jump occurs after rare mutation depends on the current microstate of the site containing the mutant individual.
(iii) the integrated value after a rare mutation can depend on the initial probability μ_0.

To investigate the emergence regime we use $\tilde{\mathcal{G}}^{++,N}$, K_t^N, $U^N(t,\cdot)$. We now summarize the ingredients used to obtain the asymptotics of the moments of $x_{E_1}(\frac{\log N}{\beta} + t)$ as $t \to -\infty$ and to prove the main results in the case $M \geq 2$. Let

T_N^{rare} time of the first rare mutation event, for the dual started with $(1,\ldots,1,0)$. \hfill (9.607)

Then we calculate the hazard function and $Prob(T_N^{\text{rare}} \geq t)$ as follows.

9.11 The Transition Regime and Emergence: Mutation-Dominant Case

Lemma 9.63 (Rare mutation transition: hazard function).
We consider the dual process $\tilde{\mathcal{G}}^{++,N}$ starting with a single factor $(1,\ldots,1,0)$ at time 0.

(a) The probability that a rare mutation event has occurred in the dual configuration by time $\frac{\log N}{\beta} + t$ is given by:

$$P(T_N^{\text{rare}} \leq \beta^{-1} \log N + t) = 1 - e^{-\frac{mM+1,\text{up}}{N} R^N(\frac{\log N}{\beta}+t)}, \qquad (9.608)$$

where (recall the state $\bar{\mathfrak{t}}$ contains the tableau $\mathfrak{t}$ at the site and $|\bar{\mathfrak{t}}|$ is the number of ranks in $\mathfrak{t}$.)

$$R^N(t) := \int_0^{\frac{\log N}{\beta}+t} \left(\bar{K}_s^N \int_{\mathfrak{T}} |\bar{\mathfrak{t}}| \bar{U}^N(s, d\bar{\mathfrak{t}}) \right) ds. \qquad (9.609)$$

Furthermore setting

$$\bar{u}^N(s) = \bar{K}_{\frac{\log N}{\beta}+s}^N, \qquad (9.610)$$

we get as $N \to \infty$, $t \to -\infty$,

$$\frac{\bar{u}^N(t)}{N} \to u(t) = \bar{W} e^{\beta t} - \bar{\kappa} \bar{W}^2 e^{2\beta t} + O(e^{3\beta t}), \qquad (9.611)$$

$$P(T_N^{\text{rare}} \leq \beta^{-1} \log N + t) \xrightarrow[N \to \infty]{} r(t) \in (0,1), \qquad (9.612)$$

$$r(t) \xrightarrow[t \to \infty]{} 1, \quad r(t) \xrightarrow[t \to -\infty]{} 0. \qquad (9.613)$$

(b) Given a rare mutation event, the conditional (given that occurence) probability that this happens at site j and to local rank i is denoted $p(j,i)$ and is a function of the current state:

$$p(j,k) = \frac{1}{\sum_\ell |\mathfrak{t}_\ell|}, \quad k = 1,\ldots,|\mathfrak{t}_j|, \quad j \in \{1,\cdots,N\} \text{ with } j \text{ occupied}. \qquad (9.614)$$

(c) Assume that for $(\tilde{\mathcal{G}}_s^{++,N})_{s \in \mathbb{R}_+}$ the following is satisfied:

a rare mutation event has occurred at time u at site j and local rank k,   (9.615)

a second rare mutation event has not occurred and   (9.616)

the site j does not experience a collision event on the interval $[u,t]$.   (9.617)

Then the possible outcomes at time t are:

(i) a decoupling event occurs at an occupied site j and the rank i is either removed before time t or the rank i has resolved to $(1, \ldots, 1, 0)$,

(ii) the rank i has not yet resolved at time t and the state at j at time t is $\bar{\mathfrak{t}}'$ and it has not been removed by a decoupling event by this time,

(iii) the local rank k which experienced a rare mutation at time u has by that rare mutation resolved to $(0, \ldots, 0)$ by time t and has not been removed by decoupling before.

(d) the (conditional) probabilities of the three outcomes (i), (ii), (iii) in (c) are measurable functions of the state $\bar{\mathfrak{t}}$ at time u at site j, namely,

$$h_i^N(\bar{\mathfrak{t}}, k, t-u), \quad h_{II}^N(\bar{\mathfrak{t}}, k, t-u, \bar{\mathfrak{t}}'), \quad h_{III}^N(\bar{\mathfrak{t}}, k, t-u). \tag{9.618}$$

We now let $N \to \infty$ and t to $-\infty$, and get as limits

$$h_i(\bar{\mathfrak{t}}, k, u), h_{II}(\bar{\mathfrak{t}}, k, u, \bar{\mathfrak{t}}'), h_{III}(\bar{\mathfrak{t}}, k, u), \tag{9.619}$$

all strictly positive.

(e) Starting with two dual particles one at site j_1 at time 0 and one at j_2 at time s, we replace (9.610) by $\bar{u}^N(t) = \bar{K}^{N, j_1, j_2}_{\frac{\log N}{\beta} + t}$ as in (9.559). □

Remark 183. We note that if (c)(i) occurs then the integrated value becomes 1. If (c)(iii) occurs, then the integrated value becomes 0. If (c)(ii) occurs, then the integrated valued depends on the initial probability $(x_1, \ldots, x_M, 0)$. Therefore in contrast to the case $M = 1$ the moments of $x_{E_0}(\frac{\log N}{\beta} + t)$ as $N \to \infty$ can depend on the initial probability on E_0. This means that the law of $*\mathcal{W}$ can depend on the initial distribution of types of E_0.

Proof of Lemma 9.63. (a),(b) This follows since the rare mutation occurs with rate $\frac{m_{M+1,up}}{N}$ times the total number of active individuals since rare mutation acts on each of them with the same rate. The expression (9.611) is given by Proposition 9.59. Then we get for $N \to \infty$ and then $t \to -\infty$ explicit positive expressions which converge to a positive number. The formula gives then immediately $r(t) \to 0$ as $t \to -\infty$. In order to get then $r(t) \to 1$ as $t \to \infty$, we apply the duality between timepoints $\beta^{-1} \log N + t, t$ very small $9 \to -\infty$) and $\beta^{-1} \log N + s$ as $(s \to \infty)$. Then the intensity of E_1-types is $\geq \varepsilon$ at time $\beta^{-1} \log N + t$ and at time $\beta^{-1} \log N + s$, s the intensity of E_0-types then converges to zero as $s \to \infty$, due to the fitness difference.

(c),(d) Under the assumptions the microstate at site j evolves via the mutation between types in E_0 and selection. As we have shown above the probability of decoupling by time v is a measurable function of the mesostate. To verify (9.619) we use the measurability and boundedness of the functions h_i, h_{ii} and h_{iii} and the convergence of the age-type distribution in the $N \to \infty$ limit as $t \to -\infty$ given by Proposition 9.58(b).

(e) This is proved in Proposition 9.61. q.e.d.

9.12 Emergence and Fixation: The Selection-Dominant Case

We have to deal also with the general case with any parameters $\mathbf{m} = (m_{i,j})_{i,j \in \{1,\cdots,M\}}$, c, d, s and $(e_1, \ldots, e_M)$. In contrast to Subsections 9.10 and 9.11 we consider therefore throughout this Subsection 9.12 emergence and fixation in the *selection-dominant* case defined in Definition 9.47 rather than in the *mutation-dominant* regime (recall Definition 9.46) treated in Subsections 9.10 and 9.11. In fact the strategy here could in principle work in some of the intermediate cases, however to avoid treating several cases in the proofs, we focus on the selection-dominant regime and comment on some of the intermediate cases shortly.

The objective of this subsection is to outline the various steps in the proof of emergence and the identification of the random initial condition for the entrance law of the McKean–Vlasov equation following the same basic strategy as in Section 8.

Of course to establish emergence for the rare mutant population we want to work with the *duality* again. We have to return to the dual dynamics presented in Subsection 9.6, which give representations of the set-valued dual process $\tilde{\mathcal{G}}^{++}$ for which we can analyse the growth behaviour. We now use the process $\mathfrak{s}_{I,\text{enr}}$ in the collision-free regime for which the reader may recall Step 2 in Subsection 9.6.3. Later on we work with a suitable analog in the collision regime on $\{1, \cdots, N\}$.

However in the case of selection dominance we can no longer work directly with the *empirical measure of microstates* (as in Subsections 9.10.2 and 9.10.3), we have to work with the *empirical measure of mesostates* associated to the population of labels (each of which consists of a multisite tableau located at a direct clan of sites which start from the founding site the unique associated *Pe*-site, see (9.294) and the sequel).

The concrete analysis involves three basic parts:

1. Consider the dual process and show that the limit as $N \to \infty$ of the empirical measure of mesostates at time $\beta^{-1} \log N + t$, $t \in \mathbb{R}$ satisfies the McKean–Vlasov equation - Proposition 9.3
2. Determining for the dual process the malthusian parameters for the number of occupied sites β and number of mesostates β_{Pe}, establishing $\beta = \beta_{\text{Pe}}$ for ergodic mutation on E_0 and to carry out the $\mathcal{W}^* =^* \mathcal{W}$ calculation (9.950) - this will follow from (9.965).
3. Proof that (1) and (2) define a unique solution (random entrance law) of the McKean–Vlasov equation for the dual proess with time parameter $\mathbb{R}$ and establish then for the original process emergence - Proposition 10.2(a)(ii).

In this subsection we focus first on (2) and then we come to the two other points.

However to carry this out there are a number of technical complications in the selection-dominant multitype case. For example, we will see that:

(I) We need to work with a class of CMJ-processes with *countably* many types if we want to study the population of labels with their mesostates. However the nice features allowing to cope with this problem is that these processes have a *two level* branching structure, a topic of interest in its own. Namely the

population of labels with their mesostates has the structure of a (countably) multitype CMJ-branching population, but also the evolution of types (i.e. the internal mesostate dynamics) follows a "generalized" branching dynamic, namely one with additional *catastrophic events*, which of course destroys the classical branching property.
(II) we cannot always assume a priori that the size of a mesostate associated with a label (given by a *Pe* site) has first and second moments which remain bounded in the $\frac{\log N}{\beta}$ time scale (used for example above in (9.546)) and we have to derive this for our range of parameters from the ergodicity of mutation on E_0.

Remark 184. The objective of this subsection is to lay out a program in some detail to obtain the necessary results in (2). However there are some steps for which we only sketch a strategy of proof and do not provide the full proof itself.

In particular two open question are:

- In the general case of ergodic mutation on E_0, i.e. case $M \geq 2, d \geq 0, m_{i,j} > 0 \; \forall \; i, j \in t_0$, we prove that the $(1 + \varepsilon)$-moment of the mesostate size at decoupling is finite for some $\varepsilon > 0$ and the second moment of the number of Pe-offspring at decoupling is finite. If we could replace the $(1 + \varepsilon)$-moment by a second moment some proofs would become much easier. However we obtain a lower bound on mesostate growth that has a finite second moment for the size but at most finitely many finite moments at decoupling.
- In the non-ergodic case, for example $d = 2$, $m_{12} = 0$, the existence of the limiting joint distribution of microstates and their lifetimes is not proven. Recall that in our scenario described in Section 1–6 this case is excluded anyway, however this case is interesting for other questions.

In order to formulate a possible approach in this case we present an Ansatz and discuss the difficulties and a possible program to establish it at the end of this Subsection 9.12. We conjecture that these steps can be fully justified.

between droplet and emergence which is complemented in 8.11.11 with the final emergence analysis.

Outline of Subsection 9.12

The whole Subsection 9.12 involves the analysis of the set-valued dual in the time window $\beta^{-1} \log N + t$, $t \in \mathbb{R}$. We begin in Subsubsections 9.12.1–9.12.8 with the description of the branching mesostate dynamics in the collision-free regime, i.e. in the limit $N \to \infty$ for times $o(\log N)$, hence with $S = \mathbb{N}$, calculate the associated *Malthusian parameter* and derive the mesostate analogue of the *stable type distribution* for sites from Section 8 (for two types) respectively Subsections 9.10–9.11 for microstates and the $M \geq 2$ case. Here Subsections 9.12.2–9.12.6 prepare the ground as we explain in 9.12.1 and 9.12.7 and 9.12.8 carry out the analysis of the exponential growth and stable age-type distribution. Then the empirical process of mesostates is introduced in Subsubsection 9.12.9 and we then come to the evolution equation in the collision regime and in 9.12.10 we turn to the transition regime between droplet and emergence which is complemented in 9.12.11 with the final convergence analysis.

More precisely we consider next in Subsections 9.12.9–9.12.10 the case $S = \{1, \ldots, N\}$ in which labels can *collide* and two of their ranks can coalesce and form a *compound label*. Here we consider the *empirical distribution of mesostates* and its limiting dynamics in the $N \to \infty$ limit both for $O(1)$-time scales and $\beta^{-1} \log N + t$ ($t \in \mathbb{R}$) time scales.

This analysis of these two models is then used in Subsections 9.12.11–9.12.13 to verify for the rare mutant population emergence and to identify the random initial condition in the McKean–Vlasov limit, using covariance calculations and L_2-arguments as in Section 8.

Finally in Subsections 9.12.14–9.12.15 we discuss aspects of the case of non-ergodic mutation on E_0 which is relevant for further research.

9.12.1 Mesostate Branching Dynamics in the Collision-Free Case: Overview

We focus now on the structure of the *dual process*. In this subsubsection we consider the collision-free case, $S = \mathbb{N}$ and the dynamic $\mathfrak{s}_{I,\mathrm{enr}}$, see (9.293). We begin by identifying the *exponential growth rate* of the *number of labels* ($\in \mathbb{N}_\ell$) identified by their (founding) P_e sites and their mesostates. Recall that the countable set of mesostates is denoted by $\hat{\mathfrak{T}}$, see (9.288). In the collision-free case (i.e. $S = \mathbb{N}$) there is a one-to-one correspondence between P_e sites and labels associated with a particular mesostate. Furthermore we establish the existence and properties of the *stable type distribution* of mesostates in the dual population.

Recall that a mesostate at a label is an indecomposable marked multisite tableau as in (9.288) in which each rank consists of a column of subsets of $\mathbb{I}_M$ (disjoint or replicas) and is assigned a mark, such that in particular ranks with the same mark occupy the same site. Given a state of the dual $\tilde{\mathcal{G}}^{++}$, i.e. an element $G \in \mathsf{I}_S$ (recall 9.181), it can be uniquely factored into a *product* with *finitely many nontrivial factors*

$$G = \prod_{j \in \mathbb{N}_\ell} \sigma(\hat{\mathfrak{t}}_j) \in (\mathbb{I}^\mathbb{N} \times \Upsilon)^{\mathbb{N}_\ell}, \tag{9.620}$$

of *marked sets* associated to a multisite tableau where $\sigma(\hat{\mathfrak{t}}_j)$ is a subset of $\mathbb{I}^\mathbb{N}$ marked with the migration information $v \in \Upsilon$ and defined by the mapping

$$\sigma : \hat{\mathfrak{T}} \to \mathcal{T} \times \Upsilon, \qquad \sigma((\tau, v, \tilde{v})) = (\tau, v), \tag{9.621}$$

where $\hat{\mathfrak{T}}$ is as defined in (9.288).

Beginning with one initial label with mesostate $\hat{\mathfrak{t}}_*$ such that $\sigma(\hat{\mathfrak{t}}_*) = ((1,\ldots,1,0),\emptyset)$, which often we refer to shortly as mesostate $(1,\ldots,1,0)$, after a finite time and evolving via Option I dynamics we have a random finite collection of labels and associated mesostates.

Assumption: Unless otherwise indicated, we work under the Assumption (9.219) of *exchangeable initial conditions* which implies that the dual expression is invariant under permutations of $\mathbb{N}_g$. For our analysis of emergence we assume even that the initial measure is of the form $\mu_0^{\otimes}$. Then the resulting integrals are invariant under permutations on $\mathbb{N}_g$. In this case we do not need the actual locations of labels but only the information specifying which ranks have the *same* location and can coalesce. Therefore it suffices to work with the empirical measure. $\square$

We will consider the *empirical measure process of mesostates* denoted $(\Psi_t)_{t\geq 0}$ associated to the process $\mathfrak{s}_{l,\text{enr}}$ as a finite counting measure on $\sigma(\hat{\mathfrak{T}})$, namely,

$$\Psi_t(\cdot) := \sum_{j\in\mathbb{N}_\ell} 1(\sigma(\hat{\mathfrak{t}}_j(t)) \neq (\mathbb{I}^{\mathbb{N}},\emptyset)) \cdot \delta_{\sigma(\hat{\mathfrak{t}}_j(t))}(\cdot) \in \mathcal{M}_{\text{fin}}(\sigma(\hat{\mathfrak{T}})). \quad (9.622)$$

The resulting process is a counting measure-valued Markov branching process with *countable* type space $\sigma(\hat{\mathfrak{T}})$. (In order to simplify the notation subsequently in this subsection we will often write simply $\hat{\mathfrak{t}}$ instead of $\sigma(\hat{\mathfrak{t}})$.)

Between decoupling times the set of mesostates forms an *independent collection* of Markov processes. At a decoupling time (i.e. times where the factor at a Pe-site, i.e. the τ-component in (9.621) resolves to the state $(11\cdots 10)$) at a Pe site its associated mesostate *reverts* to $((1,\ldots,1,0),\emptyset,*)$ (for some admissible $\tilde{v}$ as $*$) and a random number of *new offspring labels* with their initial mesostates are produced.

Since a Pe site never dies, the above transition results in a *supercritical multitype CMJ-process* in which the *individuals* in the CMJ are our *labels* and the *types* in the CMJ-process are the possible *mesostates*. Hence we have a CMJ-process with *countably* many types. However we have more structure, namely, the types evolve independently themselves by a *generalized multitype branching process*, generalized since *catastrophic events*, removing several ranks (and sites) at once, have to be built in, which destroy the classical branching property both for the process of ranks and for the process of sites. However the population of ranks can still be viewed as pruning a branching tree.

In subsection 9.7.4 we have identified the malthusian parameter β describing the growth of the number of Pe and UPe sites. In this subsection we focus on the malthusian parameter for the growth in the number of (irreducible) mesostates, β_{Pe}. It is immediate that $\beta \geq \beta_{Pe}$ but it is not apriori clear whether or not $\beta = \beta_{Pe}$. This question is related to the question as to whether the expected mesostate size (the number of sites it contains) remains bounded as $t \to \infty$.

We prove the following key results.

9.12 Emergence and Fixation: The Selection-Dominant Case

Proposition 9.64 (Finite $(1 + \varepsilon)$-moment mesostate length and mesostate offspring numbers).

We start the mesostate process in $\hat{t}_0$ with $\sigma(\hat{t}_0) = ((1, \ldots, 1, 0), \emptyset)$ and we assume that the mutation on E_0 is ergodic. Assume that we are in the selection-dominant case, at least $s > m$. Let T_{dec} be the first decoupling time of the process.

(a) The population of mesostates is a supercritical branching process.
(b) The random number of new Pe ranks produced at T_{dec} has a finite $(1 + \varepsilon)$-moment for some $\varepsilon > 0$.
(c) We have

$$\beta_{Pe} = \beta, \quad E[\|\hat{t}(t)\|] < \infty, \quad \forall t \geq 0, \tag{9.623}$$

and there exists an $\varepsilon > 0$, depending on the parameters, such that

$$E[(\|\hat{t}(T_{\text{dec}})\|)^{1+\varepsilon}] < \infty. \qquad \square \tag{9.624}$$

The above result of course raises the question, whether we can replace in the statement the $(1 + \varepsilon)$-moment by the second moment. It turns out that this is not true in general but holds for large enough selection rate s. We shall show

Proposition 9.65 (Selection-dominant case: finite second moments).

Given $m_{12}, m_{21}, m > 0$, there exists s_* such that the number of offspring mesostates produced at T_{dec} at a mesostate has a finite second moment for $s > s_*$. $\square$

This is rounded off by an example where $s > m$, but sufficiently close to m, such that mutation and selection are of comparable strength, in which in fact the second moment of the offspring number of new mesostates at T_{dec} when started in $\hat{t}_*$ is *infinite*, see Remark 192. This means we cannot in general work with L^2-techniques even though there are parameter ranges where this is possible as we saw in the above proposition and which then allows for simpler arguments. How can we prove these results?

A number of new technical issues are required in our analysis:

- the population of branching mesostates involves a *countably* many type CMJ process,
- the mesostate dynamics involves a new class of *multilevel birth and death* processes with *catastrophes*.

In order to introduce the main ideas and tools we proceed as follows:

- We first consider the special case in which $M = 2$, $d = 0$ and $m_{12} > 0, m_{21} > 0$. Most of the essential new aspects of mesostate dynamics already arise in this case. Then we extend the argument to $d > 0$.
- We then outline without full details the program to extend the results to the irreducible case with $M \geq 2, d > 0$.

We therefore begin in Subsubsection 9.12.2 with the description of the features of the mesostate processes, explained in the case $M = 2$ and $d = 0$. This leads to a generalized branching process the MBDC process (multitype birth and death process with catastrophes), which is after the discussion of concrete examples formally introduced in Definition 9.66. In Subsubsection 9.12.3 we then prove the properties of the mesostate length and the offspring distribution for the case $M = 2, d = 0$ which is then generalized in Subsubsection 9.12.4 to the case $M = 2, d > 0$. In Subsubsection 9.12.5 we then come to the questions of second moments (or higher) of mesostate length and mesostate offspring numbers. Finally in Subsubsection 9.12.6 we discuss the extension to $M > 2$.

We then at then end of Section 9.12 return to the material and discuss some aspects of the case where the mutation on E_0 is non-ergodic in order to provide some additional insights. Although our main results are formulated under the hypothesis that the mutation semigroup on E_0 is irreducible (which implies ergodicity), it is also instructive to consider this question in the non-ergodic case since it demonstrates the significance of the case in which $\beta > \beta_{Pe}$.

9.12.2 Mesostate Population Dynamics in the Case $M = 2, d = 0$

We now continue with the analysis of the population of mesostates and focus in this subsection on the case $M = 2, d = 0$. The simplification in the case $d = 0$ is that the location of factors and migration information need not be considered since now no coalescence occurs so that sites play no role in the dynamics. As mentioned earlier we have a *two-level branching dynamic*, on the population level a multitype supercritical CMJ-process of labels and their associated mesostates and on the individual mesostate level an internal system of independent branching type processes determining the current type (mesostate) which are coupled through catastrophic events when several ranks are removed from the mesostate at once.

We proceed in four steps. In Step 1 we analyse the population level, namely the probabilistic structure in the appearance of new mesostates, in Step 2 we consider the mesostate level and analyse the structure of a typical mesostate, in Step 3 the structure of the decoupling cycle, and in step 4 the offspring distribution of new labels and their associated mesostates.

This provides the basis to determine the length process of a mesostate in the simpler case $M = 2, d = 0$ in Subsubsection 9.12.3, then in Subsubsection 9.12.4 the case $M \geq 2, d \geq 0$ and the growth rate of the population of all active mesostates in Subsubsection 9.12.7.

Step 1: Mesostate Dynamics and Production of New Pe-Sites

To introduce the basic ideas we begin by recalling the mesostate dynamics, introduced in subsubsection 9.6.3, in particular the transitions that lead to a

9.12 Emergence and Fixation: The Selection-Dominant Case

decomposition of a mesostate into two or more mesostates, that is when a specific transition, namely the decoupling, results in a tableau that can be factored into two or more tableau.

Starting with a single (110) factor to model the effect of the *selection* operator consider the *pure birth* process in which the *births* take place according to a rate s Poisson process. To build in the effect of the *mutation* operator in a convenient form for our present purpose we introduce marks S, F, U for columns of the tableau and we also have a *marking* process at rate $m = m_{12} + m_{21}$ that marks ranks with S (mutation $2 \to 1$), F (mutation $1 \to 2$) or U (no mutation) which for a rank being created by one a selection event from (110) corresponds for a tableau (010) $\otimes$ (111) + (100) $\otimes$ (110) to resolving in (110) $\otimes$ (110) for S and (110) for F and unresolved for U. The probability that any given rank is marked S or F in an interval with length having an exponential distribution with mean s^{-1}, when the rank produces a new birth-selection event, (independent of the marking) is $\frac{m}{m+s}$.

Note here the importance of order - the operation of selection and mutation do not commute. We will use the concept of generation to describe this. We say a rank is of generation k, which is given roughly by the number of m_{21}-mutations on the ancestral line, we make this precise in Step 2. Namely a failure of a rank (i.e. collecting the mark F) of generation k eliminates all earlier ranks of the same generation and their descendants. However the failure of a post-successful resolution descendant of a successful rank cannot remove a rank of the same generation as the successful resolving rank. Further offspring of rank 1 are produced as a Poisson process with rate s.

Associated with this picture is a *marked branching tree* (pure birth) which describes the populations of individuals alive and their genealogical relations.

Namely consider the various individuals produced with birth times $\{T_j^b, \; j \in \text{tree}\}$ and the corresponding resolution times $\{T_j^r : j \in \text{tree}\}$. These resolution times have independent exponential distributions with mean $\frac{1}{m}$, while the birth times occur independently in a rate s binary branching process. Therefore

$$P(T_j^r \leq T_1^r | T_j^b < T_1^r) = \frac{1}{2}. \tag{9.625}$$

Now consider the *marked branching tree* which is defined via our mesostate process evolving up to a fixed time t. An individual in the tree corresponds to a rank in our multisite tableau-valued process. Each individual j in this branching tree will be given the following triple of marks:

$$T_j^b, T_j^r \text{ and } R \in \{S, F, U\} \text{ (success, failure, unresolved)}. \tag{9.626}$$

The probability that a resolution is successful, i.e. resolving in S is $\frac{m_{2,1}}{m}$ and in F, $\frac{m_{1,2}}{m}$.

An individual in our branching tree (rank) which has birth time less than the resolution time of an ancestor which is marked F is also marked F.

The mesostate which develops from one single initial factor $(1\cdots,1,0)$ looks potentially different from the ones, which arise in a decoupling event, since the new *Pe*-sites may carry at the moment of the decoupling event a mesostate having a different tableau which arises from the history of the mesostate which was giving birth to the new ones. We formalize below in Step 2 these effects further.

Remark 185. Consider the first offspring j of rank 1 that has not resolved and all of its offspring before its resolution time have resolved. Then the region to the right of j consisting of descendants of earlier births grows like the original mutation selection system. The leftmost point becomes the new first rank after the earlier offspring of rank 1 have resolved. The resulting new mesostate can be represented as a copy of the mesostate process started with (110) but at positive age where the age is given by the time between the birth of its initial rank and the parent mesostate decoupling time.

Now consider the *fragmentation* of a mesostate that takes place at a decoupling event. The result is a random number of new offspring *Pe* sites at different locations in $\mathbb{N}$ and one remaining mesostate at the original site. Those irreducible subtableau of a mesostates which are at decoupling not identical to $(1,\ldots,1,0)$ form the "*residual*" *mesostates*. We will discuss below the possible growth of the *lengths* of these "residual" mesostates as they develop from one decoupling cycle to the next.

Example 17. Assume that we have three births from rank 1 leading to the tableau:

$$\begin{matrix}(010)\\ (100)\ (010)\\ (100)\ (100)\ (010)\\ (100)\ (100)\ (100)\ (110).\end{matrix} \qquad (9.627)$$

We now consider the event that a decoupling at rank 1 occurs. At such a time the last birth from 1 occurred at rate s (exponential) time in the past we call this rank individual 1. During this time the individual 1 can in principle produce offspring itself. Now consider the branching tree describing the descendants of rank 1. The result of a F, resp. S, mutation at rank 1 results in (110) in the first case and in the second in

$$\begin{matrix}(000)\\ (110)\ (010)\\ (110)\ (100)\ (010)\\ (110)\ (100)\ (100)\ (110)\end{matrix} = (110) \otimes \begin{matrix}(010)\\ (100)\ (010)\\ (100)\ (100)\ (110).\end{matrix} \qquad (9.628)$$

In the first case only the original mesostate remains and in the second case one new offspring label and mesostate are produced in addition to the resolved parent mesostate at the decoupling event.

9.12 Emergence and Fixation: The Selection-Dominant Case

We see now that the collection of labels with their mesostates from a multitype CMJ-process where the set of possible types is countable since at the birth event of a new label and during the evolution the mesostate moves through a countable number of possible states.

Step 2: The Internal Dynamics of a Mesostate, the MBDC-Process

We have shown above that the production of new *Pe* sites and the corresponding mesostates are described by a *multitype CMJ-process*. We now consider the question concerning the growth of the individual mesostates and will resolve later on the question whether the expected mesostate size grows unboundedly with its age.

For this purpose we now look further into the *internal dynamics* of the mesostate which also has after careful inspection a *multitype branching tree structure* of its own which we extract formally below. However the difference is that in addition to birth and deaths is that some deaths of an individual result in the simultaneous death of all of its descendants which we call the *catastrophic* event. This leads to a *new* class of *branching process with catastrophes* which we will formally define below. Note that the population of ranks increases by the birth process at rate s, resolutions occur at rate m which may reduce the population. For $d = 0$ and $s > m$ this would yield a supercritical process if we exclude m_{12}-mutations resulting from a F individual and leading potentially to multiple deaths of ranks.

In order to describe the internal dynamics of the mesostate process which depends on the genealogical relations between the ranks we must introduce a new type of branching process, namely, a *multilevel birth-death-catastrophe process*, abbreviated *MBDC-process*. At this point it is important to remember that a newborn rank is placed immediately to the right of its parent rank.

Informal Description of the MBDC Process

Assume $m_{12} > 0, m_{21} > 0$. When a m_{21} transition occurs, this means that the corresponding rank and its descendants can now longer eliminate ranks born earlier and their descendants. To analyse this we introduce a *colouring system* with elements of $\mathbb{N}$ as follows. We start initially with colour 1. Consider first m_{21} and then the m_{12}-mutation for the evolution.

(1) For the $m_{2,1}$-*mutation* the colouring works as follows:

- Offspring of rank 1 and their descendants that have not experienced any m_{21} mutations are assigned colour 1.
- Ranks experiencing a m_{21} mutation and their subsequent descendants are assigned colour 2, i.e. the m_{21} mutation causes a death in the colour 1 population and immigration into the colour-2 population.
- In turn descendants of a newly created colour 2 ranks can experience a m_{21} mutation and these are assigned colour 3, etc.

- In general, when an m_{21} transition occurs at a rank of colour k it changes to colour $k+1$ (representing a *death* of a colour k rank and an *immigration* for the colour $k+1$ ranks).

Therefore the probability that the individual produced is of colour $k+1$ is proportional to the number of colour k ranks. This gives us a multicolour model in which at each transition we can either introduce a new colour or choose an existing colour proportional to the number present.

(2) Consider next m_{12}-*mutations* at a rank causing the cutpoint *catastrophes*. Here a catastrophe event is the event that a m_{12} transition occurs at a rank thus truncating the tableau to the right of this rank. Similarly a m_{12} resolution at a colour k rank truncates the subtableau corresponding to the colour k interval to which it belongs.

The combination of the two mutation transitions results in a *multilevel birth-death-catastrophe process*.

Pruned Marked Tree for MBDC-Process

Here is another viewpoint for the MBDC-process leading to a *pruned marked tree*. A positive proportion of births correspond to ranks marked F (that is, undergo $1 \rightarrow 2$ transition when mutation occurs) which now define a multilevel birth-immigration-death-catastrophe process. The level of a cutpoint catastrophe corresponds to the level at which it truncates, that is, which level (colour) it acts on. This leads to a *marked-branching-tree-valued process with multitype pruning* where the marks correspond to the rank orders. The transitions correspond to the creation of a new vertex (birth), change of colour of a vertex from k to $k+1$ and pruning which corresponds to resolving a vertex and removing all descendants of the vertex.

However note that the mapping from the genealogical tree to the tableau is not 1-1. For example the following tableau could also arise by two births at rank 1 or a birth at rank 1 and then a birth at rank 2:

$$\begin{pmatrix} (010)_1 & (111)_{12} & (111)_{11} \\ (100)_1 & (010)_{12} & (111)_{11} \\ (100)_1 & (100)_{12} & (110)_{11} \end{pmatrix} \qquad (9.629)$$

$$\begin{pmatrix} (010)_1 & (111)_{11} & (111)_{111} \\ (100)_1 & (010)_{11} & (111)_{111} \\ (100)_1 & (100)_{11} & (110)_{111} \end{pmatrix}. \qquad (9.630)$$

Formal Definitions and Examples

We now formally introduce the MBDC-process and then explain the features in Example 18.

9.12 Emergence and Fixation: The Selection-Dominant Case

Definition 9.66 (The MBDC-process with $M = 2, d = 0$).
The MBDC-process is a continuous time Markov process with the following state space, dynamic and associated marked tree.

(i) The state space.

The state of the MBDC-process is a marked-tableau-valued process in which marks, tripels (rank, colour, type) are given by:

$$\text{ranks} \in \mathbb{N}, \text{ colours: indexed by } \mathbb{N} \tag{9.631}$$

and

$$\text{type: indexed by } CT \text{ (cutpoint)}, RE \text{ (resolved} = (110)) \text{ or } DE \text{ (} = (111)) \tag{9.632}$$

which are assigned to the ranks as follows.

The colours will be assigned dynamically (see (ii)) starting with colour 1 for the initial rank 1. Next to the types. A rank is *resolved* if it corresponds to a column containing at least one factor (110) and otherwise (111) factors. A rank is a *cutpoint* (unresolved) if it corresponds to a column containing both (100) and (010) as entries. A rank is DE if its column consists of factors (111).

(ii) The dynamic: transitions and rates.

Recall that a selection operation acting on (110) located at rank ℓ produces the tableau
$$\begin{pmatrix} 010 & 111 \\ 100 & 110 \end{pmatrix} \text{ at ranks } \ell \text{ and } \ell + 1 \text{ and all the original ranks to the right of } \ell$$
are shifted one place to the right. The following transitions occur.

- A birth at a colour k rank produces an offspring colour k rank (at rate s). When an individual at rank ℓ gives birth its *offspring* is given rank $\ell + 1$.
- A resolution of the type $(100) \to (110)$ and $(010) \to (000)$ at a colour k rank changes the colour of the rank to $k + 1$ (at rate m_{21}). This is viewed as a death of a colour k rank and the immigration of a colour $k + 1$ rank.
- A resolution of the type $(100) \to (000)$ and $(010) \to (110)$ (at rate m_{12}) at a colour k rank removes all descendants (of colours $\ell \geq k$) of these ranks, that is, their type becomes DE. This is called a colour-k catastrophe. (Note therefore a colour k catastrophe can only remove ranks of colours $\ell \geq k$).
- At the resolution time at rank 1 (corresponding to a decoupling event where a number of offspring mesostates can be produced), the tableau is truncated and the state of rank 1 jumps to (110) and the whole MBDC-process now continues (this occurs with rate $m = m_{12} + m_{21}$) from this element (110) anew.

(iii) The birth process described in (ii) observed from time 0 up to t produces a genealogical tree associated with time t which is then marked with marks given by the ranks, colours and types. The marks are determined by a mapping

$$\text{genealogical tree} \to \mathbb{N} \times \mathbb{N} \times \{CT, RE, DE\} \tag{9.633}$$

corresponding to the rank, colour and type of each individual of the genealogical tree. □

We next illustrate the behaviour of this process with a concrete example below. Note however that an MBDC-process starting in a residual mesostate can always be viewed as one created in the same initial state (110) but with observation time starting from a later random time on.

Example 18. We indicate the colour of the rank by a subscript (this does not refer to space here !). We start with (110) at rank 1. Consider the case in which four births have already occurred at rank 1.

Next the second rank resolves by mutation $2 \to 1$: $(010) \to (000)$ and $(100) \to (110)$. We get:

$$\hat{t}_1 = \begin{array}{l} (010)_1 \\ (100)_1 \ (000)_2 \\ (100)_1 \ (110)_2 \ (010)_1 \\ (100)_1 \ (110)_2 \ (100)_1 \ (010)_1 \\ (100)_1 \ (110)_2 \ (100)_1 \ (100)_1 \ (110)_1. \end{array} \tag{9.634}$$

Now let a birth occur at rank 2 to get:

$$\hat{t}_2 = \begin{array}{l} (010)_1 \\ (100)_1 \ (010)_2 \ (111)_2 \ (010)_1 \\ (100)_1 \ (100)_2 \ (110)_2 \ (010)_1 \\ (100)_1 \ (010)_2 \ (111)_2 \ (100)_1 \ (010)_1 \\ (100)_1 \ (100)_2 \ (110)_2 \ (100)_1 \ (010)_1 \\ (100)_1 \ (010)_2 \ (111)_2 \ (100)_1 \ (100)_1 \ (110)_1 \\ (100)_1 \ (100)_2 \ (110)_2 \ (100)_1 \ (100)_1 \ (110)_1. \end{array} \tag{9.635}$$

We now consider here first three among all the possible resolution outcomes which are not decoupling transitions which would involve a resolution at rank 1:

(i) the present rank 2 of $\hat{t}_2$ resolves $(010) \to (000)$ and $(100) \to (110)$ $(2 \to 1$ mutation):

$$\hat{t}_3 = \begin{array}{l} (010)_1 \\ (100)_1 \ (000)_3 \ (111)_2 \ (010)_1 \\ (100)_1 \ (110)_3 \ (110)_2 \ (010)_1 \\ (100)_1 \ (000)_3 \ (111)_2 \ (100)_1 \ (010)_1 \\ (100)_1 \ (110)_3 \ (110)_2 \ (100)_1 \ (010)_1 \\ (100)_1 \ (000)_3 \ (111)_2 \ (100)_1 \ (100)_1 \ (110)_1 \\ (100)_1 \ (110)_3 \ (110)_2 \ (100)_1 \ (100)_1 \ (110)_1 \end{array}$$

9.12 Emergence and Fixation: The Selection-Dominant Case

$$= \frac{(010)_1}{(100)_1 \ (110)_3 \ (110)_2 \ (010)_1} \atop {(100)_1 \ (110)_3 \ (110)_2 \ (100)_1 \ (010)_1} \atop {(100)_1 \ (110)_3 \ (110)_2 \ (100)_1 \ (100)_1 \ (110)_1.} \quad (9.636)$$

(ii) the present rank 2 of $\hat{t}_1$ resolves $(010) \to (110)$ and $(100) \to (000)$ $(1 \to 2$ mutation):

$$\hat{t}_4 = \begin{array}{l} (010)_1 \\ (100)_1 \ (110)_2 \ (111)_2 \ (010)_1 \\ (100)_1 \ (000)_2 \ (110)_2 \ (010)_1 \\ (100)_1 \ (110)_2 \ (111)_2 \ (100)_1 \ (010)_1 \\ (100)_1 \ (000)_2 \ (110)_2 \ (100)_1 \ (010)_1 \\ (100)_1 \ (110)_2 \ (111)_2 \ (100)_1 \ (100)_1 \ (110)_1 \\ (100)_1 \ (000)_2 \ (110)_2 \ (100)_1 \ (100)_1 \ (110)_1 \end{array}$$

$$= \frac{(010)_1}{(100)_1 \ (110)_2 \ (010)_1} \atop {(100)_1 \ (110)_2 \ (100)_1 \ (010)_1} \atop {(100)_1 \ (110)_2 \ (100)_1 \ (100)_1 \ (110)_1.} \quad (9.637)$$

(iii) The present rank 4 of $\hat{t}_2$ resolves $(010) \to (110)$ *and* $(100) \to (000)$ ($1 \to 2$ mutation):

$$\hat{t}_5 = \begin{array}{l} (010)_1 \\ (100)_1 \ (010)_2 \ (111)_2 \ (110)_1 \\ (100)_1 \ (100)_2 \ (110)_2 \ (110)_1 \\ (100)_1 \ (010)_2 \ (111)_2 \ (000)_1 \ (010)_1 \\ (100)_1 \ (100)_2 \ (110)_2 \ (000)_1 \ (010)_1 \\ (100)_1 \ (010)_2 \ (111)_2 \ (000)_1 \ (100)_1 \ (110)_1 \\ (100)_1 \ (100)_2 \ (110)_2 \ (000)_1 \ (100)_1 \ (110)_1 \end{array} \quad (9.638)$$

$$= \begin{array}{l} (010)_1 \\ (100)_1 \ (010)_2 \ (111)_2 \ (110)_2 \\ (100)_1 \ (100)_2 \ (110)_2 \ (110)_2. \end{array}$$

Note that (ii) and (iii) are examples of *catastrophic transitions* in which some columns are removed.

Now consider the effect of decoupling in the case of $\hat{t}_3$ by the resolution of rank 1: we either get (110) or

$$\begin{array}{l} (110)_1 \ (110)_3 \ (110)_2 \ (010)_1 \\ (110)_1 \ (110)_3 \ (110)_2 \ (100)_1 \ (010)_1 \\ (110)_1 \ (110)_3 \ (110)_2 \ (100)_1 \ (100)_1 \ (110)_1. \end{array}$$

$$= (110) \otimes (110) \otimes (110) \otimes \begin{matrix} (010)_1 \\ (100)_1 \ (010)_1 \\ (100)_1 \ (100)_1 \ (110)_1 \end{matrix}.$$

In this case we have three new offspring mesostates (where one is a residual mesostate) in addition to the original parent mesostate.

Next consider the decoupling in case $\hat{t}_5$, we get two new offspring mesostates (where one is a residual mesostate) in addition to the original parent mesostate.

$$\begin{matrix} (000)_1 \\ (110)_1 \ (010)_2 \ (111)_2 \ (110)_2 \\ (110)_1 \ (100)_2 \ (110)_2 \ (110)_2 \end{matrix} = (110) \otimes \begin{matrix} (010) \ (111) \\ (100) \ (110) \end{matrix} \otimes (110). \qquad (9.639)$$

Note that without the catastrophic jumps, involving the death of *more* than one rank, in the case $s > m$ this mesostate evolution is a *supercritical* "branching" process. On the other hand the cutpoint process without births (immigration) at rank 1 goes to extinction. In contrast the process with immigration (via the birth at rank 1) and catastrophes has a nontrivial equilibrium. This will be described in the next subsection.

Step 3: Growth of the Mesostate During the Decoupling Cycle: Preliminary Observations

We will now consider the growth and genealogical structure of a mesostate during the decoupling cycle. A more complete analysis of the MBDC process will be given in the next subsection to obtain moment bounds on the *size of the mesostates up to the decoupling time* and this will provide moment bounds on the number of the offspring of new *Pe*-sites at a decoupling event

(i) Structure of the decoupling cycle

We now consider in detail the decoupling cycle for a mesostate with given founding *Pe* site. The distribution of the random number and types of offspring at the decoupling time depends on the mesostate at $T_{\text{dec}-}$ (decoupling time) at the *Pe* site.

The production of offspring is determined as follows. Recall that moving through the *natural tableau order* means going *backwards in time of appearance* of new ranks since the additional ranks are placed next to the right of the rank giving birth. In addition we can keep track of the genealogical tree.

Observe that:

- for one or more new *Pe* sites to be produced by a given *Pe* site at a decoupling time, the first rank (rank 1) at this site must have one or more births (due to the selection operator) and the first rank must have resolved successfully (that is, a $2 \to 1$ mutation occurred at rank 1) (cf. Example 19(b));

9.12 Emergence and Fixation: The Selection-Dominant Case 601

- for two or more to be produced at least two births at rank 1 or one at 1 and another at rank 2 are needed and both the first and the second rank (last offspring of the first rank) must have previously resolved successfully (cf. (9.406)) (cf. Example 19(b)).
- The case for three or more is more complex - see Example 18 above.

(ii) Multigeneration decomposition of a mesostate

We now focus on the internal development of the mesostate during the decoupling cycle. We do this first for $d \equiv 0$, the process with $d > 0$ can later be embedded into the $d = 0$ branching tree. First note that births occur at rank 1 as a Poisson process with rate s and that a random number Z_0 of births at rank 1 occur up to the decoupling time with distribution

$$P(Z_0 = k) = \left(\frac{s}{m+s}\right)^{k-1} \frac{m}{m+s}, \quad k \in \mathbb{N}. \tag{9.640}$$

We can then decompose the active ranks in the mesostate at time $T_{\text{dec}}-$ into the families produced by each of these Z_0 offspring of rank 1. That is, the *descendants of rank 1 at a Pe-site up to a decoupling time* can be divided into *generations*, the children of rank 1 before decoupling form generation 1, the children of these children later decoupling form generation 2, etc.

We determine the set of descendants of rank 1 as follows in two steps. Consider first the effect of the action of the selection operator and then consider the effect of mutation. We have the following scenario.

- The selection operator acts at a rank at times given by a pure birth process with rate s.
- Consider the random time T_{dec} at which rank 1 resolves resulting in a decoupling. At time T_{dec}, the *last* birth (individual 1, i.e. the first generation 1 individual) at rank 1 occurred at an exponential time with mean $1/s$ in the past, the second last birth (first generation, individual 2) occurs at an exponential time (with mean $1/s$) further back, etc. till we have reached the decoupling time of rank 1.
- The births of offspring from individual 1 during this interval (from the birth of individual 1 and to the decoupling time of rank 1) are called *second generation* individuals descending from individual 1, the ones descending from these third generation individuals, etc. At this point we focus on the complete branching tree and only later *prune* to keep only ranks which are marked S.

  The number of second generation individuals produced by individual 1 *before rank 1 produces a second offspring* is denoted as Z_1^{11}.

  Then

$$P(Z_1^{11} = k) = \frac{1}{2^k} \cdot \frac{1}{2}, \quad k = 1, 2, \ldots, \tag{9.641}$$

$$E[z^{Z_1^{11}}] = \frac{1}{2}\sum_{k=1}^{\infty} z^k \frac{1}{2^k} = \frac{1}{2}\frac{\frac{z}{2}}{1-\frac{z}{2}} = \frac{1}{2}\frac{z}{2-z},$$

$$E[Z_1^{11}] = \frac{1}{2}\frac{\partial}{\partial z}\frac{z}{2-z}|_{z=1} = 1.$$

This follows by considering two Poisson processes both of rate s and asking for k to be produced by the second and only then 1 produced by the first.

- Now consider generation k individuals. These are the offspring of the generation $(k-1)$ individuals. Consider the ith generation $k-1$ individual born *before* rank 1 produces a second offspring. Let Z_{k-1}^{i1} denote the number of generation k offspring this individual produces *before rank 1 has a second birth*. Then Z_{k-1}^{i1} also has distribution (9.641). However note that Z_{k-1}^{il} and Z_{k-1}^{jl} are not independent if $i \neq j$ (because both denote the number of births occurring before the *same* event occurs, namely, the second birth at rank 1).
- Starting with the first generation individuals produced by individual 1 before rank 1 has a second birth, this produces a genealogical tree in which the mean offspring size of an individual is 1. However we *cannot* conclude that this is a critical Galton-Watson tree due to the lack of independence of the offspring of different individuals of the same generation.
- The above genealogical tree describes the multigeneration decomposition of all descendants of individual 1 produced *before* rank 1 produces a second birth. We will verify below that the total progeny in this multigeneration decomposition does have *infinite* mean.

We note that this genealogical tree corresponds to the set of individuals in a pure birth process with rate s during a independent exponential time with mean $1/s$. That is, this tree corresponds to a Yule process tree with birth rate s observed at an independent exponential time with mean $\frac{1}{s}$. As mentioned above the tree produced in this way is not the same as the Galton-Watson branching tree with offspring distribution (9.641) due to lack of independence of offspring of different individuals. To verify this note we first compute the the generating function F of the number of leaves in the Yule tree at the random time. Recalling that the generating function of the Yule process at time t is (see [Bai] (8.14))

$$F_{Yule}(t,z) = \frac{1}{1-e^{st}(1-\frac{1}{z})}, \qquad (9.642)$$

it follows that the generating function at the random time is given by

$$F(z) = \int_0^{\infty} \frac{1}{1-e^{st}(1-\frac{1}{z})} se^{-st} dt = 1 - \frac{1-z}{z}\log\frac{1}{1-z}. \qquad (9.643)$$

On the other hand using [H], (13.2), we see that the total progeny of a Galton-Watson tree with offspring distribution (9.641) has generating function G:

9.12 Emergence and Fixation: The Selection-Dominant Case

$$G(z) = 1 - (1-z)^{1/2} \tag{9.644}$$

and therefore the distribution of the number of leaves in the two cases are different.

We can construct the Yule tree produced up to the $exp(s)$-random time recursively as follows. Starting with individual 0 corresponding to rank 1 (with factor given by $(1,\ldots,1,0)$) the first birth at rank 1 produces individual 1. Now let Z_∞ denote the total number of descendants of individual 1 produced before the second birth at rank 1. Noting that the time to the next birth produced by any individual is exponential with mean $1/s$ we have

$$P(Z_\infty \geq 1) = \frac{1}{2}, \tag{9.645}$$

$$P(Z_\infty \geq k) = P(Z_\infty \geq k-1) \tag{9.646}$$
$$\cdot P(k\text{-th birth is a descendant of individual } 1 | Z_\infty \geq k-1)$$
$$= \frac{k}{k+1} \cdot P(Z_\infty \geq k-1).$$

Then $P(Z_\infty \geq k) = \frac{1}{k+1}$ and

$$P(Z_\infty = k) = \frac{1}{(k+1)(k+2)}, \quad k = 0, 1, 2, \ldots. \tag{9.647}$$

The expected total number Z_∞ (produced during the exponential time interval with mean $1/s$) is then

$$E[Z_\infty] = \sum_{k=1}^{\infty} \frac{k}{(k+1)(k+2)} = \infty. \tag{9.648}$$

(iii) The effect of mutation.

Consider the effect of mutation with mutation rates $m_{1,2}, m_{2,1}$, $m = m_{1,2} + m_{2,1}$ on the population Z_∞ we constructed above. We introduced the two *marks* S (for success) and F (for failure). We can use an independent marking process at rate m where we mark S at rate $m_{2,1}$ or F at rate $m_{1,2}$, recall $(100) \to (110)$ in the first case and $(100) \to (000)$ in the second case.

Namely note that if a rank is marked F then when the mutation occurs we say the rank resolves in F and all ranks to the right of this rank are removed since the rows with (100) then become (000) and are removed and in those ranks to the right we get in the remaining rows to the right of this "cut" point (111) so that these ranks become inactive. If the rank is marked S, (100) becomes (110) and its offspring have the chance to decouple, see Example 19.

We now consider the effect of a $1 \to 2$ mutation on the *total number of relevant progeny* in the MBDC-process, namely $\tilde{Z}_\infty$. That is

$$\tilde{Z}_\infty \qquad (9.649)$$

is the Z_∞-*progeny without an ancestor marked* F which means the new population arises by pruning the Z_∞-population. In particular we consider the effect of ranks marked F with resolution time $\leq t_0$. In this case there is a positive probability $q > 0$ that the next transition is the death of a rank so that (recall (9.646)):

$$P(\tilde{Z}_\infty \geq k) \leq P(\tilde{Z}_\infty \geq k-1) \qquad (9.650)$$
$$\cdot (1-q) \cdot P(k\text{-th birth is a descendant of individual } 1 | Z_\infty \geq k-1)$$
$$= \frac{(1-q)k}{k+1} \cdot P(\tilde{Z}_\infty \geq k-1).$$

Therefore

$$P(\tilde{Z}_\infty \geq k) \leq (1-q)^k \frac{1}{k+1}, \qquad (9.651)$$

which implies that $\tilde{Z}_\infty$ has a finite second moment.

(iv) Population produced by all generation 1 individuals up to the decoupling time

We have obtained a multigeneration decomposition of the offspring of individual 1 up to the time of the second birth at rank 1. However in order to determine the size of the mesostate at time $T_{\text{dec}}-$ we must consider the offspring of individual 1 up to the decoupling time as well as the descendants of all the generation 1 individuals. In order to do this we must take into account that a $1 \to 2$ mutation in a descendant of individual 1 can result in the removal of the earlier generation 1 individuals and their descendants. This involves a deeper analysis of the MBDC process and will be carried out in the next subsubsection.

Step 4: Offspring Distribution of New Pe-Sites

We are now interested in the *set of new labels and associated mesostates* which arise at the decoupling time of an existing mesostate. We describe this for the case $M = 2$ and $d = 0$. This occurs when rank 1 of the founding site of a tagged label undergoes a mutation event. In the case of a mutation $1 \to 2$ at rank 1 no new offspring are produced and the mesostate returns to (110). When the mutation $2 \to 1$ occurs at least one new mesostate is produced (recall Example 18) and if more than one, then the resulting offspring mesostates correspond to a product of tableau. Each offspring mesostate consists of a tableau corresponding to a tableau

9.12 Emergence and Fixation: The Selection-Dominant Case

obtained by the action of selection and mutation starting with its first rank which can also be identified with a rank in the original mesostate at time $T_{\text{dec}}-$. We will identify the offspring labels with their founding rank and *Pe*-site.

Example 19. Here for $M = 2$ we illustrate the cases in which 1 or 2 new offspring *Pe* labels are produced after the occurrence of one birth, respectively 2 births had occurred first at rank 1, then 2 and where we started with a single factor (110).

(a) $\begin{pmatrix} 010 \ 111 \\ 100 \ 110 \end{pmatrix} \to (110) \otimes (110)$, by mutation $2 \to 1$ at rank 1.

(b) $\begin{pmatrix} 010 \ 111 \ 111 \\ 100 \ 010 \ 111 \\ 100 \ 100 \ 110 \end{pmatrix} \to (110) \otimes (110) \otimes (110)$, by mutation $2 \to 1$ at rank 2, then 1.

In case (a) rank 1 resolves S, in (b) rank 2 resolves S and then rank 1 resolves S. If we had the mutation $1 \to 2$ in case (a) rank 1 would resolve F and we would get (110) only.

Lemma 9.67 (Structure of mesostates).
Consider the mesostate process starting with a single factor $(1, \ldots, 1, 0)$ at the first decoupling time.

Then at the decoupling time the resulting tableau can be decomposed into the product of a finite random number of indecomposable tableau. Each of these tableau are given by the state of a mesostate process which started with a single factor $(1, \ldots, 1, 0)$ but is observed at a positive random age. □

Proof of Lemma 9.67.
Note that all ranks to the right of the first rank in the new mesostate can be viewed as descendants of this new first rank and the resulting tableau subsequently developed following the basic mesostate dynamics. If at a decoupling time we consider such an indecomposable tableau with a leftmost rank k then the tableau was the result of selection-mutation operations acting on ranks equal to or to the right of this rank. But this is exactly the mesostate dynamics starting with $(1, \ldots, 1, 0)$ and in which either the size of the tableau is 1 or the rank k is not resolved. q.e.d.

9.12.3 Long-Time Behaviour of the MBDC-Process: $M = 2, d = 0$

We now consider the random number of offspring *Pe* sites produced at a decoupling time and note that this can be bounded by the size of the mesostate at time $T_{\text{dec}}-$. We consider the growth of the number of mesostates in more detail in the next subsubsection but we note here a result for the case $s > m$, (the case $s \le m$, using a different approach has been treated in Subsections 9.10 and 9.11).

Lemma 9.68 (Production of Pe ranks).

(a) *Consider the case $M = 2$, $d = 0$, $s > m$ and $m_{1,2} > 0$. Assume that the mesostate process begins with $\hat{t}_*$, i.e. $\sigma(\hat{t}_*) = ((1,1,0),\emptyset)$. Then the the random number of new Pe ranks produced at the decoupling time has a finite $(1+\varepsilon)$-moment for some $\varepsilon > 0$.*
(b) *The population of mesostates is a supercritical branching process.* □

Proof of Lemma 9.68.

(a) We prove below in (9.653), that up to the decoupling time the size of the mesostate has a finite $(1+\varepsilon) - moment$. Then at fragmentation the maximum number of offspring mesostates also has a finite $(1+\varepsilon) - moment$.
(b) The branching property follows since the different mesostates do not interact and reproduce independently. Since mesostates do not die, supercriticality follows. This completes the proof of Lemma 9.68. q.e.d.

We have identified the internal mesostate dynamics and we now turn to the question of the growth of the *length* of a tagged mesostate. We first note that due to the birth process at rank 1 the mesostate process may grow, but due to the mutations at rank 1 the MBDC (recall definition 9.66) has $\hat{t}_* = (1,\ldots,1,0)$ as a *recurrent point* with mean recurrence time bounded by $\frac{1}{m}$. Therefore the MBCD process has a nontrivial stationary measure.

We now consider the growth of the number of mesostates. We will show in subsubsection 9.12.7 that this grows exponentially with malthusian parameter β_{Pe}. On the other hand we have established in subsection 9.7 that the number of Pe, UPe sites grows with malthusian parameter β (Lemma 9.75). We note that if the expected number of occupied sites in a mesostate between decoupling events remains bounded then $\beta = \beta_{Pe}$. Therefore the goal of this subsubsection is to determine under what conditions the latter is satisfied and in particular that this is the case for $m_{2,1}, m_{1,2} > 0$. However we first give an example in which $\beta > \beta_{Pe}$ occurs.

Example 20 (Process with $\beta > \beta_{Pe}$).

Consider the case $M = 2, d = 0, s > 0, m_{12} = 0, m = m_{21} > 0$ with s sufficiently large (selection-dominance). Then $\beta = s$.

If $s > 2m_{21}$ then the expected number of colour 1 individuals at decoupling is infinite. Moreover one can verify using the bound (9.714) for α, the expected number of offspring mesostates at decoupling,

$$\beta_{Pe} = m \cdot (\alpha - 1) < \beta \qquad (9.652)$$

for sufficiently large s.

The analysis of the "selection-dominant" case for non-ergodic mutation is further developed later on in the non-ergodic case for the mutation on E_0 and this results in the dependence of the emergence time scale on the initial measure facts which described in the discussion in subsubsection 9.12.15.

9.12 Emergence and Fixation: The Selection-Dominant Case

We have verified that the expected size of the mesostate remains bounded (under the assumption that $m_{ij} > 0$ for $1 \leq i, j \leq M$) and therefore $\beta = \beta_{Pe}$ in the *mutation-dominant case* (this has been verified in subsubsection 9.7.8, Lemma 9.45). We now consider the *selection dominant case* again under the assumption $m_{ij} > 0$ for $1 \leq i, j \leq M$, that is, a mutation kernel on E_0 which is ergodic.

We discuss separately the cases $d = 0$ and $d > 0$. The case $d = 0$ allows us to ignore migration, since the dynamics of the number and state of ranks is then independent of the location.

Here we work with the structure developed in the previous subsubsection and prove that following key bound.

Lemma 9.69 (Bounded $(1 + \varepsilon)$-moment for mesostate length: $M = 2, d = 0$).
Assume that $M = 2$, $d = 0$, $s > m$, $m_{12} > 0$ and $m_{21} > 0$. Furthermore start $\hat{\mathfrak{t}}(0) = \hat{\mathfrak{t}}_*$.

Then $\beta = \beta_{Pe}$ and the first moment of the number of active ranks in the mesostate process is finite and the number of active ranks satisfies in addition, for some $\varepsilon > 0$:

$$E\left[\|\hat{\mathfrak{t}}(T_{\text{dec}}-)\|^{1+\varepsilon}\right] < \infty. \qquad \Box \qquad (9.653)$$

Remark 186. The $(1 + \varepsilon)$ moment implies the usual $E[X \log X]$ condition in the theory of supercritical branching processes and therefore should be sufficient to carry out the CMJ program for this countable type CMJ process.

This raises the question of the existence of second moments. Recall that in Lemma 9.45(b) we proved the existence of second moments in the case $\hat{\lambda} > 2s$. We have proved above the existence of $(1 + \varepsilon)$-moments for some $\varepsilon > 0$ in Proposition 9.69 in the ergodic case. Moreover we have obtained a decomposition into coloured individuals such that all moments of the number of individuals of each colour is finite as we shall show below. However we shall also identify a subpopulation of the growing mesostate which has a finite second moment but at most finitely many moments at the decoupling time. This suggests that the question of second moments is delicate and the more general problem of determining the number of finite moments of the mesostate at decoupling as a function of the various parameters in the ergodic case is open.

In Lemma 9.45 we have proved that the size of the mesostate has a finite second moment in the mutation-dominant case. It also follows from Lemma 9.52 that for $d > 0$ the number of Pe, UPe sites satisfy the second moment condition at times of the form $\frac{\log N}{\beta} + T$ and the associated random constant W^{III} defined in Proposition 9.41 has a finite second moment. See subsubsection 9.12.5 for a discussion of higher moments of the offspring number and size of the mesostate process and corresponding random constant in the general case.

Proof of Lemma 9.69.
For the assertion $\beta_{Pe} = \beta$ and finite first moment we have a simpler proof (Proof 1 below) however this cannot be used to prove all claims which Proof 2 however does.

Proof 1.
We give here an indirect proof that $\beta = \beta_{Pe}$ and the first moment is finite. The proof is by *contradiction* assuming $\beta > \beta_{Pe}$.

We consider the McKean–Vlasov system with geographic space $S = \mathbb{N}$. Assume that $X_0^N = \mu_0^\otimes$ with $\mu_0(1) = q$, $\mu(2) = 1 - q$, $q \in [0, 1]$ and the number of active sites grows exponentially with malthusian parameter $\beta > \beta_{Pe}$ where β_{Pe} is the malthusian parameter for the growth in the number of mesostates.

Under our assumptions the dual at time $\frac{\log N}{\beta} + T$ with $T \in \mathbb{R}$ will be decomposed into $N^{\frac{\beta_{Pe}}{\beta}} e^{\beta_{Pe} T}$-many factors each given by tableau each of the form

$$\begin{array}{l}(010)\\(100)\,(010)\\(100)\,(100)\,(010)\\(100)\,(100)\,(100)\,(110),\end{array} \tag{9.654}$$

but with random sizes such that the total number of active ranks is $Ne^{\beta T}$. Since the expected number of mesostates at time $\frac{\log N}{\beta} + T$ is given by $N^{\beta_{Pe}} e^{\beta_{Pe} T}$, then the expected size (number of occupied sites) of typical mesostates at times $\frac{\log N}{\beta} + T$ must grow as $N^{\frac{\beta - \beta_{Pe}}{\beta}} e^{(\beta - \beta_{Pe}) T}$.

Let us now compare the emergence time of type 3 in the three cases $q = 0, q = 1$ and $q \in (0, 1)$, using the dual with $\tilde{\mathcal{G}}_0^{++} = (110)$.

First note that if $q = 0$ then $\int \tilde{\mathcal{G}}_t^{++} d\mu_0^\otimes$ jumps to 0 *only if* a rare mutation $(010) \to (000)$ takes place in the *first rank* of one of the mesostates and therefore

$$\lim_{N\to\infty} E[\bar{x}_3^N(\frac{\log N}{\beta} + T)] = \lim_{N\to\infty} E[1 - \int \tilde{\mathcal{G}}_{\frac{\log N}{\beta} + T}^{++} d\mu_0^\otimes] = 0, \tag{9.655}$$

since $\frac{m_{up}}{N} N^{\beta_{Pe}} e^{\beta_{Pe} T} \to 0$ as $N \to \infty$.

On the other hand if $q = 1$, then $\int \tilde{\mathcal{G}}_t^{++} d\mu_0^\otimes$ has jumps to 0 if a rare mutation $(100) \to (000)$ takes place in *any rank 2 or higher* at any of the mesostates and if $0 < q < 1$ it has a negative jump if a rare mutation takes place at any rank (except the last rank in a mesostate). Since there are $N - N^{\frac{\beta_{Pe}}{\beta}}$ such ranks, therefore if $q = 1$

$$\liminf_{N\to\infty} E[\bar{x}_3^N(\frac{\log N}{\beta} + T)] > 0. \tag{9.656}$$

If $0 < q < 1$

$$\liminf_{N\to\infty} E[\bar{x}_3^N(\frac{\log N}{\beta} + T)] = \liminf_{N\to\infty} E[1 - \int \tilde{\mathcal{G}}_{\frac{\log N}{\beta} + T}^{++,N} d\mu_0^\otimes] = 0. \tag{9.657}$$

9.12 Emergence and Fixation: The Selection-Dominant Case

In order to verify this let $p_k^N(t)$ denote the probability of mesostate size k at time t in the size N system. Now consider the effect of a rare mutation. This will occur at a randomly chosen rank and there the size of the mesostate to which it belongs is given by the Palm probabilities $\{\frac{kp_k(t)}{\sum kp_k(t)}, \ k \in \mathbb{N}\}$. If the rare mutation occurs in a mesostate of size k the rank at which it occurs is random. Note that if the rare mutation occurs at rank ℓ in a mesostate of size k then the the contribution of the rows $\ell+1,\ldots,k$ in the tableau associated to the mesostate to the integral $\int \tilde{\mathcal{G}}^{++,N}_{\frac{\log N}{\beta}+T} d\mu_0^\otimes$ become zero. Noting that the jth row in such a tableau contains $j-1$ factors of the form (100) the decrease in the integral term jump caused by this rare mutation has size

$$(1-q)[q^\ell + q^{\ell+1} + \cdots + q^{k-1}] \leq q^\ell \tag{9.658}$$

and therefore the expected jump (with respect to the Palm distribution of mesostate sizes) has size

$$\leq \frac{1}{\sum_k kp_k^N(t)} \sum_k p_k^N(t)[q + q^2 + \cdots + q^{k-1}]. \tag{9.659}$$

Therefore,

$$\liminf_{N\to\infty} E[\bar{x}_3^N(\frac{\log N}{\beta}+T)] \leq \frac{1}{\sum_k kp_k^N(t)} \sum_k p_k^N(t)[q + q^2 + \cdots + q^{k-1}]. \tag{9.660}$$

But under the hypothesis $\beta > \beta_{Pe}$, $\sum_k kp_k(t) \to \infty$ as $t \to \infty$ which implies that the left side of (9.660) converges 0 as $N \to \infty$ which proves (9.657).

Now recall that if $m_{12} > 0$, $m_{21} > 0$ then the system has positive mass for both types 1 and 2 at time $T_0 > 0$. This then yields a contradiction since if we recalculate the type-3 mean mass with the dual at times $\frac{\log N}{\beta} + T - T_0$ and using the Markov property, namely using (9.183),

$$\liminf_{N\to\infty} E\left[\langle \hat{X}_{\frac{\log N}{\beta}+T}, \tilde{\mathcal{G}}_0^{++,N}\rangle\right] = \lim_{N\to\infty} E\left[\langle \hat{X}_{T_0}, \tilde{\mathcal{G}}^{++,N}_{\frac{\log N}{\beta}+T-T_0}\rangle\right] > 0 \tag{9.661}$$

which contradicts (9.657).

A similar argument implies that (in the case $m_{21} > 0$, $m_{12} > 0$)

$$f(t) := E[|\hat{\mathfrak{t}}(t)|] \tag{9.662}$$

cannot satisfy $f(t) \uparrow \infty$. If this were the case then we can find a function g with $g(N) \to \infty$ such that (for a given $T \in \mathbb{R}$):

$$\liminf_{N\to\infty} \frac{1}{N} e^{\beta_{Pe}(\frac{\log N}{\beta_{Pe}}+T-g(N))} f(\frac{\log N}{\beta_{Pe}} + T - g(N)) > 0, \tag{9.663}$$

$$\limsup_{N\to\infty} \frac{1}{N} e^{\beta_{P_e}(\frac{\log N}{\beta_{P_e}} + T - g(N))} = 0, \qquad (9.664)$$

which would again produce a contradiction.

To complete the first moment result note that if $E(|\hat{t}(T_{\text{dec}}-) = \infty|)$, then $f(t) \uparrow \infty$. q.e.d.

Proof 2. We look first at the colour 1 population and later on turn to subsequent colours $2, 3, \cdots$.

Recall that the colour 1 population is a *birth-death-immigration process (BDI)* with *catastrophes*. Here immigration is given by births at rank 1 with rate s, individuals produce births at rate s, die (and produce a colour 2 individual) at rate m_{21} and undergo a catastrophe transition at rate m_{12} at ranks 2 and higher (recall Definition 9.66). In this case a catastrophe transition results in the removal of all ranks to the right of the rank at which the transition occurs.

Starting at the time of the first immigrant produced at rank 1 our objective is now to obtain information on the distribution of the time of the first catastrophe transition. Since each individual undergoes birth, death and catastrophe transitions independently, we can construct the process by assigning each individual a type (death or catastrophe) with probabilities $\frac{m_{21}}{m}$, $\frac{m_{12}}{m}$, respectively, and corresponding independent exponential transition times with means $1/m$. We now note that up to the production of the first individual of the catastrophe type, the process is a BDI process. Now fix some arbitrary $t_0 > 0$. In order to have a catastrophe occur before time t it suffices to have an individual born of catastrophe type (i.e. rank marked F) with transition time less than t_0 and this individual has to be born before time $t - t_0$. We now determine the probability of such an event.

Note that the probability that an individual has this property with probability

$$q = \frac{m_{12}}{m_{12} + m_{21}} \cdot (1 - e^{-mt_0}). \qquad (9.665)$$

Next recall that a BDI process, $\{n(t)\}_{t \geq 0}$ started with $n(0) = 0$ (with birth rate b of each individual, death rate d of each individual and immigration rate b) has probability generating function (see [Bai] (8.72)):

$$P(z,t) = E[z^{n(t)}] = \left(\frac{b-d}{be^{(b-d)t} - d}\right)\left[1 - \frac{b(e^{(b-d)t} - 1)}{be^{(b-d)t} - d} z\right]^{-1} \qquad (9.666)$$

$$= \frac{b-d}{b(1-z)e^{(b-d)t} + (bz - d)}.$$

In order for the event that a catastrophe not to have occurred by time t it is necessary that *none* of the individuals in the BDI process up to time $t - t_0$ are of catastrophe type (i.e. rank marked F) with transition time less than t_0.

Now we apply this to our setup where *immigration* of a new colour 1 individual occurs through a birth at rank 1, which occurs at rate s. In addition individuals give

9.12 Emergence and Fixation: The Selection-Dominant Case

birth at rate s, that is, a new colour 1 rank is born from a rank to the right of 1 and death of a colour 1 individual occurs when the individual mutates at rate $m_{2,1}$ from $2 \to 1$ at i.e. the death rate $d = m_{21}$ since the m_{21}-mutation changes the colour to 2. A catastrophe occurs when a resolution occurs at rate m_{12}, which is killing all ranks to the right.

Let T_C be the time of the first catastrophe transition. Since the probability that no catastrophe has occurred by time t is less than or equal to $(1-q)^{n(t-t_0)}$, using (9.666) with $b = s$, $d = m_{21}$:

$$P(T_C > t) \leq P(1-q, t-t_0) \sim \text{const} \cdot e^{-(s-m_{21})(t-t_0)}, \text{ as } t \to \infty. \quad (9.667)$$

Let T_{dec} be the resolution time at rank 1, and T_C the time of the first colour 1 catastrophe transition.

Then since resolution at rank 1 occurs at rate $m_{12} + m_{21}$ and birth occurs at rate s for each rank we have with (9.667) that:

$$P(T_{\text{dec}} \wedge T_C > t) \leq \text{const} \cdot e^{-(m_{21}+m_{12})t} e^{-(s-m_{21})t} = \text{const} \cdot e^{-(s+m_{12})t}. \quad (9.668)$$

But since the mean population growth (including all colours) is bounded by a pure birth process with birth rate s, we have that the mean mesostate size satisfies

$$E\left[|\hat{t}((T_{\text{dec}} \wedge T_C)-)|\right] \leq \text{const} \cdot \int_0^\infty e^{st} e^{-(s+m_{12})t} dt < \infty. \quad (9.669)$$

Using the method of Subsubsection 9.7.8 in the sequel of (9.394) we can now refine this argument to get a finite $1 + \varepsilon$ moment for the subpopulation of colours ≥ 2 for some $\varepsilon > 0$. We begin by estimating the length at the first catastrophe event before decoupling and then we move up to T_{dec} itself. Namely, using (9.669) and (9.412)(for the generating function and probability mass function of the pure birth process at time t) we obtain:

$$P(|\hat{t}((T_C \wedge T_{\text{dec}})-)| \geq k) \leq \text{Const} \cdot \int_0^\infty (s+m_{12}) e^{-(s+m_{12})t}(1-e^{-st})^{(k-1)} dt$$
$$\leq \frac{(m_{12}+s)}{s} \int_0^1 u^{\frac{m_{12}}{s}}(1-u)^{k-1} du. \quad (9.670)$$

Next we bound the probability of being equal to k by being $\geq k$ and note that

$$\sum_{k=1}^\infty k^{1+\varepsilon}(1-u)^k \sim \frac{1}{|\log(1-u)|^{2+\varepsilon}} \Gamma(\varepsilon+2). \quad (9.671)$$

Combining this with (9.670) gives the following bound for $(1+\varepsilon)$-moment of the mesostate size at the time of the first catastrophe given that we started with the $\hat{t}_*$:

$$E[|\hat{t}((T_C \wedge T_{\text{dec}})-)|^{1+\varepsilon}] \leq \text{Const} \cdot \int_0^1 u^{\frac{s+m_{12}}{s}} \frac{1}{|\log(1-u)|^{2+\varepsilon}} du < \infty \text{ if } \varepsilon < \frac{m_{12}}{s}.$$
$$(9.672)$$

In order to consider the continued development we now rename the first catastrophe time as $T_{C,1}$. The idea now is that we must continue the process up to the decoupling time T_{dec} but keep track of the set of intervening catastrophe times $T_{C,j}$ where $T_{C,j+1}$ is the time of the first catastrophe produced by a descendant of a birth at rank 1 which occurs after $T_{C,j}$. In this way the the remnants left at time $T_{C,j}$ (and their descendants) are removed at time $T_{C,j+1}$ (because the cut-point is to the left of the latter set of ranks). We also note that the $T_{C,j+1} - T_{C,j}$ are independent and have the same distribution at $T_{C,1}$.

If $T_{C,1} \wedge T_{\text{dec}} = T_{C,1}$, then we continue starting the process but now with initial condition given by the non-removed subpopulation, its ranks are called residual ranks and we denote by N_1 the number of these ranks. We now consider the process up to the next colour 1 catastrophe occurring in the subpopulation formed by new immigrants and their descendants. The time at which this occurs again is denoted by $T_{C,2}$. In order to obtain an upper bound on $(1 + \varepsilon)$-moment of the size of the population at time $T_{C,2} \wedge T_{\text{dec}}-$ we assume that each of the N_1 ranks are given by (110) and each of these produces descendants according to independent birth processes with birth rate s.

Note that this subpopulation is completely removed at the next colour 1 catastrophe time produced by the offspring of a new immigrant from rank 1 after the catastrophe event (because the associated subtableau is completely to its right). Define N_1 the number of residual ranks, and $\{n_i(t)\}_{t \geq 0}$ the growth of the ith residual rank up to the next catastrophe time (the dominating pure birth process with birth rate s).

Noting that the N_1, $T_{C,2} \wedge T_{\text{dec}}$ and the $\{n_i(\cdot)\}$ are independent, then then first conditioning on N_1 and $T_{C,2} \wedge T_{\text{dec}}$, then we obtain

$$E[E((\sum_{j=1}^{N_1} n_i(T_{C,2} \wedge T_{\text{dec}})^{\alpha})|N_1]] \leq E[N_1^{\alpha}] E[n_1(T_{C,2} \wedge T_{\text{dec}})^{\alpha}] \quad (9.673)$$

$$< \infty,$$

using (9.672) first for N_1 and then for the second term.

(Here we used the inequality for the sum of a random number of iid random variables:

$$(E[E((\sum_{i=1}^{N} X_i)^{\alpha})|N])^{1/\alpha} = \left(E_N [E[\left\| \sum_{i=1}^{N} X_i \right\|_{\alpha}^{\alpha} |N]]\right)^{1/\alpha} \quad \text{with } \alpha > 1$$

$$\leq \left(E_N E[(\sum_{i=1}^{N} \|X_i\|_{\alpha})^{\alpha}|N]\right)^{1/\alpha}$$

$$= E_N [(N \|X_1\|_{\alpha})^{\alpha}]^{1/\alpha}$$

$$= (E[N^{\alpha}])^{1/\alpha} \|X_1\|_{\alpha}, \quad (9.674)$$

with $\alpha = 1 + \varepsilon$, $N = N_1$ and $X_i = n_i(T_{C,2} \wedge T_{\text{dec}})$.)

This implies that the population at the $(T_{C,2} \wedge T_{\text{dec}})-$ has a finite $(1+\varepsilon)$ moment. If $(T_{C,2} \wedge T_{\text{dec}}) = T_{C,2}$, then we must again continue. Noting that all of the N_1 remnant population and their descendants are removed at $(T_{C,2} \wedge T_{\text{dec}})$ it follows that the size of the new remnant population has the same distribution as N_1. We can then continue and for some finite k we will have the event $T_{C,k} \wedge T_{\text{dec}} = T_{\text{dec}}$ and again the $(1+\varepsilon)$-moment of the population just prior to this time is finite. ■

Remark 187. The same argument shows that the $(1+\varepsilon)$-moment of a mesostate started as a residual mesostate created by a decoupling event at the parent mesostate remains bounded.

Remark 188. We can modify this argument to show that in the case $m_{12} = 0$, the number of offspring produced at decoupling has a finite ε-moment for some $\varepsilon < 1$ - to do this consider the first colour 1 individual born with lifetime larger than the decoupling time.

Offspring Size at Decoupling

We first review the history of a typical mesostate. We begin with (110) at rank 1. After the first birth (selection event) at rank 1, the *renewal time* (which results from a mutation event at rank 1) is an exponential random variable with mean $1/m$. The evolution of the mesostate at ranks $2, 3, \ldots$ is independent of the renewal time (decoupling time). At the decoupling time the state of the mesostate returns to (110) and if a $2 \to 1$ mutation event occurs a random number of offspring mesostates (factors) are created. These correspond to the internal factors existing just prior to the decoupling time. Each such factor corresponds to a mesostate started at (110) at its birth time. For example a colour 2 factor is created when a $2 \to 1$ mutation event occurs at a colour 1 rank or when an existing colour 2 factor undergoes a mutation event at its initial rank and thus produces a random number of colour 3 factors. In other words, at the renewal point the state returns to (110) and can also produce a random number of offspring each of which then evolves as an independent copy of the mesostate (which at the decoupling time is the state of the mesostate process at a random time given by the time passed since its initial rank was born). Once created a mesostate is immortal and the distribution of its state approaches an equilibrium distribution. This will then produces a supercritical branching process with some malthusian parameter β_{Pe} and random constant W_{Pe} – see Lemma 9.75.

9.12.4 The Case $M = 2$ and $d > 0$

We now generalize our results to $d > 0$ where coalescence reduces the number of ranks per site but where on the other hand migration plays an important role.

Proof of Proposition 9.64 for $M = 2, d > 0$.

We first note that when the coalescence rate $d > 0$ then the microstate at a site is recurrent so that the growth can only occur in the case $c > 0$. More precisely, recall that due to coalescence the number of ranks at a site reduces to one after a random time having an exponential moment (recall Lemma 9.25).

In this case new sites are occupied at migration events and the occupied sites now play the role played by ranks in the case $d = 0$. During the sojourn between times at which the microstate at a site becomes one factor (110), it can produce new sites by migration and the number produced in this interval has a finite second moment (Lemma 9.52 (b)).

We again introduce the *colouring* system:

- The founding site of the mesostate has colour 1 and the m_{21}-offspring of a colour 1 site is colour-2, that is, if the first rank at a site (other than the founding site) has a $2 \to 1$ mutation then the site changes to colour-2 (this occurs at rate m_{21}). Similarly if the first rank at a colour-2 site has a $2 \to 1$ mutation the site changes to colour 3, etc.
- A $1 \to 2$ mutation at the first rank of a site is a catastrophe event and this results at removing all descendants of this site (this occurs at rate m_{12}). Note that in the absence of $1 \to 2$ mutations the set of occupied sites forms a CMJ process. We denote the malthusian parameter of this process by $\tilde{s}$.

We would like now to proceed along the lines of the $d = 0$ use above. We must therefore now show that in (9.680) we can replace s by $\tilde{s}$, similarly in (9.668) and then in (9.680), (9.669)–(9.672).

We must now determine the probability of no catastrophe occuring by time t. In the case $d = 0$ this was done using the generating function of the distribution for an auxiliary process, a birth-death-immigration process $X(t)$, the number of the colour 1 ranks at time t. Since we no longer have the exact distribution at time t of the number of sites in the CMJ process we must now proceed instead as follows.

Again we work with an auxiliary process indicated by a wiggle, a process where we put $m_{12} = 0$. Let $\tilde{X}(t)$ now be the number of occupied sites in the process, which is a CMJ process, for which we know that for some $\tilde{s} \in (0, s)$ we have $\lim_{t \to \infty} e^{-\tilde{s}t} \tilde{X}(t) = \tilde{W}$ with $0 < \tilde{W} < \infty$ a.s. It is known that $\tilde{W}$ has a continuous density $f_{\tilde{W}}(.)$ (see Doney [Do], Theorem B). Let q be chosen as in (9.665). Then we have that T_C (time of first catastrophe event) is an exponential and one variable with parameter m_{21} and

$$\lim_{t \to \infty} e^{(\tilde{s}-m_{21})t} P(T_C > t) = \lim_{t \to \infty} e^{(\tilde{s}-m_{21})t} E[q^{\tilde{W}(t)e^{(\tilde{s}-m_{21})t}}] \quad (9.675)$$

$$= \lim_{t \to \infty} e^{(\tilde{s}-m_{21})t} \int_0^\infty e^{-\log(q^{-1})xe^{(\tilde{s}-m_{21})t}} dF_{\tilde{W}(t)}(x)$$

$$= \int_0^\infty e^{-\log(q^{-1})u} f_{\tilde{W}}(u) du < \infty,$$

by the continuity of $f_{\tilde{W}}(\cdot)$ at 0.

9.12 Emergence and Fixation: The Selection-Dominant Case

Then instead of using the exact distribution as in (9.670) for the case $d = 0$ we proceed now for the case $d > 0$ as follows:

$$E[(X(T_{\text{dec}} \wedge T_C)_-)^{1+\varepsilon}] \tag{9.676}$$

$$= \text{const} \cdot \int_0^\infty e^{-mt} e^{-(\tilde{s}-m_{21})} \sum_k k^{1+\varepsilon} P(\tilde{X}(t) = k) \quad (\text{using } (9.675))$$

$$= \text{const} \cdot \int_0^\infty e^{-(\tilde{s}+m_{12})t} \sum_k k^{1+\varepsilon} P(\tilde{W}(t) = ke^{-\tilde{s}t}).$$

Recall that the total number of migrants during a decoupling cycle has a finite second moment (Lemma 9.52(b)). Therefore by Bingham and Doney [BD] Theorem 3, $E[\tilde{W}^{1+\varepsilon}] < \infty$.

We next note that with $u = e^{-\tilde{s}t}$

$$\lim_{u \downarrow 0} u^{2+\varepsilon} \sum_{\ell=1}^\infty k^{1+\varepsilon} P\left(\frac{\tilde{W}}{u} = k\right) = \int_0^\infty x^{1+\varepsilon} f_{\tilde{W}}(x) dx = E[\tilde{W}^{1+\varepsilon}] < \infty. \tag{9.677}$$

We can now proceed as in the case $d = 0$ to get the bound provided that $\varepsilon < \frac{m_{12}}{\tilde{s}}$:

$$E[\tilde{X}((T_C \wedge T_{\text{dec}})_-^{1+\varepsilon})] \tag{9.678}$$

$$\leq \text{const} \int_0^\infty e^{-(\tilde{s}+m_{12})t} \sum_k k^{1+\varepsilon} P(\tilde{W} = ke^{-\tilde{s}t})$$

$$= \text{const} \int_0^1 u^{\frac{\tilde{s}+m_{12}}{\tilde{s}}} \frac{1}{u^{2+\varepsilon}} E[\tilde{W}^{1+\varepsilon}] < \infty. \quad \text{q.e.d.}$$

9.12.5 Higher Moments in the Case $M = 2$, $d = 0$

We now turn to the question of higher moments. We first show that the number of colour k mesostates has finite higher moments for each k. We then exhibit an internal supercritical branching subpopulation for the mesostate process during an excursion from (110). This implies that not all moments of the total mesostate size at the decoupling time can be finite. This raises an open question as to the number of finite moments as a function of the parameters.

Second Moments for the Colour-k-Populations: $M = 2, d = 0$

We now consider higher and in particular second moments. Through the dynamic the colours 2,3, etc. depend on the colour-1 population on which we need more information since it produces the immigration stream for colour-2. Starting with a simple factor (110) our process has a colour 1 population which is a birth-death-catastrophe process with immigration at rank 1 with rate s.

Now consider the population produced by individual 1 up to the time of the first colour 1 catastrophe. The population of descendants of individual 1 form a supercritical branching process. We can tag individuals S or F at creation and assign an exponential clock to the F's. We then consider the process in which the production of an F with clock $\leq t_0 = 1$ (for example) is considered as a death (say with rate $\delta m_{12}, 0 < \delta < 1$).

We will prove below that the colour 1 process $C_1(t)$ has an equilibrium distribution with finite first and second moments,

$$E_{\mathrm{eq}}[C_1^2] < \infty. \tag{9.679}$$

Colour 1 individuals die at rate m_{21} and become immigrants to the colour 2 population. The colour 2 population is then a birth, death, immigration, catastrophe process. Moreover the growth of the population of colours ≥ 2 is dominated by a pure birth process with rate s. Recall that the probability that a colour 1 catastrophe has not occurred by time t is

$$P(T_C > t) \leq \mathrm{const} \cdot e^{-(s-m_{21})t}. \tag{9.680}$$

Therefore the expected number of deaths (i.e. immigrants to colour 2) to occur before the next catastrophe has a finite second moment. (To directly obtain the generating function for the number of deaths of colour 1 individuals we can combine the generating function formula of Puri [PSP] with the result of Brockwell et al. [BGR].)

To obtain higher moments of the number of ranks we look at the structure of the process more closely with the goal to get moment bounds for the colour-k population. To illustrate the role of the catastrophe mechanism in this game we first consider the case $m_{21} = 0$, this means a coloured individual does not experience a simple death. Noting that the mutation $1 \to 2$ truncates the tableau, it follows that in this case the mesostate process corresponds to a birth-immigration-catastrophe process with birth and immigration rates equal to s and uniform catastrophe rate m_{12}. Then by the result of Brockwell, Gani and Resnick [BGR], equation (5.8), the stationary distribution $(p_n)_{n \in \mathbb{N}}$ is given by

$$p_n = \frac{(n+1)m_{12}}{s+m_{12}} \prod_{0 \leq j \leq n-1} \frac{(j+1)s}{(j+2)(s+m_{12})} \tag{9.681}$$

and the first and second moments are finite.

Using these ideas we next establish that the number of colour k individuals in the MBDC process in equilibrium has a finite second moment.

Proposition 9.70 (MBDC-process: colour-k populations).
Consider the MBDC-process from Definition 9.66. Assume $M = 2, d = 0, m_{12} \cdot m_{21} > 0$.

9.12 Emergence and Fixation: The Selection-Dominant Case

(a) *The colour 1 population is a birth-death-catastrophe process with immigration where the birthrate = immigration rate = s, death rate m_{21} and catastrophies occuring at rate $m_{12} > 0$. The invariant measure has finite first and second moments.*
(b) *Under the invariant measure, the total population of each colour k has a finite second moment for each colour $k \in \{2, 3, \cdots, \}$.* □

Proof of Proposition 9.70.

(a) The length process $L(t)$ of our mesostate is a non-negative integer-valued Markov process with generator $Q = \{q_{ij}\}_{i,j=0,1,2,\ldots}$, where

$$q_{ij} = is1_{i+1}(j) + im_{12}c_{ij}1_{[0,i)}(j) + im_{21}1_{i-1}(j) + s1_{i+1}(j), \quad i \neq j$$
$$q_{ii} = -s - i(s + m_{12} + m_{21}) + im_{12}c_{ii}, \quad (9.682)$$

where $c_{ij} = \frac{1}{i}$, $j = 0, \ldots, i-1$. If $\mathbf{p}(t)$ is the distribution of $L(t)$, then it satisfies the forward Kolmogorov equation

$$\mathbf{p}'(t) = \mathbf{p}(t)Q. \quad (9.683)$$

We can adapt the generating function argument of Brockwell, Gani and Resnick [BGR], Section 5 to include deaths induced by m_{21}-mutations.

Consider for every time $t \geq 0$ the generating function

$$\phi(z, t) = \sum_{j=0}^{\infty} P(L(t) = j)z^j, \quad (9.684)$$

and derive the differential equation for ϕ. We assume that $p_0(0) = 1$ so that $\phi(z, 0) = 1$. The immigration, birth and catastrophe terms arise exactly as in [BGR], equation (2.6c), and in our case we have an additional death term with coefficient m_{21}. This yields

$$\frac{\partial \phi(z,t)}{\partial t} = (sz^2 + m_{21} - sz - zm_{12} - zm_{21})\frac{\partial \phi(z,t)}{\partial z} \quad (9.685)$$
$$+ \left[s(z-1) - \frac{m_{12}}{1-z}\right]\phi(z,t) + \frac{m_{12}}{1-z}.$$

Now take Laplace transforms *with respect to* t $\Psi(z, \theta) \int_0^\infty e^{-\theta t} \psi(z, t) dt$ to obtain:

$$-\phi(z, 0) - \frac{1}{\theta}\frac{m_{12}}{1-z} = (sz^2 + m_{21} - sz - zm_{12} - zm_{21})\frac{\partial \Psi(z, \theta)}{\partial z} \quad (9.686)$$
$$+ \left[s(z-1) - \frac{m_{12}}{1-z} - \theta\right]\Psi(z, \theta).$$

Given $\theta > 0$ and $\phi(z, 0) \equiv 1$ we now solve the above linear first order differential equation for $z \in [0, 1]$:

$$e^{f(z,\theta)}\Psi(z,\theta) = -\int \frac{e^{f(z,\theta)}}{[sz(z-1) + m_{21}(1-z) - zm_{12}]}[1 + \frac{1}{\theta}\frac{m_{12}}{1-z}]dz, \quad (9.687)$$

where

$$e^{f(z,\theta)} = (z-a_1)^{(\frac{1-a_1}{a_2-a_1} - \frac{\theta}{s(a_1-a_2)} - \frac{Bm_{12}}{s})}(z-a_2)^{(\frac{a_2-1}{a_2-a_1} + \frac{\theta}{s(a_1-a_2)} - \frac{Cm_{12}}{s})}(1-z)^{\frac{m_{12}A}{s}} \quad (9.688)$$

and the constants A, B, C, a_1, a_2 depend only on s, m_{12}, m_{21}, where

$$C = \frac{1}{(a_2+a_1)(a_2-1)}, \quad B = \frac{1}{(a_2+a_1)(a_1-1)} \quad (9.689)$$

$$A = \frac{a_2 + a_1 - 2}{(a_2+a_1)(a_2-1)(a_1-1)}), \quad (9.690)$$

$$a_1 + a_2 = \frac{m+s}{s} \quad a_1 a_2 = \frac{m_{21}}{s}, \quad m = m_{12} + m_{21}, \quad a_1 < 1 < a_2, \quad (9.691)$$

(where for the latter we use $s > m$). Furthermore we have

$$A = \frac{(s-m)s}{(m+s)m_{12}}, \quad (9.692)$$

$$0 < \frac{m_{12}}{s}A = \frac{s-m}{m+s} < 1. \quad (9.693)$$

We can then determine the stationary distribution via its generating function ϕ by taking the limit

$$\phi_{eq}(z) = \lim_{t\to\infty} \phi(z,t) = \lim_{\theta\to 0} \theta\Psi(z,\theta). \quad (9.694)$$

Finally, to verify that the second moment in equilibrium is finite in the case $m_{12} > 0$ it suffices to show that $\phi_{eq}(z)$ is twice differentiable at $z = 1$, which we obtain by explicit calculation from the representation of ϕ_{eq} by Ψ as follows.

Using (9.693) we focus on the behaviour of $\phi_{eq}(z)$ at $z = 1$ after taking the limit $\theta \to 0$ in (9.694). The behaviour at $z = 1$ is determined by an expression of the form

$$\frac{1}{(1-z)^\alpha}\int_z^1 (1-z)^{\alpha-1}g(z)dz \text{ where } 0 < \alpha < 1 \text{ and } g \text{ is analytic at } z = 1. \quad (9.695)$$

Therefore the first and second derivatives of $\phi_{eq}(z)$ are finite at $z = 1$. This proves the claim that the length of a mesostate has finite second moments.

9.12 Emergence and Fixation: The Selection-Dominant Case

(b) Having now discussed the colour 1 population, we must now consider the *hierarchical generalization* of the birth-death-catastrophe process. Note that for a colour ℓ rank there are $\ell - 1$ catastrophe mechanisms that can kill it! To see this consider a rank of colour ℓ. Then it is a descendant of a founding rank of colour $\ell - 1$ that in turn is a descendant of founding rank of colour $\ell - 2$. We say a coloured rank is a $m_{1,2}$-descendent of the rank of the previous colour if a $m_{1,2}$-mutation changed this rank from colour ℓ to $\ell + 1$. Then any resolution of a m_{12}-descendant of the same founding ranks of colour $\ell - 1, \ell - 2$ (but born later) will eliminate the rank of colour ℓ, etc.

Now consider colour 2 ranks descending from a specific colour 1 rank (referred to as *segment*). These correspond to "dead" colour-1 particles. But the first and second moments of the number of dead colour 1 particles (segment founders) remains bounded since the number of live colour-1 particles does (see (9.679) which bounds the immigration rate in colour 2). Moreover then in particular the first and second moments of the number of colour-2 ranks in a segment remain bounded. Then by the branching property we can conclude that the first and second moments of the number of colour-2 particles remain bounded.

Similarly proceeding inductively this is true for $\ell = 1, \ldots, k$. q.e.d.

Remark 189. We can modify the above to compute the joint distribution of the number of colour-1 individuals and the number of deaths, that is, the number of colour-2 immigrants, as in [PSP]. This is formulated as follows.

The process $L(t) = (X_1(t), X_2(t))$ of our colour-1, colour-2 mesostate is a $\mathbb{N}^2$-valued Markov process with generator $Q = \{q_{i_1, i_2; j_1, j_2}\}$, where

$$q_{i_1, i_2; j_1, j_2} = i_1 m_{21} 1_{i_1 - 1}(j_1) 1_{i_2 + 1}(j_2), \quad (i_2 \neq j_2, i_1 \neq j_1) \tag{9.696}$$

$$q_{i_1, i_2; j_1, j_2} = i_1 s 1_{i_1 + 1}(j_1) + i_1 m_{12} c_{i_1, i_2; j_1, j_2} 1_{[0, i_1)}(j_1) + s 1_{i_1 + 1}(j_1), \quad (i_1 \neq j_1, i_2 = j_2)$$

$$q_{i_1, i_2; j_1, j_2} = -s - i_1(s + m_{12} + m_{21}) + i_1 m_{12} c_{i_1; j_1}, (i_1 = j_1, i_2 = j_2),$$

where $c_{i_j, i_2; j_1, j_2} = c_{i_1 j_1} = \frac{1}{i_1}$, $j_1 = 0, \ldots, i_1 - 1$. Let $\mathbf{p}(t)$ be the distribution of $L(t)$, then it satisfies the forward Kolmogorov equation

$$\mathbf{p}'(t) = \mathbf{p}(t) Q. \tag{9.697}$$

We can adapt the generating function argument of Brockwell, Gani and Resnick [BGR], Section 5 to include deaths.

Consider the generating function

$$\phi(z_1, z_2, t) = \sum_{j=0}^{\infty} P(L(t) = (j_1, j_2)) z_1^{j_1} z_2^{j_2}, \tag{9.698}$$

and solving the differential equation for ϕ. We assume that $p_0(0,0) = 1$ so that $\phi(z_1, z_2, 0) = 1$. We get

$$\frac{\partial \phi(z_1, z_2, t)}{\partial t} = (sz_1^2 + \frac{m_{21}}{z_2} - sz_1 - z_1 m_{12} - z_1 m_{21}) \frac{\partial \phi(z_1, z_2, t)}{\partial z_1} \quad (9.699)$$

$$+ \left[s(z_1 - 1) - \frac{m_{12}}{1 - z_1} \right] \phi(z_1, z_2, t) + \frac{m_{12}}{1 - z_1}.$$

The solution can be obtained to show that the second moment is finite but we will not give the detailed calculations here.

Remark 190 (Moments for equilibrium of the MBDC-process).
In addition to the above proof that $\beta = \beta_{Pe}$ under the conditions which guarantee ergodicity under mutation among the lower types it is of interest to prove that the mesostate process has stationary measure with number of active sites having *finite first and second* moments in equilibrium. This is equivalent to showing that the random number of individuals in the MBDC-process has finite first and second moments in equilibrium. The rigorous details will not be given here but we will now explain the main ideas.

This marked tree-valued process has a *stationary measure* concentrated on *finite trees*. The remaining problem is to show that the *expected number of vertices is finite* under the stationary measure. The pruning of a fixed Galton-Watson tree has been studied by Aldous and Pitman [AP98]. However to complete the argument we require the corresponding development of the dynamical tree pruning process described above.

Internal Factors and the Multicolour Growth Process

Recall that the growth of colour 1 ranks is a birth and death process with birth rate s and death rate m_{21}, immigration at rank 1 at rate s and catastrophe rate m_{12}. Each new colour-1 rank created has the potential to make a colour 1 catastrophe transition and this was used above to obtain the $(1 + \varepsilon)$-moment. The idea is to now to use the catastrophe transitions and multicolour structure to obtain bounds on higher moments.

Consider a colour 2 individual and its descendants up to the colour 1 decoupling time or first colour 1 catastrophe time that eliminates it. In this interval we can view this as a copy of the original system and it produces a random number of factors (the internal factors of the mesostate). The internal factors become offspring mesostates at the colour 1 decoupling time.

Now consider the growth of all descendants of a newly produced colour 2 individual (rank) up to the resolution time or colour 2 catastrophe jump in a typical subtableau (say the ith, which can only be removed at a decoupling event), namely, $T_{*,2,i} := (T_{C,1} \wedge T_{\text{dec},1}) \wedge (T_{C,2} \wedge T_{\text{dec},2})$. Recall that the colour two process up to $T_{\text{dec},2}$ is a birth and death process with immigration at the first rank and birth, death

9.12 Emergence and Fixation: The Selection-Dominant Case

and immigration rates s, m, s, respectively. Then using the same argument as for the $(1+\varepsilon)$ moment we can obtain a $(2+\varepsilon)$ moment for the size of the colour 2 tableau at $(T_{C,1} \wedge T_{\text{dec},1}) \wedge (T_{C,2} \wedge T_{\text{dec},2})$— because there are two independent times with distribution $e^{-(s+m_{12})t}$. If this occurs at the colour 2 resolution time it produces the product of a random number (with finite second moment) of tableau each of which can be removed only by a colour 1 decoupling or catastrophe. If it occurs at colour 1, the colour 2 family is removed and if it occurs at a colour 2 catastrophe it is truncated and continues growth as in the first moment development (recall that this tail of the tableau is removed at the next new catastrophe jump).

In turn each of these develops in the same way so we have a CMJ process of internal factors which grows until the colour 1 catastrophe *or* decoupling. Define

$$T_{*,1} = (T_{C,1} \wedge T_{\text{dec},1}), \; T_{*,2,i} = (T_{C,2,i} \wedge T_{\text{dec},2,i}) \tag{9.700}$$

with the catastrophe respectively decoupling times of the i-th colour-2 individual.

Let $\|\hat{t}_i\|$ denote the size of the ith tableau (corresponding to an internal factor) at the decoupling time $T_{*,1,i}$. Then since the hazard rate for $\|\hat{t}_i\|$ is $2(s+m_{21})$ (due to catastrophe at either level 1 or 2) we have

$$E[\|\hat{t}_i\|^2] \leq E[e^{2sT_{*,2,i}}] \leq \text{const} \cdot \int_0^\infty 2(s+m_{12})e^{-2(s+m_{12})t} e^{2st} dt < \infty. \tag{9.701}$$

This implies that the growth of the number of internal factors is given by a CMJ process in which the number of offspring produced at a birth event has a finite second moment. Letting $N_{\text{int}}(t)$ denote the number of internal factors at time t, we have

$$N_{\text{int}}(t) \sim We^{\tilde{\beta}t}, \tag{9.702}$$

where $E[W^2] < \infty$ and $\tilde{\beta}$ is the corresponding malthusian parameter.

We now consider this process in more detail in order to obtain an upper bound for $\tilde{\beta}$. Note that internal factors produced by a colour 2 internal factor is colour 3, internal factors produced by a colour 3 factor are colour 4, etc. Next note that during its lifetime a rank of colour k produces ranks of colour $k+1$ - this requires the colour k particle to have a birth and then for a mutation event to occur (either rank 1 for the rank 2 particle of the rank ℓ particle if $\ell > 2$). This produces a Bellman-Harris continuous time branching process in which new individuals are created after times given by the sum of two independent exponential random variables, one with mean $\frac{1}{s}$ and the other with mean $\frac{1}{m}$.)Note that if we include $1 \to 2$ mutation transitions then some of these ranks are removed by catastrophe transitions.)

Lemma 9.71. *Consider a Bellman-Harris process with times between births given by the sum of two independent exponential with mean $\frac{1}{s}$, $\frac{1}{m}$ and mean offspring size α. Then the malthusian parameter $\tilde{\beta}$ satisfies*

$$\tilde{\beta}(m,s) = \frac{-(m+s) + \sqrt{(m+s)^2 + 4(\alpha-1)ms}}{2} \leq (m+s)\frac{\sqrt{\alpha}-1}{2}. \quad (9.703)$$

□

Proof of Lemma 9.71. The time required for the production of a new individual is the sum of independent exponential with means $\frac{1}{s}$, $\frac{1}{m}$. This produces a supercritical Bellman-Harris branching process.

Note that for independent exponentials E_1, E_2 with means $\frac{1}{s}$, $\frac{1}{m}$, respectively,

$$E[e^{-\beta(E_1+E_2)}] = \frac{s}{\beta+s}\frac{m}{\beta+m} \quad (9.704)$$

Therefore by [H], Theorem 17.1, the malthusian parameter is given by the root β_{meso} of

$$\alpha\frac{s}{\beta+s}\frac{m}{\beta+m} = 1, \quad \text{that is,} \quad (9.705)$$

$$(\beta^2 + (s+m)\beta + (1-\alpha)sm) = 0$$

$$\beta(m,s) = \frac{-(m+s) + \sqrt{(m+s)^2 + 4(\alpha-1)ms}}{2} \leq (m+s)\frac{\sqrt{\alpha}-1}{2}. \quad (9.706)$$

q.e.d.

We now consider the growth of a subpopulation of ranks which leave behind 0 unresolved ancestors immediately after their birth time (except for those of colour 1). More precisely we consider the subpopulation of mesostates that are produced when a birth at the first rank of the parent mesostate is immediately followed by the resolution of the first rank at the parent site, that is before a second birth at the parent site (this has probability $\frac{m}{m+s}$).

There are two possible outcomes at resolution at the first rank: either a $1 \to 2$ mutation occurs and we are left with a colour k tableau (110) or a $2 \to 1$ mutation occurs and we have a colour $k+1$ tableau (110) and a non-empty colour k tableau left. For this to happen we must have rank 1 mutate before the next birth at rank 1. For example

$$(110)_k \to \begin{array}{c}(010)_k \\ (100)_k \ (010)_k\end{array} \to (110)_{k+1} \otimes (110)_k. \quad (9.707)$$

Then the expected offspring size is $\alpha = \frac{m_{12}+2m_{21}}{m} < 2$. Then in turn either of these two particles can produce an offspring that mutates before the next birth at the corresponding rank, etc. thus producing a branching process.

9.12 Emergence and Fixation: The Selection-Dominant Case   623

Lemma 9.72. *Even if the effect of* $1 \to 2$ *transitions are included, there is a supercritical branching subpopulation of factors with no deaths up to the decoupling time at colour 1.* □

Proof. We now note that the first rank in each colour 2 family produced by a $2 \to 1$ mutation of a colour 1 rank is immortal (up to the colour 1 decoupling time). Now consider the first time that a colour 2 subtableau has exactly two active rank and the rank 1 has a $2 \to 1$ mutation transition. This produces a colour 2 individual that is immortal (up to the colour 1 decoupling time). This supercritical branching process has malthusian parameter

$$0 < \beta_{im} \leq \tilde{\beta}. \tag{9.708}$$

Remark 191. Note that this provides a lower bound on the size of the mesostate at the time $T_C \wedge T_{\text{dec}}$. This implies that not all higher moments are finite since

$$\int_0^\infty E[(We^{\beta_{im}t})^k]me^{-mt}dt] < \infty \text{ only if } m > k\beta_{im} \tag{9.709}$$

even if W has moments of all orders.

Note that if at $T_{*,1}$ the mutation occurs at the colour 1 rank decoupling occurs and the production of internal factors terminates. On the other hand if the mutation occurs at an internal factor one or more new internal factors are produced and the process continues.

Proof of Proposition 9.65.

We now consider the branching process in which the individuals are interior factors and these become offspring mesostates at the colour 1 decoupling time $T_{\text{dec},1}$. Let $N(T_{*,1}-)$ denote the number of such interior factors at $T_{*,1}$, that is, the number of offspring produced at $T_{*,1}$. It will suffice to prove that $N(T_{*,1})$ has a finite second moment.

First, note that the number of offspring produced at $T_{*,1}$ is given by the number of internal factors, $N_{\text{int}}(T_{*,1}-)$, at $T_{*,1}$.

Recall that the growth of the set of internal factors of a mesostate between decoupling events is a multitype CMJ branching process with malthusian parameter $\tilde{\beta}$ (see (9.702)). We want to determine the second moment of the number internal factors up to the decoupling time $T_{*,1}-$, that is, $N_{\text{int}}(T_{*,1}-)$. Since the distribution of $T_{*,1}-$ is dominated by an exponential tail, const $\cdot e^{-(s+m_{12})t}$, in order to verify that the second moment is finite it suffices to show that $\tilde{\beta} < \frac{s}{2}$.

We next obtain a lower bound for the malthusian parameter. For the moment we assume that when internal factor has a decoupling event it produces a random number α of offspring internal factors (of the form (110)) and the time between the birth of an internal factor and it decoupling time is given by a birth time (selection event) and resolution at its first rank (mutation event), that is independent exponentials with means $\frac{1}{s}$, $\frac{1}{m}$, respectively.

Therefore by (9.706)

$$\beta(m,s) = \frac{-(s+m)+\sqrt{(s+m)^2-4(1-\alpha)sm}}{2} = \frac{s+m}{2}\left(-1+\sqrt{1+4(\alpha-1)\frac{sm}{(s+m)^2}}\right). \tag{9.710}$$

To see why this is a lower bound note that for the actual mesostate dynamics some of the internal factors produced will not be single (110) factors but will already form tableaus of length larger than one at the birth time. In order to handle this it would be necessary to consider the production of factors with their initial tableaus as a multitype branching process and calculate the corresponding malthusian parameter. However it is clear from this argument that $\tilde{\beta} \geq \beta(m,s)$.

On the other hand to obtain an upper bound we can consider only the time required for the mutation step. Then as a special case of [H], Theorem 17.1 corresponding to $s = \infty, \alpha = 2$ we define $\beta_1(\alpha, m)$ as the root of

$$\alpha \int_0^\infty e^{-\beta t} m e^{-mt} dt = \frac{\alpha m}{\beta + m} = 1 \tag{9.711}$$

$$\beta_1(\alpha, m) = (\alpha - 1)m$$

and

$$\tilde{\beta} \leq \beta_1(\alpha, m) = (\alpha - 1)m. \tag{9.712}$$

We must now determine α. Consider a newly created factor. When it resolves it will produce a random number of new factors with mean α. We first note that when the first rank of a factor mutates the following are the possibilities:

(i) Mutation is $1 \to 2$ and no new factors are created: prob $\frac{m_{12}}{m}$

$$\frac{m_{12}}{m} \cdot 1$$

(ii) Mutation is $2 \to 1$ and the last event at rank 2 before decoupling is a selection event - this has probability

$$\frac{m_{21}}{m} \frac{s}{s+m}$$

and in this case the offspring size is 2

(iv) Mutation is $2 \to 1$ and the last event at rank 2 is a mutation event. This has probability

$$\frac{m_{21}}{m} \frac{m}{m+s}$$

and the offspring size is bounded by the expected mesostate size, say $C < \infty$.

9.12 Emergence and Fixation: The Selection-Dominant Case

The expected size is then

$$\alpha \leq \frac{m_{12}}{m} + \frac{m_{21}}{m}[\frac{2s}{s+m} + \frac{Cm}{s+m}]. \tag{9.713}$$

We obtain a more precise upper bound for α as follows. We note that the subtableau produced during the time since the last mutation at rank 2 is dominated by a copy of the initial tableau and therefore

$$\alpha \leq \frac{m_{12}}{m} + \frac{m_{21}}{m}[\frac{2s}{s+m} + \frac{m}{s+m}(1+\alpha)] \text{ so that}$$

$$\alpha \leq \frac{2s+m}{s} = 2 + \frac{m}{s} = 2 + x, \quad x = \frac{m}{s}. \tag{9.714}$$

Then we obtain

$$\lim_{s \to \infty} \alpha = 2. \tag{9.715}$$

We then have using (9.712)

$$\lim_{s \to \infty} \tilde{\beta} = m \tag{9.716}$$

and therefore for sufficiently large s

$$E[(N_{\text{int}}(T_{*,1}-))^2] \leq \text{const} \cdot \int_0^\infty e^{2\tilde{\beta}t} e^{-(s+m_{12})t} dt < \infty. \tag{9.717}$$

q.e.d.

Remark 192. We also have the upper bound

$$\beta(s, m, \alpha) = \frac{s+m}{2}(-1 + \sqrt{1 + 4(\alpha-1)\frac{sm}{(s+m)^2}}) \tag{9.718}$$

$$\leq \frac{s+m}{2}(-1 + \sqrt{\alpha})$$

so that $\beta(s, m, \alpha) < \frac{s}{2}$ if $\alpha < \left(1 + \frac{1}{1+x}\right)^2$. By numerical calculation and using (9.714) we see that this is satisfied, for example, if $x < 0.6$.

We now show that (9.717) cannot be extended to all values of the parameters. Consider the case $m = s = 1$. We first note that the probability that exactly two offspring are produced when a factor branches is:

$$P(\{\text{no births}\} \cup \{\text{at least one birth and last event at rank 2 is a selection}\})$$

$$\tag{9.719}$$

But

$$P(\{\text{no births}\}) = \sum_{k=1}^{\infty} \frac{1}{4^k} = \frac{1}{3}, \qquad (9.720)$$

and

$$P(\{\text{at least one birth and last event at rank 2 is a selection}\}) = \frac{2}{3} \cdot \frac{1}{2} = \frac{1}{3}. \qquad (9.721)$$

Therefore

$$P(\{\text{exactly 2 offspring}\}) = \frac{2}{3}. \qquad (9.722)$$

We then obtain a lower bound on the expected offspring size α, namely,

$$\alpha \geq 2 \cdot \frac{2}{3} + 3 \cdot \frac{1}{3} = \frac{7}{3}. \qquad (9.723)$$

But recall that $\beta(1, 1, \alpha) = \sqrt{\alpha} - 1$ and this is less than $\frac{1}{2}$ only if $\alpha < \frac{9}{4}$. This means that we can pick $s > m$ (but sufficiently close to m) and m_{12} sufficiently small so that

$$2\tilde{\beta} > (s + m_{12}) \qquad (9.724)$$

and therefore the second moment of $N_{\text{int}}(T_{*,1}-)$ is infinite.

Open Problem 9.73. *We leave the question of the determination of the range of values of the parameters for which the second moment of $N_{\text{int}}(T_{*,1}-))$ is finite as an open problem.*

9.12.6 Extension to the Case $M > 2$: Conclusion of the Proof of Proposition 9.64

Now consider the general case $M \geq 2$, i.e. we are now focussing on $M > 2$ and $d > 0$. In the case $M > 2$ the action of selection and mutation is more complex because more than just one mutation event at the founding rank can be needed before decoupling occurs, so that selection and coalescence intervene and a whole cycle of mutation, selection and coalescence transitions is needed to obtain a decoupling event.

When $d > 0$, due to coalescence the number of ranks at a fixed site has a finite exponential moment (Lemma 9.25). However ranks at different sites cannot coalesce

9.12 Emergence and Fixation: The Selection-Dominant Case

so that migration plays an important role and occupied sites in a mesostate now play a role similar to that of ranks in the case $d = 0$. In addition, in general a catastrophic death (which now removes a set of occupied sites) can occur only after a *whole chain* of mutation-selection-coalescence events. We illustrate with the following example.

Example 21. Consider (in the case $M = 3$), the multisite tableau associated with label 1 arising from (1110) by a chain of 3 birth events the first two with subsequent migration. Then mutation $3 \to 2$ at rank 2 and then at rank 1:

$$
\begin{array}{l}
(0010)_1 \\
(1100)_1 \ (0010)_2 \\
(1100)_1 \ (1100)_2 \ (0010)_3 \\
(1100)_1 \ (1100)_2 \ (1100)_3 \ (1110)_3
\end{array}
\tag{9.725}
$$

$$
\to
\begin{array}{l}
(0010)_1 \\
(1100)_1 \ (0000)_2 \\
(1100)_1 \ (1110)_2 \ (0010)_3 \\
(1100)_1 \ (1110)_2 \ (1100)_3 \ (1110)_3
\end{array}
\to
\begin{array}{l}
(0000)_1 \\
(1110)_1 \ (0000)_2 \\
(1110)_1 \ (1110)_2 \ (0010)_3 \\
(1110)_1 \ (1110)_2 \ (1100)_3 \ (1110)_3.
\end{array}
$$

The last transition leads to a decoupling of label 1 and two new labels indexed by 2,3 (and located at new founding sites) are produced (note the longer tail mesostate at the end):

$$
(1110)_1 \otimes (1110)_2 \otimes \begin{array}{l}(0010)_3 \\ (1100)_3 \ (1110)_3\end{array}. \tag{9.726}
$$

The main step in extended the results of the previous subsubsection to the general case is to establish the following extension of Lemma 9.69 and Proposition 9.64.

Proposition 9.74 (New *Pe*-sites at decoupling times: general selection-dominant case).
Consider the general ergodic case with $M \geq 2$, $d > 0$ and $s > m$ and $m_{i,j} > 0$ for $i, j \in \{1, \ldots, M\}$. Consider the growth of a mesostate starting from $\hat{t}_*$.

(a) Then $\beta = \beta_{Pe}$ and the size of the mesostate at a decoupling time has a finite $(1 + \varepsilon)$-moment.
(b) The collection of Pe sites is a supercritical CMJ-branching process.
(c) In the selection dominant case, that is, with sufficiently high selection rate, the number of new Pe sites produced due to a mutation event at the decoupling time has a finite second moment. □

Proof of Proposition 9.74. The proof in the case $M > 2$ follows the same strategy as the case $M = 2$ above. We now specify the additional complications and the additional arguments required. However not all details are provided -for example the full development of the Laplace transform calculations for the analogue of the birth-death-immigration-catastrophe process is not given here.

We first prove (b) assuming (a), and then prove (a) and (c).

(b) Since we are considering the case with no collisions and no rare mutation, each label (mesostate at a *Pe* site) which is created is immortal, different mesostates evolve independently and therefore the resulting CMJ-branching process Ψ_t is supercritical.

(a) (i) We first note that the Proof 1 (of finiteness of the first moment) is essentially the same in the general case. Once again under ergodicity the emergence rate must not depend on the initial measure μ_0 and therefore we must have $\beta = \beta_{Pe}$ and finiteness of the first moment of the mesostate size.

(ii) We now outline the main steps to extend the Proof (2) of Lemma 9.69 to the case $M > 2, d > 0$ with irreducible mutation matrix in the selection-dominant case, which allows to show that there is a finite $(1 + \varepsilon)$-moment.

The main modification is that now resolution does not occur after one mutation but is a more complex process, namely a whole *chain* of subsequent selection, mutation and coalescence events. As a result the main difference from the case $M = 2$ is that time for the mutation process to hit a trap is no longer exponential. Also the process of births from rank one is now not a rate s Poisson process but for every selection operator to act we have rate $s(e_{j+1} - e_j)$ resulting in a different type of birth.

However recalling Lemmas 9.24, 9.25, 9.26 we note that the time between decoupling events at a site has a finite exponential moment, say $\hat{\lambda} > 0$. Moreover the total number of migrants produced during a decoupling cycle has a finite second moment (Lemma 9.52(b)). Therefore the mesostate process is again a CMJ branching process in which the individuals are occupied sites. We again denote the malthusian parameter of this process by $\tilde{s}$. If $\hat{\lambda} > \tilde{s}$ we again are in the mutation-dominant case which we have treated before and we focus now on the case $\hat{\lambda} \leq \tilde{s}$.

We have to define what resolution in S, F means for $M > 2$. Here success (S) for a *site* means that at resolution the first row in the tableau resolves to to $(0, \ldots, 0)$ at its first rank and failure (F) means resolution to $(1, \ldots, 1, 0)$ of that row. Recall all other rows are then either $(00 \cdots 0)$ or $(11 \cdots 10)$. Note that if F occurs all descendant sites are removed and this is a catastrophe transition. Let p_F, p_S denote the respective probabilities of resolution to F, S, respectively. (Note that these probabilities depend on the different selection events which occur with probabilities $e_j : j = 1, \ldots, M$.)

Remark 193. In order to turn the above outline into a detailed proof we have to observe that the rate-s Poisson stream of birth is now replaced by a multitype object and it is necessary to replace the probabilities p_S by $\sum_{j=1}^{M} e_j p_S(j)$ where $p_S(j)$ is the probability of resolving to S given that it was the jth selection event taking place at the birth by the founding rank.

To carry this out we must develop in detail the class of generalized birth, death and catastrophe processes (GMBDC) in which the birth and death times have distributions coming from hitting times of the mutation-selection Markov chain. These are exponential in the case $M = 2$ but in general have exponential tails

9.12 Emergence and Fixation: The Selection-Dominant Case

but are not pure by exponential. Since the finiteness of the integrals involved in our bounds depend on the tails of the distributions involved, the main results on the moments should follow as stated. The detailed development of this class of GMBDC will not be given here.

We again introduce a colouring scheme but now the *sites* occupied by a mesostate rather than ranks are coloured. A colour 1 site is one in which its *founding rank has not resolved*. A colour 1 site changes to colour 2 when the *first row* in the subtableau corresponding to that site resolves to $(0,\ldots,0)$, etc.

Then the production of colour 1 sites (occupied by a given mesostate) forms a "birth and death process" but one in which the birth and death times are no longer exponentially distributed but only have an exponential moment. However since sites evolve independently (except for catastrophe transitions), between catastrophe transitions, the colour 1 sites form a CMJ-process with malthusian parameter $\tilde{s}$.

As a result of the above properties active sites now play the role played by the factor (110) in Lemma 9.69 and the *combined* effect of coalescence-mutation now plays the role of the m_{12}, m_{21} transitions in the proof of Lemma 9.69. As before selection produces new active ranks and then migration produces new active sites. The resulting mesostate process and the corresponding CMJ-catastrophe process now play the same role as in the case $M = 2$.

We now outline the idea to obtain a $1 + \varepsilon$ moment. We first consider the time interval up to a catastrophic event at a colour 1 site. Then we obtain an upper bound on the growth of the sites with colours ≥ 2 up to $(T_C \wedge T_{\text{dec}})_-$. A key point is that the death rate for colour 1 sites plus the catastrophe rate corresponds to the decoupling rate at the founding site.

Consider (9.675), then the analog estimation runs as follows, where we use the notation $\tilde{X}$ and $\tilde{W}$ for the analogue of the processes we had for $M = 2$ in Subsubsection 9.12.4:

$$\lim_{t\to\infty} e^{(\tilde{s}-p_S\hat{\lambda})t} P(T_C > t) = \lim_{t\to\infty} e^{(\tilde{s}-p_S\hat{\lambda})t} \int_0^\infty q^{xe^{(\tilde{s}-p_S\hat{\lambda})t}} f_{\tilde{W}}(x)dx, \quad (9.727)$$

$$= \lim_{t\to\infty} e^{(\tilde{s}-p_S\hat{\lambda})t} \int_0^\infty e^{-\log(q^{-1})xe^{(\tilde{s}-p_S\hat{\lambda})t}} f_{\tilde{W}}(x)dx$$

$$= \int_0^\infty e^{-\log(q^{-1})u} f_{\tilde{W}}(u)du < \infty.$$

In the case $d > 0$, instead of using the exact distribution as in (9.670) for the pure birth process we proceed using the CMJ process as follows:

$$E[(\tilde{X}(T_{\text{dec}} \wedge T_C)_-)^{1+\varepsilon}] \quad (9.728)$$

$$= \text{const} \cdot \int_0^\infty e^{-\hat{\lambda}t} e^{-(\tilde{s}-p_S\hat{\lambda})} \sum_k k^{1+\varepsilon} P(\tilde{X}(t) = k) \quad \text{(using (9.727))}$$

$$= \text{const} \cdot \int_0^\infty e^{-(\tilde{s}+p_F\hat{\lambda})t} \sum_k k^{1+\varepsilon} P(\tilde{W} = ke^{-\tilde{s}t}).$$

We can now proceed as in the case $d = 0$ to get the bound provided that $\varepsilon < \frac{p_F \hat{\lambda}}{\tilde{s}}$

$$E[\tilde{X}((T_C \wedge T_{\text{dec}})_-^{1+\varepsilon})] \tag{9.729}$$

$$\leq \text{const} \int_0^\infty e^{-(\tilde{s}+p_F\hat{\lambda})t} \sum_k k^{1+\varepsilon} P(\tilde{W} = ke^{-\tilde{s}t})$$

$$= \int_0^1 u^{\frac{\tilde{s}+p_F}{\tilde{s}}} \frac{1}{u^{2+\varepsilon}} E[\tilde{W}^{1+\varepsilon}] < \infty. \text{ q.e.d.}$$

(c) follows from (a) combining the above argument with the argument as in Proposition 9.65 as follows. We again show that for sufficiently large selection rate, that is, there exists ς such that for $\frac{s}{m} > \varsigma$ the malthusian parameter for the internal factors satisfies $\tilde{\beta} < \frac{\tilde{s}}{2}$. Therefore we obtain

$$E[(N(T_{*,1}))^2] \leq \text{const} \cdot \int_0^\infty e^{2\tilde{\beta}t}(\tilde{s} + p_F\hat{\lambda})e^{-(\tilde{s}+p_F\hat{\lambda})t} dt < \infty. \tag{9.730}$$

9.12.7 Exponential Growth and Type Distribution of the CMJ-Mesostate Process

In this subsubsection we consider the growth of the population of labels and the associated mesostates that form a supercritical CMJ branching process with countable state space $\hat{\mathfrak{T}}$. Our objective is to determine the malthusian parameter and *stable type distribution* (i.e. the frequencies of particular mesostates associated with a label) and then using these tools to give the proof of Proposition 9.44.

We proceed in two parts. We first prepare the tools, state the results and then give the proof for the result.

Part 1: The Growth Rate and Stable Type Distribution of the Mesostate Process

(i) Basic objects

In the case $S = \mathbb{N}$ the mesostates which are offspring of different *Pe* sites yield random sets of sites that are completely decoupled and therefore product sets, which then evolve independently. Therefore the number of labels (or permanent *Pe* sites) is a *CMJ-process with countably many types*. Namely between decoupling events at a *Pe* site the internal state of an individual in the CMJ-process is the mesostate which evolves as a Markov process independently from other mesostates and produces at decoupling time potentially further new mesostates.

9.12 Emergence and Fixation: The Selection-Dominant Case

The point

$$\hat{t}_* := ((1,\ldots,1,0), \emptyset, *) \text{ is a } \textit{recurrent point of the dynamic of a } \text{single } \textit{mesostate}. \tag{9.731}$$

Each return to this point corresponds to a decoupling event.

More specifically the development of a typical mesostate is a Markov process on $\hat{\mathfrak{T}}$ and at their decoupling times a random number of new labels with new mesostates are produced, with mesostates dependent on the state of the parent mesostate at time $T_{\text{dec}}-$ (recall (9.281). We described this as a *two*-level branching system in Subsections 9.12.1–9.12.6. In particular, on the first level this includes the dynamic of a collection of labels carrying their mesostates, which we now describe in detail.

We start our system in the state $\hat{t}_*$. However we can also start in a single label with another mesostate, this will be mentioned explicitly at the occasion. Given a *Pe* site let $\mathfrak{a}$ denote the time since the birth time, or if decouplings have already occurred at this site, the time since the last decoupling at this *Pe* site. Let

$$\tilde{K}_t^{Pe}(ds, \hat{t}) \text{ denote the number of labels (}\textit{Pe}\text{ sites) with } \mathfrak{a} \in ds \text{ and mesostate } \hat{t}, \tag{9.732}$$

$$K_t^{Pe}(\hat{t}) := \tilde{K}_t^{Pe}([0, \infty), \hat{t}) \in \mathcal{M}(\hat{\mathfrak{T}}). \tag{9.733}$$

Since each offspring *Pe* site (and the associated label with its mesostate) produce further offspring labels with mesostates in an independent fashion and *Pe* sites do not die (since we consider here the McKean–Vlasov dynamic on the geographic space $\mathbb{N}$), $\{K_t^{Pe}\}_{t \geq 0}$ *is the size of a supercritical CMJ-process* with *countable type space* $\hat{\mathfrak{T}}$ and for emergence plays the role of the one-type CMJ process considered in Section 8.

Parallel to the procedure in Section 8 we can consider $(\bar{u}, \bar{U})$ the total number and the type distribution of the process of labels with mesostates as internal state. However we start out here with the *expected* quantities which we indicate using frak-letters. Consider the *expected* and *un*normalized distribution of age and type of *Pe* sites:

$$\tilde{\mathfrak{U}}^{Pe}(t, ds, \hat{t}) := E[\tilde{K}_t^{Pe}(ds, \hat{t})]. \tag{9.734}$$

Note here that since we consider the system after a long time, i.e. $t \to \infty$, and hence with a diverging number of labels with mesostates, the expected frequency distribution is the right quantity, as we shall see later on, due to a L_2-law of large numbers which we verify.

We next define as key ingredients of our description of the evolution:

$$\bar{\mathfrak{u}}(t) = E[K_t^{Pe}(\hat{\mathfrak{T}})] = \tilde{\mathfrak{U}}^{Pe}(t, [0, \infty), \hat{\mathfrak{T}}) \tag{9.735}$$

and the *normalized* expected type distribution

$$\bar{\mathfrak{U}}^{Pe}(t, ds, \hat{\mathfrak{t}}) := \frac{\tilde{\mathfrak{U}}^{Pe}(t, ds, \hat{\mathfrak{t}})}{\tilde{\mathfrak{U}}^{Pe}(t, [0, \infty), \hat{\mathfrak{T}})}. \tag{9.736}$$

The next task is now to derive for the noncollision regime both the *linear equations for* $(\mathfrak{U}^{Pe}(t, \cdot, \cdot))_{t \geq 0}$, and then the *nonlinear equations for* $(\bar{u}, \bar{\mathfrak{U}}^{Pe})$ for the *normalized* system. Finally we have to derive the growth behaviour of $(\tilde{K}_t^{Pe}(\cdot, \cdot))_{t \geq 0}$ from the growth of $\mathfrak{U}^{Pe}(t, \cdot, \cdot)$.

(ii) Evolution equations for $(\bar{u}, \bar{\mathfrak{U}}^{Pe})$.

We need some preparations, in particular we need now a list of ingredients which will enter into our equations, in particular all transition rates with which we complete the present item (ii) by writing down the evolution equations for a *semigroup* $(Q^t)_{t \geq 0}$ *of positive matrices*.

Define the matrix

$$\tilde{Q} = \{\tilde{Q}_{\hat{\mathfrak{t}}, \hat{\mathfrak{t}}'}\}, \text{ where } \tilde{Q}_{\hat{\mathfrak{t}}, \hat{\mathfrak{t}}'} \geq 0, \hat{\mathfrak{t}} \neq \hat{\mathfrak{t}}' \tag{9.737}$$

as the mean rate at which the mesostate $\hat{\mathfrak{t}}$ produces an offspring with mesostate $\hat{\mathfrak{t}}'$ at its decoupling time or changes from type $\hat{\mathfrak{t}}$ to type $\hat{\mathfrak{t}}'$ due to the Markovian dynamics between decoupling events.

More specifically we set $\tilde{Q} = \tilde{Q}^D + \tilde{Q}^I$ and define $\tilde{Q}^D_{\hat{\mathfrak{t}}, \hat{\mathfrak{t}}'}$ as the rate at which the mesostate $\hat{\mathfrak{t}}$ decouples and produces the offspring mesostate $\hat{\mathfrak{t}}'$ at a new *Pe* site and $\tilde{Q}^I_{\hat{\mathfrak{t}}, \hat{\mathfrak{t}}'}$ determines the mean transition rates for the a given mesostate between decoupling events. We set correspondingly

$$\tilde{Q}_{\hat{\mathfrak{t}}, \hat{\mathfrak{t}}} = -[\sum_{\hat{\mathfrak{t}}' \neq \hat{\mathfrak{t}}} \tilde{Q}^I_{\hat{\mathfrak{t}}, \hat{\mathfrak{t}}'} + \sum_{\hat{\mathfrak{t}}' \neq \hat{\mathfrak{t}}} \tilde{Q}^D_{\hat{\mathfrak{t}}, \hat{\mathfrak{t}}'}], \tag{9.738}$$

where the first sum is over the transitions between decoupling events and the second sum denotes the rates at a decoupling event.

Define furthermore the following aggregated *mean jump rates* in the whole population:

$$\bar{\beta}_{\hat{\mathfrak{t}}}(t) = \int_{\mathcal{P}(\hat{\mathfrak{T}}^*)} \int_0^\infty \tilde{Q}^D(\hat{\mathfrak{t}}', \hat{\mathfrak{t}}) \bar{\mathfrak{U}}(t, ds, d\hat{\mathfrak{t}}'),$$

(mean rate of production of new mesostates $\hat{\mathfrak{t}}$), (9.739)

$$\bar{\beta}(t) = \sum_{\hat{\mathfrak{t}}} \bar{\beta}_{\hat{\mathfrak{t}}}(t), \text{ (mean total rate of production of new mesostates)} \tag{9.740}$$

$$\gamma_D(\hat{\mathfrak{t}}) = \text{ decoupling rate of mesostate } \hat{\mathfrak{t}}, \tag{9.741}$$

9.12 Emergence and Fixation: The Selection-Dominant Case

$$\bar{\gamma}(t) = \int_{\mathcal{P}(\hat{\mathfrak{T}}^*)} \int_0^\infty \gamma_D(\hat{t}') \bar{\mathfrak{U}}(t, ds, d\hat{t}'), \quad \text{(mean total decoupling rate)}. \quad (9.742)$$

With (9.737–9.742) we have all ingredients to next write down the $\mathfrak{U}^{Pe}$ and $(\bar{u}, \bar{\mathfrak{U}})$ evolution equations.

We have the *linear equation* (cf. (8.271)):

$$\frac{\partial \tilde{\mathfrak{U}}^{Pe}(t, ds, \cdot)}{\partial t} = -\frac{\partial}{\partial s} \tilde{\mathfrak{U}}^{Pe}(t, ds, \cdot) + \tilde{\mathfrak{U}}^{Pe}(t, ds, \cdot) \tilde{Q}, \quad \tilde{\mathfrak{U}}^{Pe}(0, \cdot, \cdot) := \delta_{0, \hat{t}_*}. \quad (9.743)$$

To verify that there exists a solution with

$$\sum_{\hat{t}} |\hat{t}|^2 \tilde{\mathfrak{U}}^{Pe}(t, [0, \infty), \hat{t}) < \infty, \quad (9.744)$$

for all $t < \infty$, which guarantees that all terms on the r.h.s. of (9.743) are welldefined, we note that the l.h.s. in (9.744) is bounded by the second moment of the pure birth process with birth rate s.

Then the *nonlinear $(\bar{u}, \bar{\mathfrak{U}})$-equation* reads (the terms are explained below in brackets):

$$\frac{\partial \bar{u}}{\partial t} = \bar{\beta}(t) \bar{u}(t) - \bar{\gamma}(t) \bar{u}(t), \quad t \in \mathbb{R}, \quad (9.745)$$

$$\frac{\partial \bar{\mathfrak{U}}(t, ds, \hat{t})}{\partial t} =$$

$$-\frac{\partial \bar{\mathfrak{U}}(t, ds, \hat{t})}{\partial s} \quad \text{(aging)}$$

$$+ \sum_{\hat{t}' \in \hat{\mathfrak{T}}^*} \bar{\mathfrak{U}}(t, ds, \hat{t}') \tilde{Q}^I_{\hat{t}', \hat{t}} \quad \text{(mesostate dynamics)}$$

$$+ \sum_{\hat{t}' \in \hat{\mathfrak{T}}^*} \bar{\mathfrak{U}}(t, ds, \hat{t}') \tilde{Q}^D(\hat{t}', \hat{t}) 1_{(s, \hat{t})}(0, \hat{t}')) - \bar{\gamma}_D(\hat{t}) \bar{\mathfrak{U}}(t, ds, \hat{t}) \quad \text{(decoupling)}$$

$$-[\bar{\beta}(t) - \gamma_D(t)] \bar{\mathfrak{U}}(t, ds, \hat{t}) \quad \text{(normalization)}. \quad (9.746)$$

(iii) The growth of K_t^{Pe}.

Our next goal is now to determine the growth behaviour of $\bar{u}(t) = E[K_t^{Pe}(\hat{\mathfrak{T}})]$ and the longtime behaviour of $E[K_t^{Pe}(\cdot)]$ and then of K_t^{Pe} and of $K_t^{Pe}(\cdot)$ themselves. The growth is determined by the evolution equation for $(\bar{u}, \mathfrak{U})$ since $\bar{u}$ depends on $\bar{\mathfrak{U}}$. The evolution of $\bar{\mathfrak{U}}(t, [0, \infty), \cdot)$ is an evolution of probability measures on $\hat{\mathfrak{T}}$ but that of $\tilde{\mathfrak{U}}(t, [0, \infty), \cdot)$ of *finite* measures on $\hat{\mathfrak{T}}$ and $\hat{\mathfrak{T}}$ is countable, both are given by the respective evolution equations above.

To study the evolution equation for $\bar{\mathfrak{U}}(t, [0, \infty), \cdot)$ on $\hat{\mathfrak{T}}$ as $t \to \infty$ we need to use the theory of *semigroups* $(Q^t)_{t \geq 0}$ *of positive matrices* with countably many entries and as we will see in Remark 194 below this reduces in our case to the theory of powers Q^n, $n \in \mathbb{N}$ of such matrices.

Some theory of $\mathbb{N} \times \mathbb{N}$ positive matrices (a review)

We need the following facts about countable infinite matrices Q, for which $(Q^n)_{n \in \mathbb{N}}$ is well-defined for every $n \in \mathbb{N}$, which are given in Vere-Jones [VJ] I, Theorem A.

If $(Q^n)_{i,j} \geq 0$ for some $n \in \mathbb{N}$ and $Q_{i,i} > 0$ $\forall\, i \in \mathbb{N}$ (irreducibility and aperiodicity is then guaranteed), then the following limit exists

$$r = \lim_{n \to \infty} ((Q^n)_{i,j})^{\frac{1}{n}}, \quad \forall\, i, j \in \mathbb{N} \tag{9.747}$$

and is called *radius of convergence* of Q.

A positive matrix Q with radius of convergence r is called *r-recurrent* if

$$\sum_n \tilde{r}^n Q^n_{i,j}, \; < \; \text{resp.} \; = +\infty, \; \text{if } \tilde{r} < \frac{1}{r} \text{respectively}, \tilde{r} \geq \frac{1}{r}. \tag{9.748}$$

A r-recurrent matrix Q is *positive r recurrent* if for some i:

$$\lim_{n \to \infty} (r^n Q^n_{i,i}) > 0. \tag{9.749}$$

By [VJ], Theorem D, a positive 1-recurrent matrix has *left* and *right eigenvectors* v, h such that

$$Q^n_{ij} \to \frac{v_j h_i}{\sum v_i h_i}, \text{ and } \langle v, h \rangle < \infty, \text{ (shortly } Q^n \to \frac{h \otimes v}{\langle v, h \rangle}). \tag{9.750}$$

Therefore to establish the existence of positive (left) eigenvectors for Q, it suffices to show that the normalized Q, called $\hat{Q} = r^{-1} Q$ is *positive 1-recurrent* and then we have an invariant measure for $\hat{Q}$ and Q^n therefore really grows like r^n.

Remark 194. Note furthermore that if we have a continuous semigroup Q^t of positive matrices with a generator $\tilde{Q}$ with positive left and right eigenvectors (v, h) for the eigenvalue $e^{\beta t_0}$ at $t = t_0$ satisfying $\langle v, h \rangle < \infty$, then we get also these as left and right eigenvectors for the eigenvalue $e^{\beta(t_0/2^n)}$ with $n \in \mathbb{N}$. Do we get this way a (left) eigenvector for Q^t? Note that we get

$$\int_0^t v Q^s ds = \lim_{n \to \infty} \Big(\frac{1}{\lfloor 2^n t / t_0 \rfloor} \sum_{k=0}^{\lfloor 2^n t / t_0 \rfloor} v Q^{k t_0 2^{-n}} \Big)$$

9.12 Emergence and Fixation: The Selection-Dominant Case

$$= \left(\lim_{n \to \infty} \frac{1}{\lfloor 2^n t / t_0 \rfloor} \sum_{k=0}^{\lfloor 2^n t / t_0 \rfloor} v e^{\beta \frac{k t_0}{2^n}} \right) \tag{9.751}$$

$$= \frac{1}{\beta} (e^{\beta t} - 1) v. \tag{9.752}$$

Therefore v is in the domain of the generator ([EK2], Chapt. 1, Proposition 1.5), recall we have here countable state space so that we can apply the quoted result on semigroups acting on functions here for the ones acting on measures. Next writing $(vQ^t - v) = (\int_0^t vQ^s ds)\tilde{Q}$ we get by taking $t \downarrow 0$:

$$v\tilde{Q} = \beta v. \tag{9.753}$$

In other words we have found the invariant vector v and identified the growth behaviour for $(Q^t)_{t \geq 0}$ as $const \cdot e^{\beta t}$, what is what we are looking for.

Now after this excursion into positive matrix theory, we can state the main result where we use the notation $\mathfrak{U}_{\hat{t}}^{Pe}$ for the quantities for the initial state starting with one label in a mesostate $\hat{t}$.

Lemma 9.75 (Growth of mesostate population).
Fix $t_0 > 0$ and consider the infinite non-negative matrix

$$Q \text{ with entries } Q_{\hat{t}', \hat{t}} := \mathfrak{U}_{\hat{t}'}^{Pe}(t_0, [0, \infty), \mathfrak{t}). \tag{9.754}$$

(a) *There exists $\beta_{Pe} > 0$ such that $e^{-\beta_{Pe} t_0}$ is the radius of convergence of Q.*
(b) *Consider the non-negative rescaled matrix*

$$\hat{Q}_{\hat{t}, \hat{t}'} := e^{-\beta_{Pe} t_0} Q_{\hat{t}, \hat{t}'}, \tag{9.755}$$

which then has radius of convergence equal to 1.
    Then there exists a unique invariant measure (left eigenvector) for $\hat{Q}$ with eigenvalue 1.
(c) *The mesostate $\hat{t}_* = ((1, \ldots, 1, 0), \emptyset, *)$ is positive 1-recurrent for $\hat{Q}$ and the invariant measure $\mathfrak{U}^{Pe}(\infty, \cdot)$ of the dynamics given in (9.746) can be represented by*

$$\bar{\mathfrak{U}}^{Pe}(\infty, \cdot) = \frac{\nu(\cdot)}{\nu(\hat{\mathfrak{T}})} \tag{9.756}$$

where ν is given explicitly by (9.765) below.
(d) *Consider the supercritical branching process of labels $(K_t^{Pe}(\cdot))_{t \geq 0}$ with the countably infinite type (mesostate) space $\hat{\mathfrak{T}}$.*

Then for $s > m$, $\beta_{Pe} \leq \beta$ is the malthusian parameter and $\bar{\mathfrak{U}}^{Pe}(\infty, \cdot) \in \mathcal{P}(\hat{\mathfrak{T}})$ is the stable type distribution, that is,

$$\mathcal{L}[e^{-\beta_{Pe}t} K_t^{Pe}(\cdot)] \Rightarrow \mathcal{L}[W_{Pe}\bar{\mathfrak{U}}^{Pe}(\infty, \cdot)], \quad \text{as } t \to \infty, \qquad (9.757)$$

$$0 < W_{Pe} < \infty, \text{ a.s. and for some } \varepsilon > 0 : 0 < E[W_{Pe}^{1+\varepsilon}] < \infty. \qquad (9.758)$$

(e) Under the hypothesis of Proposition 9.65 ($M = 2$) or Proposition 9.74 ($M > 2$) the convergence in (9.757) holds in L^2 for $t \to \infty$. □

Corollary 9.76 (Growth, stable type distribution of $|\tilde{\mathcal{G}}^{++}|$).
There is an enriched version of the set-valued dual process $\tilde{\mathcal{G}}_t^{++}$ that asymptotically (convergence in law, L_2) can be factored into the irreducible product of

$$W_{Pe}e^{\beta_{Pe}t} \text{ many subsets of } (\mathbb{I}_M^\mathbb{N})^{\mathbb{N}_\ell}, \text{ with factors different from } \mathbb{I}_M^\mathbb{N}, \qquad (9.759)$$

where

$$\beta_{Pe} \leq \beta \qquad (9.760)$$

is the Malthusian parameter for the set of Pe sites. Each subset of ranks of an irreducible factor is supported on a random number of sites.

The limiting empirical distribution of the form of the factors (mesostates of the labels) has type distribution as given by (9.756). □

Then Proposition 9.44 also follows, namely from Corollary 9.76.

Remark 195. In the general case the argument of Nerman [N2], Theorem 6.6 (using the $(1 + \varepsilon) - moment$ of Proposition 9.64 to verify the "$X \log X$" condition) together with our tightness result should establish the *almost sure convergence* in place of L^2 convergence.

Part II: Proof of Lemma 9.75

$K_t^{Pe}(\cdot)$ is the occupancy number process of a supercritical CMJ branching process with countably many types. Each *Pe* site produces a random number of offspring after a random time having an exponential moment. We have also verified (Propositions 9.64 and 9.74) that the mean size of new *Pe* sites at decoupling is finite (and has a finite $(1 + \varepsilon)$-moment) and in the selection-dominant case the offspring size at decoupling has a finite second moment, and Proposition 9.74).

(a) We now consider for K^{Pe} and the chosen time t_0 the embedded discrete time CMJ-size process $(K_{k \cdot t_0}^{Pe})_{k \in \mathbb{N}_0}$ with *mean offspring matrix* denoted by Q. Let $\hat{\mathfrak{t}}_* := ((1, \ldots, 1, 0), \emptyset, *)$. Then given any $\hat{\mathfrak{t}}$ there exists n such that $Q_{\hat{\mathfrak{t}}_*, \hat{\mathfrak{t}}}^n > 0$ by the definition of the state space, furthermore because of positive rate of

9.12 Emergence and Fixation: The Selection-Dominant Case

resolution we have also $Q_{\hat{t}_*,\hat{t}_*} > 0$. Therefore Q is *nonnegative, irreducible* and *aperiodic* matrix with countably many entries. Since the matrix Q is *supercritical*, we can conclude from the results by Vere-Jones we quoted above, that the quantities below exist and have a common value which we denote by:

$$\lim_{k \to \infty} (Q^k_{\hat{t},\hat{t}'})^{1/k} := e^{\beta_{Pe^{t_0}}} \quad \text{for some } \beta_{Pe} > 0 \text{ for all } \hat{t}, \hat{t}' \in \hat{\mathfrak{T}}. \tag{9.761}$$

Then $e^{\beta_{Pe^{t_0}}}$, the inverse of the common radius of convergence of the powers $\{Q^n\}$ and is the potential malthusian parameter.

Since the *Pe* sites are included in the *Pe* ∪ *UPe* so that we can embed the process of labels with their mesostates into the h-transform process defining the growth constant β (of the CMJ-process $\tilde{\mathcal{G}}^{++,III}$) we have altogether $0 < \beta_{Pe} \leq \beta$. Then as in Section 7 (Lemma 8.14), β_{Pe} also gives the mean production rate of new *Pe*-sites.

(b) The question remains from (a) to show the existence of a *left eigenvector* with eigenvalue $e^{\beta_{Pe^{t_0}}}$. Consider the "normalized" problem and define:

$$\hat{Q}_{\hat{t},\hat{t}'} := e^{-\beta_{Pe^{t_0}}} Q_{\hat{t},\hat{t}'}, \tag{9.762}$$

which then has radius of convergence equal to 1. But note that $\hat{Q}$ is in general *not* a (sub)-Markov transition probability and therefore it is *not* immediate to obtain an invariant measure from the theory of Markov chains. We prove therefore with different methods the following result, which completes the proof of the claim.

Lemma 9.77 (Existence of an invariant measure).
There exists for the eigenvalue $e^{\beta_{Pe^{t_0}}}$ for the matrix Q, left and right nonnegative eigenvectors v and h such that as $n \to \infty$ and for f and μ with finite support:

$$e^{-\beta_{Pe^{t_0}}n} \sum_{\hat{t}} (\mu Q^n) f(\hat{t}) \to \sum_{\hat{t}} \mu(\hat{t}) h(\hat{t}) \cdot \sum_{\hat{t}'} f(\hat{t}') v(\hat{t}') \tag{9.763}$$

and $\sum_{\hat{t}} h(\hat{t}) v(\hat{t}) < \infty$. □

Proof of Lemma 9.77.
From (9.750) and (9.749) we know that it suffices to obtain the positive invariant measure v for $\hat{Q}$ to be able to conclude the claim. We construct such a v in the sequel in several steps.

Step 1 *(Some general theory)*

The first observation is that if we could conclude that since $\hat{t}_*$ is recurrent for the mesostate process, it is also a 1-recurrence point of $\hat{Q}$ in the sense of Vere-Jones, i.e. the series $\sum_{r=0}^{\infty} r^n \hat{Q}^n$ diverges at $r = 1 =$ (spectral radius) for the entry $(\hat{t}_*, \hat{t}_*)$, v

the socalled superinvariant measure of Althreya-Ney would be an *invariant* measure using a key result of [AN]. This would *identify* the left eigenvector for the eigenvalue 1 and guarantee the existence of the right eigenvector as we see above. This argument runs as follows.

Following [AN] (3.3)–(3.5), we can construct a *super-invariant measure* v for the kernel $\hat{Q}_{\hat{t},\hat{t}'}$ based on the recurrence point $\hat{t}_*$ of the single mesostate dynamic. Define first the "potential of the kernel stopped at $\hat{t}_*$". We set

$$v_0(\hat{t}, E) := 1_E(\hat{t}), \quad v_1(\hat{t}, E) = \hat{Q}_{\hat{t}, E-(\hat{t}_*)} \qquad (9.764)$$

$$v_r(\hat{t}_*, E) = \int_{\hat{\mathfrak{T}}-(\hat{t}_*)} \cdots \int_{\hat{\mathfrak{T}}-(\hat{t}_*)} \hat{Q}_{\hat{t}_{r-1}, E-(\hat{t}_*)} \hat{Q}_{\hat{t}_{r-2}, d\hat{t}_{r-1}} \cdots \hat{Q}_{\hat{t}, d\hat{t}_1}.$$

Then we set:

$$v(E) = \sum_{r=0}^{\infty} v_r(\hat{t}_*, E) \text{ for } E \text{ with } \hat{t}_* \notin E. \qquad (9.765)$$

We define

$$v(\hat{t}_*) := \int_{\hat{\mathfrak{T}}\setminus \hat{t}_*} \hat{Q}_{\hat{t}, \hat{t}_*} v(d\hat{t}). \qquad (9.766)$$

Then automatically

$$v \geq v\hat{Q}. \qquad (9.767)$$

Note that (due to decoupling) there exists $\delta > 0$ such that

$$\hat{Q}_{\hat{t}, \hat{t}_*} > \delta \quad \text{for any } \hat{t} \neq \hat{t}_*, \qquad (9.768)$$

therefore

$$v(\hat{\mathfrak{T}}) < \infty, \qquad (9.769)$$

that is, v is a *finite super-invariant measure*.

If $\hat{t}_*$ would be a 1-recurrence point, then v would be invariant according to Theorem 3.1 in [AN]. The problem is that this is *not* immediate from the recurrence of the point $\hat{t}_*$ in the single mesostate process that we have this 1-recurrence for the $\hat{Q}$-matrix connected with the total population of mesostates.

In order to determine the detailed limiting behaviour, in particular the existence of left and right eigenvectors the key question is whether we have a $e^{\beta p_e t_0}$-positive recurrent matrix (i.e. power series diverges at $e^{\beta p_e t_0}$) in the terminology of Vere-Jones, recall (9.749).

9.12 Emergence and Fixation: The Selection-Dominant Case

Step 2 *(Excursion: Banach lattice formulation)*

In this step we explain at which point the functional analytic spectral approach does not work. We want to show that the matrix $Q = (Q_{\hat{\mathfrak{t}},\hat{\mathfrak{t}}'})_{\hat{\mathfrak{t}},\hat{\mathfrak{t}}' \in \hat{\mathfrak{T}}}$ is $e^{\beta p_c t_0}$-*recurrent* or equivalently $\hat{Q}$ is 1-recurrent and furthermore it is *positive* 1-recurrent. For this purpose we next view the matrix Q as an operator in a *Banach space* which is a subset of $\ell^1(\mathbb{N})$, more specifically in a *Banach lattice* and consider a reformulation of this question in this context. This is given by the following result.

For a positive linear operator P with spectral radius r in a Banach lattice $\mathbb{B}$, positive r-recurrence is equivalent to showing that the operator

$$\hat{P} = (\frac{1}{r} P) \tag{9.770}$$

with spectral radius 1 is *quasi-compact* (see e.g. Martinez and Mazón [MJM], Cor. 2.3).

In order to verify the quasi-compactness it suffices to show:

$$\exists n, \text{ compact operator } K : \|(\hat{P})^n - K\| < 1, \tag{9.771}$$

where $\|\cdot\|$ is the operator norm.

We now return to our matrix $\hat{Q}_{\hat{\mathfrak{t}},\hat{\mathfrak{t}}'}$ with radius of convergence 1. We will now construct a Banach space on which $\hat{Q}$ acts as a bounded linear operator. Our Banach space will be a weighted ℓ_1-space, with weight function $\alpha(\cdot) \geq 0$ of the form:

$$\mathbb{B} = \{x \in \ell_1(\mathbb{N}) | \sum_{i=1}^{\infty} |x(i)| \alpha(i) < \infty\}, \quad \|x\|_\alpha = \sum_{l=1}^{\infty} |x(i)| \alpha(i). \tag{9.772}$$

This problem is related to the following observation. The rows of the matrix $\hat{Q}$ can be ill-behaved even if the chosen t_0 is small. We cannot uniformly bound the row sums, since large row sums can occur for small t_0 from initial states having a very large offspring at decoupling. This means that the new state has a large number of ranks. In order to verify the positive recurrence condition we need to control both the expected numbers of offspring (row sums $\sum_{\hat{\mathfrak{t}}'} \hat{Q}_{\hat{\mathfrak{t}},\hat{\mathfrak{t}}'}$) and to obtain a tightness condition of the distribution of the number of ranks $\sum_{\hat{\mathfrak{t}}'} |\hat{\mathfrak{t}}'|$ created under the rates $\hat{Q}_{\hat{\mathfrak{t}},\cdot}$.

The problem with the operator is hence the potentially unbounded row sums, hence the idea is to temper it. For $0 < z < 1$ let $\mathbb{B}_z$ be the Banach space with

$$\alpha_z(\hat{\mathfrak{t}}) = \sum_{n=0}^{\infty} \sum_{\hat{\mathfrak{t}}' \in \hat{\mathfrak{T}}} z^n \hat{Q}_{\hat{\mathfrak{t}},\hat{\mathfrak{t}}'}^n, \tag{9.773}$$

that is,

$$\mathbb{B}_z = \{x \in \ell_1(\hat{\mathfrak{T}}) : \sum_{\hat{\mathfrak{t}}\in\hat{\mathfrak{T}}}^{\infty} |x(\hat{\mathfrak{t}})|\alpha_z(\hat{\mathfrak{t}}) < \infty\}, \quad \|x\|_{LS(z)} := \sum_{\hat{\mathfrak{t}}\in\hat{\mathfrak{T}}}^{\infty} |x(\hat{\mathfrak{t}})|\alpha_z(\hat{\mathfrak{t}}). \quad (9.774)$$

On this Banach space the operator $\hat{Q}$ has the property (by its very construction)

$$\hat{Q}\alpha \leq z^{-1}\alpha. \quad (9.775)$$

Therefore for every $z \in (0,1)$:

$$\|v\hat{Q}\|_{LS(z)} = \langle v\hat{Q}, \alpha_z\rangle = \sum_{\hat{\mathfrak{t}},\hat{\mathfrak{t}}'\in\hat{\mathfrak{T}}} v(\hat{\mathfrak{t}})\hat{Q}_{\hat{\mathfrak{t}},\hat{\mathfrak{t}}'}\alpha_z(\hat{\mathfrak{t}}') = \sum_{\hat{\mathfrak{t}}\in\hat{\mathfrak{T}}} v(\hat{\mathfrak{t}}) \sum_{\hat{\mathfrak{t}}'\in\hat{\mathfrak{T}}} \hat{Q}_{\hat{\mathfrak{t}},\hat{\mathfrak{t}}'}\alpha_z(\hat{\mathfrak{t}}')$$

$$\leq z^{-1} \sum_{\hat{\mathfrak{t}}\in\hat{\mathfrak{T}}}^{\infty} v(\hat{\mathfrak{t}})\alpha_z(\hat{\mathfrak{t}}) \leq \frac{1}{z} \cdot \|v\|_{LS(z)}. \quad (9.776)$$

Therefore we have

$$\|v\hat{Q}\|_{LS(z)}\| \leq z^{-1}\|v\|_{LS(z)} \quad , \quad z \in [0,1). \quad (9.777)$$

Then we have the following conclusion.

Lemma 9.78 (Banach space properties).

If Q is r-recurrent then for all $z < r$ we have:

(i) *$\mathbb{B}_z$ is a Banach space,*
(ii) *$\hat{Q} : \mathbb{B}_z \longrightarrow \mathbb{B}_z$,*
(iii) *$\|\hat{Q}\| \leq z^{-1}$.* □

The problem here is that we cannot extend this to $z = r$ to capture the spectral radius and we would have to find a *smaller* Banach space to obtain a bounded operator with spectral radius r of which we can show $\hat{Q}$ is quasi-compact.

Therefore rather than directly proving the quasi-compactness of this operator $\hat{Q}$ on the Banach space $\mathbb{B}_z$ we will work with the probabilistic structure of the mesostate process to obtain a probabilistic version of the quasi-compactness property.

To do this, the key idea to be used in the next step is to exploit the property that the mesostate process is positive recurrent and in any state we return to the state consisting of one rank at one site (i.e. $\hat{\mathfrak{t}}_*$, recall that $\hat{\mathfrak{t}}_* := ((1,\ldots,1,0), \emptyset, *)$ to then go through the cycle again. This provides the analogue of the operator K in (9.771) and the process of replacement of the initial mesostates through decoupling and production of mesostates $\hat{\mathfrak{t}}_*$ corresponds to the contraction in (9.771).

9.12 Emergence and Fixation: The Selection-Dominant Case

Step 3 *(Bringing in the concrete branching structure)*

We now continue the analysis by bringing in the special structure of the matrix $\hat{Q}$ which arises as the expected sizes of mesostates produced at time t_0 from the two-level multitype branching type process with catastrophic events. We will proceed by verifying equation (9.749) for $i = \hat{t}_*$ by constructing a normalized eigenvector of $\hat{Q}$ by a *fixed point argument*. To do this we need two things, a *tightness property* of the mesostate length process and second a *contraction* property of the map $\hat{Q}$ which we derive after that.

To do this we use the following features of $\hat{Q}$. In the process underlying the matrix $\hat{Q}$ there are given entities which evolve by a *positive* recurrent Markov process giving the evolution of the internal states, the so-called mesostates. Then at catastrophic (decoupling) events the entity is replaced by a random number of those entities whose internal states depend on the original internal state. We know that the number of new entities produced at such an event has a finite second moment (Proposition 9.65, Proposition 9.74) and the distributions of the length of the mesostates are tight as we shall prove below. The new mesostates produced then evolve independently according to their internal dynamics up to the next catastrophic event. The rows of the matrix $\hat{Q}$ are given by the mean state after time t_0 given as function of the associate initial state. We complete this step in four parts.

(i) A tightness property

To verify that there exists a *stable type eigenvector* for the production rate matrix we need a stable type distribution producing as rates the stable type eigenvector. For this we shall need as a first ingredient a tightness property of the *single mesostate process* and begin by verifying such a *tightness condition*. Here the point is that there are only finitely many possible states of a mesostate if the length of the mesostate is $\leq L$ and only at most L such new mesostates can be produced at decoupling time. This means that large row sums can only occur by having many new mesostates produced at decoupling time for which long mesostates have to form over time. Ultimately we want to show that the family of finite measures $\{\hat{Q}^t_{\hat{t}}, t \geq 0\}$ is a tight family as $t \to \infty$, where we first consider $\{\hat{Q}^t_{\hat{t}_*}, t \geq 0\}$. Here we consider discrete times $t = k t_0$ and $k \to \infty$. Consider the recurrent mesostate process which is positive recurrent with stationary measure since the recurrence time T_{dec} of the simple mesostate $\hat{t}_*$ has a finite first moment.

Let $l(\hat{t}) = \|\hat{t}\|$ denote the number of active ranks in the mesostate $\hat{t}$ and let

$$\{\hat{t}(t)\}_{t \geq 0} \text{ be the mesostate process with } \hat{t}(0) = \hat{t}_*. \tag{9.778}$$

Then $l(\hat{t}(t \wedge T_{\text{dec}}-))$ is *dominated* by a pure birth process $\zeta(\cdot)$ with birth rate s evaluated at time $T_{\text{dec}}-$. Then recalling the definition of $\hat{\lambda}$ in (9.400) (which depends on s, m and d) and (9.399) we have

$$P[l(\hat{t}(k t_0 \wedge T_{\text{dec}}-)) > L] \leq P[\zeta(T_{\text{dec}}-) > L] = \frac{\hat{\lambda}}{s} \int_0^1 u^{\frac{\hat{\lambda}+s}{s}} (1-u)^L du. \tag{9.779}$$

Tightness of $\{\mathcal{L}[\hat{\mathfrak{t}}(t \wedge T_{\text{dec}}-)], \ t \geq 0\}$ follows since the righthand side tends to zero as $L \to \infty$.

Moreover we have verified in Lemma 9.45 that:

$$E[\mathfrak{l}(\hat{\mathfrak{t}}(kt_0 \wedge T_{\text{dec}}-))^\varepsilon] < \infty \quad \text{for } 0 < \varepsilon < \frac{\hat{\lambda}}{s} \tag{9.780}$$

and this is strengthened in Proposition 9.74 to

$$E[\mathfrak{l}(\hat{\mathfrak{t}}(kt_0 \wedge T_{\text{dec}}-)^{(1+\varepsilon)}) < \infty \quad \text{for } 0 < \varepsilon < \frac{\hat{\lambda}}{s} \tag{9.781}$$

Now we have to show the tightness for $\hat{Q}_{\hat{\mathfrak{t}}_*}^t$ as $t \to \infty$. Noting that the new labels and corresponding mesostates are always given by the descendants of a single site mesostate and then coupling a mesostate process with single site starting point with the equilibrium distribution of the single mesostates we can conclude that the normalized distributions $\tilde{\mathfrak{U}}^{Pe}(kt_0, \cdot)(\tilde{\mathfrak{U}}^{Pe}(kt_0, \hat{\mathfrak{T}}))^{-1}$ are tight on $\mathcal{M}_{\text{fin}}(\hat{\mathfrak{T}})$ (cf. Lemma 9.45) (even though we normalize with the $\tilde{\mathfrak{U}}^{Pe}$ associated with $\hat{\mathfrak{t}}_*$). This guarantees that there is a weak limit point in the space of finite measures. However we do not yet have that they are *probability* measures since the offspring rates of the weak limit can lose part of its mass if the weight rests with states with high production rate and furthermore we don't know that we have *convergence* as $k \to \infty$, both necessary conditions to get positive 1-recurrence of $\hat{Q}$.

(ii) A fixed point equation.

We want to establish now that in fact $\hat{Q}$ has a strictly positive eigenvector in $\mathcal{P}(\hat{\mathfrak{T}})$ to then obtain via (9.749) the positive 1-recurrence of $\hat{Q}$. We want to apply the *Schauder fixed point theorem* to that effect. This theorem states that if we have a *continuous* map of a *compact* and *convex* subset of a *topological vector space* into the subset *itself*, then this map has a fixed point in this space. We now want to show that this is applicable here by constructing successively all the needed ingredients.

As *vector space* we choose the space $\mathcal{M}(\hat{\mathfrak{T}})$ of *signed measures on* $\hat{\mathfrak{T}}$, which is a vector space. We observe that the topology is needed only on the subset which the map maps onto itself and will be defined only once we have that set.

Now consider the *nonlinear operator* arising as the normalized $\hat{Q}$ which is defined by:

$$\mathfrak{Q} : \mu \to (\mu \hat{Q})/\langle \mu \hat{Q}, 1 \rangle) =: \mu \mathfrak{Q}, \quad \mu \in \mathfrak{M} \cap \mathcal{M}_{\text{fin}}(\hat{\mathfrak{T}}), \ \mu \neq 0. \tag{9.782}$$

Note that because of the linearity of $\hat{Q}$:

$$\mu \mathfrak{Q}^k = \frac{\mu \hat{Q}^k}{\langle \mu \hat{Q}^k, 1 \rangle}, \quad k = 2, 3, \cdots. \tag{9.783}$$

9.12 Emergence and Fixation: The Selection-Dominant Case

We start by constructing the convex set on which we can consider the map $\mathfrak{Q}$ associated with the matrix $\hat{Q} = (\hat{Q}_{\hat{t},\hat{t}'})$. This will be a set on which the production of new mesostates $\hat{t}_*$ can be bounded.

The basic convex set will be constructed in two steps, first we get a set $\mathbb{M}$ which we then restrict to the target set $\mathbb{K}$.

A key is the following observation. We have that

$$\hat{Q}_{\hat{t},\hat{t}_*} > \delta \tag{9.784}$$

for some $\delta > 0$ for all $\hat{t}$ since this occurs if the first rank in the mesostate $\hat{t}$ decouples and this occurs after a random time with distribution having a positive density. Therefore

$$\hat{Q}^n_{\hat{t}_*,\hat{t}_*} \geq \delta \sum_{\hat{t}} \hat{Q}^{n-1}_{\hat{t}_*,\hat{t}}. \tag{9.785}$$

Then by [VJ], Theorem C(ii), $\lim_{n\to\infty} \hat{Q}^n_{\hat{t}_*,\hat{t}}$ exists and is finite. So we can assume $\sup_n \hat{Q}^n_{\hat{t}_*,\hat{t}_*} = K$ for some $K < \infty$ which implies by (9.784) $\sup_n \sum_{\hat{t}} \hat{Q}^n_{\hat{t}_*,\hat{t}} < K/\delta$.

We now consider the *convex* set $\mathfrak{M}$ (bigger than the target convex set) of measures μ in the vector space of signed measure. The set $\mathfrak{M}$ is given by:

$$\mathfrak{M} = \{\mu \in \mathcal{M}(\hat{\mathfrak{T}}) | \sup_n \sum_{\hat{t}} \mu(\hat{t}) \hat{Q}^n_{\hat{t},\hat{t}_*} \leq L\}. \tag{9.786}$$

Note that we can choose L sufficiently large so that

$$\hat{Q} : \mathfrak{M} \to \mathfrak{M} \text{ and } \mathfrak{M} \neq \emptyset, \tag{9.787}$$

the first by our definition of $\mathfrak{M}$ and the second since it contains $\delta_{\hat{t}_*}$, for L big enough, for example $\geq K/\delta$. We observe that also $\mathfrak{M} \cap \mathcal{P}(\hat{\mathfrak{T}})$ is a convex subset of our vector space $\mathcal{M}(\hat{\mathfrak{T}})$. Hence we can view $\mathfrak{Q}$ acting on the convex subset of $\mathcal{M}(\hat{\mathfrak{T}})$ given by $\mathfrak{M} \cap \mathcal{P}(\hat{\mathfrak{T}})$.

Moreover we can establish an important property of $\mathfrak{M}$, namely if $\mu \in \mathfrak{M}$, then since

$$L \geq \sum_{t_1,t_2} \mu_{t_1} \hat{Q}_{t_1,t_2} \hat{Q}_{t_2,\hat{t}_*} \geq \delta \sum_{t_1,t_2} \mu_{t_1} \hat{Q}_{t_1,t_2} : \tag{9.788}$$

we have

$$\langle \mu \hat{Q}, 1 \rangle = \sum_{\hat{t}} \mu(\hat{t}) \alpha(\hat{t}) \leq L/\delta, \tag{9.789}$$

where

$$\alpha(\hat{t}) = E[\xi(\hat{t}(t_0))|\hat{t}(0)) = \hat{t}] = \sum_{\hat{t}'} \hat{Q}_{\hat{t},\hat{t}'}, \qquad (9.790)$$

here $\xi(\hat{t}(t_0))$ denotes the number of offspring produced by a mesostate started at $(1, \ldots, 1, 0)$ during the interval $(0, t_0]$.

The function $\alpha(\cdot)$ will be crucial both to define the convex set and to define the topology. We need some additional information. We can dominate $\alpha(\hat{t})$ by stochastically dominating $\xi(\hat{t}(t_0))$ in distribution by the sum of $|\hat{t}(0)|$ i.i.d. random variables $\tilde{\xi}_i$ given by the random number of offspring produced in $(0, t_0]$ by the mesostate process starting at $(1, \ldots, 1, 0)$. To see this note that a rank in $\hat{t}$ that forms the first rank of an offspring mesostate is either one of the original ranks in $\hat{t}$ or a descendant of such an original rank. But there is a corresponding rank in the independent set - the only difference is that in $\hat{t}$ there are additional losses of ranks (and offspring) due to mutations or coalescence due to other original ranks but these do not occur in the independent family. By comparison with the Yule process with rate s, $\tilde{\xi}_i$ has a finite $(1 + \varepsilon)$-moment. Therefore

$$\alpha(\hat{t})^{(1+\varepsilon)} \leq E[(\xi(\hat{t}(t_0))^{(1+\varepsilon)}|\hat{t}(0))] \leq E\left[\sum_{i=1}^{|\hat{t}(0)|} \tilde{\xi}_i\right]^{(1+\varepsilon)} \leq \text{const} \cdot |\hat{t}(0)|^{(1+\varepsilon)}.$$
(9.791)

To define the convex set of probability measures in $\mathfrak{M}$ we want to use the characteristic properties of the frequencies of types in our mesostate. The mesostates we encounter start in initial states which can be viewed as mesostate starting in $\hat{t}^*$ but running a random time, where the random time has certain tail properties. In order to obtain the target convex set, we study in more detail the process of the internal state at the rank 1 of our mesostate, in particular its *recurrence properties*, to obtain the characteristic properties coming into the picture via the age shift of the mesosates not starting in $\hat{t}_*$.

Recall that the mutation process at rank 1 is a Markov chain with recurrence point $(1, \ldots, 1, 0)$ (recall Lemma 9.25). The time between sojourns from this point has distribution

$$F, \qquad (9.792)$$

which is tail-dominated by an exponential (see Remark 156) with parameter $\hat{\lambda}$ and on each return either 0 or a non-zero random number of offspring mesostates are produced depending on which of the types $1, \ldots, M$ takes over at the return time (cf. (9.191)). The times of return to $(1, \ldots, 1, 0)$ then form a *classical renewal process with distribution F* (where for $M = 2$, F is exponential and $\hat{\lambda} = m$). Define

9.12 Emergence and Fixation: The Selection-Dominant Case

$$\bar{F}^*(x) := \frac{1}{\mu} \int_x^\infty (1 - F(y))dy, \text{ where } \mu = \int y dF(y). \quad (9.793)$$

Note that $\bar{F}^*$ is also tail-dominated by an exponential with parameter $\hat{\lambda}$. Let $a(t)$ denote the current age at time t. Note that ([KT] (6.5)),

$$\lim_{t \to \infty} P(a(t) \geq x) = \bar{F}^*(x). \quad (9.794)$$

Note that starting at their birth times the mesostate transitions at the first rank of the different mesostates are independent of all other mesostates and form a renewal process with interval times having cumulative distribution function F.

For example when $M = 2$ then in equilibrium the age A (time since last renewal) satisfies

$$\mu_A((t, \infty)) = P(A > t) = e^{-mt}. \quad (9.795)$$

Moreover if we initiate the population with an age distribution $\mu(0)$ which is stochastically dominated by (9.795), then we can construct a probability space and random variables having marginal distributions $\mu_A, \mu(0)$ and a Poisson process with rate m such that the law of $\hat{A}(t)$ is dominated by A for all t, a.s. This follows since both age together up to the next renewal and then the two processes couple. In the general case we must replace (9.795) by the stationary age distribution (see [KT], Ch. 5, section 6) and note that this is dominated by the same exponential tail (see [KT], (6.4)).

Let $P_a \in \mathcal{P}(\hat{\mathfrak{T}})$ denote the probability law of the mesostate process started at $(1, \ldots, 1, 0)$ and current age (time from last return) a. Now consider the convex set of probability measures

$$\mathcal{P}_F(\hat{\mathfrak{T}}) \quad (9.796)$$

of the form

$$\hat{P} = \int_0^\infty P_a \tilde{P}(da), \quad \tilde{P} \in \mathcal{P}([0, \infty)), \quad \tilde{P}([x, \infty)) \leq \bar{F}^*(x), \text{ for } x \geq 0. \quad (9.797)$$

The set $\mathcal{P}_F(\hat{\mathfrak{T}})$ has the following properties. Since $\bar{F}^*$ is tail dominated by an exponential with parameter $\hat{\lambda}$, it follows by the argument of Lemma 9.69 ($M = 2$), Proposition 9.74 ($M > 2$), for $0 < \varepsilon < \hat{\lambda}s^{-1}$ there exists (for fixed ε) $L < \infty$ such that for $\{p_{\hat{\mathfrak{t}}}\}_{\hat{\mathfrak{t}} \in \hat{\mathfrak{T}}} \in \mathcal{P}_F(\hat{\mathfrak{T}})$, we have the *two* bounds (giving tightness conditions):

$$\sum_{\hat{\mathfrak{t}}} (l(\hat{\mathfrak{t}}))^{1+\varepsilon} p_{\hat{\mathfrak{t}}} \leq L, \quad (9.798)$$

$$\sum_{\hat{\mathfrak{t}}} (\alpha(\hat{\mathfrak{t}}))^{(1+\varepsilon)} p_{\hat{\mathfrak{t}}} < \text{const} \cdot L. \quad (9.799)$$

646 9 Emergence with $M \geq 2$ Lower Order Types (Phases 0,1,2)

Moreover, starting with $\zeta(0) = (1,\ldots,1,0)$, using (9.794), and noting that $1 - F(y)$ is also dominated by a multiple of the same exponential tail we have

$$\lim_{t\to\infty} E[[\mathfrak{l}(\zeta(a(t)))]^{1+\varepsilon}] = \frac{1}{\mu}\int E[[\mathfrak{l}(\zeta(u))]^{1+\varepsilon}](1 - F(u))du < \infty, \quad (9.800)$$

where $\mu = \int u dF(u)$ so we can choose L such that

$$\sup_t E\left[(\mathfrak{l}(\zeta(a(t))))^{1+\varepsilon}\right] \leq L. \quad (9.801)$$

We now introduce the basic compact *convex* set $\mathfrak{K}$ on which we analyse the nonlinear map $\mathfrak{Q}$,

$$\mathfrak{K} := \mathfrak{M} \cap \mathcal{P}_F(\hat{\mathfrak{T}}). \quad (9.802)$$

We now have a map $\mathfrak{Q}$ on a convex subset of a vector space. The next step is to introduce a *topology* so that both become $\mathfrak{K}$ *compact* and $\mathfrak{Q}$ *continuous*.

We define topology of α-weak convergence, namely as $n \to \infty$ we define

$$p_n \Rightarrow_\alpha p_\infty, \text{ iff } p_n(\hat{\mathfrak{t}}) \to p_\infty(\hat{\mathfrak{t}}) \text{ for all } \hat{\mathfrak{t}} \text{ and } \sum_{\hat{\mathfrak{t}}} \alpha(\hat{\mathfrak{t}}) p_n(\hat{\mathfrak{t}}) \to \sum_{\hat{\mathfrak{t}}} \alpha(\hat{\mathfrak{t}}) p_\infty(\hat{\mathfrak{t}}).$$

$$(9.803)$$

To verify that $\mathfrak{K}$ is compact in this topology, consider a sequence $\{p_n\} \in \mathfrak{K}$. Then $\sum_{\hat{\mathfrak{t}}} \alpha(\hat{\mathfrak{t}}) p_n(\hat{\mathfrak{t}}) < L/\delta$ by (9.789), so we can find a subsequence for which $\sum_{\hat{\mathfrak{t}}} \alpha(\hat{\mathfrak{t}}) p_n(\hat{\mathfrak{t}})$ converges. Then using the first tightness condition in (9.798) we can find a further subsequence $\{p_{n_k}\}$ that converges to p_∞ in $\mathcal{P}(\hat{\mathfrak{T}})$. Using (9.799) it follows that

$$\sum_{\hat{\mathfrak{t}}} \alpha(\hat{\mathfrak{t}}) p_{n_k}(\hat{\mathfrak{t}}) \to \sum_{\hat{\mathfrak{t}}} \alpha(\hat{\mathfrak{t}}) p_\infty(\hat{\mathfrak{t}}). \quad (9.804)$$

Furthermore we shall show below that $\mathfrak{Q}$ maps $\mathcal{P}_F(\hat{\mathfrak{T}})$ into itself and therefore $\mathfrak{Q}$ maps $\mathfrak{K}$ to $\mathfrak{K}$.

We can then verify that $\mathfrak{Q}$ is a α-weakly *continuous* map from $\mathfrak{K}$ to $\mathfrak{K}$. This follows since given $p_n \Rightarrow_\alpha p_\infty$, by definition (9.782) the denominators converge and we can verify that for all finite sets K

$$\lim_{n\to\infty} \sum_{\hat{\mathfrak{t}}} \hat{Q}_{\hat{\mathfrak{t}},\hat{\mathfrak{t}}'} p_n(\hat{\mathfrak{t}}) \geq \lim_{n\to\infty} \sum_{\hat{\mathfrak{t}}\in K} \hat{Q}_{\hat{\mathfrak{t}},\hat{\mathfrak{t}}'} p_n(\hat{\mathfrak{t}})$$

$$= \sum_{\hat{\mathfrak{t}}\in K} \hat{Q}_{\hat{\mathfrak{t}},\hat{\mathfrak{t}}'} p_\infty(\hat{\mathfrak{t}}), \quad \forall \hat{\mathfrak{t}}' \in \hat{\mathfrak{T}},$$

9.12 Emergence and Fixation: The Selection-Dominant Case

and consequently

$$\lim_{n\to\infty} \sum_{\hat{\mathfrak{t}}} \hat{Q}_{\hat{\mathfrak{t}},\hat{\mathfrak{t}}'} p_n(\hat{\mathfrak{t}}) \geq \sum_{\hat{\mathfrak{t}}} \hat{Q}_{\hat{\mathfrak{t}},\hat{\mathfrak{t}}'} p_\infty(\hat{\mathfrak{t}}), \quad \forall \, \hat{\mathfrak{t}}' \in \hat{\mathfrak{T}}. \tag{9.805}$$

But since $\sum_{\hat{\mathfrak{t}},\hat{\mathfrak{t}}'} \hat{Q}_{\hat{\mathfrak{t}},\hat{\mathfrak{t}}'} p_n(\hat{\mathfrak{t}})$, converges to $\sum_{\hat{\mathfrak{t}},\hat{\mathfrak{t}}'} \hat{Q}_{\hat{\mathfrak{t}},\hat{\mathfrak{t}}'} p_\infty(\hat{\mathfrak{t}})$ by construction of p_∞ (recall (9.804)), we have equality and the proof of continuity is complete.

We have constructed a continuous map $\mathfrak{Q}$ which is defined on a convex, compact subset of a topological vector space and so to apply Schauder's theorem it remains to show that indeed $\mathfrak{Q}$ maps $\mathfrak{K}$ into $\mathfrak{K}$.

(iii) The invariance property of $\mathfrak{Q}$ on $\mathfrak{K}$
The objective of this item is to verify that

$$\mathfrak{K}\mathfrak{Q} \subseteq \mathfrak{K}, \tag{9.806}$$

that is, if $P \in \mathfrak{K}$, then $P\mathfrak{Q} \in \mathfrak{K}$. Recalling (9.787), it suffices to show that $\mathcal{P}_F(\hat{\mathfrak{T}})$ is preserved under $\mathfrak{Q}$.

To verify this we return to the branching mesostate process. Consider a mesostate $\hat{\mathfrak{t}}$ at time t at a label that began at time zero with the mesostate $\hat{\mathfrak{t}}_*$. Recall that migration events occur when a rank from one of the existing sites in a label, i.e. in its support which we denoted by $\mathfrak{R}(\hat{\mathfrak{t}})$ migrates to an unoccupied site. The corresponding column is a partition of the set $\mathbb{I}_M$ but if a freshly colonized site is at this time decoupled from its parent its dynamics corresponds again to the evolution of a mesostate starting with $\hat{\mathfrak{t}}_*$ and otherwise being of the same form but being of *positive age* at the migration time. Thus we can view the creation of such new mesostates $\zeta_i(t), i = 1, \cdots, k$ (corresponding to internal factors prior to decoupling) as always producing mesostates starting at $\hat{\mathfrak{t}}_*$ but with possibly positive age at the decoupling time (and in addition the original mesostate has now returned to $\hat{\mathfrak{t}}_*$).

Thus at constant rate mesostates are decoupled producing offspring that are initiated by such a new mesostate in state $\hat{\mathfrak{t}}_*$ at a new site. Since mesostates never die, the growth in the number of mesostates, call it $M(t)$, is then *super*critical. Starting with any finite initial population of mesostates the population at time t can be decomposed into three groups, mesostates alive at time 0 and not yet decoupled, remnants (residuals of initial individuals) and those that have been created at decoupling times of those initial mesostates and their descendants which are having positive age but started in $\hat{\mathfrak{t}}_*$. (Note that the number of the sum of the first two is non-increasing.) The counting measure of the type of mesostate present in the time t population is then given by:

$$\Sigma(t,\cdot) = \left[\sum_{i=1}^{M(0)} \sum_{j=1}^{n_i(t)} \delta_{\zeta(a_{i,j}(t))}(\cdot) \right] \tag{9.807}$$

where $n_i(t)$ denotes the number of descendants of the ith initial mesostate at time t and $a_{i,j}(t)$ denotes the age (time since last renewal) of the jth descendant of the ith mesostate at time t.

In order to show that $\mathfrak{Q}$ satisfies the required invariance property we will use the fact that $\hat{Q}$ is the mean matrix for the branching mesostate process. Let $\hat{P} \in \mathcal{P}_F(\hat{t})$ with corresponding mixture measure $\tilde{P}$. To relate this to the branching mesostate process, let $M(0) \in \mathbb{N}$ and consider the random counting measure

$$\Sigma_{M(0)}(0) := \frac{1}{M(0)} \sum_{i=1}^{M(0)} \delta_{\zeta_i(a_i(0))} \tag{9.808}$$

where $a_i(0)$ are i.i.d. $\tilde{P}$ and the ζ_i are independent mesostate processes. Then $\frac{E[\Sigma_{M(0)}]}{M(0)} = \hat{P}$ and by the law of large numbers $\frac{\Sigma_{M(0)}}{M(0)} \to \hat{P}$ as $M(0) \to \infty$.

We then consider the population of branching mesostates started at time $t = 0$ with population $\Sigma_{M(0)}$ and observed at time t_0 and consider the occupation measure of the mesostates:

$$\left\{ \sum_{i=1}^{M(t_0)} \delta_{\zeta(a_i(t_0))}(\hat{t}) \right\}_{\hat{i} \in \hat{\mathfrak{i}}} \in \mathcal{N}(\hat{\mathfrak{T}}). \tag{9.809}$$

Now note that all offspring of a mesostates created at a decoupling event in the interval $[0, t_0]$ can have positive age at their birth times but this must be dominated by the age of the parent just prior to decoupling and therefore are dominated by $\bar{F}^*$. Therefore we have

$$\sum_{\hat{t}} \hat{P}(\hat{t}) \mathfrak{Q}_{\hat{t}, \cdot} = \lim_{M(0) \to \infty} \frac{E[\sum_{i=1}^{M(0)} \sum_{j=1}^{n_i(t_0)} \delta_{\zeta(a_{i,j}(t_0))}(\cdot)]}{E[M(t_0)]} \tag{9.810}$$

$$= \lim_{M(0) \to \infty} \frac{[\sum_{i=1}^{M(0)} \sum_{j=1}^{n_i(t_0)} \delta_{\zeta(a_{i,j}(t_0))}(\cdot)]}{\sum_{i=1}^{M(0)} n_i(t_0)},$$

and since the ages of the original mesostates $\{a_i(\cdot)\}$ are dominated by $\bar{F}^*$,

$$\frac{\sum_{i=1}^{M(0)} E(\sum_{j=1}^{n_i(t_0)} 1(a_{i,j}(t_0) > x))}{E[M(t_0)]} \leq \frac{\sum_{i=1}^{M(0)} E[n_i(t_0)] P(a_i(t_0 \wedge T_{\text{dec}}-) > x)}{\sum_{i=1}^{M(0)} E[n_i(t_0)]}$$

$$\leq \frac{\sum_{i=1}^{M(0)} E[n_i(t_0)] \bar{F}^*(x)}{\sum_{i=1}^{M(0)} E[n_i(t_0)]} \leq \bar{F}^*(x). \tag{9.811}$$

This means that the law at time t_0 is the law of a population of mesostates with random ages with distributions again dominated by $\bar{F}^*$. Therefore recalling the definition (9.797), this means that

9.12 Emergence and Fixation: The Selection-Dominant Case

$$\mathfrak{Q} : \mathcal{P}_F(\hat{\mathfrak{T}}) \to \mathcal{P}_F(\hat{\mathfrak{T}}). \tag{9.812}$$

Remark 196. Here we note that the new mesostates created by branching has a bias downwards in the distribution of ages. It is expected that the limiting population age distribution exists as in [N] but we do not elaborate on this here.

(iv) Applying the Schauder fixed point theorem

We have now verified that $\mathfrak{Q}$ is a map satisfying the assumptions of the Schauder-fixed point theorem. Therefore there is a *fixed point p^**, i.e. $p^*\mathfrak{Q} = p^*$, ([Con], 9.5) which is in particular in $\mathcal{P}(\hat{\mathfrak{T}})$. Furthermore by irreducibility (recall (9.783)). This yields a normalized eigenvector for $\hat{Q}$, namely p^* with eigenvalue $\sum_{\hat{\mathfrak{t}}} (\sum_{\hat{\mathfrak{t}}'} p_{\hat{\mathfrak{t}}}^* \hat{Q}_{\hat{\mathfrak{t}},\hat{\mathfrak{t}}'})]$ as follows. Namely the fixed point relation

$$p^*\mathfrak{Q} = p^*, \tag{9.813}$$

implies that by the definition of $\mathfrak{Q}$:

$$p^*\hat{Q} = [\sum_{\hat{\mathfrak{t}}'}(\sum_{\hat{\mathfrak{t}}} p_{\hat{\mathfrak{t}}}^* \hat{Q}_{\hat{\mathfrak{t}},\hat{\mathfrak{t}}'})]p^*. \tag{9.814}$$

Since by definition (9.762) the inverse radius of convergence of $\hat{Q}$ is 1, it follows that the eigenvalue $[\sum_{\hat{\mathfrak{t}}'}(\sum_{\hat{\mathfrak{t}}} p_{\hat{\mathfrak{t}}}^* \hat{Q}_{\hat{\mathfrak{t}},\hat{\mathfrak{t}}'})]$ must be 1. Therefore,

$$\sum_{\hat{\mathfrak{t}} \in \hat{\mathfrak{T}}} p_{\hat{\mathfrak{t}}}^* \hat{Q}_{\hat{\mathfrak{t}},\hat{\mathfrak{t}}_*} = p_{\hat{\mathfrak{t}}_*}^*. \tag{9.815}$$

This implies that (since $p_{\hat{\mathfrak{t}}}^*$ has strictly positive entries)

$$\lim_{n \to \infty} \hat{Q}_{\hat{\mathfrak{t}}_*,\hat{\mathfrak{t}}_*}^n > 0 \tag{9.816}$$

and therefore $\hat{\mathfrak{t}}_*$ is by (9.749) *positive 1-recurrent*.

We can then apply the results of Vere-Jones [VJ], Cor. to Theorem 6.2 and Athreya-Ney [AN] Theorem 3.2 and in particular there exists and invariant measure v and invariant function h (unique up to constant factors) such that as $n \to \infty$:

$$(\mu \hat{Q}^n) f(\hat{\mathfrak{t}}) \to \sum_{\mathfrak{t}} \mu(\hat{\mathfrak{t}}) h(\hat{\mathfrak{t}}) \cdot \sum_{\hat{\mathfrak{t}}'} f(\hat{\mathfrak{t}}') v(\hat{\mathfrak{t}}') \text{ and } \sum_{\mathfrak{t}} h(\hat{\mathfrak{t}}) v(\hat{\mathfrak{t}}) < \infty. \tag{9.817}$$

This completes the proof of Lemma 9.77.

In order to continue the proof of part (b) of Lemma 9.75 we finally have to argue that the growth rate and the invariant measure we found for the discrete times $(nt_0)_{n \in \mathbb{N}}$ dynamic induced by $(\tilde{\mathfrak{U}}(t,\cdot))_{t \geq 0}$ extends to the full continuous object $Q^t := e^{-\beta \rho_e t} \tilde{\mathfrak{U}}(t)$, that is, $vQ^t = v$ with $v = p_*$ and

$$(Q^t)_{\hat{t},\hat{t}'} e^{-\beta_{\text{Pe}} t} \longrightarrow \frac{v(\hat{t}') h(\hat{t}')}{\sum v(\hat{t}') h(\hat{t}')} \quad \text{for } t \to \infty, \tag{9.818}$$

compare Remark 194.

To do this we repeat the above analysis with t_0 replaced by $\frac{t_0}{2^n}$, $n \in \mathbb{N}$. Then for each n we obtain a fixed point but for consistency these fixed points are identical. Therefore

$$\lim_{k \to \infty} (Q^{\frac{kt_0}{2^n}})_{\hat{t},\hat{t}'} e^{-\beta \frac{kt_0}{2^n}} = \frac{p_*(\hat{t}') h(\hat{t}')}{\sum p_*(\hat{t}') h(\hat{t}')}, \quad \forall\, n \in \mathbb{N}. \tag{9.819}$$

We now have to show first $p^* Q^t = p^*$ and then second the convergence statement for Q^t as $t \to \infty$.

Given any sequence $(t_n)_{n \in \mathbb{N}}$ we consider

$$Q^{t_j} = Q^{t_j - \tilde{t}_j} Q^{\tilde{t}_j}, \tag{9.820}$$

where $\tilde{t}_j = \tilde{t}_j(n) = \frac{k_j t_0}{2^n}$ and $0 \leq t_j - \tilde{t}_j \leq \frac{t_0}{2^n}$. Then

$$\lim_{n \to \infty} \lim_{j \to \infty} \| p_* Q^{t_j} - p_* \| \leq \lim_{n \to \infty} \lim_{j \to \infty} (\| p_* Q^{t_j - \tilde{t}_j(n)} - p_* \|) = 0, \tag{9.821}$$

where here we use the ℓ_1-norm so that p_* is a fixed point for Q^t for every $t \geq 0$ as claimed.

Next turn to the convergence of Q^t as $t \to \infty$. We then use the strong continuity of Q^t, that is $\| p Q^s - p \| \to 0$ as $s \downarrow 0$ if $p \in \mathfrak{K}$, to show (9.818) for this sequence, namely using (9.819) on the first term and strong continuity on the second term, we get

$$\| p Q^{t_j} - p_* \| \leq \| Q^{t_j - \tilde{t}_j} \| \| p Q^{\tilde{t}_j} - p^* \| + \| p Q^{t_j - \tilde{t}_j} - p_* \| \to 0 \text{ as } j \to \infty, n \to \infty. \tag{9.822}$$

Therefore we have found a fixed point $p^* Q^t = p^*$ as claimed and we have the claimed convergence for $t \to \infty$.

(c) The Theorem 3.2 in [AN] provides the identification of the invariant measure.

(d) We first argue that the left and right eigenvectors we found are left and right eigenvectors to the eigenvalue $e^{\beta_{\text{Pe}} t}$ for the mean matrix corresponding to $t \in (0, \infty)$ and therefore we know from (9.818) that

$$E[e^{-\beta_{\text{Pe}} t} K_t^{\text{Pe}}(\cdot)] \Longrightarrow \bar{\mathfrak{U}}^{\text{Pe}}(\infty, \mathbb{R}^+, \cdot), \quad \text{as } t \to \infty. \tag{9.823}$$

The next point is then to establish the convergence in law of $e^{-\beta_{\text{Pe}} t} K_t^{\text{Pe}}$ to $W_{\text{Pe}} \bar{\mathfrak{U}}^{\text{Pe}}(\infty, \mathbb{R}^+, \cdot)$.

9.12 Emergence and Fixation: The Selection-Dominant Case

In the selection-dominant case we can calculate the second moment of the scaled random variable. Then we need to show that this converges to

$$(E[W_{Pe}])^2 + Var[W_{Pe}] \text{ as } t \to \infty, \text{ where } Var[W_{Pe}] > 0. \qquad (9.824)$$

This we show in (e). In the case of only $(1+\varepsilon)$-moments we use the tightness to reduce the finitely many types and then use Doney [Do2].

(e) We now have a multitype branching process with countable type space. By Proposition 9.65 (if $M = 2$) and Proposition 9.74(c) ($M \geq 2$), the total number of offspring produced when a mesostate branches has a finite second moment in the selection-dominant case and by (c) the mean matrix is positive 1-recurrent. The L^2-convergence follows by the second moment argument of Moy (see Moy [Mo], Theorem 1 and [Mo2]), namely this implies that there exists a real-valued random variable W_{Pe} with $E[W_{Pe}^2] < \infty$ such that:

$$\sum_{\hat{t}} e^{-\beta_{Pe} t} K_t^{Pe}(\hat{t}) f(\hat{t}) \longrightarrow W_{Pe} \sum_{\hat{t}} p^*(\hat{t}) f(\hat{t}), \text{ as } t \to \infty \text{ in } L^2, \qquad (9.825)$$

for any function f such that $\frac{f(\hat{t})}{h(\hat{t})}$ is bounded where $h(\cdot)$ is the positive right eigenvector.

This completes the proof of Lemma 9.75. q.e.d.

Remark 197. The proof of the analogous result needed in the proof of Lemma 9.55 is carried out in the same way but replacing the compact set $\mathcal{P}_F(\tilde{\mathfrak{T}})$ used above by the compact subset of $\mathcal{P}(\tilde{\mathfrak{T}})$ satisfying $\sum |\bar{t}|^2 p_{\bar{t}} \leq \mathcal{L}$ with $\mathcal{L} < \infty$.

9.12.8 Growth of $|\tilde{\mathcal{G}}_t^{++}|$: The Two Level Branching Structure

In order to study emergence we need to know at what rate the number of active ranks in the dual process grows and what the frequencies of the internal states associated with these ranks are. We investigate this in more detail and consider first $\tilde{\mathcal{G}}^{++}$ and turn then to $\tilde{\mathcal{G}}^{++,N}$ in the non-collision regime, i.e. time of order $o(\log N)$ which is asymptotically however simply the same as $\tilde{\mathcal{G}}^{++}$.

We define (note that different from Subsubsection 9.12.7 we now consider the empirical and not the expected quantities)

$$\bar{U}^{Pe}(t, \cdot) = \frac{K_t^{Pe}(\hat{t})}{K_t^{Pe}(\hat{\mathfrak{T}})} = \text{empirical distribution of mesostates}, \qquad (9.826)$$

where $K_t^{Pe}(\cdot)$ is given by (9.733), and

$$\bar{u}(t) = K^{Pe}(\hat{\mathfrak{T}}). \qquad (9.827)$$

We introduce two basic quantities which are needed for the study of the effect of a rare mutation event in the dual process, namely the functions of time $\tilde{v}$ and $\bar{v}$. Let $|\hat{t}|$ denotes the number of sites in the support of $\hat{t}$. Then set

$$\tilde{v}(t) := \int |\hat{t}| \bar{U}^{Pe}(t, d\hat{t}) \quad = \text{ empirical mean number} \tag{9.828}$$

of occupied sites per mesostate

and the *total number of occupied sites* is given by

$$\bar{v}(t) := \bar{u}(t) \cdot \tilde{v}(t). \tag{9.829}$$

Remark 198. We know from Subsubsection 9.12.7 that in the selection-dominant case the expected number of mesostates, $\bar{u}(t)$, grows with malthusian parameter β_{Pe} and from Subsubsection 9.7.5, (9.361), that the expected number of *UPe* (and $UPe \cup Tr$) sites grows with malthusian parameter $\beta \geq \beta_{Pe}$.

In the non-ergodic case we may have $\beta > \beta_{Pe}$. Since each *UPe* site belongs to a mesostate (exactly one *Pe* site belongs to a new state), by consistency it follows in this case that the expected number of sites per mesostate grows with malthusian parameter $(\beta - \beta_{Pe})$, that is,

$$\lim_{t \to \infty} e^{-(\beta - \beta_{Pe})t} \tilde{v}(t) \in (0, \infty). \tag{9.830}$$

We now consider the associated collection of *microstates* at occupied (active) sites and look for the empirical *age and type distribution* over *all* occupied active sites (recall that we did this only for *UPe* and *Pe* sites).

Define the *microstate at site j* here denoted

$$\hat{t}(j) \text{ as the element of } T \text{ obtained by the restriction of } \hat{t} \text{ to the site } j. \tag{9.831}$$

The *lifetime of a microstate*, that is the time until it is removed by a resolution at an ancestral site, depends on its position in the mesostate to which it belongs. The lifetime of a *Pe* or *UPe* site is ∞. More precisely, the lifetime distribution of a microstate at a site in a mesostate is a *measurable function of the mesostate*. Note that the microstate at an occupied site is obtained from the mesostate to which it belongs by ignoring the states of other sites in the mesostate on which of course the microstate evolution depends. We can then assign a random lifetime

$$L_j \text{ to each } \hat{t}(j), \tag{9.832}$$

with distribution given by a measurable function of $\hat{t}$, $j = 1, \ldots, |\hat{t}|$, if j indexes the sites occupied by the mesostate (which of course need not be $1, 2, \cdots$).

We next introduce the map

9.12 Emergence and Fixation: The Selection-Dominant Case

$$\hat{\theta} : \hat{\mathfrak{T}} \to \mathcal{P}(\mathcal{T} \times (0, \infty]), \tag{9.833}$$

producing the *empirical measure* of *microstates and their lifetimes* in a mesostate $\hat{\mathfrak{t}}$. Given the mesostate t this is defined as

$$\hat{\theta}(\hat{\mathfrak{t}}, \tau, dx) := \frac{1}{|\hat{\mathfrak{t}}|} \sum_{j=1}^{|\hat{\mathfrak{t}}|} 1(\hat{\mathfrak{t}}(j) = \tau, L_j \in dx). \tag{9.834}$$

Then we define the normalized size-biased empirical measure per site of microstate type and lifetime at time t as (recall (9.826)–(9.829)):

$$\bar{V}(t, \tau, dx) := \frac{\bar{u}(t)}{\bar{v}(t)} \int |\hat{\mathfrak{t}}| \hat{\theta}(\hat{\mathfrak{t}}, \tau, dx) \bar{U}^{Pe}(t, d\hat{\mathfrak{t}}) \in \mathcal{P}(\mathcal{T} \times (0, \infty]). \tag{9.835}$$

Note that here the *size-biasing term* is needed since typical occupied sites tend to lie in larger mesostates.

We can also represent and asymptotically evaluate ($t \to \infty$) the total number of active ranks at time t, $\Pi(t)$, as

$$\Pi(t) = \bar{v}(t) \sum_{\tau \in \mathcal{T}} \|\tau\| \bar{V}(t, \tau, (0, \infty]) = (1 + o(1)) W_{Pe} e^{\beta_{Pe} t} \sum_{\hat{\mathfrak{t}}} \bar{U}^{Pe}(t, \hat{\mathfrak{t}}) \sum_{i=1}^{|\hat{\mathfrak{t}}|} \|\hat{\mathfrak{t}}(i)\|, \tag{9.836}$$

where $\|\tau\|$ denotes the number of active ranks in $\tau \in \mathcal{T}$.

The next objective is to determine how the number of active sites in $\tilde{\mathcal{G}}^{++}$ grows. In order to prepare for the collision regime we can at the same time consider $\tilde{\mathcal{G}}^{++,N}$ and the corresponding mesostate process when the space $\mathbb{N}$ is replaced by the finite set $\{1, \ldots, N\}$. In the finite N case mesostates can fuse, since a rank can collide and coalesce with a rank from another mesostate. However recall that as long as we consider times of order $o(\log N)$ such collisions occur only with probability tending to zero in the considered complete time period. Hence collisions can be neglected and we have essentially as $N \to \infty$ the same branching structure. In this setting we denote the respective quantities for the process on $\{1, \cdots, N\}$ as $\bar{u}^N$, $\bar{U}^N$, $\bar{v}^N$, $\tilde{v}^N$, $\bar{V}^N$ and $\bar{\Pi}^N$.

Proposition 9.79 (Growth of number of sites in $\tilde{\mathcal{G}}^{++,N}$ in the non-collision regime).

Assume that $t_N \to \infty$ with $t_N = o(\frac{\log N}{\beta})$.

(a) Then in L^2:

$$e^{-\beta t_N} \bar{v}(t_N) \longrightarrow W_{\text{site}}, \quad e^{-\beta t_N} \bar{v}^N(t_N) \to W_{\text{site}}, \quad \text{as } N \to \infty, \tag{9.837}$$

where W_{site} is a strictly positive random variable with finite first and second moments, with

$$W_{\text{site}} = \bar{c} W_{Pe} \qquad (9.838)$$

with (this is a deterministic number)

$$\bar{c} = \sum_{\hat{\text{t}}} |\hat{\text{t}}| \bar{U}(\infty, \hat{\text{t}}). \qquad (9.839)$$

(b) Furthermore as $N \to \infty$ in L^2:

$$\bar{V}(t_N, \cdot, \cdot) \Longrightarrow \bar{V}(\infty, \cdot, \cdot) \text{ similarly } \bar{V}^N(t_N, \cdot, \cdot) \Longrightarrow \bar{V}(\infty, \cdot, \cdot), \qquad (9.840)$$

that is, for any continuous function f on $\mathcal{T} \times [0, \infty]$ with finite support in $\mathcal{T}$ and limits at ∞,

$$\sum_{\tau} \int_0^\infty f(\tau, x) \bar{V}^N(t_N, \tau, dx) \to \sum_{\tau} \int_0^\infty f(\tau, x) \bar{V}(\infty, \tau, dx) \quad \text{in } L^2. \quad \square \qquad (9.841)$$

Remark 199. An alternative formulation would be strengthen the L^2 convergence to almost sure convergence as in the result of Nerman [N2] on the CMJ process with finitely many types. The "$x \log x$" condition (9.325) is satisfied since we have a $1 + \varepsilon$ moment (see Proposition 9.74) for some $\varepsilon > 0$. Nerman's proof of the a.s. convergence of the type and age distribution [N2] (Theorems 6.6, 6.7) should remain valid in the countably many type case using our tightness condition and Lemma 9.75.

Proof of Proposition 9.79.

(a) To verify this, recall that in the case of ergodic mutation on E_0 we have that (the latter by Lemma 9.75):

$$\mathcal{L}[\bar{U}^N(t, \cdot)] \underset{N \to \infty}{\Longrightarrow} \mathcal{L}[\bar{U}(t, \cdot)] \text{ and } \mathcal{L}[\bar{U}(t, \cdot)] \underset{t \to \infty}{\Longrightarrow} \delta_{\bar{\mathfrak{U}}(\infty, \cdot)}, \qquad (9.842)$$

where $\bar{\mathfrak{U}}(\infty, \cdot)$ is the stable type distribution given by (9.756). Since $(t_N)_{N \in \mathbb{N}}$ is in the non-collision regime we see using that migration to a particular site occurs at rate N^{-1} that:

$$\bar{U}^N(t_N, d\hat{\text{t}}) \Rightarrow \bar{\mathfrak{U}}(\infty, d\hat{\text{t}}) \text{ as } N \to \infty. \qquad (9.843)$$

We claim furthermore

$$\int \bar{\mathfrak{U}}(\infty, d\hat{\text{t}}) |\hat{\text{t}}|^{(1+\varepsilon)} < \infty, \qquad (9.844)$$

9.12 Emergence and Fixation: The Selection-Dominant Case

$$\int \bar{\mathfrak{U}}(\infty, d\hat{t}) \|\hat{t}\|^{(1+\varepsilon)} < \infty. \qquad (9.845)$$

This follows from Lemma 9.69

Also by Lemma 9.69, in the case of an ergodic mutation matrix on E_0, there exists a number $\bar{c}$ given by

$$\bar{c} := \lim_{N \to \infty} \sum_{\hat{t}} |\hat{t}| \bar{U}^N(t_N, d\hat{t}) = \lim_{N \to \infty} \sum_{\hat{t}} |\hat{t}| \bar{U}(t_N, d\hat{t}) = \sum_{\hat{t}} |\hat{t}| \bar{\mathfrak{U}}(\infty, \hat{t}) < \infty. \qquad (9.846)$$

Then (9.837) holds with $W_{\text{site}} = \bar{c} W_{Pe}$. It particular, the randomness in the growth sits in W_{Pe}.

(b) Consider the empirical measures of microstates and lifetimes $\bar{V}^N(t_N, d\tau, dx)$ in the complete dual population by setting for a randomly chosen (among all labels) mesostates

$$\bar{V}^N(t_N, \tau, x) \qquad (9.847)$$

$$= \frac{1}{\sum_{|\hat{t}|} \bar{U}^N(t_N, d\hat{t}) |\hat{t}|} \sum_{|\hat{t}|} \frac{1}{|\hat{t}|} \sum_{i=1}^{|\hat{t}|} 1(\hat{t}_{(i)} = \tau, L_i \in (x, \infty)) \bar{U}^N(t_N, d\hat{t}) |\hat{t}|$$

$$= \frac{1}{\sum_{|\hat{t}|} \bar{U}^N(t_N, d\hat{t}) |\hat{t}|} \int \hat{\theta}(\hat{t}, \tau, (x, \infty)) |\hat{t}| \bar{U}^N(t_N, d\hat{t})$$

$$\to \frac{1}{\sum_{\hat{t}} |\hat{t}| \bar{\mathfrak{U}}(\infty, d\hat{t})} \int \hat{\theta}(\hat{t}, \tau, (x, \infty)) |\hat{t}| \bar{\mathfrak{U}}(\infty, d\hat{t}), \quad \text{as } N \to \infty,$$

in probability because we can argue as follows.

Using (9.842) and (9.844) it follows that the size-biased measures satisfy

$$\mathcal{L}\left[\frac{|\hat{t}| \bar{U}^N(t_N, d\hat{t})}{\sum |\hat{t}| \bar{U}^N(t_N, \hat{t})}\right] \Rightarrow \mathcal{L}\left[\frac{|\hat{t}| \bar{\mathfrak{U}}(\infty, d\hat{t})}{\sum |\hat{t}| \bar{\mathfrak{U}}(\infty, \hat{t})}\right] \quad \text{as } N \to \infty, \qquad (9.848)$$

in the sense of weak convergence of laws on probability measures on the countable set $\hat{\mathfrak{T}}$ with the r.h.s. being actually *deterministic*. In particular the tightness of the size-biased measure follows from the bound on the $(1 + \varepsilon)$ moment in the ergodic case (9.844). For any occupied site in a mesostate the local dynamics of the microstate at the site involves a linear birth, linear death and quadratic death rates. This implies that the second moments of $\|\tau\|$ are uniformly bounded by (9.845). For the convergence we argue as follows.

Given any bounded continuous function f on $\mathcal{T} \times [0, \infty]$ the mapping $\hat{t} \to \int f(\tau, x) \hat{\theta}(\hat{t}, d\tau, dx)$ is a bounded function on the countable set $\hat{\mathfrak{T}}$ and by (9.842) we have that in probability:

$$\int V^N(t_N, d\tau, dx) f(\tau, x) \rightarrow \frac{1}{\sum |\hat{\mathfrak{t}}| \bar{\mathfrak{U}}(\infty, d\hat{\mathfrak{t}})} \int \int f(\tau, x) \hat{\theta}(\hat{\mathfrak{t}}, d\tau, dx) |\hat{\mathfrak{t}}| \bar{\mathfrak{U}}(\infty, d\hat{\mathfrak{t}}).$$

q.e.d. (9.849)

9.12.9 *Empirical Process of Mesostates: Collision Case*

In the collision regime, that is times of order $\beta^{-1} \log N$ we have to work with the dual process $\tilde{\mathcal{G}}^{++,N}$ which can be viewed as a functional of the process $\mathfrak{s}^{****,N}$, which we introduced in Subsubsections 9.9.1–9.9.3 as a basis for Subsection 9.11 with focus on microstates. Now the focus is on mesostates and we need an appropriate functional, which in fact is simpler to handle as $\mathfrak{s}^{****,N}$ and will be called $\mathfrak{s}^N_{l,\text{enr}}$ and which we introduce here.

We must consider for $\tilde{\mathcal{G}}^{++,N}$ in times of order $\beta^{-1} \log N$ the additional effects in the mesostate evolution that arise in the collision regime. In this regime we must work on $S = \{1, \ldots, N\}$ and include the possibilities of the *collision and subsequent coalescence* of labels with their mesostates which then produce a *compound* label (ℓ_1, ℓ_2) with a *compound mesostate* whose original two founding *Pe* sites we call *roots* so that we now have *two Pe* roots. In fact we can get compound mesostates from the fusion of more than two mesostates and hence with more than two roots. Note *Pe*-sites are no longer immortal since they can coalesce with another Pe-site and die. Hence running the dynamics we then get multiple *Pe* roots and label $(\ell_1, \ell_2, \cdots, \ell_i)$ with multiple compound mesostates.

In particular note the following difference to the $\tilde{\mathcal{G}}^{++}$-case analysed so far. Now a *compound mesostate* is the irreducible factor in the dual and it fragments into mesostates which arose as irreducible factors and then fused. These mesostate components of a compound mesostate are for us the important functionals of the process.

We now consider the effect of a collision between a migrating rank from one label with mesostate $\hat{\mathfrak{t}}_{j_1}$ with root at site j_1 and a second label with mesostate $\hat{\mathfrak{t}}_{j_2}$ with root at site j_2 and and the migrant now sits at site $\tilde{j}$, that is after migration we have $\hat{j}$ is in the support of both labels, formally $\tilde{j} \in \mathfrak{R}(Geo)(\hat{\mathfrak{t}}_{j_1}) \cap \mathfrak{R}(Geo)(\hat{\mathfrak{t}}_{j_2})$ and then a coalescence occurs between the migrant rank and a rank from the first mesostate at the site $\tilde{j}$. At this time the two mesostates fuse and produce a new *compound label* which has a *compound mesostate* $\hat{\mathfrak{t}}_{j_1, j_2}$ with two roots at sites, $j_1, j_2 \in \{1, \ldots, N\}$.

For convenience we summarize the spaces involves and indicate where their definitions appear in the following Table 9.4.

We will describe in two steps first the basic setup and then write down the nonlinear equations.

Step 1 *(Basic ingredients)*

Recall (9.288) that the state space of the *mesostate* at label j in the collision-free case is the countable set

9.12 Emergence and Fixation: The Selection-Dominant Case

Table 9.4 Basic state spaces selection-dominant regime

State space	Process	Eqn. no.
$\widehat{\mathfrak{T}}^*$	Compound mesostate process	(9.851)
$\widetilde{\mathfrak{T}}_{\text{mark}}$	Marked mesostate process	(9.880)
$\mathcal{M}_{\text{fin}}(\widetilde{\mathfrak{T}}_{\text{mark}})$	Empirical process $\bar{\Psi}^{\text{mark},N}$	(9.873)

$$\widehat{\mathfrak{T}} = \{\hat{t} = (\tau, v, \tilde{v}) : \tau \in \mathcal{T} \text{ with } n(\tau) < \infty, \, v \in \Upsilon, \tilde{v} = Geo(j, v)\}, \quad (9.850)$$

where $\hat{t}$ is a marked multisite tableau where $\tau \in \mathcal{T}$ is an irreducible tableau, v is defined as in (9.241) and with marks $\tilde{v}$ given by the multiroot tree (recall (9.286) in the single root case). Then $(\tau, \mathcal{Y})$, the tableau and the tuple of nested subintervals describing the migration is determined from v.

Let

$$\widehat{\mathfrak{T}}^* \quad (9.851)$$

denote the set of all possible *compound mesostates with possibly multiple roots* $(j_1, \ldots, j_k)$, the latter taking care of collisions. The elements have the form

$$(\tau, v, \tilde{v}_1, \ldots, \tilde{v}_k), \quad \tau \in \mathfrak{T}, \quad v \in \Upsilon, \tilde{v}_i = Geo(j_i, v) \quad (9.852)$$

where $\tilde{v}_i = Geo(j_i, v)$ denotes the set of vertices (and subgraph) which are descendants of the root j_i. In the next subsubsection the multiroot mesostate will be decomposed into single root marked mesostates and the mappings $\tilde{v}_i, i = 1, \ldots, k$ will be replaced by the mapping $\daleth$ (see (9.874)).

Now we have to define the Markov jump dynamics of

$$\mathfrak{s}^N_{l,\text{enr}} \text{ on } (\widehat{\mathfrak{T}}^*)^{\mathbb{N}_\ell}. \quad (9.853)$$

The transitions involving collision and coalescence include:

- As in the collision-free case at a decoupling event the mesostate at a *Pe* site can fragment into one or more new mesostates with different *Pe* founding sites.
- *Collision of two labels* with their *mesostates*, say with labels ℓ_1, ℓ_2, followed by a *coalescence* between a rank from each label at the collision site. When this occurs we can introduce a *compound label* (ℓ_1, ℓ_2) mesostate. Similarly higher order compound mesostates form.
- If the site at which the collision occurs is later removed by resolution at a rank at one of the *Pe* roots, then the *compound mesostate fragments* into two separate labels and mesostates.
- The first rank at a *Pe* site can migrate and hit another *Pe* site. If coalescence occurs this means that the *Pe* site is eliminated, that is, the *Pe sites are no longer immortal*, but can *die in a collision-coalescence event*.
- Consider ranks which due to collision and coalescence belong to *two or more* labels and mesostates. In this case a transition at this rank can result in a

simultaneous change in the corresponding mesostates, i.e. in the compound mesostate.

The transitions and rates arise as before in the case on $\mathbb{N}$ from the evolution mechanism of $\tilde{\mathcal{G}}^{++,N}$ and we don't write this out explicitly again. In this case when migration occurs the new site is chosen at random in $\{1,\ldots,N\}$ and we now have $\mathfrak{R}(Geo) \subset \{1,\ldots,N\}$.

Note that from Lemma 9.53(a) when the density of sites occupied by mesostates is bounded by some $u_0 > 0$ we can show that the first two moments of the number of roots of compound mesostates remain bounded. Thus the system of interacting mesostates can then be viewed as a *coagulation-fragmentation process* in which the objects are mesostates with possibly multiple roots.

We next consider the empirical measure process of mesostates associated to this *clan coagulation-fragmentation process* and identify the limiting nonlinear equations which are valid for the limiting $N \to \infty$ empirical proportions. The feature of this limit is that two tagged mesostates will not overlap. Hence we can drop some information in the limiting equation and we only have to reintroduce this information when we study at time $\beta^{-1} \log N + t$ the $N \to \infty, t \to -\infty$ asymptotics.

Define

$$\sigma : \hat{\mathfrak{T}}^* \to \mathcal{T} \times \Upsilon, \qquad \sigma((\tau, v, \tilde{v})) = (\tau, v). \tag{9.854}$$

Convention *We now reuse the symbols from Subsections 9.10 and 9.11 for the analogous objects but of course they are different in that the state space is now $\hat{\mathfrak{T}}^*$, etc.*

The empirical process is given by

$$\bar{\Psi}_t^N(\cdot) = \bar{\Psi}_t^N(\mathfrak{s}_{l,\mathrm{enr}}^N)(\cdot) := \sum_{j \in \mathbb{N}_\ell} 1(\sigma(\hat{\mathfrak{t}}_j(t)) \neq (\mathbb{I}^\mathbb{N}, \emptyset)) \cdot \delta_{\sigma(\hat{\mathfrak{t}}_j(t))}(\cdot) \in \mathcal{M}_{\mathrm{fin}}(\sigma(\hat{\mathfrak{T}}^*)), \tag{9.855}$$

but the empirical measure is now (different from the $N = \infty$ case) over the set $\sigma(\hat{\mathfrak{T}}^*)$ of mesostates with possibly multiple roots.

Note that for fixed N the process $\Psi_t^N(\cdot)$ is *not Markov* since we must include the information coded by $\tilde{v}$ to have a *markovian* description. We will proceed as before and consider the martingale problem associated to $\mathfrak{s}_{l,\mathrm{enr}}^N(\cdot)$ and then derive limiting equations which arise via a propagation of chaos property. However we only consider the existence and uniqueness of these equations in the emergence regime and leave open the possibly complex behaviour that can occur as the proportion of sites occupied increases beyond a threshold, since it is not needed for our arguments.

For $\mathfrak{s}_{l,\mathrm{enr}}^N \in (\hat{\mathfrak{T}}^*)^{\mathbb{N}_\ell}$,

$$\bar{u}^N(t) = |\mathfrak{s}_{l,\mathrm{enr}}^N(t)| = |\tilde{\mathcal{G}}_t^{N,++}| = \Psi_t^N(\mathfrak{s}_{l,\mathrm{enr}})(\sigma(\hat{\mathfrak{T}}^*)) \quad (\text{number of (active) labels}), \tag{9.856}$$

9.12 Emergence and Fixation: The Selection-Dominant Case

$$\bar{v}^N(t) = \sum_{i=1}^{|\mathfrak{s}_{l,\mathrm{enr}}^N(t)|} |\hat{\mathfrak{t}}_i| \quad (\text{number of active sites}), \tag{9.857}$$

$$\bar{U}^N(t)(\cdot) = \frac{\Psi_t(\cdot)}{\Psi_t(\sigma(\hat{\mathfrak{T}}^*))} \quad (\text{normalized empirical measure of compound mesostates}). \tag{9.858}$$

These quantities are the ones we need to study in the $N \to \infty$ limit to analyse emergence. Hence we need to obtain limiting equations.

Step 2 *(The nonlinear equations in the collision regime)*

We must deal with the terms involving the collision and subsequent coalescence of mesostates. The first observation is that the rate at which a label with mesostate $\hat{\mathfrak{t}}_1$ can produce a migrant which collides with a second label with mesostate $\hat{\mathfrak{t}}_2$ (and consequently add a point to the support of the first label which is taken from the support of the second label) is proportional to $\|\hat{\mathfrak{t}}_1\| \cdot \frac{1}{N}$ and that if this occurs the rank which moves is randomly chosen and the site in the support of $\hat{\mathfrak{t}}_2$ to which it moves is randomly chosen.

Due to the collision-coalescence behaviour the first question is whether the number of mesostates continues to grow or even the entire dual population coagulates into a bounded number of highly complex mesostates so that there is no propagation of chaos, which we need to establish in order to get convergence to a *deterministic* limiting equation. Our objective here is to show that for as long as the density of occupied active sites in $\tilde{\mathcal{G}}^{++,N}$ remains bounded by some $u_0 > 0$ the limit as $N \to \infty$, $(\bar{u}^{N,Pe}, \bar{U}^{N,Pe}) \to (\bar{u}^{Pe}, \bar{U}^{Pe})$ exists and satisfies deterministic limiting equations to which the limit is a random entrance law.

Here we formulate the nonlinear equations for our limit dynamics in the *collision regime* below in (9.871) and (9.872). In order to do so we need to introduce some quantities, more specifically the form of transitions and corresponding rates to occur.

The rate of change due to collision coalescence events between mesostates giving *coagulation* producing from mesostates $\hat{\mathfrak{t}}', \hat{\mathfrak{t}}''$ the new mesostate $\hat{\mathfrak{t}}$ is given by a coagulation kernel R^+:

$$R^+(\hat{\mathfrak{t}}', \hat{\mathfrak{t}}''; \hat{\mathfrak{t}}), \text{ on } \sigma(\hat{\mathfrak{T}}^*) \times \sigma(\hat{\mathfrak{T}}^*) \longrightarrow \sigma(\hat{\mathfrak{T}}^*), \tag{9.859}$$

where $\hat{\mathfrak{t}}'$ is a mesostate of a ℓ_1-root label and $\hat{\mathfrak{t}}''$ a mesostate ℓ_2-root label which coagulate to form a (ℓ_1, ℓ_2)-root label with mesostate $\hat{\mathfrak{t}}$.

The related *fragmentation* operator R^- producing from the multi-root mesostate $\hat{\mathfrak{t}}'$ the new mesostate $\hat{\mathfrak{t}}$

$$R^-(\hat{\mathfrak{t}}', \hat{\mathfrak{t}}) \quad , \text{ from } \sigma(\hat{\mathfrak{T}}^*) \longrightarrow \sigma(\hat{\mathfrak{T}}^*), \tag{9.860}$$

gives the rate at which a multiroot label with mesostate $\hat{\mathfrak{t}}'$ produces (among others) a label with mesostate $\hat{\mathfrak{t}}$ by fragmentation.

Recall that the rate of production of new *Pe* sites is given by the aggregated (over all ages and mesostates) production rate in the total population at time $t \in \mathbb{R}$ arising from decoupling or through fragmentation. Here is the list of rates (the explanation in brackets):

$$\bar{\beta}_{\hat{t}}(t) = \int_{\mathcal{P}(\hat{\mathfrak{T}}*)} \int_0^\infty Q^D(\hat{t}', \hat{t}) \bar{U}(t, ds, d\hat{t}'), \quad t \in \mathbb{R}, \tag{9.861}$$

(rate of production of label with mesostate $\hat{t}$ at a *decoupling* event).

$$\bar{\beta}(t) = \sum_{\hat{t}} \bar{\beta}_{\hat{t}}(t) \quad \text{(total production rate due to } \textit{decoupling}\text{).} \tag{9.862}$$

Next we have in addition (due to compound mesostates) production by fragmentation rates:

$$\bar{\beta}_{\hat{t},2}(t) = \int \int R^-(\hat{t}', \hat{t}) \bar{U}(t, ds, d\hat{t}') \quad (\hat{t}\text{-production due to } \textit{fragmentation}), \tag{9.863}$$

$$\bar{\beta}_2(t) = \sum_{\hat{t}} \bar{\beta}_{\hat{t},2}(t) \quad \text{(total production rate due to } \textit{fragmentation}\text{).} \tag{9.864}$$

The rate at which labels with mesostates $\hat{t}$ are *removed* is given by:

$$\bar{\gamma}(\hat{t}) = \bar{\gamma}_D(\hat{t}) + \bar{\gamma}_F(\hat{t}) + \bar{\gamma}_C(\hat{t})\bar{u}(t), \tag{9.865}$$

where

$$\bar{\gamma}_D(\hat{t}) = \sum_{i,j} m_{ij} 1(\hat{t}_{ij} = (1, \ldots, 1, 0)), \tag{9.866}$$

with

$\hat{t}_{ij}$ denotes the result of a *decoupling event* due to a mutation from i to j, (9.867)

$\bar{\gamma}_C(\hat{t}) =$ rate at which $\hat{t}$ is removed due to *coagulation* at a *collision event*, (9.868)

$\bar{\gamma}_F(\hat{t})$ rate at which $\hat{t}$ is removed due to a *fragmentation event*. (9.869)

We set finally

$$\bar{\gamma}_D(t) = \int \int_0^\infty \bar{\gamma}_D(\hat{t}) \bar{U}(t, ds, d\hat{t}), \text{ similarly } \bar{\gamma}_C(t), \quad \bar{\gamma}_F(t). \tag{9.870}$$

Then the $(\bar{u}, \bar{U})$-equation reads (the terms are explained below in brackets and the components of these terms are explained in further detail after the equation):

9.12 Emergence and Fixation: The Selection-Dominant Case

$$\frac{\partial \bar{u}}{\partial t} = \bar{\beta}(t)\bar{u}(t) + \bar{\beta}_2(t)\bar{u}(t) - \bar{\gamma}_D(t)\bar{u}(t) - \gamma_F(t)\bar{u}(t) - \gamma_C(t)\bar{u}^2(t), \quad t \in \mathbb{R}, \tag{9.871}$$

$$\frac{\partial \bar{U}(t, ds, \hat{\mathfrak{t}})}{\partial t} =$$

$$-\frac{\partial \bar{U}(t, ds, \hat{\mathfrak{t}})}{\partial s} \quad \text{(aging)}$$

$$+ \sum_{\hat{\mathfrak{t}}' \in \hat{\mathfrak{T}}^*} \bar{U}(t, ds, \hat{\mathfrak{t}}') Q^I_{\hat{\mathfrak{t}}', \hat{\mathfrak{t}}} \quad \text{(mesostate dynamics)}$$

$$+ \sum_{\hat{\mathfrak{t}}' \in \hat{\mathfrak{T}}^*} U(t, ds, \hat{\mathfrak{t}}') Q^D(\hat{\mathfrak{t}}', \hat{\mathfrak{t}}) 1_{(s, \hat{\mathfrak{t}})}(0, \hat{\mathfrak{t}}') - \bar{\gamma}_D(\hat{\mathfrak{t}}) \bar{U}(t, ds, \hat{\mathfrak{t}}) \quad \text{(decoupling)}$$

$$+ \bar{u}(t) \left[\sum_{\hat{\mathfrak{t}}', \hat{\mathfrak{t}}''} \int \int \bar{U}(t, ds_1, \hat{\mathfrak{t}}') \bar{U}(t, ds_2, \hat{\mathfrak{t}}'') R^+(\hat{\mathfrak{t}}', \hat{\mathfrak{t}}''; \hat{\mathfrak{t}}) 1_{s=(s_1, s_2)} ds_1 ds_2 \right]$$

(coagulation)

$$-\bar{u}(t)\bar{\gamma}_C(\hat{\mathfrak{t}})U(t, ds, \hat{\mathfrak{t}}) \quad \text{(loss of mesostates due to coagulation)}$$

$$+ \sum_{\hat{\mathfrak{t}}' \in \hat{\mathfrak{T}}^*} \bar{U}(t, ds, \hat{\mathfrak{t}}') R^-(\hat{\mathfrak{t}}', \hat{\mathfrak{t}}) - \bar{\gamma}_F(\hat{\mathfrak{t}}) \bar{U}(t, ds, \hat{\mathfrak{t}}) \quad \text{(fragmentation)}$$

$$-[\bar{\beta}(t) + \bar{\beta}_2(t) - \bar{\gamma}_D(t) - \gamma_F(t) - \bar{\gamma}_C(t)\bar{u}(t)]\bar{U}(t, ds, \hat{\mathfrak{t}}) \quad \text{(normalization)}. \tag{9.872}$$

Explanation of terms in the nonlinear equation.

- On the r.h.s. in (9.871):
  - The $\bar{\beta}(t)\bar{u}(t)$ represents the average rate at which the current set of *Pe* sites produce new *Pe* sites by a decoupling event.
  - The $\bar{\beta}_2\bar{u}(t)$ represents the rate of creation of *Pe* sites due to a fragmentation of a multiroot label and its mesostate.
  - The $(\bar{u}(t))^2 \gamma_C$ represents the rate of removal of labels and their mesostates by coagulation, when labels are lost if two ranks in different mesostates located at the same site coalesce and form a new two multiroot label mesostate.
  - Finally $\bar{\gamma}_F$ is the rate of removal when a multiroot label with mesostate fragments produces two or more simpler mesostates.

- On the r.h.s. in (9.872):
  - the first term indicates that the age of the site (time since current founder arrived at the site) increases at the same constant rate.

- The second term corresponds to the changes due to selection, coalescence, mutation and migration and involve only one label more precisely its mesostate.
- The third term represents the creation of new *Pe* sites produced at the decoupling time of an existing *Pe* site and the loss the rate at which a mesostate decouples.
- The fourth term corresponds to the effect of production of a new multiroot labels with their mesostates at a coagulation event.
- The fifth gives the loss of labels with their mesostates due to coagulation.
- The sixth term give the rate at which *Pe* sites are produced when a multiroot label and its mesostate fragments and the corresponding loss rate of multiroot labels.
- The last term comes from the changes due to the normalization when the total number $\bar{u}(t)$ increases.

With these equations one can determine the $t \to -\infty$ behaviour.

We will not consider a formal convergence statement and proof of the system (9.871), (9.872) in this subsubsection but prove a stronger result and for that and other purposes introduce in the next subsubsection a much more refined approach. This will be based on an estimate of the order of the number of ranks belonging to more than two labels and to show it is negligible. To do so we analyse the number of ranks belonging to two labels using coloured ranks again. As before green will stand for the difference of the collision-free and collision case. For the green coloured ranks we then have to store in addition the information about the other involved labels with their mesostate. These colours then give us labels with *marked* mesostates, whose asymptotics we study in the next subsubsection.

9.12.10 *The Marked (Coloured) Mesostate Process*

To carry out the analysis of the transition regime between droplet and emergence we need to determine the asymptotic behaviour of the quantities $E[x_{E_1}^N(\beta^{-1}\log N + t)]$ and $E[(x_{E_1}^N(\beta^{-1}\log N + t))^2]$ as $N \to \infty, t \to -\infty$. This analysis will be carried out using the dual representation and we must now consider the effect of collisions in the dual dynamics in the transition regime when the rate of collision becomes small. This requires us to consider the process $\mathfrak{s}_{l,\text{enr}}^N$ in the time scale $\beta^{-1}\log N + t$ for $N \to \infty$ and the asymptotics $t \to -\infty$ later on. Therefore we must first construct more formally the appropriate *dual process with collisions* and give an appropriate description of the state of the system as well as the form of the micro-, meso- and macro-states in this case.

The principal asymptotic term in the calculation of $E[(x_{E_1}^N(\frac{\log N}{\beta} + t))^\ell]$ for $\ell = 1, 2$ and as first, then $N \to \infty, t \to -\infty$ is the result of the *additional rare mutation opportunities* lost due to the effects of collision with subsequent coalescence. The latter are described following the ideas of Subsection 8 by the

9.12 Emergence and Fixation: The Selection-Dominant Case

green ranks introduced below which grow as const $\cdot e^{2\beta t}$. We can show that the set of ranks in compound labels that belong to *more than* two mesostates are in the $N \to \infty$, $t \to -\infty$ limit of order $\leq O(e^{3\beta t})$ and are negligible in the $t \to -\infty$ limit. Similarly the set of mesostates having collisions between internal sites or two or more collisions with the *same mesostate* are therefore negligible in the limit. As a result to describe the effect of collisions on the asymptotic expressions we can focus on collisions between pairs of mesostates (but a given mesostate can have collisions with more than one other mesostate).

Using the set-valued dual, the transition regime again involves the asymptotic behaviour of the limiting (as $N \to \infty$) *nonlinear equations* as $t \to -\infty$. As in Section 8 we isolate the leading asymptotic terms in powers of $e^{\beta t}$ reintroducing the information about the *spatial overlap* of mesostates in the finite N system in the form of a *random map* ⊓ and by using an analogue of the *coloured particle system* to control the effect of collisions between mesostates or to control the effect of collisions between two dual clouds (needed with second moment expressions).

The map ⊓ identifies common ranks of overlapping mesostates and their locations in $\{1, \ldots, N\}$. We also introduce the analog of the coloured particle system of Section 7 but now identify coloured ranks and mesostates since now particles are replaced by entities having an internal state, namely, the ranks with their partitions of $\mathbb{I}_M$ and migration information. In particular the proof of (9.949) requires the analogue of the "green particles" introduced in Section 8, which is now given by *transitions in compound mesostates* which form upon collisions. We will see that only *single collisions* between mesostates are (asymptotically) relevant for us.

We construct a modification of the label-mesostate process by introducing the process of labels with *coloured mesostates* and *coloured ranks* with colours *orange, red, green* and *uncoloured*. What is hence new (compared to Section 8) is the use of the colour orange. In particular, the green marks allow us to compare the finite N system with the collision-free system on $\mathbb{N}$ discussed earlier in Subsubsections 9.12.1–9.12.8 and the corresponding empirical measures of marked mesostates in the $N \to \infty$ limit leads to nonlinear dynamics in the collision regime. We first sketch here the form, a more detailed and formalized description appears further below.

When a rank at site i_1 from one mesostate $\hat{t}$ with founding Pe site j_1 migrates to a site i_2 and collides with a second mesostate $\hat{t}'$ with founding Pe site j_2 so that $\Re(Geo)(\hat{t}) \cap \Re(Geo)(\hat{t}')$ is non-empty, a copy of this rank now at site i_2 is kept at its parent site i_1 and is coloured *orange* to indicate it can lead to a fusion of labels. If this rank at the new site then subsequently coalesces with a rank of the j_2-mesostate at an overlap site, a compound mesostate with two roots (j_1, j_2) is formed, the ghost rank is coloured *red* at the parent site i_1 (in the (j_1, j_2)-compound mesostate) and an additional *green ghost copy* is started at this rank and site which evolves with the *collision-free dynamics* on an *extra copy of* $\mathbb{N}$. The green rank and its descendants are removed if this rank is eliminated by resolution at *either parent label*. When a green rank migrates, its offspring ranks are also green and this produces a cloud of green sites and ranks. If a green rank becomes permanent after decoupling in both parent mesostates, it forms a *green mesostate* which grows at rate β.

The map $\daleth^N$ augments the coloured process and identifies ranks in pairs of mesostates which have the same location in space and therefore evolve in a coupled fashion. Note at this point compound mesostates are still irreducible factors of $\tilde{\mathcal{G}}^{++,N}$ but the components can only be characterized following the history of the evolution of the dual process, which is then stored in $\daleth^N$.

In our analysis below we use the asymptotic term given by the green ranks and mesostates as $t \to -\infty$ in the time scale $\frac{\log N}{\beta} + t$. To do this we must consider:

- Green mesostates and ranks and the asymptotics of the nonlinear equation up to terms of order $O(e^{3\beta t})$ as $t \to -\infty$. The key idea is that we again focus only on ranks involved in at most *single collisions* and then obtain an upper bound on the error resulting from omitting ranks involved in multiple collisions.
- Asymptotic growth of green mesostates and ranks and limiting constant in front of the exponential.

Outline. In the sequel we proceed in seven steps, we give the dynamics of a marked mesostate process, discuss the $N \to \infty$ limit of the dynamic, introduce the empirical distributions of microstates from the one for mesostates, work out the limiting equations, consider the dynamic in the time scale $\beta^{-1} \log N + t$, $t \in \mathbb{R}$ for $N \to \infty$ and then $t \to -\infty$ and finally study quantitatively the effect of collisions, first to compare finite N and the $N = \infty$ dynamic and second to do this for two dual clouds preparing second moment calculations.

Step 1: *The empirical process of marked (coloured) mesostates*

Given a compound mesostate $\hat{\mathfrak{t}}^* \in \hat{\mathfrak{T}}^*$ having k roots we can decompose it uniquely into k single root marked mesostates $\{\tilde{\mathfrak{t}}_1, \ldots, \tilde{\mathfrak{t}}_k\}$ with marks given by coloured ranks (that indicate that they are coupled to ranks in other single root mesostates) and the internal states of the corresponding single root mesostates - but *without* the identification of the geographic locations in $\{1, \ldots, N\}$. To completely specify the set-valued process we must also specify the set of ranks which belong to two (or more) mesostates (and which couple the dynamics of the pairs of single root mesostates). Then starting with the empirical measure of compound mesostates $\Psi_t^N(\cdot)$ defined in (9.855) (recall that this is a process with state space $\mathcal{M}_{\mathrm{fin}}(\sigma(\hat{\mathfrak{T}}^*))$) the set-valued dual is then described by a pair of objects:

- an empirical process

$$\Psi^{N,\mathrm{mark}} = \sum_j \left(\sum_{i=1}^k \delta_{\sigma(\hat{\mathfrak{t}}_{j,i})} \right) \qquad (9.873)$$

(where the sum on j is over the compound mesostates) of single root marked mesostates ($\Psi^{N,\mathrm{mark}} \in \mathcal{M}_{\mathrm{fin}}(\tilde{\mathfrak{T}}_{\mathrm{mark}})$ where $\tilde{\mathfrak{T}}_{\mathrm{mark}}$ is defined below in (9.880)),
- a mapping

$$\daleth^N \qquad (9.874)$$

9.12 Emergence and Fixation: The Selection-Dominant Case

that identifies the labels of the pairs of coupled single root mesostates, the ranks involved in the coupling and their geographic locations in $\{1, \ldots, N\}$.

Remark 200. Here we consider only the *pairwise interaction* of two mesostates, the set of the rare interactions listed above produce an error term which is negligible in the $N \to \infty$, $t \to -\infty$ limit.

Step 2: *State space and Markovian dynamics of* $(\Psi_t^{N,\text{mark}}, \daleth_t^N)$
The pair

$$(\Psi_t^{N,\text{mark}}, \daleth_t^N)_{t \geq 0}. \tag{9.875}$$

has Markovian dynamics with transitions given by (a more formal description of states and state space follows below (9.876)).

- When a rank migrates to a site occupied by a rank of another mesostate, the corresponding marked mesostate is assigned a ghost copy at the parent site marked *orange* and the migrant gets the next free local rank at the new site. Transitions occurring at this rank at the new site and its offspring are coupled to the corresponding transitions in the orange ghost. Offspring of orange sites are orange.
- If one of the resulting coupled migrant ranks then coalesces with an active rank at the new site, then the corresponding orange rank to which it is coupled at the parent site is *remarked red*, and an additional green ghost rank is added at the parent site with the same local rank as the red ghost copy but with independent dynamics. The offspring of a green rank is green and the offspring of red ghosts are red ghost ranks.
- When a green marked rank is added one potential migrant with a certain internal state has been lost due to coagulation. Once a label arises with a mesostate containing a green rank it will develop a green family of ranks attached to the mesostate of the concerned label. A decoupled green family produces then a green *Pe* mesostate.
- We follow green ranks until they are decoupled from their parent sites - then they are immortal.
- When a green rank migrates it moves to a copy of $\mathbb{N}$ and follows the normal collision-free dynamics. If *both* the green founding rank as well as the associated (at collision site) rank is *decoupled S*, the green family becomes *immortal*.
- Effect of collision-coalescence - See Fig 9.1.
    - Rank from site j_1 migrates and lands on site i_1
    - Rank at site j_1 becomes orange (and mirrors the corresponding rank at site i_1)
    - After coalescence at site i_1 orange rank becomes red and a green rank (starting with identical column) is produced on another copy of $\mathbb{N}$
    - Cut by mutation (i.e. when the mutation dynamics at the rank hits a trap) at site i_1 (to the left of the coupled rank) red rank removed and green rank becomes clear

- Cut by mutation at site j_1 (to the left of the coupled rank) green and red ranks removed
- Decoupled successfully from both sites (i.e. rank corresponds to a successful (S) selection event) green rank becomes permanent

Next we observe some effects occurring in this coupled system:

- In the $N \to \infty$ and then $t \to -\infty$ limit we view the collision with one or more additional migrants from other Pe sites as a *higher order error term*.
- Multiple simultaneous transitions: this means a rank coupled in 3 or more mesostates (Pe sites), that is, more than one green particle is created by a migrating rank. In the $N \to \infty, t \to -\infty$ limit this involves a *higher order term* and doesn't contribute to the covariance calculation we need to perform to study $\mathcal{L}[^*\mathcal{W}]$.
- At a collision event a second mesostate is effectively selected at random and the pair is followed until the collision is resolved by decoupling, this includes the creation of a green family at a coalescence event,
- In the *propagation of chaos limit* this results in choosing at random the second mesostate involved and running both up to resolution - in fact it suffices to choose a single site at random with microstate type and lifetime chosen from $\bar{V}(t)$.
- A mesostate can have have more than one green family of ranks, but the contribution of the green ranks arising from a second collision involving the mesostate produces a correction term which is *negligible* in the $N \to \infty, t \to -\infty$ limit.
- A key point is that green sites (and ranks) are produced in the $N \to \infty$ limit at rate proportional to $W_{Pe}^2 e^{2\beta t}$, for $t \in \mathbb{R}$ and $t \to -\infty$ (recall Proposition 9.44).

To formalize the construction of the *marked mesostate process* we proceed as follows. We associate with a *ghost rank* at a site in a mesostate colours from the set

$$\mathcal{C} = \{ \text{orange, green, red, non-coloured}\}. \tag{9.876}$$

The green ranks will after decoupling follow a collision-free dynamic on $\mathbb{N}$. We add orange-red-green marks according to the previously given rules. Furthermore a *green mesostate* is given by an additional new label starting from a decoupled green rank on the extra copy of $\mathbb{N}$.

Remark 201. Note that now also red (or orange ranks) could collide with another orange/red rank and in principle as in Section 8 we should need to complement orange/red by purple and green by blue ranks, etc. This however results in higher order terms we do not consider at this point.

The set of mesostates with in addition coloured ranks is denoted

$$\hat{\tilde{\mathfrak{T}}}_{\text{col}}. \tag{9.877}$$

Therefore we have now a dynamics of a collection of mesostates involving ranks marked with colours from $\mathcal{C}$ and with a collection of *"green mesostates"*. However

9.12 Emergence and Fixation: The Selection-Dominant Case

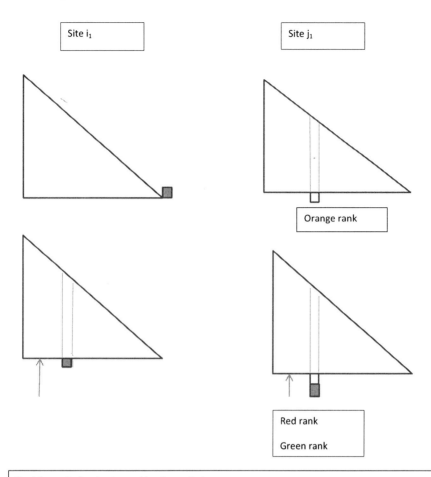

Fig. 9.1 Formation of a new permanent green rank

since the mesostates may be part of compound mesostates, we need a mesostate state space which includes the *collision and fusion information*. This is needed to obtain a Markov process and to describe the green mesostates since the green ranks and mesostates measure the effects which are important for the limiting (in time scale $\beta^{-1} \log N + t$, as $N \to \infty$, $t \to -\infty$) dynamics.

We now formally describe the state of the system.

A *marked mesostate* is given by

$$\tilde{\mathfrak{t}} = (\hat{\mathfrak{t}}, \text{mark}), \text{ where } \hat{\mathfrak{t}} \in \hat{\mathfrak{T}}_{\text{col}} \text{ and:} \quad (9.878)$$

and

$$\text{mark} = \quad (9.879)$$

- $\emptyset$ if $\hat{\mathfrak{t}}$ contains no coloured sites,
- gr if the mesostate is green,
- a list of tuples each of the form $(\ell_1, j_1, (\hat{\mathfrak{t}}', \ell_2), \tau_{or})$ if rank j_1 at site ℓ_1 has migrated and landed on site ℓ_2 in mesostate $\hat{\mathfrak{t}}'$ and has subsequently produced the orange ghost subtableau τ_{or},
- a list of tuples each of the form $(\ell_1, j_1, j_1', (\hat{\mathfrak{t}}', \ell_2, j_2), \tau_{or}, \tau_{gr})$ if rank j_1 at site ℓ_1 has migrated and landed on site ℓ_2 and rank j_1' associated with τ_{or} coalesced with rank j_2 in mesostate $\hat{\mathfrak{t}}'$,

changed the orange rank j_1' and tableau τ_{or} to τ_r to red and

produce the green subtableau τ_{gr} or $(\ell_1, j_1, j_1', j_1'', (\hat{\mathfrak{t}}', \ell_2, j_2, j_2'), \tau_r, \tau_{gr}, \tau_{gr}') \cdots$,

etc.

Note that a mesostate can have a mark referring to more than one rank if it is coupled to more than one other mesostate. The set of marks is denoted $\mathfrak{M}$.

The *green subtableau* represent sites and ranks which would have been present if collisions had not occurred. The dual system can then be represented embedded in a copy of the collision free system with the difference being the green sites and ranks.

The complete coloured system with marks and with green mesostates has as state space

$$\tilde{\mathfrak{T}}_{\text{mark}} = \{\tilde{\mathfrak{t}} = (\hat{\mathfrak{t}}, \text{mark}) | \hat{\mathfrak{t}} \in \hat{\mathfrak{T}}_{\text{col}}, \text{ mark} \in \mathfrak{M}\}. \quad (9.880)$$

To complete the dynamics we must specify $\daleth^N$. Given a finite set $\tilde{\mathfrak{t}}_{\text{mark},1}, \ldots, \tilde{\mathfrak{t}}_{\text{mark},k}$ of elements in $\tilde{\mathfrak{T}}_{\text{mark}}$ the mapping $\daleth^N$

- assigns a location in $\{1, \ldots, N\}$ to the each of the vertices of each $\tilde{\mathfrak{t}}_{\text{mark},i}$, $i = 1, \ldots, k$.

9.12 Emergence and Fixation: The Selection-Dominant Case

This corresponds to the mapping *Geo* in the collision-free case but in which the $\Re(Geo)(\tilde{t}_{\text{mark},i})$ are not disjoint and must satisfy a consistency requirement for each pair $(\tilde{t}_{\text{mark},i_1}, \tilde{t}_{\text{mark},i_2})$ which are coupled via an overlap.

Step 3: *The empirical measure of microstates*

It turns out that if we want to determine the limiting evolution equations of the empirical distribution of the marked mesostates, it is necessary to deduce from the mesostates information about the empirical distribution of marked microstates. Indeed we can now extract from the marked system the *empirical measure* of microstates at sites and their lifetimes. We therefore need the marked tableau associated with a site. Define first

$$\mathcal{T}_{\text{mark}} = \{(\tau, mark')\}, \tag{9.881}$$

where $\tau \in \mathcal{T}_{\text{col}}$ and

$$mark' = \bullet \; \emptyset \quad \text{if the site contains no coloured ranks,} \tag{9.882}$$

- gr if the site is green,

- $((j_1, \tau'), \tau_{or})$ if rank j_1 has migrated and landed on a site with state τ' and produced the orange ghost subtableau τ_{or},

- $(j_1, j'_1, (\tau', j_2), \tau_{or}, \tau_{gr})$ if rank j_1 has migrated and landed on rank $j'_1 \in \tau_{or}$ coalesced with rank j_2 in mesostate $\hat{t}'$, at a site with subtableau τ', changed the orange rank to red and produce the green subtableau τ_{gr}.

Given the empirical process of marked mesostates $\Psi_t^{N,\text{mark}}$ we next consider the *empirical measure* of *microstates and their random lifetimes*. To do this we introduce a random mapping

$$\theta_{\text{mark}} : \tilde{\mathfrak{T}}_{\text{mark}} \to \mathcal{N}(\mathcal{T}_{\text{mark}} \times (0, \infty]). \tag{9.883}$$

In order to specify the random mapping θ_{mark} recall that a site occupied by a mesostate can be removed by mutation at an ancestor site and therefore the site and its microstate can have a finite lifetime. Assuming for the moment that this mesostate evolves independently from other mesostates, then the distribution of the lifetime of the considered site (and microstate) is a measurable function of the mesostate. In general, this remains true if we replace the mesostate by the marked mesostate. A microstate at an occupied site from a mesostate at a *Pe* site is obtained by suppressing the information on the other sites in the mesostate. Then we give it a random lifetime with lifetime distribution conditioned on the full mesostate to which it belongs.

We can now define the our new object, the empirical counting measure of microstates and their lifetime:

$$\tilde{\Psi}_t^N(\cdot,\cdot) := \theta_{\text{mark}}(\Psi_t^{N,\text{mark}}) \in \mathcal{M}_{\text{fin}}(\mathcal{T}_{\text{mark}} \times (0,\infty]). \tag{9.884}$$

In order to identify the limiting empirical measures as $N \to \infty$ and the limiting nonlinear equations (9.898) and (9.899) below for $(\bar{u}, \bar{U})$ we next introduce the relevant objects. We first introduce here the finite N objects which are given by functionals of $\tilde{\Psi}_t^N(\cdot,\cdot)$. The corresponding objects in the $N \to \infty$ scaling limit are denoted by dropping the N. Let

$\bar{u}^N(t) =$ number of marked mesostates *excluding green* mesostates at time t.
$\tag{9.885}$

$\bar{u}_g^N(t) =$ number of *green* mesostates at time t. $\tag{9.886}$

$\bar{U}^N(t) =$ the empirical type distribution of the set of marked mesostates,
but *excluding green* mesostates and dropping *green ranks*.
$\tag{9.887}$

$\bar{U}_g^N(t) =$ empirical type distribution of *green* mesostates. $\tag{9.888}$

$\bar{v}^N(t) =$ number of sites *not* containing containing *orange, red* or *green* ranks
at time t. $\tag{9.889}$

$\bar{v}_{og}^N(t) =$ number of sites containing *orange, red* or *green* ranks at time t. $\tag{9.890}$

$\bar{V}^N(t) =$ the *empirical type* and *lifetime distribution* of the set of occupied sites,
not containing *orange-red-green* ranks. $\tag{9.891}$

$\bar{V}_{og}^N(t) =$ empirical type and lifetime distribution $\tag{9.892}$
of sites containing orange, red or green ranks.

Recall that we know already from Proposition 9.44 that ignoring collisions we would have

$$\bar{u}^N(t) \sim W_{P_e} e^{\beta_{P_e} t} \tag{9.893}$$

so that the time scale we want to look at is indeed $\beta^{-1} \log N + t$.

9.12 Emergence and Fixation: The Selection-Dominant Case

Step 4: *Limiting equations*

We now need to specify the equations satisfied by the limit object

$$(\bar{u}^N(\beta^{-1}\log N + t), \bar{U}^N(\beta^{-1}\log N + t))_{t\in\mathbb{R}} \Longrightarrow (\bar{u}(t), \bar{U}(t)), \text{ as } N \to \infty. \tag{9.894}$$

However as before, our primary objective is to identify the first and second order asymptotic terms of $(\bar{u}, \bar{U})$ as $t \to -\infty$.

We first describe the *propagation of chaos* effect in this setting. The idea is that in the limiting case a collision initiated by a migration event from a given mesostate can be described by choosing an *associated site* at random from the current distribution of microstates and lifetimes and then considering the future development when the migrating rank hits the associated site. The probability that a migration event results in such a collision is given by the current proportion of occupied sites $u(t) = O(e^{\beta t})$ as $t \to -\infty$. Then let the dynamics proceed for this mesostate. Create the coupled green copy and track migrations of green ranks that don't belong to the current mesostate. The number of descendant green sites grows as $e^{\beta t}$. As a result the number of green sites grows as $e^{2\beta t}$. Consider empirical measure of microstates and lifetimes at all green sites.

Let $\tilde{\gamma}_C(t)$ represents the loss of a mesostate due to collision-coalescence of the founding rank and site of the mesostate with another mesostate and $\tilde{\gamma}_D(t)$ the change due to decoupling, furthermore

$$Q^C_{\tilde{t}',j,(\tau,x);\tilde{t}}, \quad j \in 1,\ldots,\|\tilde{t}'\| \tag{9.895}$$

= rate at which a rank j in mesostate $\tilde{t}'$ migrates and collides with a site

having microstate and lifetime (τ, x) producing the marked mesostate

$\tilde{t} = (\tilde{t}', j, (\tau, x))$.

We must also obtain the limiting $\bar{v}(t), \bar{v}_g(t), \bar{V}(t), \bar{V}_g(t)$ which we have to extract from $(\bar{u}, \bar{U})$.

$$\bar{v}(t) = \bar{u}(t) \sum |\tilde{t}| \bar{U}(t, \tilde{t}) \tag{9.896}$$

and

$$\bar{V}(t, \cdot, \cdot) := \int \theta(\tilde{t}, \cdot, \cdot) \bar{U}(t, d\tilde{t}). \tag{9.897}$$

The equation for $(\bar{u}(t), \bar{U}(t, \cdot))_{t\in\mathbb{R}}$ reads as follows, compare the equations (9.746), (9.745) (in brackets we record the mechanism corresponding to that term):

$$\frac{\partial \bar{u}}{\partial t} = \tilde{\beta}_{Pe}(t)\bar{u} - \tilde{\gamma}_D(t)\bar{u} - \tilde{\gamma}_C(t)\bar{u}^2, \tag{9.898}$$

$$\frac{\partial \bar{U}(t,\tilde{t})}{\partial t} =$$

$$\sum_{\tilde{t}' \in \tilde{\mathfrak{T}}_{\text{mark}}} \bar{U}(t,\tilde{t}') Q^I_{\tilde{t}',\tilde{t}} \qquad \text{(marked mesostate dynamics –}$$

transitions not involving decoupling or collision)

$$+ \sum_{\tilde{t}' \in \tilde{\mathfrak{T}}_{\text{mark}}} \bar{U}(t,\tilde{t}') Q^D(\tilde{t}',\tilde{t}) - \tilde{\gamma}_D(\tilde{t}) \bar{U}(t,\tilde{t}) \quad \text{(decoupling)}$$

$$+ \bar{v}(t) \sum_{\tilde{t}' \in \tilde{\mathfrak{T}}_{\text{mark}}} \int_{\mathcal{T}_{\text{col}} \times [0,\infty)} \bar{V}(t,\tau,x) \bar{U}(t,\tilde{t}') \sum_{j=1}^{\|\tilde{t}'\|} Q^C_{\tilde{t}',j,(\tau,x);\tilde{t}}$$

(add new orange marks + associated mesostate)

$$-[\tilde{\beta}(t) - \tilde{\gamma}_D(t) - \tilde{\gamma}_C(t) \bar{u}(t)] \bar{U}(t,\tilde{t}) \quad \text{(normalization).} \tag{9.899}$$

Step 5: *(Solutions to the nonlinear equations)*

We must now define the sense in which equations (9.898),(9.899) have solutions and if they are unique. In order to do so we proceed as in Section 8 and we introduce an appropriate *Banach space* in order to formulate this in the setting of *nonlinear semigroups*. Appropriate means here that we can define the generator as a linear operator in that space. This requires as main restriction to not have too many transitions at once by the decoupling transitions meaning that in the limit $N \to \infty$ certain "moments" of $\bar{U}(t)(\cdot)$ must remain finite, so that we need the appropriate restrictions on the configuration $\hat{t}$ and their frequencies to prevent this.

Consider the linear space $\mathbb{B} := \mathbb{R} \times \mathcal{M}_{f,\pm}([0,\infty) \times \tilde{\mathfrak{T}}_{\text{mark}})$ with the norm

$$\|(\bar{u},\bar{U}) - (\bar{u}',\bar{U}')\| = |\bar{u} - \bar{u}'| + \sum_{j=1}^{\infty} \left\{ \left(\sum_{\{\tilde{t} \in \tilde{\mathfrak{T}}_{\text{mark}}: |\tilde{t}|=j\}} |\bar{U}(.,\tilde{t}) - \bar{U}'(.,\tilde{t})|_{\text{var}} \right) (1+j)^2 \right\}, \tag{9.900}$$

where $|\cdot|_{\text{var}}$ denotes total variation of signed measures on $\mathbb{R}$ and $|\tilde{t}|$ denotes the number of active sites in the mesostate $\tilde{t}$. Note that therefore the mesostates cannot occupy too many sites if the configuration has to stay in $\mathbb{B}$. It is therefore necessary to bound the frequency of mesostates large in size (measured in sites) as time evolves. This will mean that we cannot get easily *global* (in time) solutions.

Remark 202. Although a mesostate rank can be coupled by collision-coalescence to a random number of other mesostates, due to Lemma 9.53, this number has a finite second moment as long as the density of occupied sites has not reached u^*. Moreover in the analysis of the transition regime we focus on the case in which the contribution of mesostates having more than one collision is negligible.

9.12 Emergence and Fixation: The Selection-Dominant Case

Then we can show the following statements on solutions $(\bar{u}, \bar{U})$, respectively convergence of $(\bar{u}^N, \bar{U}^N)$ as $N \to \infty$.

Proposition 9.80 (Wellposedness of nonlinear evolution equation).
Consider the selection-dominant case with mutation ergodic on E_0. Consider a finite time horizon T. Define

$$\tau_{v^*} = \sup\{t : \bar{u}(t) \int |\tilde{t}| \bar{U}(t, d\tilde{t}) \leq v^*\}. \tag{9.901}$$

(a) *Then there exists v^* such that if we fix a time $t_0 \in \mathbb{R}$ and an initial condition in $\mathbb{B}$ with $t_0 < \tau_{v^*}$, the nonlinear system (9.898),(9.899) has a unique solution with given initial conditions up to the time $T \wedge \tau_{v^*}$.*

The state of $(\bar{u}, \bar{U})$ at time $t \in [t_0, T \wedge \tau_{v^*}]$ depends continuously on the initial value.

(b) *The linear system (9.745), (9.746) is obtained by setting $\bar{u} \equiv 0$ and starting in any admissible state (i.e. an element in $\mathbb{B}$) at some time t_0 has a solution in $[t_0, \infty)$.*

(c) *The nonlinear system (9.898),(9.899) has for given $A > 0$ a unique solution in $(-\infty, T \wedge \tau_{v^*})$, which satisfies (recall (9.842)), for $\bar{U}^0(\infty)$):*

$$\lim_{t \to -\infty} e^{-\beta t} \bar{u}(t) = A, \quad \lim_{t \to -\infty} \bar{U}(t) = \bar{U}(\infty). \quad \square \tag{9.902}$$

Proposition 9.81 (Convergence and limiting system as $N \to \infty$).
Consider the case of ergodic E_0-mutation with selection-dominance, hence $\beta = \beta_{Pe}$, fix a time horizon $T \in \mathbb{R}$ and consider the events: $\{r \leq T \wedge \tau_{v^*}^N\}$ respectively $\{r \leq T \wedge \tau_{v^*}\}$. These two events have positive probability which converge to 1 as $r \to -\infty$.

(a) *Then for all $t \leq r$, we have convergence where the laws $\tilde{\mathcal{L}}$ are taken after restricting to the above events:*

$$\tilde{\mathcal{L}}\left[\left(\frac{\bar{u}^N(\frac{\log N}{\beta} + t)}{N}, \bar{U}^N(\frac{\log N}{\beta} + t)\right)\right] \Rightarrow \tilde{\mathcal{L}}\left[(\bar{u}(t), \bar{U}(t))\right], \text{ as } N \to \infty,$$

where $(\bar{u}(t), \bar{U}(t))$ satisfy the system of equations (9.898), (9.899).

(b) *In addition the empirical (microstate, lifetime)-distribution with $\tilde{\mathcal{L}}$ as in (a) satisfies for all $t \leq r$:*

$$\tilde{\mathcal{L}}\left[\bar{V}^N(\frac{\log N}{\beta} + t)\right] \Rightarrow \tilde{\mathcal{L}}\left[\bar{V}(t, \cdot, \cdot)\right], \text{ as } N \to \infty \tag{9.903}$$

$$\text{where } \bar{V}(t, \cdot, \cdot), := \int \theta(\tilde{t}, \cdot, \cdot) \bar{U}(t, d\tilde{t}). \quad \square$$

Proof of Proposition 9.81. The proof of weak convergence of the empirical measures to the nonlinear system as $N \to \infty$ follows along the same basic steps as that of Proposition 9.57. As before we start with the martingale problem which determines the empirical measure process $\tilde{\Psi}_t^N$ and we then establish the convergence of the operators in the martingale problem to the limiting ones on a class of test functions. Let us now explain what is new here.

The first key difference is that Lemma 9.53(a) implies that for small u the number of mesostates involved in a simultaneous transition has finite second moments which remain bounded for $u < v_*$ (see (9.478)). This makes possible the main step in the derivation of the limiting nonlinear equations (at least for small values of u), that is, the control of the number of mesostates (i.e. the size of a compound label) involved in a simultaneous transition which implies that the limiting dynamics is deterministic. (Recall that we used the mutation-dominant assumption in the proof of the limiting equations for the empirical process of mircrostates and it is this point that for selection-dominance we have to switch to the mesostate approach.)

The other main point is the *propagation of chaos effect*, namely, in the limit coagulation occurs between randomly chosen mesostates (weighted by the number of occupied sites in the mesostate). This gives rise to the coagulation term

$$\bar{v}(t) \sum_{\tilde{t}' \in \tilde{\mathfrak{T}}_{\text{mark}}} \sum_j \int_{\tau,x} \bar{V}(t,\tau,x)\bar{U}(t,\tilde{t}')Q^C_{\tilde{t}',j,(\tau,x);\tilde{t}} \qquad (9.904)$$

in (9.899). This describes the change resulting from the migration of a rank from the mesostate $\tilde{t}'$ which collides with an occupied site with microstate τ resulting in the marked mesostate $\tilde{t}$, where τ is chosen at random from the limiting empirical measure on microstates, $\bar{V}$.

Proof of Proposition 9.80. Having introduced the Banach space $\mathbb{B}$ the method of proof is essentially the same as in Lemma 8.6 in Section 7 and we do not repeat all the details here. The main change is that it is harder to guarantee we stay in $\mathbb{B}$ as time evolves. This is related to the possibility that the number of compound mesostates does not continue to be of order N due to coagulation. The existence of v^* such that this does not occur before τ_{v^*} is established in Lemma 9.53(a) (note here that we have to transfer from sites and direct clans to mesostates which are special direct clans).

(a) The proof that $(\bar{u}, \bar{U})$ are uniquely determined through the system of coupled equations (9.898),(9.899) and condition (9.902) can be established as in Section 8 on the two-type case, more precisely in Subsubsection 8.3.7, at least as long as the solution stays in the Banach space $\mathbb{B}$ to guarantee this we use that $t \leq T \wedge \tau_{v^*}$, via Lemma 9.53.

As before in Section 8 (after (8.289)) and Subsection 9.10 (for the microstate description) we use a theorem from the theory on nonlinear semigroups, to get continuity in the initial state. We note that this can be viewed as a nonlinear perturbation (described by a vector field) of a linear semigroup. Since the rates

9.12 Emergence and Fixation: The Selection-Dominant Case

grow at most linearly (with weights $\leq (1+j)$) we can easily verify that the vector field is C^2 with first and second derivatives uniformly bounded in norm on bounded sets. We can then apply Marsden [MA] Theorem 4.17 (see Appendix B) to verify that the solution of equations (9.898),(9.899) defines a strongly continuous nonlinear semigroup at least for t satisfying the restriction.

(b) Note that the linear system corresponds exactly to the mean equations for the CMJ process. Then the existence of the pair $\beta, \bar{U}(\infty, \cdot)$ is given by Lemma 9.75.

(c) The proof follows along the same lines as Lemma 8.6 as follows. We can again show that for any solution and $t \in \mathbb{R}$ and $t \leq T \wedge \tau_{v^*}$ we have

$$\|(\bar{u}(t), \bar{U}(t)) - (Ae^{\beta t}, \bar{U}^{**}(t))\| \leq C_1 e^{2\beta t} \text{ for some } 0 < C_1 < \infty, \quad (9.905)$$

provided that $\bar{U}^{**}(t)$ is the solution of the linear equation arising from (9.898), (9.899) by setting $\bar{u} \equiv 0$.

Then (9.905) implies that for any two solutions we have

$$\|(\bar{u}_1(t_0), \bar{U}_1(t_0)) - (\bar{u}_2(t_0), \bar{U}_2(t_0))\| \leq 2C_1 e^{2\beta t_0} \quad (9.906)$$

But then we can solve the equation uniquely for $t \geq t_0$ with the initial state at $t = t_0$ given by $(\bar{u}_\ell, t_0), \bar{U}_\ell(t_0))$, $\ell = 1,2$ and estimate the growth of the distance by $C_2 e^{\beta(t-t_0)}$ with appropriate $C_2 < \infty$ depending on the two chosen initial states at time t_0. This implies altogether that:

$$\|(\bar{u}_1(t), \bar{U}_1(t)) - (\bar{u}_2(t), \bar{U}_2(t))\| \leq C_2 e^{\beta(t-t_0)} \cdot 2C_1 e^{2\beta t_0} = 2C_1 C_2 e^{\beta t} e^{\beta t_0},$$
$$\text{for } t \geq t_0 \in \mathbb{R}. \quad (9.907)$$

Uniqueness follows by taking $t_0 \to -\infty$.   q.e.d.

Step 6: *(The dual in the collision regime: microstates)*

We now consider the *type and lifetime empirical distribution* of microstates at occupied sites. The pair $(\bar{v}^N(t), \bar{V}^N(t))$ defined as in (9.829), (9.835) but with space $\{1, \ldots, N\}$ and with θ replaced by $\hat{\theta}_{\text{mark}}$, namely

$$\bar{V}^N(t, \tau, L > x) := \frac{1}{\bar{v}^N(t)} \int |\tilde{t}| \theta_{\text{mark}}(\tilde{t}, \tau, x) \bar{u}(t) \bar{U}^N(t, d\tilde{t}) \in \mathcal{P}(\mathcal{T}_{\text{mark}} \times (0, \infty]). \quad (9.908)$$

In order establish emergence we need the time window $(\beta^{-1} \log N + t)_{t \in \mathbb{R}}$. The key results we need in this regime are the following Propositions 9.82, 9.83 which we prove in the sequel in this step.

Proposition 9.82 (Convergence of number of rank and type and lifetime distribution).
In the case of selection-dominance and ergodic mutation on E_0 we have :

$$\mathcal{L}[e^{-\beta t}\frac{\bar{v}^N(\frac{\log N}{\beta}+t)}{N} - W_{P^c} \cdot \sum_{\tilde{\mathfrak{t}}} |\tilde{\mathfrak{t}}|\bar{U}(\infty,\tilde{\mathfrak{t}})] \Rightarrow \delta_0, \text{ as } N \to \infty, t \to -\infty,$$
(9.909)

$$e^{-\beta t}\|\bar{V}^N(\frac{\log N}{\beta}+t) - \bar{V}(\infty))\|_{\text{var}} \Rightarrow 0, \text{ as } N \to \infty, t \to -\infty,$$ (9.910)

where $\bar{V}(\infty)$ is also given by the limit (9.840). □

We have so far determined the growth of the number of sites (and their types) for $N \to \infty, t \to -\infty$ up to the first order in $e^{\beta t}$ and we now turn to *second order* asymptotic term for $\bar{v}^N$ and $\bar{V}^N$ as $N \to \infty, t \to -\infty$. This term arises as a result of migration of a rank which lands on an occupied site (i.e. a collision occurs) and then a coalescence occurs between this migrant rank and one of the ranks of the occupying family. Therefore we need now the corresponding asymptotic expansion of $\bar{v}_g^N$ and the limiting behaviour of $\bar{V}_g^N$ which describes exactly the effect of the described event.

To identify the second order asymptotic term we again follow the strategy used in Section 7 and obtain bounds between the dominating collision free system and the actual system in terms of the *green ranks, green sites* and *green labels* with their mesostates (which play the role of the green particles in Section 8). Since this has been carried out in considerable detail in Section 8, here we will focus on what is different in the case $M \geq 2$ and omit some of the detailed arguments which are parallel to those of Section 8.

We first point out the additional complications which arise in the multitype case. As we have described above we now have instead of collisions of particles at certain sites, more collisions of labels with mesostates which then involve *two or more mesostates*. In the case of ergodic E_0-mutation in which mesostates involve a random number of sites having finite second moments, the basic idea is that we can focus on single collisions of two mesostates in much the same way as was done for particles in the case $M = 1$.

What is more complicated is the coalescence and decoupling effects from such a collision. Consider the event that a rank from a site in a label migrates and hits a site contained in the support of another label. Assuming that this site does not experience further collisions what is important for the further development is a coalescence with one of the ranks at the new site, the further transitions at this site and the lifetime of this site and their effect on the green family initiated by this migration event. The green family includes green ranks at the original site of the migrant and green mesostates and has a lifetime depending on both the mesostate and the microstate and lifetime at the site where the collisions occurred. These play a role in determining the second order asymptotics of the hazard rate for rare mutations. In particular in order to determine the behaviour of the hazard rate for rare mutations we study the empirical distribution of the microstates and lifetimes of the set of sites containing green ranks.

9.12 Emergence and Fixation: The Selection-Dominant Case

We note that in the limit we consider, the contribution of a *self-collision* of a mesostate is negligible. Therefore we can focus on *collisions between two different mesostates*. As we have seen above the effect of a collision and coalescence can reduce the number of ranks at an occupied site or the number of occupied sites. Define

$$\bar{v}_g^N(t) = \sum_{\tilde{t}} |\tilde{t}|_g [\bar{u}^N(t)\bar{U}^N(t,\tilde{t}) + \bar{u}_g^N(t)\bar{U}_g^N(t,\tilde{t})] \qquad (9.911)$$

(number of sites containing green ranks),

where now $|\tilde{t}|_g$ is the number of sites containing green ranks in the label with mesostate $\tilde{t}$ and θ_g is a mapping from the marked mesostates to empirical measures of microstates and lifetimes of sites *containing green ranks*.

The effects of collisions are estimated by the green ranks and sites in the marked mesostate process, described by the analogues of (9.908), (9.829), namely,

$$\bar{V}_g^N(t,\tau,dL) = \frac{\sum_{\tilde{t}} \theta_g(\tilde{t},\tau,dL)|\tilde{t}|_g[\bar{u}^N(t)\bar{U}^N(t,\tilde{t}) + \bar{u}_g^N(t)\bar{U}_g^N(t,\tilde{t})]}{\bar{v}_g^N(t)} \in \mathcal{P}(\mathcal{T}_{\text{mark}} \times (0,\infty]), \qquad (9.912)$$

where the first summand consists of sites containing both green and non-green ranks and the second sites containing only green ranks.

We next consider the *growth constant of green ranks*. One issue here is the fact that now the lifetime of a green rank or site may be *finite* and this must be incorporated into the picture. Following the same approach as in Section 8, Proposition 8.24 we consider the growth of the set of *green ranks* $\bar{v}_g^N(s,t,\tau,L)$ starting from the creation of a green rank at a randomly chosen rank at a site having microstate τ and lifetime L calculated with the stable type distribution of the nongreen population of sites. Recall that green sites do not collide since they follow the collision-free dynamic on $\mathbb{N}$. They grow without collisions and therefore with malthusian parameter β. More over if we define

$$W^g(s,t,\tau,t-s) = e^{-\beta(t-s)}\bar{v}_g^N(s,t,\tau,L \in [t-s,\infty)), \qquad (9.913)$$
$$= 0 \quad \text{if } L < t-s,$$

then there exists the random limit

$$W^{g,*}(u,\tau) = \lim_{t \to \infty} W^g(t-u,t,\tau,u)1_{L \geq u}. \qquad (9.914)$$

It remains to explain the way the constant in the growth of the set of green sites denoted $\hat{\kappa}_2$ arises. We now state the result.

Proposition 9.83 (Emergence regime: Asymptotics of the dual for ergodic E_0-mutation).
Assume that mutation on E_0 is ergodic in the selection-dominant case. Then

(a) The asymptotic number of green sites satisfies:

$$\lim_{t \to -\infty} e^{-2\beta t} \lim_{N \to \infty} \frac{1}{N} E[\bar{v}_g^N(\frac{\log N}{\beta} + t)] = \hat{\kappa}_2 E[W_{Pe}^2], \quad (9.915)$$

where

$$\hat{\kappa}_2 = \int_0^\infty e^{-\beta u} \sum_{\tau} \bar{V}^N(\infty, \tau, [u, \infty)) E[W^{g,*}(u, \tau)] \sum_{\tilde{t}} |\tilde{t}| \bar{U}(\infty, \tilde{t}) \sum_{\tilde{t}} \|\tilde{t}\| \bar{U}(\infty, \tilde{t}) du. \quad (9.916)$$

(b) The limiting type and lifetime distribution of the green sites satisfies:

$$\bar{V}_g(\infty, \tau, dL) = \lim_{t \to -\infty} \lim_{N \to \infty} \bar{V}_g^N(\frac{\log N}{\beta} + t, \tau, dL) \quad (9.917)$$

$$= \int_0^\infty e^{-\beta u} \int \theta^{g,*}(u, \tilde{t}, \tau, dL) \|\tilde{t}\| \bar{U}(\infty, d\tilde{t}) du,$$

where $\theta^{g,*}(u, \tilde{t}, \tau, L)$ represents the empirical distribution of the microstates and lifetimes of sites, which contain green ranks at a time u after the creation of a green rank due to collision of a migrant from a randomly chosen rank in mesostate $\tilde{t}$ with a randomly chosen site, assuming that the non-green population is in stable age-type distribution. □

We next have to prove our two key propositions on the growth of mesostate numbers and number of occupied sites.

Proof of Proposition 9.82. We can choose $t_0 \in \mathbb{R}$ such that $\bar{v}(t_0) < v^*$. Then by Proposition 9.81 we know that:

$$\mathcal{L}\Big[\frac{\bar{v}^N(\frac{\log N}{\beta} + t)}{N}\Big] \Rightarrow \mathcal{L}\Big[\bar{u}(t) \sum_{\tilde{t}} |\tilde{t}| \bar{U}(t, \tilde{t}, (0, \infty])\Big], \text{ as } N \to \infty \text{ for } t \le t_0. \quad (9.918)$$

We also have

$$\bar{u}(t) \le W_{Pe} e^{\beta t}. \quad (9.919)$$

Using the fact that the contribution of the green sites produces a term bounded by a constant multiple of $e^{2\beta t}$ (see Proposition 9.83 above) it follows that

$$\lim_{t \to -\infty} e^{-\beta t}(W_{Pe} e^{\beta t} - \bar{u}(t)) = 0, \quad (9.920)$$

and

$$\lim_{t \to -\infty} e^{-\beta t} \|\bar{U}(t) - \bar{U}(\infty)\|_{\text{var}} = 0. \quad (9.921)$$

9.12 Emergence and Fixation: The Selection-Dominant Case

Then using the bound on the second moments of the size of the mesostates (uniform in N), we have

$$\lim_{t \to -\infty} e^{-\beta t} |\sum_{\tilde{t}} |\tilde{t}| \bar{U}(t, \tilde{t}) - \sum_{\tilde{t}} |\tilde{t}| \bar{U}(\infty, \tilde{t})| = 0. \qquad (9.922)$$

Together (9.920), (9.921) and (9.922) yield the result. q.e.d.

Proof of Proposition 9.83. The proof follows the same steps as Proposition 7.39 but now including the microstate and lifetime distribution. The production of green sites can be viewed as a branching process with immigration. Immigrants (the creation of new green ranks) are produced by the migration of a rank from a randomly chosen rank in a randomly chosen mesostate and collision with a randomly chosen rank from the non-green population with subsequent growth of the resulting green family. The collisions occur at rate proportional to $(\bar{u}^N(t))^2 \cdot \left(\int |\tilde{t}| \bar{U}(t, \tilde{t}) \int \|\tilde{t}\| \bar{U}(t, \tilde{t}) \right)$ and the immigrant families subsequently grow as the collision-free branching process with malthusian parameter β (up to their lifetimes). Therefore we have

$$\lim_{t \to -\infty} \lim_{N \to \infty} \frac{e^{-2\beta t}}{N} \bar{v}_g^N(\frac{\log N}{\beta} + t) \qquad (9.923)$$

$$= \lim_{t \to \infty} \lim_{N \to \infty} e^{-2\beta t}$$

$$\int_0^t \frac{W_{Pe}^2 e^{2\beta s}}{N} \sum_{\tilde{t}} |\tilde{t}| \bar{U}(s, \tilde{t}) \sum_{\tilde{t}} \|\tilde{t}\| \bar{U}(s, \tilde{t}) \sum_{\tau} \bar{V}_g^N(s, \tau, [t-s, \infty))) W^g(s, t, \tau, t-s) e^{\beta(t-s)} ds$$

$$= W_{Pe}^2 \int_0^\infty e^{-\beta u} \sum_{\tau} \bar{V}^N(\infty, \tau, [u, \infty))) W^{g,*}(u, \tau) \sum_{\tilde{t}} |\tilde{t}| \bar{U}(\infty, \tilde{t}) \sum_{\tilde{t}} \|\tilde{t}\| \bar{U}(\infty, \tilde{t}) du.$$

q.e.d.

Step 7: *(Collisions between two dual clouds)*

For second moment calculations we have, as we saw in Section 8, to compare the dual started with two single factors with two independent copies of the dual starting with one single factor. More precisely in order to evaluate dual expressions for second moments we need the following generalization of Proposition 8.24.

We first consider the two independent copies of the time-space dual process one starting in (t_1, j_1), denoted $\tilde{\mathcal{G}}_{j_1, t_1}^{++, N}$ and the other starting in (t_2, j_2), denoted $\tilde{\mathcal{G}}_{j_2, t_2}^{++, N}$ where there is no interaction between the two clouds even at sites where both clouds are present.

We can construct on the same probability space the dual $\tilde{\mathcal{G}}_{j_1, t_1; j_2, t_2}^{++, N}$ with $t_1 < t_2$ in which collisions between the descendants of the two initial individuals at j_1, j_2 do occur as in the original process. This is defined as follows:

- For $t_1 \leq t < t_2$,

$$\tilde{\mathcal{G}}^{++,N}_{j_1,t_1;j_2,j_2}(t) = \tilde{\mathcal{G}}^{++,N}_{j_1,t_1}(t), \quad \tilde{\mathcal{G}}^{++,N}_{j_1,t_1}(t_1) = (1,\ldots,1,0) \tag{9.924}$$

and for $t \geq t_2$, $\tilde{\mathcal{G}}^{++,N}_{j_1,t_1;j_2,j_2}(t)$ follows the dual dynamics with initial condition at t_2 given by

$$\tilde{\mathcal{G}}^{++,N}_{j_1,j_2;j_2,t_2}(t_2) = \tilde{\mathcal{G}}^{++,N}_{j_1,t_1}(t_2) \times (1,\ldots,1,0)_{j_2}. \tag{9.925}$$

We define $\bar{\Pi}^{N,j_i}_{t_i}(s) := \|\tilde{\mathcal{G}}^{++,N}_{j_i,t_i}(t)\|$, that is the number of active ranks in $\tilde{\mathcal{G}}^{++,N}_{j_i,t_i}(t)$ and $\bar{\Pi}^{N,j_1,j_2}_{t_1,t_2}(s) = \|\tilde{\mathcal{G}}^{++,N}_{j_i,t_i;j_2,t_2}(t)\|$.
By the argument of Propositions 8.24 and 9.83,

$$\bar{\Pi}^{N,j_i}_{t_i}(s) = W_*(s)e^{\beta s} - \frac{\hat{\kappa}_2}{N}(W^i_{Pe})^2 e^{2\beta s} + \mathcal{E}(N,s) \tag{9.926}$$

where $W_*(s) := W^i_{Pe}(s) \cdot \int \|\tilde{t}\| U^N(s,d\tilde{t})$ and $\mathcal{E}(N,s) = O\left(\frac{e^{3\beta s}}{N^2}\right)$. The second term corresponds to the number of green ranks in the respective populations. To distinguish these we denote them as green G_1, G_2 individuals, respectively. We now consider $\bar{\Pi}^{N,j_1,j_2}_{t_1,t_2}(s)$. We again obtain a set of green particles, some resulting from self collisions of the cloud starting at j_1,t_1, namely G_1-green individuals, some resulting from self collisions of the cloud starting at j_2,t_2, namely G_2-green individuals. In addition green individuals are created by the collision (and then coalescence) between the two clouds, that is, by a migrant from the (j_1,t_1) cloud landing on a site occupied by the (j_2,t_2) cloud, or vice versa. The resulting ranks and their descendants are denoted G_{12}-green sites and ranks and we denote by $\mathfrak{g}^{N,j_1,j_2}_{t_1,t_2}(s)$ the total number of green G_{12}-ranks at time s.

We are now able to compare the time-space dual process starting in (t_1, j_1) and (t_1, j_2) with one factor $(1, 1 \cdots, 1, 0)$ with two independent versions of the (t_1, j_1) and (t_2, j_2) clouds.

Proposition 9.84 (Collisions between dual clouds - production of G_{12} families).

Assume the selection-dominant case. Let $\tilde{\mathcal{G}}^{++,N}_{j_1,t_1;j_2,j_2}(t)$, $\bar{\Pi}^{N,j_1,j_2}_{t_1,t_2}(t)$ and $\mathfrak{g}^{N,j_1,j_2}_{t_1,t_2}(t)$ be defined as above. Then

(a) The total number of active ranks increases as:

$$\bar{\Pi}^{N,j_1,j_2}_{t_1,t_2}(s) = \bar{\Pi}^{N,j_1}_{t_1}(s) + \bar{\Pi}^{N,j_2}_{t_2}(s) - \mathfrak{g}^{N,j_1,j_2}_{t_1,t_2}(s) + \mathcal{E}(N,s) \tag{9.927}$$

where

$$E(N,s) = O\left(\frac{e^{3\beta s}}{N^2}\right) \text{ as } N \to \infty, s \to -\infty \tag{9.928}$$

9.12 Emergence and Fixation: The Selection-Dominant Case 681

(b) Let $t_1 = 0$, and $t_2 = t_N(1) - t_N(\delta) + t$ where $t_N(\delta) = \frac{\delta \log N}{\beta}$, $0 < \delta \le 1$. Then $\mathfrak{g}_{t_1,t_2}^{N,j_1,j_2}(s)$ satisfies the condition

$$\lim_{t \to -\infty} \lim_{N \to \infty} N \cdot e^{-\beta(t_N(1) + t_N(\delta) + t)} E[\mathfrak{g}_{0,t_N(1)-t_N(\delta)+t}^{N,j_1,j_2}(t_N + t)] = 2\hat{\kappa}_2 E[W_{Pe}^1] E[W_{Pe}^2], \tag{9.929}$$

where $\hat{\kappa}_2$ is given as in Proposition 9.83. □

Proof of Proposition 9.84. The proof follows the same lines as that of Proposition 8.24 and in particular the coupling construction between two independent duals starting in j_1 respectively j_2 and the one starting with two, one at j_1 and one at j_2. The reader is invited to recall this material. Now however we are using the result of Proposition 9.83 to determine the growth of the set of G_{12} ranks. Note that the G_{12} families grow in exactly the same fashion as the green families in Proposition 9.83(a) and therefore is given in terms of $\hat{\kappa}_2$. In particular, the proof of the existence of the limit $\hat{\kappa}_2$ follows exactly as in Proposition 9.83(a). The error term which rises by ignoring the G_1, G_2 ranks in calculating the rate of collisions between the two clouds in the expression (9.929) is again of higher order and can be neglected. q.e.d.

9.12.11 Rare Mutations and Emergence Time: Lower Bound

Assume that $\mu(E_0) = 1$ and $\mu(j) > 0$ for $j \in \{1, \ldots, M\}$. Note the latter holds in the case where the mutation E_0 is ergodic, in any positive time so this does not conflict with our starting in the type of lowest fitness.

We want to calculate $E[x_{E_1}^N(\beta^{-1} \log N + t)]$ for $N \to \infty$ as a function of t. Recall that this quantity can be written as follows in terms of the dual process (abbreviate $t_N = \beta^{-1} \log N + t$):

$$E[x_{E_0}^N(t_N)] = E[\mu_0^{\otimes}(\tilde{\mathcal{G}}_{t_N}^{++,N})], \text{ with } \tilde{\mathcal{G}}_0^{++,N} = \{1, \ldots, M\}_1. \tag{9.930}$$

In order to demonstrate emergence we must show that the limit of the right side of (9.930) is less than 1 as $N \to \infty$ for $t \in \mathbb{R}$, by the occurrence of a rare mutation. Since the initial state μ_0 satisfies $\mu_0(E_0) = 1$ the result only depends on the microstate and the lifetime of the single site. For this reason we can determine the result in terms of the limiting distribution of *microstates and lifetimes* at occupied sites, $\bar{V}(t, \cdot, \cdot)$ and do not need the full information of the mesostates to which they belong.

We begin with an overview of the strategy to prove the main results on *emergence* using the basic objects $(\bar{u}^N(t), \bar{U}^N(t))$. Emergence involves rare mutations taking place at a rank at one of the occupied sites. Hence the *number of ranks* at time t is a key quantity. A difference from the case $M = 1$ is that a single rare mutation at time t no longer makes the integral $\mu_0^\otimes(\tilde{\mathcal{G}}_t^{++,N})$ equal to 0 but only changes it to a value in $[0,1)$ and then fluctuates until it hits 0 or 1 provided that $\mu(j) > 0$ for $j \in \{1,\ldots,M\}$. Note that some rare mutations may not change the value if $\mu(j) = 0$ for some j - see subsubsection 9.12.15.

A key idea is that in the transition regime ($N \to \infty, t \to -\infty$) we can focus on a single rare mutation or possibly two in some cases (as in the second moment calculation for $\mathcal{L}[^*\mathcal{W}] = \mathcal{L}[\mathcal{W}^*]$) since in the limit $N \to \infty$, $t \to -\infty$, the contribution of one respectively two or more rare mutations is negligible for first respectively second moment calculations.

The next observation will be that then in particular the contribution of two rare mutations in the same mesostate is negligible. Thus we can focus on the event of a single rare mutation in a mesostate and the subsequent random fluctuations until the integral $\mu_0^\otimes(\tilde{\mathcal{G}}_t^{N,++})$ hits 0 or 1 - here we use the fact that the integral over a mesostate with only one occupied site containing a strict subset of $(\mathbb{I})^{\tilde{N}}$ *strictly* reduces the integral over this site.

However we have to be careful and we must take into account that an occupied site can be *removed* by a resolution at a random time and thereby the factor less than one is replaced by one again. This can be handled by assigning a lifetime distribution to an occupied site depending on its position in a mesostate (where we assign lifetime $+\infty$ to a *UPe* site; recall definition (9.884) of $\tilde{\Psi}_t^N$). We now derive in Step 1 the hazard function, in Step 2 the first moment of the E_1-mass and exploit this in Step 3 to prove emergence.

Step 1(*Hazard function*)

We first turn to the *hazard function* for rare mutations. Recall that the rare mutation in the dual evolution depends only on the total number of active ranks up to now and can be constructed from a hazard function and an exponential random variable independent of the dual evolution exponential. Recall that the *hazard rate* for a rare mutation from type $\ell \in E_0$ to E_1 at time t which is not removed by a resolution up to time x later is given by

$$\frac{m_{\ell,E_1}}{N} \cdot \text{total number of active ranks at time } t \text{ with lifetime in } dx \quad (9.931)$$

$$= \frac{m_{\ell,E_1}}{N} \cdot \bar{\Pi}^N(t,x) = \frac{m_{\ell,E_1}}{N} \cdot \bar{v}^N(t) \int_{\mathcal{T}} \|\tau\| \bar{V}^N(t, d\tau, dx)\|.$$

Now let us also introduce the number of green ranks resulting from green mesostates and green ranks in mesostates (recall (9.885), (9.886)):

$$\bar{\Pi}_g^N(t) := \int \|\tilde{t}\|_g [\bar{u}^N(t)\bar{U}^N(t,d\tilde{t}) + \bar{u}_g^N(t)\bar{U}_g^N(t,d\tilde{t})], \quad (9.932)$$

where $\|\tilde{t}\|_g$ counts the number of green ranks in the mesostate $\tilde{t}$.

9.12 Emergence and Fixation: The Selection-Dominant Case

The probability that a rare mutation has taken place before time t *given the path of $\bar{\Pi}^N$ up to time t* is given, using the *conditional hazard function*, by

$$p^N(t) = 1 - e^{-\sum_{i=1}^M \frac{m_i \cdot E_1}{N} \int_{-\infty}^t \|\tilde{\mathcal{G}}_u^{N,++}\| du} = 1 - e^{-\sum_{i=1}^M \frac{m_i \cdot E_1}{N} \int_{-\infty}^t \bar{\Pi}^N(u) du} \tag{9.933}$$

Expanding the random exponential as in Subsection 8 (see Lemma 8.26 and the sequel) we obtain

$$\lim_{t \to -\infty} \lim_{N \to \infty} e^{-\beta t} E[p^N(\frac{\log N}{\beta} + t)] = E[W_{Pe}] \sum_{\hat{t}} \|\hat{t}\| \bar{U}(\infty, \hat{t}). \tag{9.934}$$

We now consider the changes in $\tilde{\mathcal{G}}_t^{++,N}$ due to rare mutation at time $\beta^{-1} \log N + t$. For that we need the effects of a rare mutation acting on a rank at a site belonging to a mesostate. In contrast to the case $M = 1$, a rare mutation transition of the dual started with $(1, \ldots, 0)$ does not result in a jump to 0 for the integrated dual with $\mu(E_0) = 1$. Note that when such a rare mutation occurs the state of the set-valued process corresponds to a *strict subset* of $\mathbb{I}_M^N$. However due to the subsequent mutation-selection process, assuming no additional rare mutation, this will eventually return to $\mathbb{I}^N$ or become $\emptyset$ (cf. proof of ergodicity in Section 9). In particular a resolution at a rank to the left of the rank at which the rare mutation occurs can eliminate the effect of the rare mutation. Therefore we have to study both when rare mutation *occurs* and whether it *succeeds*.

To determine what happens when a rare mutation occurs we need some preparation. Set for $u > 0$:

$$A_u^i(t, \tau) \text{ is the random subset of } (E_0)^{\|\tau\|}, \tag{9.935}$$

resulting at time t from a rare mutation from type $i \to E_1, i \in E_0$ at time $u \in [0, t]$ at an occupied site with micro-state τ at a randomly chosen rank at that site and then developing up to time t following the mutation-selection-coalescence dynamics.

We observe that if a rare mutation takes place at time u it occurs at *a randomly chosen site* in supp($\tilde{\mathcal{G}}_u^{++,N}$).

Recall that it will eventually hit $\emptyset$ or $(\mathbb{I})^N$ after a finite random time with exponential moment (recall Lemma 9.25). A rare mutation event in which A hits $\emptyset$ is said to be a *successful rare mutation*. The rare mutation at time u produces at time t a random factor with values in $[0, 1]$.

Step 2 (*First moment formula*)

We now want to obtain the first moment formula for times $t_N + t$ where $t_N = \frac{\log N}{\beta}$. We first note that the contribution from two or more rare mutations by time $t_N + t$ to the calculation of the first moment is negligible in the limiting case we consider, namely, $N \to \infty, t \to -\infty$.

Now we are ready to write down a formula for the first moment of the fitter type population at time $\beta^{-1} \log N + t$ in terms of the dual. The expression has to be evaluated as $N \to \infty$. For that purpose we first write $E[x_{E_1}^N(t)] = 1 - E[x_{E_0}^N(t)]$ and then we recall the representation of the moments from Proposition 9.20 and specialize to the first moment. The dual expression is zero until the first rare mutation occurs, when this occurs we might catch a contribution. We have to rewrite the dual expression.

First we realise $\mathcal{G}^{++}$ and $\tilde{\mathcal{G}}^{++,N}$ on one probability space (recall the coloured ranks construction) so that as in Section 7 we can also use $\|\mathcal{G}^{++}\|$ as an a.s. bound for $\|\tilde{\mathcal{G}}^{++,N}\|$, but we can define the following.

We need the *conditional expectation over the lifetime L and the microstate given W_{Pe}* (here L is the lifetime of the *site i* at which the rare mutation occurs). Then consider $\tilde{\mathcal{G}}^{++,N}$ at time $\beta^{-1} \log N + t$ and set:

$$A_{(u,t)}^{N,i} = 1 - \mathbb{E}_N[\mu_0^\otimes(A_u^i(t))|W_{Pe}]$$

$$= 1 - \mathbb{E}_N\left[\sum_{\tau \in \mathcal{T}} \mu_0^\otimes(A_u^i(t,\tau))\|\tau\|\bar{V}^N(u,\tau,[t-u,\infty))|W_{Pe}\right].$$

(9.936)

We write down the expanded dual expression by expanding the random exponential in (9.933) (again as in Subsection 8 - see Lemma 8.26 and the sequel). Using this we obtain as $N \to \infty$ and with $t_N = \beta^{-1} \log N$:

$$E[x_{E_1}^N(t_N + t)] \tag{9.937}$$

$$= E\left[\sum_{i=1}^M \frac{m_{i,E_1}}{N} \int_0^{t_N+t} \bar{\Pi}^N(u) e^{-\frac{m_{i,E_1}}{N} \int_0^u \bar{\Pi}^N(r)dr} \mathbb{E}_N[1 - \mu_0^\otimes(A_u^i(t))|W_{Pe}]du\right]$$

$$+ \mathcal{E}(N,t).$$

where $\bar{\Pi}^N$ is given by (9.926) with $t_i = 0$ and arbitrary j_i and

$$\lim_{t \to -\infty} \lim_{N \to \infty} e^{-\beta t} \mathcal{E}(N,t) = 0. \tag{9.938}$$

(To handle the random exponential in the limit, we can first integrate over the event $W_{Pe} \leq K$, then take $N \to \infty$, $t \to -\infty$ and finally $K \to \infty$.)

Step 3 *(Emergence in time window $(\frac{\log N}{\beta} + t)_{t \in \mathbb{R}}$: lower bound)*

Assume that $\mu(E_0) = 1$ and $\mu(j) > 0$ for $j \in \{1, \ldots, M\}$. To estimate the hazard rate for rare mutation we must insert the growth behaviour of $\Pi^N(t)$ in (9.937). In Proposition 9.39 we have proved that for the $\tilde{\mathcal{G}}_t^{++,III}$ dynamics we have $e^{\beta t}$ occupied $Pe \cup UPe$ sites and in Proposition 9.40 we have verified that this growth rate also holds for the $Pe \cup UPe \cup URe$ sites. This gives based on (9.940) below an upper bound on the emergence time.

9.12 Emergence and Fixation: The Selection-Dominant Case

Lemma 9.85 (Expression after rare mutation).
Assume that $\mu_0(E_0) = 1$ and $\mu_0(j) \geq \delta > 0$ for $j \in \{1, \ldots, M\}$. Consider $\tilde{\mathcal{G}}_{\beta \log N + t}^{++,N}$. There exists $\varepsilon > 0$ such that for all $t \geq t_0$ for some $t_0 \in \mathbb{R}$ and $u \in [-\infty, t]$:

$$\liminf_{N \to \infty} \mathcal{A}_{(u,t)}^{N,i} \geq \varepsilon. \qquad \Box \qquad (9.939)$$

Proof of Lemma 9.85. Note since we consider $N \to \infty$ we look at everything after a large time. We verified that the growth of the number of transient sites is bounded by the growth of the Pe and UPe sites. Therefore any randomly chosen active rank has a positive probability of being Pe or UPe. Moreover by the tightness property of the mesostate length process, the probability that the rank belongs to a mesostate having more than k active ranks goes to 0 as $k \to \infty$. But a UPe active rank in a mesostate with k active ranks produces a contribution $\geq \delta^k$. q.e.d.

Remark 203. In the ergodic E_0-mutation case we can remove the assumption $\mu_0(j) \geq \delta > 0$ for $j \in \{1, \ldots, M\}$ since in this case after a rare mutation event the E_0-mutation process will produce a positive contribution for any initial measure μ_0.

Then from the equation (9.937) together with Lemma 9.85 we have (by explicit integration and an elementary estimate and W_* (see (9.926)) denoting the growth constant of $\bar{\Pi}$, the total number of active ranks) and the fact that $\bar{\Pi}^N \leq \bar{\Pi}$:

$$\liminf_{N \to \infty} E[x_{E_1}^N(t_N + t)] \geq \varepsilon \cdot E\left[\sum_{i=1}^M \frac{m_{i,E_1}}{N} \int_0^{t_N + t} \bar{\Pi}^N(u) e^{-\frac{m_{i,E_1}}{N} \int_0^u \bar{\Pi}^N(r) dr} du\right]$$

$$\geq \varepsilon \sum_i \beta^{-1}(1 - E[e^{-m_{i,E_1} W_* e^{\beta t}}]) > 0. \qquad (9.940)$$

Hence we have by the following statement a lower bound on the emergence time.

Proposition 9.86 (Lower bound on emergence time).
We have for every $t \in \mathbb{R}$ that:

$$\liminf_{N \to \infty} E[x_{E_1}(\beta^{-1} \log N + t)] > 0, \quad \forall t \in \mathbb{R}. \qquad \Box \qquad (9.941)$$

Remark 204. Does $\mathbb{E}[(1 - \mu_0^{\otimes}(A_u^i(t)))]$ converge to a constant as $t - u \to \infty$? This can be proved if $\tilde{\mathcal{G}}_s^{++}$ can be decomposed into a large number of independent pieces using a l.l.n. argument. In the ergodic E_0-mutation case this works because we have mesostates that remain of order $O(1)$ and only $O(\log N)$ are linked by the collision-coalescence process.

The outcome of a rare mutation for a Pe or UPe site only depends on the local typical state. The outcome for a Tr site depends on the type of the Pe or UPe parent and the age. Therefore, overall, the outcome depends on the stable type distribution for the UPe and Pe sites.

9.12.12 Second Moment Calculations and L^2 Convergence of the Droplet Process

In order to establish the equality (in law) of $^*\mathcal{W}$ and $\mathcal{W}^*$ in the selection-dominant case in the next subsubsection, we need to compute the second moment of the E_1-masses at times $\frac{\log N + t}{\beta}$ and $\frac{\delta \log N}{\beta}$ as $N \to \infty$ and then $t \to -\infty$. We want to do this using the dual. Since the first moments are obtained using (9.937) it suffices now to determine *covariances*. The variance (or covariance) term arises by comparing the expression for the first moment squared and the second moment expression.

We need to compute these moments assuming that the initial condition is given by

$$x_i^N(j, 0) = \mu_0(i) > 0 \quad \forall\, i \in \{1, \ldots, M\},\ \forall\, j \in \{1, \ldots, N\}, \quad (9.942)$$

$$\mu_0 \in \mathcal{P}(E_0 \cup E_1), \quad \text{and } \mu_0(E_1) = 0.$$

To prepare for this recall that given two times $t_1(N)$ and $t_2(N) > t_1(N)$ the second joint moment at two distinct sites $\{1\}, \{2\}$ can be calculated using the set-valued dual, $\{\tilde{\mathcal{G}}_t^{++,N}\} : t \geq 0$, with

$$\tilde{\mathcal{G}}_0^{++,N} = \{1, \ldots, M\}_1, \quad \text{run for times } 0 \leq t < t_2(N) - t_1(N), \quad (9.943)$$

$$\tilde{\mathcal{G}}_{(t_2(N)-t_1(N))}^{++,N} = \tilde{\mathcal{G}}_{(t_2(N)-t_1(N))-}^{++,N} \otimes \{1, \ldots, M\}_2,$$

and then evaluated at time $t = t_2(N)$. We then have with the same reasoning as for equation (9.937), that

$$E[x_{E_0}^N(1, t_1(N)) x_{E_0}^N(2, t_2(N))] \quad (9.944)$$

$$= 1 - E\Big[\sum_{i=1}^M \frac{m_{i,E_1}}{N} \int_0^{t_2(N)} \bar{\Pi}^{N;(1,t_1),(2,t_s)}(u) e^{-\frac{m_{i,E_1}}{N} \int_0^u \bar{\Pi}^{N;(1,t_1),(2,t_s)}(r) dr}$$

$$\mathbb{E}[1 - \mu_0^\otimes(A_u^i(t))] du\Big] + o_N(1),$$

where $\bar{\Pi}^{N;1,2}(r) = \|\tilde{\mathcal{G}}_r^{++,N}\|$ is given by the analogue (9.932).

The key covariance calculation depends on the green ranks and sites involving the collision and coalescence between two growing clouds of labels with their mesostates which we discussed in Proposition 9.84. We can view this as the collision of two randomly chosen occupied sites sampled from the empirical distribution of microstates and lifetimes of occupied sites and the subsequent growth of the number of green ranks and sites. We now consider this in more detail.

Note that the difference between the square of first moments and the second moment arises due to the ranks lost, due to collision and coalescence of the two clouds corresponding to the two different initial states since the square of the first moment can be interpreted in terms of the two clouds without interaction

9.12 Emergence and Fixation: The Selection-Dominant Case

between them. Recall the concept of G_{12}-green ranks discussed in Step 6 of Subsubsection 9.12.10 to quantify these effects generalizing the technique from Section 8 in the sequel of equation (8.765). We then note that the collisions occur at rate $\frac{1}{N} W_1 e^{\beta(t-t_1)} \cdot W_2 e^{\beta(t-t_2)}$ respectively. We describe the ranks lost again by a green rank created by a collision between ranks coming from two different clouds, that is, $G_{1,2}$ *green ranks*, to be distinguished from the colour green ranks marking the losses due to collision-coalescence events within one dual population (starting from one initial factor). Recall here Step 6 in Subsubsection 9.12.10.

We now sketch the required calculations for the second joint moments of $x_{E_0}^N(1,s) \, x_{E_0}^N(2,t)$ and the analogue of Proposition 8.27.

Proposition 9.87 (**Asymptotically deterministic droplet growth**)**.**

(a) We have

$$\lim_{t \to \infty} \lim_{N \to \infty} E\left[\left([e^{-\beta(t+s)} x_{E_0}^N(t+s)] - [e^{-\beta(t)} x_{E_0}^N(t)] \right)^2 \right] = 0,$$

uniformly for $s \geq 0$. (9.945)

(b) The scaled droplet process $\{e^{-\beta t} \mathbf{J}_t^{m,1}([0,1])\}_{t \geq 0}$ converges in L^2 as $t \to \infty$ to a random variable $\mathcal{W}^*$ and if $t_N(\delta) = \frac{\delta \log N}{\beta}$ with $0 < \delta < 1$, then

$$\mathcal{L}\left[e^{-\beta t_N(\delta)} \hat{x}_{E_1}^N(t_N(\delta)) \right] \Longrightarrow \mathcal{L}[\mathcal{W}^*]. \qquad \Box \qquad (9.946)$$

Proof of Proposition 9.87. (a) The proof of (9.945) is of the same type but simpler than the proof of (9.949) and is not repeated here.

(b) Recall that $\{\hat{x}_{E_1}^N(t)\}_{t \geq 0} \Longrightarrow \{\mathbf{J}_t^{m,1}([0,1])\}_{t \geq 0}$ as $N \to \infty$. Then using formula (9.945) for the second moments of increments it follows that $\{e^{-\beta t} \mathbf{J}_{t_n}^{m,1}([0,1])\}$ is Cauchy in L^2 if $t_n \to \infty$ (the boundedness of third moments can be proved as in Section 8, Lemma 8.32). We denote the limit by $\mathcal{W}^*$ (a simple argument shows that it is unique). The relation (9.946) then follows by modifying the proof of (a) to show that

$$\lim_{t \to \infty} \lim_{N \to \infty} E\left[\left([e^{-\beta t} x_{E_0}^N(t)] - [e^{-\beta t_N(\delta)} x_{E_0}^N(t_N(\delta))] \right)^2 \right] = 0. \qquad (9.947)$$

Remark 205. Note that in determining the contribution of green ranks in the rare mutation calculation we must consider in (9.951) only rare mutations acting on a green rank when considering sites having both green and non-green ranks.

9.12.13 The Random Entrance Law

The remaining task is to relate the random time shift, respectively asymptotic growth constant *$\mathcal{W}$ of the random entrance law for the nonlinear system (9.871), (9.872) and the *random growth constant for the droplet process* $\mathcal{W}^*$. This requires relating the leading asymptotic terms in the time window $(\frac{\log N}{\beta} + t)_{t \in \mathbb{R}}$ to the system of marked mesostates. In order to have the full set-valued dual and to identify the entrance law we need to include the set determined by the *Pe*, *UPe* and *URe* sites. These are however automatically included in the mesostate description.

As in the case $M = 1$, the transition between the growing droplet and the macroscopic growth described by the McKean–Vlasov equation is described by a random entrance law. Assume that the initial condition is given by

$$x_i^N(j,0) = \mu_0(i) > 0 \quad \forall\, i \in \{1,\ldots,M\},\ \forall\, j \in \{1,\ldots,N\}, \quad (9.948)$$

$$\mu_0 \in \mathcal{P}(E_0 \cup E_1), \quad \text{and } \mu_0(E_1) = 0.$$

We next identify the random entrance law by determining the random time shift of the standard (deterministic) solution derived from $^*\mathcal{W}$ and to relate it to the random growth constant for the droplet, $\mathcal{W}^*$ (see also Section 10, Proposition 10.7).

Proposition 9.88 (Entrance - Exit).

(a) *In the selection-dominant case (that is, under the hypotheses of Proposition 9.74) we have*

$$\lim_{t \to -\infty} \lim_{N \to \infty} E\left[\left(e^{-\beta t_N(\delta)} \hat{x}_{E_1}^N(t_N(\delta)) - e^{-\beta(\frac{\log N}{\beta}+t)} \hat{x}_{E_1}^N(\frac{\log N}{\beta}+t))\right)^2\right] = 0. \quad (9.949)$$

(b) *The $\bar{x}_{E_0}^N(\frac{\log N}{\beta}+t)$ converges in law to the solution of the nonlinear system with random initial condition $^*\mathcal{W}$ with*

$$\mathcal{L}[^*\mathcal{W}] = \mathcal{L}[\mathcal{W}^*]. \quad (9.950)$$

Proof of Proposition 9.88. (a) We use the dual to compute the necessary first and second moments. We have calculated first moments in Subsubsection 9.12.11 in (9.937), second moments in Subsubsection 9.12.12 in (9.944) which we shall rewrite below for our purposes.

Now to the second moments. As in Section 8, for times of the form $\frac{\log N}{\beta} + t$ we must determine terms of order $O(e^{2\beta t})$ but terms of order $O(e^{3\beta t})$ are negligible. We then note that the main contribution arises from at most two rare mutation events in the dual process since the probability of more than two rare mutations by time $\frac{\log N}{\beta} + t$ is $O(e^{3\beta t} N^{-3})$ and in the case of *two rare mutation events* they take place in *distinct mesostates*. In the case $M = 2$ we only need to consider the first rare mutation since the dual then jumps to 0 but this is not always true if $M > 2$. We now give the additional step needed to handle this added complexity.

9.12 Emergence and Fixation: The Selection-Dominant Case

Recall that as $N \to \infty$

$$E[(x_{E_0}^N(j_1, t_N(\delta)))x_{E_0}^N(j_2, t_N(1)+t)] = 1 - Z_1 - Z_2 + \mathcal{E}(N,t), \quad (9.951)$$

where $Z_1 = Z_1(N)$ is the contribution over the event that exactly one rare mutation has occurred before time t and $Z_2 = Z_2(N)$ is the contribution over the event that exactly two rare mutations have occurred before time t and $\mathcal{E}(N,t)$ is the error terms coming from more than two rare mutations. We omit the N in the notation here for convenience. We now derive formulas of Z_1, Z_2.

The probability that a rare mutation has occurred before time t is for *given* dual process just $(n(t) = n(t, N)!)$:

$$1 - e^{-n(t)}, \quad (9.952)$$

where

$$n(t) := \frac{m}{N} \int_0^t \bar{\Pi}_{t_1(N),t_2(N)}^{N,j_1,j_2}(s)ds, \text{ where } m = \sum_i m_{i,E_1}, \quad (9.953)$$

with $t_1(N) = t_N(\delta)$ and $t_2(N) = t_N(1)+t$. Define $\|\tilde{t}\|_{G_{12}}$ counts the number of G_{12} ranks in the mesostate $\tilde{t}$ where a G_{12} rank (mesostate) is a green rank (mesostate) produced by a collision-coalescence event between a rank from the family starting at j_1 and a rank from the family starting at $j_2 \neq j_1$. Furthermore $\bar{u}_{G_{12}}$ is the number of green mesostates and $\bar{U}_{G_{12}}^N$ its empirical distribution. We write the integrand in (9.953) in the form:

$$\bar{\Pi}_{t_1,t_2}^{N,j_1,j_2}(s) = \bar{\Pi}_{t_1}^{N,j_1}(s) + \bar{\Pi}_{t_2}^{N,j_2}(s) + \mathfrak{g}_{t_1,t_2}^{N,j_1,j_2}(s) + E(N,s). \quad (9.954)$$

Here

$$\bar{\mathfrak{g}}_{t_1,t_2}^{N,j_1,j_2}(s) := \int \|\tilde{t}\|_{G_{12}} [\bar{u}^N(s)\bar{U}^N(s,d\tilde{t}) + \bar{u}_{G_{12}}^N(s)\bar{U}_{G_{12}}^N(s,d\tilde{t})]. \quad (9.955)$$

Now let ($\mathbb{E}_N$ is defined in (9.936))

$$\mathcal{A}_{(u,t)}^{N,i} := 1 - \mathbb{E}_N[\mu_0^{\otimes}(A_u^{N,i}(t))] \quad (9.956)$$

$$= 1 - \mathbb{E}_N\left[\sum_{\tau \in \mathcal{T}} \mu_0^{\otimes}(A_u^i(t,\tau))\|\tau\|\bar{V}^N(u,\tau, L > t-u)\right].$$

Here $\mathcal{A}_{(u,t)}^{N,i}$ represents the effect of a rare mutation at a randomly chosen site in a randomly chosen mesostate from type i to E_1 which occurs time u evaluated at time t. If two rare mutations are involved, as noted above, we can assume that they occur in different mesostates so that they occur independently.

The contribution of exactly one rare mutation is then given by the analogous argument as in (9.937) as:

$$Z_1 := E\Big[\sum_i \int_0^t n'(t_1) e^{-n(t_1)} e^{-(n(t)-n(t_1))} \frac{m_{i,E_1}}{m} \mathcal{A}_{(t_1,t)}^{N,i} dt_1\Big] \quad (9.957)$$

$$= E\Big[e^{-n(t)} \sum_i \Big(\frac{m_{i,E_1}}{m} \int_0^t \mathcal{A}_{(t_1,t)}^{N,i} dn(t_1)\Big)\Big].$$

The contribution of exactly two rare mutations is given by:

$$Z_2 := \sum_{i,j} E\Big[\int_0^t \int_{t_1}^t n'(t_1) e^{-n(t_1)} n'(t_2) e^{-(n(t_2)-n(t_1))} e^{-(n(t)-n(t_2))} \quad (9.958)$$

$$\mathcal{A}_{(t_1,t)}^{N,i} \mathcal{A}_{(t_2,t)}^{N,j} dt_2 dt_1\Big]$$

$$= E\Big[\sum_{i,j} \frac{m_{i,E_1} m_{j,E_j}}{m} e^{-n(t)} \int_0^t \int_{t_1}^t \mathcal{A}_{(t_1,t)}^{N,i} \mathcal{A}_{(t_2,t)}^{N,j} dn(t_2) dn(dt_1)\Big].$$

Now we return (having (9.951) in mind) to the calculation of moments.

We must now evaluate these terms times of the form $\frac{\log N}{\beta} + t$ and then consider the limit as $t \to -\infty$. For that purpose insert for $n(\cdot)$ the expression in (9.953) to express everything in terms of $\bar{\Pi}^{N,j_1}$, Π^{N,j_2}, $\mathfrak{g}_{t_1,t_2}^{N,j_1,j_2}$ and $E(\cdot,\cdot)$. Then we again have to exploit the expansion of random exponentials as carried out in the Subsubsection 9.12.11 at the end of Step 2.

In order to prove (9.949) it suffices to show that the following three expressions are equal:

$$\lim_{N\to\infty} E\Big[\big(e^{-\beta t_N(\delta)} \hat{x}_{E_1}^N(t_N(\delta))\big)^2\Big] \quad (9.959)$$

$$= \lim_{t\to-\infty} \lim_{N\to\infty} E\Big[\Big(e^{-\beta(\frac{\log N}{\beta}+t)} \hat{x}_{E_1}^N\big(\frac{\log N}{\beta}+t\big)\Big)^2\Big]$$

$$= \lim_{t\to-\infty} \lim_{N\to\infty} E\Big[\Big(e^{-\beta(\frac{\log N}{\beta}+t)} \hat{x}_{E_1}^N\big(\frac{\log N}{\beta}+t\big) \cdot e^{-\beta t_N(\delta)} \hat{x}_{E_1}^N(t_N(\delta))\Big)\Big].$$

It therefore suffices to show that the following two groups of identities hold:

$$\lim_{N\to\infty} NE[e^{-\beta t_N(\delta)} x_{E_1}^N(1,t_N(\delta))] = \lim_{t\to-\infty} \lim_{N\to\infty} NE[e^{-\beta(\frac{\log N}{\beta}+t)} x_{E_1}^N(1, \frac{\log N}{\beta}+t)]$$

$$(9.960)$$

9.12 Emergence and Fixation: The Selection-Dominant Case

and

$$\lim_{t\to-\infty}\lim_{N\to\infty} N^2 E[e^{-\beta t_N(\delta)} x_{E_1}(1,t_N(\delta))e^{-\beta(\frac{\log N}{\beta}+t)} x_{E_1}(2,\frac{\log N}{\beta}+t)] \quad (9.961)$$

$$= \lim_{t\to-\infty}\lim_{N\to\infty} N^2 E[e^{-2\beta t_N(\delta)}(x_{E_1}^N(1,t_N(\delta)))(x_{E_1}^N(2,t_N(\delta)))]$$

$$= \lim_{t\to-\infty}\lim_{N\to\infty} N^2 E[e^{-2\beta(\frac{\log N}{\beta}+t)} x_{E_1}^N(1,\frac{\log N}{\beta}+t) x_{E_1}^N(2,\frac{\log N}{\beta}+t)].$$

In view of (9.960) the latter can be reformulated in terms of covariances,

$$\lim_{t\to-\infty}\lim_{N\to\infty} N^2 e^{-(\beta(2\frac{\log N}{\beta}+t-\delta\frac{\log N}{\beta}))} \text{Cov}(x_{E_0}^N(1,\frac{\log N}{\beta}-\delta\frac{\log N}{\beta}), x_{E_0}^N(2,\frac{\log N}{\beta}+t))$$

$$= \lim_{t\to-\infty}\lim_{N\to\infty} N^2 [e^{-2\beta(\frac{\log N}{\beta}+t)} \text{Cov}(x_{E_0}^N(1,\frac{\log N}{\beta}+t), x_{E_0}^N(2,\frac{\log N}{\beta}+t))], \quad etc. \quad (9.962)$$

To verify (9.960) note that using the dual expression we can show that both sides are given by

$$E[W^{g,*}]\int_{-\infty}^{0} e^{\beta s} \mathcal{A}(s,0)ds, \quad (9.963)$$

where $W^{g,*}$ is given by (9.838), recall $\mathcal{A}_{(u,t)}^{N,i}$ is defined for the time horizon $\beta^{-1}\log N + t$, and $\mathcal{A}_{(u,t)}^{N,i}$ is given by (9.956)

$$\mathcal{A}(u,t) = \lim_{N\to\infty} \mathcal{A}_{(\frac{\log N}{\beta}+u,\frac{\log N}{\beta}+t)}^{N,i}. \quad (9.964)$$

To complete the proof of (9.962) recall that the covariance arises from the contribution of the green ranks created by collisions between the two clouds descending from each of the two initial factors and sites (recall Step 3 of the proof of Proposition 8.29). Then using (9.951) to compute second moments we can then verify that the terms remaining in the covariance expression in (9.962) are given by the G_{12} ranks and therefore given by (9.955). Then (noting that the Z_2 terms correspond to a higher order correction) using (9.929) we obtain that this expression equals:

$$\lim_{t\to-\infty}\lim_{N\to\infty} \text{Cov}(e^{-\beta(t_N(1)+t)} x_{E_1}^N(t_N(1)+t), e^{-\beta t_N(\delta)} x_{E_1}(t_N(\delta))) \quad (9.965)$$

$$= \lim_{t\to-\infty}\lim_{N\to\infty} \text{Cov}(e^{-\beta(t_N(1)+t)} x_{E_1}^N(t_N(1)+t), e^{-\beta(t_N(1)+t)} x_{E_1}^N(t_N(1)+t))$$

$$= \lim_{t \to -\infty} \lim_{N \to \infty} \text{Cov}(e^{-\beta t_N(\delta)} x_{E_1}(t_N(\delta)), e^{-\beta t_N(\delta)} x_{E_1}(t_N(\delta)))$$

$$= 2\hat{\kappa}_2 E[W_1] E[W_2] \int_{-\infty}^{0} e^{2\beta s} \mathcal{A}(s,0) ds.$$

(b)

Given $t_n \to \infty$ $e^{-\beta(\frac{\log N}{\beta} + t_n)} \hat{x}_{E_1}^N(\frac{\log N}{\beta} + t_n)$ is Cauchy in L^2 if $N \to \infty, t_n \to -\infty$ with the limit point denoted by $^*\mathcal{W}$. Moreover the limit is unique a.s. and

$$\lim_{t \to -\infty} \lim_{N \to \infty} E\left[\left(^*\mathcal{W} - e^{-\beta(\frac{\log N}{\beta} + t)} \hat{x}_{E_1}^N(\frac{\log N}{\beta} + t))\right)^2\right] = 0. \quad (9.966)$$

Therefore to verify that $e^{-\beta(\frac{\log N}{\beta} + t)} \hat{x}_{E_1}^N(\frac{\log N}{\beta} + t))$ converges to $\mathcal{W}^*$ in L^2 as $t \to -\infty$ and

$$\mathcal{L}[^*\mathcal{W}] = \mathcal{L}[\mathcal{W}^*], \quad (9.967)$$

it suffices by (9.946) and (9.966) to prove (9.949). This has been done in part (a). q.e.d.

Remark 206. We have now proved $\mathcal{L}[^*\mathcal{W}] = \mathcal{L}[\mathcal{W}^*]$ for both the mutation-dominant and selection-dominant cases. We conjecture that it is also true for the parameter regime where $E[(W_{P_e})^2] = \infty$. Although $E[(W_{P_e})^2]$ does not appear in the final calculations, for example (9.965), in order to prove this result it would be necessary justify the cancelations of this term that lead to the latter.

9.12.14 Further Perspectives: Discussion of the Non-ergodic Mutation Case

We first consider the collision-free case and then the collision regime.

Collision-Free Regime

In order carry over our analysis to the case where the mutation on E_0 is nonergodic, we need the following properties. We state this as an Ansatz concerning the limiting type and lifetime distribution, The result was proved in the ergodic case in Proposition 9.79 and in their subsubsection we give heuristic arguments to support it and explain the additional technical ingredient needed to establish it in the *non-ergodic* case but this will not be developed here in full detail.

9.12 Emergence and Fixation: The Selection-Dominant Case

Ansatz 1. *Assume that $t_N \to \infty$ and $t_N = o(\frac{\log N}{\beta})$. Then (9.837) and (9.840) are satisfied.* □

Even in the non-ergodic case when $\beta > \beta_{Pe}$ case the tightness of the measures $V^N(t_N)$ remains true. However since

$$\lim_{t \to \infty} \sum |\hat{t}| U^N(t, |\hat{t}|) = \infty \text{ in the case } \beta > \beta_{Pe}, \tag{9.968}$$

we do not have the tightness of the size-biased measures and we cannot use the argument we gave in Proposition 9.82. Nevertheless we conjecture that the Ansatz is true in the general case. Although we will not provide the proof we will present some heuristic arguments to support it and explain the additional technical steps which would be needed to establish it. We will then sketch the proofs of the main results based on the Ansatz and its extension to the collision regime in next subsubsection.

In order to obtain some insight into the role of the large mesostates, we now introduce a caricature of the mesostate process that captures some features of the former. Consider a process that produces mesostates with rate β_{Pe} and with mesostates that grow with rate $\beta > \beta_{Pe}$. For this caricature we give an informal calculation to indicate that the randomness in $v(t)$, the total number of occupied sites arises from the growth of the number of mesostates Pe sites. For that purpose we determine the asymptotics of first and second moments of the number of sites as $t \to \infty$.

We calculate the *first moment* of the number of occupied sites. Note $\frac{\beta_{Pe} e^{-\beta_{Pe}(t-s)} ds}{1-e^{-\beta_{Pe}t}}, 0 \le s \le t$, is the distribution of the age of Pe sites and the expected size of a Pe site of age $t - s$ is $e^{\beta(t-s)}$.

$$E[v(t)] = E[W_{Pe} e^{\beta_{Pe}t} \int_0^t e^{\beta(t-s)} \frac{\beta_{Pe} e^{-\beta_{Pe}(t-s)} ds}{1 - e^{-\beta_{Pe}t}} ds] \tag{9.969}$$

$$\sim E[W_{Pe} \int_0^t e^{\beta(t-s)} e^{\beta_{Pe}s} ds] \sim \text{const} \cdot e^{\beta t}, \text{ as } t \to \infty,$$

Note that asymptotically as $t \to \infty$ the main contribution comes from individuals born in $[0, \varepsilon t]$ for any $\varepsilon > 0$. Furthermore if mesostate survives for a intermediate time then the successor tail offspring mesostate of this mesostate continues to grow at rate β.

Now consider the *second moment*. Let W_i be the growth constant of the population of the i-th Pe-site born at time s_i. First conditioning on W_{Pe} and using the independence of the $\{W_i, \ i \in \mathbb{N}\}$ we can show that

$$E[v^2(t)] = E\left[\sum_{i=1}^{W_{Pe} e^{\beta_{Pe}t}} W_i e^{\beta(t-s_i)}\right]^2$$

$$= \frac{1}{(1 - e^{-\beta_{Pe}t})^2} E\left[W_{Pe} e^{\beta_{Pe}t} (W_{Pe} e^{\beta_{Pe}t} - 1)\right]$$

$$\cdot E\left[\left(\int_0^t W_j e^{\beta(t-s_j)} e^{-\beta_{Pe}(t-s_j)} ds_j \int_0^t W_i e^{\beta(t-s_i)} e^{-\beta_{Pe}(t-s_i)} ds_i\right)\right]$$

$$+ \frac{1}{(1-e^{-\beta_{Pe}t})^2} E[W_{Pe} e^{\beta_{Pe}t}(\int_0^t W_j e^{\beta(t-s_j)} e^{-\beta_{Pe}(t-s_j)} ds_j)^2]$$

$$\sim \frac{1}{(1-e^{-\beta_{Pe}t})^2} \left(E[W_{Pe}^2 e^{2\beta_{Pe}t}](E[W_i])^2 e^{2(\beta-\beta_{Pe})t}\right)$$

$$+ E[W_{Pe}]E[W_i^2] e^{\beta_{Pe}t} e^{2(\beta-\beta_{Pe})t}]. \tag{9.970}$$

A similar calculation gives

$$\lim_{t\to\infty} e^{-2\beta t} \mathrm{Var}[v(t)] = (E[W_1])^2 \mathrm{Var}[W_{Pe}], \tag{9.971}$$

$$\lim_{t\to\infty} e^{-\beta t} v(t) = W_{Pe} E[W_1] \quad \text{in } L^2. \tag{9.972}$$

This suggests that the randomness arises from the mesostate CMJ-process, namely their number while for the size of them some averaging occurs.

Remark 207 (Mesostate growth process for nonergodic E_0-mutation).

We now refine this picture taking into account the finite lifetime of mesostates and following the growth of the family initiated when a new mesostate starting with the single factor $(1,\ldots,1,0)$ at one site is created. We assume that s is sufficiently large to make β large. Then β will represent the growth rate of the *tail-mesostate* corresponding to the right tail of the tableau, that is, the descendant mesostate containing the *first* offspring of the founding site and rank of the label. We can then follow the tail-mesostate containing this ancestor. The tail-mesostate will lose ranks and produce offspring, in particular "small offspring mesostates" at rate m_{21} during its lifetime. Therefore it has linear birth rate and constant death rate except when its length is 1.

We now consider the distribution of microstates and lifetimes in the case of non-ergodic E_0-mutation. If $\beta > m_{21}$, it follows from (9.969) that the main contribution to the empirical measure of microstates at times $O(\frac{\log N}{\beta})$ comes from tail-mesostates with age $O(\frac{\log N}{\beta})$. Therefore in order to verify that $V^N(t_N, \tau, x)$ converges, as $N \to \infty$, it would suffice to show that the empirical distribution of local single site states and lifetimes of a tail-mesostate converges as the age of the tail-mesostate goes to ∞.

Recall that we have already proved that there exists a limiting *age, lifetime* and *type* distribution for $Pe \cup UPe$ sites (in this case the lifetime is ∞). We must however also consider the role of the Tr sites, that is, the set of sites with *finite lifetime*. Note that the Tr sites are offspring of UPe and Pe sites. Moreover the proportion of Tr sites at time $t - T$ that have Tr offspring alive at time t decays as $e^{-\lambda T}$ and the number of Tr sites at time t older than T is $e^{\beta t} \cdot e^{-\lambda T}$. This means that we can choose T so that only a proportion $< \varepsilon$ are Tr sites older than T. But we can then add in the Tr trees produced by the typical Pe and UPe sites during the interval $(t-T, t]$ and

9.12 Emergence and Fixation: The Selection-Dominant Case

hence obtain an approximate stable type distribution at time t as $t \to \infty$. The type of a *Tr* site then depends on the *UPe* or *Pe* type of its parent and its age. Since each *UPe* or *Pe* site can produce such a *Tr* site it follows that the number of such sites also grows as $e^{\beta t}$.

This suggests that there is a stable type distribution for (active) sites and the total number of sites grows as $e^{\beta t}$. However the lifetime distribution of a Tr site depends on the *genealogy* of the site. This requires the analysis of the limiting distribution of microstate and lifetime distribution as the age of a tail-mesostate goes to infinity. The determination of the asymptotic structure of the mesostate Markovian dynamics conditioned on age increasing to infinity will not be developed here but the Ansatz asserts the existence of a limiting distribution. To establish the Ansatz it would be sufficient to prove that the distribution of microstates and lifetimes of a mesostate starting at $\hat{t}_*$ and conditioned to live up to time T converges as $T \to \infty$. The idea is that a tail-mesostate has an increasing number of internal factors that grow independently. In this way the system has a branching-type structure and this suggests that the type distribution should converge.

Working just with mean production of new sites, the mean calculations should give convergence to an invariant mean distribution of microstates, lifetimes and ages. However the full branching property is not satisfied because there are dependencies in the lifetime of related sites. This raises the question of convergence in L^2 which is unresolved. This is needed for the calculation of second moments since collision calculations involve interaction of two or more sites.

Remark 208 (Conclusion). To establish (9.840) in the case $\beta > \beta_{Pe}$ a more complete analysis of the limiting tree structure of tail-mesostates of age T, as $T \to \infty$, resulting from the Markov dynamics is required and will not be carried out in this monograph.

Asymptotics in the Collision Regime

We now consider the asymptotic behaviour in the collision regime (times $\beta^{-1} \log N + t$) for the case of non-ergodic E_0-mutation. We start with some observations and open points.

Remark 209. Consider the case $\beta > \beta_{Pe}$ and times $\beta^{-1} \log N + t$. Then there are $f(N) = uN^{\frac{\beta_{Pe}}{\beta}}$ *Pe* mesostates and uN occupied sites. Let $k^*(N) = E[|\hat{t}|] = N^{\frac{\beta - \beta_{Pe}}{\beta}}$ be the expected size of the mesostate. The probability that the mesostates have a point in common is $\frac{(k^*(N))^2}{N}$. Then the total number of linked pairs linked in this way is

$$f(N)^2 \frac{(k^*(N))^2}{N}. \tag{9.973}$$

Then the probability that a randomly chosen pair of sites in $\{1, \ldots, N\}$ are connected by an edge is $\frac{u^2}{N}$ since at time $\frac{\log N}{\beta} + t$, $f(N)k^*(N) = \text{constant} \cdot u^N(t)N$.

We can then consider the role of coalescence between mesostates that contain a common point in $\{1, \ldots, N\}$ in their *Geo* maps. This would produce merging between these mesostates which we can denote by the creation of an edge between the mesostates. Noting the possibility of multiple mergers with different pairs this produces a random graph in which vertices are mesostates and edges represent coupled joint ranks. Recall that an Erdős-Rényi random graph with $f(N) = ue^{\beta_e t} N^{\delta_1}$ vertices ($\delta_1 = \frac{\beta_{Pe}}{\beta}, \delta_2 = \frac{\beta - \beta_{Pe}}{\beta}$) and connection probability $\frac{e^{2(\beta - \beta_e)t} N^{2\delta_2}}{N}$ has a giant component if $\beta_{Pe} < \beta$ and $u = O(1)$. However we also have fragmentation which occurs when a resolution decouples one of the pair of mesostates so this looks like a multiplicative coalescent with fragmentation process (cf. [Ber, Dia]).

Open Problem 9.89. *The question as to the combined effect of coagulation and fragmentation in the case $\beta > \beta_{Pe}$ is open. In particular is there a critical value u^* such that for densities below it the coagulation-fragmentation does not exhibit a giant component?*

In contrast, if $\delta_1 = 1$ then we have an Erdős-Rényi random graph with uN vertices and connection probability $\frac{(k^*)^2}{N}$ so there is no giant component for small u. Then in this case the system (9.871), (9.872) provides and alternative to the system used in Subsection 9.11.

In the non-ergodic case we must work directly with sites rather than mesostates. We first note that the number of sites is bounded above by $We^{\beta t}$. Recall that $\bar{V}(t)$, $\bar{V}_g(t)$ arise as the limits as $N \to \infty$ of $V^N(\frac{\log N}{\beta} + t)$, $V_g^N(\frac{\log N}{\beta} + t)$. As before the number of collisions between sites is of order $O(e^{2\beta t})$ and we conclude that $v_g(t)$ grows at rate $e^{2\beta t}$. However the same difficulty arises in the analysis of $\bar{V}(t)$, $\bar{V}_g(t)$ in that we cannot use size-biasing in the non-ergodic case. However we expect that further analysis of the tree structure of the mesostate process as outlined in subsubsection 8.11.4 will justify the following extension of our basic Ansatz to the time regime $\frac{\log N}{\beta} + t$ in the limit $t \to -\infty$. Under the Ansatz the analysis below of rare mutations and emergence then proceeds as in the ergodic case.

Ansatz 2. *In the non-ergodic case the limiting distribution of types and lifetimes satisfy*

$$\lim_{t \to -\infty} \bar{V}(t) = \bar{V}(\infty), \qquad (9.974)$$

where $\bar{V}(\infty)$ is given by Ansatz 1, and the limits

$$\lim_{t \to -\infty} e^{-2\beta t} \bar{v}_g(t) = \hat{\kappa}_2 E[W^2], \qquad (9.975)$$

and

$$\bar{V}_g(\infty, \tau, L) := \lim_{t \to -\infty} \lim_{N \to \infty} \bar{V}_g^N(\frac{\log N}{\beta} + t, \tau, L) \qquad (9.976)$$

exist. □

9.12 Emergence and Fixation: The Selection-Dominant Case   697

Remark 210. We now briefly sketch the idea as to how to verify this Ansatz. Note that asymptotically as $t \to \infty$ the main contribution to the number of occupied sites comes from individuals born in $[0, \varepsilon t]$ for $\varepsilon > 0$ (see (9.969)). Therefore to prove (9.976) it would suffice to show that the size biased empirical distribution of local single site states of a mesostate converges as the age of the mesostate goes to ∞. But recall that we have proved that there exists a limiting age and type distribution for $Pe \cup UPe$ sites. We must also consider the role of the Tr sites. The idea is then to note that the Tr sites are offspring of UPe and Pe sites and depend on the stable age and type distribution of the latter types.

9.12.15 Dependence of the Emergence Time on the Initial Measure if $\beta > \beta_{Pe}$

It is natural to ask what are the implications of the dual process dichotomy, $\beta > \beta_{Pe}$ vs. $\beta = \beta_{Pe}$, for the behaviour of the original interacting systems. We now briefly indicate an important way in which the emergence behaviour differs in these two cases, namely, how the emergence time is influenced by the initial composition of the population.

To illustrate what can happen consider first the simple case $M = 2$, $s > 0$, $m_{12} = 0$, $m_{21} = 0$, $d = 0$ (recall Example 20). Now consider the case in which the initial measure μ_0 satisfies $\mu_0(1) = 1$. Then the first rare mutation (and therefore emergence) takes place after a time of order $\beta^{-1} \log N$. On the other hand if $\mu_0(2) = 1$ then the first rare mutation (and therefore emergence) takes place after a time of order $(\beta_{Pe})^{-1} \log N$. To see this recall that the dual at time $\frac{\log N}{\beta} + t$ will be decomposed into $N^{\frac{\beta_{Pe}}{\beta}} e^{\beta_{Pe} t}$-many factors given by tableau each of the form

$$\begin{array}{l}(010)\\(100)\ (010)\\(100)\ (100)\ (010)\\(100)\ (100)\ (100)\ (110)\end{array} \qquad (9.977)$$

but with random sizes such that the total number of active ranks is $Ne^{\beta t}$.

Now consider the case where $\mu_0(1) = 1$. Then a rare mutation $(100) \to (000)$ at any of the $Ne^{\beta t}$ ranks will result in a jump to zero in the dual expression for the expected mass in E_0. On the other hand if $\mu_0(2) = 1$, then the only rare mutation transition that results in a jump to zero in the dual expression for the expected mass in E_0 is the rare mutation $(010) \to (000)$ at a rank in a row not containing a factor (100). But there are only $N^{\frac{\beta_{Pe}}{\beta}} e^{\beta_{Pe} t}$ such ranks, one for each mesostate.

A modification of the above argument gives the same result for the case $m_{21} > 0$, $m_{12} = 0$, $d > 0$, $c > 0$ and s sufficiently large. However it is not depending on the initial state in the mutation-dominant case (m_{21} sufficiently large).

9.13 Proof of Propositions 9.1, 9.2, 9.4

The proof of the three key results formulated in the three propositions 9.1, 9.2, 9.4 will be given in three respective subsubsections based on the duality theory developed in the previous Subsections 9.5–9.12. We first prove emergence, fixation, uniqueness of the entrance law and finally the droplet growth.

9.13.1 Proof of Proposition 9.1, Part 1: Emergence Time

We derive now upper and lower bounds on the emergence time showing emergence happens at times $\beta^{-1} \log N + O(1)$ as $N \to \infty$ and later Subsubsection 9.13.2 we shall show that at times $\beta^{-1} \log N + t_N$ fixation occurs for any $t_N \to \infty$. The basic tool is the dual $\tilde{\mathcal{G}}^{++,N}$ as well as the limiting (as $N \to \infty$) collision-free process $\tilde{\mathcal{G}}^{++}$ which provides bounds.

Proposition 9.90 (Lower bound on rare mutation time in dual).
Consider the dual process $\tilde{\mathcal{G}}^{++}$. *Let* T_N^{rare} *denote the time of the first rare mutation event if the dual is started with a single factor* $(1,\ldots,1,0)$ *at one site, that is, the first time that a mutation jump from a type* $\{1,\ldots,M\}$ *to type* $\{M+1\}$ *occurs at an active rank.*
Then

$$\lim_{t \to -\infty} \limsup_{N \to \infty} P[T_N^{\text{rare}} - \frac{\log N}{\beta} < t] = 0. \qquad \Box \qquad (9.978)$$

Corollary 9.91 (Lower bound on emergence time for X^N).
Consider the process $(X^N(t))_{t \geq 0}$ *and assume that the initial measure is deterministic and at each site at time* 0 *the state is given by the* δ-*measure on* $(x_1,\ldots,x_M,0)$, $x_i > 0$ *for* $i = 1,\ldots,M$.
Then emergence of type $M+1$ *does not occur in times* $o(\frac{\log N}{\beta})$, *namely,*

$$\lim_{t \to -\infty} \limsup_{N \to \infty} E[x_{E_0}(\frac{\log N}{\beta} + t)] = 1. \qquad \Box \qquad (9.979)$$

We now consider the upper bound on the emergence time of type $M + 1$.

Proposition 9.92 (Upper bound on emergence time).
Consider the process X^N *and assume that the initial state at each site at time* 0 *is given by* $(x_1,\ldots,x_M,0)$, $x_i > 0$ *for* $i = 1,\ldots,M$.
Then for $t \in \mathbb{R}$:

$$\limsup_{N \to \infty} E[x_{E_0}^N(\frac{\log N}{\beta} + t)] > 0. \qquad \Box \qquad (9.980)$$

So far we now only know that at times $\beta^{-1}\log N + t$ the density of $(M+1)$-types is positive with positive probability, but we want to actually claim that it is positive with probability 1.

Proposition 9.93 (Upper bound on emergence time in the strong sense). *Under the assumptions of Proposition 9.92 we have for every $t \in \mathbb{R}$, that*

$$\lim_{\varepsilon \to 0} \liminf_{N \to \infty} P[\bar{x}_{M+1}^N(\beta^{-1}\log N + t) > \varepsilon] = 1. \quad \square \quad (9.981)$$

Corollary 9.94 (Limiting expression for mean of E_1-mass at emergence).
We have in the mutation dominance regime, that

$$E[x_{E_1}^N(\beta^{-1}\log N + t)] \xrightarrow[N \to \infty]{} 1 - E\left[\int_{-\infty}^{t} r'(u)e^{-r(u)} H_{ii,iii}(u,t)du\right] \quad (9.982)$$

where

$$r(u) = m_{M+1,up} W e^{\beta u} \quad (9.983)$$

and $H_{ii,iii}$ is given by (9.991). Moreover Then

$$\lim_{t \to -\infty} \lim_{N \to \infty} e^{-\beta t} E[x_{E_1}^N(\frac{\log N}{\beta} + t)] > 0. \quad (9.984)$$

$\square$

For the proofs of the above propositions we have to distinguish the mutation-dominant and the selection-dominant case.

Case 1: Mutation-Dominance
Consider first the mutation-dominance case, i.e. in this whole Part 1 **we assume that our parameters are in the mutation-dominance regime**.

In Lemma 9.63 we have identified the hazard rate for the rare mutation transition in the dual process and the probability that a rare mutation has occurred before time $\frac{\log N}{\beta} + t$ as well as the asymptotics as $t \to -\infty$.

Proof of Proposition 9.90. Using Proposition 9.40 and the $\tilde{\mathcal{G}}^{++}$-versions of the expression (9.608), together with (9.609) and (9.611) it follows that

$$\lim_{t \to -\infty} \limsup_{N \to \infty} P(T_N^{\text{rare}} \le \frac{\log N}{\beta} + t) \le \lim_{t \to -\infty} (\text{const} \cdot e^{\beta t}) = 0. \quad \text{q.e.d.} \quad (9.985)$$

Proof of Corollary 9.91. We use the dual representation of $E[x_{E_0}^N(t)]$ by $\tilde{\mathcal{G}}^{++,N}$. Then the claim follows since before a rare mutation event occurs the integrated value of the dual with initial state has a.s. mass 0 on type $M+1$, then the dual starting with a single factor $(11 \cdots 10)$ satisfies $\tilde{\mathcal{G}}^{++,N} \supseteq (E_0^{\mathbb{N}})^{\mathbb{N}}$ (each unresolved pair of factors arising from a selection event lies between $(1 \cdots 10) \otimes 1$ and $(1 \cdots 10) \otimes (1 \cdots 10)$) and hence contributes 1 in the expectation. The hazard function for a rare mutation to occur can be bounded since we bound $\tilde{\mathcal{G}}^{++,N}$ from above by $\tilde{\mathcal{G}}^{++}$. Then the claim follows from (9.985).   q.e.d.

Proof of Proposition 9.92. We now have to consider $\tilde{T}_N^{\text{rare}}$ the time till the first rare mutation in $\tilde{\mathcal{G}}^{++,N}$ rather than $\tilde{\mathcal{G}}^{++}$. To do this we have to control the difference between $\tilde{\mathcal{G}}^{++}$ and $\tilde{\mathcal{G}}^{++,N}$, the latter may have fewer sites and ranks due to collision. We argue as follows. We know that the number of potential sites (and their offspring) lost to collision can be bounded by the green ranks and green mesostates and therefore can be bounded by a term of order $O(e^{2\beta t})$ as shown in (9.915).

Now using the expression (9.608), together with (9.609) and (9.611) it follows that for every $t \in \mathbb{R}$:

$$\liminf_{N \to \infty} P(\tilde{T}_N^{\text{rare}} \leq \frac{\log N}{\beta} + t) > 0, \tag{9.986}$$

and

$$\lim_{t \to -\infty} e^{-\beta t} \liminf_{N \to \infty} P(T_N^{\text{rare}} \leq \frac{\log N}{\beta} + t) > 0. \tag{9.987}$$

We now have to use this information to evaluate the dual expression for the expectation in (9.980). In Lemma 9.63(b),(c),(d) we have described the possible outcomes after such a rare mutation event and their probabilities. As a result we can conclude that after such a rare mutation event there is a positive probability that a zero factor is produced after a finite random time. Then it follows that:

$$\limsup_{N \to \infty} E[x_{E_0}^N(\frac{\log N}{\beta} + t)] < 1. \quad q.e.d. \tag{9.988}$$

Proof of Corollary 9.94. We can derive an "explicit" expression for the limiting expected type E_0-mass. The reader should review at this point Subsubsection 9.11.4. From Lemma 9.63 we take a number of quantities to define for $-\infty < u < t$ we get

$$H_{ii}^N(u,t,\mu_0) = \int_{\mathbb{T}^E} \frac{1}{|\bar{t}|} \sum_{i=1}^{|\bar{t}|} \left(\int_{\mathbb{I}|\bar{t}'|} (1 - h_{ii}^N(\bar{t},i,t-u,\bar{t}'))d\mu_0^{\otimes} \right) U^N(u,d\bar{t}) > 0, \tag{9.989}$$

and for $-\infty < u < t$, we set

$$H_{iii}^N(u,t) = \int_{\mathbb{T}^E} \frac{1}{|\bar{t}|} \sum_{i=1}^{|\bar{t}|} h_{iii}^N(\bar{t},i,t-u) U^N(u,d\bar{t}) > 0. \tag{9.990}$$

9.13 Proof of Propositions 9.1, 9.2, 9.4

Furthermore define:

$$H_{ii,iii}(u,t) = \int_{\mathfrak{T}^E} \frac{1}{|\bar{t}|} \sum_{i=1}^{|\bar{t}|} h_{iii}(\bar{t},i,t-u) U(u,d\bar{t}) \qquad (9.991)$$

$$+ \int_{\mathfrak{T}^E} \frac{1}{|\bar{t}|} \sum_{i=1}^{|\bar{t}|} \left(\int_{\mathbb{I}|\bar{t}'|} (1 - h_{ii}(\bar{t},i,t-u,\bar{t}')) d\mu_0^{\otimes} \right) U(u,d\bar{t}).$$

From Lemma 9.63 it follows next that

$$\lim_{t_1 \to -\infty} e^{-\beta t_1} \lim_{N \to \infty} P(\text{two or more rare mutation events before time } \frac{\log N}{\beta} + t_1) = 0. \qquad (9.992)$$

This implies that as $t \to -\infty$ the main contribution to the integrated value at time $\frac{\log N}{\beta} + t$ is given by the occurrence of the first rare mutation event before this time and that the rank that underwent rare mutation has not been removed by a resolution. More precisely, using the expression (9.608), together with (9.609) and (9.611) it follows that there exists a sequence $o_N(1)$ converging to zero such that:

$$E[x_{E_1}^N(\frac{\log N}{\beta} + t)] \qquad (9.993)$$

$$= E\left[\left(1 - \frac{1}{N} \int_0^{\frac{\log N}{\beta} + t} \left[m_{M+1,\text{up}} \bar{W}^N(u) \beta e^{\beta u} e^{-m_{M+1,\text{up}} \bar{W}^N(u) e^{\beta u}} \right. \right. \right.$$
$$\left. \left. \left. (H_{ii}^N(u, \frac{\log N}{\beta} + t, \mu_0) + H_{iii}^N(u, \frac{\log N}{\beta} + t)) \right] du \right) \right] + o_N(1)(e^{\beta t}).$$

We have by (9.611) that

$$\lim_{t \to -\infty} \limsup_{N \to \infty} e^{-\beta t} o_N(e^{\beta t}) = 0. \qquad (9.994)$$

But since as $N \to \infty$, $\bar{W}^N(u) \to \bar{W} \in (0,\infty)$ a.s. and $U^N(u,\cdot) \longrightarrow U(u,\cdot)$ and using (9.619) it follows that:

$$E[x_{E_1}^N(\frac{\log N}{\beta} + t)] \to 1 - E[\int_{-\infty}^t r'(u) e^{-r(u)} H_{ii,iii}(u,t) du], \qquad (9.995)$$

where

$$r(u) = m_{M+1,up} W e^{\beta u}. \qquad (9.996)$$

Then

$$\lim_{t\to-\infty}\lim_{N\to\infty} e^{-\beta t} E[x_{E_1}^N(\frac{\log N}{\beta}+t)] > 0. \qquad (9.997)$$

Proof of Proposition 9.93. To obtain the strong emergence (9.981) we compute all higher moments exactly as in Section 7 - we do not give the details here.

Case 2: Selection-Dominant Case

The proof of emergence in the case of selection dominance follows the same lines as the above but uses the results of Subsection 9.12. For example, from (9.940) we have

$$\liminf_{N\to\infty} E[x_{E_1}^N(t_N+t)] \geq \varepsilon \sum_i \beta^{-1}(1 - E[e^{-m_{i,E_1} W_* e^{\beta t}}]) > 0$$

which then proves the analogue of (9.984) in this case. This proves Proposition 9.92.

We can then get a lower bound on the emergence time from (9.937). This proves Proposition 9.90 and Corollary 9.91.

9.13.2 Proof of Proposition 9.1, Part 2: Fixation

In this subsubsection the distinction between mutation dominance or selection dominance plays *no role*, essentially since once emergence has created a positive spatial intensity of E_1-types which is $\geq \varepsilon > 0$, then these types take over as time proceeds.

Next we want to determine the time scale of fixation. We need more information about the behaviour in the collision regime for that purpose we need to derive that the evolution of the dual in times $t_N = \beta^{-1}\log N + t$, $t \in \mathbb{R}$ as $N \to \infty$ be described by a limiting evolution in t and leads as $t \to \infty$ to fixation.

More precisely we next consider the phase of fixation and in the next proposition we use the dual process $\tilde{\mathcal{G}}_t^{++,N}$ to calculate the expected values of the type $M+1$ in order to establish fixation in time $\beta^{-1}\log N + t$ if $N \to \infty$, $t \to \infty$, or $\beta^{-1}\log N + s_N$ as $N \to \infty$ with $s_N \to \infty$.

Proposition 9.95 (Fixation in limiting dynamics).

(a) Assume that $(x_1^N(j,0),\dots,x_{M+1}^N(j,0))$, $j = 1,\dots,N$ are exchangeable and

$$\frac{1}{N}\sum_{j=1}^N 1_{(\varepsilon,1]}(x_{M+1}^N(j,0)) > \varepsilon > 0. \qquad (9.998)$$

Then

$$\lim_{t\to\infty}\lim_{N\to\infty} E[x_{M+1}^N(1,t)] = 1. \qquad (9.999)$$

(b) Assume that $x_{M+1}^N(j,0) = 0$, $j = 1, \cdots, N$. Then

$$\lim_{t\to\infty} \lim_{N\to\infty} E[x_{M+1}^N(1, \beta^{-1}\log N + t)] = 1. \qquad \Box \qquad (9.1000)$$

Corollary 9.96 (Fixation).
Consider the assumptions as above. Let $t_N = \beta^{-1}\log N + s_N$. Then for $s_N \to \infty$

$$\lim_{N\to\infty} E[x_{M+1}^N(1, t_N)] = 1. \qquad (9.1001)$$

$$\lim_{t\to\infty} \lim_{N\to\infty} \sup E[x_{E_0}^N(\frac{\log N}{\beta} + t)] = 0. \qquad \Box \qquad (9.1002)$$

Proof of Proposition 9.95. (a) Recall that

$$E[x_1^N(1,t) + \cdots + x_M^N(1,t)] = E\left[\int \tilde{\mathcal{G}}_t^{++,N} d(\mu_N^{\otimes \mathbb{N}})^{\otimes N}\right], \qquad (9.1003)$$

where $\mu_N = \mathcal{L}[(x_1^N(i,0),\ldots,x_{M+1}^N(i,0))]$ and $\mathfrak{s}(0)$ is given by (9.278). Note that then each resolved factor corresponds to $(1,\ldots,1,0)$ and each unresolved factor lies between $(1,\ldots,1,0)$ and $(1,\ldots,1,0) \otimes (1,\ldots,1,0)$. This means that each resolved or unresolved factor at a site j gives factor less than or equal to $x_1^N(j,0) + \cdots + x_M^N(j,0)$. The r.h.s. of (9.1003) has as $N \to \infty$ limit points of the form (here μ is a limit point of μ_N):

$$E[\int \tilde{\mathcal{G}}_t^{++} d((\mu^{\otimes \mathbb{N}})^{\otimes \mathbb{N}})]. \qquad (9.1004)$$

This expression goes by (9.998) to zero as $t \to \infty$, since the number of resolved factors $(1\cdots 10)$ tends to ∞ as $t \to \infty$.

(b) Furthermore applying (9.1003) to the time horizon $\beta^{-1}\log N + t_0 + t_0$, it follows that for any t_0 a (unbounded) series of rare mutation events takes place after time $\frac{\log N}{\beta} + t_0$ since there is a positive rate of rare mutation once K_t^N reaches εN for any $\varepsilon > 0$. In particular with $T_{N,L}$ being the waiting time for the first L rare mutation jumps in the dual process $\tilde{\mathcal{G}}^{++,N}$:

$$\lim_{t\to\infty} \lim_{N\to\infty} \inf P(T_{N,L} \leq \frac{\log N}{\beta} + t) = 1. \qquad (9.1005)$$

But since after each rare mutation there is a positive probability that a transition to a zero factor occurs it follows that after at most finitely many such events the dual becomes 0. This implies (9.999). q.e.d.

Remark 211. Note that the assumption of now down mutation from type $M + 1$ is irrelevant here, namely a rare mutation from type $M + 1$ to type M can also change $(1,\ldots,1,0)$ to $(1,\ldots,1,1)$ and therefore decrease by one the number of factors $(1,\ldots,1,0)$ which can become $(0,\cdots,0)$ in times of order 1, however in time horizon of interest we will have at most $O(N)$ such factors and since the rate of rare mutation is N^{-1}, there are only $O(1) = o(N)$ of these that are killed by rare mutations. Therefore this effect does not change the macroscopic dynamics of the number of factors.

Proof of Corollary 9.96. Now we have to work with the dual $\tilde{\mathcal{G}}_{t_N}^{++,N}$ and $N \to \infty$ instead of getting first with $N \to \infty, \tilde{\mathcal{G}}_t^{++}$ and then take $t \to \infty$. However since we have established emergence at times $\beta^{-1}\log N$ we can take this time as the initial time and consider the evolution for time s_N and use that $\tilde{\mathcal{G}}_{s_N}^{++,N}$ with $s_N = o(N)$ can be approximated by $\tilde{\mathcal{G}}_{s_N}^{++}$ and then (9.999) can be applied.

9.13.3 Proof of Proposition 9.2: Uniqueness for McKean–Vlasov Entrance Law

Here in this subsubsection again the distinction between mutation dominance and selection dominance plays absolutely no role.

The existence and uniqueness of the solution $\{\mathcal{L}_t\}_{t \geq t_0}$ of the McKean–Vlasov equation (9.18) starting at time t_0 in a given state $\mathcal{L}_{t_0} \in \mathcal{P}(\Delta_M)$ is classical (compare [DG99]) for any number of types. Moreover, the law of the corresponding process at a tagged site is characterized via the set-valued dual of the McKean–Vlasov process. Therefore the following two problems to solve remain.

(1) What we need first is existence and uniqueness of the *entrance law* from $-\infty$. This means that we have to identify the rate of growth of the solution as it comes in from $-\infty$. In other words we need the $M \geq 2$-type version of the assertion in (2.27) for $M = 1$ together with the argument showing that a unique mean curve of the entrance law from $-\infty$ of the McKean–Vlasov equation results in a unique entrance law from $-\infty$. Hence next one first has to show that one has a unique mean curve and then second given the mean curve a unique solution. An essential part of showing that indeed we have a unique mean curve is to verify this for the mean of type $(M+1)$.

(2) Secondly we consider the question of the existence of $^*\mathcal{W}$, i.e. the question whether

$$\mathcal{L}\big[e^{-\beta t} \int_{\Delta_M} \mu(\{M+1\})\mathcal{L}_t(d\mu)\big] = \lim_{N \to \infty} \mathcal{L}\big[e^{-\beta t}\int_{\Delta_M} x_{M+1} \Xi_N^{\log,\beta}(t,dx)\big]$$

(9.1006)

converges as $t \to -\infty$, to the law of a random variable $^*\mathcal{W}$ and verify that the distribution of the limit $^*\mathcal{W}$ is concentrated on $(0, \infty)$.

9.13 Proof of Propositions 9.1, 9.2, 9.4

The argument follows closely the argument given in Section 8 for the two-type case and the reader should recall this argument at this point.

Problem (1)

Step 1 For the first problem we can follow the same strategy using estimates on the growth of $\bar{K}_t$ to argue as in the proof of Proposition 2.3 that in fact every solution of the McKean–Vlasov equation satisfying

$$\limsup_{t \to -\infty} e^{\beta|t|} \int \mu(\{M+1\}) \mathcal{L}_t(d\mu) < \infty \tag{9.1007}$$

has a limit

$$A = \lim_{t \to -\infty} e^{\beta|t|} \int \mu(\{M+1\}) \mathcal{L}_t(d\mu). \tag{9.1008}$$

The proof is similar to that of Proposition 2.3 (see Subsubsection 8.3.2) where the key lemma giving (9.1008) under (9.1007) for the two-type case, i.e. $M = 1$ was Lemma 8.6 whose line of argument carries over but starting the dual with the indicator of $(1, \ldots, 1, 0)$.

Then using the $e^{\beta t}$ growth of the number of dual factors (i.e. ranks) following the discussion of the dual $\tilde{\mathcal{G}}^{++}$ and the result (9.354) we can conclude that indeed (9.1007) is satisfied for a potential limit dynamics. Then we get that the solutions of the McKean–Vlasov equation we are interested in, does satisfy (9.1008) with some random $A > 0$.

Step 2

Next we have to argue that given a mean-curve $\bar{x}_{M+1}(t)$ satisfying that

$$\lim_{t \to -\infty} e^{-\beta t} \bar{x}_{M+1}(t) \in (0, \infty), \tag{9.1009}$$

the law of the corresponding process at a tagged site is unique.

Recall that the solution of the McKean–Vlasov equation corresponds to the law of a multi-dimensional diffusion process with a prescribed mean curve. In Section 8 with one type on the lower level uniqueness for a given mean-curve was a problem for a one-dimensional diffusion with a driftterm varying in time, which we solved by coupling. Now we have for a given mean curve a multi-dimensional diffusion.

Recall the coupling technique for the case $M = 1$ in Section 8 which was based on the fact that the higher level type $(M + 1)$ hits zero along a sequence $(t_n)_{n \in \mathbb{N}}$ with $t_n \to -\infty$ and then comparing with the solutions started at these points.

The basic idea for the multitype case is as follows. Assume that we are given two solutions (with the same mean curve) $(x_1(t), \ldots, x_M(t), x_{M+1}(t))_{t \in \mathbb{R}}$ and $(\tilde{x}_1(t), \ldots, \tilde{x}_M(t), \tilde{x}_{M+1}(t))_{t \in \mathbb{R}}$. We will verify that we can couple the processes so that the total masses of the type $(M + 1)$ have simultaneous zeros $\{t_n\}_{n \in \mathbb{N}}$ with $t_n \to -\infty$ and we can couple the processes $(x_{M+1}(t))_{t \in \mathbb{R}}$ and $(\tilde{x}_{M+1}(t))_{t \in \mathbb{R}}$ and the processes $(x_1(t) + \cdots + x_M(t))_{t \in \mathbb{R}}$ and $(\tilde{x}_1(t) + \cdots + \tilde{x}_M(t))_{t \in \mathbb{R}}$. But the problem is that if we have coupled $\tilde{x}_{M+1}$ and x_M at the joint zeros, then there are M degrees of freedom left, namely the relative proportions of the types $x_1, \ldots, x_M$ at this moment of coupling.

The key idea to handle this problem is to show that for $s_N \to \infty$, $s_N = o(\log N)$, these proportions converge in distribution to the *mutation-selection equilibrium* π_{loweq} on the types $\{1, \ldots, M\}$ for the McKean–Vlasov limit on this reduced type space. We will now show more precisely that if $t_N = \frac{\log N}{\beta}$, then

$$\lim_{t \to -\infty} \lim_{N \to \infty} \mathcal{L}[(x_1^N(t_N + t), \ldots, x_M^N(t_N + t))] = \pi_{\text{loweq}}, \quad (9.1010)$$

so that we can, as shown below, couple solutions for all components.

To prove (9.1010) we will show that every solution to the McKean–Vlasov equation satisfying

$$\limsup_{t \to -\infty} e^{\beta|t|} x_{M+1}(t) < B < \infty, \quad (9.1011)$$

has the property that as $t \to -\infty$ the distribution of types $x_1, \ldots, x_M$ converges to the M-type McKean–Vlasov mutation-selection equilibrium π_{loweq}.

We prove this by comparing the $M + 1$ type McKean–Vlasov system starting with mass of type $M + 1$ of size $x_{M+1}(t_0) \leq Be^{\beta t_0}$ at time t_0, with a system starting with types concentrated on E_0. Suppose we were having mutation from E_0 to E_1 and back. Then using the dual representation we can compute

$$E\left[\int f(u_1, \ldots, u_n) \prod_{i=1}^{n} X_t(du_i)\right], \quad (9.1012)$$

with functions of the form

$$f(u_1, \ldots, u_n) = \bigotimes_{i=1}^{n} 1_{A_i}(u_i). \quad (9.1013)$$

Theorem 2 (proved in subsection 10.3) establishes the convergence of $\mathcal{L}[X_t]$ as $t \to \infty$ to a unique equilibrium law. Using a comparison argument we will show that these limits as $t \to -\infty$ agree with those of the mutation-selection equilibrium.

We next temporarily combine

$$\text{types } M \text{ and } M + 1 \text{ into a single type } M^* \quad (9.1014)$$

with mass x_{M^*} and make an appropriate modification of the dynamics of the resulting two types $1, \ldots, M - 1, M^*$ with masses $x_1, \ldots, x_{M^*}$.

We then have a system essentially identical to the M type system except that the mutation rate from this new type to types $j \leq M - 1$ are slightly smaller than

9.13 Proof of Propositions 9.1, 9.2, 9.4

$m_{M,j}$ because part of the mass belongs to the original type M+1, an effect which is arbitrarily small for values $t \leq t_0$ with t_0 negative enough. More precisely we must replace the mutation rates $m_{M,j}$, $j = 1, \ldots, M - 1$ by the time inhomogeneous mutation rates

$$\tilde{m}_{M,j}(t) = \frac{(m_{M,j}(x_{M^*}(t) - x_{M+1}(t)))}{x_{M^*}(t)}. \tag{9.1015}$$

Note that this varying rate will converge to a constant eventually, namely

$$\tilde{m}_{M,j}(t) \to m_{M,j} \text{ as } t \to -\infty. \tag{9.1016}$$

If we assume that $x_{M+1}(t) \leq \varepsilon$ for $t \leq t_0(\varepsilon)$, then the mutation rate $\tilde{m}_{M,j}(t)$ is bounded above by $m_{M,j}$ and below by $m_{M,j} \frac{(x^{M^*}(t)-\varepsilon)^+}{x^{M^*}}$. Now let

$$\pi_{\text{loweq}} \text{ resp. } \pi_{\text{loweq}.\varepsilon} \tag{9.1017}$$

denote the respective equilibrium measures.

Now using a standard coupling it is easy to verify that the masses of types $1, \ldots, M - 1$ are stochastically monotone increasing in $m_{M,j}$ so that we can use the modified system and the M-type systems to provide upper and lower bounds in the stochastic order. In order words our system is sandwiched in stochastic order between two systems we understand well.

By a continuity argument (based on the duality for example) we obtain

$$\lim_{\varepsilon \to 0} \pi_{\text{loweq}.\varepsilon} = \pi_{\text{loweq}}. \tag{9.1018}$$

We can then choose a starting time $t_{00} \ll t_0$ and conclude that the Prohorov distance between π_{loweq} and $\pi_{\text{loweq}.\varepsilon}$ up to time time t_0 decreases to zero as $\varepsilon \downarrow 0$. Now choose t_0 so that $e^{\beta t_0} \sup_{t \leq t_0} x_{M+1}(t_0) < \varepsilon$, to conclude that the two systems bound our system from below and above with a ε arbitrarily small if we start at time t_{00} small enough.

Putting everything together we can show that for the entrance law the joint distribution of $x_1, \ldots, x_M$ converges to the M type mutation-selection equilibrium in the limit as $t_0 \to -\infty$, that is

$$\mathcal{L}[(x_1(t_0), \ldots, x_M(t_0))] \Rightarrow \pi_{\text{loweq}} \quad \text{as } t_0 \to -\infty. \tag{9.1019}$$

Step 3

We now complete the proof of uniqueness in law of the process $(x_1(t), \ldots, x_{M+1}(t))_{t \geq \mathbb{R}}$ given a mean curve $(m_{M+1}(t))_{t \in \mathbb{R}}$ using the coupling argument as follows.

Given any two solutions x and $\tilde{x}$ we can take independent realizations on one probability space. Introduce

$$y(t) := x_{M+1}(t) + \tilde{x}_{M+1}(t). \tag{9.1020}$$

We show next, that we can find a sequence $t_n \to -\infty$ at which $y(t_n) = x_{M+1}(t_n) + \tilde{x}_{M+1}(t_n) = 0$.

We can verify that y can be represented in the form

$$dy(t) = 2m_{M+1}(t)dt + \sqrt{g(t)y(t)}d\tilde{w}(t), \tag{9.1021}$$

where $\tilde{w}(t)$ is a Brownian motion and $g(t)$ is a predictable function on the joint probability space and bounded below by a positive constant on the interval on which $y(t) \leq \varepsilon < 1$.

By a time change this is the square of a Bessel process with time-varying dimension. Then by a result of P. Carmona (1994), [CPY], (Proposition 3.3) (also see H. Shirakawa (2002) [HS02] Theorem 2.1) the zero set contains the zero set of a Bessel process with dimension $\delta < 1$ which is a perfect set with empty interior (see Revuz-Yor p. 419). The hitting time of zero for the Bessel process with dimension δ has density [GJY],

$$f_T(t) = \frac{1}{t\Gamma(\hat{v})} \left(\frac{x}{2t}\right)^{\hat{v}} \exp(-\frac{x}{2t}), \quad \hat{v} = \frac{4-\delta}{2} - 1. \tag{9.1022}$$

Using the recurrence of the process $y(t)$ we can then obtain the required sequence $t_n \to -\infty$ such $t_n - t_{n+1} \geq T$ any $T > 0$.

Step 4

We next use the fact that the excursions of type $M+1$ starting at t_n in a random environment of types $1, \ldots, M$ have distribution arbitrarily close to the two type equilibrium π_{loweq} as $n \to \infty$. By the uniform ergodicity of the lower-level mutation-selection process and the continuity of the equilibrium measure as a function of the parameters (when bounded away from 0) and our assumption on the sequence $(t_n)_{n\in\mathbb{N}}$ we know that $(x_1(t_n), \ldots, x_M(t_n))$ converges to the M-type equilibrium as $n \to \infty$.

We consider now the Markov chain obtained by observing the diffusion process $(x(t))_{t\geq 0}$ at time $(t_n)_{n\in\mathbb{N}}$. Then the transition probabilities for $\{(x_1(t_n), \ldots, x_M(t_n))\}_{n\in\mathbb{N}}$ satisfy for large enough n that

$$\frac{1}{2} \sup_{x'_1, x'_2, x''_1, x''_2} \int_{[0,1]\times[0,1]} |P_n((x'_1, x'_2), (x_1, x_2)) - P_n((x''_1, x''_2), (x_1, x_2))| dx_1 dx_2 < 1. \tag{9.1023}$$

Hence we can then couple the masses of the types $(1, \ldots, M)$ at times t_n so that

$$P(x_i(t_n) = \tilde{x}_i(t_n); \; i = 1, \ldots, M) \geq \delta > 0 \text{ for all large enough } n. \tag{9.1024}$$

This then implies that the two solutions couple with probability 1 and therefore the laws of x and $\tilde{x}$ are identical and therefore the entrance law for the McKean–Vlasov equation is unique.   q.e.d.

Problem 2:

Here we have to establish first that the limit on the r.h.s. of (9.1006) exists and leads to an entrance law of the McKean–Vlasov equation. Once we have that we know that the entrance law grows like $e^{\beta t}$ as $t \to -\infty$ and that for every realisation the scaled object in (9.1006) has a limit, which is in $(0, \infty)$, since we know that (9.1007) is satisfied and therefore we now need only the convergence result, which we prove in Subsection 9.14.

Proof of Proposition 9.4. This follows as a special case of Propositions (10.6), (10.7) and (10.8) in Section 10.   q.e.d.

9.14 Proof of Propositions 9.3 and 9.5

In this subsection we collect the proof of statements which need some more serious modification compared to the Section 8 since we cannot work with the empirical measure of microstates which are indexed by sites, but we have to consider the empirical measure of mesostates associated with labels which are involving multisite tableaus and the clan structure enters into the picture.

9.14.1 Proof of Proposition 9.3: Convergence to Limiting Nonlinear Dynamics

We prove piece by piece the different assertions in Part (a)–(f) in the proposition combining parts whenever convenient.

In order to prove the convergence to the McKean–Vlasov limit we follow a program similar to the $M = 1$ case.

(a), (b).

We must prove

1. Convergence on finite time intervals $[0, T]$ with initial upper level mass - here no collisions are involved and rare mutations play no role - this is standard starting with the martingale problem and uses the mean field dual constructed above (compare [DG99]).
2. Assuming no higher level initial mass - convergence of moments of the higher level mass at times $\frac{\log N}{\beta} + t_0$ for some t_0. This is obtained using the results on the limiting dual process (Proposition 9.58(b)).
3. Identification of the limiting distribution of $x^N(1, t)$ at time $\frac{\log N}{\beta} + t_0$ in terms of a random entrance law and unique entrance solution. Existence of the limit

$N \to \infty$ of $e^{-\beta t_N} \bar{x}^N(t_N)$ with $t_N = \beta^{-1} \log N + s_N, s_N \to -\infty, s_N = o(\log N)$ (using (1)) and identification with the limiting droplet $\mathcal{W}^*$ and of β with the droplet Malthusian parameter which are obtained (as a special case with one upper level type) in Proposition 10.7. This involves the L^2 argument using the dual in the emergence regime using the first two asymptotic terms that was developed in the last sections using the coloured particle scheme.

We show directly the convergence as $N \to \infty$ of the empirical measure at time $\beta^{-1} \log N + t_0$ for every t_0, by showing the convergence of the moments

$$E\left[\left(\int_{\mathcal{P}(\mathbb{K})} F(p) \Xi_N(t)(dp) \right)^k \right] \quad ; \quad F(p) = \langle f, p \rangle^j, \text{ with } f = 1_A, \quad A \subseteq \mathbb{K};$$

$$k, j \in \mathbb{N}. \qquad (9.1025)$$

In order to obtain emergence we must show that for $A = \{1, \cdots, M\}$ the limits (9.1025) as $N \to \infty$ for $k = 1, 2, \cdots$ defines a probability measure on $[0, 1]$ which has full measure on $(0, 1)$ for some $t_0 \in \mathbb{R}$. The moments in (9.1025) are calculated using the dual starting with j individuals located at each of sites ℓ different sites.

We now proceed to verify that:
(i) for $t_0 > 0$, $\Xi_N^{\log \cdot \beta}(t_0)$ converges to a non-degenerate (non-deterministic) random measure and (ii) after time t_0 the limit satisfies the McKean–Vlasov dynamics and this dynamic is continuous in the initial state.

First we consider the proof of (ii) assuming that (i) is true and that the random variable is strictly positive (see (c)). We can study the limiting system for $t \geq t_0$ by applying the dual starting at time $t_N - t_0$ and evolving for time $t_0 + t$ further.

Here we can again use the same argument as in the case of two types in order to identify the limiting dynamics for $t \geq t_0$, as the McKean–Vlasov dynamics, recall that this is for a *finite* time horizon. From the dual one reads of the Feller property immediately since it is a finite system.

We now turn to the proof of (i). Here we must prove that we have convergence of the t_0-marginal and that the limiting solution is random. This is done in two steps.

(1) First, to prove the convergence of the marginal distribution, we prove the convergence of all joint moments as $N \to \infty$.
(2) To prove that the limiting solution is random we consider again

$$m_{k,\ell}^N(t) = E\left[\left(\int_{[0,1]} (1-x)^k \; \Xi_N^{\log \cdot \beta}(t, M+1)(dx) \right)^\ell \right], \quad k, \ell \in \mathbb{N} \quad (9.1026)$$

and show that

$$\liminf_{N \to \infty} [m_{1,2}^N(t) - (m_{1,1}^N(t))^2] > 0. \qquad (9.1027)$$

9.14 Proof of Propositions 9.3 and 9.5

For the easier point (2), we need positivity of the limiting variance. This follows exactly as before from the fact that one of the two particles initially at the same site coalesce before migration with positive probability.

For the point (1) we have must carry out calculations of the various joint moments of all order using the dual representation. To do this we calculate the moments (see (9.1025)) using the dual in the time scale $\beta^{-1} \log N + t_0$ to conclude the proof of emergence by showing that these moments converge as $N \to \infty$ to a limit such that they determine the law of the limiting random measure. Recall that the moments corresponding to $k_1, \ldots, k_{M+1}, \ell \in \mathbb{N}$ are of the form

$$m^N_{k_1,\ldots,k_{M+1},\ell}(t) = \qquad (9.1028)$$

$$E\Big[\Big(\int_{[0,1]} (x_1^{k_1} \ldots x_M^{k_M} (1-x_{M+1})^{k_{M+1}}) \; \Xi_N^{\log,\beta}(t)(dx_1,\ldots,dx_{M+1})\Big)^\ell\Big],$$

and hence are obtained by integrating the set-valued dual process starting with the appropriate dual factors at each of ℓ distinct sites. We then obtain the set-valued dual process involving ℓ independent clouds prior to the collision regime each in the stable age-type distribution. The total dual cloud then has the same form as (9.550) except that we must replace W in (9.551) by

$$W_1 + \cdots + W_\ell \qquad (9.1029)$$

where $W_1, \ldots, W_\ell$ are independent copies of W.

The convergence of the joint moments then follows in the mutation-dominant regime from Proposition 9.58 which gives in particular a formula for the limit. This then implies the convergence (9.21) as claimed.

In order to deal with the selection-dominant case we need the convergence of the empirical distribution of the mesostates, which is obtained in Proposition 9.81.

(c) This follows from the convergence (a),(b) and Proposition 9.1 (a). The proof of (9.23) follows in the mutation-dominant regime from the mean calculation that shows that the mean mass of type $M + 1$ goes to zero as $t \to -\infty$. The proof of (9.24) follows from the McKean–Vlasov dynamics as in the case of two types.

(d) To prove (9.25) it suffices again as in the case of two types in (8.619) to consider the asymptotics as the number of initial particles goes to infinity. This however does not change since it concerns only the dual particle system which is the same regardless of the number of types which only influences the form of $\mathcal{G}_t^{++}$.

(e) This we showed in Subsubsection 9.13.3.

(f) This follows from the uniqueness of the entrance law with (9.26).

9.14.2 Proof of Proposition 9.5

We have to distinguish here the mutation-dominant case where we get through with an argument based on the empirical measure of microstates and the selection-dominant case, where we have to work with the mesostates and the clan structure.

(1) Consider first the mutation-dominance regime. We must now prove the equality of $^*\mathcal{W}$ and $\mathcal{W}^*$ following the method of the proof of Proposition 8.29. As before the proof is based on second moment calculations (and bounds on third moments to have uniform square integrability).

The proof follows in the mutation-dominant case the same lines as the $M = 1$ case (see Subsection (8.4.6) e.g. (8.872)) in order to prove the L_2 convergence for the interacting Wright Fisher. The essential difference is that now we must carry out the L_2 calculations using the set-valued dual $\tilde{\mathcal{G}}_t^{++}$. The required asymptotic result is obtained using Proposition 9.61 and in particular the asymptotic formula (9.559) for $\tilde{K}_{t_N+s}^{N,j_1,j_2}$ and the second moment calculation as in the proof of Proposition 9.62 (d).

(2) Next to the selection-dominant case. We proved the claim in Subsubsection 9.12.13 in Proposition 9.88.

9.15 Asymptotic Elimination of Downward Mutations

In the arguments above we suppressed down mutation, i.e. we assumed $m_{M+1,\text{down}} = 0$. In order to verify that even for $m_{M+1,\text{down}} > 0$ down-mutation is *asymptotically* negligible for emergence, fixation and for the evolution from emergence towards fixation (recall the latter takes place over times $O(1)$), it suffices to prove the following result.

For simplicity we state and prove the next proposition in the special case $M = 2$ with $m_{1,2}, m_{2,1}$ the lower level, i.e. type 1,2, mutation rates and $m_{3,\text{up}}, m_{3,\text{down}}$ the up to respectively down mutation from the superior type 3. The corresponding statement and proof for $M \geq 2$ are given by a simple (essentially notational) modification.

Proposition 9.97 (Effects of rare downward mutation).
Assume $d > 0$, $s > 0$, $c > 0$, $m_{1,2} \neq 0, m_{2,1} \neq 0, m_{3,\text{up}} = 0, \frac{m_{3,\text{down}}}{N} > 0$. The evolution starts in the state $\mu_0^\otimes$ and is denoted $(X^N(t))_{t \geq 0}$.

(a) Let $(\tilde{X}^N(t))_{t \in [0,T]}$ be the dynamic with $m_{3,\text{down}} > 0$ and $(X^N(t))_{t \in [0,T]}$ with $m_{3,\text{down}} = 0$ and both starting in the same initial state. Then for all $T > 0$ we have:

$$(\mathcal{L}[(\tilde{X}^N(t))_{t \in [0,T]}] - \mathcal{L}[(X^N(t))_{t \in [0,T]}]) \Longrightarrow 0, \text{ as } N \to \infty. \quad (9.1030)$$

9.15 Asymptotic Elimination of Downward Mutations   713

(b) If $\mu_0(3) > 0$, then

$$\lim_{t \to \infty} \lim_{N \to \infty} E[x_3^N(t)] = 1. \tag{9.1031}$$

(c) If $\mu_0(3) > 0$ and $t_N \to \infty$ as $N \to \infty$, then

$$\lim_{N \to \infty} E[x_3^N(t_N + t)] = 1 \ \forall \ t > 0. \tag{9.1032}$$

(d) If $\mu_0(3) = 1$, then

$$\mathcal{L}[(x_3^N(t))_{t \geq 0}] \underset{N \to \infty}{\Longrightarrow} \delta_1. \tag{9.1033}$$

(e) The results (a)–(d) remain true if $m_{3,\mathrm{up}} > 0$. □

Proof of Proposition 9.97. (a) This is immediate from the dual representation since uniformly in $t \in [0, T]$ the probability to see at least one mutation event $\{3\} \to \{2\}$ or $\{3\} \to \{1\}$ is tending to zero as $N \to \infty$.
(b) We consider

$$\mathcal{F}_0^{++} = (110) \text{ and assume that } \mu_0(3) > 0. \tag{9.1034}$$

The key idea is that in the dual we get only terms which contain a diverging number of factors (110). To make this precise we proceed in two steps, we first assume that

$$d = 0, \ c = 0, \ m_{3,\mathrm{down}} = 0 \tag{9.1035}$$

and then we treat the general case.
Step (1): Then consider the following chain of transitions, first by selection

$$(110) \to (010) + (100) \otimes (110) \tag{9.1036}$$

and then next by either of the two mutation transitions:

$$\begin{aligned}
&\to (110) + (110) \otimes (000) \quad (\text{with prob } = \tfrac{m_{1,2}}{m_{1,2}+m_{2,1}}) \\
&\longrightarrow (000) + (110) \otimes (110) \quad (\text{with prob } = \tfrac{m_{2,1}}{m_{1,2}+m_{2,1}}).
\end{aligned} \tag{9.1037}$$

This chain of events occurs with positive rate ($m_{1,2} + m_{2,1} > 0$). We then repeat this chain of events so that after a finite number of repetitions the single factor term dies and we are left with two or more factors.

Therefore the evolution of the dual has the property that the above process either (II) comes to an end when all factors have died or (I) we have a sum of products in which there is an increasing minimal number of factors. Moreover among nonzero factors a positive proportion is (110) or (100).

Note next all terms are nonnegative and the expression (i.e. the sum of all products), even if we factor out the first $n(t)$ factors (110) in each sum is less than 1 or equal to 1 if the integral of (110) is 1 (i.e. $\mu_0(3) = 0$). We get therefore that there is a random number $n(t)$ such that:

$$\mathcal{F}_t^{++} \leq \delta(110)^{\otimes n(t)} \text{ with } \delta \leq 1 \text{ and } n(t) \to \infty, \qquad (9.1038)$$

so that (since the $\mu_0(3) > 0$)

$$\int \mathcal{F}_t^{++} d\mu_0^\otimes \leq (1 - \mu_0(3))^{n(t)} \longrightarrow 0, \text{ as } t \to \infty. \qquad (9.1039)$$

Next we have to remove the restriction in (9.1035).

Step 2: First we handle downward mutation, i.e. $m_{3,\text{down}} > 0$. The effect of down-mutation is to replace at rate $\frac{m_{3,\text{down}}}{N}$ factors (110) by (111) in (9.1038). However in every interval $[0, t]$ as $N \to \infty$ this effect becomes negligible and the resulting expression for (9.1039) still goes to zero as the time stretch diverges.

If $d > 0$, $c = 0$, then the number of factors at each site is limited and since there is no migration instead we get fixation of types 1, 2 with positive probability. However if also $c > 0$ then again we get an increasing number of (110) or (100) factors so that $E[x_3(t)] \to 1$ as $t \to \infty$.

(c) Here the only change is that migration can lead to a collision, however that is not possible in times $o(\log N)$ and then the claim follows in the same way.

(d) Assume that $\mu_0(3) = 1$ and $m_{3,\text{down}} > 0$ (where $\frac{m_{3,\text{down}}}{N}$ is the downward mutation rate). Start the dual in (110) again and recall (9.1038). Then we see that all the $n(t)$-factors of (110) would have to mutate to (111) in order to prevent that $\int \mathcal{F}_t^{++} d\mu^\otimes$ is 0. This has probability tending to 0 as $N \to \infty$. Hence $\lim_{N \to \infty} E(x_3^N(t)) = 1$, for all $t > 0$ so that types 1, 2 never appear, that is, "downward mutations are instantaneously removed".

Then in order to obtain (9.1033) it suffices to check tightness of the law of $\{(x_3^N(t))_{t \in [0,T]}, N \in \mathbb{N}\}$. These processes are by definition semi-martingales with path in $C(\mathbb{R}_+, \mathbb{K})$. To get tightness in such a situation is standard. Namely we know that the local characteristics, i.e. local drift and diffusion constants are bounded uniformly in N so that this follows from the criterion in Joffe and Métivier given in the appendix.

(e) Additional upward mutation makes $x_3^N(t)$ stochastically larger, so the claim is immediate. q.e.d.

Chapter 10
The General (M, M)-Type Mean-Field Model: Emergence, Fixation and Droplets

The purpose of this section is to establish the main results for the general case of a finite number of distinct types also on the *higher* levels of fitness (i.e. $|E_1| > 1$) which brings in some *new* features we deal with in Subsection 10.5 (results) and requiring new duality techniques in Subsection 10.7.

What we need are four types of results, (1) emergence of the E_1-types at the macroscopic level, (2) fixation at the higher level E_1-types, (3) the early formation of microscopic droplets of rare advantageous mutant types from E_1 and finally (4) the growth of the microscopic droplets as time gets large. In addition we prove (5) an ergodic theorem for the (M, M) finite N and McKean–Vlasov system.

Recall the results that we have obtained in Sections 8 and 9 for the system with 1 type at the higher level and with 1, respectively M, types on the lower level. Our goal in this section is to present the results on the $N \to \infty$ asymptotics, for the time to emergence, the resulting limiting random entrance law, the dynamic of fixation, the early droplet formation for the time-evolution of the general (M, M)-type case, i.e. with M types in E_0 and M types in E_1, and to provide proofs using the tools and ideas developed in earlier sections completed by some new ideas needed to handle the richer structure on the upper level. (*The assumption that the numbers of types at the two levels are the same plays no role and is used to avoid extra notation.*)

The good news is that the growth rate for emergence is not affected by the number of types on the upper level. But there are still essentially two (related) new points:

- The droplet now is described by an atomic measure on $[0, 1] \times \{M+1, \cdots, 2M\}$ and has therefore a *multi-dimensional* state (M-dimensional to be precise).
- Upon emergence from E_0 to E_1 we have besides the random time shift of the entrance law given by an $\mathbb{R}$-valued random variable $\mathcal{W}^*$ now in addition a random $|E_1|$-dimensional vector, the *random emergence-frequency* $\vec{\mathcal{W}}^*$ on E_1 types.

Note in that context that merging the upper M-types, i.e.

$$(x_1(t, i), \cdots, x_{2M}(t, i))_{i \in \{1, \cdots, N\}} \longrightarrow (\tilde{x}_1(t, i), \cdots, \tilde{x}_{M+1}(t, i))_{i \in \{1, \cdots, N\}} \quad (10.1)$$

with

$$\tilde{x}_\ell(t,i) = x_\ell(t,i), \quad \ell = 1,\cdots,M; \quad \tilde{x}_{M+1}(t,i) = x_{M+1}(t,i) + \cdots + x_{2M}(t,i) \tag{10.2}$$

we obtain a system $\tilde{X}^N$ to which the theory of Section 9 applies if we suppress the downward mutation, which we saw in Subsection 9.15 plays no role. Therefore we have to prove only those parts of the statements concerning relative proportions of the upper types.

Since the main tools have been introduced in Section 8 and Section 9 and many of the arguments are minor modifications of those in these sections, we concentrate here on those two above points where new features content-wise or on the technical level are needed to complete the proofs. This means that we will focus to a large extent on how to handle the above two new effects in detail but we omit details on the more routine aspects of M upper types.

Outline of Section 10. We introduce the precise model in Subsection 10.1. In Subsection 10.2 we formulate in a short overview the five phases for the (M, M)-model. In Subsection 10.3, we state as preparation an ergodic theorem for the (M, M)-model. In Subsection 10.4 we discuss the results on emergence, the random entrance law and fixation. In Subsection 10.5 we formulate the results on droplet formation both the early time window and the subsequent growth. The sections on proofs are Subsections 10.6–10.13 with Subsection 10.6 proving the ergodic theorem which also is a good preparation for Subsection 10.7 providing the *new dual calculations* which are then employed in Subsections 10.8–10.13 to prove the various propositions from Subsection 10.4 and 10.5.

10.1 The Interacting System on $\{1,\ldots,N\}$ with (M, M)-Types

We begin with our basic (M, M)-model, which has the form of N exchangeably interacting multitype Fisher-Wright diffusions with $2M$ types, where we have M types $\{1,\ldots,M\} = E_0$ at the lower level and M types $\{M+1,\ldots,2M\} = E_1$ at the higher level. The state space of each spatial component is then the simplex Δ_{2M-1}. We will interpret here the states of the process X^N mostly as a collection in the index $\{1,2,\cdots,N\}$ of 2M-vectors, and then write $X_t^N = (\mathbf{x}^N(j,t), j = 1,\cdots,N)$ as a collection of random vectors, or alternatively as a collection of random probability measures on $E_0 \cup E_1$, whichever is more convenient. Therefore our diffusion process has the form

$$X^N = (X_t^N)_{t \geq 0} = ((x_i^N(j,t), \quad i = 1,\cdots,2M), \quad j = 1,\cdots,N)_{t \geq 0}. \tag{10.3}$$

10.1 The Interacting System on $\{1,\ldots,N\}$ with (M,M)-Types

We assume the following about the parameters of the dynamics

selection rate is s
types $M+1,\ldots,2M$ have fitness: $\quad 1-\frac{e_i}{N}, \quad 0 \leq e_i < 1, \quad i = M+1,\cdots,2M$
types $i = 1,\ldots,M-1$ have fitness: $\quad 0 \leq e_i < 1 - \frac{1}{N}$ and
type M has fitness: $\quad 1 - \frac{1}{N}$,
$$\tag{10.4}$$

the mutation rates, m_{ij} between types $1,\ldots,M$, are strictly positive, $\quad$ (10.5)

the mutation rate from lower type i to higher type j is $\dfrac{m_{ij}}{N}$. $\quad$ (10.6)

The formal description of the system is in terms of the generator of the martingale problem as explained in Section 3 which we now write out for the case of the geographic space $\{1,2,\cdots,N\}$ (see (10.12) below) as a *sum* of the *single site operators* which in turn we decompose suitably for our purposes in various parts depending on their $N \to \infty$ asymptotics.

The single site dynamics is given by the linear operator $G^{0,1,N}$ on $C_b(\Delta_{2M-1}, \mathbb{R})$ defined below in (10.10) with domain $C_b^2(\Delta_{2M-1}, \mathbb{R})$. Define first as ingredients for this operator the following linear operators on $C_b(\Delta_{2M-1}, \mathbb{R})$ with domain $C_b^2(\Delta_{2M-1}, \mathbb{R})$. We split here the single-site generator in *three parts* playing different roles in the asymptotics ($N \to \infty$) of the emergence process. For that purpose define the following three *drift terms*.

Let $\mathbf{x} \in \Delta_{2M-1}$, $\mathbf{x} = (x_i)_{i=1,\cdots,2M}$. Then we set

$$G^0 f(\mathbf{x}) = \sum_{i=1}^{M}\left(\sum_{j=1}^{M}(m_{ji}x_j - m_{ij}x_i)\right)\frac{\partial f(\mathbf{x})}{\partial x_i} \tag{10.7}$$

$$+ \sum_{i=1}^{M} sx_i \left(e_i - \sum_{k=1}^{M} e_k x_k - \sum_{k=M+1}^{2M} x_k\right)\frac{\partial f(\mathbf{x})}{\partial x_i}$$

$$+ \sum_{i=M+1}^{2M} sx_i \left(1 - \sum_{k=1}^{M} e_k x_k - \sum_{k=M+1}^{2M} x_k\right)\frac{\partial f(\mathbf{x})}{\partial x_i},$$

$$G^{1,N} f(\mathbf{x}) \tag{10.8}$$

$$= \frac{1}{N}\sum_{i=M+1}^{2M}\left[\left(\sum_{j=M+1}^{2M}(m_{ji}x_j - m_{ij}x_i)\right) + sx_i\left(e_i - \sum_{k=M+1}^{2M} e_k x_k\right)\right]\frac{\partial f(\mathbf{x})}{\partial x_i}$$

$$+ \frac{1}{N} \sum_{i=1}^{M} \left(\sum_{j=M+1}^{2M} m_{ji} x_j \right) \frac{\partial f(\mathbf{x})}{\partial x_i} - \frac{1}{N} \sum_{i=M+1}^{2M} \left(\sum_{j=1}^{M} m_{ij} x_i \right) \frac{\partial f(\mathbf{x})}{\partial x_i},$$

$$G^{2,N} f(\mathbf{x}) = \frac{1}{N} \sum_{i=M+1}^{2M} \left(\sum_{j=1}^{M} m_{ji} x_j \right) \frac{\partial f(\mathbf{x})}{\partial x_i} - \frac{1}{N} \sum_{j=M+1}^{2M} \left(\sum_{i=1}^{M} m_{ji} x_j \right) \frac{\partial f(\mathbf{x})}{\partial x_j}. \tag{10.9}$$

Here G^0 corresponds to the $O(1)$, as $N \to \infty$, driftterms (in fact this part does not depend on N), $G^{1,N}$ corresponds to the $O(\frac{1}{N})$ driftterms (from selection and mutation) that play no role in the emergence of the sparse set of E_1-mutant sites and $G^{2,N}$ has $O(\frac{1}{N})$-terms that correspond to the rare mutation mechanism that creates the sparse set of E_1-mutant sites.

Then the linear operator for the martingale problem corresponding to the single site dynamic acts on C_b^2-functions f on Δ_{2M-1} as follows:

$$G^{0,1,N} f(\mathbf{x}) = \frac{1}{2} \sum_{i,j=1}^{2M} x_i (\delta_{ij} - x_j) \frac{\partial^2 f(\mathbf{x})}{\partial x_i \partial x_j} + G^0 f(\mathbf{x}) + G^{1,N} f(x) + G^{2,N} f(\mathbf{x}). \tag{10.10}$$

We now consider the spatially interacting system on N sites $\{1, \ldots, N\}$. Define then for the spatial system linear operators on $C_b((\Delta_{2M-1})^N, \mathbb{R})$ with domain $C_b^2((\Delta_{2M-1})^N, \mathbb{R})$:

$$G_k^{0,1,N}, \tag{10.11}$$

where the subscript k means that $G^{0,1,N}$ is applied to the state of the k-th spatial component of the system as variable.

The generator of the martingale problem for the N exchangeably interacting diffusions is then the following linear operator on $C_b((\Delta_{2M-1})^N, \mathbb{R})$ with domain $C_b^2((\Delta_{2M-1})^N, \mathbb{R})$:

$$G_{int}^{0,1,N} f(\mathbf{x}(1), \ldots, \mathbf{x}(N)) = c \sum_{k=1}^{N} \left[\sum_{j=1}^{2M} \left(\frac{1}{N} \sum_{\ell=1}^{N} x_j(\ell) - x_j(k) \right) \frac{\partial f}{\partial x_j(k)} \right] + \sum_{k=1}^{N} G_k^{0,1,N} f, \tag{10.12}$$

which defines uniquely a solution and this solution then has *continuous path*, the *strong Markov property* and is associated with a *Feller semi-group* on $C_b((\Delta_{2M-1})^N, \mathbb{R})$.

We assume here for our purposes that the random initial state satisfies:

$$(\mathbf{x}(1,0), \ldots, \mathbf{x}(N,0)) \text{ is exchangeable.} \tag{10.13}$$

10.1 The Interacting System on $\{1,\ldots,N\}$ with (M,M)-Types

The dynamic in (10.12) has a McKean–Vlasov limit dynamic as $N \to \infty$ which is a *nonlinear Markov process*. We will after some preparation write out the defining equation explicitly later on in (10.16).

We need furthermore the following objects. If we remove $G^{1,N}$ and $G^{2,N}$ from the generator (10.12) and start with configurations concentrated on E_0, then the exchangeable system can be viewed as a process with state space $(\Delta_{M-1})^N = (\mathcal{P}(E_0))^N$ and has a unique stationary measure and corresponding stationary process with law (see Subsection 10.3, Theorem 14) which we denote by:

$$\mu_{eq}^N \text{ respectively } P_{eq}^{0,N}, \tag{10.14}$$

the latter with marginal μ_{eq}^N.

We can then define the *out of equilibrium rate of mutation* to type j, namely $m_j (= m_j^N)$, as:

$$m_j = \frac{1}{N} \sum_{i=1}^N \mu_{eq}^N(i) m_{i,j}, \quad j = M+1,\cdots,2M. \tag{10.15}$$

The limiting process as $N \to \infty$ of the process defined by (10.12) is the law of a nonlinear Markov process which can be characterized either by a nonlinear martingale problem as in the sequel of (4.27) or by the forward equation for the probability density. The general expressions of Subsubsection 4.2.2 take here the following concrete and more explicit form.

In order to write down the limiting martingale problem respectively McKean–Vlasov equation arising for the process in (10.33) we introduce the following ingredient. Consider the one site linear operator G on $C_b(\Delta_{2M-1}, \mathbb{R})$ with domain $C_b^2(\Delta_{2M-1}, \mathbb{R})$:

$$Gf(\mathbf{x}) = \frac{d}{2} \sum_{i,j}^{2M} x_i (\delta_{ij} - x_j) \frac{\partial^2 f(\mathbf{x})}{\partial x_i \partial x_j} + \sum_{i=1}^M \left(\sum_{j=1}^M (m_{ji} x_j - m_{ij} x_i) \right) \frac{\partial f(\mathbf{x})}{\partial x_i}$$

$$+ s \sum_{i=1}^M x_i \left(e_i - \sum_{k=1}^M e_k x_k - \sum_{k=M+1}^{2M} x_k \right) \frac{\partial f(\mathbf{x})}{\partial x_i}$$

$$+ s \sum_{i=M+1}^{2M} x_i \left(1 - \sum_{k=1}^M e_k x_k - \sum_{k=M+1}^{2M} x_k \right) \frac{\partial f(\mathbf{x})}{\partial x_i}. \tag{10.16}$$

The $(G, C_b^2(\Delta_{2M-1}, \mathbb{R}), \delta_{\mathbf{x}})$-martingale problem is wellposed and defines a Markov process on Δ_{2M-1}.

The adjoint operator of G is a linear operator G^* on $L^1(\Delta_{2M-1}, \lambda^{(2M-1)})$, with $\lambda^{(2M-1)}$ the $(2M-1)$-dimensional Lebesgue measure on the simplex Δ_{2M-1}, satisfying $< g, Gf > = < G^*g, f >$ for all $f \in \mathcal{D}(G), g \in L^1(\Delta_{2M-1}, \lambda^{(2M-1)})$ and hence maps

$$G^* : \mathcal{D}(G^*) \longrightarrow L^1(\Delta_{2M-1}, \lambda^{(2M-1)}). \tag{10.17}$$

This map can be extended to $\mathcal{M}_{\text{fin}}(\Delta_{2M-1})$ immediately.

We can introduce a nonlinear Markov process if we consider the operator

$$G + c[E[\mathbf{x}(t)] - \mathbf{x}(t)] \cdot \nabla. \tag{10.18}$$

We obtain again a wellposed martingale problem (see [DG99]). This process is the limit of the process given by a single component of the process X^N.

Now the McKean–Vlasov entrance law will be the marginal distribution of a nonlinear Markov process given via (10.18) later an dis denoted $\mathcal{L} = (\mathcal{L}_t)_{t \in \mathbb{R}}$ with $\mathcal{L}_t \in C([t_0, \infty), \mathcal{P}(\Delta_{2M-1}))$ with

$$\mathcal{L}_t(d\mathbf{x}) = u(t, \mathbf{x})d\mathbf{x} \tag{10.19}$$

and which satisfies the *McKean–Vlasov equation* given by:

$$\begin{aligned}\frac{\partial u(t, \mathbf{x})}{\partial t} &= (G^*u)(t, \mathbf{x}) - c \sum_{i=1}^{2M} \frac{\partial}{\partial x_i} \left(\left[\int_{\Delta_{2M-1}} \tilde{x}_i u(t, d\tilde{\mathbf{x}}) - x_i \right] u(t, \mathbf{x}) \right) \\ &= (G^*u)(t, \mathbf{x}) - c \sum_{i=1}^{2M} \frac{\partial}{\partial x_i} \left([\pi_t(i) - x_i] u(t, \mathbf{x}) \right), \end{aligned} \tag{10.20}$$

where we denote $\pi_t(i) = \int_{\Delta_{2M-1}} \tilde{x}_i u(t, d\tilde{\mathbf{x}})$.

This entrance law $\mathcal{L}$ will arise below as limit as $N \to \infty$ of the process in (10.33), which is the time-scaled process of empirical measures of the X^N-population at a late time.

10.2 Overview of Phases 0–3 for the (M, M) System on $\{1, \ldots, N\}$

We now outline the formulation of the main results on Phases 0–3 with M types in both E_0 and E_1 for the interacting system indexed by $\{1, \ldots, N\}$ and list where this is stated. We proceed here in the order of time in the model which the actual statements are ordered from the point of view of methodology used, namely first looking at the short, then large time scale for the E_0 respectively

$(E_0 \cup E_1)$-population and then at early times for the E_1-population. New features occur concerning the droplet formation in phase 0 and emergence, phase 1.

As a starting point we assume that at time $t = 0$ only the least fit type is present, that is,

$$x_1^N(j, 0) = 1, \ j = 1, \ldots, N. \tag{10.21}$$

- *Phase 0: E_0 population*: For times $t_N \to \infty$, $t_N = o(\log N)$ as $N \to \infty$, k tagged sites approach a product equilibrium measure concentrated on $(E_0)^k$.— Subsection 10.3.
- *Phase 0: E_1 population*: Microscopic emergence of rare mutants in the form of a mutant droplet formation in times of the same order as above: exponential growth in time and limiting Palm measure - Subsection 10.5
- *Phase 1*: Macroscopic emergence of rare mutants - Subsubsections 10.4.1, 10.4.2. This involves in time scales $\beta^{-1} \log N$ the emergence of essentially disjoint *islands* consisting of sites containing a single higher level type whose distribution is given by the limiting Palm measure. This requires two steps:
  - McKean–Vlasov limit in macroscopic time scale, i.e. at times $\beta^{-1} \log N + t, t \in \mathbb{R}$ the macroscopic time scale as $N \to \infty$ on Δ_{2M-1} and existence and uniqueness of solutions with entrance law at $t = -\infty$—Subsubsection 10.4.1
  - Identification of the random entrance law—Subsubsection 10.4.2.
- *Linking Phase 0 and Phase 1:* Subsection 10.5.2.
- *Phase 2*: Fixation - Subsubsection 10.4.3 in times $\beta^{-1} \log N + t_N$, $t_N \to \infty$ with $t_N = o(N)$.
- *Phase 3*: Neutral evolution after fixation and equilibrium in the E_1-equilibrium is treated with different methods in Section 11 and Section 12.

10.3 Ergodic Theorem for the (M, M) Spatial and Mean-Field Systems

We are first looking at the $t \to \infty$ limit of the law of a system for fixed N and having only the E_0 population. In Section 4 we proved the ergodic theorem (Theorem 2) in some special cases using the refined dual with a Feynman-Kac term. In this subsection we formulate an ergodic theorem and we use the set-valued dual to prove this in Subsection 10.7 and which gives Theorem 2 in the general case and to investigate the ergodic behaviour for multitype *interacting finite site mean-field* and *infinite site real spatial systems*.

In particular we formulate and prove the ergodic theorem for the spatial and mean-field systems in the time regimes in which the rare mutation is suppressed, which gives us the needed ingredient for Phase 0.

In addition we obtain a new result on survival of fit types in Corollary 10.1 for the $N \to \infty$ limiting McKean–Vlasov system (compare Remark 212).

In the limit $N \to \infty$ rare mutations are negligible for time of order $o(\log N)$ and we will see that the system approaches equilibrium on the lower level types in times of order $O(1)$ so that as $N \to \infty$ the time scales for the approach of low-level equilibrium and emergence via rare mutations *separate*. Recall that the exchangeable system X^N from (10.3) converges as $N \to \infty$ to the McKean–Vlasov process (the defining equations are in (10.16)–(10.20) the martingale problem is specified via (10.18)).

Theorem 14 (Ergodic theorem for (M, M)-system). *Assume that the rare mutation is absent, i.e.*

$$m_{j,j'} = 0 \text{ if } j = 1, \ldots, M, \; j' = M+1, \ldots, 2M, \qquad (10.22)$$

the initial configuration satisfies

$$\sum_{i \in E_0} x_i^N(j, 0) = 1, \; j = 1, \ldots, N \qquad (10.23)$$

and since this property is preserved under (10.22) we can consider $(X_t^N)_{t \geq 0}$ to be a system restricted to the type space $\{1, \ldots, M\}$ and we use this point of view below in (a)–(c).

(a) Let $N < \infty$ and consider the exchangeably interacting system (recall (10.3))

$$X_t^N = (\mathbf{x}^N(1,t), \ldots, \mathbf{x}^N(N,t)) \in (\Delta_{M-1})^N,$$
$$\text{with } c > 0, d \geq 0 \text{ and } m_{i,j} > 0 \text{ for all } (i,j) \in \{1, \ldots, M\}. \qquad (10.24)$$

Consider the distribution of ℓ tagged sites $\mathcal{L}[(\mathbf{x}^N(1,t), \ldots, \mathbf{x}^N(\ell,t))]$. Then for $\ell \leq N$

$$\mathcal{L}[(\mathbf{x}^N(1,t), \ldots, \mathbf{x}^N(\ell,t))] \Rightarrow \mu_{eq}^{N,\ell} \in \mathcal{P}((\Delta_{M-1})^\ell) \text{ as } t \to \infty, \qquad (10.25)$$

where $\mu_{eq}^{N,\ell}$ is the ℓ-dimensional marginal of

$$P_{eq}^{0,N} \text{ the unique invariant measure of the process } X^N, \qquad (10.26)$$

which is in particular independent of $X_0^N \in (\Delta_{M-1})^N$.

(b) Consider the McKean–Vlasov dynamic $(\mathcal{L}_t)_{t \geq 0}$ (the limit of $N \to \infty$ of the law of a component in (10.24)) corresponding to above set-up on $\mathbb{I} = \{1, \ldots, M\}$ and defined as in (10.20)). Then

$$\mathcal{L}_t \Rightarrow \mathcal{L}_\infty = \mu_{eq}^\infty \;, \text{ as } t \to \infty, \qquad (10.27)$$

10.3 Ergodic Theorem for the (M, M) Spatial and Mean-Field Systems

where $\mu_{eq}^\infty \in \mathcal{P}(\Delta_{M-1})$ does not depend on $\mathcal{L}_0$ and is the marginal distribution of the unique equilibrium of the McKean–Vlasov process given by the martingale problem for the operator (10.18).

(c) Now consider the sequence of the ℓ-dimensional marginals of the equilibrium measure $P_{eq}^{0,N}$, namely $\{\mu_{eq}^{N,\ell}\}_{N \in \mathbb{N}}$ from Part (a). Then for every $\ell \in \mathbb{N}$, $(\ell \leq N)$, as $N \to \infty$,

$$\mu_{eq}^{N,\ell}(dx_1,\ldots,dx_\ell) \Rightarrow \prod_{i=1}^{\ell} (\mu_{eq}^\infty(dx_i)), \qquad (10.28)$$

where μ_{eq}^∞ is the stationary distribution given in (10.27).

(d) Consider the geographic space Ω_N and the system $X^N(t) := \{x_i^N(\xi,t)\}_{i \in \{1,\ldots,M\}, \xi \in \Omega_N}$ with $d > 0$, under the above assumptions on the $\{m(i,j); i,j \in \{1,\cdots,2M\}\}$ and with $c_j > 0$ for all j. Assume we have a spatially homogeneous and shift ergodic initial condition.

In this case

$$\mathcal{L}[X_N(t)] \Rightarrow \mathcal{L}_{eq}^{\Omega_N}, \text{ as } t \to \infty, \qquad (10.29)$$

where $\mathcal{L}_{eq}^{\Omega_N} \in \mathcal{P}((\Delta_{M-1})^{\Omega_N})$ is the law of a spatially homogeneous shift-ergodic random field, which is an invariant measure of the evolution and which is unique under all invariant measures which are translation-invariant. □

The dual argument used in the proof of Theorem 14 also establishes the following generalization of a result originally due to Kimura [Kim2]. This involves the question of the survival of a *one-time mutation* of an advantageous type which can undergo mutation to a lower level type. The main idea is that if the relative fitness of the advantageous type is sufficiently large compared to its deleterious mutation rate, then in the mean-field limit model, i.e. the McKean–Vlasov process, this type survives in the limit as $t \to \infty$ and if it is too small, it may get extinct stochastically. This implies non-ergodicity even though the mutation Markov chain has only one ergodic class (selection-dominance).

Corollary 10.1 (Survival of fittest type - non-ergodicity). *Now consider the McKean–Vlasov system (given by a process solving the martingale based on (10.18), in the case in which type M has fitness 1, type i has fitness e_i, $i = 1,\ldots,M-1$, furthermore $c > 0$, $d > 0$. Define*

$$e^* := \max\{e_i, i = 1,\cdots, M-1\} < 1, \qquad (10.30)$$

$$\sum_{i=1}^{M-1} m_{M,i} \leq m^*, \text{ but } m_{i,M} = 0 \text{ for } 1 \leq i < M. \qquad (10.31)$$

Then given positive initial mass of type M, and fixed total deleterious mutation rate m^* there exists s^* such that for $s > s^*$

$$\liminf_{t\to\infty} E[x_M(t)] > 0, \qquad (10.32)$$

i.e. type M survives forever. In this case there are two stationary measures, one with $E[x_M(\infty)] = 0$ and one with $E[x_M(\infty)] > 0$. □

Remark 212 (Survival of advantageous type in spatial system). Consider the analogue of the Corollary in the spatial case on Ω_N. Then using a coupling argument as developed in Sections 8 and 9 using the coloured particle system it can be shown that survival of the rare mutant in the spatial case does not occur if it does not occur for the same parameters for selection, mutation, resampling and migration for the McKean–Vlasov case. In fact the conditions on the parameters for survival for the mean-field limit case provides a necessary condition for survival for any analogous system with migration given by a random walk on a countable abelian group. However we do not give the details here.

10.4 Results 1: Macroscopic Emergence, Fixation and Random Entrance Law of Higher Level Multiple Types

In this subsection we generalize our emergence results to the case where instead of one higher level type we have M such types and we study the limiting behaviour of the random empirical process in a new time scale:

$$\Xi_N^{\log,\beta}(t) = \frac{1}{N} \sum_{i=1}^{N} \delta_{x^N(i,\frac{\log N}{\beta}+t)} \in \mathcal{P}(\Delta_{2M-1}), \quad t \geq -\frac{\log N}{\beta}, \qquad (10.33)$$

as $N \to \infty$. We can extend the definition of $\Xi_N^{\log,\beta}(t)$ to all $t \in \mathbb{R}$ by setting $\Xi_N^{\log,\beta}(t) \equiv \Xi_N^{\log,\beta}(-\frac{\log N}{\beta})$ for all $t \leq -\frac{\log N}{\beta}$.

Using the dual in the multitype case at the lower level (see Section 9) with associated Malthusian parameter β and with initial element 1_{E_0} one proves that asymptotically as $N \to \infty$ there is strictly positive mean mass of type E_1 at times of the form $\frac{\log N}{\beta} + t$ in essentially the same way as in the 2-type case. Moreover we will verify that the resulting limiting mean mass of the types E_1 has non-trivial variance and therefore the state is random.

The results and concepts needed to describe phases 1 and 2, random entrance laws, emergence and fixation are presented in Subsubsections 10.4.1–10.4.3 respectively and the proofs in Subsections 10.7 to 10.13.

10.4.1 The Multitype McKean–Vlasov Equation and Its Entrance Laws

In this subsubsection we treat the multitype McKean–Vlasov entrance law introduced in (10.20). We need some more ingredients.

The *mean measure* $m(t, \cdot) \in \mathcal{M}_f(E_o \cup E_1)$ is defined by

$$m(t, A) = m(u, t, A) = \int \mathbf{x}(A) u(t, \mathbf{x}) d\mathbf{x}, \quad A \in \mathcal{B}(E_0 \cup E_1). \tag{10.34}$$

Then

$$(m(t, E_1))_{t \geq 0} \tag{10.35}$$

is the mean curve of the total mass of the higher level. Define also the set where the mean curve restricted to E_1 will take its values after emergence on the higher level (here δ_i is the vector which is 1 at the i-th position, 0 elsewhere):

$$\tilde{\Delta}^+_{M-1} := \{ w_{M+1} \delta_{M+1} + \cdots + w_{2M} \delta_{2M}, \quad w_i \geq 0, \quad \sum_{i=M+1}^{2M} w_i > 0 \}. \tag{10.36}$$

The system in (10.16) has the following important property which corresponds to (10.14) and the sequel. There is a unique McKean–Vlasov mutation-selection equilibrium process on the set of lower types with law

$$P^0_{eq} = (\mu^\infty_{eq})^{\otimes \mathbb{N}}. \tag{10.37}$$

Recall that the mutation rate from the lower order types in the lower-type selection-mutation equilibrium μ^∞_{eq} to higher level type $j \in E_1$ is given by

$$m_j = \sum_{i \in E_0} \mu^\infty_{eq}(i) m_{i,j}. \tag{10.38}$$

The key statement on the existence and uniqueness of entrance laws from $-\infty$ for the McKean–Vlasov equation reads as follows:

Proposition 10.2 (The McKean–Vlasov random entrance law from $-\infty$).

(a) *Given* $\mathbf{A} \in \tilde{\Delta}^+_{M-1} \subseteq \mathcal{M}_f(E_1)$ *of the form*

$$\mathbf{A} = A_{M+1} \delta_{M+1} + \cdots + A_{2M} \delta_{2M}, \tag{10.39}$$

there exists a unique solution $\{u(t) : -\infty < t < \infty\}$ *of (10.20) and a unique $\beta > 0$ such that*

$$\lim_{t \to -\infty} e^{-\beta t} m(u, t, \cdot \cap E_1) = \mathbf{A}. \tag{10.40}$$

(b) Moreover every deterministic solution of (10.20) satisfying

$$\limsup_{t \to -\infty} e^{-\beta t} m(t, E_1) < \infty, \quad \liminf_{t \to -\infty} e^{-\beta t} m(t, E_1) > 0, \tag{10.41}$$

that is in particular $e^{-\beta t} \int \mathbf{x}(E_1) u(t, \mathbf{x}) d\mathbf{x}$ is tight, satisfies (10.40) for some $\mathbf{A}$ of the form given in (10.39).

(c) Given any solution u of (10.20) satisfying the tightness condition (10.41), we know the backward limits exist and satisfy

$$\lim_{t \to -\infty} e^{-\beta t} m(u, t, E_1) = \bar{A} \tag{10.42}$$

$$\lim_{t \to -\infty} e^{-\beta t} m(u, t, \{i\}) = A_i, \quad i = M+1, \cdots, 2M, \tag{10.43}$$

$$\bar{A} = A_{M+1} + \cdots + A_{2M}, \tag{10.44}$$

$$\lim_{t \to -\infty} m(u, t, \{i\}) = q_i, \quad i = 1, \cdots, M, \tag{10.45}$$

with

$$q_i = \int x_i \mu_{eq}(d\mathbf{x}). \tag{10.46}$$

In addition with $\hat{\pi}_A$ denoting the image measure under the projection π_A on the set A:

$$\hat{\pi}_{E_0} \circ \mathcal{L}_t \underset{t \to -\infty}{\Longrightarrow} \mu_{eq}. \tag{10.47}$$

(d) If we have a solution u satisfying (10.41) and for which (10.43) holds, then the following forward limits as $t \to \infty$ for the corresponding mean curve $(m(t, u, \cdot))_{t \in \mathbb{R}}$ exist and satisfy:

$$\lim_{t \to \infty} m(u, t, E_1) = 1, \tag{10.48}$$

$$\lim_{t \to \infty} m(u, t, \{i\}) = \frac{A_i}{\sum_{i=M+1}^{2M} A_i}, \quad \text{for } i = M+1, \cdots, 2M. \quad \square \tag{10.49}$$

10.4.2 Emergence: The Multitype Case

We now return to the investigation of the random empirical measures $\Xi_N^{\log,\beta}(t) \in \mathcal{P}(\Delta_{2M-1})$ in the limit as $N \to \infty$. We observe that if we work with the total mass

10.4 Results 1: Macroscopic Emergence, Fixation and Random Entrance Law ...

of E_1 types, then this total mass evolves exactly as in a model with only *one* superior type since the fitness difference in E_1 is of $O(N^{-1})$ and hence have no effect in time scale $o(N)$. Therefore the time scale of emergence is

$$\beta^{-1} \log N + t \text{ with } \beta \text{ as given in Subsection 9.} \tag{10.50}$$

What remains to be determined is the fine structure of the emerging process, i.e. the *relative proportions* of the higher level types upon emergence and later on fixation. In other words, we must determine the distribution of *random* relative frequencies of the higher level types at emergence - this will be determined by the respective growth constants and is given by the sum of independent contributions of the different upper level types. This means that we have to show that $(\Xi_N^{\log,\beta}(t))_{t \in \mathbb{R}}$ converges to an entrance law of the form we found in Proposition 10.2 and we have to identify the characteristic vector **A** appearing in (10.40).

Proposition 10.3 (Convergence and Emergence). *Consider the model given by (10.3)–(10.6), (12.83).*

(a) *As $N \to \infty$ the $\mathcal{P}(\Delta_{2M-1})$-valued empirical processes*

$$\left((\Xi_N^{\log,\beta}(t)) \right)_{t \in \mathbb{R}} \tag{10.51}$$

converge (in the sense of weak convergence of laws on path of measure-valued processes) to a random solution $(\mathcal{L}_t)_{t \in \mathbb{R}}$ of the McKean–Vlasov equation, that is

$$\mathcal{L}_t = u(t, \mathbf{x}) d\mathbf{x}, \tag{10.52}$$

with $u(t, \cdot)$ solving the equation, given in (10.20) with corresponding (random) mean process $m(u, t, \cdot)$ (see (10.34)).

The limit is random in the sense that the mean process satisfies:

$$\mathrm{Var}[m(u, t_0, E_1)] > 0, \ \forall\, t_0 \in \mathbb{R}. \tag{10.53}$$

(b) *We have the following asymptotics for the mean curve of the limit of (10.51) which then characterizes (recall Proposition 10.2 part b) the limiting (as $N \to \infty$) entrance law uniquely:*

$$\lim_{t \to -\infty} (e^{-\beta t} m(u, t, \cdot)) = {}^*\mathcal{W}_{M+1} \delta_{M+1} + \cdots + {}^*\mathcal{W}_{2M} \delta_{2M}, \text{ in probability} \tag{10.54}$$

with random variables ${}^*\vec{\mathcal{W}} = ({}^*\mathcal{W}_i, \ i = M+1, \cdots, 2M)$, satisfying:

$$\{{}^*\mathcal{W}_i, i = M+1, \cdots, 2M\} \text{ are independent.} \tag{10.55}$$

If we start with the distribution $(\mu_{\mathrm{eq}})^{\otimes N}$ initially, then furthermore

$$E[{}^*\mathcal{W}_i] = m_i \quad (given\ by\ (10.38)).\tag{10.56}$$

(c) Define the hitting time of the higher level types, (here $\bar{x}^N$ is the empirical mean measure):

$$\bar{T}_\varepsilon^N = \inf_{t\geq 0}(\bar{x}^N(t)[\{M+1,\cdots,2M\}] \geq \varepsilon).\tag{10.57}$$

We have

$$\mathcal{L}[\bar{T}_\varepsilon^N - (\frac{1}{\beta}\log N)] \underset{N\to\infty}{\Longrightarrow} \nu_\varepsilon, \quad \nu_\varepsilon \in \mathcal{P}((-\infty,\infty)),\tag{10.58}$$

where ν_ε is nontrivial. $\square$

Remark 213 (Random shift). We see that the random entrance law is now again a simple random time shift of a deterministic entrance law. Namely by introducing the random variable $^*\bar{\mathcal{W}} = \sum_{i=M+1}^{2M} {}^*\mathcal{W}_i$ which is the growth factor of the (total mass on E_1)-process we obtain the random shift $\beta^{-1}\log {}^*\bar{\mathcal{W}}$ of a standard solution as before in Section 9. However the solution involves now in addition another characteristic feature, namely a random "relative weight" of the higher level types.

Remark 214. The the second part of (b) says in particular that the different types emerge in disjoint parts of the geographic space and do so independently. The interaction between these types sets in only after emergence.

Remark 215 (Palm entrance laws). Compare here with Remark 12. The Palm entrance law is defined by

$$\hat{\mathcal{L}}_t(dx_{M+1},\cdots,dx_{2M}) = \frac{x_{M+1}+\cdots+x_{2M}}{m(u,t,E_1)}\mathcal{L}_t(dx_{M+1},\cdots,dx_{2M}).\tag{10.59}$$

Let

$$\mu_u^*(dx_{M+1},\ldots,x_{2M}) = \text{distribution of an excursion into } E_1 \text{ of age } u.\tag{10.60}$$

At time t let

$$\Gamma_t([a,b)) = \text{number of atoms of age in } [a,b) \text{ at time } t, \quad 0 < a < b \leq t\tag{10.61}$$

and then define the corresponding size-biased distribution on the product space of age and size given by

10.5 Results 2: Mutant Droplet Formation in (M, M)-Case

$$\hat{\Gamma}_t(du, dx_{M+1}, \ldots, dx_{2M}) = \frac{\Gamma_t(du)\ (x_{M+1}, \ldots, x_{2M})\mu_u^*(dx_{M+1}, \ldots, x_{2M})}{\int_0^t \Gamma_t(du)\int_0^1 (x_{M+1}, \ldots, x_{2M})\mu_u^*(dx_{M+1}, \ldots, x_{2M})}. \tag{10.62}$$

As in the two type case the asymptotic behaviour of $\hat{\mathcal{L}}_t$ as $t \to -\infty$ satisfies:

$$\lim_{t \to -\infty} \hat{\mathcal{L}}_t(dx_M, \ldots, dx_{2M}) = \lim_{t \to \infty} \hat{\Gamma}_t((0, \infty), dx_{M+1}, \ldots, dx_{2M}). \tag{10.63}$$

10.4.3 Fixation

After the emergence of the E_1-types at time $\beta^{-1} \log N + t_0$ they fixate on E_1 in time of order $O(1)$ as $N \to \infty$, namely we show the following:

Proposition 10.4 (Fixation). *Consider the model given in (10.3)–(10.6), (10.21).*

(a) The system fixates asymptotically in the emergence frequencies:

$$\lim_{t \to \infty} \lim_{N \to \infty} (\mathcal{L}[\int \sum_{i=1}^M x_i \, \Xi_N^{\log,\beta}(t)(d\mathbf{x})]) = \delta_0, \tag{10.64}$$

$$\lim_{t \to \infty} \lim_{N \to \infty} \mathcal{L}[(\int x_i \, \Xi_N^{\log,\beta}(t)(d\mathbf{x}))_{i=M+1,\cdots,2M}] = \mathcal{L}[{}^*\vec{\mathcal{W}}/{}^*\mathcal{W}(E_1)]. \tag{10.65}$$

(b) If we choose $t_N \to \infty$ $t_N = o(\log N)$ as $N \to \infty$, then the above holds in a joint (of time and N) limit:

$$\lim_{N \to \infty} \mathcal{L}([\int \sum_{i=1}^M x_i \, \Xi_N^{\log,\beta}(t_N)(d\mathbf{x})]) = \delta_0, \tag{10.66}$$

$$\lim_{N \to \infty} \mathcal{L}[(\int x_i \, \Xi_N^{\log,\beta}(t_N)(d\mathbf{x}))_{i=M+1,\cdots,2M}] = \mathcal{L}[{}^*\vec{\mathcal{W}}/{}^*\vec{\mathcal{W}}(E_1)]. \quad \square \tag{10.67}$$

10.5 Results 2: Mutant Droplet Formation in (M, M)-Case

In this section we want to relate the phase 1 of emergence with the phase 0 of the where a E_0-equilibrium develops. For this purpose we have to first study the formation of E_1-droplets in Subsubsection 10.5.1 and then relate their longtime time evolution with the emergence in Subsubsection 10.5.2.

10.5.1 Formation of Droplets and Their Longtime Behaviour

Our goal is to analyse the time evolution of the droplet of E_1-types and in particular to establish the exponential growth of the droplet in the (M,M)-type mean-field model. We present here the concepts and results and the proofs are in Subsection 10.13. We now have to analyse the behaviour of the droplet of advantageous types in two time windows and as $N \to \infty$, namely in the time of $O(1)$ but first starting from time 0 and second from time t_N, with $t_N \to \infty$ but $t_N = o(\log N)$.

We proceed in steps, first we give the finite N-sites model and the description of droplets therein, then we give in Step 2 the candidate for the $N \to \infty$ limit and in Step 3 the actual convergence results.

Step 1: *Droplet description.*

To carry out the analysis of the droplet of the $N \to \infty$ limit we need to give a suitable description of the droplet of upper types. To do this we introduce an atomic measure-valued process on $[0,1] \times \{M+1, \cdots, 2M\}$ (for an alternative description on $[0,1] \times \Delta_{2M-1}$), see Remark 216 below. Let

$$\mathbf{x}^N(j,t) = ((x_i^N(j,t)))_{i=1,\cdots,2M}, \quad \mathbf{x}^N(j,t) \in \mathcal{P}(E_0 \cup E_1), \qquad (10.68)$$

denote the proportion of type i at site j at time t. Let now $\{a(j) \in [0,1], j \in \mathbb{N}\}$ be independently randomly chosen points in $[0,1]$ (which are also chosen independently of the further dynamic), that serve as a labels for site $j \in \{1, \cdots, N\}$. Then by exchangeability we can describe the geographic distribution of the types in E_1 by the object:

$$\mathfrak{z}_t^{N,M} = \sum_{i=M+1}^{2M} \sum_{j=1}^{N} x_i^N(j,t)(\delta_{a(j)} \otimes \delta_i), \qquad (10.69)$$

which is an *atomic finite measure* on $[0,1] \times \{M+1, \cdots, 2M\}$.

To describe the droplet we need the abbreviations:

$$\Delta_{2M-1}^{i+} = \{\mathbf{x} \in \Delta_{2M-1} : x(\{i\}) > 0, \mathbf{x}(\{i'\}) = 0, i' \neq i, i' \in \{M+1, \ldots, 2M\}\}, \qquad (10.70)$$

$$\Delta_{2M-1}^{E_1+} = \{\mathbf{x} \in \Delta_{2M-1} : \sum_{i=M+1}^{2M} \mathbf{x}(\{i\}) > 0\}.$$

Remark 216. Often a different point of view on (10.69) is useful.

(a) If we extend the first sum in (10.69) also over $i \in \{1, \cdots, M\}$,s we have also coded the complete configuration of the system and the measure extends to the complete type space $E_0 \cup E_1$.

10.5 Results 2: Mutant Droplet Formation in (M, M)-Case

(b) Since the $\mathbf{x}(j,t)$ for every j is an element of Δ_{2M-1} we can code the configuration of the system for a given labelling of the sites by $a(j)$, also as an element of

$$\mathcal{M}_a([0,1] \times \Delta_{2M-1}^{E_1+}) \tag{10.71}$$

by writing

$$\mathbf{J}_t^{N,M} = \sum_{j=1}^{N} 1_{(\mathbf{x}^N(j,t) \in \Delta_{2M-1}^{E_1+})} \delta_{\left(a(j), \{(x_i^N(j,t), \ i=1,\cdots,2M)\}\right)}. \tag{10.72}$$

This means that we can view $\mathbf{J}_t^{N,m}$ as an element of $\mathcal{M}_a([0,1] \times \Delta_{2M-1}^{E_1+})$ and we will use both these representations whichever is more convenient.

Now emergence begins as a collection of droplets of types $M+1,\ldots,2M$ whose aggregated masses eventually reach a positive intensity in the collection of N sites. This arises since we have after some $O(1)$-time a sparse set of components with a nontrivial mass of higher types and then the number of such components grows exponentially with a random factor as time gets large. The sparse sites of $O(1)$-mass of types $i = M+1,\ldots,2M$ are now described by the atomic measure $\mathbf{J}_t^{N,m}$ on $[0,1] \times \{M+1,\ldots,2M\}$ (or $[0,1] \times \Delta_{2M-1}^{E_1+}$). Note however that the evolution of this atomic measure is like a *process in a random medium* where the medium is given by the process of relative frequencies of the types $\{1,\ldots,M\}$ at each of the geographic sites which generates a (small) flow of rare mutations.

Step 2: *The limiting dynamic for the droplet*

We next construct a measure-valued process, $(\mathbf{J}_t^M)_{t \geq 0}$ on $[0,1] \times \{M+1,\ldots,2M\}$ with infinitely divisible marginal distributions. We will then later establish that the measure-valued processes $(\mathbf{J}_t^{N,M})_{t \geq 0}$ converge to this process $(\mathbf{J}_t^M)_{t \geq 0}$ as $N \to \infty$ and we identify the Palm measures with realizations $(\hat{\mathbf{J}}_t^M)_{t \geq 0}$ which we will identify later on.

The construction is based on the *excursions* to E_1 of the McKean–Vlasov process and we next introduce the necessary tools. We introduce the following diffusion on Δ_{2M-1} defined via the martingale problem associated with the operator $G^{\mathbf{q}}$ which we introduce below in (10.73).

Given $\mathbf{q} = (q_1,\ldots,q_M,0,\ldots,0) \in \Delta_{2M-1}$ we define the linear operator $G^{\mathbf{q}}$ on $C_b(\Delta_{2M-1}, \mathbb{R})$ with domain $C_b^2(\Delta_{2M-1}, \mathbb{R})$:

$$G^{\mathbf{q}} f(\mathbf{x}) = G^{0,1} f(\mathbf{x}) + c \sum_{j=1}^{2M} (q_j - x_j) \frac{\partial f}{\partial x_j}, \tag{10.73}$$

where $G^{0,1}$ arises from $\sum_{k=1}^{N} G_k^{0,1,N}$ by letting $N \to \infty$, which amounts to drop in $G_k^{0,1,N}$ in (10.10) the last two terms (i.e. (10.8) and (10.9)). The

$(G^{\mathbf{q}}, C_b^2(\Delta_{2M-1}, \mathbb{R}), \delta_{\mathbf{x}})$-martingale problem is for every $\mathbf{x} \in \Delta_{2M-1}$ well-posed and its solution is the law of a diffusion process on Δ_{2M-1}.

Now consider the case in which $\mathbf{q} = (q_1, q_2, \cdots, q_{2M})$ is determined as the mean vector of the mean-field mutation-selection equilibrium, $\mu_{eq}^\infty(dx_1, \ldots, dx_M)$, for the McKean–Vlasov dynamic on E_0 that is extended to a measure $\tilde{\mu}_{eq}^\infty$ on Δ_{2M-1}, such that $\tilde{\mu}_{eq}^\infty$ is concentrated on $\{x : x \in \Delta_{2M-1}, x_{\{M+1,\ldots,2M\}} = 0\}$ and agrees there with μ_{eq}^∞. We set:

$$q_i = \int_0^1 x_i \tilde{\mu}_{eq}^\infty(dx_1, \ldots, dx_{2M}), \quad i = 1, \cdots 2M. \tag{10.74}$$

Note with this choice for $\mathbf{q}$ in (10.73) we denote the generator $G^{\mathbf{q}}$ by:

$$G^{eq}. \tag{10.75}$$

Then the diffusion with generator G^{eq} describes the dynamics at a tagged site in mean-field equilibrium and $\tilde{\mu}_{eq}$ gives the stationary measure for the G^{eq} diffusion.

Remark 217. Recall that assuming that initially $x_{M+1}, \ldots, x_{2M} = 0$, this law $\tilde{\mu}_{eq}$ arises if we consider the state at time t_N in the limit $N \to \infty$, $t_N \to \infty$, $t_N = o(\log N)$ for the dynamic on $E_0 \cup E_1$ and the limiting dynamic arises if we drop terms in the generator containing rates of order $\frac{1}{N}$.

Remark 218. We can also consider the McKean–Vlasov dynamics starting with an arbitrary initial law $\mathcal{L}_0$ concentrated on types $\{1, \ldots, M\}$ and the analogous excursion process for this diffusion. If we start with another configuration in the M lower types, then the one given by μ_{eq}^∞, then we have to write everything out in a *time-inhomogeneous* fashion. This poses no problems, but notation becomes a bit more complex.

In order to use the excursion theory for the G^{eq}-diffusion we first introduce the set of possible excursions in $\Delta_{2M-1}^{E_1+}$ starting at time 0:

$$W_0^{2M} = \{w \in C([0,\infty), \Delta_{2M-1}^+)), w(0) \in \{\mathbf{x} : x_i = 0; i = M+1, \ldots, 2M\},$$
$$\exists \zeta < \infty \text{ with } \sum_{i=M+1}^{2M} w(t,i) = 0 \; \forall t \geq \zeta, \quad w(t) \in \Delta_{2M-1}^{E_1+} \text{ for } t \in (0, \zeta)\}.$$
$$\tag{10.76}$$

We now construct a set of *excursion laws* for every $\mathbf{x} = (x_1, \cdots, x_{2M}) \in (\Delta_{2M-1}^{E_1+})^c$ with excursions into $\Delta_{2M-1}^{E_1+}$.

Consider the diffusion starting in a point $\mathbf{x} \in \Delta_{2M-1}$ with the dynamic given by the generator G^{eq} from (10.73) and denote the law on path space by $P_{\mathbf{x}}$. The collection $\mathbb{Q}_{(x_i,\cdots,x_m)}^i, i = M+1, \cdots, 2M$ of *excursion laws*, are defined by the following limits.

10.5 Results 2: Mutant Droplet Formation in (M, M)-Case

Here the convergence is defined as weak convergence of those finite measures on the path space $C([0, \infty), \Delta_{2M-1})$, which are obtained by restricting to the sets of paths where $\sup_{t \geq 0} w(t, i) > \delta$ for the specified i for all $\delta > 0$. We set:

$$\mathbb{Q}^i_{(x_1, \ldots, x_M)} := \lim_{\varepsilon \to 0} \frac{1}{\varepsilon} P_{(1-\varepsilon)(x_1, \ldots, x_M, 0, \ldots, 0) + \varepsilon \delta_i}, \quad i = M+1, \cdots 2M. \tag{10.77}$$

Furthermore we define the excursion law of the total E_1-mass process by:

$$\mathbb{Q}_{(x_1, \cdots, x_M)} := \lim_{\varepsilon \to 0} \frac{1}{\varepsilon} P_{(1-\hat{\varepsilon})(x_1, \ldots, x_M, 0, \cdots, 0) + \hat{\varepsilon} 1}, \quad \hat{\varepsilon} = \varepsilon/M. \tag{10.78}$$

Proposition 10.5 (Excursions on upper level).

(a) The limit in (10.77) exists.
(b) The excursion law for the total E_1-mass defined in (10.78) is given by

$$\mathbb{Q}_{(x_1, \cdots, x_M)} = \sum_{i=M+1}^{2M} \mathbb{Q}^i_{(x_1, \cdots, x_M)}. \tag{10.79}$$

(c) For any $(x_1, \ldots, x_M) \in \Delta_{M-1}$ the excursion law $\mathbb{Q}^i_{(x_1, \ldots, x_M)}$ satisfies:

$$\mathbb{Q}^i_{(x_1, \ldots, x_M)} \text{ is a } \sigma\text{-finite measure on } W_0^{2M}, \tag{10.80}$$

namely for every $i \in \{M+1, \cdots, 2M\}$ we have that

$$\mathbb{Q}^i_{(x_1, \ldots, x_M)}(\{w : w(t, \{i\}) > \delta\}) < \infty \text{ for all } \delta > 0, \ t > 0. \quad \square \tag{10.81}$$

Based on the excursion law we can introduce now a process of $\tilde{\mu}_s$-driven random measure as follows. Let

$$\{\tilde{\mu}_s\}_{s \geq 0} \text{ be a measurable function from } [0, \infty) \text{ to } \mathcal{P}(E_0) \tag{10.82}$$

and consider for every $i \in \{M+1, \cdots, 2M\}$, the *Poisson random measure*

$$N^i(ds, da, du, dw) \text{ on } [0, \infty) \times [0, 1] \times [0, \infty) \times W_0^{2M}, \tag{10.83}$$

such that for different i, the N^i are *independent* and the *intensity measure* of the random measure with index i is given by

$$ds \, da \, du \left(\int_{\Delta_{M-1}} (\mathbb{Q}^i_{x_1, \ldots, x_M}(dw)) \tilde{\mu}_s(dx_1, \ldots, dx_M) \right). \tag{10.84}$$

A system may start already in a nontrivial atomic measure. Here the initial state $(\mathbb{J}_*^0(u))_{u \geq 0}$ is assumed to have the form

$$\sum_{j=M+1}^{M} \sum_{i \in \mathbb{N}} y_j(i, u) \delta_{a_i}, \qquad (10.85)$$

where $\{y_j(i, \cdot) \in C([0, \infty), \Delta_{2M-1}), \; j = M+1, \ldots, 2M, \; i = 1, 2, \ldots\}$ are given by independent solutions of the mutation-selection system with generator (10.73) for which the initial conditions satisfy $y_j(i, 0) > 0$ for some $j \in \{M+1, \ldots, 2M\}$ for each i.

With these ingredients we can construct the atomic measure-valued dynamic generalizing the construction for two and three types discussed in previous Sections 8 and 9.

Proposition 10.6 (A continuous atomic-measure-valued Markov process: limit droplet dynamic). *Given the ingredients from (10.83)–(10.85) we have the following:*

(a) Define $q^i(s, a)$ as the non-negative predictable function

$$q^i(s, a) := (\tilde{m}_i(s) + c \mathbb{J}_{s-}^M([0, 1] \times \{i\})), \quad i = M+1, \cdots, 2M, \quad (10.86)$$

where

$$\tilde{m}_j(s) := \sum_{i=1}^{M} \tilde{\mu}_s(i) m_{ij}, \quad j \in \{M+11, \cdots, 2M\}. \qquad (10.87)$$

Then the stochastic integral equation

$$\mathbb{J}_t^M(dv, d\mathbf{x}) = \mathbb{J}_*^0(t)$$

$$+ \int_0^t \int_{[0,1]} \sum_{i=M+1}^{2M} \int_0^{q^i(s,a)} \int_{W_0} \delta_{w(t-s)}(d\mathbf{x}) \delta_a(dv) N^i(ds, da, du, dw),$$

$$(10.88)$$

has a unique continuous solution, $(\mathbb{J}_t^M)_{t \geq 0}$.

(b) $(\mathbb{J}_t^M)_{t \geq 0}$ is a continuous $\mathcal{M}_a([0, 1] \times \{M+1, \cdots, 2M\})$ respectively $\mathcal{M}_a([0, 1] \times \Delta_{M-1}^+)$-valued strong Markov process.

(c) The process $(\mathbb{J}_t^M)_{t \geq 0}$ has the following properties:

- *the mass of each atom follows an excursion from zero, which is obtained at the intensity given by the excursion law $\mathbb{Q}_{x_1, \ldots, x_M}$,*
- *new Δ_{M-1}^{i+}-valued excursions are produced at time t at rate*

$$\tilde{m}_i(t) + c \mathbb{J}_t^M([0, 1] \times \{i\}), \qquad (10.89)$$

10.5 Results 2: Mutant Droplet Formation in (M, M)-Case

- each new Δ_{M-1}^+-valued excursion for a type $i \in \{M+1, \cdots, 2M\}$ produces an atom located at a point in $[0,1] \times \{i\}$, the first component chosen according to the uniform distribution on $[0,1]$,
- at each t and $\delta > 0$ there are at most finitely many atoms $\{a_i\} \times \{i\}$ with mass $\geq \delta$, for some $i \in M+1, \ldots, 2M$,
- $t \to \mathbf{J}_t([0,1] \times \{i\})$ is a.s. continuous, $i = M+1, \ldots, 2M$. □

Remark 219. If we assume that the process has reached the McKean–Vlasov mutation-selection equilibrium on level E_0, μ_{eq}, we can take $\tilde{\mu}_s = \mu_{\text{eq}}$ for all $s \geq 0$.

The large time behaviour of $\mathbf{J}_t^M$ is similar to the previous case of two types in Section 8, but in addition we now have a distribution of the droplet mass on the M different types in E_1, whose mean vector and law we determine in the next result. First we introduce the main ingredient to calculate the exponential growth rate, namely $w(s, x_1(s), \cdots, x_{2M}(s))$ by

$$w(s, \Delta_{2M-1}^{E_1+}) = \sum_{i=M+1}^{2M} x_i(s) \qquad (10.90)$$

and then set:

$$f(s) = \int_{\Delta_{M-1}} \int_{W_0} w(s, \Delta_{2M-1}^{E_1+})(\mathbb{Q}_{(p_1,\ldots,p_M)}(dw)) \mu_{eq}(dp_1, \ldots, dp_M). \qquad (10.91)$$

Then the growth behaviour of the limit $(N \to \infty)$ droplet can be completely determined.

Proposition 10.7 (Exponential growth of droplet total mass, limiting droplet frequencies). *Assume that $\mathbf{J}_0^M$ satisfies that we start in the lowest type only at every site.*

(a) *There exists $\beta^* > 0$, the Malthusian parameter, such that*

$$e^{-\beta^* t} E[\mathbf{J}_t^M([0,1] \times \{M+1, \ldots, 2M\})] \to A \in (0, \infty), \text{ as } t \to \infty. \qquad (10.92)$$

The parameter β^* is given by the unique positive solution β^* of the equation

$$c \int_0^\infty e^{-\beta^* s} f(s) ds = 1. \qquad (10.93)$$

We have that (β as in Proposition 9.39 in Section 9):

$$\beta^* = \beta. \qquad (10.94)$$

(b) *Furthermore there exist $A_i \in (0, \infty)$ such that as $t \to \infty$,*

$$e^{-\beta t}(E[\mathbf{J}_t^M([0,1] \times \{i\})])_{i=M+1,\cdots,2M} \longrightarrow (A_i)_{i=M+1,\cdots,2M}. \qquad (10.95)$$

(c) *There exist non-degenerate random variables* $\vec{W}^* = (W_i^*)_{i=M+1,\cdots,2M}$ *such that*

$$\mathcal{L}[e^{-\beta t}(\mathbf{J}_t^M([0,1]\times\{i\}])_{i=M+1,\cdots,2M})] \underset{t\to\infty}{\Longrightarrow} \mathcal{L}[\vec{W}^*]. \qquad \Box \qquad (10.96)$$

Remark 220. The parameter β^* does not depend on the initial state if we consider only such states where types are concentrated on E_0. However the vector $\vec{W}^*$ does depend strongly on this initial state.

Remark 221 (Stable size distribution). Now let $i \in \{M+1,\ldots,2M\}$ and define for $b \in [0,1]$

$$\tilde{U}_i(t,[b,1]) := \text{number of type } i \text{ atoms of size } \in [b,1], \qquad (10.97)$$

and the above measure normalized by the total droplet size, the size distribution:

$$\hat{U}_i(t,dx) := \frac{\tilde{U}_i(t,dx)}{\int y\tilde{U}_i(t,dy)}. \qquad (10.98)$$

Note that $\tilde{U}_i(t,dx)$ is a sigma-finite random measure on $(0,1)$.

We can view $\mathbf{J}_t^M$ as a generalized CMJ process and the analogue of the stable size distribution for the CMJ process would be the statement

$$\lim_{t\to\infty}\hat{U}_i(t,dx) \to \hat{U}_i(\infty,dx) \text{ as } t \to \infty, \qquad (10.99)$$

where $\hat{U}_i(\infty,dx)$ is *non*-random.

Step 3: *Convergence and droplet growth*

Now we can describe the droplet formation in a finite time horizon in the limit $N \to \infty$ indeed by the limit dynamic $\mathbf{J}^M$. Let $\mathcal{M}_a([0,1]\times\Delta_{2M-1}^{E_1+})$ denote the set of atomic measures on $[0,1]\times\Delta_{2M-1}^{E_1+}$ with the weak atomic topology (recall Section 8 for a review of the definition).

Proposition 10.8 (Convergence of droplet $\mathbf{J}^{N,M}$ as $N \to \infty$).

(a) *Assume that the N-site system starts in the equilibrium μ_{eq}^N of the E_0-dynamic corresponding to the mean vector $(q_1,\ldots,q_M)$. Then*

$$\mathcal{L}[(\mathbf{J}_t^{N,M})_{t\geq 0}] \underset{N\to\infty}{\Longrightarrow} \mathcal{L}[(\mathbf{J}_t^M)_{t\geq 0}], \qquad (10.100)$$

where the r.h.s. satisfies (10.88) with $\mu_s \equiv \mu_{eq}^\infty$ and $\mathbf{J}_0^* = 0$.

(b) *Assume that the system starts with only the type 1 present. Then*

$$\mathcal{L}[(\mathbf{J}_t^{N,M})_{t\geq 0}] \underset{N\to\infty}{\Longrightarrow} \mathcal{L}[(\mathbf{J}_t^{M,\text{td}})_{t\geq 0}], \qquad (10.101)$$

10.5 Results 2: Mutant Droplet Formation in (M, M)-Case

where $\mathbf{J}_t^{M,\text{td}}$ is defined as in (10.88) but with (10.84) replaced by (here $\tilde{\mu}_t^{\delta_1}$ gives the current frequency on E_0 of the diffusion process defined by the solution of the McKean–Vlasov equation starting in the state δ_1):

$$ds\, da\, du \left(\int_{\Delta_{M-1}} (\mathbb{Q}^i_{x_1,\ldots,x_M}(dw)) \tilde{\mu}_t^{\delta_1}(dx_1,\ldots,dx_M) \right). \quad \square \tag{10.102}$$

We not only have the above convergence for $N \to \infty$ in the finite time horizon and for $t \to \infty$ of the ($N \to \infty$) limit dynamics as in Proposition 10.7, but we get this convergence if N and t go both to infinity and we find again exponential growth of the droplet at rate β and a growth constant $\vec{\mathcal{W}}^*$. We recall that $\hat{x}_i^N(t) = \sum_{j=1}^{N} x_i^N(j,t)$.

Proposition 10.9 (Droplet growth). *Assume that at time* $t = 0: \hat{x}_1^N(i,0) = 1$ *for* $i = 1,\cdots,N$ *and that* $t_N \to \infty, t_N = o(\log N)$ *as* $N \to \infty$.
Then

$$\mathcal{L}[e^{-\beta t_N}(\hat{x}_{M+1}^N(t_N),\ldots,\hat{x}_{2M}^N(t_N))] \tag{10.103}$$

$$= \mathcal{L}\left[e^{-\beta t_N} \left(\mathbf{J}_{t_N}^{N,m}([0,1] \times \{M+1\}),\ldots,\mathbf{J}_{t_N}^{N,m}([0,1] \times \{2M\}) \right) \right]$$

$$\underset{N \to \infty}{\Longrightarrow} \mathcal{L}\left[\sum_{i=M+1}^{2M} \mathcal{W}_i^* \delta_i \right],$$

where $\mathcal{W}_i^*$, $i = M+1,\ldots,2M$ *are non-degenerate* $\mathbb{R}^+$*-valued independent random variables.* $\square$

One can say more about the fine structure of the E_1-droplet even in the general case of the (M, M)-model. We shall need the measures

$$\mu_t^{N,i} = \mathcal{L}[x_i^N(t,1)] \in \mathcal{P}([0,1]), \quad i = M+1,\cdots,2M \tag{10.104}$$

and the following notation for the Palm measures of a measure ν on $[0,1]$:

$$\hat{\nu}(dx) = \frac{x\nu(dx)}{\int_0^1 x\nu(dx)}. \tag{10.105}$$

Then we introduce (existence of limits will be part of the statement below)

$$\hat{\mu}_t^i = \lim_{N \to \infty} \hat{\mu}_t^{N,i}, \tag{10.106}$$

and

$$\hat{\mu}_\infty^i \in \mathcal{P}([0,1]) \quad,\quad \hat{\mu}_\infty^i := \lim_{t \to \infty} \lim_{N \to \infty} \hat{\mu}_t^{N,i}. \tag{10.107}$$

Then we can formulate a result on the finer structure of the E_1-droplet in terms of $\hat{\mu}_\infty^i$.

Proposition 10.10 (Fine structure of advantageous droplet). *We have the following structure of the E_1-type population on emergence.*

(a) The limiting empirical distribution of emerging types satisfies

$$\lim_{t \to -\infty} e^{-\beta t} \lim_{N \to \infty} \frac{1}{N} \sum_{i=1}^N \delta_{(x_{M+1}^N(\frac{\log N}{\beta}+t,i),\ldots,x_{2M}^N(\frac{\log N}{\beta}+t,i))} \quad (10.108)$$

$$= \int_0^1 {}^*\mathcal{W}_{M+1} U_{M+1}(\infty, da_{M+1}) (\delta_{a_{M+1}(1,0,\ldots,0)}) + \cdots +$$

$${}^*\mathcal{W}_{2M} U_{2M}(\infty, da_{2M}) (\delta_{a_{2M}(0,\ldots,0,1)}).$$

Here $U_i(\infty, dx)$ (cf (10.99)) is the stable size distribution of type i excursions. This means that as $t \to -\infty$, the number of sites with mass of type $i \in \{M+1, M+2, \cdots, 2M\}$ in $(b, 1]$, $b > 0$, grows like $e^{\beta t} \, {}^*\mathcal{W}_i \cdot U_i(\infty, (b, 1])$.

(b) The following limits exist:

$$\lim_{t \to \infty} \lim_{N \to \infty} \hat{\mu}_t^{N,i}(\cdot) = \hat{\mu}_\infty^i(\cdot). \quad (10.109)$$

(c) Moreover

$$\frac{xU_i(\infty, dx)}{\int_0^1 xU_i(\infty, dx)} = \hat{\mu}_\infty^i(dx) \in \mathcal{P}([0,1]), \quad (10.110)$$

where

$$\hat{\mu}_\infty^i(dx) = \frac{x\tilde{\mathbb{Q}}_\infty^i(dx)}{\int x\tilde{\mathbb{Q}}_\infty^i(dx)} \quad (10.111)$$

and with ζ_b and ζ_d denoting birth time and extinction time of an excursion:

$$\tilde{\mathbb{Q}}_\infty^i(dy) = \int_{-\infty}^0 \int_{\Delta_{M-1}} e^{\beta s} \tilde{\mathbb{Q}}_{\vec{x}}(\zeta_b \in ds, \zeta_d > 0, w(0) \in dy) \mu_{\mathrm{eq}}(d\vec{x}).$$

$$(10.112)$$

This is the analogue of the limiting single site Palm distribution determined in (8.635). □

10.5.2 The Equality of $\mathcal{W}^*$ and $^*\mathcal{W}$ (Linkage Phase 0 and Phase 1)

We have considered above in Proposition 10.9 the droplet growth. Then the droplet formation is described by a simple picture, namely exponential growth of the droplet at rate β with a growth vector $\vec{\mathcal{W}}^*$. In formulas, for $t_N \to \infty$ with $t_N = o(\log N)$ and $\hat{x}$ denoting the sum overall N spatial components:

$$\mathcal{L}[e^{-\beta t_N}(\hat{x}_{M+1}(t_N), \cdots, \hat{x}_{2M}(t_N))] \underset{N \to \infty}{\Longrightarrow} (\mathcal{W}_1^*, \cdots, \mathcal{W}_M^*), \qquad (10.113)$$

where the $(\mathcal{W}_i^*)_{i=1,\cdots,M}$ are independent. We next turn to the question of how these relate to the $(^*\mathcal{W}_i)_{i=M+1,\cdots,2M}$ given in (10.54) and describing the factor in the exponential emergence in the entrance law from $-\infty$ which occurs with the same rate β. We want to show that also for the (M, M)-type model:

Proposition 10.11 (Exit = Entrance: equality of growth constants). *We have for the initial state starting all in type 1:*

$$\mathcal{L}[^*\vec{\mathcal{W}}] = \mathcal{L}[\vec{\mathcal{W}}^*]. \qquad \square \qquad (10.114)$$

In particular we conclude from (10.114) that

$$\mathcal{L}[(\mathcal{W}^*(E_1), \vec{\mathcal{W}}^*/\mathcal{W}^*(E_1))] = \mathcal{L}[(^*\mathcal{W}(E_1), ^*\vec{\mathcal{W}}/^*\mathcal{W}(E_1))]. \qquad (10.115)$$

That is we have the same random time shift for the droplet growth dynamic and relative weights on E_1 in times of $O(1)$, then for the entrance law arising around time $\beta^{-1} \log N$. Hence even though the time scales do separate in the $N \to \infty$ limit they fit together in the exit-entrance behaviour.

10.6 Dual Calculations: Proof of Theorem 14

The proof of Theorem 14 is based on the set-valued dual and provides a gentle introduction into the type of calculations we need later on in Subsection 10.7. We proceed with the proof step by step through the parts a)-d) of the theorem.

(a) It suffices that the joint moments of the field $\{x^N(j,t), \ j = 1, \ldots, N\}$ converge. To show this we use the set-valued dual $\tilde{\mathcal{G}}_t^{++,N}$. It suffices to take the initial state for the dual given by

$$\tilde{\mathcal{G}}_0^{++,N} = \prod_{j=1}^{\ell} \prod_{i=1}^{n_j} 1_{A_{i,j}}(u_{i,j}), \ A_{i,j} \subsetneq E_0, \qquad (10.116)$$

where $1 \leq \sum_j n_j = n_* < \infty$ where n_* is independent of N. This means at a collection of sites $j \in \{1, \cdots, \ell\}$ we have n_j factors at the j-th site. The behaviour depends on whether $d = 0$ or $d > 0$, $N = 1$ or $N > 1$ and we discuss these possible cases.

Case i: $d = 0, N = 1$.

In this deterministic case it suffices to consider only $n_* = 1$ since different variables in the dual do not interact by coalescence and hence evolve independently so that the product structure is preserved.

First consider the case $M = 2$ where type 2 has fitness 1 and type 1 fitness 0 which clarifies the principle.

Start the dual with (01). (Note that this suffices in this case). Consider the tableau which results from three selection events before the first mutation event occurs (here the special form results in the fact that selection creates only four nonzero rows rather than eight):

$$
\begin{array}{llll}
(01) & \otimes 1 & \otimes 1 & \otimes 1 \\
(10) & \otimes (01) & \otimes 1 & \otimes 1 \\
(10) & \otimes (10) & \otimes (01) & \otimes 1 \\
(10) & \otimes (10) & \otimes (10) & \otimes (01).
\end{array}
\qquad (10.117)
$$

Observe next that the mutation $2 \to 1$ replaces (01) by (00) and (10) by (11) and this removes one row and column. However if there is a single row, that is, the tableau (01), then the result is a trap and the value is 0. On the other hand a mutation $1 \to 2$ replaces (01) by (11) and (10) by (00) and this immediately produces the value 1 at the column where it occurs. Therefore note that the first $m_{1,2}$ mutation terminates the process, since columns to the left of the column ℓ at which the mutation occurs form $\{(01) \cup (10)\}$ and we end up with only inactive columns.

We can then view the number of active rows and columns as a birth and death process with birth rate s and death rate $m_{2,1}$ and which in addition is terminated forever at rate $m_{1,2}$.

We must now consider two cases $s > m_{2,1}$ and $s \leq m_{2,1}$.

In the latter case the birth and death process is recurrent and returns to (01) infinitely often if no mutation occurs on each return. Since on each such return there is a positive probability of reaching a *trap* on the next transition, the probability that the tableau has not reached a trap by time t goes to 0 as $t \to \infty$.

In the first case if $s > m_{2,1}$ then there are an increasing number of rows each containing a factor (01). The probability that the next transition at this factor produces a (11) and terminates the process is positive. So again the probability that the tableau has not reached a trap by time t goes to 0 as $t \to \infty$.

Hence combining the two cases the process reaches the trap with probability tending to 1 as $t \to \infty$. The value of the dual expressions is 0 or 1 irrespective of the initial value of the process X^N. Hence we have convergence of moments to a value independently of the initial state.

Now consider the case $M > 2$.

10.6 Dual Calculations: Proof of Theorem 14

Again if $d = 0$, at each birth event a column is created by selection. This column is eventually resolved to a factor which is either $(1,\ldots,1)$ or $(0,\ldots,0)$. Since the probability of the latter is strictly positive, even in the limit $t \to \infty$ there the number of those which are not $(0,\ldots,0)$ is stochastically bounded. Also the initial column will finally by mutation reach one of the states $(1,\cdots,1)$ or $(0,\cdots,0)$. Therefore after a finite random time we have $\tilde{\mathcal{G}}_0^{++,N}$ is empty or the full set $\{1,2,\cdots,M\}^{\mathbb{N}}$.

Denote the probabilities, that the process reaches the trap $(\emptyset)$ or $\{1,\ldots,M\}^{\mathbb{N}}$ from the initial factor $(0\cdots 10\cdots), 1$ at the i-the position by q_0^i, q_1^i respectively. Then

$$\lim_{t\to\infty} E[x_i^N(t)] = q_1^i \quad \text{independent of } \mathbf{x}^N(0) \text{ for every } i \in \{1,\cdots,M\}. \quad (10.118)$$

Remark 222. The mutation among the first $(M-1)$ types can kill extra rows from the action of selection but not the primary row added by a selection event. We illustrate this as follows in the case $M = 5$. Start with the tableau (00001). Let selection act with 1_A on the column number 1 first with $A = \{4,5\}$ which gives:

$$\begin{array}{llll}(00001) & \otimes 1 & \otimes 1 & \otimes 1 \\ (11100) & \otimes (00001) & \otimes 1 & \otimes 1.\end{array} \quad (10.119)$$

Then selection acts with birth from the first column and with $A = \{3,4,5\}$, which yields the new tableau:

$$\begin{array}{llll}(00001) & \otimes 1 & \otimes 1 & \otimes 1 \\ (11000) & \otimes (00001) & \otimes 1 & \otimes 1 \\ (00100) & \otimes (00001) & \otimes 1 & \otimes 1 \\ (11000) & \otimes (11100) & \otimes (00001) & \otimes 1.\end{array} \quad (10.120)$$

If we have transitions such that in a column $(00001) \to (11111)$, this no longer automatically terminates the process but does cut the tableau at the level of this factor.

Case ii: $d > 0$, $N = 1$.

In this case we must consider the initial condition (10.116) for $\ell = 1$ and arbitrary n_*.

Again the selection process creates births (i.e. new columns and rows in the tableau). One new feature is that a selection operation can produce more than one new row since the number of rows doubles at each step in which $1_A f \otimes 1$ or $(1 - 1_A) \otimes f$ does not contain a $1_A f$ which is identically zero.

Since $d > 0$, the process $|\tilde{\mathcal{G}}_t^{++,N}|$, the number of active columns, is always *positive recurrent*. Each time the number of factors reduces to 1, there is a positive probability that the mutation process will hit a trap before the next selection event. Therefore $\tilde{\mathcal{G}}_t^{++,N}$ will reach a trap at a finite random time with probability 1 and then satisfies $\tilde{\mathcal{G}}_t^{++,N} = \emptyset$ or $\{1, \cdots, M\}^{\mathbb{N}}$ (see Lemma 9.24 and Lemma 9.25). Again denote the probabilities that the process reaches the trap $\emptyset$, $\{1, \cdots, M\}^{\mathbb{N}}$ by q_0, q_1 (note that $q_0 = q_0(\ell, n_1, \cdots, n_\ell)$, similarly q_1). Then

$$\lim_{t \to \infty} E[\prod_{j=1}^{\ell} (x_j^N(t))^{n_j(t)}] = q_1 \quad \text{independent of } \mathbf{x}^N(0). \tag{10.121}$$

This again shows convergence of all mixed moments and the claim follows.

Case iii: $d > 0$, $1 < N < \infty$.

In this case we must consider initial conditions given by (10.116) with initial factors at ℓ different sites. In this case the dynamics is again positive recurrent and returns to a state with one occupied factor and the argument is exactly as above and the analysis at a site is as above and all mixed moments converge as $t \to \infty$ to a value which is independent of the initial state.

(b) Now consider the McKean–Vlasov process in which we now use the set-valued dual with $S = \mathbb{N}$. The new feature is that the dual is described by a set of occupied sites and the corresponding CMJ process describing the cloud of occupied (non-resolved sites) can be supercritical. First of all all dual clouds starting from different sites are independent. Furthermore since each site has a positive probability of resolving to $\emptyset$, in a growing cloud of unresolved sites, eventually one of these will resolve to $\emptyset$ and kill the whole expression.

Since the probability that this has not yet happened is monotone decreasing in time all moments converge to a limit which proves (10.27).

(c) In order to prove the convergence fix ℓ and consider (10.116). As before $\tilde{\mathcal{G}}^{++,N}$ will hit a trap at a finite random time. The point is then that as $N \to \infty$ the probability that the time to hit the trap occurs before any collisions occur tends to 1 and hence for $N \to \infty$ the expectation over the dual process in the duality relation converges to the same value as for the case $N = \infty$, that is, the McKean–Vlasov case which identifies the factor in (10.28) as μ_{eq}^{∞}.

(d) Now consider the ergodic theorem for the M-type interacting system on the hierarchical group Ω_N in the case $m_{i,j} > 0$ for all $i, j \in \{1, \ldots, M\}$. We must now prove the convergence of the mixed moments. To do this we start n_ℓ factors at each of ℓ different sites.

The main point for the convergence of the mixed moments is again that the dual system eventually hits a trap and as before in b) this will guarantee convergence of all moments. In fact we see that the system with collisions will tend to hit a trap faster than the one for the McKean–Vlasov dual.

The *spatial homogeneity* of the limit law follows from the one of the initial state if this has this property, but since all initial states lead to the same limit we are done.

The next point is the shift-ergodicity of the limit law. Furthermore we note that two initial clouds of factors starting at two sites which are at hierarchical distance distance ℓ will have a probability to ever meet before they resolve to 0 or 1 which

10.6 Dual Calculations: Proof of Theorem 14

tends to 0 as $\ell \to \infty$. More precisely, if λ is the parameter in the exponential tail of the trapping time, the probability of a collision between two clouds starting at two sites at hierarchical distance ℓ before trapping is

$$\leq \text{const} \cdot \frac{\frac{c_\ell}{N^{2\ell}}}{\lambda + \frac{c_\ell}{N^{2\ell}}}. \tag{10.122}$$

Therefore the *all* the mixed moments connected with two sites at distance ℓ factor asymptotically as $\ell \to \infty$. Hence the limiting state is *shift-ergodic*.   q.e.d.

Proof of Corollary 10.1. We first consider the case $M = 2$ and then $M \geq 2$.

In the case $M = 2$, we start the dual with the single factor (01). Then the subsequent action of selection without other type of transitions occuring leads to

$$\begin{array}{llll} (01) & \otimes 1 & \otimes 1 & \otimes 1 \\ (10) & \otimes (01) & \otimes 1 & \otimes 1 \\ (10) & \otimes (10) & \otimes (01) & \otimes 1 \\ (10) & \otimes (10) & \otimes (10) & \otimes (01). \end{array} \tag{10.123}$$

In this case the mutation $2 \to 1$ replaces (01) by (00) and this removes one row and column. However if there is a single row, that is, the tableau (01), then the result is a trap and the value is 0. Now there are two cases. (1) If the birth and death process is critical or subcritical it will return to (01) and on one of these returns it will become (00). (2) If the birth and death process is supercritical, then there will be a number of rows increasing to infinity and each of these has positive integral if there is positive initial mass on $\{2\}$. The condition for criticality is as usual dependent on the expected number of offspring sites produced during the lifetime of a site. Once again the dual hits 0 when the last active site hits the trap (00).

Recall that the dynamics at a fixed site are recurrent (due to coalescence) and the local state will hit the trap (00) after a finite random time. However before it hits the trap it can produce a random number (≥ 0) of new sites with migrant columns, which carry in a row (01) factors.

We illustrate this as follows: if the third column in the above tableau migrates, then in the first column the mutation transition $2 \to 1$ occurs and hence $(01) \to (00)$, $(10) \to (11)$ and finally the second and fourth columns coalesce, then the result is

$$\begin{array}{l} (01)_1 \otimes (11)_2 \\ (10)_1 \otimes (01)_2, \end{array} \tag{10.124}$$

where now the subscript indicates the site at which the tableau is located. Here the birth rate of new sites from a given site depends on the internal state at this site which is determined by the birth rate s of new factors, the death rate d induced by coalescence, the migration rate c and the mutation rates which determine the probability of death due to resolution by the selection-mutation process.

Noting that the birth rate of new sites increases linearly as s increases (for fixed values of the other parameters), for sufficiently large s, this produces a *supercritical* birth and death process of sites of such columns. Since this results in an increasing number of summands as in (10.123) each having non-zero integral if the initial mass of type 2 is positive, the limiting expected mass of type 2 as $t \to \infty$ is positive.

Now consider the case $M > 2$. The selection operation $1_{\{M\}}$ at rate s and fitness 1 plays the same role as the selection operation $1_{\{2\}}$ in the case $M = 2$. Again the selection process creates births (i.e. rows in the tableau). If due to mutation the last factor in a row (0000001) becomes (0000000) then one row and column are removed.

To show that survival can still occur we compare this to the case $M = 2$. We illustrate this in the case $M = 5$. One new feature is that a selection operation 1_{A_i}, $i < M$ can also produce new rows as illustrated in (10.120). Mutation among the first $(M - 1)$ components can then remove these extra rows, for example mutation $3 \to 2$ applied to

$$\begin{array}{llll} (00001) \otimes 1 & \otimes 1 & \otimes 1 & \\ (11000) \otimes (00001) \otimes 1 & & \otimes 1 & \\ (00100) \otimes (00001) \otimes 1 & & \otimes 1 & \\ (11000) \otimes (11100) \otimes (00001) \otimes 1. & & & \end{array} \quad (10.125)$$

yields

$$\begin{array}{llll} (00001) \otimes 1 & \otimes 1 & \otimes 1 & \\ (11100) \otimes (00001) \otimes 1 & & \otimes 1 & \\ (11100) \otimes (11100) \otimes (00001) \otimes 1. & & & \end{array} \quad (10.126)$$

It is easy to check that the additional births (even is some are removed) results in a faster growth of the number of sites compared to the $M = 2$ case and therefore is supercritical if this is true for the $M = 2$ case. This implies the result but the determination of the critical parameters in this case is complicated and we do not consider this question here.   q.e.d.

10.7 Dual Calculations with M Higher Level Types

This section develops the necessary *new* dual calculations and techniques to cope with M upper level types if $M > 1$. The point is to calculate moments and *mixed moments* of our processes efficiently.

We start our system $X^N = (x_j(t, i); \ i = 1, \cdots, N$ and $j \in \{1, \cdots, 2M\})$ with

$$x_j^N(0, i) = 0, \text{ for all } i = 1, \ldots, N \text{ and } j = M + 1, \cdots, 2M. \quad (10.127)$$

10.7 Dual Calculations with M Higher Level Types

The key task we now address is to apply duality to study the joint distribution of the emergence frequencies of the higher level types starting with all mass on the lowest type. Recall the notation $(0, 0 \cdots, 1, 0, \cdots, 0)$ with the 1 on the ith position for the indicator of type i.

In particular in order to prove the independence in (10.55) we must choose $(k_{M+1}, \cdots, k_{2M}) \in \mathbb{N}^M$ and verify that the corresponding limiting joint moments satisfy:

$$\lim_{t \to -\infty} \lim_{N \to \infty} E \left[\prod_{\ell=M+1}^{2M} \left(e^{-\beta t} \bar{x}_\ell^N (\frac{\log N}{\beta} + t) \right)^{k_\ell} \Big| \{x_1^N(0, i) = 1, \ i = 1, \ldots, N\} \right]$$

$$= \prod_{\ell=M+1}^{2M} \left(\lim_{t \to -\infty} \lim_{N \to \infty} E \left[\left(e^{-\beta t} \bar{x}_\ell^N (\frac{\log N}{\beta} + t) \right)^{k_\ell} \Big| \{x_1^N(0, i) = 1, \ i = 1, \ldots, N\} \right] \right) \tag{10.128}$$

and then to verify that the limiting joint moments uniquely characterize the limiting distribution.

Remark 223. We recall from the $M = 1$ case treated in Section 8 that we need to distinguish the following three different calculations:

1. The time regime in which no collisions occur (for example times $o(\log N)$). In this case to compute the higher moments of $\bar{x}_t^N$ we must include the second and higher order corrections (of order $\frac{1}{N^{k-1}}$ $k \geq 2$) when we start with one particle at each of two or more different sites.
2. The time regime $\frac{\log N}{\beta} + t$ with collisions. In this case we need to compute the kth moments of $\bar{x}^N(\frac{\log N}{\beta} + t)$. To do this we must start k particles at different sites.
3. To compute the kth moment of the ℓ-th empirical moment of $x^N(\frac{\log N}{\beta} + t, i)$, that is, for $i \in \{M + 1, \ldots, 2M\}$ we have to calculate asymptotically as $N \to \infty$, the quantity

$$E[(\frac{1}{N} \sum_{j=1}^N (x_i^N(t, j))^\ell)^k], \tag{10.129}$$

hence we must place ℓ particles at each of k different sites.

The evaluation of the moments will again be done using the set-valued dual $\tilde{\mathcal{G}}^{++,N}$ of X^N and $\tilde{\mathcal{G}}_t^{++}$ of the McKean–Vlasov limit X of X^N developed in Section 9 using that the moments of $\bar{x}_\ell^N$ can be calculated based on the moments of $x^N(i, t)$ for $i = 1, \cdots, k; k \in \mathbb{N}$.

In order to compute the expressions in (10.128) we take the dual with η_0 consisting for each $\ell = M + 1, \ldots, 2M$ of k_ℓ dual particles all at different sites associated with factors $1_\ell(\cdot)$ at each of the k_ℓ sites attached to ℓ, that is, we have $\tilde{\mathcal{G}}^{++,N}$ is determined by $\mathcal{G}^{++,N}$,

$$\mathcal{G}_0^{++,N} := \prod_{\ell=M+1}^{2M} \prod_{j=1}^{k_\ell} 1_\ell(u_{j,\ell}), \tag{10.130}$$

where $u_{j,\ell} \in \Delta_{2M-1}$ denotes the variable at distinct sites $\xi_{j,\ell} \in \{1,\ldots,N\}$ for $j = 1,\cdots,k_\ell$ and $\ell = M+1,\cdots,2M$ with $\sum_\ell k_\ell < N$.

We can identify $\tilde{\mathcal{G}}^{++,N}$ with an indicator function $\mathcal{G}^{++,N}$ and we can extend $\mathcal{G}^{++,N}$ to a function on $(\mathbb{I}_M^\mathbb{N})^N$, which we use next. We use for the set-valued dual the initial state

$$\tilde{\mathcal{G}}_0^{++,N} = \{u \in (\mathbb{I}_M^\mathbb{N})^{\{1,\cdots,N\}} | \mathcal{G}_0^{++,N}(u) = 1\}. \tag{10.131}$$

We then obtain with $T_N = \frac{\log N}{\beta}$ and choosing for $X(0)$ in each component the δ-measure on the lowest type 1 (recall (9.182) for $\hat{X}$-notation)

$$\begin{aligned}&\lim_{N\to\infty} E[\prod_{\ell=M+1}^{2M}(\bar{x}_\ell^N(T_N+t))^{k_\ell}|x_1^N(0,i)=1), \forall i = 1,\cdots,N]\\ &= \lim_{N\to\infty} E_{1_{M+1}^{\hat{\otimes}k_1}\otimes\cdots 1_{2M}^{\hat{\otimes}k_{2M}}}[\int \tilde{\mathcal{G}}_{T_N+t}^{++,N} d\hat{X}^N(0)].\end{aligned} \tag{10.132}$$

Here $\hat{\otimes}$ denotes that the factors appear at distinct sites.

We shall develop a scheme to efficiently evaluate the r.h.s. of the relation (10.132) and to verify the factorization. To introduce the main point we first restrict our attention to the case $S = \mathbb{N}$, that is, *without collisions* and *no rare mutation*. We will then after understanding the collision-free case deal with collisions between dual individuals as well as with rare mutations, i.e. we always focus first on the evaluation of the expression in (10.132) when we replace $\tilde{\mathcal{G}}^{++,N}$ by $\tilde{\mathcal{G}}^{++}$ and then subsequently we pass to $\tilde{\mathcal{G}}^{++,N}$.

In order to make the *new* effects from more types on the upper level more transparent we first prove the results for a simplified $(1, M)$-case in Subsubsection 10.7.1 (first moments) and 10.7.2 (higher moments) and then in Subsubsection 10.7.3 we introduce the modifications needed to prove that the results hold for the general (M, M)-case.

To be able to focus on the key features, we first work with some simplifications which we then remove at the end of the argument. We make precisely the following simplifying assumptions:

- There is only *one* lower level type.
- Assume that the fitness on the higher level types is the *same* for all and that there is *no mutation* between these higher level types.
- We assume that there is *no downward mutation* from E_1 to E_0.
- Rare mutation occurs at rate $m_{1,j} = \frac{m_j}{N}$, $j \in E_1$.

10.7 Dual Calculations with M Higher Level Types

Summarized in formulas:

$$\begin{array}{l} \text{fitness } 0, \text{ for type } 1 \\ \text{fitness } 1, \text{ for types } 2, \ldots, M+1 \\ m_{i,j} = 0, \text{ for } i, j \in 2, \ldots, M+1 \\ m_{i,j} = \frac{m_j}{N}, \text{ for } i = 1, j \in \{2, \cdots, M+1\}. \end{array} \quad (10.133)$$

10.7.1 Dual Calculations in the Case $(1, M)$: First Moment

We consider the set-valued dual $\tilde{\mathcal{G}}_t^{++}$ starting with one dual individual and with the initial state one of the following (by making a choice for j)

$$\tilde{\mathcal{G}}_0^{++} = (\{j\} \times ((\mathbb{I}_M)^{\mathbb{N}}) \times ((\mathbb{I}_M)^{\mathbb{N}})^{\{2,\cdots,N\}}, \text{ for } j = 2, 3, \cdots, M. \quad (10.134)$$

and we write in the corresponding tableau:

$$(0, 0, \ldots, 1, 0, \ldots, 0) \quad (\text{the 1 in position } j). \quad (10.135)$$

To describe the calculation and keep the notation simple we let $M = 2$ for the moment and at the end the case $M > 2$ will be immediate.

Step 1. $M=2$

We begin by considering the first moment, that is, we begin with $\tilde{\mathcal{G}}_0^{++}$ respectively $\tilde{\mathcal{G}}_0^{++,N}$ being set-induced by

$$\text{one factor } (010) \text{ or } (001). \quad (10.136)$$

This means we have to study the effect of the action of selection, coalescence and migration on $(\tilde{\mathcal{G}}_t^{++})_{t \geq 0}$. We do this step by step and look separately at the case for $\tilde{\mathcal{G}}^{++}$ and $\tilde{\mathcal{G}}^{++,N}$ and $d = 0, d > 0$.

Case : $S = \mathbb{N}$ and no coalescence

For the moment we consider only the effect of selection and suppress coalescence (this corresponds to the deterministic (infinite population) case and we can then w.l.o.g. also ignore migration).

If the selection operator corresponding to $A_2 = (011)$ acts we can calculate the outcome using

$$1_{A_2} \cdot (010) = (010), \, 1_{A_2} \cdot (001) = (001), \, 1_{A_2} \cdot (100) = (000) \quad (10.137)$$

$$\text{and } 1 - 1_{A_2} = (100). \quad (10.138)$$

Therefore the selection operator produces in $\tilde{\mathcal{G}}^{++}$ the transitions:

$$(010) \to (010) \otimes (111) + (100) \otimes (010),$$
$$(001) \to (001) \otimes (111) + (100) \otimes (001)), \qquad (10.139)$$

$$(100) \to (000) \otimes (111) + (100) \otimes (100) = (100) \otimes (100), \qquad (10.140)$$

$$(111) \to (011) \otimes (111) + (100) \otimes (111) = (111) \otimes (111). \qquad (10.141)$$

Hence in the two latter cases no new row is produced by the selection action and in the first case exactly one new row is produced. Of course always a new column is produced.

From these rules we can now write down the tableau we get if selection acts several times. If we start with (010) (the structure is the same starting in (001)) and selection acts in a time interval $[0, t]$ and exactly $N(t) = 4$ selection events do occur during this interval we obtain *always* the following tableau with $N(t)$ rows and columns, where we leave out the entries (111) in the figure to enhance readability and we also put down only the active ranks:

$$\begin{array}{l} (010) \\ (100)\ (010) \\ (100)\ (100)\ (010) \\ (100)\ (100)\ (100)\ (010) \\ (100)\ (100)\ (100)\ (100)\ (010). \end{array} \qquad (10.142)$$

This follows since a new column is produced when selection acts on any active rank in this tableau and from (10.139)–(10.141) we conclude that only *one* new row is created. Therefore the number of rows (and columns) grows like a pure birth process, $N(t)$ with $N(0) = 1$ and with birth rate s.

We now consider the effect of a rare mutation at a rank 3 in the above tableau. If at time t we have the tableau (10.142) then a *rare mutation from type 1 to type 2* gives $(010) \to (110), (100) \to (000)$ and then we have as outcome:

$$\begin{array}{ll} (010) & (010) \\ (100)\ (010) & (100)\ (010) \\ (100)\ (100)\ (110) & = (100)\ (100)\ (110). \\ (100)\ (100)\ (000)\ (010) & \\ (100)\ (100)\ (000)\ (100)\ (010) & \end{array} \qquad (10.143)$$

The same effect occurs if mutation acts at other ranks, the tableau is cut to the right and below the (010)-factor.

Then considering the initial state with mass 1 on type 1 and integrating with respect to $\delta_1^\otimes$ we obtain before the rare mutation in (10.142) the value 0, but after rare mutation $1 \to 2$ at *any* rank that the integral gets equal to 1, since the remaining columns yield $(010) + (100) = (110)$. That is, a rare mutation at any rank produces an instantaneous transition to 1. Therefore the instantaneous rate at

10.7 Dual Calculations with M Higher Level Types

which the integral jumps from 0 to 1 is $\frac{m_{1,2}N(t)}{N}$, with $N(t)$ the number of ranks at time t.

Then (integrating with respect to $(\delta_1)^\otimes$) the conditional probability given the process of birth events in the dual that a rare mutation from type 1 to type 2 occurs by time t is

$$1 - e^{-\frac{m_{1,2}}{N}\int_0^t N(u)du}. \tag{10.144}$$

We then have (recall here we are in the case $d = 0$)

$$E[\bar{x}_2^N(t)] = E[1 - e^{-\frac{m_{1,2}}{N}\int_0^t N(u)du}]. \tag{10.145}$$

Recall that this allows us to prove emergence in times $\frac{\log N}{s} + t$ for the $d = 0$, i.e. deterministic case. Similarly we can proceed with type 3 and we get the corresponding formula ($m_{1,3}$ replacing $m_{1,2}$).

Case $S = \mathbb{N}$: Effect of coalescence and migration

We now consider for the dual of the McKean–Vlasov process (as above) the effect of *coalescence* and *migration*. Two factors occupying the same site can *coalesce* before either one makes a migration step. However once one of them migrates to a new site they cannot coalesce anymore in $\tilde{G}^{++}$.

The coalescence between (100)-factors just reduces their number but otherwise does not change the form of the expression. On the other hand coalescence of factors (100) and (010) produces (000). It is then clear that the coalescence of two columns in the tableau (10.142) results in the removal of *one row* and *one column*. Therefore if there is no migration the number of active ranks approaches an equilibrium with finite mean and as before this implies that the probability of a rare mutation occurring in time of order $O(\log N)$ is negligible and it is the migration which makes the emergence possible.

We now turn to migration and again start with one factor (010) at site 1. Then each rank present can jump to a new site 2 at rate c.

For example, let rank 2 migrate. We then have the tableau (starting with (10.142))

$$\begin{array}{l}(010)_1 \\ (100)_1 \ (010)_2 \\ (100)_1 \ (100)_2 \ (010)_1 \\ (100)_1 \ (100)_2 \ (100)_1 \ (010)_1 \\ (100)_1 \ (100)_2 \ (100)_1 \ (100)_1 \ (010)_1.\end{array} \tag{10.146}$$

Here the subscript denotes the site at which the rank is located. This means that migration does not effect the structure of the tableau but the site-2 columns can no longer coalesce (and be removed) with the columns at site 1.

Remark 224. The tableau describes a sum of products of indicators located at $K(t)$ sites. We can also decompose the corresponding event in $\{1, 2, 3\}^N$ into disjoint events (based on the location of the (010) factor) as follows:

$$\bigcup_{j=1}^{K(t)} \left((010)_j \otimes (100)_j^{n_j} \otimes \prod_{i \neq j=1}^{K(t)} (100)_i^{n_i} \right). \tag{10.147}$$

We note that selection, coalescence and migration will continue to act at both sites 1 and 2. Due to coalescence which balances the linear birth rate of factors at a site this produces at each site a positive recurrent Markov chain. This is precisely a system of the type studied in Section 9 but starting initially instead of with (010) or (001) as we do here with (100) (better (10) as we there had only one upper type). We also note that the number of occupied sites $K(t)$ and number of active ranks Π_t at time t is precisely the same as that described in Section 8

Consider the configuration at time t. Note that the rows of the tableau correspond to *disjoint* events and no replication of rows occurs. There will be some number $K(t)$ of occupied sites and Π_t of active ranks.

Then as above the instantaneous rate of transition to 1 at time t is Π_t and the integrating with respect to $\delta_1^\otimes$ we obtain for the conditional probability of the rare mutation given the process of tableaus:

$$1 - e^{-\frac{m_{1,2}}{N} \int_0^t \Pi_u du}. \tag{10.148}$$

We then have again an explicit formula for the mean in the McKean–Vlasov process:

$$E[\bar{x}_2(t)] = E[1 - e^{-\frac{m_{1,2}}{N} \int_0^t \Pi_u du}]. \tag{10.149}$$

We can proceed exactly the same way with type 3 and we get the same formula with $m_{1,2}$ replaced by $m_{1,3}$.

Case $S = \{1, \cdots, N\}$.

We next consider the dual process $\tilde{\mathcal{G}}^{++,N}$, which means that now migration can lead to an occupied site, which means we have to discuss the effect occuring where two columns collide which came from different sites. We first note that again the form of the tableau in (10.146) does not change only the chance of collisions and subsequent coalescence reduces further the number of ranks. Hence we have

$$E[\bar{x}_2^N(t)] = E[1 - \exp(-\frac{m_{1,2}}{N} \int_0^t \Pi_u^N du)]. \tag{10.150}$$

The behaviour of Π_t^N however is exactly the same as we studied in Section 8 and hence we have all the means from Section 8 to derive the growth behaviour of $\Pi_{t_N}^N$

as $N \to \infty$ and $t_N \to \infty$ and determine the behaviour of the mean of the fitter type in the emergence regime $\alpha^{-1} \log N + t$.

Step 2: $M > 2$

We now have to see how this works for $M > 2$-types on the upper level. Here we have however then still only one selection operator, since all higher types have one fitness level, by our assumption. Then with the indicator of any higher-level type results in a tableau of the same structure and evolving exactly the same way giving then the formula (10.150) with 2 or 3 replaced by $\ell \in \{2, 3, \cdots, M\}$. Then for the analysis of Π_t^N exactly the same analysis of Section 8 holds as it did for $M = 2$.

10.7.2 Dual Calculation for the $(1, M)$ System: Higher Moments

We next need the behaviour of the higher moments of $\bar{x}_\ell^N(T_N + t)$ for $\ell = 2, \cdots, M+1$. We want to show the following result.

Proposition 10.12 (Factorization and growth of mixed moments).

(a) The first moment satisfies

$$\lim_{t \to -\infty} \lim_{N \to \infty} e^{-\alpha t} E[\bar{x}_i^N(\frac{\log N}{\alpha} + t)] = m_{1,i} E[W], \quad i = 2, \cdots, M. \quad (10.151)$$

(b) The scaled k-th moment of type i converges as $N \to \infty, t \to -\infty$ and satisfies

$$m_k = \lim_{t \to -\infty} \lim_{N \to \infty} e^{-k\alpha t} E[(\bar{x}_i^N(\frac{\log N}{\alpha} + t))^k] \leq B^k k!,$$

for some $B < \infty$, for $i = 2, \cdots, M$. \quad (10.152)

(c) All limiting moments and mixed moments of $(\bar{x}_\ell^N)_{i=2,\cdots,M+1}$ (as $N \to \infty, t \to -\infty$) are finite and the limiting mixed moments of the $(M-1)$ components

$$(e^{-\alpha t} \bar{x}_2^N(\frac{\log N}{\alpha} + t)), e^{-\alpha t} \bar{x}_3^N(\frac{\log N}{\alpha} + t), \cdots, e^{-\alpha t} \bar{x}_M^N(\frac{\log N}{\alpha} + t)) \quad (10.153)$$

factor into the product of the corresponding moments of the components.

(d) The Laplace transform of the limiting law of $\mathcal{L}[\{e^{-\alpha t} \bar{x}_i^N(\frac{\log N}{\alpha} + t), i = 2, \cdots, M\}]$ as $N \to \infty, t \to -\infty$ is determined by the joint moments. This Laplace transform factors into the transforms of the limiting $(\mathcal{L}[e^{-\alpha t} \bar{x}_i^N(\frac{\log N}{\alpha} + t)])_{i=2,\cdots,M+1}$. $\square$

Proof of Proposition 10.12. **(a)** follows from (8.718) of Proposition 8.21 with the reasoning we gave in Subsubsection 10.7.1.

(b) Recall (see (8.594)), for $k \in \mathbb{N}$, $i_\ell \in \{2, \ldots, M+1\}$, $T_N = \frac{\log N}{\alpha}$:

$$\lim_{N \to \infty} E\left[\prod_{\ell=1}^{k} \left(\frac{1}{N} \sum_{j=1}^{N} x_{i_\ell}^N(T_N + t, j)\right)\right]$$
$$= \lim_{N \to \infty} E[x_{i_1}^N(T_N + t, j_1) \ldots x_{i_k}^N(T_N + t, j_k)], \quad (10.154)$$
$$j_1 \neq j_2 \neq \cdots \neq j_k.$$

We begin to set up the stage for the formal proof of (10.152). We shall see that this is easy for the McKean–Vlasov process, i.e. $\tilde{\mathcal{G}}^{++}$, but since we deal with $\tilde{\mathcal{G}}^{++,N}$, where collisions of factors arising from different initial sites occurs, this is more complicated. We begin therefore by explaining the role of collisions in the dual for the assertion.

For notational convenience we again discuss first the case $M = 2$. To obtain now the right hand side of (10.154) we start the dual with a single factor (010) at each of k distinct sites. Then for times of order $o(\log N)$ we have the *product of k tableaus* of the form (10.146) which do not interfere with probability tending to 1 as $N \to \infty$, that is, are located at disjoint subsets of $\{1, \ldots, N\}$.

We have to evaluate next the r.h.s. of the duality relation. Observe for that purpose that integrating $\tilde{\mathcal{G}}_t^{++,N}$ with respect to $\delta_1^{\otimes}$ the integral remains 0 until *all* the factors (010) undergo rare mutation and change to (110) since only then there is a row which contributes. Indeed if we write the tableau corresponding to $\tilde{\mathcal{G}}^{++,N}$ respectively $\tilde{\mathcal{G}}^{++}$ as a product of the tableaus descending from the initial factor at the sites $j_1, \cdots, j_k$, which we can do always in $\tilde{\mathcal{G}}^{++}$ and in $\tilde{\mathcal{G}}^{++,N}$ till the first collision we see that then we need in each such"factor" one row where all entries 010 have undergone rare mutation to (110). (Recall here (10.142) for the structure of the part of the tableau descending from one initial factor which shows that in each row descending from one factor is exactly one (010).) This shows that this decomposition in subfamilies is crucial.

We denote by

$$\Pi_{j_\ell}^N, \Pi_{j_\ell}, \quad \ell = 1, \cdots, k \quad (10.155)$$

the number of factors descending from the factor at site j_ℓ. Note that *after collision* between these subfamilies and subsequent *coalescence* a column can belong to more than one of the subfamilies descending from $j_\ell, \ell = 1, \cdots, k$.

Suppose we would look at $\tilde{\mathcal{G}}^{++}$ instead of $\tilde{\mathcal{G}}^{++,N}$, then the conditional probability (given the processes Π_{j_ℓ} for $\ell = 1, \cdots, k$) that all factors in one row undergo rare mutation by time $T_N + t$ is given by

10.7 Dual Calculations with M Higher Level Types

$$\prod_{\ell=1}^{k}(1 - e^{-\frac{m_{1,\ell}}{N}\int_0^{T_N+t}\Pi_{j_\ell}(u)du}). \tag{10.156}$$

In the system without collisions (i.e. the McKean–Vlasov limit system) we obtain simply

$$E[\prod_{\ell=1}^{k}\bar{x}_\ell(T_N+t)] = E[\prod_{\ell=1}^{k}(1-e^{-\frac{m_{1,i_\ell}}{N}\int_0^{T_N+t}\Pi_{j_\ell}(u)du})]$$

$$= \prod_{\ell=1}^{k}E[1-e^{-\frac{m_{1,i_\ell}}{N}\int_0^{T_N+t}\Pi_{j_\ell}(u)du}], \tag{10.157}$$

which factor.

However once we consider times $\alpha^{-1}\log N + t$ we want to calculate this for the $\bar{x}^N$ rather for the $\bar{x}$ process, hence we calculate with the dual $\tilde{\mathcal{G}}^{++,N}$ rather than $\tilde{\mathcal{G}}^{++}$ but *collision occurs* in the dual $\tilde{\mathcal{G}}^{++,N}$ between the different i_ℓ-clouds and subsequent coalescence causes the terms *not* to factor at time $T_N + t$ as $N \to \infty$ but with t fixed. However we are interested in the case $t \to -\infty$. For time $\frac{\log N}{\alpha}+t$ in the limit $N \to \infty$, the proportion of interacting sites goes to 0 as $t \to -\infty$.

We can split the dual population starting with one factor now in the two parts $\tilde{\Pi}_{j_1}^N, \tilde{\Pi}_{j_2}^N$ which have not been (till time t) affected by *interfamily collision with subsequent coalescence* and then we have a coupled part, we denote by $\tilde{\Pi}_{j_1,j_2}^N$. In particular we have that $\tilde{\Pi}_{j_k}^N + \tilde{\Pi}_{j_1,j_2}^N = \Pi_{j_k}^N$, $k = 1, 2$.

We need two properties of the dual. Collision and coalescence *within* each growing family has the effect of slowing the growth of Π^N but this produces only a higher order correction in the limit as $t \to -\infty$. However collisions *between* the two families produces a contribution to the limiting second moment. These two facts we show next.

If the two families meet and the corresponding ranks coalesce, they and their descendants are coupled and the formula for the occurrence of a rare mutation event for all involved factors has now a different form than (10.156). To describe the effect of collisions we first consider the case $k = 2$.

Recall that on arrival we obtain the product of the current tableau and the newly arrived tableau both belonging to partitions of the product of the type space. Coalescence of the new rank with one of the current ranks can then occur at this site with rate proportional to the number of current other ranks at this site. Once this occurs the two involved rows become *coupled*.

We distinguish two cases when in (10.154) $k = 2$, namely (denoting with i_1, i_2 the type of the two factors, i.e. either (010) or (001), involved in the coalescence)

- $i_1 = i_2$ and then we have replaced $(010) \otimes (010)$ by a single (010),
- $i_1 \neq i_2$, that is we have coalescence of (010) and (001) which produces (000).

The key observation is that after such a coalescence in the first case *only one* rare mutation is required to have a jump in the integrated expression to 1, while in the second case we need two rare mutations to get the jump to 1. We now have to argue that this means that both these cases contribute in the $N \to \infty, t \to -\infty$ limit.

The rate at which collision between the two families occur is (the asymptotic equivalence is established below)

$$\frac{c\tilde{\Pi}^N_{j_1}(T_N+t)K^N_{j_2}(T_N+t)}{N} \sim \frac{c\Pi^N_{j_1}(T_N+t)K^N_{j_2}(T_N+t)}{N}, \text{ as } N \to \infty, t \to -\infty, \quad (10.158)$$

where $\tilde{\Pi}^N_{j_1}$ denotes the number of ranks in family starting in j_1 and $\tilde{K}^N_{j_2}$ denotes the number of sites occupied by the family starting in j_2, which are both not yet affected by collisions with subsequent coalescence.

Therefore the event collision-coalescence with subsequent collision occurs by (10.158) at the same order as the two independent rare mutations required before coalescence in order to produce a non-negligible contribution to the second moment, since a rank from the first family can migrate to a site occupied by the second family at a high enough rate.

We illustrate this in the two cases by an example (we leave out here the (111) factors for better readability of the table).

Example 22 (Case 1). Consider first $i_1 = i_2$ for the following tableaus located at sites 1 and 2:

$$\begin{pmatrix} (010)_1 & & & & \\ (100)_1 & (010)_1 & & & \\ (100)_1 & (100)_1 & (010)_1 & & \\ (100)_1 & (100)_1 & (100)_1 & (010)_1 & \\ (100)_1 & (100)_1 & (100)_1 & (100)_1 & (010)_1 \end{pmatrix} \otimes \begin{pmatrix} (010)_2 & & & & \\ (100)_2 & (010)_2 & & & \\ (100)_2 & (100)_2 & (010)_{1*} & & \\ (100)_2 & (100)_2 & (100)_{1*} & (010)_2 & \\ (100)_2 & (100)_2 & (100)_{1*} & (100)_2 & (010)_2 \end{pmatrix} \quad (10.159)$$

In the case $(i_1 = i_2)$ coalescence between the $*$-column and the 3rd column at site 1 followed by a rare mutation $(010) \to (110)$ would produce a jump in the integrated dual to 1. Note the picture does not depend on the position of the ranks in the respective clouds. We therefore obtain two contributions to the second moment, one involving two rare mutations and the other collision-coalescence with subsequently one rare mutation. Both of these contributions are of order $\frac{1}{N^2}e^{2\alpha(\frac{\log N}{\alpha}+t)}$ asymptotically as $N \to \infty, t \to -\infty$ giving nontrivial values.

10.7 Dual Calculations with M Higher Level Types

We make this now precise. In the case $i_1 = i_2$ we focus on the event that a collision occurs between the two families of occupied sites starting from the two factors (010) at two different sites j_1 and j_2. Recall that the number sites in the j_1 family grows like $W_{j_1} e^{\alpha s}$ in the collision-free regime and we have obtained in section 8 with multi-colour particle systems the expansion of this size including collisions in times s of the from $\alpha^{-1} \log N + t$, to obtain the behaviour as $N \to \infty$ and then $t \to -\infty$. This gives that the rate of collisions of the j_1-family with the j_2-family occurs then at time $\alpha^{-1} \log N + t$ at rate (for $N \to \infty, r \to \infty$)

$$\frac{c}{N} \alpha W_{j_1} e^{\alpha r} W_{j_2} e^{\alpha r}. \tag{10.160}$$

Coalescence of the columns of equal rank then occurs at rate d so that the number of ranks where this has occurred by time $\frac{\log N}{\alpha} + t$ is by explicit integration (the actual constants are by the way not important for our argument):

$$(\frac{c}{N} \alpha W_{j_1} W_{j_2} \int_0^{\frac{\log N}{\alpha} + t} e^{2\alpha s} (1 - e^{-d(T_N + t - s)}) ds) \tag{10.161}$$

$$\sim c W_{j_1} W_{j_2} \frac{d}{2(2\alpha + d)} e^{2\alpha t} N, \text{ as } N \to \infty, t \to -\infty.$$

Since the probability of a subsequent rare mutation by time u later is

$$\left(1 - \exp(-\frac{m_{1,2}}{N} u)\right) \sim u \cdot \frac{m_{1,2}}{N}, \tag{10.162}$$

the probability of a jump by time $\alpha^{-1} \log N + t + u$ converges for $u \in (0, \infty)$ to a *positive* constant as $N \to \infty$.

Example 23 (Case 2).
Turn now to the case $i_1 \neq i_2$ for the following tableaus located at sites 1 and 2.

$$\begin{pmatrix} (010)_1 \\ (100)_1 \ (010)_1 \\ (100)_1 \ (100)_1 \ (010)_1 \\ (100)_1 \ (100)_1 \ (100)_1 \ (010)_1 \\ (100)_1 \ (100)_1 \ (100)_1 \ (100)_1 \ (010)_1 \end{pmatrix} \otimes \begin{pmatrix} (001)_2 \\ (100)_2 \ (001)_2 \\ (100)_2 \ (100)_2 \ (001)_{1*} \\ (100)_2 \ (100)_2 \ (100)_{1*} \ (001)_2 \\ (100)_2 \ (100)_2 \ (100)_{1*} \ (100)_2 \ (001)_2 \end{pmatrix} \tag{10.163}$$

Coalescence between the *-column and the 3rd column at site 1 would remove the third row in the first tableau thus reducing by 2 the number of possible rank at which a rare mutation event can occur but it would also still require *two* rare mutations to change the integrated dual to 1 which has rate of order N^{-2}. Therefore altogether we have a term of order N^{-3} and this does not change the limiting second moment which arises from contributions corresponding to two rare mutations, one in each family and the contribution in the limit is not changed in this case and hence gives still factorization.

To calculate the second moment in the limit $N \to \infty$ in the two cases, explicitly we have to distinguish them for the final formula for their contribution. We have to first write down the analogue of (10.157).

The expression in (10.156) has to be replaced (for $k = 2$) if we have $i_1 = i_2$ by (we include the simpler $i_1 \neq i_2$ case in the formula):

$$\left(1 - \exp\left(\frac{m_{1,i_1}}{N}\int_0^{\alpha^{-1}\log Nr+t} \tilde{\Pi}_{j_1}^N(u)du\right)\right)\left(1 - \exp\left(-\frac{m_{1,i_2}}{N}\int_0^{\alpha^{-1}\log N+t} \tilde{\Pi}_{j_1}^N(u)du\right)\right)$$

$$+\delta_{i_1,i_2}\left(1 - \exp\left(-\frac{m_{1,i_1}}{N}\int_0^{\alpha^{-1}\log N+t} \tilde{\Pi}_{j_1,j_2}^N(u)du\right)\right). \tag{10.164}$$

Then we continue by taking expectations

$$E[x_{i_1}^N(\alpha^{-1}\log N + t, 0)x_{i_2}^N(\alpha^{-1}\log N + t, 1)]$$
$$= E\left[\left(1 - \exp\left(-\frac{m_{1,i_2}}{N}\int_0^{\alpha^{-1}\log N+t} \tilde{\Pi}_{j_1}^N(du)\right)\right)\left(1 - \exp\left(-\frac{m_{1,i_1}}{N}\int_0^{\alpha^{-1}\log N+t} \tilde{\Pi}_{j_2}^N(u)du\right)\right)\right.$$
$$\left.+\delta_{i_1,i_2}\left(1 - \exp\left(-\frac{m_{1,i_1}}{N}\int_0^{\alpha^{-1}\log N+t} \tilde{\Pi}_{j_1,j_2}^N(u)du\right)\right)\right]. \tag{10.165}$$

Case $i_1 = i_2$. We consider random times

$$T_1^N, T_2^N \text{ and } \tilde{T}_1^N, \tilde{T}_2^N, \tilde{T}^{*,N}, \tag{10.166}$$

defined as the ordered times $T_1^N < T_2^N$ of the first rare mutation events in the two families $\Pi_{j_1}^N$ and $\Pi_{j_2}^N$, with $\tilde{T}_1^N, \tilde{T}_2^N$ we denote the corresponding random times for $\tilde{\Pi}_{j_1}^N, \tilde{\Pi}_{j_2}^N$ and with $\tilde{T}^{*,N}$ the time of the first rare mutation event in $\tilde{\Pi}_{j_1,j_2}^N$. We have to establish the behaviour as $N \to \infty$ and then $t \to -\infty$.

Recall that if we start with k dual particles at k-different sites $\{\frac{1}{N}\Pi^N(\frac{\log N}{\alpha} + r)\}_{r\geq 0} \Rightarrow \{u(r)(\alpha(s) + \gamma(r))\}_{r\geq 0}$ as $N \to \infty$ in law and $u(t)(\alpha(t)) + \gamma(t)) \sim e^{-k\alpha t}$ as $t \to -\infty$. Since in law $T_1^N = \inf\{t : \int_{-\infty}^t \frac{1}{N}\Pi^N((\frac{\log N}{\alpha} + r))dr \geq \mathcal{E}\}$

10.7 Dual Calculations with M Higher Level Types

where $\mathcal{E}$ is an independent exponential random variable with parameter m_{12}, it follows that the random time T_1^N has a weak limit as $N \to \infty$ and similarly T_2^N (note we have implicitly centered the waiting times here by $\alpha^{-1} \log N$). We want to argue now that this behaviour is also exhibited by $\tilde{T}_1^N$ and $\tilde{T}_2^N$. Here we have to deal with the fact that $\tilde{\Pi}_{j_1}$ behaves different than the dual just started with a factor at site j_1 since we now have to handle collisions with the j_2-family.

A similar representation then for $\tilde{T}_1^N$ holds for $\tilde{T}^{*,N}$, where we consider in the hazard function the occupation integral of pairs of dual particles consisting of a particle of each family $\Pi_{j_1}^N$, $\Pi_{j_2}^N$, such that one at least belongs to $\tilde{\Pi}_{j_1}^N$, $\tilde{\Pi}_{j_2}^N$, respectively the waiting time of the first rare mutation in $\tilde{\Pi}_{j_1,j_2}^N$. Again we have to establish the $N \to \infty, t \to -\infty$ behaviour. This requires in particular that we show the convergence and verify that $\tilde{\Pi}_{j_1,j_2}$ is for $N \to \infty$ and $t \to -\infty$ of lower order.

First we have to establish the convergence of the occupation integral
$$\int_{-\infty}^{t} \frac{1}{N} \tilde{\Pi}_{j_1}^N ((\alpha^{-1} \log N + r)^+) dr \text{ as } N \to \infty.$$

What we know is that this holds if we replace $\tilde{\Pi}_{j_1}^N$ by $\Pi^{N, \{j_1, j_2\}}$, Π^{N, j_ℓ}, $\ell = 1, 2$ the sizes of the dual populations if we *start* with a factor at j_1 and j_2 respectively j_ℓ. On the other hand we know

$$\Pi^{N, \{j_1, j_2\}} = \Pi_{j_1}^N + \Pi_{j_2}^N - \tilde{\Pi}_{j_1, j_2}^N, \quad \Pi^{N, j_\ell} = \tilde{\Pi}_{j_\ell}^N + \tilde{\Pi}_{j_1, j_2}^N. \quad (10.167)$$

By a coupling construction we then see that the claimed convergence as $N \to \infty$ and the claimed behaviour as $t \to -\infty$ holds for $\tilde{\Pi}_{j_1, j_2}^N$, the latter using formula (10.158) and the following fact.

Since the term:

$$\left(\frac{m_{1, i_1}}{N} \int_0^{\alpha^{-1} \log N + t} \tilde{\Pi}_{j_1, j_2}^N (u) du \right) = S_t \quad (10.168)$$

can be stochastically bounded by a term such that

$$S_t \leq \frac{\text{Const}}{N^2} e^{-2\alpha t} \quad (10.169)$$

since we combine (10.161) and (10.162).

Furthermore we have that at times $\alpha^{-1} \log N + t$ $\tilde{\Pi}_{j_1}^N$ and $\tilde{\Pi}_{j_2}^N$ behave as $\Pi_{i_1}^N$ and $\Pi_{j_2}^N$ as $N \to \infty$ and then $t \to -\infty$ and only taking the limit $N \to \infty$ we have to take a correction based on a multi-colour construction within an expansion giving first $\tilde{\Pi}_{j_1, j_2}^N$.

Since we can represent in fact all three random times jointly on one probability space as $(\tilde{T}_1^N, \tilde{T}_2^N, \tilde{T}^{*,N})$, then if we center them by $\alpha^{-1} \log N$ we have for $N \to \infty$ weak limits which we call $(\tilde{T}_1, \tilde{T}_2, T^*)$.

Then by inspection the limiting second moment at time $\frac{\log N}{\alpha} + t$ as $N \to \infty$ is given by

$$P(\tilde{T}^* \leq t, \tilde{T}^* \leq \tilde{T}_2) + P(\tilde{T}_2 \leq t, \tilde{T}^* > \tilde{T}_2). \tag{10.170}$$

Note that this expression is $\sim \text{const} \cdot e^{2\alpha t}$ as $t \to -\infty$ due to the asymptotics of the hazard function following from the limiting evolution of $\tilde{\Pi}^N(\alpha^{-1} \log N + t)$ for $A = j_1, j_2$ or (j_1, j_2), which we established above.

Case $i_1 \neq i_2$. If $i_1 \neq i_2$ let $\hat{T}_1, \hat{T}_2$ denote the times of the first rare mutation events in the j_1, j_2 families $\tilde{\Pi}_{i_1}^N, \tilde{\Pi}_{j_2}^N$ respectively. Again we get a weak limit $\mathcal{L}[(\hat{T}_1, \hat{T}_2)]$ of $\mathcal{L}[(\hat{T}_1^N, \hat{T}_2^N)]$.

Then the second cross moment is given in the limit $N \to \infty$ by

$$P(\max(\hat{T}_1, \hat{T}_2) \leq t) = P(\hat{T}_1 \leq t) P(\hat{T}_2 \leq t) \tag{10.171}$$

via the independence of $\hat{T}_1, \hat{T}_2$. Again as in the previous case this expression behaves $\sim e^{2\alpha t}$ as $t \to -\infty$.

The two cases together prove the claim that the second moments scale with $e^{2\alpha t}$ as $t \to -\infty$.

We can now use these same ideas for the calculation for higher moments to prove the result for arbitrary moments as follows.

In order to calculate k-th moments, we begin with the factor (010) at k distinct locations in $\{1, \ldots, N\}$. Now a series of collisions with subsequent coalescence can take place so that there may be in the k-subpopulations only ℓ of the (010)-factors left which could then each undergo rare mutations, so that in this event $(k-\ell)$-collision-coalescent events combined with ℓ rare mutations could produce the value 1. In other words for each $1 \leq \ell \leq k-1$ there is a contribution to the integrated dual arising from each of the possibilities leading to an event of the form:

- the k families are involved in a series of collision-coalescence events resulting in exactly ℓ remaining (010) factors resulting from $k - \ell$ coalescence events, respectively,
- each of these ℓ factors undergoes a rare mutation event before further coalescence.

This leads to a decomposition of the dual population in

$$\tilde{\Pi}_{j_1}^N, \ldots, \tilde{\Pi}_{j_k}^N, \tilde{\Pi}_{j_1, j_2}^N, \ldots, \tilde{\Pi}_{j_1, j_2, j_2}^N, \ldots, \tilde{\Pi}_{j_1, \cdots, j_k}^N \tag{10.172}$$

10.7 Dual Calculations with M Higher Level Types   759

which belong to exactly one, two, $\cdots$, k families after the respective collision-coalescence events. Now similar formulas as for the $k = 2$ case can be written down.

Since each of these above events defining these subpopulations involves $k-\ell$ coalescence events and we then wait for exactly ℓ rare mutation events. They have the combined rate $O(\frac{1}{N^k})$ to occur and consequently lead to contributions at time $\frac{\log N}{\alpha} + t \sim \text{const} \cdot e^{k\alpha t}$ as $N \to \infty$ and then $t \to -\infty$.

From this scenario we can obtain the convergence of the scaled moments by calculating the contributions in the various cases based on the analysis following the route given in the case $k = 2$ above with the only difficulty of the higher notational complexity. We omit further details.

The extension to the case with $M > 2$ many types on the upper level is now immediate, we only have to replace $(010), (001)$ by $(010, \cdots), \cdots (00\cdots 1)$ and can carry out the same argument.

To obtain an upper bound on the growth in k for $k \to \infty$ of the limiting k-th moments (as $N \to \infty, t \to -\infty$) we proceed as follows. Recall that the number of ways to divide a set of k elements into ℓ nonempty subsets is given by

$$S(k, \ell), \text{ the Stirling number of the second kind (cf. [Rio]).} \quad (10.173)$$

This produces a contribution bounded by

$$\sum_{\ell=1}^{k} S(k, \ell) z^\ell = P(k, z), \quad \forall k \in \mathbb{N}, \quad (10.174)$$

for some $0 < z < \infty$ where we can bound for every $t \geq 0$ the variable z by

$$z \leq \text{const} \cdot e^{\alpha t} \quad (10.175)$$

and

$P(k, z)$ is the kth moment of the Poisson distribution with parameter z.
$\quad (10.176)$

This follows as indicated in (10.161) using in addition the independence of the W_ℓ and $E[W_\ell] < \infty$ together with a bound on the constants arising from the integration over time.

We then have

$$m_k \leq \sum_{\ell=1}^{k-1} S(k, \ell) z^\ell = P(k, z) \quad (10.177)$$

for some $0 < z < \infty$.

Then (see e.g. [Rio], eqn. (24), p. 76),

$$\sum \lambda^k \frac{m_k}{k!} \leq \sum_{k=1}^{\infty} \left(\sum_{\ell=1}^{k} S(k,\ell) z^\ell \right) \frac{\lambda^k}{k!} = \sum_{k=1}^{\infty} \lambda^k \frac{P(k,z)}{k!} = e^{z(e^\lambda - 1)} < \infty.$$
(10.178)

(c) The finiteness of the moments follows from (b). To verify that the mixed moments factor we generalize the observation used above for the second joint moments, namely, we use the fact that the contributions due to coalescence of rows corresponding to the different components of the scaled vector only occur when two members of two clouds of the same type coalesce but not when two members of clouds of different type coalesce, recall here (10.163).

The extension of these arguments to the case $(1, M)$ for $M > 2$ follows in exactly the same way the only simplification arises as we see from the argument above in the notation, (010), (001) simply have to be replaced by $(0, 0, \cdots, 1, 0, \cdots)$ at the i-th position with $i = 2, \cdots, M + 1$. Hence we have shown the Proposition 10.12 in the case for $(1, M)$ as claimed.

(d) Part (b) yields a growth condition for the moments which implies that the joint Laplace transform function is analytic (cf. for example [DU], Chap. 2, Theorem 3.11) and is then determined by the joint moments. Denote by $*\mathcal{W}_\ell$ the growth constant of the type-ℓ population, for $\ell = 2, \cdots, M + 1$. Since the joint moments factor this implies

$$E[\Pi_{\ell=2}^{M+1} e^{-(\lambda_\ell \, *\mathcal{W}_\ell)}] = \Pi_{\ell=2}^{M+1} E[e^{(-\lambda_\ell \, *\mathcal{W}_\ell)}],$$
(10.179)

which is equivalent to the independence of $\{*\mathcal{W}_i\}_{i \in 2,5,\cdots,M}$.

q.e.d.

10.7.3 The Dual Calculation in the Case of (M, M)-Types

Here we point out the changes needed in the calculations with the dual which we introduced for one type on the lower level in Subsubsections 10.7.1 and 10.7.2. These techniques will then also play a role subsequently in the arguments of Subsection 10.8, to handle there more types on the lower level. We focus on the new features arising from this situation compared to 10.7.1 and 10.7.2.

In Subsubsection 10.7.2 we established the factorization of the joint moments of the different higher level types for the $(1, M)$ system with only one lower level type.

We now consider M types $1, \ldots, M$ on the lower level and M on the higher level with upward mutation only at order N^{-1}, the lower level types have fitness $0 < e_1 < \cdots < e_{M-1} < e_M = 1$ and higher level types $M + 1, \ldots, 2M$ have fitness 1 and there is mutation between the lower level types at positive rates of

10.7 Dual Calculations with M Higher Level Types

order 1 and *no* mutation on the upper level. In 10.12 we remove the simplifying assumptions we have used here on the parameters.

Proposition 10.13. *The conclusions of Proposition 10.12 remain valid for the (M, M) system given above with the Malthusian parameter α replaced by β.*
□

Proof of Proposition 10.13. We again use the the set-valued dual $(\tilde{\mathcal{G}}_t^{++})_{t \geq 0}$, using the representation we developed in Section 9.

The dual calculations now change, since the lower-level selection and mutation acting requires to work instead of with $(0; 1, 0, \cdots), (1; 0, \cdots), (1; 1, 0, \cdots)$, $(0; 0, 1, 0, \cdots)$ etc., $(0; 0, 0 \cdots 0)$ we had in Subsubsection 10.7.2 now with $(\underline{*}; 1, 0, \ldots, 0), (\underline{*}; 0, 1, \ldots, 0)$, etc., where $\underline{*}$ is a vector consisting at each position of 0 or 1's of length M. This we saw already moving from Section 8 to Section 9, but there we did not have the richer structure on the upper level.

We first note that in the collision free regime the lower level mutation-selection-migration process proceeds as in the analysis of Section 8 resulting in a growing number of occupied sites with the richer internal structure associated with the sites. This leads to the Malthusian parameter β (which was smaller than α).

We first observe that during the time evolution we still get a tableau of the form we had in (10.142), only the first entry 0 or 1 is now replaced by $\underline{*}$, where $\underline{*}$ evolves as analysed in Section 9 and not as we had in Subsubsection 10.7.2 where we simply had 0 or 1 and could use the analysis of Section 8. To carry out the adaption to $\underline{*}$ we must use the set-valued dual again but with $M \geq 2$ which we illustrate with an example.

Example 24. We now illustrate the dual for the case of the $(3, 3)$ system. To compute the mean of the different E_1 types we again use the dual. For example, if $M = 3$, the mean of type 5 is obtained by starting the dual with the factor $(000|010)$. Then, for example, after four selection operations we have

$$
\begin{array}{l}
(000|010) \\
(110|000)\ (000|010) \\
(110|000)\ (110|000)\ (000|010) \\
(110|000)\ (110|000)\ (110|000)\ (000|010) \\
(110|000)\ (110|000)\ (110|000)\ (110|000)\ (000|010)
\end{array}
\tag{10.180}
$$

To see this effect more clearly we start next with one factor $(111|010)$ and apply the selection operators in the following order, first 1_{A_2} to column 1 followed by 1_{A_2} to column 1 followed by 1_{A_3} to column 2.

$$
\left(\begin{array}{l}
(011|010) \\
(100|000)\ (001|010) \\
(100|000)\ (110|000)\ (011|010) \\
(100|000)\ (110|000)\ (100|000)\ (011|010)
\end{array}\right)
\tag{10.181}
$$

Note that the above is the tableau if we begin with $(111|010)$ so that we can identify the process of occupied sites with that of Section 9, that is, a CMJ process with Malthusian parameter β.

We now briefly review the effect of the mutation and coalescence operations on the tableau:

- coalescence of two columns - this removes one column and one row,
- lower level mutation and resolution can remove a number of columns and rows - for example the mutation $(011|010) \to (111|010)$ in the third column removes the last row and column.

In order to exhibit the structure for k-th moments we write the tableau again as a k-fold product (compare (10.159)) and then we see again that we need to remove (000—010) factors by rare mutations and subsequent revolution on the lower level to (111—010).

The basic strategy of proof in the (M, M)-case is exactly the same but we must note that in the calculations using the set-valued dual with $M > 2$ some new effects have occurred. The differences are

- lower order selection operators now act in addition, they result in birth events, which can lead to new migrant sites,
- also lower order mutation occurs now, with the effect that some of these above additional new sites can be removed by lower order mutation,
- the above two mechanisms lead to the Malthusian parameter β (rather than α),
- some but possibly not all rare mutations are *successful* so that more than one rare mutation might be needed (see Subsubsection 9.11.4). However the main contribution to the moments in the limit $t \to -\infty$ are then again the initial rare mutations. The hazard rate for a successful rare mutation is given in Lemma 9.63 for the mutation-dominant case and in (9.933) for the general case.

We come now to the argument. Start with $(\underline{*}; 0, \cdots, 1, 0, \cdots, 0)_i$ with the one at the i-th position among the upper half of the factor. Next we observe that since selection does not affect the last M positions corresponding to types $M+1, \cdots, 2M$ and neither does mutation, the dual $\tilde{\mathcal{G}}^{++}$ (without collision) then has the form

$$\sum_{j=1}^{N_t^*} \prod_{i=1}^{K_t} (\mathfrak{s}_{ji} \otimes (0, \ldots, 1, \ldots, 0)_i), \qquad (10.182)$$

where N_t^* is the number of summands, K_t is the number of occupied sites and for each i, j, the factor denoted $\mathfrak{s}_{j,i}$ is a product of subsets of $\{1, \ldots, M\}$ at the active ranks at the site i which we analysed in Section 9 in detail.

The key point is again as in Subsubsection 10.7.2 the fact that if 1_{M+i} and 1_{M+j}, $i \neq j$ coalesce, then they *annihilate* whereas if two 1_{M+i} coalesce, then this allows for a substantial contribution since afterwards only one rare mutation is required. Therefore up to the first rare mutation we can follow closely the argument we had with one lower level type, but what is new here is that a rare mutation does

10.7 Dual Calculations with M Higher Level Types

not always lead to a jump in the integral (depending on resolution) and the rate of rare mutation differs for different lower level types, that is, we must now also take into account the possibly different rare mutation rates of the lower level types, which lead to effective mutation rates from the lower level to the upper type j which was before occuring at a given rate m_j/N.

The proof of the factorization of the joint moments of $^*\mathcal{W}_{M+1}, \ldots, {}^*\mathcal{W}_{2M}$ works otherwise exactly as before. (Note that we do not need to determine the actual value of the moment but only need to verify that the respective mixed moments factor in the limit $t \to -\infty$.)

Now we get *effective rates* determined from the the stable type distribution in the collision-free dual $\tilde{\mathcal{G}}^{++}$ on the internal states of the columns for the part of the factors referring to the E_0-types, which we call

$$\frac{\tilde{m}_j}{N}, \quad j \in \{M+1, \cdots, 2M\}. \tag{10.183}$$

If we now consider the dual $\tilde{\mathcal{G}}^{++}$ (ignoring collisions, which becomes valid for $\tilde{\mathcal{G}}^{++,N}$ as $t \to -\infty$) starting with factors $(*|0\cdots 10\cdots)_{i_\ell}$, $i_\ell \in \{1,\cdots,M\}$, $\ell = 1,\cdots k$ at positions $j_1,\cdots,j_k$ all distinct, in particular we can take $(00\cdots 0|0\cdots 10\cdots)_i$, then this leads to the conditional hazard function for all factors undergoing rare mutation :

$$\prod_{\ell=1}^{k}(1 - e^{-\tilde{m}_{i_\ell} \int_0^{T_N+t} \Pi_{j_\ell}(u)du}). \tag{10.184}$$

Here we can use the results (e.g. Lemma 9.63 in Subsubsection 8.10.4, respectively (9.933) in Subsubsection 9.12) since asymptotically as $t \to -\infty$ this growth behaviour of the higher order types does not influence the lower order distribution.

The next step is to now pass again (as in Subsubsection 10.7.2) to $\tilde{\mathcal{G}}^{++,N}$ and the effect of the collisions. The effect occurring is exactly as before that we get contributions due to collisions if we consider initial states where the same $(0\cdots 0|0\cdots 10\cdots)_{i_\ell}$, $\ell = 1,\cdots,k$ do appear and no such contribution if they are all different. We omit here further (standard but notationally involved) details.

Corollary 10.14. *In the case (M, M) the components of $(^*\mathcal{W}_{M+1},\ldots,{}^*\mathcal{W}_{2M})$ are independent and the distribution is determined by the mixed moments.* □

Proof of Corollary 10.14. Again for the sum of the frequencies of type $M+1,\cdots,2M$ we get a growth constant $^*\mathcal{W}$ as before. The independence of the components follows immediately from the factorization of the joint moments. q.e.d.

10.8 Proof of Proposition 10.2

We shall prove separately the four claims (a)–(d).

(a) We have to verify both the existence of the entrance law with time index in $\mathbb{R}$ as well as the uniqueness, which are separate arguments given in (i) and (ii) below.

(i) Existence

The proof follows the same strategy as in the two-type case in Subsubsection 8.3.2 which the reader may recall at this point.

Recall that we can identify $\mathcal{P}(E_0 \cup E_1) = \Delta_{2M-1}$. Let $\mathcal{L}_t \in \mathcal{P}(\Delta_{2M-1})$ denote the solution to the McKean–Vlasov dynamics (10.20). We shall use μ, ν as typical elements of Δ_{2M-1}. We write as before $\mathcal{L}_t(d\nu) = u(t,\nu)d\nu$. Given $A_i \geq 0$, $i = M+1,\ldots,2M$, with $\sum A_i > 0$ we want to prove that there exists a solution satisfying

$$\lim_{t \to -\infty} e^{-\beta t} \int_{\Delta_{2M-1}} \nu(i) u(t,\nu) d\nu = A_i, \quad i = M+1,\ldots,2M. \qquad (10.185)$$

We first sketch the ideas for the case $(1, M)$ to introduce one of the new features and then treat the general (M, M) case in detail.

Case $(1, M)$

To prove existence we take a sequence $t_n \to -\infty$ and $u(t_n, \cdot) = \delta_{\mu(t_n)}(\cdot)$, with $\mu(t_n) \in \Delta_{2M-1}$ such that with

$$e(n) = e^{-\alpha t_n}, \qquad (10.186)$$

$$\mu(t_n) = (1 - \frac{1}{e(n)})\delta_1 + \frac{1}{e(n)}(\sum_{i=2}^{M+1} A_i \delta_i). \qquad (10.187)$$

It is easy to follow the line of argument in Section 8, in particular in Subsubsection 8.3.2 for $M = 1$ subsequently to (8.46) (note we are in the case of one type on the lower level), and to verify tightness and to obtain a convergent subsequence and to verify that the corresponding limit point (u, U) satisfies (10.20). But we can then use the McKean–Vlasov dual to compute the mean at a time t as a function of n and verify (10.43), that is show that the mean curve has the desired asymptotic behaviour as $t \to -\infty$.

We wish to show that we obtain the same limit if we have any sequence $u(t_n, \cdot)$ of initial states having the same asymptotic mean $m(t_n)[i]$, $i \in E_1$. As in (8.63) and (8.65), to do this we again consider a sequence $t_n \to -\infty$ and $\mu_0 \in \mathcal{P}(E_1 \cup E_0)$, $\mathrm{supp}(\mu_0) \subseteq E_1$, $p_{t_n} \in \mathcal{P}(\mathcal{P}(E_0 \cup E_1))$, $q_1 > 0$, such that:

10.8 Proof of Proposition 10.2

$$\mathcal{L}_{t_n}(dv) = (1 - \frac{q_1}{e(n)})\delta_{(1-\frac{1}{e(n)})\delta_1 + \frac{1}{e(n)}\mu_0}(dv) + \frac{q_1}{e(n)}p_{t_n}(dv), \qquad (10.188)$$

$$\lim_{n \to \infty} \int v(i) p_{t_n}(dv) = A'_i, \quad i = 2, 3, \cdots, M+1, \qquad (10.189)$$

where we let

$$e(n) = e^{-\alpha t_n}, A_i = \mu_0(i) + q_1 A'_i \text{ for } i = 2, 3, \cdots, M+1. \qquad (10.190)$$

Then we can show that:

$$e^{\alpha|t_n|} \int_{\mathcal{P}(\mathcal{P}(E_0 \cup E_1))} v(i) \mathcal{L}_{t_n}(dv) \to A_i, \ i \in E_1, \text{ as } n \to \infty. \qquad (10.191)$$

Namely in order to verify that the mean curve has the required limiting form as $t \to -\infty$ we argue again with the dual with one initial factor $(0, \ldots, 1, 0 \ldots)$ with the 1 in the i-th position, that is the indicator function of type i and then generalize immediately the argument of Case 3 in Subsubsection 8.3.2, since we have only one type on the lower level.

Case (M, M).

The essential difference for the general (M, M) case is that now the initial measure at time t_n which we let tend to $-\infty$ must be specified on both the lower types and higher level types and is not as above on the lower level just a number. Recalling the McKean–Vlasov mutation-selection equilibrium μ_{eq}^∞ on the lower level types (see Theorem 14 (b)) we can assume that as $t_n \to -\infty$ the limiting law of μ_{t_n} is supported by the lower level types $\{1, \ldots, M\}$ and given by $\tilde{\mu}_{eq}^\infty$. We then take $\mathcal{L}_{t_n} \in \mathcal{P}(\Delta_{2M-1})$ to be a perturbation of this law of the form (recall (10.188)–(10.191))

$$\mathcal{L}_{t_n}(d\tilde{v}) = (1 - \frac{q_1}{e(n)})\delta_{(1-\frac{1}{e(n)})\tilde{\mu}_{eq}^\infty + \frac{1}{e(n)}\mu_0}(dv) + \frac{q_1}{e(n)}p_{t_n}(dv), \qquad (10.192)$$

where $\mu_0 \in \mathcal{P}(E_0 \cup E_1), supp(\mu_0) \subseteq E_1$, and $p_{t_n} \in \mathcal{P}(\mathcal{P}(E_0 \cup E_1))$ satisfying (10.189), $q_1 > 0$.

The existence result is then obtained in the same way as in the proof of Proposition 2.3 but with the additional difficulty that we must now use the set-valued dual to describe the dynamics on the lower level. We point out next the major changes needed to carry this out.

Case 1: *(Mutation dominance)*

To compute the mean E_0 mass we use the dual process with one initial factor at one site, namely $(1, \ldots, 1; 0, \ldots, 0)$. We then know that the dual process at time t_0 is in the limit $N \to \infty$ is the McKean–Vlasov dual $\tilde{\mathcal{G}}^{++}$, which is for $t_0 \to \infty$ a collection of $K_{t_0} = We^{\beta t_0}$ occupied sites and at each site there are set-valued states arising from the selection-mutation-coalescence and migration driven dynamic of the dual, which in the large time limit $t_0 \to \infty$ satisfy the *stable type distribution* $\bar{U}^0(\infty, \bar{t})$, $\bar{t} \in \bar{\mathfrak{T}}$ given by (9.553).

The argument then proceeds as in Subsubsection 8.3.2 with α replaced by β and the stable size distribution, a probability measure on $\mathbb{N}$, replaced by the stable distribution $\bar{U}^0(\infty, \cdot)$.

The main steps are then as in Subsubsection 8.3.2 but we must replace $G(z)$ in (8.53) by (here the expectation E_{stable} is with respect to the random sets A_k of the dual and taken over the stable distribution of set-valued states at a site, which is induced by $\hat{U}^0(\infty) \in \mathcal{P}(\mathcal{T})$, recall $\mathcal{T}$ is given by (9.178) and which is induced by $\bar{U}^0(\infty)$):

$$G(\nu) := E_{stable}\left[\sum_{k=1}^{\infty} \nu^{\otimes k}(A_k)\right], \quad \nu \in \mathcal{M}_{\leq 1}(\{1,\cdots,M\}), A_k \subseteq (\{1,\ldots,M\}^k)$$

(10.193)

$$= \sum_{k=1}^{\infty} \sum_{A_k \in \{1,\ldots,M\}^k} \nu^{\otimes k}(A_k)\hat{U}^0(\infty, A_k).$$

As a consequence we have to replace (8.82) by

$$\tilde{E}\left[\int (1 - \nu(\{1,\ldots,M\}))\tilde{\mathcal{L}}_t^n(d\nu)\right] \to e^{\int (G(\nu)-1)\Gamma(d\nu)We^{\beta t}}, \text{ as } n \to \infty, \quad (10.194)$$

where Γ is given by the analogue of (8.73).

Case 2 *(Selection dominance)*

Here we have to replace $\bar{U}^0(\infty)$ from above by the stable distribution of mesostates (9.902) and then the argument follows along the same lines as above in case 1.

(ii) Uniqueness

The proof of the uniqueness of the entrance law proceeds in two steps, first the uniqueness of the mean curve is established and then the uniqueness of the solution $\{\mathcal{L}_t(\cdot)\}_{t \in \mathbb{R}}$ given this curve is shown.

Step 1

The proof of *uniqueness of the mean curve* of the entrance law follows the same lines as the proof of Proposition 2.3 except that we must use the dual with M types at the lower level as developed in Section 9—cf. Proposition 9.2. The key point in the proof of 2.3 was the exponential growth of the number of sites together with the stable type-distribution for the configuration at the sites and this property also holds with M types at the lower level but now with the appropriate Malthusian parameter β, and exponential growth $W \cdot e^{\beta t}$ instead of α and also the law of W is now more complicated since it results from a multitype CMJ-process.

10.8 Proof of Proposition 10.2

Given two solutions

$$(m^1(t_n, i), \quad i \in \{M+1, \cdots, 2M\}), \qquad (m^2(t_{n,i}) \quad i \in \{M+1, \cdots, 2M\}) \tag{10.195}$$

evaluated at a sequence $t_n \to -\infty$ we shall see below: (1) that the separation of trajectories of the mean curve for two different solutions would asymptotically be of order $O(e^{2\beta(t-t_n)})$ and (2) that therefore they then must actually agree.

(1) We first consider (1) in the case where only one type $i \in E_1$ has positive initial mass which means at the reference time t_n from above since here we already know exactly what happens. We denote a corresponding solution $(\mathcal{L}_t^i)_{t \in \mathbb{R}}$ and its mean curve $(m_i(t))_{t \in \mathbb{R}}$, the corresponding solutions are indexed by $i \in \{M+1, \cdots, 2M\}$. We then have with the analysis for only one upper type which we already have carried out in Section 9 that:

$$\lim_{t \to -\infty} e^{-\beta t} m_i(t) = a_i \delta_i, \quad i \in \{M+1, \ldots, 2M\}, \tag{10.196}$$

and

$$[m_i(t) - a_i e^{\beta t}] \sim O(e^{2\beta t}). \tag{10.197}$$

We have to see next why this holds also for the general case. Return to the calculation we did in the example in Subsubsection 10.7.3, in particular Eq. (10.180) for $M = 3$. We argue for this case to simplify notation but everything is absolutely general. Start with a 1 in the i-th position and otherwise having 0, for $i = 4, 5, 6$. We recall that the occupation by a single factor a site is a recurrent state. Then using the dynamics we can see that each site will be trapped eventually as follows. Each summand ends in, depending on the initial factor, $(000|100)$, $(000|010)$, $(000|001)$ respectively. It follows after integrating with respect to the initial measure the corresponding means are proportional to the initial measure of types $4, 5, 6$, respectively. Therefore we have for general M that:

$$\frac{(m(t, M+1), \ldots, m(t, 2M))}{m(t, E_1)} \to \frac{(a_{M+1}, \ldots, a_{2M})}{\sum_{i=M+1}^{2M} a_i}, \text{ as } t \to -\infty. \tag{10.198}$$

Hence we can concentrate on the *total mass of E_1* or equivalently as before the total mass of E_0 types.

To do this we begin with the factor $(111|000)$. Then after a selection operation we have

$$\frac{(001|000)}{(110|000) \ (111|000)}. \tag{10.199}$$

After resolution of the first column we have

$$
\begin{matrix}(001|000)\\(110|000)\ (111|000)\end{matrix} \rightarrow \begin{matrix}(000|000)\\(111|000)\ (111|000)\end{matrix} \text{ or } \begin{matrix}(111|000)\\(000|000)\ (111|000)\end{matrix}. \tag{10.200}
$$

In the first case we have a new permanent $(111|000)$ factor (i.e. a "success") and in the second case the effect of the selection operation is canceled. But as discussed in Section 9 the number of Pe sites (and therefore effectively the number of $(111|000)$ factors) grows like $K_t = W(t)e^{\beta t}$ and as we saw above with local states asymptotically given by the stable type distribution $\bar{U}^0(t,\cdot)$. We next note that when integrated with respect to the initial mass distribution this introduces a factor less than one at those sites containing $(111|000)$ factors and for which the initial type E_0 mass is less than 1. This implies that the mean total type E_0-mass behaves as $1 - \text{const} \cdot e^{\beta t}$ as $t \to -\infty$.

From the analysis in Subsubsection 8.3.2 (adapted to M types at the lower level) and the analogue of (8.86) it follows that as $t \to -\infty$:

$$
0 \leq \sum_{i=M+1}^{2M} a_i e^{\beta t} - m(t, E_1) \sim O(e^{2\beta t}). \tag{10.201}
$$

Therefore we must have for every $i \in \{M+1, \cdots, 2M\}$ that as $t \to -\infty$:

$$
[m(t, \{i\}) - a_i e^{\beta t}] \sim O(e^{2\beta t}). \tag{10.202}
$$

(2) To carry out the next part of the argument we assume that there are two solutions $m^{(1)}, m^{(2)}$ satisfying for $\ell = 1, 2$ that

$$
e^{-\beta t} m^{(\ell)}(t, \{i\}) \longrightarrow a_i, \text{ as } t \to -\infty, \quad i = M+1, \cdots, 2. \tag{10.203}
$$

Then we must have by (10.202) for $t_n \to -\infty$ and a sequence $z(n) \to 0$ as $n \to \infty$ that

$$
|m^{(1)}(t, \{i\}) - m^{(2)}(t, \{i\})| \leq e^{\beta(t-t_n)} z(n) e^{\beta t_n}, \tag{10.204}
$$

by using for times $t \geq t_n$ the duality relation and the $e^{\beta t}$-growth of the number of dual particles. Next let in (10.204) $n \to \infty$, which gives $m^{(1)}(t, \{i\}) = m^{(2)}(t, \{i\})$ for every i. This gives the uniqueness for $m(t, \cdot)$ given (10.203) holds as claimed.

Step 2

To prove *uniqueness for the law of a tagged site given the mean curve* $m(t, \cdot)$ we adapt the argument in the proof of Proposition 9.2. Consider a tagged component denoted $(\mathbf{x}(t))_{t \geq t_0}$, $\mathbf{x}(t) \in \Delta_{2M-1}^+$) in the McKean–Vlasov system, given a realization of the mean curve of total mass at the upper level $(\bar{x}_{M+1}(t), \ldots, \bar{x}_{2M}(t))_{t \in \mathbb{R}}$.

10.8 Proof of Proposition 10.2

Then $\{\mathcal{L}_t\}_{t \geq t_0}$ is the law of the tagged component $(\mathbf{x}(t))_{t \geq t_0}$ which is a Fisher-Wright diffusion with time dependent immigration depending on the total mass curve and therefore has a unique weak solution given an *initial* value at time t_0. However we must also prove the existence and uniqueness of an *entrance law*, i.e. a solution with time running in $\mathbb{R}$ and satisfying the appropriate limiting behaviour as $t \to -\infty$.

The proof of the uniqueness for the total upper level mass process given the mean curve of upper level mass is as in Section 8 by a coupling argument. We now use that coupling to complete the proof that the weights of the multiple types at the higher level are unique. To do this we return to the coupling in the proof of the uniqueness of the entrance law in Section 9 (see Subsection 9.13.3, Step 2). Recall that the proportions of lower order types couple at some joint zero of the E_1-mass for two coupled solutions (cf. 9.1024). Since at such a joint zero at time t_0 the distribution of E_1-types arises as an excursion with rates given by the mean curves which agree at time t_0 (recall that excursions with rates $\{\tilde{m}_i(\cdot) : i = M+1, \ldots, 2M\}$ are described by Proposition 10.6). In particular at a tagged site the next excursion starting at a joint zero is chosen with proportions from the mean curve, i.e.

$$\frac{\bar{x}_{M+1}(t_0), \ldots, \bar{x}_{2M}(t_0)}{\bar{x}_{E_1}(t_0)}, \tag{10.205}$$

so that we can also couple the choice of proportions in the excursion then giving us a complete coupling of the $2M$ types which in turn proves the uniqueness of the entrance law.

(b) By tightness there exists a subsequence $t_n \to -\infty$ such that

$$\lim_{n \to \infty} e^{-\beta t_n} m(t_n, \{i\}) = \mu(\{i\}). \tag{10.206}$$

But then as in the proof of Propositions 2.3 and 9.2 (based on Proposition 9.13 in Subsubsection 9.7) we get

$$\lim_{t \to -\infty} e^{-\beta t} m(t, \{i\}) = \mu(\{i\}). \tag{10.207}$$

(c) Given the tightness we can choose a subsequence $t_n \to -\infty$ such that the convergence claimed in (10.42) and (10.43) occurs. But then the solution must be the corresponding limit of the solution we get using the initial conditions at t_n and for those solutions the limit as $t \to -\infty$ exists due to (10.202), compare Section 8.

(d) The first statement follows from the results of [DG99], Theorem 1. To verify the second statement we consider the solution $\mathcal{L}$ of the McKean–Vlasov equation on $[t_0, \infty)$. We can replace E_1 by $[0, 1]$ where all types in $[0, 1]$ have the same (positive) fitness. If we subdivide $[0, 1]$ into K equal parts, then the limiting element of Δ_{K-1} must be $(\frac{1}{K}, \ldots, \frac{1}{K})$ by symmetry. It then follows that the limiting proportions of the types in E_1 must equal their starting values at time t_0. The result then follows by letting $t_0 \to -\infty$ using (10.43).

10.9 Proof of of Proposition 10.3

We proceed for the proof assertion by assertion.

(a) If we have the convergence of $\Xi_N^{\log.\beta}(t)$ at time $t = t_0$, then the convergence to the McKean Vlasov equation is the standard mean-field limit combined with the Feller property of the McKean–Vlasov dynamic (which follows easily using the duality as we pointed out already in Section 8). Hence we have to get the convergence for a fixed (macroscopic) time t_0.

The tightness of the sequence

$$m_N(t, E_1) = \left(\int_{\Delta_{2M-1}} \mathbf{x}(E_1) \Xi_N^{\log.\beta}(t_0, d\mathbf{x}) \right), \quad N \in \mathbb{N} \qquad (10.208)$$

as well as the tightness condition on $m(t_0, E_1)$ for any limit point follows from the first moment calculation we gave. It remains to prove convergence in distribution of $m_N(t_0, \cdot)$ for a fixed t_0.

The convergence for $N \to \infty$ of the distribution at time t_0 will be established as in Section 8 by verifying the convergence of the moments for $N \to \infty$. Namely we calculate the moments of $1 - m_N(t, E_1)$ by means of the modified dual $(\tilde{\mathcal{G}}_t^{++,N})_{t \geq 0}$ (developed in Subsection 9.4). The proof of the convergence of all moments of $1 - m_N(t_0, E_1)$ and the non-triviality of the limiting variance follows in the exactly the same way as in the proof of Proposition 9.3. This completes the proof of weak convergence of the total mass $m_N(t_0, E_1)$ and the randomness of the limit.

To extend this to the weak convergence of $\{m_N(t_0, i) : i \in E_1\}$ we compute the joint moments again using the dual. This has been done for second moments in (10.170), (10.171). The convergence of the higher moments follows in the same way.

(b) The a.s. existence of the limit in (10.54) follows from Proposition 10.2. In order to actually identify the distribution of

$$[^*\mathcal{W}_{M+1}\delta_{M+1} + \cdots +^* \mathcal{W}_{2M}\delta_{2M}], \qquad (10.209)$$

we shall use the dual. Hence we now turn to the proof of (10.55), namely, to obtain the joint weak convergence of $\{m(t, \{i\}), i = M+1, \ldots, 2M\}$ and to show that the limiting random coefficients as $t \to -\infty$

$$^*\mathcal{W}_{M+1}, \ldots, ^*\mathcal{W}_{2M}, \qquad (10.210)$$

are independent and $E[^*\mathcal{W}_i] = m_i$.

The analysis is again based on calculations of moments using the dual, as in Subsubsections 10.7 and 10.7.3.

To prove independence we will prove that (1) all joint moments are finite, (2) all the joint moments factor, that is, we prove that for $k_1, k_2 \in \mathbb{N}$, $t_N = \frac{\log N}{\beta}$ and $\ell_1, \ell_2 \in \{M+1, \cdots, 2M\}$ with $\ell_1 \neq \ell_2$

$$\lim_{t\to-\infty}\lim_{N\to\infty} E[(\bar{x}^N_{\ell_1}(t_N+t))^{k_1}(\bar{x}^N_{\ell_2}(t_N+t))^{k_2}] \tag{10.211}$$

$$= \lim_{t\to-\infty}\lim_{N\to\infty} E[(\bar{x}^N_{\ell_1}(t_N+t))^{k_1}] \cdot \lim_{t\to-\infty}\lim_{N\to\infty} E[(\bar{x}^N_{\ell_2}(t_N+t))^{k_2}]$$

and (3) that the limiting moments determine the joint distribution. This we proved in Subsection 10.7.2 in Proposition 10.12.

To prove that asymptotically the random measure on $\{M+1,\dots,2M\}$,

$$e^{-\beta t}\left(\frac{1}{N}\sum_{j=1}^{N}(x^N_{M+1}(t_N+t,j),\dots,x^N_{M+1}(t_N+t,j))\right) \tag{10.212}$$

consists of single type atoms with random masses, we use (10.242).

(c) We now have to prove the convergence of the hitting time of total mass ε on the upper level. This is however identical to the argument given in Section 9, since the distribution among the higher level is here irrelevant.

Remark 225. The intuitive reason behind the concentration on single atom measures on Δ^+_{M-1} is that the different types occupy disjoint sets of sites. To make this precise note that we have essentially the same situation as that in the discussion of the droplet formation in which the macroscopic excursions of the Wright-Fisher diffusion (with vanishing immigration rate) are single type due to the standard properties of the Poisson distribution.

10.10 Proof of Proposition 10.11

First recall that we proved this for the case of two types in Proposition 2.14 and in the case $(M,1)$ in Subsection 9.14.1. By Proposition 10.3 the limiting values $^*W_{M+1},\dots,^*W_{2M}$ are independent. Therefore for every $j \in \{M+1,\cdots,2M\}$ we can condition on $\sum_{i\neq j}{}^*W_i < \varepsilon$ and let $\varepsilon \to 0$ to get in the limit the law of $(0,\dots,^*W_j,\dots,0)$, but under this condition we are in the $M+1$ type case $(M,1)$, since the laws depend continuously on the parameter $(m_1,\cdots,m_M)$, as is seen immediately from the dual. Therefore we can conclude that $\mathcal{L}(^*W_j) = \mathcal{L}(W^*_j)$ by Proposition 2.14 in the case $(1,1)$ and in Proposition 9.3 in the $(M+1)$-type case $(M,1)$. Hence we get from this the general (M,M)-case.

10.11 Proof of Proposition 10.4

First consider the dynamics of $\mathcal{L}_t(d\mathbf{x}) = u(t,\mathbf{x})d\mathbf{x}$ given by the McKean–Vlasov equation for the density $u(t,\cdot)$ starting at time 0 in the case in which there is positive mean mass in E_1 that is $u(t,\Delta^+_{M-1}) > 0$:

$$\frac{\partial u(t,\mathbf{x})}{\partial t} = G^* u(t,\mathbf{x}) - c \sum_{i=1}^{2M} \frac{\partial}{\partial x_i} \left(\left[\int \tilde{x}_i u(t, d\tilde{\mathbf{x}}) - x_i \right] u(t,\mathbf{x}) \right) \quad (10.213)$$

It then follows by a simple extension of the arguments in the proof of the ergodic theorem, Theorem 13(a) in Section 6 that the expected proportion of type E_1 goes to 1 as $t \to \infty$. Notice that the duality gives us that the ergodic theorem holds uniformly in the initial state.

Since from the emergence results we know that positive E_1 mass exists w.p.1 at times of the form $\frac{\log N}{\beta} + t$ in the limit as $N \to \infty$, we conclude that at times $\frac{\log N}{\beta} + t_N + t$ there is no mass on E_0 in the $N \to \infty$ limit and we fixate on types in E_1.

The identification of the limiting proportions on the right side of (10.65) follows by combining Proposition 10.3(b) and Proposition 10.2(d).

10.12 Removing the Simplifying Assumption

Next to remove the simplifications still used above, namely (1) no mutation on E_1, (2) fitness equal 1 on E_1 rather than varying by terms of order N^{-1}.

We observe that the selective difference on the higher level and the mutation on that level are of order N^{-1}. We want to claim that this is irrelevant in the time scale we consider which is of order $\log N$. Turn to the dual process. The rates of selection operators or mutation operators acting between higher-level types correspond to transitions of order N^{-1}. But in our duality calculations carried out in the proofs there are only $O(1)$ factors on which this mutation operator or this selection operator could act other than the identity operator and produce a different integral. Therefore the probability of such an event to happen in times of order $\beta^{-1} \log N + O(1)$ goes to zero as $N \to \infty$ and hence no effect occurs. The standard details, which were explained in Subsection 9.15 and need hardly a modification are left to the reader.

10.13 Droplets: Proofs of Propositions 10.5, 10.6, 10.7, 10.8, 10.9, 10.10

Proof of Proposition 10.5. We must extend the construction of excursions for the one dimensional case developed in Pitman-Yor [PY].

(a) To keeping notation simpler we illustrate the principle in the case of the $(2, M)$. The for $i \in \{3, \ldots, M+2\}$ we consider the system which satisfies the system of SDE

$$dx_i(t) = -cx_i(t)dt + sx_i(t)x_1(t)dt \quad (10.214)$$
$$+ \sqrt{x_i(t)x_1(t)}dw_i(t) + \sqrt{x_i(t)x_2(t)}dw_2$$

10.13 Droplets: Proofs of Propositions 10.5, 10.6, 10.7, 10.8, 10.9, 10.10

$$dx_1(t) = c(E[x_1(t)] - x_1(t))dt + (m_{2,1}x_2(t) - m_{1,2}x_1(t))dt - sx_1(t)(x_2(t) + x_i(t))dt$$
$$+ \sqrt{x_1(t)x_2(t)}dw_1(t) - \sqrt{x_1(t)x_i(t)}dw_i(t)$$
$$dx_2(t) = c(E[x_2(t)] - x_2(t))dt + (m_{1,2}x_1(t) - m_{2,1}x_2(t))dt + sx_2(t)x_1(t)dt$$
$$- \sqrt{x_1(t)x_2(t)}dw_1(t) - \sqrt{x_2(t)x_i(t)}dw_2(t),$$
(10.215)

where w_1, w_2, w_i are independent standard Brownian motions. In order to use the Girsanov theorem to transform to the neutral case, we exhibit the selection term in the notation, namely, we write P^s to denote the law of the process, when the the selection rate is $s \geq 0$. For $i \in \{3, \ldots, M + 2\}$ we want to prove the existence of the limit

$$\mathbb{Q}^{s,i}_{p_1, p_2, 0}(\cdot) = \lim_{\varepsilon \to 0} \frac{P^s_{(1-\varepsilon)p_1, (1-\varepsilon)p_2, \varepsilon\delta_i}(\cdot)}{\varepsilon},$$
(10.216)

and to show that this limit defines a nontrivial σ-finite measure.

We use the Girsanov formula to obtain the probability in question. Namely denoting with $\mathcal{F}_t$ the σ-algebra generated by the paths up to time t, we have with R^s_t denoting the *Girsanov density* with respect to the neutral case:

$$P^s_{(1-\varepsilon)p_1, (1-\varepsilon)p_2, \varepsilon\delta_i}|_{\mathcal{F}_t} = R^s_t \cdot (P^0_{(1-\varepsilon)p_1, (1-\varepsilon)p_2, \varepsilon\delta_i}|_{\mathcal{F}_t}).$$
(10.217)

Recall that in this case R^s_t is a bounded $\mathcal{F}_t$-measurable continuous function and $R^s_t \to 1$, a.s. as $t \downarrow 0$ (see [D], Section 10.1.2 with $r(\mu, y) = s1_{\{2,i\}}(y)$). Therefore it suffices to study the neutral case and to consider

$$\lim_{\varepsilon \to 0} \frac{P^0_{(1-\varepsilon)p_1, (1-\varepsilon)p_2, \varepsilon\delta_i}(\cdot)}{\varepsilon}.$$
(10.218)

Note next that under P^0 the pair $((x_1(t) + x_2(t)), x_i(t))$ is a neutral 2-type Fisher-Wright diffusion which was discussed already in Section 8. The existence of the limit can be obtained by a *modification* of the argument in Lemma 2.4 and the analogues of (2.40), (2.41) follow. What is new and modification is needed, is that the mutation rate from types 1,2 to type i depend on the proportions of types $1, 2$ which are not constant under the law $P^0_{(1-\varepsilon)p_1, (1-\varepsilon)p_2, \varepsilon\delta_i}$ (recall (10.86)). To handle this and identify the limit in (10.216) we first freeze $(x_1(t), x_2(t)) = ((1 - x_i(t))p_1, (1 - x_i(t))p_2)$, $0 \leq t \leq \delta$, but use the full 3-type dynamics for $t \geq \delta$. For fixed δ the existence of the limit as $\varepsilon \downarrow 0$ then follows from the case $M = 1$. Then letting $\delta \to 0$ and noting the continuity of the mutation rate to type 3 as a function of the proportions of types 1, 2 we obtain the required excursion measure. The generalization to M lower types is immediate.

(b) Note that for $i_1, i_2 \in \{M+1, \ldots, 2M\}$ and $\eta > 0$, $t > 0$ we have that:

$$\lim_{\varepsilon \to 0} \left(\frac{1}{\varepsilon} P_{(1-\varepsilon)(p_1, \ldots, p_M, 0, \ldots, 0) + \frac{\varepsilon}{2}(\delta_{i_1} + \delta_{i_2})}(w(t, i_1) > \eta \text{ and } w(t, i_2) > \eta) \right) = 0. \tag{10.219}$$

This can be checked calculating mixed moments of the type i_1 and type i_2 masses using the duality.

Therefore the excursion measure for the complete E_1 mass is given by

$$\mathbb{Q}_{(p_1, \ldots, p_M)} = \sum_{i=1}^{M} Q^i_{(p_1, \ldots, p_M)}. \tag{10.220}$$

(c) This follows as in the one-dimensional case. q.e.d.

Proof of Proposition 10.6. We first prove existence of a solution. We first note that by a simple first moment inequality (recalling that the fitness of types $\{M+1, \ldots, 2M\}$ is 1, selection rate is s)

$$E[\mathbf{J}_t^M([0,1] \times \{M+1, \ldots, 2M\})] \leq (e^{st} - 1) \sum_{i=M+1}^{2M} \tilde{m}_i. \tag{10.221}$$

This guarantees tightness of finite dimensional marginals.

We next consider the increasing sequence of *approximations*, where we here abbreviate with $\mathbf{J}_{t,k}(\{i\})$ for $\mathbf{J}_{t,k}([0,1] \times \{i\})$:

$$\mathbf{J}_{t,0} = \int_0^t \int_{[0,1]} \sum_{i=M+1}^{2M} \int_0^{\tilde{m}_i(s)} \int_{W_0} \delta_{w(t-s)}(dp) \delta_a(dv) N_0^i(ds, da, du, dw)$$

$$\mathbf{J}_{t,1} = \mathbf{J}_{t,0} + \int_0^t \int_{[0,1]} \sum_{i=M+1}^{2M} \int_0^{c\mathbf{J}_{s-,0}(\{i\})} \int_{W_0} \delta_{w(t-s)}(dp) \delta_a(dv) N_1^i(ds, da, du, dw)$$

$$\mathbf{J}_{t,k+1} = \mathbf{J}_{t,k} + \tag{10.222}$$

$$\int_0^t \int_{[0,1]} \sum_{i=M+1}^{2M} \int_0^{c(\mathbf{J}_{s-,k}(\{i\}) - \mathbf{J}_{s-,k-1}(\{i\}))} \int_{W_0} \delta_{w(t-s)}(dp) \delta_a(dv) N_k^i(ds, da, du, dw),$$

where the $(N_k^i)_{k \in \mathbb{N}}$ are for every i independent Poisson random measures and furthermore the sequences are independent in i.

Since the sequence $\mathbf{J}_{t,k}$ is stochastically increasing and tight, the existence of a limit follows. The proof that the limit satisfies the equation is now straightforward.

10.13 Droplets: Proofs of Propositions 10.5, 10.6, 10.7, 10.8, 10.9, 10.10

To get uniqueness observe first that the limit we constructed must be a lower bound for any solution. Since the first moment measure of the limit is bounded (10.221) and is given by the unique solution to the linear first moment equation uniqueness of the solution of (10.88) then follows immediately.   q.e.d.

Proof of Proposition 10.7. The part (a) follows from Proposition 9.62(a) (resp., Subsubsection 9.11.3), since it concerns only the total mass on the upper level.

The proof of the remaining parts involving verifying that the droplets of different types do not interact in the collision free regime of time. This follows since essentially the different type droplets grow in different parts of space. To verify this, note that by Proposition 10.10 (b),(c), given any $\varepsilon > 0$, there exists random subsets $B_N^{i_1}(\varepsilon), B_N^{i_2}(\varepsilon)$ of $\{1,\ldots,N\}$ of size $O(e^{\beta t_N})$ such that $e^{-\beta t_N} \sum_{j \in B^{i_k}(\varepsilon)^c} x_{i_k}^N(t_N, j) < \varepsilon$, $k = 1, 2$. But then if $i_1 \neq i_2$, $t_M \to \infty$ and $t_N = o(\frac{\log N}{\beta})$ then the expected mass of either type in the intersection of $B^{i_1} \cap B^{i_2}$ tends to 0 as $N \to \infty$.   q.e.d.

Remark 226. The intuitive reason behind the concentration on single atom measures on Δ_{M-1}^+ is that the different types occupy disjoint sets of sites. Note that the mutation of each type $i = M+1,\ldots,2M$ and the corresponding new excursion occurs at a rate of order $O(\frac{1}{N})$.

Now consider the excursion processes:

$$\mathbb{Q}^i_{(x_1,\ldots,x_M)} := \lim_{\varepsilon \to 0} \frac{1}{\varepsilon} P_{(1-\varepsilon)(x_1,\ldots,x_M,0,\ldots,0)+\varepsilon \delta_i}, \quad i = M+1, \cdots 2M. \quad (10.223)$$

$$\mathbb{Q}^{i,j}_{(x_1,\ldots,x_M)} := \lim_{\varepsilon \to 0} \frac{1}{\varepsilon} P_{(1-\varepsilon)(x_1,\ldots,x_M,0,\ldots,0)+\frac{\varepsilon}{2}(\delta_i+\delta_j)}, \quad i \neq j \in \{M+1,\cdots 2M\}. \quad (10.224)$$

Note that for $i_1, i_2 \in \{M+1,\ldots,2M\}$ and $\eta > 0$, $t > 0$

$$\lim_{\varepsilon \to 0} \left(\frac{1}{\varepsilon} P_{(1-\varepsilon)(p_1,\ldots,p_M,0,\ldots,0)+\frac{\varepsilon}{2}(\delta_{i_1}+\delta_{i_2})}[w(t,i_1) > \eta \text{ and } w(t,i_2) > \eta] \right) = 0. \quad (10.225)$$

Therefore we can consider only one type excursions given by $\mathbb{Q}^i$. Namely recall that the probability that a given site has an excursion of type i of size $> \eta$ at a fixed time $t > 0$ is given by a Poisson distribution with mean $\frac{\tilde{m}_i}{N}\mathbb{Q}^i_\mathbf{x}(\{w : w(t,\{i\}) > \eta\}) < \frac{\text{const}}{N}$. Therefore the probability that such a site has excursions of both type i and $j \neq i$ is of order $\frac{\text{const}}{N^2}$. Since there are only N sites, in the limit as $N \to \infty$, no two-type excursions are created. We then note that each one type excursion only produces new excursions by migration of the same type. Since in the $N \to \infty$ only single type excursions are created by mutation and the migration process cannot create a two-type droplet from a single type excursion, it follows that all droplets (up to collision times) are single type.

Also see Eq. (10.55) which establishes the same property in the macroscopic emergence regime.

Proof of Proposition 10.8. The proof is obtained by a modification of the proofs of Proposition 2.9 and Lemma 8.20. We now explain the necessary supplementary arguments.

(a) This follows essentially using that the dynamics of the dual process $\tilde{\mathcal{G}}^{++,N}$ converges to $\tilde{\mathcal{G}}^{++}$ for fixed time horizon, which is working exactly as before, and then the argument is as before showing that all moments converge.

(b) In order to describe the development starting from the N-site interacting system with only the least fit type present at time 0 we proceed as follows. We observe the evolution of a tagged site say 1. Here the rate of production of type $i \in E_1$ at time t is given by $\frac{1}{N}\left(\tilde{m}_i^N(t) + c q^{N,i}(t)\right)$ where

$$q^{N,i}(t) := \sum_{j=1}^{N} x_i^N(j,t), \quad i \in \{M+1, \ldots, 2M\}, \tag{10.226}$$

$$\tilde{m}_i^N(t) = \sum_{j=1}^{M} m_{ji} x_j^N(1,t), \quad i \in \{M+1, \cdots, 2M\}. \tag{10.227}$$

During the period $[0, T]$ some early $O(1)$ excursions arise at a sparse set of sites which must be described by the time-dependent version of the excursion process, where the intensity measure for the excursions is given by (see (10.84)):

$$ds \left(\frac{1}{N} \sum_{i=1}^{N} \delta_{a_i}\right)(da)\, du \left(\int_{\Delta_{M-1}} (\mathbb{Q}^i_{x_1,\ldots,x_M}(dw)) \tilde{\mu}_t^N(dx_1,\ldots,dx_M)\right), \tag{10.228}$$

where $\{a_i\}$ are iid uniform $[0, 1]$ and

$$\tilde{\mu}_t^N(dx_1,\ldots,dx_M) = \frac{1}{N} \sum_{j=1}^{N} \delta_{(x_1^N(j,t),\ldots,x_M^N(j,t))}(dx_1,\ldots,dx_M). \tag{10.229}$$

Let $\tilde{\mathcal{L}}_t$ be the solution of the McKean–Vlasov equation on the type set E_0. Then in the limit as $N \to \infty$ we get as claimed:

$$ds\, da\, du \left(\int_{\Delta_{M-1}} (\mathbb{Q}^i_{x_1,\ldots,x_M}(dw)) \tilde{\mathcal{L}}_t(dx_1,\ldots,dx_M)\right), \tag{10.230}$$

The following consequence we will need below in the next proof. Using the arguments of Theorem 14 it follows that as $N \to \infty$ and $t \to \infty$

10.13 Droplets: Proofs of Propositions 10.5, 10.6, 10.7, 10.8, 10.9, 10.10

$$\mathcal{L}[\tilde{m}_i^N(t)] \Longrightarrow \sum_{j=1}^{M} \mu_{\text{eq}}^{\infty}(j) m_{ji}, \qquad (10.231)$$

in the following sense. Given $\varepsilon > 0$ there exists $T_\varepsilon < \infty$ and N_ε such that for $t \geq T_\varepsilon$ and $N \geq N_\varepsilon$

$$P(|\tilde{m}_i^N(t) - \sum_{j=1}^{M} \mu_{\text{eq}}^{\infty}(j) m_{ji}| > \varepsilon) < \varepsilon. \qquad (10.232)$$

To verify this, first find T_ε so that

$$P(|\tilde{m}_i(T_\varepsilon)) - \sum_{j=1}^{M} \mu_{\text{eq}}^{\infty}(j) m_{ji}| > \varepsilon) < \frac{\varepsilon}{2} \qquad (10.233)$$

for the McKean–Vlasov dynamics and secondly choose N_ε sufficiently large so that the N-site interacting system is sufficiently close to the McKean–Vlasov system.

Proof of Proposition 10.9. Because the types $M+1, \ldots, 2M$ all have asymptotically as $N \to \infty$ the fitness 1, the proof that there exists $\mathcal{W}^*$ such that

$$\mathcal{L}[e^{-\beta t_N} \mathbf{1}_{t_N}^{N,M}([0,1] \times \{M+1, \ldots, 2M\})] \underset{N \to \infty}{\Longrightarrow} \mathcal{W}^*, \qquad (10.234)$$

involves the same argument as that for the proof of Proposition 8.27 but now for the case of M types at the lower level using the dual developed in Subsections 9.4. The additional point here is that if we do not start in μ_{eq} the growth rate during a fixed time interval, say $[0, T]$, is not equal to β. However the point is, that this will be the case after some large time later on, which is what is important. We make this precise.

Consider a time T_ε, as introduced above (10.232). Then using the fact that we ε-approximate the equilibrium, see (10.232), it follows that the growth taking or not taking place up to T_ε has limited effect, namely is of size $e^{(\beta+\text{error})T_\varepsilon}$ and hence does not change the exponential growth up to time t_N but only influences the constant $\mathcal{W}^*$, that is we still have

$$\lim_{N \to \infty} e^{-\beta t_N} \mathbf{1}_{t_N}^{N,M}([0,1] \times \{M+1, \ldots, 2M\}) = \mathcal{W}^*, \qquad (10.235)$$

but where $\mathcal{W}^*$ now depends on the distribution at time $t = 0$.

Similarly if we suppress mutation to all but one upper level type we obtain the corresponding $\mathcal{W}_i^*$. Since the different types occupy disjoint sets and do not interact, the result follows. q.e.d.

Proof of Proposition 10.10. We prove step by step the parts (a)–(c) of the proposition.

(a) It remains to verify that the r.h.s. is a random combination of E_1-monotype configurations. Consider the development at a tagged site i. Let $0 < \varepsilon < 1$. Recall that for $j \in \{M+1,\ldots,2M\}$

$$e^{-\beta t} \lim_{N \to \infty} E[(x_j^N(\frac{\log N}{\beta} + t, i))] \to E[{}^*\mathcal{W}_j] > 0 \quad \text{as } t \to -\infty \quad (10.236)$$

and (recall coalescence of a pair occurs with positive probability before a migration step occurs)

$$\lim_{t \to -\infty} e^{-\beta t} \lim_{N \to \infty} E[(x_j^N(\frac{\log N}{\beta} + t, i))^2] \geq C \cdot E[{}^*\mathcal{W}_j] > 0, \quad C > 0. \quad (10.237)$$

Note that

$$E[x_j^N(i,t)^2] = \int_0^1 x^2 \mu_t^N(dx) \leq \varepsilon \int_0^\varepsilon x \mu_t^N(dx) + \int_\varepsilon^1 \mu_t^N(dx), \quad (10.238)$$

where μ_t^N is the law of $x_j^N(i,t)$ so that

$$P[x_j^N(i,t) > \varepsilon] \geq E[x_j^N(i,t)^2] - \varepsilon E[x_j^N(i,t)]. \quad (10.239)$$

Therefore there exists $\varepsilon > 0$ such that

$$\lim_{t \to -\infty} \lim_{N \to \infty} e^{-\beta t} P[x_j^N(\frac{\log N}{\beta} + t, i) \geq \varepsilon] \geq \text{const} > 0. \quad (10.240)$$

On the other hand if $j_1 \neq j_2$, we have

$$e^{-2\beta t} \lim_{N \to \infty} E[(x_{j_1}^N(\frac{\log N}{\beta} + t, i)) x_{j_2}^N(\frac{\log N}{\beta} + t, i))] \to E[{}^*\mathcal{W}_{j_1}][E[{}^*\mathcal{W}_{j_2}] \quad \text{as } t \to -\infty \quad (10.241)$$

so that by Chebyshev

$$\lim_{t \to -\infty} e^{-\beta t} \lim_{N \to \infty} P[\{x_{j_1}^N(\frac{\log N}{\beta} + t, i) > \varepsilon\} \cap \{x_{j_2}^N(\frac{\log N}{\beta} + t, i) > \varepsilon\}] = 0. \quad (10.242)$$

Then (10.240) and (10.242) implies the form of the limit in the r.h.s. of (10.108).

(b) To verify the existence of the limiting Palm distribution $\hat{\mu}_\infty^i(\cdot)$, by (10.104), (10.105) we compute the higher moments of $x_j^N(\frac{\log N}{\beta} + t, i)$ and and then take the limits $N \to \infty$ and then $t \to -\infty$. Recall that to compute the kth moment

10.13 Droplets: Proofs of Propositions 10.5, 10.6, 10.7, 10.8, 10.9, 10.10

of $x_j^N(i, \frac{\log N}{\beta} + t)$ we start the dual with k factors $(0, \ldots, 1, 0, \ldots)$ with 1 at j-th position in the tuple and located at site i. We then consider the first time at which site i contains only one factor of this type. Note that if ℓ of these factors have migrated before this time, then ℓ rare mutations are required before the integral becomes non-zero. Therefore, asymptotically the only non-zero contribution to the higher moments of the single site $x_j^N(\frac{\log N}{\beta} + t, i)$ as $N \to \infty$ comes from the event that all initial factors at the site coalesce and do not migrate. From that time of reaching one factor at the initial site after the k-coalescence events we then continue with the first moment calculation carried out in Subsection 10.7.1. Then recalling (10.105) this implies the convergence of the moments.

(c) First recall (10.97) which sets

$$\tilde{U}_i^N(t, dy) := \sum_{j=1}^{N} 1_{\{x_i^N(j,t) \in dy\}}. \tag{10.243}$$

From Proposition 10.8 we have

$$\{\tilde{U}^N(t, dy)\}_{t \geq 0} \Rightarrow \{\tilde{U}_i(t, dy)\}_{t \geq 0} \text{ as } N \to \infty. \tag{10.244}$$

Then choosing $t_0(N) \to \infty$, $t_0(N) = o(\log N)$ we have by Proposition 10.9 that

$$\lim_{N \to \infty} \mathcal{L}[e^{-\beta t_0(N)} \sum_{j=1}^{N} x_i^N(j, t_0(N))] = \lim_{t \to \infty} \mathcal{L}[\int_0^1 \mathfrak{z}_t^i(da)] = \mathcal{L}[\mathcal{W}_i^*], \ \mathcal{W}_i^* > 0, \text{a.s.}. \tag{10.245}$$

Therefore, defining $\hat{U}_i^N(t, \cdot)$ as in (10.98), we have

$$\lim_{N \to \infty} \hat{U}_i^N(t_0(N), dy) \to \hat{U}_i(\infty, dy). \tag{10.246}$$

Next define ζ_b as the birth time and ζ_d is the death time of an excursion starting from $\vec{x} = (x_1, \ldots, x_M)$. Then using the definition (10.110) we have

$$\hat{\mu}_t^i(dy) = \frac{y \int_0^1 1_{\{a \in dy\}} \mathfrak{z}_t^i(da)}{\int_0^1 \mathfrak{z}_t^i(da)} = \frac{y \tilde{N}_t^i(dy)}{\int_0^1 y \tilde{N}_t^i(dy)}, \tag{10.247}$$

where $\tilde{N}_t^i(\cdot)$ is a Poisson measure with intensity $\text{Pois}(e^{\beta t} \tilde{Q}^i(t, dy))$, with

$$\tilde{Q}^i(t, dy)) = e^{-\beta t} \int_0^t e^{\beta s} \mathcal{W}^*(s) \int_{\Delta_{M-1}} \tilde{Q}_{\vec{x}}(\zeta_b \in ds, \zeta_d > t, w(t) \in dy) \mu_{\text{eq}}(d\vec{x}). \tag{10.248}$$

Recalling that $\mathcal{W}^*(s) \to \mathcal{W}^*$ as $s \to \infty$, we obtain (10.111) and (10.112) by conditioning on $\mathcal{W}^*$, letting $t \to \infty$ and using the law of large numbers for the Poisson variables.   q.e.d.

Remark 227. We can also compute the joint moments of the lower types and the higher types for the limiting Palm measure on Δ_{2M-1}. In this case the non-zero contributions come from the event that no coalescence between upper level and lower level types occurs.

Chapter 11
Neutral Evolution on E_1 After Fixation (Phase 3)

Recall the five phases for the transition from E_0-types to E_1-types, namely 0,1,2,3,4, which we introduced in Subsubsection 7.1.1 for the mean-field model. So far we have discussed in Sections 8–10 the Phases 0, 1 and 2 and we now have to deal here in this section with the next Phase 3 (and in the next Section 12 with Phase 4). In Subsection 11.1 we describe the result, then in 11.2 give the proof.

11.1 Evolution in Neutral Equilibrium: Results

Consider the mean-field model with the two levels of fitness E_0 and E_1 and N spatial components, which is denoted $(X^N(t))_{t \geq 0}$ defined by (7.1). If we observe the system at times $\beta^{-1} \log N + t_N$ for $t_N \to \infty, t_N = o(N)$ all types of level zero, i.e. in E_0, are (asymptotically as $N \to \infty$) extinct and we have reached fixation in a tagged colony on level-one types, i.e. elements of E_1, as we proved in Section 10 and for later times we come into the *Phase 3*. In this phase we see as limit dynamic an exchangeable collection of neutral evolutions on level E_1 without any selection and mutation over times of order $o(N)$. If we start such a limit dynamic in a frequency vector θ on E_1 (everything in the limit $N \to \infty$), then we finally reach as time proceeds further a corresponding *neutral McKean–Vlasov equilibrium* earlier denoted $\Gamma_\theta^{c,d}$ among the types in E_1 (recalling due to migration monotype configurations do not arise). Recall that the initial state of the limiting neutral evolution is random since the proportions among the emerging fitter types in E_1 are *random* as we saw in the Section 10 and hence we see a *mixture* of the equilibria $\Gamma_\theta^{c,d}$ over the parameter θ representing the frequencies of the E_1-types at fixation on E_1.

The proof of the approach of the neutral equilibrium has two parts. First we verify that we get indeed a limiting dynamics in the considered time scale and second that in later times we approach the unique equilibrium of this dynamics for the given realization of the type intensities on E_1. We recall that the random frequencies at fixation are given in law by the random proportions between E_1-types upon emergence.

To begin we now formulate the fact that in a shifted time parameter t of the form

$$\frac{1}{\beta} \log N + t_N + t, \text{ with } t_N \to \infty \text{ but } t_N << N, \tag{11.1}$$

for a fixed realization of the proportions at fixation asymptotically as $N \to \infty$ in the time index t, we obtain roughly speaking, a $\mathbb{N}$-fold collection of (neutral) multitype Fisher-Wright diffusions with immigration-emigration all taking place on types of level-one fitness, i.e. types in E_1. The immigration source is given by a *random* distribution on E_1 with weights proportional to frequencies at fixation.

This limiting dynamic arises formally from the McKean–Vlasov dynamic on $E_0 \cup E_1$ by dropping all mutation and selection terms from the generator and then restricting the state to an element in $(\mathcal{P}(E_1))^\mathbb{N}$. In fact we obtain for *given* fixation intensities on E_1 the corresponding *neutral* equilibrium state which is the equilibrium of a multitype Fisher-Wright diffusion with immigration-emigration living on the higher level types in E_1 only. Recall here that the emerged types in E_1 have a *random* proportion λ on E_1 with distribution Q and its mean is ρ_1 (the rare mutation distribution on E_1), these random proportions are the ones we have at fixation and for given realization of λ there is a *unique* equilibrium measure of the McKean–Vlasov dynamic.

Precisely we have the following statement. Denote by

$$\tilde{\pi}_1 : (\mathcal{P}(E_0 \cup E_1))^N \longrightarrow (\mathcal{M}(E_1))^N, \quad \tilde{\pi}_1(\mu)(\cdot) := \bigotimes_{i=1}^{N} \mu_i(\cdot \cap E_1) \tag{11.2}$$

the map on measures given by $(\tilde{\pi}_1 \mu)(A) = \mu(A \cap E_1)$, $A \in \mathcal{B}(E_0 \cup E_1)$ induced by $\pi_1 : A \to A \cap E_1$. (Recall that $\mathcal{M}(E_1)$ is the space of subprobability measures.)

Proposition 11.1 (Phase 3: Neutral evolution). *Consider the system $(X^N(t))_{t \geq 0}$ defined in Section 7.1.1. Let $(Z(t))_{t \in \mathbb{R}}$ be an $\mathbb{N}$-indexed exchangeable collection of stationary multiple type Fisher-Wright diffusions on $\Delta_{|E_1|-1}$ with resampling rate d, with immigration, emigration both at rate c and immigration for each component from the same random (but constant in time) source $\hat{\mathcal{V}}^*(\cdot) = {}^*\mathcal{V}(\cdot)/{}^*\mathcal{V}(E_1)$ which is given independently of the dynamic. Recall (4.11) for the generator of Z.*

For given value of ${}^*\mathcal{V} = \theta$ at that immigration source each component of the process Z has a unique equilibrium defined in (4.25) and denoted as usual by $\Gamma_{\hat{\theta}}^{c,d}$ with $\hat{\theta}(\cdot) = \theta(\cdot)/\theta(E_1)$ and we get as t-marginal of the process Z the law

$$\int_{\mathcal{P}(E_1)} \Gamma_{\hat{\theta}}^{c,d}(\mathcal{L}[{}^*\hat{\mathcal{V}}])(d\hat{\theta}). \tag{11.3}$$

11.2 Proof of Proposition 11.1

Then assuming that $t_N \to \infty$ as $N \to \infty$ with $t_N = o(N)$:

$$\mathcal{L}\left[(\tilde{\pi}_1 \circ X^N\left((\frac{1}{\beta}\log N + t_N + t)^+\right)\right)_{t \in \mathbb{R}}\right] \underset{N \to \infty}{\Longrightarrow} \mathcal{L}[(Z(t))_{t \in \mathbb{R}}], \quad (11.4)$$

where the marginal distributions of the limit process are given by

$$\mathcal{L}[Z(t)] = \int Q(d\hat{\theta})(\Gamma_{\hat{\theta}}^{c,d})^{\otimes \mathbb{N}}, \quad t \in \mathbb{R}, \quad (11.5)$$

with

$$Q = \mathcal{L}[{}^*\mathcal{W}(\cdot)/{}^*\mathcal{W}(E_1)]. \quad (11.6)$$

In particular for given value of ${}^*\mathcal{W} = \theta$, the process Z is stationary with marginal $\Gamma_\theta^{c,d}$. □

Remark 228. What is new here compared to the analysis of hierarchical mean-field models in [DG99] is that we must show in spite of rare deleterious mutations from E_1 to E_0, the lower level, that the lower level types can be ignored in the evolution at higher levels at the specified age of the system and observed over short time intervals.

11.2 Proof of Proposition 11.1

We proceed in three steps. First verify that the dynamics in t lives on measures with support E_1. The next point is twofold, in a first step show that the dynamics in t converges to the limit dynamic as claimed, then in the next step show that we are indeed in equilibrium. For notational convenience we consider first the time parameter t running in $[0, \infty)$.

Step 1 We have to show (using the exchangeability) that for every $t \in (0, \infty)$:

$$\int_0^t x^N(1, \frac{1}{\beta}\log N + t_N + s)(E_0) ds \underset{N \to \infty}{\longrightarrow} 0 \text{ in probability}. \quad (11.7)$$

For this we show that if $X^N(0)(E_1) = 1$, then for fixed t:

$$E[\int_0^t x^N(1, s)(E_0) ds] \underset{N \to \infty}{\longrightarrow} 0. \quad (11.8)$$

This follows from the fact that the rate of downward mutation form E_1 to E_0 is $O(\frac{1}{N})$ in probability in a time interval of length t. The very same arguments as in the proof of Proposition 9.97 work here again. Then we note that by the Feller property of the evolution we know that the quantity in (11.8) is continuous in the initial law. We conclude the proof of (11.7) by observing that at time $T_N = \frac{1}{\beta} \log N + t_N$ as $t_N \to \infty$, the quantity $E[x^N(1, T_N)(E_1)]$ converges to 1 so that the rest follows via dominated convergence.

This proves that the weak limit points of

$$\mathcal{L}[(\tilde{\pi}_1 \circ X^N(\beta^{-1} \log N + t_N + t)_{t \in [0,T]}) \in \mathcal{P}(C([0,T], (\mathcal{M}(E_1))^N)) \quad (11.9)$$

are necessarily concentrated on the closed subspace $C([0, T], \mathcal{P}(E_1))^{\mathbb{N}})$ which we wanted to prove in this step.

Step 2 Note first that the process $(Z(t))_{t \geq 0}$ has the Feller property since each of the independent components has, so that convergence of distribution at time $\beta^{-1} \log N + t_N$ together with uniform (in all initial states) convergence of the dynamic give the claim.

In order to prove first the claimed convergence of the dynamics as t varies, we use the dual process to represent the state at time $(\frac{1}{\beta} \log N + t_N) + t$ in terms of the state at time $\frac{1}{\beta} \log N + t_N$. In order to verify that in time scale t mutation and selection between the level-one types disappears, we first note that these terms have coefficients which are of order N^{-1} and we observe the process only over a time interval of finite length, namely t. Therefore the result that the dynamic converges as claimed is an immediate consequence of the continuity of the model in the parameters s and m, in the sense of processes which we prove in the rest of this step.

In fact we next show that the model considered in (3.24) is for the case $\mathbb{I}$ finite indeed continuous in all the parameters:

Lemma 11.2 (Continuity in parameters). *Assume that $|\mathbb{I}| < \infty$ and $|\Omega| < \infty$.*

(a) Then

$$(m, s, c, d; M, a, \chi, X_0) \longrightarrow \mathcal{L}[(X(t))_{t \geq 0} | X(0) = X_0] \in \mathcal{P}(C([0, \infty), (\mathcal{P}(\mathbb{I}))^\Omega)) \quad (11.10)$$

is a continuous function. The function is uniformly continuous in X_0 for varying m, s, c, d.

(b) If we consider the mean-field dynamics Y instead of the process X then (11.10) holds.

(c) If we drop the assumption $|\Omega| < \infty$ in (a) then (11.10) holds if we consider the projection of paths onto finitely many sites. □

Proof of Lemma 11.2:. (a) We proceed in two parts, first we show that any collection of finite dimensional distributions is a continuous function of the parameters

11.2 Proof of Proposition 11.1

(uniformly in the initial state) and then that we have continuity in path space (uniformly in the initial state).

Part 1 Consider first the one-dimensional marginals. Then it suffices to show that all mixed moments are a continuous function of the parameters. This translates by the duality into the statement that the map (involving only the dynamical parameters)

$$(m, s, c, d; M, a, \chi) \longrightarrow \mathcal{L}((\eta_t, \mathcal{F}_t^+)) \tag{11.11}$$

is continuous. Since $(\eta_t)_{t \geq 0}$ is a pure Markov jump process, this is true considering the first component of the dual process $(\eta_t, \mathcal{F}_t^+)_{t \geq 0}$. The second component is changing at the jump times of η and the transitions of $\mathcal{F}_t^+$ are continuous in the parameters χ and M as is seen by inspection. This proves the claim for the dual process.

Since the processes as we consider are all Markov, we have proved the convergence of the transition kernels over an arbitrary time interval. This gives the continuity of the finite dimensional distributions in the dynamical parameters.

The continuity in the initial state follows again by applying the duality to show the Feller property of all the processes.

Part 2 It remains to verify tightness in path space for a converging sequence of parameters. To do this we use that $X(t) = (x_\xi(t))_{\xi \in \Omega}$ and each of the component processes $(x_\xi(t))_{t \geq 0}$ is a *semimartingale*. Then we can use tightness criteria for semimartingales on $C([0, \infty), \Delta_{|\mathbb{I}|-1})$, with Δ_{k-1} denoting the k-simplex. We use here the tightness criteria of Joffe and Métivier [JM] (see appendix) in terms of the local characteristics, which depend continuously on the parameters uniformly on the state space. Since all rates are bounded the claim follows immediately and Lemma 11.2 is proved.

To prove part (b), (c) of the Lemma we work with the obvious modification of the above argument. q.e.d.

Step 3 We return to the proof of Proposition 11.1. In order to see that the process Z is in equilibrium we use the ergodicity of the neutral McKean–Vlasov evolution which was established in [DGV] (Theorem 0.4), together with the fact that $t_N \to \infty$, but $t_N = o(N)$. Namely, observe that on E_1 we have state-independent mutation and therefore the ergodic theorem holds *uniformly* in the initial condition, i.e. we have that $\mathcal{L}_0[Z_t] \Longrightarrow \Gamma$ as $t \to \infty$ uniformly in Z_0 if Γ denotes the equilibrium state.

Now we can use a restart argument, namely it suffices to consider $t_N = t_N^1 + t_N^2$ with $t_N^\ell \uparrow \infty$ and to choose a subsequence such that the law at time t_N^1 converges along the subsequence. Then consider the process at time t_N^2 later and apply the uniform ergodicity, the Feller property and the convergence of the dynamic, which we showed above in Step 2. Namely note that we can *metrize* the weak topology and then use the triangle inequality to show that

$$\nu^N S^N(t_N^2) \text{ as } N \to \infty \text{ has the same limit as } \nu S^\infty(t) \text{ as } t \to \infty \tag{11.12}$$

if $v^N \to v$ and S^N, S^∞ are the semigroups of the Markov processes for the finite N respectively the McKean–Vlasov process. This continuity is uniform in all initial states as we read of immediately from the duality.

Finally we observe that the arguments above run for every time horizon $t \in [t_0, \infty)$ with $t_0 \in \mathbb{R}$ and hence we can obtain all the statements indeed for $t \in \mathbb{R}$ as claimed.   q.e.d.

Chapter 12
Re-equilibration on Higher Level E_1 (Phase 4)

Having now handled phases 0,1,2,3, in this section we are concerned with Phase 4 (recall Subsubsection 4.2.1 for the five phases scenario), that is, the time regime in which the mean-field model on spatial level 2 is concentrated on the second fitness level (E_1 types) and approaches a McKean–Vlasov *mutation-selection* equilibrium but *before* the emergence of the third fitness level (E_2-types) in a typical spatial 1-ball. This will then also serve as Phase 0 for the transition $E_1 \to E_2$ in a model at spatial scale 2 which will be considered in the next section.

For phase 4 of our basic scenario we consider the macroscopic time scale Nt (instead of t) and instead of components we consider blockaverages of N components and then we get a E_1-mutation-selection dynamic in t and if we replace t by $t_N \to \infty$ with $t_N = o(N)$ we get a new mutation-selection equilibrium on level E_1 rather than E_0 which occurred in time scale t. Since we want to study finite systems which can serve in the specified time scale, here Nt, as approximations to the infinite hierarchical system, we need to introduce systems with *two* levels of spatial organisation and hence in particular N^2 many components divided into N subsets of the type considered in the previous sections.

We state in Subsection 12.1 the results and Subsections 12.2 and 12.4 contain the proofs of the three propositions while Subsection 12.4 contains material on the suitable topology on path space.

12.1 Statement of Results: From Neutral Equilibrium to Mutation-Selection on E_1

Recall that in the previous section we proved that the population in a ball of radius 1, that is, interacting Fisher-Wright processes at N sites at distance 1, is asymptotically concentrated on types in E_1 and types in E_0 disappear by times of order $\beta^{-1} \cdot \log N + t_N$ for sufficiently large N. Furthermore we proved in the previous section that we then reach the neutral equilibrium on types in E_1 if $t_N \to \infty$ with $t_N = o(N)$.

Moreover if we turn to our infinite hierarchical system then in this time scale two balls of radius 1, a collection of states of the form $\{x^N(i,j;t); i,j \in \{1,\cdots,N\}$ at distance 2 are still asymptotically independent as $N \to \infty$ if we start with this property.

The *phase 4* takes place in time scale Nt in which two changes occur. First mutation and selection of types in E_1, the higher level, and second *migration at distance 2* becomes relevant which requires now to consider N copies of the mean-field model, i.e. N^2-sites altogether. We now give an overview of the program we must carry out to handle these two effects:

- we first define the system of N interacting balls of radius 1 and the corresponding *blockaverages* we denote by $\{y^N(j,t), j = 1, \ldots, N\}$,
- the N different 1-balls have independent initial conditions converging as $N \to \infty$ to the neutral equilibria $(\Gamma_{\lambda_i}^{c,d})_{i=1,\ldots,N} \in (\mathcal{P}(E_1))^N$ defined in Section 11 where the $\{\lambda_i\}$ are i.i.d. $\mathcal{P}(E_1)$-valued random variables with law Q given by (11.6),
- we consider the empirical distributions $\mathcal{Y}^N(t)$ (see (12.6)) of the N blockaverages of measures restricted to E_1 and its stochastic dynamics with trajectories in $D([0,\infty), \mathcal{P}(\mathcal{M}(E_1)))$,
- prove that it converges in the sense of weak convergence of processes to a McKean–Vlasov dynamics on $\mathcal{P}(\mathcal{P}(E_1))$, for this
  - we must formulate an appropriate topology on trajectories,
  - verify tightness,
  - prove that the limiting process has statespace $\mathcal{P}(\mathcal{P}(E_1))$,
  - prove that any limit point on $D([0,\infty), \mathcal{P}(\mathcal{M}(E_1)))$ is given by the unique solution of the nonlinear McKean–Vlasov dynamics,
- verify the propagation of chaos for the $\{y^N(j,t), j = 1,\cdots,N\}$ and that the limit dynamics of each $y^N(j,t)$ is the mean-field McKean–Vlasov equation now on E_1,
- apply the ergodic theorem to the resulting McKean–Vlasov dynamics to show that the in times of order $O(Nt_N)$, $t_N = o(\log N)$, the limiting empirical measure of blockaverages converges to the mutation-selection equilibrium on E_1.

We proceed in four subsubsections, three preparing the ground to state the result in the fourth.

12.1.1 The Finite Two-Level System: Construction

We consider a system (two spatial levels, two fitness levels)

$$(X^{N,2}(t))_{t \geq 0} \tag{12.1}$$

12.1 Statement of Results: From Neutral Equilibrium to Mutation-Selection on E_1

of size N^2 after times of order N in time scales of order $O(1)$, which has the form:

$$(\{x_k^N(i,j;Nt+u)|k \in \mathbb{K}, i=1,\ldots,N;\ j=1,\ldots,N\})_{u \geq 0}, \quad \mathbb{K} = E_0 \cup E_1. \tag{12.2}$$

Here (i,j) denotes the site i in the j-th ball of radius 1. We view this geographic space as a *subgroup of the hierarchical group*. We must study the evolving random field in (12.2) as a function of both the macroscopic time variable t and the microscopic time variable u.

The type space is $\mathbb{K} = E_0 \cup E_1$ with mutation, selection and resampling in each colony as before in the mean-field model. The dynamics is modified from the mean-field dynamic as follows changing only the mechanism of migration. Now the migration includes besides the jumps of distance 1 at rate c_0 now also

$$\text{uniformly distributed jumps of distance 2 at rate } \frac{c_1}{N}. \tag{12.3}$$

This means that in the Ito-equation setting we get an additional drift term in the equation for $x_k^N(i,j;t)$, namely

$$\frac{c_1}{N^2} \sum_{i,j=1}^{N} x_k(i,j;t), \quad k \in \mathbb{K}. \tag{12.4}$$

We start this system in an *i.i.d. initial state* with the expected type frequencies being $\hat{\theta} \in \mathcal{P}(\mathbb{K})$ which are concentrated on E_0.

We can now observe the system with N^2 sites on three spatial levels for the appropriate time scales, (1) individual sites, (2) blockaverages of sites over balls of radius 1 with N sites and (3) empirical distributions of these N blockaverages.

Note that in the *macroscopic* time scale Nt we mainly focus on the empirical distribution of blockaverages in each of N balls of radius 1 separated by distance 2. However single components will also be considered for times $Nt + u$ as the *microscopic* time u varies and the macroscopic time t is fixed.

To proceed we define three functionals of the process $X^{N,2}$ as follows. We first define the *collection of blockaverage processes* $\{(y_k^N(j;t))_{t \geq 0,\ k \in \mathbb{K}}\}_{j=1,\ldots,N}$ in *macroscopic time* by setting:

$$y_k^N(j;t) = \frac{1}{N} \sum_{i=1}^{N} x_k^N(i,j;Nt),\ k \in \mathbb{K},$$

$$y^N(j,t) = (y_k^N(j,t))_{k \in \mathbb{K}} \in \mathcal{P}(\mathbb{K}),\ j=1,\ldots,N \tag{12.5}$$

and the *scaled empirical process* $(\mathcal{Y}^N(t))_{t \geq 0}$ by setting:

$$\mathcal{Y}^N(t) = \frac{1}{N} \sum_{j=1}^{N} \delta_{y^N(j,t)} \in \mathcal{P}(\mathcal{P}(\mathbb{K})). \tag{12.6}$$

Finally, define the *scaled empirical mean over the ball of radius* 2, $(\bar{y}_k^N(t), k \in \mathbb{K})_{t \geq 0}$ by setting:

$$\bar{y}_k^N(t) = \frac{1}{N} \sum_{j=1}^{N} y_k^N(j,t), \quad k \in \mathbb{K}, \quad \bar{y}^N = (\bar{y}_k^N, k \in \mathbb{K}). \tag{12.7}$$

In order to consider the restriction of the measure to E_1 we next introduce the map

$$\tilde{\pi}_1 : \mathcal{P}(\mathbb{K}) \longrightarrow \mathcal{M}(E_1) \tag{12.8}$$

induced by π_1 mapping a subset of $\mathbb{K}$ to E_1 by the restriction map to types in E_1 as follows:

$$\pi_1(A) = A \cap E_1, \quad \tilde{\pi}_1(\mu)(B) = \mu(\pi_1^{-1}(B)), \quad B \subset E_1. \tag{12.9}$$

Similarly on the next level the induced map is

$$\tilde{\tilde{\pi}}_1 : \mathcal{P}(\mathcal{P}(\mathbb{K})) \longrightarrow \mathcal{P}(\mathcal{M}(E_1)). \tag{12.10}$$

Based on the martingale problem defining the dynamic of $X^{N,2}$, we now want to prove that the empirical distribution of blockaverages has a limiting dynamic $\mathcal{Y}(t)$ given by a Wright-Fisher system with types in E_1 and with selection - mutation, and finally a nonlinear term arising from the migration. Under the assumptions on the initial conditions described above (below (12.4)) the limiting initial condition $\mathcal{Y}(0)$ is given by (12.14 below).

12.1.2 Limit System: Construction

We introduce next the candidate for the limiting (as $N \to \infty$) dynamics for the process defined above with the initial condition for $\{\mathcal{Y}^N(t) \in \mathcal{P}((\mathcal{M}(E_1)) : t \geq 0\}$ given via the initial state of $X^{N,2}$ described in Subsubsection 12.1.1.

Our objective is to establish that the limiting dynamics is given by the deterministic McKean–Vlasov dynamics $(\mathcal{Y}(t))_{t \geq 0}$ on $\mathcal{P}(\mathcal{P}(E_1))$ characterized as the unique solution to the nonlinear martingale problem for a $\mathcal{P}(\mathcal{P}(E_1))$-valued stochastic process corresponding to the McKean–Vlasov equation (recall (2.19)),

12.1 Statement of Results: From Neutral Equilibrium to Mutation-Selection on E_1

$$\frac{d\mathcal{L}_t}{dt} = (L^{\mathcal{L}_t})^* \mathcal{L}_t, \qquad (12.11)$$

which has for given initial state at a time t_0 a unique solution for $t \geq t_0$ ([DG99]). Here L_t^π is defined as in (10.20) where the parameters are given by the renormalized parameters as follows:

$$c = c_1, s = s_1, d = d_1, \chi = \chi_1, M(\ell_1, \ell_2) \equiv m_{1,1}(\ell_1, \ell_2) \text{ with } \ell_1, \ell_2 \in E_1, \qquad (12.12)$$

and d_1, s_1 are recalled from Subsection 12.2 below in (12.65), (12.66), respectively, and

$$\pi_t = \int_{\mathcal{P}(\mathbb{K})} \mu \mathcal{Y}^\infty(t, d\mu) \in \mathcal{P}(\mathbb{K}). \qquad (12.13)$$

It remains to specify the initial distribution of the dynamic described above. Since we want to consider the dynamics of $\mathcal{Y}^\infty = \mathcal{Y}^\infty(t, dx)$ after fixation, we must take as initial state the limiting empirical distribution produced by the *neutral equilibrium* corresponding to the intensity measure at fixation. In this time scale the equilibrium associated with the means in the N balls of radius 1 are independent. Therefore by the Glivenko-Cantelli theorem the initial state is given by the law of the mean in a ball of radius 1 in the neutral equilibrium, namely,

$$\mathcal{Y}(0) = \mathcal{L}[\Gamma_\lambda^{c,d}], \qquad (12.14)$$

where the *random* measure λ is given by

$$\lambda(\cdot) = \frac{{}^*\mathcal{W}(\cdot)}{{}^*\mathcal{W}(E_1)} \qquad (12.15)$$

and for $\mu \in \mathcal{P}(E_1)$, $\Gamma_\mu^{c,d}$ is defined as in (4.25).

We introduce the corresponding collection of independent realizations

$$\{(y^\infty(j,t))_{t\geq 0}, \quad j \in \mathbb{N}\}, \qquad (12.16)$$

of the evolutions on E_1 given by the process $\mathcal{Y}$ solving the $L_{M,\chi}^{\pi_t}$-martingale problem from (12.11), (12.12) above.

Furthermore we have for the single colonies observed at late times over time spans of order 1, i.e. times $tN + u$ a limiting dynamic in u. Define for every *fixed* macroscopic time t and each $i \in \mathbb{N}$ independently for *fixed* values of $\{y^\infty(j,t), j \in \mathbb{N}\}$, the evolutions in microscopic time u

$$(\{x_k^\infty(i,j,u); i \in \mathbb{N}\})_{u \geq 0, k \in \mathbb{K}}, \qquad (12.17)$$

as independent collections of realizations of solutions of the $L_{\pi^j}^{c,d}$-nonlinear martingale problem on E_1 (recall (2.19)) with the following choice of the parameters:

$$c = c_0, \pi^j \equiv y^\infty(j,t), d = d_0, \chi \equiv 0, M \equiv 0 \qquad (12.18)$$

and the initial state being the *equilibrium* state $\Gamma_\theta^{c,d}$ with $\theta = y^\infty(j,t)$.

Note that the evolution of the collection in (12.17) is an evolution in a random medium, where the medium is the configuration in (12.16) at a fixed time t. In particular the law of the collection in (12.17) depends on the choice of t. For fixed j we have:

$$\mathcal{L}[\{(x^\infty(i,j,u))_{u \geq 0}, \quad i \in \mathbb{N}\}|y(j,t)] = \bigotimes_{i \in \mathbb{N}} \mathcal{L}[(x^\infty(i,j,u))_{u \geq 0} | y(j,t)].$$
$$(12.19)$$

We want to state convergence results in path space and not just f.d.d. convergence. Note that in the fixed N-model downward mutations from level E_1 to E_0 do occur (at rate $O(1)$) in time scale Nt in every component but are wiped out at rate $O(N)$ as we shall prove further on, but first in the next subsubsection we have to specify the topologies in which we can make this claim.

12.1.3 Topologies on Path Space

Since we want to establish convergence from a dynamic on $\mathcal{P}(\mathcal{P}(\mathbb{K}))$ to one on $\mathcal{P}(\mathcal{P}(E_1))$, it is necessary to discuss the appropriate *topology on path space* $D([0,\infty), \mathcal{P}(\mathcal{P}(\mathbb{K})))$ (the space of càdlàg paths) or continuous path $C([0,\infty), \mathcal{P}(\mathcal{P}(\mathbb{K})))$.

The usual topology for weak convergence for processes is the Skorohod topology which yields a Polish space. We expect that our limiting process is concentrated on types in E_1. However we will show below in Proposition 12.9 in Subsection 12.3 that for $0 < \varepsilon < 1$, $\lim_{N \to \infty} P[\sup_{0 \leq s \leq NT} x_1^N(s) > \varepsilon] > 0$. In view of this we will begin by introducing a weaker topology on the measure-valued paths analogous to the Meyer-Zheng topology (see [MZ]) which allows us to formulate the desired results. This involves proving that the occupation integral of mass on E_0 over finite macroscopic times vanishes in the limit so that E_0-types are "not observable". This involves replacing the pathwise supnorm on compacta (based on the variational distance in the space of values) by the L_1-norm of the path considered on compacta.

Recall that our paths, both for finite N and for $N \to \infty$, are almost surely continuous. Since rare downward mutations occur in time scale Nt at order $O(1)$

12.1 Statement of Results: From Neutral Equilibrium to Mutation-Selection on E_1

and selection acts at rate $O(N)$ we want to determine if excursions to E_0-types result in small occupation mass-time integrals.

Denote by $\|\cdot\|_{\text{var}}$ the variational distance between two probability measures. Define the seminorms:

$$\|f\|_{I,1} = \int_I \|f(s)\|_{\text{var}} ds, \quad I = (a,b), \ a,b \in \mathbb{R}. \tag{12.20}$$

Then we consider the topology in which convergence is defined by the *Meyer-Zheng topology*

$$f_n \longrightarrow f \iff \|f - f_n\|_{I,1} \xrightarrow[n\to\infty]{} 0 \quad \forall \text{ finite intervals } I. \tag{12.21}$$

If we complete $C(\mathbb{R}, \mathcal{P}(\mathcal{P}(\mathbb{K})))$ under a metric derived from the seminorms

$$d(f,g) = \sum_{m=0}^{\infty} 2^{-m} \|f - g\|_{[-m,m],1}, \tag{12.22}$$

then we end up with a larger space than just the space of continuous functions and we therefore have to check an appropriate tightness criterion in a bigger space. We can pass to the space of pseudopaths and adapt the tightness criterion of Meyer-Zheng [MZ], Theorem 4 to our case.

We first introduce a topology on the set of measure-valued pseudopaths on $\mathcal{P}(\mathcal{P}(\mathbb{K}))$. Following Meyer-Zheng [MZ], a *pseudopath* is a probability law on $[0,\infty] \times \mathcal{P}(\mathcal{P}(\mathbb{K}))$: namely the image measure ν of $\lambda(dt) = e^{-t} dt$ under the mapping $t \to (t, w(t))$ where $w(t)$ is a $\mathcal{P}(\mathcal{P}(\mathbb{K}))$-valued Borel measurable function on $\mathbb{R}_+$, that is for $a, b \in \mathbb{R}^+$ and $B \in \mathcal{B}(\mathcal{P}(\mathcal{P}(\mathbb{K})))$:

$$\nu((a,b) \times B) = \int_a^b e^{-s} 1_B(w(s)) ds. \tag{12.23}$$

Some notation will be needed later on. Furthermore consider the spaces

$$C(\mathbb{R}, \mathcal{P}(\mathcal{P}(\mathbb{K}))), \ \mathbf{D}(\mathbb{R}, \mathcal{P}(\mathcal{P}(\mathbb{K}))), \tag{12.24}$$

of continuous measure-valued paths, respectively $\mathcal{P}(\mathcal{P}(\mathbb{K}))$-valued pseudopaths. Let

$$\mathcal{M}_0(\mathbb{K}) = \{\mu \in \mathcal{P}(\mathbb{K}) : \mu(E_0) = 0\}. \tag{12.25}$$

We conclude giving a tightness criterion in the Meyer-Zheng topology. Next we define the conditional variation as follows. For a subdivision $\tau: 0 = t_0 < t_1 < \cdots < t_n = \infty$ (define $X_\infty = 0$) set (here $(\mathcal{F}_t)_{t \geq 0}$ is the natural filtration)

$$V_\tau(X) = \sum_{0 \le i < n} E[|E[X_{t_{i+1}} - X_{t_i}|\mathcal{F}_{t_i}]|] \qquad (12.26)$$

and

$$V(X) = \sup_\tau V_\tau(X). \qquad (12.27)$$

If $V(X) < \infty$ then X is called a *quasi-martingale*. Note that we can stop the processes at T.

The Meyer-Zheng result is:

Lemma 12.1 (Tightness in Meyer-Zheng topology).

(a) If P_n is a sequence of probability laws on **D** such that under P_n, the coordinate process (X_t) is a quasi-martingale with conditional variation $V_n(X)$ uniformly bounded in n, then there exists a subsequence P_{n_k} which converges weakly on **D** to a law P and (X_t) is a quasi-martingale under P.

(b) Assume

$$X_t^n = \liminf_{h \downarrow 0} \frac{1}{h} \int_t^{t+h} X_s^n ds, \; a.s. \qquad (12.28)$$

In this case there exists a subsequence $(X_t^{n_k})$ and a set I of full Lebesgue measure, such that the finite dimensional distributions of $(X_t^n)_{t \in I}$ converge weakly to those of $(X_t)_{t \in I}$. □

(Also see [MZ], Theorem 7 for identification of a limiting semimartingale.)

Remark 229. Note that the space of pseudopaths with the Meyer-Zheng topology does not lead to a Polish space (see [MZ], page 372).

12.1.4 Statement of Three Results on Scaling Limit

We begin with two results on the limiting dynamic on level-one types of the collection of blockaverages over 1-balls and of the average over the whole system of N^2 sites both in macroscopic time.

Proposition 12.2 (Phase 4: Approach to selection-mutation dynamics on E_1).
Let $\mathcal{Y}(0) \in \mathcal{P}(\mathcal{P}(\mathbb{K}))$ be given by (12.11)–(12.13). Under the above assumptions

$$\mathcal{L}[\widetilde{\widetilde{\pi}}_1 \circ \mathcal{Y}^N(t))_{t \ge 0}] \underset{N \to \infty}{\Longrightarrow} \mathcal{L}[(\mathcal{Y}^\infty(t))_{t \ge 0}], \qquad (12.29)$$

where $\mathcal{Y}^\infty(t)$ is a solution to the McKean–Vlasov equation (12.11). The convergence is in the sense of weak convergence in the Meyer-Zheng topology. Moreover,

12.1 Statement of Results: From Neutral Equilibrium to Mutation-Selection on E_1

$$\mathcal{L}[(\tilde{\pi}^1 \circ \bar{y}^N(t))_{t \geq 0}] \underset{N \to \infty}{\Longrightarrow} \mathcal{L}[(\int_{\mathcal{P}(E_1)} \mu \mathcal{Y}^\infty(t, d\mu)))_{t \geq 0}]. \qquad \Box \qquad (12.30)$$

It is a natural question to ask whether or not the convergence above holds also in the *uniform topology* on path space, since we have downward mutation at rate $O(1)$ in macroscopic time, but the selection favours E_1- over E_0-types by terms $O(N)$. To verify this is quite technically challenging, but we shall present partial results and arguments in Subsection 12.3 to this effect.

The next result shows that the random proportions arising in emergence and persisting in fixation do not persist in time scales $O(N)$ and are washed out and a unique equilibrium point is reached.

Proposition 12.3 (Phase 4: Mutation-selection equilibrium on E_1). *Let $t_N \to \infty$, $\lim_{N \to \infty} \frac{t_N}{\log N} = 0$. Then*

$$\mathcal{L}[\tilde{\tilde{\pi}}_1 \circ \mathcal{Y}^N(t_N)] \Rightarrow \delta_{\Gamma_{M, \chi_1}^{s_1, c_1, d_1}} \text{ as } N \to \infty, \qquad (12.31)$$

where $\Gamma_{M, \chi_1}^{s_1, c_1, d_1} \in \mathcal{P}(\mathcal{P}(E_1))$ *is the unique equilibrium for the nonlinear McKean–Vlasov mutation-selection dynamics on E_1.* $\Box$

Remark 230. In the case in which the mutation process at the second level has two or more ergodic classes (see Remark 38) it is expected that the emergence is described by a set of random variables, one for each ergodic class. If these have the same fitness at the second level then both will appear in the equilibrium on this level (by the mean-field mechanism) but with random proportions depending on the values of the respective $*\mathcal{W}$ variables. For successive levels this results in a tree structure with random proportions associated to the different branches. These results can be obtained by extensions of the methods of this section but will not be developed in this monograph.

Next consider averages of N-blocks in macroscopic time $O(N)$ and single sites in $O(1)$-time, the latter moving in the random medium provided by the N-blockaverages:

Proposition 12.4 (Multi-scale analysis phase 4: Propagation of Chaos).

(a) *Under the above assumptions we have for blockaverages in (macroscopic) Nt-time scale propagation of chaos, i.e. for given $L \in \mathbb{N}$,*

$$\mathcal{L}[\{(\tilde{\pi}_1 \circ y^N(j;t))_{t>0}, \quad j = 1, \cdots, L\}] \underset{N \to \infty}{\Longrightarrow} \bigotimes_{j=1}^L \mathcal{L}[(y^\infty(j,t))_{t>0}]. \qquad (12.32)$$

(b) *For fixed $t \in (0, \infty)$ and $j \in \mathbb{N}$, let $\tilde{\mathcal{L}}_t$ denoting the law of $\{y^\infty(j,t)\}$, we have for $L \in \mathbb{N}$, that the tagged components in microscopic time scale satisfy:*

$$\mathcal{L}[\{\tilde{\pi}_1 \circ x^N(i,j;Nt+u)_{i,j=1,2,\cdots,L;u\geq 0}\}] \tag{12.33}$$

$$\underset{N\to\infty}{\Longrightarrow} \int_{(\mathcal{P}(E_1))^L} \left(\bigotimes_{i=1}^{L} \mathcal{L}[(x^\infty(j;u))_{u\geq 0}|\mu_j]\right) \bigotimes_{j=1,\cdots,L} \prod_{j=1}^{L} \tilde{\mathcal{L}}_t(d\mu_j). \qquad \Box$$

12.2 Proof of Proposition 12.2

We proceed in three main steps, showing elimination of E_0-mass, tightness (which we do in Step 2) and finally identify the limit process.

Step 1: Asymptotic elimination of E_0 types

Assume that initially $x_k^N(i,j;0) = 0 \; \forall k \in E_0, \; i,j = 1,\ldots,N$. We then want to show that at a tagged ball of radius 1 labeled by j,

$$\sum_{\ell \in E_0} y_\ell^N(j,\cdot) \xrightarrow[N\to\infty]{} 0 \tag{12.34}$$

in the Meyer-Zheng topology and for this it suffices to show that

$$\lim_{N\to\infty} \int_a^b \sum_{\ell \in E_0} E[y_\ell^N(j,t)]dt = 0, \text{ for all } 0 \leq a < b < \infty. \tag{12.35}$$

This guarantees low occupation integral on types in E_0 and convergence to a continuous path with states concentrated on types in E_1 provided we have shown tightness (which we do in Step 2) and this is what we want.

We will also show that at a tagged site i in a tagged ball j we have:

$$\lim_{N\to\infty} \int_0^T \left(\sum_{\ell \in E_0} E[x_\ell^N(i,j,Nt)]\right) dt \tag{12.36}$$

$$= \lim_{N\to\infty} \frac{1}{N} \int_0^{NT} \left(\sum_{\ell \in E_0} E[x_\ell^N(i,j,t)]\right) dt = 0.$$

For this purpose we have to include the downward mutations from E_1 to E_0 into the picture. First note that a downward mutation appears at rate N^{-1} and hence in time scale Nt at rate $O(1)$. However this is compensated by selection acting at rate $O(1)$ and hence in time scale Nt at rate $\sim N$. Consequently a downward mutation lives for a time of order $O(\frac{1}{N})$ and in a fixed time window of width $O(N)$ it should not be visible in the limit $N \to \infty$. The details are given in this step.

12.2 Proof of Proposition 12.2

We establish here the effective absence of lower level types in the two scenarios described above which we will be needed in all the propositions. Focus first on three types again, i.e. assuming

$$K = 3, |E_0| = 2, |E_1| = 1. \tag{12.37}$$

(1) Consider first times of order $O(1)$ as $N \to \infty$. It follows from the previous section, Eq. (11.7) (see also Proposition 9.97), that over time periods of length $O(1)$ the inferior types are absent at tagged sites since the cumulative downward mutation rate converges to zero. Then using that $x_1^N(t) + x_2^N(t)$ is a semimartingale with driftterm bounded by a multiple of the downward mutation rate, we obtain that the path does not exceed ε over finite time horizons, that is, abbreviating $x_k(i, j, t)$ by $x_k(t)$ for some fixed i, j, if $(x_1^N(0) + x_2^N(0)) = 0$, then:

$$\lim_{N \to \infty} P[\sup_{0 \le t \le T} (x_1^N(t) + x_2^N(t)) > \varepsilon] = 0. \tag{12.38}$$

This implies in the limit $N \to \infty$ the absence of E_0-types for times of order $O(1)$.

(2) We now verify that the probability that the weighted occupation time of types 1 and 2 in an interval of length $O(N)$ (i.e. $O(1)$ in the Nt scale) can affect the level-1 blockaverage is also 0. We will return to the question of convergence in the stronger topology in Subsection 12.3.

Abbreviate $y_k(j, t)$ by $y_k(t)$ for fixed j and recall that $y_k(t)$ involves the macroscopic time scale Nt. Then we claim:

Lemma 12.5 (No mass on lower types). *Assume that* $(y_1^N(0) + y_2^N(0)) = 0$. *Then*

$$\lim_{N \to \infty} \sup_{0 \le t \le T} P[y_1^N(t) + y_2^N(t)) > \varepsilon] = 0, \tag{12.39}$$

and

$$\lim_{N \to \infty} P[\int_0^T (y_1^N(t) + y_2^N(t))dt > \varepsilon] = 0. \quad \square \tag{12.40}$$

Proof. This follows immediately from Lemma 12.6 (a) below using Chebyshev's inquality.   q.e.d.

Remark 231. This implies that $(y_1^N(t) + y_2^N(t))_{0 \le t \le T}$ converges to the 0 as process in the Meyer-Zheng topology (see [MZ]).

We now prove the key estimate:

Lemma 12.6 (Estimates on E_0-mass). *Assume that* $|E_0| = 2, |E_1| = 1, m_{3,\text{up}} = 0, m_{1,2} > 0, m_{2,1} > 0$. *Assume that the fitness levels are* $0, 1, 1$, *respectively. Then abbreviating a generic* $y^N(j, t)(E_0)$ *by* $y_{E_0}^N(t)$ *we have the following estimates.*

(a) If $y_{E_0}^N(0) = 0$ and $m_{3,\text{down}} > 0$, then there exists a constant $\tilde{\kappa}_1 < \infty$ such that

$$E[y_{E_0}^N(t)] \leq \frac{\tilde{\kappa}_1}{N} \quad \forall\, t \geq 0. \tag{12.41}$$

(b) If $y_{E_0}^N(0) \in (0, 1)$, and $m_{3,\text{down}} = 0$, then there exists a constant $\tilde{\kappa}_2$ such that

$$E[y_{E_0}^N(t)] \leq e^{-\tilde{\kappa}_2 N t} y_{E_0}^N(0). \quad \square \tag{12.42}$$

Proof of Lemma 12.6. Note that in our model in time scale Nt the down-mutation rate is $m_{3,\text{down}} > 0$ and the fitness difference of types 1 and 3 is sN. As a warm-up we discuss first the deterministic case and then come to the stochastic case, for which the argument is then easily adapted.

Case $d = 0$ *(Deterministic case)*

Let $y_{100}(t) = E[y_{100}^N(t)]$, etc. Then for the deterministic system the quantities satisfy the system of ode:

$$\frac{dy_{010}}{dt} = m_{21} N (y_{110} - y_{010}) - N m_{12} y_{010} + s N y_{010} y_{100} + m_{3,\text{down}}(y_{011} - y_{010}), \tag{12.43}$$

$$\frac{dy_{110}}{dt} = s N (y_{010} + y_{110} y_{100} - y_{110}) + m_{3,\text{down}}(1 - y_{110}). \tag{12.44}$$

By an analysis of this system of differential equations it can be proved that if $m_{3,\text{down}} = 0$ and $y_{110}(0) < 1$, then $y_{110}(t) \to 0$ exponentially fast for any N. If $m_{3,\text{down}} > 0$, then as $N \to \infty$ the equilibrium value of $y_{(110)}(t)$ goes to zero at speed N^{-1}.

However as an introduction to the stochastic case we will prove this using the dual representation. Let

$$y_{100}(t) := E_{(100)}\left[\int \mathcal{F}_t^{++} d\mu^\otimes\right] \tag{12.45}$$

starting the dual $\mathcal{F}_t^{++}$ at $\mathcal{F}_0^{++} = (100)$. Then by independence, due to the fact that no coalescence takes place, we have for the dual started in $(100) \otimes (100)$:

$$E_{(100)\otimes(100)}\left[\int \mathcal{F}_t^{++} d\mu_0^\otimes\right] = \left(E_{(100)}\left[\int \mathcal{F}_t^{++} d\mu_0^\otimes\right]\right)^2. \tag{12.46}$$

with the analog relation, if we start in $(100)^{\otimes k}$, $k = 3, 4, \cdots$.

Using this we have the backward equation (recall $(001)=(111)-(110)$):

$$\tfrac{dy_{100}}{dt} = -N m_{12} 1 y_{100} + N m_{21}(y_{110} - y_{100}) + N s (y_{100}^2 - y_{100}), \tag{12.47}$$

12.2 Proof of Proposition 12.2

Fix a value of T (recall t, T are macroscopic variables, i.e. Nt is the real time). Note that this takes care of the speeded-up time.

Assume that the $\mu_0 = (0, 0, 1)$ at each site (as in the neutral equilibrium), that is

$$y_3(0) = 1. \qquad (12.48)$$

To get an *upper bound* on the mass that *comes* from the downward mutation we will show that if we start the dual process with the indicator function

$$f_0 = (110) \qquad (12.49)$$

and with initial state μ_0, then for $t > 0$:

$$E_{\mu_0}[y_2^N(t) + y_1^N(t)] = E[\int (\tilde{\mathcal{G}}_{Nt}^{++}) d\mu_0^{\otimes |\pi_t|}] \to 0 \text{ as } N \to \infty. \qquad (12.50)$$

The transitions of the dual process $\mathcal{G}_{Nt}^{++}$ by selection, mutation and migration are as follows:

- $(110) \to (010) \otimes (111) + (100) \otimes (110)$ at rate sN,
- $(010) \to (010) \otimes (111) + (100) \otimes (010)$, at rate sN
- $(100) \to (100) \otimes (100)$, at rate sN
- $(100) \to (000)$ and $(010) \to (110)$ rate Nm_{12},
- $(100) \to (110)$ and $(010) \to (000)$ at rate Nm_{21}
- $(110) \to (111)$ at rate $2m_{3,\text{down}}$, and $(100) \to (101)$ and $(010) \to (011)$ at rate $m_{3,\text{down}}$
- $(***)_i \longrightarrow (***)_j$, where $*$ is 0 or 1 at rate cN.

We first note that for the given initial state the factors $(110), (100), (010)$, do not result in a contribution, this occurs only for factors $(001), (011), (111)$ and (101). Hence only summands which consist of factors from the second set *exclusively* can contribute. Hence the positive contributions require a rare downward mutation to create terms $(**1)$.

Then beginning with (110) we consider the first time τ_1 the state of the dual is bounded by $(110) \otimes (110)$, that is a second factor of (110) is produced. Note that this occurs, for example, when the transitions $(110) \to (010) + (100) \otimes (110) \to (110) \otimes (110)$ that is one selection operation followed by one mutation $(010) \to 0$, $(100) \to (110)$ (recall that these are coupled). Note that the probability that a rare downward mutation occurs before the first selection operation is

$$\frac{m_{3,\text{down}}}{sN + m_{3,\text{down}}}. \qquad (12.51)$$

At the time of the first selection event we have

$$(110) \to (010) + (100) \otimes (110). \qquad (12.52)$$

We must also consider the effects of the mutations $1 \to 2$, $2 \to 1$ on the first column. A mutation $1 \to 2$ produces (110), that is, the selection event is erased. A mutation $2 \to 1$ produces $(110) \otimes (110)$, that is, the selection event is successful. This decision is made after an exponential $N(m_{12} + m_{21})$ random time and the probability of success is $q = \frac{m_{21}}{m_{21}+m_{12}}$. Therefore the waiting time $\mathfrak{W}$ before the first successful selection occurs is the sum of a geometric number $\mathfrak{N}$ of random times having distribution given by the sum of an $\text{Exp}(sN)$ and an $\text{Exp}(N(m_{1,2} + m_{2,1}))$, that is the sum of an Erlang $(\mathfrak{N}, sN)$ and an Erlang $(\mathfrak{N}, N(m_{1,2} + m_{2,1})$. The results are exponentials with parameters sqN, $(Nq(m_{1,2} + m_{2,1}))$. Therefore there exists $\kappa > 0$ such that

$$P(\mathfrak{W} > t) \leq e^{-\frac{Nt}{\kappa}}. \tag{12.53}$$

Therefore using (12.50) we see that the contributions to $E_{\mu_0}[y_2^N(t) + y_1^N(t)]$ come from the following dual events:

- one rare mutation down before the second (110) is produced (and before time t)
- two rare mutations down before the third (110) factor (and before time t) is produced,
- three rare mutations are produced before the third (110) is produced, etc.

We can then show that the probability of the first event is bounded above by

$$q_N := \int_0^\infty m_{3,\text{down}} e^{-m_{3,\text{down}} t} e^{-Nt/\kappa} dt = \frac{\tilde{\kappa}}{N} \quad \text{for some } \tilde{\kappa} < \infty. \tag{12.54}$$

Similarly the probability of the second event is bounded by

$$\left(\frac{\tilde{\kappa}}{N}\right)^2, \text{ etc.} \tag{12.55}$$

Therefore we conclude that if $\mu_0(E_0) = 0$, then for any $t > 0$,

$$E_{\mu_0}[y_2^N(t) + y_1^N(t)] = \sum_{k=1}^{\infty} \left(\frac{\tilde{\kappa}}{N}\right)^k \leq \text{const} \cdot \left(\frac{\tilde{\kappa}}{N}\right). \tag{12.56}$$

Similarly we can show using the dual that if $y_{E_0}(0) \in (0, 1)$ and $m_{3,\text{down}} = 0$, then

$$\lim_{N \to \infty} E[y_{E_0}^N(t)] \leq e^{-m_* Nt} y_{E_0}(0) \tag{12.57}$$

where $m_* = \sum_{j=1,\ldots,M-1} m_{M,j}$. This immediately gives the claim in (12.42) and completes the proof of the lemma. q.e.d.

12.2 Proof of Proposition 12.2

(If the initial state $\mu_0 = (p_1, p_2, p_3)$ at each site satisfies $p_3 > 0$, then excluding rare mutations and using the dual it is easy to verify that $E(y_2^N(t) + y_1^N(t))]$ goes to 0 exponentially fast.)

Case $d > 0$.

Now consider the case $d > 0$. We now start with the factor (110) at a fixed site and consider the resolution cycles, that is, the times at this site after the first selection operation until the resolution time T_{R*}. During each such cycle there is a positive probability of producing a new permanent site occupied by at least one (110) factor. Moreover by Proposition 9.32 we know that the resolution T_{R*} has an exponential moment. Since a new permanent site is produced after at most a geometrically distributed number of cycles, we also have that the time to produce the first new permanent site $T_{1,**}$ has an exponential bound. Therefore we obtain an upper bound on the E_0-mass by summing the probabilities of the events that a rare mutation occurs before $T_{1,**}$ or two rare mutations occur before $T_{2,**}$, etc. where $T_{2,**}$ denotes the first time there are 2 new permanent sites, etc. The argument then proceeds as in the deterministic case.

This concludes the proof for $d \geq 0$. q.e.d.

Corollary 12.7 (No mass on lower types: general case).

(a) The relation (12.39) can be generalized if we have M types in E_0 and M-types in E_1 to

$$\lim_{N\to\infty} \sup_{0\leq t\leq T} P[y^N(1,t)(E_0) > \varepsilon] = 0. \tag{12.58}$$

(b) Define

$$\bar{Y}_{E_0}^N(t) = \sum_{k\in E_0} \bar{y}_k^N(t). \tag{12.59}$$

Then we have

$$\mathcal{L}[(\bar{Y}_{E_0}^N(t))_{t\geq 0}] \underset{N\to\infty}{\Longrightarrow} \delta_{\underline{0}} \text{ in the Meyer-Zheng topology.} \qquad \square \tag{12.60}$$

Proof of Corollary 12.7. We now have to argue that this argument also works for $|E_0| = |E_1| = M$. There is no effect from having M types in E_1, since the internal structure of the higher level did not enter in the argument. What is needed is to check M types in E_0.

Here the new point is that types in E_1 have selective advantage $O(1)$ over all those in E_0 except the maximal one in E_0, that is type M. We can however make a comparison to a system where all lower types have the second highest fitness level and all types mutate to this maximal type in E_0 with the maximal rate. For this system we can use the result for the resulting effective system with $M = 2$ types since we can efficiently merge the types $1, \cdots, M-1$ into one. Our real system has by coupling of the duals less mass on E_0. q.e.d.

Step 2: Tightness

Next we want to show that the sequence of $\mathcal{P}(\mathcal{P}(\mathbb{K}))$-valued processes $\{y^N(t)\}$ is tight in the time scale Nt and that a limiting process is given by the solution of the nonlinear selection-mutation martingale problem now on the higher level of fitness, i.e. types in E_1. The tightness can be verified using the criterion of Meyer-Zheng [MZ], Theorem 4, adapted to our case as discussed in Subsubsection 12.1.3.

To do this we need to compute

$$V_\tau(y^N) = \sum_{0 \leq i < n} E[\|E[y^N(t_{i+1}) - y^N(t_i)|\mathcal{F}_{t_i}]\|] \quad (12.61)$$

$$= \sum_{0 \leq i < n} \sum_{k \in E_1} |E[y_k^N(t_{i+1}) - y_k^N(t_i)|\mathcal{F}_{t_i}]|$$

We first observe that from the martingale problem for the process we have:

$$y_j^N(t_{i+1}) - y_j^N(t_i)$$
$$= \int_{t_i}^{t_{i+1}} (f_{j,1}(u) + Nsf_{j,2}(u)y_{E_0 \setminus \{M\}}^N(s))du + (M_j(t_{i+1}) - M_j(t_i)), \quad (12.62)$$
$$\text{if } j \in E_1$$

$$y_{E_0}^N(t_{i+1}) - y_{E_0}^N(t_i)$$
$$= \int_{t_i}^{t_{i+1}} (f_1(u) - Nsf_2(u)y_{E_0 \setminus M}^N(u))du + (M_j(t_{i+1}) - M_j(t_i)), \quad \text{if } M \geq 2,$$
$$(12.63)$$

where $(M_j(t))_{t \geq 0}$ are martingales and $\sup_{0 \leq s \leq T}(|f_1(s)| + |f_2(s)|) < \infty$, $f_1(s)$, $f_2(s) \geq 0$. Taking expectations we obtain

$$|E[(y_{E_0}^N(t_{i+1}) - y_{E_0}^N(t_i))]|\mathcal{F}_{t_i}]| \leq K(t_{i+1} - t_i) \quad (12.64)$$

for some constant K independent of N. Therefore $V(y^N) \leq KT$, completing the proof.

Step 3: Identification of the limiting dynamics

We establish first the convergence of the $\{\mathcal{Y}^N\}$ to a limiting dynamic, second convergence of a tagged N-block (i.e. ball of radius 1) $\{y^N(j;t)_{t \geq 0}\}$ to the selection-mutation process on types in E_1 (instead of E_0 as in phase 0), and third convergence to the mutation-selection mean-field equilibrium of this system in the time scale $t_N N$ as $N \to \infty$ with $t_N \uparrow \infty$, $t_N = o(\log N)$.

12.2 Proof of Proposition 12.2

The key point here is to see that the averaging over N-blocks and the change of the time scale to Nt together results in a *new system* which is asymptotically as $N \to \infty$ a selection, mutation, migration and resampling dynamics as were the single sites in scale $O(1)$ before, only that now we are on the level of the N-blockaverages not single sites and because of the new time scale tN we have to choose the appropriate *new rates* for migration, resampling, selection and mutation.

In order to verify this picture we proceed in two steps, establishing first the convergence needed for (12.29) and (12.30) *suppressing mutation from E_1 to E_0* and second incorporating the downward mutation.

We focus first on the convergence of the dynamics over a fixed time horizon and use the dual representation for the process to identify the moment measures of the empirical measure process. We must show that asymptotically the k-th moment measure converges to the one of the nonlinear process with mutation, selection and resampling as given in (12.11) and (12.12). This is done by showing in Step 1 below that the dual system in time scale Nt converges to the dual of the nonlinear McKean–Vlasov process with parameters chosen as claimed in (12.12).

In order to show the convergence of empirical moment measures we consider the dual process in the Nt time scale and show that we obtain in the limit $N \to \infty$ a system of the same type, but now with sites in distance 1 identified and mutation and selection events only taking place on E_1 with χ_1 as fitness function and $m_{1,1}$ as mutation matrix.

First of all we modify our system by removing mutations $E_1 \to E_0$. Consider therefore the dual process starting in functions *supported by E_1*. We then observe first that all selection and mutation events of order $O(1)$ in the new time scale do not change the form of the dual $\mathcal{F}^{++}$. This means that we can replace the selection and mutation transition in the macroscopic time scale by a new transition of this kind. The selection and mutation operators are the same as before but are now taken w.r.t. fitness function χ_1 and mutation matrix $m_{1,1}$. In other words we have to show that the original dual modified by removing the $O(N^{-1})$ rates in the new time scale is the one just described above as $N \to \infty$. This means we can now forget about the $O(1)$ rates in the dual.

For that purpose we have to analyse the particle system and then the function-valued evolution. We begin by looking at the evolution of the particle process and the neutral part of the dynamic, i.e. consider first the mechanism of migration and coalescence. In the limiting dynamic of the dual in time scale Nt migration at distance 1 becomes infinitely fast as $N \to \infty$ and coalescence of particles at the same site instantaneous if they don't jump first and therefore we *locate the particles not by sites but by balls of radius 1* - with the convention that if there are k particles in a ball they are internally located at random sites within the ball according to the uniform distribution.

Particles will migrate between balls at rate c_1 and two particles that are in the same ball coalesce at a rate d_1 depending on c_0 and d_0 (proportional to $c_0/(c_0 + d_0)$) based on *instantaneous* coalescence at a site and the fact that two particles in a ball will meet in time scale Nt at the same site at a certain rate of order $O(1)$ with (asymptotically as $N \to \infty$) exponential times between collisions. This can be found in detail in [DGV], Corollary 2.1, [DG99], Theorem 2.) Therefore the coalescence between particles in the same ball will occur with *effective rate d_1* as $N \to \infty$ where

$$d_1 = c_0 \left(\frac{d_0}{c_0 + d_0} \right) = d_0 \left(\frac{c_0}{c_0 + d_0} \right). \qquad (12.65)$$

To verify this note that the rate at which two particles meet (in time scale Nt) at a site (i.e. one particle jumps on the site occupied by the other particle) due to migration is $N \frac{c_0}{N} = c_0$. Therefore asymptotically, the particles instantaneously undergo coalescence with probability $\frac{d_0}{d_0 + c_0}$ or migration with probability $\frac{c_0}{d_0 + c_0}$.

Now we come to the transitions in the dual due to selection and mutation events. First consider selection and in particular the birth of particles. The effect of the selection operator is the identity if a newborn particle at a site coalesces with its ancestor since then $\chi(u_i) + (1 - \chi(u_{n+1})) \longrightarrow 1$. This reduces the effective birthrate exactly at the rate at which distance 1 migration does not occur before coalescence, i.e.

$$s_1 = s_0(c_0/(c_0 + d_0)). \qquad (12.66)$$

Therefore we can now work with the *effective birth rate* above. This leads altogether to a correction of birth rates corresponding to selection and coalescing rates corresponding to resampling rates with exactly the same factor.

Concerning mutation jumps in the dual, we observe that mutations for types from E_1 occur at rates of order N^{-1} and hence the change of time scale turns them into rates constant in macroscopic time of the form as claimed. Other mutation events can only occur after a $E_1 \to E_0$ downward mutation from E_0 types. But we have proved in Step 1 that such types do not persist in the macroscopic time scale.

To this asymptotically equivalent dual system we apply now the argument leading to the McKean–Vlasov limit and we get the claim on the level of f.d.d. convergence (and convergence of the transition measures) immediately if we simply *suppress* downward mutation from level E_1 to level E_0. However we have shown in Step 1 that the types in E_0 play no role, so we are done.

12.3 Convergence in the Uniform Topology

It is natural to ask if convergence to zero of the E_0-mass actually occurs in the uniform topology. We will show that uniform convergence to zero *does not occur* for the evolution at *fixed sites* over the longtime interval $[t_N, NT]$ with $t_N \to \infty$ but $t_N = o(N)$ in Subsubsection 12.3.2 but *does occur* for the blockaverages, that is, the averages over balls of radius 1 in Subsubsection 12.3.1. To do this we consider the processes $x_{E_0}^N(j, \cdot), y_{E_0}^N(\cdot)$ given by the masses of types in E_0 in a single component j, respectively, blockaverage and we start with all the mass on the type E_1. We shall

12.3.1 Microscopic Viewpoint: Uniform Convergence of Blockaverages

We first sketch the idea behind the proof that $y_{E_0}^N(\cdot)$ converges to 0 in the uniform topology. Consider the excursions of height $\geq \varepsilon > 0$ occurring successive intervals $[k, k+1)$. Then since at a given site j there are a Poisson number of excursions x_k with mean proportional to $\frac{m_{3,\text{down}} Q([\varepsilon,1])}{N}$. (Note that by subcriticality each such excursion has length with exponential moment.) Consider one such interval. Then the normalized sum over a typical interval k satisfies,

$$\frac{X_k(\varepsilon)}{N}, \tag{12.67}$$

where X_k has an exponential moment.

Using a large deviation argument we then get that the max of these over the N intervals is no larger than $O(\frac{\log N}{N})$. This implies that there exists $0 < C(\varepsilon) < \infty$ such that the contribution of the type E_0 excursions larger than or equal to ε denoted $y_{E_0,\varepsilon}^N$ is given by

$$\lim_{N\to\infty} P[\sup_{0\leq t\leq T} y_{E_0,\varepsilon}^N(t) > \frac{C(\varepsilon)\log N}{N}] = 0. \tag{12.68}$$

In fact below we include all E_0-excursions and obtain a more complete picture of the total type E_0 mass in the N sites in the following proposition. To do this we introduce as in the supercritical case a process $\mathbf{J}_{E_0}^N(t)$ with values in the set of atomic measures on $[0,1] \otimes \{1,\ldots,M\}$, denoted here $M_a([0,1] \otimes \{1,\ldots,M\})$. The atom sizes are given by the E_0-excursions and their locations in $[0,1]$ are i.i.d. $U(0,1)$. Then

$$\hat{x}_{E_0}^N(t) = \mathbf{J}_{E_0}^N(t, [0,1] \otimes \{1,\ldots,M\}). \tag{12.69}$$

Proposition 12.8 (Uniform convergence for blockaverages). *Start the system with all the mass concentrated on E_1. Let*

$$\hat{x}_{E_0}^N(t) := \sum_{i=1}^{N}\sum_{\ell \in E_0} x_\ell^N(i,t),\ 0 \leq t \leq NT, \quad y_{E_0}^N(t) = \frac{1}{N}\hat{x}_{E_0}^N(Nt),\ 0 \leq t \leq T. \tag{12.70}$$

Then the following conclusions hold.

(a) As $N \to \infty$,

$$\mathcal{L}[\{\hat{x}^N_{E_0}(t)\}_{t \geq 0}] \Longrightarrow \mathcal{L}[\{\hat{x}^\infty_{E_0}(t)\}_{t \geq 0}] \text{ and } \mathcal{L}[\{\mathfrak{I}^N_{E_0}(t)\}_{t \geq 0}] \Longrightarrow \mathcal{L}[\{\mathfrak{I}_{E_0}(t)\}_{t \geq 0}]. \quad (12.71)$$

(b) If $\hat{x}^\infty_{E_0}(0) = 0$, then as $t \to \infty$

$$\mathcal{L}[\hat{x}^\infty_{E_0}(t)] \Longrightarrow \mathcal{L}[\hat{x}^\infty_{E_0, \text{eq}}], \quad \mathcal{L}[x^\infty_{E_0}(t)] \leq \mathcal{L}[\hat{x}^\infty_{E_0, \text{eq}}]. \quad (12.72)$$

(c) There exists $C \in (0, \infty)$ such that

$$\lim_{N \to \infty} P[\sup_{0 \leq t \leq NT} \hat{x}^\infty_{E_0}(t) > C \log N] = 0. \quad (12.73)$$

(d) The blockaverages in time scale NT, $y^N_1(\cdot)$, converge as $N \to \infty$ to 0 in the uniform topology, that is, there exists $C \in (0, \infty)$ such that

$$\lim_{N \to \infty} P[\sup_{0 \leq t \leq T} y^N_{E_0}(t) > \frac{C \log N}{N}] = 0. \quad \square \quad (12.74)$$

Remark 232. Essentially the same proof can be used to extend (d) in the above result to time intervals of length Tt_N where $t_N \to \infty$, $t_N = o(N)$, namely, there exists $C \in (0, \infty)$ such that

$$\lim_{N \to \infty} P(\sup_{0 \leq t \leq N t_N T} \hat{x}^\infty_{E_0}(t) > C \log N) = 0. \quad (12.75)$$

Proof of Proposition 12.8. The key idea is that as $N \to \infty$ we can consider the total lower order mass, $\hat{x}_{E_0}(t)$, produced by the downward mutation and show that it converges to a stationary process and then consider the maximum of this process over the interval $[0, NT]$.

Let $\hat{x}^N_{E_0}(t) := \sum_{i=1}^N \sum_{j \in E_0} x_j(i, t) = \mathfrak{I}(t, [0, 1])$. Note that by (12.56) that

$$\lim_{N \to \infty} E[\hat{x}^N_{E_0}(t)] \leq \text{const} \cdot \tilde{\kappa}. \quad (12.76)$$

We can verify the weak convergence of $\{\mathfrak{I}^N_{E_0}(t)\}_{t \geq 0}$ to a process $\{\mathfrak{I}(t)\}_{t \geq 0}$. The definition of $\mathfrak{I}(t)$ is analogous to (10.69) but with types given by E_0. The proof follows along the same lines as the proof of Proposition 2.9. Tightness follows using Joffe-Métivier and the joint moments of the finite dimensional distributions follows using the dual representation.

Remark 233. For fixed N the process $\mathfrak{I}^N_{E_0}(t)$ converges as $t \to \infty$ to the zero measure because with finite N the system (in the absence of upward mutation) becomes extinct (but the mean extinction time goes superexponentially to ∞ as $N \to \infty$). Of course if we include upward mutation this does not occur.

12.3 Convergence in the Uniform Topology

Next, note that

$$d\hat{x}_{E_0}(t) = [c_1(E(\hat{x}_{E_0}(t)) - \hat{x}_{E_0}(t)) - s(\sum_i x_1(i)(1 - x_{E_0}(i)) + m_{3,\text{down}}]dt$$

$$+ \sqrt{\hat{x}_{E_0} - \sum_i x_{E_0}^2(i)}\, dw(t). \qquad (12.77)$$

Let $X_k := \sup_{k \leq t < k+1} \hat{x}^\infty(t)$. We obtain upper bounds for the moments of X_k using the Burkholder-Davis-Gundy inequality $c_p E([M]_t^{p/2}) \leq E((M_t^*)^p) \leq C_p E([M]_t^{p/2})$ or by comparison with Feller branching and then verifying that it has a finite exponential moment. Therefore

$$P(X_k > x) \leq \text{const} \cdot e^{-\lambda x} \qquad (12.78)$$

for large x for some $\lambda > 0$.

The existence of a limiting stationary measure for $\hat{x}_{E_0}(t)$ as $t \to \infty$ follows since we have a Markov process with tight marginals and we can also verify that there is a unique limit $\hat{x}_{E_0,\text{eq}}$ point by coupling.

We then obtain the corresponding stationary process X_k. Recall that a stationary process is called strongly mixing if: for $A \in \mathcal{B}(X_1, \ldots, X_m)$ and $B \in \mathcal{B}(X_{m+k}, X_{m+k+1}, \ldots)$, we have

$$|P(A \cap B) - P(A)P(B)| \leq g(k) \text{ and } g(k) \to 0 \text{ as } k \to \infty. \qquad (12.79)$$

To verify that $\{X_k\}$ is strong mixing we use the fact that by subcriticality the probability that all excursions starting in $[0, t_1]$ have become extinct by time t_2 goes to 1 as $t_2 \to \infty$. Let $1 - g(k)$ denote the probability of the event G_k that all mutant families generated in $(-\infty, t_1]$ have become extinct by time $t_1 + k$. By extinction, $g(k) \to 0$ as $k \to \infty$. Then if $A \in \mathcal{F}_{t_1}$ and $B \in \mathcal{F}_{t_1+k}$, then

$$P(A \cap B) = P(A \cap B|G_k)P(G_k) + P(A \cap B \cap G_k^c) \qquad (12.80)$$

$$= P(A|G_k)P(B|G_k)P(G_k) + P(A \cap B \cap G_k^c)$$

$$P(A)P(B) = (P(A|G_k)P(G_k) + P(A \cap G_k^c))(P(B|G_k)P(G_k) + P(B \cap G_k^c))$$

$$= P(A|G_k)P(B|G_k)(P(G_k))^2$$

$$+ P(B \cap G_k)P(A \cap G_k^c) + P(B \cap G_k)P(A \cap G_k^c) + P(A \cap G_k^c)P(B \cap G_k^c).$$

Therefore

$$|P(A \cap B) - P(A)P(B)| \leq P(A|G_k)P(B|G_k)(P(G_k) - P(G_k)^2) + \text{const} \cdot g_k \leq \text{const} \cdot g_k. \tag{12.81}$$

Note that (12.78) implies that

$$P\left(X_k > \frac{\log N - \log x}{\lambda}\right) \leq \frac{x}{N}. \tag{12.82}$$

Therefore applying the results of (cf. [LO] Loynes (1965), Theorem 2) to $\{X_k\}_{k \in \mathbb{Z}}$ we obtain

$$\lim_{N \to \infty} P\left(\max_{1 \leq k \leq N} X_k \leq \frac{\log N - \log x}{\lambda}\right) = e^{-\gamma x} \tag{12.83}$$

for some $0 < \gamma \leq 1$. This implies that for $C > \frac{1}{\lambda}$

$$\lim_{N \to \infty} P(\sup_{0 \leq t \leq NT} \hat{x}_{E_0}^{\infty}(t) > C \log N) = \lim_{N \to \infty} P(\sup_{1 \leq k \leq N} X_k > C \log N) = 0. \tag{12.84}$$

(d) follows immediately from (c).   q.e.d.

12.3.2 The Microscopic Viewpoint: Tagged Sites

In this subsection we prove that convergence in the uniform topology does not occur at the single site level if we start with all mass on the fitter types in the case $M = 1$, i.e. types 1 and 2. The general case follows by a similar argument.

We start with the mass of the lower level types at a fixed site j:

$$dx_1^N(j,t) = (c(z^N(t) - x_1^N(j,t))dt - s(x_1^N(j,t) - x_1^N(j,t)^2)dt \tag{12.85}$$
$$+ d\sqrt{x_1^N(j,t)(1 - x_1^N(j,t))}dw_j(t) + a_N m_{3,\text{down}}(1 - x_1^N(j,t))dt,$$

$$z^N(t) = \bar{x}_1^N(t) = \frac{1}{N}\sum_{\ell=1}^{N} x_1^N(\ell, t), \quad \bar{x}_1^N(0) = 0, \quad a_N = \frac{1}{N}, \tag{12.86}$$

on the time interval $[0, NT]$ with $T > 0$.

Proposition 12.9 (Non-convergence in uniform topology for components). Given $0 < \varepsilon < 1, T > 0$,

$$\liminf_{N \to \infty} P[\sup_{0 \leq s \leq NT} x_1^N(s) > \varepsilon] > 0, \tag{12.87}$$

that is, $x_1^N(\cdot)$ does not converge to 0 in the uniform topology.   □

Proof of Proposition 12.9. We consider the intervals $[k, k+1)$, $0 \leq k \leq \lfloor NT \rfloor$ and set $z(t) \equiv 0$ to get a lower bound for x_1^N in (12.85). Note that there are $\lfloor NT \rfloor$ such intervals and there is a probability of order $O(\frac{1}{N})$ that a E_0-type excursion of $\mathbf{J}_t^{m_{3,\text{down}}}$ of height $\geq \varepsilon$ will begin in such an interval. Hence in the limit as $N \to \infty$ there is a Poisson number of such excursions and we *cannot* get uniform convergence to zero in probability for the single component $x_1^N(j, t)$.   q.e.d.

12.4 Proofs of Propositions 12.3 and 12.4

Proof of Proposition 12.3:. As in Section 11 the result follows from three facts. First since $\mathcal{Y}^\infty(t)$ is a nonlinear mean-field process the ergodic theorem for it follows from Theorem 14. Secondly we have the weak convergence as $N \to \infty$ of $\mathcal{Y}^N$ to $\mathcal{Y}^\infty$ for fixed macroscopic time t *uniformly* in the initial state, see [DG99]. (This is an immediate consequence of the convergence of the dual process to a limiting dual involving for finite time horizon a finite particle system.) Thirdly we observe that due to the state-independent mutation component on E_1 the ergodic theorem for $\mathcal{Y}^\infty$ holds *uniform* in the initial state. Combining these observations as in the proof of Proposition 11.1 (see in particular Step 3) gives the claim.   q.e.d.

Proof of Proposition 12.4:. (1) We treat first the distance 1 blockaverages. Here the main work has been done above in the proof of Proposition 12.2. Therefore we only need to verify propagation of chaos. This means that we have to show that mixed moments for the collection $\{y^N(j,t), \quad j = 1, \cdots, N\}$ given a state $s < t$ factorize in moments for the different j, since then the transition kernels are asymptotically the product of transition kernels for the different j and for each factor we have already identified the limit. The same type of argument will give that the mixed moments at time $t_N \uparrow \infty, t_N = o(N)$ factorize, which will take care of the initial state and its product form. Verifying this factorization is by using the dual process.

The key fact about the dual we need is the following. One first shows that particles initially at distance 2 do not meet in times of order N, this is the same argument as for distance 1 in times $O(1)$. Particles initially at distance one meet in time scale Nt in the limit after an exponential waiting time and then coalesce at rate $(d_0/c_0 + d_0)$. Selection and mutation on E_1 occur in time scale Nt at exponential scales in times of order 1. Hence it remains to consider the downward mutations occurring at rate $O(1)$ and to verify that they are eliminated by the selection operation. But this has already been demonstrated in (12.50).

(2) Next turn to the single components in the small time scale $O(1)$. Note that here we consider for the components a small time scale (at a very late time) so that the downward mutations do not pose any problem since their rates are of order N^{-1}. Therefore the program is that we show the propagation of chaos and relaxation of a single component using duality between the state at time Nt and $Nt + u$ and conditioning on a value of the empirical mean process at time Nt for some $t \in$

$(0, \infty)$. The process in the u-scale is in equilibrium which is proven with a restart argument.

This piece of the program does not require too much *new* work since we can use existing results. First, we have shown above that effectively the dynamics take place on types E_1. In addition selection and mutation have rates of order N^{-1} and act therefore only in time scales $O(N)$ on E_1. Therefore by a standard argument using generator convergence on test functions we are left in time scale $O(1)$ with a *neutral* model. However for the neutral model all the work for the program outlined above was done in [DGV], namely, establishing the socalled *mean-field finite system scheme*. The key idea there was to use certain coupling arguments to get the uniformity in the estimates. We refer the reader for the details of this step to the references mentioned above.   q.e.d.

Chapter 13
Iteration of the Cycle I: Emergence and Fixation on E_2

We have now completed the analysis of the transition from $E_0 \to E_1$ and the equilibration on level E_1 in its various phases 0–4. The question is now if we can iterate and continue now with the transition from E_1 to E_2 with now two or more spatial levels. This problem we address in the present section. Later on we have to pass to the general $E_{j-1} \to E_j$ transition. We begin with an overview where we stand after Section 7–12 and how we plan to complete the program before we come to $E_1 \to E_2$ transition.

13.1 How to Proceed to Higher Levels

In the last sections (namely Sections 8–12) we established starting from a population of E_0-types tht they form an (quasi)-equilibrium on that level and at the same time we see the forming of droplets which expand and lead to the emergence of the types in E_1 in time scale $\beta^{-1} \log N$ as $N \to \infty$, then follows the fixation dynamic (in t) on these higher level types in times $\beta^{-1} \log N + t$ as $N \to \infty$, next the development of a neutral equilibrium for times $\beta^{-1} \log N + t_N$ where $t_N \to \infty$, $\lim_{N \to \infty} \frac{t_N}{N} = 0$ and finally the development of a mutation-selection equilibrium concentrated on E_1-types by replacing components by blockaverages over N components and observing in time scale Nt as $N \to \infty$ and $t \to \infty$. For the latter we already considered two levels of spatial organisation. In other words we verified our *complete scenario* for two levels of fitness and two levels of spatial organization in a *finite* system. This part of the analysis requires most of the new mathematical techniques. From now on we can use more results and techniques already in the literature.

Our task is now to increase the number of fitness levels and the number of levels of spatial organization to finally treat the full model we consider in Section 3 and 4. It is necessary to proceed in three main steps. In this section we continue by first analysing systems with a *finite* number of fitness levels and spatial organisation (hence systems involving finitely many interacting sites) to obtain the tool to analyse the infinite hierarchy and therefore infinitely many interacting sites model in the $N \to \infty$ limit.

We start by demonstrating in this Section 13, that we can extend the previous analysis here for a system with two levels of spatial organization and three levels of fitness which requires an additional step we explain at the end of this introduction. Namely we show for $N \to \infty$ the emergence of types in E_2 in time scale $\beta^{-1}(N \log N)$, then show that the process of fixation (time parameter t) occurs in times $(\beta^{-1} N \log N) + tN$, neutral equilibrium on E_2 forms at time $\beta^{-1} N \log N + t_N$ (with $t_N >> N$ but $t_N << N^2$) and finally mutation-selection equilibrium on E_2 in spatial scale given by blockaverages over N^2-many sites and in time scale $N^2 t$ occurs as both $N \to \infty$ and $t \to \infty$, where for the latter we then need to treat three levels of spatial organisation and hence things follow the same pattern as earlier.

We now continue the program in two sections. In Section 13 we carry out this step described above from two to three levels of fitness and from two to three levels of spatial organisations. Then later in Section 14 we continue and we explain in Subsection 14.1 how this can be extended to any finite number, j of fitness levels and $(j - 1)$-levels of spatial organisation. Finally in Subsection 14.2 we extend this analysis of the transition of a population on E_{j-1}-types to one on E_j-types in the corresponding time scales in a system of finitely many levels to a system with *infinitely* many levels of fitness and a system with *infinitely* many spatial components.

Finally we can use these results in a final step to analyse the system with the infinite hierarchy both on the level of fitness and on the level of spatial organisation, i.e. the original spatially infinite model on the hierarchical group with a whole sequence of fitness levels, which then gives us all we need to prove in Section 15 our Theorems 3–5, 6 and 7.

In this section we will again use the duality relation with the refined dual to analyse the behaviour of the process in various time scales. For the proofs we can follow closely the strategy developed in [DGV] for the neutral case and [DG99] in the case with selection. Namely compared to the case of two fitness level, essential new features arise in the case of three fitness levels only through *lower order perturbations* arising - if we pass to blockaverages and time scales Nt - from rates of transitions between sites respectively on smaller time scales. But in contrast to the references mentioned above they occur not only because of migration and selection but also due to mutation. Therefore throughout this section we can rely on modifications of techniques which were developed in [DGV] and [DG99]. What must be added to the analysis of Sections 8–12 and the quoted literature are new methods to handle the effects of the presence of *downward mutation*. Therefore we carefully state all facts needed and then only outline how to adapt existing techniques to the models under consideration here to take care of the downward mutation.

13.2 The Case of Two Levels of Spatial Organisation and Three Fitness Levels

In the last section we considered two levels of fitness denoted E_0 and E_1 and the phases leading to the transition from E_0-valued states to E_1-valued states and we now need to describe the emergence of a third level of fitness namely the types in E_2. We considered in the previous section the mean-field system in order to get approximations which are valid in certain time scales namely $O(1), O(\log N), O(\log N + t), T_N$ with $\log N \ll T_N \ll N, O(N)$. In the last time scale we equilibrated on E_1-types. For that we already considered two levels of spatial organisation. Now we need to see when the emergence of types in the third level E_2 occurs, which is expected in times of order $N \log N$.

In order to analyse the behaviour in time scales of order $N \log N$ we cannot use the mean-field system and get good approximations of $(X^N(t))_{t \geq 0}$ for $N \to \infty$. We have to embed our mean-field system in a system with two spatial levels since we need to carry out an analysis of rare mutants appearing in a "level-E_2-fitness McKean–Vlasov" limit, exhibiting the equilibration into the level-E_2-fitness selection-mutation equilibria of the level-E_2 nonlinear Markov process. In the time scale where this happens the system feels the migration on the hierarchical group at distance two.

Therefore we consider again as in the second part of Section 12 an array of N of the size N mean-field models and again we allow a weak interaction between these size N systems. This requires then to argue that this does not perturb the $E_0 \to E_1$ transition. Later when we discuss the time scale $N^2 t$ where we expect equilibration on E_2 we will then need even three levels of spatial organisation, a change we already needed to perform passing from Sections 8–10 to Section 12.

The key idea of this section is twofold. (1) First to study the transition from E_0 to E_1 and show that we can ignore all transitions of order $O(\frac{1}{N^2})$ and apply the previous analysis to each of the N different 1-balls of N sites each. (2) Then for the second step to study the transition from E_1 to E_2 we observe that we can introduce the time scale Nt and in space replace components by blockaverages over N sites and then if we ignore types in E_0, study the transition from E_1 to E_2 by the same techniques we used to study the transition from E_0 to E_1 with the additional problem to handle *lower order perturbations* since only in the limit $N \to \infty$ we obtain the dynamic we had before already exactly for finite N.

13.2.1 The (3,2)-Model and Outline

We now let (three level of types, two levels of spatial organisation)

$$(X_t^{N,(3,2)})_{t \geq 0}, \tag{13.1}$$

be a system with N^2 colonies grouped in N blocks of N components and only E_0, E_1, E_2 as levels of fitness. We obtain this system formally if we set in the definition of the system $(X^N(t))_{t \geq 0}$:

$$c_2 = c_3 = \cdots = 0, \quad \chi_N : \chi_N(k, \cdot) \equiv 0 \text{ for } k \geq 3, \\ M : m_k, m_k^+, m_k^- = 0 \quad k \geq 3, \tag{13.2}$$

where

$$\chi_N(k, \cdot) \text{ as in } (3.29), k = 0, 1, 2 \tag{13.3}$$

and then *project* the obtained system with values in $(\mathcal{P}(\mathbb{I}))^{\Omega_N}$, with $\mathbb{I} = \bigcup_{i=0}^{\infty} E_i$ onto one indexed by $B_2(0)$, the sites in hierarchical distance at most 2 from zero (i.e. the 2-ball around 0) and furthermore *project* the state in each of the remaining sites from a measure on $\bigcup_0^{\infty} E_i$ to a measure on $\mathbb{K} = E_0 \cup E_1 \cup E_2$.

Convention on notation. *We distinguished in Section 12 the map mapping $E_1 \cup E_0 \to E_0$ and the induced maps on $\mathcal{P}(E_1 \cup E_0)$ and $\mathcal{P}(\mathcal{P}(E_0 \cup E_1))$ by writing $\pi_1, \tilde{\pi}_1$ and $\tilde{\tilde{\pi}}_1$. Here we will drop this and write a bit more informally just one symbol.* □

We always use as initial state an i.i.d. state, but here one with:

$$X_0^{N,(3.2)} \text{ concentrated on } E_0. \tag{13.4}$$

Then we define (only the type space is now compared to the objects in Section 11 and 12 different, mainly $E_0 \cup E_1 \cup E_2$ instead of $E_0 \cup E_1$) the objects

$$y_k^N(j, t), \quad \mathcal{Y}^N(t), \quad \bar{y}_k^N(t), \tag{13.5}$$

as in (12.5)–(12.7) and in addition we introduce the analog of (9.7), ($\bar{\mathcal{P}}$ denotes here subprobability measures)

$$\Xi_N^{\log, \beta} = \frac{1}{N} \left(\sum_{i=1}^N \delta_{(y_k^N(i, \frac{N \log N}{\beta} + \cdot); k \in E_1 \cup E_2)} \right) \in \mathcal{P}(\bar{\mathcal{P}}(E_1 \cup E_2)), \tag{13.6}$$

where we have made three changes
(1) for the larger time scale $\beta^{-1} N \log N$ instead of $\beta^{-1} \log N$ and
(2) replacing components by blockaverages of the N sites at distance 1,
(3) types in $E_1 \cup E_2$ instead of E_1.

Outline Subsection 13.2. First we have to argue that the transition $E_0 \to E_1$ takes place in $X^{N,(3.2)}$ the same way as before in the mean-field system by dropping all terms of N^{-2} in the generator, this is in Subsubsection 13.2.2. Next we have to carry out the asymptotic analysis of the five phases again for the transition $E_1 \to E_2$, now starting with a phase 0 involving times of order larger than tN but less than $N \log N$ up to the last phase 4 taking place in times of order tN^2 and $t_N N^2$ with $t_N = o(\log N)$. Each phase gets its own subsubsection.

13.2.2 A Two-Level System with Three Fitness Levels: Time Scale $O(N)$

We deal with the phases 0–4 of the transition from E_0 to E_1 in the new system and then with the *phase 0* of the transition $E_1 \to E_2$ here in two steps. Hence we first have to show that for times and time scales which are $o(N \log N)$ we get exactly the results we had in Section 7, 8, 9, 10, 11 and 12 for each of the N balls of radius 1 evolving independently.

We therefore state that the systems in the N balls of radius 1 become independent and that the projection of the states onto types of $E_0 \cup E_1$ and restriction to the N sites which are in distance 1 to the "origin" $(0,0)$ converges as $N \to \infty$ to the mean-field system on $E_0 \cup E_1$. Denote this map of restricting the system on the levels E_0, E_1 of types and spatially restricting to just one N-block by

$$\pi_{2,1}^{3,2} : \quad (\mathcal{P}(E_0 \cup E_1 \cup E_2))^{\{1,2,\cdots,N\}^2} \longrightarrow (\bar{\mathcal{P}}(E_0 \cup E_1))^{\{1,\cdots,N\}}. \tag{13.7}$$

Let $X^{N,(3,2)}$ be the system with three levels of fitness and two levels of spatial organisation and correspondingly $X^{N,(2,1)}$ be the mean-field system with types in $E_0 \cup E_1$. Consider on paths the topology induced by the seminorms

$$\|f\|_{a,b} = \sup_{a \leq x \leq b} |f(x)|, \tag{13.8}$$

where we choose $a = 0$ and $b = \infty$ for our purposes of a rather strong statement.

Then we can conclude if we continue path beyond t_N as *constant and equal to the t_N value*:

Proposition 13.1 (Higher order terms are negligible, 1-blocks independent). Assume $t_N = o(N \log N)$. Then:

$$(\{x_t^N(i,j), \quad i = 1, \cdots, N\})_{t \leq t_N}, \quad j = 1, \cdots, L, \tag{13.9}$$

are asymptotically as $N \to \infty$ independent for every fixed $L \in \mathbb{N}$. Furthermore in the weak topology:

$$(\mathcal{L}[(\pi_{2,1}^{3,2} \circ X_t^{N,(3,2)})_{t \leq t_N}] - \mathcal{L}[(X_t^{N,(2,1)})_{t \leq t_N}]) \underset{N \to \infty}{\Longrightarrow} 0. \quad \square \tag{13.10}$$

From the statement in Proposition 13.1 we can conclude that in time scales $o(N \log N)$ and for components we can simply apply the results we obtained in Section 8, 9 and 10 on the one hand and Section 11, 12 on the other hand, i.e. the complete scenario of phase 0 to phase 4 in the transition from E_0 to E_1 with all its features follows for $X^{N,(3,2)}$ from that for $X^{N,(2,1)}$.

Proof of Proposition 13.1. We can neglect in the time scale $o(N^2)$ the higher order mutation, selection and migration terms (i.e. term with N^{-2}) occurring in the model. This means that we can neglect jumps in distance two and mutations to level E_2 entirely as well as the selection transitions corresponding to the fitness levels of types in E_2 versus those in $E_0 \cup E_1$ ending up with N independent dynamics on $E_0 \cup E_1$ in the N different N-balls of radius 1.

This can be proved using standard methods namely using a comparison-duality argument. Here we compare the process with respectively without the $O(N^{-2})$-terms. To do this we simply compare the two dual processes and see they agree asymptotically running them over a time stretch $[0, t_N]$ with $t_N = o(N^2)$ since the probability that any transition occurs in $[0, t_N]$ corresponding to a rate associated with a term containing N^{-2} tends to zero. See [DGV] for such type of arguments, we omit here further details. q.e.d.

13.2.3 A Two-Level System with Three Fitness Levels in Time Scale $o(N \log N)$ (Phase 0)

To turn phase 4 of the transition $E_0 \to E_1$ into phase 0 for the transition $E_1 \to E_2$ we need one more step. We want to show that if we take blockaverages over the balls of radius 1 and scale time to Nt then we obtain asymptotically in macroscopic time t a mean-field system of N-components, only with changed type space $E_1 \cup E_2$ instead of $E_0 \cup E_1$ and changed parameters for migration, resampling and selection.

Proposition 13.2 (Renormalised system). *Define the following two processes*

$$Y^N(t) = (y^N(j,t))_{j=1,\cdots,N}, \ (recall \ (12.5) \ for \ y^N(j,t)), \qquad (13.11)$$

$X^N(t)$ is given in (7.1) with c_0 and d replaced by c_1 and d_1 and $\mathbb{K}$ by $E_1 \cup E_2$.
$\qquad (13.12)$

We get the following.

$$(\mathcal{L}[(Y^N(t))_{t>0}] - \mathcal{L}[(X^N(t))_{t>0}]) \underset{N \to \infty}{\Longrightarrow} 0 - measure. \qquad (13.13)$$

Furthermore

$$\lim_{t \to \infty} \lim_{N \to \infty} \mathcal{L}[Y^N(t)] = \bigotimes_{\mathbb{N}} \Gamma_{M_1, \chi_1}^{c_1, d_1, m_1, s_1}. \qquad \square \qquad (13.14)$$

Remark 234. The convergence can be strengthened. Consider $(\tilde{Y}^N(t))_{t>0}$ which equals $Y^N(t)$ for $t \leq t_N$ and $\tilde{Y}^N(t) = \tilde{Y}^N(t_N)$ for $t > t_N$ similarly $\tilde{X}^N$. Here $t_N = o(N^2)$. Then (13.13) holds for $\tilde{Y}^N, \tilde{X}^N$ in path space topologized by the sup-norm.

Here we can use for the analysis that both processes are for every N Markov processes and it therefore suffices to prove that they have asymptotically the same transition semigroup. Note that both processes can be considered on the same state space. Hence we have to verify the asymptotic equivalence of the semigroup.

Proof of Proposition 13.2. This follows from the fact (which was derived in [DG99] without mutation) that the averages of N components and the empirical distribution in the N^2-system in such a system in time scale Nt are as $N \to \infty$ asymptotically equivalent to a new such system with a corrected resampling and selection rate. This argument was presented in Section 12.   q.e.d.

Hence we can now turn to the next transition from E_1 to E_2 in our system starting with phase 1 (phase 0 of the transition $E_1 \to E_2$ we have already which is according to the above phase 4 of the previous regime).

Remark 235. Note that now we must start in the equilibrium on E_1 rather than in the value with the lowest fitness in E_1 as we did in the analysis of Section 9.

13.2.4 Droplet Formation (Transition Phase)

As before we can now observe in the time-space rescaled system droplets of the types taken from E_2 begin to form. These arise from 1-blocks where in the time scale Nt already E_2-types have established mass of positive intensity. We then have to define a $(\mathfrak{J}_t^{m,N})_{t\geq 0}$-process of E_2-droplets among the N-blockaverages over 1-balls, where this droplet process is considered in time scale Nt. Then we again have to establish that as $N \to \infty$ we obtain the $(\mathfrak{J}_t^m)_{t\geq 0}$-process, where just the parameters change, namely $c_0, d_0, s_0, m_0, \cdots$ replaced by c_1, d_1, s_1, m_1, etc. The keypoint is that once we have this object we can define $\mathcal{W}_2^*$ (2 indicating the level of the types). With the same machinery we can then address the problem to establish $\mathcal{L}[\vec{\mathcal{W}}_2^*] = \mathcal{L}[{}^*\vec{\mathcal{W}}_2]$, where we discuss $\mathcal{L}[{}^*\vec{\mathcal{W}}]$ in Subsubsection 13.2.5. If we can show that we can replace the system $\{(y^N(i, NT))_{t\geq 0}, \ i = 1, \cdots, N\}$ by a system of interacting Fisher-Wright diffusions, where only parameters are changed $c_0 \to c_1, s \to s_1, d \to d_1$, etc. than immediately our theory of Sections 8, 9, 10 with droplet results holds. Hence we only have to verify this asymptotic equivalence.

Define for $k \in E_2$:

$$\mathfrak{J}_{t,k}^{m,N,2} = \sum_{i=1}^{N} y_k^N(i, Nt)\delta_{a_i} \quad , \quad (a_i)_{i\in\mathbb{N}} \text{ i.i.d. uniform on } [0, 1]. \qquad (13.15)$$

Then we have

Proposition 13.3 (Droplet formation).

(a) We have

$$\mathcal{L}[(\mathbf{J}_t^{m_2,N,2})_{t \geq 0}] \Longrightarrow \mathcal{L}[(\mathbf{J}_t^{m_2,2})_{t \geq 0}], \qquad (13.16)$$

where the r.h.s. is obtained from $\mathbf{J}_t^{m_2}$ by replacing c_0, d_0, s_0 by c_1, d_1, s_1.

(b) Furthermore there exists a $\beta \in (0, s_1)$ such that

$$\mathcal{L}\left[e^{-\beta t_N} y_{t_N}^{m_2,N,2}\right] \underset{N \to \infty}{\Longrightarrow} \mathcal{L}\left[\vec{\mathcal{W}}_2^*\right]. \qquad (13.17)$$

where β and the law of $\vec{\mathcal{W}}^*$ now depend on the parameters m_2, c_1, d_1, s_1 instead of m_1, c_0, d_0, s_0. □

13.2.5 A Two-Level System with Three Fitness Levels in Time Scale $N \log N$ (Phase 1)

Here we focus on the emergence of level E_2 types, i.e. *phase 1* of the transition $E_1 \to E_2$. This means we would now like to pass to a system which has only types of level E_1 and of level E_2 and to pass to random fields consisting only of the collection of the N-different N-blockaverages and the N^2-blockaverage, leaving out single components from the picture, where we observe in time scales based on Nt instead of t. If indeed we can show that we can ignore the level E_0 types we can then proceed after scaling time $t \to Nt$ exactly as we did before passing from E_0 to E_1 in our two-level spatial system with two levels of fitness. Here we have again the problem that the mutations downward from E_1 to E_0 do occur so that we have to show that they are wiped out very quickly again. This we shall see by similar arguments as before. One point here is that we want to use this approximation not only in a finite (macroscopically) time horizon but in a time horizon $[0, t_N]$ where t_N tends to ∞ with $t_N = o(N)$. This involves a mild extension of the result we had so far - see Remark 232.

Recall the definition in (12.5) and (12.7) of $(\{y_k^N(j,t), j \in \{1, \cdots, N\}\})_{t \geq 0}$ and $(\bar{y}_k^N(t))_{t \geq 0}$ defined for $k \in \mathbb{K}$. Here we have $\mathbb{K} = E_0 \cup E_1 \cup E_2$. We now restrict these quantities to $E_1 \cup E_2$. Then define (here $\bar{\mathcal{P}}$ denotes the possibly defective probability measures) the empirical measure of the restricted blockaverages over 1-balls:

$$\Xi_N^{\log,\beta}(t) = \frac{1}{N}\left(\sum_{i=1}^N \delta_{(y_k^N(i, \frac{\log N}{\beta}+t); k \in E_1 \cup E_2)}\right) \in \mathcal{P}(\bar{\mathcal{P}}(E_1 \cup E_2)). \qquad (13.18)$$

13.2 The Case of Two Levels of Spatial Organisation and Three Fitness Levels

One point here is that we want to use this approximation not only in a finite (macroscopically) time horizon but in a horizon $[0, t_N]$ where t_N tends to ∞.

We obtain here:

Proposition 13.4 (Phase 1: Emergence of E_2). *With the constant $\beta \in (0, s_1)$ from (13.17) we have*

$$\lim_{t \to \infty} \liminf_{N \to \infty} P(\{\Xi_N^{\log,\beta}(t)(E_2) \geq 1 - \varepsilon\}) = 1 \quad \forall \varepsilon \in (0, 1), \tag{13.19}$$

$$\lim_{t \to -\infty} \limsup_{N \to \infty} P(\{\Xi_N^{\log,\beta}(t)(E_2) \geq \varepsilon\}) = 0 \quad \forall \varepsilon \in (0, 1). \tag{13.20}$$

The constant β is determined as in Proposition 9.1 in Section 9 but now the E_1-parameters replaced by the E_2 parameters and the E_0 parameters by E_1-parameters. □

Proof of Proposition 13.4. As mentioned above two steps have to be carried out. First we have to argue that we can ignore the types in E_0 in this time scale. For the second step suppose we can do so and consider the type space $E_1 \cup E_2$ instead of $E_0 \cup E_1 \cup E_2$. Then we scale time $t \to Nt$ and replace single components by block averages over N components and obtain in the limit $N \to \infty$ a new system for these averages as before the components and which follows the same dynamic as before except that the parameters c_0, s_0, d_0, m_0 are replaced by c_1, s_1, d_1 and m_1 and mutation matrix M_0 by M_1 and the selection function χ_0 by χ_1. Then apply the analysis we did for the transition from E_0 to E_1 now to the transition from E_1 to E_2. That this is possible was verified at the end of Section 11 in Proposition 11.1 based on [DGV] and [DG99]. The only difference is that in the new time scale we have to start the system in the E_1-equilibrium rather than in the type of lowest fitness in E_1. For some intuitive insight into the emergence process for the reader who does not have [DGV] or [DG99] at hand, we sketch below an argument for three types based on the dual.

Excursion on Coalescence in Fast Time Scale

We now turn to the announced explanation of the reduction of the collection of blockaverages to a new system. Consider the dual calculations. To explain the new features assume that $E_0 = \{1\}$, $E_1 = \{2\}$, $E_2 = \{3\}$. We again start the system with all sites only occupied by type $\{2\}$ and consider it in time scale Nt. As we have argued above we can ignore mutations down to type $\{1\}$. We then take as initial condition for the dual $1_{\{2\}}$. Then it will be subject to the 1_{A_3} operator which will produce copies of $1_{\{2\}}$ at rate $O(1)$ in the new time scale and mutation to type $\{3\}$ at rate $\frac{m}{N}$. Note that the migration rates between sites in 1-balls is Nc and the coalescence rate of particles at the same site is Nd. Therefore collisions between a given pair of particles occur in time scale $O(1)$ and there is a positive probability of a coalescence at each collision. The rate at which a given particle coalesces is proportional to the number of particles in the 1-ball. In addition particles migrate to a different 1-ball at rate $O(1)$. Again once a 1-ball is occupied it can only become empty by the migration of the occupying particle to a different 1-ball in the 2-ball. Occupied 1-balls will reach equilibrium (but fluctuating) in time $0+$ and have an average of $O(N)$ particles. In this way the behaviour in the one ball is analogous to the behaviour at a single site in the case of one spatial level. We then have a supercritical branching process of occupied 1-balls and emergence when $O(N)$ 1-balls are occupied - this occurs at time $N \log N$.

We can carry out the same argument for a system of N different 2-balls. In this case in time scale $N^2 t$ a 2-ball reaches equilibrium in time $O(1)$ with $O(N^2)$ particles and produces particles to migrate to a different 2-ball at rate $O(1)$. Emergence then occurs at time $N^2 \log N$.

We now have to establish that we can indeed drop the types in E_0 and transitions to E_0 in time scales $N \log N + Nt$ in the sense that in the topology of path in the time variable t we have as $N \to \infty$ convergence to the system with values in $E_1 \cup E_2$. Next we have to see that also in the larger time scales $CN \log N + t$ the types in E_0 can be neglected. These have disappeared in the time scale $O(N)$, so that (recall the Feller property of the limiting dynamics) the only way they could be reintroduced is by deleterious mutation from $(E_1 \cup E_2)$-types. Note that we have passed to averages over N-sites (respectively N^2 sites) and scaled time by $t \to Nt$. Therefore we are back to the same problem we had previously in scales $C \log N$ and deleterious mutation at rate $\frac{1}{N}$. However the fact that types in E_0 are negligible was proved in Lemma 12.5, Proposition 12.8 and Remark 232. q.e.d.

13.2.6 A Two-Level System with Three Fitness Levels in Time Scale $N \log N + t$ (Phase 2)

Next we come to phase 2 of our basic scenario and prove that after the emergence the system follows a dynamic (on $E_1 \cup E_2$) leading to fixation on E_2 if we observe the size N blockaverages in time scale Nt and their empirical measure.

We note that due to the fact that for finite N we have transitions to E_0 so that the state of $\Xi_N^{\log,\beta}$ is a defective probability measure on $E_1 \cup E_2$. Therefore we must use the topology on path space introduced in Section 12 in order to get the convergence in (13.21) below.

Proposition 13.5 (Phase 2: Fixation on E_2). *We start $X^{N,(3,2)}$ in an i.i.d. state concentrated on the lowest type.*

(a) There exists a process $(\mathcal{L}_t)_{t \geq 0}$ with values in $\mathcal{P}(\mathcal{P}(E_1 \cup E_2))$ such that

$$\mathcal{L}[(\Xi_N^{\log,\beta}(t))_{t \in \mathbb{R}}] \underset{N \to \infty}{\Longrightarrow} \mathcal{L}[(\mathcal{L}_t)_{t \in \mathbb{R}}] \tag{13.21}$$

13.2 The Case of Two Levels of Spatial Organisation and Three Fitness Levels

and $\mathcal{L}[(\mathcal{L}_t)_{t\in\mathbb{R}}]$ is a true random solution to the McKean–Vlasov equation

$$\frac{d}{dt}(\mathcal{L}_t) = (L_{M,\chi}^{\pi_t})^*(\mathcal{L}_t) \qquad (13.22)$$

with parameters given as follows.

The resampling, selection rates d_2, s_2 are given by

$$d_2 = \frac{c_1}{d_2 + c_1} d_1, \quad s_2 = \frac{c_1}{d_2 + c_1} s_1 \qquad (13.23)$$

and mutation matrix, fitness function and immigration source are given by

$$M \text{ via } m_2^+, m_2^-, m_2 \text{ and } \rho_2; \quad \chi = \hat{\chi}_2, \qquad (13.24)$$

$$\pi_t = \int_{\mathcal{P}(E_1 \cup E_2)} x \mathcal{L}_t(dx). \qquad (13.25)$$

(b) We have

$$\mathcal{L}[(\mathcal{L}_t)_{t\in\mathbb{R}}] = \mathcal{L}[(\tilde{\mathcal{L}}_{t+{}^*\mathcal{E}})_{t\in\mathbb{R}}] \qquad (13.26)$$

where ${}^*\mathcal{E}$ is an $\mathbb{R}$-valued random variable and $(\tilde{\mathcal{L}}_t)_{t\in\mathbb{R}}$ the unique solution to the McKean–Vlasov equation satisfying

$$e^{\beta t} \tilde{\mathcal{L}}_t(E_2) \longrightarrow 1 \text{ in probability as } t \to -\infty, \qquad (13.27)$$

$$e^{\beta t} \tilde{\mathcal{L}}_t(\cdot \bigcup E_2) \longrightarrow^* \vec{\mathcal{W}}/{}^*\mathcal{W}(E_2) \text{ as } t \to -\infty \qquad (13.28)$$

and

$${}^*\mathcal{E} = \frac{1}{s_2} \log({}^*\mathcal{W}), \qquad (13.29)$$

where the r.h.s., i.e. $\mathcal{L}[{}^*\mathcal{W}], \mathcal{L}[{}^*\vec{\mathcal{W}}]$ is given in terms of the parameters via the dual process as in Lemma 10.3 but c_1, d_1, m_1, etc. replaced by c_2, d_2, m_2, etc. □

Remark 236. If necessary we denote ${}^*\vec{\mathcal{W}}$ with the new parameters as ${}^*\mathcal{W}_2$.

Proof of Proposition 13.5. Here again we recall that in the time scale Nt mutations from E_1 to E_0 types do not play a role in the limit since even though they occur at rate $O(1)$ are wiped out by selection at rate $O(N)$. If we have established this, then we can consider a system where the state space is $E_1 \cup E_2$ and mutations to E_0 are removed.

Then we apply the results from the transition E_0 to E_1 to a system with types E_1, E_2 and size-N blockaverages in scale Nt. This system is asymptotically equivalent to a system with N components and types in $E_1 \cup E_2$, where parameters are adapted as above in the proof of Proposition 13.4. This asymptotics equivalence we established in Proposition 13.1 based on [DGV], [DG99]. Then the proof follows applying Section 10 to this new system.

The fact that we can ignore the types in E_0 arises now as in the equivalent system with two levels of spatial organisation and two levels of fitness which we discussed already in Subsubsection 13.2.5.   q.e.d.

We now have to argue that again the $^*\vec{\mathcal{W}}$ equals in law $\vec{\mathcal{W}}^*$ from the droplet formation of E_2-droplets and claim

Proposition 13.6 (Exit and entrance constant). *The random variables $\vec{\mathcal{W}}^*, {}^*\vec{\mathcal{W}}$ defined in (13.17) and (13.21)–(13.29) satisfy:*

$$\mathcal{L}[\vec{\mathcal{W}}^*] = \mathcal{L}[^*\vec{\mathcal{W}}]. \qquad \square \qquad (13.30)$$

Proof Proposition 13.6. The proof follows the same basic steps as in Proposition 8.29 in Section 7. Recall that the key steps involve the covariance calculation and the disappearance of the correction terms given by the "green" particles in the limit $t \to -\infty$. We do not give the details but briefly outline the second moment calculation.

Recall (13.11). Let

$$\bar{y}^N_{E_2}(t) = \sum_{j=1}^N y^N_{E_2}(t,j), \quad \hat{y}^N_{E_2}(t) = \sum_{k \in E_2} \sum_{j=1}^N \sum_{i=1}^N x_k^N(i,j,t) \qquad (13.31)$$

where

$$y^N_{E_2}(t,j) = \frac{1}{N^2} \sum_{k \in E_2} \sum_{i=1}^N x_k^N(i,j; N \log N + Nt). \qquad (13.32)$$

To compute the second moment of $\bar{y}^N_{E_2}(t)$ it suffices to compute

$$E[y^N_{E_2}(t, j_1) y^N_{E_2}(t, j_2)] \text{ with } j_1 \neq j_2. \qquad (13.33)$$

For simplicity we now assume that $E_0 = \{1\}$, $E_1 = \{2\}$, $E_2 = \{3\}$. We use the dual with two initial factors $1_{E_0 \cup E_1}$ located in two distinct 1-balls. We then have a CMJ growth process in the pre-collision regime. The variance is produced by the particles lost due to a 1-collision, that is, entering the same 1-ball, and subsequent coalescence when they meet at the same site within this 1-ball (recall the "excursion on coalescence" in Subsubsection 13.2.5). Let

13.2 The Case of Two Levels of Spatial Organisation and Three Fitness Levels

$$t_N(\delta, t) = \frac{\delta}{\beta} N \log N + Nt, \quad 0 < \delta \leq 1. \tag{13.34}$$

The parallel computation to that in Proposition 8.29 then yields the following analogue of (8.915):

$$\lim_{t \to -\infty} \lim_{N \to \infty} \mathrm{cov}\left(e^{-\beta t_N(1,t)} \hat{y}_{E_2}^N(t_N(1,t)), e^{-\beta t_N(\delta,0)} \hat{y}_{E_2}^N(t_N(\delta,0))\right) \tag{13.35}$$

is constant (as a function of δ). This yields the result. q.e.d.

13.2.7 A Two-Level System with Three Fitness Levels After Fixation (Phase 3)

Here we look at times of the order $N(\beta^{-1} \log N + t_N + t)$, with $t_N \to \infty$ but $t_N \mathbb{N} \to 0_2$ and $t \in \mathbb{R}$ as the macroscopic time variable, where we see the *phase 3* of our basic scenario. The dynamic here is the neutral evolution on the set E_2 of types if we observe the size N block-averages *after* the time of emergence in time scale Nt in the macroscopic time parameter t. Furthermore denote (the 1 indicating that we have an empirical mean of level-1 spatial averages and the corresponding time scale Nt).

$$Y_1^N(t) = \{\bar{y}_{k,1}^N(t), k \in E_2\}. \tag{13.36}$$

Here again we have to remember that after fixation in phase 2 the intensity is a random measure, which we denote (the 2 indicating that we had fixation on E_2):

$$\lambda_2 \text{ and } \mathcal{L}[\lambda_2] := Q_2, \text{ where } Q_2 = \mathcal{L}[{}^*\vec{W}/{}^*\mathcal{W}(E_2)]. \tag{13.37}$$

Proposition 13.7 (Phase 3: Neutral evolution on E_2). *Consider the system defined in (13.36) and let $(Z^\lambda(t))_{t \geq 0}$ be an independent countable collection of equilibrium multiple type Fisher-Wright diffusions on $\Delta_{|E_2|-1}$, with immigration, emigration both at rate c_2 and immigration in every component from the same random, but constant in time source $\lambda = \lambda_2$. We denote by π_2 the restriction on the type space $E_0 \cup E_1 \cup E_2$ to E_2 and by $\tilde{\pi}_2$ the induced map on measures.*

Then we get for $t_N \to \infty$ as $N \to \infty$ with $t_N = o(N)$ but $t_N - \beta^{-1} \log N \to \infty$, that

$$\mathcal{L}[(\tilde{\pi}_2 \circ Y_1^N(t_N + t)^+)_{t \in \mathbb{R}}] \underset{N \to \infty}{\Longrightarrow} \int Q_2(d\lambda)(\mathcal{L}[(Z^\lambda(t))_{t \in \mathbb{R}}]). \quad \square \tag{13.38}$$

Proof of Proposition 13.7. Again as before, if we exclude the effect of downward mutation to types in E_0 (which we established in Subsection 13.2.5), we can apply the same scaling technique as before (i.e. replacing N-1-ball blockaverages in time scale Nt asymptotically by a system of N components in time scale t with adapted parameters) to reduce everything to a system with N components, types in $E_1 \cup E_2$ and new parameters, by using Proposition 13.2. Then we are literally in the situation in Subsection 11. q.e.d.

13.2.8 A Three-Level System with Three Fitness Levels in Time Scale $N^2 t$ (Phase 4)

After completion of Phase 3 (neutral evolution among level E_2 types), in the final *phase 4* of our basic scenario the system reaches a new mutation-selection equilibrium on E_2-types in times of order $N^2 t_N$ with $t = t_N \uparrow \infty, t_N = o(\log N)$ which is given by the equilibrium of a nonlinear Markov process with mutation and selection on the types in E_2.

Here we have now to take into account migration steps in distance 3. This means that instead of the $x_k^N(i, j; t), k \in \mathbb{K}, i = 1, \cdots, N; j = 1, \cdots, N$ of (13.1), we now have to deal with

$$X^{N,(3,3)}(t) = \{x_\ell^N(i, j, k; t)\}, \quad i, j, k \in \{1, \cdots, N\}; \quad \ell \in \mathbb{K} = E_0 \cup E_1 \cup E_2, \tag{13.39}$$

where we keep all mechanism but we add migration steps at distance 3 at rate c_2/N^2, in addition to the distance 1 jumps at rate c_0 and distance 2 jumps at rate c_1/N which we already had.

As before we need some functionals of the process. Here we observe a system with types in $E_0 \cup E_1 \cup E_2$ but we expect that the intensity on the types in $E_0 \cup E_1$ goes to zero in the time scale we consider. We therefore observe only the restrictions on E_2 which gives then defective probability vectors for states of components for finite N:

$$\tilde{\pi}_2 \circ X^{N,(3,3)}(t) \in (\bar{\mathcal{P}}(E_2))^{N^3}, \tag{13.40}$$

($\bar{\mathcal{P}}$ denotes as before defective probability measures). We use for the restriction on E_2 the same notation for the components $x_\ell^N(i, j, k; t)$ to keep the equation transparent.

There is now the need to incorporate the level of spatial organisation as an index. Define the components (level-0), spatial level-1 and level-2 averages in their respective natural time scale after the long time $N^2 t_N$ with $t_N \to \infty$ but $t_N = o(\log N)$ as follows:

13.2 The Case of Two Levels of Spatial Organisation and Three Fitness Levels

$$y_{\ell,0}^N(i,j,k;u,s,t) = x_\ell^N(i,j,k;N^2u+Ns+t),$$
$$\bar{y}_0^N = \{y_{\ell,0}; \ell \in E_2\}, \quad (i,j,k) \in \{1,\cdots,N\}^3, \tag{13.41}$$

$$y_{\ell,1}^N(i,j,s,t) = \frac{1}{N} \sum_{k=1}^N x_\ell^N(i,j,k;N^2s+Nt),$$
$$\bar{y}_1^N = \{y_{\ell,1}; \ell \in E_2\}, \quad (i,j) \in \{1,\cdots,N\}^2, \tag{13.42}$$

$$y_{\ell,2}^N(i,t) = \frac{1}{N^2} \sum_{k=1}^N \sum_{j=1}^N x_\ell^N(i,j,k;N^2t), \quad \bar{y}_2^N = \{y_{\ell,2}; \ell \in E_2\}, \quad i \in \{1,\cdots,N\}. \tag{13.43}$$

We write as further abbreviations for the respective complete spatial configuration,

$$y_0^N(u,s,t), \quad y_1^N(s,t), \quad y_2^N(t), \tag{13.44}$$

i.e. for the collections of $\bar{y}_0^N, \bar{y}_1^N, \bar{y}_2^N$ in the spatial indices.

Define the level-3 empirical mean of level-2 averages (the index 2 indicating we have an empirical mean of averages over N^2 sites):

$$\mathcal{Y}_2^N(t) = \frac{1}{N} \sum_{i=1}^N \delta_{y_{\ell,2}^N(i,t)}, \quad \ell \in E_2. \tag{13.45}$$

We have to define the *limiting dynamics* corresponding to these four functionals defined above, which are denoted

$$(\mathcal{Y}_2^\infty(t))_{t \geq 0} \text{ with values in } \mathcal{P}(\mathcal{P}(E_2)), \tag{13.46}$$

$$(\{(y_2^\infty(i,t); i \in \mathbb{N}\})_{t \geq 0}, \text{ with values in } \mathcal{P}(E_2), \tag{13.47}$$

$$(\{y_1^\infty(i,j;s,t); i,j \in \mathbb{N}\})_{t \geq 0}, \text{ with values in } \mathcal{P}(E_2), \tag{13.48}$$

$$(\{y_0^\infty(i,j,k;u,s,t); \; i,j,k \in \mathbb{N})_{t \geq 0} \text{ with values in } \mathcal{P}(E_2). \tag{13.49}$$

Here $\mathcal{Y}_2^\infty$ arises as law of the $(L_{M,\chi}^{\pi_t}, \delta_{\lambda_2})$-martingale problem, y_2^∞ as independent realizations of it, while y_0^∞, y_1^∞ arise as realizations of solutions from certain $(L_\theta^{c,d})$-martingale problems in equilibrium. Namely we chose the parameters for $\mathcal{Y}_2^\infty, y_2^\infty$:

$$c = c_2, d = d_2, s = s_2; \; M \text{ given by } m_2^+, m_2^-, m_2 \text{ and } \rho_2; \quad \chi \text{ by } \hat{\chi}_2 \tag{13.50}$$

and

$$\pi_t = \int_{\mathcal{P}(E_2)} \mu \mathcal{Y}_2^\infty(t)(d\mu) \tag{13.51}$$

respectively for y_0^∞, y_1^∞:

$$c = c_0, d = d_0 \text{ respectively } c = c_1, d = d_1, \tag{13.52}$$

while θ is the current value of the limiting next higher level average at the macroscopic time u, s respectively s.

The initial state $\mathcal{Y}_2(0)$ has the law

$$\mathcal{L}[\Gamma_{\lambda_2}^{c,d}] \tag{13.53}$$

and the *random* measure λ_2 is the (random) emergence proportion on E_2 given by $^*\mathcal{W}(\cdot)/^*\mathcal{W}(E_2)$, as specified in (13.29).

Now we are ready to establish the scenario of phase 4 for our three-level system as follows.

Proposition 13.8 (Phase 4: Convergence to mutation-selection dynamic, equilibrium on E_2)). *With the definitions (13.39)–(13.52) we have the following convergence relations. (Recall that here our observables are already projected to E_2).*

a) *The empirical measures in macroscopic scale behaves as follows:*

$$\mathcal{L}[(\mathcal{Y}_2^N(t))_{t\geq 0}] \underset{N\to\infty}{\Longrightarrow} \mathcal{L}[(\mathcal{Y}_2^\infty(t))_{t\geq 0}] \tag{13.54}$$

and for $t_N \to \infty$, $\lim(t_N/\log N) = 0$ we get (recall (4.33)):

$$\mathcal{L}[\mathcal{Y}_2^N(t_N N^2)] \underset{N\to\infty}{\Longrightarrow} \delta_{\Gamma_{\rho_2,\tilde{\lambda}_2}^{m,s_2,c_2,d_2}}. \tag{13.55}$$

b) *Furthermore level-2, level-1 blockaverages and single components behave as follows with $\tilde{\mathcal{L}}$ the law of λ_2 in (13.53):*

$$\mathcal{L}[(\{\bar{y}_2^N(i, t_N + t); i \in \{1, \cdots, N\}\})_{t\geq 0}] \underset{N\to\infty}{\Longrightarrow} \tilde{\mathcal{L}}[\bigotimes_{i=1}^\infty \mathcal{L}[(y_2^\infty(i,t))_{t\geq 0}]], \tag{13.56}$$

with $(\{y_1^\infty((i, j; s, t), j \in \mathbb{N}\})_{t\geq 0}$ defined for given random field

$$\{y_2^\infty(i, s); i \in \mathbb{N}\}, \tag{13.57}$$

13.2 The Case of Two Levels of Spatial Organisation and Three Fitness Levels

whose law is denoted $\tilde{\mathcal{L}}_2$, then for every $i \in \{1, \cdots, N\}, s > 0$,

$$\mathcal{L}[(\{\bar{y}_1^N(i, j, s, t); j \in \{1, \cdots, N\}\})_{t \geq 0}] \underset{N \to \infty}{\Longrightarrow} \tilde{\mathcal{L}}_2[\bigotimes_{j=1}^{\infty} \mathcal{L}[(y_1^\infty(i, j; s, t))_{t \geq 0}]]. \tag{13.58}$$

analogously if $y_0^\infty(i, j, k; u, s, t)$ is defined for given collection ($\tilde{\mathcal{L}}$ denotes again its law)

$$\{y_1^\infty(i, j; u, s), (i, j) \in \mathbb{N}^2\}, \tag{13.59}$$

we get for every $i, j \in \{1, \cdots, N\}; u, s > 0$,

$$\mathcal{L}[(\{\bar{y}_0^N(i, j, k; u, s, t); k \in \{1, \cdots, N\}\})_{t \geq 0})] \underset{N \to \infty}{\Longrightarrow} \tilde{\mathcal{L}}[\bigotimes_{k=1}^{\infty} \mathcal{L}[(y_0^\infty(i, j, k; t))_{t \geq 0}]]. \tag{13.60}$$

The

$$\{\bar{y}_1^N(i, j; s, \cdot); (i, j) \in \{1, \cdots, N\}^2\} \quad , \quad \{\bar{y}_0(i, j, k; u, s); (i, j, k) \in \{1, \cdots, N\}^3\} \tag{13.61}$$

are as $N \to \infty$ asymptotically independent for different i respectively different i, j if s respectively u, s is fixed. □

Proof of Proposition 13.8. Exactly as before we have to show that (1) we can ignore the types in E_0 and and E_1, (2) by forming level-1 and level-2 block averages together with a time-rescaling $t \to Nt$ and $t \to N^2 t$ we obtain in the limit $N \to \infty$ again an asymptotically equivalent system of the form we had before for components and level-1 averages, with the only difference that we use different values for the parameters and (3) the components on the lowest level behave as claimed.

For point (1) of course we again have to show that the effect of downward mutations to E_0 or E_1 can be neglected. We don't repeat the details here which are of the same nature as before showing for the $E_0 \cup E_1$ system, that we can neglect E_0 types.

The point (2) is handled again by showing that the time rescaled dual process converges to the dual process of the claimed limit. For that purpose we have to show that the lower level transitions can be ignored, which we see for mutation, selection terms from point (1), since on E_2 these transitions with terms of order 1 simply do not act. What remains are the migration steps between components and their interaction with resampling. This has been discussed in the "excursion on coalescence" in Subsubsection 13.2.5. This problem however appears already in the neutral case. For the analogous detailed treatment of lower order terms without mutation we refer to [DGV], Section 5.

Therefore we can apply the results of the Sections 11 and 12 if we choose the parameters properly and we obtain the statements (13.54)–(13.56).

The two remaining statements (13.58), (13.60) and (13.61) are statements about lower level averages in an asymptotically *neutral* dynamic (since only E_2-types appear with mutation and selection terms acting in time scales of order at least N^2 and which therefore can be neglected for evolutions in time scales $O(N)$ or $O(1)$). Then the *neutral* theory of Section 5 in [DGV] applies and gives immediately the claim.   q.e.d.

Chapter 14
Iteration of the Cycle II: Extension to the General Multilevel Hierarchy

In the previous section we have followed phases 0–4 for up to 3 fitness levels and 2 spatial levels, i.e. we treated the transition $E_0 \to E_1$ and then $E_1 \to E_2$. The latter case of 3 fitness levels leading to fixation and neutral equilibrium on E_2 includes all the complexity of higher level systems, namely, the necessity to deal treating the McKean–Vlasov dynamic on types in E_2 the *perturbations* arising with the McKean–Vlasov dynamics at the next lower lower level in smaller space-time scales. The program can otherwise then be continued in essentially the same way to describe the emergence, fixation and neutral equilibrium at levels $E_3, E_4, E_5, \ldots$ in *finite* spatial systems with *finitely* many fitness levels. In the next subsection we briefly indicate how the general transition $E_{j-1} \to E_j$ is carried out. In the second subsection we then explain how these transitions for the *finite* (spatial and number of fitness levels) systems is used to obtain the transitions for the original *in*finite systems.

14.1 Combination of Three Time Scales for $(j + 1, j)$-Level Systems

Here we have to return to the analysis we did in the previous subsection for $j = 2$ but now we want to describe the transition from types in E_{j-1} to types in E_j respectively the equilibration on types of E_j. This also automatically requires the consideration of the corresponding j-levels of spatial organisation. This means we are starting with a model of $(j + 1)$ levels of fitness, namely, $E_0, E_1, \cdots, E_j$, and level 1 to j of spatial organisation, i.e. the set of locations is $\{1, \cdots, N\}^j$.

The $(j + 1, j)$-level systems have the form:

$$X^{N,(j+1,j)}(t) = \{x_k^N(i,t)\}; \quad k \in E_0 \cup \cdots E_j, i \in \{1, \cdots, N\}^j \qquad (14.1)$$

and are defined by three operations, namely *choosing* in the definition of $X^N(t)$ the parameters satisfying

$$c_j = c_{j+1} = \cdots = 0, \quad m_k^+, m_k^-, m_k \equiv 0 \text{ for } k \geq j+1, \chi_N(k,\cdot) \equiv 0 \text{ for } k \geq j+1, \tag{14.2}$$

then *restricting space* from $(\mathcal{P}(\mathbb{K}))^{\Omega_N}$ to sites at distance j from the origin (i.e. $B_j(0)$) and finally restricting *types* for all the components of the process from the set of types $\bigcup_0^\infty E_i$ to the set $E_0 \cup \cdots \cup E_j$.

We study these arising *new* systems in time scales

$$uN^{j-1}, \quad \beta_j^{-1} N^{j-1} \log N, \quad \beta^{-1} N^{j-1} \log N + sN^{j-1}, \quad tN^j, \tag{14.3}$$

and in spatial levels of averaging corresponding to spatial levels $(j-2), j-1$ and j which means we consider the block averages over $(j-2)$-balls, as the basic components to align with the $(3,2)$ model of types in $E_0 \cup E_1 \cup E_2$ on $\{1, \cdots, N\}^2$ as geographic space.

This construction now allows us to make visible the transition from types in E_{j-1} to types in E_j. As before we want to exhibit the phases 0–4 in this transition. Again if one studies approach to equilibrium on E_j, one needs $j+1$ levels of spatial organisation, i.e.

$$X^{N,(j+1,j+1)} = \{x_k^N(i,t); \quad k \in E_0 \cup E_1 \cdots E_j, \quad i \in \{1,\cdots,N\}^{j+1}, \tag{14.4}$$

according to the same principle as before in Section 12 and Subsection 13.2 but adding now migration steps in distance $j+1$ at rate c_j.

The argument should work in principle as before in Section 13.2. More precisely if we would

- only have types of fitness in E_{j-1}, E_j,
- no mutations to lower levels $E_0, \cdots, E_{j-2}$,
- no migration jumps in distances $1, 2, \cdots, j-2$,

then by a *time rescaling*

$$t \to tN^{j-2} \tag{14.5}$$

and *block-averaging* aggregating all $(j-2)$-blocks

$$x_\xi \longrightarrow \sum_{d(\xi,\xi') \leq j-2} x_{\xi'}, \tag{14.6}$$

we would arrive with a simple modification of the averaging results for the neutral model in [DGV] *exactly* at the situation in Subsection 13.2 by just replacing E_0, E_1, E_2 by E_{j-2}, E_{j-1}, E_j and components at sites by $(j-2)$ blockaverages, and adapting the parameters to the new situation.

14.1 Combination of Three Time Scales for $(j+1,j)$-Level Systems

In conclusion, we mainly have to deal here with the point that in our real system we have *lower order perturbations* which are not visible if either we consider types in E_{j-2}, E_{j-1}, E_j only, or migration steps in distance $j-1$ and j only. These perturbations arise on the one hand from mutations leading to types in $E_0, \cdots, E_{j-2}$ and lower order selection resulting from this and on the other hand from migration in the small distances $1, \cdots, j-2$.

Nevertheless the key point to establish is that these lower order perturbations are negligible and that we go through phases 0,1,2,3,4 as described before for two-level systems with three levels of fitness which we already analysed in detail and so the main result of this section is that everything we proved in Subsection 13.2 can be lifted to level j.

This means in particular given $c_{j-1}, d_{j-1}, s_{j-1}$ for the evolution of block-averages over sites in distance $j-1$ and in time scale N^{j-1} there exists $\gamma(c_{j-1}, d_{j-1}, s_{j-1})$ such that we again have the basic scenario. Namely with $T(\cdot)$ is increasing and $T(N)$ satisfying $\lim_{N \to \infty} T(N) = \infty$, $\lim_{N \to \infty} \frac{T(N)}{\log N} = 0$:

- Time scale $N^{j-1} T(N)$ (phase 0: quasi-equilibrium on E_{j-1}-types of averages in blocks of radius $j-1$),
- time scale $N^{j-1}(\frac{\log N}{\gamma} + t)$, (phase 1: Emergence of E_j-types in j-blocks),
- transition regime with random time shift and emergence proportions given via $^*\vec{\mathcal{W}}_j$ and droplet formation with growth behaviour given by $\vec{\mathcal{W}}_j^*$ such that $\mathcal{L}[^*\vec{\mathcal{W}}_j] = \mathcal{L}[\vec{\mathcal{W}}_j^*]$,
- time scale $N^{j-1}(\frac{\log N}{\gamma} + T(N))$, (phase 2: Fixation of E_j-types that is, the absence of E_{j-1} and lower order types),
- time scale $N^{j-1}(\frac{\log N}{\gamma} + T(N) + t), t \in \mathbb{R}$ (phase 3: Neutral evolution on E_j-types),
- time scale $N^j t T(N)$ with $t > 0$, (phase 4: evolution towards level-j selection-mutation, i.e. the equilibrium on E_j-types is reached in blockaverages of radius j).

For the purpose of the analysis in the above time scales we define the analogs of the quantities defined in Subsection 13.2 to describe components, level-1 blockaverages up to level-j blockaverages and finally empirical distributions for the whole space now for j levels of spatial organisations. We get:

Proposition 14.1 (Scenario for $(j+1,j)$-systems in phases 0–4). *Consider the system $X^{N,(j+1,j)}$ respectively $X^{N,(j+1,j+1)}$ and consider for $\gamma \in (0, \infty)$ the time scales (here $S(N), T(N) \to \infty, S(N)/\log N \to 0, T(N)/N \to 0$)*

$$N^{j-1}S(N), \quad N^{j-1}(\tfrac{\log N}{\gamma}+t), \quad N^{j-1}(\tfrac{\log N}{\gamma}+T(N)), \quad N^{j-1}(\tfrac{\log N}{\gamma}+T(N)+t),$$

$$N^j t \text{ and } T(N) N^j. \tag{14.7}$$

Turn to the statements in Proposition 13.4-Proposition 13.8. Then replace there

$$E_0 \text{ by } E_0 \cup E_1 \cup \cdots E_{j-2}, E_1 \text{ by } E_{j-1} \text{ and } E_2 \text{ by } E_j, \qquad (14.8)$$

furthermore replace

$$X^{N,(3,2)} \text{ by } X^{N,(j+1,j)} \text{ and } X^{N,(3,3)} \text{ by } X^{N,(j+1,j+1)}, \qquad (14.9)$$

the time scales

$$NS(N), N(\log N + t), \quad N \log N + T(N), \quad N(\log N + T(N) + t), \quad tN^2 \text{ and } T(N)N^2 \qquad (14.10)$$

by the ones in (14.7), where γ is chosen depending on the parameters on level j in the same way β was chosen as function of the parameters on level-2 in Subsection 13.2.

Then the Proposition 13.4-Proposition 13.8 hold for the new objects for the appropriate values of the parameters in the topology on path space we gave in Section 12. □

Proof Proposition 14.1. If mutations to the much lower levels $(E_0, \cdots, E_{j-2})$ are suppressed and if all migration steps in distance at most $j - 1$ are suppressed and time is scaled by $t \to N^{j-1}t$, then the transition from types in E_{j-1} to E_j is identical to the transition from types in E_1 to types in E_2 in the model with three fitness levels studied in Section 13. However we do have *deleterious mutations* to lower levels and we do have *small migration steps* which we have to handle.

The problem how to handle the perturbation by the lower order *migration* steps is well understood from [DG99] and [DGV] (in particular Section 5 therein) and we refer the reader to the latter for details. Here the main new step is therefore to show that in spite of the possible mutations to lower levels, in the $N \to \infty$ limit there is no mass on the set of lower types. This is based on the following lemma which then immediately allows to lift the results of Section 13 to transitions from E_{j-1} to E_j in the spatially system with j-levels of spatial structure.   q.e.d.

Lemma 14.2 (No low types are present). *In all time scales listed above, asymptotically as $N \to \infty$, there is no mass on types in $E_0 \cup \cdots \cup E_{j-2}$ in the sense of the Meyer-Zhang topology.* □

This now completes the proof of Proposition 14.1 and it remains to verify the lemma above.

Proof of Lemma 14.2. It suffices to prove that the expected mass of the types in $E_0, \cdots, E_{j-2}$ is asymptotically zero at fixed times and that this can be extended to a statement in path space in the time scale $N^{j-1}t$ respectively $N^j t$.

The proof uses the fact that in time scale $N^{j-1}t$ or tN^j there is a fitness difference of size $O(N)$ between some E_{j-2}-types and E_{j-1}-types (and even of bigger order for E_k types with $k < j - 2$), there is mutation among the

E_{j-2}-types at rate $O(N)$ and there is mutation between E_{j-2}- and E_{j-1}-types at rate $O(1)$. The point is to show that despite the fact that E_{j-2}-types (from E_{j-1}-types) are produced at rate $O(1)$ in rescaled time, this mass is essentially "instantaneously" eliminated by the mutation-selection process occurring at rate $O(N)$ namely selection between E_{j-1} and E_{j-2} respectively deleterious mutation of the latter. This should eliminate the E_{j-2}-types (and of course the even lower ones). This is exactly the same phenomenon as we had in Lemma 12.5 and there it was only used what difference between the two scales exist rather than their absolute size and therefore we refer the reader to the proof of this lemma for detail. q.e.d.

14.2 The Infinite Hierarchy in Multiple Time-Space Scales

The final step is to approximate our original system with infinitely many levels both of fitness and spatial organization for the purpose of studying particular time and space scales corresponding to the transition $E_{j-1} \to E_j$ and in the limit $N \to \infty$ by finite systems, more precisely systems with $(j+1)$ levels of fitness and $(j+1)$ levels of spatial organization. This works for time scales up to order $t_N N^j$ with $t_N \uparrow \infty$, $t_N = o(N)$ and for blockaverages of all orders up to j.

In order to establish this approximation, we have to prove that the process obtained from $X^N(t)$ by *restricting* on the sites in distance at most j from 0 and by *restricting* the state of the components onto measures on types in $\bigcup_0^j E_k$ is approximated uniformly in times $o(N^{(j+1)})$ by the finite system with N^j sites and $(j+1)$ levels of fitness, which was treated in Subsection 14.1. This means precisely the following.

Let $(X^{N,(j+1,j)}(t))_{t \geq 0}$ denote the system defined in Subsection 14.1 in (14.1) and (14.2). Let

$$\pi_{\infty,k}^{\infty,j}(X^N) \tag{14.11}$$

denote the system $X^N(t)$ where the components are restricted to types in $E_0 \cup \cdots \cup E_j$ and then the system is restricted to sites in the k-ball $B(k)$ around 0 as described above. Let

$$d(\cdot, \cdot) \text{ be the Prohorov distance on } C([0, \infty), (\mathcal{P}(\mathbb{K}))^{B(k)}), \text{ resp. } C([0, T], \mathcal{P}(\mathbb{K})^{B(k)}) \tag{14.12}$$

for some k and recall that this metric generates the weak topology. We note here that we do not have to switch to a weaker topology since we cut off only mutations upward to high levels (which are rare in all the considered time scales) and migration steps to sites far away which again are rare in all considered time scales.

Then we get the key approximation result:

Proposition 14.3 (Approximation of infinite by finite systems).

a) *For $t_N = O(N^{j+1})$:*

$$d(\mathcal{L}[(\pi_{\infty,j+1}^{\infty,j+1}(X^N(t))_{t\in[0,t_N]}], \mathcal{L}[(X^{N,(j+1,j+1)}(t))_{t\in[0,t_N]}]) \underset{N\to\infty}{\Longrightarrow} 0. \quad (14.13)$$

b) *If we consider at a late time $O(N^{j+1})$ the collection of k-blockaverages with $k \leq j+1$ in time scales tN^k, then the restricted infinite system and the $(j+1, j+1)$-system have as $N \to \infty$ asymptotically the same distribution in path space for these functionals.* □

Proof. (a) Since all involved processes are Markov processes with the same initial states, it suffices to show that all transition kernels from time s to t with $s < t$ and $s, t \in [0, t_N]$ of the X^N and $\pi_{\infty,j+1}^{\infty,j+1} \circ X^N$ process are asymptotically equal uniformly in s, t.

The proof of this result is easy, again we can use the refined dual process $(\mathcal{F}_t^{++})_{t \geq 0}$. We show that the dual process of the spatially infinite system and types in $\bigcup E_i$ and of the system with $(j+1)$-levels of spatial organization, i.e. sites indexed by $B(j+1)$, and types in $E_0 \cup \cdots \cup E_j$ yield as $N \to \infty$ the same expressions in the duality relations if we evaluate the duality at times $t_N = o(N^{j+1})$ as $N \to \infty$. This way we get the same transition probabilities uniformly for all times satisfying the restriction.

We show for that purpose that in the dual process no mutations occur to levels of fitness $j+2$ and higher, no selection operators act corresponding to sets A_k with $k \geq j+2$ and no migration jumps in distance $j+2$ or more occur in the time interval $[0, t_N]$. This fact is implied since the rates for these transitions are of order $O(N^{-(j+2)})$ hence in the time interval $[0, t_N]$ their total intensity is $O(N^{-1})$, which gives the claim. This will imply the approximation of the marginal distributions at fixed times up to times $O(N^{j+1})$ *uniformly* in the initial state and *uniformly* in time below the threshold t_N.

This argument gives via the Markov property the approximation of the finite dimensional distributions. Therefore we have f.d.d. convergence. What remains is to lift this to a statement about the law on path space, the path set constant beyond time t_N.

The latter point is handled as follows. We can couple the two dynamics by introducing a bivariate dynamic where both components evolve jointly with the common part of the generator. This process has a dual and we use the argument above to show that the *difference process* converges to the 0-process f.d.d. and satisfies the tightness criteria in the appendix (this is immediate using the terms are of order $N^{-(j+2)}$) so that we get the statement in path space as wanted.

(b) Immediate from the above.   q.e.d.

14.3 Local Genealogy and Biodiversity During Emergence and Fixation

The main new aspect here will be to bring the historical process aspect back into the picture and to prove our assertions concerning the historical process.

14.3.1 Proof of Theorem 11

Here we have to use the asymptotic results from Sections 5–13 and Subsubsections 14.1, 14.2 in order to identify the asymptotic behaviour of the ancestral lines and resulting from this the statement on biodiversity. The key point is that we are interested in times $t = t(N)$ of the order such that fixation on E_j has taken place, which allows to use results on the neutral case in [DGV] by applying our approximation techniques in Section 13 and Subsections 14.1, 14.2. Furthermore recall that we do not work with the full but the reduced historical process, but only the functional which allows use to distinguish *families*.

First recall that *countably* many different rare mutants contribute to the growth of the superior type (see Remark 4) in the spatial box of size N^j corresponding to the process of fixation on the level j of fitness which takes time $\gamma^{-1}(\log N)N^{j-1}$. However we are interested in what happens in a small box or even one component of observation if the level j of the transition $E_{j-1} \to E_j$ becomes very large.

Consider the case in which there are many types of equal fitness in the region of radius j. We call a dynamic a *level-k dynamic* if for $k \in \mathbb{N}_0$:

- we remove all contributions to the fitness function and the mutation matrix which are associated with terms $N^{-(k+1)}$ and higher,
- we start the evolution in $\bigcup_{k}^{\infty} E_i$ and suppress all mutations to $E_0 \cup E_1 \cdots \cup E_{k-1}$,
- we use time scale tN^k,
- we identify locations within k-blocks.

Note that in the level k dynamics all level $\ell \geq k$ types have equal fitness so we are in the *neutral* case, which allows us to apply the results which were obtained for the historical process in this case in [DGV] which we recall next.

Consider the level-j dynamic and the limit $N \to \infty$. Assume that $c_j = c^j$ for $c \in (0, \infty)$. Then if j goes to ∞, and $c \leq 1$ (recurrent case) we see for a fixed $\ell \in \mathbb{N}$ only one type in a ball of radius ℓ (local fixation) whereas if $c > 1$ (transient case) we will see coexistence of types in the limit (see [DGV], Cor. 0.1). In other words for such a j-level dynamic we have a dichotomy of *local fixation* versus *local coexistence* corresponding $c \leq 1$ and $c > 1$ as claimed in the theorem.

Therefore we only have to argue that we can compare our dynamic at times of fixation on level j with a j-level dynamic. We know that in our dynamic we have fixation before time N^j. Therefore we want to approximate our system by an appropriate finite system (in type and space). This exactly was shown however in Proposition 14.3 and hence the theorem follows from the results on the neutral system quoted above.

14.3.2 The Geographic Source of E_j-Valued Mutant Families: Proof of Proposition 4.14

(a) This follows immediately from the dual process. Namely, no jumps to $CB_j(0)$ occur in times $o(N^{j+1})$ in the dual process. Therefore the configuration in $B_j(0)$ up to this time does only depend on the initial configuration in $B_j(0)$.

(b) To prove (b) we now enrich the space of E_{j+1}-types by including the locations of the founding fathers of families in which a successful rare mutation occurs which then contributes a non-trivial part of the emerging E_{j+1}-type population. We denote by $(\bar{Z}^{*,j}(t))^i$ the part of the E_{j+1} mass arising from such families with founding fathers in $A_N(i)$ for $i = 1, \ldots, k$. To do this consider the enriched type space $\{1, \ldots, k\} \times (E_0 \cup E_1)$, $E_0 = \{1\}$, $E_1 = \{2, \ldots, M+1\}$. We assume that the fitness of types (j, ℓ) depends only on ℓ and mutation from (j, ℓ) to j', ℓ' with $j \neq j'$ does not occur.

Assume that rates of a rare mutation $(j, 1) \to (j, \ell)$ are given by

$$\frac{m_{(j,1),(j,\ell)}}{N} = \frac{m_{1,\ell}}{N}, \quad \ell = 2, \ldots, M+1,$$

and assume that the initial state of the system satisfies

$$\mu(i, (j, 1)) = 1 \quad \text{if } i \in A_{N,j}, \; j = 1, \ldots, k.$$

The proof uses the dual calculations of Subsection 10.7 as in the proof of Proposition 10.3 starting the dual at a tagged site with a single (nontrivial) rank given by the indicator function

$$1_{(j,\ell)}.$$

We use this to first compute the mean of type (j, ℓ) at a tagged site at times $\frac{\log N}{\beta} + t$. The instantaneous rate that the dual has a rare mutation jump is $\frac{m_{(j,1),(j,\ell)} K(t)}{N}$. Then the integrated dual is 1 at time $\frac{\log N}{\beta} + t$ if a dual rare mutation jump has occurred and the rank at which the dual rare mutation jump occurred is located in $A_{N,j}$ at time $\frac{\log N}{\beta} + t$ and otherwise is 0. But by uniform migration rates between j-blocks in the $(j+1)$-block the probability that this rank is at a site in $A_{N,j}$ is $\frac{1}{k}$ so that we obtain

14.3 Local Genealogy and Biodiversity During Emergence and Fixation

$$\lim_{t \to -\infty} e^{-\beta t} E[\bar{x}^N_{(j,\ell)}(\frac{\log N}{\beta} + t)] = \frac{1}{k} \lim_{t \to -\infty} e^{-\beta t} E[\bar{x}^N_{\ell}(\frac{\log N}{\beta} + t)]. \quad (14.14)$$

It then follows that the distribution of expected mass of E_{j+1}-types based on the location of the founding mutation is uniform.

Finally, in order to prove independence of the contributions from different sets $A_{N,j}$, $A_{N,k}$, $j \neq k$, we follow the same argument as in the proof of Proposition 10.12, that is, to prove that the joint moments factor. The reason for the latter is exactly the same as in Proposition 10.12 since the collision and coalescence of $1_{(j,\ell)}$ and $1_{(k,\ell)}$ results in their mutual annihilation.

(c) Recall that the rare mutant families firsts arise as excursions for the single site dynamics. The position of these initial rare mutations is decided on the basis of exchangeable experiments with (as $N \to \infty$) small success probability. It follows from (b) that as $N \to \infty$ the ancestral positions are asymptotically uniformly distributed.

(d) Recall that the probability that distance between a tagged site and a randomly chosen point in $B_j(0)$ in Ω_N is equal to j converges to 1 as $N \to \infty$. Together with (b) this implies that as $N \to \infty$ the probability that the origin of a E_j-valued family in time scale $N^j t$ at a tagged site originated at a site at hierarchical distance j converges to 1.

Chapter 15
Winding-Up: Proofs of the Theorems 3–11

In this section we complete the proofs of the Theorems 3–11 by putting all the pieces together. First recall Theorem 11 was proved in Subsection 14.3.

Here we first use the result of Subsection 14.2 to *reduce* all assertions on phases 0–4 in the transition from types in E_{j-1} to E_j to the corresponding scenario for finite multi-level systems with $(j+1)$-levels of fitness and j-levels (or $(j+1)$-levels) of spatial organization by Proposition 14.3. Then we conclude with the proposition from Subsection 14.1 which were based on the arguments in Section 13.2, the statement of the Theorems 3–10 immediately. (The basis for these results were statements in Section 10 up to Section 12 on phases 0–4).

Appendix A
Tightness

For convenience, in this section we review the tightness criteria that are needed in this monograph.

We first recall the Joffe–Métivier criterion ([JM]) for weak convergence on $D([0,\infty), \mathbb{R}^d)$. This criterion is concerned with a collection $(X^{(n)}(t))_{t \geq 0}$ of semi-martingales with values in $\mathbb{R}^d$ with càdlàg paths. First observe that by forming

$$(< X^{(n)}(t), \lambda >)_{t \geq 0} \quad , \quad \lambda \in \mathbb{R}^d \tag{A.1}$$

we obtain $\mathbb{R}$-valued semi-martingales. If for every $\lambda \in \mathbb{R}^d$ the laws of these projections are tight on $D([0,\infty), \mathbb{R})$ then this is true for $\{[\mathcal{L}[(X^{(n)}(t))_{t \geq 0}], n \in \mathbb{N}\}$.

The tightness criterion for $\mathbb{R}$-valued semimartingales which is needed after the above reduction, is in terms of the socalled *local characteristics* of the semimartingales.

A.1 The Joffe–Métivier criteria for tightness of D-semimartingales

We recall the Joffe Métivier criterion ([JM]) for tightness of locally square integrable processes.

A càdlàg adapted process X, defined on $(\Omega, \mathcal{F}, \mathcal{F}_t, P)$ with values in $\mathbb{R}$ is called a *D-semimartingale* if there exists a càdlàg function $A(t)$, a linear subspace $D(L) \subset C(\mathbb{R})$ and a mapping $L : (D(L) \times \mathbb{R} \times [0, \infty) \times \Omega) \to \mathbb{R}$ with the following properties:

1. for every $(x, t, \omega) \in \mathbb{R} \times [0, \infty) \times \Omega$ the mapping $\phi \to L(\phi, x, t, \omega)$ is a linear functional on $D(L)$ and $L(\phi, \cdot, t, \omega) \in D(L)$,
2. for every $\phi \in D(L)$, $(x, t, \omega) \to L(\phi, x, t, \omega)$ is $\mathcal{B}(\mathbb{R}) \times \mathcal{P}$-measurable, where $\mathcal{P}$ is the predictable σ-algebra on $[0, \infty) \times \Omega$, ($\mathcal{P}$ is generated by sets of the form $(s, t] \times F$ where $F \in \mathcal{F}_s$ and s, t are arbitrary),

3. for every $\phi \in D(L)$ the process M^ϕ defined by

$$M^\phi(t,\omega) := \phi(X_t(\omega)) - \phi(X_0(\omega)) - \int_0^t L(\phi, X_{s-}(\omega), s, \omega) dA_s, \quad (A.2)$$

is a locally square integrable martingale on $(\Omega, \mathcal{F}, \mathcal{F}_t, P)$,

4. the functions $\psi(x) := x$ and ψ^2 belong to $D(L)$.

The functions (recall here $\Psi(x) = x$)

$$\beta(x,t,\omega) := L(\psi, x, t, \omega) \quad (A.3)$$

$$\alpha(x,t,\omega) := L((\psi)^2, x, t, \omega) - 2x\beta(x,t,\omega) \quad (A.4)$$

are called the *local characteristics of the first and second order*.

Theorem 15 (Tightness criterion). *Let $X^m = (\Omega^m, \mathcal{F}^m, \mathcal{F}_t^M, P^m)$ be a sequence of D-semimartingales with common $D(L)$ and associated operators L^m, functions A^m, α^m, β^m.*

Then the sequence $\{X^m : m \in \mathbb{N}\}$ is tight in $D_\mathbb{R}([0, \infty)$ provided the following conditions are satisfied:

1. $\sup_m E|X_0^m|^2 < \infty$,
2. there is a $K > 0$ and a sequence of positive adapted processes $\{\{C_t^m : t \geq 0\}$ on $\Omega^m\}_{m \in \mathbb{N}}$ such that for every $m \in \mathbb{N}, x \in \mathbb{R}, \omega \in \Omega^m$,

$$|\beta_m(x,t,\omega)|^2 + \alpha_m(x,t,\omega) \leq K(C_t^m(\omega) + x^2) \quad (A.5)$$

and for every $T > 0$,

$$\sup_m \sup_{t \in [0,T]} E|C_t^m| < \infty, \text{ and } \lim_{k \to \infty} \sup_m P^m[\sup_{t \in [0,T]} C_t^m \geq k] = 0, \quad (A.6)$$

3. there exists a positive function γ on $[0, \infty)$ and a decreasing sequence of numbers (δ_m) such that $\lim_{t \to 0} \gamma(t) = 0$, $\lim_{m \to \infty} \delta_m = 0$ and for all $0 < s < t$ and all m,

$$|(A^m(t) - A^m(s))| \leq \gamma(t-s) + \delta_m, \quad (A.7)$$

4. *if we set*

$$M_t^m := X_t^m - X_0^m - \int_0^t \beta_m(X_{s-}^m, s, \cdot) dA_s^m, \quad (A.8)$$

then for each $T > 0$ there is a constant K_T and m_0 such that for all $m \geq m_0$, then

$$E[\sup_{t \in [0,T]} |X_t^m|^2] \leq K_T(1 + E|X_0^m|^2), \quad (A.9)$$

and
$$E[\sup_{t\in[0,T]} |M_t^m|^2] \le K_T(1 + E|X_0^m|^2). \qquad \Box \qquad (A.10)$$

Corollary A.1. *Assume that for $T > 0$ there is a constant K_T such that*

$$\sup_m \sup_{t\le T, x\in\mathbb{R}} (|\alpha_m(t,x)| + |\beta_m(t,x)|) \le K_T, \text{ a.s.} \qquad (A.11)$$

$$\sum_m (A^m(t) - A^m(s)) \le K_T(t-s) \text{ if } 0 \le s \le t \le T, \qquad (A.12)$$

and

$$\sup_m E|X_0^m|^2 < \infty, \qquad (A.13)$$

and M_t^m is a square integrable martingale with $\sup_m E(|M_T^m|^2) \le K_T$. Then the $\{X^m : m \in \mathbb{N}\}$ are tight in $D_{\mathbb{R}}([0,\infty))$. $\Box$

A.2 Tightness criteria for continuous processes

Now consider the special case of probability measures on $C([0,\infty), \mathbb{R}^d)$. This criterion is concerned with a collection $(X^{(n)}(t))_{t\ge 0}$ of semimartingales with values in $\mathbb{R}^d$ with continuous paths. First observe that by forming

$$(<X^{(n)}(t), \lambda>)_{t\ge 0} \quad , \quad \lambda \in \mathbb{R}^d \qquad (A.14)$$

we obtain $\mathbb{R}$-valued semi-martingales. If for every $\lambda \in \mathbb{R}^d$ the laws of these projections are tight on $C([0,\infty), \mathbb{R})$ then this is true for $\{[\mathcal{L}[(X^{(n)}(t))_{t\ge 0}], n \in \mathbb{N}\}$. The tightness criterion for $\mathbb{R}$-valued semimartingales is in terms of the socalled local characteristics of the semimartingales.

For Itô processes the local characteristics can be calculated directly from the coefficients. For example, if we have a sequence of semimartingales X^n that are also a Markov processes with generators:

$$L^{(n)} f = \Big(\sum_{i=1}^d a_i^n(x) \frac{\partial}{\partial x_i} + \sum_{i=1}^d \sum_{j=1}^d b_{i,j}^n(x) \frac{\partial^2}{\partial x_i \partial x_j}\Big) f, \qquad (A.15)$$

then the local characteristics are given by

$$a^n = (a_i^n)_{i=1,\cdots,d}, \quad b^n = (b_{i,j}^n)_{i,j,=1,\cdots,d}. \qquad (A.16)$$

The Joffe–Métivier criterion implies that if

$$\sup_{n} \sup_{0 \leq t \leq T} E[(|a^n(X^{(n)}(t)| + |b^n(X^{(n)}(t)|)^2] < \infty,$$

$$\lim_{k \to \infty} \sup_{n} P[\sup_{0 \leq t \leq T} (|a^n(X^{(n)})(t)| + |b^n(X^{(n)})(t)|) \geq k] = 0 \qquad (A.17)$$

then $\{\mathcal{L}[(X^{(n)}(t))_{t \geq 0}], n \in \mathbb{N}\}$ are tight in $C([0, \infty), \mathbb{R})$. See [JM] for details.

Theorem 16 (Ethier-Kurtz [EK2] Chapt. 3, Theorem 10.2). *Let*

$$J(x) = \int_0^\infty e^{-u}[J(x, u) \wedge 1]du, \quad J(x, u) = \sup_{0 \leq t \leq u} d(x(t), x(t-)). \qquad (A.18)$$

Assume that a sequence of processes $X_n \Rightarrow X$ in for a Polish space $ED([0, \infty), E)$. Then X is a.s. continuous if and only if $J(X_n) \Rightarrow 0$. □

Appendix B
Nonlinear Semigroup Perturbations

We use the following result of Marsden [MA], (4.17).

Theorem 17 (Perturbation). *Let $\mathbb{B}$ be a Banach space and let A_S be the infinitesimal generator of a strongly continuous semigroup, with $\|S_t\| \leq Me^{Ct}$ for some C. Let $A_T : \mathbb{B} \to \mathbb{B}$ be a vector field on $\mathbb{B}$ such that A_T is of class C^2 with its first and second derivatives uniformly bounded on bounded subsets and let $\{T_t\}$ be the flow of A_T.*

Then $A_S + A_T$ has a unique flow which is Lipschitz for each t, $0 \leq t \leq T$, and

$$V_t x = \lim_{n \to \infty} (S_{t/n} \cdot T_{t/n})^n x \tag{B.1}$$

uniformly in t for each x on bounded sets of t. If $x \in \mathcal{D}(A_S + A_T)$, then

$$\frac{d}{dt} V_t x = (A_S + A_T) U_t \tag{B.2}$$

on $[0, \tau)$ where τ is the exit time from $\mathbb{B}$. □

References

[AP98] D. Aldous, J. Pitman, Tree-valued Markov chains derived from Galton-Watson processes. Ann. Inst. Henri Poincare **34**, 637–686 (1998)

[AN] K.B. Athreya, P. Ney, A renewal approach to Perron-Frobenius theory of non-negative kernels on general state spaces. Math. Zeit. **179**, 507–529 (1982)

[AS] S.R. Athreya, J.M. Swart, Branching-coalescing systems. Probab. Theory Relat. Fields **131**(3), 376–414. Electronic, 39 p. (2005). doi:10.1007/s00440-004-0377-4

[Bai] N.T.J. Bailey, *The Elements of Stochastic Processes* (Wiley, New York, 1964)

[BEV] N. Barton, A. Etheridge, A. Véber, A new model for evolution in a spatial continuum. Electron. J. Probab. **15**, 162–216 (2010)

[Ber] J. Bertoin, *Random Fragmentation and Coagulation Processes* (Cambridge University Press, Cambridge, 2006)

[BD] N.H. Bingham, R.A. Doney, Asmptotic properties of super-critical branching processes II: Crump-mode and Jirina processes. Adv. Appl. Probab. **7**, 66–82 (1975)

[BGR] P.J. Brockwell, J.Gani, S.I. Resnick, Birth, immigration and catastrophe processes. Adv. Appl. Probab. **14**, 709–731 (1982)

[Bu] R. Bürger, *The Mathematical Theory of Selection, Recombination, and Mutation* (Wiley, New York, 2001)

[Bu] R. Bürger, Mathematical properties of mutation-selection models. Genetica **102/103**, 279–298 (1998)

[CPY] P. Carmona, F. Petit, M. Yor, Some extensions of the arc sine law as (partial) consequences of the scaling property of Brownian motion. Probab. Theory Relat. Fields **100**, 1–29 (1994)

[CFM] N. Champagnat, R. Ferrière, S. Méléard, Unifying evolutionary dynamics: From individual stochastic processes to macroscopic models via timescale separation. Theor. Popul. Biol. **69**, 297–321 (2006)

[CM] N. Champagnat, S. Méléard, Invasion and adaptive evolution for individual based spatially structured population. J. Math. Biol. **55**, 147–188 (2007)

[CC] B. Charlesworth, D. Charlesworth, Some evolutionary consequences of deleterious mutations. Genetica **102/103**, 2–19 (1998)

[CD] J.G. Conlon, C.R. Doering, On travelling waves for the stochastic Fisher-Kolmogorov-Petrovsky-Piscunov equation. J. Stat. Phys. **120**(3), 421–477 (2005)

[Con] J.B. Conway, *A Course in Functional Analysis*, 2nd edn. (Springer, Berlin, 1990)

[D] D.A. Dawson, Measure-valued Markov processes, in *École d'Été de Probabilités de Saint Flour XXI*. Lecture Notes in Mathematics **1541** (Springer, Berlin, 1993), pp. 1–261

[D2013]   D.A. Dawson, *Multilevel mutation-selection systems and set-valued duals* (in preparation)
[DG1]   D.A. Dawson, A. Greven, Multiple time scale analysis of hierarchically interacting systems, in *A Festschrift to Honor G. Kallianpur* (Springer, Berlin, 1993a), pp. 41–50
[DG2]   D.A. Dawson, A. Greven, Multiple time scale analysis of interacting diffusions. Probab. Theory Relat. Fields **95**, 467–508 (1993b)
[DG3]   D.A. Dawson, A. Greven, Hierarchical models of interacting diffusions: multiple time scale phenomena. Phase transition and pattern of cluster-formation. Probab. Theory Rel. Fields **96**, 435–473 (1993c)
[DG96]   D.A. Dawson, A. Greven, Multiple space-time scale analysis for interacting branching models. Electron. J. Probab. **1**, paper no. 14, 1–84 (1996)
[DG99]   D.A. Dawson, A. Greven, Hierarchically interacting Fleming-Viot processes with selection and mutation: Multiple space time scale analysis and quasi equilibria. Electron. J. Probab. **4**, paper no. 4, 1–81 (1999)
[DG12]   D.A. Dawson, A. Greven, Multiscale analysis: Fisher-Wright diffusions with rare mutations and selection, Logistic branching system, in *Probability in Complex Physical Systems*. Springer Proceedings in Mathematics, vol. 11 (Springer, Heidelberg, 2012) pp. 373–408
[DGV]   D.A. Dawson, A. Greven, J. Vaillancourt, Equilibria and Quasi-equilibria for Infinite Collections of Interacting Fleming-Viot processes. Trans. Am. Math. Soc. **347**(7), 2277–2360 (1995)
[DGW01]   D.A. Dawson, L.G. Gorostiza, A. Wakolbinger, Occupation time fluctuations in branching systems. J. Theor. Probab. **14**, 729–796 (2001)
[D-Li]   D.A. Dawson, Z. Li, Construction of immigration superprocesses with dependent spatial motion from one-dimensional excursions. Probab. Theory Relat. Fields **127**, 37–61 (2003)
[DGP11]   A. Depperschmidt, A. Greven, P. Pfaffelhuber, Marked metric measure spaces. ECP **16**, 174–188 (2011)
[DGP]   A. Depperschmidt, A. Greven, P. Pfaffelhuber, Tree-valued Fleming-Viot dynamics with mutation and selection. Ann. Appl. Probab. **22**, 2560–2615 (2012)
[Dia]   P. Diaconis, E. Mayer-Wolf, O. Zeitouni, M.P.W. Zerner, The Poisson-Dirichlet la is the unique invariant distribution for uniform split-merge transformations. Ann. Probab. **32**, 915–938 (2004)
[Do]   R.A. Doney, A limit theorem for a class of supercritical branching processes. J. Appl. Probab. **9**, 707–724 (1972)
[Do2]   R.A. Doney, On single and multi-type age-dependent branching processes. J. Appl. Probab. **13**, 239–246 (1976)
[DMS]   C.R. Doering, C. Mueller, P. Smereka, Interacting particles, the stochastic Fisher-Kolmogorov-Petrovsky-Piscounov equation, and duality. Phys. A **325**, 243–259 (2003)
[DU]   R. Durrett, *Probability, Theory and Examples*, 2nd edn. (Duxbury, Wadsworth Publ. Co., Belmont, 1996)
[EG1]   N. Eldridge, S.J. Gould, in *Models in Paleobiology*, ed. by T.J.M. Schopf (Freeman, San Francisco, 1972), pp. 82–115
[EG2]   N. Eldridge, S.J. Gould, Punctuated equilibria: The tempo and mode of evolution reconsidered. Paleobiology, **3**, 115–151 (1977)
[Eth00]   A.M. Etheridge, in *An Introduction to Superprocesses*. (English summary). University Lecture Series, vol. 20 (American Mathematical Society, Providence, 2000)
[EG09]   A.M. Etheridge, R.C. Griffiths, A coalescent dual process in a Moran model with genic selection. Theor. Popul. Biol. **75**, 320–330 (2009)

References

[EPW] A. Etheridge, P. Pfaffelhuber, A. Wakolbinger, How often does the ratchet click? Facts, heuristics, asymptotics, in *Trends in Stochastic Analysis*, vol. LMS 353 (Cambridge University Press, Cambridge, 2008)

[ECL] S.F. Elena, V. Cooper, R. Lenski, Punctuated equilibrium caused by selection of rare beneficial mutation. Science **272**, 1802–1804 (1996)

[EG] S.N. Ethier, B. Griffiths, The infinitely-many-sites model as a measure-valued diffusion. Ann. Probab. **15**, 515–545 (1981)

[EK1] S.N. Ethier, T.G. Kurtz, The infinitely-many-neutral-alleles diffusion model. Adv. Appl. Probab. **13**, 429–452 (1981)

[EK2] S.N. Ethier, T.G. Kurtz, *Markov Processes, Characterization and Convergence* (Wiley, New York, 1986)

[EK3] S.N. Ethier, T.G. Kurtz, The infinitely-many-alleles-model with selection as a measure-valued diffusion, in *Lecture Notes in Biomathematics*, vol. 70 (Springer, Berlin, 1987), pp. 72–86

[EK4] S.N. Ethier, T.G. Kurtz, Convergence to Fleming-Viot processes in the weak atomic topology. Stoch. Process. Appl. **54**, 1–27 (1994)

[EK5] S.N. Ethier, T.G. Kurtz, Coupling and ergodic theorems for Fleming-Viot processes. Ann. Probab. **26**, 533–561 (1998)

[Ew] W.J. Ewens, *Mathematical Population Genetics*, 2nd edn. (Springer, Berlin, 2004)

[F] P. Fernhead, Perfect simulation from population genetic models with selection. Theor. Popul. Biol. **59**, 263–279 (2001)

[FL] Z. Fu, Z. Li, Measure-valued diffusions and stochastic equations with Poisson processes. Osaka J. Math. **41**, 727–744 (2004)

[Gar] J. Gärtner, On the McKean-Vlasov limit for interacting diffusions. Math. Nachr. **137**, 197–248 (1988)

[Gav] S. Gavrilets, Evolution and speciation in a hyperspace: the roles of neutrality, selection, mutation and random drift, in *Towards a Comprehensive Dynamics of Evolution - Exploring the Interplay of Selection, Neutrality, Accident, and Function*, ed. by J. Crutchfield, P. Schuster (Oxford University Press, Oxford, 1999)

[GRD] N.S. Goel, N. Richter-Dyn, *Stochastic Models in Biology* (Academic, New York, 1974)

[GJY] A. Göing-Jaeschke, M. Yor, A survey and some generalizations of Bessel process (1999), http://www.risklab.ch/papers.htm.

[GLW] A. Greven, V. Limic, A. Winter, Representation theorems for interacting Moran models, interacting Fisher–Wright diffusions and applications. Electron. J. Probab. **10**(39), 1286–1358 (2005)

[GPWmp12] A. Greven, P. Pfaffelhuber, A. Winter, Tree-valued resampling dynamics: Martingale Problems and applications. Probab. Theor. Relat. Fields **155**, 787–838 (2013)

[GPWmetric09] A. Greven, P. Pfaffelhuber, A. Winter, Convergence in distribution of random metric measure spaces (The Λ-coalescent measure tree). Probab. Theor. Relat. Fields **145**, 285–322 (2009)

[GHKK] A. Greven, F. den Hollander, S. Kliem, A. Klimovsky, Renormalization of hierarchically interacting Cannings processes (2012)

[H] T.E. Harris, *On the Theory of Branching Processes* (Springer, Berlin, 1963)

[HS] J. Hofbauer, K. Sigmund, *The Theory of Evolution and Dynamical Systems* (Cambridge University Press, Cambridge, 1988)

[Hu] M. Hutzenthaler, The virgin island model. Electron. J. Probab. **14**, 1117–1161 (2009)

[Hu2] M. Hutzenthaler, Interacting diffusions and trees of excursions: Convergence and comparison. Electron. J. Probab. **17**, (71), 1–49 (2012)

[J92]	P. Jagers, Stability and instability in population dynamics. J. Appl. Probab. **29**, 770–780 (1992)
[JN]	P. Jagers, O. Nerman, The growth and composition of branching populations. Adv. Appl. Probab. **16**, 221–259 (1984)
[JN2]	P. Jagers, O. Nerman, in *The Asymptotic Composition of Supercritical, Multitype Branching Populations*. Séminaire de probabilités (Strasbourg), tome 30 (1996), pp. 40–54
[JM]	A. Joffe, M. Métivier, Weak convergence of sequences of semi-martingales with application to multitype branching processes. Adv. Appl. Probab. **18**, 20–65 (1986)
[KT]	S. Karlin, H.M. Taylor, *A First Course in Stochastic Processes*, 2nd edn. (Academic, New York, 1975)
[Kim1]	M. Kimura, Diffusion models in population genetics. J. Appl. Probab. **1**, 177–232 (1964)
[Kim2]	M. Kimura, Diffusion model of population genetics incorporating group selection, with special reference to an altruistic trait, in *Lecture Notes in Mathematics*, vol. 1203 (Springer, Berlin, 1986), pp. 101–118
[KN]	C. Kipnis, C.M. Newman, The metastable behavior of infrequently observed weakly random one dimensional diffusion processes. SIAM J. Appl. Math. **45**, 972–982 (1985)
[KN97]	S. Krone, C. Neuhauser, Ancestral processes with selection. Theor. Popul. Biol. **51**, 210–237 (1997)
[Lig85]	T.M. Liggett, *Interacting Particle Systems* (Springer, New York, 1985)
[LS]	V. Limic, A. Sturm, The spatial Λ-coalescent. Electron. J. Probab. **11**, 363–393 (2006)
[L1]	G.D. Lin, On weak convergence within the $\mathcal{L}$-like classes of life distributions. Sankhyā Indian J. Stat. **60**, 176–183 (1998)
[LO]	R.M. Loynes, Extreme values in uniformly mixing stationary stochastic processes. Ann. Math. Stat. **36**, 993–999 (1965)
[MA]	J.E. Marsden, On product formulas for nonlinear semigroups. J. Funct. Anal. **13**, 51–74 (1973)
[MJM]	J. Martinez, J.M. Mazón, Quasi-compactness of dominated positive operators and C_0-semigroups. Math. Z. **207**, 109–120 (1991)
[MZ]	P.A. Meyer, W.A. Zheng, Tightness criteria for laws of semimartingales. Ann. l'Institut Henri Poincaré **20**, 353–372 (1984)
[Mo]	S.-T.C. Moy, Extensions of a limit theorem of Everett, Ulm and Harris on multitype branching processes to a branching process with countably many types. Ann. Math. Stat. **38**, 992–999 (1967)
[Mo2]	S.-T.C. Moy, Ergodic properties of expectation matrices of a branching process with countably many types. J. Math. Mech. **16**, 1207–1225 (1967)
[N]	O. Nerman, On the convergence of supercritical general (C-M-J) branching processes. Zeitschrift f. Wahrscheinlichkeitsth. verw. Gebiete **57**, 365–395 (1981)
[N2]	O. Nerman, On the Convergence of Supercritical General Branching Processes. Thesis, Department of Mathematics, Chalmers University of Technology and the University of Göteborg, 1979
[NCK]	C.M. Newman, J.E. Cohen, C. Kipnis, Neo-Darwinian evolution implies punctuated equilibria. Nature **315**, 400–401 (1985)
[Paz]	A. Pazy, *Semigroups of Linear Operators and Applications to Partial Differential Equations* (Springer, Berlin, 1983)
[PY]	J. Pitman, M. Yor, A decomposition of Bessel bridges. Z. Wahr. verw. Geb. **59**, 425–457 (1982)
[PSP]	P.S. Puri, Some further results on the birth-and-death process and its integral. Proc. Camb. Philos. Soc. **64**, 141–154 (1968)
[Rio]	J. Riordan, *Introduction to Combinatorial Analysis* (Dover, New York, 1958)

References

[RW]	L.C.G. Rogers, D. Williams, *Diffusions, Markov Processes and Martingales*, vol. 2 (Wiley, New York, 1987)
[Saw]	S. Sawyer, Branching diffusion processes in population genetics. Adv. Appl. Probab. **8**, 659–689 (1976)
[SF]	S. Sawyer, J. Felsenstein, Isolation by distance in a hierarchically clustered population. J. Appl. Prob. **20**, 1–10 (1983)
[Seidel08]	P. Seidel, The historical process of the spatial Moran model with selection and mutation. Master thesis, Department of Math., Erlangen, Germany, 2008
[Seidel]	P. Seidel, *The historical process of interacting Fleming-Viot processes with selection* (in preparation)
[S1]	T. Shiga, Diffusion processes in population genetics. J. Math. Kyoto Univ. **21-1**, 133–151 (1981)
[S2]	T. Shiga, Continuous time multi-allelic stepping stone models in population genetics. J. Math. Kyoto Univ. **22-1**, 1–40 (1982)
[HS02]	H. Shirakawa, Squared Bessel processes and their applications to the square root interest rate model. Asia-Pac. Financ. Mark. **9**, 169–190 (2002)
[SU]	T. Shiga, Uchiyama, Stationary states and the stability of the stepping stone model involving mutation and selection. Probab. Theory Relat. Fields **73**, 87–117 (1986)
[W]	M.J. Wade, Sewall Wright, gene interaction and the shifting balance theory. Oxf. Surv. Evol. Biol. **8**, 35–62 (1992)
[VJ]	D. Vere-Jones, Ergodic properties of non-negative matrices, I, II. Pac. J. Math. **22**, 361–386; **26**, 601–620 (1967/1968)
[Wr1]	S. Wright, The roles of mutation, inbreeding, crossbreeding and selection in evolution, in *Proceedings of the. Sixth International Congress of Genetics*, vol. 1, (1932), pp. 356–366
[Wr2]	S. Wright, *Evolution and the Genetics of Populations*, vol. 3 (1977), pp. 443–473; vol. 4 (1978), pp. 460–476 (University of Chicago Press, Chicago, 1977)

Index of Notation and Tables of Basic Objects

- $\mathbb{Z}$—the set of integers
- $\mathbb{N} = \{1, 2, 3, \ldots\}$
- $\Omega_N = \otimes_{\mathbb{N}} Z_N$, Z_N cyclical group of order N
- $\mathcal{M}(E)$ denotes the space of finite Borel measures on a Polish space E
- $\mathcal{P}(E)$ denotes the space of probability measures on the Borel field on a Polish space E.
- Table 9.1: List of Basic Dual Objects
- Table 9.2: State Spaces of Basic Dual Objects
- Table 9.3: Dual State Spaces—Mutation-dominant regime
- Table 9.4: Dual State Spaces—Selection-dominant regime

Index

basic scenario, 161
biodiversity, 835
blockaverages, 805

Crump–Mode–Jagers (CMJ), 174, 188, 195, 486

direct clan, 534
droplet formation, 28, 30, 387, 817
dual clouds, 324
dual particle system, 111
dual process, 107
duality relation, 116

emergence characteristic, 88
emergence time, 79
empirical process, 541, 542
enriched ordered dual, 410
entrance law, 18, 20
ergodic theorem, 148, 157, 721
excursion laws, 732

factorization dynamics, 499
family decomposition, 98
Feynman–Kac dual, 116
Fisher–Wright dynamics, 14
fitness function, 46
five phases of transition, 63
fixation, 729
fixation of rare mutants, 161
fixation time, 28

hierarchical mean-field limit, 63
hierarchical mean-field model, 161
historical interpretation, 126

interaction chains, 73

Joffe–Métivier conditions, 841

local genealogy, 835

macroscopic emergence, 27
Malthusian parameter, 195, 483
marked set-valued dual process, 426, 435
martingale problem, 43
MBDC process, 597
McKean–Vlasov dual process, 142
McKean–Vlasov dynamics, 18
mean-field emergence, 161
mean-field model, 161
mesostate dynamics, 592
mesostates, 474
microscopic emergence, 30
microstate process, 546
microstates, 475
modified dual, 124
multicolour systems, 569
mutation dominance condition, 504

quasi-equilibria, 76

random solution of McKean–Vlasov, 21
rare mutations, 681

refined dual, 135
renormalization, 58

selection dominance condition, 505
set-valued dual, 391
successive invasions, 56

Table 1, 395
Table 2, 396
Table 3, 547
Table 4, 657

weak atomic topology, 37

LECTURE NOTES IN MATHEMATICS

Edited by J.-M. Morel, B. Teissier; P.K. Maini

Editorial Policy (for the publication of monographs)

1. Lecture Notes aim to report new developments in all areas of mathematics and their applications - quickly, informally and at a high level. Mathematical texts analysing new developments in modelling and numerical simulation are welcome.

   Monograph manuscripts should be reasonably self-contained and rounded off. Thus they may, and often will, present not only results of the author but also related work by other people. They may be based on specialised lecture courses. Furthermore, the manuscripts should provide sufficient motivation, examples and applications. This clearly distinguishes Lecture Notes from journal articles or technical reports which normally are very concise. Articles intended for a journal but too long to be accepted by most journals, usually do not have this "lecture notes" character. For similar reasons it is unusual for doctoral theses to be accepted for the Lecture Notes series, though habilitation theses may be appropriate.

2. Manuscripts should be submitted either online at www.editorialmanager.com/lnm to Springer's mathematics editorial in Heidelberg, or to one of the series editors. In general, manuscripts will be sent out to 2 external referees for evaluation. If a decision cannot yet be reached on the basis of the first 2 reports, further referees may be contacted: The author will be informed of this. A final decision to publish can be made only on the basis of the complete manuscript, however a refereeing process leading to a preliminary decision can be based on a pre-final or incomplete manuscript. The strict minimum amount of material that will be considered should include a detailed outline describing the planned contents of each chapter, a bibliography and several sample chapters.

   Authors should be aware that incomplete or insufficiently close to final manuscripts almost always result in longer refereeing times and nevertheless unclear referees' recommendations, making further refereeing of a final draft necessary.

   Authors should also be aware that parallel submission of their manuscript to another publisher while under consideration for LNM will in general lead to immediate rejection.

3. Manuscripts should in general be submitted in English. Final manuscripts should contain at least 100 pages of mathematical text and should always include

   - a table of contents;
   - an informative introduction, with adequate motivation and perhaps some historical remarks: it should be accessible to a reader not intimately familiar with the topic treated;
   - a subject index: as a rule this is genuinely helpful for the reader.

   For evaluation purposes, manuscripts may be submitted in print or electronic form (print form is still preferred by most referees), in the latter case preferably as pdf- or zipped ps-files. Lecture Notes volumes are, as a rule, printed digitally from the authors' files. To ensure best results, authors are asked to use the LaTeX2e style files available from Springer's web-server at:

   ftp://ftp.springer.de/pub/tex/latex/svmonot1/ (for monographs) and
   ftp://ftp.springer.de/pub/tex/latex/svmultt1/ (for summer schools/tutorials).

Additional technical instructions, if necessary, are available on request from lnm@springer.com.

4. Careful preparation of the manuscripts will help keep production time short besides ensuring satisfactory appearance of the finished book in print and online. After acceptance of the manuscript authors will be asked to prepare the final LaTeX source files and also the corresponding dvi-, pdf- or zipped ps-file. The LaTeX source files are essential for producing the full-text online version of the book (see http://www.springerlink.com/openurl.asp?genre=journal&issn=0075-8434 for the existing online volumes of LNM). The actual production of a Lecture Notes volume takes approximately 12 weeks.

5. Authors receive a total of 50 free copies of their volume, but no royalties. They are entitled to a discount of 33.3 % on the price of Springer books purchased for their personal use, if ordering directly from Springer.

6. Commitment to publish is made by letter of intent rather than by signing a formal contract. Springer-Verlag secures the copyright for each volume. Authors are free to reuse material contained in their LNM volumes in later publications: a brief written (or e-mail) request for formal permission is sufficient.

Addresses:
Professor J.-M. Morel, CMLA,
École Normale Supérieure de Cachan,
61 Avenue du Président Wilson, 94235 Cachan Cedex, France
E-mail: morel@cmla.ens-cachan.fr

Professor B. Teissier, Institut Mathématique de Jussieu,
UMR 7586 du CNRS, Équipe "Géométrie et Dynamique",
175 rue du Chevaleret
75013 Paris, France
E-mail: teissier@math.jussieu.fr

For the "Mathematical Biosciences Subseries" of LNM:

Professor P. K. Maini, Center for Mathematical Biology,
Mathematical Institute, 24-29 St Giles,
Oxford OX1 3LP, UK
E-mail: maini@maths.ox.ac.uk

Springer, Mathematics Editorial, Tiergartenstr. 17,
69121 Heidelberg, Germany,
Tel.: +49 (6221) 4876-8259

Fax: +49 (6221) 4876-8259
E-mail: lnm@springer.com

CPSIA information can be obtained at www.ICGtesting.com
Printed in the USA
LVOW10s1514221213

366440LV00002B/4/P